CONVERSION FACTORS

Length
1 m = 39.37 in. = 3.281 ft
1 in. = 2.54 cm (exact)
1 km = 0.621 mi
1 mi = 5 280 ft = 1.609 km
1 lightyear (ly) = 9.461 × 10^{15} m
1 angstrom (Å) = 10^{-10} m

Mass
1 kg = 10^3 g = 6.85 × 10^{-2} slug
1 slug = 14.59 kg
1 u = 1.66 × 10^{-27} kg = 931.5 MeV/c^2

Time
1 min = 60 s
1 h = 3 600 s
1 day = 24 h = 1.44 × 10^3 min = 8.64 × 10^4 s
1 yr = 365.242 days = 3.156 × 10^7 s

Volume
1 L = 1 000 cm^3 = 0.035 3 ft^3
1 ft^3 = 2.832 × 10^{-2} m^3
1 gal = 3.786 L = 231 $in.^3$

Angle
180° = π rad
1 rad = 57.30°
1° = 60 min = 1.745 × 10^{-2} rad

Speed
1 km/h = 0.278 m/s = 0.621 mi/h
1 m/s = 2.237 mi/h = 3.281 ft/s
1 mi/h = 1.61 km/h = 0.447 m/s = 1.47 ft/s

Force
1 N = 0.224 8 lb = 10^5 dynes
1 lb = 4.448 N
1 dyne = 10^{-5} N = 2.248 × 10^{-6} lb

Work and energy
1 J = 10^7 erg = 0.738 ft·lb = 0.239 cal
1 cal = 4.186 J
1 ft·lb = 1.356 J
1 Btu = 1.054 × 10^3 J = 252 cal
1 J = 6.24 × 10^{18} eV
1 eV = 1.602 × 10^{-19} J
1 kWh = 3.60 × 10^6 J

Pressure
1 atm = 1.013 × 10^5 N/m^2 (or Pa) = 14.70 lb/$in.^2$
1 Pa = 1 N/m^2 = 1.45 × 10^{-4} lb/$in.^2$
1 lb/$in.^2$ = 6.895 × 10^3 N/m^2

Power
1 hp = 550 ft·lb/s = 0.746 kW
1 W = 1 J/s = 0.738 ft·lb/s
1 Btu/h = 0.293 W

5 REASONS
to buy your textbooks and course materials at
CENGAGEbrain.com

1 SAVINGS:
Prices up to 65% off, daily coupons, and free shipping on orders over $25

2 CHOICE:
Multiple format options including textbook, eBook and eChapter rentals

3 CONVENIENCE:
Anytime, anywhere access of eBooks or eChapters via mobile devices

4 SERVICE:
Free eBook access while your text ships, and instant access to online homework products

5 STUDY TOOLS:
Free study tools* for your text, plus writing, research, career and job search resources
*availability varies

Find your course materials and start saving at:
www.cengagebrain.com

Physics 1101 / 1102

Algebra - based, UNCC Custom

Raymond A. Serway | Chris Vuille

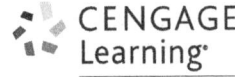

Australia • Brazil • Japan • Korea • Mexico • Singapore • Spain • United Kingdom • United States

**Physics 1101 / 1102:
Algebra - based, UNCC Custom**

College Physics, 10th Edition
Raymond A. Serway | Chris Vuille

© 2015, 2012, 2008 Cengage Learning. All rights reserved.

Senior Manager, Student Engagement:
Linda deStefano
Janey Moeller

Manager, Student Engagement:
Julie Dierig

Marketing Manager:
Rachael Kloos

Manager, Production Editorial:
Kim Fry

Manager, Intellectual Property Project Manager:
Brian Methe

Senior Manager, Production and Manufacturing:
Donna M. Brown

Manager, Production:
Terri Daley

ALL RIGHTS RESERVED. No part of this work covered by the copyright herein may be reproduced, transmitted, stored or used in any form or by any means graphic, electronic, or mechanical, including but not limited to photocopying, recording, scanning, digitizing, taping, Web distribution, information networks, or information storage and retrieval systems, except as permitted under Section 107 or 108 of the 1976 United States Copyright Act, without the prior written permission of the publisher.

> For product information and technology assistance, contact us at
> **Cengage Learning Customer & Sales Support, 1-800-354-9706**
> For permission to use material from this text or product,
> submit all requests online at **cengage.com/permissions**
> Further permissions questions can be emailed to
> **permissionrequest@cengage.com**

This book contains select works from existing Cengage Learning resources and was produced by Cengage Learning Custom Solutions for collegiate use. As such, those adopting and/or contributing to this work are responsible for editorial content accuracy, continuity and completeness.

Compilation © 2014 Cengage Learning

ISBN-13: 978-1-305-30532-8
ISBN-10: 1-305-30532-9

WCN: 01-100-101

Cengage Learning
5191 Natorp Boulevard
Mason, Ohio 45040
USA

Cengage Learning is a leading provider of customized learning solutions with office locations around the globe, including Singapore, the United Kingdom, Australia, Mexico, Brazil, and Japan. Locate your local office at:
international.cengage.com/region.

Cengage Learning products are represented in Canada by Nelson Education, Ltd.
For your lifelong learning solutions, visit **www.cengage.com/custom.**
Visit our corporate website at **www.cengage.com.**

Brief Contents

About the Authors viii
Chapter 1   Introduction 1
Chapter 2   Motion in One Dimension 26
Chapter 3   Vectors and Two-Dimensional Motion... 57
Chapter 4   The Laws of Motion 88
Chapter 5   Energy 127
Chapter 6   Momentum and Collisions 170
Chapter 7   Rotational Motion and the Law of Gravity 202
Chapter 8   Rotational Equilibrium and Rotational Dynamics 240
Chapter 13   Vibrations and Waves 445
Chapter 15   Electric Forces and Electric Fields 523
Chapter 16   Electrical Energy and Capacitance 558
Chapter 17   Current and Resistance 600
Chapter 18   Direct-Current Circuits 626
Chapter 19   Magnetism 659
Chapter 20   Induced Voltages and Inductance 699
Chapter 21   Alternating-Current Circuits and Electromagnetic Waves 733
Chapter 22   Reflection and Refraction of Light 772
Chapter 23   Mirrors and Lenses 801
Chapter 24   Wave Optics 835
Chapter 25   Optical Instruments 870
Appendix A   Mathematics Review A.1
Appendix B   An Abbreviated Table of Isotopes A.14
Appendix C   Some Useful Tables A.19
Appendix D   SI Units A.21
Answers to Quick Quizzes, Example Questions, Odd-Numbered Warm-Up
Exercises, Conceptual Questions, and Problems A.23
Index I.1

About the Authors

Raymond A. Serway received his doctorate at Illinois Institute of Technology and is Professor Emeritus at James Madison University. In 2011, he was awarded an honorary doctorate degree from his alma mater, Utica College. He received the 1990 Madison Scholar Award at James Madison University, where he taught for 17 years. Dr. Serway began his teaching career at Clarkson University, where he conducted research and taught from 1967 to 1980. He was the recipient of the Distinguished Teaching Award at Clarkson University in 1977 and the Alumni Achievement Award from Utica College in 1985. As Guest Scientist at the IBM Research Laboratory in Zurich, Switzerland, he worked with K. Alex Müller, 1987 Nobel Prize recipient. Dr. Serway was also a visiting scientist at Argonne National Laboratory, where he collaborated with his mentor and friend, the late Sam Marshall. Early in his career, he was employed as a research scientist at the Rome Air Development Center from 1961 to 1963 and at the IIT Research Institute from 1963 to 1967. Dr. Serway is also the coauthor of *Physics for Scientists and Engineers*, ninth edition; *Principles of Physics: A Calculus-Based Text*, fifth edition; *Essentials of College Physics*, *Modern Physics*, third edition; and the high school textbook *Physics*, published by Holt, Rinehart and Winston. In addition, Dr. Serway has published more than 40 research papers in the field of condensed matter physics and has given more than 60 presentations at professional meetings. Dr. Serway and his wife Elizabeth enjoy traveling, playing golf, fishing, gardening, singing in the church choir, and especially spending quality time with their four children, nine grandchildren, and a recent great grandson.

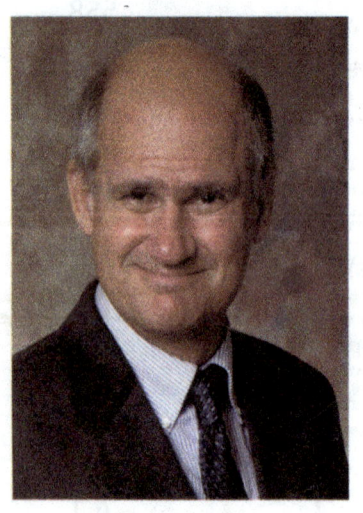

Chris Vuille is an associate professor of physics at Embry-Riddle Aeronautical University (ERAU), Daytona Beach, Florida, the world's premier institution for aviation higher education. He received his doctorate in physics from the University of Florida in 1989 and moved to Daytona after a year at ERAU's Prescott, Arizona, campus. Although he has taught courses at all levels, including postgraduate, his primary interest has been instruction at the level of introductory physics. He has received several awards for teaching excellence, including the Senior Class Appreciation Award (three times). He conducts research in general relativity and quantum theory and was a participant in the JOVE program, a special three-year NASA grant program during which he studied neutron stars. His work has appeared in a number of scientific journals, and he has been a featured science writer in Analog Science Fiction/Science Fact magazine. In addition to this textbook, he is coauthor of *Essentials of College Physics*. Dr. Vuille enjoys tennis, swimming, and playing classical piano, and he is a former chess champion of St. Petersburg and Atlanta. In his spare time he writes fiction and goes to the beach. His wife, Dianne Kowing, is Chief of Optometry for a local Veterans' Administration clinic. They have a daughter, Kira, and two sons, Christopher and James.

In the eighteenth century, navigators of ocean-going ships could obtain their latitude by observations of the north star, but there was no reliable way of determining longitude. The H1 clock was invented by John Harrison in 1736 in an attempt to address that need. His clock had to remain highly accurate for months at sea while withstanding constant motion, dampness, and changes of temperature. To determine longitude, navigators had only to compare local noon, when the sun was highest in the sky, with the time on the clock, which was Greenwich time. The difference in the number of hours then revealed their longitude.

Introduction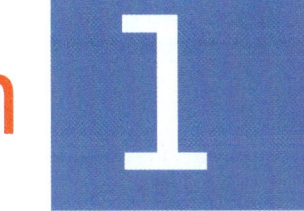

The goal of physics is to provide an understanding of the physical world by developing theories based on experiments. A physical theory, usually expressed mathematically, describes how a given physical system works. The theory makes certain predictions about the physical system which can then be checked by observations and experiments. If the predictions turn out to correspond closely to what is actually observed, then the theory stands, although it remains provisional. No theory to date has given a complete description of all physical phenomena, even within a given subdiscipline of physics. Every theory is a work in progress.

The basic laws of physics involve such physical quantities as force, velocity, volume, and acceleration, all of which can be described in terms of more fundamental quantities. In mechanics, it is conventional to use the quantities of **length** (L), **mass** (M), and **time** (T); all other physical quantities can be constructed from these three.

1.1 Standards of Length, Mass, and Time
1.2 The Building Blocks of Matter
1.3 Dimensional Analysis
1.4 Uncertainty in Measurement and Significant Figures
1.5 Conversion of Units
1.6 Estimates and Order-of-Magnitude Calculations
1.7 Coordinate Systems
1.8 Trigonometry
1.9 Problem-Solving Strategy

1.1 Standards of Length, Mass, and Time

LEARNING OBJECTIVES

1. State and use the SI units for length, mass, and time.
2. Give examples of the approximate magnitudes of common measurements.

To communicate the result of a measurement of a certain physical quantity, a *unit* for the quantity must be defined. If our fundamental unit of length is defined to be 1.0 meter, for example, and someone familiar with our system of measurement reports that a wall is 2.0 meters high, we know that the height of the wall is twice the fundamental unit of length. Likewise, if our fundamental unit of mass is defined as 1.0 kilogram and we are told that a person has a mass of 75 kilograms, then that person has a mass 75 times as great as the fundamental unit of mass.

In 1960 an international committee agreed on a standard system of units for the fundamental quantities of science, called **SI** (Système International). Its units of length, mass, and time are the meter, kilogram, and second, respectively.

Length

In 1799 the legal standard of length in France became the meter, defined as one ten-millionth of the distance from the equator to the North Pole. Until 1960, the official length of the meter was the distance between two lines on a specific bar of platinum-iridium alloy stored under controlled conditions. This standard was abandoned for several reasons, the principal one being that measurements of the separation between the lines were not precise enough. In 1960 the meter was defined as 1 650 763.73 wavelengths of orange-red light emitted from a krypton-86 lamp. In October 1983 this definition was abandoned also, and **the meter was redefined as the distance traveled by light in vacuum during a time interval of 1/299 792 458 second**. This latest definition establishes the speed of light at 299 792 458 meters per second.

▶ Definition of the meter

Mass

▶ Definition of the kilogram

The SI unit of mass, the kilogram, is defined as the mass of a specific platinum–iridium alloy cylinder kept at the International Bureau of Weights and Measures at Sèvres, France (similar to that shown in Fig. 1.1a). As we'll see in Chapter 4, mass is a quantity used to measure the resistance to a change in the motion of an object. It's more difficult to cause a change in the motion of an object with a large mass than an object with a small mass.

> **Tip 1.1 No Commas in Numbers with Many Digits**
>
> In science, numbers with more than three digits are written in groups of three digits separated by spaces rather than commas; so that 10 000 is the same as the common American notation 10,000. Similarly, $\pi = 3.14159265$ is written as 3.141 592 65.

Time

Before 1960, the time standard was defined in terms of the average length of a solar day in the year 1900. (A solar day is the time between successive appearances of the Sun at the highest point it reaches in the sky each day.) The basic unit of

Figure 1.1 (a) International Prototype of the Kilogram, an accurate copy of the International Standard Kilogram kept at Sèvres, France, is housed under a double bell jar in a vault at the National Institute of Standards and Technology. (b) A cesium fountain atomic clock. The clock will neither gain nor lose a second in 20 million years.

1.1 | Standards of Length, Mass, and Time

Table 1.1 Approximate Values of Some Measured Lengths

	Length (m)
Distance from Earth to most remote known quasar	1×10^{26}
Distance from Earth to most remote known normal galaxies	4×10^{25}
Distance from Earth to nearest large galaxy (M31, the Andromeda galaxy)	2×10^{22}
Distance from Earth to nearest star (Proxima Centauri)	4×10^{16}
One light year	9×10^{15}
Mean orbit radius of Earth about Sun	2×10^{11}
Mean distance from Earth to Moon	4×10^{8}
Mean radius of Earth	6×10^{6}
Typical altitude of satellite orbiting Earth	2×10^{5}
Length of football field	9×10^{1}
Length of housefly	5×10^{-3}
Size of smallest dust particles	1×10^{-4}
Size of cells in most living organisms	1×10^{-5}
Diameter of hydrogen atom	1×10^{-10}
Diameter of atomic nucleus	1×10^{-14}
Diameter of proton	1×10^{-15}

time, the second, was defined to be $(1/60)(1/60)(1/24) = 1/86\,400$ of the average solar day. In 1967 the second was redefined to take advantage of the high precision attainable with an atomic clock, which uses the characteristic frequency of the light emitted from the cesium-133 atom as its "reference clock." **The second is now defined as 9 192 631 700 times the period of oscillation of radiation from the cesium atom.** The newest type of cesium atomic clock is shown in Figure 1.1b.

◀ Definition of the second

Approximate Values for Length, Mass, and Time Intervals

Approximate values of some lengths, masses, and time intervals are presented in Tables 1.1, 1.2, and 1.3, respectively. Note the wide ranges of values. Study these tables to get a feel for a kilogram of mass (this book has a mass of about 2 kilograms), a time interval of 10^{10} seconds (one century is about 3×10^{9} seconds), or 2 meters of length (the approximate height of a forward on a basketball team). Appendix A reviews the notation for powers of 10, such as the expression of the number 50 000 in the form 5×10^{4}.

Systems of units commonly used in physics are the Système International, in which the units of length, mass, and time are the meter (m), kilogram (kg), and second (s); the cgs, or Gaussian, system, in which the units of length, mass, and time

Table 1.2 Approximate Values of Some Masses

	Mass (kg)
Observable Universe	1×10^{52}
Milky Way galaxy	7×10^{41}
Sun	2×10^{30}
Earth	6×10^{24}
Moon	7×10^{22}
Shark	1×10^{2}
Human	7×10^{1}
Frog	1×10^{-1}
Mosquito	1×10^{-5}
Bacterium	1×10^{-15}
Hydrogen atom	2×10^{-27}
Electron	9×10^{-31}

Table 1.3 Approximate Values of Some Time Intervals

	Time Interval (s)
Age of Universe	5×10^{17}
Age of Earth	1×10^{17}
Average age of college student	6×10^{8}
One year	3×10^{7}
One day	9×10^{4}
Time between normal heartbeats	8×10^{-1}
Period[a] of audible sound waves	1×10^{-3}
Period[a] of typical radio waves	1×10^{-6}
Period[a] of vibration of atom in solid	1×10^{-13}
Period[a] of visible light waves	2×10^{-15}
Duration of nuclear collision	1×10^{-22}
Time required for light to travel across a proton	3×10^{-24}

[a]A *period* is defined as the time required for one complete vibration.

Table 1.4 Some Prefixes for Powers of Ten Used with "Metric" (SI and cgs) Units

Power	Prefix	Abbreviation
10^{-18}	atto-	a
10^{-15}	femto-	f
10^{-12}	pico-	p
10^{-9}	nano-	n
10^{-6}	micro-	μ
10^{-3}	milli-	m
10^{-2}	centi-	c
10^{-1}	deci-	d
10^{1}	deka-	da
10^{3}	kilo-	k
10^{6}	mega-	M
10^{9}	giga-	G
10^{12}	tera-	T
10^{15}	peta-	P
10^{18}	exa-	E

are the centimeter (cm), gram (g), and second; and the U.S. customary system, in which the units of length, mass, and time are the foot (ft), slug, and second. SI units are almost universally accepted in science and industry, and will be used throughout the book. Limited use will be made of Gaussian and U.S. customary units.

Some of the most frequently used "metric" (SI and cgs) prefixes representing powers of 10 and their abbreviations are listed in Table 1.4. For example, 10^{-3} m is equivalent to 1 millimeter (mm), and 10^3 m is 1 kilometer (km). Likewise, 1 kg is equal to 10^3 g, and 1 megavolt (MV) is 10^6 volts (V). It's a good idea to memorize the more common prefixes early on: femto- to centi-, and kilo- to giga- are used routinely by most physicists.

1.2 The Building Blocks of Matter

LEARNING OBJECTIVES

1. State the fundamental components of matter.
2. Describe qualitatively the levels of structure of matter.

A 1-kg ($\approx$ 2-lb) cube of solid gold has a length of about 3.73 cm ($\approx$ 1.5 in.) on a side. If the cube is cut in half, the two resulting pieces retain their chemical identity. But what happens if the pieces of the cube are cut again and again, indefinitely? The Greek philosophers Leucippus and Democritus couldn't accept the idea that such cutting could go on forever. They speculated that the process ultimately would end when it produced a particle that could no longer be cut. In Greek, *atomos* means "not sliceable." From this term comes our English word *atom*, once believed to be the smallest particle of matter but since found to be a composite of more elementary particles.

The atom can be naively visualized as a miniature solar system, with a dense, positively charged nucleus occupying the position of the Sun and negatively charged electrons orbiting like planets. This model of the atom, first developed by the great Danish physicist Niels Bohr nearly a century ago, led to the understanding of certain properties of the simpler atoms such as hydrogen but failed to explain many fine details of atomic structure.

Notice the size of a hydrogen atom, listed in Table 1.1, and the size of a proton—the nucleus of a hydrogen atom—one hundred thousand times smaller. If the proton were the size of a ping-pong ball, the electron would be a tiny speck about the size of a bacterium, orbiting the proton a kilometer away! Other atoms are similarly constructed. So there is a surprising amount of empty space in ordinary matter.

After the discovery of the nucleus in the early 1900s, questions arose concerning its structure. Although the structure of the nucleus remains an area of active research even today, by the early 1930s scientists determined that two basic entities—protons and neutrons—occupy the nucleus. The *proton* is nature's most common carrier of positive charge, equal in magnitude but opposite in sign to the charge on the electron. The number of protons in a nucleus determines what the element is. For instance, a nucleus containing only one proton is the nucleus of an atom of hydrogen, regardless of how many neutrons may be present. Extra neutrons correspond to different isotopes of hydrogen—deuterium and tritium—which react chemically in exactly the same way as hydrogen, but are more massive. An atom having two protons in its nucleus, similarly, is always helium, although again, differing numbers of neutrons are possible.

The existence of *neutrons* was verified conclusively in 1932. A neutron has no charge and has a mass about equal to that of a proton. Except for hydrogen, all atomic nuclei contain neutrons, which, together with the protons, interact through the strong nuclear force. That force opposes the strongly repulsive electrical force of the protons, which otherwise would cause the nucleus to disintegrate.

The division doesn't stop here; strong evidence collected over many years indicates that protons, neutrons, and a zoo of other exotic particles are composed of six particles called **quarks** (rhymes with "sharks" though some rhyme it with "forks"). These particles have been given the names *up, down, strange, charm, bottom*, and *top*. The up, charm, and top quarks each carry a charge equal to $+\frac{2}{3}$ that of the proton, whereas the down, strange, and bottom quarks each carry a charge equal to $-\frac{1}{3}$ the proton charge. The proton consists of two up quarks and one down quark (see Fig. 1.2), giving the correct charge for the proton, +1. The neutron is composed of two down quarks and one up quark and has a net charge of zero.

The up and down quarks are sufficient to describe all normal matter, so the existence of the other four quarks, indirectly observed in high-energy experiments, is something of a mystery. Despite strong indirect evidence, no isolated quark has ever been observed. Consequently, the possible existence of yet more fundamental particles remains purely speculative.

1.3 Dimensional Analysis

LEARNING OBJECTIVES

1. State the definition of a dimension and give examples of the dimensions of some basic physical quantities.
2. Use dimensions to check equations for consistency.
3. Use dimensions to derive relationships between physical quantities.

In physics the word *dimension* denotes the physical nature of a quantity. The distance between two points, for example, can be measured in feet, meters, or furlongs, which are different ways of expressing the dimension of *length*.

The symbols used in this section to specify the dimensions of length, mass, and time are L, M, and T, respectively. Brackets [ ] will often be used to denote the dimensions of a physical quantity. In this notation, for example, the dimensions of velocity v are written $[v] = L/T$, and the dimensions of area A are $[A] = L^2$. The dimensions of area, volume, velocity, and acceleration are listed in Table 1.5, along with their units in the three common systems. The dimensions of other quantities, such as force and energy, will be described later as they are introduced.

In physics it's often necessary to deal with mathematical expressions that relate different physical quantities. One way to analyze such expressions, called **dimensional analysis,** makes use of the fact that **dimensions can be treated as algebraic quantities.** Adding masses to lengths, for example, makes no sense, so it follows that quantities can be added or subtracted only if they have the same dimensions. If the terms on the opposite sides of an equation have the same dimensions, then that equation may be correct, although correctness can't be guaranteed on the basis of dimensions alone. Nonetheless, dimensional analysis has value as a partial check of an equation and can also be used to develop insight into the relationships between physical quantities.

The procedure can be illustrated by developing some relationships between acceleration, velocity, time, and distance. Distance x has the dimension of length: $[x] = L$. Time t has dimension $[t] = T$. Velocity v has the dimensions length over

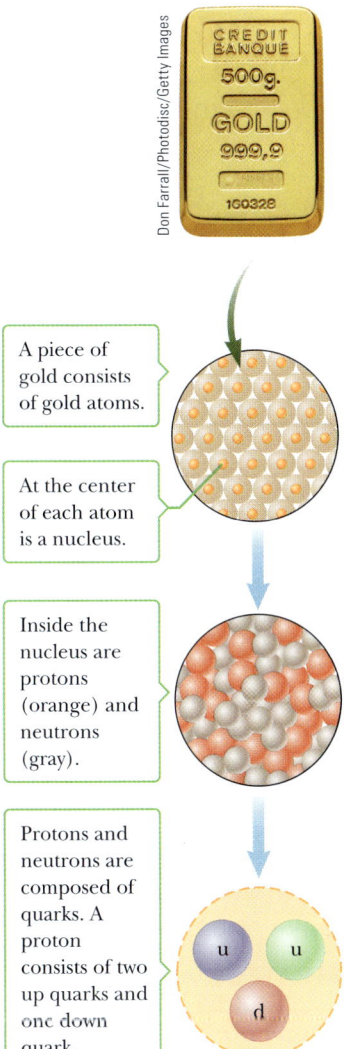

A piece of gold consists of gold atoms.

At the center of each atom is a nucleus.

Inside the nucleus are protons (orange) and neutrons (gray).

Protons and neutrons are composed of quarks. A proton consists of two up quarks and one down quark.

Figure 1.2 Levels of organization in matter.

Table 1.5 Dimensions and Some Units of Area, Volume, Velocity, and Acceleration

System	Area (L^2)	Volume (L^3)	Velocity (L/T)	Acceleration (L/T^2)
SI	m^2	m^3	m/s	m/s^2
cgs	cm^2	cm^3	cm/s	cm/s^2
U.S. customary	ft^2	ft^3	ft/s	ft/s^2

6 **CHAPTER 1** | Introduction

time: $[v] = L/T$, and acceleration the dimensions length divided by time squared: $[a] = L/T^2$. Notice that velocity and acceleration have similar dimensions, except for an extra dimension of time in the denominator of acceleration. It follows that

$$[v] = \frac{L}{T} = \frac{L}{T^2}T = [a][t]$$

From this it might be guessed that velocity equals acceleration multiplied by time, $v = at$, and that is true for the special case of motion with constant acceleration starting at rest. Noticing that velocity has dimensions of length divided by time and distance has dimensions of length, it's reasonable to guess that

$$[x] = L = L\frac{T}{T} = \frac{L}{T}T = [v][t] = [a][t]^2$$

Here it appears that $x = at^2$ might correctly relate the distance traveled to acceleration and time; however, that equation is not even correct in the case of constant acceleration starting from rest. The correct expression in that case is $x = \frac{1}{2}at^2$. These examples serve to show the inherent limitations in using dimensional analysis to discover relationships between physical quantities. Nonetheless, such simple procedures can still be of value in developing a preliminary mathematical model for a given physical system. Further, because it's easy to make errors when solving problems, dimensional analysis can be used to check the consistency of the results. When the dimensions in an equation are not consistent, it indicates an error has been made in a prior step.

■ EXAMPLE 1.1 | Analysis of an Equation

GOAL Check an equation using dimensional analysis.

PROBLEM Show that the expression $v = v_0 + at$ is dimensionally correct, where v and v_0 represent velocities, a is acceleration, and t is a time interval.

STRATEGY Analyze each term, finding its dimensions, and then check to see if all the terms agree with each other.

SOLUTION
Find dimensions for v and v_0.

$$[v] = [v_0] = \frac{L}{T}$$

Find the dimensions of at.

$$[at] = [a][t] = \frac{L}{T^2}(T) = \frac{L}{T}$$

REMARKS All the terms agree, so the equation is dimensionally correct.

QUESTION 1.1 True or False. An equation that is dimensionally correct is always physically correct, up to a constant of proportionality.

EXERCISE 1.1 Determine whether the equation $x = vt^2$ is dimensionally correct. If not, provide a correct expression, up to an overall constant of proportionality.

ANSWER Incorrect. The expression $x = vt$ is dimensionally correct.

■ EXAMPLE 1.2 | Find an Equation

GOAL Derive an equation by using dimensional analysis.

PROBLEM Find a relationship between an acceleration of constant magnitude a, speed v, and distance r from the origin for a particle traveling in a circle.

STRATEGY Start with the term having the most dimensionality, a. Find its dimensions, and then rewrite those dimensions in terms of the dimensions of v and r. The dimensions of time will have to be eliminated with v, because that's the only quantity (other than a, itself) in which the dimension of time appears.

SOLUTION

Write down the dimensions of a: $\quad [a] = \dfrac{L}{T^2}$

Solve the dimensions of speed for T: $\quad [v] = \dfrac{L}{T} \quad \rightarrow \quad T = \dfrac{L}{[v]}$

Substitute the expression for T into the equation for $[a]$: $\quad [a] = \dfrac{L}{T^2} = \dfrac{L}{(L/[v])^2} = \dfrac{[v]^2}{L}$

Substitute $L = [r]$, and guess at the equation: $\quad [a] = \dfrac{[v]^2}{[r]} \quad \rightarrow \quad \boxed{a = \dfrac{v^2}{r}}$

REMARKS This is the correct equation for the magnitude of the centripetal acceleration—acceleration toward the center of motion—to be discussed in Chapter 7. In this case it isn't necessary to introduce a numerical factor. Such a factor is often displayed explicitly as a constant k in front of the right-hand side; for example, $a = kv^2/r$. As it turns out, $k = 1$ gives the correct expression. A good technique sometimes introduced in calculus-based textbooks involves using unknown powers of the dimensions. This problem would then be set up as $[a] = [v]^b[r]^c$. Writing out the dimensions and equating powers of each dimension on both sides of the equation would result in $b = 2$ and $c = -1$.

QUESTION 1.2 True or False: Replacing v by r/t in the final answer also gives a dimensionally correct equation.

EXERCISE 1.2 In physics, energy E carries dimensions of mass times length squared divided by time squared. Use dimensional analysis to derive a relationship for energy in terms of mass m and speed v, up to a constant of proportionality. Set the speed equal to c, the speed of light, and the constant of proportionality equal to 1 to get the most famous equation in physics. (Note, however, that the first relationship is associated with energy of motion, and the second with energy of mass. See Chapter 26.)

ANSWER $E = kmv^2 \quad \rightarrow \quad E = mc^2$ when $k = 1$ and $v = c$.

1.4 Uncertainty in Measurement and Significant Figures

LEARNING OBJECTIVES

1. Identify the number of significant figures in a given physical measurement.
2. Apply significant figures to estimate the proper accuracy of a combination of physical measurements.

Physics is a science in which mathematical laws are tested by experiment. No physical quantity can be determined with complete accuracy because our senses are physically limited, even when extended with microscopes, cyclotrons, and other instruments. Consequently, it's important to develop methods of determining the accuracy of measurements.

All measurements have uncertainties associated with them, whether or not they are explicitly stated. The accuracy of a measurement depends on the sensitivity of the apparatus, the skill of the person carrying out the measurement, and the number of times the measurement is repeated. Once the measurements, along with their uncertainties, are known, it's often the case that calculations must be carried out using those measurements. Suppose two such measurements are multiplied. When a calculator is used to obtain this product, there may be eight digits in the calculator window, but often only two or three of those numbers have any significance. The rest have no value because they imply greater accuracy than was actually achieved in the original measurements. In experimental work, determining how many numbers to retain requires the application of statistics and the mathematical propagation of uncertainties. In a textbook it isn't practical to apply those sophisticated tools in the numerous calculations, so instead a simple method, called *significant figures*, is used to indicate the approximate number of digits that should be retained at the end of a calculation. Although that method is not mathematically rigorous, it's easy to apply and works fairly well.

Suppose that in a laboratory experiment we measure the area of a rectangular plate with a meter stick. Let's assume that the accuracy to which we can measure a particular dimension of the plate is ±0.1 cm. If the length of the plate is measured to be 16.3 cm, we can only claim that it lies somewhere between 16.2 cm and 16.4 cm. In this case, we say the measured value has three significant figures. Likewise, if the plate's width is measured to be 4.5 cm, the actual value lies between 4.4 cm and 4.6 cm. This measured value has only two significant figures. We could write the measured values as 16.3 ± 0.1 cm and 4.5 ± 0.1 cm. In general, **a significant figure is a reliably known digit** (other than a zero used to locate a decimal point). Note that in each case, the final number has some uncertainty associated with it, and is therefore not 100% reliable. Despite the uncertainty, that number is retained and considered significant because it does convey some information.

Suppose we would like to find the area of the plate by multiplying the two measured values together. The final value can range between (16.3 − 0.1 cm)(4.5 − 0.1 cm) = (16.2 cm)(4.4 cm) = 71.28 cm^2 and (16.3 + 0.1 cm)(4.5 + 0.1 cm) = (16.4 cm)(4.6 cm) = 75.44 cm^2. Claiming to know anything about the hundredths place, or even the tenths place, doesn't make any sense, because it's clear we can't even be certain of the units place, whether it's the 1 in 71, the 5 in 75, or somewhere in between. The tenths and the hundredths places are clearly not significant. We have *some* information about the units place, so that number is significant. Multiplying the numbers at the middle of the uncertainty ranges gives (16.3 cm)(4.5 cm) = 73.35 cm^2, which is also in the middle of the area's uncertainty range. Because the hundredths and tenths are not significant, we drop them and take the answer to be 73 cm^2, with an uncertainty of ±2 cm^2. Note that the answer has two significant figures, the same number of figures as the least accurately known quantity being multiplied, the 4.5-cm width.

Calculations as carried out in the preceding paragraph can indicate the proper number of significant figures, but those calculations are time-consuming. Instead, two rules of thumb can be applied. The first, concerning multiplication and division, is as follows: **In multiplying (dividing) two or more quantities, the number of significant figures in the final product (quotient) is the same as the number of significant figures in the *least accurate* of the factors being combined, where *least accurate* means *having the lowest number of significant figures.***

To get the final number of significant figures, it's usually necessary to do some rounding. If the last digit dropped is less than 5, simply drop the digit. If the last digit dropped is greater than or equal to 5, raise the last retained digit by one.[1]

Zeros may or may not be significant figures. Zeros used to position the decimal point in such numbers as 0.03 and 0.007 5 are not considered significant figures. Hence, 0.03 has one significant figure, and 0.007 5 has two.

When zeros are placed after other digits in a whole number, there is a possibility of misinterpretation. For example, suppose the mass of an object is given as 1 500 g. This value is ambiguous, because we don't know whether the last two zeros are being used to locate the decimal point or whether they represent significant figures in the measurement.

Using scientific notation to indicate the number of significant figures removes this ambiguity. In this case, we express the mass as 1.5×10^3 g if there are two significant figures in the measured value, 1.50×10^3 g if there are three significant figures, and 1.500×10^3 g if there are four. Likewise, 0.000 15 is expressed in scientific notation as 1.5×10^{-4} if it has two significant figures or as 1.50×10^{-4} if it

> **Tip 1.2** Using Calculators
> Calculators are designed by engineers to yield as many digits as the memory of the calculator chip permits, so be sure to round the final answer down to the correct number of significant figures.

---

[1] Some prefer to round to the nearest even digit when the last dropped digit is 5, which has the advantage of rounding 5 up half the time and down half the time. For example, 1.55 would round to 1.6, but 1.45 would round to 1.4. Because the final significant figure is only one representative of a range of values given by the uncertainty, this very slight refinement will not be used in this text.

three significant figures. The three zeros between the decimal point and the digit 1 in the number 0.000 15 are not counted as significant figures because they only locate the decimal point. Similarly, trailing zeros are not considered significant. However, any zeros written after a decimal point are considered significant. For example, 3.00, 30.0, and 300. have three significant figures, whereas 300 has only one. In this book, **most of the numerical examples and end-of-chapter problems will yield answers having two or three significant figures.**

For addition and subtraction, it's best to focus on the number of decimal places in the quantities involved rather than on the number of significant figures. **When numbers are added (subtracted), the number of decimal places in the result should equal the smallest number of decimal places of any term in the sum (difference).** For example, if we wish to compute 123 (zero decimal places) + 5.35 (two decimal places), the answer is 128 (zero decimal places) and not 128.35. If we compute the sum 1.000 1 (four decimal places) + 0.000 3 (four decimal places) = 1.000 4, the result has the correct number of decimal places, namely four. Observe that the rules for multiplying significant figures don't work here because the answer has five significant figures even though one of the terms in the sum, 0.000 3, has only one significant figure. Likewise, if we perform the subtraction 1.002 − 0.998 = 0.004, the result has three decimal places because each term in the subtraction has three decimal places.

To show why this rule should hold, we return to the first example in which we added 123 and 5.35, and rewrite these numbers as 123.*xxx* and 5.35*x*. Digits written with an *x* are completely unknown and can be any digit from 0 to 9. Now we line up 123.*xxx* and 5.35*x* relative to the decimal point and perform the addition, using the rule that an unknown digit added to a known or unknown digit yields an unknown:

$$
\begin{array}{r}
123.xxx \\
+5.35x \\
\hline
128.xxx
\end{array}
$$

The answer of 128.*xxx* means that we are justified only in keeping the number 128 because everything after the decimal point in the sum is actually unknown. The example shows that the controlling uncertainty is introduced into an addition or subtraction by the term with the smallest number of decimal places.

■ EXAMPLE 1.3 | Carpet Calculations

GOAL Apply the rules for significant figures.

PROBLEM Several carpet installers make measurements for carpet installation in the different rooms of a restaurant, reporting their measurements with inconsistent accuracy, as compiled in Table 1.6. Compute the areas for **(a)** the banquet hall, **(b)** the meeting room, and **(c)** the dining room, taking into account significant figures. **(d)** What total area of carpet is required for these rooms?

Table 1.6 Dimensions of Rooms in Example 1.3

	Length (m)	Width (m)
Banquet hall	14.71	7.46
Meeting room	4.822	5.1
Dining room	13.8	9

STRATEGY For the multiplication problems in parts (a)–(c), count the significant figures in each number. The smaller result is the number of significant figures in the answer. Part (d) requires a sum, where the area with the least accurately known decimal place determines the overall number of significant figures in the answer.

SOLUTION

(a) Compute the area of the banquet hall.

Count significant figures:

$\quad$ 14.71 m $\rightarrow$ 4 significant figures

$\quad$ 7.46 m $\rightarrow$ 3 significant figures

To find the area, multiply the numbers keeping only three digits:

$\quad$ 14.71 m × 7.46 m = 109.74 m² $\rightarrow$ 1.10×10^2 m²

(*Continued*)

(b) Compute the area of the meeting room.

Count significant figures:

$4.822 \text{ m} \rightarrow 4$ significant figures

$5.1 \text{ m} \rightarrow 2$ significant figures

To find the area, multiply the numbers keeping only two digits:

$4.822 \text{ m} \times 5.1 \text{ m} = 24.59 \text{ m}^2 \rightarrow \boxed{25 \text{ m}^2}$

(c) Compute the area of the dining room.

Count significant figures:

$13.8 \text{ m} \rightarrow 3$ significant figures

$9 \text{ m} \rightarrow 1$ significant figure

To find the area, multiply the numbers keeping only one digit:

$13.8 \text{ m} \times 9 \text{ m} = 124.2 \text{ m}^2 \rightarrow \boxed{100 \text{ m}^2}$

(d) Calculate the total area of carpet required, with the proper number of significant figures.

Sum all three answers without regard to significant figures:

$1.10 \times 10^2 \text{ m}^2 + 25 \text{ m}^2 + 100 \text{ m}^2 = 235 \text{ m}^2$

The least accurate number is 100 m², with one significant figure in the hundred's decimal place:

$235 \text{ m}^2 \rightarrow \boxed{2 \times 10^2 \text{ m}^2}$

REMARKS Notice that the final answer in part (d) has only one significant figure, in the hundred's place, resulting in an answer that had to be rounded down by a sizable fraction of its total value. That's the consequence of having insufficient information. The value of 9 m, without any further information, represents a true value that could be anywhere in the interval [8.5 m, 9.5 m), all of which round to 9 when only one digit is retained.

QUESTION 1.3 How would the final answer change if the width of the dining room were given as 9.0 m?

EXERCISE 1.3 A ranch has two fenced rectangular areas. Area A has a length of 750 m and width 125 m, and area B has length 400 m and width 150 m. Find (a) area A, (b) area B, and (c) the total area, with attention to the rules of significant figures. Assume trailing zeros are not significant.

ANSWERS (a) $9.4 \times 10^4 \text{ m}^2$ (b) $6 \times 10^4 \text{ m}^2$ (c) $1.5 \times 10^5 \text{ m}^2$

In performing any calculation, especially one involving a number of steps, there will always be slight discrepancies introduced by both the rounding process and the algebraic order in which steps are carried out. For example, consider $2.35 \times 5.89/1.57$. This computation can be performed in three different orders. First, we have $2.35 \times 5.89 = 13.842$, which rounds to 13.8, followed by $13.8/1.57 = 8.789\,8$, rounding to 8.79. Second, $5.89/1.57 = 3.751\,6$, which rounds to 3.75, resulting in $2.35 \times 3.75 = 8.812\,5$, rounding to 8.81. Finally, $2.35/1.57 = 1.496\,8$ rounds to 1.50, and $1.50 \times 5.89 = 8.835$ rounds to 8.84. So three different algebraic orders, following the rules of rounding, lead to answers of 8.79, 8.81, and 8.84, respectively. Such minor discrepancies are to be expected, because the last significant digit is only one representative from a range of possible values, depending on experimental uncertainty. To avoid such discrepancies, some carry one or more extra digits during the calculation, although it isn't conceptually consistent to do so because those extra digits are not significant. As a practical matter, in the worked examples in this text, intermediate reported results will be rounded to the proper number of significant figures, and only those digits will be carried forward. In the problem sets, however, given data will usually be assumed accurate to two or three digits, even when there are trailing zeros. **In solving the problems, the student should be aware that slight differences in rounding practices can result in answers varying from the text in the last significant digit, which is normal and not cause for**

concern. The method of significant figures has its limitations in determining accuracy, but it's easy to apply. In experimental work, however, statistics and the mathematical propagation of uncertainty must be used to determine the accuracy of an experimental result.

1.5 Conversion of Units

LEARNING OBJECTIVE
1. Convert physical quantities from one system of units to another.

Sometimes it's necessary to convert units from one system to another. Conversion factors between the SI and U.S. customary systems for units of length are as follows:

$$1 \text{ mi} = 1\,609 \text{ m} = 1.609 \text{ km} \qquad 1 \text{ ft} = 0.304\,8 \text{ m} = 30.48 \text{ cm}$$
$$1 \text{ m} = 39.37 \text{ in.} = 3.281 \text{ ft} \qquad 1 \text{ in.} = 0.025\,4 \text{ m} = 2.54 \text{ cm}$$

A more extensive list of conversion factors can be found on the front endsheets of this book. In all the given conversion equations, the "1" on the left is assumed to have the same number of significant figures as the quantity given on the right of the equation.

Units can be treated as algebraic quantities that can "cancel" each other. We can make a fraction with the conversion that will cancel the units we don't want, and multiply that fraction by the quantity in question. For example, suppose we want to convert 15.0 in. to centimeters. Because 1 in. = 2.54 cm, we find that

$$15.0 \text{ in.} = 15.0 \text{ in.} \times \left(\frac{2.54 \text{ cm}}{1.00 \text{ in.}}\right) = 38.1 \text{ cm}$$

The next two examples show how to deal with problems involving more than one conversion and with powers.

The speed limit is given in both kilometers per hour and miles per hour on this road sign. How accurate is the conversion?

EXAMPLE 1.4 | Pull Over, Buddy!

GOAL Convert units using several conversion factors.

PROBLEM If a car is traveling at a speed of 28.0 m/s, is the driver exceeding the speed limit of 55.0 mi/h?

STRATEGY Meters must be converted to miles and seconds to hours, using the conversion factors listed on the front endsheets of the book. Here, three factors will be used.

SOLUTION

Convert meters to miles:

$$28.0 \text{ m/s} = \left(28.0 \frac{\text{m}}{\text{s}}\right)\left(\frac{1.00 \text{ mi}}{1\,609 \text{ m}}\right) = 1.74 \times 10^{-2} \text{ mi/s}$$

Convert seconds to hours:

$$1.74 \times 10^{-2} \text{ mi/s} = \left(1.74 \times 10^{-2} \frac{\text{mi}}{\text{s}}\right)\left(60.0 \frac{\text{s}}{\text{min}}\right)\left(60.0 \frac{\text{min}}{\text{h}}\right)$$
$$= \boxed{62.6 \text{ mi/h}}$$

REMARKS The driver should slow down because he's exceeding the speed limit.

QUESTION 1.4 Repeat the conversion, using the relationship 1.00 m/s = 2.24 mi/h. Why is the answer slightly different?

EXERCISE 1.4 Convert 152 mi/h to m/s.

ANSWER 67.9 m/s

EXAMPLE 1.5 Press the Pedal to the Metal

GOAL Convert a quantity featuring powers of a unit.

PROBLEM The traffic light turns green, and the driver of a high-performance car slams the accelerator to the floor. The accelerometer registers 22.0 m/s². Convert this reading to km/min².

STRATEGY Here we need one factor to convert meters to kilometers and another two factors to convert seconds squared to minutes squared.

SOLUTION
Multiply by the three factors:

$$\frac{22.0 \text{ m}}{1.00 \text{ s}^2}\left(\frac{1.00 \text{ km}}{1.00 \times 10^3 \text{ m}}\right)\left(\frac{60.0 \text{ s}}{1.00 \text{ min}}\right)^2 = \boxed{79.2 \frac{\text{km}}{\text{min}^2}}$$

REMARKS Notice that in each conversion factor the numerator equals the denominator when units are taken into account. A common error in dealing with squares is to square the units inside the parentheses while forgetting to square the numbers!

QUESTION 1.5 What time conversion factor or factors would be used to further convert the answer to km/h²?

EXERCISE 1.5 Convert 4.50×10^3 kg/m³ to g/cm³.

ANSWER 4.50 g/cm³

1.6 Estimates and Order-of-Magnitude Calculations

LEARNING OBJECTIVE
1. Create estimates for physical quantities using approximations and educated guesses.

Getting an exact answer to a calculation may often be difficult or impossible, either for mathematical reasons or because limited information is available. In these cases, estimates can yield useful approximate answers that can determine whether a more precise calculation is necessary. Estimates also serve as a partial check if the exact calculations are actually carried out. If a large answer is expected but a small exact answer is obtained, there's an error somewhere.

For many problems, knowing the approximate value of a quantity—within a factor of 10 or so—is sufficient. This approximate value is called an **order-of-magnitude** estimate, and requires finding the power of 10 that is closest to the actual value of the quantity. For example, 75 kg ~ 10² kg, where the symbol ~ means "is on the order of" or "is approximately". Increasing a quantity by three orders of magnitude means that its value increases by a factor of $10^3 = 1\ 000$.

Occasionally the process of making such estimates results in fairly crude answers, but answers ten times or more too large or small are still useful. For example, suppose you're interested in how many people have contracted a certain disease. Any estimates under ten thousand are small compared with Earth's total population, but a million or more would be alarming. So even relatively imprecise information can provide valuable guidance.

In developing these estimates, you can take considerable liberties with the numbers. For example, $\pi \sim 1$, $27 \sim 10$, and $65 \sim 100$. To get a less crude estimate, it's permissible to use slightly more accurate numbers (e.g., $\pi \sim 3$, $27 \sim 30$, $65 \sim 70$). Better accuracy can also be obtained by systematically underestimating as many numbers as you overestimate. Some quantities may be completely unknown, but it's standard to make reasonable guesses, as the examples show.

EXAMPLE 1.6 Brain Cells Estimate

GOAL Develop a simple estimate.

PROBLEM Estimate the number of cells in the human brain.

STRATEGY Estimate the volume of a human brain and divide by the estimated volume of one cell. The brain is located in the upper portion of the head, with a volume that could be approximated by a cube $\ell = 20$ cm on a side. Brain cells, consisting of about 10% neurons and 90% glia, vary greatly in size, with dimensions ranging from a few microns to a meter or so. As a guess, take $d = 10$ microns as a typical dimension and consider a cell to be a cube with each side having that length.

SOLUTION

Estimate of the volume of a human brain:
$$V_{\text{brain}} = \ell^3 \approx (0.2 \text{ m})^3 = 8 \times 10^{-3} \text{ m}^3 \approx 1 \times 10^{-2} \text{ m}^3$$

Estimate the volume of a cell:
$$V_{\text{cell}} = d^3 \approx (10 \times 10^{-6} \text{ m})^3 = 1 \times 10^{-15} \text{ m}^3$$

Divide the volume of a brain by the volume of a cell:
$$\text{number of cells} = \frac{V_{\text{brain}}}{V_{\text{cell}}} = \frac{0.01 \text{ m}^3}{1 \times 10^{-15} \text{ m}^3} = \boxed{1 \times 10^{13} \text{ cells}}$$

REMARKS Notice how little attention was paid to obtaining precise values. Some general information about a problem is required if the estimate is to be within an order of magnitude of the actual value. Here, knowledge of the approximate dimensions of brain cells and the human brain were essential to developing the estimate.

QUESTION 1.6 Would 10^{12} cells also be a reasonable estimate? What about 10^9 cells? Explain.

EXERCISE 1.6 Estimate the total number of cells in the human body.

ANSWER 10^{14} (Answers may vary.)

EXAMPLE 1.7 Stack One-Dollar Bills to the Moon

GOAL Estimate the number of stacked objects required to reach a given height.

PROBLEM How many one-dollar bills, stacked one on top of the other, would reach the Moon?

STRATEGY The distance to the Moon is about 400 000 km. Guess at the number of dollar bills in a millimeter, and multiply the distance by this number, after converting to consistent units.

SOLUTION

We estimate that ten stacked bills form a layer of 1 mm. Convert mm to km:
$$\frac{10 \text{ bills}}{1 \text{ mm}} \left(\frac{10^3 \text{ mm}}{1 \text{ m}}\right)\left(\frac{10^3 \text{ m}}{1 \text{ km}}\right) = \frac{10^7 \text{ bills}}{1 \text{ km}}$$

Multiply this value by the approximate lunar distance:
$$\text{\# of dollar bills} \sim (4 \times 10^5 \text{ km})\left(\frac{10^7 \text{ bills}}{1 \text{ km}}\right) = \boxed{4 \times 10^{12} \text{ bills}}$$

REMARKS That's within an order of magnitude of the U.S. national debt!

QUESTION 1.7 Based on the answer, about how many stacked pennies would reach the Moon?

EXERCISE 1.7 How many pieces of cardboard, typically found at the back of a bound pad of paper, would you have to stack up to match the height of the Washington monument, about 170 m tall?

ANSWER $\sim 10^5$ (Answers may vary.)

EXAMPLE 1.8 Number of Galaxies in the Universe

GOAL Estimate a volume and a number density, and combine.

PROBLEM Given that astronomers can see about 10 billion light years into space and that there are 14 galaxies in our local group, 2 million light years from the next local group, estimate the number of galaxies in the observable universe. (Note: One light year is the distance traveled by light in one year, about 9.5×10^{15} m.) (See Fig. 1.3.)

STRATEGY From the known information, we can estimate the number of galaxies per unit volume. The local group of 14 galaxies is contained in a sphere a million light years in radius, with the Andromeda group in a similar sphere, so there are about 10 galaxies within a volume of radius 1 million light years. Multiply that number density by the volume of the observable universe.

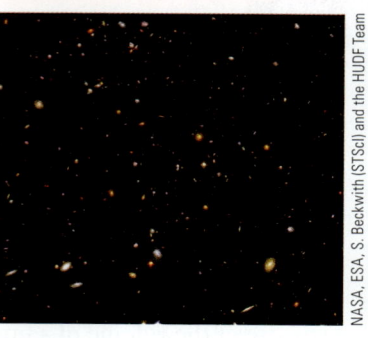

Figure 1.3 In this deep-space photograph, there are few stars—just galaxies without end.

SOLUTION

Compute the approximate volume V_{lg} of the local group of galaxies:

$$V_{lg} = \tfrac{4}{3}\pi r^3 \sim (10^6 \text{ ly})^3 = 10^{18} \text{ ly}^3$$

Estimate the density of galaxies:

$$\text{density of galaxies} = \frac{\text{\# of galaxies}}{V_{lg}}$$

$$\sim \frac{10 \text{ galaxies}}{10^{18} \text{ ly}^3} = 10^{-17} \frac{\text{galaxies}}{\text{ly}^3}$$

Compute the approximate volume of the observable universe:

$$V_u = \tfrac{4}{3}\pi r^3 \sim (10^{10} \text{ ly})^3 = 10^{30} \text{ ly}^3$$

Multiply the density of galaxies by V_u:

$$\text{\# of galaxies} \sim (\text{density of galaxies})V_u$$

$$= \left(10^{-17} \frac{\text{galaxies}}{\text{ly}^3}\right)(10^{30} \text{ ly}^3)$$

$$= \boxed{10^{13} \text{ galaxies}}$$

REMARKS Notice the approximate nature of the computation, which uses $4\pi/3 \sim 1$ on two occasions and $14 \sim 10$ for the number of galaxies in the local group. This is completely justified: Using the actual numbers would be pointless, because the other assumptions in the problem—the size of the observable universe and the idea that the local galaxy density is representative of the density everywhere—are also very rough approximations. Further, there was nothing in the problem that required using volumes of spheres rather than volumes of cubes. Despite all these arbitrary choices, the answer still gives useful information, because it rules out a lot of reasonable possible answers. Before doing the calculation, a guess of a billion galaxies might have seemed plausible.

QUESTION 1.8 About one in ten galaxies in the local group are not dwarf galaxies. Estimate the number of galaxies in the universe that are not dwarfs.

EXERCISE 1.8 (a) Given that the nearest star is about 4 light years away, develop an estimate of the density of stars per cubic light year in our galaxy. (b) Estimate the number of stars in the Milky Way galaxy, given that it's roughly a disk 100 000 light years across and a thousand light years thick.

ANSWER (a) 0.02 stars/ly^3 (b) 2×10^{11} stars (Estimates will vary. The actual answer is probably about twice that number.)

1.7 Coordinate Systems

LEARNING OBJECTIVES

1. Describe and locate points in a plane using a Cartesian coordinate system.
2. Describe and locate points in a plane using a polar coordinate system.

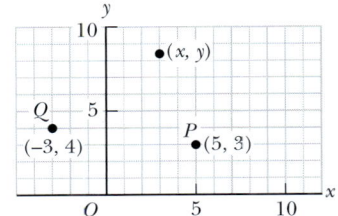

Figure 1.4 Designation of points in a two-dimensional Cartesian coordinate system. Every point is labeled with coordinates (x, y).

Many aspects of physics deal with locations in space, which require the definition of a coordinate system. A point on a line can be located with one coordinate, a point in a plane with two coordinates, and a point in space with three.

A coordinate system used to specify locations in space consists of the following:

- A fixed reference point O, called the *origin*
- A set of specified axes, or directions, with an appropriate scale and labels on the axes
- Instructions on labeling a point in space relative to the origin and axes

One convenient and commonly used coordinate system is the **Cartesian coordinate system,** sometimes called the **rectangular coordinate system.** Such a system in two dimensions is illustrated in Figure 1.4. An arbitrary point in this system is labeled with the coordinates (x, y). For example, the point P in the figure has coordinates $(5, 3)$. If we start at the origin O, we can reach P by moving 5 meters horizontally to the right and then 3 meters vertically upward. In the same way, the point Q has coordinates $(-3, 4)$, which corresponds to going 3 meters horizontally to the left of the origin and 4 meters vertically upward from there.

Positive x is usually selected as right of the origin and positive y upward from the origin, but in two dimensions this choice is largely a matter of taste. (In three dimensions, however, there are "right-handed" and "left-handed" coordinates, which lead to minus sign differences in certain operations. These will be addressed as needed.)

Sometimes it's more convenient to locate a point in space by its **plane polar coordinates** (r, θ), as in Figure 1.5. In this coordinate system, an origin O and a reference line are selected as shown. A point is then specified by the distance r from the origin to the point and by the angle θ between the reference line and a line drawn from the origin to the point. The standard reference line is usually selected to be the positive x-axis of a Cartesian coordinate system. The angle θ is considered positive when measured counterclockwise from the reference line and negative when measured clockwise. For example, if a point is specified by the polar coordinates 3 m and 60°, we locate this point by moving out 3 m from the origin at an angle of 60° above (counterclockwise from) the reference line. A point specified by polar coordinates 3 m and $-60°$ is located 3 m out from the origin and 60° below (clockwise from) the reference line.

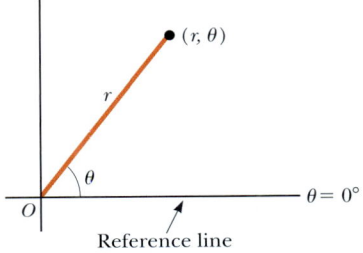

Figure 1.5 The plane polar coordinates of a point are represented by the distance r and the angle θ, where θ is measured counterclockwise from the positive x-axis.

1.8 Trigonometry

LEARNING OBJECTIVES

1. Convert between Cartesian and polar coordinates using the basic trigonometric functions and the Pythagorean theorem.
2. Apply the basic trigonometric functions and the Pythagorean theorem in simple physical contexts.

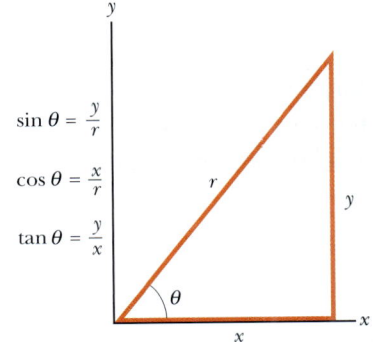

Figure 1.6 Certain trigonometric functions of a right triangle.

Consider the right triangle shown in Figure 1.6, where side y is opposite the angle θ, side x is adjacent to the angle θ, and side r is the hypotenuse of the triangle. The basic trigonometric functions defined by such a triangle are the ratios of the lengths of the sides of the triangle. These relationships are called the sine (sin), cosine (cos), and tangent (tan) functions. In terms of θ, the basic trigonometric functions are as follows[2]:

$$\sin \theta = \frac{\text{side opposite } \theta}{\text{hypotenuse}} = \frac{y}{r}$$

$$\cos \theta = \frac{\text{side adjacent to } \theta}{\text{hypotenuse}} = \frac{x}{r} \quad [1.1]$$

$$\tan \theta = \frac{\text{side opposite } \theta}{\text{side adjacent to } \theta} = \frac{y}{x}$$

For example, if the angle θ is equal to 30°, then the ratio of y to r is always 0.50; that is, $\sin 30° = 0.50$. Note that the sine, cosine, and tangent functions are quantities without units because each represents the ratio of two lengths.

Another important relationship, called the **Pythagorean theorem,** exists between the lengths of the sides of a right triangle:

$$r^2 = x^2 + y^2 \quad [1.2]$$

Finally, it will often be necessary to find the values of inverse relationships. For example, suppose you know that the sine of an angle is 0.866, but you need to know the value of the angle itself. The inverse sine function may be expressed as $\sin^{-1}(0.866)$, which is a shorthand way of asking the question "What angle has a sine of 0.866?" Punching a couple of buttons on your calculator reveals that this angle is 60.0°. Try it for yourself and show that $\tan^{-1}(0.400) = 21.8°$. Be sure that your calculator is set for degrees and not radians. In addition, the inverse tangent function can return only values between −90° and +90°, so when an angle is in the second or third quadrant, it's necessary to add 180° to the answer in the calculator window.

The definitions of the trigonometric functions and the inverse trigonometric functions, as well as the Pythagorean theorem, can be applied to *any* right triangle, regardless of whether its sides correspond to x- and y-coordinates.

These results from trigonometry are useful in converting from rectangular coordinates to polar coordinates, or vice versa, as the next example shows.

> **Tip 1.3 Degrees vs. Radians**
> When calculating trigonometric functions, make sure your calculator setting—degrees or radians—is consistent with the angular measure you're using in a given problem.

■ EXAMPLE 1.9 | Cartesian and Polar Coordinates

GOAL Understand how to convert from plane rectangular coordinates to plane polar coordinates and vice versa.

PROBLEM (a) The Cartesian coordinates of a point in the xy-plane are $(x, y) = (-3.50 \text{ m}, -2.50 \text{ m})$, as shown in Figure 1.7. Find the polar coordinates of this point. (b) Convert $(r, \theta) = (5.00 \text{ m}, 37.0°)$ to rectangular coordinates.

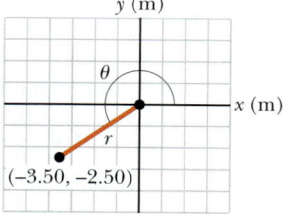

Figure 1.7 (Example 1.9) Converting from Cartesian coordinates to polar coordinates.

STRATEGY Apply the trigonometric functions and their inverses, together with the Pythagorean theorem.

SOLUTION

(a) Cartesian to Polar conversion

Take the square root of both sides of Equation 1.2 to find the radial coordinate:

$$r = \sqrt{x^2 + y^2} = \sqrt{(-3.50 \text{ m})^2 + (-2.50 \text{ m})^2} = \boxed{4.30 \text{ m}}$$

Use Equation 1.1 for the tangent function to find the angle with the inverse tangent, adding 180° because the angle is actually in third quadrant:

$$\tan \theta = \frac{y}{x} = \frac{-2.50 \text{ m}}{-3.50 \text{ m}} = 0.714$$

$$\theta = \tan^{-1}(0.714) = 35.5° + 180° = \boxed{216°}$$

---

[2]Many people use the mnemonic *SOHCAHTOA* to remember the basic trigonometric formulas: Sine = Opposite/Hypotenuse, Cosine = Adjacent/Hypotenuse, and Tangent = Opposite/Adjacent. (Thanks go to Professor Don Chodrow for pointing this out.)

(b) Polar to Cartesian conversion

Use the trigonometric definitions, Equation 1.1.

$$x = r\cos\theta = (5.00 \text{ m}) \cos 37.0° = \boxed{3.99 \text{ m}}$$

$$y = r\sin\theta = (5.00 \text{ m}) \sin 37.0° = \boxed{3.01 \text{ m}}$$

REMARKS When we take up vectors in two dimensions in Chapter 3, we will routinely use a similar process to find the direction and magnitude of a given vector from its components, or, conversely, to find the components from the vector's magnitude and direction.

QUESTION 1.9 Starting with the answers to part (b), work backwards to recover the given radius and angle. Why are there slight differences from the original quantities?

EXERCISE 1.9 (a) Find the polar coordinates corresponding to $(x, y) = (-3.25 \text{ m}, 1.50 \text{ m})$. (b) Find the Cartesian coordinates corresponding to $(r, \theta) = (4.00 \text{ m}, 53.0°)$.

ANSWERS (a) $(r, \theta) = (3.58 \text{ m}, 155°)$ (b) $(x, y) = (2.41 \text{ m}, 3.19 \text{ m})$

EXAMPLE 1.10 | How High Is the Building?

GOAL Apply basic results of trigonometry.

PROBLEM A person measures the height of a building by walking out a distance of 46.0 m from its base and shining a flashlight beam toward the top. When the beam is elevated at an angle of 39.0° with respect to the horizontal, as shown in Figure 1.8, the beam just strikes the top of the building. (a) If the flashlight is held at a height of 2.00 m, find the height of the building. (b) Calculate the length of the light beam.

STRATEGY Refer to the right triangle shown in the figure. We know the angle, 39.0°, and the length of the side adjacent to it. Because the height of the building is the side opposite the angle, we can use the tangent function. With the adjacent and opposite sides known, we can then find the hypotenuse with the Pythagorean theorem.

Figure 1.8 (Example 1.10)

SOLUTION

(a) Find the height of the building.

Use the tangent of the given angle:

$$\tan 39.0° = \frac{\Delta y}{46.0 \text{ m}}$$

Solve for the height:

$$\Delta y = (\tan 39.0°)(46.0 \text{ m}) = (0.810)(46.0 \text{ m}) = 37.3 \text{ m}$$

Add 2.00 m to Δy to obtain the height:

height = $\boxed{39.3 \text{ m}}$

(b) Calculate the length of the light beam.

Use the Pythagorean theorem:

$$r = \sqrt{x^2 + y^2} = \sqrt{(37.3 \text{ m})^2 + (46.0 \text{ m})^2} = \boxed{59.2 \text{ m}}$$

REMARKS In a later chapter, right-triangle trigonometry is often used when working with vectors.

QUESTION 1.10 Could the distance traveled by the light beam be found without using the Pythagorean theorem? How?

EXERCISE 1.10 While standing atop a building 50.0 m tall, you spot a friend standing on a street corner. Using a protractor and dangling a plumb bob, you find that the angle between the horizontal and the direction to the spot on the sidewalk where your friend is standing is 25.0°. Your eyes are located 1.75 m above the top of the building. How far away from the foot of the building is your friend?

ANSWER 111 m

1.9 Problem-Solving Strategy

LEARNING OBJECTIVE

1. Systematically organize and solve a general physics problem.

Most courses in general physics require the student to learn the skills used in solving problems, and examinations usually include problems that test such skills. This brief section presents some useful suggestions to help increase your success in solving problems. An organized approach to problem solving will also enhance your understanding of physical concepts and reduce exam stress. Throughout the book, there will be a number of sections labeled "Problem-Solving Strategy," many of them just a specializing of the list given below (and illustrated in Fig. 1.9).

General Problem-Solving Strategy

Problem
1. **Read** the problem carefully at least twice. Be sure you understand the nature of the problem before proceeding further.
2. **Draw** a diagram while rereading the problem.
3. **Label** all physical quantities in the diagram, using letters that remind you what the quantity is (e.g., m for mass). Choose a coordinate system and label it.

Strategy
4. **Identify** physical principles, the knowns and unknowns, and list them. Put circles around the unknowns. There must be as many equations as there are unknowns.
5. **Equations**, the relationships between the labeled physical quantities, should be written down next. Naturally, the selected equations should be consistent with the physical principles identified in the previous step.

Solution
6. **Solve** the set of equations for the unknown quantities in terms of the known. Do this algebraically, without substituting values until the next step, except where terms are zero.
7. **Substitute** the known values, together with their units. Obtain a numerical value with units for each unknown.

Check Answer
8. **Check** your answer. Do the units match? Is the answer reasonable? Does the plus or minus sign make sense? Is your answer consistent with an order of magnitude estimate?

Figure 1.9 A guide to problem solving.

This same procedure, with minor variations, should be followed throughout the course. The first three steps are extremely important, because they get you mentally oriented. Identifying the proper concepts and physical principles assists you in choosing the correct equations. The equations themselves are essential, because when you understand them, you also understand the relationships between the physical quantities. This understanding comes through a lot of daily practice.

Equations are the tools of physics: To solve problems, you have to have them at hand, like a plumber and his wrenches. Know the equations, and understand what they mean and how to use them. Just as you can't have a conversation without knowing the local language, you can't solve physics problems without knowing and understanding the equations. This understanding grows as you study and apply the concepts and the equations relating them.

Carrying through the algebra for as long as possible (substituting numbers only at the end) is also important, because it helps you think in terms of the physical quantities involved, not merely the numbers that represent them. Many beginning physics students are eager to substitute, but once numbers are substituted it's harder to understand relationships and easier to make mistakes.

The physical layout and organization of your work will make the final product more understandable and easier to follow. Although physics is a challenging discipline, your chances of success are excellent if you maintain a positive attitude and keep trying.

> **Tip 1.4 Get Used to Symbolic Algebra**
> Whenever possible, solve problems symbolically and then substitute known values. This process helps prevent errors and clarifies the relationships between physical quantities.

■ EXAMPLE 1.11 A Round Trip by Air

GOAL Illustrate the Problem-Solving Strategy.

PROBLEM An airplane travels $x = 4.50 \times 10^2$ km due east and then travels an unknown distance y due north. Finally, it returns to its starting point by traveling a distance of $r = 525$ km. How far did the airplane travel in the northerly direction?

STRATEGY We've finished reading the problem (step 1), and have drawn a diagram (step 2) in Figure 1.10 and labeled it (step 3). From the diagram, we recognize a right triangle and identify (step 4) the principle involved: the Pythagorean theorem. Side y is the unknown quantity, and the other sides are known.

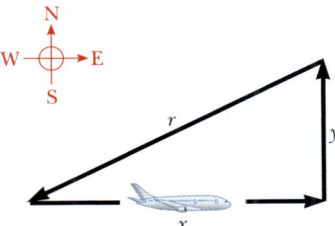

Figure 1.10 (Example 1.11)

SOLUTION

Write the Pythagorean theorem (step 5):

$$r^2 = x^2 + y^2$$

Solve symbolically for y (step 6):

$$y^2 = r^2 - x^2 \quad \rightarrow \quad y = +\sqrt{r^2 - x^2}$$

Substitute the numbers, with units (step 7):

$$y = \sqrt{(525 \text{ km})^2 - (4.50 \times 10^2 \text{ km})^2} = \boxed{2.70 \times 10^2 \text{ km}}$$

REMARKS Note that the negative solution has been disregarded, because it's not physically meaningful. In checking (step 8), note that the units are correct and that an approximate answer can be obtained by using the easier quantities, 500 km and 400 km. Doing so gives an answer of 300 km, which is approximately the same as our calculated answer of 270 km.

QUESTION 1.11 What is the answer if both the distance traveled due east and the direct return distance are both doubled?

EXERCISE 1.11 A plane flies 345 km due south, then turns and flies 615 km at a heading north of east, until it's due east of its starting point. If the plane now turns and heads for home, how far will it have to go?

ANSWER 509 km

■ SUMMARY

1.1 Standards of Length, Mass, and Time
The physical quantities in the study of mechanics can be expressed in terms of three fundamental quantities: length, mass, and time, which have the SI units meters (m), kilograms (kg), and seconds (s), respectively.

1.2 The Building Blocks of Matter
Matter is made of atoms, which in turn are made up of a relatively small nucleus of protons and neutrons within a cloud of electrons. Protons and neutrons are composed of still smaller particles, called quarks.

1.3 Dimensional Analysis
Dimensional analysis can be used to check equations and to assist in deriving them. When the dimensions on both sides of the equation agree, the equation is often correct up to a numerical factor. When the dimensions don't agree, the equation must be wrong.

1.4 Uncertainty in Measurement and Significant Figures
No physical quantity can be determined with complete accuracy. The concept of significant figures affords a basic

method of handling these uncertainties. A significant figure is a reliably known digit, other than a zero used to locate the decimal point. The two rules of significant figures are as follows:

1. When multiplying or dividing using two or more quantities, the result should have the same number of significant figures as the quantity having the fewest significant figures.
2. When quantities are added or subtracted, the number of decimal places in the result should be the same as in the quantity with the fewest decimal places.

Use of scientific notation can avoid ambiguity in significant figures. In rounding, if the last digit dropped is less than 5, simply drop the digit; otherwise, raise the last retained digit by one.

1.5 Conversion of Units

Units in physics equations must always be consistent. In solving a physics problem, it's best to start with consistent units, using the table of conversion factors on the front endsheets as necessary.

Converting units is a matter of multiplying the given quantity by a fraction, with one unit in the numerator and its equivalent in the other units in the denominator, arranged so the unwanted units in the given quantity are canceled out in favor of the desired units.

1.6 Estimates and Order-of-Magnitude Calculations

Sometimes it's useful to find an approximate answer to a question, either because the math is difficult or because information is incomplete. A quick estimate can also be used to check a more detailed calculation. In an order-of-magnitude calculation, each value is replaced by the closest power of ten, which sometimes must be guessed or estimated when the value is unknown. The computation is then carried out. For quick estimates involving known values, each value can first be rounded to one significant figure.

1.7 Coordinate Systems

The Cartesian coordinate system consists of two perpendicular axes, usually called the x-axis and y-axis, with each axis labeled with all numbers from negative infinity to positive infinity. Points are located by specifying the x- and y-values. Polar coordinates consist of a radial coordinate r, which is the distance from the origin, and an angular coordinate θ, which is the angular displacement from the positive x-axis.

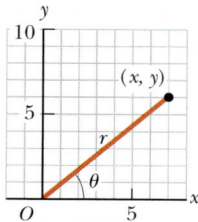

A point in the plane can be described with Cartesian coordinates (x, y) or with the polar coordinates (r, θ).

1.8 Trigonometry

The three most basic trigonometric functions of a right triangle are the sine, cosine, and tangent, defined as follows:

$$\sin \theta = \frac{\text{side opposite } \theta}{\text{hypotenuse}} = \frac{y}{r}$$

$$\cos \theta = \frac{\text{side adjacent to } \theta}{\text{hypotenuse}} = \frac{x}{r} \quad [1.1]$$

$$\tan \theta = \frac{\text{side opposite } \theta}{\text{side adjacent to } \theta} = \frac{y}{x}$$

The **Pythagorean theorem** is an important relationship between the lengths of the sides of a right triangle:

$$r^2 = x^2 + y^2 \quad [1.2]$$

where r is the hypotenuse of the triangle and x and y are the other two sides.

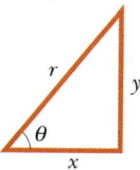

■ WARM-UP EXERCISES

WebAssign The warm-up exercises in this chapter may be assigned online in Enhanced WebAssign.

1. **Math Review** Convert the following numbers to scientific notation. (a) 568 017 (b) 0.000 309

2. **Math Review** Simplify the following expression in terms of the dimensions mass, length, and time given by [M], [L], and [T]. (See Section 1.3.)

$$\frac{[M][L]^2}{[T]^3} \cdot \frac{[T]}{[L]} \cdot [T] = ?$$

3. Simplify the following expression, combining terms as appropriate and combining and canceling units. (See Section 1.5.)

4. The Roman cubitus is an ancient unit of measure equivalent to about 0.445 m. Convert the 2.00-m height of a basketball forward to cubiti. (See Section 1.5.)

5. A house is advertised as having 1 420 square feet under roof. What is the area of this house in square meters? (See Section 1.5.)

6. A rectangular airstrip measures 32.30 m by 210 m, with the width measured more accurately than the length. Find the area, taking into account significant figures. (See Section 1.4.)

7. Use the rules for significant figures to find the answer to the addition problem 21.4 + 15 + 17.17 + 4.003. (See Section 1.4).

8. Find the polar coordinates corresponding to a point located at (−5.00, 12.00) in Cartesian coordinates. (See Section 1.7.)

9. At a horizontal distance of 45 m from the bottom of a tree, the angle of elevation to the top of the tree is 26°. How tall is the tree? (See Section 1.8.)

CONCEPTUAL QUESTIONS

The conceptual questions in this chapter may be assigned online in Enhanced WebAssign.

1. Estimate the order of magnitude of the length, in meters, of each of the following: (a) a mouse, (b) a pool cue, (c) a basketball court, (d) an elephant, (e) a city block.

2. What types of natural phenomena could serve as time standards?

3. Find the order of magnitude of your age in seconds.

4. An object with a mass of 1 kg weighs approximately 2 lb. Use this information to estimate the mass of the following objects: (a) a baseball; (b) your physics textbook; (c) a pickup truck.

5. **BIO** (a) Estimate the number of times your heart beats in a month. (b) Estimate the number of human heartbeats in an average lifetime.

6. Estimate the number of atoms in 1 cm³ of a solid. (Note that the diameter of an atom is about 10^{-10} m.)

7. The height of a horse is sometimes given in units of "hands." Why is this a poor standard of length?

8. How many of the lengths or time intervals given in Tables 1.2 and 1.3 could you verify, using only equipment found in a typical dormitory room?

9. (a) If an equation is dimensionally correct, does this mean that the equation must be true? (b) If an equation is not dimensionally correct, does this mean that the equation can't be true? Explain your answers.

10. Why is the metric system of units considered superior to most other systems of units?

11. How can an estimate be of value even when it is off by an order of magnitude? Explain and give an example.

12. Suppose two quantities, A and B, have different dimensions. Determine which of the following arithmetic operations *could* be physically meaningful. (a) $A + B$ (b) $B - A$ (c) $A - B$ (d) A/B (e) AB

13. Answer each question yes or no. Must two quantities have the same dimensions (a) if you are adding them? (b) If you are multiplying them? (c) If you are subtracting them? (d) If you are dividing them? (e) If you are equating them?

PROBLEMS

The problems in this chapter may be assigned online in Enhanced WebAssign.

1. denotes straightforward problem; 2. denotes intermediate problem;
3. denotes challenging problem
1. denotes full solution available in *Student Solutions Manual/Study Guide*
1. denotes problems most often assigned in Enhanced WebAssign

BIO denotes biomedical problems
GP denotes guided problems
M denotes Master It tutorial available in Enhanced WebAssign
Q|C denotes asking for quantitative and conceptual reasoning
S denotes symbolic reasoning problem
W denotes Watch It video solution available in Enhanced WebAssign

1.3 Dimensional Analysis

1. The period of a simple pendulum, defined as the time necessary for one complete oscillation, is measured in time units and is given by

$$T = 2\pi\sqrt{\frac{\ell}{g}}$$

where ℓ is the length of the pendulum and g is the acceleration due to gravity, in units of length divided by time squared. Show that this equation is dimensionally consistent. (You might want to check the formula using your keys at the end of a string and a stopwatch.)

2. (a) Suppose that the displacement of an object is related to time according to the expression $x = Bt^2$. What are the dimensions of B? (b) A displacement is related to time as $x = A \sin(2\pi ft)$, where A and f are constants. Find the dimensions of A. (*Hint:* A trigonometric function appearing in an equation must be dimensionless.)

3. **S** A shape that covers an area A and has a uniform height h has a volume $V = Ah$. (a) Show that $V = Ah$ is dimensionally correct. (b) Show that the volumes of a cylinder and of a rectangular box can be written in the form $V = Ah$, identifying A in each case.

(Note that A, sometimes called the "footprint" of the object, can have any shape and that the height can, in general, be replaced by the average thickness of the object.)

4. Each of the following equations was given by a student during an examination: (a) $\frac{1}{2}mv^2 = \frac{1}{2}mv_0^2 + \sqrt{mgh}$ (b) $v = v_0 + at^2$ (c) $ma = v^2$. Do a dimensional analysis of each equation and explain why the equation can't be correct.

5. Newton's law of universal gravitation is represented by

$$F = G\frac{Mm}{r^2}$$

where F is the gravitational force, M and m are masses, and r is a length. Force has the SI units kg · m/s². What are the SI units of the proportionality constant G?

6. **Q|C** Kinetic energy KE (Chapter 5) has dimensions kg · m²/s². It can be written in terms of the momentum p (Chapter 6) and mass m as

$$KE = \frac{p^2}{2m}$$

(a) Determine the proper units for momentum using dimensional analysis. (b) Refer to Problem 5. Given the units of force, write a simple equation relating a constant force F exerted on an object, an interval of time t during which the force is applied, and the resulting momentum of the object, p.

1.4 Uncertainty in Measurement and Significant Figures

7. **W** A carpet is to be installed in a room of length 9.72 m and width 5.3 m. Find the area of the room retaining the proper number of significant figures.

8. **Q|C** Use your calculator to determine $(\sqrt{8})^3$ to three significant figures in two ways: (a) Find $\sqrt{8}$ to four significant figures; then cube this number and round to three significant figures. (b) Find $\sqrt{8}$ to three significant figures; then cube this number and round to three significant figures. (c) Which answer is more accurate? Explain.

9. How many significant figures are there in (a) 78.9 ± 0.2, (b) 3.788×10^9, (c) 2.46×10^{-6}, (d) 0.003 2

10. The speed of light is now defined to be $2.997\,924\,58 \times 10^8$ m/s. Express the speed of light to (a) three significant figures, (b) five significant figures, and (c) seven significant figures.

11. **Q|C** A block of gold has length 5.62 cm, width 6.35 cm, and height 2.78 cm. (a) Calculate the length times the width and round the answer to the appropriate number of significant figures. (b) Now multiply the rounded result of part (a) by the height and again round, obtaining the volume. (c) Repeat the process, first finding the width times the height, rounding it, and then obtaining the volume by multiplying by the length. (d) Explain why the answers don't agree in the third significant figure.

12. The radius of a circle is measured to be (10.5 ± 0.2) m. Calculate (a) the area and (b) the circumference of the circle, and give the uncertainty in each value.

13. The edges of a shoebox are measured to be 11.4 cm, 17.8 cm, and 29 cm. Determine the volume of the box retaining the proper number of significant figures in your answer.

14. Carry out the following arithmetic operations: (a) the sum of the measured values 756, 37.2, 0.83, and 2.5; (b) the product $0.003\,2 \times 356.3$; (c) the product $5.620 \times \pi$.

1.5 Conversion of Units

15. A fathom is a unit of length, usually reserved for measuring the depth of water. A fathom is approximately 6 ft in length. Take the distance from Earth to the Moon to be 250 000 miles, and use the given approximation to find the distance in fathoms.

16. A small turtle moves at a speed of 186 furlongs per fortnight. Find the speed of the turtle in centimeters per second. Note that 1 furlong = 220 yards and 1 fortnight = 14 days.

17. A firkin is an old British unit of volume equal to 9 gallons. How many cubic meters are there in 6.00 firkins?

18. Find the height or length of these natural wonders in kilometers, meters, and centimeters: (a) The longest cave system in the world is the Mammoth Cave system in Central Kentucky, with a mapped length of 348 miles. (b) In the United States, the waterfall with the greatest single drop is Ribbon Falls in California, which drops 1 612 ft. (c) At 20 320 feet, Mount McKinley in Alaska is America's highest mountain. (d) The deepest canyon in the United States is King's Canyon in California, with a depth of 8 200 ft.

19. A car is traveling at a speed of 38.0 m/s on an interstate highway where the speed limit is 75.0 mi/h. Is the driver exceeding the speed limit? Justify your answer.

20. A certain car has a fuel efficiency of 25.0 miles per gallon (mi/gal). Express this efficiency in kilometers per liter (km/L).

21. The diameter of a sphere is measured to be 5.36 in. Find (a) the radius of the sphere in centimeters, (b) the surface area of the sphere in square centimeters, and (c) the volume of the sphere in cubic centimeters.

22. **W** **BIO** Suppose your hair grows at the rate of 1/32 inch per day. Find the rate at which it grows in nanometers per second. Because the distance between atoms in a molecule is on the order of 0.1 nm, your answer suggests how rapidly atoms are assembled in this protein synthesis.

23. The speed of light is about 3.00×10^8 m/s. Convert this figure to miles per hour.

24. **M** A house is 50.0 ft long and 26 ft wide and has 8.0-ft-high ceilings. What is the volume of the interior of the house in cubic meters and in cubic centimeters?

25. The amount of water in reservoirs is often measured in acre-ft. One acre-ft is a volume that covers an area of one acre to a depth of one foot. An acre is 43 560 ft². Find the volume in SI units of a reservoir containing 25.0 acre-ft of water.

26. The base of a pyramid covers an area of 13.0 acres (1 acre = 43 560 ft²) and has a height of 481 ft (Fig. P1.26). If the volume of a pyramid is given by the expression $V = bh/3$, where b is the area of the base and h is the height, find the volume of this pyramid in cubic meters.

Figure P1.26

27. A quart container of ice cream is to be made in the form of a cube. What should be the length of a side, in centimeters? (Use the conversion 1 gallon = 3.786 liter.)

1.6 Estimates and Order-of-Magnitude Calculations

Note: In developing answers to the problems in this section, you should state your important assumptions, including the numerical values assigned to parameters used in the solution.

28. Estimate the number of steps you would have to take to walk a distance equal to the circumference of the Earth.

29. **BIO** Estimate the number of breaths taken by a human being during an average lifetime.

30. **BIO** Estimate the number of people in the world who are suffering from the common cold on any given day. (Answers may vary. Remember that a person suffers from a cold for about a week.)

31. **BIO** **Q|C** (a) About how many microorganisms are found in the human intestinal tract? (A typical bacterial length scale is one micron = 10^{-6} m. Estimate the intestinal volume and assume bacteria occupy one hundredth of it.) (b) Discuss your answer to part (a). Are these bacteria beneficial, dangerous, or neutral? What functions could they serve?

32. **BIO** Treat a cell in a human as a sphere of radius 1.0 μm. (a) Determine the volume of a cell. (b) Estimate the volume of your body. (c) Estimate the number of cells in your body.

33. An automobile tire is rated to last for 50 000 miles. Estimate the number of revolutions the tire will make in its lifetime.

34. **BIO** **Q|C** Bacteria and other prokaryotes are found deep underground, in water, and in the air. One micron (10^{-6} m) is a typical length scale associated with these microbes. (a) Estimate the total number of bacteria and other prokaryotes in the biosphere of the Earth. (b) Estimate the total mass of all such microbes. (c) Discuss the relative importance of humans and microbes to the ecology of planet Earth. Can *Homo sapiens* survive without them?

1.7 Coordinate Systems

35. **M** A point is located in a polar coordinate system by the coordinates $r = 2.5$ m and $\theta = 35°$. Find the x- and y-coordinates of this point, assuming that the two coordinate systems have the same origin.

36. A certain corner of a room is selected as the origin of a rectangular coordinate system. If a fly is crawling on an adjacent wall at a point having coordinates (2.0, 1.0), where the units are meters, what is the distance of the fly from the corner of the room?

37. Express the location of the fly in Problem 36 in polar coordinates.

38. **W** Two points in a rectangular coordinate system have the coordinates (5.0, 3.0) and (−3.0, 4.0), where the units are centimeters. Determine the distance between these points.

39. Two points are given in polar coordinates by (r, θ) = (2.00 m, 50.0°) and (r, θ) = (5.00 m, −50.0°), respectively. What is the distance between them?

40. **S** Given points (r_1, θ_1) and (r_2, θ_2) in polar coordinates, obtain a general formula for the distance between them. Simplify it as much as possible using the identity $\cos^2 \theta + \sin^2 \theta = 1$. *Hint:* Write the expressions for the two points in Cartesian coordinates and substitute into the usual distance formula.

1.8 Trigonometry

41. **M** For the triangle shown in Figure P1.41, what are (a) the length of the unknown side, (b) the tangent of θ, and (c) the sine of ϕ?

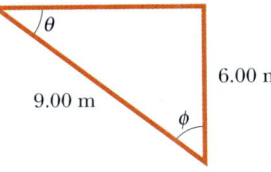

Figure P1.41

42. A ladder 9.00 m long leans against the side of a building. If the ladder is inclined at an angle of 75.0° to the horizontal, what is the horizontal distance from the bottom of the ladder to the building?

43. A high fountain of water is located at the center of a circular pool as shown in Figure P1.43. Not wishing to get his feet wet, a student walks around the pool and measures its circumference to be 15.0 m. Next, the student stands

Figure P1.43

at the edge of the pool and uses a protractor to gauge the angle of elevation at the bottom of the fountain to be 55.0°. How high is the fountain?

44. **W** A right triangle has a hypotenuse of length 3.00 m, and one of its angles is 30.0°. What are the lengths of (a) the side opposite the 30.0° angle and (b) the side adjacent to the 30.0° angle?

45. In Figure P1.45, find (a) the side opposite θ, (b) the side adjacent to ϕ, (c) $\cos \theta$, (d) $\sin \phi$, and (e) $\tan \phi$.

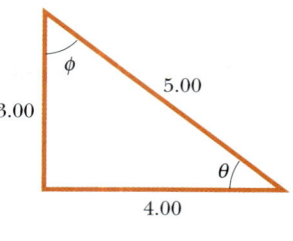

Figure P1.45

46. In a certain right triangle, the two sides that are perpendicular to each other are 5.00 m and 7.00 m long. What is the length of the third side of the triangle?

47. In Problem 46, what is the tangent of the angle for which 5.00 m is the opposite side?

48. **GP** **S** A woman measures the angle of elevation of a mountaintop as 12.0°. After walking 1.00 km closer to the mountain on level ground, she finds the angle to be 14.0°. (a) Draw a picture of the problem, neglecting the height of the woman's eyes above the ground. *Hint:* Use two triangles. (b) Select variable names for the mountain height (suggestion: y) and the woman's original distance from the mountain (suggestion: x) and label the picture. (c) Using the labeled picture and the tangent function, write two trigonometric equations relating the two selected variables. (d) Find the height y of the mountain by first solving one equation for x and substituting the result into the other equation.

49. A surveyor measures the distance across a straight river by the following method: Starting directly across from a tree on the opposite bank, he walks $x = 100$ m along the riverbank to establish a baseline. Then he sights across to the tree. The angle from his baseline to the tree is $\theta = 35.0°$ (Fig. P1.49). How wide is the river?

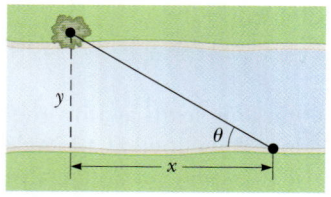

Figure P1.49

50. **S** Refer to Problem 48. Suppose the mountain height is y, the woman's original distance from the mountain is x, and the angle of elevation she measures from the horizontal to the top of the mountain is θ. If she moves a distance d closer to the mountain and measures an angle of elevation ϕ, find a general equation for the height of the mountain y in terms of d, ϕ, and θ, neglecting the height of her eyes above the ground.

Additional Problems

51. (a) One of the fundamental laws of motion states that the acceleration of an object is directly proportional to the resultant force on it and inversely proportional to its mass. If the proportionality constant is defined to have no dimensions, determine the dimensions of force. (b) The newton is the SI unit of force. According to the results for (a), how can you express a force having units of newtons by using the fundamental units of mass, length, and time?

52. (a) Find a conversion factor to convert from miles per hour to kilometers per hour. (b) For a while, federal law mandated that the maximum highway speed would be 55 mi/h. Use the conversion factor from part (a) to find the speed in kilometers per hour. (c) The maximum highway speed has been raised to 65 mi/h in some places. In kilometers per hour, how much of an increase is this over the 55-mi/h limit?

53. **BIO** One cubic centimeter (1.0 cm³) of water has a mass of 1.0×10^{-3} kg. (a) Determine the mass of 1.0 m³ of water. Assuming that biological substances are 98% water, estimate the masses of (b) a cell with a diameter of 1.0 μm, (c) a human kidney, and (d) a fly. Take a kidney to be roughly a sphere with a radius of 4.0 cm and a fly to be roughly a cylinder 4.0 mm long and 2.0 mm in diameter.

54. Soft drinks are commonly sold in aluminum containers. (a) To an order of magnitude, how many such containers are thrown away or recycled each year by U.S. consumers? (b) How many tons of aluminum does this represent? In your solution, state the quantities you measure or estimate and the values you take for them.

55. The displacement of an object moving under uniform acceleration is some function of time and the acceleration. Suppose we write this displacement as $s = ka^m t^n$, where k is a dimensionless constant. Show by dimensional analysis that this expression is satisfied if $m = 1$ and $n = 2$. Can the analysis give the value of k?

56. Assume that it takes 7.00 minutes to fill a 30.0-gal gasoline tank. (a) Calculate the rate at which the tank is filled in gallons per second. (b) Calculate the rate at which the tank is filled in cubic meters per second. (c) Determine the time interval, in hours, required to fill a 1.00-m³ volume at the same rate. (1 U.S. gal = 231 in.³)

57. **M** One gallon of paint (volume = 3.79×10^{-3} m³) covers an area of 25.0 m². What is the thickness of the fresh paint on the wall?

58. **S** Sphere 1 has surface area A_1 and volume V_1, and sphere 2 has surface area A_2 and volume V_2. If the radius of sphere 2 is double the radius of sphere 1, what is the ratio of (a) the areas, A_2/A_1 and (b) the volumes, V_2/V_1?

59. **M** Assume that there are 100 million passenger cars in the United States and that the average fuel consumption is 20 mi/gal of gasoline. If the average distance

traveled by each car is 10 000 mi/yr, how much gasoline would be saved per year if average fuel consumption could be increased to 25 mi/gal?

60. In 2013, the U.S. national debt was about $17 trillion. (a) If payments were made at the rate of $1 000 per second, how many years would it take to pay off the debt, assuming that no interest were charged? (b) A dollar bill is about 15.5 cm long. If seventeen trillion dollar bills were laid end to end around the Earth's equator, how many times would they encircle the planet? Take the radius of the Earth at the equator to be 6 378 km. (*Note:* Before doing any of these calculations, try to guess at the answers. You may be very surprised.)

61. (a) How many seconds are there in a year? (b) If one micrometeorite (a sphere with a diameter on the order of 10^{-6} m) struck each square meter of the Moon each second, estimate the number of years it would take to cover the Moon with micrometeorites to a depth of one meter. (*Hint:* Consider a cubic box, 1 m on a side, on the Moon, and find how long it would take to fill the box.)

62. Imagine that you are the equipment manager of a professional baseball team. One of your jobs is to keep baseballs on hand for games. Balls are sometimes lost when players hit them into the stands as either home runs or foul balls. Estimate how many baseballs you have to buy per season in order to make up for such losses. Assume that your team plays an 81-game home schedule in a season.

63. The nearest neutron star (a collapsed star made primarily of neutrons) is about 3.00×10^{18} m away from Earth. Given that the Milky Way galaxy (Fig. P1.63) is roughly a disk of diameter $\sim 10^{21}$ m and thickness $\sim 10^{19}$ m, estimate the number of neutron stars in the Milky Way to the nearest order of magnitude.

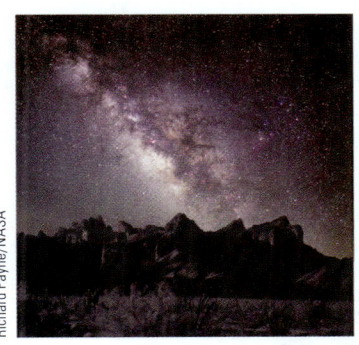

Figure P1.63

The current absolute land speed record holder is the British designed ThrustSSC, a twin turbofan-powered car which achieved 763 miles per hour (1,228 km/h) for the measured mile (1.6 km), breaking the sound barrier. The car was driven by Andy Green (UK) on 10/15/1997 in the Black Rock Desert in Gerlach, Nevada.

2 Motion in One Dimension

2.1 Displacement
2.2 Velocity
2.3 Acceleration
2.4 Motion Diagrams
2.5 One-Dimensional Motion with Constant Acceleration
2.6 Freely Falling Objects

Life is motion. Our muscles coordinate motion microscopically to enable us to walk and jog. Our hearts pump tirelessly for decades, moving blood through our bodies. Cell wall mechanisms move select atoms and molecules in and out of cells. From the prehistoric chase of antelopes across the savanna to the pursuit of satellites in space, mastery of motion has been critical to our survival and success as a species.

The study of motion and of physical concepts such as force and mass is called **dynamics.** The part of dynamics that describes motion without regard to its causes is called **kinematics.** In this chapter the focus is on kinematics in one dimension: motion along a straight line. This kind of motion—and, indeed, *any* motion—involves the concepts of displacement, velocity, and acceleration. Here, we use these concepts to study the motion of objects undergoing constant acceleration. In Chapter 3 we will repeat this discussion for objects moving in two dimensions.

The first recorded evidence of the study of mechanics can be traced to the people of ancient Sumeria and Egypt, who were interested primarily in understanding the motions of heavenly bodies. The most systematic and detailed early studies of the heavens were conducted by the Greeks from about 300 B.C. to A.D. 300. Ancient scientists and laypeople regarded the Earth as the center of the Universe. This **geocentric model** was accepted by such notables as Aristotle (384–322 B.C.) and Claudius Ptolemy (about A.D. 140). Largely because of the authority of Aristotle, the geocentric model became the accepted theory of the Universe until the seventeenth century.

About 250 B.C., the Greek philosopher Aristarchus worked out the details of a model of the solar system based on a spherical Earth that rotated on its axis and revolved around the Sun. He proposed that the sky appeared to turn westward because the Earth was turning eastward. This model wasn't given much consideration because it was believed that a turning

Earth would generate powerful winds as it moved through the air. We now know that the Earth carries the air and everything else with it as it rotates.

The Polish astronomer Nicolaus Copernicus (1473–1543) is credited with initiating the revolution that finally replaced the geocentric model. In his system, called the **heliocentric model,** Earth and the other planets revolve in circular orbits around the Sun.

This early knowledge formed the foundation for the work of Galileo Galilei (1564–1642), who stands out as the dominant facilitator of the entrance of physics into the modern era. In 1609 he became one of the first to make astronomical observations with a telescope. He observed mountains on the Moon, the larger satellites of Jupiter, spots on the Sun, and the phases of Venus. Galileo's observations convinced him of the correctness of the Copernican theory. His quantitative study of motion formed the foundation of Newton's revolutionary work in the next century.

2.1 Displacement

LEARNING OBJECTIVES

1. Calculate displacements in one dimension.
2. Explain the difference between scalars and vectors.

Motion involves the displacement of an object from one place in space and time to another. Describing motion requires some convenient coordinate system and a specified origin. A **frame of reference** is a choice of coordinate axes that defines the starting point for measuring any quantity, an essential first step in solving virtually any problem in mechanics (Fig. 2.1). In Figure 2.2a, for example, a car moves along the *x*-axis. The coordinates of the car at any time describe its position in space and, more importantly, its *displacement* at some given time of interest.

◂ Definition of displacement

The **displacement** Δx of an object is defined as its *change in position* and is given by

$$\Delta x \equiv x_f - x_i \quad [2.1]$$

where x_i is the coordinate of the initial position of the car and x_f is the coordinate of the car's final position. (The indices *i* and *f* stand for initial and final, respectively.)

SI unit: meter (m)

Figure 2.1 (a) How large is the canyon? Without a frame of reference, it's hard to tell. (b) The canyon is Valles Marineris on Mars, and with a frame of reference provided by a superposed outline of the United States, its size is easily grasped.

28 CHAPTER 2 | Motion in One Dimension

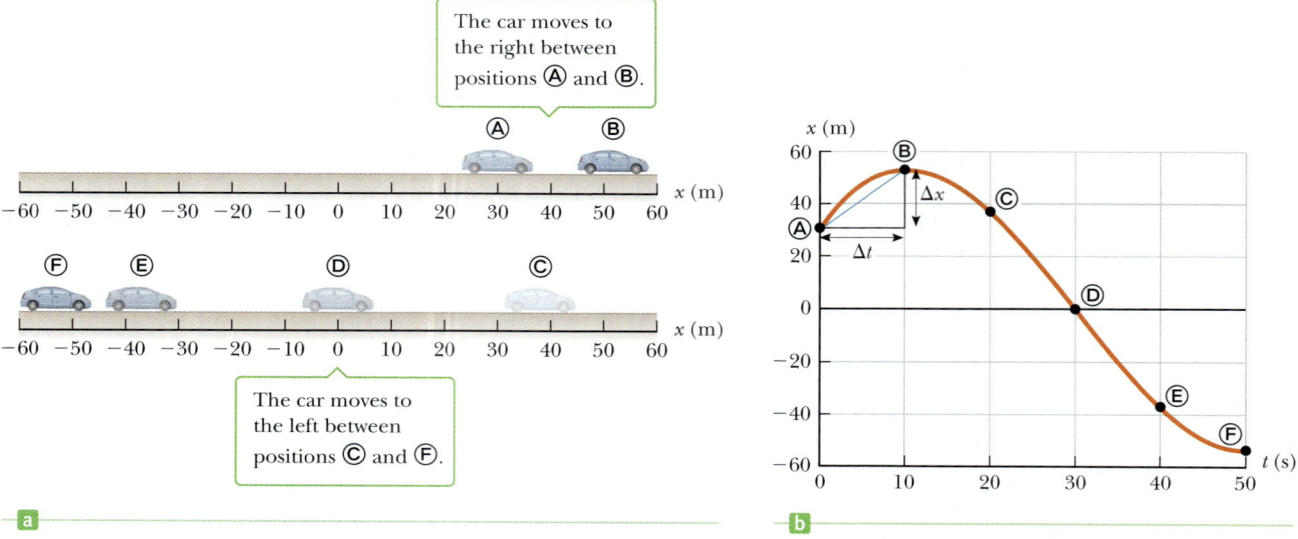

Figure 2.2
(a) A car moves back and forth along a straight line taken to be the x-axis. Because we are interested only in the car's translational motion, we can model it as a particle. (b) Graph of position vs. time for the motion of the "particle."

Tip 2.1 A Displacement Isn't a Distance!
The displacement of an object is *not* the same as the distance it travels. Toss a tennis ball up and catch it. The ball travels a *distance* equal to twice the maximum height reached, but its *displacement* is zero.

Tip 2.2 Vectors Have Both a Magnitude and a Direction
Scalars have size. Vectors, too, have size, but they also indicate a direction.

We will use the Greek letter delta, Δ, to denote a change in any physical quantity. From the definition of displacement, we see that Δx (read "delta ex") is positive if x_f is greater than x_i and negative if x_f is less than x_i. For example, if the car moves from point Ⓐ to point Ⓑ so that the initial position is $x_i = 30$ m and the final position is $x_f = 52$ m, the displacement is $\Delta x = x_f - x_i = 52$ m $- 30$ m $= +22$ m. However, if the car moves from point Ⓒ to point Ⓕ, then the initial position is $x_i = 38$ m and the final position is $x_f = -53$ m, and the displacement is $\Delta x = x_f - x_i = -53$ m $- 38$ m $= -91$ m. A positive answer indicates a displacement in the positive x-direction, whereas a negative answer indicates a displacement in the negative x-direction. Figure 2.2b displays the graph of the car's position as a function of time.

Because displacement has both a magnitude (size) and a direction, it's a vector quantity, as are velocity and acceleration. In general, **a vector quantity is characterized by having both a magnitude and a direction.** By contrast, **a scalar quantity has magnitude, but no direction.** Scalar quantities such as mass and temperature are completely specified by a numeric value with appropriate units; no direction is involved.

Vector quantities will usually be denoted in boldface type with an arrow over the top of the letter. For example, $\vec{\mathbf{v}}$ represents velocity and $\vec{\mathbf{a}}$ denotes an acceleration, both vector quantities. In this chapter, however, it won't be necessary to use that notation because in one-dimensional motion an object can only move in one of two directions, and these directions are easily specified by plus and minus signs.

2.2 Velocity

LEARNING OBJECTIVES

1. Calculate the average speed of an object.
2. Calculate the average velocity of an object.
3. Show by example that an object's average speed can differ from the magnitude of its average velocity.
4. Analyze a graph of position vs. time to obtain average and instantaneous velocities.

In everyday usage the terms *speed* and *velocity* are interchangeable. In physics, however, there's a clear distinction between them: speed is a scalar quantity, having only magnitude, whereas velocity is a vector, having both magnitude and direction.

Why must velocity be a vector? If you want to get to a town 70 km away in an hour's time, it's not enough to drive at a speed of 70 km/h; you must travel in the correct direction as well. That's obvious, but it shows that velocity gives considerably more information than speed, as will be made more precise in the formal definitions.

> The **average speed** of an object over a given time interval is the length of the path it travels divided by the total elapsed time:
>
> $$\text{Average speed} \equiv \frac{\text{path length}}{\text{elapsed time}}$$
>
> **SI unit: meter per second (m/s)**

◀ Definition of average speed

This equation might be written with symbols as $v = d/t$, where v represents the average speed (not average velocity), d represents the path length, and t represents the elapsed time during the motion. The path length is often called the "total distance," but that can be misleading because distance has a different, precise mathematical meaning based on differences in the coordinates between the initial and final points. Distance (neglecting any curvature of the surface) is given by the Pythagorean theorem, $\Delta s = \sqrt{(x_f - x_i)^2 + (y_f - y_i)^2}$, which depends only on the endpoints, (x_i, y_i) and (x_f, y_f), and not on what happens in between. The same equation gives the magnitude of a displacement. The straight-line distance from Atlanta, Georgia, to St. Petersburg, Florida, for example, is about 500 miles. If someone drives a car that distance in 10 h, the car's average speed is 500 mi/10 h = 50 mi/h, even if the car's speed varies greatly during the trip. If the driver takes scenic detours off the direct route along the way, however, or doubles back for a while, the path length increases while the distance between the two cities remains the same. A side trip to Jacksonville, Florida, for example, might add 100 miles to the path length, so the car's average speed would then be 600 mi/10 h = 60 mi/h. The magnitude of the average velocity, however, would remain 50 mi/h.

■ EXAMPLE 2.1 | The Tortoise and the Hare

GOAL Apply the concept of average speed.

PROBLEM A turtle and a rabbit engage in a footrace over a distance of 4.00 km. The rabbit runs 0.500 km and then stops for a 90.0-min nap. Upon awakening, he remembers the race and runs twice as fast. Finishing the course in a total time of 1.75 h, the rabbit wins the race. **(a)** Calculate the average speed of the rabbit. **(b)** What was his average speed before he stopped for a nap? Assume no detours or doubling back.

STRATEGY Finding the overall average speed in part (a) is just a matter of dividing the path length by the elapsed time. Part (b) requires two equations and two unknowns, the latter turning out to be the two different average speeds: v_1 before the nap and v_2 after the nap. One equation is given in the statement of the problem ($v_2 = 2v_1$), whereas the other comes from the fact the rabbit ran for only 15 minutes because he napped for 90 minutes.

SOLUTION

(a) Find the rabbit's overall average speed.

Apply the equation for average speed:

$$\text{Average speed} \equiv \frac{\text{path length}}{\text{elapsed time}} = \frac{4.00 \text{ km}}{1.75 \text{ h}}$$
$$= \boxed{2.29 \text{ km/h}}$$

(b) Find the rabbit's average speed before his nap.

Sum the running times, and set the sum equal to 0.25 h: $t_1 + t_2 = 0.25$ h

(Continued)

Substitute $t_1 = d_1/v_1$ and $t_2 = d_2/v_2$:

(1) $\quad \dfrac{d_1}{v_1} + \dfrac{d_2}{v_2} = 0.25 \text{ h}$

Substitute $v_2 = 2v_1$ and the values of d_1 and d_2 into Equation (1):

(2) $\quad \dfrac{0.500 \text{ km}}{v_1} + \dfrac{3.50 \text{ km}}{2v_1} = 0.25 \text{ h}$

Solve Equation (2) for v_1:

$v_1 = $ 9.0 km/h

REMARKS As seen in this example, average speed can be calculated regardless of any variation in speed over the given time interval.

QUESTION 2.1 Does a doubling of an object's average speed always double the magnitude of its displacement in a given amount of time? Explain.

EXERCISE 2.1 Estimate the average speed of the *Apollo* spacecraft in meters per second, given that the craft took five days to reach the Moon from Earth. (The Moon is 3.8×10^8 m from Earth.)

ANSWER ~ 900 m/s

Unlike average speed, **average velocity** is a vector quantity, having both a magnitude and a direction. Consider again the car of Figure 2.2, moving along the road (the *x*-axis). Let the car's position be x_i at some time t_i and x_f at a later time t_f. In the time interval $\Delta t = t_f - t_i$, the displacement of the car is $\Delta x = x_f - x_i$.

▶ Definition of average velocity

The average velocity $\bar{v}$ during a time interval Δt is the displacement Δx divided by Δt:

$$\bar{v} \equiv \dfrac{\Delta x}{\Delta t} = \dfrac{x_f - x_i}{t_f - t_i} \qquad [2.2]$$

SI unit: meter per second (m/s)

Table 2.1 Position of the Car at Various Times

Position	t (s)	x (m)
Ⓐ	0	30
Ⓑ	10	52
Ⓒ	20	38
Ⓓ	30	0
Ⓔ	40	−37
Ⓕ	50	−53

Unlike the average speed, which is always positive, the average velocity of an object in one dimension can be either positive or negative, depending on the sign of the displacement. (The time interval Δt is always positive.) In Figure 2.2a, for example, the average velocity of the car is positive in the upper illustration, a positive sign indicating motion to the right along the *x*-axis. Similarly, a negative average velocity for the car in the lower illustration of the figure indicates that it moves to the left along the *x*-axis.

As an example, we can use the data in Table 2.1 to find the average velocity in the time interval from point Ⓐ to point Ⓑ (assume two digits are significant):

$$\bar{v} = \dfrac{\Delta x}{\Delta t} = \dfrac{52 \text{ m} - 30 \text{ m}}{10 \text{ s} - 0 \text{ s}} = 2.2 \text{ m/s}$$

Aside from meters per second, other common units for average velocity are feet per second (ft/s) in the U.S. customary system and centimeters per second (cm/s) in the cgs system.

To further illustrate the distinction between speed and velocity, suppose we're watching a drag race from a stationary blimp. In one run we see a car follow the straight-line path from Ⓟ to Ⓠ shown in Figure 2.3 during the time interval Δt, and in a second run a car follows the curved path during the same interval. From the definition in Equation 2.2, the two cars had the same average velocity because they had the same displacement $\Delta x = x_f - x_i$ during the same time interval Δt. The car taking the curved route, however, traveled a greater path length and had the higher average speed.

Figure 2.3 A drag race viewed from a stationary blimp. One car follows the rust-colored straight-line path from Ⓟ to Ⓠ, and a second car follows the blue curved path.

Quick Quiz

2.1 Figure 2.4 shows the unusual path of a confused football player. After receiving a kickoff at his own goal, he runs downfield to within inches of a touchdown, then reverses direction and races back until he's tackled at the exact location where he first caught the ball. During this run, which took 25 s, what is (a) the path length he travels, (b) his displacement, and (c) his average velocity in the x-direction? (d) What is his average speed?

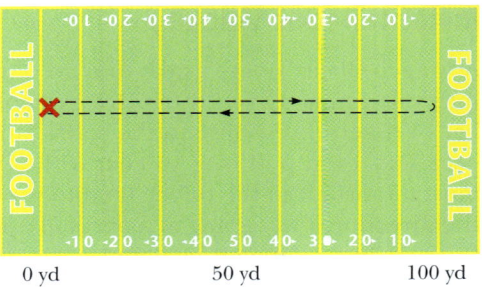

Figure 2.4 (Quick Quiz 2.1) The path followed by a confused football player.

Graphical Interpretation of Velocity

If a car moves along the x-axis from Ⓐ to Ⓑ to Ⓒ, and so forth, we can plot the positions of these points as a function of the time elapsed since the start of the motion. The result is a **position vs. time graph** like those of Figure 2.5. In Figure 2.5a, the graph is a straight line because the car is moving at constant velocity. The same displacement Δx occurs in each time interval Δt. In this case, the average velocity is always the same and is equal to $\Delta x/\Delta t$. Figure 2.5b is a graph of the data in Table 2.1. Here, the position vs. time graph is not a straight line because the velocity of the car is changing. Between any two points, however, we can draw a straight line just as in Figure 2.5a, and the slope of that line is the average velocity $\Delta x/\Delta t$ in that time interval. In general, **the average velocity of an object during the time interval Δt is equal to the slope of the straight line joining the initial and final points on a graph of the object's position versus time.**

From the data in Table 2.1 and the graph in Figure 2.5b, we see that the car first moves in the positive x-direction as it travels from Ⓐ to Ⓑ, reaches a position of 52 m at time $t = 10$ s, then reverses direction and heads backwards. In the first 10 s of its motion, as the car travels from Ⓐ to Ⓑ, its average velocity is 2.2 m/s, as previously calculated. In the first 40 seconds, as the car goes from Ⓐ to Ⓔ, its displacement is $\Delta x = -37$ m $-$ (30 m) $= -67$ m. So the average velocity in this interval, which equals the slope of the blue line in Figure 2.5b from Ⓐ to Ⓔ, is $\bar{v} = \Delta x/\Delta t = (-67$ m$)/(40$ s$) = -1.7$ m/s. In general, there will be a different average velocity between any distinct pair of points.

> **Tip 2.3 Slopes of Graphs**
> The word *slope* is often used in reference to the graphs of physical data. Regardless of the type of data, the *slope* is given by
> $$\text{Slope} = \frac{\text{change in vertical axis}}{\text{change in horizontal axis}}$$
> Slope carries units.

> **Tip 2.4 Average Velocity vs. Average Speed**
> Average velocity is *not* the same as average speed. If you run from $x = 0$ m to $x = 25$ m and back to your starting point in a time interval of 5 s, the average velocity is zero, whereas the average speed is 10 m/s.

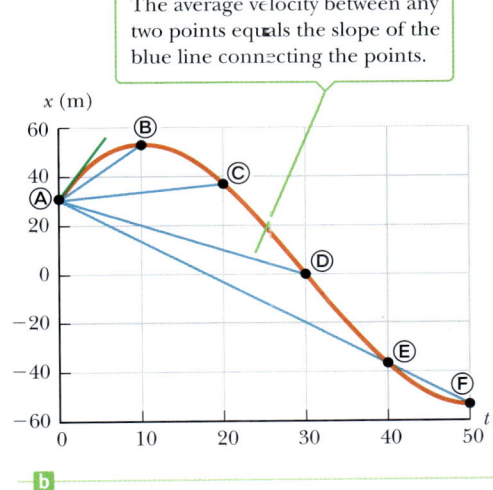

Figure 2.5 (a) Position vs. time graph for the motion of a car moving along the x-axis at constant velocity. (b) Position vs. time graph for the motion of a car with changing velocity, using the data in Table 2.1.

The average velocity between any two points equals the slope of the blue line connecting the points.

Instantaneous Velocity

Average velocity doesn't take into account the details of what happens during an interval of time. On a car trip, for example, you may speed up or slow down a number of times in response to the traffic and the condition of the road, and on rare occasions even pull over to chat with a police officer about your speed. What is most important to the police (and to your own safety) is the speed of your car and the direction it was going at a particular instant in time, which together determine the car's **instantaneous velocity.**

So in driving a car between two points, the average velocity must be computed over an interval of time, but the magnitude of instantaneous velocity can be read on the car's speedometer.

▶ Definition of instantaneous velocity

The instantaneous velocity v is the limit of the average velocity as the time interval Δt becomes infinitesimally small:

$$v \equiv \lim_{\Delta t \to 0} \frac{\Delta x}{\Delta t} \qquad [2.3]$$

SI unit: meter per second (m/s)

The notation $\lim_{\Delta t \to 0}$ means that the ratio $\Delta x/\Delta t$ is repeatedly evaluated for smaller and smaller time intervals Δt. As Δt gets extremely close to zero, the ratio $\Delta x/\Delta t$ gets closer and closer to a fixed number, which is defined as the instantaneous velocity.

To better understand the formal definition, consider data obtained on our vehicle via radar (Table 2.2). At $t = 1.00$ s, the car is at $x = 5.00$ m, and at $t = 3.00$ s, it's at $x = 52.5$ m. The average velocity computed for this interval $\Delta x/\Delta t = (52.5$ m $- 5.00$ m$)/(3.00$ s $- 1.00$ s$) = 23.8$ m/s. This result could be used as an estimate for the velocity at $t = 1.00$ s, but it wouldn't be very accurate because the speed changes considerably in the 2-second time interval. Using the rest of the data, we can construct Table 2.3. As the time interval gets smaller, the average velocity more closely approaches the instantaneous velocity. Using the final interval of only 0.010 0 s, we find that the average velocity is $\bar{v} = \Delta x/\Delta t = 0.470$ m$/0.010$ 0 s $= 47.0$ m/s. Because 0.010 0 s is a very short time interval, the actual instantaneous velocity is probably very close to this latter average velocity, given the limits on the car's ability to accelerate. Finally using the conversion factor on the front endsheets of the book, we see that this is 105 mi/h, a likely violation of the speed limit.

As can be seen in Figure 2.6, the chords formed by the blue lines gradually approach a tangent line as the time interval becomes smaller. **The slope of the line tangent to the position vs. time curve at "a given time" is defined to be the instantaneous velocity at that time.**

Table 2.2 Positions of a Car at Specific Instants of Time

t (s)	x (m)
1.00	5.00
1.01	5.47
1.10	9.67
1.20	14.3
1.50	26.3
2.00	34.7
3.00	52.5

Table 2.3 Calculated Values of the Time Intervals, Displacements, and Average Velocities of the Car of Table 2.2

Time Interval (s)	Δt (s)	Δx (m)	$\bar{v}$ (m/s)
1.00 to 3.00	2.00	47.5	23.8
1.00 to 2.00	1.00	29.7	29.7
1.00 to 1.50	0.50	21.3	42.6
1.00 to 1.20	0.20	9.30	46.5
1.00 to 1.10	0.10	4.67	46.7
1.00 to 1.01	0.01	0.470	47.0

> The slopes of the blue lines are average velocities which approach the slope of the green tangent line, an instantaneous velocity.

Figure 2.6 Graph representing the motion of the car from the data in Table 2.2.

The instantaneous speed of an object, which is a scalar quantity, is defined as the magnitude of the instantaneous velocity. Like average speed, instantaneous speed (which we will usually call, simply, "speed") has no direction associated with it and hence carries no algebraic sign. For example, if one object has an instantaneous velocity of +15 m/s along a given line and another object has an instantaneous velocity of −15 m/s along the same line, both have an instantaneous speed of 15 m/s.

◀ Definition of instantaneous speed

■ EXAMPLE 2.2 | Slowly Moving Train

GOAL Obtain average and instantaneous velocities from a graph.

PROBLEM A train moves slowly along a straight portion of track according to the graph of position versus time in Figure 2.7a. Find (a) the average velocity for the total trip, (b) the average velocity during the first 4.00 s of motion, (c) the average velocity during the next 4.00 s of motion, (d) the instantaneous velocity at $t = 2.00$ s, and (e) the instantaneous velocity at $t = 9.00$ s.

Figure 2.7 (a) (Example 2.2) (b) (Exercise 2.2)

STRATEGY The average velocities can be obtained by substituting the data into the definition. The instantaneous velocity at $t = 2.00$ s is the same as the average velocity at that point because the position vs. time graph is a straight line, indicating constant velocity. Finding the instantaneous velocity when $t = 9.00$ s requires sketching a line tangent to the curve at that point and finding its slope.

SOLUTION
(a) Find the average velocity from Ⓞ to Ⓒ.

Calculate the slope of the dashed blue line:

$$\bar{v} = \frac{\Delta x}{\Delta t} = \frac{10.0 \text{ m}}{12.0 \text{ s}} = +0.833 \text{ m/s}$$

(b) Find the average velocity during the first 4 seconds of the train's motion.

Again, find the slope:

$$\bar{v} = \frac{\Delta x}{\Delta t} = \frac{4.00 \text{ m}}{4.00 \text{ s}} = +1.00 \text{ m/s}$$

(c) Find the average velocity during the next 4 seconds.

Here, there is no change in position as the train moves from Ⓐ to Ⓑ, so the displacement Δx is zero:

$$\bar{v} = \frac{\Delta x}{\Delta t} = \frac{0 \text{ m}}{4.00 \text{ s}} = 0 \text{ m/s}$$

(Continued)

(d) Find the instantaneous velocity at $t = 2.00$ s.

This is the same as the average velocity found in (b), because the graph is a straight line:

$v = $ 1.00 m/s

(e) Find the instantaneous velocity at $t = 9.00$ s.

The tangent line appears to intercept the x-axis at (3.0 s, 0 m) and graze the curve at (9.0 s, 4.5 m). The instantaneous velocity at $t = 9.00$ s equals the slope of the tangent line through these points:

$$v = \frac{\Delta x}{\Delta t} = \frac{4.5 \text{ m} - 0 \text{ m}}{9.0 \text{ s} - 3.0 \text{ s}} = 0.75 \text{ m/s}$$

REMARKS From the origin to Ⓐ, the train moves at constant speed in the positive x-direction for the first 4.00 s, because the position vs. time curve is rising steadily toward positive values. From Ⓐ to Ⓑ, the train stops at $x = 4.00$ m for 4.00 s. From Ⓑ to Ⓒ, the train travels at increasing speed in the positive x-direction.

QUESTION 2.2 Would a vertical line in a graph of position versus time make sense? Explain.

EXERCISE 2.2 Figure 2.7b graphs another run of the train. Find (a) the average velocity from Ⓞ to Ⓒ; (b) the average velocity from Ⓞ to Ⓐ and the instantaneous velocity at any given point between Ⓞ and Ⓐ; (c) the approximate instantaneous velocity at $t = 6.0$ s; and (d) the average velocity on the open interval from Ⓑ to Ⓒ and instantaneous velocity at $t = 9.0$ s.

ANSWERS (a) 0 m/s (b) both are +0.5 m/s (c) 2 m/s (d) both are −2.5 m/s

---

Figure 2.8 A car moving to the right accelerates from a velocity of v_i to a velocity of v_f in the time interval $\Delta t = t_f - t_i$.

2.3 Acceleration

LEARNING OBJECTIVES

1. Calculate an object's average acceleration.
2. Analyze an object's velocity vs. time graph to obtain average and instantaneous accelerations.

Going from place to place in your car, you rarely travel long distances at constant velocity. The velocity of the car increases when you step harder on the gas pedal and decreases when you apply the brakes. The velocity also changes when you round a curve, altering your direction of motion. The changing of an object's velocity with time is called **acceleration.**

Average Acceleration

A car moves along a straight highway as in Figure 2.8. At time t_i it has a velocity of v_i, and at time t_f its velocity is v_f, with $\Delta v = v_f - v_i$ and $\Delta t = t_f - t_i$.

▶ Definition of average acceleration

The average acceleration $\bar{a}$ during the time interval Δt is the change in velocity Δv divided by Δt:

$$\bar{a} \equiv \frac{\Delta v}{\Delta t} = \frac{v_f - v_i}{t_f - t_i} \qquad [2.4]$$

SI unit: meter per second per second (m/s²)

For example, suppose the car shown in Figure 2.8 accelerates from an initial velocity of $v_i = +10$ m/s to a final velocity of $v_f = +20$ m/s in a time interval of 2 s.

(Both velocities are toward the right, selected as the positive direction.) These values can be inserted into Equation 2.4 to find the average acceleration:

$$\bar{a} = \frac{\Delta v}{\Delta t} = \frac{20 \text{ m/s} - 10 \text{ m/s}}{2 \text{ s}} = +5 \text{ m/s}^2$$

Acceleration is a vector quantity having dimensions of length divided by the time squared. Common units of acceleration are meters per second per second ((m/s)/s, which is usually written m/s^2) and feet per second per second (ft/s^2). An average acceleration of +5 m/s^2 means that, on average, the car increases its velocity by 5 m/s every second in the positive x-direction.

For the case of motion in a straight line, the direction of the velocity of an object and the direction of its acceleration are related as follows: **When the object's velocity and acceleration are in the same direction, the speed of the object increases with time. When the object's velocity and acceleration are in opposite directions, the speed of the object decreases with time.**

To clarify this point, suppose the velocity of a car changes from −10 m/s to −20 m/s in a time interval of 2 s. The minus signs indicate that the velocities of the car are in the negative x-direction; they do *not* mean that the car is slowing down! The average acceleration of the car in this time interval is

$$\bar{a} = \frac{\Delta v}{\Delta t} = \frac{-20 \text{ m/s} - (-10 \text{ m/s})}{2 \text{ s}} = -5 \text{ m/s}^2$$

The minus sign indicates that the acceleration vector is also in the negative x-direction. Because the velocity and acceleration vectors are in the same direction, the speed of the car must increase as the car moves to the left. Positive and negative accelerations specify directions relative to chosen axes, not "speeding up" or "slowing down." The terms *speeding up* or *slowing down* refer to an increase and a decrease in speed, respectively.

> **Tip 2.5 Negative Acceleration**
> Negative acceleration doesn't necessarily mean an object is slowing down. If the acceleration is negative and the velocity is also negative, the object is speeding up!

> **Tip 2.6 Deceleration**
> The word *deceleration* means a reduction in speed, a slowing down. Some confuse it with a negative acceleration, which can speed something up. (See Tip 2.5.)

■ Quick Quiz

2.2 True or False? (a) A car must always have an acceleration in the same direction as its velocity. (b) It's possible for a slowing car to have a positive acceleration. (c) An object with constant nonzero acceleration can never stop and remain at rest.

An object with nonzero acceleration can have a velocity of zero, but only instantaneously. When a ball is tossed straight up, its velocity is zero when it reaches its maximum height. Gravity still accelerates the ball at that point, however; otherwise, it wouldn't fall down.

Instantaneous Acceleration

The value of the average acceleration often differs in different time intervals, so it's useful to define the **instantaneous acceleration,** which is analogous to the instantaneous velocity discussed in Section 2.2.

The instantaneous acceleration a is the limit of the average acceleration as the time interval Δt goes to zero:

$$a \equiv \lim_{\Delta t \to 0} \frac{\Delta v}{\Delta t} \quad [2.5]$$

SI unit: meter per second per second (m/s^2)

◀ Definition of instantaneous acceleration

Here again, the notation $\lim_{\Delta t \to 0}$ means that the ratio $\Delta v/\Delta t$ is evaluated for smaller and smaller values of Δt. The closer Δt gets to zero, the closer the ratio gets to a fixed number, which is the instantaneous acceleration.

36 CHAPTER 2 | Motion in One Dimension

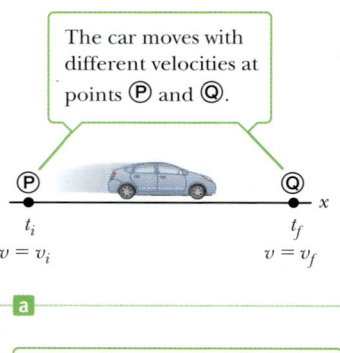

The car moves with different velocities at points Ⓟ and Ⓠ.

a

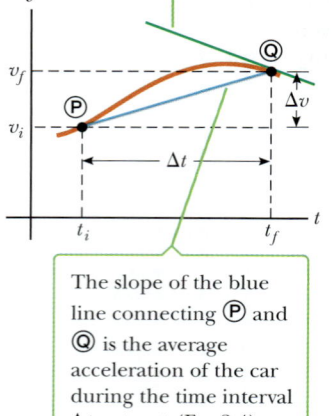

The slope of the green line is the instantaneous acceleration of the car at point Ⓠ (Eq. 2.5).

The slope of the blue line connecting Ⓟ and Ⓠ is the average acceleration of the car during the time interval $\Delta t = t_f - t_i$ (Eq. 2.4).

b

Figure 2.9 (a) A car, modeled as a particle, moving along the x-axis from Ⓟ to Ⓠ, has velocity v_{xi} at $t = t_i$ and velocity v_{xf} at $t = t_f$. (b) Velocity vs. time graph for an object moving in a straight line.

Figure 2.9, a **velocity vs. time graph,** plots the velocity of an object against time. The graph could represent, for example, the motion of a car along a busy street. The average acceleration of the car between times t_i and t_f can be found by determining the slope of the line joining points Ⓟ and Ⓠ. If we imagine that point Ⓠ is brought closer and closer to point Ⓟ, the line comes closer and closer to becoming tangent at Ⓟ. The **instantaneous acceleration of an object at a given time equals the slope of the tangent to the velocity vs. time graph at that time.** From now on, we will use the term *acceleration* to mean "instantaneous acceleration."

In the special case where the velocity vs. time graph of an object's motion is a straight line, the instantaneous acceleration of the object at any point is equal to its average acceleration. That also means the tangent line to the graph overlaps the graph itself. In that case, the object's acceleration is said to be *uniform*, which means that it has a constant value. Constant acceleration problems are important in kinematics and will be studied extensively in this and the next chapter.

■ **Quick Quiz**

2.3 Parts (a), (b), and (c) of Figure 2.10 represent three graphs of the velocities of different objects moving in straight-line paths as functions of time. The possible accelerations of each object as functions of time are shown in parts (d), (e), and (f). Match each velocity vs. time graph with the acceleration vs. time graph that best describes the motion.

Figure 2.10 (Quick Quiz 2.3) Match each velocity vs. time graph to its corresponding acceleration vs. time graph.

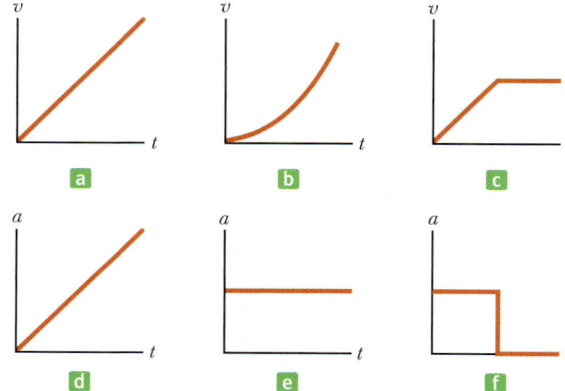

■ EXAMPLE 2.3 Catching a Fly Ball

GOAL Apply the definition of instantaneous acceleration.

PROBLEM A baseball player moves in a straight-line path in order to catch a fly ball hit to the outfield. His velocity as a function of time is shown in Figure 2.11a. Find his instantaneous acceleration at points Ⓐ, Ⓑ, and Ⓒ.

STRATEGY At each point, the velocity vs. time graph is a straight line segment, so the instantaneous acceleration will be the slope of that segment. Select two points on each segment and use them to calculate the slope.

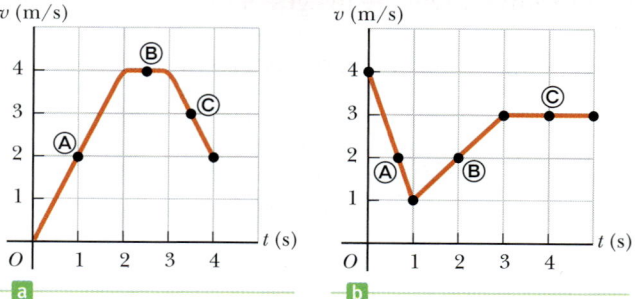

Figure 2.11 (a) (Example 2.3) (b) (Exercise 2.3)

SOLUTION

Acceleration at Ⓐ.

The acceleration at Ⓐ equals the slope of the line connecting the points (0 s, 0 m/s) and (2.0 s, 4.0 m/s):

$$a = \frac{\Delta v}{\Delta t} = \frac{4.0 \text{ m/s} - 0}{2.0 \text{ s} - 0} = \boxed{+2.0 \text{ m/s}^2}$$

Acceleration at Ⓑ.

$\Delta v = 0$, because the segment is horizontal:

$$a = \frac{\Delta v}{\Delta t} = \frac{4.0 \text{ m/s} - 4.0 \text{ m/s}}{3.0 \text{ s} - 2.0 \text{ s}} = 0 \text{ m/s}^2$$

Acceleration at Ⓒ.

The acceleration at Ⓒ equals the slope of the line connecting the points (3.0 s, 4.0 m/s) and (4.0 s, 2.0 m/s):

$$a = \frac{\Delta v}{\Delta t} = \frac{2.0 \text{ m/s} - 4.0 \text{ m/s}}{4.0 \text{ s} - 3.0 \text{ s}} = -2.0 \text{ m/s}^2$$

REMARKS For the first 2.0 s, the ballplayer moves in the positive x-direction (the velocity is positive) and steadily accelerates (the curve is steadily rising) to a maximum speed of 4.0 m/s. He moves for 1.0 s at a steady speed of 4.0 m/s and then slows down in the last second (the v vs. t curve is falling), still moving in the positive x-direction (v is always positive).

QUESTION 2.3 Can the tangent line to a velocity vs. time graph ever be vertical? Explain.

EXERCISE 2.3 Repeat the problem, using Figure 2.11b.

ANSWER The accelerations at Ⓐ, Ⓑ, and Ⓒ are -3.0 m/s^2, 1.0 m/s^2, and 0 m/s^2, respectively.

2.4 Motion Diagrams

LEARNING OBJECTIVE

1. Analyze an object's motion using a motion diagram.

Velocity and acceleration are sometimes confused with each other, but they're very different concepts, as can be illustrated with the help of motion diagrams. A **motion diagram** is a representation of a moving object at successive time intervals, with velocity and acceleration vectors sketched at each position, red for velocity vectors and violet for acceleration vectors, as in Figure 2.12. The time intervals between adjacent positions in the motion diagram are assumed equal.

A motion diagram is analogous to images resulting from a stroboscopic photograph of a moving object. Each image is made as the strobe light flashes. Figure 2.12 represents three sets of strobe photographs of cars moving along a straight roadway from left to right. The time intervals between flashes of the stroboscope are equal in each diagram.

In Figure 2.12a, the images of the car are equally spaced: The car moves the same distance in each time interval. This means that the car moves with *constant positive velocity* and has *zero acceleration*. The red arrows are all the same length (constant velocity) and there are no violet arrows (zero acceleration).

In Figure 2.12b, the images of the car become farther apart as time progresses and the velocity vector increases with time, because the car's displacement between adjacent positions increases as time progresses. The car is moving with a *positive*

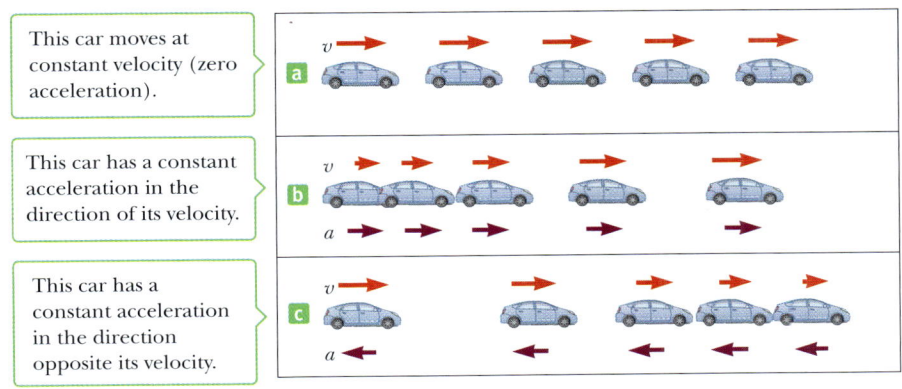

Figure 2.12
Motion diagrams of a car moving along a straight roadway in a single direction. The velocity at each instant is indicated by a red arrow, and the constant acceleration is indicated by a purple arrow.

This car moves at constant velocity (zero acceleration).

This car has a constant acceleration in the direction of its velocity.

This car has a constant acceleration in the direction opposite its velocity.

CHAPTER 2 | Motion in One Dimension

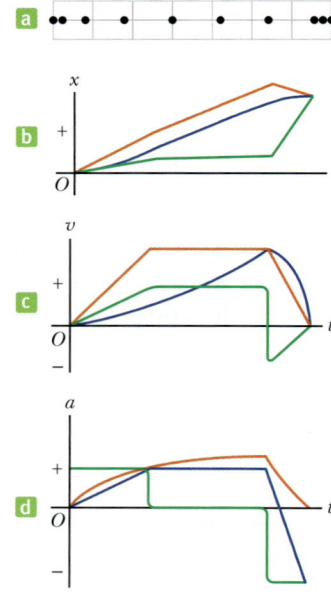

Figure 2.14 (Quick Quiz 2.5) Choose the correct graphs.

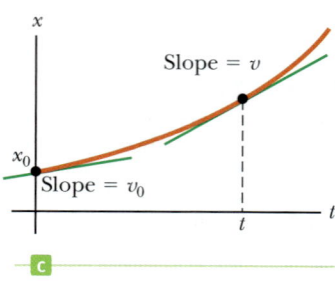

Figure 2.15
A particle moving along the x-axis with constant acceleration a.
(a) the acceleration vs. time graph,
(b) the velocity vs. time graph, and
(c) the position vs. time graph.

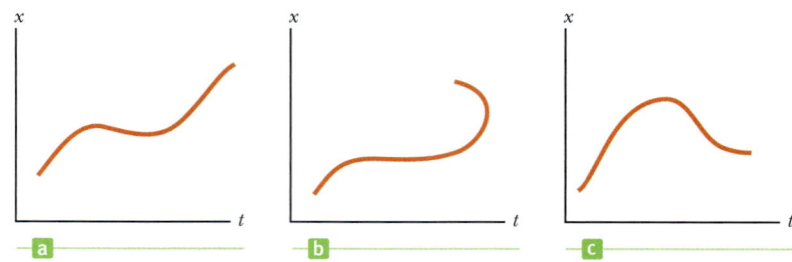

Figure 2.13 (Quick Quiz 2.4) Which position vs. time curve is impossible?

velocity and a constant *positive acceleration*. The red arrows are successively longer in each image, and the violet arrows point to the right.

In Figure 2.12c, the car slows as it moves to the right because its displacement between adjacent positions decreases with time. In this case, the car moves initially to the right with a constant negative acceleration. The velocity vector decreases in time (the red arrows get shorter) and eventually reaches zero, as would happen when the brakes are applied. Note that the acceleration and velocity vectors are *not* in the same direction. The car is moving with a *positive velocity*, but with a *negative acceleration*.

Try constructing your own diagrams for various problems involving kinematics.

■ Quick Quiz

2.4 The three graphs in Figure 2.13 represent the position vs. time for objects moving along the x-axis. Which, if any, of these graphs is not physically possible?

2.5 Figure 2.14a is a diagram of a multiflash image of an air puck moving to the right on a horizontal surface. The images sketched are separated by equal time intervals, and the first and last images show the puck at rest. **(a)** In Figure 2.14b, which color graph best shows the puck's position as a function of time? **(b)** In Figure 2.14c, which color graph best shows the puck's velocity as a function of time? **(c)** In Figure 2.14d, which color graph best shows the puck's acceleration as a function of time?

2.5 One-Dimensional Motion with Constant Acceleration

LEARNING OBJECTIVES

1. Apply the kinematics equations for objects moving at constant acceleration.
2. Find accelerations and displacements by analyzing a velocity vs. time graph.

Many applications of mechanics involve objects moving with *constant acceleration*. This type of motion is important because it applies to numerous objects in nature, such as an object in free fall near Earth's surface (assuming air resistance can be neglected). A graph of acceleration versus time for motion with constant acceleration is shown in Figure 2.15a. **When an object moves with constant acceleration, the instantaneous acceleration at any point in a time interval is equal to the value of the average acceleration over the entire time interval.** Consequently, the velocity increases or decreases at the same rate throughout the motion, and a plot of v versus t gives a straight line with either positive, zero, or negative slope.

Because the average acceleration equals the instantaneous acceleration when a is constant, we can eliminate the bar used to denote average values from our defining equation for acceleration, writing $\bar{a} = a$, so that Equation 2.4 becomes

$$a = \frac{v_f - v_i}{t_f - t_i}$$

The observer timing the motion is always at liberty to choose the initial time, so for convenience, let $t_i = 0$ and t_f be any arbitrary time t. Also, let $v_i = v_0$ (the initial velocity at $t = 0$) and $v_f = v$ (the velocity at any arbitrary time t). With this notation, we can express the acceleration as

$$a = \frac{v - v_0}{t}$$

or

$$v = v_0 + at \quad \text{(for constant } a\text{)} \quad [2.6]$$

Equation 2.6 states that the acceleration a steadily changes the initial velocity v_0 by an amount at. For example, if a car starts with a velocity of $+2.0$ m/s to the right and accelerates to the right with $a = +6.0$ m/s^2, it will have a velocity of $+14$ m/s after 2.0 s have elapsed:

$$v = v_0 + at = +2.0 \text{ m/s} + (6.0 \text{ m/s}^2)(2.0 \text{ s}) = +14 \text{ m/s}$$

The graphical interpretation of v is shown in Figure 2.15b. The velocity varies linearly with time according to Equation 2.6, as it should for constant acceleration.

Because the velocity is increasing or decreasing *uniformly* with time, we can express the average velocity in any time interval as the arithmetic average of the initial velocity v_0 and the final velocity v:

$$\bar{v} = \frac{v_0 + v}{2} \quad \text{(for constant } a\text{)} \quad [2.7]$$

Remember that this expression is valid only when the acceleration is constant, in which case the velocity increases uniformly.

We can now use this result along with the defining equation for average velocity, Equation 2.2, to obtain an expression for the displacement of an object as a function of time. Again, we choose $t_i = 0$ and $t_f = t$, and for convenience, we write $\Delta x = x_f - x_i = x - x_0$. This results in

$$\Delta x = \bar{v}t = \left(\frac{v_0 + v}{2}\right)t$$

$$\Delta x = \tfrac{1}{2}(v_0 + v)t \quad \text{(for constant } a\text{)} \quad [2.8]$$

We can obtain another useful expression for displacement by substituting the equation for v (Eq. 2.6) into Equation 2.8:

$$\Delta x = \tfrac{1}{2}(v_0 + v_0 + at)t$$

$$\Delta x = v_0 t + \tfrac{1}{2}at^2 \quad \text{(for constant } a\text{)} \quad [2.9]$$

This equation can also be written in terms of the position x, since $\Delta x = x - x_0$. Figure 2.15c shows a plot of x versus t for Equation 2.9, which is related to the graph of velocity vs. time: The area under the curve in Figure 2.15b is equal to $v_0 t + \tfrac{1}{2}at^2$, which is equal to the displacement Δx. In fact, **the area under the graph of v versus t for any object is equal to the displacement Δx of the object.**

Finally, we can obtain an expression that doesn't contain time by solving Equation 2.6 for t and substituting into Equation 2.8, resulting in

$$\Delta x = \tfrac{1}{2}(v + v_0)\left(\frac{v - v_0}{a}\right) = \frac{v^2 - v_0^2}{2a}$$

$$v^2 = v_0^2 + 2a\Delta x \quad \text{(for constant } a\text{)} \quad [2.10]$$

Equations 2.6 and 2.9 together can solve any problem in one-dimensional motion with constant acceleration, but Equations 2.7, 2.8, and, especially, 2.10 are sometimes

40 CHAPTER 2 | Motion in One Dimension

Table 2.4 Equations for Motion in a Straight Line Under Constant Acceleration

Equation	Information Given by Equation
$v = v_0 + at$	Velocity as a function of time
$\Delta x = v_0 t + \frac{1}{2}at^2$	Displacement as a function of time
$v^2 = v_0^2 + 2a\Delta x$	Velocity as a function of displacement

Note: Motion is along the x-axis. At $t = 0$, the velocity of the particle is v_0.

convenient. The three most useful equations—Equations 2.6, 2.9, and 2.10—are listed in Table 2.4.

The best way to gain confidence in the use of these equations is to work a number of problems. There is usually more than one way to solve a given problem, depending on which equations are selected and what quantities are given. The difference lies mainly in the algebra.

■ PROBLEM-SOLVING STRATEGY

Motion in One Dimension at Constant Acceleration
The following procedure is recommended for solving problems involving accelerated motion.

1. **Read** the problem.
2. **Draw** a diagram, choosing a coordinate system, labeling initial and final points, and indicating directions of velocities and accelerations with arrows.
3. **Label** all quantities, circling the unknowns. Convert units as needed.
4. **Equations** from Table 2.4 should be selected next. All kinematics problems in this chapter can be solved with the first two equations, and the third is often convenient.
5. **Solve** for the unknowns. Doing so often involves solving two equations for two unknowns.
6. **Check** your answer, using common sense and estimates.

> **Tip 2.7 Pigs Don't Fly**
> After solving a problem, you should think about your answer and decide whether it seems reasonable. If it isn't, look for your mistake!

Most of these problems reduce to writing the kinematic equations from Table 2.4 and then substituting the correct values into the constants a, v_0, and x_0 from the given information. Doing this produces two equations—one linear and one quadratic—for two unknown quantities.

■ EXAMPLE 2.4 The Daytona 500

GOAL Apply the basic kinematic equations.

PROBLEM (a) A race car starting from rest accelerates at a constant rate of 5.00 m/s². What is the velocity of the car after it has traveled 1.00×10^2 ft? (b) How much time has elapsed? (c) Calculate the average velocity two different ways.

Figure 2.16 (Example 2.4)

STRATEGY We've read the problem, drawn the diagram in Figure 2.16, and chosen a coordinate system (steps 1 and 2). We'd like to find the velocity v after a certain known displacement Δx. The acceleration a is also known, as is the initial velocity v_0 (step 3, labeling, is complete), so the third equation in Table 2.4 looks most useful for solving part (a). Given the velocity, the first equation in Table 2.4 can then be used to find the time in part (b). Part (c) requires substitution into Equations 2.2 and 2.7, respectively.

SOLUTION

(a) Convert units of Δx to SI, using the information in the inside front cover.

$$1.00 \times 10^2 \text{ ft} = (1.00 \times 10^2 \text{ ft})\left(\frac{1 \text{ m}}{3.28 \text{ ft}}\right) = 30.5 \text{ m}$$

Write the kinematics equation for v^2 (step 4):

$$v^2 = v_0^2 + 2a\,\Delta x$$

2.5 One-Dimensional Motion with Constant Acceleration

Solve for v, taking the positive square root because the car moves to the right (step 5):

$$v = \sqrt{v_0^2 + 2a\,\Delta x}$$

Substitute $v_0 = 0$, $a = 5.00$ m/s², and $\Delta x = 30.5$ m:

$$v = \sqrt{v_0^2 + 2a\,\Delta x} = \sqrt{(0)^2 + 2(5.00 \text{ m/s}^2)(30.5 \text{ m})}$$
$$= \boxed{17.5 \text{ m/s}}$$

(b) How much time has elapsed?

Apply the first equation of Table 2.4:

$$v = at + v_0$$

Substitute values and solve for time t:

$$17.5 \text{ m/s} = (5.00 \text{ m/s}^2)t$$

$$t = \frac{17.5 \text{ m/s}}{5.00 \text{ m/s}^2} = \boxed{3.50 \text{ s}}$$

(c) Calculate the average velocity in two different ways.

Apply the definition of average velocity, Equation 2.2:

$$\bar{v} = \frac{x_f - x_i}{t_f - t_i} = \frac{30.5 \text{ m}}{3.50 \text{ s}} = \boxed{8.71 \text{ m/s}}$$

Apply the definition of average velocity in Equation 2.7:

$$\bar{v} = \frac{v_0 + v}{2} = \frac{0 + 17.5 \text{ m/s}}{2} = \boxed{8.75 \text{ m/s}}$$

REMARKS The answers are easy to check. An alternate technique is to use $\Delta x = v_0 t + \frac{1}{2}at^2$ to find t and then use the equation $v = v_0 + at$ to find v. Notice that the two different equations for calculating the average velocity, due to rounding, give slightly different answers.

QUESTION 2.4 What is the final speed if the displacement is increased by a factor of 4?

EXERCISE 2.4 Suppose the driver in this example now slams on the brakes, stopping the car in 4.00 s. Find (a) the acceleration, (b) the distance the car travels while braking, assuming the acceleration is constant, and (c) the average velocity.

ANSWERS (a) -4.38 m/s² (b) 35.0 m (c) 8.75 m/s

■ EXAMPLE 2.5 | Car Chase

GOAL Solve a problem involving two objects, one moving at constant acceleration and the other at constant velocity.

PROBLEM A car traveling at a constant speed of 24.0 m/s passes a trooper hidden behind a billboard, as in Figure 2.17. One second after the speeding car passes the billboard, the trooper sets off in chase with a constant acceleration of 3.00 m/s². **(a)** How long does it take the trooper to overtake the speeding car? **(b)** How fast is the trooper going at that time?

STRATEGY Solving this problem involves two simultaneous kinematics equations of position, one for the trooper and the other for the car. Choose $t = 0$ to correspond to the time the trooper takes up the chase, when the car is at $x_{car} = 24.0$ m because of its head start (24.0 m/s × 1.00 s). The trooper catches up with the car when their positions are the same, which suggests setting $x_{trooper} = x_{car}$ and solving for time, which can then be used to find the trooper's speed in part (b).

Figure 2.17 (Example 2.5) A speeding car passes a hidden trooper. When does the trooper catch up to the car?

SOLUTION

(a) How long does it take the trooper to overtake the car?

Write the equation for the car's displacement:

$$\Delta x_{car} = x_{car} - x_0 = v_0 t + \tfrac{1}{2}a_{car}t^2$$

Take $x_0 = 24.0$ m, $v_0 = 24.0$ m/s, and $a_{car} = 0$. Solve for x_{car}:

$$x_{car} = x_0 + vt = 24.0 \text{ m} + (24.0 \text{ m/s})t$$

(Continued)

Write the equation for the trooper's position, taking $x_0 = 0$, $v_0 = 0$, and $a_{trooper} = 3.00 \text{ m/s}^2$:

$$x_{trooper} = \tfrac{1}{2} a_{trooper} t^2 = \tfrac{1}{2}(3.00 \text{ m/s}^2) t^2 = (1.50 \text{ m/s}^2) t^2$$

Set $x_{trooper} = x_{car}$, and solve the quadratic equation. (The quadratic formula appears in Appendix A, Equation A.8.) Only the positive root is meaningful.

$$(1.50 \text{ m/s}^2) t^2 = 24.0 \text{ m} + (24.0 \text{ m/s}) t$$
$$(1.50 \text{ m/s}^2) t^2 - (24.0 \text{ m/s}) t - 24.0 \text{ m} = 0$$
$$t = \boxed{16.9 \text{ s}}$$

(b) Find the trooper's speed at that time.

Substitute the time into the trooper's velocity equation:

$$v_{trooper} = v_0 + a_{trooper} t = 0 + (3.00 \text{ m/s}^2)(16.9 \text{ s})$$
$$= \boxed{50.7 \text{ m/s}}$$

REMARKS The trooper, traveling about twice as fast as the car, must swerve or apply his brakes strongly to avoid a collision! This problem can also be solved graphically by plotting position versus time for each vehicle on the same graph. The intersection of the two graphs corresponds to the time and position at which the trooper overtakes the car.

QUESTION 2.5 The graphical solution corresponds to finding the intersection of what two types of curves in the tx-plane?

EXERCISE 2.5 A motorist with an expired license tag is traveling at 10.0 m/s down a street, and a policeman on a motorcycle, taking another 5.00 s to finish his donut, gives chase at an acceleration of 2.00 m/s². Find (a) the time required to catch the car and (b) the distance the trooper travels while overtaking the motorist.

ANSWERS (a) 13.7 s (b) 188 m

■ EXAMPLE 2.6 | Runway Length

GOAL Apply kinematics to horizontal motion with two phases.

PROBLEM A typical jetliner lands at a speed of 1.60×10^2 mi/h and decelerates at the rate of (10.0 mi/h)/s. If the plane travels at a constant speed of 1.60×10^2 mi/h for 1.00 s after landing before applying the brakes, what is the total displacement of the aircraft between touchdown on the runway and coming to rest?

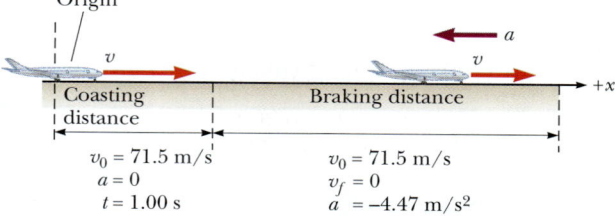

Figure 2.18 (Example 2.6) Coasting and braking distances for a landing jetliner.

STRATEGY See Figure 2.18. First, convert all quantities to SI units. The problem must be solved in two parts, or phases, corresponding to the initial coast after touchdown, followed by braking. Using the kinematic equations, find the displacement during each part and add the two displacements.

SOLUTION

Convert units of speed and acceleration to SI:

$$v_0 = (1.60 \times 10^2 \text{ mi/h})\left(\frac{0.447 \text{ m/s}}{1.00 \text{ mi/h}}\right) = 71.5 \text{ m/s}$$

$$a = (-10.0 \text{ (mi/h)/s})\left(\frac{0.447 \text{ m/s}}{1.00 \text{ mi/h}}\right) = -4.47 \text{ m/s}^2$$

Taking $a = 0$, $v_0 = 71.5$ m/s, and $t = 1.00$ s, find the displacement while the plane is coasting:

$$\Delta x_{coasting} = v_0 t + \tfrac{1}{2} a t^2 = (71.5 \text{ m/s})(1.00 \text{ s}) + 0 = 71.5 \text{ m}$$

Use the time-independent kinematic equation to find the displacement while the plane is braking.

$$v^2 = v_0^2 + 2 a \Delta x_{braking}$$

Take $a = -4.47$ m/s² and $v_0 = 71.5$ m/s. The negative sign on a means that the plane is slowing down.

$$\Delta x_{braking} = \frac{v^2 - v_0^2}{2a} = \frac{0 - (71.5 \text{ m/s})^2}{2.00(-4.47 \text{ m/s}^2)} = 572 \text{ m}$$

Sum the two results to find the total displacement:

$$\Delta x_{coasting} + \Delta x_{braking} = 71.5 \text{ m} + 572 \text{ m} = \boxed{644 \text{ m}}$$

REMARKS To find the displacement while braking, we could have used the two kinematics equations involving time, namely, $\Delta x = v_0 t + \tfrac{1}{2} a t^2$ and $v = v_0 + at$, but because we weren't interested in time, the time-independent equation was easier to use.

QUESTION 2.6 How would the answer change if the plane coasted for 2.00 s before the pilot applied the brakes?

EXERCISE 2.6 A jet lands at 80.0 m/s, the pilot applying the brakes 2.00 s after landing. Find the acceleration needed to stop the jet within 5.00×10^2 m after touchdown.

ANSWER -9.41 m/s^2

■ EXAMPLE 2.7 | The Acela: The Porsche of American Trains

GOAL Find accelerations and displacements from a velocity vs. time graph.

PROBLEM The sleek high-speed electric train known as the Acela (pronounced ahh-sell-ah) is currently in service on the Washington-New York-Boston run. The Acela consists of two power cars and six coaches and can carry 304 passengers at speeds up to 170 mi/h. In order to negotiate curves comfortably at high speeds, the train carriages tilt as much as 6° from the vertical, preventing passengers from being pushed to the side. A velocity vs. time graph for the Acela is shown in Figure 2.19a. **(a)** Describe the motion of the Acela. **(b)** Find the peak acceleration of the Acela in miles per hour per second ((mi/h)/s) as the train speeds up from 45 mi/h to 170 mi/h. **(c)** Find the train's displacement in miles between $t = 0$ and $t = 200$ s. **(d)** Find the average acceleration of the Acela and its displacement in miles in the interval from 200 s to 300 s. (The train has regenerative braking, which means that it feeds energy back into the utility lines each time it stops!) **(e)** Find the total displacement in the interval from 0 to 400 s. *Note:* Assume that all given quantities and estimates are good to two significant figures. (Estimates by different individuals may vary, and result in slightly different answers.)

STRATEGY For part (a), remember that the slope of the tangent line at any point of the velocity vs. time graph gives the acceleration at that time. To find the peak acceleration in part (b), study the graph and locate the point at which the slope is steepest. In parts (c) through (e), estimating the area under the curve gives the displacement during a given period, with areas below the time axis, as in part (e), subtracted from the total. The average acceleration in part (d) can be obtained by substituting numbers taken from the graph into the definition of average acceleration, $\bar{a} = \Delta v / \Delta t$.

SOLUTION

(a) Describe the motion.

From about −50 s to 50 s, the Acela cruises at a constant velocity in the +x-direction. Then the train accelerates in the +x-direction from 50 s to 200 s, reaching a top speed of about 170 mi/h, whereupon it brakes to rest at 350 s and reverses, steadily gaining speed in the −x-direction.

Figure 2.19 (Example 2.7) (a) Velocity vs. time graph for the Acela. (b) The slope of the steepest tangent blue line gives the peak acceleration, and the slope of the green line is the average acceleration between 200 s and 300 s. (c) The area under the velocity vs. time graph in some time interval gives the displacement of the Acela in that time interval. (d) (Exercise 2.7).

(Continued)

(b) Find the peak acceleration.

Calculate the slope of the steepest tangent line, which connects the points (50 s, 50 mi/h) and (100 s, 150 mi/h) (the light blue line in Figure 2.19b):

$$a = \text{slope} = \frac{\Delta v}{\Delta t} = \frac{(1.5 \times 10^2 - 5.0 \times 10^1) \text{ mi/h}}{(1.0 \times 10^2 - 5.0 \times 10^1) \text{s}}$$
$$= \boxed{2.0 \text{ (mi/h)/s}}$$

(c) Find the displacement between 0 s and 200 s.

Using triangles and rectangles, approximate the area in Figure 2.19c:

$$\Delta x_{0 \to 200 \text{ s}} = \text{area}_1 + \text{area}_2 + \text{area}_3 + \text{area}_4 + \text{area}_5$$
$$\approx (5.0 \times 10^1 \text{ mi/h})(5.0 \times 10^1 \text{ s})$$
$$+ (5.0 \times 10^1 \text{ mi/h})(5.0 \times 10^1 \text{ s})$$
$$+ (1.6 \times 10^2 \text{ mi/h})(1.0 \times 10^2 \text{ s})$$
$$+ \tfrac{1}{2}(5.0 \times 10^1 \text{ s})(1.0 \times 10^2 \text{ mi/h})$$
$$+ \tfrac{1}{2}(1.0 \times 10^2 \text{ s})(1.7 \times 10^2 \text{ mi/h} - 1.6 \times 10^2 \text{ mi/h})$$
$$= 2.4 \times 10^4 \text{ (mi/h)s}$$

Convert units to miles by converting hours to seconds:

$$\Delta x_{0 \to 200 \text{ s}} \approx 2.4 \times 10^4 \frac{\text{mi} \cdot \text{s}}{\text{h}} \left(\frac{1 \text{ h}}{3600 \text{ s}}\right) = \boxed{6.7 \text{ mi}}$$

(d) Find the average acceleration from 200 s to 300 s, and find the displacement.

The slope of the green line is the average acceleration from 200 s to 300 s (Fig. 2.19b):

$$\bar{a} = \text{slope} = \frac{\Delta v}{\Delta t} = \frac{(1.0 \times 10^1 - 1.7 \times 10^2) \text{ mi/h}}{1.0 \times 10^2 \text{ s}}$$
$$= \boxed{-1.6 \text{ (mi/h)/s}}$$

The displacement from 200 s to 300 s is equal to area₆, which is the area of a triangle plus the area of a very narrow rectangle beneath the triangle:

$$\Delta x_{200 \to 300 \text{ s}} \approx \tfrac{1}{2}(1.0 \times 10^2 \text{ s})(1.7 \times 10^2 - 1.0 \times 10^1) \text{ mi/h}$$
$$+ (1.0 \times 10^1 \text{ mi/h})(1.0 \times 10^2 \text{ s})$$
$$= 9.0 \times 10^3 \text{(mi/h)(s)} = \boxed{2.5 \text{ mi}}$$

(e) Find the total displacement from 0 s to 400 s.

The total displacement is the sum of all the individual displacements. We still need to calculate the displacements for the time intervals from 300 s to 350 s and from 350 s to 400 s. The latter is negative because it's below the time axis.

$$\Delta x_{300 \to 350 \text{ s}} \approx \tfrac{1}{2}(5.0 \times 10^1 \text{ s})(1.0 \times 10^1 \text{ mi/h})$$
$$= 2.5 \times 10^2 \text{(mi/h)(s)}$$
$$\Delta x_{350 \to 400 \text{ s}} \approx \tfrac{1}{2}(5.0 \times 10^1 \text{ s})(-5.0 \times 10^1 \text{ mi/h})$$
$$= -1.3 \times 10^3 \text{(mi/h)(s)}$$

Find the total displacement by summing the parts:

$$\Delta x_{0 \to 400 \text{ s}} \approx (2.4 \times 10^4 + 9.0 \times 10^3 + 2.5 \times 10^2$$
$$- 1.3 \times 10^3)\text{(mi/h)(s)} = \boxed{8.9 \text{ mi}}$$

REMARKS There are a number of ways to find the approximate area under a graph. Choice of technique is a personal preference.

QUESTION 2.7 According to the graph in Figure 2.19a, at what different times is the acceleration zero?

EXERCISE 2.7 Suppose the velocity vs. time graph of another train is given in Figure 2.19d. Find (a) the maximum instantaneous acceleration and (b) the total displacement in the interval from 0 s to 4.00×10^2 s.

ANSWERS (a) 1.0 (mi/h)/s (b) 4.7 mi

2.6 Freely Falling Objects

LEARNING OBJECTIVES

1. Apply the kinematics equations for constant acceleration to freely falling objects near Earth's surface.
2. Construct and solve the kinematics equations for motion involving two distinct phases of acceleration.

2.6 | Freely Falling Objects

When air resistance is negligible, all objects dropped under the influence of gravity near Earth's surface fall toward Earth with the same constant acceleration. This idea may seem obvious today, but it wasn't until about 1600 that it was accepted. Prior to that time, the teachings of the great philosopher Aristotle (384–322 B.C.) had held that heavier objects fell faster than lighter ones.

According to legend, Galileo discovered the law of falling objects by observing that two different weights dropped simultaneously from the Leaning Tower of Pisa hit the ground at approximately the same time. Although it's unlikely that this particular experiment was carried out, we know that Galileo performed many systematic experiments with objects moving on inclined planes. In his experiments he rolled balls down a slight incline and measured the distances they covered in successive time intervals. The purpose of the incline was to reduce the acceleration and enable Galileo to make accurate measurements of the intervals. (Some people refer to this experiment as "diluting gravity.") By gradually increasing the slope of the incline he was finally able to draw mathematical conclusions about freely falling objects, because a falling ball is equivalent to a ball going down a vertical incline. Galileo's achievements in the science of mechanics paved the way for Newton in his development of the laws of motion, which we will study in Chapter 4.

Try the following experiment: Drop a hammer and a feather simultaneously from the same height. The hammer hits the floor first because air drag has a greater effect on the much lighter feather. On August 2, 1971, this same experiment was conducted on the Moon by astronaut David Scott, and the hammer and feather fell with exactly the same acceleration, as expected, hitting the lunar surface at the same time. In the idealized case where air resistance is negligible, such motion is called *free fall*.

The expression *freely falling object* doesn't necessarily refer to an object dropped from rest. **A freely falling object is any object moving freely under the influence of gravity alone, regardless of its initial motion.** Objects thrown upward or downward and those released from rest are all considered freely falling.

We denote the magnitude of the **free-fall acceleration** by the symbol g. The value of g decreases with increasing altitude, and varies slightly with latitude as well. At Earth's surface, the value of g is approximately 9.80 m/s^2. Unless stated otherwise, we will use this value for g in doing calculations. For quick estimates, use $g \approx 10 \text{ m/s}^2$.

If we neglect air resistance and assume that the free-fall acceleration doesn't vary with altitude over short vertical distances, then the motion of a freely falling object is the same as motion in one dimension under constant acceleration. This means that the kinematics equations developed in Section 2.5 can be applied. It's conventional to define "up" as the $+y$-direction and to use y as the position variable. In that case the acceleration is $a = -g = -9.80 \text{ m/s}^2$. In Chapter 7, we study the variation in g with altitude.

Galileo Galilei
Italian Physicist and Astronomer
(1564–1642)
Galileo formulated the laws that govern the motion of objects in free fall. He also investigated the motion of an object on an inclined plane, established the concept of relative motion, invented the thermometer, and discovered that the motion of a swinging pendulum could be used to measure time intervals. After designing and constructing his own telescope, he discovered four of Jupiter's moons, found that our own Moon's surface is rough, discovered sunspots and the phases of Venus, and showed that the Milky Way consists of an enormous number of stars. Galileo publicly defended Nicolaus Copernicus's assertion that the Sun is at the center of the Universe (the heliocentric system). He published *Dialogue Concerning Two New World Systems* to support the Copernican model, a view the Church declared to be heretical. After being taken to Rome in 1633 on a charge of heresy, he was sentenced to life imprisonment and later was confined to his villa at Arcetri, near Florence, where he died in 1642.

■ Quick Quiz

2.6 A tennis player on serve tosses a ball straight up. While the ball is in free fall, does its acceleration (a) increase, (b) decrease, (c) increase and then decrease, (d) decrease and then increase, or (e) remain constant?

2.7 As the tennis ball of Quick Quiz 2.6 travels through the air, does its speed (a) increase, (b) decrease, (c) decrease and then increase, (d) increase and then decrease, or (e) remain the same?

2.8 A skydiver jumps out of a hovering helicopter. A few seconds later, another skydiver jumps out, so they both fall along the same vertical line relative to the helicopter. Assume both skydivers fall with the same acceleration. Does the vertical distance between them (a) increase, (b) decrease, or (c) stay the same? Does the difference in their velocities (d) increase, (e) decrease, or (f) stay the same?

EXAMPLE 2.8 — Not a Bad Throw for a Rookie!

GOAL Apply the kinematic equations to a freely falling object with a nonzero initial velocity.

PROBLEM A ball is thrown from the top of a building with an initial velocity of 20.0 m/s straight upward, at an initial height of 50.0 m above the ground. The ball just misses the edge of the roof on its way down, as shown in Figure 2.20. Determine (a) the time needed for the ball to reach its maximum height, (b) the maximum height, (c) the time needed for the ball to return to the height from which it was thrown and the velocity of the ball at that instant, (d) the time needed for the ball to reach the ground, and (e) the velocity and position of the ball at $t = 5.00$ s. Neglect air drag.

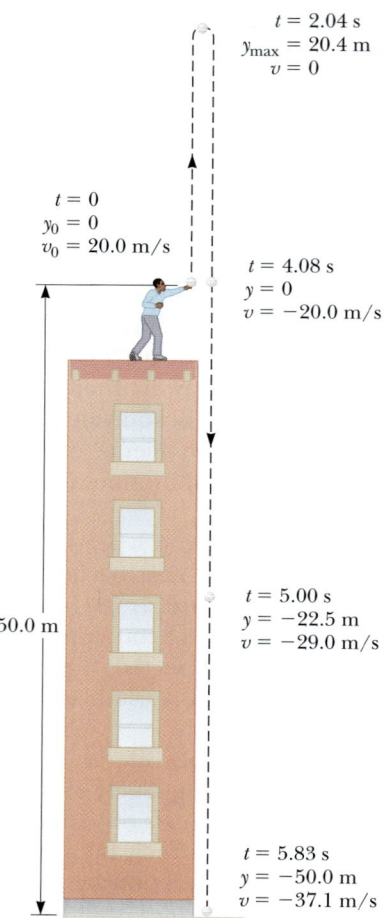

Figure 2.20 (Example 2.8) A ball is thrown upward with an initial velocity of $v_0 = +20.0$ m/s. Positions and velocities are given for several times.

STRATEGY The diagram in Figure 2.20 establishes a coordinate system with $y_0 = 0$ at the level at which the ball is released from the thrower's hand, with y positive upward. Write the velocity and position kinematic equations for the ball, and substitute the given information. All the answers come from these two equations by using simple algebra or by just substituting the time. In part (a), for example, the ball comes to rest for an instant at its maximum height, so set $v = 0$ at this point and solve for time. Then substitute the time into the displacement equation, obtaining the maximum height.

SOLUTION

(a) Find the time when the ball reaches its maximum height.

Write the velocity and position kinematic equations:

$$v = at + v_0$$
$$\Delta y = y - y_0 = v_0 t + \tfrac{1}{2}at^2$$

Substitute $a = -9.80$ m/s^2, $v_0 = 20.0$ m/s, and $y_0 = 0$ into the preceding two equations:

(1) $\quad v = (-9.80 \text{ m/s}^2)t + 20.0 \text{ m/s}$

(2) $\quad y = (20.0 \text{ m/s})t - (4.90 \text{ m/s}^2)t^2$

Substitute $v = 0$, the velocity at maximum height, into Equation (1) and solve for time:

$$0 = (-9.80 \text{ m/s}^2)t + 20.0 \text{ m/s}$$
$$t = \frac{-20.0 \text{ m/s}}{-9.80 \text{ m/s}^2} = \boxed{2.04 \text{ s}}$$

(b) Determine the ball's maximum height.

Substitute the time $t = 2.04$ s into Equation (2):

$$y_{max} = (20.0 \text{ m/s})(2.04 \text{ s}) - (4.90 \text{ m/s}^2)(2.04 \text{ s})^2 = \boxed{20.4 \text{ m}}$$

(c) Find the time the ball takes to return to its initial position, and find the velocity of the ball at that time.

Set $y = 0$ in Equation (2) and solve for t:

$$0 = (20.0 \text{ m/s})t - (4.90 \text{ m/s}^2)t^2$$
$$= t(20.0 \text{ m/s} - 4.90 \text{ m/s}^2 t)$$
$$t = \boxed{4.08 \text{ s}}$$

Substitute the time into Equation (1) to get the velocity:

$v = 20.0 \text{ m/s} + (-9.80 \text{ m/s}^2)(4.08 \text{ s}) = \boxed{-20.0 \text{ m/s}}$

(d) Find the time required for the ball to reach the ground.

In Equation (2), set $y = -50.0$ m:

$-50.0 \text{ m} = (20.0 \text{ m/s})t - (4.90 \text{ m/s}^2)t^2$

Apply the quadratic formula and take the positive root:

$t = \boxed{5.83 \text{ s}}$

(e) Find the velocity and position of the ball at $t = 5.00$ s.

Substitute values into Equations (1) and (2):

$v = (-9.80 \text{ m/s}^2)(5.00 \text{ s}) + 20.0 \text{ m/s} = \boxed{-29.0 \text{ m/s}}$

$y = (20.0 \text{ m/s})(5.00 \text{ s}) - (4.90 \text{ m/s}^2)(5.00 \text{ s})^2 = \boxed{-22.5 \text{ m}}$

REMARKS Notice how everything follows from the two kinematic equations. Once they are written down and the constants correctly identified as in Equations (1) and (2), the rest is relatively easy. If the ball were thrown downward, the initial velocity would have been negative.

QUESTION 2.8 How would the answer to part (b), the maximum height, change if the person throwing the ball was jumping upward at the instant he released the ball?

EXERCISE 2.8 A projectile is launched straight up at 60.0 m/s from a height of 80.0 m, at the edge of a sheer cliff. The projectile falls, just missing the cliff and hitting the ground below. Find (a) the maximum height of the projectile above the point of firing, (b) the time it takes to hit the ground at the base of the cliff, and (c) its velocity at impact.

ANSWERS (a) 184 m (b) 13.5 s (c) −72.3 m/s

■ EXAMPLE 2.9 | Maximum Height Derived

GOAL Find the maximum height of a thrown projectile using symbols.

PROBLEM Refer to Example 2.8. Use symbolic manipulation to find **(a)** the time t_{max} it takes the ball to reach its maximum height and **(b)** an expression for the maximum height that doesn't depend on time. Answers should be expressed in terms of the quantities v_0, g, and y_0 only.

STRATEGY When the ball reaches its maximum height, its velocity is zero, so for part (a) solve the kinematics velocity equation for time t and set $v = 0$. For part (b), substitute the expression for time found in part (a) into the displacement equation, solving it for the maximum height.

SOLUTION

(a) Find the time it takes the ball to reach its maximum height.

Write the velocity kinematics equation:

$v = at + v_0$

Move v_0 to the left side of the equation:

$v - v_0 = at$

Divide both sides by a:

$\dfrac{v - v_0}{a} = \dfrac{at}{a} = t$

Turn the equation around so that t is on the left and substitute $v = 0$, corresponding to the velocity at maximum height:

(1) $\quad t = \dfrac{-v_0}{a}$

Replace t by t_{max} and substitute $a = -g$:

(2) $\quad t_{max} = \boxed{\dfrac{v_0}{g}}$

(b) Find the maximum height.

Write the equation for the position y at any time:

$y = y_0 + v_0 t + \tfrac{1}{2}at^2$

(Continued)

Substitute $t = -v_0/a$, which corresponds to the time it takes to reach y_{max}, the maximum height:

$$y_{max} = y_0 + v_0\left(\frac{-v_0}{a}\right) + \frac{1}{2}a\left(\frac{-v_0}{a}\right)^2$$

$$= y_0 - \frac{v_0^2}{a} + \frac{1}{2}\frac{v_0^2}{a}$$

Combine the last two terms and substitute $a = -g$:

(3) $\quad y_{max} = y_0 + \dfrac{v_0^2}{2g}$

REMARKS Notice that $g = +9.8$ m/s^2, so the second term is positive overall. Equations (1)–(3) are much more useful than a numerical answer because the effect of changing one value can be seen immediately. For example, doubling the initial velocity v_0 quadruples the displacement above the point of release. Notice also that y_{max} could be obtained more readily from the time-independent equation, $v^2 - v_0^2 = 2a\Delta y$.

QUESTION 2.9 By what factor would the maximum displacement above the rooftop be increased if the building were transported to the Moon, where $a = -\frac{1}{6}g$?

EXERCISE 2.9 (a) Using symbols, find the time t_E it takes for a ball to reach the ground on Earth if released from rest at height y_0. (b) In terms of t_E, how much time t_M would be required if the building were on Mars, where $a = -0.385g$?

ANSWERS (a) $t_E = \sqrt{\dfrac{2y_0}{g}}$ (b) $t_M = 1.61 t_E$

EXAMPLE 2.10 | A Rocket Goes Ballistic

GOAL Solve a problem involving a powered ascent followed by free-fall motion.

PROBLEM A rocket moves straight upward, starting from rest with an acceleration of $+29.4$ m/s^2. It runs out of fuel at the end of 4.00 s and continues to coast upward, reaching a maximum height before falling back to Earth. **(a)** Find the rocket's velocity and position at the end of 4.00 s. **(b)** Find the maximum height the rocket reaches. **(c)** Find the velocity the instant before the rocket crashes on the ground.

STRATEGY Take $y = 0$ at the launch point and y positive upward, as in Figure 2.21. The problem consists of two phases. In phase 1 the rocket has a net *upward* acceleration of 29.4 m/s^2, and we can use the kinematic equations with constant acceleration a to find the height and velocity of the rocket at the end of phase 1, when the fuel is burned up. In phase 2 the rocket is in free fall and has an acceleration of -9.80 m/s^2, with initial velocity and position given by the results of phase 1. Apply the kinematic equations for free fall.

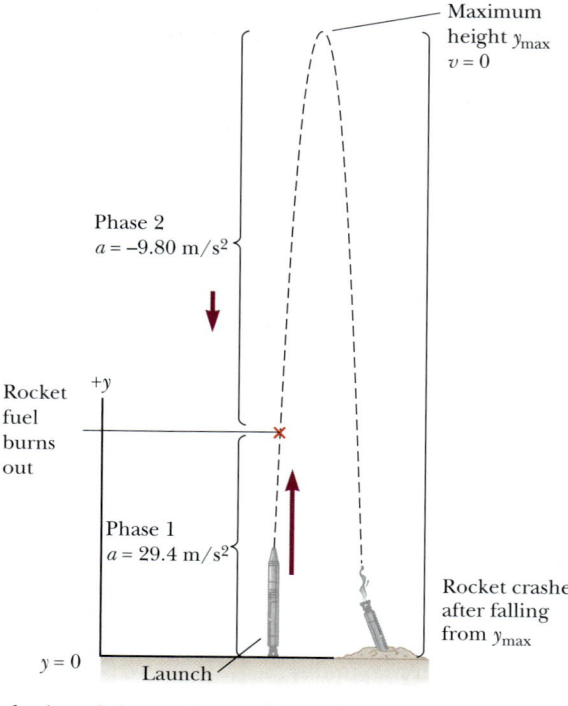

Figure 2.21 (Example 2.10) Two linked phases of motion for a rocket that is launched, uses up its fuel, and crashes.

SOLUTION
(a) Phase 1: Find the rocket's velocity and position after 4.00 s.

Write the velocity and position kinematic equations:

(1) $\quad v = v_0 + at$

(2) $\quad \Delta y = y - y_0 = v_0 t + \frac{1}{2}at^2$

Adapt these equations to phase 1, substituting $a = 29.4 \text{ m/s}^2$, $v_0 = 0$, and $y_0 = 0$:

(3) $v = (29.4 \text{ m/s}^2)t$
(4) $y = \frac{1}{2}(29.4 \text{ m/s}^2)t^2 = (14.7 \text{ m/s}^2)t^2$

Substitute $t = 4.00$ s into Equations (3) and (4) to find the rocket's velocity v and position y at the time of burnout. These will be called v_b and y_b, respectively.

$\boxed{v_b = 118 \text{ m/s}}$ and $\boxed{y_b = 235 \text{ m}}$

(b) Phase 2: Find the maximum height the rocket attains.

Adapt Equations (1) and (2) to phase 2, substituting $a = -9.8 \text{ m/s}^2$, $v_0 = v_b = 118 \text{ m/s}$, and $y_0 = y_b = 235 \text{ m}$:

(5) $v = (-9.8 \text{ m/s}^2)t + 118 \text{ m/s}$
(6) $y = 235 \text{ m} + (118 \text{ m/s})t - (4.90 \text{ m/s}^2)t^2$

Substitute $v = 0$ (the rocket's velocity at maximum height) in Equation (5) to get the time it takes the rocket to reach its maximum height:

$0 = (-9.8 \text{ m/s}^2)t + 118 \text{ m/s} \rightarrow t = \dfrac{118 \text{ m/s}}{9.80 \text{ m/s}^2} = 12.0 \text{ s}$

Substitute $t = 12.0$ s into Equation (6) to find the rocket's maximum height:

$y_{max} = 235 \text{ m} + (118 \text{ m/s})(12.0 \text{ s}) - (4.90 \text{ m/s}^2)(12.0 \text{ s})^2$
$= \boxed{945 \text{ m}}$

(c) Phase 2: Find the velocity of the rocket just prior to impact.

Find the time to impact by setting $y = 0$ in Equation (6) and using the quadratic formula:

$0 = 235 \text{ m} + (118 \text{ m/s})t - (4.90 \text{ m/s})t^2$
$t = 25.9 \text{ s}$

Substitute this value of t into Equation (5):

$v = (-9.80 \text{ m/s}^2)(25.9 \text{ s}) + 118 \text{ m/s} = \boxed{-136 \text{ m/s}}$

REMARKS You may think that it is more natural to break this problem into three phases, with the second phase ending at the maximum height and the third phase a free fall from maximum height to the ground. Although this approach gives the correct answer, it's an unnecessary complication. Two phases are sufficient, one for each different acceleration.

QUESTION 2.10 If, instead, some fuel remains, at what height should the engines be fired again to brake the rocket's fall and allow a perfectly soft landing? (Assume the same acceleration as during the initial ascent.)

EXERCISE 2.10 An experimental rocket designed to land upright falls freely from a height of 2.00×10^2 m, starting at rest. At a height of 80.0 m, the rocket's engines start and provide constant upward acceleration until the rocket lands. What acceleration is required if the speed on touchdown is to be zero? (Neglect air resistance.)

ANSWER 14.7 m/s^2

SUMMARY

2.1 Displacement

The **displacement** of an object moving along the x-axis is defined as the change in position of the object,

$$\Delta x \equiv x_f - x_i \qquad [2.1]$$

where x_i is the initial position of the object and x_f is its final position.

A **vector** quantity is characterized by both a magnitude and a direction. A **scalar** quantity has a magnitude only.

2.2 Velocity

The **average speed** of an object is given by

$$\text{Average speed} \equiv \dfrac{\text{path length}}{\text{elapsed time}}$$

The **average velocity** $\bar{v}$ during a time interval Δt is the displacement Δx divided by Δt.

$$\bar{v} \equiv \dfrac{\Delta x}{\Delta t} = \dfrac{x_f - x_i}{t_f - t_i} \qquad [2.2]$$

The average velocity is equal to the slope of the straight line joining the initial and final points on a graph of the position of the object versus time.

The slope of the line tangent to the position vs. time curve at some point is equal to the **instantaneous velocity** at that time. The **instantaneous speed** of an object is defined as the magnitude of the instantaneous velocity.

2.3 Acceleration

The **average acceleration** $\bar{a}$ of an object undergoing a change in velocity Δv during a time interval Δt is

$$\bar{a} \equiv \frac{\Delta v}{\Delta t} = \frac{v_f - v_i}{t_f - t_i} \quad [2.4]$$

The **instantaneous acceleration** of an object at a certain time equals the slope of a velocity vs. time graph at that instant.

2.5 One-Dimensional Motion with Constant Acceleration

The most useful equations that describe the motion of an object moving with constant acceleration along the x-axis are as follows:

$$v = v_0 + at \quad [2.6]$$

$$\Delta x = v_0 t + \tfrac{1}{2} a t^2 \quad [2.9]$$

$$v^2 = v_0^2 + 2a\Delta x \quad [2.10]$$

All problems can be solved with the first two equations alone, the last being convenient when time doesn't explicitly enter the problem. After the constants are properly identified, most problems reduce to one or two equations in as many unknowns.

2.6 Freely Falling Objects

An object falling in the presence of Earth's gravity exhibits a free-fall acceleration directed toward Earth's center. If air friction is neglected and if the altitude of the falling object is small compared with Earth's radius, then we can assume that the free-fall acceleration $g = 9.8$ m/s² is constant over the range of motion. Equations 2.6, 2.9, and 2.10 apply, with $a = -g$.

■ WARM-UP EXERCISES

The warm-up exercises in this chapter may be assigned online in Enhanced WebAssign.

1. **Math Review** Solve the quadratic equation $2.00t^2 - 6.00t - 9.00 = 0$ using the quadratic formula, finding both solutions.

2. **Math Review** Solve the following two equations for (a) the time t, and (b) the position, x. Assume SI units.

    $-9.8t + 49 = 0$ and $x = -4.9t^2 + 49t + 16$

3. **Math Review** Solve the following two equations for (a) the (positive) time t, and (b) the position x. Assume SI units.

    $x = 3.00t^2 \qquad x = 24.0t + 72.0$

4. A football player runs from his own goal line to the opposing team's goal line, returning to the fifty-yard line, all in 18.0 s. Calculate (a) his average speed, and (b) the magnitude of his average velocity. (See Section 2.2.)

5. A ball is thrown downward from the top of a 40.0 m tower with an initial speed of 12.0 m/s. Assuming negligible air resistance, what is the speed of the ball just before hitting the ground? (See Section 2.6.)

6. An arrow is shot straight up in the air at an initial speed of 15.0 m/s. After how much time is the arrow heading downward at a speed of 8.00 m/s? (See Section 2.6.)

7. A red ball is dropped from rest at a height of 6.00 m. A blue ball at a height of 10.0 m is thrown down at the same instant at 4.00 m/s. How long does it take the blue ball to catch up with the red ball? (See Sections 2.5 and 2.6.)

■ CONCEPTUAL QUESTIONS

The conceptual questions in this chapter may be assigned online in Enhanced WebAssign.

1. If the velocity of a particle is nonzero, can the particle's acceleration be zero? Explain.

2. If the velocity of a particle is zero, can the particle's acceleration be nonzero? Explain.

3. If a car is traveling eastward, can its acceleration be westward? Explain.

4. (a) Can the equations in Table 2.4 be used in a situation where the acceleration varies with time? (b) Can they be used when the acceleration is zero?

5. Two cars are moving in the same direction in parallel lanes along a highway. At some instant, the velocity of car A exceeds the velocity of car B. Does that mean that the acceleration of A is greater than that of B at that instant? Explain.

6. Figure CQ2.6 shows strobe photographs taken of a disk moving from left to right under different conditions. The time interval between images is constant. Taking the direction to the right to be positive, describe the motion of the disk in each case. For which case is (a) the acceleration positive? (b) the acceleration negative? (c) the velocity constant?

Figure CQ2.6

7. (a) Can the instantaneous velocity of an object at an instant of time ever be greater in magnitude than the average velocity over a time interval containing that instant? (b) Can it ever be less?

8. A ball is thrown vertically upward. (a) What are its velocity and acceleration when it reaches its maximum altitude? (b) What is the acceleration of the ball just before it hits the ground?

9. Consider the following combinations of signs and values for the velocity and acceleration of a particle with respect to a one-dimensional x-axis:

	Velocity	Acceleration
a.	Positive	Positive
b.	Positive	Negative
c.	Positive	Zero
d.	Negative	Positive
e.	Negative	Negative
f.	Negative	Zero
g.	Zero	Positive
h.	Zero	Negative

Describe what the particle is doing in each case and give a real-life example for an automobile on an east–west one-dimensional axis, with east considered the positive direction.

10. A ball rolls in a straight line along the horizontal direction. Using motion diagrams (or multiflash photographs), describe the velocity and acceleration of the ball for each of the following situations: (a) The ball moves to the right at a constant speed. (b) The ball moves from right to left and continually slows down. (c) The ball moves from right to left and continually speeds up. (d) The ball moves to the right, first speeding up at a constant rate and then slowing down at a constant rate.

11. An object moves along the x-axis, its position measured at each instant of time. The data are organized into an accurate graph of x vs. t. Which of the following quantities cannot be obtained from this graph? (a) The velocity at any instant (b) the acceleration at any instant (c) the displacement during some time interval (d) the average velocity during some time interval (e) the speed of the particle at any instant.

12. A ball is thrown straight up in the air. For which situation are both the instantaneous velocity and the acceleration zero? (a) On the way up (b) at the top of the flight path (c) on the way down (d) halfway up and halfway down (e) none of these.

13. A juggler throws a bowling pin straight up in the air. After the pin leaves his hand and while it is in the air, which statement is true? (a) The velocity of the pin is always in the same direction as its acceleration. (b) The velocity of the pin is never in the same direction as its acceleration. (c) The acceleration of the pin is zero. (d) The velocity of the pin is opposite its acceleration on the way up. (e) The velocity of the pin is in the same direction as its acceleration on the way up.

14. A racing car starts from rest and reaches a final speed v in a time t. If the acceleration of the car is constant during this time, which of the following statements must be true? (a) The car travels a distance vt. (b) The average speed of the car is $v/2$. (c) The acceleration of the car is v/t. (d) The velocity of the car remains constant. (e) None of these.

■ PROBLEMS

WebAssign The problems in this chapter may be assigned online in Enhanced WebAssign.

1. denotes straightforward problem; 2. denotes intermediate problem;
3. denotes challenging problem

1. denotes full solution available in *Student Solutions Manual/Study Guide*

1. denotes problems most often assigned in Enhanced WebAssign

BIO denotes biomedical problems
GP denotes guided problems
M denotes Master It tutorial available in Enhanced WebAssign
Q|C denotes asking for quantitative and conceptual reasoning
S denotes symbolic reasoning problem
W denotes Watch It video solution available in Enhanced WebAssign

2.1 Displacement
2.2 Velocity

1. **BIO** The speed of a nerve impulse in the human body is about 100 m/s. If you accidentally stub your toe in the dark, estimate the time it takes the nerve impulse to travel to your brain.

2. Light travels at a speed of about 3×10^8 m/s. (a) How many miles does a pulse of light travel in a time interval of 0.1 s, which is about the blink of an eye? (b) Compare this distance to the diameter of Earth.

3. A person travels by car from one city to another with different constant speeds between pairs of cities. She drives for 30.0 min at 80.0 km/h, 12.0 min at 100 km/h, and 45.0 min at 40.0 km/h and spends 15.0 min eating lunch and buying gas. (a) Determine the average speed for the trip. (b) Determine the distance between the initial and final cities along the route.

4. The current indoor world record time in the 200-m race is 19.92 s, held by Frank Fredericks of Namibia (1996), while the indoor record time in the one-mile race is 228.5 s, held by Hicham El Guerrouj of Morroco (1997). Find the mean speed in meters per second corresponding to these record times for (a) the 200-m event and (b) the one-mile event.

5. Two boats start together and race across a 60-km-wide lake and back. Boat A goes across at 60 km/h and returns at 60 km/h. Boat B goes across at 30 km/h, and its crew, realizing how far behind it is getting, returns at 90 km/h. Turnaround times are negligible, and the boat that completes the round trip first wins. (a) Which boat wins and by how much? (Or is it a tie?) (b) What is the average velocity of the winning boat?

6. A graph of position versus time for a certain particle moving along the x-axis is shown in Figure P2.6. Find the average velocity in the time intervals from (a) 0 to 2.00 s, (b) 0 to 4.00 s, (c) 2.00 s to 4.00 s, (d) 4.00 s to 7.00 s, and (e) 0 to 8.00 s.

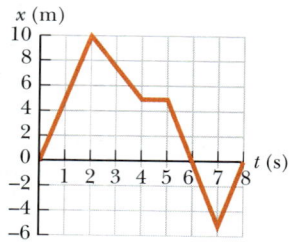

Figure P2.6 Problems 6 and 17

7. **W** A motorist drives north for 35.0 minutes at 85.0 km/h and then stops for 15.0 minutes. He then continues north, traveling 130 km in 2.00 h. (a) What is his total displacement? (b) What is his average velocity?

8. A tennis player moves in a straight-line path as shown in Figure P2.8. Find her average velocity in the time intervals from (a) 0 to 1.0 s, (b) 0 to 4.0 s, (c) 1.0 s to 5.0 s, and (d) 0 to 5.0 s.

Figure P2.8

9. A jet plane has a takeoff speed of $v_{to} = 75$ m/s and can move along the runway at an average acceleration of 1.3 m/s². If the length of the runway is 2.5 km, will the plane be able to use this runway safely? Defend your answer.

10. Two cars travel in the same direction along a straight highway, one at a constant speed of 55 mi/h and the other at 70 mi/h. (a) Assuming they start at the same point, how much sooner does the faster car arrive at a destination 10 mi away? (b) How far must the faster car travel before it has a 15-min lead on the slower car?

11. The cheetah can reach a top speed of 114 km/h (71 mi/h). While chasing its prey in a short sprint, a cheetah starts from rest and runs 45 m in a straight line, reaching a final speed of 72 km/h. (a) Determine the cheetah's average acceleration during the short sprint, and (b) find its displacement at $t = 3.5$ s.

12. **S** An athlete swims the length L of a pool in a time t_1 and makes the return trip to the starting position in a time t_2. If she is swimming initially in the positive x-direction, determine her average velocities symbolically in (a) the first half of the swim, (b) the second half of the swim, and (c) the round trip. (d) What is her average speed for the round trip?

13. **M** A person takes a trip, driving with a constant speed of 89.5 km/h, except for a 22.0-min rest stop. If the person's average speed is 77.8 km/h, (a) how much time is spent on the trip and (b) how far does the person travel?

14. A tortoise can run with a speed of 0.10 m/s, and a hare can run 20 times as fast. In a race, they both start at the same time, but the hare stops to rest for 2.0 minutes. The tortoise wins by a shell (20 cm). (a) How long does the race take? (b) What is the length of the race?

15. To qualify for the finals in a racing event, a race car must achieve an average speed of 250 km/h on a track with a total length of 1 600 m. If a particular car covers the first half of the track at an average speed of 230 km/h, what minimum average speed must it have in the second half of the event in order to qualify?

16. **Q|C S** One athlete in a race running on a long, straight track with a constant speed v_1 is a distance d behind a second athlete running with a constant speed v_2. (a) Under what circumstances is the first athlete able to overtake the second athlete? (b) Find the time t it takes the first athlete to overtake the second athlete, in terms of d, v_1, and v_2. (c) At what minimum distance d_2 from the leading athlete must the finish line be located so that the trailing athlete can at least tie for first place? Express d_2 in terms of d, v_1, and v_2 by using the result of part (b).

17. A graph of position versus time for a certain particle moving along the x-axis is shown in Figure P2.6. Find the instantaneous velocity at the instants (a) $t = 1.00$ s, (b) $t = 3.00$ s, (c) $t = 4.50$ s, and (d) $t = 7.50$ s.

18. A race car moves such that its position fits the relationship

$$x = (5.0 \text{ m/s})t + (0.75 \text{ m/s}^3)t^3$$

where x is measured in meters and t in seconds. (a) Plot a graph of the car's position versus time. (b) Determine the instantaneous velocity of the car at $t = 4.0$ s, using time intervals of 0.40 s, 0.20 s, and 0.10 s. (c) Compare the average velocity during the first 4.0 s with the results of part (b).

19. Runner A is initially 4.0 mi west of a flagpole and is running with a constant velocity of 6.0 mi/h due east. Runner B is initially 3.0 mi east of the flagpole and is running with a constant velocity of 5.0 mi/h due west. How far are the runners from the flagpole when they meet?

2.3 Acceleration

20. A particle starts from rest and accelerates as shown in Figure P2.20. Determine (a) the particle's speed at $t = 10.0$ s and at $t = 20.0$ s, and (b) the distance traveled in the first 20.0 s.

21. A 50.0-g Super Ball traveling at 25.0 m/s bounces off a brick wall and rebounds at 22.0 m/s. A high-speed camera records this event. If the ball is in contact with the wall for 3.50 ms, what is the magnitude of the average acceleration of the ball during this time interval?

22. **BIO** The average person passes out at an acceleration of $7g$ (that is, seven times the gravitational acceleration on Earth). Suppose a car is designed to accelerate at this rate. How much time would be required for the car to accelerate from rest to 60.0 miles per hour? (The car would need rocket boosters!)

23. **W** A certain car is capable of accelerating at a rate of 0.60 m/s². How long does it take for this car to go from a speed of 55 mi/h to a speed of 60 mi/h?

24. The velocity vs. time graph for an object moving along a straight path is shown in Figure P2.24. (i) Find the average acceleration of the object during the time intervals (a) 0 to 5.0 s, (b) 5.0 s to 15 s, and (c) 0 to 20 s. (ii) Find the instantaneous acceleration at (a) 2.0 s, (b) 10 s, and (c) 18 s.

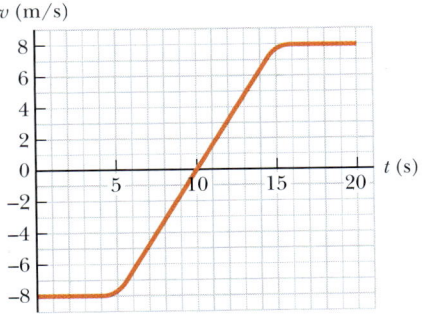

Figure P2.24

25. A steam catapult launches a jet aircraft from the aircraft carrier *John C. Stennis*, giving it a speed of 175 mi/h in 2.50 s. (a) Find the average acceleration of the plane. (b) Assuming the acceleration is constant, find the distance the plane moves.

2.5 One-Dimensional Motion with Constant Acceleration

26. Solve Example 2.5, "Car Chase" by a graphical method. On the same graph, plot position versus time for the car and the trooper. From the intersection of the two curves, read the time at which the trooper overtakes the car.

27. **M** An object moving with uniform acceleration has a velocity of 12.0 cm/s in the positive x-direction when its x-coordinate is 3.00 cm. If its x-coordinate 2.00 s later is -5.00 cm, what is its acceleration?

28. **W** In 1865 Jules Verne proposed sending men to the Moon by firing a space capsule from a 220-m-long cannon with final speed of 10.97 km/s. What would have been the unrealistically large acceleration experienced by the space travelers during their launch? (A human can stand an acceleration of $15g$ for a short time.) Compare your answer with the free-fall acceleration, 9.80 m/s².

29. A truck covers 40.0 m in 8.50 s while uniformly slowing down to a final velocity of 2.80 m/s. (a) Find the truck's original speed. (b) Find its acceleration.

30. **GP** A speedboat increases its speed uniformly from $v_i = 20.0$ m/s to $v_f = 30.0$ m/s in a distance of 2.00×10^2 m. (a) Draw a coordinate system for this situation and label the relevant quantities, including vectors. (b) For the given information, what single equation is most appropriate for finding the acceleration? (c) Solve the equation selected in part (b) symbolically for the boat's acceleration in terms of v_f, v_i, and Δx. (d) Substitute given values, obtaining that acceleration. (e) Find the time it takes the boat to travel the given distance.

31. A Cessna aircraft has a liftoff speed of 120 km/h. (a) What minimum constant acceleration does the aircraft require if it is to be airborne after a takeoff run of 240 m? (b) How long does it take the aircraft to become airborne?

32. An object moves with constant acceleration 4.00 m/s² and over a time interval reaches a final velocity of 12.0 m/s. (a) If its original velocity is 6.00 m/s, what is its displacement during the time interval? (b) What is the distance it travels during this interval? (c) If its original velocity is -6.00 m/s, what is its displacement during this interval? (d) What is the total distance it travels during the interval in part (c)?

33. **Q|C** In a test run, a certain car accelerates uniformly from zero to 24.0 m/s in 2.95 s. (a) What is the magnitude of the car's acceleration? (b) How long does it take the car to change its speed from 10.0 m/s to 20.0 m/s? (c) Will doubling the time always double the change in speed? Why?

54 CHAPTER 2 | Motion in One Dimension

34. **Q|C** A jet plane lands with a speed of 100 m/s and can accelerate at a maximum rate of -5.00 m/s^2 as it comes to rest. (a) From the instant the plane touches the runway, what is the minimum time needed before it can come to rest? (b) Can this plane land on a small tropical island airport where the runway is 0.800 km long?

35. **Q|C** Speedy Sue, driving at 30.0 m/s, enters a one-lane tunnel. She then observes a slow-moving van 155 m ahead traveling at 5.00 m/s. Sue applies her brakes but can accelerate only at -2.00 m/s^2 because the road is wet. Will there be a collision? State how you decide. If yes, determine how far into the tunnel and at what time the collision occurs. If no, determine the distance of closest approach between Sue's car and the van.

36. A record of travel along a straight path is as follows:
    1. Start from rest with a constant acceleration of 2.77 m/s^2 for 15.0 s.
    2. Maintain a constant velocity for the next 2.05 min.
    3. Apply a constant negative acceleration of -9.47 m/s^2 for 4.39 s.

    (a) What was the total displacement for the trip? (b) What were the average speeds for legs 1, 2, and 3 of the trip, as well as for the complete trip?

37. A train is traveling down a straight track at 20 m/s when the engineer applies the brakes, resulting in an acceleration of -1.0 m/s^2 as long as the train is in motion. How far does the train move during a 40-s time interval starting at the instant the brakes are applied?

38. A car accelerates uniformly from rest to a speed of 40.0 mi/h in 12.0 s. Find (a) the distance the car travels during this time and (b) the constant acceleration of the car.

39. A car starts from rest and travels for 5.0 s with a uniform acceleration of $+1.5$ m/s^2. The driver then applies the brakes, causing a uniform acceleration of -2.0 m/s^2. If the brakes are applied for 3.0 s, (a) how fast is the car going at the end of the braking period, and (b) how far has the car gone?

40. **S** A car starts from rest and travels for t_1 seconds with a uniform acceleration a_1. The driver then applies the brakes, causing a uniform acceleration a_2. If the brakes are applied for t_2 seconds, (a) how fast is the car going just before the beginning of the braking period? (b) How far does the car go before the driver begins to brake? (c) Using the answers to parts (a) and (b) as the initial velocity and position for the motion of the car during braking, what total distance does the car travel? Answers are in terms of the variables a_1, a_2, t_1, and t_2.

41. In the Daytona 500 auto race, a Ford Thunderbird and a Mercedes Benz are moving side by side down a straightaway at 71.5 m/s. The driver of the Thunderbird realizes that she must make a pit stop, and she smoothly slows to a stop over a distance of 250 m. She spends 5.00 s in the pit and then accelerates out, reaching her previous speed of 71.5 m/s after a distance of 350 m. At this point, how far has the Thunderbird fallen behind the Mercedes Benz, which has continued at a constant speed?

42. A certain cable car in San Francisco can stop in 10 s when traveling at maximum speed. On one occasion, the driver sees a dog a distance d in front of the car and slams on the brakes instantly. The car reaches the dog 8.0 s later, and the dog jumps off the track just in time. If the car travels 4.0 m beyond the position of the dog before coming to a stop, how far was the car from the dog? (Hint: You will need three equations.)

43. **M** A hockey player is standing on his skates on a frozen pond when an opposing player, moving with a uniform speed of 12 m/s, skates by with the puck. After 3.0 s, the first player makes up his mind to chase his opponent. If he accelerates uniformly at 4.0 m/s^2, (a) how long does it take him to catch his opponent, and (b) how far has he traveled in that time? (Assume the player with the puck remains in motion at constant speed.)

44. A train 400 m long is moving on a straight track with a speed of 82.4 km/h. The engineer applies the brakes at a crossing, and later the last car passes the crossing with a speed of 16.4 km/h. Assuming constant acceleration, determine how long the train blocked the crossing. Disregard the width of the crossing.

2.6 Freely Falling Objects

45. A ball is thrown vertically upward with a speed of 25.0 m/s. (a) How high does it rise? (b) How long does it take to reach its highest point? (c) How long does the ball take to hit the ground after it reaches its highest point? (d) What is its velocity when it returns to the level from which it started?

46. A ball is thrown directly downward with an initial speed of 8.00 m/s, from a height of 30.0 m. After what time interval does it strike the ground?

47. A certain freely falling object, released from rest, requires 1.50 s to travel the last 30.0 m before it hits the ground. (a) Find the velocity of the object when it is 30.0 m above the ground. (b) Find the total distance the object travels during the fall.

48. **Q|C** An attacker at the base of a castle wall 3.65 m high throws a rock straight up with speed 7.40 m/s at a height of 1.55 m above the ground. (a) Will the rock reach the top of the wall? (b) If so, what is the rock's speed at the top? If not, what initial speed must the rock have to reach the top? (c) Find the change in the speed of a rock thrown straight down from the top of the wall at an initial speed of 7.40 m/s and moving between the same two points. (d) Does the change in speed of the downward-moving rock agree with the magnitude of the speed change of the rock moving upward between the same elevations? Explain physically why or why not.

49. **BIO** Traumatic brain injury such as concussion results when the head undergoes a very large acceleration. Generally, an acceleration less than 800 m/s^2 lasting for any length of time will not cause injury, whereas an

acceleration greater than 1 000 m/s² lasting for at least 1 ms will cause injury. Suppose a small child rolls off a bed that is 0.40 m above the floor. If the floor is hardwood, the child's head is brought to rest in approximately 2.0 mm. If the floor is carpeted, this stopping distance is increased to about 1.0 cm. Calculate the magnitude and duration of the deceleration in both cases, to determine the risk of injury. Assume the child remains horizontal during the fall to the floor. Note that a more complicated fall could result in a head velocity greater or less than the speed you calculate.

50. A small mailbag is released from a helicopter that is descending steadily at 1.50 m/s. After 2.00 s, (a) what is the speed of the mailbag, and (b) how far is it below the helicopter? (c) What are your answers to parts (a) and (b) if the helicopter is rising steadily at 1.50 m/s?

51. A tennis player tosses a tennis ball straight up and then catches it after 2.00 s at the same height as the point of release. (a) What is the acceleration of the ball while it is in flight? (b) What is the velocity of the ball when it reaches its maximum height? Find (c) the initial velocity of the ball and (d) the maximum height it reaches.

52. **S** A package is dropped from a helicopter that is descending steadily at a speed v_0. After t seconds have elapsed, (a) what is the speed of the package in terms of v_0, g, and t? (b) What distance d is it from the helicopter in terms of g and t? (c) What are the answers to parts (a) and (b) if the helicopter is rising steadily at the same speed?

53. **Q|C** A model rocket is launched straight upward with an initial speed of 50.0 m/s. It accelerates with a constant upward acceleration of 2.00 m/s² until its engines stop at an altitude of 150 m. (a) What can you say about the motion of the rocket after its engines stop? (b) What is the maximum height reached by the rocket? (c) How long after liftoff does the rocket reach its maximum height? (d) How long is the rocket in the air?

54. **W** A baseball is hit so that it travels straight upward after being struck by the bat. A fan observes that it takes 3.00 s for the ball to reach its maximum height. Find (a) the ball's initial velocity and (b) the height it reaches.

Additional Problems

55. A truck tractor pulls two trailers, one behind the other, at a constant speed of 100 km/h. It takes 0.600 s for the big rig to completely pass onto a bridge 400 m long. For what duration of time is all or part of the truck–trailer combination on the bridge?

56. **M BIO** Colonel John P. Stapp, USAF, participated in studying whether a jet pilot could survive emergency ejection. On March 19, 1954, he rode a rocket-propelled sled that moved down a track at a speed of 632 mi/h (see Fig. P2.56). He and the sled were safely brought to rest in 1.40 s. Determine in SI units (a) the negative acceleration he experienced and (b) the distance he traveled during this negative acceleration.

Figure P2.56 (left) Col. John Stapp and his rocket sled are brought to rest in a very short time interval. (right) Stapp's face is contorted by the stress of rapid negative acceleration.

57. A bullet is fired through a board 10.0 cm thick in such a way that the bullet's line of motion is perpendicular to the face of the board. If the initial speed of the bullet is 400 m/s and it emerges from the other side of the board with a speed of 300 m/s, find (a) the acceleration of the bullet as it passes through the board and (b) the total time the bullet is in contact with the board.

58. A speedboat moving at 30.0 m/s approaches a no-wake buoy marker 100 m ahead. The pilot slows the boat with a constant acceleration of −3.50 m/s² by reducing the throttle. (a) How long does it take the boat to reach the buoy? (b) What is the velocity of the boat when it reaches the buoy?

59. A student throws a set of keys vertically upward to his fraternity brother, who is in a window 4.00 m above. The brother's outstretched hand catches the keys 1.50 s later. (a) With what initial velocity were the keys thrown? (b)? What was the velocity of the keys just before they were caught?

60. **S** A student throws a set of keys vertically upward to his fraternity brother, who is in a window a distance h above. The brother's outstretched hand catches the keys on their way up a time t later. (a) With what initial velocity were the keys thrown? (b) What was the velocity of the keys just before they were caught? (Answers should be in terms of h, g, and t.)

61. **BIO** It has been claimed that an insect called the froghopper (*Philaenus spumarius*) is the best jumper in the animal kingdom. This insect can accelerate at 4 000 m/s² over a distance of 2.0 mm as it straightens its specially designed "jumping legs." (a) Assuming a uniform acceleration, what is the velocity of the insect after it has accelerated through this short distance, and (b) how long did it take to reach that velocity? (c) How high would the insect jump if air resistance could be ignored? Note that the actual height obtained is about 0.7 m, so air resistance is important here.

62. **Q|C** Draw motion diagrams (see Section 2.5) for (a) an object moving to the right at constant speed, (b) an object moving to the right and speeding up at a constant rate, (c) an object moving to the right and slowing down at a constant rate, (d) an object moving to the left and speeding up at a constant rate, and (e) an object moving to the left and slowing down at a constant rate. (f) How would your drawings change if the changes in speed were not uniform; that is, if the speed were not changing at a constant rate?

63. **M** A ball is thrown upward from the ground with an initial speed of 25 m/s; at the same instant, another ball is dropped from a building 15 m high. After how long will the balls be at the same height?

64. To pass a physical education class at a university, a student must run 1.0 mi in 12 min. After running for 10 min, she still has 500 yd to go. If her maximum acceleration is 0.15 m/s^2, can she make it? If the answer is no, determine what acceleration she would need to be successful.

65. In Chapter 5 we will define the center of mass of an object. The center of mass moves with constant acceleration when constant forces act on the object. A gymnast jumps straight up, with her center of mass moving at 2.80 m/s as she leaves the ground. How high above this point is her center of mass (a) 0.100 s, (b) 0.200 s, (c) 0.300 s, and (d) 0.500 s thereafter?

66. **S** Two students are on a balcony a distance h above the street. One student throws a ball vertically downward at a speed v_0; at the same time, the other student throws a ball vertically upward at the same speed. Answer the following symbolically in terms of v_0, g, h, and t. (a) Write the kinematic equation for the y-coordinate of each ball. (b) Set the equations found in part (a) equal to height 0 and solve each for t symbolically using the quadratic formula. What is the difference in the two balls' time in the air? (c) Use the time-independent kinematics equation to find the velocity of each ball as it strikes the ground. (d) How far apart are the balls at a time t after they are released and before they strike the ground?

67. You drop a ball from a window on an upper floor of a building and it is caught by a friend on the ground when the ball is moving with speed v_f. You now repeat the drop, but you have a friend on the street below throw another ball upward at speed v_f exactly at the same time that you drop your ball from the window. The two balls are initially separated by 28.7 m. (a) At what time do they pass each other? (b) At what location do they pass each other relative the window?

68. The driver of a truck slams on the brakes when he sees a tree blocking the road. The truck slows down uniformly with an acceleration of −5.60 m/s^2 for 4.20 s, making skid marks 62.4 m long that end at the tree. With what speed does the truck then strike the tree?

69. **QC** Emily challenges her husband, David, to catch a $1 bill as follows. She holds the bill vertically as in Figure P2.69, with the center of the bill between David's index finger and thumb. David must catch the bill after Emily releases it without moving his hand downward. If his reaction time is 0.2 s, will he succeed?

Figure P2.69

Explain your reasoning. (This challenge is a good trick you might want to try with your friends.)

70. A mountain climber stands at the top of a 50.0-m cliff that overhangs a calm pool of water. She throws two stones vertically downward 1.00 s apart and observes that they cause a single splash. The first stone had an initial velocity of −2.00 m/s. (a) How long after release of the first stone did the two stones hit the water? (b) What initial velocity must the second stone have had, given that they hit the water simultaneously? (c) What was the velocity of each stone at the instant it hit the water?

71. An ice sled powered by a rocket engine starts from rest on a large frozen lake and accelerates at +40 ft/s^2. After some time t_1, the rocket engine is shut down and the sled moves with constant velocity v for a time t_2. If the total distance traveled by the sled is 17 500 ft and the total time is 90 s, find (a) the times t_1 and t_2 and (b) the velocity v. At the 17 500-ft mark, the sled begins to accelerate at −20 ft/s^2. (c) What is the final position of the sled when it comes to rest? (d) How long does it take to come to rest?

72. In Bosnia, the ultimate test of a young man's courage used to be to jump off a 400-year-old bridge (destroyed in 1993; rebuilt in 2004) into the River Neretva, 23 m below the bridge. (a) How long did the jump last? (b) How fast was the jumper traveling upon impact with the river? (c) If the speed of sound in air is 340 m/s, how long after the jumper took off did a spectator on the bridge hear the splash?

73. A person sees a lightning bolt pass close to an airplane that is flying in the distance. The person hears thunder 5.0 s after seeing the bolt and sees the airplane overhead 10 s after hearing the thunder. The speed of sound in air is 1 100 ft/s. (a) Find the distance of the airplane from the person at the instant of the bolt. (Neglect the time it takes the light to travel from the bolt to the eye.) (b) Assuming the plane travels with a constant speed toward the person, find the velocity of the airplane. (c) Look up the speed of light in air and defend the approximation used in part (a).

74. **QC** A glider on an air track carries a flag of length ℓ through a stationary photogate, which measures the time interval Δt_d during which the flag blocks a beam of infrared light passing across the photogate. The ratio $v_d = \ell/\Delta t_d$ is the average velocity of the glider over this part of its motion. Suppose the glider moves with constant acceleration. (a) Is v_d necessarily equal to the instantaneous velocity of the glider when it is halfway through the photogate in space? Explain. (b) Is v_d equal to the instantaneous velocity of the glider when it is halfway through the photogate in time? Explain.

75. A stuntman sitting on a tree limb wishes to drop vertically onto a horse galloping under the tree. The constant speed of the horse is 10.0 m/s, and the man is initially 3.00 m above the level of the saddle. (a) What must be the horizontal distance between the saddle and the limb when the man makes his move? (b) How long is he in the air?

Dave "Human Cannonball" Smith hurtles through the air along a parabolic path, depending on the correct initial velocity and cannon angle to send him safely over the U.S.-Mexico border fence and into a net.

Vectors and Two-Dimensional Motion   3

In our discussion of one-dimensional motion in Chapter 2, we used the concept of vectors only to a limited extent. In our further study of motion, manipulating vector quantities will become increasingly important, so much of this chapter is devoted to vector techniques. We'll then apply these mathematical tools to two-dimensional motion, especially that of projectiles, and to the understanding of relative motion.

3.1 Vectors and Their Properties
3.2 Components of a Vector
3.3 Displacement, Velocity, and Acceleration in Two Dimensions
3.4 Motion in Two Dimensions
3.5 Relative Velocity

3.1 Vectors and Their Properties

LEARNING OBJECTIVES

1. Apply the definitions of scalar and vector to categorize different physical quantities.
2. Use the geometric interpretation of vector addition, subtraction and multiplication to find the resultant vectors of those operations.

Each of the physical quantities we will encounter in this book can be categorized as either a *vector quantity* or a *scalar quantity*. As noted in Chapter 2, a vector has both direction and magnitude (size). A scalar can be completely specified by its magnitude with appropriate units; it has no direction. An example of each kind of quantity is shown in Figure 3.1.

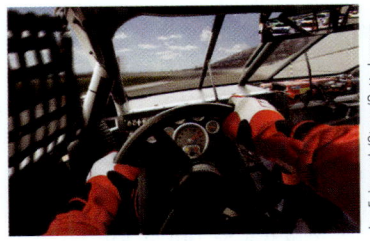

Figure 3.1 A vector such as velocity has a magnitude, shown on the race car's speedometer, and a direction, straight out through the race car's front windshield. The mass of the car is a scalar quantity, as is the volume of gasoline in its fuel tank.

58 CHAPTER 3 | Vectors and Two-Dimensional Motion

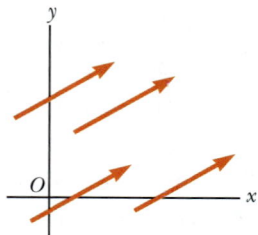

Figure 3.2 These four vectors are equal because they have equal lengths and point in the same direction.

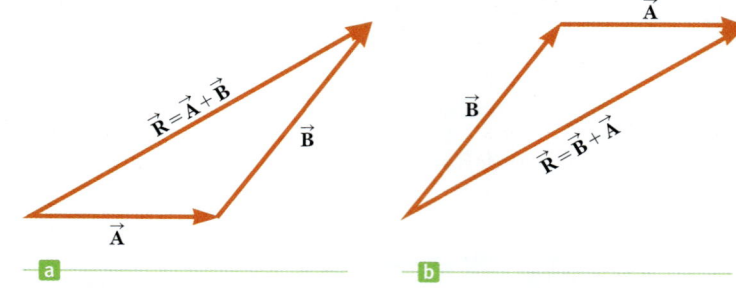

Figure 3.3 (a) When vector $\vec{B}$ is added to vector $\vec{A}$, the vector sum $\vec{R}$ is the vector that runs from the tail of $\vec{A}$ to the tip of $\vec{B}$. (b) Here the resultant runs from the tail of $\vec{B}$ to the tip of $\vec{A}$. These constructions prove that $\vec{A} + \vec{B} = \vec{B} + \vec{A}$.

Tip 3.1 Vector Addition vs. Scalar Addition
$\vec{A} + \vec{B} = \vec{C}$ differs significantly from $A + B = C$. The first is a vector sum, which must be handled graphically or with components, whereas the second is a simple arithmetic sum of numbers.

As described in Chapter 2, displacement, velocity, and acceleration are vector quantities. Temperature is an example of a scalar quantity. If the temperature of an object is −5°C, that information completely specifies the temperature of the object; no direction is required. Masses, time intervals, and volumes are scalars as well. Scalar quantities can be manipulated with the rules of ordinary arithmetic. Vectors can also be added and subtracted from each other, and multiplied, but there are a number of important differences, as will be seen in the following sections.

When a vector quantity is handwritten, it is often represented with an arrow over the letter ($\vec{A}$). As mentioned in Section 2.1, a vector quantity in this book will be represented by boldface type with an arrow on top (for example, $\vec{A}$). The magnitude of the vector $\vec{A}$ will be represented by italic type, as A. Italic type will also be used to represent scalars.

Equality of Two Vectors Two vectors $\vec{A}$ and $\vec{B}$ are equal if they have the same magnitude and the same direction. This property allows us to translate a vector parallel to itself in a diagram without affecting the vector. In fact, for most purposes, any vector can be moved parallel to itself without being affected. (See Fig. 3.2.)

Adding Vectors When two or more vectors are added, they must all have the same units. For example, it doesn't make sense to add a velocity vector, carrying units of meters per second, to a displacement vector, carrying units of meters. Scalars obey the same rule: It would be similarly meaningless to add temperatures to volumes or masses to time intervals.

Vectors can be added geometrically or algebraically. (The latter is discussed at the end of the next section.) To add vector $\vec{B}$ to vector $\vec{A}$ geometrically, first draw $\vec{A}$ on a piece of graph paper to some scale, such as 1 cm = 1 m, so that its direction is specified relative to a coordinate system. Then draw vector $\vec{B}$ to the same scale with the tail of $\vec{B}$ starting at the tip of $\vec{A}$, as in Figure 3.3a. Vector $\vec{B}$ must be drawn along the direction that makes the proper angle relative vector $\vec{A}$. The **resultant vector** $\vec{R} = \vec{A} + \vec{B}$ is the vector drawn from the tail of $\vec{A}$ to the tip of $\vec{B}$. This procedure is known as the **triangle method of addition.**

When two vectors are added, their sum is independent of the order of the addition: $\vec{A} + \vec{B} = \vec{B} + \vec{A}$. This relationship can be seen from the geometric construction in Figure 3.3b, and is called the **commutative law of addition.**

This same general approach can also be used to add more than two vectors, as is done in Figure 3.4 for four vectors. The resultant vector sum $\vec{R} = \vec{A} + \vec{B} + \vec{C} + \vec{D}$ is the vector drawn from the tail of the first vector to the tip of the last. Again, the order in which the vectors are added is unimportant.

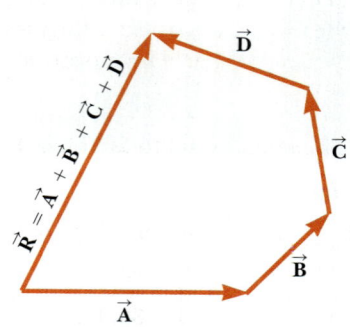

Figure 3.4 A geometric construction for summing four vectors. The resultant vector $\vec{R}$ is the vector that completes the polygon.

3.1 | Vectors and Their Properties

Negative of a Vector The negative of the vector $\vec{A}$ is defined as the vector that gives zero when added to $\vec{A}$. This means that $\vec{A}$ and $-\vec{A}$ have the same magnitude but opposite directions.

Subtracting Vectors Vector subtraction makes use of the definition of the negative of a vector. We define the operation $\vec{A} - \vec{B}$ as the vector $-\vec{B}$ added to the vector $\vec{A}$:

$$\vec{A} - \vec{B} = \vec{A} + (-\vec{B}) \quad [3.1]$$

Vector subtraction is really a special case of vector addition. The geometric construction for subtracting two vectors is shown in Figure 3.5.

Multiplying or Dividing a Vector by a Scalar Multiplying or dividing a vector by a scalar gives a vector. For example, if vector $\vec{A}$ is multiplied by the scalar number 3, the result, written $3\vec{A}$, is a vector with a magnitude three times that of $\vec{A}$ and pointing in the same direction. If we multiply vector $\vec{A}$ by the scalar -3, the result is $-3\vec{A}$, a vector with a magnitude three times that of $\vec{A}$ and pointing in the opposite direction (because of the negative sign).

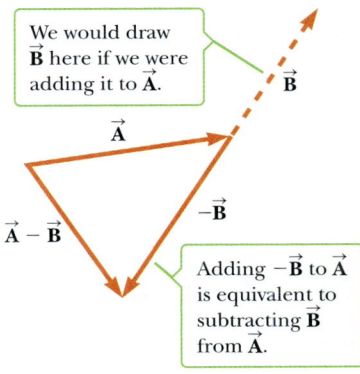

Figure 3.5 This construction shows how to subtract vector $\vec{B}$ from vector $\vec{A}$. The vector $-\vec{B}$ has the same magnitude as the vector $\vec{B}$, but points in the opposite direction.

■ Quick Quiz

3.1 The magnitudes of two vectors $\vec{A}$ and $\vec{B}$ are 12 units and 8 units, respectively. What are the largest and smallest possible values for the magnitude of the resultant vector $\vec{R} = \vec{A} + \vec{B}$? (a) 14.4 and 4 (b) 12 and 8 (c) 20 and 4 (d) none of these.

■ EXAMPLE 3.1 | Taking a Trip

GOAL Find the sum of two vectors by using a graph.

PROBLEM A car travels 20.0 km due north and then 35.0 km in a direction 60.0° west of north, as in Figure 3.6. Using a graph, find the magnitude and direction of a single vector that gives the net effect of the car's trip. This vector is called the car's *resultant displacement*.

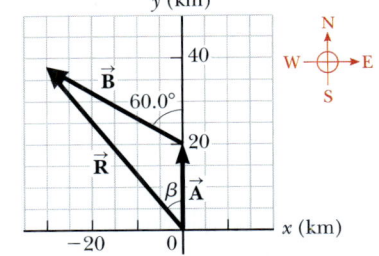

Figure 3.6 (Example 3.1) A graphical method for finding the resultant displacement vector $\vec{R} = \vec{A} + \vec{B}$.

STRATEGY Draw a graph and represent the displacement vectors as arrows. Graphically locate the vector resulting from the sum of the two displacement vectors. Measure its length and angle with respect to the vertical.

SOLUTION
Let $\vec{A}$ represent the first displacement vector, 20.0 km north, and $\vec{B}$ the second displacement vector, extending west of north. Carefully graph the two vectors, drawing a resultant vector $\vec{R}$ with its base touching the base of $\vec{A}$ and extending to the tip of $\vec{B}$. Measure the length of this vector, which turns out to be about 48 km. The angle β, measured with a protractor, is about 39° west of north.

REMARKS Notice that ordinary arithmetic doesn't work here: the correct answer of 48 km is not equal to 20.0 km + 35.0 km = 55.0 km!

QUESTION 3.1 Suppose two vectors are added. Under what conditions would the sum of the magnitudes of the vectors equal the magnitude of the resultant vector?

EXERCISE 3.1 Graphically determine the magnitude and direction of the displacement if a man walks 30.0 km 45° north of east and then walks due east 20.0 km.

ANSWER 46 km, 27° north of east

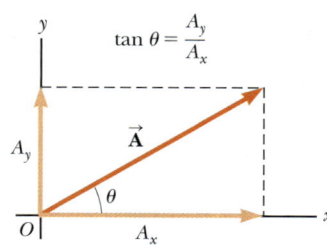

Figure 3.7 Any vector $\vec{A}$ lying in the xy-plane can be represented by its rectangular components A_x and A_y.

3.2 Components of a Vector

LEARNING OBJECTIVES

1. Represent vectors in terms of their magnitudes and directions.
2. Represent vectors in terms of x- and y-components.
3. Perform arithmetic operations with vectors using their components.

One method of adding vectors makes use of the projections of a vector along the axes of a rectangular coordinate system. These projections are called **components**. Any vector can be completely described by its components.

Consider a vector $\vec{A}$ in a rectangular coordinate system, as shown in Figure 3.7. $\vec{A}$ can be expressed as the sum of two vectors: $\vec{A}_x$, parallel to the x-axis, and $\vec{A}_y$, parallel to the y-axis. Mathematically,

$$\vec{A} = \vec{A}_x + \vec{A}_y$$

where $\vec{A}_x$ and $\vec{A}_y$ are the component vectors of $\vec{A}$. The projection of $\vec{A}$ along the x-axis, A_x, is called the x-component of $\vec{A}$, and the projection of $\vec{A}$ along the y-axis, A_y, is called the y-component of $\vec{A}$. These components can be either positive or negative numbers with units. From the definitions of sine and cosine, we see that $\cos \theta = A_x/A$ and $\sin \theta = A_y/A$, so the components of $\vec{A}$ are

$$A_x = A \cos \theta \quad [3.2a]$$
$$A_y = A \sin \theta \quad [3.2b]$$

Tip 3.2 x- and y-components
Equation 3.2 for the x- and y-components of a vector associates cosine with the x-component and sine with the y-component, as in Figure 3.8a. This association is due *solely* to the fact that we chose to measure the angle θ with respect to the positive x-axis. If the angle were measured with respect to the y-axis, as in Figure 3.8b, the components would be given by $A_x = A \sin \theta$ and $A_y = A \cos \theta$.

These components form two sides of a right triangle having a hypotenuse with magnitude A. It follows that $\vec{A}$'s magnitude and direction are related to its components through the Pythagorean theorem and the definition of the tangent:

$$A = \sqrt{A_x^2 + A_y^2} \quad [3.3]$$

$$\tan \theta = \frac{A_y}{A_x} \quad [3.4]$$

To solve for the angle θ, which is measured counterclockwise from the positive x-axis by convention, the inverse tangent can be taken of both sides of Equation 3.4:

$$\theta = \tan^{-1}\left(\frac{A_y}{A_x}\right)$$

Tip 3.3 Inverse Tangents on Calculators: Right Half the Time
The inverse tangent function on calculators returns an angle between $-90°$ and $+90°$. If the vector lies in the second or third quadrant, the angle, as measured from the positive x-axis, will be the angle returned by your calculator plus 180°.

This formula gives the right answer for θ only half the time! The inverse tangent function returns values only from $-90°$ to $+90°$, so the answer in your calculator window will only be correct if the vector happens to lie in the first or fourth quadrant. If it lies in the second or third quadrant, adding 180° to the number in the calculator window will always give the right answer. The angle in Equations 3.2 and 3.4 must be measured from the positive x-axis. Other choices of reference line are possible, but certain adjustments must then be made. (See Tip 3.2 and Fig. 3.8.)

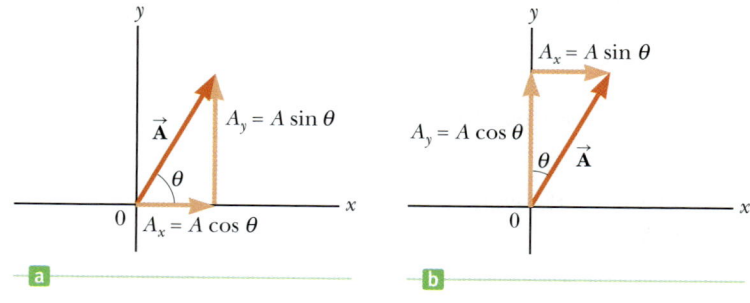

Figure 3.8 The angle θ need not always be defined from the positive x-axis.

3.2 | Components of a Vector

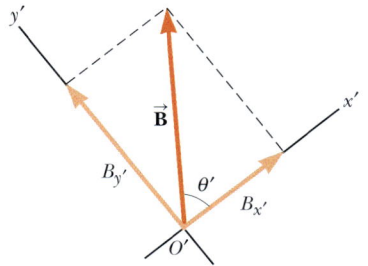

Figure 3.9 The components of vector $\vec{B}$ in a tilted coordinate system.

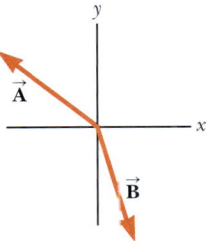

Figure 3.10 (Quick Quizzes 3.2 and 3.3)

If a coordinate system other than the one shown in Figure 3.7 is chosen, the components of the vector must be modified accordingly. In many applications it's more convenient to express the components of a vector in a coordinate system having axes that are not horizontal and vertical, but are still perpendicular to each other. Suppose a vector $\vec{B}$ makes an angle θ' with the x'-axis defined in Figure 3.9. The rectangular components of $\vec{B}$ along the axes of the figure are given by $B_{x'} = B \cos \theta'$ and $B_{y'} = B \sin \theta'$, as in Equations 3.2. The magnitude and direction of $\vec{B}$ are then obtained from expressions equivalent to Equations 3.3 and 3.4.

Quick Quiz

3.2 Figure 3.10 shows two vectors lying in the xy-plane. Determine the signs of the x- and y-components of $\vec{A}$, $\vec{B}$, and $\vec{A} + \vec{B}$.

3.3 Which vector has an angle with respect to the positive x-axis that is in the range of the inverse tangent function?

■ EXAMPLE 3.2 | Help Is on the Way!

GOAL Find vector components, given a magnitude and direction, and vice versa.

PROBLEM (a) Find the horizontal and vertical components of the $d = 1.00 \times 10^2$ m displacement of a superhero who flies from the top of a tall building along the path shown in Figure 3.11a. (b) Suppose instead the superhero leaps in the other direction along a displacement vector $\vec{B}$ to the top of a flagpole where the displacement components are given by $B_x = -25.0$ m and $B_y = 10.0$ m. Find the magnitude and direction of the displacement vector.

STRATEGY (a) The triangle formed by the displacement and its components is shown in Figure 3.11b. Simple trigonometry gives the components relative to the standard xy-coordinate system: $A_x = A \cos \theta$ and $A_y = A \sin \theta$ (Eqs. 3.2). Note that $\theta = -30.0°$, negative because it's measured clockwise from the positive x-axis. (b) Apply Equations 3.3 and 3.4 to find the magnitude and direction of the vector.

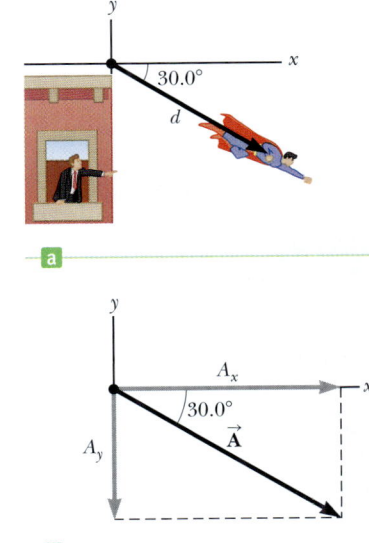

Figure 3.11 (Example 3.2)

SOLUTION

(a) Find the vector components of $\vec{A}$ from its magnitude and direction.

Use Equations 3.2 to find the components of the displacement vector $\vec{A}$:

$A_x = A \cos \theta = (1.00 \times 10^2 \text{ m}) \cos(-30.0°) = \boxed{+86.6 \text{ m}}$

$A_y = A \sin \theta = (1.00 \times 10^2 \text{ m}) \sin(-30.0°) = \boxed{-50.0 \text{ m}}$

(Continued)

(b) Find the magnitude and direction of the displacement vector $\vec{B}$ from its components.

Compute the magnitude of $\vec{B}$ from the Pythagorean theorem:

$$B = \sqrt{B_x^2 + B_y^2} = \sqrt{(-25.0 \text{ m})^2 + (10.0 \text{ m})^2} = \boxed{26.9 \text{ m}}$$

Calculate the direction of $\vec{B}$ using the inverse tangent, remembering to add 180° to the answer in your calculator window, because the vector lies in the second quadrant:

$$\theta = \tan^{-1}\left(\frac{B_y}{B_x}\right) = \tan^{-1}\left(\frac{10.0}{-25.0}\right) = -21.8°$$

$$\theta = \boxed{158°}$$

REMARKS In part (a), note that $\cos(-\theta) = \cos\theta$; however, $\sin(-\theta) = -\sin\theta$. The negative sign of A_y reflects the fact that displacement in the y-direction is *downward*.

QUESTION 3.2 What other functions, if any, can be used to find the angle in part (b)?

EXERCISE 3.2 (a) Suppose the superhero had flown 150 m at a 120° angle with respect to the positive x-axis. Find the components of the displacement vector. (b) Suppose instead the superhero had leaped with a displacement having an x-component of 32.5 m and a y-component of 24.3 m. Find the magnitude and direction of the displacement vector.

ANSWERS (a) $A_x = -75$ m, $A_y = 130$ m (b) 40.6 m, 36.8°

Adding Vectors Algebraically

The graphical method of adding vectors is valuable in understanding how vectors can be manipulated, but most of the time vectors are added algebraically in terms of their components. Suppose $\vec{R} = \vec{A} + \vec{B}$. Then the components of the resultant vector $\vec{R}$ are given by

$$R_x = A_x + B_x \quad \text{[3.5a]}$$

$$R_y = A_y + B_y \quad \text{[3.5b]}$$

So x-components are added only to x-components, and y-components only to y-components. The magnitude and direction of $\vec{R}$ can subsequently be found with Equations 3.3 and 3.4.

Subtracting two vectors works the same way because it's a matter of adding the negative of one vector to another vector. You should make a rough sketch when adding or subtracting vectors, in order to get an approximate geometric solution as a check.

EXAMPLE 3.3 | Take a Hike

GOAL Add vectors algebraically and find the resultant vector.

PROBLEM A hiker begins a trip by first walking 25.0 km 45.0° south of east from her base camp. On the second day she walks 40.0 km in a direction 60.0° north of east, at which point she discovers a forest ranger's tower. **(a)** Determine the components of the hiker's displacements in the first and second days. **(b)** Determine the components of the hiker's total displacement for the trip. **(c)** Find the magnitude and direction of the displacement from base camp.

STRATEGY This problem is just an application of vector addition using components, Equations 3.5. We denote the displacement vectors on the first and second days by $\vec{A}$ and $\vec{B}$, respectively.

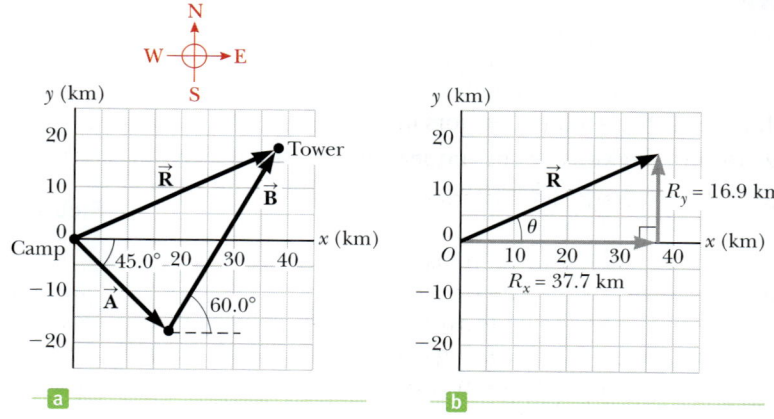

Figure 3.12 (Example 3.3) (a) Hiker's path and the resultant vector. (b) Components of the hiker's total displacement from camp.

Using the camp as the origin of the coordinates, we get the vectors shown in Figure 3.12a. After finding x- and y-components for each vector, we add them "componentwise." Finally, we determine the magnitude and direction of the resultant vector $\vec{R}$, using the Pythagorean theorem and the inverse tangent function.

SOLUTION

(a) Find the components of $\vec{A}$.

Use Equations 3.2 to find the components of $\vec{A}$:

$A_x = A \cos(-45.0°) = (25.0 \text{ km})(0.707) = $ 17.7 km

$A_y = A \sin(-45.0°) = -(25.0 \text{ km})(0.707) = $ −17.7 km

Find the components of $\vec{B}$:

$B_x = B \cos 30.0° = (40.0 \text{ km})(0.500) = $ 20.0 km

$B_y = B \sin 60.0° = (40.0 \text{ km})(0.866) = $ 34.6 km

(b) Find the components of the resultant vector, $\vec{R} = \vec{A} + \vec{B}$.

To find R_x, add the x-components of $\vec{A}$ and $\vec{B}$:

$R_x = A_x + B_x = 17.7 \text{ km} + 20.0 \text{ km} = $ 37.7 km

To find R_y, add the y-components of $\vec{A}$ and $\vec{B}$:

$R_y = A_y + B_y = -17.7 \text{ km} + 34.6 \text{ km} = $ 16.9 km

(c) Find the magnitude and direction of $\vec{R}$.

Use the Pythagorean theorem to get the magnitude:

$R = \sqrt{R_x^2 + R_y^2} = \sqrt{(37.7 \text{ km})^2 + (16.9 \text{ km})^2} = $ 41.3 km

Calculate the direction of $\vec{R}$ using the inverse tangent function:

$\theta = \tan^{-1}\left(\dfrac{16.9 \text{ km}}{37.7 \text{ km}}\right) = $ 24.1°

REMARKS Figure 3.12b shows a sketch of the components of $\vec{R}$ and their directions in space. The magnitude and direction of the resultant can also be determined from such a sketch.

QUESTION 3.3 A second hiker follows the same path the first day, but then walks 15.0 km east on the second day before turning and reaching the ranger's tower. Is the second hiker's resultant displacement vector the same as the first hiker's, or different?

EXERCISE 3.3 A cruise ship leaving port travels 50.0 km 45.0° north of west and then 70.0 km at a heading 30.0° north of east. Find (a) the components of the ship's displacement vector and (b) the displacement vector's magnitude and direction.

ANSWER (a) $R_x = 25.3$ km, $R_y = 70.4$ km (b) 74.8 km, 70.2° north of east

---

3.3 Displacement, Velocity, and Acceleration in Two Dimensions

LEARNING OBJECTIVES

1. Define displacement vectors in two dimensions.
2. Define average and instantaneous velocity vectors in two dimensions.
3. Define average and instantaneous acceleration vectors in two dimensions.

In one-dimensional motion, as discussed in Chapter 2, the direction of a vector quantity such as a velocity or acceleration can be taken into account by specifying whether the quantity is positive or negative. The velocity of a rocket, for example, is positive if the rocket is going up and negative if it's going down. This simple solution is no longer available in two or three dimensions. Instead, we must make full use of the vector concept.

Consider an object moving through space as shown in Figure 3.13. When the object is at some point ⓟ at time t_i, its position is described by the position vector

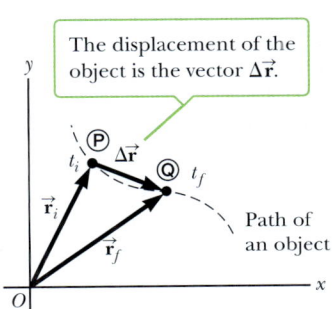

Figure 3.13 An object moving along some curved path between points ⓟ and ⓠ. The displacement vector $\Delta\vec{r}$ is the difference in the position vectors: $\Delta\vec{r} = \vec{r}_f - \vec{r}_i$.

$\vec{\mathbf{r}}_i$, drawn from the origin to ⓟ. When the object has moved to some other point ⓠ at time t_f, its position vector is $\vec{\mathbf{r}}_f$. From the vector diagram in Figure 3.13, the final position vector is the sum of the initial position vector and the displacement $\Delta\vec{\mathbf{r}}$: $\vec{\mathbf{r}}_f = \vec{\mathbf{r}}_i + \Delta\vec{\mathbf{r}}$. From this relationship, we obtain the following one:

An object's **displacement** is defined as the change in its position vector, or

$$\Delta\vec{\mathbf{r}} \equiv \vec{\mathbf{r}}_f - \vec{\mathbf{r}}_i \qquad [3.6]$$

SI unit: meter (m)

We now present several generalizations of the definitions of velocity and acceleration given in Chapter 2.

Average velocity ▶

An object's **average velocity** during a time interval Δt is its displacement divided by Δt:

$$\vec{\mathbf{v}}_{av} \equiv \frac{\Delta\vec{\mathbf{r}}}{\Delta t} \qquad [3.7]$$

SI unit: meter per second (m/s)

Because the displacement is a vector quantity and the time interval is a scalar quantity, we conclude that the average velocity is a *vector* quantity directed along $\Delta\vec{\mathbf{r}}$.

Instantaneous velocity ▶

An object's **instantaneous velocity** $\vec{\mathbf{v}}$ is the limit of its average velocity as Δt goes to zero:

$$\vec{\mathbf{v}} \equiv \lim_{\Delta t \to 0} \frac{\Delta\vec{\mathbf{r}}}{\Delta t} \qquad [3.8]$$

SI unit: meter per second (m/s)

The direction of the instantaneous velocity vector is along a line that is tangent to the object's path and in the direction of its motion.

Average acceleration ▶

An object's **average acceleration** during a time interval Δt is the change in its velocity $\Delta\vec{\mathbf{v}}$ divided by Δt, or

$$\vec{\mathbf{a}}_{av} \equiv \frac{\Delta\vec{\mathbf{v}}}{\Delta t} \qquad [3.9]$$

SI unit: meter per second squared (m/s²)

Instantaneous acceleration ▶

An object's **instantaneous acceleration** vector $\vec{\mathbf{a}}$ is the limit of its average acceleration vector as Δt goes to zero:

$$\vec{\mathbf{a}} \equiv \lim_{\Delta t \to 0} \frac{\Delta\vec{\mathbf{v}}}{\Delta t} \qquad [3.10]$$

SI unit: meter per second squared (m/s²)

It's important to recognize that an object can accelerate in several ways. First, the magnitude of the velocity vector (the speed) may change with time. Second, the direction of the velocity vector may change with time, even though the speed is constant, as can happen along a curved path. Third, both the magnitude and the direction of the velocity vector may change at the same time.

Quick Quiz

3.4 Which of the following objects can't be accelerating? (a) An object moving with a constant speed; (b) an object moving with a constant velocity; (c) an object moving along a curve.

3.5 Consider the following controls in an automobile: gas pedal, brake, steering wheel. The controls in this list that can cause an acceleration of the car are (a) all three controls, (b) the gas pedal and the brake, (c) only the brake, or (d) only the gas pedal.

3.4 Motion in Two Dimensions

LEARNING OBJECTIVES

1. Describe projectile motion in two dimensions graphically.
2. Apply the two-dimensional kinematics equations to motion with constant acceleration near the surface of the Earth.

In Chapter 2 we studied objects moving along straight-line paths, such as the x-axis. In this chapter, we look at objects that move in both the x- and y-directions simultaneously under constant acceleration. An important special case of this two-dimensional motion is called **projectile motion.**

◀ Projectile motion

Anyone who has tossed any kind of object into the air has observed projectile motion. If the effects of air resistance and the rotation of Earth are neglected, the path of a projectile in Earth's gravity field is curved in the shape of a parabola, as shown in Figure 3.14.

The positive x-direction is horizontal and to the right, and the y-direction is vertical and positive upward. The most important experimental fact about projectile motion in two dimensions is that **the horizontal and vertical motions are completely independent of each other.** This means that motion in one direction has no effect on motion in the other direction. If a baseball is tossed in a parabolic path, as in Figure 3.14, the motion in the y-direction will look just like a ball tossed straight up under the influence of gravity. Figure 3.15 shows the effect of various initial angles; note that complementary angles give the same horizontal range.

Figure 3.16 is an experiment illustrating the independence of horizontal and vertical motion. The gun is aimed directly at the target ball and fired at the instant the target is released. In the absence of gravity, the projectile would hit the target because the target wouldn't move. However, the projectile still hits the target in the presence of gravity. That means

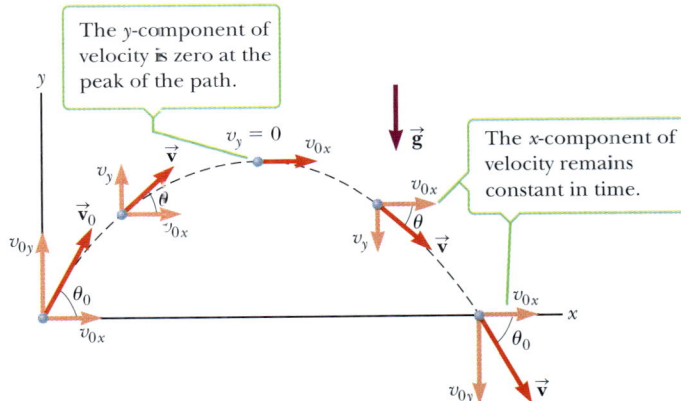

Figure 3.14
The parabolic trajectory of a particle that leaves the origin with a velocity of $\vec{v}_0$. Note that $\vec{v}$ changes with time. However, the x-component of the velocity, v_x, remains constant in time, equal to its initial velocity, v_{0x}. Also, $v_y = 0$ at the peak of the trajectory, but the acceleration is always equal to the free-fall acceleration and acts vertically downward.

Figure 3.15
A projectile launched from the origin with an initial speed of 50 m/s at various angles of projection.

Figure 3.16 A ball is fired at a target at the same instant the target is released. Both fall vertically at the same rate and collide.

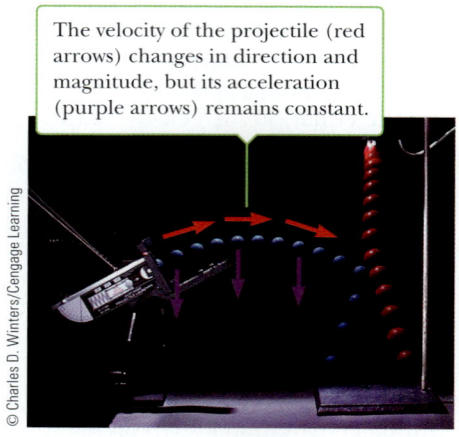

Figure 3.17 Multiflash photograph of the projectile–target demonstration. If the gun is aimed directly at the target and is fired at the same instant the target begins to fall, the projectile will hit the target.

> **Tip 3.4 Acceleration at the Highest Point**
>
> The acceleration in the y-direction is *not* zero at the top of a projectile's trajectory. Only the y-component of the velocity is zero there. If the acceleration were zero, too, the projectile would never come down!

the projectile is falling through the same vertical displacement as the target despite its horizontal motion. The experiment also works when set up as in Figure 3.17, when the initial velocity has a vertical component.

In general, the equations of constant acceleration developed in Chapter 2 follow separately for both the x-direction and the y-direction. An important difference is that the initial velocity now has two components, not just one as in that chapter. We assume that at $t = 0$ the projectile leaves the origin with an initial velocity $\vec{v}_0$. If the velocity vector makes an angle θ_0 with the horizontal, where θ_0 is called the *projection angle*, then from the definitions of the cosine and sine functions and Figure 3.14 we have

$$v_{0x} = v_0 \cos \theta_0 \quad \text{and} \quad v_{0y} = v_0 \sin \theta_0$$

where v_{0x} is the initial velocity (at $t = 0$) in the x-direction and v_{0y} is the initial velocity in the y-direction.

Now, Equations 2.6, 2.9, and 2.10 developed in Chapter 2 for motion with constant acceleration in one dimension carry over to the two-dimensional case; there is one set of three equations for each direction, with the initial velocities modified as just discussed. In the x-direction, with a_x constant, we have

$$v_x = v_{0x} + a_x t \quad [3.11a]$$

$$\Delta x = v_{0x} t + \tfrac{1}{2} a_x t^2 \quad [3.11b]$$

$$v_x^2 = v_{0x}^2 + 2 a_x \Delta x \quad [3.11c]$$

where $v_{0x} = v_0 \cos \theta_0$. In the y-direction, we have

$$v_y = v_{0y} + a_y t \quad [3.12a]$$

$$\Delta y = v_{0y} t + \tfrac{1}{2} a_y t^2 \quad [3.12b]$$

$$v_y^2 = v_{0y}^2 + 2 a_y \Delta y \quad [3.12c]$$

where $v_{0y} = v_0 \sin \theta_0$ and a_y is constant. The object's speed v can be calculated from the components of the velocity using the Pythagorean theorem:

$$v = \sqrt{v_x^2 + v_y^2}$$

The angle that the velocity vector makes with the x-axis is given by

$$\theta = \tan^{-1} \left(\frac{v_y}{v_x} \right)$$

This formula for θ, as previously stated, must be used with care, because the inverse tangent function returns values only between $-90°$ and $+90°$. Adding $180°$ is necessary for vectors lying in the second or third quadrant.

The kinematic equations are easily adapted and simplified for projectiles close to the surface of the Earth. In this case, assuming air friction is negligible, the acceleration in the x-direction is 0 (because air resistance is neglected). **This means that $a_x = 0$, and the projectile's velocity component along the x-direction remains constant.** If the initial value of the velocity component in the x-direction is $v_{0x} = v_0 \cos \theta_0$, then this is also the value of v_x at any later time, so

$$v_x = v_{0x} = v_0 \cos \theta_0 = \text{constant} \quad [3.13a]$$

whereas the horizontal displacement is simply

$$\Delta x = v_{0x} t = (v_0 \cos \theta_0) t \quad [3.13b]$$

For the motion in the y-direction, we make the substitution $a_y = -g$ and $v_{0y} = v_0 \sin \theta_0$ in Equations 3.12, giving

$$v_y = v_0 \sin \theta_0 - gt \quad [3.14a]$$

$$\Delta y = (v_0 \sin \theta_0) t - \tfrac{1}{2} g t^2 \quad [3.14b]$$

$$v_y^2 = (v_0 \sin \theta_0)^2 - 2g \Delta y \quad [3.14c]$$

The important facts of projectile motion can be summarized as follows:

1. Provided air resistance is negligible, the horizontal component of the velocity v_x remains constant because there is no horizontal component of acceleration.
2. The vertical component of the acceleration is equal to the free-fall acceleration $-g$.
3. The vertical component of the velocity v_y and the displacement in the y-direction are identical to those of a freely falling body.
4. Projectile motion can be described as a superposition of two independent motions in the x- and y-directions.

EXAMPLE 3.4 Projectile Motion with Diagrams

GOAL Approximate answers in projectile motion using a motion diagram.

PROBLEM A ball is thrown so that its initial vertical and horizontal components of velocity are 40 m/s and 20 m/s, respectively. Use a motion diagram to estimate the ball's total time of flight and the distance it traverses before hitting the ground.

STRATEGY Use the diagram, estimating the acceleration of gravity as -10 m/s². By symmetry, the ball goes up and comes back down to the ground at the same y-velocity as when it left, except with opposite sign. With this fact and the fact that the acceleration of gravity decreases the velocity in the y-direction by 10 m/s every second, we can find the total time of flight and then the horizontal range.

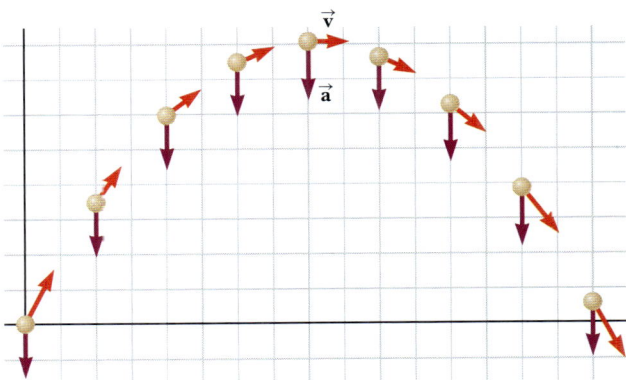

Figure 3.18 (Example 3.4) Motion diagram for a projectile.

SOLUTION

In the motion diagram shown in Figure 3.18, the acceleration vectors are all the same, pointing downward with magnitude of nearly 10 m/s². By symmetry, we know that the ball will hit the ground at the same speed in the y-direction as when it was thrown, so the velocity in the y-direction goes from 40 m/s to -40 m/s in steps of -10 m/s every second; hence, approximately 8 seconds elapse during the motion.

(Continued)

The velocity vector constantly changes direction, but the horizontal velocity never changes because the acceleration in the horizontal direction is zero. Therefore, the displacement of the ball in the x-direction is given by Equation 3.13b, $\Delta x \approx v_{0x}t = (20 \text{ m/s})(8 \text{ s}) = 160 \text{ m}$.

REMARKS This example emphasizes the independence of the x- and y-components in projectile motion problems.

QUESTION 3.4 Is the magnitude of the velocity vector at impact greater than, less than, or equal to the magnitude of the initial velocity vector? Why?

EXERCISE 3.4 Estimate the maximum height in this same problem.

ANSWER 80 m

■ Quick Quiz

3.6 Suppose you are carrying a ball and running at constant speed, and wish to throw the ball and catch it as it comes back down. Neglecting air resistance, should you (a) throw the ball at an angle of about 45° above the horizontal and maintain the same speed, (b) throw the ball straight up in the air and slow down to catch it, or (c) throw the ball straight up in the air and maintain the same speed?

3.7 As a projectile moves in its parabolic path, the velocity and acceleration vectors are perpendicular to each other (a) everywhere along the projectile's path, (b) at the peak of its path, (c) nowhere along its path, or (d) not enough information is given.

■ PROBLEM-SOLVING STRATEGY

Projectile Motion

1. **Select** a coordinate system and sketch the path of the projectile, including initial and final positions, velocities, and accelerations.
2. **Resolve** the initial velocity vector into x- and y-components.
3. **Treat** the horizontal motion and the vertical motion independently.
4. **Follow** the techniques for solving problems with constant velocity to analyze the horizontal motion of the projectile.
5. **Follow** the techniques for solving problems with constant acceleration to analyze the vertical motion of the projectile.

■ EXAMPLE 3.5 | Stranded Explorers

GOAL Solve a two-dimensional projectile motion problem in which an object has an initial horizontal velocity.

PROBLEM An Alaskan rescue plane drops a package of emergency rations to stranded hikers, as shown in Figure 3.19. The plane is traveling horizontally at 40.0 m/s at a height of 1.00×10^2 m above the ground. Neglect air resistance. **(a)** Where does the package strike the ground relative to the point at which it was released? **(b)** What are the horizontal and vertical components of the velocity of the package just before it hits the ground? **(c)** Find the angle of the impact.

STRATEGY Here, we're just taking Equations 3.13 and 3.14, filling in known quantities, and solving for the remaining unknown quantities. Sketch the problem using a coordinate system as in Figure 3.19. In part (a), set the y-component of the displacement equations equal to -1.00×10^2 m—the ground level where the package lands—and solve for the time it takes the package to reach the ground. Substitute this time into the displacement equation for the x-component to find the range. In part (b), substitute the time found in part (a) into the velocity components. Notice that the initial velocity has only an x-component, which simplifies the math. Solving part (c) requires the inverse tangent function.

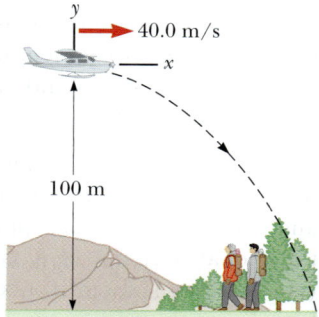

Figure 3.19 (Example 3.5) From the point of view of an observer on the ground, a package released from the rescue plane travels along the path shown.

SOLUTION

(a) Find the range of the package.

Use Equation 3.14b to find the y-displacement:

$$\Delta y = y - y_0 = v_{0y}t - \tfrac{1}{2}gt^2$$

Substitute $y_0 = 0$ and $v_{0y} = 0$, and set $y = -1.00 \times 10^2$ m, the final vertical position of the package relative to the airplane. Solve for time:

$$y = -(4.90 \text{ m/s}^2)t^2 = -1.00 \times 10^2 \text{ m}$$
$$t = 4.52 \text{ s}$$

Use Equation 3.13b to find the x-displacement:

$$\Delta x = x - x_0 = v_{0x}t$$

Substitute $x_0 = 0$, $v_{0x} = 40.0$ m/s, and the time:

$$x = (40.0 \text{ m/s})(4.52 \text{ s}) = \boxed{181 \text{ m}}$$

(b) Find the components of the package's velocity at impact:

Find the x-component of the velocity at the time of impact:

$$v_x = v_0 \cos\theta = (40.0 \text{ m/s}) \cos 0° = \boxed{40.0 \text{ m/s}}$$

Find the y-component of the velocity at the time of impact:

$$v_y = v_0 \sin\theta - gt = 0 - (9.80 \text{ m/s}^2)(4.52 \text{ s}) = \boxed{-44.3 \text{ m/s}}$$

(c) Find the angle of the impact.

Write Equation 3.4 and substitute values:

$$\tan\theta = \frac{v_y}{v_x} = \frac{-44.3 \text{ m/s}}{40.0 \text{ m/s}} = -1.11$$

Apply the inverse tangent functions to both sides:

$$\theta = \tan^{-1}(-1.11) = \boxed{-48.0°}$$

REMARKS Notice how motion in the x-direction and motion in the y-direction are handled separately.

QUESTION 3.5 Neglecting air resistance effects, what path does the package travel as observed by the pilot? Explain.

EXERCISE 3.5 A bartender slides a beer mug at 1.50 m/s toward a customer at the end of a frictionless bar that is 1.20 m tall. The customer makes a grab for the mug and misses, and the mug sails off the end of the bar. (a) How far away from the end of the bar does the mug hit the floor? (b) What are the speed and direction of the mug at impact?

ANSWERS (a) 0.742 m (b) 5.08 m/s, $\theta = -72.8°$

■ EXAMPLE 3.6 The Long Jump

GOAL Solve a two-dimensional projectile motion problem involving an object starting and ending at the same height.

PROBLEM A long jumper (Fig. 3.20) leaves the ground at an angle of 20.0° to the horizontal and at a speed of 11.0 m/s. (a) How long does it take for her to reach maximum height? (b) What is the maximum height? (c) How far does she jump? (Assume her motion is equivalent to that of a particle, disregarding the motion of her arms and legs.) (d) Use Equation 3.14c to find the maximum height she reaches.

Figure 3.20 (Example 3.6) This multiple-exposure shot of a long jumper shows that in reality, the jumper's motion is not the equivalent of the motion of a particle. The center of mass of the jumper follows a parabola, but to extend the length of the jump before impact, the jumper pulls her feet up so she strikes the ground later than she otherwise would have.

(Continued)

70 CHAPTER 3 | Vectors and Two-Dimensional Motion

STRATEGY Again, we take the projectile equations, fill in the known quantities, and solve for the unknowns. At the maximum height, the velocity in the y-direction is zero, so setting Equation 3.14a equal to zero and solving gives the time it takes her to reach her maximum height. By symmetry, given that her trajectory starts and ends at the same height, doubling this time gives the total time of the jump.

SOLUTION

(a) Find the time t_{max} taken to reach maximum height.

Set $v_y = 0$ in Equation 3.14a and solve for t_{max}:

$$v_y = v_0 \sin \theta_0 - g t_{max} = 0$$

$$(1) \quad t_{max} = \frac{v_0 \sin \theta_0}{g}$$

$$= \frac{(11.0 \text{ m/s})(\sin 20.0°)}{9.80 \text{ m/s}^2} = \boxed{0.384 \text{ s}}$$

(b) Find the maximum height she reaches.

Substitute the time t_{max} into the equation for the y-displacement, equation 3.14b:

$$y_{max} = (v_0 \sin \theta_0) t_{max} - \tfrac{1}{2} g (t_{max})^2$$

$$y_{max} = (11.0 \text{ m/s})(\sin 20.0°)(0.384 \text{ s})$$

$$- \tfrac{1}{2}(9.80 \text{ m/s}^2)(0.384 \text{ s})^2$$

$$y_{max} = \boxed{0.722 \text{ m}}$$

(c) Find the horizontal distance she jumps.

First find the time for the jump, which is twice t_{max}:

$$t = 2t_{max} = 2(0.384 \text{ s}) = 0.768 \text{ s}$$

Substitute this result into the equation for the x-displacement:

$$(2) \quad \Delta x = (v_0 \cos \theta_0) t$$

$$= (11.0 \text{ m/s})(\cos 20.0°)(0.768 \text{ s}) = \boxed{7.94 \text{ m}}$$

(d) Use an alternate method to find the maximum height.

Use Equation 3.14c, solving for Δy:

$$v_y^2 - v_{0y}^2 = -2g \Delta y$$

$$\Delta y = \frac{v_y^2 - v_{0y}^2}{-2g}$$

Substitute $v_y = 0$ at maximum height, and the fact that $v_{0y} = (11.0 \text{ m/s}) \sin 20.0°$:

$$\Delta y = \frac{0 - [(11.0 \text{ m/s}) \sin 20.0°]^2}{-2(9.80 \text{ m/s}^2)} = \boxed{0.722 \text{ m}}$$

REMARKS Although modeling the long jumper's motion as that of a projectile is an oversimplification, the values obtained are reasonable.

QUESTION 3.6 True or False: Because the x-component of the displacement doesn't depend explicitly on g, the horizontal distance traveled doesn't depend on the acceleration of gravity.

EXERCISE 3.6 A grasshopper jumps a horizontal distance of 1.00 m from rest, with an initial velocity at a 45.0° angle with respect to the horizontal. Find (a) the initial speed of the grasshopper and (b) the maximum height reached.

ANSWERS (a) 3.13 m/s (b) 0.250 m

■ **EXAMPLE 3.7** | **The Range Equation**

GOAL Find an equation for the maximum horizontal displacement of a projectile fired from ground level.

PROBLEM An athlete participates in a long-jump competition, leaping into the air with a velocity v_0 at an angle θ_0 with the horizontal. Obtain an expression for the length of the jump in terms of v_0, θ_0, and g.

STRATEGY Use the results of Example 3.6, eliminating the time t from Equations (1) and (2).

SOLUTION

Use Equation (1) of Example 3.6 to find the time of flight, t:

$$t = 2t_{max} = \frac{2v_0 \sin \theta_0}{g}$$

Substitute that expression for t into Equation (2) of Example 3.6:

$$\Delta x = (v_0 \cos \theta_0)t = (v_0 \cos \theta_0)\left(\frac{2v_0 \sin \theta_0}{g}\right)$$

Simplify:

$$\Delta x = \frac{2v_0^2 \cos \theta_0 \sin \theta_0}{g}$$

Substitute the identity $2 \cos \theta_0 \sin \theta_0 = \sin 2\theta_0$ to reduce the foregoing expression to a single trigonometric function:

$$(1) \quad \Delta x = \frac{v_0^2 \sin 2\theta_0}{g}$$

REMARKS The use of a trigonometric identity in the final step isn't necessary, but it makes Question 3.7 easier to answer.

QUESTION 3.7 What angle θ_0 produces the longest jump?

EXERCISE 3.7 Obtain an expression for the athlete's maximum displacement in the vertical direction, Δy_{max} in terms of v_0, θ_0, and g.

ANSWER $\Delta y_{max} = \dfrac{v_0^2 \sin^2 \theta_0}{2g}$

EXAMPLE 3.8 | That's Quite an Arm

GOAL Solve a two-dimensional kinematics problem with a nonhorizontal initial velocity, starting and ending at different heights.

PROBLEM A ball is thrown upward from the top of a building at an angle of 30.0° above the horizontal and with an initial speed of 20.0 m/s, as in Figure 3.21. The point of release is 45.0 m above the ground. **(a)** How long does it take for the ball to hit the ground? **(b)** Find the ball's speed at impact. **(c)** Find the horizontal range of the stone. Neglect air resistance.

STRATEGY Choose coordinates as in the figure, with the origin at the point of release. **(a)** Fill in the constants of Equation 3.14b for the y-displacement and set the displacement equal to -45.0 m, the y-displacement when the ball hits the ground. Using the quadratic formula, solve for the time. To solve part **(b)**, substitute the time from part (a) into the components of the velocity, and substitute the same time into the equation for the x-displacement to solve part **(c)**.

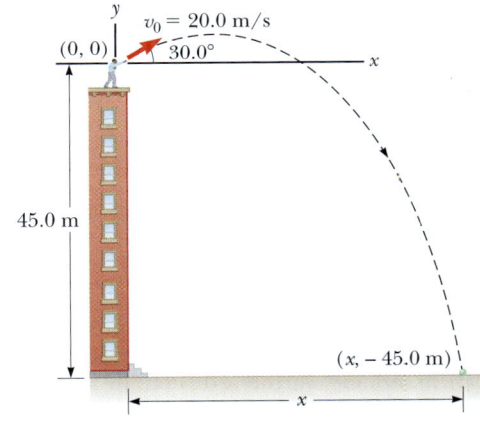

Figure 3.21 (Example 3.8)

SOLUTION

(a) Find the ball's time of flight.

Find the initial x- and y-components of the velocity:

$$v_{0x} = v_0 \cos \theta_0 = (20.0 \text{ m/s})(\cos 30.0°) = +17.3 \text{ m/s}$$
$$v_{0y} = v_0 \sin \theta_0 = (20.0 \text{ m/s})(\sin 30.0°) = +10.0 \text{ m/s}$$

Find the y-displacement, taking $y_0 = 0$, $y = -45.0$ m, and $v_{0y} = 10.0$ m/s:

$$\Delta y = y - y_0 = v_{0y}t - \tfrac{1}{2}gt^2$$
$$-45.0 \text{ m} = (10.0 \text{ m/s})t - (4.90 \text{ m/s}^2)t^2$$

Reorganize the equation into standard form and use the quadratic formula (see Appendix A) to find the positive root:

$$t = \boxed{4.22 \text{ s}}$$

(b) Find the ball's speed at impact.

Substitute the value of t found in part (a) into Equation 3.14a to find the y-component of the velocity at impact:

$$v_y = v_{0y} - gt = 10.0 \text{ m/s} - (9.80 \text{ m/s}^2)(4.22 \text{ s})$$
$$= -31.4 \text{ m/s}$$

(Continued)

Use this value of v_y, the Pythagorean theorem, and the fact that $v_x = v_{0x} = 17.3$ m/s to find the speed of the ball at impact:

$$v = \sqrt{v_x^2 + v_y^2} = \sqrt{(17.3 \text{ m/s})^2 + (-31.4 \text{ m/s})^2}$$
$$= \boxed{35.9 \text{ m/s}}$$

(c) Find the horizontal range of the ball.

Substitute the time of flight into the range equation:

$$\Delta x = x - x_0 = (v_0 \cos \theta)t = (20.0 \text{ m/s})(\cos 30.0°)(4.22 \text{ s})$$
$$= \boxed{73.1 \text{ m}}$$

REMARKS The angle at which the ball is thrown affects the velocity vector throughout its subsequent motion, but doesn't affect the speed at a given height. This is a consequence of the conservation of energy, described in Chapter 5.

QUESTION 3.8 True or False: All other things being equal, if the ball is thrown at half the given speed it will travel half as far.

EXERCISE 3.8 Suppose the ball is thrown from the same height as in the example at an angle of 30.0° below the horizontal. If it strikes the ground 57.0 m away, find (a) the time of flight, (b) the initial speed, and (c) the speed and the angle of the velocity vector with respect to the horizontal at impact. (*Hint:* For part (a), use the equation for the *x*-displacement to eliminate $v_0 t$ from the equation for the *y*-displacement.)

ANSWERS (a) 1.57 s (b) 41.9 m/s (c) 51.3 m/s, −45.0°

Two-Dimensional Constant Acceleration

So far we have studied only problems in which an object with an initial velocity follows a trajectory determined by the acceleration of gravity alone. In the more general case, other agents, such as air drag, surface friction, or engines, can cause accelerations. These accelerations, taken together, form a vector quantity with components a_x and a_y. When both components are constant, we can use Equations 3.11 and 3.12 to study the motion, as in the next example.

■ EXAMPLE 3.9 | The Rocket

GOAL Solve a problem involving accelerations in two directions.

PROBLEM A jet plane traveling horizontally at 1.00×10^2 m/s drops a rocket from a considerable height. (See Fig. 3.22.) The rocket immediately fires its engines, accelerating at 20.0 m/s² in the *x*-direction while falling under the influence of gravity in the *y*-direction. When the rocket has fallen 1.00 km, find **(a)** its velocity in the *y*-direction, **(b)** its velocity in the *x*-direction, and **(c)** the magnitude and direction of its velocity. Neglect air drag and aerodynamic lift.

STRATEGY Because the rocket maintains a horizontal orientation (say, through gyroscopes), the *x*- and *y*-components of acceleration are independent of each other. Use the time-independent equation for the velocity in the *y*-direction to find the *y*-component of the velocity after the rocket falls 1.00 km. Then calculate the time of the fall and use that time to find the velocity in the *x*-direction.

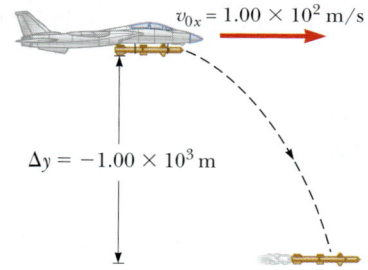

Figure 3.22 (Example 3.9)

SOLUTION

(a) Find the velocity in the *y*-direction.

Use Equation 3.14c:

$$v_y^2 = v_{0y}^2 - 2g\Delta y$$

Substitute $v_{0y} = 0$, $g = -9.80$ m/s², and $\Delta y = -1.00 \times 10^3$ m, and solve for v_y:

$$v_y^2 - 0 = 2(-9.8 \text{ m/s}^2)(-1.00 \times 10^3 \text{ m})$$
$$v_y = \boxed{-1.40 \times 10^2 \text{ m/s}}$$

(b) Find the velocity in the *x*-direction.

Find the time it takes the rocket to drop 1.00×10^3 m, using the *y*-component of the velocity:

$$v_y = v_{0y} + a_y t$$
$$-1.40 \times 10^2 \text{ m/s} = 0 - (9.80 \text{ m/s}^2)t \quad \rightarrow \quad t = 14.3 \text{ s}$$

Substitute t, v_{0x}, and a_x into Equation 3.11a to find the velocity in the x-direction:

$v_x = v_{0x} + a_x t = 1.00 \times 10^2$ m/s $+ (20.0$ m/s$^2)(14.3$ s$)$
$= 386$ m/s

(c) Find the magnitude and direction of the velocity.

Find the magnitude using the Pythagorean theorem and the results of parts (a) and (b):

$v = \sqrt{v_x^2 + v_y^2} = \sqrt{(-1.40 \times 10^2 \text{ m/s})^2 + (386 \text{ m/s})^2}$
$= 411$ m/s

Use the inverse tangent function to find the angle:

$\theta = \tan^{-1}\left(\dfrac{v_y}{v_x}\right) = \tan^{-1}\left(\dfrac{-1.40 \times 10^2 \text{ m/s}}{386 \text{ m/s}}\right) = -19.9°$

REMARKS Notice the similarity: The kinematic equations for the x- and y-directions are handled in exactly the same way. Having a nonzero acceleration in the x-direction doesn't greatly increase the difficulty of the problem.

QUESTION 3.9 True or False: Neglecting air friction and lift effects, a projectile with a horizontal acceleration always stays in the air longer than a projectile that is freely falling.

EXERCISE 3.9 Suppose a rocket-propelled motorcycle is fired from rest horizontally across a canyon 1.00 km wide. (a) What minimum constant acceleration in the x-direction must be provided by the engines so the cycle crosses safely if the opposite side is 0.750 km lower than the starting point? (b) At what speed does the motorcycle land if it maintains this constant horizontal component of acceleration? Neglect air drag, but remember that gravity is still acting in the negative y-direction.

ANSWERS (a) 13.1 m/s^2 (b) 202 m/s

In a stunt similar to that described in Exercise 3.9, motorcycle daredevil Evel Knievel tried to vault across Hells Canyon, part of the Snake River system in Idaho, on his rocket-powered Harley-Davidson X-2 "Skycycle." He lost consciousness at takeoff and released a lever, prematurely deploying his parachute and falling short of the other side. He landed safely in the canyon.

3.5 Relative Velocity

LEARNING OBJECTIVES

1. Derive the relative velocity equation.
2. Solve problems involving relative velocity.

Relative velocity is all about relating the measurements of two different observers, one moving with respect to the other. The measured velocity of an object depends on the velocity of the observer with respect to the object. On highways, for example, cars moving in the same direction are often moving at high speed relative to Earth, but relative to each other they hardly move at all. To an observer at rest at the side of the road, a car might be traveling at 60 mi/h, but to an observer in a truck traveling in the same direction at 50 mi/h, the car would appear to be traveling only 10 mi/h.

So measurements of velocity depend on the **reference frame** of the observer. Reference frames are just coordinate systems. Most of the time, we use a **stationary frame of reference** relative to Earth, but occasionally we use a **moving frame of reference** associated with a bus, car, or plane moving with constant velocity relative to Earth.

In two dimensions relative velocity calculations can be confusing, so a systematic approach is important and useful. Let E be an observer, assumed stationary

74 CHAPTER 3 | Vectors and Two-Dimensional Motion

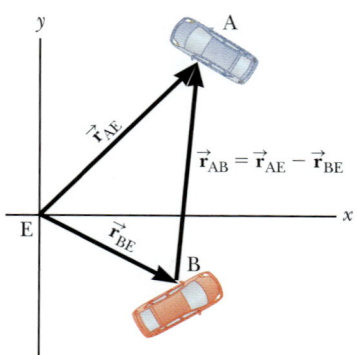

Figure 3.23 The position of Car A relative to Car B can be found by vector subtraction. The rate of change of the resultant vector with respect to time is the relative velocity equation.

with respect to Earth. Let two cars be labeled A and B, and introduce the following notation (see Fig. 3.23):

$\vec{r}_{AE}$ = the position of Car A as measured by E (in a coordinate system fixed with respect to Earth).

$\vec{r}_{BE}$ = the position of Car B as measured by E.

$\vec{r}_{AB}$ = the position of Car A as measured by an observer in Car B.

According to the preceding notation, the first letter tells us what the vector is pointing at and the second letter tells us where the position vector starts. The position vectors of Car A and Car B relative to E, $\vec{r}_{AE}$ and $\vec{r}_{BE}$, are given in the figure. How do we find $\vec{r}_{AB}$, the position of Car A as measured by an observer in Car B? We simply draw an arrow pointing from Car B to Car A, which can be obtained by subtracting $\vec{r}_{BE}$ from $\vec{r}_{AE}$:

$$\vec{r}_{AB} = \vec{r}_{AE} - \vec{r}_{BE} \qquad [3.15]$$

Now, the rate of change of these quantities with time gives us the relationship between the associated velocities:

$$\vec{v}_{AB} = \vec{v}_{AE} - \vec{v}_{BE} \qquad [3.16]$$

The coordinate system of observer E need not be fixed to Earth, although it often is. Take careful note of the pattern of subscripts; rather than memorize Equation 3.16, it's better to study the short derivation based on Figure 3.23. Note also that the equation doesn't work for observers traveling a sizable fraction of the speed of light, when Einstein's theory of special relativity comes into play.

■ PROBLEM-SOLVING STRATEGY

Relative Velocity

1. **Label** each object involved (usually three) with a letter that reminds you of what it is (for example, E for Earth).
2. **Look** through the problem for phrases such as "The velocity of A relative to B" and write the velocities as $\vec{v}_{AB}$. When a velocity is mentioned but it isn't explicitly stated as relative to something, it's almost always relative to Earth.
3. **Take** the three velocities you've found and assemble them into an equation just like Equation 3.16, with subscripts in an analogous order.
4. **There will be** two unknown components. Solve for them with the *x*- and *y*-components of the equation developed in step 3.

■ EXAMPLE 3.10 | Pitching Practice on the Train

GOAL Solve a one-dimensional relative velocity problem.

PROBLEM A train is traveling with a speed of 15 m/s relative to Earth. A passenger standing at the rear of the train pitches a baseball with a speed of 15 m/s relative to the train off the back end, in the direction opposite the motion of the train. **(a)** What is the velocity of the baseball relative to Earth? **(b)** What is the velocity of the baseball relative to the Earth if thrown in the opposite direction at the same speed?

STRATEGY Solving these problems involves putting the proper subscripts on the velocities and arranging them as in Equation 3.16. In the first sentence of the problem statement, we are informed that the train travels at "15 m/s relative to Earth." This quantity is $\vec{v}_{TE}$, with T for train and E for Earth. The passenger throws the baseball at "15 m/s relative to the train," so this quantity is $\vec{v}_{BT}$, where B stands for baseball. The second sentence asks for the velocity of the baseball relative to Earth, $\vec{v}_{BE}$. The rest of the problem can be solved by identifying the correct components of the known quantities and solving for the unknowns, using an analog of Equation 3.16. Part(b) just requires a change of sign.

SOLUTION

(a) What is the velocity of the baseball relative to the Earth?

Write the x-components of the known quantities:

$(\vec{v}_{TE})_x = +15 \text{ m/s}$
$(\vec{v}_{BT})_x = -15 \text{ m/s}$

Follow Equation 3.16:

(1) $(\vec{v}_{BT})_x = (\vec{v}_{BE})_x - (\vec{v}_{TE})_x$

Insert the given values and solve:

$-15 \text{ m/s} = (\vec{v}_{BE})_x - 15 \text{ m/s}$
$(\vec{v}_{BE})_x = 0$

(b) What is the velocity of the baseball relative the to Earth if thrown in the opposite direction at the same speed?

Substitute $(\vec{v}_{BT})_x = +15 \text{ m/s}$ into Equation (1):

$+15 \text{ m/s} = (\vec{v}_{BT})_x - 15 \text{ m/s}$

Solve for $(\vec{v}_{BT})_x$:

$(\vec{v}_{BT})_x = +3.0 \times 10^1 \text{ m/s}$

QUESTION 3.10 Describe the motion of the ball in part (a) as related by an observer on the ground.

EXERCISE 3.10 A train is traveling at 27 m/s relative to Earth in the positive x-direction. A passenger standing on the ground throws a ball at 15 m/s relative to Earth in the same direction as the train's motion. **(a)** Find the speed of the ball relative to an observer on the train. **(b)** Repeat the exercise if the ball is thrown in the opposite direction.

ANSWERS (a) −12 m/s (b) −42 m/s

■ EXAMPLE 3.11 | Crossing a River

GOAL Solve a simple two-dimensional relative motion problem.

PROBLEM The boat in Figure 3.24 is heading due north as it crosses a wide river with a velocity of 10.0 km/h relative to the water. The river has a uniform velocity of 5.00 km/h due east. Determine the magnitude and direction of the boat's velocity with respect to an observer on the riverbank.

STRATEGY Again, we look for key phrases. "The boat . . . (has) a velocity of 10.0 km/h relative to the water" gives $\vec{v}_{BR}$. "The river has a uniform velocity of 5.00 km/h due east" gives $\vec{v}_{RE}$, because this implies velocity with respect to Earth. The observer on the riverbank is in a reference frame at rest with respect to Earth. Because we're looking for the velocity of the boat with respect to that observer, this last velocity is designated $\vec{v}_{BE}$. Take east to be the $+x$-direction, north the $+y$-direction.

Figure 3.24 (Example 3.10)

SOLUTION

Arrange the three quantities into the proper relative velocity equation:

$\vec{v}_{BR} = \vec{v}_{BE} - \vec{v}_{RE}$

Write the velocity vectors in terms of their components. For convenience, these are organized in the following table:

Vector	x-Component (km/h)	y-Component (km/h)
$\vec{v}_{BR}$	0	10.0
$\vec{v}_{BE}$	v_x	v_y
$\vec{v}_{RE}$	5.00	0

Find the x-component of velocity:

$0 = v_x - 5.00 \text{ km/h} \rightarrow v_x = 5.00 \text{ km/h}$

Find the y-component of velocity:

$10.0 \text{ km/h} = v_y - 0 \rightarrow v_y = 10.0 \text{ km/h}$

(Continued)

Find the magnitude of $\vec{v}_{BE}$:

$$v_{BE} = \sqrt{v_x^2 + v_y^2}$$
$$= \sqrt{(5.00 \text{ km/h})^2 + (10.0 \text{ km/h})^2} = \boxed{11.2 \text{ km/h}}$$

Find the direction of $\vec{v}_{BE}$:

$$\theta = \tan^{-1}\left(\frac{v_x}{v_y}\right) = \tan^{-1}\left(\frac{5.00 \text{ m/s}}{10.0 \text{ m/s}}\right) = \boxed{26.6°}$$

REMARKS The boat travels at a speed of 11.2 km/h in the direction 26.6° east of north with respect to Earth.

QUESTION 3.11 If the speed of the boat relative to the water is increased, what happens to the angle?

EXERCISE 3.11 Suppose the river is flowing east at 3.00 m/s and the boat is traveling south at 4.00 m/s with respect to the river. Find the speed and direction of the boat relative to Earth.

ANSWER 5.00 m/s, 53.1° south of east

EXAMPLE 3.12 | Bucking the Current

GOAL Solve a complex two-dimensional relative motion problem.

PROBLEM If the skipper of the boat of Example 3.11 moves with the same speed of 10.0 km/h relative to the water but now wants to travel due north, as in Figure 3.25a, in what direction should he head? What is the speed of the boat, according to an observer on the shore? The river is flowing east at 5.00 km/h.

Figure 3.25
(a) (Example 3.12)
(b) (Exercise 3.12)

STRATEGY Proceed as in the previous example. In this situation, we must find the heading of the boat and its velocity with respect to the water, using the fact that the boat travels due north.

SOLUTION

Arrange the three quantities, as before:

$$\vec{v}_{BR} = \vec{v}_{BE} - \vec{v}_{RE}$$

Organize a table of velocity components:

Vector	x-Component (km/h)	y-Component (km/h)
$\vec{v}_{BR}$	$-(10.0 \text{ km/h}) \sin\theta$	$(10.0 \text{ km/h}) \cos\theta$
$\vec{v}_{BE}$	0	v
$\vec{v}_{RE}$	5.00 km/h	0

The x-component of the relative velocity equation can be used to find θ:

$$-(10.0 \text{ m/s}) \sin\theta = 0 - 5.00 \text{ km/h}$$
$$\sin\theta = \frac{5.00 \text{ km/h}}{10.0 \text{ km/h}} = \frac{1.00}{2.00}$$

Apply the inverse sine function and find θ, which is the boat's heading, west of north:

$$\theta = \sin^{-1}\left(\frac{1.00}{2.00}\right) = \boxed{30.0°}$$

The y-component of the relative velocity equation can be used to find v:

$$(10.0 \text{ km/h}) \cos\theta = v \rightarrow v = \boxed{8.66 \text{ km/h}}$$

REMARKS From Figure 3.25, we see that this problem can be solved with the Pythagorean theorem, because the problem involves a right triangle: the boat's x-component of velocity exactly cancels the river's velocity. When this is not the case, a more general technique is necessary, as shown in the following exercise. Notice that in the x-component of the relative velocity equation a minus sign had to be included in the term $-(10.0 \text{ km/h})\sin\theta$ because the x-component of the boat's velocity with respect to the river is negative.

QUESTION 3.12 The speeds in this example are the same as in Example 3.11. Why isn't the angle the same as before?

EXERCISE 3.12 Suppose the river is moving east at 5.00 km/h and the boat is traveling 45.0° south of east with respect to Earth. Find (a) the speed of the boat with respect to Earth and (b) the speed of the boat with respect to the river if the boat's heading in the water is 60.0° south of east. (See Fig. 3.25b.) You will have to solve two equations with two unknowns. (As an alternative, the law of sines can be used.)

ANSWERS (a) 16.7 km/h (b) 13.7 km/h

■ SUMMARY

3.1 Vectors and Their Properties

Two vectors $\vec{A}$ and $\vec{B}$ can be added geometrically with the **triangle method.** The two vectors are drawn to scale on graph paper, with the tail of the second vector located at the tip of the first. The **resultant** vector is the vector drawn from the tail of the first vector to the tip of the second.

The negative of a vector $\vec{A}$ is a vector with the same magnitude as $\vec{A}$, but pointing in the opposite direction. A vector can be multiplied by a scalar, changing its magnitude, and its direction if the scalar is negative.

3.2 Components of a Vector

A vector $\vec{A}$ can be split into two components, one pointing in the x-direction and the other in the y-direction. These components form two sides of a right triangle having a hypotenuse with magnitude A and are given by

$$A_x = A \cos\theta \quad [3.2a]$$

$$A_y = A \sin\theta \quad [3.2b]$$

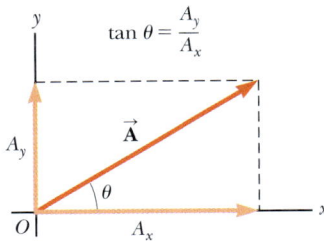

A vector can be written in terms of components in the x- and y-directions.

The magnitude and direction of $\vec{A}$ are related to its components through the Pythagorean theorem and the definition of the tangent:

$$A = \sqrt{A_x^2 + A_y^2} \quad [3.3]$$

$$\tan\theta = \frac{A_y}{A_x} \quad [3.4]$$

In Equation (3.4), $\theta = \tan^{-1}(A_y/A_x)$ gives the correct vector angle only for vectors with $-90° < \theta < 90°$. If the vector has a negative x-component, 180° must be added to the answer in the calculator window.

If $\vec{R} = \vec{A} + \vec{B}$, then the components of the resultant vector $\vec{R}$ are

$$R_x = A_x + B_x \quad [3.5a]$$

$$R_y = A_y + B_y \quad [3.5b]$$

3.3 Displacement, Velocity, and Acceleration in Two Dimensions

The displacement of an object in two dimensions is defined as the change in the object's position vector:

$$\Delta\vec{r} \equiv \vec{r}_f - \vec{r}_i \quad [3.6]$$

The average velocity of an object during the time interval Δt is

$$\vec{v}_{av} \equiv \frac{\Delta\vec{r}}{\Delta t} \quad [3.7]$$

Taking the limit of this expression as Δt gets arbitrarily small gives the instantaneous velocity $\vec{v}$:

$$\vec{v} \equiv \lim_{\Delta t \to 0} \frac{\Delta\vec{r}}{\Delta t} \quad [3.8]$$

The direction of the instantaneous velocity vector is along a line that is tangent to the path of the object and in the direction of its motion.

The average acceleration of an object with a velocity changing by $\Delta\vec{v}$ in the time interval Δt is

$$\vec{a}_{av} \equiv \frac{\Delta\vec{v}}{\Delta t} \quad [3.9]$$

Taking the limit of this expression as Δt gets arbitrarily small gives the instantaneous acceleration vector $\vec{a}$:

$$\vec{a} \equiv \lim_{\Delta t \to 0} \frac{\Delta\vec{v}}{\Delta t} \quad [3.10]$$

3.4 Motion in Two Dimensions

The general kinematic equations in two dimensions for objects with constant acceleration are, for the x-direction,

$$v_x = v_{0x} + a_x t \quad [3.11a]$$

$$\Delta x = v_{0x}t + \tfrac{1}{2}a_x t^2 \quad [3.11b]$$

$$v_x^2 = v_{0x}^2 + 2a_x \Delta x \quad [3.11c]$$

where $v_{0x} = v_0 \cos\theta_0$, and, for the y-direction,

$$v_y = v_{0y} + a_y t \quad [3.12a]$$

$$\Delta y = v_{0y}t + \tfrac{1}{2}a_y t^2 \quad [3.12b]$$

$$v_y^2 = v_{0y}^2 + 2a_y \Delta y \quad [3.12c]$$

where $v_{0y} = v_0 \sin\theta_0$. The speed v of the object at any instant can be calculated from the components of velocity at that instant using the Pythagorean theorem:

$$v = \sqrt{v_x^2 + v_y^2}$$

The angle that the velocity vector makes with the x-axis is given by

$$\theta = \tan^{-1}\left(\frac{v_y}{v_x}\right)$$

The horizontal and vertical motions of a projectile are completely independent of each other.

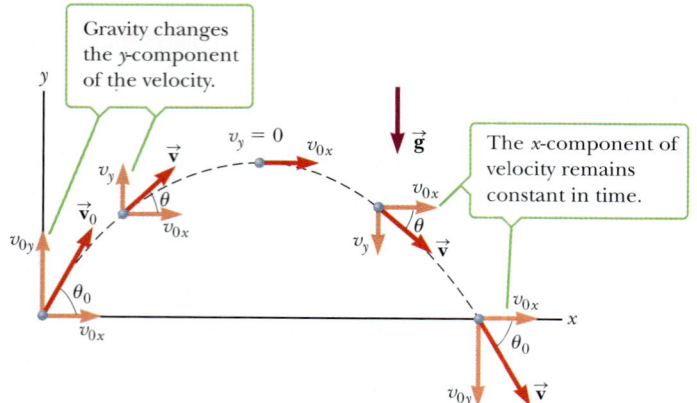

Gravity acts on the y-component of the velocity and has no effect on the x-component, illustrating the independence of horizontal and vertical projectile motion.

The kinematic equations are easily adapted and simplified for projectiles close to the surface of the Earth. The equations for the motion in the horizontal or x-direction are

$$v_x = v_{0x} = v_0 \cos \theta_0 = \text{constant} \quad [3.13a]$$

$$\Delta x = v_{0x}t = (v_0 \cos \theta_0)t \quad [3.13b]$$

while the equations for the motion in the vertical or y-direction are

$$v_y = v_0 \sin \theta_0 - gt \quad [3.14a]$$

$$\Delta y = (v_0 \sin \theta_0)t - \tfrac{1}{2}gt^2 \quad [3.14b]$$

$$v_y^2 = (v_0 \sin \theta_0)^2 - 2g\Delta y \quad [3.14c]$$

Problems are solved by algebraically manipulating one or more of these equations, which often reduces the system to two equations and two unknowns.

3.5 Relative Velocity

Let E be an observer, and B a second observer traveling with velocity $\vec{v}_{BE}$ as measured by E. If E measures the velocity of an object A as $\vec{v}_{AE}$, then B will measure A's velocity as

$$\vec{v}_{AB} = \vec{v}_{AE} - \vec{v}_{BE} \quad [3.16]$$

Equation 3.16 can be derived from Figure 3.21 by dividing the relative position equation by the Δt and taking the limit as Δt goes to zero.

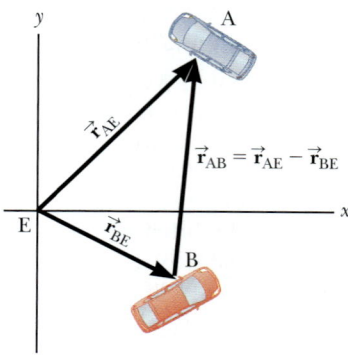

The time rate of change of the difference of the two position vectors $\vec{r}_{AE}$ and $\vec{r}_{BE}$ gives the relative velocity equation, Equation 3.16.

Solving relative velocity problems involves identifying the velocities properly and labeling them correctly, substituting into Equation 3.16, and then solving for unknown quantities.

■ WARM-UP EXERCISES

WebAssign The warm-up exercises in this chapter may be assigned online in Enhanced WebAssign.

1. A vector $\vec{A}$ has components $A_x = -5.00$ m and $A_y = 9.00$ m. Find (a) the magnitude and (b) the direction of the vector. (See Section 3.2.)

2. Calculate (a) the x- and (b) y-components of the vector with magnitude 24.0 m and direction 56.0°. (See Section 3.2.)

3. Find (a) the x- and (b) y-components of $\vec{R} = 2\vec{A} - \vec{B}$ if $\vec{A}$ has components $A_x = 15.0$ m and $A_y = 12.0$ m whereas $\vec{B}$ has components $B_x = 24.0$ m and $B_y = 8.00$ m. (See Section 3.2.)

4. A hiker walks from $(x_1, y_1) = (-4.00$ km, 3.00 km) to $(x_2, y_2) = (3.00$ km, 6.00 km). (a) What distance has he traveled? (b) The hiker desires to return to his starting point. In what direction should he go? (Give the angle with respect to due east.) (See Sections 3.2 and 3.3.)

5. A hiker walks 3.00 km north and then 4.00 km west, all in one hour and forty minutes. (a) Calculate his average speed in km/h. (b) Calculate the magnitude of his average velocity. (See Section 3.2 and 3.3.)

6. A car is traveling east at 25.0 m/s when it turns north and accelerates to 35.0 m/s, all during a time of 6.00 s. Calculate the magnitude of the car's average acceleration. (See Section 3.3.)

7. A skier leaves the end of a horizontal ski jump at 22.0 m/s and falls through a vertical distance of 3.20 m before landing. Neglecting air resistance, (a) how long

does it take the skier to reach the ground? (b) How far horizontally does the skier travel in the air before landing? (See Section 3.4.)

8. A catapult launches a large stone from ground level at a speed of 45.0 m/s at an angle of 55.0° with the horizontal. The stone returns to ground level shortly thereafter. (a) How long is it in the air? (b) What maximum height does the stone reach? (Neglect air friction.) (See Section 3.4.)

9. A cruise ship sails due north at 4.50 m/s while a coast guard patrol boat heads 45.0° north of west at 5.20 m/s. What are (a) the x- and (b) y-components of the velocity of the cruise ship relative to the patrol boat? (See Section 3.5.)

CONCEPTUAL QUESTIONS

WebAssign The conceptual questions in this chapter may be assigned online in Enhanced WebAssign.

1. If $\vec{B}$ is added to $\vec{A}$, under what conditions does the resultant vector have a magnitude equal to $A + B$? Under what conditions is the resultant vector equal to zero?

2. Under what circumstances would a vector have components that are equal in magnitude?

3. As a projectile moves in its path, is there any point along the path where the velocity and acceleration vectors are (a) perpendicular to each other? (b) Parallel to each other?

4. Construct motion diagrams showing the velocity and acceleration of a projectile at several points along its path, assuming (a) the projectile is launched horizontally and (b) the projectile is launched at an angle θ with the horizontal.

5. Explain whether the following particles do or do not have an acceleration: (a) a particle moving in a straight line with constant speed and (b) a particle moving around a curve with constant speed.

6. A ball is projected horizontally from the top of a building. One second later, another ball is projected horizontally from the same point with the same velocity. (a) At what point in the motion will the balls be closest to each other? (b) Will the first ball always be traveling faster than the second? (c) What will be the time difference between them when the balls hit the ground? (d) Can the horizontal projection velocity of the second ball be changed so that the balls arrive at the ground at the same time?

7. A spacecraft drifts through space at a constant velocity. Suddenly, a gas leak in the side of the spacecraft causes it to constantly accelerate in a direction perpendicular to the initial velocity. The orientation of the spacecraft does not change, so the acceleration remains perpendicular to the original direction of the velocity. What is the shape of the path followed by the spacecraft?

8. Determine which of the following moving objects obey the equations of projectile motion developed in this chapter. (a) A ball is thrown in an arbitrary direction. (b) A jet airplane crosses the sky with its engines thrusting the plane forward. (c) A rocket leaves the launch pad. (d) A rocket moves through the sky after its engines have failed. (e) A stone is thrown under water.

9. Two projectiles are thrown with the same initial speed, one at an angle θ with respect to the level ground and the other at angle $90° - \theta$. Both projectiles strike the ground at the same distance from the projection point. Are both projectiles in the air for the same length of time?

10. A ball is thrown upward in the air by a passenger on a train that is moving with constant velocity. (a) Describe the path of the ball as seen by the passenger. Describe the path as seen by a stationary observer outside the train. (b) How would these observations change if the train were accelerating along the track?

11. A projectile is launched at some angle to the horizontal with some initial speed v_i, and air resistance is negligible. (a) Is the projectile a freely falling body? (b) What is its acceleration in the vertical direction? (c) What is its acceleration in the horizontal direction?

12. A baseball is thrown from the outfield toward the catcher. When the ball reaches its highest point, which statement is true? (a) Its velocity and its acceleration are both zero. (b) Its velocity is not zero, but its acceleration is zero. (c) Its velocity is perpendicular to its acceleration. (d) Its acceleration depends on the angle at which the ball was thrown. (e) None of statements (a) through (d) is true.

13. A student throws a heavy red ball horizontally from a balcony of a tall building with an initial speed v_0. At the same time, a second student drops a lighter blue ball from the same balcony. Neglecting air resistance, which statement is true? (a) The blue ball reaches the ground first. (b) The balls reach the ground at the same instant. (c) The red ball reaches the ground first. (d) Both balls hit the ground with the same speed. (e) None of statements (a) through (d) is true.

14. A car moving around a circular track with constant speed (a) has zero acceleration, (b) has an acceleration component in the direction of its velocity, (c) has an acceleration directed away from the center of its path, (d) has an acceleration directed toward the center of its path, or (e) has an acceleration with

a direction that cannot be determined from the information given.

15. As an apple tree is transported by a truck moving to the right with a constant velocity, one of its apples shakes loose and falls toward the bed of the truck. Of the curves shown in Figure CQ3.15, (i) which best describes the path followed by the apple as seen by a stationary observer on the ground, who observes the truck moving from his left to his right? (ii) Which best describes the path as seen by an observer sitting in the truck?

Figure CQ3.15

PROBLEMS

WebAssign The problems in this chapter may be assigned online in Enhanced WebAssign.

1. denotes straightforward problem; 2. denotes intermediate problem;
3. denotes challenging problem
1. denotes full solution available in *Student Solutions Manual/Study Guide*
1. denotes problems most often assigned in Enhanced WebAssign

BIO denotes biomedical problems
GP denotes guided problems
M denotes Master It tutorial available in Enhanced WebAssign
Q|C denotes asking for quantitative and conceptual reasoning
S denotes symbolic reasoning problem
W denotes Watch It video solution available in Enhanced WebAssign

3.1 Vectors and Their Properties

1. Vector $\vec{A}$ has a magnitude of 29 units and points in the positive y-direction. When vector $\vec{B}$ is added to $\vec{A}$, the resultant vector $\vec{A} + \vec{B}$ points in the negative y-direction with a magnitude of 14 units. Find the magnitude and direction of $\vec{B}$.

2. Vector $\vec{A}$ has a magnitude of 8.00 units and makes an angle of 45.0° with the positive x-axis. Vector $\vec{B}$ also has a magnitude of 8.00 units and is directed along the negative x-axis. Using graphical methods, find (a) the vector sum $\vec{A} + \vec{B}$ and (b) the vector difference $\vec{A} - \vec{B}$.

3. Vector $\vec{A}$ is 3.00 units in length and points along the positive x-axis. Vector $\vec{B}$ is 4.00 units in length and points along the negative y-axis. Use graphical methods to find the magnitude and direction of the vectors (a) $\vec{A} + \vec{B}$ and (b) $\vec{A} - \vec{B}$.

4. **Q|C** Three displacements are $\vec{A} = 200$ m due south, $\vec{B} = 250$ m due west, and $\vec{C} = 150$ m at 30.0° east of north. (a) Construct a separate diagram for each of the following possible ways of adding these vectors: $\vec{R}_1 = \vec{A} + \vec{B} + \vec{C}$; $\vec{R}_2 = \vec{B} + \vec{C} + \vec{A}$; $\vec{R}_3 = \vec{C} + \vec{B} + \vec{A}$. (b) Explain what you can conclude from comparing the diagrams.

5. A roller coaster moves 200 ft horizontally and then rises 135 ft at an angle of 30.0° above the horizontal. Next, it travels 135 ft at an angle of 40.0° below the horizontal. Use graphical techniques to find the roller coaster's displacement from its starting point to the end of this movement.

6. **Q|C** An airplane flies 200 km due west from city A to city B and then 300 km in the direction of 30.0° north of west from city B to city C. (a) In straight-line distance, how far is city C from city A? (b) Relative to city A, in what direction is city C? (c) Why is the answer only approximately correct?

7. A plane flies from base camp to lake A, a distance of 280 km at a direction of 20.0° north of east. After dropping off supplies, the plane flies to lake B, which is 190 km and 30.0° west of north from lake A. Graphically determine the distance and direction from lake B to the base camp.

8. A force $\vec{F}_1$ of magnitude 6.00 units acts on an object at the origin in a direction $\theta = 30.0°$ above the positive x-axis (Fig. P3.8). A second force $\vec{F}_2$ of magnitude 5.00 units acts on the object in the direction of the positive y-axis. Find graphically the magnitude and direction of the resultant force $\vec{F}_1 + \vec{F}_2$.

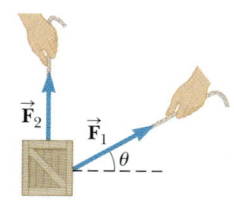

Figure P3.8

9. A man in a maze makes three consecutive displacements. His first displacement is 8.00 m westward, and the second is 13.0 m northward. At the end of his third displacement he is back to where he started. Use the graphical method to find the magnitude and direction of his third displacement.

3.2 Components of a Vector

10. A person walks 25.0° north of east for 3.10 km. How far due north and how far due east would she have to walk to arrive at the same location?

11. The magnitude of vector $\vec{A}$ is 35.0 units and points in the direction 325° counterclockwise from the positive x-axis. Calculate the x- and y-components of this vector.

12. A figure skater glides along a circular path of radius 5.00 m. If she coasts around one half of the circle, find (a) the magnitude of the displacement vector and

(b) what distance she skated. (c) What is the magnitude of the displacement if she skates all the way around the circle?

13. A girl delivering newspapers covers her route by traveling 3.00 blocks west, 4.00 blocks north, and then 6.00 blocks east. (a) What is her resultant displacement? (b) What is the total distance she travels?

14. **QC** A hiker starts at his camp and moves the following distances while exploring his surroundings: 75.0 m north, 2.50×10^2 m east, 125 m at an angle 30.0° north of east, and 1.50×10^2 m south. (a) Find his resultant displacement from camp. (Take east as the positive x-direction and north as the positive y-direction.) (b) Would changes in the order in which the hiker makes the given displacements alter his final position? Explain.

15. A vector has an x-component of -25.0 units and a y-component of 40.0 units. Find the magnitude and direction of the vector.

16. A quarterback takes the ball from the line of scrimmage, runs backwards for 10.0 yards, then runs sideways parallel to the line of scrimmage for 15.0 yards. At this point, he throws a 50.0-yard forward pass straight downfield, perpendicular to the line of scrimmage. What is the magnitude of the football's resultant displacement?

17. **M** The eye of a hurricane passes over Grand Bahama Island in a direction 60.0° north of west with a speed of 41.0 km/h. Three hours later the course of the hurricane suddenly shifts due north, and its speed slows to 25.0 km/h. How far from Grand Bahama is the hurricane 4.50 h after it passes over the island?

18. A map suggests that Atlanta is 730 miles in a direction 5.00° north of east from Dallas. The same map shows that Chicago is 560 miles in a direction 21.0° west of north from Atlanta. Figure P3.18 shows the location of these three cities. Modeling the Earth as flat, use this information to find the displacement from Dallas to Chicago.

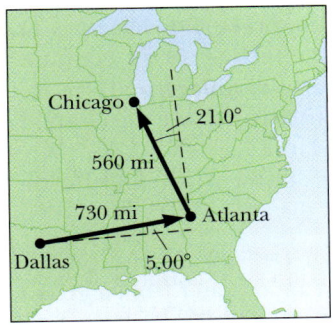

Figure P3.18

19. A commuter airplane starts from an airport and takes the route shown in Figure P3.19. The plane first flies to city A. located 175 km away in a direction 30.0° north of east. Next, it flies for 150 km 20.0° west of north, to city B. Finally, the plane flies 190 km due west, to city C. Find the location of city C relative to the location of the starting point.

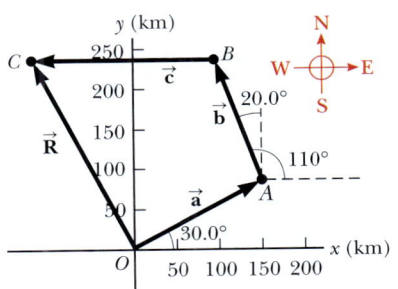

Figure P3.19

20. The helicopter view in Figure P3.20 shows two people pulling on a stubborn mule. Find (a) the single force that is equivalent to the two forces shown and (b) the force a third person would have to exert on the mule to make the net force equal to zero. The forces are measured in units of newtons (N).

Figure P3.20

21. **W** A novice golfer on the green takes three strokes to sink the ball. The successive displacements of the ball are 4.00 m to the north, 2.00 m northeast, and 1.00 m at 30.0° west of south (Fig. P3.21). Starting at the same initial point, an expert golfer could make the hole in what single displacement?

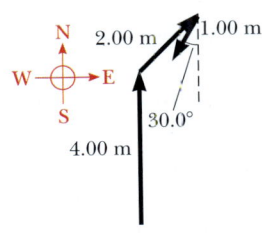

Figure P3.21

3.3 Displacement, Velocity, and Acceleration in Two Dimensions

3.4 Motion in Two Dimensions

22. One of the fastest recorded pitches in major-league baseball, thrown by Tim Lincecum in 2009, was clocked

at 101.0 mi/h (Fig. P3.22). If a pitch were thrown horizontally with this velocity, how far would the ball fall vertically by the time it reached home plate, 60.5 ft away?

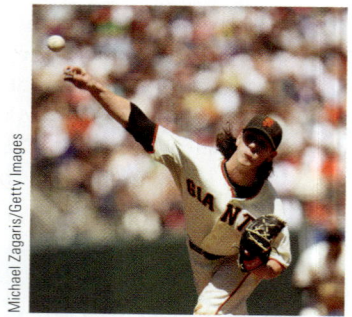

Figure P3.22 Tim Lincecum throws a baseball.

23. **GP** A student stands at the edge of a cliff and throws a stone horizontally over the edge with a speed of 18.0 m/s. The cliff is 50.0 m above a flat, horizontal beach as shown in Figure P3.23. (a) What are the coordinates of the initial position of the stone? (b) What are the components of the initial velocity? (c) Write the equations for the x- and y-components of the velocity of the stone with time. (d) Write the equations for the position of the stone with time, using the coordinates in Figure P3.23. (e) How long after being released does the stone strike the beach below the cliff? (f) With what speed and angle of impact does the stone land?

Figure P3.23

24. **Q|C S** A rock is thrown upward from the level ground in such a way that the maximum height of its flight is equal to its horizontal range R. (a) At what angle θ is the rock thrown? (b) In terms of the original range R, what is the range R_{max} the rock can attain if it is launched at the same speed but at the optimal angle for maximum range? (c) Would your answer to part (a) be different if the rock is thrown with the same speed on a different planet? Explain.

25. **W** The best leaper in the animal kingdom is the puma, which can jump to a height of 3.7 m when leaving the ground at an angle of 45°. With what speed must the animal leave the ground to reach that height?

26. **Q|C** The record distance in the sport of throwing cowpats is 81.1 m. This record toss was set by Steve Urner of the United States in 1981. Assuming the initial launch angle was 45° and neglecting air resistance, determine (a) the initial speed of the projectile and (b) the total time the projectile was in flight. (c) Qualitatively, how would the answers change if the launch angle were greater than 45°? Explain.

27. **M** A place-kicker must kick a football from a point 36.0 m (about 40 yards) from the goal. Half the crowd hopes the ball will clear the crossbar, which is 3.05 m high. When kicked, the ball leaves the ground with a speed of 20.0 m/s at an angle of 53.0° to the horizontal. (a) By how much does the ball clear or fall short of clearing the crossbar? (b) Does the ball approach the crossbar while still rising or while falling?

28. **GP** From the window of a building, a ball is tossed from a height y_0 above the ground with an initial velocity of 8.00 m/s and angle of 20.0° below the horizontal. It strikes the ground 3.00 s later. (a) If the base of the building is taken to be the origin of the coordinates, with upward the positive y-direction, what are the initial coordinates of the ball? (b) With the positive x-direction chosen to be out the window, find the x- and y-components of the initial velocity. (c) Find the equations for the x- and y-components of the position as functions of time. (d) How far horizontally from the base of the building does the ball strike the ground? (e) Find the height from which the ball was thrown. (f) How long does it take the ball to reach a point 10.0 m below the level of launching?

29. A brick is thrown upward from the top of a building at an angle of 25° to the horizontal and with an initial speed of 15 m/s. If the brick is in flight for 3.0 s, how tall is the building?

30. An artillery shell is fired with an initial velocity of 300 m/s at 55.0° above the horizontal. To clear an avalanche, it explodes on a mountainside 42.0 s after firing. What are the x- and y-coordinates of the shell where it explodes, relative to its firing point?

31. A car is parked on a cliff overlooking the ocean on an incline that makes an angle of 24.0° below the horizontal. The negligent driver leaves the car in neutral, and the emergency brakes are defective. The car rolls from rest down the incline with a constant acceleration of 4.00 m/s² for a distance of 50.0 m to the edge of the cliff, which is 30.0 m above the ocean. Find (a) the car's position relative to the base of the cliff when the car lands in the ocean and (b) the length of time the car is in the air.

32. **W** A fireman $d = 50.0$ m away from a burning building directs a stream of water from a ground-level fire hose at an angle of $\theta_i = 30.0°$ above the horizontal as shown in Figure P3.32. If the speed of the stream as it leaves the hose is $v_i = 40.0$ m/s, at what height will the stream of water strike the building?

Figure P3.32

33. A projectile is launched with an initial speed of 60.0 m/s at an angle of 30.0° above the horizontal. The projectile lands on a hillside 4.00 s later. Neglect air friction. (a) What is the projectile's velocity at the highest point of its trajectory? (b) What is the straight-line distance from where the projectile was launched to where it hits its target?

34. A playground is on the flat roof of a city school, 6.00 m above the street below (Fig. P3.34). The vertical wall of the building is $h = 7.00$ m high, to form a 1-m-high railing around the playground. A ball has fallen to the street below, and a passerby returns it by launching it at an angle of $\theta = 53.0°$ above the horizontal at a point $d = 24.0$ m from the base of the building wall. The ball takes 2.20 s to reach a point vertically above the wall. (a) Find the speed at which the ball was launched. (b) Find the vertical distance by which the ball clears the wall. (c) Find the horizontal distance from the wall to the point on the roof where the ball lands.

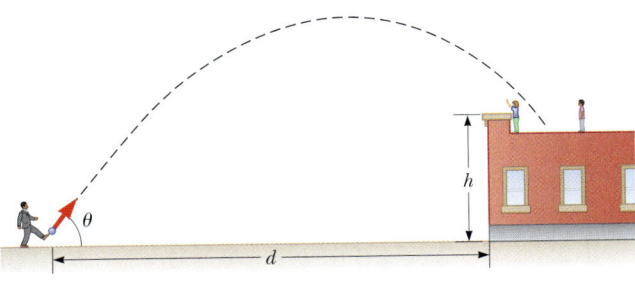

Figure P3.34

3.5 Relative Velocity

35. **GP** A jet airliner moving initially at 3.00×10^2 mi/h due east enters a region where the wind is blowing 1.00×10^2 mi/h in a direction 30.0° north of east. (a) Find the components of the velocity of the jet airliner relative to the air, $\vec{v}_{JA}$. (b) Find the components of the velocity of the air relative to Earth, $\vec{v}_{AE}$. (c) Write an equation analogous to Equation 3.16 for the velocities $\vec{v}_{JA}$, $\vec{v}_{AE}$, and $\vec{v}_{JE}$. (d) What are the speed and direction of the aircraft relative to the ground?

36. **M** A car travels due east with a speed of 50.0 km/h. Raindrops are falling at a constant speed vertically with respect to the Earth. The traces of the rain on the side windows of the car make an angle of 60.0° with the vertical. Find the velocity of the rain with respect to (a) the car and (b) the Earth.

37. **QC** A bolt drops from the ceiling of a moving train car that is accelerating northward at a rate of 2.50 m/s². (a) What is the acceleration of the bolt relative to the train car? (b) What is the acceleration of the bolt relative to the Earth? (c) Describe the trajectory of the bolt as seen by an observer fixed on the Earth.

38. A Coast Guard cutter detects an unidentified ship at a distance of 20.0 km in the direction 15.0° east of north. The ship is traveling at 26.0 km/h on a course at 40.0° east of north. The Coast Guard wishes to send a speedboat to intercept and investigate the vessel. (a) If the speedboat travels at 50.0 km/h, in what direction should it head? Express the direction as a compass bearing with respect to due north. (b) Find the time required for the cutter to intercept the ship.

39. An airplane maintains a speed of 630 km/h relative to the air it is flying through, as it makes a trip to a city 750 km away to the north. (a) What time interval is required for the trip if the plane flies through a headwind blowing at 35.0 km/h toward the south? (b) What time interval is required if there is a tailwind with the same speed? (c) What time interval is required if there is a crosswind blowing at 35.0 km/h to the east relative to the ground?

40. **BIO** Suppose a chinook salmon needs to jump a waterfall that is 1.50 m high. If the fish starts from a distance 1.00 m from the base of the ledge over which the waterfall flows, (a) find the x- and y-components of the initial velocity the salmon would need to just reach the ledge at the top of its trajectory. (b) Can the fish make this jump? (Note that a chinook salmon can jump out of the water with an initial speed of 6.26 m/s.)

41. **QC** A river has a steady speed of 0.500 m/s. A student swims upstream a distance of 1.00 km and swims back to the starting point. (a) If the student can swim at a speed of 1.20 m/s in still water, how long does the trip take? (b) How much time is required in still water for the same length swim? (c) Intuitively, why does the swim take longer when there is a current?

42. **QC S** *This is a symbolic version of Problem 41.* A river has a steady speed of v_s. A student swims upstream a distance d and back to the starting point. (a) If the student can swim at a speed of v in still water, how much time t_{up} does it take the student to swim upstream a distance d? Express the answer in terms of d, v, and v_s. (b) Using the same variables, how much time t_{down} does it take to swim back downstream to the starting point? (c) Sum the answers found in parts (a) and (b) and show that the time t_a required for the whole trip can be written as

$$t_a = \frac{2d/v}{1 - v_s^2/v^2}$$

(d) How much time t_b does the trip take in still water? (e) Which is larger, t_a or t_b? Is it always larger?

43. A bomber is flying horizontally over level terrain at a speed of 275 m/s relative to the ground and at an altitude of 3.00 km. (a) The bombardier releases one bomb. How far does the bomb travel horizontally between its release and its impact on the ground? Ignore the effects of air resistance. (b) Firing from the people on the ground suddenly incapacitates the bombardier before he can call, "Bombs away!" Consequently, the pilot maintains the plane's original course, altitude, and speed through a storm of flak. Where is the plane relative to the bomb's point of impact when the bomb hits the ground? (c) The plane has a telescopic bombsight set so that the bomb hits the target seen in the sight at the moment of release. At what angle from the vertical was the bombsight set?

Additional Problems

44. **S** A moving walkway at an airport has a speed v_1 and a length L. A woman stands on the walkway as it moves from one end to the other, while a man in a hurry to reach his flight walks on the walkway with a speed of v_2 relative to the moving walkway. (a) How long does it take the woman to travel the distance L? (b) How long does it take the man to travel this distance?

45. How long does it take an automobile traveling in the left lane of a highway at 60.0 km/h to overtake (become even with) another car that is traveling in the right lane at 40.0 km/h when the cars' front bumpers are initially 100 m apart?

46. You can use any coordinate system you like to solve a projectile motion problem. To demonstrate the truth of this statement, consider a ball thrown off the top of a building with a velocity $\vec{v}$ at an angle θ with respect to the horizontal. Let the building be 50.0 m tall, the initial horizontal velocity be 9.00 m/s, and the initial vertical velocity be 12.0 m/s. Choose your coordinates such that the positive y-axis is upward, the x-axis is to the right, and the origin is at the point where the ball is released. (a) With these choices, find the ball's maximum height above the ground and the time it takes to reach the maximum height. (b) Repeat your calculations choosing the origin at the base of the building.

47. **QC** A Nordic jumper goes off a ski jump at an angle of 10.0° below the horizontal, traveling 108 m horizontally and 55.0 m vertically before landing. (a) Ignoring friction and aerodynamic effects, calculate the speed needed by the skier on leaving the ramp. (b) Olympic Nordic jumpers can make such jumps with a jump speed of 23.0 m/s, which is considerably less than the answer found in part (a). Explain how that is possible.

48. **S** In a local diner, a customer slides an empty coffee cup down the counter for a refill. The cup slides off the counter and strikes the floor at distance d from the base of the counter. If the height of the counter is h, (a) find an expression for the time t it takes the cup to fall to the floor in terms of the variables h and g. (b) With what speed does the mug leave the counter? Answer in terms of the variables d, g, and h. (c) In the same terms, what is the speed of the cup immediately before it hits the floor? (d) In terms of h and d, what is the direction of the cup's velocity immediately before it hits the floor?

49. **M** Towns A and B in Figure P3.49 are 80.0 km apart. A couple arranges to drive from town A and meet a couple driving from town B at the lake, L. The two couples leave simultaneously and drive for 2.50 h in the directions shown. Car 1 has a speed of 90.0 km/h. If the cars arrive simultaneously at the lake, what is the speed of car 2?

Figure P3.49

50. **BIO** A chinook salmon has a maximum underwater speed of 3.58 m/s, but it can jump out of water with a speed of 6.26 m/s. To move upstream past a waterfall, the salmon does not need to jump to the top of the fall, but only to a point in the fall where the water speed is less than 3.58 m/s; it can then swim up the fall for the remaining distance. Because the salmon must make forward progress in the water, let's assume it can swim to the top if the water speed is 3.00 m/s. If water has a speed of 1.50 m/s as it passes over a ledge, (a) how far below the ledge will the water be moving with a speed of 3.00 m/s? (Note that water undergoes projectile motion once it leaves the ledge.) (b) If the salmon

is able to jump vertically upward from the base of the fall, what is the maximum height of waterfall that the salmon can clear?

51. A rocket is launched at an angle of 53.0° above the horizontal with an initial speed of 100 m/s. The rocket moves for 3.00 s along its initial line of motion with an acceleration of 30.0 m/s². At this time, its engines fail and the rocket proceeds to move as a projectile. Find (a) the maximum altitude reached by the rocket, (b) its total time of flight, and (c) its horizontal range.

52. **W** Two canoeists in identical canoes exert the same effort paddling and hence maintain the same speed relative to the water. One paddles directly upstream (and moves upstream), whereas the other paddles directly downstream. With downstream as the positive direction, an observer on shore determines the velocities of the two canoes to be −1.2 m/s and +2.9 m/s, respectively. (a) What is the speed of the water relative to the shore? (b) What is the speed of each canoe relative to the water?

53. (a) If a person can jump a maximum horizontal distance (by using a 45° projection angle) of 3.0 m on Earth, what would be his maximum range on the Moon, where the free-fall acceleration is g/6 and g = 9.80 m/s²? (b) Repeat for Mars, where the acceleration due to gravity is 0.38g.

54. **QIC** A farm truck moves due east with a constant velocity of 9.50 m/s on a limitless, horizontal stretch of road. A boy riding on the back of the truck throws a can of soda upward (Fig. P3.54) and catches the projectile at the same location on the truck bed, but 16.0 m farther down the road. (a) In the frame of reference of the truck, at what angle to the vertical does the boy throw the can? (b) What is the initial speed of the can relative to the truck? (c) What is the shape of the can's trajectory as seen by the boy? An observer on the ground watches the boy throw the can and catch it. In this observer's frame of reference, (d) describe the shape of the can's path and (e) determine the initial velocity of the can.

Figure P3.54

55. **M** A home run is hit in such a way that the baseball just clears a wall 21 m high, located 130 m from home plate. The ball is hit at an angle of 35° to the horizontal, and air resistance is negligible. Find (a) the initial speed of the ball, (b) the time it takes the ball to reach the wall, and (c) the velocity components and the speed of the ball when it reaches the wall. (Assume the ball is hit at a height of 1.0 m above the ground.)

56. A ball is thrown straight upward and returns to the thrower's hand after 3.00 s in the air. A second ball thrown at an angle of 30.0° with the horizontal reaches the same maximum height as the first ball. (a) At what speed was the first ball thrown? (b) At what speed was the second ball thrown?

57. A quarterback throws a football toward a receiver with an initial speed of 20 m/s at an angle of 30° above the horizontal. At that instant the receiver is 20 m from the quarterback. In (a) what direction and (b) with what constant speed should the receiver run in order to catch the football at the level at which it was thrown?

58. A 2.00-m-tall basketball player is standing on the floor 10.0 m from the basket, as in Figure P3.58. If he shoots the ball at a 40.0° angle with the horizontal, at what initial speed must he throw the basketball so that it goes through the hoop without striking the backboard? The height of the basket is 3.05 m.

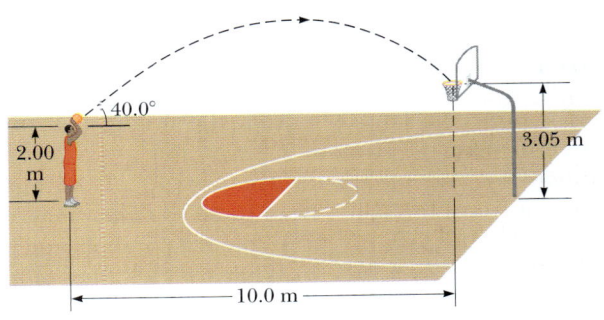

Figure P3.58

59. **QIC S** In a very popular lecture demonstration, a projectile is fired at a falling target as in Figure P3.59. The projectile leaves the gun at the same instant the target is dropped from rest. Assuming the gun is initially aimed at the target, show that the projectile will hit the target. (One restriction of this experiment is that the projectile must reach the target before the target strikes the floor.)

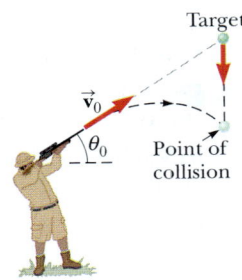

Figure P3.59

60. **BIO** Figure P3.60 illustrates the difference in proportions between the male (m) and female (f) anatomies. The displacements $\vec{d}_{1m}$ and $\vec{d}_{1f}$ from the bottom of the feet to the navel have magnitudes of 104 cm and 84.0 cm, respectively. The displacements $\vec{d}_{2m}$ and $\vec{d}_{2f}$ have magnitudes of 50.0 cm and 43.0 cm, respectively. (a) Find the vector sum of the displacements $\vec{d}_{d1}$ and $\vec{d}_{d2}$ in each case. (b) The male figure is 180 cm tall, the female 168 cm. Normalize the displacements of each

figure to a common height of 200 cm and re-form the vector sums as in part (a). Then find the vector difference between the two sums.

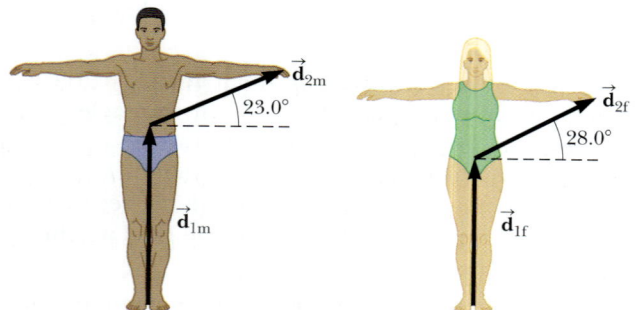

Figure P3.60

61. **Q|C** By throwing a ball at an angle of 45°, a girl can throw the ball a maximum horizontal distance R on a level field. How far can she throw the same ball vertically upward? Assume her muscles give the ball the same speed in each case. (Is this assumption valid?)

62. **S** The equation of a parabola is $y = ax^2 + bx + c$, where a, b, and c are constants. The x- and y-coordinates of a projectile launched from the origin as a function of time are given by $x = v_{0x}t$ and $y = v_{0y}t - \frac{1}{2}gt^2$, where v_{0x} and v_{0y} are the components of the initial velocity. (a) Eliminate t from these two equations and show that the path of a projectile is a parabola and has the form $y = ax + bx^2$. (b) What are the values of a, b, and c for the projectile?

63. A hunter wishes to cross a river that is 1.5 km wide and flows with a speed of 5.0 km/h parallel to its banks. The hunter uses a small powerboat that moves at a maximum speed of 12 km/h with respect to the water. What is the minimum time necessary for crossing?

64. **S** When baseball outfielders throw the ball, they usually allow it to take one bounce, on the theory that the ball arrives at its target sooner that way. Suppose that, after the bounce, the ball rebounds at the same angle θ that it had when it was released (as in Fig. P3.64), but loses half its speed. (a) Assuming that the ball is always thrown with the same initial speed, at what angle θ should the ball be thrown in order to go the same distance D with one bounce as a ball thrown upward at 45.0° with no bounce? (b) Determine the ratio of the times for the one-bounce and no-bounce throws.

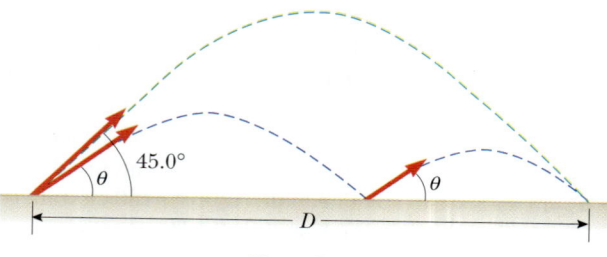

Figure P3.64

65. A daredevil is shot out of a cannon at 45.0° to the horizontal with an initial speed of 25.0 m/s. A net is positioned a horizontal distance of 50.0 m from the cannon. At what height above the cannon should the net be placed in order to catch the daredevil?

66. **BIO** Chinook salmon are able to move upstream faster by jumping out of the water periodically; this behavior is called *porpoising*. Suppose a salmon swimming in still water jumps out of the water with a speed of 6.26 m/s at an angle of 45°, sails through the air a distance L before returning to the water, and then swims a distance L underwater at a speed of 3.58 m/s before beginning another porpoising maneuver. Determine the average speed of the fish.

67. **Q|C** A student decides to measure the muzzle velocity of a pellet shot from his gun. He points the gun horizontally. He places a target on a vertical wall a distance x away from the gun. The pellet hits the target a vertical distance y below the gun. (a) Show that the position of the pellet when traveling through the air is given by $y = Ax^2$, where A is a constant. (b) Express the constant A in terms of the initial (muzzle) velocity and the free-fall acceleration. (c) If $x = 3.00$ m and $y = 0.210$ m, what is the initial speed of the pellet?

68. A sailboat is heading directly north at a speed of 20 knots (1 knot = 0.514 m/s). The wind is blowing toward the east with a speed of 17 knots. (a) Determine the magnitude and direction of the wind velocity as measured on the boat. (b) What is the component of the wind velocity in the direction parallel to the motion of the boat? (See Problem 58 in Chapter 4 for an explanation of how a sailboat can move "into the wind.")

69. A golf ball with an initial speed of 50.0 m/s lands exactly 240 m downrange on a level course. (a) Neglecting air friction, what *two* projection angles would achieve this result? (b) What is the maximum height reached by the ball, using the two angles determined in part (a)?

70. **Q|C** A landscape architect is planning an artificial waterfall in a city park. Water flowing at 0.750 m/s leaves the end of a horizontal channel at the top of a vertical wall $h = 2.35$ m high and falls into a pool (Fig. P3.70). (a) How far from the wall will the water land? Will the space behind the waterfall be wide enough for a pedestrian walkway? (b) To sell her plan to the city council, the architect wants to build a model to standard scale, one-twelfth actual size. How fast should the water flow in the channel in the model?

Figure P3.70

71. One strategy in a snowball fight is to throw a snowball at a high angle over level ground. Then, while your opponent is watching that snowball, you throw a second one

at a low angle timed to arrive before or at the same time as the first one. Assume both snowballs are thrown with a speed of 25.0 m/s. The first is thrown at an angle of 70.0° with respect to the horizontal. (a) At what angle should the second snowball be thrown to arrive at the same point as the first? (b) How many seconds later should the second snowball be thrown after the first in order for both to arrive at the same time?

72. A dart gun is fired while being held horizontally at a height of 1.00 m above ground level and while it is at rest relative to the ground. The dart from the gun travels a horizontal distance of 5.00 m. A college student holds the same gun in a horizontal position while sliding down a 45.0° incline at a constant speed of 2.00 m/s. How far will the dart travel if the student fires the gun when it is 1.00 m above the ground?

73. The determined Wile E. Coyote is out once more to try to capture the elusive roadrunner. The coyote wears a new pair of power roller skates, which provide a constant horizontal acceleration of 15 m/s², as shown in Figure P3.73. The coyote starts off at rest 70 m from the edge of a cliff at the instant the roadrunner zips by in the direction of the cliff. (a) If the roadrunner moves with constant speed, find the minimum speed the roadrunner must have to reach the cliff before the coyote. (b) If the cliff is 100 m above the base of a canyon, find where the coyote lands in the canyon. (Assume his skates are still in operation when he is in "flight" and that his horizontal component of acceleration remains constant at 15 m/s².)

Figure P3.73

74. A truck loaded with cannonball watermelons stops suddenly to avoid running over the edge of a washed-out bridge (Fig. P3.74). The quick stop causes a number of melons to fly off the truck. One melon rolls over the edge with an initial speed $v_i = 10.0$ m/s in the horizontal direction. A cross section of the bank has the shape of the bottom half of a parabola with its vertex at the edge of the road, and with the equation $y^2 = (16.0\text{ m})\, x$, where x and y are measured in meters. What are the x- and y-coordinates of the melon when it splatters on the bank?

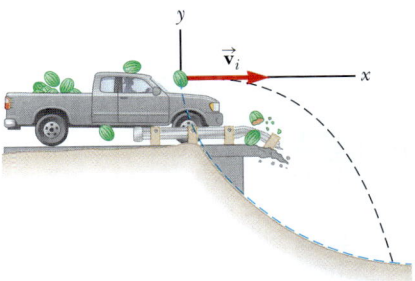

Figure P3.74 The blue dashed curve shows the parabolic shape of the bank.

A rock climber depends on a number of different forces to overcome the force of gravity and reach a summit. Her muscles apply forces to the rocks and to her own body, static friction forces helping her maintain her grip and enabling motion. Safety lines and the tension forces they can exert offer insurance against a fall. The climber uses the third law of motion continually: by exerting a downward force on the rock, the rock exerts an equal and opposite force on her, propelling her up the sheer cliff.

4 The Laws of Motion

4.1 Forces
4.2 Newton's First Law
4.3 Newton's Second Law
4.4 Newton's Third Law
4.5 Applications of Newton's Laws
4.6 Forces of Friction

Classical mechanics describes the relationship between the motion of objects found in our everyday world and the forces acting on them. As long as the system under study doesn't involve objects comparable in size to an atom or traveling close to the speed of light, classical mechanics provides an excellent description of nature.

This chapter introduces Newton's three laws of motion and his law of gravity. The three laws are simple and sensible. The first law states that a force must be applied to an object in order to change its velocity. Changing an object's velocity means accelerating it, which implies a relationship between force and acceleration. This relationship, the second law, states that the net force on an object equals the object's mass times its acceleration. Finally, the third law says that whenever we push on something, it pushes back with equal force in the opposite direction. Those are the three laws in a nutshell.

Newton's three laws, together with his invention of calculus, opened avenues of inquiry and discovery that are used routinely today in virtually all areas of mathematics, science, engineering, and technology. Newton's theory of universal gravitation had a similar impact, starting a revolution in celestial mechanics and astronomy that continues to this day. With the advent of this theory, the orbits of all the planets could be calculated to high precision and the tides understood. The theory even led to the prediction of "dark stars," now called black holes, more than two centuries before any evidence for their existence was observed.[1] Newton's three laws of motion, together with his law of gravitation, are considered among the greatest achievements of the human mind.

---

[1]In 1783, John Michell combined Newton's theory of light and theory of gravitation, predicting the existence of "dark stars" from which light itself couldn't escape.

4.1 Forces

LEARNING OBJECTIVES
1. Describe the distinction between contact forces and field forces.
2. Identify the four field forces and describe their roles in the interactions of matter.

A **force** is commonly imagined as a push or a pull on some object, perhaps rapidly, as when we hit a tennis ball with a racket. (See Fig. 4.1.) We can hit the ball at different speeds and direct it into different parts of the opponent's court. That means we can control the magnitude of the applied force and also its direction, so force is a vector quantity, just like velocity and acceleration.

If you pull on a spring (Fig. 4.2a), the spring stretches. If you pull hard enough on a wagon (Fig. 4.2b), the wagon moves. When you kick a football (Fig. 4.2c), it deforms briefly and is set in motion. These are all examples of **contact forces,** so named because they result from physical contact between two objects.

Another class of forces doesn't involve any direct physical contact. Early scientists, including Newton, were uneasy with the concept of forces that act between two disconnected objects. Nonetheless, Newton used this "action-at-a-distance" concept in his law of gravity, whereby a mass at one location, such as the Sun, affects the motion of a distant object such as Earth despite no evident physical connection between the two objects. To overcome the conceptual difficulty associated with action at a distance, Michael Faraday (1791–1867) introduced the concept of a *field*. The corresponding forces are called **field forces.** According to this approach, an object of mass M, such as the Sun, creates an invisible influence that stretches throughout space. A second object of mass m, such as Earth, interacts with the *field* of the Sun, not directly with the Sun itself. So the force of gravitational attraction between two objects, illustrated in Figure 4.2d, is an example of a field force. The force of gravity keeps objects bound to Earth and also gives rise to what we call the *weight* of those objects.

Another common example of a field force is the electric force that one electric charge exerts on another (Fig. 4.2e). A third example is the force exerted by a bar magnet on a piece of iron (Fig. 4.2f).

The known fundamental forces in nature are all field forces. These are, in order of decreasing strength, (1) the strong nuclear force between subatomic particles, (2) the electromagnetic forces between electric charges, (3) the weak nuclear force, which arises in certain radioactive decay processes, and (4) the gravitational force between objects. The strong force keeps the nucleus of an atom from flying apart due to the repulsive electric force of the protons. The weak force is involved in most radioactive processes and plays an important role in the nuclear reactions

Figure 4.1 A tennis player applies a contact force to the ball with her racket, accelerating and directing the ball toward the open court.

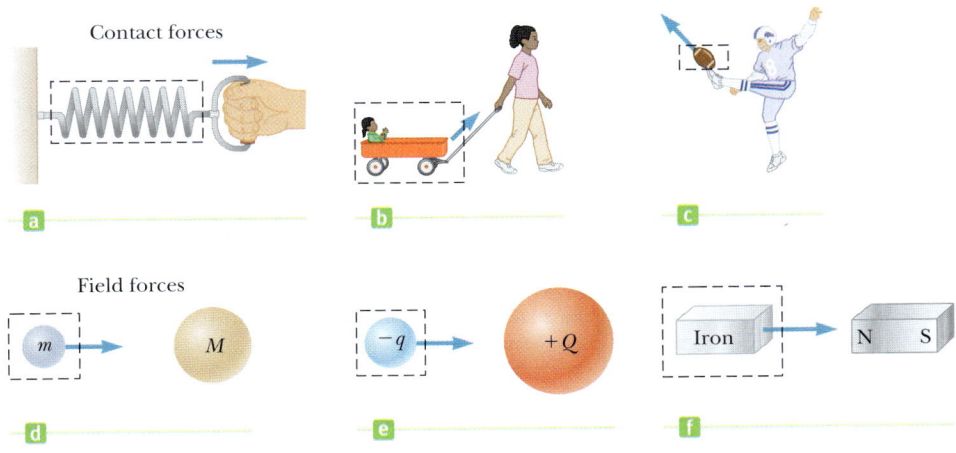

Figure 4.2 Examples of forces applied to various objects. In each case, a force acts on the object surrounded by the dashed lines. Something in the environment external to the boxed area exerts the force.

that generate the Sun's energy output. The strong and weak forces operate only on the nuclear scale, with a very short range on the order of 10^{-15} m. Outside this range they have no influence. Classical physics, however, deals only with gravitational and electromagnetic forces, which have infinite range.

Forces exerted on an object can change the object's shape. For example, striking a tennis ball with a racket, as in Figure 4.1, deforms the ball to some extent. Even objects we usually consider rigid and inflexible are deformed under the action of external forces. Often the deformations are permanent, as in the case of a collision between automobiles.

4.2 Newton's First Law

LEARNING OBJECTIVES

1. Explain what the first law of motion implies about an object's motion and the forces acting on it.
2. Explain the concepts of mass and inertia and the relationship between them.

Consider a book lying on a table. Obviously, the book remains at rest if left alone. Now imagine pushing the book with a horizontal force great enough to overcome the force of friction between the book and the table, setting the book in motion. Because the magnitude of the applied force exceeds the magnitude of the friction force, the book accelerates. When the applied force is withdrawn (Fig. 4.3a), friction soon slows the book to a stop.

Now imagine pushing the book across a smooth, waxed floor. The book again comes to rest once the force is no longer applied, but not as quickly as before. Finally, if the book is moving on a horizontal frictionless surface (Fig. 4.3b), it continues to move in a straight line with constant velocity until it hits a wall or some other obstruction.

Before about 1600, scientists felt that the natural state of matter was the state of rest. Galileo, however, devised thought experiments—such as an object moving on a frictionless surface, as just described—and concluded that **it's not the nature of an object to stop once set in motion, but rather to continue in its original state of motion.** This approach was later formalized as **Newton's first law of motion:**

▶ Newton's first law

An object moves with a velocity that is constant in magnitude and direction unless a non-zero net force acts on it.

Figure 4.3 The first law of motion. (a) A book moves at an initial velocity of $\vec{v}$ on a surface with friction. Because there is a friction force acting horizontally, the book slows to rest. (b) A book moves at velocity $\vec{v}$ on a frictionless surface. In the absence of a net force, the book keeps moving at velocity v.

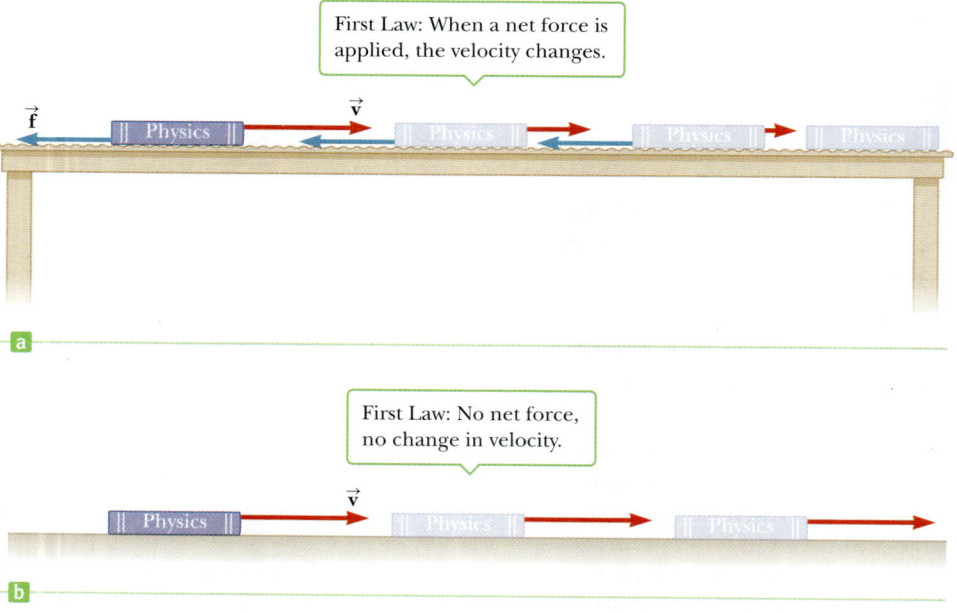

The net force on an object is defined as the vector sum of all external forces exerted on the object. External forces come from the object's environment. If an object's velocity isn't changing in either magnitude or direction, then its acceleration and the net force acting on it must both be zero.

Internal forces originate within the object itself and can't change the object's velocity (although they can change the object's rate of rotation, as described in Chapter 8). As a result, internal forces aren't included in Newton's second law. It's not really possible to "pull yourself up by your own bootstraps."

A consequence of the first law is the feasibility of space travel. After just a few moments of powerful thrust, the spacecraft coasts for months or years, its velocity only slowly changing with time under the relatively faint influence of the distant Sun and planets.

Mass and Inertia

Imagine hitting a golf ball off a tee with a driver. If you're a good golfer, the ball will sail over two hundred yards down the fairway. Now imagine teeing up a bowling ball and striking it with the same club (an experiment we don't recommend). Your club would probably break, you might sprain your wrist, and the bowling ball, at best, would fall off the tee, take half a roll, and come to rest.

From this thought experiment, we conclude that although both balls resist changes in their state of motion, the bowling ball offers much more effective resistance. The tendency of an object to continue in its original state of motion is called **inertia.**

Although inertia is the tendency of an object to continue its motion in the absence of a force, **mass** is a measure of the object's resistance to changes in its motion due to a force. The greater the mass of a body, the less it accelerates under the action of a given applied force. The SI unit of mass is the kilogram. Mass is a scalar quantity that obeys the rules of ordinary arithmetic.

Inertia can be used to explain the operation of one type of seat belt mechanism. The purpose of the seat belt is to hold the passenger firmly in place relative to the car, to prevent serious injury in the event of an accident. Figure 4.4 illustrates how one type of shoulder harness operates. Under normal conditions, the ratchet turns freely to allow the harness to wind on or unwind from the pulley as the passenger moves. In an accident, the car undergoes a large acceleration and rapidly comes to rest. Because of its inertia, the large block under the seat continues to slide forward along the tracks. The pin connection between the block and the rod causes the rod to pivot about its center and engage the ratchet wheel. At this point, the ratchet wheel locks in place and the harness no longer unwinds.

Unless acted on by an external force, an object at rest will remain at rest and an object in motion will continue in motion with constant velocity. In this case, the wall of the building did not exert a large enough external force on the moving train to stop it.

APPLICATION
Seat Belts

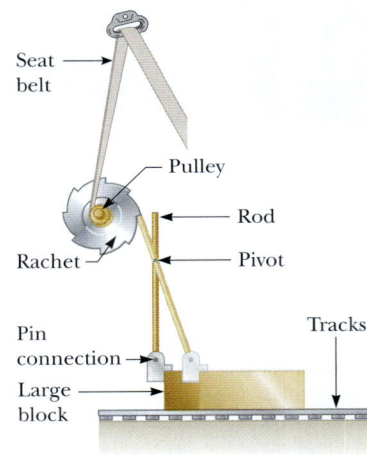

Figure 4.4 A mechanical arrangement for an automobile seat belt.

Tip 4.1 Force Causes Changes in Motion

Motion can occur even in the absence of forces. Force causes *changes* in motion.

4.3 Newton's Second Law

LEARNING OBJECTIVES

1. Relate accelerations to forces with the second law of motion.
2. Convert forces between SI and U.S. customary units.
3. Use the second law to study the motion of an object in elementary applications.
4. Apply Newton's Universal Law of gravitation to elementary systems.
5. Contrast the concepts of mass and weight.

Newton's first law explains what happens to an object that has no net force acting on it: The object either remains at rest or continues moving in a straight line with constant speed. Newton's second law answers the question of what happens to an object that *does* have a net force acting on it.

Imagine pushing a block of ice across a frictionless horizontal surface. When you exert some horizontal force on the block, it moves with an acceleration of,

Figure 4.5 The second law of motion. For the block of mass m, the net force $\sum \vec{F}$ acting on the block equals the mass m times the acceleration vector $\vec{a}$.

say, 2 m/s². If you apply a force twice as large, the acceleration doubles to 4 m/s². Pushing three times as hard triples the acceleration, and so on. From such observations, we conclude that **the acceleration of an object is directly proportional to the net force acting on it.**

Mass also affects acceleration. Suppose you stack identical blocks of ice on top of each other while pushing the stack with constant force. If the force applied to one block produces an acceleration of 2 m/s², then the acceleration drops to half that value, 1 m/s², when two blocks are pushed, to one-third the initial value when three blocks are pushed, and so on. We conclude that **the acceleration of an object is inversely proportional to its mass.** These observations are summarized in **Newton's second law:**

▶ Newton's second law

> The acceleration $\vec{a}$ of an object is directly proportional to the net force acting on it and inversely proportional to its mass.

The constant of proportionality is equal to one, so in mathematical terms the preceding statement can be written

$$\vec{a} = \frac{\sum \vec{F}}{m}$$

where $\vec{a}$ is the acceleration of the object, m is its mass, and $\sum \vec{F}$ is the vector sum of all forces acting on it. Multiplying through by m, we have

$$\sum \vec{F} = m\vec{a} \qquad [4.1]$$

Isaac Newton
English Physicist and Mathematician (1642–1727)
Newton was one of the most brilliant scientists in history. Before he was 30, he formulated the basic concepts and laws of mechanics, discovered the law of universal gravitation, and invented the mathematical methods of the calculus. As a consequence of his theories, Newton was able to explain the motions of the planets, the ebb and flow of the tides, and many special features of the motions of the Moon and Earth. He also interpreted many fundamental observations concerning the nature of light. His contributions to physical theories dominated scientific thought for two centuries and remain important today.

Physicists commonly refer to this equation as '$F = ma$.' Figure 4.5 illustrates the relationship between the mass, acceleration, and the net force. The second law is a vector equation, equivalent to the following three component equations:

$$\sum F_x = ma_x \qquad \sum F_y = ma_y \qquad \sum F_z = ma_z \qquad [4.2]$$

When there is no net force on an object, its acceleration is zero, which means the velocity is constant.

Units of Force and Mass

The SI unit of force is the **newton.** When 1 newton of force acts on an object that has a mass of 1 kg, it produces an acceleration of 1 m/s² in the object. From this definition and Newton's second law, we see that the newton can be expressed in terms of the fundamental units of mass, length, and time as

$$1 \text{ N} \equiv 1 \text{ kg} \cdot \text{m/s}^2 \qquad [4.3]$$

In the U.S. customary system, the unit of force is the **pound.** The conversion from newtons to pounds is given by

$$1 \text{ N} = 0.225 \text{ lb} \qquad [4.4]$$

The units of mass, acceleration, and force in the SI and U.S. customary systems are summarized in Table 4.1.

Tip 4.2 $m\vec{a}$ **Is Not a Force**
Equation 4.1 does *not* say that the product $m\vec{a}$ is a force. All forces exerted on an object are summed as vectors to generate the net force on the left side of the equation. This net force is then equated to the product of the mass and resulting acceleration of the object. Do *not* include an "$m\vec{a}$ force" in your analysis.

■ Quick Quiz

4.1 Which of the following statements are true? (a) An object can move even when no force acts on it. (b) If an object isn't moving, no external forces act on it. (c) If a single force acts on an object, the object accelerates. (d) If an object accelerates, a force is acting on it. (e) If an object isn't accelerating, no external force is acting on it. (f) If the net force acting on an object is in the positive x-direction, the object moves only in the positive x-direction.

Table 4.1 Units of Mass, Acceleration, and Force

System	Mass	Acceleration	Force
SI	kg	m/s^2	N = kg·m/s^2
U.S. customary	slug	ft/s^2	lb = slug·ft/s^2

■ EXAMPLE 4.1 | Airboat

GOAL Apply Newton's second law in one dimension, together with the equations of kinematics.

PROBLEM An airboat with mass 3.50×10^2 kg, including the passenger, has an engine that produces a net horizontal force of 7.70×10^2 N, after accounting for forces of resistance (see Fig. 4.6). **(a)** Find the acceleration of the airboat. **(b)** Starting from rest, how long does it take the airboat to reach a speed of 12.0 m/s? **(c)** After reaching that speed, the pilot turns off the engine and drifts to a stop over a distance of 50.0 m. Find the resistance force, assuming it's constant.

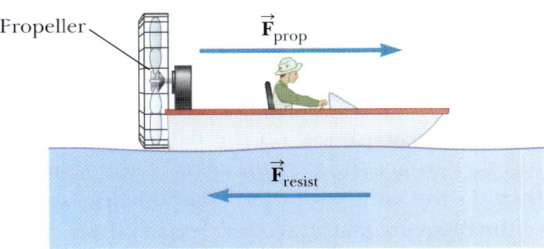

Figure 4.6 (Example 4.1)

STRATEGY In part (a), apply Newton's second law to find the acceleration, and in part (b) use that acceleration in the one-dimensional kinematics equation for the velocity. When the engine is turned off in part (c), only the resistance forces act on the boat in the x-direction, so the net acceleration can be found from $v^2 - v_0^2 = 2a\,\Delta x$. Then Newton's second law gives the resistance force.

SOLUTION

(a) Find the acceleration of the airboat.

Apply Newton's second law and solve for the acceleration:

$$ma = F_{net} \quad \rightarrow \quad a = \frac{F_{net}}{m} = \frac{7.70 \times 10^2 \text{ N}}{3.50 \times 10^2 \text{ kg}}$$

$$= \boxed{2.20 \text{ m/s}^2}$$

(b) Find the time necessary to reach a speed of 12.0 m/s.

Apply the kinematics velocity equation:

$$v = at + v_0 = (2.20 \text{ m/s}^2)t = 12.0 \text{ m/s} \quad \rightarrow \quad t = \boxed{5.45 \text{ s}}$$

(c) Find the resistance force after the engine is turned off.

Using kinematics, find the net acceleration due to resistance forces:

$$v^2 - v_0^2 = 2a\,\Delta x$$
$$0 - (12.0 \text{ m/s})^2 = 2a(50.0 \text{ m}) \quad \rightarrow \quad a = -1.44 \text{ m/s}^2$$

Substitute the acceleration into Newton's second law, finding the resistance force:

$$F_{resist} = ma = (3.50 \times 10^2 \text{ kg})(-1.44 \text{ m/s}^2) = \boxed{-504 \text{ N}}$$

REMARKS The propeller exerts a force on the air, pushing it backwards behind the boat. At the same time, the air exerts a force on the propellers and consequently on the airboat. Forces always come in pairs of this kind, which are formalized in the next section as Newton's third law of motion. The negative answer for the acceleration in part (c) means that the airboat is slowing down.

QUESTION 4.1 What other forces act on the airboat? Describe them.

EXERCISE 4.1 Suppose the pilot, starting again from rest, opens the throttle partway. At a constant acceleration, the airboat then covers a distance of 60.0 m in 10.0 s. Find the net force acting on the boat.

ANSWER 4.20×10^2 N

> **Tip 4.3 Newton's Second Law Is a Vector Equation**
>
> In applying Newton's second law, add all of the forces on the object as vectors and then find the resultant vector acceleration by dividing by m. Don't find the individual magnitudes of the forces and add them like scalars.

EXAMPLE 4.2 | Horses Pulling a Barge

GOAL Apply Newton's second law in a two-dimensional problem.

PROBLEM Two horses are pulling a barge with mass 2.00×10^3 kg along a canal, as shown in Figure 4.7. The cable connected to the first horse makes an angle of $\theta_1 = 30.0°$ with respect to the direction of the canal, while the cable connected to the second horse makes an angle of $\theta_2 = -45.0°$. Find the initial acceleration of the barge, starting at rest, if each horse exerts a force of magnitude 6.00×10^2 N on the barge. Ignore forces of resistance on the barge.

STRATEGY Using trigonometry, find the vector force exerted by each horse on the barge. Add the x-components together to get the x-component of the resultant force, and then do the same with the y-components. Divide by the mass of the barge to get the accelerations in the x- and y-directions.

Figure 4.7 (Example 4.2)

SOLUTION

Compute the x-components of the forces exerted by the horses.

$$F_{1x} = F_1 \cos \theta_1 = (6.00 \times 10^2 \text{ N}) \cos (30.0°) = 5.20 \times 10^2 \text{ N}$$
$$F_{2x} = F_2 \cos \theta_2 = (6.00 \times 10^2 \text{ N}) \cos (-45.0°) = 4.24 \times 10^2 \text{ N}$$

Find the total force in the x-direction by adding the x-components:

$$F_x = F_{1x} + F_{2x} = 5.20 \times 10^2 \text{ N} + 4.24 \times 10^2 \text{ N}$$
$$= 9.44 \times 10^2 \text{ N}$$

Compute the y-components of the forces exerted by the horses:

$$F_{1y} = F_1 \sin \theta_1 = (6.00 \times 10^2 \text{ N}) \sin 30.0° = 3.00 \times 10^2 \text{ N}$$
$$F_{2y} = F_2 \sin \theta_2 = (6.00 \times 10^2 \text{ N}) \sin (-45.0°)$$
$$= -4.24 \times 10^2 \text{ N}$$

Find the total force in the y-direction by adding the y-components:

$$F_y = F_{1y} + F_{2y} = 3.00 \times 10^2 \text{ N} - 4.24 \times 10^2 \text{ N}$$
$$= -1.24 \times 10^2 \text{ N}$$

Obtain the components of the acceleration by dividing each of the force components by the mass:

$$a_x = \frac{F_x}{m} = \frac{9.44 \times 10^2 \text{ N}}{2.00 \times 10^3 \text{ kg}} = 0.472 \text{ m/s}^2$$

$$a_y = \frac{F_y}{m} = \frac{-1.24 \times 10^2 \text{ N}}{2.00 \times 10^3 \text{ kg}} = -0.062\,0 \text{ m/s}^2$$

Calculate the magnitude of the acceleration:

$$a = \sqrt{a_x^2 + a_y^2} = \sqrt{(0.472 \text{ m/s}^2)^2 + (-0.062\,0 \text{ m/s}^2)^2}$$
$$= \boxed{0.476 \text{ m/s}^2}$$

Calculate the direction of the acceleration using the tangent function:

$$\tan \theta = \frac{a_y}{a_x} = \frac{-0.062\,0 \text{ m/s}^2}{0.472 \text{ m/s}^2} = -0.131$$
$$\theta = \tan^{-1}(-0.131) = \boxed{-7.46°}$$

REMARKS Notice that the angle is in fourth quadrant, in the range of the inverse tangent function, so it is not necessary to add 180° to the answer. The horses exert a force on the barge through the tension in the cables, while the barge exerts an equal and opposite force on the horses, again through the cables. If that were not true, the horses would easily move forward, as if unburdened. This example is another illustration of forces acting in pairs.

QUESTION 4.2 True or False: In general, the magnitude of the acceleration of an object is determined by the magnitudes of the forces acting on it.

EXERCISE 4.2 Repeat Example 4.2, but assume the first horse pulls at a 40.0° angle, the second horse at −20.0°.

ANSWER 0.520 m/s^2, 10.0°

The Gravitational Force

The **gravitational force** is the mutual force of attraction between any two objects in the Universe. Although the gravitational force can be very strong between very large objects, it's the weakest of the fundamental forces. A good demonstration of how weak it is can be carried out with a small balloon. Rubbing the balloon in your hair gives the balloon a tiny electric charge. Through electric forces, the balloon then adheres to a wall, resisting the gravitational pull of the entire Earth!

In addition to contributing to the understanding of motion, Newton studied gravity extensively. **Newton's law of universal gravitation states that every particle in the Universe attracts every other particle with a force that is directly proportional to the product of the masses of the particles and inversely proportional to the square of the distance between them.** If the particles have masses m_1 and m_2 and are separated by a distance r, as in Figure 4.8, the magnitude of the gravitational force F_g is

$$F_g = G \frac{m_1 m_2}{r^2} \qquad [4.5]$$

where $G = 6.67 \times 10^{-11} \text{ N} \cdot \text{m}^2/\text{kg}^2$ is the **universal gravitation constant.** We examine the gravitational force in more detail in Chapter 7.

◀ Law of universal gravitation

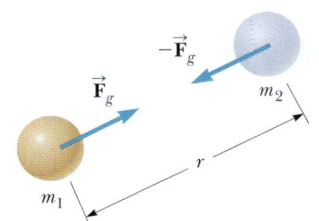

Figure 4.8 The gravitational force between two particles is attractive.

Weight

The magnitude of the gravitational force acting on an object of mass m is called the **weight** w of the object, given by

$$w = mg \qquad [4.6]$$

where g is the acceleration of gravity.
SI unit: newton (N)

From Equation 4.5, an alternate definition of the weight of an object with mass m can be written as

$$w = G \frac{M_E m}{r^2} \qquad [4.7]$$

where M_E is the mass of Earth and r is the distance from the object to Earth's center. If the object is at rest on Earth's surface, then r is equal to Earth's radius R_E. Because r^2 is in the denominator of Equation 4.7, the weight decreases with increasing r. So the weight of an object on a mountaintop is less than the weight of the same object at sea level.

Comparing Equations 4.6 and 4.7, it follows that

$$g = G \frac{M_E}{r^2} \qquad [4.8]$$

Unlike mass, weight is not an inherent property of an object because it can take different values, depending on the value of g in a given location. If an object has a mass of 70.0 kg, for example, then its weight at a location where $g = 9.80 \text{ m/s}^2$ is $mg = 686$ N. In a high-altitude balloon, where g might be 9.76 m/s², the object's weight would be 683 N. The value of g also varies slightly due to the density of matter in a given locality. **In this text, unless stated otherwise, the value of g will be understood to be 9.80 m/s², its value near the surface of the Earth.**

Equation 4.8 is a general result that can be used to calculate the acceleration of an object falling near the surface of any massive object if the more massive object's radius and mass are known. Using the values in Table 7.3 (p. 228), you should be able to show that $g_{Sun} = 274 \text{ m/s}^2$ and $g_{Moon} = 1.62 \text{ m/s}^2$. An important fact is that for spherical bodies, distances are calculated from the centers of the objects, a consequence of Gauss's law (explained in Chapter 15), which holds for both gravitational and electric forces.

The life-support unit strapped to the back of astronaut Harrison Schmitt weighed 300 lb on Earth and had a mass of 136 kg. During his training, a 50-lb mock-up with a mass of 23 kg was used. Although the mock-up had the same weight as the actual unit would have on the Moon, the smaller mass meant it also had a lower inertia. The weight of the unit is caused by the acceleration of the local gravity field, but the astronaut must also accelerate anything he's carrying in order to move it around. Consequently, the actual unit used on the Moon, with the same weight but greater inertia, was harder for the astronaut to handle than the mock-up unit on Earth.

96 | **CHAPTER 4** | The Laws of Motion

■ Quick Quiz

4.2 Which has greater value, a newton of gold on Earth or a newton of gold on the Moon? (a) The newton of gold on the Earth. (b) The newton of gold on the Moon. (c) The value is the same, regardless.

4.3 Respond to each statement, true or false: (a) No force of gravity acts on an astronaut in an orbiting space station. (b) At three Earth radii from the center of Earth, the acceleration of gravity is 1/9 its surface value. (c) If two identical planets, each with surface gravity g and volume V, coalesce into one planet with volume $2V$, the surface gravity of the new planet is $2g$. (d) One kilogram of gold would have greater value on Earth than on the Moon.

■ EXAMPLE 4.3 | Forces of Distant Worlds

GOAL Calculate the magnitude of a gravitational force using Newton's law of gravitation.

PROBLEM (a) Find the gravitational force exerted by the Sun on a 70.0-kg man located at the Earth's equator at noon, when the man is closest to the Sun. (b) Calculate the gravitational force of the Sun on the man at midnight, when he is farthest from the Sun. (c) Calculate the difference in the acceleration due to the Sun between noon and midnight. (For values, see Table 7.3 on page 228.)

STRATEGY To obtain the distance of the Sun from the man at noon, subtract the Earth's radius from the solar distance. At midnight, add the Earth's radius. Retain enough digits so that rounding doesn't remove the small difference between the two answers. For part (c), subtract the answer for (b) from (a) and divide by the man's mass.

SOLUTION

(a) Find the gravitational force exerted by the Sun on the man at the Earth's equator at noon.

Write the law of gravitation, Equation 4.5, in terms of the distance from the Sun to the Earth, r_S, and Earth's radius, R_E:

$$(1) \quad F_{\text{Sun}}^{\text{noon}} = \frac{mM_SG}{r^2} = \frac{mM_SG}{(r_S - R_E)^2}$$

Substitute values into (1) and retain two extra digits:

$$F_{\text{Sun}}^{\text{noon}} = \frac{(70.0 \text{ kg})(1.991 \times 10^{30} \text{ kg})(6.67 \times 10^{-11} \text{ kg}^{-1}\text{m}^3/\text{s}^2)}{(1.496 \times 10^{11} \text{ m} - 6.38 \times 10^6 \text{ m})^2}$$

$$= \boxed{0.415\ 40 \text{ N}}$$

(b) Calculate the gravitational force of the Sun on the man at midnight.

Write the law of gravitation, adding Earth's radius this time:

$$(2) \quad F_{\text{Sun}}^{\text{mid}} = \frac{mM_SG}{r^2} = \frac{mM_SG}{(r_S + R_E)^2}$$

Substitute values into (2):

$$F_{\text{Sun}}^{\text{mid}} = \frac{(70.0 \text{ kg})(1.991 \times 10^{30} \text{ kg})(6.67 \times 10^{-11} \text{ kg}^{-1}\text{m}^3/\text{s}^2)}{(1.496 \times 10^{11} \text{ m} + 6.38 \times 10^6 \text{ m})^2}$$

$$= \boxed{0.415\ 33 \text{ N}}$$

(c) Calculate the difference in the man's solar acceleration between noon and midnight.

Write an expression for the difference in acceleration and substitute values:

$$a = \frac{F_{\text{Sun}}^{\text{noon}} - F_{\text{Sun}}^{\text{mid}}}{m} = \frac{0.415\ 19 \text{ N} - 0.415\ 12 \text{ N}}{70.0 \text{ kg}}$$

$$\cong \boxed{1 \times 10^{-6} \text{ m/s}^2}$$

REMARKS The gravitational attraction between the Sun and objects on Earth is easily measurable and has been exploited in experiments to determine whether gravitational attraction depends on the composition of the object. The gravitational force on Earth due to the Moon is much weaker than the gravitational force on Earth due to the Sun. Paradoxically, the Moon's effect on the tides is over twice that of the Sun because the tides depend on *differences* in the gravitational force across the Earth, and those differences are greater for the Moon's gravitational force because the Moon is much closer to Earth than the Sun.

QUESTION 4.3 Mars is about one and a half times as far from the Sun as Earth. Without doing an explicit calculation, estimate to one significant digit the gravitational force of the Sun on a 70.0 kg man standing on Mars.

EXERCISE 4.3 During a new Moon, the Moon is directly overhead in the middle of the day. (a) Find the gravitational force exerted by the Moon on a 70.0-kg man at the Earth's equator at noon. (b) Calculate the gravitational force of the Moon on the man at midnight. (c) Calculate the difference in the man's acceleration due to the Moon between noon and midnight. *Note:* The distance from the Earth to the Moon is 3.84×10^8 m. The mass of the Moon is 7.36×10^{22} kg.

ANSWERS (a) 2.41×10^{-3} N (b) 2.25×10^{-3} N (c) 2.3×10^{-6} m/s^2

■ EXAMPLE 4.4   Weight on Planet X

GOAL Understand the effect of a planet's mass and radius on the weight of an object on the planet's surface.

PROBLEM An astronaut on a space mission lands on a planet with three times the mass and twice the radius of Earth. What is her weight w_X on this planet as a multiple of her Earth weight w_E?

STRATEGY Write M_X and r_X, the mass and radius of the planet, in terms of M_E and R_E, the mass and radius of Earth, respectively, and substitute into the law of gravitation.

SOLUTION

From the statement of the problem, we have the following relationships:

$$M_X = 3M_E \quad r_X = 2R_E$$

Substitute the preceding expressions into Equation 4.5 and simplify, algebraically associating the terms giving the weight on Earth:

$$w_X = G\frac{M_X m}{r_X^2} = G\frac{3M_E m}{(2R_E)^2} = \frac{3}{4} G\frac{M_E m}{R_E^2} = \frac{3}{4} w_E$$

REMARKS This problem shows the interplay between a planet's mass and radius in determining the weight of objects on its surface. Although Jupiter has about three hundred times the mass of the Earth, the weight of an object at Jupiter's planetary radius is only a little over two and a half times the weight of the same object on Earth's surface.

QUESTION 4.4 A volume of rock has a mass roughly three times a similar volume of ice. Suppose one world is made of ice whereas another world with the same radius is made of rock. If g is the acceleration of gravity on the surface of the ice world, what is the approximate acceleration of gravity on the rock world?

EXERCISE 4.4 An astronaut lands on Ganymede, a giant moon of Jupiter that is larger than the planet Mercury. Ganymede has one-fortieth the mass of Earth and two-fifths the radius. Find the weight of the astronaut standing on Ganymede in terms of his Earth weight w_E.

ANSWER $w_G = (5/32)w_E$

4.4 Newton's Third Law

LEARNING OBJECTIVE

1. Apply the third law of motion to simple systems, for each force identifying the proper reaction force.

In Section 4.1 we found that a force is exerted on an object when it comes into contact with some other object. Consider the task of driving a nail into a block of wood, for example, as illustrated in Figure 4.9a (page 98). To accelerate the nail and drive it into the block, the hammer must exert a net force on the nail.

Figure 4.9 Newton's third law. (a) The force exerted by the hammer on the nail is equal in magnitude and opposite in direction to the force exerted by the nail on the hammer. (b) The force $\vec{F}_{12}$ exerted by object 1 on object 2 is equal in magnitude and opposite in direction to the force $\vec{F}_{21}$ exerted by object 2 on object 1.

Newton recognized, however, that a single isolated force couldn't exist. Instead, **forces in nature always exist in pairs.** According to Newton, as the nail is driven into the block by the force exerted by the hammer, the hammer is slowed down and stopped by the force exerted by the nail.

Newton described such paired forces with his **third law:**

▶ Newton's third law

> If object 1 and object 2 interact, the force $\vec{F}_{12}$ exerted by object 1 on object 2 is equal in magnitude but opposite in direction to the force $\vec{F}_{21}$ exerted by object 2 on object 1.

This law, which is illustrated in Figure 4.9b, states that **a single isolated force can't exist.** The force $\vec{F}_{12}$ exerted by object 1 on object 2 is sometimes called the *action force,* and the force $\vec{F}_{21}$ exerted by object 2 on object 1 is called the *reaction force.* In reality, either force can be labeled the action or reaction force. **The action force is equal in magnitude to the reaction force and opposite in direction. In all cases, the action and reaction forces act on different objects.** For example, the force acting on a freely falling projectile is the force of gravity exerted by Earth on the projectile, $\vec{F}_g$, and the magnitude of this force is its weight mg. The reaction to force $\vec{F}_g$ is the gravitational force exerted by the projectile on Earth, $\vec{F}_g{'} = -2\vec{F}_g$. The reaction force $\vec{F}_g{'}$ must accelerate the Earth towards the projectile, just as the action force $\vec{F}_g$ accelerates the projectile towards Earth. Because Earth has such a large mass, however, its acceleration due to this reaction force is negligibly small.

Newton's third law constantly affects our activities in everyday life. Without it, no locomotion of any kind would be possible, whether on foot, on a bicycle, or in a motorized vehicle. When walking, for example, we exert a frictional force against the ground. The reaction force of the ground against our foot propels us forward. In the same way, the tires on a bicycle exert a frictional force against the ground, and the reaction of the ground pushes the bicycle forward. As we'll see shortly, friction plays a large role in such reaction forces.

APPLICATION
Helicopter Flight

For another example of Newton's third law, consider the helicopter. Most helicopters have a large set of blades rotating in a horizontal plane above the body of the vehicle and another, smaller set rotating in a vertical plane at the back. Other helicopters have two large sets of blades above the body rotating in opposite directions. Why do helicopters always have two sets of blades? In the first type of helicopter, the engine applies a force to the blades, causing them to change their rotational motion. According to Newton's third law, however, the blades must exert a force on the engine of equal magnitude and in the opposite direction. This force would cause the body of the helicopter to rotate in the direction opposite the blades. A rotating helicopter would be impossible to control, so a second set of blades is used. The small blades in the back provide a force opposite to that tending to rotate the body of the helicopter, keeping the body oriented in a stable position. In helicopters with two sets of large counterrotating blades,

Tip 4.4 Action–Reaction Pairs

In applying Newton's third law, remember that an action and its reaction force always act on *different* objects. Two external forces acting on the same object, even if they are equal in magnitude and opposite in direction, *can't* be an action–reaction pair.

Figure 4.10 When a monitor is sitting on a table, the forces acting on the monitor are the normal force $\vec{\mathbf{n}}$ exerted by the table and the force of gravity, $\vec{\mathbf{F}}_g$, as illustrated in (b). The reaction to $\vec{\mathbf{n}}$ is the force exerted by the monitor on the table, $\vec{\mathbf{n}}'$. The reaction to $\vec{\mathbf{F}}_g$ is the force exerted by the monitor on Earth, $\vec{\mathbf{F}}_g'$.

engines apply forces in opposite directions, so there is no net force rotating the helicopter.

As mentioned earlier, Earth exerts a gravitational force $\vec{\mathbf{F}}_g$ on any object. If the object is a monitor at rest on a table, as in Figure 4.10a, the reaction force to $\vec{\mathbf{F}}_g$ is the gravitational force the monitor exerts on the Earth, $\vec{\mathbf{F}}_g'$. The monitor doesn't accelerate downward because it's held up by the table. The table therefore exerts an upward force $\vec{\mathbf{n}}$, called the **normal force,** on the monitor. (*Normal*, a technical term from mathematics, means "perpendicular" in this context.) The normal force is an elastic force arising from the cohesion of matter and is electromagnetic in origin. It balances the gravitational force acting on the monitor, preventing the monitor from falling through the table, and can have any value needed, up to the point of breaking the table. The reaction to $\vec{\mathbf{n}}$ is the force exerted by the monitor on the table, $\vec{\mathbf{n}}'$. Therefore,

$$\vec{\mathbf{F}}_g = -\vec{\mathbf{F}}_g' \quad \text{and} \quad \vec{\mathbf{n}} = -\vec{\mathbf{n}}'$$

The forces $\vec{\mathbf{n}}$ and $\vec{\mathbf{n}}'$ both have the same magnitude as $\vec{\mathbf{F}}_g$. Note that the forces acting on the monitor are $\vec{\mathbf{F}}_g$ and $\vec{\mathbf{n}}$, as shown in Figure 4.10b. The two reaction forces, $\vec{\mathbf{F}}_g'$ and $\vec{\mathbf{n}}'$, are exerted by the monitor on objects other than the monitor. Remember that the two forces in an action–reaction pair always act on two different objects.

Because the monitor is not accelerating in any direction ($\vec{\mathbf{a}} = 0$), it follows from Newton's second law that $m\vec{\mathbf{a}} = 0 = \vec{\mathbf{F}}_g + \vec{\mathbf{n}}$. However, $F_g = -mg$, so $n = mg$, a useful result.

> **Tip 4.5 Equal and Opposite but Not a Reaction Force**
> A common error in Figure 4.10b is to consider the normal force on the object to be the reaction force to the gravity force, because in this case these two forces are equal in magnitude and opposite in direction. That is impossible, however, because they act on the same object!

Quick Quiz

4.4 A small sports car collides head-on with a massive truck. The greater impact force (in magnitude) acts on (a) the car, (b) the truck, (c) neither, the force is the same on both. Which vehicle undergoes the greater magnitude acceleration? (d) the car, (e) the truck, (f) the accelerations are the same.

APPLICATION
Colliding Vehicles

■ EXAMPLE 4.5 | Action–Reaction and the Ice Skaters

GOAL Illustrate Newton's third law of motion.

PROBLEM A man of mass $M = 75.0$ kg and woman of mass $m = 55.0$ kg stand facing each other on an ice rink, both wearing ice skates. The woman pushes the man with a horizontal force of $F = 85.0$ N in the positive x-direction. Assume the ice is frictionless. **(a)** What is the man's acceleration? **(b)** What is the reaction force acting on the woman? **(c)** Calculate the woman's acceleration.

STRATEGY Parts (a) and (c) are simple applications of the second law. An application of the third law solves part (b).

(Continued)

SOLUTION

(a) What is the man's acceleration?

Write the second law for the man:
$$Ma_M = F$$

Solve for the man's acceleration and substitute values:
$$a_M = \frac{F}{M} = \frac{85.0 \text{ N}}{75.0 \text{ kg}} = \boxed{1.13 \text{ m/s}^2}$$

(b) What is the reaction force acting on the woman?

Apply Newton's third law of motion, finding that the reaction force R acting on the woman has the same magnitude and opposite direction:
$$R = -F = \boxed{-85.0 \text{ N}}$$

(c) Calculate the woman's acceleration.

Write Newton's second law for the woman:
$$ma_W = R = -F$$

Solve for the woman's acceleration and substitute values:
$$a_W = \frac{-F}{m} = \frac{-85.0 \text{ N}}{55.0 \text{ kg}} = \boxed{-1.55 \text{ m/s}^2}$$

REMARKS Notice that the forces are equal and opposite each other, but the accelerations are not because the two masses differ from each other.

QUESTION 4.5 Name two other forces acting on the man and the two reaction forces that are paired with them.

EXERCISE 4.5 A space-walking astronaut of total mass 148 kg exerts a force of 265 N on a free-floating satellite of mass 635 kg, pushing it in the positive x-direction. (a) What is the reaction force exerted by the satellite on the astronaut? Calculate the accelerations of (b) the astronaut, and (c) the satellite.

ANSWERS (a) −265 N (b) −1.79 m/s² (c) 0.417 m/s²

4.5 Applications of Newton's Laws

LEARNING OBJECTIVES

1. Draw free-body diagrams for physical systems.
2. Apply the second law to an object in equilibrium.
3. Apply the second law to an object under acceleration.
4. Apply the second law to systems of two objects.

Figure 4.11 Newton's second law applied to a rope gives $T - T' = ma$. However, if $m = 0$, then $T = T'$. Thus, the tension in a massless rope is the same at all points in the rope.

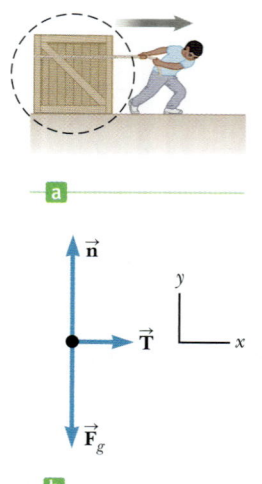

Figure 4.12 (a) A crate being pulled to the right on a frictionless surface. (b) The free-body diagram that represents the forces exerted on the crate.

This section applies Newton's laws to objects moving under the influence of constant external forces. We assume that objects behave as particles, so we need not consider the possibility of rotational motion. We also neglect any friction effects and the masses of any ropes or strings involved. With these approximations, the magnitude of the force exerted along a rope, called the **tension**, is the same at all points in the rope. This is illustrated by the rope in Figure 4.11, showing the forces $\vec{T}$ and $\vec{T}'$ acting on it. If the rope has mass m, then Newton's second law applied to the rope gives $T - T' = ma$. If the mass m is taken to be negligible, however, as in the upcoming examples, then $T = T'$.

When we apply Newton's law to an object, we are interested only in those forces which act *on the object*. For example, in Figure 4.10b, the only external forces acting on the monitor are $\vec{n}$ and $\vec{F}_g$. The reactions to these forces, $\vec{n}'$ and $\vec{F}_g'$, act on the table and on Earth, respectively, and don't appear in Newton's second law applied to the monitor.

Consider a crate being pulled to the right on a frictionless, horizontal surface, as in Figure 4.12a. Suppose you wish to find the acceleration of the crate and the force the surface exerts on it. The horizontal force exerted on the crate acts through the rope. The force that the rope exerts on the crate is denoted by $\vec{T}$ (because it's a tension force). The magnitude of $\vec{T}$ is equal to the tension in the rope. What we mean by the words "tension in the rope" is just the force read by a spring scale when the rope in question has been cut and the scale inserted between the cut ends. A dashed circle is drawn around the crate in Figure 4.12a to emphasize the importance of isolating the crate from its surroundings.

Because we are interested only in the motion of the crate, we must be able to identify all forces acting on it. These forces are illustrated in Figure 4.12b. In addition to displaying the force $\vec{T}$, the force diagram for the crate includes the force of gravity $\vec{F}_g$ exerted by Earth and the normal force $\vec{n}$ exerted by the floor. Such a force diagram is called a **free-body diagram** because the environment is replaced by a series of forces on an otherwise free body. The construction of a correct free-body diagram is an essential step in applying Newton's laws. An incorrect diagram will most likely lead to incorrect answers!

The *reactions* to the forces we have listed—namely, the force exerted by the rope on the hand doing the pulling, the force exerted by the crate on Earth, and the force exerted by the crate on the floor—aren't included in the free-body diagram because they act on other objects and not on the crate. Consequently, they don't directly influence the crate's motion. Only forces acting directly on the crate are included.

Now let's apply Newton's second law to the crate. First we choose an appropriate coordinate system. In this case it's convenient to use the one shown in Figure 4.12b, with the x-axis horizontal and the y-axis vertical. We can apply Newton's second law in the x-direction, y-direction, or both, depending on what we're asked to find in a problem. Newton's second law applied to the crate in the x- and y-directions yields the following two equations:

$$ma_x = T \qquad ma_y = n - mg = 0$$

From these equations, we find that the acceleration in the x-direction is constant, given by $a_x = T/m$, and that the normal force is given by $n = mg$. Because the acceleration is constant, the equations of kinematics can be applied to obtain further information about the velocity and displacement of the object.

■ PROBLEM-SOLVING STRATEGY

Newton's Second Law
Problems involving Newton's second law can be very complex. The following protocol breaks the solution process down into smaller, intermediate goals:

1. **Read** the problem carefully at least once.
2. **Draw** a picture of the system, identify the object of primary interest, and indicate forces with arrows.
3. **Label** each force in the picture in a way that will bring to mind what physical quantity the label stands for (e.g., *T* for tension).
4. **Draw** a free-body diagram of the object of interest, based on the labeled picture. If additional objects are involved, draw separate free-body diagrams for them. Choose convenient coordinates for each object.
5. **Apply Newton's second law.** The x- and y-components of Newton's second law should be taken from the vector equation and written individually. This usually results in two equations and two unknowns.
6. **Solve** for the desired unknown quantity, and substitute the numbers.

In the special case of equilibrium, the foregoing process is simplified because the acceleration is zero.

Objects in Equilibrium

Objects that are either at rest or moving with constant velocity are said to be in equilibrium. Because $\vec{a} = 0$, Newton's second law applied to an object in equilibrium gives

$$\sum \vec{F} = 0 \qquad [4.9]$$

Tip 4.6 Free-Body Diagrams
The *most important step* in solving a problem by means of Newton's second law is to draw the correct free-body diagram. Include only those forces that act directly on the object of interest.

Tip 4.7 A Particle in Equilibrium
A zero net force on a particle does *not* mean that the particle isn't moving. It means that the particle isn't *accelerating*. If the particle has a nonzero initial velocity and is acted upon by a zero net force, it continues to move with the same velocity.

Figure 4.13 (Quick Quiz 4.5) (i) A person pulls with a force of magnitude F on a spring scale attached to a wall. (ii) Two people pull with forces of magnitude F in opposite directions on a spring scale attached between two ropes.

This statement signifies that the *vector* sum of all the forces (the net force) acting on an object in equilibrium is zero. Equation 4.9 is equivalent to the set of component equations given by

$$\sum F_x = 0 \quad \text{and} \quad \sum F_y = 0 \qquad [4.10]$$

We won't consider three-dimensional problems in this book, but the extension of Equation 4.10 to a three-dimensional problem can be made by adding a third equation: $\sum F_z = 0$.

Quick Quiz

4.5 Consider the two situations shown in Figure 4.13, in which there is no acceleration. In both cases the men pull with a force of magnitude F. Is the reading on the scale in part (i) of the figure (a) greater than, (b) less than, or (c) equal to the reading in part (ii)?

EXAMPLE 4.6 A Traffic Light at Rest

GOAL Use the second law in an equilibrium problem requiring two free-body diagrams.

PROBLEM A traffic light weighing 1.00×10^2 N hangs from a vertical cable tied to two other cables that are fastened to a support, as in Figure 4.14a. The upper cables make angles of 37.0° and 53.0° with the horizontal. Find the tension in each of the three cables.

STRATEGY There are three unknowns, so we need to generate three equations relating them, which can then be solved. One equation can be obtained by applying Newton's second law to the traffic light, which has forces in the y-direction only. Two more equations can be obtained by applying the second law to the knot joining the cables—one equation from the x-component and one equation from the y-component.

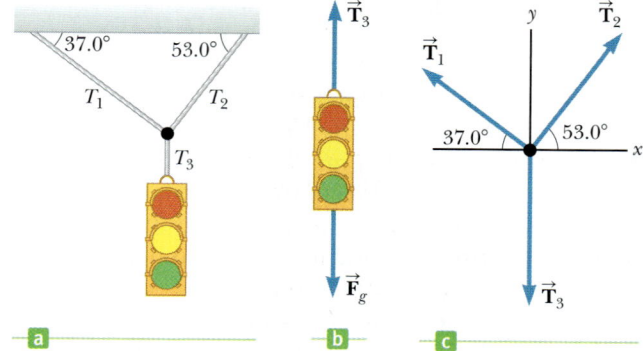

Figure 4.14 (Example 4.6) (a) A traffic light suspended by cables. (b) The forces acting on the traffic light. (c) A free-body diagram for the knot joining the cables.

SOLUTION

Find T_3 from Figure 4.14b, using the condition of equilibrium:

$$\sum F_y = 0 \quad \rightarrow \quad T_3 - F_g = 0$$
$$T_3 = F_g = 1.00 \times 10^2 \text{ N}$$

Using Figure 4.14c, resolve all three tension forces into components and construct a table for convenience:

Force	x-Component	y-Component
$\vec{T}_1$	$-T_1 \cos 37.0°$	$T_1 \sin 37.0°$
$\vec{T}_2$	$T_2 \cos 53.0°$	$T_2 \sin 53.0°$
$\vec{T}_3$	0	-1.00×10^2 N

Apply the conditions for equilibrium to the knot, using the components in the table:

(1) $\sum F_x = -T_1 \cos 37.0° + T_2 \cos 53.0° = 0$

(2) $\sum F_y = T_1 \sin 37.0° + T_2 \sin 53.0° - 1.00 \times 10^2 \text{ N} = 0$

There are two equations and two remaining unknowns. Solve Equation (1) for T_2:

$$T_2 = T_1 \left(\frac{\cos 37.0°}{\cos 53.0°} \right) = T_1 \left(\frac{0.799}{0.602} \right) = 1.33 T_1$$

Substitute the result for T_2 into Equation (2):

$T_1 \sin 37.0° + (1.33 T_1)(\sin 53.0°) - 1.00 \times 10^2 \text{ N} = 0$

$T_1 = $ 60.1 N

$T_2 = 1.33 T_1 = 1.33(60.1 \text{ N}) = $ 79.9 N

REMARKS It's very easy to make sign errors in this kind of problem. One way to avoid them is to always measure the angle of a vector from the positive x-direction. The trigonometric functions of the angle will then automatically give the correct signs for the components. For example, T_1 makes an angle of $180° - 37° = 143°$ with respect to the positive x-axis, and its x-component, $T_1 \cos 143°$, is negative, as it should be.

QUESTION 4.6 How would the answers change if a second traffic light were attached beneath the first?

EXERCISE 4.6 Suppose the traffic light is hung so that the tensions T_1 and T_2 are both equal to 80.0 N. Find the new angles they make with respect to the x-axis. (By symmetry, these angles will be the same.)

ANSWER Both angles are 38.7°.

■ **EXAMPLE 4.7** Sled on a Frictionless Hill

GOAL Use the second law and the normal force in an equilibrium problem.

PROBLEM A sled is tied to a tree on a frictionless, snow-covered hill, as shown in Figure 4.15a. If the sled weighs 77.0 N, find the magnitude of the tension force $\vec{T}$ exerted by the rope on the sled and that of the normal force $\vec{n}$ exerted by the hill on the sled.

STRATEGY When an object is on a slope, it's convenient to use tilted coordinates, as in Figure 4.15b, so that the normal force $\vec{n}$ is in the y-direction and the tension force $\vec{T}$ is in the x-direction. In the absence of friction, the hill exerts no force on the sled in the x-direction. Because the sled is at rest, the conditions for equilibrium, $\sum F_x = 0$ and $\sum F_y = 0$, apply, giving two equations for the two unknowns—the tension and the normal force.

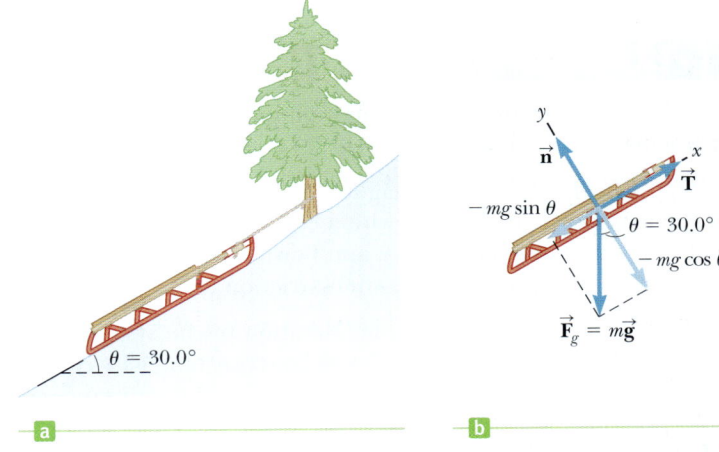

Figure 4.15 (Example 4.7) (a) A sled tied to a tree on a frictionless hill. (b) A diagram of forces acting on the sled.

SOLUTION

Apply Newton's second law to the sled, with $\vec{a} = 0$:

$$\sum \vec{F} = \vec{T} + \vec{n} + \vec{F}_g = 0$$

(Continued)

Extract the x-component from this equation to find T. The x-component of the normal force is zero, and the sled's weight is given by $mg = 77.0$ N.

$$\sum F_x = T + 0 - mg \sin \theta = T - (77.0 \text{ N}) \sin 30.0° = 0$$

$$T = \boxed{38.5 \text{ N}}$$

Write the y-component of Newton's second law. The y-component of the tension is zero, so this equation will give the normal force.

$$\sum F_y = 0 + n - mg \cos \theta = n - (77.0 \text{ N})(\cos 30.0°) = 0$$

$$n = \boxed{66.7 \text{ N}}$$

REMARKS Unlike its value on a horizontal surface, n is less than the weight of the sled when the sled is on the slope. This is because only part of the force of gravity (the x-component) is acting to pull the sled down the slope. The y-component of the force of gravity balances the normal force.

QUESTION 4.7 Consider the same scenario on a hill with a steeper slope. Would the magnitude of the tension in the rope get larger, smaller, or remain the same as before? How would the normal force be affected?

EXERCISE 4.7 Suppose a child of weight w climbs onto the sled. If the tension force is measured to be 60.0 N, find the weight of the child and the magnitude of the normal force acting on the sled.

ANSWERS $w = 43.0$ N, $n = 104$ N

■ *Quick Quiz*

4.6 For the woman being pulled forward on the toboggan in Figure 4.16, is the magnitude of the normal force exerted by the ground on the toboggan (a) equal to the total weight of the woman plus the toboggan, (b) greater than the total weight, (c) less than the total weight, or (d) possibly greater than or less than the total weight, depending on the size of the weight relative to the tension in the rope?

Accelerating Objects and Newton's Second Law

When a net force acts on an object, the object accelerates, and we use Newton's second law to analyze the motion.

Figure 4.16 (Quick Quiz 4.6)

■ EXAMPLE 4.8 Moving a Crate

GOAL Apply the second law of motion for a system not in equilibrium, together with a kinematics equation.

PROBLEM The combined weight of the crate and dolly in Figure 4.17 is 3.00×10^2 N. If the man pulls on the rope with a constant force of 20.0 N, what is the acceleration of the system (crate plus dolly), and how far will it move in 2.00 s? Assume the system starts from rest and that there are no friction forces opposing the motion.

STRATEGY We can find the acceleration of the system from Newton's second law. Because the force exerted on the system is constant, its acceleration is constant. Therefore, we can apply a kinematics equation to find the distance traveled in 2.00 s.

Figure 4.17 (Example 4.8)

SOLUTION

Find the mass of the system from the definition of weight, $w = mg$:

$$m = \frac{w}{g} = \frac{3.00 \times 10^2 \text{ N}}{9.80 \text{ m/s}^2} = 30.6 \text{ kg}$$

Find the acceleration of the system from the second law:

$$a_x = \frac{F_x}{m} = \frac{20.0 \text{ N}}{30.6 \text{ kg}} = \boxed{0.654 \text{ m/s}^2}$$

Use kinematics to find the distance moved in 2.00 s, with $v_0 = 0$:

$$\Delta x = \tfrac{1}{2} a_x t^2 = \tfrac{1}{2}(0.654 \text{ m/s}^2)(2.00 \text{ s})^2 = \boxed{1.31 \text{ m}}$$

REMARKS Note that the constant applied force of 20.0 N is assumed to act on the system at all times during its motion. If the force were removed at some instant, the system would continue to move with constant velocity and hence zero acceleration. The rollers have an effect that was neglected, here.

QUESTION 4.8 What effect does doubling the weight have on the acceleration and the displacement?

EXERCISE 4.8 A man pulls a 50.0-kg box horizontally from rest while exerting a constant horizontal force, displacing the box 3.00 m in 2.00 s. Find the force the man exerts on the box. (Ignore friction.)

ANSWER 75.0 N

EXAMPLE 4.9 The Runaway Car

GOAL Apply the second law and kinematic equations to a problem involving an object moving on an incline.

PROBLEM (a) A car of mass m is on an icy driveway inclined at an angle $\theta = 20.0°$, as in Figure 4.18a. Determine the acceleration of the car, assuming the incline is frictionless. (b) If the length of the driveway is 25.0 m and the car starts from rest at the top, how long does it take to travel to the bottom? (c) What is the car's speed at the bottom?

STRATEGY Choose tilted coordinates as in Figure 4.18b so that the normal force $\vec{n}$ is in the positive y-direction, perpendicular to the driveway, and the positive x-axis is down the slope. The force of gravity $\vec{F}_g$ then has an x-component, $mg\sin\theta$, and a y-component, $2mg\cos\theta$. The components of Newton's second law form a system of two equations and two unknowns for the acceleration down the slope, a_x, and the normal force. Parts (b) and (c) can be solved with the kinematics equations.

Figure 4.18 (Example 4.9)

SOLUTION

(a) Find the acceleration of the car.

Apply Newton's second law to the car:
$$m\vec{a} = \sum \vec{F} = \vec{F}_g + \vec{n}$$

Extract the x- and y-components from the second law:
(1) $\quad ma_x = \sum F_x = mg\sin\theta$
(2) $\quad 0 = \sum F_y = -mg\cos\theta + n$

Divide Equation (1) by m and substitute the given values:
$$a_x = g\sin\theta = (9.80 \text{ m/s}^2)\sin 20.0° = \boxed{3.35 \text{ m/s}^2}$$

(b) Find the time taken for the car to reach the bottom.

Use Equation 3.11b for displacement, with $v_{0x} = 0$:
$$\Delta x = \tfrac{1}{2}a_x t^2 \quad \rightarrow \quad \tfrac{1}{2}(3.35 \text{ m/s}^2)t^2 = 25.0 \text{ m}$$
$$t = \boxed{3.86 \text{ s}}$$

(c) Find the speed of the car at the bottom of the driveway.

Use Equation 3.11a for velocity, again with $v_{0x} = 0$:
$$v_x = a_x t = (3.35 \text{ m/s}^2)(3.86 \text{ s}) = \boxed{12.9 \text{ m/s}}$$

REMARKS Notice that the final answer for the acceleration depends only on g and the angle θ, not the mass. Equation (2), which gives the normal force, isn't useful here, but is essential when friction plays a role.

QUESTION 4.9 If the car is parked on a more gentle slope, how will the time required for it to slide to the bottom of the hill be affected? Explain.

EXERCISE 4.9 (a) Suppose a hockey puck slides down a frictionless ramp with an acceleration of 5.00 m/s². What angle does the ramp make with respect to the horizontal? (b) If the ramp has a length of 6.00 m, how long does it take
(Continued)

the puck to reach the bottom? **(c)** Now suppose the mass of the puck is doubled. What's the puck's new acceleration down the ramp?

ANSWER (a) 30.7° (b) 1.55 s (c) unchanged, 5.00 m/s²

■ EXAMPLE 4.10 | Weighing a Fish in an Elevator

GOAL Explore the effect of acceleration on the apparent weight of an object.

PROBLEM A woman weighs a fish with a spring scale attached to the ceiling of an elevator, as shown in Figures 4.19a and 4.19b. While the elevator is at rest, she measures a weight of 40.0 N. **(a)** What weight does the scale read if the elevator accelerates upward at 2.00 m/s²? **(b)** What does the scale read if the elevator accelerates downward at 2.00 m/s², as in Figure 4.19b? **(c)** If the elevator cable breaks, what does the scale read?

STRATEGY Write down Newton's second law for the fish, including the force $\vec{T}$ exerted by the spring scale and the force of gravity, $m\vec{g}$. The scale doesn't measure the true weight, it measures the force T that it exerts on the fish, so in each case solve for this force, which is the apparent weight as measured by the scale.

When the elevator accelerates upward, the spring scale reads a value greater than the weight of the fish.

When the elevator accelerates downward, the spring scale reads a value less than the weight of the fish.

Figure 4.19 (Example 4.10)

SOLUTION

(a) Find the scale reading as the elevator accelerates upward, as in Figure 4.19a.

Apply Newton's second law to the fish, taking upward as the positive direction:

$$ma = \sum F = T - mg$$

Solve for T:

$$T = ma + mg = m(a + g)$$

Find the mass of the fish from its weight of 40.0 N:

$$m = \frac{w}{g} = \frac{40.0 \text{ N}}{9.80 \text{ m/s}^2} = 4.08 \text{ kg}$$

Compute the value of T, substituting $a = +2.00$ m/s²:

$$T = m(a + g) = (4.08 \text{ kg})(2.00 \text{ m/s}^2 + 9.80 \text{ m/s}^2)$$
$$= \boxed{48.1 \text{ N}}$$

(b) Find the scale reading as the elevator accelerates downward, as in Figure 4.19b.

The analysis is the same, the only change being the acceleration, which is now negative: $a = -2.00$ m/s².

$$T = m(a + g) = (4.08 \text{ kg})(-2.00 \text{ m/s}^2 + 9.80 \text{ m/s}^2)$$
$$= \boxed{31.8 \text{ N}}$$

(c) Find the scale reading after the elevator cable breaks.

Now $a = -9.80$ m/s², the acceleration due to gravity:

$$T = m(a + g) = (4.08 \text{ kg})(-9.80 \text{ m/s}^2 + 9.80 \text{ m/s}^2)$$
$$= \boxed{0 \text{ N}}$$

REMARKS Notice how important it is to have correct signs in this problem! Accelerations can increase or decrease the apparent weight of an object. Astronauts experience very large changes in apparent weight, from several times normal weight during ascent to weightlessness in free fall.

QUESTION 4.10 Starting from rest, an elevator accelerates upward, reaching and maintaining a constant velocity thereafter until it reaches the desired floor, when it begins to slow down. Describe the scale reading during this time.

EXERCISE 4.10 Find the initial acceleration of a rocket if the astronauts on board experience eight times their normal weight during an initial vertical ascent. (*Hint:* In this exercise, the scale force is replaced by the normal force.)

ANSWER 68.6 m/s²

EXAMPLE 4.11 Atwood's Machine

GOAL Use the second law to solve a simple two-body problem symbolically.

PROBLEM Two objects of mass m_1 and m_2, with $m_2 > m_1$, are connected by a light, inextensible cord and hung over a frictionless pulley, as in Figure 4.20a. Both cord and pulley have negligible mass. Find the magnitude of the acceleration of the system and the tension in the cord.

STRATEGY The heavier mass, m_2, accelerates downward, in the negative y-direction. Because the cord can't be stretched, the accelerations of the two masses are equal in magnitude, but *opposite* in direction, so that a_1 is positive and a_2 is negative, and $a_2 = -a_1$. Each mass is acted on by a force of tension $\vec{T}$ in the upward direction and a force of gravity in the downward direction. Figure 4.20b shows free-body diagrams for the two masses. Newton's second law for each mass, together with the equation relating the accelerations, constitutes a set of three equations for the three unknowns—a_1, a_2, and T.

Figure 4.20 (Example 4.11) Atwood's machine. (a) Two hanging objects connected by a light string that passes over a frictionless pulley. (b) Free-body diagrams for the objects.

SOLUTION

Apply the second law to each of the two objects individually:

(1) $m_1 a_1 = T - m_1 g$ (2) $m_2 a_2 = T - m_2 g$

Substitute $a_2 = -a_1$ into Equation (2) and multiply both sides by -1:

(3) $m_2 a_1 = -T + m_2 g$

Add Equations (1) and (3), and solve for a_1:

$(m_1 + m_2) a_1 = m_2 g - m_1 g$

$$a_1 = \left(\frac{m_2 - m_1}{m_1 + m_2}\right) g$$

Substitute this result into Equation (1) to find T:

$$T = \left(\frac{2 m_1 m_2}{m_1 + m_2}\right) g$$

REMARKS The acceleration of the second object is the same as that of the first, but negative. When m_2 gets very large compared with m_1, the acceleration of the system approaches g, as expected, because m_2 is falling nearly freely under the influence of gravity. Indeed, m_2 is only slightly restrained by the much lighter m_1.

QUESTION 4.11 How could this simple machine be used to raise objects too heavy for a person to lift?

EXERCISE 4.11 Suppose in the same Atwood setup another string is attached to the bottom of m_1 and a constant force f is applied, retarding the upward motion of m_1. If $m_1 = 5.00$ kg and $m_2 = 10.00$ kg, what value of f will reduce the acceleration of the system by 50%?

ANSWER 24.5 N

4.6 Forces of Friction

LEARNING OBJECTIVES

1. Explain the physical origins of friction forces.
2. Apply the concept of the kinetic friction force.
3. Apply the concept of the static friction force.
4. Apply the system approach to multiple-body problems.

An object moving on a surface or through a viscous medium such as air or water encounters resistance as it interacts with its surroundings. This resistance is called **friction.** Forces of friction are essential in our everyday lives. Friction makes it possible to grip and hold things, drive a car, walk, and run. Even standing in one spot would be impossible without friction, as the slightest shift would instantly cause you to slip and fall.

Imagine that you've filled a plastic trash can with yard clippings and want to drag the can across the surface of your concrete patio. If you apply an external horizontal force $\vec{F}$ to the can, acting to the right as shown in Figure 4.21a, the can remains stationary if $\vec{F}$ is small. The force that counteracts $\vec{F}$ and keeps the can from moving acts to the left, opposite the direction of $\vec{F}$, and is called the **force of static friction,** $\vec{f}_s$. As long as the can isn't moving, $\vec{f}_s = -\vec{F}$. If $\vec{F}$ is increased, $\vec{f}_s$ also increases. Likewise, if $\vec{F}$ decreases, $\vec{f}_s$ decreases. Experiments show that the friction force arises from the nature of the two surfaces: Because of their roughness, contact is made at only a few points.

If we increase the magnitude of $\vec{F}$, as in Figure 4.21b, the trash can eventually slips. When the can is on the verge of slipping, f_s is a maximum, as shown in Figure 4.21c. When F exceeds $f_{s,\text{max}}$, the can accelerates to the right. When the can is in motion, the friction force is less than $f_{s,\text{max}}$ (Fig. 4.21c). We call the

Figure 4.21 (a) and (b) When pulling on a trash can, the direction of the force of friction ($\vec{f}_s$ in part (a) and $\vec{f}_k$ in part (b)) between the can and a rough surface is opposite the direction of the applied force $\vec{F}$. (c) A graph of the magnitude of the friction force versus applied force. Note that $f_{s,\text{max}} > f_k$.

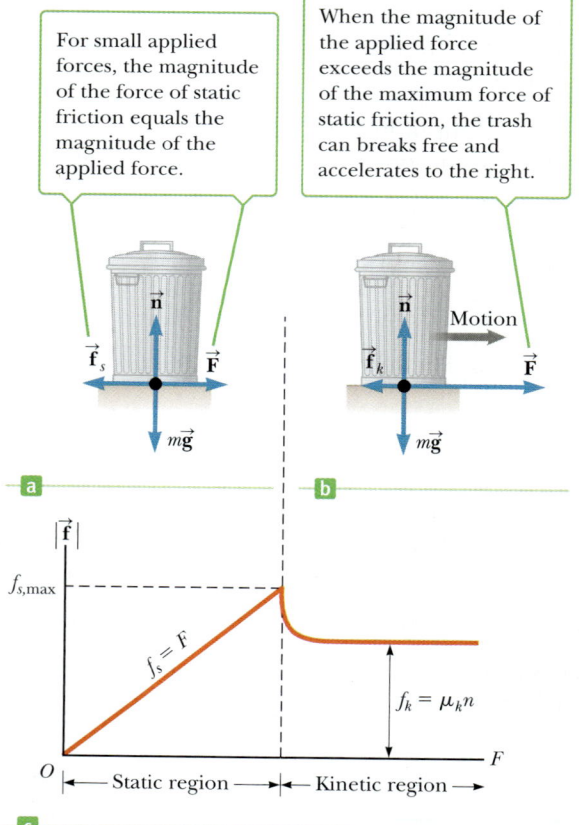

friction force for an object in motion the **force of kinetic friction**, $\vec{f}_k$. The net force $F - f_k$ in the x-direction produces an acceleration to the right, according to Newton's second law. If $F = f_k$, the acceleration is zero, and the can moves to the right with constant speed. If the applied force is removed, the friction force acting to the left provides an acceleration of the can in the $-x$-direction and eventually brings it to rest, again consistent with Newton's second law.

Experimentally, to a good approximation, both $f_{s,\text{max}}$ and f_k for an object on a surface are proportional to the normal force exerted by the surface on the object. The experimental observations can be summarized as follows:

- The magnitude of the force of static friction between any two surfaces in contact can have the values

$$f_s \leq \mu_s n \qquad [4.11]$$

where the dimensionless constant μ_s is called the **coefficient of static friction** and n is the magnitude of the normal force exerted by one surface on the other. Equation 4.11 also holds for $f_s = f_{s,\text{max}} \equiv \mu_s n$ when an object is on the verge of slipping. This situation is called *impending motion*. The strict inequality holds when the component of the applied force parallel to the surfaces is less than $\mu_s n$.

- The magnitude of the force of kinetic friction acting between two surfaces is

$$f_k = \mu_k n \qquad [4.12]$$

where μ_k is the **coefficient of kinetic friction.**

- The values of μ_k and μ_s depend on the nature of the surfaces, but μ_k is generally less than μ_s. Table 4.2 lists some reported values.
- The direction of the friction force exerted by a surface on an object is opposite the actual motion (kinetic friction) or the impending motion (static friction) of the object relative to the surface.
- The coefficients of friction are nearly independent of the area of contact between the surfaces.

Although the coefficient of kinetic friction varies with the speed of the object, we will neglect any such variations. The approximate nature of Equations 4.11 and 4.12 is easily demonstrated by trying to get an object to slide down an incline at constant acceleration. Especially at low speeds, the motion is likely to be characterized by alternating stick and slip episodes.

> **Tip 4.8 Use the Equals Sign in Limited Situations**
>
> In Equation 4.11, the equals sign is used *only* when the surfaces are just about to break free and begin sliding. Don't fall into the common trap of using $f_s = \mu_s n$ in *any* static situation.

Table 4.2 Coefficients of Friction[a]

	μ_s	μ_k
Steel on steel	0.74	0.57
Aluminum on steel	0.61	0.47
Copper on steel	0.53	0.36
Rubber on concrete	1.0	0.8
Wood on wood	0.25–0.5	0.2
Glass on glass	0.94	0.4
Waxed wood on wet snow	0.14	0.1
Waxed wood on dry snow	—	0.04
Metal on metal (lubricated)	0.15	0.06
Ice on ice	0.1	0.03
Teflon on Teflon	0.04	0.04
Synovial joints in humans	0.01	0.003

[a]All values are approximate.

Figure 4.22 (Quick Quiz 4.9)

Quick Quiz

4.7 If you press a book flat against a vertical wall with your hand, in what direction is the friction force exerted by the wall on the book? (a) downward (b) upward (c) out from the wall (d) into the wall

4.8 A crate is sitting in the center of a flatbed truck. As the truck accelerates to the east, the crate moves with it, not sliding on the bed of the truck. In what direction is the friction force exerted by the bed of the truck on the crate? (a) To the west. (b) To the east. (c) There is no friction force, because the crate isn't sliding.

4.9 Suppose your friend is sitting on a sled and asks you to move her across a flat, horizontal field. You have a choice of (a) pushing her from behind by applying a force downward on her shoulders at 30° below the horizontal (Fig. 4.22a) or (b) attaching a rope to the front of the sled and pulling with a force at 30° above the horizontal (Fig 4.22b). Which option would be easier and why?

■ EXAMPLE 4.12 | A Block on a Ramp

GOAL Apply the concept of static friction to an object resting on an incline.

PROBLEM Suppose a block with a mass of 2.50 kg is resting on a ramp. If the coefficient of static friction between the block and ramp is 0.350, what maximum angle can the ramp make with the horizontal before the block starts to slip down?

STRATEGY This is an application of Newton's second law involving an object in equilibrium. Choose tilted coordinates, as in Figure 4.23. Use the fact that the block is just about to slip when the force of static friction takes its maximum value, $f_s = \mu_s n$.

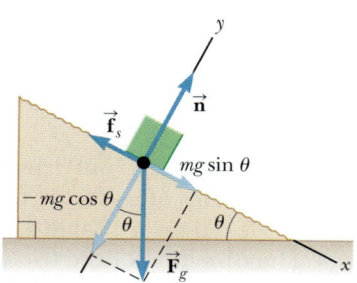

Figure 4.23 (Example 4.12)

SOLUTION

Write Newton's laws for a static system in component form. The gravity force has two components, just as in Examples 4.7 and 4.9.

(1) $\sum F_x = mg \sin \theta - \mu_s n = 0$

(2) $\sum F_y = n - mg \cos \theta = 0$

Rearrange Equation (2) to get an expression for the normal force n:

$n = mg \cos \theta$

Substitute the expression for n into Equation (1) and solve for $\tan \theta$:

$\sum F_x = mg \sin \theta - \mu_s mg \cos \theta = 0 \quad \rightarrow \quad \tan \theta = \mu_s$

Apply the inverse tangent function to get the answer:

$\tan \theta = 0.350 \quad \rightarrow \quad \theta = \tan^{-1}(0.350) = \boxed{19.3°}$

REMARKS It's interesting that the final result depends only on the coefficient of static friction. Notice also how similar Equations (1) and (2) are to the equations developed in Examples 4.7 and 4.9. Recognizing such patterns is key to solving problems successfully.

QUESTION 4.12 How would a larger static friction coefficient affect the maximum angle?

EXERCISE 4.12 The ramp in Example 4.12 is roughed up and the experiment repeated. (a) What is the new coefficient of static friction if the maximum angle turns out to be 30.0°? (b) Find the maximum static friction force that acts on the block.

ANSWER (a) 0.577 (b) 12.2 N

EXAMPLE 4.13 The Sliding Hockey Puck

GOAL Apply the concept of kinetic friction.

PROBLEM The hockey puck in Figure 4.24, struck by a hockey stick, is given an initial speed of 20.0 m/s on a frozen pond. The puck remains on the ice and slides 1.20×10^2 m, slowing down steadily until it comes to rest. Determine the coefficient of kinetic friction between the puck and the ice.

STRATEGY The puck slows "steadily," which means that the acceleration is constant. Consequently, we can use the kinematic equation $v^2 = v_0^2 + 2a\Delta x$ to find a, the acceleration in the x-direction. The x- and y-components of Newton's second law then give two equations and two unknowns for the coefficient of kinetic friction, μ_k, and the normal force n.

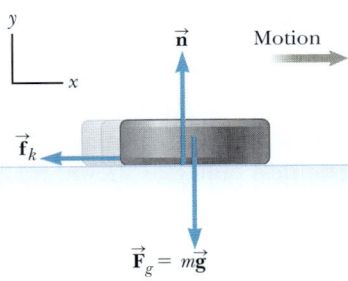

Figure 4.24 (Example 4.13) *After the puck is given an initial velocity to the right, the external forces acting on it are the force of gravity $\vec{F}_g$, the normal force $\vec{n}$, and the force of kinetic friction, $\vec{f}_k$.*

SOLUTION

Solve the time-independent kinematic equation for the acceleration a:

$$v^2 = v_0^2 + 2a\Delta x$$
$$a = \frac{v^2 - v_0^2}{2\Delta x}$$

Substitute $v = 0$, $v_0 = 20.0$ m/s, and $\Delta x = 1.20 \times 10^2$ m. Note the negative sign in the answer: $\vec{a}$ is opposite $\vec{v}$:

$$a = \frac{0 - (20.0 \text{ m/s})^2}{2(1.20 \times 10^2 \text{ m})} = -1.67 \text{ m/s}^2$$

Find the normal force from the y-component of the second law:

$$\sum F_y = n - F_g = n - mg = 0$$
$$n = mg$$

Obtain an expression for the force of kinetic friction, and substitute it into the x-component of the second law:

$$f_k = \mu_k n = \mu_k mg$$
$$ma = \sum F_x = -f_k = -\mu_k mg$$

Solve for μ_k and substitute values:

$$\mu_k = -\frac{a}{g} = \frac{1.67 \text{ m/s}^2}{9.80 \text{ m/s}^2} = \boxed{0.170}$$

REMARKS Notice how the problem breaks down into three parts: kinematics, Newton's second law in the y-direction, and then Newton's law in the x-direction.

QUESTION 4.13 How would the answer be affected if the puck were struck by an astronaut on a patch of ice on Mars, where the acceleration of gravity is $0.35g$, with all other given quantities remaining the same?

EXERCISE 4.13 An experimental rocket plane lands on skids on a dry lake bed. If it's traveling at 80.0 m/s when it touches down, how far does it slide before coming to rest? Assume the coefficient of kinetic friction between the skids and the lake bed is 0.600.

ANSWER 544 m

The System Approach

Two-body problems can often be treated as single objects and solved with a system approach. When the objects are rigidly connected—say, by a string of negligible mass that doesn't stretch—this approach can greatly simplify the analysis. When the two bodies are considered together, one or more of the forces end up becoming forces that are internal to the system, rather than external forces affecting each of the individual bodies. Both approaches will be used in Example 4.14.

EXAMPLE 4.14 | Connected Objects

GOAL Use both the general method and the system approach to solve a connected two-body problem involving gravity and friction.

PROBLEM (a) A block with mass $m_1 = 4.00$ kg and a ball with mass $m_2 = 7.00$ kg are connected by a light string that passes over a massless, frictionless pulley, as shown in Figure 4.25a. The coefficient of kinetic friction between the block and the surface is 0.300. Find the acceleration of the two objects and the tension in the string. (b) Check the answer for the acceleration by using the system approach.

STRATEGY Connected objects are handled by applying Newton's second law separately to each object. The force diagrams for the block and the ball are shown in Figure 4.25b, with the +x-direction to the right and the +y-direction upwards. The magnitude of the acceleration for both objects has the same value, $|a_1| = |a_2| = a$. The block with mass m_1 moves in the positive x-direction, and the ball with mass m_2 moves in the negative y-direction, so $a_1 = -a_2$. Using Newton's second law, we can develop two equations involving the unknowns T and a that can be solved simultaneously. In part (b), treat the two masses as a single object, with the gravity force on the ball increasing the combined object's speed and the friction force on the block retarding it. The tension forces then become internal and don't appear in the second law.

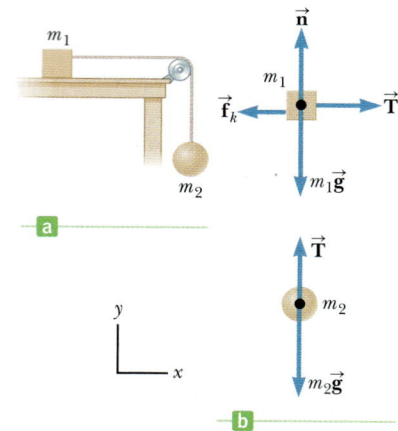

Figure 4.25 (Example 4.14) (a) Two objects connected by a light string that passes over a frictionless pulley. (b) Force diagrams for the objects.

SOLUTION

(a) Find the acceleration of the objects and the tension in the string.

Write the components of Newton's second law for the block of mass m_1:

$$\sum F_x = T - f_k = m_1 a_1 \quad \sum F_y = n - m_1 g = 0$$

The equation for the y-component gives $n = m_1 g$. Substitute this value for n and $f_k = \mu_k n$ into the equation for the x-component:

$$(1) \quad T - \mu_k m_1 g = m_1 a_1$$

Apply Newton's second law to the ball, recalling that $a_2 = -a_1$:

$$\sum F_y = T - m_2 g = m_2 a_2 = -m_2 a_1$$
$$(2) \quad T - m_2 g = -m_2 a_1$$

Subtract Equation (2) from Equation (1), eliminating T and leaving an equation that can be solved for a_1:

$$m_2 g - \mu_k m_1 g = (m_1 + m_2) a_1$$
$$a_1 = \frac{m_2 g - \mu_k m_1 g}{m_1 + m_2}$$

Substitute the given values to obtain the acceleration:

$$a_1 = \frac{(7.00 \text{ kg})(9.80 \text{ m/s}^2) - (0.300)(4.00 \text{ kg})(9.80 \text{ m/s}^2)}{(4.00 \text{ kg} + 7.00 \text{ kg})}$$
$$= \boxed{5.17 \text{ m/s}^2}$$

Substitute the value for a_1 into Equation (1) to find the tension T:

$$T = \boxed{32.4 \text{ N}}$$

(b) Find the acceleration using the system approach, where the system consists of the two blocks.

Apply Newton's second law to the system and solve for a:

$$(m_1 + m_2) a = m_2 g - \mu_k n = m_2 g - \mu_k m_1 g$$
$$a = \frac{m_2 g - \mu_k m_1 g}{m_1 + m_2}$$

REMARKS Although the system approach appears quick and easy, it can be applied only in special cases and can't give any information about the internal forces, such as the tension. To find the tension, you must consider the free-body diagram of one of the blocks separately as was done in part (a) of Example 4.14.

QUESTION 4.14 If mass m_2 is increased, does the acceleration of the system increase, decrease, or remain the same? Does the tension increase, decrease, or remain the same?

EXERCISE 4.14 What if an additional mass is attached to the ball in Example 4.14? How large must this mass be to increase the downward acceleration by 50%? Why isn't it possible to add enough mass to double the acceleration?

ANSWER 14.0 kg. Doubling the acceleration to 10.3 m/s² isn't possible simply by suspending more mass because all objects, regardless of their mass, fall freely at 9.8 m/s² near Earth's surface.

■ EXAMPLE 4.15 Two Blocks and a Cord

GOAL Apply Newton's second law and static friction to a two-body system.

PROBLEM A block of mass m = 5.00 kg rides on top of a second block of mass M = 10.0 kg. A person attaches a string to the bottom block and pulls the system horizontally across a frictionless surface, as in Figure 4.26a. Friction between the two blocks keeps the 5.00-kg block from slipping off. If the coefficient of static friction is 0.350, **(a)** what maximum force can be exerted by the string on the 10.0-kg block without causing the 5.00-kg block to slip? **(b)** Use the system approach to calculate the acceleration.

Figure 4.26 (a) (Example 4.15) (b) (Exercise 4.15)

STRATEGY Draw a free-body diagram for each block. The static friction force causes the top block to move horizontally, and the maximum such force corresponds to $f_s = \mu_s n$. That same static friction retards the motion of the bottom block. As long as the top block isn't slipping, the acceleration of both blocks is the same. Write Newton's second law for each block, and eliminate the acceleration a by substitution, solving for the tension T. Once the tension is known, use the system approach to calculate the acceleration.

SOLUTION

(a) Find the maximum force that can be exerted by the string.

Write the two components of Newton's second law for the top block:

x-component: $ma = \mu_s n_1$
y-component: $0 = n_1 - mg$

Solve the y-component for n_1, substitute the result into the x-component, and then solve for a:

$n_1 = mg \;\rightarrow\; ma = \mu_s mg \;\rightarrow\; a = \mu_s g$

Write the x-component of Newton's second law for the bottom block:

(1) $\quad Ma = -\mu_s mg + T$

Substitute the expression for $a = \mu_s g$ into Equation (1) and solve for the tension T:

$M\mu_s g = T - \mu_s mg \;\rightarrow\; T = (m + M)\mu_s g$

Now evaluate to get the answer:

$T = (5.00 \text{ kg} + 10.0 \text{ kg})(0.350)(9.80 \text{ m/s}^2) = \boxed{51.5 \text{ N}}$

(b) Use the system approach to calculate the acceleration.

Write the second law for the x-component of the force on the system:

$(m + M)a = T$

Solve for the acceleration and substitute values:

$a = \dfrac{T}{m + M} = \dfrac{51.5 \text{ N}}{5.00 \text{ kg} + 10.0 \text{ kg}} = \boxed{3.43 \text{ m/s}^2}$

(Continued)

REMARKS Notice that the y-component for the 10.0-kg block wasn't needed because there was no friction between that block and the underlying surface. It's also interesting to note that the top block was accelerated by the force of static friction. The system acceleration could also have been calculated with $a = \mu_s g$. Does the result agree with the answer found by the system approach?

QUESTION 4.15 What would happen if the tension force exceeded 51.5 N?

EXERCISE 4.15 Suppose instead the string is attached to the top block in Example 4.15 (see Fig. 4.26b). Find the maximum force that can be exerted by the string on the block without causing the top block to slip.

ANSWER 25.7 N

APPLYING PHYSICS 4.1 | Cars and Friction

Forces of friction are important in the analysis of the motion of cars and other wheeled vehicles. How do such forces both help and hinder the motion of a car?

EXPLANATION There are several types of friction forces to consider, the main ones being the force of friction between the tires and the road surface and the retarding force produced by air resistance.

Assuming the car is a four-wheel-drive vehicle of mass m, as each wheel turns to propel the car forward, the tire exerts a rearward force on the road. The reaction to this rearward force is a forward force $\vec{f}$ exerted by the road on the tire (Fig. 4.27). If we assume the same forward force $\vec{f}$ is exerted on each tire, the net forward force on the car is $4\vec{f}$, and the car's acceleration is therefore $\vec{a} = 4\vec{f}/m$.

The friction between the moving car's wheels and the road is normally static friction, unless the car is skidding.

When the car is in motion, we must also consider the force of air resistance, $\vec{R}$, which acts in the direction opposite the velocity of the car. The net force exerted

Figure 4.27 (Applying Physics 4.1) The horizontal forces acting on the car are the *forward* forces $\vec{f}$ exerted by the road on each tire and the force of air resistance $\vec{R}$, which acts *opposite* the car's velocity. (The car's tires exert a rearward force on the road, not shown in the diagram.)

on the car is therefore $4\vec{f} - \vec{R}$, so the car's acceleration is $\vec{a} = (4\vec{f} - \vec{R})/m$. At normal driving speeds, the magnitude of $\vec{R}$ is proportional to the first power of the speed, $R = bv$, where b is a constant, so the force of air resistance increases with increasing speed. When R is equal to $4f$, the acceleration is zero and the car moves at a constant speed. To minimize this resistive force, race cars often have very low profiles and streamlined contours. ∎

APPLYING PHYSICS 4.2 | Air Drag

Air resistance isn't always undesirable. What are some applications that depend on it?

EXPLANATION Consider a skydiver plunging through the air, as in Figure 4.28. Despite falling from a height of several thousand meters, she never exceeds a speed of around 120 miles per hour. This is because, aside from the downward force of gravity $m\vec{g}$, there is also an upward force of air resistance, $\vec{R}$. Before she reaches a final constant speed, the magnitude of $\vec{R}$ is less than her weight. As her downward speed increases, the force of air resistance increases. The vector sum of the force of gravity and the force of air resistance gives a total force that decreases with time, so her acceleration decreases. Once the two forces balance each other, the net force is zero, so the acceleration is zero, and she reaches a **terminal speed**.

Terminal speed is generally still high enough to be fatal on impact, although there have been amazing stories of survival. In one case, a man fell flat on his back in a freshly plowed field and survived. (He did, however, break

Figure 4.28 (Applying Physics 4.2)

virtually every bone in his body.) In another case, a flight attendant survived a fall from thirty thousand feet into a snowbank. In neither case would the person have had any chance of surviving without the effects of air drag.

Parachutes and paragliders create a much larger drag force due to their large area and can reduce the terminal speed to a few meters per second. Some sports enthusiasts have even developed special suits with wings, allowing a long glide to the ground. In each case, a larger cross-sectional area intercepts more air, creating greater air drag, so the terminal speed is lower.

Air drag is also important in space travel. Without it, returning to Earth would require a considerable amount of fuel. Air drag helps slow capsules and spaceships, and aerocapture techniques have been proposed for trips to other planets. These techniques significantly reduce fuel requirements by using air drag to reduce the speed of the spacecraft. ■

SUMMARY

4.1 Forces
There are four known fundamental forces of nature: (1) the strong nuclear force between subatomic particles; (2) the electromagnetic forces between electric charges; (3) the weak nuclear forces, which arise in certain radioactive decay processes; and (4) the gravitational force between objects. These are collectively called field forces. Classical physics deals only with the gravitational and electromagnetic forces.

Forces such as friction or the force of a bat hitting a ball are called contact forces. On a more fundamental level, contact forces have an electromagnetic nature.

4.2 Newton's First Law
Newton's first law states that an object moves at constant velocity unless acted on by a force.

The tendency for an object to maintain its original state of motion is called **inertia**. **Mass** is the physical quantity that measures the resistance of an object to changes in its velocity.

4.3 Newton's Second Law
Newton's second law states that the acceleration of an object is directly proportional to the net force acting on it and inversely proportional to its mass. The net force acting on an object equals the product of its mass and acceleration:

A net force $\sum \vec{F}$ acting on a mass m creates an acceleration proportional to the force and inversely proportional to the mass.

$$\sum \vec{F} = m\vec{a} \quad [4.1]$$

Newton's universal law of gravitation is

$$F_g = G\frac{m_1 m_2}{r^2} \quad [4.5]$$

The **weight** w of an object is the magnitude of the force of gravity exerted on that object and is given by

$$w = mg \quad [4.6]$$

where $g = F_g/m$ is the acceleration of gravity.

Solving problems with Newton's second law involves finding all the forces acting on a system and writing Equation 4.1 for the x-component and y-component separately.

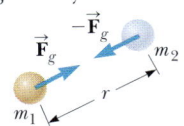

The gravity force between any two objects is proportional to their masses and inversely proportional to the square of the distance between them.

These two equations are then solved algebraically for the unknown quantities.

4.4 Newton's Third Law
Newton's third law states that if two objects interact, the force $\vec{F}_{12}$ exerted by object 1 on object 2 is equal in magnitude and opposite in direction to the force $\vec{F}_{21}$ exerted by object 2 on object 1:

$$\vec{F}_{12} = -\vec{F}_{21}$$

Newton's third law in action: the hammer drives the nail forward into the wall, and the nail slows the head of the hammer down to rest with an equal and opposite force.

An isolated force can never occur in nature.

4.5 Applications of Newton's Laws
An **object in equilibrium** has no net external force acting on it, and the second law, in component form, implies that $\sum F_x = 0$ and $\sum F_y = 0$ for such an object. These two equations are useful for solving problems in statics, in which the object is at rest or moving at constant velocity.

An object under acceleration requires the same two equations, but with the acceleration terms included: $\sum F_x = ma_x$ and $\sum F_y = ma_y$. When the acceleration is constant, the equations of kinematics can supplement Newton's second law.

4.6 Forces of Friction
The magnitude of the maximum force of static friction, $f_{s,\max}$, between an object and a surface is proportional to the magnitude of the normal force acting on the object. This maximum force occurs when the object is on the verge of slipping. In general,

$$f_s \leq \mu_s n \quad [4.11]$$

where μ_s is the **coefficient of static friction**. When an object slides over a surface, the direction of the force of kinetic friction, $\vec{f}_k$, on the object is opposite the direction of the motion of the object relative to the surface and proportional to the magnitude of the normal force. The magnitude of $\vec{f}_k$ is

$$f_k = \mu_k n \quad [4.12]$$

where μ_k is the **coefficient of kinetic friction**. In general, $\mu_k < \mu_s$.

Solving problems that involve friction is a matter of using these two friction forces in Newton's second law. The static friction force must be handled carefully because it refers to a maximum force, which is not always called upon in a given problem.

CHAPTER 4 | The Laws of Motion

WARM-UP EXERCISES

The warm-up exercises in this chapter may be assigned online in Enhanced WebAssign.

1. **Physics Review** A hockey player strikes a puck, giving it an initial velocity of 10.0 m/s in the positive x-direction. The puck slows uniformly to 6.00 m/s when it has traveled 40.0 m. (a) What is the puck's acceleration? (b) At what velocity is it traveling after 2.00 s? (c) How long does it take to travel 40.0 m? (See Section 2.5.)

2. Four forces act on an object, given by $\vec{A} = 40.0$ N east, $\vec{B} = 50.0$ N north, $\vec{C} = 70.0$ N west, and $\vec{D} = 90.0$ N south. (a) What is the magnitude of the net force on the object? (b) What is the direction of the force? (See Sections 3.2 and 4.3.)

3. A force of 30.0 N is applied in the positive x-direction to a block of mass 8.00 kg, at rest on a frictionless surface. (a) What is the block's acceleration? (b) How fast is it going after 6.00 s? (See Sections 2.5 and 4.3.)

4. What would be the acceleration of gravity at the surface of a world with twice Earth's mass and twice its radius? (See Section 4.3.)

5. Two monkeys are holding onto a single vine of negligible mass that hangs vertically from a tree, with one monkey a few meters higher than the other. The upper monkey has mass 20.0 kg and the lower monkey mass 10.0 kg. What is the ratio of the tension in the vine above the upper monkey to the tension in the vine between the two monkeys? (See Section 4.5.)

6. Two identical strings making an angle of $\theta = 30.0°$ with respect to the vertical support a block of mass $m = 15.0$ kg (Figure WU4.6). What is the tension in each of the strings? (See Section 4.5.)

Figure WU4.6

7. Calculate the normal force on a 15.0 kg block in the following circumstances: (a) The block is resting on a level surface. (b) The block is resting on a surface tilted up at a 30.0° angle with respect to the horizontal. (c) The block is resting on the floor of an elevator that is accelerating upwards at 3.00 m/s². (d) The block is on a level surface and a force of 125 N is exerted on it at an angle of 30.0° above the horizontal. (See Section 4.5.)

8. A horizontal force of 95.0 N is applied to a 60.0-kg crate on a rough, level surface. If the crate accelerates at 1.20 m/s², what is the magnitude of the force of kinetic friction acting on the crate? (See Section 4.5.)

9. A car of mass 875 kg is traveling 30.0 m/s when the driver applies the brakes, which lock the wheels. The car skids for 5.60 s in the positive x-direction before coming to rest. (a) What is the car's acceleration? (b) What magnitude force acted on the car during this time? (c) How far did the car travel? (See Sections 2.5 and 4.5.)

10. A block of mass 12.0 kg is sliding at an initial velocity of 8.00 m/s in the positive x-direction. The surface has a coefficient of kinetic friction of 0.300. (a) What is the force of kinetic friction acting on the block? (b) What is the block's acceleration? (c) How far will it slide before coming to rest? (See Sections 2.5 and 4.6.)

11. A man exerts a horizontal force of 112 N on a refrigerator of mass 42.0 kg. If the refrigerator doesn't move, what is the minimum coefficient of static friction between the refrigerator and the floor? (See Section 4.6.)

12. An Atwood's machine (Figure 4.20) consists of two masses, one of mass 3.00 kg and the other of mass 8.00 kg. When released from rest, what is the acceleration of the system? (See Section 4.6.)

13. A block of mass $m_1 = 16$ kg is on a frictionless table to the left of a second block of mass $m_2 = 24$ kg, attached by a horizontal string (Figure WU4.13). If a horizontal force of 120 N is exerted on the block m_2 in the positive x-direction, (a) use the system approach to find the acceleration of the two blocks. (b) What is the tension in the string connecting the blocks? (See Section 4.6.)

Figure WU4.13

CONCEPTUAL QUESTIONS

The conceptual questions in this chapter may be assigned online in Enhanced WebAssign.

1. A passenger sitting in the rear of a bus claims that she was injured as the driver slammed on the brakes, causing a suitcase to come flying toward her from the front of the bus. If you were the judge in this case, what disposition would you make? Explain.

2. A space explorer is moving through space far from any planet or star. He notices a large rock, taken as a specimen from an alien planet, floating around the cabin of the ship. Should he push it gently, or should he kick it toward the storage compartment? Explain.

3. (a) If gold were sold by weight, would you rather buy it in Denver or in Death Valley? (b) If it were sold by mass, in which of the two locations would you prefer to buy it? Why?

4. If you push on a heavy box that is at rest, you must exert some force to start its motion. Once the box is sliding, why does a smaller force maintain its motion?

5. A ball is held in a person's hand. (a) Identify all the external forces acting on the ball and the reaction to each. (b) If the ball is dropped, what force is exerted on it while it is falling? Identify the reaction force in this case. (Neglect air resistance.)

6. A weight lifter stands on a bathroom scale. (a) As she pumps a barbell up and down, what happens to the reading on the scale? (b) Suppose she is strong enough to actually *throw* the barbell upward. How does the reading on the scale vary now?

7. (a) What force causes an automobile to move? (b) A propeller-driven airplane? (c) A rowboat?

8. If only one force acts on an object, can it be in equilibrium? Explain.

9. In the motion picture *It Happened One Night* (Columbia Pictures, 1934), Clark Gable is standing inside a stationary bus in front of Claudette Colbert, who is seated. The bus suddenly starts moving forward and Clark falls into Claudette's lap. Why did this happen?

10. Analyze the motion of a rock dropped in water in terms of its speed and acceleration as it falls. Assume a resistive force is acting on the rock that increases as the velocity of the rock increases.

11. Identify the action–reaction pairs in the following situations: (a) a man takes a step, (b) a snowball hits a girl in the back, (c) a baseball player catches a ball, (d) a gust of wind strikes a window.

12. Draw a free-body diagram for each of the following objects: (a) a projectile in motion in the presence of air resistance, (b) a rocket leaving the launch pad with its engines operating, (c) an athlete running along a horizontal track.

13. In a tug-of-war between two athletes, each pulls on the rope with a force of 200 N. What is the tension in the rope? If the rope doesn't move, what horizontal force does each athlete exert against the ground?

14. Suppose you are driving a car at a high speed. Why should you avoid slamming on your brakes when you want to stop in the shortest possible distance? (Newer cars have antilock brakes that avoid this problem.)

15. As a block slides down a frictionless incline, which of the following statements is true? (a) Both its speed and acceleration increase. (b) Its speed and acceleration remain constant. (c) Its speed increases and its acceleration remains constant. (d) Both its speed and acceleration decrease. (e) Its speed increases and its acceleration decreases.

16. A crate remains stationary after it has been placed on a ramp inclined at an angle with the horizontal. Which of the following statements must be true about the magnitude of the frictional force that acts on the crate? (a) It is larger than the weight of the crate. (b) It is at least equal to the weight of the crate. (c) It is equal to $\mu_s n$. (d) It is greater than the component of the gravitational force acting down the ramp. (e) It is equal to the component of the gravitational force acting down the ramp.

17. In the photo on page 91, a locomotive has broken through the wall of a train station. During the collision, what can be said about the force exerted by the locomotive on the wall? (a) The force exerted by the locomotive on the wall was larger than the force the wall could exert on the locomotive. (b) The force exerted by the locomotive on the wall was the same in magnitude as the force exerted by the wall on the locomotive. (c) The force exerted by the locomotive on the wall was less than the force exerted by the wall on the locomotive. (d) The wall cannot be said to "exert" a force; after all, it broke.

18. If an object is in equilibrium, which of the following statements is not true? (a) The speed of the object remains constant. (b) The acceleration of the object is zero. (c) The net force acting on the object is zero. (d) The object must be at rest. (e) The velocity is constant.

19. A truck loaded with sand accelerates along a highway. The driving force on the truck remains constant. What happens to the acceleration of the truck as its trailer leaks sand at a constant rate through a hole in its bottom? (a) It decreases at a steady rate. (b) It increases at a steady rate. (c) It increases and then decreases. (d) It decreases and then increases. (e) It remains constant.

20. A large crate of mass m is placed on the back of a truck but not tied down. As the truck accelerates forward with an acceleration a, the crate remains at rest relative to the truck. What force causes the crate to accelerate forward? (a) the normal force (b) the force of gravity (c) the force of friction between the crate and the floor of the truck (d) the "ma" force (e) none of these

21. Which of the following statements are true? (a) An astronaut's weight is the same on the Moon as on Earth. (b) An astronaut's mass is the same on the International Space Station as it is on Earth. (c) Earth's gravity has no effect on astronauts inside the International Space Station. (d) An astronaut's mass is greater on Earth than on the Moon. (e) None of these statements are true.

PROBLEMS

WebAssign The problems in this chapter may be assigned online in Enhanced WebAssign.

1. denotes straightforward problem; 2. denotes intermediate problem;
3. denotes challenging problem
1. denotes full solution available in *Student Solutions Manual/Study Guide*
1. denotes problems most often assigned in Enhanced WebAssign

BIO denotes biomedical problems
GP denotes guided problems
M denotes Master It tutorial available in Enhanced WebAssign
Q|C denotes asking for quantitative and conceptual reasoning
S denotes symbolic reasoning problem
W denotes Watch It video solution available in Enhanced WebAssign

4.1 Forces
4.2 Newton's First Law
4.3 Newton's Second Law
4.4 Newton's Third Law

1. The heaviest invertebrate is the giant squid, which is estimated to have a weight of about 2 tons spread out over its length of 70 feet. What is its weight in newtons?

2. A football punter accelerates a football from rest to a speed of 10 m/s during the time in which his toe is in contact with the ball (about 0.20 s). If the football has a mass of 0.50 kg, what average force does the punter exert on the ball?

3. A 6.0-kg object undergoes an acceleration of 2.0 m/s². (a) What is the magnitude of the resultant force acting on it? (b) If this same force is applied to a 4.0-kg object, what acceleration is produced?

4. **Q|C** One or more external forces are exerted on each object enclosed in a dashed box shown in Figure 4.2. Identify the reaction to each of these forces.

5. A bag of sugar weighs 5.00 lb on Earth. What would it weigh in newtons on the Moon, where the free-fall acceleration is one-sixth that on Earth? Repeat for Jupiter, where g is 2.64 times that on Earth. Find the mass of the bag of sugar in kilograms at each of the three locations.

6. A freight train has a mass of 1.5×10^7 kg. If the locomotive can exert a constant pull of 7.5×10^5 N, how long does it take to increase the speed of the train from rest to 80 km/h?

7. A 75-kg man standing on a scale in an elevator notes that as the elevator rises, the scale reads 825 N. What is the acceleration of the elevator?

8. **Q|C** Consider a solid metal sphere (S) a few centimeters in diameter and a feather (F). For each quantity in the list that follows, indicate whether the quantity is the same, greater, or lesser in the case of S or in that of F. Explain in each case why you gave the answer you did. Here is the list: (a) the gravitational force, (b) the time it will take to fall a given distance in air, (c) the time it will take to fall a given distance in vacuum, (d) the total force on the object when falling in vacuum.

9. **BIO** As a fish jumps vertically out of the water, assume that only two significant forces act on it: an upward force F exerted by the tail fin and the downward force due to gravity. A record Chinook salmon has a length of 1.50 m and a mass of 61.0 kg. If this fish is moving upward at 3.00 m/s as its head first breaks the surface and has an upward speed of 6.00 m/s after two-thirds of its length has left the surface, assume constant acceleration and determine (a) the salmon's acceleration and (b) the magnitude of the force F during this interval.

10. **W** A 5.0-g bullet leaves the muzzle of a rifle with a speed of 320 m/s. What force (assumed constant) is exerted on the bullet while it is traveling down the 0.82-m-long barrel of the rifle?

11. A boat moves through the water with two forces acting on it. One is a 2 000-N forward push by the water on the propeller, and the other is a 1 800-N resistive force due to the water around the bow. (a) What is the acceleration of the 1 000-kg boat? (b) If it starts from rest, how far will the boat move in 10.0 s? (c) What will its velocity be at the end of that time?

12. Two forces are applied to a car in an effort to move it, as shown in Figure P4.12. (a) What is the resultant vector of these two forces? (b) If the car has a mass of 3 000 kg, what acceleration does it have? Ignore friction.

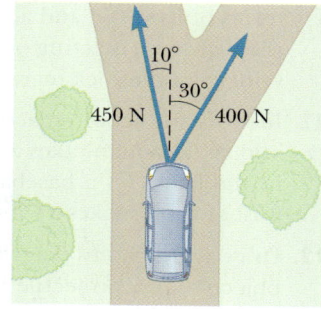

Figure P4.12

13. A 970-kg car starts from rest on a horizontal roadway and accelerates eastward for 5.00 s when it reaches a speed of 25.0 m/s. What is the average force exerted on the car during this time?

14. **S** An object of mass m is dropped from the roof of a building of height h. While the object is falling, a wind blowing parallel to the face of the building exerts a constant horizontal force F on the object. (a) How long does it take the object to strike the ground? Express the time t in terms of g and h. (b) Find an expression in terms of m and F for the acceleration a_x of the object in the horizontal direction (taken as the positive x-direction). (c) How far is the object displaced horizontally before hitting the ground? Answer in terms of m, g, F, and h. (d) Find the magnitude of the object's acceleration while it is falling, using the variables F, m, and g.

15. After falling from rest from a height of 30 m, a 0.50-kg ball rebounds upward, reaching a height of 20 m. If the contact between ball and ground lasted 2.0 ms, what average force was exerted on the ball?

16. **M** The force exerted by the wind on the sails of a sailboat is 390 N north. The water exerts a force of 180 N east. If the boat (including its crew) has a mass of 270 kg, what are the magnitude and direction of its acceleration?

4.5 Applications of Newton's Laws

17. **Q|C** (a) Find the tension in each cable supporting the 600-N cat burglar in Figure P4.17. (b) Suppose the horizontal cable were reattached higher up on the wall. Would the tension in the other cable increase, decrease, or stay the same? Why?

18. **BIO** A certain orthodontist uses a wire brace to align a patient's crooked tooth as in Figure P4.18. The tension in the wire is adjusted to have a magnitude of 18.0 N. Find the magnitude of the net force exerted by the wire on the crooked tooth.

Figure P4.17

Figure P4.18

19. A 150-N bird feeder is supported by three cables as shown in Figure P4.19. Find the tension in each cable.

20. **BIO** The leg and cast in Figure P4.20 weigh 220 N (w_1). Determine the weight w_2 and the angle α needed so that no force is exerted on the hip joint by the leg plus the cast.

Figure P4.19

Figure P4.20

21. Two blocks each of mass $m = 3.50$ kg are fastened to the top of an elevator as in Figure P4.21. (a) If the elevator has an upward acceleration $a = 1.60$ m/s², find the tensions T_1 and T_2 in the upper and lower strings. (b) If the strings can withstand a maximum tension of 85.0 N, what maximum acceleration can the elevator have before the upper string breaks?

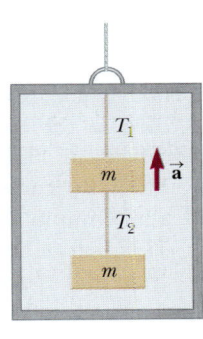

Figure P4.21
(Problems 21 and 22)

22. **S** Two blocks each of mass m are fastened to the top of an elevator as in Figure P4.21. The elevator has an upward acceleration a. The strings have negligible mass. (a) Find the tensions T_1 and T_2 in the upper and lower strings in terms of m, a, and g. (b) Compare the two tensions and determine which string would break first if a is made sufficiently large. (c) What are the tensions if the cable supporting the elevator breaks?

23. The distance between two telephone poles is 50.0 m. When a 1.00-kg bird lands on the telephone wire midway between the poles, the wire sags 0.200 m. Draw a free-body diagram of the bird. How much tension does the bird produce in the wire? Ignore the weight of the wire.

24. The systems shown in Figure P4.24 are in equilibrium. If the spring scales are calibrated in newtons, what do they read? Ignore the masses of the pulleys and strings and assume the pulleys and the incline in Figure P4.24d are frictionless.

Figure P4.24

25. **W** A 5.0-kg bucket of water is raised from a well by a rope. If the upward acceleration of the bucket is 3.0 m/s², find the force exerted by the rope on the bucket.

26. A crate of mass $m = 32$ kg rides on the bed of a truck attached by a cord to the back of the cab as in Figure P4.26. The cord can withstand a maximum tension of 68 N before breaking. Neglecting friction between the crate and truck bed, find the maximum acceleration the truck can have before the cord breaks.

Figure P4.26

27. **S** Two blocks of masses m and $2m$ are held in equilibrium on a frictionless incline as in Figure P4.27. In terms of m and θ, find (a) the magnitude of the tension T_1 in the upper cord and (b) the magnitude of the tension T_2 in the lower cord connecting the two blocks.

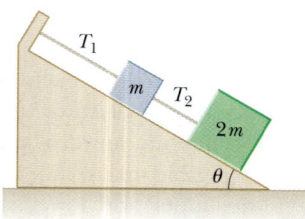

Figure P4.27

28. Two packing crates of masses 10.0 kg and 5.00 kg are connected by a light string that passes over a frictionless pulley as in Figure P4.28. The 5.00-kg crate lies on a smooth incline of angle 40.0°. Find (a) the acceleration of the 5.00-kg crate and (b) the tension in the string.

Figure P4.28

29. **M** Assume the three blocks portrayed in Figure P4.29 move on a frictionless surface and a 42-N force acts as shown on the 3.0-kg block. Determine (a) the acceleration given this system, (b) the tension in the cord connecting the 3.0-kg and the 1.0-kg blocks, and (c) the force exerted by the 1.0-kg block on the 2.0-kg block.

Figure P4.29

30. A block of mass $m = 5.8$ kg is pulled up a $\theta = 25°$ incline as in Figure P4.30 with a force of magnitude $F = 32$ N. (a) Find the acceleration of the block if the incline is frictionless. (b) Find the acceleration of the block if the coefficient of kinetic friction between the block and incline is 0.10.

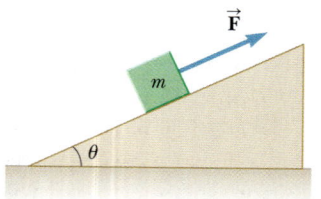

Figure P4.30

31. **BIO** A setup similar to the one shown in Figure P4.31 is often used in hospitals to support and apply a traction force to an injured leg. (a) Determine the force of tension in the rope supporting the leg. (b) What is the traction force exerted on the leg? Assume the traction force is horizontal.

Figure P4.31

32. **GP** Two blocks of masses m_1 and m_2 ($m_1 > m_2$) are placed on a frictionless table in contact with each other. A horizontal force of magnitude F is applied to the block of mass m_1 in Figure P4.32. (a) If P is the magnitude of the contact force between the blocks, draw the free-body diagrams for each block. (b) What is the net force on the system consisting of both blocks? (c) What is the net force acting on m_1? (d) What is the net force acting on m_2? (e) Write the x-component of Newton's second law for each block. (f) Solve the resulting system of two equations and two unknowns, expressing the acceleration a and contact force P in terms of the masses and force. (g) How would the answers change if the force had been applied to m_2 instead? (*Hint:* use symmetry; don't calculate!) Is the contact force larger, smaller, or the same in this case? Why?

Figure P4.32

33. A 276-kg glider is being pulled by a 1 950-kg jet along a horizontal runway with an acceleration of $\vec{a} = 2.20$ m/s^2 to the right as in Figure P4.33. Find (a) the thrust provided by the jet's engines and (b) the magnitude of the tension in the cable connecting the jet and glider.

Figure P4.33

34. **Q|C** In Figure P4.34, the light, taut, unstretchable cord B joins block 1 and the larger-mass block 2. Cord A exerts a force on block 1 to make it accelerate forward. (a) How does the magnitude of the force exerted by cord A on block 1 compare with the magnitude of the force exerted by cord B on block 2? (b) How does the

Figure P4.34

acceleration of block 1 compare with the acceleration of block 2? (c) Does cord B exert a force on block 1? Explain your answer.

35. **Q|C** (a) An elevator of mass m moving upward has two forces acting on it: the upward force of tension in the cable and the downward force due to gravity. When the elevator is accelerating upward, which is greater, T or w? (b) When the elevator is moving at a constant velocity upward, which is greater, T or w? (c) When the elevator is moving upward, but the acceleration is downward, which is greater, T or w? (d) Let the elevator have a mass of 1 500 kg and an upward acceleration of 2.5 m/s². Find T. Is your answer consistent with the answer to part (a)? (e) The elevator of part (d) now moves with a constant upward velocity of 10 m/s. Find T. Is your answer consistent with your answer to part (b)? (f) Having initially moved upward with a constant velocity, the elevator begins to accelerate downward at 1.50 m/s². Find T. Is your answer consistent with your answer to part (c)?

36. **W** An object with mass m_1 = 5.00 kg rests on a frictionless horizontal table and is connected to a cable that passes over a pulley and is then fastened to a hanging object with mass m_2 = 10.0 kg, as shown in Figure P4.36. Find (a) the acceleration of each object and (b) the tension in the cable.

Figure P4.36
(Problems 36, 40, and 45)

37. A 1 000-kg car is pulling a 300-kg trailer. Together, the car and trailer have an acceleration of 2.15 m/s² in the positive x-direction. Neglecting frictional forces on the trailer, determine (a) the net force on the car, (b) the net force on the trailer, (c) the magnitude and direction of the force exerted by the trailer on the car, and (d) the resultant force exerted by the car on the road.

38. Two objects with masses of 3.00 kg and 5.00 kg are connected by a light string that passes over a frictionless pulley, as in Figure P4.38. Determine (a) the tension in the string, (b) the acceleration of each object, and (c) the distance each object will move in the first second of motion if both objects start from rest.

Figure P4.38

4.6 Forces of Friction

39. A dockworker loading crates on a ship finds that a 20-kg crate, initially at rest on a horizontal surface, requires a 75-N horizontal force to set it in motion. However, after the crate is in motion, a horizontal force of 60 N is required to keep it moving with a constant speed. Find the coefficients of static and kinetic friction between crate and floor.

40. In Figure P4.36, m_1 = 10 kg and m_2 = 4.0 kg. The coefficient of static friction between m_1 and the horizontal surface is 0.50, and the coefficient of kinetic friction is 0.30. (a) If the system is released from rest, what will its acceleration be? (b) If the system is set in motion with m_2 moving downward, what will be the acceleration of the system?

41. A 1 000-N crate is being pushed across a level floor at a constant speed by a force $\vec{F}$ of 300 N at an angle of 20.0° below the horizontal, as shown in Figure P4.41a. (a) What is the coefficient of kinetic friction between the crate and the floor? (b) If the 300-N force is instead pulling the block at an angle of 20.0° above the horizontal, as shown in Figure P4.41b, what will be the acceleration of the crate? Assume that the coefficient of friction is the same as that found in part (a).

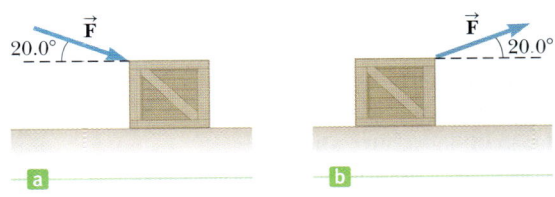

Figure P4.41

42. **Q|C** **S** A block of mass $3m$ is placed on a frictionless horizontal surface, and a second block of mass m is placed on top of the first block. The surfaces of the blocks are rough.

Figure P4.42

A constant force of magnitude F is applied to the first block as in Figure P4.42. (a) Construct free-body diagrams for each block. (b) Identify the horizontal force that causes the block of mass m to accelerate. (c) Assume that the upper block does not slip on the lower block, and find the acceleration of each block in terms of m and F.

43. Consider a large truck carrying a heavy load, such as steel beams. A significant hazard for the driver is that the load may slide forward, crushing the cab, if the truck stops suddenly in an accident or even in braking. Assume, for example, a 10 000-kg load sits on the flatbed of a 20 000-kg truck moving at 12.0 m/s. Assume the load is not tied down to the truck and has a coefficient of static friction of 0.500 with the truck bed. (a) Calculate the minimum stopping distance for which the load will not slide forward relative to the truck. (b) Is any piece of data unnecessary for the solution?

44. A crate of mass 45.0 kg is being transported on the flatbed of a pickup truck. The coefficient of static friction

between the crate and the truck's flatbed is 0.350, and the coefficient of kinetic friction is 0.320. (a) The truck accelerates forward on level ground. What is the maximum acceleration the truck can have so that the crate does not slide relative to the truck's flatbed? (b) The truck barely exceeds this acceleration and then moves with constant acceleration, with the crate sliding along its bed. What is the acceleration of the crate relative to the ground?

45. **M** Objects with masses $m_1 = 10.0$ kg and $m_2 = 5.00$ kg are connected by a light string that passes over a frictionless pulley as in Figure P4.36. If, when the system starts from rest, m_2 falls 1.00 m in 1.20 s, determine the coefficient of kinetic friction between m_1 and the table.

46. **S** A hockey puck struck by a hockey stick is given an initial speed v_0 in the positive x-direction. The coefficient of kinetic friction between the ice and the puck is μ_k. (a) Obtain an expression for the acceleration of the puck. (b) Use the result of part (a) to obtain an expression for the distance d the puck slides. The answer should be in terms of the variables v_0, μ_k, and g only.

47. **W** The coefficient of static friction between the 3.00-kg crate and the 35.0° incline of Figure P4.47 is 0.300. What minimum force $\vec{F}$ must be applied to the crate perpendicular to the incline to prevent the crate from sliding down the incline?

Figure P4.47

48. A student decides to move a box of books into her dormitory room by pulling on a rope attached to the box. She pulls with a force of 80.0 N at an angle of 25.0° above the horizontal. The box has a mass of 25.0 kg, and the coefficient of kinetic friction between box and floor is 0.300. (a) Find the acceleration of the box. (b) The student now starts moving the box up a 10.0° incline, keeping her 80.0 N force directed at 25.0° above the line of the incline. If the coefficient of friction is unchanged, what is the new acceleration of the box?

49. **Q|C** An object falling under the pull of gravity is acted upon by a frictional force of air resistance. The magnitude of this force is approximately proportional to the speed of the object, which can be written as $f = bv$. Assume $b = 15$ kg/s and $m = 50$ kg. (a) What is the terminal speed the object reaches while falling? (b) Does your answer to part (a) depend on the initial speed of the object? Explain.

50. A car is traveling at 50.0 km/h on a flat highway. (a) If the coefficient of friction between road and tires on a rainy day is 0.100, what is the minimum distance in which the car will stop? (b) What is the stopping distance when the surface is dry and the coefficient of friction is 0.600?

51. A 3.00-kg block starts from rest at the top of a 30.0° incline and slides 2.00 m down the incline in 1.50 s. Find (a) the acceleration of the block, (b) the coefficient of kinetic friction between the block and the incline, (c) the frictional force acting on the block, and (d) the speed of the block after it has slid 2.00 m.

52. A 15.0-lb block rests on a horizontal floor. (a) What force does the floor exert on the block? (b) A rope is tied to the block and is run vertically over a pulley. The other end is attached to a free-hanging 10.0-lb object. What now is the force exerted by the floor on the 15.0-lb block? (c) If the 10.0-lb object in part (b) is replaced with a 20.0-lb object, what is the force exerted by the floor on the 15.0-lb block?

53. To meet a U.S. Postal Service requirement, employees' footwear must have a coefficient of static friction of 0.500 or more on a specified tile surface. A typical athletic shoe has a coefficient of 0.800. In an emergency, what is the minimum time interval in which a person starting from rest can move 3.00 m on the tile surface if she is wearing (a) footwear meeting the Postal Service minimum and (b) a typical athletic shoe?

54. Objects of masses $m_1 = 4.00$ kg and $m_2 = 9.00$ kg are connected by a light string that passes over a frictionless pulley as in Figure P4.54. The object m_1 is held at rest on the floor, and m_2 rests on a fixed incline of $\theta = 40.0°$. The objects are released from rest, and m_2 slides 1.00 m down the incline in 4.00 s. Determine (a) the acceleration of each object, (b) the tension in the string, and (c) the coefficient of kinetic friction between m_2 and the incline.

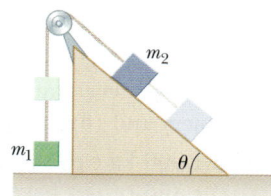

Figure P4.54

55. **BIO** The person in Figure P4.55 weighs 170 lb. Each crutch makes an angle of 22.0° with the vertical (as seen from the front). Half of the person's weight is supported by the crutches, the other half by the vertical forces exerted by the ground on his feet. Assuming he is at rest and the force exerted by the ground on the crutches acts along the crutches, determine (a) the smallest possible coefficient of friction between crutches and ground and (b) the magnitude of the compression force supported by each crutch.

Figure P4.55

Additional Problems

56. As a protest against the umpire's calls, a baseball pitcher throws a ball straight up into the air at a speed of 20.0 m/s. In the process, he moves his hand through a distance of 1.50 m. If the ball has a mass of 0.150 kg, find the force he exerts on the ball to give it this upward speed.

57. Q|C Three objects are connected on a table as shown in Figure P4.57. The coefficient of kinetic friction between the block of mass m_2 and the table is 0.350. The objects have masses of $m_1 = 4.00$ kg, $m_2 = 1.00$ kg, and $m_3 = 2.00$ kg as shown, and the pulleys are frictionless. (a) Draw a diagram of the forces on each object. (b) Determine the acceleration of each object, including its direction. (c) Determine the tensions in the two cords. (d) If the tabletop were smooth, would the tensions increase, decrease, or remain the same? Explain.

Figure P4.57

58. The force exerted by the wind on a sailboat is approximately perpendicular to the sail and proportional to the component of the wind velocity perpendicular to the sail. For the 800-kg sailboat shown in Figure P4.58, the force exerted by the wind on the sailboat is

$$F_{sail} = \left(550 \frac{N}{m/s}\right) v_{wind\perp}$$

Water exerts a force along the keel (bottom) of the boat that prevents it from moving sideways, as shown in the figure. Once the boat starts moving forward, water also exerts a drag force backwards on the boat, opposing the forward motion. If a 17-knot wind (1 knot = 0.514 m/s) is blowing to the east, what is the initial acceleration of the sailboat?

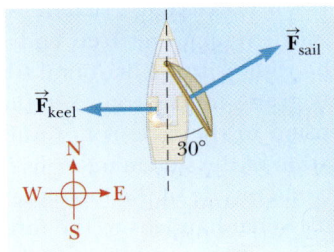

Figure P4.58

59. (a) What is the resultant force exerted by the two cables supporting the traffic light in Figure P4.59? (b) What is the weight of the light?

60. (a) What is the minimum force of friction required to hold the system of Figure P4.60 in equilibrium? (b) What coefficient of static friction between the 100-N block and the table ensures equilibrium? (c) If the coefficient of kinetic friction between the 100-N block and the table is 0.250, what hanging weight should replace the 50.0-N weight to allow the system to move at a constant speed once it is set in motion?

Figure P4.59

Figure P4.60

61. A boy coasts down a hill on a sled, reaching a level surface at the bottom with a speed of 7.0 m/s. If the coefficient of friction between the sled's runners and the snow is 0.050 and the boy and sled together weigh 600 N, how far does the sled travel on the level surface before coming to rest?

62. A woman at an airport is towing her 20.0-kg suitcase at constant speed by pulling on a strap at an angle θ above the horizontal (Fig. 4.62). She pulls on the strap with a 35.0-N force, and the friction force on the suitcase is 20.0 N. (a) Draw a free-body diagram of the suitcase. (b) What angle does the strap make with the horizontal? (c) What is the magnitude of the normal force that the ground exerts on the suitcase?

Figure P4.62

63. Q|C A box rests on the back of a truck. The coefficient of static friction between the box and the bed of the truck is 0.300. (a) When the truck accelerates forward, what force accelerates the box? (b) Find the maximum acceleration the truck can have before the box slides.

64. Three objects are connected by light strings as shown in Figure P4.64. The string connecting the 4.00-kg object and the 5.00-kg object passes over a light frictionless pulley. Determine (a) the acceleration of each object and (b) the tension in the two strings.

Figure P4.64

65. M A frictionless plane is 10.0 m long and inclined at 35.0°. A sled starts at the bottom with an initial speed of 5.00 m/s up the incline. When the sled reaches the point at which it momentarily stops, a second sled is released from the top of the incline with an initial speed v_i. Both sleds reach the bottom of the incline at the same moment. (a) Determine the distance that the first sled traveled up the incline. (b) Determine the initial speed of the second sled.

66. A high diver of mass 70.0 kg steps off a board 10.0 m above the water and falls vertical to the water, starting from rest. If her downward motion is stopped 2.00 s after her feet first touch the water, what average upward force did the water exert on her?

67. A 2.00-kg aluminum block and a 6.00-kg copper block are connected by a light string over a frictionless pulley. The two blocks are allowed to move on a fixed steel block wedge (of angle $\theta = 30.0°$) as shown in Figure P4.67. Making use of Table 4.2, determine (a) the acceleration of the two blocks and (b) the tension in the string.

Figure P4.67

68. S An object of mass m_1 hangs from a string that passes over a very light fixed pulley P_1 as shown in Figure P4.68. The string connects to a second very light pulley P_2. A second string passes around this pulley with one end attached to a wall and the other to an object of mass m_2 on a frictionless, horizontal table. (a) If a_1 and a_2 are the accelerations of m_1 and m_2, respectively, what is the relation between these accelerations? Find expressions for (b) the tensions in the strings and (c) the accelerations a_1 and a_2 in terms of the masses m_1 and m_2, and g.

Figure P4.68

69. Two boxes of fruit on a frictionless horizontal surface are connected by a light string as in Figure P4.69, where $m_1 = 10$ kg and $m_2 = 20$ kg. A force of 50 N is applied to the 20-kg box. (a) Determine the acceleration of each box and the tension in the string. (b) Repeat the problem for the case where the coefficient of kinetic friction between each box and the surface is 0.10.

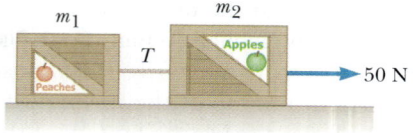

Figure P4.69

70. Q|C Measuring coefficients of friction A coin is placed near one edge of a book lying on a table, and that edge of the book is lifted until the coin just slips down the incline as shown in Figure P4.70. The angle of the incline, θ_c, called the critical angle, is measured. (a) Draw a free-body diagram for the coin when it is on the verge of slipping and identify all forces acting on it. Your free-body diagram should include a force of static friction acting up the incline. (b) Is the magnitude of the friction force equal to $\mu_s n$ for angles less than θ_c? Explain. What can you definitely say about the magnitude of the friction force for any angle $\theta \leq \theta_c$? (c) Show that the coefficient of static friction is given by $\mu_s = \tan \theta_c$. (d) Once the coin starts to slide down the incline, the angle can be adjusted to a new value $\theta_c' \leq \theta_c$ such that the coin moves down the incline with constant speed. How does observation enable you to obtain the coefficient of kinetic friction?

Figure P4.70

71. Q|C A fisherman poles a boat as he searches for his next catch. He pushes parallel to the length of the light pole, exerting a force of 240 N on the bottom of a shallow lake. The pole lies in the vertical plane containing the boat's keel. At one moment, the pole makes an angle of 35.0° with the vertical and the water exerts a horizontal drag force of 47.5 N on the boat, opposite to its forward velocity of magnitude 0.857 m/s. The mass of the boat including its cargo and the worker is 370 kg. (a) The water exerts a buoyant force vertically upward on the boat. Find the magnitude of this force. (b) Assume the forces are constant over a short interval of time. Find the velocity of the boat 0.450 s after the moment described. (c) If the angle of the pole with respect to the vertical increased but the exerted force against the bottom remained the same, what would happen to buoyant force and the acceleration of the boat?

72. Q|C A rope with mass m_r is attached to a block with mass m_b as in Figure P4.72. Both the rope and the block rest on a horizontal, frictionless surface. The rope does not stretch. The free end of the rope is pulled to the right with a horizontal force F. (a) Draw free-body diagrams for the rope and the block, noting that the tension in the rope is not uniform. (b) Find the acceleration of the system in terms of m_b, m_r, and F. (c) Find the magnitude of the force the rope exerts on the block. (d) What happens to the force on the block as the rope's mass approaches zero? What can you state about the tension in a *light* cord joining a pair of moving objects?

73. A van accelerates down a hill (Fig. P4.73), going from rest to 30.0 m/s in 6.00 s. During the acceleration, a

toy ($m = 0.100$ kg) hangs by a string from the van's ceiling. The acceleration is such that the string remains perpendicular to the ceiling. Determine (a) the angle θ and (b) the tension in the string.

Figure P4.73

74. An inquisitive physics student, wishing to combine pleasure with scientific inquiry, rides on a roller coaster sitting on a bathroom scale. (Do not try this

Figure P4.72

yourself on a roller coaster that forbids loose, heavy packages.) The bottom of the seat in the roller-coaster car is in a plane parallel to the track. The seat has a perpendicular back and a seat belt that fits around the student's chest in a plane parallel to the bottom of the seat. The student lifts his feet from the floor so that the scale reads his weight, 200 lb, when the car is horizontal. At one point during the ride, the car zooms with negligible friction down a straight slope inclined at 30.0° below the horizontal. What does the scale read at that point?

75. The parachute on a race car of weight 8 820 N opens at the end of a quarter-mile run when the car is traveling at 35 m/s. What total retarding force must be supplied by the parachute to stop the car in a distance of 1 000 m?

76. **M** On an airplane's takeoff, the combined action of the air around the engines and wings of an airplane exerts an 8 000-N force on the plane, directed upward at an angle of 65.0° above the horizontal. The plane rises with constant velocity in the vertical direction while continuing to accelerate in the horizontal direction. (a) What is the weight of the plane? (b) What is its horizontal acceleration?

77. The board sandwiched between two other boards in Figure P4.77 weighs 95.5 N. If the coefficient of friction between the boards is 0.663, what must be the magnitude of the compression forces (assumed to be horizontal) acting on both sides of the center board to keep it from slipping?

Figure P4.77

78. A sled weighing 60.0 N is pulled horizontally across snow so that the coefficient of kinetic friction between sled and snow is 0.100. A penguin weighing 70.0 N rides on the sled, as in Figure P4.78. If the coefficient of static friction between

Figure P4.78

penguin and sled is 0.700, find the maximum horizontal force that can be exerted on the sled before the penguin begins to slide off.

79. A 72-kg man stands on a spring scale in an elevator. Starting from rest, the elevator ascends, attaining its maximum speed of 1.2 m/s in 0.80 s. The elevator travels with this constant speed for 5.0 s, undergoes a uniform *negative* acceleration for 1.5 s, and then comes to rest. What does the spring scale register (a) before the elevator starts to move? (b) During the first 0.80s of the elevator's ascent? (c) While the elevator is traveling at constant speed? (d) During the elevator's negative acceleration?

80. A magician pulls a tablecloth from under a 200-g mug located 30.0 cm from the edge of the cloth. The cloth exerts a friction force of 0.100 N on the mug and is pulled with a constant acceleration of 3.00 m/s². How far does the mug move relative to the horizontal tabletop before the cloth is completely out from under it? Note that the cloth must move more than 30 cm relative to the tabletop during the process.

81. An inventive child wants to reach an apple in a tree without climbing the tree. Sitting in a chair connected to a rope that passes over a frictionless pulley (Fig. P4.81), the child pulls on the loose end of the rope with such a force that the spring scale reads 250 N. The child's true weight is 320 N, and the chair weighs 160 N. The child's feet are not touching the ground.
(a) Show that the acceleration of the system is *upward*, and find its magnitude. (b) Find the force the child exerts on the chair.

Figure P4.81

82. A fire helicopter carries a 620-kg bucket of water at the end of a 20.0-m-long cable. Flying back from a fire at a constant speed of 40.0 m/s, the cable makes an angle of 40.0° with respect to the vertical. Determine the force exerted by air resistance on the bucket.

83. **S** A crate of weight F_g is pushed by a force $\vec{P}$ on a horizontal floor as shown in Figure P4.83. The coefficient of static friction is μ_s, and $\vec{P}$ is directed at angle θ below the horizontal. (a) Show that the minimum value of P that will move the crate is given by

Figure P4.83

$$P = \frac{\mu_s F_g \sec \theta}{1 - \mu_s \tan \theta}$$

(b) Find the condition on θ in terms of μ_s for which motion of the crate is impossible for any value of P.

84. **QC** In Figure P4.84, the pulleys and the cord are light, all surfaces are frictionless, and the cord does not stretch. (a) How does the acceleration of block 1 compare with the acceleration of block 2? Explain your reasoning. (b) The mass of block 2 is $m_2 = $ 1.30 kg. Derive an expression for the acceleration of the block having mass m_2 as a function of the mass of block 1, m_1. (c) What does the result of part (b) predict if m_1 is very much less than 1.30 kg? (d) What does the result of part (b) predict if m_1 approaches infinity? (e) In this last case, what is the tension in the cord? (f) Could you anticipate the answers to parts (c), (d), and (e) without first doing part (b)? Explain.

Figure P4.84

85. **S** What horizontal force must be applied to a large block of mass M shown in Figure P4.85 so that the blocks remain stationary relative to M? Assume all surfaces and the pulley are frictionless. Notice that the force exerted by the string accelerates m_2.

Figure P4.85

By timing his jumps, a student does work on the pogo-stick-student system, increasing his height with each jump. The work is transformed into gravitational potential energy at maximum height, converted to kinetic energy while falling, and stored as spring potential energy after contact with the ground.

Energy 5

Energy is one of the most important concepts in the world of science. In everyday use energy is associated with the fuel needed for transportation and heating, with electricity for lights and appliances, and with the foods we consume. These associations, however, don't tell us what energy *is*, only what it *does*, and that producing it requires fuel. Our goal in this chapter, therefore, is to develop a better understanding of energy and how to quantify it.

Energy is present in the Universe in a variety of forms, including mechanical, chemical, electromagnetic, and nuclear energy. Even the inert mass of everyday matter contains a very large amount of energy. Although energy can be transformed from one kind to another, all observations and experiments to date suggest that the total amount of energy in the Universe never changes. That's also true for an isolated system, which is a collection of objects that can exchange energy with each other, but not with the rest of the Universe. If one form of energy in an isolated system decreases, then another form of energy in the system must increase. For example, if the system consists of a motor connected to a battery, the battery converts chemical energy to electrical energy and the motor converts electrical energy to mechanical energy. Understanding how energy changes from one form to another is essential in all the sciences.

In this chapter the focus is mainly on *mechanical energy*, which is the sum of *kinetic energy*, the energy associated with motion, and *potential energy*—the energy associated with relative position. Using an energy approach to solve certain problems is often much easier than using forces and Newton's three laws. These two very different approaches are linked through the concept of *work*.

5.1 Work

5.2 Kinetic Energy and the Work–Energy Theorem

5.3 Gravitational Potential Energy

5.4 Spring Potential Energy

5.5 Systems and Energy Conservation

5.6 Power

5.7 Work Done by a Varying Force

Figure 5.1 A constant force $\vec{F}$ in the same direction as the displacement, $\Delta \vec{x}$, does work $F\Delta x$.

5.1 Work

LEARNING OBJECTIVES

1. Contrast by example the physics concept of work with the commonly accepted concept.
2. Calculate the work done by a force on an object in basic contexts.

Work has a different meaning in physics than it does in everyday usage. In the physics definition, a physics textbook author does very little work typing away at a computer. A mason, by contrast, may do a lot of work laying concrete blocks. In physics, work is done only if an object is moved through some displacement while a force is applied to it. If either the force or displacement is doubled, the work is doubled. Double them both, and the work is quadrupled. Doing work involves applying a force to an object while moving it a given distance.

The definition for work W might be taken as

▶ Intuitive definition of work

$$W = Fd \quad [5.1]$$

where F is the magnitude of the force acting on the object and d is the magnitude of the object's displacement. That definition, however, gives only the magnitude of work done on an object when the force is constant and parallel to the displacement, which must be along a line. A more sophisticated definition is required.

Figure 5.1 shows a block undergoing a displacement $\Delta \vec{x}$ along a straight line while acted on by a constant force $\vec{F}$ in the same direction. We have the following definition:

> The work W done on an object by a constant force $\vec{F}$ during a linear displacement along the x-axis is

▶ Work by a constant force during a linear displacement

$$W = F_x \Delta x \quad [5.2]$$

> where F_x is the x-component of the force $\vec{F}$ and $\Delta x = x_f - x_i$ is the object's displacement.
>
> **SI unit: joule (J) = newton · meter (N · m) = kg · m²/s²**

Note that in one dimension, $\Delta x = x_f - x_i$ is a vector quantity, just as it was defined in Chapter 2, not a magnitude as might be inferred from definitions of a vector and its magnitude in Chapter 3. Therefore Δx can be either positive or negative. Work as defined in Equation 5.2 is rigorous for displacements of any object along the x-axis while a constant force acts on it and, therefore, is suitable for many one-dimensional problems. Work is a positive number if F_x and Δx are both positive or both negative, in which case, as discussed in the next section, the work increases the mechanical energy of the object. If F_x is positive and Δx is negative, or vice versa, then the work is negative, and the object loses mechanical energy. The definition in Equation 5.2 works even when the constant force $\vec{F}$ is not parallel to the x-axis. Work is only done by the part of the force acting parallel to the object's direction of motion.

Tip 5.1 Work Is a Scalar Quantity

Work is a simple number—a scalar, not a vector—so there is no direction associated with it. Energy and energy transfer are also scalars.

It's easy to see the difference between the physics definition and the everyday definition of work. The author exerts very little force on the keys of a keyboard, creating only small displacements, so relatively little physics work is done. The mason must exert much larger forces on concrete blocks and move them significant distances, and so performs a much greater amount of work. Even very tiring tasks, however, may not constitute work according to the physics definition. A truck driver, for example, may drive for several hours, but if he doesn't exert a force, then $F_x = 0$ in Equation 5.2 and he doesn't do any work. Similarly, a student pressing against a wall for hours in an isometric exercise also does no work, because the displacement in Equation 5.2, Δx, is zero.[1] Atlas, of Greek mythology, bore the world on his shoulders, but that, too, wouldn't qualify as work in the physics definition.

Work is a scalar quantity—a number rather than a vector—and consequently is easier to handle. No direction is associated with it. Further, work doesn't depend explicitly on time, which can be an advantage in problems involving only velocities and positions. Because the units of work are those of force and distance, the SI unit is the **newton-meter** (N · m). Another name for the newton-meter is the **joule (J)** (rhymes with "pool"). The U.S. customary unit of work is the **foot-pound**, because distances are measured in feet and forces in pounds in that system.

A useful alternate definition relates the work done on an object to the angle the displacement makes with respect to the force. This definition exploits the triangle shown in Figure 5.2. The components of the vector $\vec{F}$ can be written as $F_x = F\cos\theta$ and $F_y = F\sin\theta$. However, only the x-component, which is parallel to the direction of motion, makes a nonzero contribution to the work done on the object.

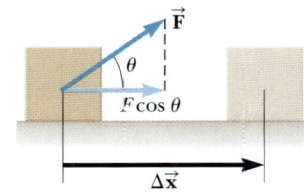

Figure 5.2 A constant force $\vec{F}$ exerted at an angle θ with respect to the displacement, $\Delta\vec{x}$, does work $(F\cos\theta)\Delta x$.

The work W done on an object by a constant force $\vec{F}$ during a linear displacement $\Delta\vec{x}$ is

$$W = (F\cos\theta)d \qquad [5.3]$$

where d is the magnitude of the displacement and θ is the angle between the vectors $\vec{F}$ and $\Delta\vec{x}$.

SI unit: joule (J)

◀ Work by a constant force at an angle to the displacement

Figure 5.3 No work is done on a bucket when it is moved horizontally because the applied force $\vec{F}$ is perpendicular to the displacement.

The definition in Equation 5.3 can also be used more generally when the displacement is not specifically along the x-axis or any other axis.

In Figure 5.3 a man carries a bucket of water horizontally at constant velocity. The upward force exerted by the man's hand on the bucket is perpendicular to the direction of motion, so it does no work on the bucket. This can also be seen from Equation 5.3 because the angle between the force exerted by the hand and the direction of motion is 90°, giving $\cos 90° = 0$ and $W = 0$. Similarly, the force of gravity does no work on the bucket.

Work always requires a system of more than just one object. A nail, for example, can't do work on itself, but a hammer can do work on the nail by driving it into a board. In general, an object may be moving under the influence of several external forces. In that case, the net work done on the object as it undergoes some displacement is just the sum of the amount of work done by each force.

Work can be either positive or negative. In the definition of work in Equation 5.3, F and d are magnitudes, which are never negative. Work is therefore positive or negative depending on whether $\cos\theta$ is positive or negative. This, in turn, depends on the direction of $\vec{F}$ relative the direction of $\Delta\vec{x}$. When these vectors are pointing in the same direction, the angle between them is 0°, so $\cos 0° = +1$ and the work is positive. For example, when a student lifts a box as in Figure 5.4, the work he does on the box is positive because the force he exerts on the box is upward, in the same direction as the displacement. In lowering the box slowly back down, however, the student still exerts an upward force on the box, but the motion of the box is downwards. Because the vectors $\vec{F}$ and $\Delta\vec{x}$ are now in opposite directions, the angle between them is 180°, and $\cos 180° = -1$ and the work done by the student is negative. In general, when the part of $\vec{F}$ parallel to $\Delta\vec{x}$ points in the same direction as $\Delta\vec{x}$, the work is positive; otherwise, it's negative.

Because Equations 5.1–5.3 assume a force constant in both direction and magnitude, they are only special cases of a more general definition of work—that done by a varying force—treated briefly in Section 5.7.

Tip 5.2 Work Is Done by Something, on Something Else

Work doesn't happen by itself. Work is done *by* something in the environment, *on* the object of interest.

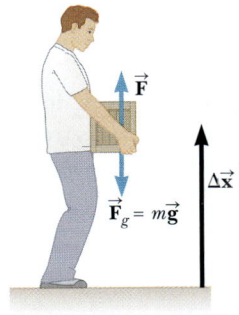

Figure 5.4 The student does positive work when he lifts the box from the floor, because the applied force $\vec{F}$ is in the same direction as the displacement. When he lowers the box to the floor, he does negative work.

---

[1] Actually, you do expend energy while doing isometric exercises because your muscles are continuously contracting and relaxing in the process. This internal muscular movement qualifies as work according to the physics definition.

Figure 5.5 (Quick Quiz 5.1) A force $\vec{F}$ is exerted on an object that undergoes a displacement $\Delta\vec{x}$ to the right. Both the magnitude of the force and the displacement are the same in all four cases.

■ Quick Quiz

5.1 In Figure 5.5 (a)–(d), a block moves to the right in the positive x-direction through the displacement $\Delta\vec{x}$ while under the influence of a force with the same magnitude $\vec{F}$. Which of the following is the correct order of the amount of work done by the force $\vec{F}$, from most positive to most negative? (a) d, c, a, b (b) c, a, b, d (c) c, a, d, b

---

■ EXAMPLE 5.1 | Sledding Through the Yukon

GOAL Apply the basic definitions of work done by a constant force.

PROBLEM An Eskimo returning from a successful fishing trip pulls a sled loaded with salmon. The total mass of the sled and salmon is 50.0 kg, and the Eskimo exerts a force of magnitude 1.20×10^2 N on the sled by pulling on the rope. **(a)** How much work does he do on the sled if the rope is horizontal to the ground ($\theta = 0°$ in Fig. 5.6) and he pulls the sled 5.00 m? **(b)** How much work does he do on the sled if $\theta = 30.0°$ and he pulls the sled the same distance? (Treat the sled as a point particle, so details such as the point of attachment of the rope make no difference.) **(c)** At a coordinate position of 12.4 m, the Eskimo lets up on the applied force. A friction force of 45.0 N between the ice and the sled brings the sled to rest at a coordinate position of 18.2 m. How much work does friction do on the sled?

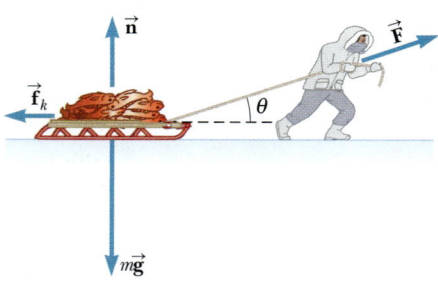

Figure 5.6 (Examples 5.1 and 5.2) An Eskimo pulling a sled with a rope at an angle θ to the horizontal.

STRATEGY Substitute the given values of F and Δx into the basic equations for work, Equations 5.2 and 5.3.

SOLUTION

(a) Find the work done when the force is horizontal.

Use Equation 5.2, substituting the given values:

$W = F_x \Delta x = (1.20 \times 10^2 \text{ N})(5.00 \text{ m}) = \boxed{6.00 \times 10^2 \text{ J}}$

(b) Find the work done when the force is exerted at a 30° angle.

Use Equation 5.3, again substituting the given values:

$W = (F \cos \theta) d = (1.20 \times 10^2 \text{ N})(\cos 30°)(5.00 \text{ m})$
$= \boxed{5.20 \times 10^2 \text{ J}}$

(c) How much work does a friction force of 45.0 N do on the sled as it travels from a coordinate position of 12.4 m to 18.2 m?

Use Equation 5.2, with F_x replaced by f_k:

$W_{\text{fric}} = F_x \Delta x = f_k (x_f - x_i)$

Substitute $f_k = -45.0$ N and the initial and final coordinate positions into x_i and x_f:

$W_{\text{fric}} = (-45.0 \text{ N})(18.2 \text{ m} - 12.4 \text{ m}) = \boxed{-2.6 \times 10^2 \text{ J}}$

REMARKS The normal force $\vec{n}$, the gravitational force $m\vec{g}$, and the upward component of the applied force do *no* work on the sled because they're perpendicular to the displacement. The mass of the sled didn't come into play here, but it is important when the effects of friction must be calculated, and in the next section, where we introduce the work–energy theorem.

QUESTION 5.1 How does the answer for the work done by the applied force change if the load is doubled? Explain.

EXERCISE 5.1 Suppose the Eskimo is pushing the same 50.0-kg sled across level terrain with a force of 50.0 N. (a) If he does 4.00×10^2 J of work on the sled while exerting the force horizontally, through what distance must he have pushed it? (b) If he exerts the same force at an angle of 45.0° with respect to the horizontal and moves the sled through the same distance, how much work does he do on the sled?

ANSWERS (a) 8.00 m (b) 283 J

Work and Dissipative Forces

Frictional work is extremely important in everyday life because doing almost any other kind of work is impossible without it. The Eskimo in the last example, for instance, depends on surface friction to pull his sled. Otherwise, the rope would slip in his hands and exert no force on the sled, while his feet slid out from underneath him and he fell flat on his face. Cars wouldn't work without friction, nor could conveyor belts, nor even our muscle tissue.

The work done by pushing or pulling an object is the application of a single force. Friction, on the other hand, is a complex process caused by numerous microscopic interactions over the entire area of the surfaces in contact. Consider a metal block sliding over a metal surface. Microscopic "teeth" in the block encounter equally microscopic irregularities in the underlying surface. Pressing against each other, the teeth deform, get hot, and weld to the opposite surface. Work must be done breaking these temporary bonds, and that comes at the expense of the energy of motion of the block, to be discussed in the next section. The energy lost by the block goes into heating both the block and its environment, with some energy converted to sound.

The friction force of two objects in contact and in relative motion to each other always dissipates energy in these relatively complex ways. For our purposes, the phrase "work done by friction" will denote the effect of these processes on mechanical energy alone.

The edge of a razor blade looks smooth to the eye, but under a microscope proves to have numerous irregularities.

■ EXAMPLE 5.2 | More Sledding

GOAL Calculate the work done by friction when an object is acted on by an applied force.

PROBLEM Suppose that in Example 5.1 the coefficient of kinetic friction between the loaded 50.0-kg sled and snow is 0.200. **(a)** The Eskimo again pulls the sled 5.00 m, exerting a force of 1.20×10^2 N at an angle of 0°. Find the work done on the sled by friction, and the net work. **(b)** Repeat the calculation if the applied force is exerted at an angle of 30.0° with the horizontal.

STRATEGY See Figure 5.6. The frictional work depends on the magnitude of the kinetic friction coefficient, the normal force, and the displacement. Use the y-component of Newton's second law to find the normal force $\vec{n}$, calculate the work done by friction using the definitions, and sum with the result of Example 5.1(a) to obtain the net work on the sled. Part (b) is solved similarly, but the normal force is smaller because it has the help of the applied force $\vec{F}_{app}$ in supporting the load.

SOLUTION
(a) Find the work done by friction on the sled and the net work, if the applied force is horizontal.

First, find the normal force from the y-component of Newton's second law, which involves only the normal force and the force of gravity:

$$\sum F_y = n - mg = 0 \quad \rightarrow \quad n = mg$$

Use the normal force to compute the work done by friction:

$$W_{\text{fric}} = -f_k \Delta x = -\mu_k n \Delta x = -\mu_k mg \Delta x$$
$$= -(0.200)(50.0 \text{ kg})(9.80 \text{ m/s}^2)(5.00 \text{ m})$$
$$= \boxed{-4.90 \times 10^2 \text{ J}}$$

(Continued)

Sum the frictional work with the work done by the applied force from Example 5.1 to get the net work (the normal and gravity forces are perpendicular to the displacement, so they don't contribute):

$$W_{net} = W_{app} + W_{fric} + W_n + W_g$$
$$= 6.00 \times 10^2 \text{ J} + (-4.90 \times 10^2 \text{ J}) + 0 + 0$$
$$= \boxed{1.10 \times 10^2 \text{ J}}$$

(b) Recalculate the frictional work and net work if the applied force is exerted at a 30.0° angle.

Find the normal force from the y-component of Newton's second law:

$$\Sigma F_y = n - mg + F_{app} \sin \theta = 0$$
$$n = mg - F_{app} \sin \theta$$

Use the normal force to calculate the work done by friction:

$$W_{fric} = -f_k \Delta x = -\mu_k n \Delta x = -\mu_k (mg - F_{app} \sin \theta) \Delta x$$
$$= -(0.200)(50.0 \text{ kg} \cdot 9.80 \text{ m/s}^2$$
$$-1.20 \times 10^2 \text{ N} \sin 30.0°)(5.00 \text{ m})$$
$$W_{fric} = \boxed{-4.30 \times 10^2 \text{ J}}$$

Sum this answer with the result of Example 5.1(b) to get the net work (again, the normal and gravity forces don't contribute):

$$W_{net} = W_{app} + W_{fric} + W_n + W_g$$
$$= 5.20 \times 10^2 \text{ J} - 4.30 \times 10^2 \text{ J} + 0 + 0 = \boxed{9.0 \times 10^1 \text{ J}}$$

REMARKS The most important thing to notice here is that exerting the applied force at different angles can dramatically affect the work done on the sled. Pulling at the optimal angle (11.3° in this case) will result in the most net work for the same applied force.

QUESTION 5.2 How does the net work change in each case if the displacement is doubled?

EXERCISE 5.2 (a) The Eskimo pushes the same 50.0-kg sled over level ground with a force of 1.75×10^2 N exerted horizontally, moving it a distance of 6.00 m over new terrain. If the net work done on the sled is 1.50×10^2 J, find the coefficient of kinetic friction. (b) Repeat the exercise with the same data, finding the coefficient of kinetic friction, but assume the applied force is upwards at a 45.0° angle with the horizontal.

ANSWERS (a) 0.306 (b) 0.270

5.2 Kinetic Energy and the Work–Energy Theorem

LEARNING OBJECTIVES

1. Define kinetic energy and derive the work–energy theorem.
2. Apply the work–energy theorem in elementary physical contexts.
3. Categorize forces as conservative or non-conservative forces.

Solving problems using Newton's second law can be difficult if the forces involved are complicated. An alternative is to relate the speed of an object to the net work done on it by external forces. If the net work can be calculated for a given displacement, the change in the object's speed is easy to evaluate.

Figure 5.7 shows an object of mass m moving to the right under the action of a constant net force $\vec{F}_{net}$, also directed to the right. Because the force is constant, we know from Newton's second law that the object moves with constant acceleration $\vec{a}$. If the object is displaced by Δx, the work done by $\vec{F}_{net}$ on the object is

$$W_{net} = F_{net} \Delta x = (ma) \Delta x \quad [5.4]$$

In Chapter 2, we found that the following relationship holds when an object undergoes constant acceleration:

$$v^2 = v_0^2 + 2a\Delta x \quad \text{or} \quad a\Delta x = \frac{v^2 - v_0^2}{2}$$

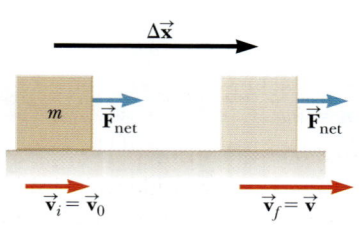

Figure 5.7 An object undergoes a displacement and a change in velocity under the action of a constant net force $\vec{F}_{net}$.

We can substitute this expression into Equation 5.4 to get

$$W_{net} = m\left(\frac{v^2 - v_0^2}{2}\right)$$

or

$$W_{net} = \tfrac{1}{2}mv^2 - \tfrac{1}{2}mv_0^2 \qquad [5.5]$$

So the net work done on an object equals a change in a quantity of the form $\tfrac{1}{2}mv^2$. Because this term carries units of energy and involves the object's speed, it can be interpreted as energy associated with the object's motion, leading to the following definition:

> The **kinetic energy** *KE* of an object of mass *m* moving with a speed *v* is
>
> $$KE \equiv \tfrac{1}{2}mv^2 \qquad [5.6]$$
>
> **SI unit: joule (J) = kg · m²/s²**

◀ Kinetic energy

Like work, kinetic energy is a scalar quantity. Using this definition in Equation 5.5, we arrive at an important result known as the **work–energy theorem**:

> The net work done on an object is equal to the change in the object's kinetic energy:
>
> $$W_{net} = KE_f - KE_i = \Delta KE \qquad [5.7]$$
>
> where the change in the kinetic energy is due entirely to the object's change in speed.

◀ Work–energy theorem

The proviso on the speed is necessary because work that deforms or causes the object to warm up invalidates Equation 5.7, although under most circumstances it remains approximately correct. From that equation, a positive net work W_{net} means that the final kinetic energy KE_f is greater than the initial kinetic energy KE_i. This, in turn, means that the object's final speed is greater than its initial speed. So positive net work increases an object's speed, and negative net work decreases its speed.

We can also turn Equation 5.7 around and think of kinetic energy as the work a moving object can do in coming to rest. For example, suppose a hammer is on the verge of striking a nail, as in Figure 5.8. The moving hammer has kinetic energy and can therefore do work on the nail. The work done on the nail is $F\Delta x$, where *F* is the average net force exerted on the nail and Δx is the distance the nail is driven into the wall. That work, plus small amounts of energy carried away by heat and sound, is equal to the change in kinetic energy of the hammer, ΔKE.

For convenience, the work–energy theorem was derived under the assumption that the net force acting on the object was constant. A more general derivation, using calculus, would show that Equation 5.7 is valid under all circumstances, including the application of a variable force.

Figure 5.8 The moving hammer has kinetic energy and can do work on the nail, driving it into the wall.

■ APPLYING PHYSICS 5.1 | Leaving Skid Marks

Suppose a car traveling at a speed *v* skids a distance *d* after its brakes lock. Estimate how far it would skid if it were traveling at speed 2*v* when its brakes locked.

EXPLANATION Assume for simplicity that the force of kinetic friction between the car and the road surface is constant and the same at both speeds. From the work–energy theorem, the net force exerted on the car times the displacement of the car, $F_{net}\Delta x$, is equal in magnitude to its initial kinetic energy, $\tfrac{1}{2}mv^2$. When the speed is doubled, the kinetic energy of the car is quadrupled. So for a given applied friction force, the distance traveled must increase fourfold when the initial speed is doubled, and the estimated distance the car skids is 4*d*. ■

EXAMPLE 5.3 Collision Analysis

GOAL Apply the work–energy theorem with a known force.

PROBLEM The driver of a 1.00×10^3 kg car traveling on the interstate at 35.0 m/s (nearly 80.0 mph) slams on his brakes to avoid hitting a second vehicle in front of him, which had come to rest because of congestion ahead (Fig. 5.9). After the brakes are applied, a constant kinetic friction force of magnitude 8.00×10^3 N acts on the car. Ignore air resistance. **(a)** At what minimum distance should the brakes be applied to avoid a collision with the other vehicle? **(b)** If the distance between the vehicles is initially only 30.0 m, at what speed would the collision occur?

Figure 5.9 (Example 5.3) A braking vehicle just prior to an accident.

STRATEGY Compute the net work, which involves just the kinetic friction, because the normal and gravity forces are perpendicular to the motion. Then set the net work equal to the change in kinetic energy. To get the minimum distance in part (a), we take the final speed v_f to be zero just as the braking vehicle reaches the rear of the vehicle at rest. Solve for the unknown, Δx. For part (b) proceed similarly, except that the unknown is the final velocity v_f.

SOLUTION

(a) Find the minimum necessary stopping distance.

Apply the work–energy theorem to the car:
$$W_{net} = \tfrac{1}{2}mv_f^2 - \tfrac{1}{2}mv_i^2$$

Substitute an expression for the frictional work and set $v_f = 0$:
$$-f_k \Delta x = 0 - \tfrac{1}{2}mv_i^2$$

Substitute $v_i = 35.0$ m/s, $f_k = 8.00 \times 10^3$ N, and $m = 1.00 \times 10^3$ kg. Solve for Δx:
$$-(8.00 \times 10^3 \text{ N})\Delta x = -\tfrac{1}{2}(1.00 \times 10^3 \text{ kg})(35.0 \text{ m/s})^2$$
$$\Delta x = \boxed{76.6 \text{ m}}$$

(b) At the given distance of 30.0 m, the car is too close to the other vehicle. Find the speed at impact.

Write down the work–energy theorem:
$$W_{net} = W_{fric} = -f_k \Delta x = \tfrac{1}{2}mv_f^2 - \tfrac{1}{2}mv_i^2$$

Multiply by $2/m$ and rearrange terms, solving for the final velocity v_f:
$$v_f^2 = v_i^2 - \frac{2}{m}f_k \Delta x$$
$$v_f^2 = (35.0 \text{ m/s})^2 - \left(\frac{2}{1.00 \times 10^3 \text{ kg}}\right)(8.00 \times 10^3 \text{ N})(30.0 \text{ m})$$
$$= 745 \text{ m}^2/\text{s}^2$$
$$v_f = \boxed{27.3 \text{ m/s}}$$

REMARKS This calculation illustrates how important it is to remain alert on the highway, allowing for an adequate stopping distance at all times. It takes about a second to react to the brake lights of the car in front of you. On a high-speed highway, your car may travel more than 30 m before you can engage the brakes. Bumper-to-bumper traffic at high speed, which often occurs on the highways near big cities, is extremely unsafe.

QUESTION 5.3 Qualitatively, how would the answer for the final velocity change in part (b) if it's raining during the incident? Explain.

EXERCISE 5.3 A police investigator measures straight skid marks 27.0 m long in an accident investigation. Assuming a friction force and car mass the same as in the previous problem, what was the minimum speed of the car when the brakes locked?

ANSWER 20.8 m/s

Conservative and Nonconservative Forces

It turns out there are two general kinds of forces. The first is called a **conservative force.** Gravity is probably the best example of a conservative force. To understand the origin of the name, think of a diver climbing to the top of a 10-meter platform. The diver has to do work against gravity in making the climb. Once at the top, however, she can recover the work as kinetic energy by taking a dive. Her speed just before hitting the water will give her a kinetic energy equal to the work she did against gravity in climbing to the top of the platform, minus the effect of some nonconservative forces, such as air drag and internal muscular friction.

A **nonconservative force** is generally dissipative, which means that it tends to randomly disperse the energy of bodies on which it acts. This dispersal of energy often takes the form of heat or sound. Kinetic friction and air drag are good examples. Propulsive forces, like the force exerted by a jet engine on a plane or by a propeller on a submarine, are also nonconservative.

Work done against a nonconservative force can't be easily recovered. Dragging objects over a rough surface requires work. When the Eskimo in Example 5.2 dragged the sled across terrain having a nonzero coefficient of friction, the net work was smaller than in the frictionless case. The missing energy went into warming the sled and its environment. As will be seen in the study of thermodynamics, such losses can't be avoided, nor all the energy recovered, so these forces are called nonconservative.

Another way to characterize conservative and nonconservative forces is to measure the work done by a force on an object traveling between two points along different paths. The work done by gravity on someone going down a frictionless slide, as in Figure 5.10, is the same as that done on someone diving into the water from the same height. This equality doesn't hold for nonconservative forces. For example, sliding a book directly from point Ⓐ to point Ⓓ in Figure 5.11 requires a certain amount of work against friction, but sliding the book along the three other legs of the square, from Ⓐ to Ⓑ, Ⓑ to Ⓒ, and finally Ⓒ to Ⓓ, requires three times as much work. This observation motivates the following definition of a conservative force:

> A force is conservative if the work it does moving an object between two points is the same no matter what path is taken.

Nonconservative forces, as we've seen, don't have this property. The work–energy theorem, Equation 5.7, can be rewritten in terms of the work done by conservative forces W_c and the work done by nonconservative forces W_{nc} because the net work is just the sum of these two:

$$W_{nc} + W_c = \Delta KE \quad [5.8]$$

It turns out that conservative forces have another useful property: The work they do can be recast as something called **potential energy,** a quantity that depends only on the beginning and end points of a curve, not the path taken.

5.3 Gravitational Potential Energy

LEARNING OBJECTIVES

1. Understand the relationship between gravitational potential energy and gravitational work.
2. Apply the conservation of mechanical energy to solving problems.
3. Extend and apply the work–energy theorem to problems involving gravity.

An object with kinetic energy (energy of motion) can do work on another object, just like a moving hammer can drive a nail into a wall. A brick on a high shelf can also do work: it can fall off the shelf, accelerate downwards, and hit a nail squarely, driving it into the floorboards. The brick is said to have **potential energy** associated with it, because from its location on the shelf it can potentially do work.

Potential energy is a property of a **system,** rather than of a single object, because it's due to the relative positions of interacting objects in the system, such as the position of the diver in Figure 5.10 relative to the Earth. In this chapter we define

Figure 5.10 Because the gravity field is conservative, the diver regains as kinetic energy the work she did against gravity in climbing the ladder. Taking the frictionless slide gives the same result.

◀ Conservative force

The work done in moving the book is greater along the rust-colored path than along the blue path.

Figure 5.11 Because friction is a nonconservative force, a book pushed along the three segments Ⓐ–Ⓑ, Ⓑ–Ⓒ, and Ⓒ–Ⓓ requires three times the work as pushing the book directly from Ⓐ to Ⓓ.

a system as a collection of objects interacting via forces or other processes that are internal to the system. It turns out that potential energy is another way of looking at the work done by conservative forces.

Gravitational Work and Potential Energy

Using the work–energy theorem in problems involving gravitation requires computing the work done by gravity. For most trajectories—say, for a ball traversing a parabolic arc—finding the gravitational work done on the ball requires sophisticated techniques from calculus. Fortunately, for conservative fields there's a simple alternative: potential energy.

Gravity is a conservative force, and for every conservative force a special expression called a potential energy function can be found. Evaluating that function at any two points in an object's path of motion and finding the difference will give the negative of the work done by that force between those two points. It's also advantageous that potential energy, like work and kinetic energy, is a scalar quantity.

Our first step is to find the work done by gravity on an object when it moves from one position to another. The negative of that work is the change in the gravitational potential energy of the system, and from that expression, we'll be able to identify the potential energy function.

In Figure 5.12, a book of mass m falls from a height y_i to a height y_f, where the positive y-coordinate represents position above the ground. We neglect the force of air friction, so the only force acting on the book is gravitation. How much work is done? The magnitude of the force is mg and that of the displacement is $\Delta y = y_i - y_f$ (a positive number), while both $\vec{F}$ and $\Delta \vec{y}$ are pointing downwards, so the angle between them is zero. We apply the definition of work in Equation 5.3, with $d = y_i - y_f$:

$$W_g = Fd \cos \theta = mg(y_i - y_f) \cos 0° = -mg(y_f - y_i) \quad [5.9]$$

Factoring out the minus sign was deliberate, to clarify the coming connection to potential energy. Equation 5.9 for gravitational work holds for any object, regardless of its trajectory in space, because the gravitational force is conservative. Now, W_g will appear as the work done by gravity in the work–energy theorem. For the rest of this section, assume for simplicity that we are dealing only with systems involving gravity and nonconservative forces. Then Equation 5.8 can be written as

$$W_{net} = W_{nc} + W_g = \Delta KE$$

where W_{nc} is the work done by the nonconservative forces. Substituting the expression for W_g from Equation 5.9, we obtain

$$W_{nc} - mg(y_f - y_i) = \Delta KE \quad [5.10a]$$

Next, we add $mg(y_f - y_i)$ to both sides:

$$W_{nc} = \Delta KE + mg(y_f - y_i) \quad [5.10b]$$

Now, by definition, we'll make the connection between gravitational work and gravitational potential energy.

▶ **Gravitational potential energy**

The gravitational potential energy of a system consisting of Earth and an object of mass m near Earth's surface is given by

$$PE \equiv mgy \quad [5.11]$$

where g is the acceleration of gravity and y is the vertical position of the mass relative the surface of Earth (or some other reference point).

SI unit: joule (J)

In this definition, $y = 0$ is usually taken to correspond to Earth's surface, but that is not strictly necessary, as discussed in the next subsection. It turns out that only *differences* in potential energy really matter.

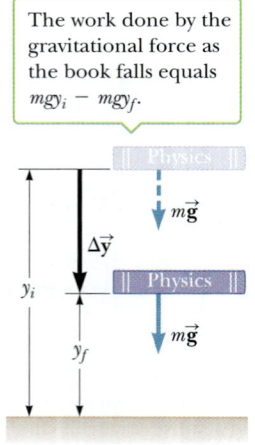

The work done by the gravitational force as the book falls equals $mgy_i - mgy_f$.

Figure 5.12 A book of mass m falls from a height y_i to a height y_f.

5.3 | Gravitational Potential Energy

So the gravitational potential energy associated with an object located near the surface of Earth is the object's weight mg times its vertical position y above Earth. From this *definition*, we have the relationship between gravitational work and gravitational potential energy:

$$W_g = -(PE_f - PE_i) = -(mgy_f - mgy_i) \quad [5.12]$$

The work done by gravity is one and the same as the negative of the change in gravitational potential energy.

Finally, using the relationship in Equation 5.12 in Equation 5.10b, we obtain an extension of the work–energy theorem:

$$W_{nc} = (KE_f - KE_i) + (PE_f - PE_i) \quad [5.13]$$

This equation says that the work done by nonconservative forces, W_{nc}, is equal to the change in the kinetic energy plus the change in the gravitational potential energy.

Equation 5.13 will turn out to be true in general, even when other conservative forces besides gravity are present. The work done by these additional conservative forces will again be recast as changes in potential energy and will appear on the right-hand side along with the expression for gravitational potential energy.

> **Tip 5.3** Potential Energy Takes Two
>
> Potential energy always takes a system of at least two interacting objects—for example, the Earth and a baseball interacting via the gravitational force.

Reference Levels for Gravitational Potential Energy

In solving problems involving gravitational potential energy, it's important to choose a location at which to set that energy equal to zero. Given the form of Equation 5.11, this is the same as choosing the place where $y = 0$. The choice is completely arbitrary because the important quantity is the *difference* in potential energy, and this difference will be the same regardless of the choice of zero level. However, once this position is chosen, it must remain fixed for a given problem.

While it's always possible to choose the surface of Earth as the reference position for zero potential energy, the statement of a problem will usually suggest a convenient position to use. As an example, consider a book at several possible locations, as in Figure 5.13. When the book is at Ⓐ, a natural zero level for potential energy is the surface of the desk. When the book is at Ⓑ, the floor might be a more convenient reference level. Finally, a location such as Ⓒ, where the book is held out a window, would suggest choosing the surface of Earth as the zero level of potential energy. The choice, however, makes no difference: Any of the three reference levels could be used as the zero level, regardless of whether the book is at Ⓐ, Ⓑ, or Ⓒ. Example 5.4 illustrates this important point.

Figure 5.13 Any reference level—the desktop, the floor of the room, or the ground outside the building—can be used to represent zero gravitational potential energy in the book–Earth system.

■ EXAMPLE 5.4 | Wax Your Skis

GOAL Calculate the change in gravitational potential energy for different choices of reference level.

PROBLEM A 60.0-kg skier is at the top of a slope, as shown in Figure 5.14. At the initial point Ⓐ, she is 10.0 m vertically above point Ⓑ. **(a)** Setting the zero level for gravitational potential energy at Ⓑ, find the gravitational potential energy of this system when the skier is at Ⓐ and then at Ⓑ. Finally, find the change in potential energy of the skier–Earth system as the skier goes from point Ⓐ to point Ⓑ. **(b)** Repeat this problem with the zero level at point Ⓐ. **(c)** Repeat again, with the zero level 2.00 m higher than point Ⓑ.

STRATEGY Follow the definition and be careful with signs. Ⓐ is the initial point, with gravitational potential energy PE_i, and Ⓑ is the final point, with gravitational potential energy PE_f. The location chosen for $y = 0$ is also the zero point for the potential energy, because $PE = mgy$.

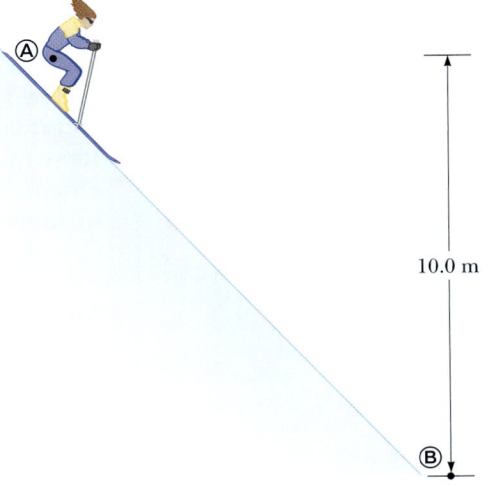

Figure 5.14 (Example 5.4)

(Continued)

SOLUTION

(a) Let $y = 0$ at Ⓑ. Calculate the potential energy at Ⓐ and at Ⓑ, and calculate the change in potential energy.

Find PE_i, the potential energy at Ⓐ, from Equation 5.11:

$$PE_i = mgy_i = (60.0 \text{ kg})(9.80 \text{ m/s}^2)(10.0 \text{ m}) = 5.88 \times 10^3 \text{ J}$$

$PE_f = 0$ at Ⓑ by choice. Find the difference in potential energy between Ⓐ and Ⓑ:

$$PE_f - PE_i = 0 - 5.88 \times 10^3 \text{ J} = \boxed{-5.88 \times 10^3 \text{ J}}$$

(b) Repeat the problem if $y = 0$ at Ⓐ, the new reference point, so that PE = 0 at Ⓐ.

Find PE_f, noting that point Ⓑ is now at $y = -10.0$ m:

$$PE_f = mgy_f = (60.0 \text{ kg})(9.80 \text{ m/s}^2)(-10.0 \text{ m})$$
$$= -5.88 \times 10^3 \text{ J}$$
$$PE_f - PE_i = -5.88 \times 10^3 \text{ J} - 0 = \boxed{-5.88 \times 10^3 \text{ J}}$$

(c) Repeat the problem, if $y = 0$ two meters above Ⓑ.

Find PE_i, the potential energy at Ⓐ:

$$PE_i = mgy_i = (60.0 \text{ kg})(9.80 \text{ m/s}^2)(8.00 \text{ m}) = 4.70 \times 10^3 \text{ J}$$

Find PE_f, the potential energy at Ⓑ:

$$PE_f = mgy_f = (60.0 \text{ kg})(9.8 \text{ m/s}^2)(-2.00 \text{ m})$$
$$= -1.18 \times 10^3 \text{ J}$$

Compute the change in potential energy:

$$PE_f - PE_i = -1.18 \times 10^3 \text{ J} - 4.70 \times 10^3 \text{ J}$$
$$= \boxed{-5.88 \times 10^3 \text{ J}}$$

REMARKS These calculations show that the change in the gravitational potential energy when the skier goes from the top of the slope to the bottom is -5.88×10^3 J, *regardless of the zero level selected.*

QUESTION 5.4 If the angle of the slope is increased, does the change of gravitational potential energy between two heights (a) increase, (b) decrease, (c) remain the same?

EXERCISE 5.4 If the zero level for gravitational potential energy is selected to be midway down the slope, 5.00 m above point Ⓑ, find the initial potential energy, the final potential energy, and the change in potential energy as the skier goes from point Ⓐ to Ⓑ in Figure 5.14.

ANSWER 2.94 kJ, −2.94 kJ, −5.88 kJ

Gravity and the Conservation of Mechanical Energy

Conservation principles play a very important role in physics. **When a physical quantity is conserved the numeric value of the quantity remains the same throughout the physical process.** Although the form of the quantity may change in some way, **its final value is the same as its initial value.**

The kinetic energy KE of an object falling only under the influence of gravity is constantly changing, as is the gravitational potential energy PE. Obviously, then, these quantities aren't conserved. Because all nonconservative forces are assumed absent, however, we can set $W_{nc} = 0$ in Equation 5.13. Rearranging the equation, we arrive at the following very interesting result:

$$KE_i + PE_i = KE_f + PE_f \quad [5.14]$$

According to this equation, **the sum of the kinetic energy and the gravitational potential energy remains constant at all times and hence is a conserved quantity.** We denote the total mechanical energy by $E = KE + PE$, and say that **the total mechanical energy is conserved.**

To show how this concept works, think of tossing a rock off a cliff, ignoring the drag forces. As the rock falls, its speed increases, so its kinetic energy increases. As the rock approaches the ground, the potential energy of the rock–Earth system decreases. Whatever potential energy is lost as the rock moves downward appears as kinetic energy, and Equation 5.14 says that in the absence of nonconservative

> **Tip 5.4 Conservation Principles**
>
> There are many conservation laws like the conservation of mechanical energy in isolated systems, as in Equation 5.14. For example, momentum, angular momentum, and electric charge are all conserved quantities, as will be seen later. Conserved quantities may change form during physical interactions, but their sum total for a system never changes.

forces like air drag, the trading of energy is exactly even. This is true for all conservative forces, not just gravity.

> In any isolated system of objects interacting only through conservative forces, the total mechanical energy $E = KE + PE$, of the system, remains the same at all times.

◀ Conservation of mechanical energy

If the force of gravity is the *only* force doing work within a system, then the principle of conservation of mechanical energy takes the form

$$\tfrac{1}{2}mv_i^2 + mgy_i = \tfrac{1}{2}mv_f^2 + mgy_f \qquad [5.15]$$

This form of the equation is particularly useful for solving problems explicitly involving only one mass and gravity. In that special case, which occurs commonly, notice that the mass cancels out of the equation. However, that is possible only because any change in kinetic energy of the Earth in response to the gravity field of the object of mass m has been (rightfully) neglected. In general, there must be kinetic energy terms for each object in the system, and gravitational potential energy terms for every pair of objects. Further terms have to be added when other conservative forces are present, as we'll soon see.

■ Quick Quiz

5.2 Three identical balls are thrown from the top of a building, all with the same initial speed. The first ball is thrown horizontally, the second at some angle above the horizontal, and the third at some angle below the horizontal, as in Figure 5.15. Neglecting air resistance, rank the speeds of the balls as they reach the ground, from fastest to slowest. (a) 1, 2, 3 (b) 2, 1, 3 (c) 3, 1, 2 (d) All three balls strike the ground at the same speed.

5.3 Bob, of mass m, drops from a tree limb at the same time that Esther, also of mass m, begins her descent down a frictionless slide. If they both start at the same height above the ground, which of the following is true about their kinetic energies as they reach the ground?

(a) Bob's kinetic energy is greater than Esther's.
(b) Esther's kinetic energy is greater than Bob's.
(c) They have the same kinetic energy.
(d) The answer depends on the shape of the slide.

Figure 5.15 (Quick Quiz 5.2) A student throws three identical balls from the top of a building, each at the same initial speed but at a different initial angle.

■ PROBLEM-SOLVING STRATEGY

Applying Conservation of Mechanical Energy

Take the following steps when applying conservation of mechanical energy to problems involving gravity:

1. **Define the system,** including all interacting bodies. Verify the absence of nonconservative forces.
2. **Choose a location for $y = 0$,** the zero point for gravitational potential energy.
3. **Select the body of interest and identify two points**—one point where you have given information and the other point where you want to find out something about the body of interest.
4. **Write down the conservation of energy equation,** Equation 5.15, for the system. **Identify the unknown quantity** of interest.
5. **Solve for the unknown quantity,** which is usually either a speed or a position, and substitute known values.

As previously stated, it's usually best to do the algebra with symbols rather than substituting known numbers first, because it's easier to check the symbols for possible errors. The exception is when a quantity is clearly zero, in which case immediate substitution greatly simplifies the ensuing algebra.

EXAMPLE 5.5 | Platform Diver

GOAL Use conservation of energy to calculate the speed of a body falling straight down in the presence of gravity.

PROBLEM A diver of mass m drops from a board 10.0 m above the water's surface, as in Figure 5.16. Neglect air resistance. **(a)** Use conservation of mechanical energy to find his speed 5.00 m above the water's surface. **(b)** Find his speed as he hits the water.

STRATEGY Refer to the problem-solving strategy. Step 1: The system consists of the diver and Earth. As the diver falls, only the force of gravity acts on him (neglecting air drag), so the mechanical energy of the system is conserved, and we can use conservation of energy for both parts (a) and (b). Step 2: Choose $y = 0$ for the water's surface. Step 3: In part (a), $y = 10.0$ m and $y = 5.00$ m are the points of interest, while in part (b), $y = 10.0$ m and $y = 0$ m are of interest.

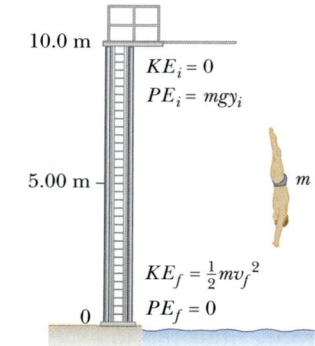

Figure 5.16 (Example 5.5) The zero of gravitational potential energy is taken to be at the water's surface.

SOLUTION

(a) Find the diver's speed halfway down, at $y = 5.00$ m.

Step 4: We write the energy conservation equation and supply the proper terms:

$$KE_i + PE_i = KE_f + PE_f$$
$$\tfrac{1}{2}mv_i^2 + mgy_i = \tfrac{1}{2}mv_f^2 + mgy_f$$

Step 5: Substitute $v_i = 0$, cancel the mass m, and solve for v_f:

$$0 + gy_i = \tfrac{1}{2}v_f^2 + gy_f$$
$$v_f = \sqrt{2g(y_i - y_f)} = \sqrt{2(9.80 \text{ m/s}^2)(10.0 \text{ m} - 5.00 \text{ m})}$$
$$v_f = \boxed{9.90 \text{ m/s}}$$

(b) Find the diver's speed at the water's surface, $y = 0$.

Use the same procedure as in part (a), taking $y_f = 0$:

$$0 + mgy_i = \tfrac{1}{2}mv_f^2 + 0$$
$$v_f = \sqrt{2gy_i} = \sqrt{2(9.80 \text{ m/s}^2)(10.0 \text{ m})} = \boxed{14.0 \text{ m/s}}$$

REMARKS Notice that the speed halfway down is not half the final speed. Another interesting point is that the final answer doesn't depend on the mass. That is really a consequence of neglecting the change in kinetic energy of Earth, which is valid when the mass of the object, the diver in this case, is much smaller than the mass of Earth. In reality, Earth also falls towards the diver, reducing the final speed, but the reduction is so minuscule it could never be measured.

QUESTION 5.5 Qualitatively, how will the answers change if the diver takes a running dive off the end of the board?

EXERCISE 5.5 Suppose the diver vaults off the springboard, leaving it with an initial speed of 3.50 m/s upward. Use energy conservation to find his speed when he strikes the water.

ANSWER 14.4 m/s

EXAMPLE 5.6 | The Jumping Bug

GOAL Use conservation of mechanical energy and concepts from ballistics in two dimensions to calculate a speed.

PROBLEM A powerful grasshopper launches itself at an angle of 45° above the horizontal and rises to a maximum height of 1.00 m during the leap. (See Fig. 5.17.) With what speed v_i did it leave the ground? Neglect air resistance.

STRATEGY This problem can be solved with conservation of energy and the relation between the initial velocity and its x-component. Aside from the origin, the other point of interest is the maximum height $y = 1.00$ m, where the grasshopper has a velocity v_x in the x-direction only. Energy conservation then gives one equation with two unknowns: the initial speed v_i and speed at maximum height, v_x. Because there are no forces in the x-direction, however, v_x is the same as the x-component of the initial velocity.

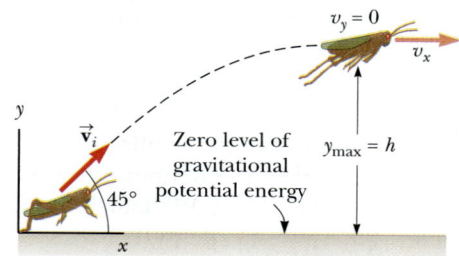

Figure 5.17 (Example 5.6)

SOLUTION

Use energy conservation:

$$\tfrac{1}{2}mv_i^2 + mgy_i = \tfrac{1}{2}mv_f^2 + mgy_f$$

Substitute $y_i = 0$, $v_f = v_x$, and $y_f = h$:

$$\tfrac{1}{2}mv_i^2 = \tfrac{1}{2}mv_x^2 + mgh$$

Multiply each side by $2/m$, obtaining one equation and two unknowns:

(1) $\quad v_i^2 = v_x^2 + 2gh$

Eliminate v_x by substituting $v_x = v_i \cos 45°$ into Equation (1), solving for v_i, and substituting known values:

$$v_i^2 = (v_i \cos 45°)^2 + 2gh = \tfrac{1}{2}v_i^2 + 2gh$$
$$v_i = 2\sqrt{gh} = 2\sqrt{(9.80 \text{ m/s}^2)(1.00 \text{ m})} = \boxed{6.26 \text{ m/s}}$$

REMARKS The final answer is a surprisingly high value and illustrates how strong insects are relative to their size.

QUESTION 5.6 All other given quantities remaining the same, how would the answer change if the initial angle were smaller? Why?

EXERCISE 5.6 A catapult launches a rock at a 30.0° angle with respect to the horizontal. Find the maximum height attained if the speed of the rock at its highest point is 30.0 m/s.

ANSWER 15.3 m

Gravity and Nonconservative Forces

When nonconservative forces are involved along with gravitation, the full work–energy theorem must be used, often with techniques from Chapter 4. Solving problems requires the basic procedure of the problem-solving strategy for conservation-of-energy problems in the previous section. The only difference lies in substituting Equation 5.13, the work–energy equation with potential energy, for Equation 5.15.

> **Tip 5.5 Don't Use Work Done by the Force of Gravity *and* Gravitational Potential Energy!**
>
> Gravitational potential energy is just another way of including the work done by the force of gravity in the work–energy theorem. Don't use both of them in the same equation or you'll count it twice!

■ EXAMPLE 5.7 Der Stuka!

GOAL Use the work–energy theorem with gravitational potential energy to calculate the work done by a nonconservative force.

PROBLEM Waterslides are nearly frictionless, hence can provide bored students with high-speed thrills (Fig. 5.18). One such slide, Der Stuka, named for the terrifying German dive bombers of World War II, is 72.0 feet high (21.9 m), found at Six Flags in Dallas, Texas, and at Wet'n Wild in Orlando, Florida. **(a)** Determine the speed of a 60.0-kg woman at the bottom of such a slide, assuming no friction is present. **(b)** If the woman is clocked at 18.0 m/s at the bottom of the slide, find the work done on the woman by friction.

STRATEGY The system consists of the woman, Earth, and the slide. The normal force, always perpendicular to the displacement, does no work. Let $y = 0$ m represent the bottom of the slide. The two points of interest are $y = 0$ m and $y = 21.9$ m. Without friction, $W_{nc} = 0$, and we can apply conservation of mechanical energy, Equation 5.15. For part (b), use Equation 5.13, substitute two velocities and heights, and solve for W_{nc}.

Figure 5.18 (Example 5.7) If the slide is frictionless, the woman's speed at the bottom depends only on the height of the slide, not on the path it takes.

(Continued)

142 CHAPTER 5 | Energy

SOLUTION

(a) Find the woman's speed at the bottom of the slide, assuming no friction.

Write down Equation 5.15, for conservation of energy:

$$\tfrac{1}{2}mv_i^2 + mgy_i = \tfrac{1}{2}mv_f^2 + mgy_f$$

Insert the values $v_i = 0$ and $v_f = 0$:

$$0 + mgy_i = \tfrac{1}{2}mv_f^2 + 0$$

Solve for v_f and substitute values for g and y_i:

$$v_f = \sqrt{2gy_i} = \sqrt{2(9.80 \text{ m/s}^2)(21.9 \text{ m})} = \boxed{20.7 \text{ m/s}}$$

(b) Find the work done on the woman by friction if $v_f = 18.0$ m/s < 20.7 m/s.

Write Equation 5.13, substituting expressions for the kinetic and potential energies:

$$W_{nc} = (KE_f - KE_i) + (PE_f - PE_i)$$
$$= (\tfrac{1}{2}mv_f^2 - \tfrac{1}{2}mv_i^2) + (mgy_f - mgy_i)$$

Substitute $m = 60.0$ kg, $v_f = 18.0$ m/s, and $v_i = 0$, and solve for W_{nc}:

$$W_{nc} = [\tfrac{1}{2} \cdot 60.0 \text{ kg} \cdot (18.0 \text{ m/s})^2 - 0]$$
$$+ [0 - 60.0 \text{ kg} \cdot (9.80 \text{ m/s}^2) \cdot 21.9 \text{ m}]$$
$$W_{nc} = \boxed{-3.16 \times 10^3 \text{ J}}$$

REMARKS The speed found in part (a) is the same as if the woman fell vertically through a distance of 21.9 m, consistent with our intuition in Quick Quiz 5.3. The result of part (b) is negative because the system loses mechanical energy. Friction transforms part of the mechanical energy into thermal energy and mechanical waves, absorbed partly by the system and partly by the environment.

QUESTION 5.7 If the slide were not frictionless, would the shape of the slide affect the final answer? Explain.

EXERCISE 5.7 Suppose a slide similar to Der Stuka is 35.0 m high, but is a straight slope, inclined at 45.0° with respect to the horizontal. (a) Find the speed of a 60.0-kg woman at the bottom of the slide, assuming no friction. (b) If the woman has a speed of 20.0 m/s at the bottom, find the change in mechanical energy due to friction and (c) the magnitude of the force of friction, assumed constant.

ANSWERS (a) 26.2 m/s (b) -8.58×10^3 J (c) 173 N

■ EXAMPLE 5.8 | Hit the Ski Slopes

GOAL Combine conservation of mechanical energy with the work–energy theorem involving friction on a horizontal surface.

PROBLEM A skier starts from rest at the top of a frictionless incline of height 20.0 m, as in Figure 5.19. At the bottom of the incline, the skier encounters a horizontal surface where the coefficient of kinetic friction between skis and snow is 0.210. **(a)** Find the skier's speed at the bottom. **(b)** How far does the skier travel on the horizontal surface before coming to rest? Neglect air resistance.

Figure 5.19 (Example 5.8) The skier slides down the slope and onto a level surface, stopping after traveling a distance d from the bottom of the hill.

STRATEGY Going down the frictionless incline is physically no different than going down the slide of the previous example and is handled the same way, using conservation of mechanical energy to find the speed $v_Ⓑ$ at the bottom. On the flat, rough surface, use the work–energy theorem, Equation 5.13, with $W_{nc} = W_{fric} = -f_k d$, where f_k is the magnitude of the force of friction and d is the distance traveled on the horizontal surface before coming to rest.

SOLUTION

(a) Find the skier's speed at the bottom.

Follow the procedure used in part (a) of the previous example as the skier moves from the top, point Ⓐ, to the bottom, point Ⓑ:

$$v_Ⓑ = \sqrt{2gh} = \sqrt{2(9.80 \text{ m/s}^2)(20.0 \text{ m})} = \boxed{19.8 \text{ m/s}}$$

(b) Find the distance traveled on the horizontal, rough surface.

Apply the work–energy theorem as the skier moves from Ⓑ to Ⓒ:

$$W_{net} = -f_k d = \Delta KE = \tfrac{1}{2}mv_{Ⓒ}^2 - \tfrac{1}{2}mv_{Ⓑ}^2$$

Substitute $v_{Ⓒ} = 0$ and $f_k = \mu_k n = \mu_k mg$:

$$-\mu_k mgd = -\tfrac{1}{2}mv_{Ⓑ}^2$$

Solve for d:

$$d = \frac{v_{Ⓑ}^2}{2\mu_k g} = \frac{(19.8 \text{ m/s})^2}{2(0.210)(9.80 \text{ m/s}^2)} = \boxed{95.2 \text{ m}}$$

REMARKS Substituting the symbolic expression $v_{Ⓑ} = \sqrt{2gh}$ into the equation for the distance d shows that d is linearly proportional to h: Doubling the height doubles the distance traveled.

QUESTION 5.8 Give two reasons why skiers typically assume a crouching position down when going down a slope.

EXERCISE 5.8 Find the horizontal distance the skier travels before coming to rest if the incline also has a coefficient of kinetic friction equal to 0.210. Assume that $\theta = 20.0°$.

ANSWER 40.3 m

5.4 Spring Potential Energy

LEARNING OBJECTIVES

1. Understand the relationship between spring potential energy and the work done by springs.
2. Extend and apply spring potential energy using the work–energy theorem.

Springs are important elements in modern technology. They are found in machines of all kinds, in watches, toys, cars, and trains. Springs will be introduced here, then studied in more detail in Chapter 13.

Work done by an applied force in stretching or compressing a spring can be recovered by removing the applied force, so like gravity, the spring force is conservative, as long as losses through internal friction of the spring can be neglected. That means a potential energy function can be found and used in the work–energy theorem.

Figure 5.20a shows a spring in its equilibrium position, where the spring is neither compressed nor stretched. Pushing a block against the spring as in Figure 5.20b compresses it a distance x. Although x appears to be merely a coordinate, for springs it also represents a displacement from the equilibrium position, which for our purposes will always be taken to be at $x = 0$. Experimentally, it turns out that doubling a given displacement requires twice the force, and tripling it takes three times the force. This means the force exerted by the spring, F_s, must be proportional to the displacement x, or

$$F_s = -kx \qquad [5.16]$$

where k is a constant of proportionality, the *spring constant*, carrying units of newtons per meter. Equation 5.16 is called **Hooke's law**, after Sir Robert Hooke, who discovered the relationship. The force F_s is often called a *restoring force* because the spring always exerts a force in a direction opposite the displacement of its end, tending to restore whatever is attached to the spring to its original position. For positive values of x, the force is negative, pointing back towards equilibrium at $x = 0$, and for negative x, the force is positive, again pointing towards $x = 0$. For a flexible spring, k is a small number (about 100 N/m), whereas for a stiff spring k is large (about 10 000 N/m). The value of the spring constant k is determined by how the spring was formed, its material composition, and the thickness of the wire. The minus sign ensures that the spring force is always directed back towards the equilibrium point.

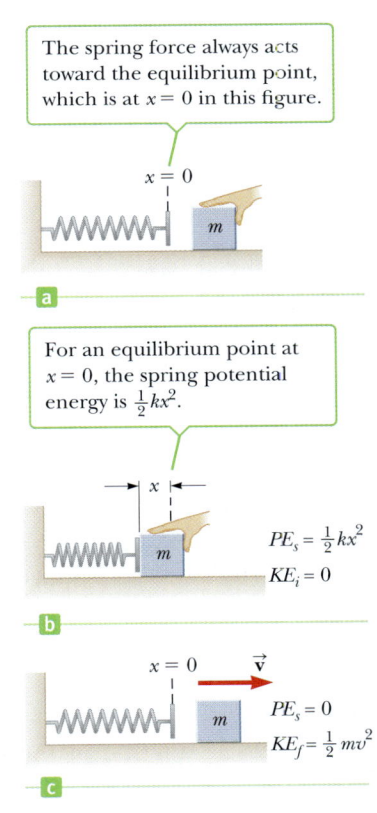

Figure 5.20 (a) A spring at equilibrium, neither compressed nor stretched. (b) A block of mass m on a frictionless surface is pushed against the spring. (c) When the block is released, the energy stored in the spring is transferred to the block in the form of kinetic energy.

As in the case of gravitation, a potential energy, called the **elastic potential energy,** can be associated with the spring force. Elastic potential energy is another way of looking at the work done by a spring during motion because it is equal to the negative of the work done by the spring. It can also be considered stored energy arising from the work done to compress or stretch the spring.

Consider a horizontal spring and mass at the equilibrium position. We determine the work done by the spring when compressed by an applied force from equilibrium to a displacement x, as in Figure 5.20b. The spring force points in the direction opposite the motion, so we expect the work to be negative. When we studied the constant force of gravity near Earth's surface, we found the work done on an object by multiplying the gravitational force by the vertical displacement of the object. However, this procedure can't be used with a varying force such as the spring force. Instead, we use the average force, $\overline{F}$:

$$\overline{F} = \frac{F_0 + F_1}{2} = \frac{0 - kx}{2} = -\frac{kx}{2}$$

Therefore, the work *done by the spring force* is

$$W_s = \overline{F}x = -\tfrac{1}{2}kx^2$$

In general, when the spring is stretched or compressed from x_i to x_f, the work done by the spring is

$$W_s = -(\tfrac{1}{2}kx_f^2 - \tfrac{1}{2}kx_i^2)$$

The work done by a spring can be included in the work–energy theorem. Assume Equation 5.13 now includes the work done by springs on the left-hand side. It then reads

$$W_{nc} - (\tfrac{1}{2}kx_f^2 - \tfrac{1}{2}kx_i^2) = \Delta KE + \Delta PE_g$$

where PE_g is the gravitational potential energy. We now define the elastic potential energy associated with the spring force, PE_s, by

$$PE_s \equiv \tfrac{1}{2}kx^2 \qquad [5.17]$$

Inserting this expression into the previous equation and rearranging gives the new form of the work–energy theorem, including both gravitational and elastic potential energy:

$$W_{nc} = (KE_f - KE_i) + (PE_{gf} - PE_{gi}) + (PE_{sf} - PE_{si}) \qquad [5.18]$$

where W_{nc} is the work done by nonconservative forces, KE is kinetic energy, PE_g is gravitational potential energy, and PE_s is the elastic potential energy. PE, formerly used to denote gravitational potential energy alone, will henceforth denote the total potential energy of a system, including potential energies due to all conservative forces acting on the system.

It's important to remember that the work done by gravity and springs in any given physical system is already included on the right-hand side of Equation 5.18 as potential energy and should not also be included on the left as work.

Figure 5.20c shows how the stored elastic potential energy can be recovered. When the block is released, the spring snaps back to its original length, and the stored elastic potential energy is converted to kinetic energy of the block. The elastic potential energy stored in the spring is zero when the spring is in the equilibrium position ($x = 0$). As given by Equation 5.17, potential energy is also stored in the spring when it's stretched. Further, the elastic potential energy is a maximum when the spring has reached its maximum compression or extension. Finally, because PE_s is proportional to x^2, the potential energy is always positive when the spring is not in the equilibrium position.

In the absence of nonconservative forces, $W_{nc} = 0$, so the left-hand side of Equation 5.18 is zero, and an extended form for conservation of mechanical energy results:

$$(KE + PE_g + PE_s)_i = (KE + PE_g + PE_s)_f \quad [5.19]$$

Problems involving springs, gravity, and other forces are handled in exactly the same way as described in the problem-solving strategy for conservation of mechanical energy, except that the equilibrium point of any spring in the problem must be defined in addition to the zero point for gravitational potential energy.

■ EXAMPLE 5.9 | A Horizontal Spring

GOAL Use conservation of energy to calculate the speed of a block on a horizontal spring with and without friction.

PROBLEM A block with mass of 5.00 kg is attached to a horizontal spring with spring constant $k = 4.00 \times 10^2$ N/m, as in Figure 5.21. The surface the block rests upon is frictionless. If the block is pulled out to $x_i = 0.050\ 0$ m and released, **(a)** find the speed of the block when it first reaches the equilibrium point, **(b)** find the speed when $x = 0.025\ 0$ m, and **(c)** repeat part (a) if friction acts on the block, with coefficient $\mu_k = 0.150$.

STRATEGY In parts (a) and (b) there are no nonconservative forces, so conservation of energy, Equation 5.19, can be applied. In part (c) the definition of work and the work–energy theorem are needed to deal with the loss of mechanical energy due to friction.

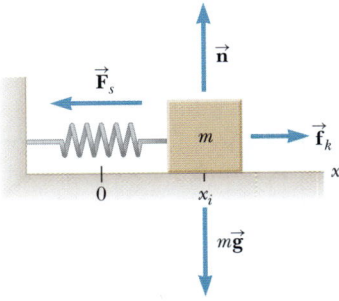

Figure 5.21 (Example 5.9) A mass attached to a spring.

SOLUTION

(a) Find the speed of the block at equilibrium point.

Start with Equation 5.19:

$$(KE + PE_g + PE_s)_i = (KE + PE_g + PE_s)_f$$

Substitute expressions for the block's kinetic energy and the potential energy, and set the gravity terms to zero:

$$(1) \quad \tfrac{1}{2}mv_i^2 + \tfrac{1}{2}kx_i^2 = \tfrac{1}{2}mv_f^2 + \tfrac{1}{2}kx_f^2$$

Substitute $v_i = 0$, $x_f = 0$, and multiply by $2/m$:

$$\frac{k}{m}x_i^2 = v_f^2$$

Solve for v_f and substitute the given values:

$$v_f = \sqrt{\frac{k}{m}}\, x_i = \sqrt{\frac{4.00 \times 10^2\ \text{N/m}}{5.00\ \text{kg}}}\,(0.050\ 0\ \text{m})$$

$$= \boxed{0.447\ \text{m/s}}$$

(b) Find the speed of the block at the halfway point.

Set $v_i = 0$ in Equation (1) and multiply by $2/m$:

$$\frac{kx_i^2}{m} = v_f^2 + \frac{kx_f^2}{m}$$

Solve for v_f and substitute the given values:

$$v_f = \sqrt{\frac{k}{m}(x_i^2 - x_f^2)}$$

$$= \sqrt{\frac{4.00 \times 10^2\ \text{N/m}}{5.00\ \text{kg}}\,[(0.050\ \text{m})^2 - (0.025\ \text{m})^2]}$$

$$= \boxed{0.387\ \text{m/s}}$$

(c) Repeat part (a), this time with friction.

Apply the work–energy theorem. The work done by the force of gravity and the normal force is zero because these forces are perpendicular to the motion.

$$W_{\text{fric}} = \tfrac{1}{2}mv_f^2 - \tfrac{1}{2}mv_i^2 + \tfrac{1}{2}kx_f^2 - \tfrac{1}{2}kx_i^2$$

Substitute $v_i = 0$, $x_f = 0$, and $W_{fric} = -\mu_k n x_i$:

$$-\mu_k n x_i = \tfrac{1}{2} m v_f^2 - \tfrac{1}{2} k x_i^2$$

Set $n = mg$ and solve for v_f:

$$\tfrac{1}{2} m v_f^2 = \tfrac{1}{2} k x_i^2 - \mu_k m g x_i$$

$$v_f = \sqrt{\frac{k}{m} x_i^2 - 2\mu_k g x_i}$$

$$v_f = \sqrt{\frac{4.00 \times 10^2 \text{ N/m}}{5.00 \text{ kg}} (0.050\ 0 \text{ m})^2 - 2(0.150)(9.80 \text{ m/s}^2)(0.050\ 0 \text{ m})}$$

$$v_f = \boxed{0.230 \text{ m/s}}$$

REMARKS Friction or drag from immersion in a fluid damps the motion of an object attached to a spring, eventually bringing the object to rest.

QUESTION 5.9 In the case of friction, what percent of the mechanical energy was lost by the time the mass first reached the equilibrium point? (*Hint:* use the answers to parts (a) and (c).)

EXERCISE 5.9 Suppose the spring system in the last example starts at $x = 0$ and the attached object is given a kick to the right, so it has an initial speed of 0.600 m/s. (a) What distance from the origin does the object travel before coming to rest, assuming the surface is frictionless? (b) How does the answer change if the coefficient of kinetic friction is $\mu_k = 0.150$? (Use the quadratic formula.)

ANSWERS (a) 0.067 1 m (b) 0.051 2 m

■ EXAMPLE 5.10 Circus Acrobat

GOAL Use conservation of mechanical energy to solve a one-dimensional problem involving gravitational potential energy and spring potential energy.

Figure 5.22 (Example 5.10) An acrobat drops onto a springboard, causing it to compress.

PROBLEM A 50.0-kg circus acrobat drops from a height of 2.00 meters straight down onto a springboard with a force constant of 8.00×10^3 N/m, as in Figure 5.22. By what maximum distance does she compress the spring?

STRATEGY Nonconservative forces are absent, so conservation of mechanical energy can be applied. At the two points of interest, the acrobat's initial position and the point of maximum spring compression, her velocity is zero, so the kinetic energy terms will be zero. Choose $y = 0$ as the point of maximum compression, so the final gravitational potential energy is zero. This choice also means that the initial position of the acrobat is $y_i = h + d$, where h is the acrobat's initial height above the platform and d is the spring's maximum compression.

SOLUTION

Use conservation of mechanical energy:

(1) $(KE + PE_g + PE_s)_i = (KE + PE_g + PE_s)_f$

The only nonzero terms are the initial gravitational potential energy and the final spring potential energy.

$$0 + mg(h + d) + 0 = 0 + 0 + \tfrac{1}{2} k d^2$$
$$mg(h + d) = \tfrac{1}{2} k d^2$$

Substitute the given quantities and rearrange the equation into standard quadratic form:

$$(50.0 \text{ kg})(9.80 \text{ m/s}^2)(2.00 \text{ m} + d) = \tfrac{1}{2}(8.00 \times 10^3 \text{ N/m}) d^2$$
$$d^2 - (0.123 \text{ m})d - 0.245 \text{ m}^2 = 0$$

Solve with the quadratic formula (Equation A.8): $d = \boxed{0.560 \text{ m}}$

REMARKS The other solution, $d = -0.437$ m, can be rejected because d was chosen to be a positive number at the outset. A change in the acrobat's center of mass, say, by crouching as she makes contact with the springboard, also affects the spring's compression, but that effect was neglected. Shock absorbers often involve springs, and this example illustrates how they work. The spring action of a shock absorber turns a dangerous jolt into a smooth deceleration, as excess kinetic energy is converted to spring potential energy.

QUESTION 5.10 Is it possible for the acrobat to rebound to a height greater than her initial height? If so, how?

EXERCISE 5.10 An 8.00-kg block drops straight down from a height of 1.00 m, striking a platform spring having force constant 1.00×10^3 N/m. Find the maximum compression of the spring.

ANSWER $d = 0.482$ m

EXAMPLE 5.11 | A Block Projected up a Frictionless Incline

GOAL Use conservation of mechanical energy to solve a problem involving gravitational potential energy, spring potential energy, and a ramp.

PROBLEM A 0.500-kg block rests on a horizontal, frictionless surface as in Figure 5.23. The block is pressed back against a spring having a constant of $k = 625$ N/m, compressing the spring by 10.0 cm to point Ⓐ. Then the block is released. (a) Find the maximum distance d the block travels up the frictionless incline if $\theta = 30.0°$. (b) How fast is the block going at half its maximum height?

Figure 5.23 (Example 5.11)

STRATEGY In the absence of other forces, conservation of mechanical energy applies to parts (a) and (b). In part (a), the block starts at rest and is also instantaneously at rest at the top of the ramp, so the kinetic energies at Ⓐ and Ⓒ are both zero. Note that the question asks for a distance d along the ramp, not the height h. In part (b), the system has both kinetic and gravitational potential energy at Ⓑ.

SOLUTION

(a) Find the distance the block travels up the ramp.

Apply conservation of mechanical energy:

$$\tfrac{1}{2}mv_i^2 + mgy_i + \tfrac{1}{2}kx_i^2 = \tfrac{1}{2}mv_f^2 + mgy_f + \tfrac{1}{2}kx_f^2$$

Substitute $v_i = v_f = 0$, $y_i = 0$, $y_f = h = d \sin \theta$, and $x_f = 0$:

$$\tfrac{1}{2}kx_i^2 = mgh = mgd \sin \theta$$

Solve for the distance d and insert the known values:

$$d = \frac{\tfrac{1}{2}kx_i^2}{mg \sin \theta} = \frac{\tfrac{1}{2}(625 \text{ N/m})(-0.100 \text{ m})^2}{(0.500 \text{ kg})(9.80 \text{ m/s}^2) \sin (30.0°)}$$

$$= \boxed{1.28 \text{ m}}$$

(b) Find the velocity at half the height, $h/2$. Note that $h = d \sin \theta = (1.28 \text{ m}) \sin 30.0° = 0.640$ m.

Use energy conservation again:

$$\tfrac{1}{2}mv_i^2 + mgy_i + \tfrac{1}{2}kx_i^2 = \tfrac{1}{2}mv_f^2 + mgy_f + \tfrac{1}{2}kx_f^2$$

Take $v_i = 0$, $y_i = 0$, $y_f = \tfrac{1}{2} h$, and $x_f = 0$, yielding

$$\tfrac{1}{2} kx_i^2 = \tfrac{1}{2}mv_f^2 + mg\left(\tfrac{1}{2}h\right)$$

Multiply by $2/m$ and solve for v_f:

$$\frac{k}{m} x_i^2 = v_f^2 + gh$$

$$v_f = \sqrt{\frac{k}{m} x_i^2 - gh}$$

$$= \sqrt{\left(\frac{625 \text{ N/m}}{0.500 \text{ kg}}\right)(-0.100 \text{ m})^2 - (9.80 \text{ m/s}^2)(0.640 \text{ m})}$$

$$v_f = \boxed{2.50 \text{ m/s}}$$

REMARKS Notice that it wasn't necessary to compute the velocity gained upon leaving the spring: only the mechanical energy at each of the two points of interest was required, where the block was at rest.

(Continued)

QUESTION 5.11 A real spring will continue to vibrate slightly after the mass has left it. How would this affect the answer to part (a), and why?

EXERCISE 5.11 A 1.00-kg block is shot horizontally from a spring, as in the previous example, and travels 0.500 m up along a frictionless ramp before coming to rest and sliding back down. If the ramp makes an angle of 45.0° with respect to the horizontal, and the spring was originally compressed by 0.120 m, find the spring constant.

ANSWER 481 N/m

APPLYING PHYSICS 5.2 | Accident Reconstruction

Sometimes people involved in automobile accidents make exaggerated claims of chronic pain due to subtle injuries to the neck or spinal column. The likelihood of injury can be determined by finding the change in velocity of a car during the accident. The larger the change in velocity, the more likely it is that the person suffered spinal injury resulting in chronic pain. How can reliable estimates for this change in velocity be found after the fact?

EXPLANATION The metal and plastic of an automobile acts much like a spring, absorbing the car's kinetic energy by flexing during a collision. When the magnitude of the difference in velocity of the two cars is under 5 mi/h, there is usually no visible damage, because bumpers are designed to absorb the impact and return to their original shape at such low speeds. At greater relative speeds there will be permanent damage to the vehicle. Despite the fact the structure of the car may not return to its original shape, a certain force per meter is still required to deform it, just as it takes a certain force per meter to compress a spring. The greater the original kinetic energy, the more the car is compressed during a collision, and the greater the damage. By using data obtained through crash tests, it's possible to obtain effective spring constants for all the different models of cars and determine reliable estimates of the change in velocity of a given vehicle during an accident. Medical research has established the likelihood of spinal injury for a given change in velocity, and the estimated velocity change can be used to help reduce insurance fraud. ∎

5.5 Systems and Energy Conservation

LEARNING OBJECTIVES

1. State the work–energy theorem in terms of the total mechanical energy.
2. Discuss the different forms of energy and energy transfer, and give examples.
3. State the general principle of conservation of energy and discuss its consequences.

Recall that the work–energy theorem can be written as

$$W_{nc} + W_c = \Delta KE$$

where W_{nc} represents the work done by nonconservative forces and W_c is the work done by conservative forces in a given physical context. As we have seen, any work done by conservative forces, such as gravity and springs, can be accounted for by changes in potential energy. The work–energy theorem can therefore be written in the following way:

$$W_{nc} = \Delta KE + \Delta PE = (KE_f - KE_i) + (PE_f - PE_i) \qquad [5.20]$$

where now, as previously stated, PE includes all potential energies. This equation is easily rearranged to

$$W_{nc} = (KE_f + PE_f) - (KE_i + PE_i) \qquad [5.21]$$

Recall, however, that the total mechanical energy is given by $E = KE + PE$. Making this substitution into Equation 5.21, we find that the work done on a system by all nonconservative forces is equal to the change in mechanical energy of that system:

$$W_{nc} = E_f - E_i = \Delta E \qquad [5.22]$$

If the mechanical energy is changing, it has to be going somewhere. The energy either leaves the system and goes into the surrounding environment, or it stays in the system and is converted into a nonmechanical form such as thermal energy.

A simple example is a block sliding along a rough surface. Friction creates thermal energy, absorbed partly by the block and partly by the surrounding environment. When the block warms up, something called *internal energy* increases. The internal energy of a system is related to its temperature, which in turn is a consequence of the activity of its parts, such as the motion of atoms in a gas or the vibration of atoms in a solid. (Internal energy will be studied in more detail in Chapters 10–12.)

Energy can be transferred between a nonisolated system and its environment. If positive work is done on the system, energy is transferred from the environment to the system. If negative work is done on the system, energy is transferred from the system to the environment.

So far, we have encountered three methods of storing energy in a system: kinetic energy, potential energy, and internal energy. On the other hand, we've seen only one way of transferring energy into or out of a system: through work. Other methods will be studied in later chapters, but are summarized here:

- **Work**, in the mechanical sense of this chapter, transfers energy to a system by displacing it with an applied force.
- **Heat** is the process of transferring energy through microscopic collisions between atoms or molecules. For example, a metal spoon resting in a cup of coffee becomes hot because some of the kinetic energy of the molecules in the liquid coffee is transferred to the spoon as internal energy.
- **Mechanical waves** transfer energy by creating a disturbance that propagates through air or another medium. For example, energy in the form of sound leaves your stereo system through the loudspeakers and enters your ears to stimulate the hearing process. Other examples of mechanical waves are seismic waves and ocean waves.
- **Electrical transmission** transfers energy through electric currents. This is how energy enters your stereo system or any other electrical device.
- **Electromagnetic radiation** transfers energy in the form of electromagnetic waves such as light, microwaves, and radio waves. Examples of this method of transfer include cooking a potato in a microwave oven and light energy traveling from the Sun to Earth through space.

Conservation of Energy in General

The most important feature of the energy approach is the idea that energy is conserved; it can't be created or destroyed, only transferred from one form into another. This is the principle of **conservation of energy.**

The principle of conservation of energy is not confined to physics. In biology, energy transformations take place in myriad ways inside all living organisms. One example is the transformation of chemical energy to mechanical energy that causes flagella to move and propel an organism. Some bacteria use chemical energy to produce light. (See Fig. 5.24.) Although the mechanisms that produce these light emissions are not well understood, living creatures often rely on this light for their existence. For example, certain fish have sacs beneath their eyes filled with light-emitting bacteria. The emitted light attracts creatures that become food for the fish.

Quick Quiz

5.4 A book of mass m is projected with a speed v across a horizontal surface. The book slides until it stops due to the friction force between the book and the surface. The surface is now tilted 30°, and the book is projected up the surface with the same initial speed v. When the book has come to rest, how does the decrease in mechanical energy of the book–Earth system compare with that when the book slid over the horizontal surface? (a) It's the same. (b) It's larger on the tilted surface. (c) It's smaller on the tilted surface. (d) More information is needed.

BIO APPLICATION
Flagellar Movement; Bioluminescence

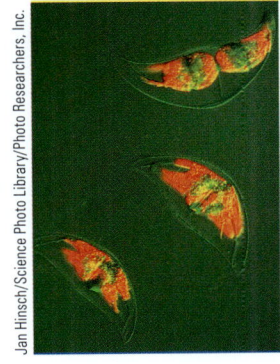

Figure 5.24 This small plant, found in warm southern waters, exhibits bioluminescence, a process in which chemical energy is converted to light. The red areas are chlorophyll, which fluoresce when irradiated with blue light.

APPLYING PHYSICS 5.3 | Asteroid Impact!

An asteroid about 10 kilometers in diameter has been blamed for the extinction of the dinosaurs 65 million years ago. How can a relatively small object, which could fit inside a college campus, inflict such injury on the vast biosphere of Earth?

EXPLANATION While such an asteroid is comparatively small, it travels at a very high speed relative to Earth, typically on the order of 40 000 m/s. A roughly spherical asteroid 10 kilometers in diameter and made mainly of rock has a mass of approximately 1 000 trillion kilograms—a mountain of matter. The kinetic energy of such an asteroid would be about 10^{24} J, or a trillion trillion joules. By contrast, the atomic bomb that devastated Hiroshima was equivalent to 15 kilotons of TNT, approximately 6×10^{13} J of energy. On striking Earth, the asteroid's enormous kinetic energy changes into other forms, such as thermal energy, sound, and light, with a total energy release greater than ten billion Hiroshima explosions! Aside from the devastation in the immediate blast area and fires across a continent, gargantuan tidal waves would scour low-lying regions around the world and dust would block the Sun for decades.

For this reason, asteroid impacts represent a threat to life on Earth. Asteroids large enough to cause widespread extinction hit Earth only every 60 million years or so.

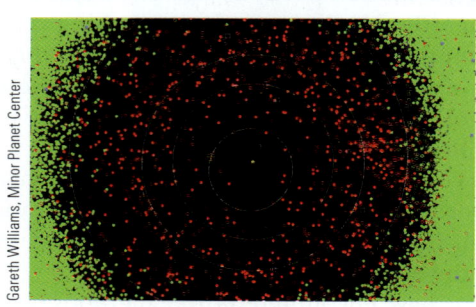

Figure 5.25 Asteroid map of the inner solar system. The violet circles represent the orbits of the inner planets. Green dots stand for asteroids not considered dangerous to Earth; those that are considered threatening are represented by red dots.

Smaller asteroids, of sufficient size to cause serious damage to civilization on a global scale, are thought to strike every five to ten thousand years. There have been several near misses by such asteroids in the last century and even in the last decade. In 1907, a small asteroid or comet fragment struck Tunguska, Siberia, annihilating a region 60 kilometers across. Had it hit northern Europe, millions of people might have perished.

Figure 5.25 is an asteroid map of the inner solar system. More asteroids are being discovered every year. ■

5.6 Power

LEARNING OBJECTIVES

1. Define average power and instantaneous power and explain their physical meaning.
2. Calculate average power in simple physical contexts.
3. Calculate instantaneous power in simple physical contexts.

Power, the rate at which energy is transferred, is important in the design and use of practical devices, such as electrical appliances and engines of all kinds. The concept of power, however, is essential whenever a transfer of any kind of energy takes place. The issue is particularly interesting for living creatures because the maximum work per second, or power output, of an animal varies greatly with output duration. Power is defined as the rate of energy transfer with time:

▶ Average power

If an external force does work W on an object in the time interval Δt, then the **average power** delivered to the object is the work done divided by the time interval, or

$$\overline{P} = \frac{W}{\Delta t} \quad [5.23]$$

SI unit: watt (W = J/s)

It's sometimes useful to rewrite Equation 5.23 by substituting $W = F \Delta x$ and noticing that $\Delta x / \Delta t$ is the average velocity of the object during the time Δt:

$$\overline{P} = \frac{W}{\Delta t} = \frac{F \Delta x}{\Delta t} = F \overline{v} \quad [5.24]$$

According to Equation 5.24, average power is a constant force times the average velocity. The force F is the component of force in the direction of the average velocity. A more general definition, called the **instantaneous power,** can be written down with a little calculus and has the same form as Equation 5.24:

$$P = Fv \qquad [5.25]$$

◀ Instantaneous power

In Equation 5.25 both the force F and the velocity v must be parallel, but can change with time. The SI unit of power is the joule per second (J/s), also called the **watt,** named after James Watt:

$$1 \text{ W} = 1 \text{ J/s} = 1 \text{ kg} \cdot \text{m}^2/\text{s}^3 \qquad [5.26\text{a}]$$

The unit of power in the U.S. customary system is the horsepower (hp), where

$$1 \text{ hp} \equiv 550 \frac{\text{ft} \cdot \text{lb}}{\text{s}} = 746 \text{ W} \qquad [5.26\text{b}]$$

The horsepower was first defined by Watt, who needed a large power unit to rate the power output of his new invention, the steam engine.

The watt is commonly used in electrical applications, but it can be used in other scientific areas as well. For example, European sports car engines are rated in kilowatts.

In electric power generation, it's customary to use the kilowatt-hour as a measure of energy. One kilowatt-hour (kWh) is the energy transferred in 1 h at the constant rate of 1 kW = 1 000 J/s. Therefore,

$$1 \text{ kWh} = (10^3 \text{ W})(3\,600 \text{ s}) = (10^3 \text{ J/s})(3\,600 \text{ s}) = 3.60 \times 10^3 \text{ J}$$

Tip 5.6 Watts the Difference?

Don't confuse the nonitalic symbol for watts, W, with the italic symbol W for work. A watt is a unit, the same as joules per second. Work is a concept, carrying units of joules.

It's important to realize that a kilowatt-hour is a unit of energy, *not* power. When you pay your electric bill, you're buying energy, and that's why your bill lists a charge for electricity of about 10 cents/kWh. The amount of electricity used by an appliance can be calculated by multiplying its power rating (usually expressed in watts and valid only for normal household electrical circuits) by the length of time the appliance is operated. For example, an electric bulb rated at 100 W (= 0.100 kW) "consumes" 3.6×10^5 J of energy in 1 h.

■ EXAMPLE 5.12 | Power Delivered by an Elevator Motor

GOAL Apply the force-times-velocity definition of power.

PROBLEM A 1.00×10^3-kg elevator car carries a maximum load of 8.00×10^2 kg. A constant frictional force of 4.00×10^3 N retards its motion upward, as in Figure 5.26. What minimum power, in kilowatts and in horsepower, must the motor deliver to lift the fully loaded elevator car at a constant speed of 3.00 m/s?

STRATEGY To solve this problem, we need to determine the force the elevator car's motor must deliver through the force of tension in the cable, $\vec{\mathbf{T}}$. Substituting this force together with the given speed v into $P = Fv$ gives the desired power. The tension in the cable, T, can be found with Newton's second law.

Figure 5.26 (a) (Example 5.12) The motor exerts an upward force $\vec{\mathbf{T}}$ on the elevator. A frictional force $\vec{\mathbf{f}}$ and the force of gravity $M\vec{\mathbf{g}}$ act downward. (b) The free-body diagram for the elevator car.

(Continued)

SOLUTION

Apply Newton's second law to the elevator car:

$$\sum \vec{F} = m\vec{a}$$

The velocity is constant, so the acceleration is zero. The forces acting on the elevator car are the force of tension in the cable, $\vec{T}$, the friction $\vec{f}$, and gravity $M\vec{g}$, where M is the mass of the elevator car.

$$\vec{T} + \vec{f} + M\vec{g} = 0$$

Write the equation in terms of its components:

$$T - f - Mg = 0$$

Solve this equation for the tension T and evaluate it:

$$T = f + Mg$$
$$= 4.00 \times 10^3 \text{ N} + (1.80 \times 10^3 \text{ kg})(9.80 \text{ m/s}^2)$$
$$T = 2.16 \times 10^4 \text{ N}$$

Substitute this value of T for F in the power equation:

$$P = Fv = (2.16 \times 10^4 \text{ N})(3.00 \text{ m/s}) = 6.48 \times 10^4 \text{ W}$$
$$P = 64.8 \text{ kW} = \boxed{86.9 \text{ hp}}$$

REMARKS The friction force acts to retard the motion, requiring more power. For a descending elevator car, the friction force can actually reduce the power requirement.

QUESTION 5.12 In general, are the minimum power requirements of an elevator car ascending at constant velocity (a) greater than, (b) less than, or (c) equal to the minimum power requirements of an elevator car descending at constant velocity?

EXERCISE 5.12 Suppose the same elevator car with the same load descends at 3.00 m/s. What minimum power is required? (Here, the motor removes energy from the elevator car by not allowing it to fall freely.)

ANSWER 4.09×10^4 W = 54.9 hp

■ EXAMPLE 5.13 | Shamu Sprint BIO

GOAL Calculate the average power needed to increase an object's kinetic energy.

PROBLEM Killer whales are known to reach 32 ft in length and have a mass of over 8 000 kg. They are also very quick, able to accelerate up to 30 mi/h in a matter of seconds. Disregarding the considerable drag force of water, calculate the average power a killer whale named Shamu with mass 8.00×10^3 kg would need to generate to reach a speed of 12.0 m/s in 6.00 s.

STRATEGY Find the change in kinetic energy of Shamu and use the work–energy theorem to obtain the minimum work Shamu has to do to effect this change. (Internal and external friction forces increase the necessary amount of energy.) Divide by the elapsed time to get the average power.

SOLUTION

Calculate the change in Shamu's kinetic energy. By the work–energy theorem, this equals the minimum work Shamu must do:

$$\Delta KE = \tfrac{1}{2}mv_f^2 - \tfrac{1}{2}mv_i^2$$
$$= \tfrac{1}{2} \cdot 8.00 \times 10^3 \text{ kg} \cdot (12.0 \text{ m/s})^2 - 0$$
$$= 5.76 \times 10^5 \text{ J}$$

Divide by the elapsed time (Eq. 5.23), noting that $W = \Delta KE$:

$$\overline{P} = \frac{W}{\Delta t} = \frac{5.76 \times 10^5 \text{ J}}{6.00 \text{ s}} = \boxed{9.60 \times 10^4 \text{ W}}$$

REMARKS This is enough power to run a moderate-sized office building! The actual requirements are larger because of friction in the water and muscular tissues. Something similar can be done with gravitational potential energy, as the exercise illustrates.

QUESTION 5.13 If Shamu could double his velocity in double the time, by what factor would the average power requirement change?

EXERCISE 5.13 What minimum average power must a 35-kg human boy generate climbing up the stairs to the top of the Washington monument? The trip up the nearly 170-m-tall building takes him 10 minutes. Include only work done against gravity, ignoring biological inefficiency.

ANSWER 97 W

EXAMPLE 5.14 | Speedboat Power

GOAL Combine power, the work–energy theorem, and nonconservative forces with one-dimensional kinematics.

PROBLEM (a) What average power would a 1.00×10^3-kg speedboat need to go from rest to 20.0 m/s in 5.00 s, assuming the water exerts a constant drag force of magnitude $f_d = 5.00 \times 10^2$ N and the acceleration is constant. (b) Find an expression for the instantaneous power in terms of the drag force f_d, the mass m, acceleration a, and time t.

STRATEGY The power is provided by the engine, which creates a nonconservative force. Use the work–energy theorem together with the work done by the engine, W_{engine}, and the work done by the drag force, W_{drag}, on the left-hand side. Use one-dimensional kinematics to find the acceleration and then the displacement Δx. Solve the work–energy theorem for W_{engine}, and divide by the elapsed time to get the average power. For part (b), use Newton's second law to obtain an example for F_E, and then substitute into the definition of instantaneous power.

SOLUTION

(a) Write the work–energy theorem:

$$W_{net} = \Delta KE = \tfrac{1}{2}mv_f^2 - \tfrac{1}{2}mv_i^2$$

Fill in the two work terms and take $v_i = 0$:

(1) $\quad W_{engine} + W_{drag} = \tfrac{1}{2}mv_f^2$

To get the displacement Δx, first find the acceleration using the velocity equation of kinematics:

$$v_f = at + v_i \rightarrow v_f = at$$
$$20.0 \text{ m/s} = a(5.00 \text{ s}) \rightarrow a = 4.00 \text{ m/s}^2$$

Substitute a into the time-independent kinematics equation and solve for Δx:

$$v_f^2 - v_i^2 = 2a\,\Delta x$$
$$(20.0 \text{ m/s})^2 - 0^2 = 2(4.00 \text{ m/s}^2)\,\Delta x$$
$$\Delta x = 50.0 \text{ m}$$

Now that we know Δx, we can find the mechanical energy lost due to the drag force:

$$W_{drag} = -f_d\,\Delta x = -(5.00 \times 10^2 \text{ N})(50.0 \text{ m}) = -2.50 \times 10^4 \text{ J}$$

Solve equation (1) for W_{engine}:

$$W_{engine} = \tfrac{1}{2}mv_f^2 - W_{drag}$$
$$= \tfrac{1}{2}(1.00 \times 10^3 \text{ kg})(20.0 \text{ m/s})^2 - (-2.50 \times 10^4 \text{ J})$$
$$W_{engine} = 2.25 \times 10^5 \text{ J}$$

Compute the average power:

$$\overline{P} = \frac{W_{engine}}{\Delta t} = \frac{2.25 \times 10^5 \text{ J}}{5.00 \text{ s}} = 4.50 \times 10^4 \text{ W} = \boxed{60.3 \text{ hp}}$$

(b) Find a symbolic expression for the instantaneous power.

Use Newton's second law:

$$ma = F_E - f_d$$

Solve for the force exerted by the engine, F_E:

$$F_E = ma + f_d$$

Substitute the expression for F_E and $v = at$ into Equation 5.25 to obtain the instantaneous power:

$$P = F_E v = (ma + f_d)(at)$$
$$\boxed{P = (ma^2 + af_d)t}$$

REMARKS In fact, drag forces generally get larger with increasing speed.

QUESTION 5.14 How does the instantaneous power at the end of 5.00 s compare to the average power?

EXERCISE 5.14 What average power must be supplied to push a 5.00-kg block from rest to 10.0 m/s in 5.00 s when the coefficient of kinetic friction between the block and surface is 0.250? Assume the acceleration is uniform.

ANSWER 111 W

Energy and Power in a Vertical Jump BIO

The stationary jump consists of two parts: extension and free flight.[2] In the extension phase the person jumps up from a crouch, straightening the legs and throwing up the arms; the free-flight phase occurs when the jumper leaves the ground.

---

[2] For more information on this topic, see E. J. Offenbacher, *American Journal of Physics*, **38**, 829 (1969).

Figure 5.27 Extension and free flight in the vertical jump.

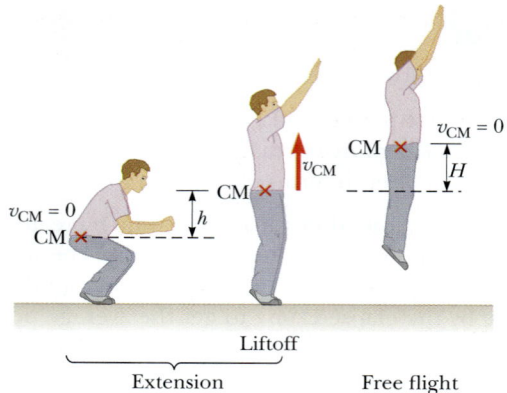

Because the body is an extended object and different parts move with different speeds, we describe the motion of the jumper in terms of the position and velocity of the **center of mass (CM)**, which is the point in the body at which all the mass may be considered to be concentrated. Figure 5.27 shows the position and velocity of the CM at different stages of the jump.

Using the principle of the conservation of mechanical energy, we can find H, the maximum increase in height of the CM, in terms of the velocity v_{CM} of the CM at liftoff. Taking PE_i, the gravitational potential energy of the jumper–Earth system just as the jumper lifts off from the ground to be zero, and noting that the kinetic energy KE_f of the jumper at the peak is zero, we have

$$PE_i + KE_i = PE_f + KE_f$$

$$\tfrac{1}{2}mv_{CM}^2 = mgH \quad \text{or} \quad H = \frac{v_{CM}^2}{2g}$$

We can estimate v_{CM} by assuming that the acceleration of the CM is constant during the extension phase. If the depth of the crouch is h and the time for extension is Δt, we find that $v_{CM} = 2\bar{v} = 2h/\Delta t$. Measurements on a group of male college students show typical values of $h = 0.40$ m and $\Delta t = 0.25$ s, the latter value being set by the fixed speed with which muscle can contract. Substituting, we obtain

$$v_{CM} = 2(0.40 \text{ m})/(0.25 \text{ s}) = 3.2 \text{ m/s}$$

and

$$H = \frac{v_{CM}^2}{2g} = \frac{(3.2 \text{ m/s})^2}{2(9.80 \text{ m/s}^2)} = 0.52 \text{ m}$$

Measurements on this same group of students found that H was between 0.45 m and 0.61 m in all cases, confirming the basic validity of our simple calculation.

To relate the abstract concepts of energy, power, and efficiency to humans, it's interesting to calculate these values for the vertical jump. The kinetic energy given to the body in a jump is $KE = \tfrac{1}{2}mv_{CM}^2$, and for a person of mass 68 kg, the kinetic energy is

$$KE = \tfrac{1}{2}(68 \text{ kg})(3.2 \text{ m/s})^2 = 3.5 \times 10^2 \text{ J}$$

Although this may seem like a large expenditure of energy, we can make a simple calculation to show that jumping and exercise in general are not good ways to lose weight, in spite of their many health benefits. Because the muscles are at most 25% efficient at producing kinetic energy from chemical energy (muscles always produce a lot of internal energy and kinetic energy as well as work—that's why you perspire when you work out), they use up four times the 350 J (about 1 400 J) of chemical energy in one jump. This chemical energy ultimately comes from the food we eat, with energy content given in units of food

BIO APPLICATION
Diet Versus Exercise in Weight-loss Programs

calories and one food calorie equal to 4 200 J. So the total energy supplied by the body as internal energy and kinetic energy in a vertical jump is only about one-third of a food calorie!

Finally, it's interesting to calculate the average mechanical power that can be generated by the body in strenuous activity for brief periods. Here we find that

$$\overline{P} = \frac{KE}{\Delta t} = \frac{3.5 \times 10^2 \text{ J}}{0.25 \text{ s}} = 1.4 \times 10^3 \text{ W}$$

or $(1\,400 \text{ W})(1 \text{ hp}/746 \text{ W}) = 1.9$ hp. So humans can produce about 2 hp of mechanical power for periods on the order of seconds. Table 5.1 shows the maximum power outputs from humans for various periods while bicycling and rowing, activities in which it is possible to measure power output accurately.

Table 5.1 Maximum Power Output from Humans over Various Periods BIO

Power	Time
2 hp, or 1 500 W	6 s
1 hp, or 750 W	60 s
0.35 hp, or 260 W	35 min
0.2 hp, or 150 W	5 h
0.1 hp, or 75 W (safe daily level)	8 h

5.7 Work Done by a Varying Force

LEARNING OBJECTIVE

1. Analyze a graph of force vs. position to find the work done on an object by a varying force.

Suppose an object is displaced along the x-axis under the action of a force F_x that acts in the x-direction and varies with position, as shown in Figure 5.28. The object is displaced in the direction of increasing x from $x = x_i$ to $x = x_f$. In such a situation, we can't use Equation 5.2 to calculate the work done by the force because this relationship applies only when $\vec{F}$ is constant in magnitude and direction. However, if we imagine that the object undergoes the *small* displacement Δx shown in Figure 5.28a, then the x-component F_x of the force is nearly constant over this interval and we can approximate the work done by the force for this small displacement as

$$W_1 \cong F_x \Delta x \quad [5.27]$$

This quantity is just the area of the shaded rectangle in Figure 5.28a. If we imagine that the curve of F_x versus x is divided into a large number of such intervals, then the total work done for the displacement from x_i to x_f is approximately equal to the sum of the areas of a large number of small rectangles:

$$W \cong F_1 \Delta x_1 + F_2 \Delta x_2 + F_3 \Delta x_3 + \cdots \quad [5.28]$$

Now imagine going through the same process with twice as many intervals, each half the size of the original Δx. The rectangles then have smaller widths and will

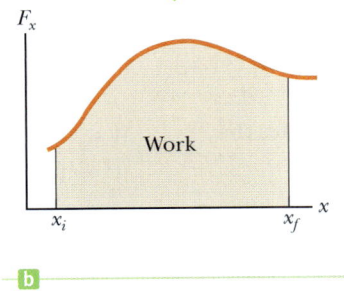

Figure 5.28 (a) The work done on a particle by the force component F_x for the small displacement Δx is approximately $F_x \Delta x$, the area of the shaded rectangle. (b) The width Δx of each rectangle is shrunk to zero.

better approximate the area under the curve. Continuing the process of increasing the number of intervals while allowing their size to approach zero, the number of terms in the sum increases without limit, but the value of the sum approaches a definite value equal to the area under the curve bounded by F_x and the x-axis in Figure 5.28b. In other words, **the work done by a variable force acting on an object that undergoes a displacement is equal to the area under the graph of F_x versus x.**

A common physical system in which force varies with position consists of a block on a horizontal, frictionless surface connected to a spring, as discussed in Section 5.4. When the spring is stretched or compressed a small distance x from its equilibrium position $x = 0$, it exerts a force on the block given by $F_x = -kx$, where k is the force constant of the spring.

Now let's determine the work done by an external agent on the block as the spring is stretched *very slowly* from $x_i = 0$ to $x_f = x_{max}$, as in Figure 5.29a. This work can be easily calculated by noting that at any value of the displacement, Newton's third law tells us that the applied force $\vec{F}_{app}$ is equal in magnitude to the spring force $\vec{F}_s$ and acts in the opposite direction, so that $F_{app} = -(-kx) = kx$. A plot of F_{app} versus x is a straight line, as shown in Figure 5.29b. Therefore, the work done by this applied force in stretching the spring from $x = 0$ to $x = x_{max}$ is the area under the straight line in that figure, which in this case is the area of the shaded triangle:

$$W_{F_{app}} = \tfrac{1}{2} k x_{max}^2$$

During this same time the spring has done exactly the same amount of work, but that work is negative, because the spring force points in the direction opposite the motion. The potential energy of the system is exactly equal to the work done by the applied force and is the same sign, which is why potential energy is thought of as stored work.

Figure 5.29 (a) A block being pulled from $x_i = 0$ to $x_f = x_{max}$ on a frictionless surface by a force $\vec{F}_{app}$. (b) A graph of F_{app} versus x.

■ EXAMPLE 5.15 | Work Required to Stretch a Spring

GOAL Apply the graphical method of finding work.

PROBLEM One end of a horizontal spring ($k = 80.0$ N/m) is held fixed while an external force is applied to the free end, stretching it slowly from $x_Ⓐ = 0$ to $x_Ⓑ = 4.00$ cm. **(a)** Find the work done by the applied force on the spring. **(b)** Find the additional work done in stretching the spring from $x_Ⓑ = 4.00$ cm to $x_Ⓒ = 7.00$ cm.

STRATEGY For part (a), simply find the area of the smaller triangle in Figure 5.30, using $A = \tfrac{1}{2}bh$, one-half the base times the height. For part (b), the easiest way to find the additional work done from $x_Ⓑ = 4.00$ cm to $x_Ⓒ = 7.00$ cm is to find the area of the new, larger triangle and subtract the area of the smaller triangle.

Figure 5.30 (Example 5.15) A graph of the external force required to stretch a spring that obeys Hooke's law versus the elongation of the spring.

SOLUTION

(a) Find the work from $x_Ⓐ = 0$ cm to $x_Ⓑ = 4.00$ cm.

Compute the area of the smaller triangle:

$$W = \tfrac{1}{2} k x_Ⓑ^2 = \tfrac{1}{2}(80.0 \text{ N/m})(0.040 \text{ m})^2 = \boxed{0.064\,0 \text{ J}}$$

(b) Find the work from $x_Ⓑ = 4.00$ cm to $x_Ⓒ = 7.00$ cm.

Compute the area of the large triangle and subtract the area of the smaller triangle:

$$W = \tfrac{1}{2} k x_Ⓒ^2 - \tfrac{1}{2} k x_Ⓑ^2$$
$$W = \tfrac{1}{2}(80.0 \text{ N/m})(0.070\,0 \text{ m})^2 - 0.064\,0 \text{ J}$$
$$= 0.196 \text{ J} - 0.064\,0 \text{ J}$$
$$= \boxed{0.132 \text{ J}}$$

REMARKS Only simple geometries (rectangles and triangles) can be solved exactly with this method. More complex shapes require calculus or the square-counting technique in the next worked example.

QUESTION 5.15 True or False: When stretching springs, half the displacement requires half as much work.

EXERCISE 5.15 How much work is required to stretch this same spring from $x_i = 5.00$ cm to $x_f = 9.00$ cm?

ANSWER 0.224 J

■ EXAMPLE 5.16 | Estimating Work by Counting Boxes

GOAL Use the graphical method and counting boxes to estimate the work done by a force.

PROBLEM Suppose the force applied to stretch a thick piece of elastic changes with position as indicated in Figure 5.31a. Estimate the work done by the applied force.

STRATEGY To find the work, simply count the number of boxes underneath the curve and multiply that number by the area of each box. The curve will pass through the middle of some boxes, in which case only an estimated fractional part should be counted.

Figure 5.31 (a) (Example 5.16) (b) (Exercise 5.16)

SOLUTION
There are 62 complete or nearly complete boxes under the curve, 6 boxes that are about half under the curve, and a triangular area from $x = 0$ m to $x = 0.10$ m that is equivalent to 1 box, for a total of about 66 boxes. Because the area of each box is 0.10 J, the total work done is approximately 66×0.10 J = 6.6 J.

REMARKS Mathematically, there are a number of other methods for creating such estimates, all involving adding up regions approximating the area. To get a better estimate, make smaller boxes.

QUESTION 5.16 In developing such an estimate, is it necessary for all boxes to have the same length and width?

EXERCISE 5.16 Suppose the applied force necessary to pull the drawstring on a bow is given by Figure 5.31b. Find the approximate work done by counting boxes.

ANSWER About 50 J. (Individual answers may vary.)

■ SUMMARY

5.1 Work
The work done on an object by a constant force is

$$W = (F \cos \theta) d \quad [5.3]$$

where F is the magnitude of the force, d is the magnitude of the object's displacement, and θ is the angle between the direction of the force $\vec{F}$ and the displacement $\Delta \vec{x}$. Solving simple problems requires substituting values into this equation. More complex problems, such as those involving friction, often require using Newton's second law, $m\vec{a} = \vec{F}_{net}$, to determine forces.

A constant force $\vec{F}$ applied during a displacement $\Delta \vec{x}$ does work ($F \cos \theta$) Δx.

5.2 Kinetic Energy and the Work–Energy Theorem
The kinetic energy of a body with mass m and speed v is given by

$$KE \equiv \tfrac{1}{2}mv^2 \quad [5.6]$$

The work–energy theorem states that the net work done on an object of mass m is equal to the change in its kinetic energy, or

$$W_{net} = KE_f - KE_i = \Delta KE \quad [5.7]$$

Work and energy of any kind carry units of joules. Solving problems involves finding the work done by each force acting

Work done by a net force $\vec{F}_{net}$ on an object changes the object's velocity.

on the object and summing them up, which is W_{net}, followed by substituting known quantities into Equation 5.7, solving for the unknown quantity.

Conservative forces are special: Work done against them can be recovered—it's conserved. An example is gravity: The work done in lifting an object through a height is effectively stored in the gravity field and can be recovered in the kinetic energy of the object simply by letting it fall. Nonconservative forces, such as surface friction and drag, dissipate energy in a form that can't be readily recovered. To account for such forces, the work–energy theorem can be rewritten as

$$W_{nc} + W_c = \Delta KE \quad [5.8]$$

where W_{nc} is the work done by nonconservative forces and W_c is the work done by conservative forces.

5.3 Gravitational Potential Energy

The gravitational force is a conservative field. Gravitational potential energy is another way of accounting for gravitational work W_g:

$$W_g = -(PE_f - PE_i)$$
$$= -(mgy_f - mgy_i) \quad [5.12]$$

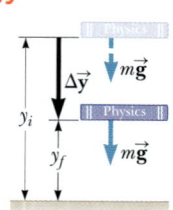

The work done by the gravitational force as the book falls equals $mgy_i - mgy_f$.

To find the change in gravitational potential energy as an object of mass m moves between two points in a gravitational field, substitute the values of the object's y-coordinates.

The work–energy theorem can be generalized to include gravitational potential energy:

$$W_{nc} = (KE_f - KE_i) + (PE_f - PE_i) \quad [5.13]$$

Gravitational work and gravitational potential energy should not both appear in the work–energy theorem at the same time, only one or the other, because they're equivalent. Setting the work due to nonconservative forces to zero and substituting the expressions for KE and PE, a form of the conservation of mechanical energy with gravitation can be obtained:

$$\tfrac{1}{2}mv_i^2 + mgy_i = \tfrac{1}{2}mv_f^2 + mgy_f \quad [5.15]$$

To solve problems with this equation, identify two points in the system—one where information is known and the other where information is desired. Substitute and solve for the unknown quantity.

The work done by other forces, as when frictional forces are present, isn't always zero. In that case, identify two points as before, calculate the work due to all other forces, and solve for the unknown in Equation 5.13.

5.4 Spring Potential Energy

The spring force is conservative, and its potential energy is given by

$$PE_s \equiv \tfrac{1}{2}kx^2 \quad [5.17]$$

Spring potential energy can be put into the work–energy theorem, which then reads

$$W_{nc} = (KE_f - KE_i) + (PE_{gf} - PE_{gi}) + (PE_{sf} - PE_{si}) \quad [5.18]$$

When nonconservative forces are absent, $W_{nc} = 0$ and mechanical energy is conserved.

5.5 Systems and Energy Conservation

The principle of the conservation of energy states that energy can't be created or destroyed. It can be transformed, but the total energy content of any isolated system is always constant. The same is true for the universe at large. The work done by all nonconservative forces acting on a system equals the change in the total mechanical energy of the system:

$$W_{nc} = (KE_f + PE_f) - (KE_i + PE_i) = E_f - E_i \quad [5.21–5.22]$$

where PE represents all potential energies present.

5.6 Power

Average power is the amount of energy transferred divided by the time taken for the transfer:

$$\overline{P} = \frac{W}{\Delta t} \quad [5.23]$$

This expression can also be written

$$\overline{P} = F\overline{v} \quad [5.24]$$

where $\overline{v}$ is the object's average velocity and F is constant and parallel to $\overline{v}$. The instantaneous power is given by.

$$P = Fv \quad [5.25]$$

where F must be parallel to the velocity v and both quantities can change with time. The unit of power is the watt (W = J/s). To solve simple problems, substitute given quantities into one of these equations. More difficult problems usually require finding the work done on the object using the work–energy theorem or the definition of work.

■ WARM-UP EXERCISES

The warm-up exercises in this chapter may be assigned online in Enhanced WebAssign.

1. **Physics Review** A crane lifts a load of bricks of mass 1 570 kg at an initial acceleration of 1.60 m/s². Calculate the tension in the cable. (See Section 4.5.)
2. **Physics Review** A crate of mass 20.0 kg rest on a level surface. If the coefficient of kinetic friction between the crate and surface is 0.400, **(a)** calculate the normal force and **(b)** the magnitude of the kinetic friction force when a horizontal applied force of 90.0 N moves the crate. **(c)** Calculate the normal force and **(d)** the magnitude of the kinetic friction force when

WU5.4 Exercises 4 & 5.

the 90.0-N applied force is exerted at an angle of 35.0° above the horizontal. (See Section 4.6.)

3. Calculate the work done by an applied force of 75.0 N on a crate if (a) the force is exerted horizontally while pushing the create 5.00 m and (b) the force is exerted at an angle of 35.0° above the horizontal. (See Section 5.1.)

4. In each of the diagrams WU5.4a-WU5.4c, calculate the work done by the graph of the force vs. position. (See Section 5.7.)

5. Suppose that in each of the diagrams WU5.4a-WU5.4c, the force is applied to a block of mass 5.00 kg at rest on a level, frictionless surface. Calculate the block's speed in each case after the work is done. (See Section 5.2.)

6. A 4.00-kg crate starting at rest slides down a rough 6.00-m-long ramp, inclined at 30.0° below the horizontal. The magnitude of the force of friction between the crate and the ramp is 8.00 N. (a) How much work is done on the crate by friction? (b) What is the change in potential energy of the crate in sliding down the ramp? (c) What is the speed of the crate at the bottom of the incline? (See Sections 5.2 and 5.3.)

7. A skier leaves a ski jump at 15.0 m/s at some angle θ. At what speed is he traveling at his maximum height of 4.50 m above the level of the end of the ski jump? (Neglect air friction.) (See Section 5.3.)

8. A block of mass 3.00 kg is placed against a horizontal spring of constant k = 875 N/m and pushed so the spring compresses by 0.070 0 m. (a) What is the spring potential energy of the block-spring system? (b) If the block is now released and the surface is frictionless, calculate the block's speed after leaving the spring. (See Section 5.4.)

9. What average mechanical power must a 70.0-kg mountain climber generate to climb to the summit of a hill of height 325 m in 45.0 min? Note: Due to inefficiencies in converting chemical energy to mechanical energy, the amount calculated here is only a fraction of the power that must be produced by the climber's body.

10. A puck of mass 0.170 kg slides across ice in the positive x-direction with a kinetic friction coefficient between the ice and puck of 0.150. If the puck is moving at an initial speed of 12.0 m/s, (a) what is the force of kinetic friction? (b) What is the acceleration of the puck? (c) How long does it take for the puck to come to rest? (d) What distance does the puck travel during that time? (e) What total work does friction do on the puck? (f) What average power does friction generate in the puck during that time? (g) What instantaneous power does friction generate in the puck when the velocity is 6.00 m/s? (See Sections 2.5, 4.6, 5.1, and 5.6.)

■ CONCEPTUAL QUESTIONS

WebAssign The conceptual questions in this chapter may be assigned online in Enhanced WebAssign.

1. Consider a tug-of-war as in Figure CQ5.1, in which two teams pulling on a rope are evenly matched so that no motion takes place. Is work done on the rope? On the pullers? On the ground? Is work done on anything?

2. **BIO** During a stress test of the cardiovascular system, a patient walks and runs on a treadmill. (a) Is the energy expended by the patient equivalent to the energy of walking and running on the ground? Explain. (b) What effect, if any, does tilting the treadmill upward have? Discuss.

3. (a) If the height of a playground slide is kept constant, will the length of the slide or whether it has bumps make any difference in the final speed of children playing on it? Assume that the slide is slick enough to be considered frictionless. (b) Repeat part (a), assuming that the slide is not frictionless.

Figure CQ5.1

4. (a) Can the kinetic energy of a system be negative? (b) Can the gravitational potential energy of a system be negative? Explain.

5. Roads going up mountains are formed into switchbacks, with the road weaving back and forth along the face of the slope such that there is only a gentle rise on any portion of the roadway. Does this configuration require any less work to be done by an automobile climbing the mountain, compared with one traveling on a roadway that is straight up the slope? Why are switchbacks used?

6. A bowling ball is suspended from the ceiling of a lecture hall by a strong cord. The ball is drawn away from its equilibrium position and released from rest at the tip of the demonstrator's nose, as shown in Figure CQ5.6. (a) If the demonstrator remains stationary, explain why the ball does not strike her on its return swing. (b) Would this demonstrator be safe if the ball were given a push from its starting position at her nose?

Figure CQ5.6

7. As a simple pendulum swings back and forth, the forces acting on the suspended object are the force of gravity, the tension in the supporting cord, and air resistance. (a) Which of these forces, if any, does no work on the pendulum? (b) Which of these forces does negative work at all times during the pendulum's motion? (c) Describe the work done by the force of gravity while the pendulum is swinging.

8. Discuss whether any work is being done by each of the following agents and, if so, whether the work is positive or negative: (a) a chicken scratching the ground, (b) a person studying, (c) a crane lifting a bucket of concrete, (d) the force of gravity on the bucket in part (c), (e) the leg muscles of a person in the act of sitting down.

9. When a punter kicks a football, is he doing any work on the ball while the toe of his foot is in contact with it? Is he doing any work on the ball after it loses contact with his toe? Are any forces doing work on the ball while it is in flight?

10. The driver of a car slams on her brakes to avoid colliding with a deer crossing the highway. What happens to the car's kinetic energy as it comes to rest?

11. A weight is connected to a spring that is suspended vertically from the ceiling. If the weight is displaced downward from its equilibrium position and released, it will oscillate up and down. (a) If air resistance is neglected, will the total mechanical energy of the system (weight plus Earth plus spring) be conserved? (b) How many forms of potential energy are there for this situation?

12. In most situations we have encountered in this chapter, frictional forces tend to reduce the kinetic energy of an object. However, frictional forces can sometimes increase an object's kinetic energy. Describe a few situations in which friction causes an increase in kinetic energy.

13. Suppose you are re-shelving books in a library. You lift a book from the floor to the top shelf. The kinetic energy of the book on the floor was zero, and the kinetic energy of the book on the top shelf is zero, so there is no change in kinetic energy. Yet you did some work in lifting the book. Is the work–energy theorem violated?

14. The feet of a standing person of mass m exert a force equal to mg on the floor, and the floor exerts an equal and opposite force upwards on the feet, which we call the normal force. During the extension phase of a vertical jump (see page 154), the feet exert a force on the floor that is greater than mg, so the normal force is greater than mg. As you learned in Chapter 4, we can use this result and Newton's second law to calculate the acceleration of the jumper:

$$a = F_{net}/m = (n - mg)/m$$

Using energy ideas, we know that work is performed on the jumper to give him or her kinetic energy. But the normal force can't perform any work here because the feet don't undergo any displacement. How is energy transferred to the jumper?

15. An Earth satellite is in a circular orbit at an altitude of 500 km. Explain why the work done by the gravitational force acting on the satellite is zero. Using the work–energy theorem, what can you say about the speed of the satellite?

16. Mark and David are loading identical cement blocks onto David's pickup truck. Mark lifts his block straight up from the ground to the truck, whereas David slides his block up a ramp on massless, frictionless rollers. Which statement is true? (a) Mark does more work than David. (b) Mark and David do the same amount of work. (c) David does more work than Mark. (d) None of these statements is necessarily true because the angle of the incline is unknown. (e) None of these statements is necessarily true because the mass of one block is not given.

17. If the speed of a particle is doubled, what happens to its kinetic energy? (a) It becomes four times larger. (b) It becomes two times larger. (c) It becomes $\sqrt{2}$ times larger. (d) It is unchanged. (e) It becomes half as large.

18. A certain truck has twice the mass of a car. Both are moving at the same speed. If the kinetic energy of the truck is K, what is the kinetic energy of the car? (a) $K/4$ (b) $K/2$ (c) $0.71K$ (d) K (e) $2K$

19. If the net work done on a particle is zero, which of the following statements must be true? (a) The velocity is

zero. (b) The velocity is decreased. (c) The velocity is unchanged. (d) The speed is unchanged. (e) More information is needed.

20. A car accelerates uniformly from rest. Ignoring air friction, when does the car require the greatest power? (a) When the car first accelerates from rest, (b) just as the car reaches its maximum speed, (c) when the car reaches half its maximum speed. (d) The question is misleading because the power required is constant. (e) More information is needed.

PROBLEMS

WebAssign The problems in this chapter may be assigned online in Enhanced WebAssign.

1. denotes straightforward problem; 2. denotes intermediate problem;
3. denotes challenging problem
1. denotes full solution available in *Student Solutions Manual/Study Guide*
1. denotes problems most often assigned in Enhanced WebAssign

BIO denotes biomedical problems
GP denotes guided problems
M denotes Master It tutorial available in Enhanced WebAssign
QC denotes asking for quantitative and conceptual reasoning
S denotes symbolic reasoning problem
W denotes Watch It video solution available in Enhanced WebAssign

5.1 Work

1. A weight lifter lifts a 350-N set of weights from ground level to a position over his head, a vertical distance of 2.00 m. How much work does the weight lifter do, assuming he moves the weights at constant speed?

2. In 1990 Walter Arfeuille of Belgium lifted a 281.5-kg object through a distance of 17.1 cm using only his teeth. (a) How much work did Arfeuille do on the object? (b) What magnitude force did he exert on the object during the lift, assuming the force was constant?

3. The record number of boat lifts, including the boat and its ten crew members, was achieved by Sami Heinonen and Juha Räsänen of Sweden in 2000. They lifted a total mass of 653.2 kg approximately 4 in. off the ground a total of 24 times. Estimate the total mechanical work done by the two men in lifting the boat 24 times, assuming they applied the same force to the boat during each lift. (Neglect any work they may have done allowing the boat to drop back to the ground.)

4. **QC** A shopper in a supermarket pushes a cart with a force of 35 N directed at an angle of 25° below the horizontal. The force is just sufficient to overcome various frictional forces, so the cart moves at constant speed. (a) Find the work done by the shopper as she moves down a 50.0-m length aisle. (b) What is the *net work* done on the cart? Why? (c) The shopper goes down the next aisle, pushing horizontally and maintaining the same speed as before. If the work done by frictional forces doesn't change, would the shopper's applied force be larger, smaller, or the same? What about the work done on the cart by the shopper?

5. **QC** Starting from rest, a 5.00-kg block slides 2.50 m down a rough 30.0° incline. The coefficient of kinetic friction between the block and the incline is $\mu_k = 0.436$. Determine (a) the work done by the force of gravity, (b) the work done by the friction force between block and incline, and (c) the work done by the normal force. (d) Qualitatively, how would the answers change if a shorter ramp at a steeper angle were used to span the same vertical height?

6. A horizontal force of 150 N is used to push a 40.0-kg packing crate a distance of 6.00 m on a rough horizontal surface. If the crate moves at constant speed, find (a) the work done by the 150-N force and (b) the coefficient of kinetic friction between the crate and surface.

7. A sledge loaded with bricks has a total mass of 18.0 kg and is pulled at constant speed by a rope inclined at 20.0° above the horizontal. The sledge moves a distance of 20.0 m on a horizontal surface. The coefficient of kinetic friction between the sledge and surface is 0.500. (a) What is the tension in the rope? (b) How much work is done by the rope on the sledge? (c) What is the mechanical energy lost due to friction?

8. **M** A block of mass $m = 2.50$ kg is pushed a distance $d = 2.20$ m along a frictionless horizontal table by a constant applied force of magnitude $F = 16.0$ N directed at an angle $\theta = 25.0°$ below the horizontal as shown in Figure P5.8. Determine the work done by (a) the applied force, (b) the normal force exerted by the table, (c) the force of gravity, and (d) the net force on the block.

Figure P5.8

5.2 Kinetic Energy and the Work–Energy Theorem

9. A mechanic pushes a 2.50×10^3-kg car from rest to a speed of v, doing 5 000 J of work in the process. During this time, the car moves 25.0 m. Neglecting friction between car and road, find (a) v and (b) the horizontal force exerted on the car.

10. A 7.00-kg bowling ball moves at 3.00 m/s. How fast must a 2.45-g Ping-Pong ball move so that the two balls have the same kinetic energy?

11. A 65.0-kg runner has a speed of 5.20 m/s at one instant during a long-distance event. (a) What is the runner's

kinetic energy at this instant? (b) If he doubles his speed to reach the finish line, by what factor does his kinetic energy change?

12. **Q|C** A worker pushing a 35.0-kg wooden crate at a constant speed for 12.0 m along a wood floor does 350 J of work by applying a constant horizontal force of magnitude F_0 on the crate. (a) Determine the value of F_0. (b) If the worker now applies a force greater than F_0, describe the subsequent motion of the crate. (c) Describe what would happen to the crate if the applied force is less than F_0.

13. **W** A 70-kg base runner begins his slide into second base when he is moving at a speed of 4.0 m/s. The coefficient of friction between his clothes and Earth is 0.70. He slides so that his speed is zero just as he reaches the base. (a) How much mechanical energy is lost due to friction acting on the runner? (b) How far does he slide?

14. A running 62-kg cheetah has a top speed of 32 m/s. (a) What is the cheetah's maximum kinetic energy? (b) Find the cheetah's speed when its kinetic energy is one half of the value found in part (a).

15. A 7.80-g bullet moving at 575 m/s penetrates a tree trunk to a depth of 5.50 cm. (a) Use work and energy considerations to find the average frictional force that stops the bullet. (b) Assuming the frictional force is constant, determine how much time elapses between the moment the bullet enters the tree and the moment it stops moving.

16. A 0.60-kg particle has a speed of 2.0 m/s at point A and a kinetic energy of 7.5 J at point B. What is (a) its kinetic energy at A? (b) Its speed at point B? (c) The total work done on the particle as it moves from A to B?

17. A large cruise ship of mass 6.50×10^7 kg has a speed of 12.0 m/s at some instant. (a) What is the ship's kinetic energy at this time? (b) How much work is required to stop it? (c) What is the magnitude of the constant force required to stop it as it undergoes a displacement of 2.50 km?

18. A man pushing a crate of mass $m = 92.0$ kg at a speed of $v = 0.850$ m/s encounters a rough horizontal surface of length $\ell = 0.65$ m as in Figure P5.18. If the coefficient of kinetic friction between the crate and rough surface is 0.358 and he exerts a constant horizontal force of 275 N on the crate, find (a) the magnitude and direction of the net force on the crate while it is on the rough surface, (b) the net work done on the crate while it is on the rough surface, and (c) the speed of the crate when it reaches the end of the rough surface.

Figure P5.18

5.3 Gravitational Potential Energy
5.4 Spring Potential Energy

19. A 0.20-kg stone is held 1.3 m above the top edge of a water well and then dropped into it. The well has a depth of 5.0 m. Taking $y = 0$ at the top edge of the well, what is the gravitational potential energy of the stone–Earth system (a) before the stone is released and (b) when it reaches the bottom of the well. (c) What is the change in gravitational potential energy of the system from release to reaching the bottom of the well?

20. When a 2.50-kg object is hung vertically on a certain light spring described by Hooke's law, the spring stretches 2.76 cm. (a) What is the force constant of the spring? (b) If the 2.50-kg object is removed, how far will the spring stretch if a 1.25-kg block is hung on it? (c) How much work must an external agent do to stretch the same spring 8.00 cm from its unstretched position?

21. In a control system, an accelerometer consists of a 4.70-g object sliding on a calibrated horizontal rail. A low-mass spring attaches the object to a flange at one end of the rail. Grease on the rail makes static friction negligible, but rapidly damps out vibrations of the sliding object. When subject to a steady acceleration of $0.800g$, the object should be at a location 0.500 cm away from its equilibrium position. Find the force constant of the spring required for the calibration to be correct.

22. **GP** A 60.0-kg athlete leaps straight up into the air from a trampoline with an initial speed of 9.0 m/s. The goal of this problem is to find the maximum height she attains and her speed at half maximum height. (a) What are the interacting objects and how do they interact? (b) Select the height at which the athlete's speed is 9.0 m/s as $y = 0$. What is her kinetic energy at this point? What is the gravitational potential energy associated with the athlete? (c) What is her kinetic energy at maximum height? What is the gravitational potential energy associated with the athlete? (d) Write a general equation for energy conservation in this case and solve for the maximum height. Substitute and obtain a numerical answer. (e) Write the general equation for energy conservation and solve for the velocity at half the maximum height. Substitute and obtain a numerical answer.

23. **M** A 2 100-kg pile driver is used to drive a steel I-beam into the ground. The pile driver falls 5.00 m before coming into contact with the top of the beam, and it drives the beam 12.0 cm farther into the ground as it comes to rest. Using energy considerations, calculate the average force the beam exerts on the pile driver while the pile driver is brought to rest.

24. **S** Two blocks are connected by a light string that passes over two frictionless pulleys as in Figure P5.24.

The block of mass m_2 is attached to a spring of force constant k and $m_1 > m_2$. If the system is released from rest, and the spring is initially not stretched or compressed, find an expression for the maximum displacement d of m_2.

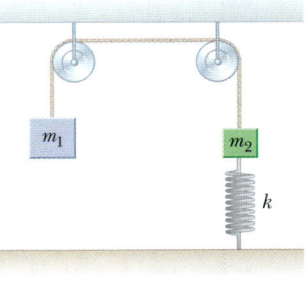

Figure P5.24

25. A daredevil on a motorcycle leaves the end of a ramp with a speed of 35.0 m/s as in Figure P5.25. If his speed is 33.0 m/s when he reaches the peak of the path, what is the maximum height that he reaches? Ignore friction and air resistance.

Figure P5.25

26. Truck suspensions often have "helper springs" that engage at high loads. One such arrangement is a leaf spring with a helper coil spring mounted on the axle, as shown in Figure P5.26. When the main leaf spring is compressed by distance y_0, the helper spring engages and then helps to support any additional load. Suppose the leaf spring constant is 5.25×10^5 N/m, the helper spring constant is 3.60×10^5 N/m, and $y_0 = 0.500$ m. (a) What is the compression of the leaf spring for a load of 5.00×10^5 N? (b) How much work is done in compressing the springs?

Figure P5.26

27. **BIO** The chin-up is one exercise that can be used to strengthen the biceps muscle. This muscle can exert a force of approximately 800 N as it contracts a distance of 7.5 cm in a 75-kg male.[3] How much work can the biceps muscles (one in each arm) perform in a single contraction? Compare this amount of work with the energy required to lift a 75-kg person 40 cm in performing a chin-up. Do you think the biceps muscle is the only muscle involved in performing a chin-up?

28. **BIO** A flea is able to jump about 0.5 m. It has been said that if a flea were as big as a human, it would be able to jump over a 100-story building! When an animal jumps, it converts work done in contracting muscles into gravitational potential energy (with some steps in between). The maximum force exerted by a muscle is proportional to its cross-sectional area, and the work done by the muscle is this force times the length of contraction. If we magnified a flea by a factor of 1 000, the cross section of its muscle would increase by $1\,000^2$ and the length of contraction would increase by 1 000. How high would this "superflea" be able to jump? (Don't forget that the mass of the "superflea" increases as well.)

29. A 50.0-kg projectile is fired at an angle of 30.0° above the horizontal with an initial speed of 1.20×10^2 m/s from the top of a cliff 142 m above level ground, where the ground is taken to be $y = 0$. (a) What is the initial total mechanical energy of the projectile? (b) Suppose the projectile is traveling 85.0 m/s at its maximum height of $y = 427$ m. How much work has been done on the projectile by air friction? (c) What is the speed of the projectile immediately before it hits the ground if air friction does one and a half times as much work on the projectile when it is going down as it did when it was going up?

30. **S** A projectile of mass m is fired horizontally with an initial speed of v_0 from a height of h above a flat, desert surface. Neglecting air friction, at the instant before the projectile hits the ground, find the following in terms of m, v_0, h, and g: (a) the work done by the force of gravity on the projectile, (b) the change in kinetic energy of the projectile since it was fired, and (c) the final kinetic energy of the projectile. (d) Are any of the answers changed if the initial angle is changed?

31. **GP** A horizontal spring attached to a wall has a force constant of 850 N/m. A block of mass 1.00 kg is attached to the spring and oscillates freely on a horizontal, frictionless surface as in Figure 5.20. The initial goal of this problem is to find the velocity at the equilibrium point after the block is released. (a) What objects constitute the system, and through what forces do they interact? (b) What are the two points of interest? (c) Find the energy stored in the spring when the mass is stretched 6.00 cm from equilibrium and again

[3] G. P. Pappas et al., "Nonuniform shortening in the biceps brachii during elbow flexion," *Journal of Applied Physiology* **92**, 2381, 2002.

when the mass passes through equilibrium after being released from rest. (d) Write the conservation of energy equation for this situation and solve it for the speed of the mass as it passes equilibrium. Substitute to obtain a numerical value. (e) What is the speed at the halfway point? Why isn't it half the speed at equilibrium?

5.5 Systems and Energy Conservation

32. A 50-kg pole vaulter running at 10 m/s vaults over the bar. Her speed when she is above the bar is 1.0 m/s. Neglect air resistance, as well as any energy absorbed by the pole, and determine her altitude as she crosses the bar.

33. **W** A child and a sled with a combined mass of 50.0 kg slide down a frictionless slope. If the sled starts from rest and has a speed of 3.00 m/s at the bottom, what is the height of the hill?

34. Hooke's law describes a certain light spring of unstretched length 35.0 cm. When one end is attached to the top of a door frame and a 7.50-kg object is hung from the other end, the length of the spring is 41.5 cm. (a) Find its spring constant. (b) The load and the spring are taken down. Two people pull in opposite directions on the ends of the spring, each with a force of 190 N. Find the length of the spring in this situation.

35. **Q|C** A 0.250-kg block along a horizontal track has a speed of 1.50 m/s immediately before colliding with a light spring of force constant 4.60 N/m located at the end of the track. (a) What is the spring's maximum compression if the track is frictionless? (b) If the track is *not* frictionless, would the spring's maximum compression be greater than, less than, or equal to the value obtained in part (a)?

36. A block of mass $m = 5.00$ kg is released from rest from point Ⓐ and slides on the frictionless track shown in Figure P5.36. Determine (a) the block's speed at points Ⓑ and Ⓒ and (b) the net work done by the gravitational force on the block as it moves from point from Ⓐ to Ⓒ.

Figure P5.36

37. Tarzan swings on a 30.0-m-long vine initially inclined at an angle of 37.0° with the vertical. What is his speed at the bottom of the swing (a) if he starts from rest? (b) If he pushes off with a speed of 4.00 m/s?

38. **S** Two blocks are connected by a light string that passes over a frictionless pulley as in Figure P5.38. The system is released from rest while m_2 is on the floor and m_1 is a distance h above the floor. (a) Assuming $m_1 > m_2$, find an expression for the speed of m_1 just as it reaches the floor. (b) Taking $m_1 = 6.5$ kg, $m_2 = 4.2$ kg, and $h = 3.2$ m, evaluate your answer to part (a), and (c) find the speed of each block when m_1 has fallen a distance of 1.6 m.

Figure P5.38

39. **M** The launching mechanism of a toy gun consists of a spring of unknown spring constant, as shown in Figure P5.39a. If the spring is compressed a distance of 0.120 m and the gun fired vertically as shown, the gun can launch a 20.0-g projectile from rest to a maximum height of 20.0 m above the starting point of the projectile. Neglecting all resistive forces, (a) describe the mechanical energy transformations that occur from the time the gun is fired until the projectile reaches its maximum height, (b) determine the spring constant, and (c) find the speed of the projectile as it moves through the equilibrium position of the spring (where $x = 0$), as shown in Figure P5.39b.

Figure P5.39

40. **GP** (a) A block with a mass m is pulled along a horizontal surface for a distance x by a constant force $\vec{F}$ at an angle θ with respect to the horizontal. The coefficient of kinetic friction between block and table is μ_k. Is the force exerted by friction equal to $\mu_k mg$? If not, what is the force exerted by friction? (b) How much work is done by the friction force and by $\vec{F}$? (Don't forget the signs.) (c) Identify all the forces that do no work on the block. (d) Let $m = 2.00$ kg, $x = 4.00$ m, $\theta = 37.0°$, $F = 15.0$ N, and $\mu_k = 0.400$, and find the answers to parts (a) and (b).

41. **Q|C** (a) A child slides down a water slide at an amusement park from an initial height h. The slide can be considered frictionless because of the water flowing down it. Can the equation for conservation of mechanical energy be used on the child? (b) Is the mass of the child a factor in determining his speed at the bottom of the slide? (c) The child drops straight down rather than following the curved ramp of the slide. In which case will he be traveling faster at ground level? (d) If friction is present, how would the conservation-of-energy equation be modified? (e) Find the maximum speed of the child when the slide is frictionless if the initial height of the slide is 12.0 m.

42. **QC** An airplane of mass 1.50×10^4 kg is moving at 60.0 m/s. The pilot then increases the engine's thrust to 7.50×10^4 N. The resistive force exerted by air on the airplane has a magnitude of 4.00×10^4 N. (a) Is the work done by the engine on the airplane equal to the change in the airplane's kinetic energy after it travels through some distance through the air? Is mechanical energy conserved? Explain. (b) Find the speed of the airplane after it has traveled 5.00×10^2 m. Assume the airplane is in level flight throughout the motion.

43. The system shown in Figure P5.43 is used to lift an object of mass $m = 76.0$ kg. A constant downward force of magnitude F is applied to the loose end of the rope such that the hanging object moves upward at constant speed. Neglecting the masses of the rope and pulleys, find (a) the required value of F, (b) the tensions T_1, T_2, and T_3, and (c) the work done by the applied force in raising the object a distance of 1.80 m.

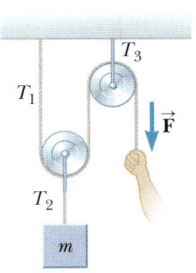

Figure P5.43

44. A 25.0-kg child on a 2.00-m-long swing is released from rest when the ropes of the swing make an angle of 30.0° with the vertical. (a) Neglecting friction, find the child's speed at the lowest position. (b) If the actual speed of the child at the lowest position is 2.00 m/s, what is the mechanical energy lost due to friction?

45. A 2.1×10^3-kg car starts from rest at the top of a 5.0-m-long driveway that is inclined at 20° with the horizontal. If an average friction force of 4.0×10^3 N impedes the motion, find the speed of the car at the bottom of the driveway.

46. **GP S** A child of mass m starts from rest and slides without friction from a height h along a curved waterslide (Fig. P5.46). She is launched from a height $h/5$ into the pool. (a) Is mechanical energy conserved? Why? (b) Give the gravitational potential energy associated with the child and her kinetic energy in terms of mgh at the following positions: the top of the waterslide, the launching point, and the point where she lands in the pool. (c) Determine her initial speed v_0 at the launch point in terms of g and h. (d) Determine her maximum airborne height y_{max} in terms of h, g, and the horizontal speed at that height, v_{0x}. (e) Use the x-component of the answer to part (c) to eliminate v_0 from the answer to part (d), giving the height y_{max} in terms of g, h, and the launch angle θ. (f) Would your answers be the same if the waterslide were not frictionless? Explain.

47. A skier starts from rest at the top of a hill that is inclined 10.5° with respect to the horizontal. The hillside is 200 m long, and the coefficient of friction between snow and skis is 0.075 0. At the bottom of the hill, the snow is level and the coefficient of friction is unchanged. How far does the skier glide along the horizontal portion of the snow before coming to rest?

48. In a circus performance, a monkey is strapped to a sled and both are given an initial speed of 4.0 m/s up a 20° inclined track. The combined mass of monkey and sled is 20 kg, and the coefficient of kinetic friction between sled and incline is 0.20. How far up the incline do the monkey and sled move?

49. An 80.0-kg skydiver jumps out of a balloon at an altitude of 1 000 m and opens the parachute at an altitude of 200.0 m. (a) Assuming that the total retarding force on the diver is constant at 50.0 N with the parachute closed and constant at 3 600 N with the parachute open, what is the speed of the diver when he lands on the ground? (b) Do you think the skydiver will get hurt? Explain. (c) At what height should the parachute be opened so that the final speed of the skydiver when he hits the ground is 5.00 m/s? (d) How realistic is the assumption that the total retarding force is constant? Explain.

5.6 Power

50. **W** A skier of mass 70 kg is pulled up a slope by a motor-driven cable. (a) How much work is required to pull him 60 m up a 30° slope (assumed frictionless) at a constant speed of 2.0 m/s? (b) What power (expressed in hp) must a motor have to perform this task?

51. A 3.50-kN piano is lifted by three workers at constant speed to an apartment 25.0 m above the street using a pulley system fastened to the roof of the building. Each worker is able to deliver 165 W of power, and the pulley system is 75.0% efficient (so that 25.0% of the mechanical energy is lost due to friction in the pulley). Neglecting the mass of the pulley, find the time required to lift the piano from the street to the apartment.

52. **BIO** While running, a person dissipates about 0.60 J of mechanical energy per step per kilogram of body mass. If a 60-kg person develops a power of 70 W during a race, how fast is the person running? (Assume a running step is 1.5 m long.)

53. The electric motor of a model train accelerates the train from rest to 0.620 m/s in 21.0 ms. The total mass of the train is 875 g. Find the average power delivered to the train during its acceleration.

54. When an automobile moves with constant speed down a highway, most of the power developed by the engine is used to compensate for the mechanical energy loss

Figure P5.46

due to frictional forces exerted on the car by the air and the road. If the power developed by an engine is 175 hp, estimate the total frictional force acting on the car when it is moving at a speed of 29 m/s. One horsepower equals 746 W.

55. An older-model car accelerates from 0 to speed v in 10 s. A newer, more powerful sports car of the same mass accelerates from 0 to $2v$ in the same time period. Assuming the energy coming from the engine appears only as kinetic energy of the cars, compare the power of the two cars.

56. A certain rain cloud at an altitude of 1.75 km contains 3.20×10^7 kg of water vapor. How long would it take for a 2.70-kW pump to raise the same amount of water from Earth's surface to the cloud's position?

57. A 1.50×10^3-kg car starts from rest and accelerates uniformly to 18.0 m/s in 12.0 s. Assume that air resistance remains constant at 400 N during this time. Find (a) the average power developed by the engine and (b) the instantaneous power output of the engine at $t = 12.0$ s, just before the car stops accelerating.

58. A 650-kg elevator starts from rest and moves upward for 3.00 s with constant acceleration until it reaches its cruising speed, 1.75 m/s. (a) What is the average power of the elevator motor during this period? (b) How does this amount of power compare with its power during an upward trip with constant speed?

5.7 Work Done by a Varying Force

59. **M** The force acting on a particle varies as in Figure P5.59. Find the work done by the force as the particle moves (a) from $x = 0$ to $x = 8.00$ m, (b) from $x = 8.00$ m to $x = 10.0$ m, and (c) from $x = 0$ to $x = 10.0$ m.

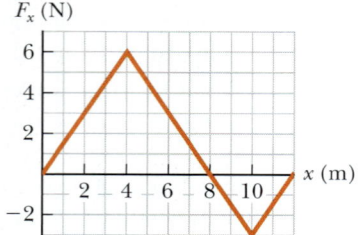

Figure P5.59

60. An object of mass 3.00 kg is subject to a force F_x that varies with position as in Figure P5.60. Find the work done by the force on the object as it moves (a) from $x = 0$ to $x = 5.00$ m, (b) from $x = 5.00$ m to $x = 10.0$ m,

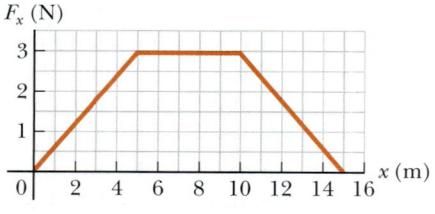

Figure P5.60

and (c) from $x = 10.0$ m to $x = 15.0$ m. (d) If the object has a speed of 0.500 m/s at $x = 0$, find its speed at $x = 5.00$ m and its speed at $x = 15.0$ m.

61. The force acting on an object is given by $F_x = (8x - 16)$ N, where x is in meters. (a) Make a plot of this force versus x from $x = 0$ to $x = 3.00$ m. (b) From your graph, find the net work done by the force as the object moves from $x = 0$ to $x = 3.00$ m.

Additional Problems

62. An outfielder throws a 0.150-kg baseball at a speed of 40.0 m/s and an initial angle of 30.0°. What is the kinetic energy of the ball at the highest point of its motion?

63. A person doing a chin-up weighs 700 N, exclusive of the arms. During the first 25.0 cm of the lift, each arm exerts an upward force of 355 N on the torso. If the upward movement starts from rest, what is the person's velocity at that point?

64. **S** A boy starts at rest and slides down a frictionless slide as in Figure P5.64. The bottom of the track is a height h above the ground. The boy then leaves the track horizontally, striking the ground a distance d as shown. Using energy methods, determine the initial height H of the boy in terms of h and d.

Figure P5.64

65. A roller-coaster car of mass 1.50×10^3 kg is initially at the top of a rise at point Ⓐ. It then moves 35.0 m at an angle of 50.0° below the horizontal to a lower point Ⓑ. (a) Find both the potential energy of the system when the car is at points Ⓐ and Ⓑ and the change in potential energy as the car moves from point Ⓐ to point Ⓑ, assuming $y = 0$ at point Ⓑ. (b) Repeat part (a), this time choosing $y = 0$ at point Ⓒ, which is another 15.0 m down the same slope from point Ⓑ.

66. A ball of mass $m = 1.80$ kg is released from rest at a height $h = 65.0$ cm above a light vertical spring of force constant k as in Figure P5.66a. The ball strikes the top of the spring and compresses it a distance $d = 9.00$ cm as in Figure P5.66b. Neglecting any energy losses

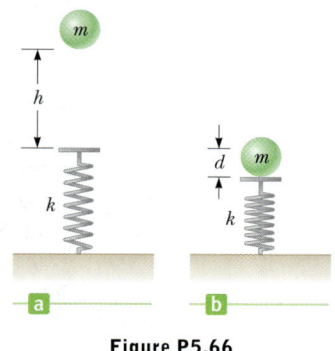

Figure P5.66

during the collision, find (a) the speed of the ball just as it touches the spring and (b) the force constant of the spring.

67. An archer pulls her bowstring back 0.400 m by exerting a force that increases uniformly from zero to 230 N. (a) What is the equivalent spring constant of the bow? (b) How much work does the archer do in pulling the bow?

68. A block of mass 12.0 kg slides from rest down a frictionless 35.0° incline and is stopped by a strong spring with $k = 3.00 \times 10^4$ N/m. The block slides 3.00 m from the point of release to the point where it comes to rest against the spring. When the block comes to rest, how far has the spring been compressed?

69. BIO (a) A 75-kg man steps out a window and falls (from rest) 1.0 m to a sidewalk. What is his speed just before his feet strike the pavement? (b) If the man falls with his knees and ankles locked, the only cushion for his fall is an approximately 0.50-cm give in the pads of his feet. Calculate the average force exerted on him by the ground during this 0.50 cm of travel. This average force is sufficient to cause damage to cartilage in the joints or to break bones.

70. A toy gun uses a spring to project a 5.3-g soft rubber sphere horizontally. The spring constant is 8.0 N/m, the barrel of the gun is 15 cm long, and a constant frictional force of 0.032 N exists between barrel and projectile. With what speed does the projectile leave the barrel if the spring was compressed 5.0 cm for this launch?

71. Two objects ($m_1 = 5.00$ kg and $m_2 = 3.00$ kg) are connected by a light string passing over a light, frictionless pulley as in Figure P5.71. The 5.00-kg object is released from rest at a point $h = 4.00$ m above the table. (a) Determine the speed of each object when the two pass each other. (b) Determine the speed of each object at the moment the 5.00-kg object hits the table. (c) How much higher does the 3.00-kg object travel after the 5.00-kg object hits the table?

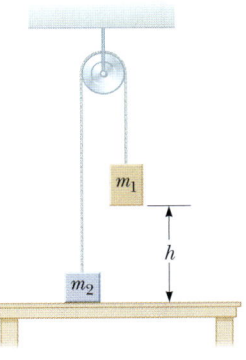

Figure P5.71

72. BIO In a needle biopsy, a narrow strip of tissue is extracted from a patient with a hollow needle. Rather than being pushed by hand, to ensure a clean cut the needle can be fired into the patient's body by a spring. Assume the needle has mass 5.60 g, the light spring has force constant 375 N/m, and the spring is originally compressed 8.10 cm to project the needle horizontally without friction. The tip of the needle then moves through 2.40 cm of skin and soft tissue, which exerts a resistive force of 7.60 N on it. Next, the needle cuts 5.50 cm into an organ, which exerts a backward force of 9.20 N on it. Find (a) the maximum speed of the needle and (b) the speed at which a flange on the back end of the needle runs into a stop, set to limit the penetration to 5.90 cm.

73. A 2.00×10^2-g particle is released from rest at point A on the inside of a smooth hemispherical bowl of radius $R = 30.0$ cm (Fig. P5.73). Calculate (a) its gravitational potential energy at A relative to B, (b) its kinetic energy at B, (c) its speed at B, (d) its potential energy at C relative to B, and (e) its kinetic energy at C.

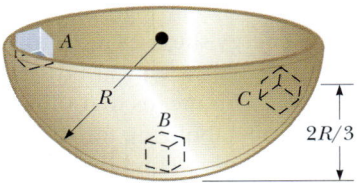

Figure P5.73 Problems 73 and 74.

74. Q|C The particle described in Problem 73 (Fig. P5.73) is released from point A at rest. Its speed at B is 1.50 m/s. (a) What is its kinetic energy at B? (b) How much mechanical energy is lost as a result of friction as the particle goes from A to B? (c) Is it possible to determine μ from these results in a simple manner? Explain.

75. A light spring with spring constant 1.20×10^3 N/m hangs from an elevated support. From its lower end hangs a second light spring, which has spring constant 1.80×10^3 N/m. A 1.50-kg object hangs at rest from the lower end of the second spring. (a) Find the total extension distance of the pair of springs. (b) Find the effective spring constant of the pair of springs as a system. We describe these springs as being *in series*. *Hint:* Consider the forces on each spring separately.

76. S *Symbolic Version of Problem 75* A light spring with spring constant k_1 hangs from an elevated support. From its lower end hangs a second light spring, which has spring constant k_2. An object of mass m hangs at rest from the lower end of the second spring. (a) Find the total extension distance x of the pair of springs in terms of the two displacements x_1 and x_2. (b) Find the effective spring constant of the pair of springs as a system. We describe these springs as being *in series*.

77. BIO In terms of saving energy, bicycling and walking are far more efficient means of transportation than is travel by automobile. For example, when riding at 10.0 mi/h, a cyclist uses food energy at a rate of about 400 kcal/h above what he would use if he were merely sitting still. (In exercise physiology, power is often measured in kcal/h rather than in watts. Here, 1 kcal = 1 nutritionist's Calorie = 4 186 J.) Walking at 3.00 mi/h requires about 220 kcal/h. It is interesting to compare these values with the energy consumption required for travel by car. Gasoline yields about

1.30 × 10⁸ J/gal. Find the fuel economy in equivalent miles per gallon for a person (a) walking and (b) bicycling.

78. **BIO** Energy is conventionally measured in Calories as well as in joules. One Calorie in nutrition is 1 kilocalorie, which we define in Chapter 11 as 1 kcal = 4 186 J. Metabolizing 1 gram of fat can release 9.00 kcal. A student decides to try to lose weight by exercising. She plans to run up and down the stairs in a football stadium as fast as she can and as many times as necessary. Is this in itself a practical way to lose weight? To evaluate the program, suppose she runs up a flight of 80 steps, each 0.150 m high, in 65.0 s. For simplicity, ignore the energy she uses in coming down (which is small). Assume that a typical efficiency for human muscles is 20.0%. This means that when your body converts 100 J from metabolizing fat, 20 J goes into doing mechanical work (here, climbing stairs). The remainder goes into internal energy. Assume the student's mass is 50.0 kg. (a) How many times must she run the flight of stairs to lose 1 pound of fat? (b) What is her average power output, in watts and in horsepower, as she is running up the stairs?

79. A ski jumper starts from rest 50.0 m above the ground on a frictionless track and flies off the track at an angle of 45.0° above the horizontal and at a height of 10.0 m above the level ground. Neglect air resistance. (a) What is her speed when she leaves the track? (b) What is the maximum altitude she attains after leaving the track? (c) Where does she land relative to the end of the track?

80. A 5.0-kg block is pushed 3.0 m up a vertical wall with constant speed by a constant force of magnitude F applied at an angle of $\theta = 30°$ with the horizontal, as shown in Figure P5.80. If the coefficient of kinetic friction between block and wall is 0.30, determine the work done by (a) $\vec{F}$, (b) the force of gravity, and (c) the normal force between block and wall. (d) By how much does the gravitational potential energy increase during the block's motion?

Figure P5.80

81. A child's pogo stick (Fig. P5.81) stores energy in a spring ($k = 2.50 \times 10^4$ N/m). At position Ⓐ ($x_1 = -0.100$ m), the spring compression is a maximum and the child is momentarily at rest. At position Ⓑ ($x = 0$), the spring is relaxed and the child is moving upward. At position Ⓒ, the child is again momentarily at rest at the top of the jump. Assuming that the combined mass of child

Figure P5.81

and pogo stick is 25.0 kg, (a) calculate the total energy of the system if both potential energies are zero at $x = 0$, (b) determine x_2, (c) calculate the speed of the child at $x = 0$, (d) determine the value of x for which the kinetic energy of the system is a maximum, and (e) obtain the child's maximum upward speed.

82. **BIO** A hummingbird is able to hover because, as the wings move downward, they exert a downward force on the air. Newton's third law tells us that the air exerts an equal and opposite force (upward) on the wings. The average of this force must be equal to the weight of the bird when it hovers. If the wings move through a distance of 3.5 cm with each stroke, and the wings beat 80 times per second, determine the work performed by the wings on the air in 1 m if the mass of the hummingbird is 3.0 g.

83. In the dangerous "sport" of bungee jumping, a daring student jumps from a hot-air balloon with a specially designed elastic cord attached to his waist. The unstretched length of the cord is 25.0 m, the student weighs 700 N, and the balloon is 36.0 m above the surface of a river below. Calculate the required force constant of the cord if the student is to stop safely 4.00 m above the river.

84. **Q|C** The masses of the javelin, discus, and shot are 0.80 kg, 2.0 kg, and 7.2 kg, respectively, and record throws in the corresponding track events are about 98 m, 74 m, and 23 m, respectively. Neglecting air resistance, (a) calculate the minimum initial kinetic energies that would produce these throws, and (b) estimate the average force exerted on each object during the throw, assuming the force acts over a distance of 2.0 m. (c) Do your results suggest that air resistance is an important factor?

85. A truck travels uphill with constant velocity on a highway with a 7.0° slope. A 50-kg package sits on the floor of the back of the truck and does not slide, due to a static frictional force. During an interval in which the truck travels 340 m, (a) what is the net work done on the package? What is the work done on the package by (b) the force of gravity, (c) the normal force, and (d) the friction force?

86. A daredevil wishes to bungee-jump from a hot-air balloon 65.0 m above a carnival midway. He will use a piece of uniform elastic cord tied to a harness around his body to stop his fall at a point 10.0 m above the ground. Model his body as a particle and the cord as having negligible mass and a tension force described by Hooke's force law. In a preliminary test, hanging at rest from a 5.00-m length of the cord, the jumper finds that his body weight stretches it by 1.50 m. He will drop from rest at the point where the top end of a longer section of the cord is attached to the stationary balloon. (a) What length of cord should he use? (b) What maximum acceleration will he experience?

87. **M** A loaded ore car has a mass of 950 kg and rolls on rails with negligible friction. It starts from rest and is pulled up a mine shaft by a cable connected to a winch. The shaft is inclined at 30.0° above the horizontal. The car accelerates uniformly to a speed of 2.20 m/s in 12.0 s and then continues at constant speed. (a) What power must the winch motor provide when the car is moving at constant speed? (b) What maximum power must the motor provide? (c) What total energy transfers out of the motor by work by the time the car moves off the end of the track, which is of length 1 250 m?

88. **S** An object of mass m is suspended from the top of a cart by a string of length L as in Figure P5.88a. The cart and object are initially moving to the right at a constant speed v_0. The cart comes to rest after colliding and sticking to a bumper, as in Figure P5.88b, and the suspended object swings through an angle θ. (a) Show that the initial speed is $v_0 = \sqrt{2gL(1 - \cos\theta)}$. (b) If $L = 1.20$ m and $\theta = 35.0°$, find the initial speed of the cart. (*Hint:* The force exerted by the string on the object does no work on the object.)

Figure P5.88

89. Three objects with masses $m_1 = 5.0$ kg, $m_2 = 10$ kg, and $m_3 = 15$ kg, respectively, are attached by strings over frictionless pulleys as indicated in Figure P5.89. The horizontal surface exerts a force of friction of 30 N on m_2. If the system is released from rest, use energy concepts to find the speed of m_3 after it moves down 4.0 m.

90. A light spring with force constant 3.85 N/m is compressed by 8.00 cm as it is held between a 0.250-kg block on the left and a 0.500-kg block on the right, both resting on a horizontal surface. The spring exerts a force on each block, tending to push them apart. The blocks are simultaneously released from rest. Find the acceleration with which each block starts to move, given that the coefficient of kinetic friction between each block and the surface is (a) 0, (b) 0.100, and (c) 0.462.

91. **BIO** In bicycling for aerobic exercise, a woman wants her heart rate to be between 136 and 166 beats per minute. Assume that her heart rate is directly proportional to her mechanical power output. Ignore all forces on the woman-plus-bicycle system, except for static friction forward on the drive wheel of the bicycle and an air resistance force proportional to the square of the bicycler's speed. When her speed is 22.0 km/h, her heart rate is 90.0 beats per minute. In what range should her speed be so that her heart rate will be in the range she wants?

92. Two blocks, A and B (with mass 50 kg and 100 kg, respectively), are connected by a string, as shown in Figure P5.92. The pulley is frictionless and of negligible mass. The coefficient of kinetic friction between block A and the incline is $\mu_k = 0.25$. Determine the change in the kinetic energy of block A as it moves from Ⓒ to Ⓓ, a distance of 20 m up the incline (and block B drops downward a distance of 20 m) if the system starts from rest.

Figure P5.92

Figure P5.89

Rockets such as the Falcon 9 transform a large part of their initial mass into hot gas through chemical reactions. The energetic gas molecules collide with the reaction chamber walls, transferring momentum to the rest of the rocket before escaping out the exhaust nozzle.

NASA/Tony Gray and Robert Murray

6 Momentum and Collisions

6.1 Momentum and Impulse
6.2 Conservation of Momentum
6.3 Collisions
6.4 Glancing Collisions
6.5 Rocket Propulsion

What happens when two automobiles collide? How does the impact affect the motion of each vehicle, and what basic physical principles determine the likelihood of serious injury? How do rockets work, and what mechanisms can be used to overcome the limitations imposed by exhaust speed? Why do we have to brace ourselves when firing small projectiles at high velocity? Finally, how can we use physics to improve our golf game?

To begin answering such questions, we introduce *momentum*. Intuitively, anyone or anything that has a lot of momentum is going to be hard to stop. In politics, the term is metaphorical. Physically, the more momentum an object has, the more force has to be applied to stop it in a given time. This concept leads to one of the most powerful principles in physics: *conservation of momentum*. Using this law, complex collision problems can be solved without knowing much about the forces involved during contact. We'll also be able to derive information about the average force delivered in an impact. With conservation of momentum, we'll have a better understanding of what choices to make when designing an automobile or a moon rocket, or when addressing a golf ball on a tee.

6.1 Momentum and Impulse

LEARNING OBJECTIVES

1. Define momentum and impulse and state the impulse–momentum theorem.
2. Apply the impulse–momentum theorem to find estimates of average forces during collisions.

In physics, momentum has a precise definition. A slowly moving brontosaurus has a lot of momentum, but so does a little hot lead shot from the muzzle of a gun. We therefore expect that momentum will depend on an object's mass and velocity.

The linear momentum $\vec{\mathbf{p}}$ of an object of mass m moving with velocity $\vec{\mathbf{v}}$ is the product of its mass and velocity:

$$\vec{\mathbf{p}} \equiv m\vec{\mathbf{v}} \quad [6.1]$$

◀ Linear momentum

SI unit: kilogram-meter per second (kg · m/s)

Doubling either the mass or the velocity of an object doubles its momentum; doubling both quantities quadruples its momentum. Momentum is a vector quantity with the same direction as the object's velocity. Its components are given in two dimensions by

$$p_x = mv_x \quad p_y = mv_y$$

where p_x is the momentum of the object in the x-direction and p_y its momentum in the y-direction.

The magnitude of the momentum p of an object of mass m can be related to its kinetic energy KE:

$$KE = \frac{p^2}{2m} \quad [6.2]$$

This relationship is easy to prove using the definitions of kinetic energy and momentum (see Problem 6) and is valid for objects traveling at speeds much less than the speed of light. Equation 6.2 is useful in grasping the interplay between the two concepts, as illustrated in Quick Quiz 6.1.

■ Quick Quiz

6.1 Two masses m_1 and m_2, with $m_1 < m_2$, have equal kinetic energy. How do the magnitudes of their momenta compare? (a) Not enough information is given. (b) $p_1 < p_2$ (c) $p_1 = p_2$ (d) $p_1 > p_2$.

Changing the momentum of an object requires the application of a force. This is, in fact, how Newton originally stated his second law of motion. Starting from the more common version of the second law, we have

$$\vec{\mathbf{F}}_{net} = m\vec{\mathbf{a}} = m\frac{\Delta \vec{\mathbf{v}}}{\Delta t} = \frac{\Delta(m\vec{\mathbf{v}})}{\Delta t}$$

where the mass m and the forces are assumed constant. The quantity in parentheses is just the momentum, so we have the following result:

The change in an object's momentum $\Delta \vec{\mathbf{p}}$ divided by the elapsed time Δt equals the constant net force $\vec{\mathbf{F}}_{net}$ acting on the object:

$$\frac{\Delta \vec{\mathbf{p}}}{\Delta t} = \frac{\text{change in momentum}}{\text{time interval}} = \vec{\mathbf{F}}_{net} \quad [6.3]$$

◀ Newton's second law and momentum

This equation is also valid when the forces are not constant, provided the limit is taken as Δt becomes infinitesimally small. Equation 6.3 says that if the net force on an object is zero, the object's momentum doesn't change. In other words, the linear momentum of an object is conserved when $\vec{\mathbf{F}}_{net} = 0$. Equation 6.3 also shows us that changing an object's momentum requires the

Figure 6.1 (a) A net force acting on a particle may vary in time. (b) The value of the constant force F_{av} (horizontal dashed line) is chosen so that the area of the rectangle $F_{av}\Delta t$ is the same as the area under the curve in (a).

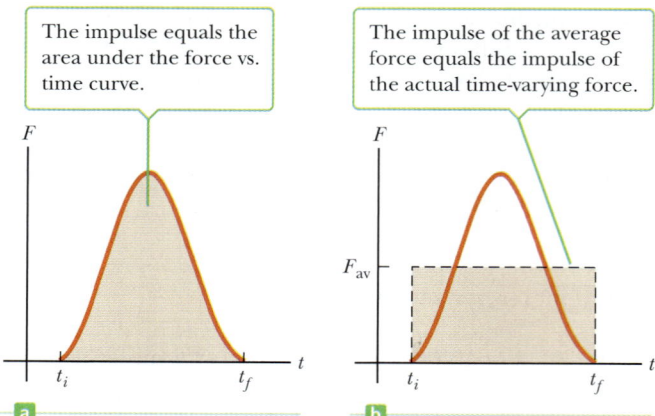

The impulse equals the area under the force vs. time curve.

The impulse of the average force equals the impulse of the actual time-varying force.

continuous application of a force over a period of time Δt, leading to the definition of *impulse*:

> If a constant force $\vec{F}$ acts on an object, the **impulse** $\vec{I}$ delivered to the object over a time interval Δt is given by
> $$\vec{I} \equiv \vec{F}\Delta t \qquad [6.4]$$
> **SI unit: kilogram meter per second (kg · m/s)**

Impulse is a vector quantity with the same direction as the constant force acting on the object. When a single constant force $\vec{F}$ acts on an object, Equation 6.3 can be written as

$$\vec{I} = \vec{F}\Delta t = \Delta\vec{p} = m\vec{v}_f - m\vec{v}_i \qquad [6.5]$$

◄ Impulse–momentum theorem

which is a special case of the **impulse–momentum theorem.** Equation 6.5 shows that **the impulse of the force acting on an object equals the change in momentum of that object.** That equality is true even if the force is not constant, as long as the time interval Δt is taken to be arbitrarily small. (The proof of the general case requires concepts from calculus.)

In real-life situations, the force on an object is only rarely constant. For example, when a bat hits a baseball, the force increases sharply, reaches some maximum value, and then decreases just as rapidly. Figure 6.1(a) shows a typical graph of force versus time for such incidents. The force starts out small as the bat comes in contact with the ball, rises to a maximum value when they are firmly in contact, and then drops off as the ball leaves the bat. In order to analyze this rather complex interaction, it's useful to define an **average force** $\vec{F}_{av}$, shown as the dashed line in Figure 6.1b. The average force is the constant force delivering the same impulse to the object in the time interval Δt as the actual time-varying force. We can then write the impulse–momentum theorem as

$$\vec{F}_{av}\Delta t = \Delta\vec{p} \qquad [6.6]$$

The magnitude of the impulse delivered by a force during the time interval Δt is equal to the area under the force vs. time graph as in Figure 6.1a or, equivalently, to $F_{av}\Delta t$ as shown in Figure 6.1b. The brief collision between a bullet and an apple is illustrated in Figure 6.2.

Figure 6.2 An apple being pierced by a 30-caliber bullet traveling at a supersonic speed of 900 m/s. This collision was photographed with a microflash stroboscope using an exposure time of 0.33 μs. Shortly after the photograph was taken, the apple disintegrated completely. Note that the points of entry and exit of the bullet are visually explosive.

APPLYING PHYSICS 6.1 | Boxing and Brain Injury BIO

Boxers in the nineteenth century used their bare fists. In modern boxing, fighters wear padded gloves. How do gloves protect the brain of the boxer from injury? Also, why do boxers often "roll with the punch"?

EXPLANATION The brain is immersed in a cushioning fluid inside the skull. If the head is struck suddenly by a bare fist, the skull accelerates rapidly. The brain matches this acceleration only because of the large impulsive force exerted by the skull on the brain. This large and sudden force (large F_{av} and small Δt) can cause severe brain injury. Padded gloves extend the time Δt over which the force is applied to the head. For a given impulse $F_{av}\Delta t$, a glove results in a longer time interval than a bare fist, decreasing the average force. Because the average force is decreased, the acceleration of the skull is decreased, reducing (but not eliminating) the chance of brain injury. The same argument can be made for "rolling with the punch": If the head is held steady while being struck, the time interval over which the force is applied is relatively short and the average force is large. If the head is allowed to move in the same direction as the punch, the time interval is lengthened and the average force reduced. ■

EXAMPLE 6.1 | Teeing Off

GOAL Use the impulse–momentum theorem to estimate the average force exerted during an impact.

PROBLEM A golf ball with mass 5.0×10^{-2} kg is struck with a club as in Figure 6.3. The force on the ball varies from zero when contact is made up to some maximum value (when the ball is maximally deformed) and then back to zero when the ball leaves the club, as in the graph of force vs. time in Figure 6.1. Assume that the ball leaves the club face with a velocity of 44 m/s. **(a)** Find the magnitude of the impulse due to the collision. **(b)** Estimate the duration of the collision and the average force acting on the ball.

STRATEGY In part (a), use the fact that the impulse is equal to the change in momentum. The mass and the initial and final velocities are known, so this change can be computed. In part (b), the average force is just the change in momentum computed in part (a) divided by an estimate of the duration of the collision. Estimate the distance the ball travels on the face of the club (about 2.0 cm, roughly the same as the radius of the ball). Divide this distance by the average velocity (half the final velocity) to get an estimate of the time of contact.

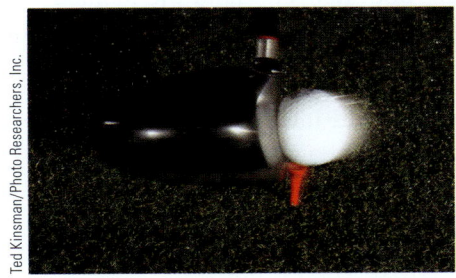

Figure 6.3 (Example 6.1) During impact, the club head momentarily flattens the side of the golf ball.

SOLUTION

(a) Find the impulse delivered to the ball.

The problem is essentially one dimensional. Note that $v_i = 0$, and calculate the change in momentum, which equals the impulse:

$$I = \Delta p = p_f - p_i = (5.0 \times 10^{-2} \text{ kg})(44 \text{ m/s}) - 0$$
$$= 2.2 \text{ kg} \cdot \text{m/s}$$

(b) Estimate the duration of the collision and the average force acting on the ball.

Estimate the time interval of the collision, Δt, using the approximate displacement (radius of the ball) and its average speed (half the maximum speed):

$$\Delta t = \frac{\Delta x}{v_{av}} = \frac{2.0 \times 10^{-2} \text{ m}}{22 \text{ m/s}} = 9.1 \times 10^{-4} \text{ s}$$

Estimate the average force from Equation 6.6:

$$F_{av} = \frac{\Delta p}{\Delta t} = \frac{2.2 \text{ kg} \cdot \text{m/s}}{9.1 \times 10^{-4} \text{ s}} = 2.4 \times 10^3 \text{ N}$$

REMARKS This estimate shows just how large such contact forces can be. A good golfer achieves maximum momentum transfer by shifting weight from the back foot to the front foot, transmitting the body's momentum through the shaft and head of the club. This timing, involving a short movement of the hips, is more effective than a shot powered exclusively by the arms and shoulders. Following through with the swing ensures that the motion isn't slowed at the critical instant of impact.

QUESTION 6.1 What average club speed would double the average force? (Assume the final velocity is unchanged.)

EXERCISE 6.1 A 0.150-kg baseball, thrown with a speed of 40.0 m/s, is hit straight back at the pitcher with a speed of 50.0 m/s. (a) What is the magnitude of the impulse delivered by the bat to the baseball? (b) Find the magnitude of the average force exerted by the bat on the ball if the two are in contact for 2.00×10^{-3} s.

ANSWERS (a) 13.5 kg · m/s (b) 6.75 kN

EXAMPLE 6.2 How Good Are the Bumpers?

GOAL Find an impulse and estimate a force in a collision of a moving object with a stationary object.

PROBLEM In a crash test, a car of mass 1.50×10^3 kg collides with a wall and rebounds as in Figure 6.4a. The initial and final velocities of the car are $v_i = -15.0$ m/s and $v_f = 2.60$ m/s, respectively. If the collision lasts for 0.150 s, find **(a)** the impulse delivered to the car due to the collision and **(b)** the magnitude and direction of the average force exerted on the car.

STRATEGY This problem is similar to the previous example, except that the initial and final momenta are both nonzero. Find the momenta and substitute into the impulse–momentum theorem, Equation 6.6, solving for F_{av}.

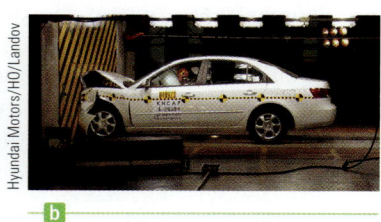

Figure 6.4 (Example 6.2) (a) This car's momentum changes as a result of its collision with the wall. (b) In a crash test (an inelastic collision), much of the car's initial kinetic energy is transformed into the energy it took to damage the vehicle.

SOLUTION

(a) Find the impulse delivered to the car.

Calculate the initial and final momenta of the car:

$$p_i = mv_i = (1.50 \times 10^3 \text{ kg})(-15.0 \text{ m/s})$$
$$= -2.25 \times 10^4 \text{ kg} \cdot \text{m/s}$$
$$p_f = mv_f = (1.50 \times 10^3 \text{ kg})(+2.60 \text{ m/s})$$
$$= +0.390 \times 10^4 \text{ kg} \cdot \text{m/s}$$

The impulse is just the difference between the final and initial momenta:

$$I = p_f - p_i$$
$$= +0.390 \times 10^4 \text{ kg} \cdot \text{m/s} - (-2.25 \times 10^4 \text{ kg} \cdot \text{m/s})$$
$$I = \boxed{2.64 \times 10^4 \text{ kg} \cdot \text{m/s}}$$

(b) Find the average force exerted on the car.

Apply Equation 6.6, the impulse–momentum theorem:

$$F_{av} = \frac{\Delta p}{\Delta t} = \frac{2.64 \times 10^4 \text{ kg} \cdot \text{m/s}}{0.150 \text{ s}} = \boxed{+1.76 \times 10^5 \text{ N}}$$

REMARKS If the car doesn't rebound off the wall, the average force exerted on the car is smaller than the value just calculated. With a final momentum of zero, the car undergoes a smaller change in momentum.

QUESTION 6.2 When a person is involved in a car accident, why is the likelihood of injury greater in a head-on collision as opposed to being hit from behind? Answer using the concepts of relative velocity, momentum, and average force.

EXERCISE 6.2 Suppose the car doesn't rebound off the wall, but the time interval of the collision remains at 0.150 s. In this case, the final velocity of the car is zero. Find the average force exerted on the car.

ANSWER $+1.50 \times 10^5$ N

Injury in Automobile Collisions

BIO APPLICATION
Injury to Passengers in Car Collisions

The main injuries that occur to a person hitting the interior of a car in a crash are brain damage, bone fracture, and trauma to the skin, blood vessels, and internal organs. Here, we compare the rather imprecisely known thresholds for human injury with typical forces and accelerations experienced in a car crash.

A force of about 90 kN (20 000 lb) compressing the tibia can cause fracture. Although the breaking force varies with the bone considered, we may take this value as the threshold force for fracture. It's well known that rapid acceleration of the head, even without skull fracture, can be fatal. Estimates show that head accelerations of $150g$ experienced for about 4 ms or $50g$ for 60 ms are fatal 50% of the time. Such injuries from rapid acceleration often result in nerve damage to the spinal cord where the nerves enter the base of the brain. The threshold for damage to skin, blood vessels, and internal organs may be estimated from whole-body impact data, where the force is uniformly distributed over the entire front surface area of 0.7 to 0.9 m². These data show that if the collision lasts for less than about 70 ms, a person will survive if the whole-body impact pressure (force per unit area) is less than 1.9×10^5 N/m² (28 lb/in.²). Death results in 50% of cases in which the whole-body impact pressure reaches 3.4×10^5 N/m² (50 lb/in.²).

Armed with the data above, we can estimate the forces and accelerations in a typical car crash and see how seat belts, air bags, and padded interiors can reduce the chance of death or serious injury in a collision. Consider a typical collision involving a 75-kg passenger not wearing a seat belt, traveling at 27 m/s (60 mi/h) who comes to rest in about 0.010 s after striking an unpadded dashboard. Using $F_{av} \Delta t = mv_f - mv_i$, we find that

$$F_{av} = \frac{mv_f - mv_i}{\Delta t} = \frac{0 - (75 \text{ kg})(27 \text{ m/s})}{0.010 \text{ s}} = -2.0 \times 10^5 \text{ N}$$

and

$$a = \left|\frac{\Delta v}{\Delta t}\right| = \frac{27 \text{ m/s}}{0.010 \text{ s}} = 2\,700 \text{ m/s}^2 = \frac{2\,700 \text{ m/s}^2}{9.8 \text{ m/s}^2}g = 280g$$

If we assume the passenger crashes into the dashboard and windshield so that the head and chest, with a combined surface area of 0.5 m², experience the force, we find a whole-body pressure of

$$\frac{F_{av}}{A} = \frac{2.0 \times 10^5 \text{ N}}{0.5 \text{ m}^2} \cong 4 \times 10^5 \text{ N/m}^2$$

We see that the force, the acceleration, and the whole-body pressure all *exceed* the threshold for fatality or broken bones and that an unprotected collision at 60 mi/h is almost certainly fatal.

What can be done to reduce or eliminate the chance of dying in a car crash? The most important factor is the collision time, or the time it takes the person to come to rest. If this time can be increased by 10 to 100 times the value of 0.01 s for a hard collision, the chances of survival in a car crash are much higher because the increase in Δt makes the contact force 10 to 100 times smaller. Seat belts restrain people so that they come to rest in about the same amount of time it takes to stop the car, typically about 0.15 s. This increases the effective collision time by an order of magnitude. Figure 6.5 shows the measured force on a car versus time for a car crash.

Air bags also increase the collision time, absorb energy from the body as they rapidly deflate, and spread the contact force over an area of the body of about 0.5 m², preventing penetration wounds and fractures. Air bags must deploy very rapidly (in less than 10 ms) in order to stop a human traveling at 27 m/s before he or she comes to rest against the steering column about 0.3 m away. To achieve this rapid deployment, accelerometers send a signal to discharge a bank of capacitors (devices that store electric charge), which then ignites an explosive, thereby filling the air bag with gas very quickly. The electrical charge for ignition is stored in capacitors to ensure that the air bag deploys in the event of damage to the battery or the car's electrical system in a severe collision.

The important reduction in potentially fatal forces, accelerations, and pressures to tolerable levels by the simultaneous use of seat belts and air bags is summarized as follows: If a 75-kg person traveling at 27 m/s is stopped by a seat belt in 0.15 s, the person experiences an average force of 9.8 kN, an average acceleration of 18g, and a whole-body pressure of 2.8×10^4 N/m² for a contact area of 0.5 m². These values are about one order of magnitude less than the values estimated earlier for an unprotected person and well below the thresholds for life-threatening injuries.

Figure 6.5 Force on a car versus time for a typical collision.

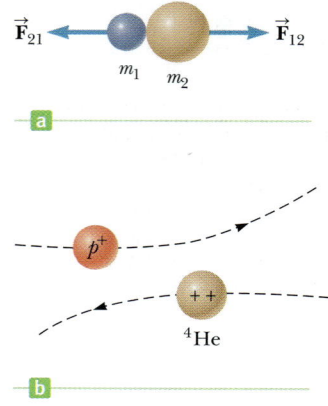

Figure 6.6 (a) A collision between two objects resulting from direct contact. (b) A collision between two charged objects (in this case, a proton and a helium nucleus).

6.2 Conservation of Momentum

LEARNING OBJECTIVES

1. Obtain the principle of conservation of momentum from the impulse–momentum theorem and the third law of motion.
2. Apply conservation of momentum to the problem of recoil.

When a collision occurs in an isolated system, the total momentum of the system doesn't change with the passage of time. Instead, it remains constant both in magnitude and in direction. The momenta of the individual objects in the system may change, but the vector sum of *all* the momenta will not change. The total momentum is therefore said to be *conserved*. In this section, we will see how the laws of motion lead us to this important conservation law.

A collision may be the result of physical contact between two objects, as illustrated in Figure 6.6a. This is a common macroscopic event, as when a pair of billiard balls or a baseball and a bat strike each other. By contrast, because contact on a submicroscopic scale is hard to define accurately, the notion of *collision* must be generalized to that scale. Forces between two objects arise from the electrostatic interaction of the electrons in the surface atoms of the objects. As will be discussed in Chapter 15, electric charges are either positive or negative. Charges with the same sign repel each other, while charges with opposite sign attract each other. To understand the distinction between macroscopic and microscopic collisions, consider the collision between two positive charges, as shown in Figure 6.6b. Because the two particles in the figure are both positively charged, they repel each other. During such a microscopic collision, particles need not touch in the normal sense in order to interact and transfer momentum.

Figure 6.7 shows an isolated system of two particles before and after they collide. By "isolated," we mean that no external forces, such as the gravitational force or friction, act on the system. Before the collision, the velocities of the two particles are $\vec{v}_{1i}$ and $\vec{v}_{2i}$; after the collision, the velocities are $\vec{v}_{1f}$ and $\vec{v}_{2f}$. The impulse–momentum theorem applied to m_1 becomes

$$\vec{F}_{21}\Delta t = m_1\vec{v}_{1f} - m_1\vec{v}_{1i}$$

Likewise, for m_2, we have

$$\vec{F}_{12}\Delta t = m_2\vec{v}_{2f} - m_2\vec{v}_{2i}$$

where $\vec{F}_{21}$ is the average force exerted by m_2 on m_1 during the collision and $\vec{F}_{12}$ is the average force exerted by m_1 on m_2 during the collision, as in Figure 6.6a.

We use average values for $\vec{F}_{21}$ and $\vec{F}_{12}$ even though the actual forces may vary in time in a complicated way, as is the case in Figure 6.8. Newton's third law states that at all times these two forces are equal in magnitude and opposite in direction: $\vec{F}_{21} = -\vec{F}_{12}$. In addition, the two forces act over the same time interval. As a result, we have

$$\vec{F}_{21}\Delta t = -\vec{F}_{12}\Delta t$$

or

$$m_1\vec{v}_{1f} - m_1\vec{v}_{1i} = -(m_2\vec{v}_{2f} - m_2\vec{v}_{2i})$$

after substituting the expressions obtained for $\vec{F}_{21}$ and $\vec{F}_{12}$. This equation can be rearranged to give the following important result:

$$m_1\vec{v}_{1i} + m_2\vec{v}_{2i} = m_1\vec{v}_{1f} + m_2\vec{v}_{2f} \qquad [6.7]$$

This result is a special case of the law of **conservation of momentum** and is true of isolated systems containing any number of interacting objects.

▶ Conservation of momentum

When no net external force acts on a system, the total momentum of the system remains constant in time.

Figure 6.7 Before and after a head-on collision between two particles. The momentum of each object changes during the collision, but the total momentum of the system is constant. Notice that the magnitude of the change of velocity of the lighter particle is greater than that of the heavier particle, which is true in general.

6.2 | Conservation of Momentum   177

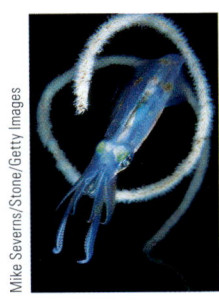

Conservation of momentum is the principle behind a squid's propulsion system. It propels itself by expelling water at a high velocity.

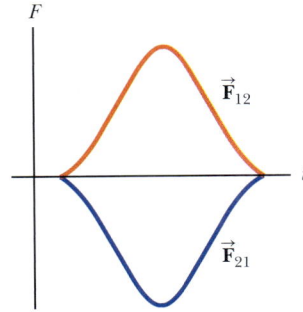

Figure 6.8 Force as a function of time for the two colliding particles in Figures 6.6a and 6.7. Note that $\vec{F}_{21} = -\vec{F}_{12}$.

> **Tip 6.1** Momentum Conservation Applies to a *System*!
>
> The momentum of an isolated system is conserved, but not necessarily the momentum of one particle within that system, because other particles in the system may be interacting with it. Apply conservation of momentum to an isolated system *only*.

BIO APPLICATION
Conservation of momentum and squid propulsion

Defining the isolated system is an important feature of applying this conservation law. A cheerleader jumping upwards from rest might appear to violate conservation of momentum, because initially her momentum is zero and suddenly she's leaving the ground with velocity $\vec{v}$. The flaw in this reasoning lies in the fact that the cheerleader isn't an isolated system. In jumping, she exerts a downward force on Earth, changing its momentum. This change in Earth's momentum isn't noticeable, however, because of Earth's gargantuan mass compared to the cheerleader's. When we define the system to be *the cheerleader and Earth,* momentum is conserved.

Action and reaction, together with the accompanying exchange of momentum between two objects, is responsible for the phenomenon known as *recoil*. Everyone knows that throwing a baseball while standing straight up, without bracing one's feet against Earth, is a good way to fall over backwards. This reaction, an example of recoil, also happens when you fire a gun or shoot an arrow. Conservation of momentum provides a straightforward way to calculate such effects, as the next example shows.

■ EXAMPLE 6.3 The Archer

GOAL Calculate recoil velocity using conservation of momentum.

PROBLEM An archer stands at rest on frictionless ice; his total mass including his bow and quiver of arrows is 60.00 kg. (See Fig. 6.9.) **(a)** If the archer fires a 0.030 0-kg arrow horizontally at 50.0 m/s in the positive *x*-direction, what is his subsequent velocity across the ice? **(b)** He then fires a second identical arrow at the same speed relative to the ground but at an angle of 30.0° above the horizontal. Find his new speed. **(c)** Estimate the average normal force acting on the archer as the second arrow is accelerated by the bowstring. Assume a draw length of 0.800 m.

STRATEGY To solve part (a), set up the conservation of momentum equation in the *x*-direction and solve for the final velocity of the archer. The system of the archer (including the bow) and the arrow is not isolated, because the gravitational and normal forces act on it. Those forces, however, are perpendicular to the motion of the system during the release of the arrow, and in addition are equal in magnitude and opposite in direction. Consequently, they produce no impulse during the arrow's release and conservation of momentum can be used. In part (b), conservation of momentum can be applied again, neglecting the tiny effect of gravitation on the arrow during its release. This time there is a non-zero initial velocity. Part (c) requires using the impulse–momentum theorem and estimating the time, which can be carried out with simple ballistics.

Figure 6.9 (Example 6.3) An archer fires an arrow horizontally to the right. Because he is standing on frictionless ice, he will begin to slide to the left across the ice.

(Continued)

SOLUTION

(a) Find the archer's subsequent velocity across the ice.

Write the conservation of momentum equation for the x-direction.

$$p_{ix} = p_{fx}$$

Let m_1 and v_{1f} be the archer's mass and velocity after firing the arrow, respectively, and m_2 and v_{2f} the arrow's mass and velocity. Both velocities are in the x-direction. Substitute $p_i = 0$ and expressions for the final momenta:

$$0 = m_1 v_{1f} + m_2 v_{2f}$$

Solve for v_{1f} and substitute $m_1 = 59.97$ kg, $m_2 = 0.0300$ kg, and $v_{2f} = 50.0$ m/s:

$$v_{1f} = -\frac{m_2}{m_1} v_{2f} = -\left(\frac{0.0300 \text{ kg}}{59.97 \text{ kg}}\right)(50.0 \text{ m/s})$$

$$v_{1f} = \boxed{-0.0250 \text{ m/s}}$$

(b) Calculate the archer's velocity after he fires a second arrow at an angle of 30.0° above the horizontal.

Write the x-component of the momentum equation with m_1 again the archer's mass after firing the first arrow as in part (a) and m_2 the mass of the next arrow:

$$m_1 v_{1i} = (m_1 - m_2) v_{1f} + m_2 v_{2f} \cos\theta$$

Solve for v_{1f}, the archer's final velocity, and substitute:

$$v_{1f} = \frac{m_1}{(m_1 - m_2)} v_{1i} - \frac{m_2}{(m_1 - m_2)} v_{2f} \cos\theta$$

$$= \left(\frac{59.97 \text{ kg}}{59.94 \text{ kg}}\right)(-0.0250 \text{ m/s}) - \left(\frac{0.0300 \text{ kg}}{59.94 \text{ kg}}\right)(50.0 \text{ m/s}) \cos(30.0°)$$

$$v_{1f} = \boxed{-0.0467 \text{ m/s}}$$

(c) Estimate the average normal force acting on the archer as the arrow is accelerated by the bowstring.

Use kinematics in one dimension to estimate the acceleration of the arrow:

$$v^2 - v_0^2 = 2a\Delta x$$

Solve for the acceleration and substitute values setting $v = v_{2f}$, the final velocity of the arrow:

$$a = \frac{v_{2f}^2 - v_0^2}{2\Delta x} = \frac{(50.0 \text{ m/s})^2 - 0}{2(0.800 \text{ m})} = 1.56 \times 10^3 \text{ m/s}^2$$

Find the time the arrow is accelerated using $v = at + v_0$:

$$t = \frac{v_{2f} - v_0}{a} = \frac{50.0 \text{ m/s} - 0}{1.56 \times 10^3 \text{ m/s}^2} = 0.0320 \text{ s}$$

Write the y-component of the impulse–momentum theorem:

$$F_{y,av} \Delta t = \Delta p_y$$

$$F_{y,av} = \frac{\Delta p_y}{\Delta t} = \frac{m_2 v_{2f} \sin\theta}{\Delta t}$$

$$F_{y,av} = \frac{(0.0300 \text{ kg})(50.0 \text{ m/s}) \sin(30.0°)}{0.0320 \text{ s}} = 23.4 \text{ N}$$

The average normal force is given by the archer's weight plus the reaction force R of the arrow on the archer:

$$\Sigma F_y = n - mg - R = 0$$

$$n = mg + R = (59.94 \text{ kg})(9.80 \text{ m/s}^2) + (23.4 \text{ N}) = \boxed{6.11 \times 10^2 \text{ N}}$$

REMARKS The negative sign on v_{1f} indicates that the archer is moving opposite the arrow's direction, in accordance with Newton's third law. Because the archer is much more massive than the arrow, his acceleration and velocity are much smaller than the acceleration and velocity of the arrow. A technical point: the second arrow was fired at the same velocity relative to the ground, but because the archer was moving backwards at the time, it was traveling slightly faster than the first arrow relative to the archer. Velocities must always be given relative to a frame of reference.

Notice that conservation of momentum was effective in leading to a solution in parts (a) and (b). The final answer for the normal force is only an average because the force exerted on the arrow is unlikely to be constant. If the ice really were frictionless, the archer would have trouble standing. In general the coefficient of static friction of ice is more than sufficient to prevent sliding in response to such small recoils.

QUESTION 6.3 Would firing a heavier arrow necessarily increase the recoil velocity? Explain, using the result of Quick Quiz 6.1.

EXERCISE 6.3 A 70.0-kg man and a 55.0-kg woman holding a 2.50-kg purse on ice skates stand facing each other. (a) If the woman pushes the man backwards so that his final speed is 1.50 m/s, with what average force did she push him, assuming they were in contact for 0.500 s? (b) What is the woman's recoil speed? (c) If she now throws her 2.50-kg purse at him at a 20.0° angle above the horizontal and at 4.20 m/s relative to the ground, what is her subsequent speed?

ANSWERS (a) 2.10×10^2 N (b) 1.83 m/s (c) 2.09 m/s

Quick Quiz

6.2 A boy standing at one end of a floating raft that is stationary relative to the shore walks to the opposite end of the raft, away from the shore. As a consequence, the raft (a) remains stationary, (b) moves away from the shore, or (c) moves toward the shore. (*Hint:* Use conservation of momentum.)

6.3 Collisions

LEARNING OBJECTIVES

1. Define inelastic, perfectly inelastic, and elastic collisions.
2. Apply conservation of momentum to inelastic and perfectly inelastic collisions in one dimension.
3. Apply conservation of momentum and energy to one-dimensional elastic collisions.

We have seen that for any type of collision, the total momentum of the system just before the collision equals the total momentum just after the collision as long as the system may be considered isolated. The total kinetic energy, on the other hand, is generally not conserved in a collision because some of the kinetic energy is converted to internal energy, sound energy, and the work needed to permanently deform the objects involved, such as cars in a car crash. **We define an inelastic collision as a collision in which momentum is conserved, but kinetic energy is not.** The collision of a rubber ball with a hard surface is inelastic, because some of the kinetic energy is lost when the ball is deformed during contact with the surface. **When two objects collide and stick together, the collision is called *perfectly inelastic.*** For example, if two pieces of putty collide, they stick together and move with some common velocity after the collision. If a meteorite collides head on with Earth, it becomes buried in Earth and the collision is considered perfectly inelastic. Only in very special circumstances is all the initial kinetic energy lost in a perfectly inelastic collision.

An elastic collision is defined as one in which both momentum and kinetic energy are conserved. Billiard ball collisions and the collisions of air molecules with the walls of a container at ordinary temperatures are highly elastic. Macroscopic collisions such as those between billiard balls are only approximately elastic, because some loss of kinetic energy takes place—for example, in the clicking sound when two balls strike each other. Perfectly elastic collisions do occur, however, between atomic and subatomic particles. Elastic and perfectly inelastic collisions are *limiting* cases; most actual collisions fall into a range in between them.

As a practical application, an inelastic collision is used to detect glaucoma, a disease in which the pressure inside the eye builds up and leads to blindness by damaging the cells of the retina. In this application, medical professionals use a device called a *tonometer* to measure the pressure inside the eye. This device releases a puff of air against the outer surface of the eye and measures the speed of the air after reflection from the eye. At normal pressure, the eye is slightly spongy, and the pulse

Tip 6.2 Momentum and Kinetic Energy in Collisions
The momentum of an isolated system is conserved in all collisions. However, the kinetic energy of an isolated system is conserved only when the collision is elastic.

Tip 6.3 Inelastic vs. Perfectly Inelastic Collisions
If the colliding particles stick together, the collision is perfectly inelastic. If they bounce off each other (and kinetic energy is not conserved), the collision is inelastic.

BIO APPLICATION
Glaucoma Testing

is reflected at low speed. As the pressure inside the eye increases, the outer surface becomes more rigid, and the speed of the reflected pulse increases. In this way, the speed of the reflected puff of air can measure the internal pressure of the eye.

We can summarize the types of collisions as follows:

Elastic collision ▶
Inelastic collision ▶

- In an elastic collision, both momentum and kinetic energy are conserved.
- In an inelastic collision, momentum is conserved but kinetic energy is not.
- In a *perfectly* inelastic collision, momentum is conserved, kinetic energy is not, and the two objects stick together after the collision, so their final velocities are the same.

In the remainder of this section, we will treat perfectly inelastic collisions and elastic collisions in one dimension.

■ Quick Quiz

6.3 A car and a large truck traveling at the same speed collide head-on and stick together. Which vehicle undergoes the larger change in the magnitude of its momentum? (a) the car (b) the truck (c) the change in the magnitude of momentum is the same for both (d) impossible to determine without more information.

Perfectly Inelastic Collisions

Consider two objects having masses m_1 and m_2 moving with known initial velocity components v_{1i} and v_{2i} along a straight line, as in Figure 6.10. If the two objects collide head-on, stick together, and move with a common velocity component v_f after the collision, then the collision is perfectly inelastic. Because the total momentum of the two-object isolated system before the collision equals the total momentum of the combined-object system after the collision, we can solve for the final velocity using conservation of momentum alone:

$$m_1 v_{1i} + m_2 v_{2i} = (m_1 + m_2) v_f \qquad [6.8]$$

$$v_f = \frac{m_1 v_{1i} + m_2 v_{2i}}{m_1 + m_2} \qquad [6.9]$$

It's important to notice that v_{1i}, v_{2i}, and v_f represent the x-components of the velocity vectors, so care is needed in entering their known values, particularly with regard to signs. For example, in Figure 6.10, v_{1i} would have a positive value (m_1 moving to the right), whereas v_{2i} would have a negative value (m_2 moving to the left). Once these values are entered, Equation 6.9 can be used to find the correct final velocity, as shown in Examples 6.4 and 6.5.

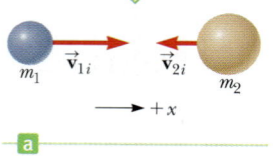

Before a perfectly inelastic collision the objects move independently.

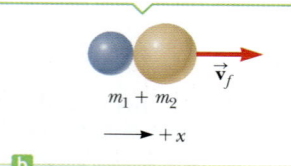

After the collision the objects remain in contact. System momentum *is* conserved, but system energy is *not* conserved.

Figure 6.10 (a) Before and (b) after a perfectly inelastic head-on collision between two objects.

■ EXAMPLE 6.4 | A Truck Versus a Compact

GOAL Apply conservation of momentum to a one-dimensional inelastic collision.

PROBLEM A pickup truck with mass 1.80×10^3 kg is traveling eastbound at +15.0 m/s, while a compact car with mass 9.00×10^2 kg is traveling westbound at −15.0 m/s. (See Fig. 6.11.) The vehicles collide head-on, becoming entangled. **(a)** Find the speed of the entangled vehicles after the collision. **(b)** Find the change in the velocity of each vehicle. **(c)** Find the change in the kinetic energy of the system consisting of both vehicles.

STRATEGY The total momentum of the vehicles before the collision, p_i, equals the total momentum of the vehicles after the collision, p_f, if we ignore friction and assume the two vehicles form an isolated system. (This is called the "impulse approximation.") Solve the momentum conservation equation for the final velocity of the entangled vehicles. Once the velocities are in hand, the other parts can be solved by substitution.

Figure 6.11 (Example 6.4)

SOLUTION

(a) Find the final speed after collision.

Let m_1 and v_{1i} represent the mass and initial velocity of the pickup truck, while m_2 and v_{2i} pertain to the compact. Apply conservation of momentum:

$$p_i = p_f$$
$$m_1 v_{1i} + m_2 v_{2i} = (m_1 + m_2) v_f$$

Substitute the values and solve for the final velocity, v_f:

$$(1.80 \times 10^3 \text{ kg})(15.0 \text{ m/s}) + (9.00 \times 10^2 \text{ kg})(-15.0 \text{ m/s})$$
$$= (1.80 \times 10^3 \text{ kg} + 9.00 \times 10^2 \text{ kg}) v_f$$
$$v_f = \boxed{+5.00 \text{ m/s}}$$

(b) Find the change in velocity for each vehicle.

Change in velocity of the pickup truck:
$$\Delta v_1 = v_f - v_{1i} = 5.00 \text{ m/s} - 15.0 \text{ m/s} = \boxed{-10.0 \text{ m/s}}$$

Change in velocity of the compact car:
$$\Delta v_2 = v_f - v_{2i} = 5.00 \text{ m/s} - (-15.0 \text{ m/s}) = \boxed{20.0 \text{ m/s}}$$

(c) Find the change in kinetic energy of the system.

Calculate the initial kinetic energy of the system:
$$KE_i = \tfrac{1}{2} m_1 v_{1i}^2 + \tfrac{1}{2} m_2 v_{2i}^2 = \tfrac{1}{2}(1.80 \times 10^3 \text{ kg})(15.0 \text{ m/s})^2$$
$$+ \tfrac{1}{2}(9.00 \times 10^2 \text{ kg})(-15.0 \text{ m/s})^2$$
$$= 3.04 \times 10^5 \text{ J}$$

Calculate the final kinetic energy of the system and the change in kinetic energy, ΔKE.

$$KE_f = \tfrac{1}{2}(m_1 + m_2) v_f^2$$
$$= \tfrac{1}{2}(1.80 \times 10^3 \text{ kg} + 9.00 \times 10^2 \text{ kg})(5.00 \text{ m/s})^2$$
$$= 3.38 \times 10^4 \text{ J}$$
$$\Delta KE = KE_f - KE_i = \boxed{-2.70 \times 10^5 \text{ J}}$$

REMARKS During the collision, the system lost almost 90% of its kinetic energy. The change in velocity of the pickup truck was only 10.0 m/s, compared to twice that for the compact car. This example underscores perhaps the most important safety feature of any car: its mass. Injury is caused by a change in velocity, and the more massive vehicle undergoes a smaller velocity change in a typical accident.

QUESTION 6.4 If the mass of both vehicles were doubled, how would the final velocity be affected? The change in kinetic energy?

EXERCISE 6.4 Suppose the same two vehicles are both traveling eastward, the compact car leading the pickup truck. The driver of the compact car slams on the brakes suddenly, slowing the vehicle to 6.00 m/s. If the pickup truck traveling at 18.0 m/s crashes into the compact car, find (a) the speed of the system right after the collision, assuming the two vehicles become entangled, (b) the change in velocity for both vehicles, and (c) the change in kinetic energy of the system, from the instant before impact (when the compact car is traveling at 6.00 m/s) to the instant right after the collision.

ANSWERS (a) 14.0 m/s (b) pickup truck: $\Delta v_1 = -4.0$ m/s, compact car: $\Delta v_2 = 8.0$ m/s (c) -4.32×10^4 J

■ EXAMPLE 6.5 The Ballistic Pendulum

GOAL Combine the concepts of conservation of energy and conservation of momentum in inelastic collisions.

PROBLEM The ballistic pendulum (Fig. 6.12a) is a device used to measure the speed of a fast-moving projectile such as a bullet. The bullet is fired into a large block of wood suspended from some light wires. The bullet embeds in the block, and the entire system swings up to a height h. It is possible to obtain the initial speed of the bullet by measuring h and the two masses. As an example of the technique, assume that the mass of the bullet, m_1, is 5.00 g, the mass of the pendulum, m_2, is 1.000 kg, and h is 5.00 cm. **(a)** Find the velocity of the system after the bullet embeds in the block. **(b)** Calculate the initial speed of the bullet.

STRATEGY Use conservation of energy to find the initial velocity of the block–bullet system, labeling it v_{sys}. Part (b) requires the conservation of momentum equation, which can be solved for the initial velocity of the bullet, v_{1i}.

(Continued)

Figure 6.12 (Example 6.5) (a) Diagram of a ballistic pendulum. Note that $\vec{v}_{sys}$ is the velocity of the system just *after* the perfectly inelastic collision. (b) Multiflash photograph of a laboratory ballistic pendulum.

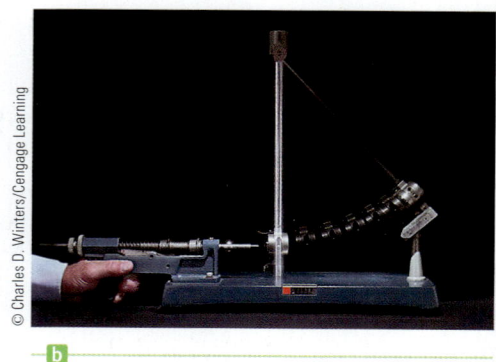

SOLUTION

(a) Find the velocity of the system after the bullet embeds in the block.

Apply conservation of energy to the block–bullet system after the collision:

$$(KE + PE)_{\text{after collision}} = (KE + PE)_{\text{top}}$$

Substitute expressions for the kinetic and potential energies. Note that both the potential energy at the bottom and the kinetic energy at the top are zero:

$$\tfrac{1}{2}(m_1 + m_2)v_{sys}^2 + 0 = 0 + (m_1 + m_2)gh$$

Solve for the final velocity of the block–bullet system, v_{sys}:

$$v_{sys}^2 = 2gh$$
$$v_{sys} = \sqrt{2gh} = \sqrt{2(9.80 \text{ m/s}^2)(5.00 \times 10^{-2} \text{ m})}$$
$$v_{sys} = \boxed{0.990 \text{ m/s}}$$

(b) Calculate the initial speed of the bullet.

Write the conservation of momentum equation and substitute expressions.

$$p_i = p_f$$
$$m_1 v_{1i} + m_2 v_{2i} = (m_1 + m_2)v_{sys}$$

Solve for the initial velocity of the bullet, and substitute values:

$$v_{1i} = \frac{(m_1 + m_2)v_{sys}}{m_1}$$
$$v_{1i} = \frac{(1.005 \text{ kg})(0.990 \text{ m/s})}{5.00 \times 10^{-3} \text{ kg}} = \boxed{199 \text{ m/s}}$$

REMARKS Because the impact is inelastic, it would be incorrect to equate the initial kinetic energy of the incoming bullet to the final gravitational potential energy associated with the bullet–block combination. The energy isn't conserved!

QUESTION 6.5 List three ways mechanical energy can be lost from the system in this experiment.

EXERCISE 6.5 A bullet with mass 5.00 g is fired horizontally into a 2.000-kg block attached to a horizontal spring. The spring has a constant 6.00×10^2 N/m and reaches a maximum compression of 6.00 cm. (a) Find the initial speed of the bullet–block system. (b) Find the speed of the bullet.

ANSWERS (a) 1.04 m/s (b) 417 m/s

■ Quick Quiz

6.4 An object of mass m moves to the right with a speed v. It collides head-on with an object of mass $3m$ moving with speed $v/3$ in the opposite direction. If the two objects stick together, what is the speed of the combined object, of mass $4m$, after the collision?

(a) 0 (b) $v/2$ (c) v (d) $2v$

6.5 A skater is using very low-friction rollerblades. A friend throws a Frisbee to her, on the straight line along which she is coasting. Describe each of the following events as an elastic, an inelastic, or a perfectly inelastic collision between the skater and the Frisbee. (a) She catches the Frisbee and holds it. (b) She tries to catch the Frisbee, but it bounces off her hands and falls to the ground in front of her. (c) She catches the Frisbee and immediately throws it back with the same speed (relative to the ground) to her friend.

6.6 In a perfectly inelastic one-dimensional collision between two objects, what initial condition alone is necessary so that *all* of the original kinetic energy of the system is gone after the collision? (a) The objects must have momenta with the same magnitude but opposite directions. (b) The objects must have the same mass. (c) The objects must have the same velocity. (d) The objects must have the same speed, with velocity vectors in opposite directions.

Elastic Collisions

Now consider two objects that undergo an **elastic head-on collision** (Fig. 6.13). In this situation, **both the momentum and the kinetic energy of the system of two objects are conserved.** We can write these conditions as

$$m_1 v_{1i} + m_2 v_{2i} = m_1 v_{1f} + m_2 v_{2f} \quad [6.10]$$

and

$$\tfrac{1}{2} m_1 v_{1i}^2 + \tfrac{1}{2} m_2 v_{2i}^2 = \tfrac{1}{2} m_1 v_{1f}^2 + \tfrac{1}{2} m_2 v_{2f}^2 \quad [6.11]$$

where v is positive if an object moves to the right and negative if it moves to the left.

In a typical problem involving elastic collisions, there are two unknown quantities, and Equations 6.10 and 6.11 can be solved simultaneously to find them. These two equations are linear and quadratic, respectively. An alternate approach simplifies the quadratic equation to another linear equation, facilitating solution. Canceling the factor $\tfrac{1}{2}$ in Equation 6.11, we rewrite the equation as

$$m_1(v_{1i}^2 - v_{1f}^2) = m_2(v_{2f}^2 - v_{2i}^2)$$

Here we have moved the terms containing m_1 to one side of the equation and those containing m_2 to the other. Next, we factor both sides of the equation:

$$m_1(v_{1i} - v_{1f})(v_{1i} + v_{1f}) = m_2(v_{2f} - v_{2i})(v_{2f} + v_{2i}) \quad [6.12]$$

Now we separate the terms containing m_1 and m_2 in the equation for the conservation of momentum (Eq. 6.10) to get

$$m_1(v_{1i} - v_{1f}) = m_2(v_{2f} - v_{2i}) \quad [6.13]$$

Next, we divide Equation 6.12 by Equation 6.13, producing

$$v_{1i} + v_{1f} = v_{2f} + v_{2i}$$

Gathering initial and final values on opposite sides of the equation gives

$$v_{1i} - v_{2i} = -(v_{1f} - v_{2f}) \quad [6.14]$$

This equation, in combination with Equation 6.10, will be used to solve problems dealing with perfectly elastic head-on collisions. According to Equation 6.14, the relative velocity of the two objects before the collision, $v_{1i} - v_{2i}$, equals the negative of the relative velocity of the two objects after the collision, $-(v_{1f} - v_{2f})$. To better understand the equation, imagine that you are riding along on one of the objects. As you measure the velocity of the other object from your vantage point, you will be measuring the relative velocity of the two objects. In your view of the collision, the other object comes toward you and bounces off, leaving the collision with the same speed, but in the opposite direction. This is just what Equation 6.14 states.

Figure 6.13 (a) Before and (b) after an elastic head-on collision between two hard spheres. Unlike an inelastic collision, both the total momentum and the total energy are conserved.

PROBLEM-SOLVING STRATEGY

One-Dimensional Collisions

The following procedure is recommended for solving one-dimensional problems involving collisions between two objects:

1. **Coordinates.** Choose a coordinate axis that lies along the direction of motion.
2. **Diagram.** Sketch the problem, representing the two objects as blocks and labeling velocity vectors and masses.
3. **Conservation of Momentum.** Write a general expression for the *total* momentum of the system of two objects *before* and *after* the collision, and equate the two, as in Equation 6.10. On the next line, fill in the known values.
4. **Conservation of Energy.** If the collision is elastic, write a general expression for the total energy before and after the collision, and equate the two quantities, as in Equation 6.11 or (preferably) Equation 6.14. Fill in the known values. (*Skip* this step if the collision is *not* perfectly elastic.)
5. **Solve** the equations simultaneously. Equations 6.10 and 6.14 form a system of two linear equations and two unknowns. If you have forgotten Equation 6.14, use Equation 6.11 instead.

Steps 1 and 2 of the problem-solving strategy are generally carried out in the process of sketching and labeling a diagram of the problem. This is clearly the case in our next example, which makes use of Figure 6.13. Other steps are pointed out as they are applied.

■ EXAMPLE 6.6   Let's Play Pool

GOAL Solve an elastic collision in one dimension.

PROBLEM Two billiard balls of identical mass move toward each other as in Figure 6.13, with the positive *x*-axis to the right (steps 1 and 2). Assume that the collision between them is perfectly elastic. If the initial velocities of the balls are +30.0 cm/s and −20.0 cm/s, what are the velocities of the balls after the collision? Assume friction and rotation are unimportant.

STRATEGY Solution of this problem is a matter of solving two equations, the conservation of momentum and conservation of energy equations, for two unknowns, the final velocities of the two balls. Instead of using Equation 6.11 for conservation of energy, use Equation 6.14, which is linear, hence easier to handle.

SOLUTION

Write the conservation of momentum equation. Because $m_1 = m_2$, we can cancel the masses, then substitute $v_{1i} = +30.0$ m/s and $v_{2i} = -20.0$ cm/s (Step 3).

$$m_1 v_{1i} + m_2 v_{2i} = m_1 v_{1f} + m_2 v_{2f}$$
$$30.0 \text{ cm/s} + (-20.0 \text{ cm/s}) = v_{1f} + v_{2f}$$
$$(1) \quad 10.0 \text{ cm/s} = v_{1f} + v_{2f}$$

Next, apply conservation of energy in the form of Equation 6.14 (Step 4):

$$(2) \quad v_{1i} - v_{2i} = -(v_{1f} - v_{2f})$$
$$30.0 \text{ cm/s} - (-20.0 \text{ cm/s}) = v_{2f} - v_{1f}$$
$$(3) \quad 50.0 \text{ cm/s} = v_{2f} - v_{1f}$$

Now solve Equations (1) and (3) simultaneously by adding them together (Step 5):

$$10.0 \text{ cm/s} + 50.0 \text{ cm/s} = (v_{1f} + v_{2f}) + (v_{2f} - v_{1f})$$
$$60.0 \text{ cm/s} = 2v_{2f} \quad \rightarrow \quad v_{2f} = \boxed{30.0 \text{ m/s}}$$

Substitute the answer for v_{2f} into Equation (1):

$$10.0 \text{ cm/s} = v_{1f} + 30.0 \text{ m/s} \quad \rightarrow \quad v_{1f} = \boxed{-20.0 \text{ m/s}}$$

REMARKS Notice the balls exchanged velocities—almost as if they'd passed through each other. This is always the case when two objects of equal mass undergo an elastic head-on collision.

QUESTION 6.6 In this example, is it possible to adjust the initial velocities of the balls so that both are at rest after the collision? Explain.

EXERCISE 6.6 Find the final velocities of the two balls if the ball with initial velocity $v_{2i} = -20.0$ cm/s has a mass equal to one-half that of the ball with initial velocity $v_{1i} = +30.0$ cm/s.

ANSWER $v_{1f} = -3.33$ cm/s; $v_{2f} = +46.7$ cm/s

■ EXAMPLE 6.7   Two Blocks and a Spring

GOAL Solve an elastic collision involving spring potential energy.

PROBLEM A block of mass $m_1 = 1.60$ kg, initially moving to the right with a velocity of $+4.00$ m/s on a frictionless horizontal track, collides with a massless spring attached to a second block of mass $m_2 = 2.10$ kg moving to the left with a velocity of -2.50 m/s, as in Figure 6.14a. The spring has a spring constant of 6.00×10^2 N/m. **(a)** Determine the velocity of block 2 at the instant when block 1 is moving to the right with a velocity of $+3.00$ m/s, as in Figure 6.14b. **(b)** Find the compression of the spring at that time.

STRATEGY We identify the system as the two blocks and the spring. Write down the conservation of momentum equations, and solve for the final velocity of block 2, v_{2f}. Then use conservation of energy to find the compression of the spring at that time.

Figure 6.14 (Example 6.7)

SOLUTION

(a) Find the velocity v_{2f} when block 1 has velocity $+3.00$ m/s.

Write the conservation of momentum equation for the system and solve for v_{2f}:

(1) $m_1 v_{1i} + m_2 v_{2i} = m_1 v_{1f} + m_2 v_{2f}$

$$v_{2f} = \frac{m_1 v_{1i} + m_2 v_{2i} - m_1 v_{1f}}{m_2}$$

$$= \frac{(1.60 \text{ kg})(4.00 \text{ m/s}) + (2.10 \text{ kg})(-2.50 \text{ m/s}) - (1.60 \text{ kg})(3.00 \text{ m/s})}{2.10 \text{ kg}}$$

$v_{2f} = \boxed{-1.74 \text{ m/s}}$

(b) Find the compression of the spring.

Use energy conservation for the system, noticing that potential energy is stored in the spring when it is compressed a distance x:

$E_i = E_f$

$\frac{1}{2}m_1 v_{1i}^2 + \frac{1}{2}m_2 v_{2i}^2 + 0 = \frac{1}{2}m_1 v_{1f}^2 + \frac{1}{2}m_2 v_{2f}^2 + \frac{1}{2}kx^2$

Substitute the given values and the result of part (a) into the preceding expression, solving for x:

$x = \boxed{0.173 \text{ m}}$

REMARKS The initial velocity component of block 2 is -2.50 m/s because the block is moving to the left. The negative value for v_{2f} means that block 2 is still moving to the left at the instant under consideration.

QUESTION 6.7 Is it possible for both blocks to come to rest while the spring is being compressed? Explain. *Hint:* Look at the momentum in Equation (1).

EXERCISE 6.7 Find (a) the velocity of block 1 and (b) the compression of the spring at the instant that block 2 is at rest.

ANSWERS (a) 0.719 m/s to the right   (b) 0.251 m

6.4 Glancing Collisions

LEARNING OBJECTIVE

1. Solve two-dimensional collisions with conservation of momentum.

In Section 6.2 we showed that the total linear momentum of a system is conserved when the system is isolated (that is, when no external forces act on the system). For a general collision of two objects in three-dimensional space, the conservation of momentum principle implies that the total momentum of the system in each direction is conserved. However, an important subset of collisions takes place in a plane. The game of billiards is a familiar example involving multiple collisions of objects moving on a two-dimensional surface. We restrict our attention to a single two-dimensional collision between two objects that takes place in a plane, and ignore any possible rotation. For such collisions, we obtain two component equations for the conservation of momentum:

$$m_1 v_{1ix} + m_2 v_{2ix} = m_1 v_{1fx} + m_2 v_{2fx}$$

$$m_1 v_{1iy} + m_2 v_{2iy} = m_1 v_{1fy} + m_2 v_{2fy}$$

We must use three subscripts in this general equation, to represent, respectively, (1) the object in question, and (2) the initial and final values of the components of velocity.

Now, consider a two-dimensional problem in which an object of mass m_1 collides with an object of mass m_2 that is initially at rest, as in Figure 6.15. After the collision, object 1 moves at an angle θ with respect to the horizontal, and object 2 moves at an angle ϕ with respect to the horizontal. This is called a *glancing* collision. Applying the law of conservation of momentum in component form, and noting that the initial y-component of momentum is zero, we have

$$x\text{-component:} \quad m_1 v_{1i} + 0 = m_1 v_{1f} \cos \theta + m_2 v_{2f} \cos \phi \qquad [6.15]$$

$$y\text{-component:} \quad 0 + 0 = m_1 v_{1f} \sin \theta + m_2 v_{2f} \sin \phi \qquad [6.16]$$

If the collision is elastic, we can write a third equation, for conservation of energy, in the form

$$\tfrac{1}{2} m_1 v_{1i}^2 = \tfrac{1}{2} m_1 v_{1f}^2 + \tfrac{1}{2} m_2 v_{2f}^2 \qquad [6.17]$$

If we know the initial velocity v_{1i} and the masses, we are left with four unknowns (v_{1f}, v_{2f}, θ, and ϕ). Because we have only three equations, one of the four remaining quantities must be given in order to determine the motion after the collision from conservation principles alone.

If the collision is inelastic, the kinetic energy of the system is *not* conserved, and Equation 6.17 does *not* apply.

Figure 6.15 A glancing collision between two objects.

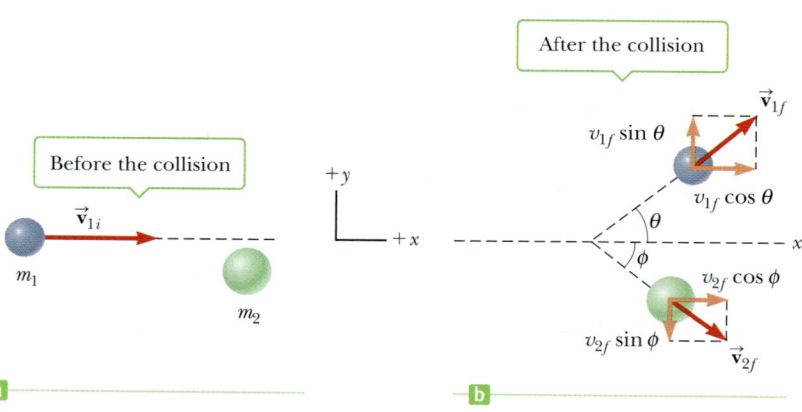

PROBLEM-SOLVING STRATEGY

Two-Dimensional Collisions

To solve two-dimensional collisions, follow this procedure:

1. **Coordinate Axes.** Use both x- and y-coordinates. It's convenient to have either the x-axis or the y-axis coincide with the direction of one of the initial velocities.
2. **Diagram.** Sketch the problem, labeling velocity vectors and masses.
3. **Conservation of Momentum.** Write a separate conservation of momentum equation for each of the x- and y-directions. In each case, the total initial momentum in a given direction equals the total final momentum in that direction.
4. **Conservation of Energy.** If the collision is elastic, write a general expression for the total energy before and after the collision, and equate the two expressions, as in Equation 6.11. Fill in the known values. (**Skip** this step if the collision is **not** perfectly elastic.) The energy equation can't be simplified as in the one-dimensional case, so a quadratic expression such as Equation 6.11 or 6.17 must be used when the collision is elastic.
5. **Solve** the equations simultaneously. There are two equations for inelastic collisions and three for elastic collisions.

EXAMPLE 6.8 | Collision at an Intersection

GOAL Analyze a two-dimensional inelastic collision.

PROBLEM A car with mass 1.50×10^3 kg traveling east at a speed of 25.0 m/s collides at an intersection with a 2.50×10^3-kg van traveling north at a speed of 20.0 m/s, as shown in Figure 6.16. Find the magnitude and direction of the velocity of the wreckage after the collision, assuming that the vehicles undergo a perfectly inelastic collision (that is, they stick together) and assuming that friction between the vehicles and the road can be neglected.

STRATEGY Use conservation of momentum in two dimensions. (Kinetic energy is not conserved.) Choose coordinates as in Figure 6.16. Before the collision, the only object having momentum in the x-direction is the car, while the van carries all the momentum in the y-direction. After the totally inelastic collision, both vehicles move together at some common speed v_f and angle θ. Solve for these two unknowns, using the two components of the conservation of momentum equation.

Figure 6.16 (Example 6.8) A top view of a perfectly inelastic collision between a car and a van.

SOLUTION

Find the x-components of the initial and final total momenta:

$\sum p_{xi} = m_{car}v_{car} = (1.50 \times 10^3 \text{ kg})(25.0 \text{ m/s})$
$= 3.75 \times 10^4 \text{ kg} \cdot \text{m/s}$

$\sum p_{xf} = (m_{car} + m_{van})v_f \cos\theta = (4.00 \times 10^3 \text{ kg})v_f \cos\theta$

Set the initial x-momentum equal to the final x-momentum:

(1) $\quad 3.75 \times 10^4 \text{ kg} \cdot \text{m/s} = (4.00 \times 10^3 \text{ kg})v_f \cos\theta$

Find the y-components of the initial and final total momenta:

$\sum p_{iy} = m_{van}v_{van} = (2.50 \times 10^3 \text{ kg})(20.0 \text{ m/s})$
$= 5.00 \times 10^4 \text{ kg} \cdot \text{m/s}$

$\sum p_{fy} = (m_{car} + m_{van})v_f \sin\theta = (4.00 \times 10^3 \text{ kg})v_f \sin\theta$

Set the initial y-momentum equal to the final y-momentum:

(2) $\quad 5.00 \times 10^4 \text{ kg} \cdot \text{m/s} = (4.00 \times 10^3 \text{ kg})v_f \sin\theta$

(Continued)

Divide Equation (2) by Equation (1) and solve for θ:

$$\tan \theta = \frac{5.00 \times 10^4 \text{ kg} \cdot \text{m/s}}{3.75 \times 10^4 \text{ kg} \cdot \text{m}} = 1.33$$

$$\theta = \boxed{53.1°}$$

Substitute this angle back into Equation (2) to find v_f:

$$v_f = \frac{5.00 \times 10^4 \text{ kg} \cdot \text{m/s}}{(4.00 \times 10^3 \text{ kg}) \sin 53.1°} = \boxed{15.6 \text{ m/s}}$$

REMARKS It's also possible to first find the x- and y-components v_{fx} and v_{fy} of the resultant velocity. The magnitude and direction of the resultant velocity can then be found with the Pythagorean theorem, $v_f = \sqrt{v_{fx}^2 + v_{fy}^2}$, and the inverse tangent function $\theta = \tan^{-1}(v_{fy}/v_{fx})$. Setting up this alternate approach is a simple matter of substituting $v_{fx} = v_f \cos \theta$ and $v_{fy} = v_f \sin \theta$ into Equations (1) and (2).

QUESTION 6.8 If the car and van had identical mass and speed, what would the resultant angle have been?

EXERCISE 6.8 A 3.00-kg object initially moving in the positive x-direction with a velocity of $+5.00$ m/s collides with and sticks to a 2.00-kg object initially moving in the negative y-direction with a velocity of -3.00 m/s. Find the final components of velocity of the composite object.

ANSWER $v_{fx} = 3.00$ m/s; $v_{fy} = -1.20$ m/s

6.5 Rocket Propulsion

LEARNING OBJECTIVE

1. Apply the physics of rocket propulsion to calculate rocket motion in elementary contexts.

When ordinary vehicles such as cars and locomotives move, the driving force of the motion is friction. In the case of the car, this driving force is exerted by the road on the car, a reaction to the force exerted by the wheels against the road. Similarly, a locomotive "pushes" against the tracks; hence, the driving force is the reaction force exerted by the tracks on the locomotive. However, a rocket moving in space has no road or tracks to push against. How can it move forward?

In fact, reaction forces also propel a rocket. (You should review Newton's third law, discussed in Chapter 4.) To illustrate this point, we model our rocket with a spherical chamber containing a combustible gas, as in Figure 6.17a. When an explosion occurs in the chamber, the hot gas expands and presses against all sides of the chamber, as indicated by the arrows. Because the sum of the forces exerted on the rocket is zero, it doesn't move. Now suppose a hole is drilled in the bottom of the chamber, as in Figure 6.17b. When the explosion occurs, the gas presses against the chamber in all directions, but can't press against anything at the hole, where it simply escapes into space. Adding the forces on the spherical chamber now results in a net force upwards. Just as in the case of cars and locomotives, this is a reaction force. A car's wheels press against the ground, and the reaction force of the ground on the car pushes it forward. The wall of the rocket's combustion chamber exerts a force on the gas expanding against it. The reaction force of the gas on the wall then pushes the rocket upward.

In a now infamous article in *The New York Times*, rocket pioneer Robert Goddard was ridiculed for thinking that rockets would work in space, where, according to the *Times*, there was nothing to push against. The *Times* retracted, rather belatedly, during the first *Apollo* moon landing mission in 1969. The hot gases are not pushing against anything external, but against the rocket itself—and ironically, rockets actually work *better* in a vacuum. In an atmosphere, the gases have to do work against the outside air pressure to escape the combustion chamber, slowing the exhaust velocity and reducing the reaction force.

A rocket reaction chamber without a nozzle has reaction forces pushing equally in all directions, so no motion results.

a

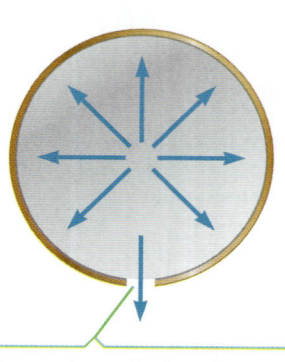

An opening at the bottom of the chamber removes the downward reaction force, resulting in a net upward reaction force.

b

Figure 6.17 A rocket reaction chamber containing a combusting gas works because it has a nozzle where gases can escape. The chamber wall acts on the expanding gas; the reaction force of the gas on the chamber wall pushes the rocket forward.

At the microscopic level, this process is complicated, but it can be simplified by applying conservation of momentum to the rocket and its ejected fuel. In principle, the solution is similar to that in Example 6.3, with the archer representing the rocket and the arrows the exhaust gases.

Suppose that at some time t, the momentum of the rocket plus the fuel is given by $(M + \Delta m)v$, where Δm is an amount of fuel about to be burned (Fig. 6.18a). This fuel is traveling at a speed v relative to, say, Earth, just like the rest of the rocket. During a short time interval Δt, the rocket ejects fuel of mass Δm, and the rocket's speed increases to $v + \Delta v$ (Fig. 6.18b). If the fuel is ejected with exhaust speed v_e *relative to the rocket,* the speed of the fuel relative to the Earth is $v - v_e$. Equating the total initial momentum of the system with the total final momentum, we have

$$(M + \Delta m)v = M(v + \Delta v) + \Delta m(v - v_e)$$

Simplifying this expression gives

$$M\Delta v = v_e \Delta m$$

The increase Δm in the mass of the exhaust corresponds to an equal decrease in the mass of the rocket, so that $\Delta m = -\Delta M$. Using this fact, we have

$$M\Delta v = -v_e \Delta M \quad [6.18]$$

This result, together with the methods of calculus, can be used to obtain the following equation:

$$v_f - v_i = v_e \ln\left(\frac{M_i}{M_f}\right) \quad [6.19]$$

where M_i is the initial mass of the rocket plus fuel and M_f is the final mass of the rocket plus its remaining fuel. This is the basic expression for rocket propulsion; it tells us that the increase in velocity is proportional to the exhaust speed v_e and to the natural logarithm of M_i/M_f. Because the maximum ratio of M_i to M_f for a single-stage rocket is about 10:1, the increase in speed can reach $v_e \ln 10 = 2.3 v_e$ or about twice the exhaust speed! For best results, therefore, the exhaust speed should be as high as possible. Currently, typical rocket exhaust speeds are several kilometers per second.

The **thrust** on the rocket is defined as the force exerted on the rocket by the ejected exhaust gases. We can obtain an expression for the instantaneous thrust by dividing Equation 6.18 by Δt:

$$\text{Instantaneous thrust} = Ma = M\frac{\Delta v}{\Delta t} = \left|v_e \frac{\Delta M}{\Delta t}\right| \quad [6.20]$$

◀ Rocket thrust

The absolute value signs are used for clarity: In Equation 6.18, $-\Delta M$ is a positive quantity (as is v_e, a speed). Here we see that the thrust increases as the exhaust velocity increases and as the rate of change of mass $\Delta M/\Delta t$ (the burn rate) increases.

Figure 6.18 Rocket propulsion. (a) The initial mass of the rocket and fuel is $M + \Delta m$ at a time t, and the rocket's speed is v. (b) At a time $t + \Delta t$, the rocket's mass has been reduced to M, and an amount of fuel Δm has been ejected. The rocket's speed increases by an amount Δv.

APPLYING PHYSICS 6.2 | Multistage Rockets

The current maximum exhaust speed of $v_e = 4\,500$ m/s can be realized with rocket engines fueled with liquid hydrogen and liquid oxygen. But this means that the maximum speed attainable for a given rocket with a mass ratio of 10 is $v_e \ln 10 \simeq 10\,000$ m/s. To reach the Moon, however, requires a change in velocity of over 11 000 m/s. Further, this change must occur while working against gravity and atmospheric friction. How can that be managed without developing better engines?

EXPLANATION The answer is the multistage rocket. By dropping stages, the spacecraft becomes lighter, so that fuel burned later in the mission doesn't have to accelerate mass that no longer serves any purpose. Strap-on boosters, as used by the space shuttle and a number of other rockets, such as the *Titan 4* or Russian *Proton*, employ a similar method. The boosters are jettisoned after their fuel is exhausted, so the rocket is no longer burdened by their weight.

EXAMPLE 6.9 | Single Stage to Orbit (SSTO)

GOAL Apply the velocity and thrust equations of a rocket.

PROBLEM A rocket has a total mass of 1.00×10^5 kg and a burnout mass of 1.00×10^4 kg, including engines, shell, and payload. The rocket blasts off from Earth and exhausts all its fuel in 4.00 min, burning the fuel at a steady rate with an exhaust velocity of $v_e = 4.50 \times 10^3$ m/s. **(a)** If air friction and gravity are neglected, what is the speed of the rocket at burnout? **(b)** What thrust does the engine develop at liftoff? **(c)** What is the initial acceleration of the rocket if gravity is not neglected? **(d)** Estimate the speed at burnout if gravity isn't neglected.

STRATEGY Although it sounds sophisticated, this problem is mainly a matter of substituting values into the appropriate equations. Part (a) requires substituting values into Equation 6.19 for the velocity. For part (b), divide the change in the rocket's mass by the total time, getting $\Delta M/\Delta t$, then substitute into Equation 6.20 to find the thrust. (c) Using Newton's second law, the force of gravity, and the result of (b), we can find the initial acceleration. For part (d), the acceleration of gravity is approximately constant over the few kilometers involved, so the velocity found in part (b) will be reduced by roughly $\Delta v_g = -gt$. Add this loss to the result of part (a).

SOLUTION

(a) Calculate the velocity at burnout, ignoring gravity and air drag.

Substitute $v_i = 0$, $v_e = 4.50 \times 10^3$ m/s, $M_i = 1.00 \times 10^5$ kg, and $M_f = 1.00 \times 10^4$ kg into Equation 6.19:

$$v_f = v_i + v_e \ln\left(\frac{M_i}{M_f}\right)$$

$$= 0 + (4.5 \times 10^3 \text{ m/s}) \ln\left(\frac{1.00 \times 10^5 \text{ kg}}{1.00 \times 10^4 \text{ kg}}\right)$$

$$v_f = \boxed{1.04 \times 10^4 \text{ m/s}}$$

(b) Find the thrust at liftoff.

Compute the change in the rocket's mass:

$$\Delta M = M_f - M_i = 1.00 \times 10^4 \text{ kg} - 1.00 \times 10^5 \text{ kg}$$
$$= -9.00 \times 10^4 \text{ kg}$$

Calculate the rate at which rocket mass changes by dividing the change in mass by the time (where the time interval equals 4.00 min = 2.40×10^2 s):

$$\frac{\Delta M}{\Delta t} = \frac{-9.00 \times 10^4 \text{ kg}}{2.40 \times 10^2 \text{ s}} = -3.75 \times 10^2 \text{ kg/s}$$

Substitute this rate into Equation 6.20, obtaining the thrust:

$$\text{Thrust} = \left|v_e \frac{\Delta M}{\Delta t}\right| = (4.50 \times 10^3 \text{ m/s})(3.75 \times 10^2 \text{ kg/s})$$
$$= \boxed{1.69 \times 10^6 \text{ N}}$$

(c) Find the initial acceleration, including the gravity force.

Write Newton's second law, where T stands for thrust, and solve for the acceleration a:

$$Ma = \Sigma F = T - Mg$$

$$a = \frac{T}{M} - g = \frac{1.69 \times 10^6 \text{ N}}{1.00 \times 10^5 \text{ kg}} - 9.80 \text{ m/s}^2$$

$$= \boxed{7.10 \text{ m/s}^2}$$

(d) Estimate the speed at burnout when gravity is not neglected.

Find the approximate loss of speed due to gravity:

$$\Delta v_g = -g\Delta t = -(9.80 \text{ m/s}^2)(2.40 \times 10^2 \text{ s})$$
$$= -2.35 \times 10^3 \text{ m/s}$$

Add this loss to the result of part (b):

$$v_f = 1.04 \times 10^4 \text{ m/s} - 2.35 \times 10^3 \text{ m/s}$$
$$= \boxed{8.05 \times 10^3 \text{ m/s}}$$

REMARKS Even taking gravity into account, the speed is sufficient to attain orbit. Some additional boost may be required to overcome air drag.

QUESTION 6.9 What initial normal force would be exerted on an astronaut of mass m in a rocket traveling vertically upward with an acceleration a? Answer symbolically in terms of the positive quantities m, g, and a.

EXERCISE 6.9 A spaceship with a mass of 5.00×10^4 kg is traveling at 6.00×10^3 m/s relative to a space station. What mass will the ship have after it fires its engines in order to reach a relative speed of 8.00×10^3 m/s, traveling in the same direction? Assume an exhaust velocity of 4.50×10^3 m/s.

ANSWER 3.21×10^4 kg

SUMMARY

6.1 Momentum and Impulse

The **linear momentum** $\vec{\mathbf{p}}$ of an object of mass m moving with velocity $\vec{\mathbf{v}}$ is defined as

$$\vec{\mathbf{p}} \equiv m\vec{\mathbf{v}} \quad [6.1]$$

Momentum carries units of kg·m/s. The **impulse** $\vec{\mathbf{I}}$ of a constant force $\vec{\mathbf{F}}$ delivered to an object is equal to the product of the force and the time interval during which the force acts:

$$\vec{\mathbf{I}} \equiv \vec{\mathbf{F}}\Delta t \quad [6.4]$$

These two concepts are unified in the **impulse–momentum theorem**, which states that the impulse of a constant force delivered to an object is equal to the change in momentum of the object:

$$\vec{\mathbf{I}} = \vec{\mathbf{F}}\Delta t = \Delta \vec{\mathbf{p}} \equiv m\vec{\mathbf{v}}_f - m\vec{\mathbf{v}}_i \quad [6.5]$$

Solving problems with this theorem often involves estimating speeds or contact times (or both), leading to an average force.

6.2 Conservation of Momentum

When no net external force acts on an isolated system, the total momentum of the system is constant. This principle is called **conservation of momentum.** In particular, if the isolated system consists of two objects undergoing a collision, the total momentum of the system is the same before and after the collision. Conservation of momentum can be written mathematically for this case as

$$m_1\vec{\mathbf{v}}_{1i} + m_2\vec{\mathbf{v}}_{2i} = m_1\vec{\mathbf{v}}_{1f} + m_2\vec{\mathbf{v}}_{2f} \quad [6.7]$$

In an isolated system of two objects undergoing a collision, the total momentum of the system remains constant.

Collision and recoil problems typically require finding unknown velocities in one or two dimensions. Each vector component gives an equation, and the resulting equations are solved simultaneously.

6.3 Collisions

In an **inelastic collision,** the momentum of the system is conserved, but kinetic energy is not. In a **perfectly inelastic collision,** the colliding objects stick together. In an **elastic collision,** both the momentum and the kinetic energy of the system are conserved.

A one-dimensional **elastic collision** between two objects can be solved by using the conservation of momentum and conservation of energy equations:

$$m_1 v_{1i} + m_2 v_{2i} = m_1 v_{1f} + m_2 v_{2f} \quad [6.10]$$

$$\tfrac{1}{2}m_1 v_{1i}^2 + \tfrac{1}{2}m_2 v_{2i}^2 = \tfrac{1}{2}m_1 v_{1f}^2 + \tfrac{1}{2}m_2 v_{2f}^2 \quad [6.11]$$

The following equation, derived from Equations 6.10 and 6.11, is usually more convenient to use than the original conservation of energy equation:

$$v_{1i} - v_{2i} = -(v_{1f} - v_{2f}) \quad [6.14]$$

These equations can be solved simultaneously for the unknown velocities. Energy is not conserved in **inelastic collisions**, so such problems must be solved with Equation 6.10 alone.

6.4 Glancing Collisions

In glancing collisions, conservation of momentum can be applied along two perpendicular directions: an x-axis and a y-axis. Problems can be solved by using the x- and y-components of Equation 6.7. Elastic two-dimensional collisions will usually require Equation 6.11 as well. (Equation 6.14 doesn't apply to two dimensions.) Generally, one of the two objects is taken to be traveling along the x-axis, undergoing a deflection at some angle θ after the collision. The final velocities and angles can be found with elementary trigonometry.

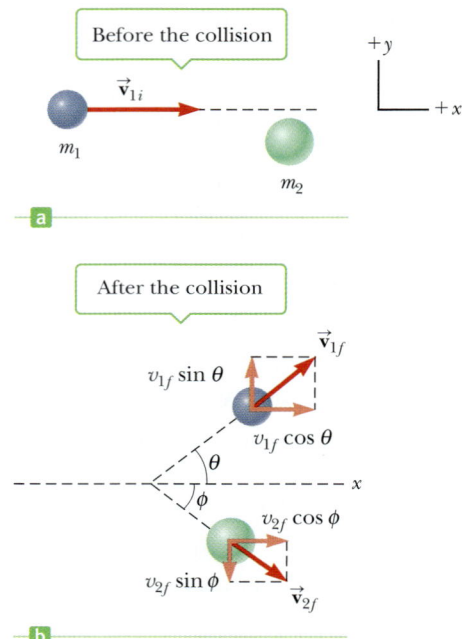

In a two-dimensional collision the system momentum is conserved, whereas the system energy is conserved only if the collision is elastic.

WARM-UP EXERCISES

WebAssign The warm-up exercises in this chapter may be assigned online in Enhanced WebAssign.

1. **Math Review** Solve the two equations $mv_i + MV_i = mv_f + MV_f$ and $v_i - V_i = -(v_f - V_f)$ for (a) v_f and (b) V_f if $m = 2.00$ kg, $v_i = 4.00$ m/s, $M = 3.00$ kg, and $V_i = 0$. (See Section 6.3.)

2. **Math Review** Given the equations $-507 = 147\, V_f \cos\theta$ and $-377 = 147\, V_f \sin\theta$, find (a) V_f by using the identity $\cos^2\theta + \sin^2\theta = 1$, and (b) θ by using the inverse tangent function. (*Note:* Some may consider it easier finding the angle θ first, and then V_f by back substitution.) (See Section 6.4.)

3. **Math Review** (a) Solve the equation 7.20×10^3 m/s $= (4.20 \times 10^3$ m/s$) \ln(M_i/M_f)$ for the fraction M_i/M_f.
   (b) If $M_i = 2.65 \times 10^4$ kg, calculate M_f.

4. A soccer player runs up behind a 0.450-kg soccer ball traveling at 3.20 m/s and kicks it in the same direction as it is moving, increasing its speed to 12.8 m/s. (a) What is the change in the magnitude of the ball's momentum? (b) What magnitude impulse did the soccer player deliver to the ball? (c) What magnitude impulse would be required to kick the ball in the opposite direction at 12.8 m/s, instead? (See Section 6.1.)

5. A 57.0-g tennis ball is traveling straight at a player at 21.0 m/s. The player volleys the ball straight back at 25.0 m/s. (a) What is the magnitude of the ball's change of momentum? (b) If the ball remains in contact with the racket for 0.060 0 s, what average force acts on the ball? (See Section 6.1.)

6. An astronaut, of total mass 85.0 kg including her suit, stands on a spherical satellite of mass 375 kg, both at rest relative a nearby space station. She jumps at a speed of 2.56 m/s directly away from the satellite, as measured by an observer in the station. At what speed does that observer measure the satellite traveling in the opposite direction? (See Section 6.2.)

7. A small china bowl of mass 0.450 kg is sliding along a frictionless countertop at speed 1.28 m/s. (a) What is the kinetic energy of the bowl? Subsequently a server, with perfect timing, places a rice ball of the same mass into the bowl as it passes him. (b) What is the subsequent speed of the system and (c) what is the system's kinetic energy? (See Section 6.3.)

8. A car of mass 750 kg traveling at a velocity of 27 m/s in the positive x-direction crashes into the rear of a truck of mass 1 500 kg that is at rest and in neutral at an intersection. If the collision is inelastic and the truck moves forward at 15.0 m/s, what is the velocity of the car after the collision? (See Section 6.3.)

9. A car of mass 1 560 kg traveling east and a truck of equal mass traveling north collide and become entangled, moving as a unit at 15.0 m/s and 60.0° north of east. Find the speed of (a) the car, and (b) the truck prior to the collision. (See Section 6.4.)

10. A rocket with total mass 3.00×10^5 kg is in circular orbit around the Earth. It begins to accelerate at

36.0 m/s² tangent to its orbit (hence doing no work against gravity). If the speed of the exhausted gases is 4.50×10^3 m/s, at what rate is the rocket initially burning fuel? (b) If the rocket were to be launched vertically from Earth's surface with the same initial acceleration, at what rate would the fuel have to be burned? (Disregard the reduction in exhaust speed due to the ambient atmospheric pressure.) (See Section 6.5.)

11. A spacecraft in circular orbit around Earth has nuclear hydrogen rocket engines with an exhaust velocity of 9.00×10^3 m/s. If the rocket has an initial mass of 6.70×10^5 kg, (a) what mass will it have after the rockets have fired and changed the spacecraft's velocity by 3.50×10^3 m/s? Assume changes in radial position during the burn are negligible. (b) What mass of fuel will the rocket use during that time? (See Section 6.5.)

CONCEPTUAL QUESTIONS

The conceptual questions in this chapter may be assigned online in Enhanced WebAssign.

1. A batter bunts a pitched baseball, blocking the ball without swinging. (a) Can the baseball deliver more kinetic energy to the bat and batter than the ball carries initially? (b) Can the baseball deliver more momentum to the bat and batter than the ball carries initially? Explain each of your answers.

2. If two objects collide and one is initially at rest, (a) is it possible for both to be at rest after the collision? (b) Is it possible for only one to be at rest after the collision? Explain.

3. In perfectly inelastic collisions between two objects, there are events in which all of the original kinetic energy is transformed to forms other than kinetic. Give an example of such an event.

4. Americans will never forget the terrorist attack on September 11, 2001. One commentator remarked that the force of the explosion at the Twin Towers of the World Trade Center was strong enough to blow glass and parts of the steel structure to small fragments. Yet the television coverage showed thousands of sheets of paper floating down, many still intact. Explain how that could be.

5. A ball of clay of mass m is thrown with a speed v against a brick wall. The clay sticks to the wall and stops. Is the principle of conservation of momentum violated in this example?

6. A skater is standing still on a frictionless ice rink. Her friend throws a Frisbee straight to her. In which of the following cases is the largest momentum transferred to the skater? (a) The skater catches the Frisbee and holds onto it. (b) The skater catches the Frisbee momentarily, but then drops it vertically downward. (c) The skater catches the Frisbee, holds it momentarily, and throws it back to her friend.

7. A more ordinary example of conservation of momentum than a rocket ship occurs in a kitchen dishwashing machine. In this device, water at high pressure is forced out of small holes on the spray arms. Use conservation of momentum to explain why the arms rotate, directing water to all the dishes.

8. (a) If two automobiles collide, they usually do not stick together. Does this mean the collision is elastic? (b) Explain why a head-on collision is likely to be more dangerous than other types of collisions.

9. Your physical education teacher throws you a tennis ball at a certain velocity, and you catch it. You are now given the following choice: The teacher can throw you a medicine ball (which is much more massive than the tennis ball) with the same velocity, the same momentum, or the same kinetic energy as the tennis ball. Which option would you choose in order to make the easiest catch, and why?

10. A large bedsheet is held vertically by two students. A third student, who happens to be the star pitcher on the baseball team, throws a raw egg at the sheet. Explain why the egg doesn't break when it hits the sheet, regardless of its initial speed. (If you try this, make sure the pitcher hits the sheet near its center, and don't allow the egg to fall on the floor after being caught.)

11. A sharpshooter fires a rifle while standing with the butt of the gun against his shoulder. If the forward momentum of a bullet is the same as the backward momentum of the gun, why isn't it as dangerous to be hit by the gun as by the bullet?

12. An air bag inflates when a collision occurs, protecting a passenger (the dummy in Figure CQ6.12) from serious injury. Why does the air bag soften the blow? Discuss the physics involved in this dramatic photograph.

Figure CQ6.12

13. In golf, novice players are often advised to be sure to "follow through" with their swing. (a) Why does this make the ball travel a longer distance? (b) If a shot is taken near the green, very little follow-through is required. Why?

14. An open box slides across a frictionless, icy surface of a frozen lake. What happens to the speed of the box as water from a rain shower falls vertically downward into the box? Explain.

15. Does a larger net force exerted on an object always produce a larger change in the momentum of the object, compared to a smaller net force? Explain.

16. Does a larger net force always produce a larger change in kinetic energy than a smaller net force? Explain.

17. If two particles have equal momenta, are their kinetic energies equal? (a) yes, always (b) no, never (c) no, except when their masses are equal (d) no, except when their speeds are the same (e) yes, as long as they move along parallel lines.

18. Two particles of different mass start from rest. The same net force acts on both of them as they move over equal distances. How do their final kinetic energies compare? (a) The particle of larger mass has more kinetic energy. (b) The particle of smaller mass has more kinetic energy. (c) The particles have equal kinetic energies. (d) Either particle might have more kinetic energy.

PROBLEMS

WebAssign The problems in this chapter may be assigned online in Enhanced WebAssign.

1. denotes straightforward problem; 2. denotes intermediate problem;
3. denotes challenging problem
1. denotes full solution available in *Student Solutions Manual/Study Guide*
1. denotes problems most often assigned in Enhanced WebAssign

BIO denotes biomedical problems
GP denotes guided problems
M denotes Master It tutorial available in Enhanced WebAssign
QC denotes asking for quantitative and conceptual reasoning
S denotes symbolic reasoning problem
W denotes Watch It video solution available in Enhanced WebAssign

6.1 Momentum and Impulse

1. Calculate the magnitude of the linear momentum for the following cases: (a) a proton with mass equal to 1.67×10^{-27} kg, moving with a speed of 5.00×10^6 m/s; (b) a 15.0-g bullet moving with a speed of 300 m/s; (c) a 75.0-kg sprinter running with a speed of 10.0 m/s; (d) the Earth (mass = 5.98×10^{24} kg) moving with an orbital speed equal to 2.98×10^4 m/s.

2. A high-speed photograph of a club hitting a golf ball is shown in Figure 6.3. The club was in contact with a ball, initially at rest, for about 0.002 0 s. If the ball has a mass of 55 g and leaves the head of the club with a speed of 2.0×10^2 ft/s, find the average force exerted on the ball by the club.

3. A pitcher claims he can throw a 0.145-kg baseball with as much momentum as a 3.00-g bullet moving with a speed of 1.50×10^3 m/s. (a) What must the baseball's speed be if the pitcher's claim is valid? (b) Which has greater kinetic energy, the ball or the bullet?

4. **S** A ball of mass m is thrown straight up into the air with an initial speed v_0. (a) Find an expression for the maximum height reached by the ball in terms of v_0 and g. (b) Using conservation of energy and the result of part (a), find the magnitude of the momentum of the ball at one-half its maximum height in terms of m and v_0.

5. **QC** Drops of rain fall perpendicular to the roof of a parked car during a rainstorm. The drops strike the roof with a speed of 12 m/s, and the mass of rain per second striking the roof is 0.035 kg/s. (a) Assuming the drops come to rest after striking the roof, find the average force exerted by the rain on the roof. (b) If hailstones having the same mass as the raindrops fall on the roof at the same rate and with the same speed, how would the average force on the roof compare to that found in part (a)?

6. **S** Show that the kinetic energy of a particle of mass m is related to the magnitude of the momentum p of that particle by $KE = p^2/2m$. (*Note:* This expression is invalid for particles traveling at speeds near that of light.)

7. An object has a kinetic energy of 275 J and a momentum of magnitude 25.0 kg · m/s. Find the (a) speed and (b) mass of the object.

8. An estimated force vs. time curve for a baseball struck by a bat is shown in Figure P6.8. From this curve, determine (a) the impulse delivered to the ball and (b) the average force exerted on the ball.

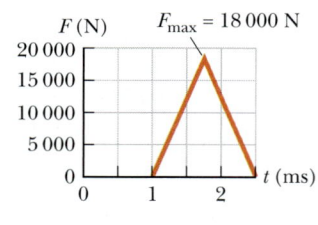

Figure P6.8

9. A 0.280-kg volleyball approaches a player horizontally with a speed of 15.0 m/s. The player strikes the ball with her fist and causes the ball to move in the opposite direction with a speed of 22.0 m/s. (a) What impulse is delivered to the ball by the player? (b) If the player's fist is in contact with the ball for 0.060 0 s, find the magnitude of the average force exerted on the player's fist.

10. **QC** A man claims he can safely hold on to a 12.0-kg child in a head-on collision with a relative speed of 120-mi/h lasting for 0.10 s as long as he has his seat belt on. (a) Find the magnitude of the average force needed

to hold onto the child. (b) Based on the result to part (a), is the man's claim valid? (c) What does the answer to this problem say about laws requiring the use of proper safety devices such as seat belts and special toddler seats?

11. A ball of mass 0.150 kg is dropped from rest from a height of 1.25 m. It rebounds from the floor to reach a height of 0.960 m. What impulse was given to the ball by the floor?

12. A tennis player receives a shot with the ball (0.060 0 kg) traveling horizontally at 50.0 m/s and returns the shot with the ball traveling horizontally at 40.0 m/s in the opposite direction. (a) What is the impulse delivered to the ball by the racket? (b) What work does the racket do on the ball?

13. A car is stopped for a traffic signal. When the light turns green, the car accelerates, increasing its speed from 0 to 5.20 m/s in 0.832 s. What are (a) the magnitudes of the linear impulse and (b) the average total force experienced by a 70.0-kg passenger in the car during the time the car accelerates?

14. A 65.0-kg basketball player jumps vertically and leaves the floor with a velocity of 1.80 m/s upward. (a) What impulse does the player experience? (b) What force does the floor exert on the player before the jump? (c) What is the total average force exerted by the floor on the player if the player is in contact with the floor for 0.450 s during the jump?

15. The force shown in the force vs. time diagram in Figure P6.15 acts on a 1.5-kg object. Find (a) the impulse of the force, (b) the final velocity of the object if it is initially at rest, and (c) the final velocity of the object if it is initially moving along the x-axis with a velocity of −2.0 m/s.

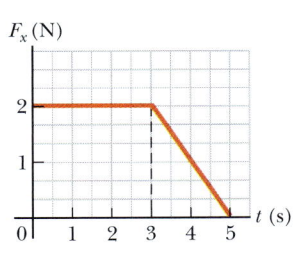

Figure P6.15

16. A force of magnitude F_x acting in the x-direction on a 2.00-kg particle varies in time as shown in Figure P6.16. Find (a) the impulse of the force, (b) the final velocity of the particle if it is initially at rest, and (c) the final velocity of the particle if it is initially moving along the x-axis with a velocity of −2.00 m/s.

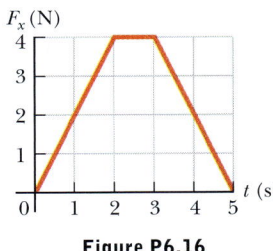

Figure P6.16

17. The forces shown in the force vs. time diagram in Figure P6.17 act on a 1.5-kg particle. Find

Figure P6.17

(a) the impulse for the interval from $t = 0$ to $t = 3.0$ s and (b) the impulse for the interval from $t = 0$ to $t = 5.0$ s. If the forces act on a 1.5-kg particle that is initially at rest, find the particle's speed (c) at $t = 3.0$ s and (d) at $t = 5.0$ s.

18. **W** A 3.00-kg steel ball strikes a massive wall at 10.0 m/s at an angle of $\theta = 60.0°$ with the plane of the wall. It bounces off the wall with the same speed and angle (Fig. P6.18). If the ball is in contact with the wall for 0.200 s, what is the average force exerted by the wall on the ball?

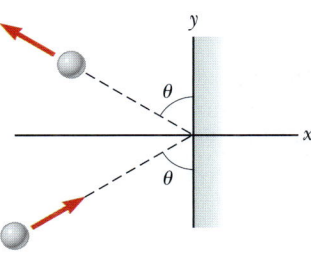

Figure P6.18

19. **M** The front 1.20 m of a 1 400-kg car is designed as a "crumple zone" that collapses to absorb the shock of a collision. If a car traveling 25.0 m/s stops uniformly in 1.20 m, (a) how long does the collision last, (b) what is the magnitude of the average force on the car, and (c) what is the acceleration of the car? Express the acceleration as a multiple of the acceleration of gravity.

20. **QC** A pitcher throws a 0.14-kg baseball toward the batter so that it crosses home plate horizontally and has a speed of 42 m/s just before it makes contact with the bat. The batter then hits the ball straight back at the pitcher with a speed of 48 m/s. Assume the ball travels along the same line leaving the bat as it followed before contacting the bat. (a) What is the magnitude of the impulse delivered by the bat to the baseball? (b) If the ball is in contact with the bat for 0.005 0 s, what is the magnitude of the average force exerted by the bat on the ball? (c) How does your answer to part (b) compare to the weight of the ball?

6.2 Conservation of Momentum

21. **W** High-speed stroboscopic photographs show that the head of a 200-g golf club is traveling at 55 m/s just before it strikes a 46-g golf ball at rest on a tee. After the collision, the club head travels (in the same direction) at 40 m/s. Find the speed of the golf ball just after impact.

22. A rifle with a weight of 30 N fires a 5.0-g bullet with a speed of 300 m/s. (a) Find the recoil speed of the rifle. (b) If a 700-N man holds the rifle firmly against his shoulder, find the recoil speed of the man and rifle.

23. A 45.0-kg girl is standing on a 150-kg plank. The plank, originally at rest, is free to slide on a frozen lake, which is a flat, frictionless surface. The girl begins to walk along the plank at a constant velocity of 1.50 m/s to the right relative to the plank. (a) What is her velocity relative to the surface of the ice? (b) What is the velocity of the plank relative to the surface of the ice?

24. **S** *This is a symbolic version of Problem 23.* A girl of mass m_G is standing on a plank of mass m_P. Both are

originally at rest on a frozen lake that constitutes a frictionless, flat surface. The girl begins to walk along the plank at a constant velocity v_{GP} to the right relative to the plank. (The subscript GP denotes the girl relative to plank.) (a) What is the velocity v_{PI} of the plank relative to the surface of the ice? (b) What is the girl's velocity v_{GI} relative to the ice surface?

25. **QC** An astronaut in her space suit has a total mass of 87.0 kg, including suit and oxygen tank. Her tether line loses its attachment to her spacecraft while she's on a spacewalk. Initially at rest with respect to her spacecraft, she throws her 12.0-kg oxygen tank away from her spacecraft with a speed of 8.00 m/s to propel herself back toward it (Fig. P6.25). (a) Determine the maximum distance she can be from the craft and still return within 2.00 min (the amount of time the air in her helmet remains breathable). (b) Explain in terms of Newton's laws of motion why this strategy works.

Figure P6.25

26. A 75-kg fisherman in a 125-kg boat throws a package of mass $m = 15$ kg horizontally toward the right with a speed of $v_i = 4.5$ m/s as in Figure P6.26. Neglecting water resistance, and assuming the boat is at rest before the package is thrown, find the velocity of the boat after the package is thrown.

Figure P6.26

27. A 65.0-kg person throws a 0.045 0-kg snowball forward with a ground speed of 30.0 m/s. A second person, with a mass of 60.0 kg, catches the snowball. Both people are on skates. The first person is initially moving forward with a speed of 2.50 m/s, and the second person is initially at rest. What are the velocities of the two people after the snowball is exchanged? Disregard friction between the skates and the ice.

28. **QC S** An amateur skater of mass M (when fully dressed) is trapped in the middle of an ice rink and is unable to return to the side where there is no ice. Every motion she makes causes her to slip on the ice and remain in the same spot. She decides to try to return to safety by removing her gloves of mass m and throwing them in the direction opposite the safe side. (a) She throws the gloves as hard as she can, and they leave her hand with a velocity $\vec{v}_{gloves}$. Explain whether or not she moves. If she does move, calculate her velocity $\vec{v}_{girl}$ relative to the Earth after she throws the gloves. (b) Discuss her motion from the point of view of the forces acting on her.

6.3 Collisions

6.4 Glancing Collisions

29. **GP** A man of mass $m_1 = 70.0$ kg is skating at $v_1 = 8.00$ m/s behind his wife of mass $m_2 = 50.0$ kg, who is skating at $v_2 = 4.00$ m/s. Instead of passing her, he inadvertently collides with her. He grabs her around the waist, and they maintain their balance. (a) Sketch the problem with before-and-after diagrams, representing the skaters as blocks. (b) Is the collision best described as elastic, inelastic, or perfectly inelastic? Why? (c) Write the general equation for conservation of momentum in terms of m_1, v_1, m_2, v_2, and final velocity v_f. (d) Solve the momentum equation for v_f. (e) Substitute values, obtaining the numerical value for v_f, their speed after the collision.

30. An archer shoots an arrow toward a 300-g target that is sliding in her direction at a speed of 2.50 m/s on a smooth, slippery surface. The 22.5-g arrow is shot with a speed of 35.0 m/s and passes through the target, which is stopped by the impact. What is the speed of the arrow after passing through the target?

31. Gayle runs at a speed of 4.00 m/s and dives on a sled, initially at rest on the top of a frictionless, snow-covered hill. After she has descended a vertical distance of 5.00 m, her brother, who is initially at rest, hops on her back, and they continue down the hill together. What is their speed at the bottom of the hill if the total vertical drop is 15.0 m? Gayle's mass is 50.0 kg, the sled has a mass of 5.00 kg, and her brother has a mass of 30.0 kg.

32. **BIO** A 75.0-kg ice skater moving at 10.0 m/s crashes into a stationary skater of equal mass. After the collision, the two skaters move as a unit at 5.00 m/s. Suppose the average force a skater can experience without breaking a bone is 4 500 N. If the impact time is 0.100 s, does a bone break?

33. A railroad car of mass 2.00×10^4 kg moving at 3.00 m/s collides and couples with two coupled railroad cars, each of the same mass as the single car and moving in the same direction at 1.20 m/s. (a) What is the speed of the three coupled cars after the collision? (b) How much kinetic energy is lost in the collision?

34. **S** *This is a symbolic version of Problem 33.* A railroad car of mass M moving at a speed v_1 collides and couples with two coupled railroad cars, each of the same mass M and moving in the same direction at a speed v_2. (a) What is the speed v_f of the three coupled cars after the collision in terms of v_1 and v_2? (b) How much kinetic energy is lost in the collision? Answer in terms of M, v_1, and v_2.

35. **S** Consider the ballistic pendulum device discussed in Example 6.5 and illustrated in Figure 6.12. (a) Determine the ratio of the momentum immediately after the collision to the momentum immediately before the collision. (b) Show that the ratio of the kinetic energy immediately after the collision to the kinetic energy immediately before the collision is $m_1/(m_1 + m_2)$.

36. **S** A car of mass m moving at a speed v_1 collides and couples with the back of a truck of mass $2m$ moving initially in the same direction as the car at a lower speed v_2. (a) What is the speed v_f of the two vehicles immediately after the collision? (b) What is the change in kinetic energy of the car–truck system in the collision?

37. In a Broadway performance, an 80.0-kg actor swings from a 3.75-m-long cable that is horizontal when he starts. At the bottom of his arc, he picks up his 55.0-kg costar in an inelastic collision. What maximum height do they reach after their upward swing?

38. **W** Two shuffleboard disks of equal mass, one orange and the other green, are involved in a perfectly elastic glancing collision. The green disk is initially at rest and is struck by the orange disk moving initially to the right at 5.00 m/s as in Figure P6.38a. After the collision, the orange disk moves in a direction that makes an angle of 37.0° with the horizontal axis while the green disk makes an angle of 53.0° with this axis as in Figure P6.38b. Determine the speed of each disk after the collision.

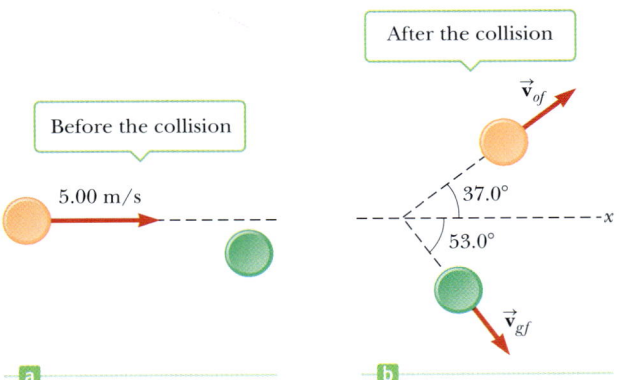

Figure P6.38

39. A 0.030-kg bullet is fired vertically at 200 m/s into a 0.15-kg baseball that is initially at rest. How high does the combined bullet and baseball rise after the collision, assuming the bullet embeds itself in the ball?

40. **M** An bullet of mass m = 8.00 g is fired into a block of mass M = 250 g that is initially at rest at the edge of a table of height h = 1.00 m (Fig. P6.40). The bullet remains in the block, and after the impact the block lands d = 2.00 m from the bottom of the table. Determine the initial speed of the bullet.

Figure P6.40

41. **W** A 12.0-g bullet is fired horizontally into a 100-g wooden block that is initially at rest on a frictionless horizontal surface and connected to a spring having spring constant 150 N/m. The bullet becomes embedded in the block. If the bullet–block system compresses the spring by a maximum of 80.0 cm, what was the speed of the bullet at impact with the block?

42. A 1 200-kg car traveling initially with a speed of 25.0 m/s in an easterly direction crashes into the rear end of a 9 000-kg truck moving in the same direction at 20.0 m/s (Fig. P6.42). The velocity of the car right after the collision is 18.0 m/s to the east. (a) What is the velocity of the truck right after the collision? (b) How much mechanical energy is lost in the collision? Account for this loss in energy.

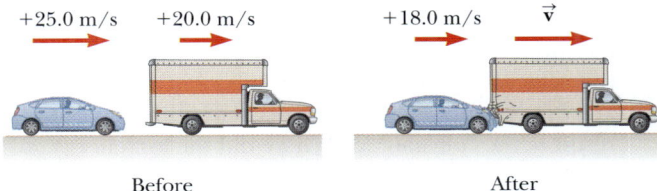

Figure P6.42

43. **QC S** A boy of mass m_b and his girlfriend of mass m_g, both wearing ice skates, face each other at rest while standing on a frictionless ice rink. The boy pushes the girl, sending her away with velocity v_g toward the east. Assume that $m_b > m_g$. (a) Describe the subsequent motion of the boy. (b) Find expressions for the final kinetic energy of the girl and the final kinetic energy of the boy, and show that the girl has greater kinetic energy than the boy. (c) The boy and girl had zero kinetic energy before the boy pushed the girl, but ended up with kinetic energy after the event. How do you account for the appearance of mechanical energy?

44. **GP QC** A space probe, initially at rest, undergoes an internal mechanical malfunction and breaks into three pieces. One piece of mass m_1 = 48.0 kg travels in the positive x-direction at 12.0 m/s, and a second piece of mass m_2 = 62.0 kg travels in the xy-plane at an angle of 105° at 15.0 m/s. The third piece has mass m_3 = 112 kg. (a) Sketch a diagram of the situation, labeling the different masses and their velocities. (b) Write the general

expression for conservation of momentum in the x- and y-directions in terms of m_1, m_2, m_3, v_1, v_2, and v_3 and the sines and cosines of the angles, taking θ to be the unknown angle. (c) Calculate the final x-components of the momenta of m_1 and m_2. (d) Calculate the final y-components of the momenta of m_1 and m_2. (e) Substitute the known momentum components into the general equations of momentum for the x- and y-directions, along with the known mass m_3. (f) Solve the two momentum equations for $v_3 \cos \theta$ and $v_3 \sin \theta$, respectively, and use the identity $\cos^2 \theta + \sin^2 \theta = 1$ to obtain v_3. (g) Divide the equation for $v_3 \sin \theta$ by that for $v_3 \cos \theta$ to obtain $\tan \theta$, then obtain the angle by taking the inverse tangent of both sides. (h) In general, would three such pieces necessarily have to move in the same plane? Why?

45. A 25.0-g object moving to the right at 20.0 cm/s overtakes and collides elastically with a 10.0-g object moving in the same direction at 15.0 cm/s. Find the velocity of each object after the collision.

46. A billiard ball rolling across a table at 1.50 m/s makes a head-on elastic collision with an identical ball. Find the speed of each ball after the collision (a) when the second ball is initially at rest, (b) when the second ball is moving toward the first at a speed of 1.00 m/s, and (c) when the second ball is moving away from the first at a speed of 1.00 m/s.

47. **QC M** A 90.0-kg fullback running east with a speed of 5.00 m/s is tackled by a 95.0-kg opponent running north with a speed of 3.00 m/s. (a) Why does the tackle constitute a perfectly inelastic collision? (b) Calculate the velocity of the players immediately after the tackle and (c) determine the mechanical energy that is lost as a result of the collision. (d) Where did the lost energy go?

48. Identical twins, each with mass 55.0 kg, are on ice skates and at rest on a frozen lake, which may be taken as frictionless. Twin A is carrying a backpack of mass 12.0 kg. She throws it horizontally at 3.00 m/s to Twin B. Neglecting any gravity effects, what are the subsequent speeds of Twin A and Twin B?

49. A 2 000-kg car moving east at 10.0 m/s collides with a 3 000-kg car moving north. The cars stick together and move as a unit after the collision, at an angle of 40.0° north of east and a speed of 5.22 m/s. Find the speed of the 3 000-kg car before the collision.

50. Two automobiles of equal mass approach an intersection. One vehicle is traveling with velocity 13.0 m/s toward the east, and the other is traveling north with velocity v_{2i}. Neither driver sees the other. The vehicles collide in the intersection and stick together, leaving parallel skid marks at an angle of 55.0° north of east. The speed limit for both roads is 35 mi/h, and the driver of the northward-moving vehicle claims he was within the limit when the collision occurred. Is he telling the truth?

51. A billiard ball moving at 5.00 m/s strikes a stationary ball of the same mass. After the collision, the first ball moves at 4.33 m/s at an angle of 30° with respect to the original line of motion. (a) Find the velocity (magnitude and direction) of the second ball after collision. (b) Was the collision inelastic or elastic?

Additional Problems

52. **BIO** In research in cardiology and exercise physiology, it is often important to know the mass of blood pumped by a person's heart in one stroke. This information can be obtained by means of a *ballistocardiograph*. The instrument works as follows: The subject lies on a horizontal pallet floating on a film of air. Friction on the pallet is negligible. Initially, the momentum of the system is zero. When the heart beats, it expels a mass m of blood into the aorta with speed v, and the body and platform move in the opposite direction with speed V. The speed of the blood can be determined independently (e.g., by observing an ultrasound Doppler shift). Assume that the blood's speed is 50.0 cm/s in one typical trial. The mass of the subject plus the pallet is 54.0 kg. The pallet moves at a speed of 6.00×10^{-5} m in 0.160 s after one heartbeat. Calculate the mass of blood that leaves the heart. Assume that the mass of blood is negligible compared with the total mass of the person. This simplified example illustrates the principle of ballistocardiography, but in practice a more sophisticated model of heart function is used.

53. **QC** Most of us know intuitively that in a head-on collision between a large dump truck and a subcompact car, you are better off being in the truck than in the car. Why is this? Many people imagine that the collision force exerted on the car is much greater than that exerted on the truck. To substantiate this view, they point out that the car is crushed, whereas the truck is only dented. This idea of unequal forces, of course, is false; Newton's third law tells us that both objects are acted upon by forces of the same magnitude. The truck suffers less damage because it is made of stronger metal. But what about the two drivers? Do they experience the same forces? To answer this question, suppose that each vehicle is initially moving at 8.00 m/s and that they undergo a perfectly inelastic head-on collision. Each driver has mass 80.0 kg. Including the masses of the drivers, the total masses of the vehicles are 800 kg for the car and 4 000 kg for the truck. If the collision time is 0.120 s, what force does the seat belt exert on each driver?

54. Consider a frictionless track as shown in Figure P6.54. A block of mass $m_1 = 5.00$ kg is released from Ⓐ.

Figure P6.54

It makes a head-on elastic collision at Ⓑ with a block of mass $m_2 = 10.0$ kg that is initially at rest. Calculate the maximum height to which m_1 rises after the collision.

55. **M** A 2.0-g particle moving at 8.0 m/s makes a perfectly elastic head-on collision with a resting 1.0-g object. (a) Find the speed of each particle after the collision. (b) Find the speed of each particle after the collision if the stationary particle has a mass of 10 g. (c) Find the final kinetic energy of the incident 2.0-g particle in the situations described in parts (a) and (b). In which case does the incident particle lose more kinetic energy?

56. **S** A bullet of mass m and speed v passes completely through a pendulum bob of mass M as shown in Figure P6.56. The bullet emerges with a speed of $v/2$. The pendulum bob is suspended by a stiff rod of length ℓ and negligible mass. What is the minimum value of v such that the bob will barely swing through a complete vertical circle?

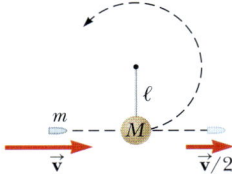

Figure P6.56

57. Two objects of masses $m_1 = 0.56$ kg and $m_2 = 0.88$ kg are placed on a horizontal frictionless surface and a compressed spring of force constant $k = 280$ N/m is placed between them as in Figure P6.57a. Neglect the mass of the spring. The spring is not attached to either object and is compressed a distance of 9.8 cm. If the objects are released from rest, find the final velocity of each object as shown in Figure P6.57b.

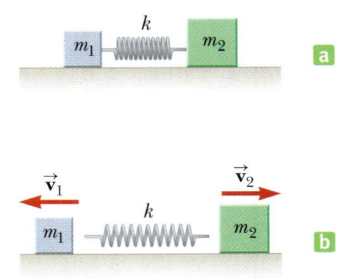

Figure P6.57

58. A 0.400-kg blue bead slides on a frictionless, curved wire, starting from rest at point Ⓐ in Figure P6.58, where $h = 1.50$ m. At point Ⓑ, the bead collides elastically with a 0.600-kg green bead at rest. Find the maximum height the green bead rises as it moves up the wire.

Figure P6.58

59. A 730-N man stands in the middle of a frozen pond of radius 5.0 m. He is unable to get to the other side because of a lack of friction between his shoes and the ice. To overcome this difficulty, he throws his 1.2-kg physics textbook horizontally toward the north shore at a speed of 5.0 m/s. How long does it take him to reach the south shore?

60. An unstable nucleus of mass 1.7×10^{-26} kg, initially at rest at the origin of a coordinate system, disintegrates into three particles. One particle, having a mass of $m_1 = 5.0 \times 10^{-27}$ kg, moves in the positive y-direction with speed $v_1 = 6.0 \times 10^6$ m/s. Another particle, of mass $m_2 = 8.4 \times 10^{-27}$ kg, moves in the positive x-direction with speed $v_2 = 4.0 \times 10^6$ m/s. Find the magnitude and direction of the velocity of the third particle.

61. **S** Two blocks of masses m_1 and m_2 approach each other on a horizontal table with the same constant speed, v_0, as measured by a laboratory observer. The blocks undergo a perfectly elastic collision, and it is observed that m_1 stops but m_2 moves opposite its original motion with some constant speed, v. (a) Determine the ratio of the two masses, m_1/m_2. (b) What is the ratio of their speeds, v/v_0?

62. Two blocks of masses $m_1 = 2.00$ kg and $m_2 = 4.00$ kg are each released from rest at a height of $h = 5.00$ m on a frictionless track, as shown in Figure P6.62, and undergo an elastic head-on collision. (a) Determine the velocity of each block just before the collision. (b) Determine the velocity of each block immediately after the collision. (c) Determine the maximum heights to which m_1 and m_2 rise after the collision.

Figure P6.62

63. A block with mass $m_1 = 0.500$ kg is released from rest on a frictionless track at a distance $h_1 = 2.50$ m above the top of a table. It then collides elastically with an object having mass $m_2 = 1.00$ kg that is initially at rest on the table, as shown in Figure P6.63. (a) Determine the velocities of the two objects just after the collision. (b) How high up the track does the 0.500-kg object travel back after the collision? (c) How far away from the bottom of the table does the 1.00-kg object land, given that the height of the table is $h_2 = 2.00$ m? (d) How far away from the bottom of the table does the 0.500-kg object eventually land?

Figure P6.63

64. **S** Two objects of masses m and $3m$ are moving toward each other along the x-axis with the same initial speed v_0. The object with mass m is traveling to the left, and the object with mass $3m$ is traveling to the right. They undergo an elastic glancing collision such that m is moving downward after the collision at right angles from its initial direction. (a) Find the final speeds of the two objects. (b) What is the angle θ at which the object with mass $3m$ is scattered?

65. A small block of mass $m_1 = 0.500$ kg is released from rest at the top of a curved wedge of mass $m_2 = 3.00$ kg, which sits on a frictionless horizontal surface as in Figure P6.65a. When the block leaves the wedge, its velocity is measured to be 4.00 m/s to the right, as in Figure P6.65b. (a) What is the velocity of the wedge after the block reaches the horizontal surface? (b) What is the height h of the wedge?

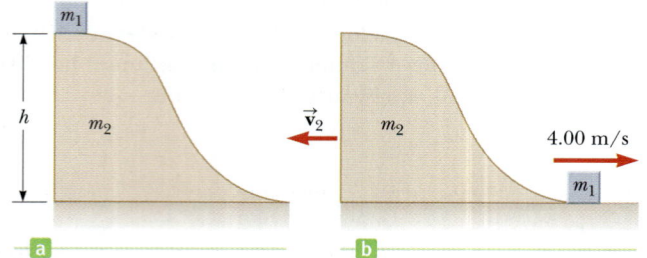

Figure P6.65

66. A cue ball traveling at 4.00 m/s makes a glancing, elastic collision with a target ball of equal mass that is initially at rest. The cue ball is deflected so that it makes an angle of 30.0° with its original direction of travel. Find (a) the angle between the velocity vectors of the two balls after the collision and (b) the speed of each ball after the collision.

67. **Q|C** A cannon is rigidly attached to a carriage, which can move along horizontal rails, but is connected to a post by a large spring, initially unstretched and with force constant $k = 2.00 \times 10^4$ N/m, as in Figure P6.67. The cannon fires a 200-kg projectile at a velocity of 125 m/s directed 45.0° above the horizontal. (a) If the mass of the cannon and its carriage is 5 000 kg, find the recoil speed of the cannon. (b) Determine the maximum extension of the spring. (c) Find the maximum force the spring exerts on the carriage. (d) Consider the system consisting of the cannon, the carriage, and the shell. Is the momentum of this system conserved during the firing? Why or why not?

Figure P6.67

68. The "force platform" is a tool that is used to analyze the performance of athletes by measuring the vertical force as a function of time that the athlete exerts on the ground in performing various activities. A simplified force vs. time graph for an athlete performing a standing high jump is shown in Figure P6.68. The athlete started the jump at $t = 0.0$ s. How high did this athlete jump?

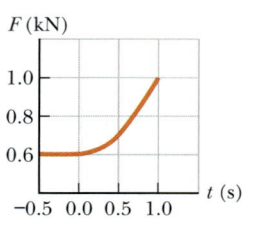

Figure P6.68

69. A neutron in a reactor makes an elastic head-on collision with a carbon atom that is initially at rest. (The mass of the carbon nucleus is about 12 times that of the neutron.) (a) What fraction of the neutron's kinetic energy is transferred to the carbon nucleus? (b) If the neutron's initial kinetic energy is 1.6×10^{-13} J, find its final kinetic energy and the kinetic energy of the carbon nucleus after the collision.

70. **Q|C S** Two blocks collide on a frictionless surface. After the collision, the blocks stick together. Block A has a mass M and is initially moving to the right at speed v. Block B has a mass $2M$ and is initially at rest. System C is composed of both blocks. (a) Draw a force diagram for each block at an instant *during* the collision. (b) Rank the magnitudes of the horizontal forces in your diagram. Explain your reasoning. (c) Calculate the change in momentum of block A, block B, and system C. (d) Is kinetic energy conserved in this collision? Explain your answer. (This problem is courtesy of Edward F. Redish. For more such problems, visit http://www.physics.umd.edu/perg.)

71. **Q|C** (a) A car traveling due east strikes a car traveling due north at an intersection, and the two move together as a unit. A property owner on the southeast corner of the intersection claims that his fence was torn down in the collision. Should he be awarded damages by the insurance company? Defend your answer. (b) Let the eastward-moving car have a mass of 1 300 kg and a speed of 30.0 km/h and the northward-moving car a mass of 1 100 kg and a speed of 20.0 km/h. Find the velocity after the collision. Are the results consistent with your answer to part (a)?

72. A 60-kg soccer player jumps vertically upwards and heads the 0.45-kg ball as it is descending vertically with a speed of 25 m/s. (a) If the player was moving upward with a speed of 4.0 m/s just before impact, what will be the speed of the ball immediately after the collision if the ball rebounds vertically upwards and the collision is elastic? (b) If the ball is in contact with the player's head for 20 ms, what is the average acceleration of the ball? (Note that the force of gravity may be ignored during the brief collision time.)

73. **M** A tennis ball of mass 57.0 g is held just above a basketball of mass 590 g. With their centers vertically

aligned, both balls are released from rest at the same time, to fall through a distance of 1.20 m, as shown in Figure P6.73. (a) Find the magnitude of the downward velocity with which the basketball reaches the ground. (b) Assume that an elastic collision with the ground instantaneously reverses the velocity of the basketball while the tennis ball is still moving down. Next, the two balls meet in an elastic collision. To what height does the tennis ball rebound?

Figure P6.73

74. A 20.0-kg toboggan with 70.0-kg driver is sliding down a frictionless chute directed 30.0° below the horizontal at 8.00 m/s when a 55.0-kg woman drops from a tree limb straight down behind the driver. If she drops through a vertical displacement of 2.00 m, what is the subsequent velocity of the toboggan immediately after impact?

75. **S** *Measuring the speed of a bullet.* A bullet of mass m is fired horizontally into a wooden block of mass M lying on a table. The bullet remains in the block after the collision. The coefficient of friction between the block and table is μ, and the block slides a distance d before stopping. Find the initial speed v_0 of the bullet in terms of M, m, μ, g, and d.

76. A flying squid (family Ommastrephidae) is able to "jump" off the surface of the sea by taking water into its body cavity and then ejecting the water vertically downward. A 0.85-kg squid is able to eject 0.30 kg of water with a speed of 20 m/s. (a) What will be the speed of the squid immediately after ejecting the water? (b) How high in the air will the squid rise?

77. A 0.30-kg puck, initially at rest on a frictionless horizontal surface, is struck by a 0.20-kg puck that is initially moving along the x-axis with a velocity of 2.0 m/s. After the collision, the 0.20-kg puck has a speed of 1.0 m/s at an angle of $\theta = 53°$ to the positive x-axis. (a) Determine the velocity of the 0.30-kg puck after the collision. (b) Find the fraction of kinetic energy lost in the collision.

78. **Q|C S** A wooden block of mass M rests on a table over a large hole as in Figure P6.78. A bullet of mass m with an initial velocity v_i is fired upward into the bottom of the block and remains in the block after the collision. The block and bullet rise to a maximum height of h. (a) Describe how you would find the initial velocity of the bullet using ideas you have learned in this chapter. (b) Find an expression for the initial velocity of the bullet.

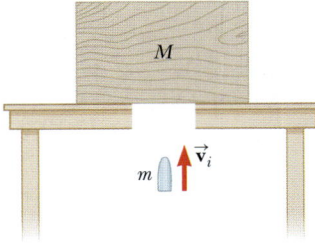

Figure P6.78 Problems 78 and 79.

79. **Q|C** A 1.25-kg wooden block rests on a table over a large hole as in Figure P6.78. A 5.00-g bullet with an initial velocity v_i is fired upward into the bottom of the block and remains in the block after the collision. The block and bullet rise to a maximum height of 22.0 cm. (a) Describe how you would find the initial velocity of the bullet using ideas you have learned in this chapter. (b) Calculate the initial velocity of the bullet from the information provided.

The International Space Station falls freely around the Earth at thousands of meters per second, held in orbit by the centripetal force provided by gravity.

7 Rotational Motion and the Law of Gravity

7.1 Angular Speed and Angular Acceleration

7.2 Rotational Motion Under Constant Angular Acceleration

7.3 Relations Between Angular and Linear Quantities

7.4 Centripetal Acceleration

7.5 Newtonian Gravitation

7.6 Kepler's Laws

Rotational motion is an important part of everyday life. The rotation of the Earth creates the cycle of day and night, the rotation of wheels enables easy vehicular motion, and modern technology depends on circular motion in a variety of contexts, from the tiny gears in a Swiss watch to the operation of lathes and other machinery. The concepts of *angular speed, angular acceleration,* and *centripetal acceleration* are central to understanding the motions of a diverse range of phenomena, from a car moving around a circular racetrack to clusters of galaxies orbiting a common center.

Rotational motion, when combined with Newton's law of universal gravitation and his laws of motion, can also explain certain facts about space travel and satellite motion, such as where to place a satellite so it will remain fixed in position over the same spot on the Earth. The generalization of gravitational potential energy and energy conservation offers an easy route to such results as planetary escape speed. Finally, we present Kepler's three laws of planetary motion, which formed the foundation of Newton's approach to gravity.

7.1 Angular Speed and Angular Acceleration

LEARNING OBJECTIVES

1. Define radian measure, angular position, and angular displacement.
2. Define average and instantaneous angular speed.
3. Define average and instantaneous angular acceleration.
4. Perform elementary calculations with angular variables.

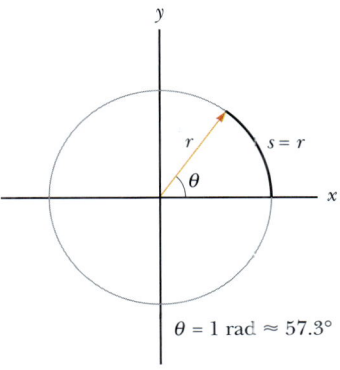

Figure 7.1 For a circle of radius r, one radian is the angle subtended by an arc length equal to r.

In the study of linear motion, the important concepts are *displacement* Δx, *velocity* v, and *acceleration* a. Each of these concepts has its analog in rotational motion: *angular displacement* $\Delta \theta$, *angular velocity* ω, and *angular acceleration* α.

The *radian*, a unit of angular measure, is essential to the understanding of these concepts. Recall that the distance s around a circle is given by $s = 2\pi r$, where r is the radius of the circle. Dividing both sides by r results in $s/r = 2\pi$. This quantity is dimensionless because both s and r have dimensions of length, but the value 2π corresponds to a displacement around a circle. A half circle would give an answer of π, a quarter circle an answer of $\pi/2$. The numbers 2π, π, and $\pi/2$ correspond to angles of 360°, 180°, and 90°, respectively, so a new unit of angular measure, the **radian,** can be introduced, with $180° = \pi$ rad relating degrees to radians.

The angle θ subtended by an arc length s along a circle of radius r, measured in radians counterclockwise from the positive x-axis, is

$$\theta = \frac{s}{r} \quad [7.1]$$

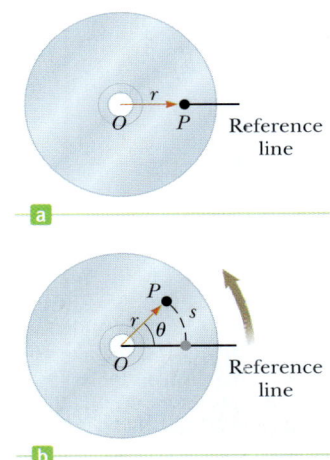

Figure 7.2 (a) The point P on a rotating compact disc at $t = 0$. (b) As the disc rotates, P moves through an arc length s.

The angle θ in Equation 7.1 is actually an angular displacement from the positive x-axis, and s the corresponding displacement along the circular arc, again measured from the positive x-axis. Figure 7.1 illustrates the size of 1 radian, which is approximately 57°. Converting from degrees to radians requires multiplying by the ratio (π rad/180°). For example, 45° (π rad/180°) = (π/4) rad.

Generally, angular quantities in physics must be expressed in radians. Be sure to set your calculator to radian mode; neglecting to do so is a common error.

Armed with the concept of radian measure, we can now discuss angular concepts in physics. Consider Figure 7.2a, a top view of a rotating compact disc. Such a disk is an example of a "rigid body," with each part of the body fixed in position relative to all other parts of the body. When a rigid body rotates through a given angle, all parts of the body rotate through the same angle at the same time. For the compact disc, the axis of rotation is at the center of the disc, O. A point P on the disc is at a distance r from the origin and moves about O in a circle of radius r. We set up a *fixed* reference line, as shown in Figure 7.2a, and assume that at time $t = 0$ the point P is on that reference line. After a time interval Δt has elapsed, P has advanced to a new position (Fig. 7.2b). In this interval, the line OP has moved through the angle θ with respect to the reference line. The angle θ, measured in radians, is called the **angular position** and is analogous to the linear position variable x. Likewise, P has moved an arc length s measured along the circumference of the circle.

In Figure 7.3, as a point on the rotating disc moves from Ⓐ to Ⓑ in a time Δt, it starts at an angle θ_i and ends at an angle θ_f. The difference $\theta_f - \theta_i$ is called the **angular displacement.**

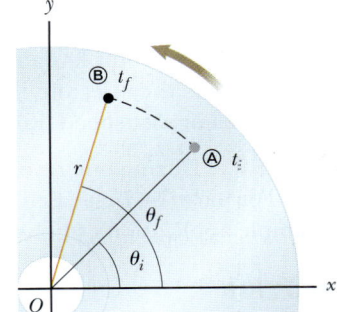

Figure 7.3 As a point on the compact disc moves from Ⓐ to Ⓑ, the disc rotates through the angle $\Delta \theta = \theta_f - \theta_i$.

An object's angular displacement, $\Delta\theta$, is the difference in its final and initial angles:

$$\Delta\theta = \theta_f - \theta_i \quad [7.2]$$

SI unit: radian (rad)

For example, if a point on a disc is at $\theta_i = 4$ rad and rotates to angular position $\theta_f = 7$ rad, the angular displacement is $\Delta\theta = \theta_f - \theta_i = 7$ rad $-$ 4 rad $=$ 3 rad. Note that we use angular variables to describe the rotating disc because **each point on the disc undergoes the same angular displacement in any given time interval.**

Using the definition in Equation 7.2, Equation 7.1 can be written more generally as $\Delta\theta = \Delta s/r$, where Δs is a displacement along the circular arc subtended by the angular displacement. Having defined angular displacements, it's natural to define an angular speed:

The average angular speed ω_{av} of a rotating rigid object during the time interval Δt is the angular displacement $\Delta\theta$ divided by Δt:

$$\omega_{av} \equiv \frac{\theta_f - \theta_i}{t_f - t_i} = \frac{\Delta\theta}{\Delta t} \quad [7.3]$$

SI unit: radian per second (rad/s)

For very short time intervals, the average angular speed approaches the instantaneous angular speed, just as in the linear case.

The **instantaneous angular speed** ω of a rotating rigid object is the limit of the average speed $\Delta\theta/\Delta t$ as the time interval Δt approaches zero:

$$\omega \equiv \lim_{\Delta t \to 0} \frac{\Delta\theta}{\Delta t} \quad [7.4]$$

SI unit: radian per second (rad/s)

We take ω to be positive when θ is increasing (counterclockwise motion) and negative when θ is decreasing (clockwise motion). When the angular speed is constant, the instantaneous angular speed is equal to the average angular speed.

> **Tip 7.1 Remember the Radian**
>
> Equation 7.1 uses angles measured in *radians*. Angles expressed in terms of degrees must first be converted to radians. Also, be sure to check whether your calculator is in degree or radian mode when solving problems involving rotation.

■ EXAMPLE 7.1 | Whirlybirds

GOAL Perform some elementary calculations with angular variables.

PROBLEM The rotor on a helicopter turns at an angular speed of 3.20×10^2 revolutions per minute. (In this book, we sometimes use the abbreviation rpm, but in most cases we use rev/min.) **(a)** Express this angular speed in radians per second. **(b)** If the rotor has a radius of 2.00 m, what arclength does the tip of the blade trace out in 3.00×10^2 s? **(c)** The pilot opens the throttle, and the angular speed of the blade increases while rotating twenty-six times in 3.60 s. Calculate the average angular speed during that time.

STRATEGY During one revolution, the rotor turns through an angle of 2π radians. Use this relationship as a conversion factor. For part (b), first calculate the angular displacement in radians by multiplying the angular speed by time. Part (c) is a simple application of Equation 7.3.

SOLUTION

(a) Express this angular speed in radians per second.

Apply the conversion factors 1 rev = 2π rad and 60.0 s = 1 min:

$$\omega = 3.20 \times 10^2 \, \frac{\text{rev}}{\text{min}}$$

$$= 3.20 \times 10^2 \, \frac{\text{rev}}{\text{min}} \left(\frac{2\pi \text{ rad}}{1 \text{ rev}}\right)\left(\frac{1.00 \text{ min}}{60.0 \text{ s}}\right)$$

$$= \boxed{33.5 \text{ rad/s}}$$

(b) Find the arclength traced out by the tip of the blade.

Multiply the angular speed by the time to obtain the angular displacement:

$$\Delta\theta = \omega t = (33.5 \text{ rad/s})(3.00 \times 10^2 \text{ s}) = 1.01 \times 10^4 \text{ rad}$$

Multiply the angular displacement by the radius to get the arc length:

$$\Delta s = r\Delta\theta = (2.00 \text{ m})(1.01 \times 10^4 \text{ rad}) = \boxed{2.02 \times 10^4 \text{ m}}$$

(c) Calculate the average angular speed of the blade while its angular speed increases.

Apply Equation 7.3, noticing that

$$\Delta\theta = (26 \text{ rev})(2\pi \text{ rad/rev}) = 52\pi \text{ rad}:$$

$$\omega_{av} = \frac{\Delta\theta}{\Delta t} = \frac{52\pi \text{ rad}}{3.60 \text{ s}} = 45 \text{ rad/s}$$

REMARKS It's best to express angular speeds in radians per second. Consistent use of radian measure minimizes errors.

QUESTION 7.1 Is it possible to express angular speed in degrees per second? If so, what's the conversion factor from radians per second?

EXERCISE 7.1 A Ferris wheel turns at a constant 185.0 revolutions per hour. **(a)** Express this rate of rotation in units of radians per second. **(b)** If the wheel has a radius of 12.0 m, what arclength does a passenger trace out during a ride lasting 5.00 min? **(c)** If the wheel then slows to rest in 9.72 s while making a quarter turn, calculate the magnitude of its average angular speed during that time.

ANSWERS (a) 0.323 rad/s (b) 1.16×10^3 m (c) 0.162 rad/s

Quick Quiz

7.1 A rigid body is rotating counterclockwise about a fixed axis. Each of the following pairs of quantities represents an initial angular position and a final angular position of the rigid body. Which of the sets can occur *only* if the rigid body rotates through more than 180°? (a) 3 rad, 6 rad; (b) −1 rad, 1 rad; (c) 1 rad, 5 rad.

7.2 Suppose the change in angular position for each of the pairs of values in Quick Quiz 7.1 occurred in 1 s. Which choice represents the lowest average angular speed?

Figure 7.4 shows a bicycle turned upside down so that a repair technician can work on the rear wheel. The bicycle pedals are turned so that at time t_i the wheel has angular speed ω_i (Fig. 7.4a) and at a later time t_f it has angular speed ω_f (Fig. 7.4b). Just as a changing speed leads to the concept of an acceleration, a changing angular speed leads to the concept of an angular acceleration.

An object's average angular acceleration α_{av} during the time interval Δt is the change in its angular speed $\Delta\omega$ divided by Δt:

$$\alpha_{av} \equiv \frac{\omega_f - \omega_i}{t_f - t_i} = \frac{\Delta\omega}{\Delta t} \quad [7.5]$$

◀ Average angular acceleration

SI unit: radian per second squared (rad/s²)

Figure 7.4 An accelerating bicycle wheel rotates with (a) angular speed ω_i at time t_i and (b) angular speed ω_f at time t_f.

As with angular velocity, positive angular accelerations are in the counterclockwise direction, negative angular accelerations in the clockwise direction. If the angular speed goes from 15 rad/s to 9.0 rad/s in 3.0 s, the average angular acceleration during that time interval is

$$\alpha_{av} = \frac{\Delta \omega}{\Delta t} = \frac{9.0 \text{ rad/s} - 15 \text{ rad/s}}{3.0 \text{ s}} = -2.0 \text{ rad/s}^2$$

The negative sign indicates that the angular acceleration is clockwise (although the angular speed, still positive but slowing down, is in the counterclockwise direction). There is also an instantaneous version of angular acceleration:

▶ Instantaneous angular acceleration

The instantaneous angular acceleration α is the limit of the average angular acceleration $\Delta\omega/\Delta t$ as the time interval Δt approaches zero:

$$\alpha \equiv \lim_{\Delta t \to 0} \frac{\Delta \omega}{\Delta t} \quad [7.6]$$

SI unit: radian per second squared (rad/s²)

When a rigid object rotates about a fixed axis, as does the bicycle wheel, every portion of the object has the same angular speed and the same angular acceleration. This fact is what makes these variables so useful for describing rotational motion. In contrast, the tangential (linear) speed and acceleration of the object take different values that depend on the distance from a given point to the axis of rotation.

7.2 Rotational Motion Under Constant Angular Acceleration

LEARNING OBJECTIVES

1. Identify the correspondence between the equations for linear motion at constant acceleration and those for angular motion.
2. Apply rotational kinematics to objects undergoing uniform angular acceleration.

A number of parallels exist between the equations for rotational motion and those for linear motion. For example, compare the defining equation for the average angular speed,

$$\omega_{av} \equiv \frac{\theta_f - \theta_i}{t_f - t_i} = \frac{\Delta \theta}{\Delta t}$$

with that of the average linear speed,

$$v_{av} \equiv \frac{x_f - x_i}{t_f - t_i} = \frac{\Delta x}{\Delta t}$$

In these equations, ω takes the place of v and θ takes the place of x, so the equations differ only in the names of the variables. In the same way, every linear quantity we have encountered so far has a corresponding "twin" in rotational motion.

The procedure used in Section 2.5 to develop the kinematic equations for linear motion under constant acceleration can be used to derive a similar set of equations for rotational motion under constant angular acceleration. The resulting equations of rotational kinematics, along with the corresponding equations for linear motion, are as follows:

Linear Motion with a Constant (Variables: x and v)	**Rotational Motion About a Fixed Axis with α Constant (Variables: θ and ω)**	
$v = v_i + at$	$\omega = \omega_i + \alpha t$	[7.7]
$\Delta x = v_i t + \frac{1}{2}at^2$	$\Delta\theta = \omega_i t + \frac{1}{2}\alpha t^2$	[7.8]
$v^2 = v_i^2 + 2a\Delta x$	$\omega^2 = \omega_i^2 + 2\alpha\Delta\theta$	[7.9]

Notice that every term in a given linear equation has a corresponding term in the analogous rotational equation.

■ Quick Quiz

7.3 Consider again the pairs of angular positions for the rigid object in Quick Quiz 7.1. If the object starts from rest at the initial angular position, moves counterclockwise with constant angular acceleration, and arrives at the final angular position with the same angular speed in all three cases, for which choice is the angular acceleration the highest?

■ EXAMPLE 7.2 | A Rotating Wheel

GOAL Apply the rotational kinematic equations.

PROBLEM A wheel rotates with a constant angular acceleration of 3.50 rad/s^2. If the angular speed of the wheel is 2.00 rad/s at $t = 0$, **(a)** through what angle does the wheel rotate between $t = 0$ and $t = 2.00$ s? Give your answer in radians and in revolutions. **(b)** What is the angular speed of the wheel at $t = 2.00$ s? **(c)** What angular displacement (in revolutions) results while the angular speed found in part (b) doubles?

STRATEGY The angular acceleration is constant, so this problem just requires substituting given values into Equations 7.7–7.9.

SOLUTION
(a) Find the angular displacement after 2.00 s, in both radians and revolutions.

Use Equation 7.8, setting $\omega_i = 2.00$ rad/s, $\alpha = 3.5$ rad/s^2, and $t = 2.00$ s:

$$\Delta\theta = \omega_i t + \frac{1}{2}\alpha t^2$$
$$= (2.00 \text{ rad/s})(2.00 \text{ s}) + \frac{1}{2}(3.50 \text{ rad/s}^2)(2.00 \text{ s})^2$$
$$= \boxed{11.0 \text{ rad}}$$

Convert radians to revolutions.

$$\Delta\theta = (11.0 \text{ rad})(1.00 \text{ rev}/2\pi \text{ rad}) = \boxed{1.75 \text{ rev}}$$

(b) What is the angular speed of the wheel at $t = 2.00$ s?

Substitute the same values into Equation 7.7:

$$\omega = \omega_i + \alpha t = 2.00 \text{ rad/s} + (3.50 \text{ rad/s}^2)(2.00 \text{ s})$$
$$= \boxed{9.00 \text{ rad/s}}$$

(c) What angular displacement (in revolutions) results during the time in which the angular speed found in part (b) doubles?

Apply the time-independent rotational kinematics equation:

$$\omega_f^2 - \omega_i^2 = 2\alpha\Delta\theta$$

Substitute values, noting that $\omega_f = 2\omega_i$:

$$(2 \times 9.00 \text{ rad/s})^2 - (9.00 \text{ rad/s})^2 = 2(3.50 \text{ rad/s}^2)\Delta\theta$$

Solve for the angular displacement and convert to revolutions:

$$\Delta\theta = (34.7 \text{ rad})\left(\frac{1 \text{ rev}}{2\pi \text{ rad}}\right) = \boxed{5.52 \text{ rev}}$$

REMARKS The result of part (b) could also be obtained from Equation 7.9 and the results of part (a).

(Continued)

QUESTION 7.2 Suppose the radius of the wheel is doubled. Are the answers affected? If so, in what way?

EXERCISE 7.2 (a) Find the angle through which the wheel rotates between $t = 2.00$ s and $t = 3.00$ s. (b) Find the angular speed when $t = 3.00$ s. (c) What is the magnitude of the angular speed two revolutions following $t = 3.00$ s?

ANSWERS (a) 10.8 rad (b) 12.5 rad/s (c) 15.6 rad/s

7.3 Relations Between Angular and Linear Quantities

LEARNING OBJECTIVE

1. Apply the relationships between angular and linear quantities.

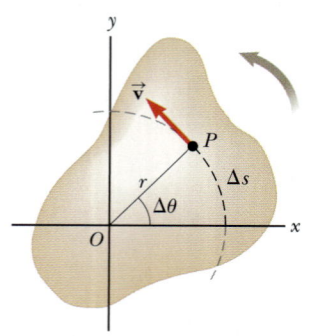

Figure 7.5 Rotation of an object about an axis through O (the z-axis) that is perpendicular to the plane of the figure. Note that a point P on the object rotates in a circle of radius r centered at O.

Angular variables are closely related to linear variables. Consider the arbitrarily shaped object in Figure 7.5 rotating about the z-axis through the point O. Assume the object rotates through the angle $\Delta\theta$, and hence point P moves through the arc length Δs, in the interval Δt. We know from the defining equation for radian measure that

$$\Delta\theta = \frac{\Delta s}{r}$$

Dividing both sides of this equation by Δt, the time interval during which the rotation occurs, yields

$$\frac{\Delta\theta}{\Delta t} = \frac{1}{r}\frac{\Delta s}{\Delta t}$$

When Δt is very small, the angle $\Delta\theta$ through which the object rotates is also small and the ratio $\Delta\theta/\Delta t$ is close to the instantaneous angular speed ω. On the other side of the equation, similarly, the ratio $\Delta s/\Delta t$ approaches the instantaneous linear speed v for small values of Δt. Hence, when Δt gets arbitrarily small, the preceding equation is equivalent to

$$\omega = \frac{v}{r}$$

In Figure 7.5, the point P traverses a distance Δs along a circular arc during the time interval Δt at a linear speed of v. The direction of P's velocity vector $\vec{v}$ is *tangent to the circular path*. The magnitude of $\vec{v}$ is the linear speed $v = v_t$, called the **tangential speed** of a particle moving in a circular path, written

▶ Tangential speed

$$v_t = r\omega \qquad [7.10]$$

The tangential speed of a point on a rotating object equals the distance of that point from the axis of rotation multiplied by the angular speed. Equation 7.10 shows that the linear speed of a point on a rotating object increases as that point is moved outward from the center of rotation toward the rim, as expected; however, **every point on the rotating object has the same *angular* speed.**

Equation 7.10, derived using the defining equation for radian measure, is valid only when ω is measured in radians per unit time. Other measures of angular speed, such as degrees per second and revolutions per second, shouldn't be used.

To find a second equation relating linear and angular quantities, refer again to Figure 7.5 and suppose the rotating object changes its angular speed by $\Delta\omega$ in the time interval Δt. At the end of this interval, the speed of a point on the object, such as P, has changed by the amount Δv_t. From Equation 7.10 we have

$$\Delta v_t = r\Delta\omega$$

Dividing by Δt gives

$$\frac{\Delta v_t}{\Delta t} = r\frac{\Delta \omega}{\Delta t}$$

As the time interval Δt is taken to be arbitrarily small, $\Delta \omega/\Delta t$ approaches the instantaneous angular acceleration. On the left-hand side of the equation, note that the ratio $\Delta v_t/\Delta t$ tends to the instantaneous linear acceleration, called the tangential acceleration of that point, given by

$$a_t = r\alpha \qquad [7.11]$$ ◀ Tangential acceleration

The tangential acceleration of a point on a rotating object equals the distance of that point from the axis of rotation multiplied by the angular acceleration. Again, radian measure must be used for the angular acceleration term in this equation.

One last equation that relates linear quantities to angular quantities will be derived in the next section.

■ Quick Quiz

7.4 Andrea and Chuck are riding on a merry-go-round. Andrea rides on a horse at the outer rim of the circular platform, twice as far from the center of the circular platform as Chuck, who rides on an inner horse. When the merry-go-round is rotating at a constant angular speed, Andrea's angular speed is (a) twice Chuck's (b) the same as Chuck's (c) half of Chuck's (d) impossible to determine.

7.5 When the merry-go-round of Quick Quiz 7.4 is rotating at a constant angular speed, Andrea's tangential speed is (a) twice Chuck's (b) the same as Chuck's (c) half of Chuck's (d) impossible to determine.

■ APPLYING PHYSICS 7.1 | ESA Launch Site

Why is the launch area for the European Space Agency in South America and not in Europe?

EXPLANATION Satellites are boosted into orbit on top of rockets, which provide the large tangential speed necessary to achieve orbit. Due to its rotation, the surface of Earth is already traveling toward the east at a tangential speed of nearly 1 700 m/s at the equator. This tangential speed is steadily reduced farther north because the distance to the axis of rotation is decreasing. It finally goes to zero at the North Pole. Launching eastward from the equator gives the satellite a starting initial tangential speed of 1 700 m/s, whereas a European launch provides roughly half that speed (depending on the exact latitude). ■

■ EXAMPLE 7.3 | Compact Discs

GOAL Apply the rotational kinematics equations in tandem with tangential acceleration and speed.

PROBLEM A compact disc rotates from rest up to an angular speed of 31.4 rad/s in a time of 0.892 s. **(a)** What is the angular acceleration of the disc, assuming the angular acceleration is uniform? **(b)** Through what angle does the disc turn while coming up to speed? **(c)** If the radius of the disc is 4.45 cm, find the tangential speed of a microbe riding on the rim of the disc when $t = 0.892$ s. **(d)** What is the magnitude of the tangential acceleration of the microbe at the given time?

STRATEGY We can solve parts (a) and (b) by applying the kinematic equations for angular speed and angular displacement (Eqs. 7.7 and 7.8). Multiplying the radius by the angular acceleration yields the tangential acceleration at the rim, whereas multiplying the radius by the angular speed gives the tangential speed at that point.

SOLUTION

(a) Find the angular acceleration of the disc.

Apply the angular velocity equation $\omega = \omega_i + \alpha t$, taking $\omega_i = 0$ at $t = 0$:

$$\alpha = \frac{\omega}{t} = \frac{31.4 \text{ rad/s}}{0.892 \text{ s}} = \boxed{35.2 \text{ rad/s}^2}$$

(Continued)

(b) Through what angle does the disc turn?

Use Equation 7.8 for angular displacement, with $t = 0.892$ s and $\omega_i = 0$:

$$\Delta\theta = \omega_i t + \tfrac{1}{2}\alpha t^2 = \tfrac{1}{2}(35.2 \text{ rad/s}^2)(0.892 \text{ s})^2 = \boxed{14.0 \text{ rad}}$$

(c) Find the final tangential speed of a microbe at $r = 4.45$ cm.

Substitute into Equation 7.10:

$$v_t = r\omega = (0.044\,5 \text{ m})(31.4 \text{ rad/s}) = \boxed{1.40 \text{ m/s}}$$

(d) Find the tangential acceleration of the microbe at $r = 4.45$ cm.

Substitute into Equation 7.11:

$$a_t = r\alpha = (0.044\,5 \text{ m})(35.2 \text{ rad/s}^2) = \boxed{1.57 \text{ m/s}^2}$$

REMARKS Because 2π rad $= 1$ rev, the angular displacement in part (b) corresponds to 2.23 rev. In general, dividing the number of radians by 6 gives a rough approximation to the number of revolutions, because $2\pi \sim 6$.

QUESTION 7.3 If the angular acceleration were doubled for the same duration, by what factor would the angular displacement change? Why is the answer true in this case but not in general?

EXERCISE 7.3 (a) What are the angular speed and angular displacement of the disc 0.300 s after it begins to rotate? (b) Find the tangential speed at the rim at this time.

ANSWERS (a) 10.6 rad/s; 1.58 rad (b) 0.472 m/s

APPLICATION
Phonograph Records and Compact Discs

Before MP3s became the medium of choice for recorded music, compact discs and phonographs were popular. There are similarities and differences between the rotational motion of phonograph records and that of compact discs. A phonograph record rotates at a constant angular speed. Popular angular speeds were $33\tfrac{1}{3}$ rev/min for long-playing albums (hence the nickname "LP"), 45 rev/min for "singles," and 78 rev/min used in very early recordings. At the outer edge of the record, the pickup needle (stylus) moves over the vinyl material at a faster tangential speed than when the needle is close to the center of the record. As a result, the sound information is compressed into a smaller length of track near the center of the record than near the outer edge.

CDs, on the other hand, are designed so that the disc moves under the laser pickup at a constant tangential speed. Because the pickup moves radially as it follows the tracks of information, the angular speed of the compact disc must vary according to the radial position of the laser. Because the tangential speed is fixed, the information density (per length of track) anywhere on the disc is the same. Example 7.4 demonstrates numerical calculations for both compact discs and phonograph records.

EXAMPLE 7.4 | Track Length of a Compact Disc

GOAL Relate angular to linear variables.

PROBLEM In a compact disc player, as the read head moves out from the center of the disc, the angular speed of the disc changes so that the linear speed at the position of the head remains at a constant value of about 1.3 m/s. **(a)** Find the angular speed of a compact disc of radius 6.00 cm when the read head is at $r = 2.0$ cm and again at $r = 5.6$ cm. **(b)** An old-fashioned record player rotates at a constant angular speed, so the linear speed of the record groove moving under the detector (the stylus) changes. Find the linear speed of a 45.0-rpm record at points 2.0 and 5.6 cm from the center. **(c)** In both the CDs and phonograph records, information is recorded in a continuous spiral track. Calculate the total length of the track for a CD designed to play for 1.0 h.

STRATEGY This problem is just a matter of substituting numbers into the appropriate equations. Part (a) requires relating angular and linear speed with Equation 7.10, $v_t = r\omega$, solving for ω and substituting given values. In part (b), convert from rev/min to rad/s and substitute straight into Equation 7.10 to obtain the linear speeds. In part (c), linear speed multiplied by time gives the total distance.

SOLUTION

(a) Find the angular speed of the disc when the read head is at $r = 2.0$ cm and $r = 5.6$ cm.

Solve $v_t = r\omega$ for ω and calculate the angular speed at $r = 2.0$ cm:

$$\omega = \frac{v_t}{r} = \frac{1.3 \text{ m/s}}{2.0 \times 10^{-2} \text{ m}} = \boxed{65 \text{ rad/s}}$$

Likewise, find the angular speed at $r = 5.6$ cm:

$$\omega = \frac{v_t}{r} = \frac{1.3 \text{ m/s}}{5.6 \times 10^{-2} \text{ m}} = \boxed{23 \text{ rad/s}}$$

(b) Find the linear speed in m/s of a 45.0-rpm record at points 2.0 cm and 5.6 cm from the center.

Convert rev/min to rad/s:

$$45.0 \frac{\text{rev}}{\text{min}} = 45.0 \frac{\text{rev}}{\text{min}} \left(\frac{2\pi \text{ rad}}{\text{rev}}\right)\left(\frac{1.00 \text{ min}}{60.0 \text{ s}}\right) = 4.71 \frac{\text{rad}}{\text{s}}$$

Calculate the linear speed at $r = 2.0$ cm:

$$v_t = r\omega = (2.0 \times 10^{-2} \text{ m})(4.71 \text{ rad/s}) = \boxed{0.094 \text{ m/s}}$$

Calculate the linear speed at $r = 5.6$ cm:

$$v_t = r\omega = (5.6 \times 10^{-2} \text{ m})(4.71 \text{ rad/s}) = \boxed{0.26 \text{ m/s}}$$

(c) Calculate the total length of the track for a CD designed to play for 1.0 h.

Multiply the linear speed of the read head by the time in seconds:

$$d = v_t t = (1.3 \text{ m/s})(3\,600 \text{ s}) = \boxed{4\,700 \text{ m}}$$

REMARKS Notice that for the record player in part (b), even though the angular speed is constant at all points along a radial line, the tangential speed steadily increases with increasing r. The calculation for a CD in part (c) is easy only because the linear (tangential) speed is constant. It would be considerably more difficult for a record player, where the tangential speed depends on the distance from the center.

QUESTION 7.4 What is the angular acceleration of a record player while it's playing a song? Can a CD player have the same angular acceleration as a record player? Explain.

EXERCISE 7.4 Compute the linear speed on a record playing at $33\frac{1}{3}$ revolutions per minute (a) at $r = 2.00$ cm and (b) at $r = 5.60$ cm.

ANSWERS (a) 0.069 8 m/s (b) 0.195 m/s

7.4 Centripetal Acceleration

LEARNING OBJECTIVES

1. Calculate the centripetal, tangential, and total accelerations of objects in circular motion.
2. Apply the second law to objects in uniform circular motion.
3. Identify forces responsible for centripetal accelerations in physical contexts.

Figure 7.6a shows a car moving in a circular path with *constant* linear speed v. **Even though the car moves at a constant speed, it still has an acceleration.** To understand this, consider the defining equation for average acceleration:

$$\vec{a}_{av} = \frac{\vec{v}_f - \vec{v}_i}{t_f - t_i} \qquad [7.12]$$

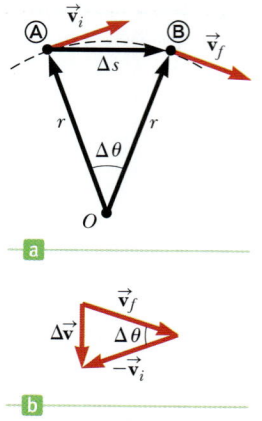

Figure 7.6 (a) Circular motion of a car moving with constant speed. (b) As the car moves along the circular path from Ⓐ to Ⓑ, the direction of its velocity vector changes, so the car undergoes a centripetal acceleration.

The numerator represents the difference between the velocity vectors $\vec{v}_f$ and $\vec{v}_i$. These vectors may have the same *magnitude*, corresponding to the same speed, but if they have different *directions*, their difference can't equal zero. The direction of the car's velocity as it moves in the circular path is continually changing, as shown in Figure 7.6b. For circular motion at constant speed, the acceleration vector always points toward the center of the circle. Such an acceleration is called a **centripetal** (center-seeking) **acceleration**. Its magnitude is given by

$$a_c = \frac{v^2}{r} \qquad [7.13]$$

To derive Equation 7.13, consider Figure 7.7a. An object is first at point Ⓐ with velocity $\vec{v}_i$ at time t_i and then at point Ⓑ with velocity $\vec{v}_f$ at a later time t_f. We assume $\vec{v}_i$ and $\vec{v}_f$ differ only in direction; their magnitudes are the same ($v_i = v_f = v$). To calculate the acceleration, we begin with Equation 7.12,

$$\vec{a}_{av} = \frac{\vec{v}_f - \vec{v}_i}{t_f - t_i} = \frac{\Delta \vec{v}}{\Delta t} \qquad [7.14]$$

where $\Delta \vec{v} = \vec{v}_f - \vec{v}_i$ is the change in velocity. When Δt is very small, Δs and $\Delta \theta$ are also very small. In Figure 7.7b $\vec{v}_f$ is almost parallel to $\vec{v}_i$, and the vector $\Delta \vec{v}$ is approximately perpendicular to them, pointing toward the center of the circle. In the limiting case when Δt becomes vanishingly small, $\Delta \vec{v}$ points exactly toward the center of the circle, and the average acceleration $\vec{a}_{av}$ becomes the instantaneous acceleration $\vec{a}$. From Equation 7.14, $\vec{a}$ and $\Delta \vec{v}$ point in the same direction (in this limit), so the instantaneous acceleration points to the center of the circle.

The triangle in Figure 7.7a, which has sides Δs and r, is similar to the one formed by the vectors in Figure 7.7b, so the ratios of their sides are equal:

$$\frac{\Delta v}{v} = \frac{\Delta s}{r}$$

or

$$\Delta v = \frac{v}{r} \Delta s \qquad [7.15]$$

Substituting the result of Equation 7.15 into $a_{av} = \Delta v/\Delta t$ gives

$$a_{av} = \frac{v}{r} \frac{\Delta s}{\Delta t} \qquad [7.16]$$

But Δs is the distance traveled along the arc of the circle in time Δt, and in the limiting case when Δt becomes very small, $\Delta s/\Delta t$ approaches the instantaneous value of the tangential speed, v. At the same time, the average acceleration a_{av} approaches a_c, the instantaneous centripetal acceleration, so Equation 7.16 reduces to Equation 7.13:

$$a_c = \frac{v^2}{r}$$

Because the tangential speed is related to the angular speed through the relation $v_t = r\omega$ (Eq. 7.10), an alternate form of Equation 7.13 is

$$a_c = \frac{r^2 \omega^2}{r} = r\omega^2 \qquad [7.17]$$

Dimensionally, $[r] = L$ and $[\omega] = 1/T$, so the units of centripetal acceleration are L/T^2, as they should be. This is a geometric result relating the centripetal acceleration to the angular speed, but physically an acceleration can occur only if some force is present. For example, if a car travels in a circle on flat ground, the force of static friction between the tires and the ground provides the necessary centripetal force.

Note that a_c in Equations 7.13 and 7.17 represents only the *magnitude* of the centripetal acceleration. The acceleration itself is always directed toward the center of rotation.

Figure 7.7 (a) As the particle moves from Ⓐ to Ⓑ, the direction of its velocity vector changes from $\vec{v}_i$ to $\vec{v}_f$. (b) The construction for determining the direction of the change in velocity $\Delta \vec{v}$, which is toward the center of the circle.

7.4 | Centripetal Acceleration

The foregoing derivations concern circular motion at constant speed. When an object moves in a circle but is speeding up or slowing down, a tangential component of acceleration, $a_t = r\alpha$, is also present. Because the tangential and centripetal components of acceleration are perpendicular to each other, we can find the magnitude of the **total acceleration** with the Pythagorean theorem:

$$a = \sqrt{a_t^2 + a_c^2} \qquad [7.18]$$ ◀ Total acceleration

■ Quick Quiz

7.6 A racetrack is constructed such that two arcs of radius 80 m at Ⓐ and 40 m at Ⓑ are joined by two stretches of straight track as in Figure 7.8. In a particular trial run, a driver travels at a constant speed of 50 m/s for one complete lap.
1. The ratio of the tangential acceleration at Ⓐ to that at Ⓑ is
   (a) $\frac{1}{2}$ (b) $\frac{1}{4}$ (c) 2 (d) 4 (e) The tangential acceleration is zero at both points.
2. The ratio of the centripetal acceleration at Ⓐ to that at Ⓑ is
   (a) $\frac{1}{2}$ (b) $\frac{1}{4}$ (c) 2 (d) 4 (e) The centripetal acceleration is zero at both points.
3. The angular speed is greatest at
   (a) Ⓐ (b) Ⓑ (c) It is equal at both Ⓐ and Ⓑ.

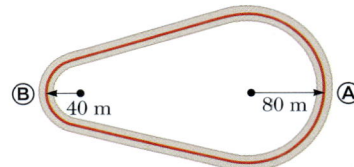

Figure 7.8 (Quick Quiz 7.6)

7.7 An object moves in a circular path with constant speed v. Which of the following statements is true concerning the object? (a) Its velocity is constant, but its acceleration is changing. (b) Its acceleration is constant, but its velocity is changing. (c) Both its velocity and acceleration are changing. (d) Its velocity and acceleration remain constant.

■ EXAMPLE 7.5 | At the Racetrack

GOAL Apply the concepts of centripetal acceleration and tangential speed.

PROBLEM A race car accelerates uniformly from a speed of 40.0 m/s to a speed of 60.0 m/s in 5.00 s while traveling counterclockwise around a circular track of radius 4.00×10^2 m. When the car reaches a speed of 50.0 m/s, calculate (a) the magnitude of the car's centripetal acceleration, (b) the angular speed, (c) the magnitude of the tangential acceleration, and (d) the magnitude of the total acceleration.

STRATEGY Substitute values into the definitions of centripetal acceleration (Eq. 7.13), tangential speed (Eq. 7.10), and total acceleration (Eq. 7.18). Dividing the change in linear speed by the time yields the tangential acceleration.

SOLUTION

(a) Calculate the magnitude of the centripetal acceleration when $v = 50.0$ m/s.

Substitute into Equation 7.13:
$$a_c = \frac{v^2}{r} = \frac{(50.0 \text{ m/s})^2}{4.00 \times 10^2 \text{ m}} = \boxed{6.25 \text{ m/s}^2}$$

(b) Calculate the angular speed.

Solve Equation 7.10 for ω and substitute:
$$\omega = \frac{v}{r} = \frac{50.0 \text{ m/s}}{4.00 \times 10^2 \text{ m}} = \boxed{0.125 \text{ rad/s}}$$

(c) Calculate the magnitude of the tangential acceleration.

Divide the change in linear speed by the time:
$$a_t = \frac{v_f - v_i}{\Delta t} = \frac{60.0 \text{ m/s} - 40.0 \text{ m/s}}{5.00 \text{ s}} = \boxed{4.00 \text{ m/s}^2}$$

(d) Calculate the magnitude of the total acceleration.

Substitute into Equation 7.18:
$$a = \sqrt{a_t^2 + a_c^2} = \sqrt{(4.00 \text{ m/s}^2)^2 + (6.25 \text{ m/s}^2)^2}$$
$$a = \boxed{7.42 \text{ m/s}^2}$$

(Continued)

REMARKS We can also find the centripetal acceleration by substituting the derived value of ω into Equation 7.17.

QUESTION 7.5 If the force causing the centripetal acceleration suddenly vanished, would the car (a) slide away along a radius, (b) proceed along a line tangent to the circular motion, or (c) proceed at an angle intermediate between the tangent and radius?

EXERCISE 7.5 Suppose the race car now slows down uniformly from 60.0 m/s to 30.0 m/s in 4.50 s to avoid an accident, while still traversing a circular path 4.00×10^2 m in radius. Calculate the car's (a) centripetal acceleration, (b) angular speed, (c) tangential acceleration, and (d) total acceleration when the speed is 40.0 m/s.

ANSWERS (a) 4.00 m/s² (b) 0.100 rad/s (c) −6.67 m/s² (d) 7.78 m/s²

Figure 7.9 (a) The right-hand rule for determining the direction of the angular velocity vector $\vec{\omega}$. (b) The direction of $\vec{\omega}$ is in the direction of advance of a right-handed screw.

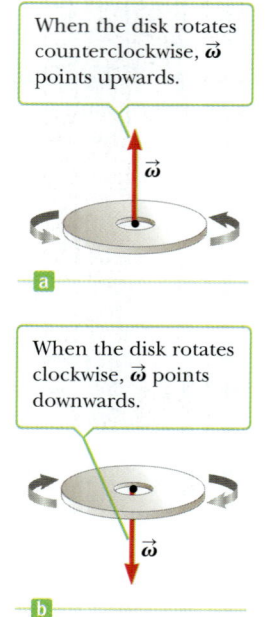

Figure 7.10 The direction of the angular velocity vector $\vec{\omega}$ depends on the direction of rotation.

When the disk rotates counterclockwise, $\vec{\omega}$ points upwards.

When the disk rotates clockwise, $\vec{\omega}$ points downwards.

Angular Quantities Are Vectors

When we discussed linear motion in Chapter 2, we emphasized that displacement, velocity, and acceleration are all vector quantities. In describing rotational motion, angular displacement, angular velocity, and angular acceleration are also vector quantities.

The direction of the angular velocity vector $\vec{\omega}$ can be found with the **right-hand rule,** as illustrated in Figure 7.9a. Grasp the axis of rotation with your right hand so that your fingers wrap in the direction of rotation. Your extended thumb then points in the direction of $\vec{\omega}$. Figure 7.9b shows that $\vec{\omega}$ is also in the direction of advance of a rotating right-handed screw.

We can apply this rule to a disk rotating about a vertical axis through its center, as in Figure 7.10. When the disk rotates counterclockwise (Fig. 7.10a), the right-hand rule shows that the direction of $\vec{\omega}$ is upward. When the disk rotates clockwise (Fig. 7.10b), the direction of $\vec{\omega}$ is downward.

Finally, the directions of the angular acceleration $\vec{\alpha}$ and the angular velocity $\vec{\omega}$ are the same if the angular speed ω (the magnitude of $\vec{\omega}$) is increasing with time, and are opposite each other if the angular speed is decreasing with time.

Forces Causing Centripetal Acceleration

An object can have a centripetal acceleration *only* if some external force acts on it. For a ball whirling in a circle at the end of a string, that force is the tension in the string. In the case of a car moving on a flat circular track, the force is friction between the car and track. A satellite in circular orbit around Earth has a centripetal acceleration due to the gravitational force between the satellite and Earth.

Some books use the term "centripetal force," which can give the mistaken impression that it is a new force of nature. This is not the case: The adjective "centripetal" in "centripetal force" simply means that the force in question acts toward a center. The force of tension in the string of a yo-yo whirling in a vertical circle is an example of a centripetal force, as is the force of gravity on a satellite circling the Earth.

Consider a puck of mass m that is tied to a string of length r and is being whirled at constant speed in a horizontal circular path, as illustrated in Figure 7.11. Its weight is supported by a frictionless table. Why does the puck move in a circle? Because of its inertia, the tendency of the puck is to move in a straight line; however, the string prevents motion along a straight line by exerting a radial force on the puck—a tension force—that makes it follow the circular path. The tension $\vec{T}$ is directed along the string toward *the center of the circle*, as shown in the figure.

In general, converting Newton's second law to polar coordinates yields an equation relating the net centripetal force, F_c, which is the sum of the radial components of all forces acting on a given object, to the centripetal acceleration. The *magnitude* of the net centripetal force equals the mass times the magnitude of the centripetal acceleration:

$$F_c = ma_c = m\frac{v^2}{r} \qquad [7.19]$$

A net force causing a centripetal acceleration acts toward the center of the circular path and effects a change in the direction of the velocity vector. If that force should vanish, the object would immediately leave its circular path and move along a straight line tangent to the circle at the point where the force vanished.

Centrifugal ('center-fleeing') forces also exist, such as the force between two particles with the same sign charge (see Chapter 15). The normal force that prevents an object from falling toward the center of the Earth is another example of a centrifugal force. Sometimes an insufficient centripetal force is mistaken for the presence of a centrifugal force (see "Fictitious Forces," page 219).

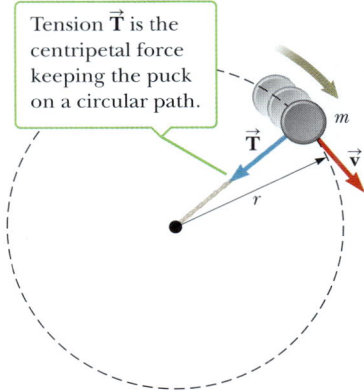

Figure 7.11 A puck attached to a string of length r rotates in a horizontal plane at constant speed.

> **Tip 7.2 Centripetal Force Is a *Type* of Force, Not a Force in Itself!**
> "Centripetal force" is a classification that includes forces acting toward a central point, like the horizontal component of the string tension on a tetherball or gravity on a satellite. A centripetal force must be *supplied* by some actual, physical force.

■ APPLYING PHYSICS 7.2 | Artificial Gravity

Astronauts spending lengthy periods of time in space experience a number of negative effects due to weightlessness, such as weakening of muscle tissue and loss of calcium in bones. These effects may make it very difficult for them to return to their usual environment on Earth. How could artificial gravity be generated in space to overcome such complications?

SOLUTION A rotating cylindrical space station creates an environment of artificial gravity. The normal force of the rigid walls provides the centripetal force, which keeps the astronauts moving in a circle (Fig. 7.12). To an astronaut, the normal force can't be easily distinguished from a gravitational force as long as the radius of the station is large compared with the astronaut's height. (Otherwise there are unpleasant inner ear effects.) This same principle is used in certain amusement park rides in which passengers are pressed against the inside of a rotating cylinder as it tilts in various directions. The visionary physicist Gerard O'Neill proposed creating a giant space colony a kilometer in radius that rotates slowly, creating Earth-normal artificial gravity for the inhabitants in its interior. These inside-out artificial worlds could enable safe transport on a several-thousand-year journey to another star system. ■

Figure 7.12 Artificial gravity inside a spinning cylinder is provided by the normal force.

■ PROBLEM-SOLVING STRATEGY

Forces That Cause Centripetal Acceleration
Use the following steps in dealing with centripetal accelerations and the forces that produce them:

1. **Draw a free-body diagram** of the object under consideration, labeling all forces that act on it.

2. **Choose a coordinate system** that has one axis perpendicular to the circular path followed by the object (the radial direction) and one axis tangent to the circular path (the tangential, or angular, direction). The normal direction, perpendicular to the plane of motion, is also often needed.
3. **Find the net force F_c toward the center** of the circular path, $F_c = \Sigma F_r$, where ΣF_r is the sum of the radial components of the forces. This net radial force causes the centripetal acceleration.
4. **Use Newton's second law for the radial, tangential, and normal directions,** as required, writing $\Sigma F_r = ma_c$, $\Sigma F_t = ma_t$, and $\Sigma F_n = ma_n$. Remember that the magnitude of the centripetal acceleration for uniform circular motion can always be written $a_c = v_t^2/r$.
5. **Solve** for the unknown quantities.

EXAMPLE 7.6 | Buckle Up for Safety

GOAL Calculate the frictional force that causes an object to have a centripetal acceleration.

PROBLEM A car travels at a constant speed of 30.0 mi/h (13.4 m/s) on a level circular turn of radius 50.0 m, as shown in the bird's-eye view in Figure 7.13a. What minimum coefficient of static friction, μ_s, between the tires and roadway will allow the car to make the circular turn without sliding?

STRATEGY In the car's free-body diagram (Fig. 7.13b) the normal direction is vertical and the tangential direction is into the page (Step 2). Use Newton's second law. The net force acting on the car in the radial direction is the force of static friction toward the center of the circular path, which causes the car to have a centripetal acceleration. Calculating the maximum static friction force requires the normal force, obtained from the normal component of the second law.

Figure 7.13 (Example 7.6) (a) The centripetal force is provided by the force of static friction, which is directed radially toward the center of the circular path. (b) Gravity, the normal force, and the static friction force act on the car.

SOLUTION

(Steps 3, 4) Write the components of Newton's second law. The radial component involves only the maximum static friction force, $f_{s,\max}$:

$$m\frac{v^2}{r} = f_{s,\max} = \mu_s n$$

In the vertical component of the second law, the gravity force and the normal force are in equilibrium:

$$n - mg = 0 \quad \rightarrow \quad n = mg$$

(Step 5) Substitute the expression for n into the first equation and solve for μ_s:

$$m\frac{v^2}{r} = \mu_s mg$$

$$\mu_s = \frac{v^2}{rg} = \frac{(13.4 \text{ m/s})^2}{(50.0 \text{ m})(9.80 \text{ m/s}^2)} = \boxed{0.366}$$

REMARKS The value of μ_s for rubber on dry concrete is very close to 1, so the car can negotiate the curve with ease. If the road were wet or icy, however, the value for μ_s could be 0.2 or lower. Under such conditions, the radial force provided by static friction wouldn't be great enough to keep the car on the circular path, and it would slide off on a tangent, leaving the roadway.

QUESTION 7.6 If the static friction coefficient were increased, would the maximum safe speed be reduced, increased, or remain the same?

EXERCISE 7.6 At what maximum speed can a car negotiate a turn on a wet road with coefficient of static friction 0.230 without sliding out of control? The radius of the turn is 25.0 m.

ANSWER 7.51 m/s

EXAMPLE 7.7 | Daytona International Speedway

GOAL Solve a centripetal force problem involving two dimensions.

PROBLEM The Daytona International Speedway in Daytona Beach, Florida, is famous for its races, especially the Daytona 500, held every February. Both of its courses feature four-story, 31.0° banked curves, with maximum radius of 316 m. If a car negotiates the curve too slowly, it tends to slip down the incline of the turn, whereas if it's going too fast, it may begin to slide up the incline. **(a)** Find the necessary centripetal acceleration on this banked curve so the car won't tend to slip down or slide up the incline. (Neglect friction.) **(b)** Calculate the speed of the race car.

Figure 7.14 (Example 7.7) As the car rounds a curve banked at an angle θ, the centripetal force that keeps it on a circular path is supplied by the radial component of the normal force. Friction also contributes, although is neglected in this example. The car is moving forward, into the page. (a) Force diagram for the car. (b) Components of the forces.

STRATEGY Two forces act on the race car: the force of gravity and the normal force $\vec{n}$. (See Fig. 7.14.) Use Newton's second law in the upward and radial directions to find the centripetal acceleration a_c. Solving $a_c = v^2/r$ for v then gives the race car's speed.

SOLUTION

(a) Find the centripetal acceleration.

Write Newton's second law for the car:
$$m\vec{a} = \sum \vec{F} = \vec{n} + m\vec{g}$$

Use the y-component of Newton's second law to solve for the normal force n:
$$n \cos \theta - mg = 0$$
$$n = \frac{mg}{\cos \theta}$$

Obtain an expression for the horizontal component of $\vec{n}$, which is the centripetal force F_c in this example:
$$F_c = n \sin \theta = \frac{mg \sin \theta}{\cos \theta} = mg \tan \theta$$

Substitute this expression for F_c into the radial component of Newton's second law and divide by m to get the centripetal acceleration:
$$ma_c = F_c$$
$$a_c = \frac{F_c}{m} = \frac{mg \tan \theta}{m} = g \tan \theta$$
$$a_c = (9.80 \text{ m/s}^2)(\tan 31.0°) = \boxed{5.89 \text{ m/s}^2}$$

(b) Find the speed of the race car.

Apply Equation 7.13:
$$\frac{v^2}{r} = a_c$$
$$v = \sqrt{ra_c} = \sqrt{(316 \text{ m})(5.89 \text{ m/s}^2)} = \boxed{43.1 \text{ m/s}}$$

(Continued)

REMARKS In fact, both banking and friction assist in keeping the race car on the track.

QUESTION 7.7 What three physical quantities determine the minimum and maximum safe speeds on a banked racetrack?

EXERCISE 7.7 A racetrack is to have a banked curve with radius of 245 m. What should be the angle of the bank if the normal force alone is to allow safe travel around the curve at 58.0 m/s?

ANSWER 54.5°

APPLICATION
Banked Roadways

EXAMPLE 7.8 Riding the Tracks

GOAL Combine centripetal force with conservation of energy. Derive results symbolically.

PROBLEM Figure 7.15a shows a roller-coaster car moving around a circular loop of radius R. **(a)** What speed must the car have at the top of the loop so that it will just make it over the top without any assistance from the track? **(b)** What speed will the car subsequently have at the bottom of the loop? **(c)** What will be the normal force on a passenger at the bottom of the loop if the loop has a radius of 10.0 m?

STRATEGY This problem requires Newton's second law and centripetal acceleration to find an expression for the car's speed at the top of the loop, followed by conservation of energy to find its speed at the bottom. If the car just makes it over the top, the force $\vec{n}$ must become zero there, so the only force exerted on the car at that point is the force of gravity, $m\vec{g}$. At the bottom of the loop, the normal force acts up toward the center and the gravity force acts down, away from the center. The difference of these two is the centripetal force. The normal force can then be calculated from Newton's second law.

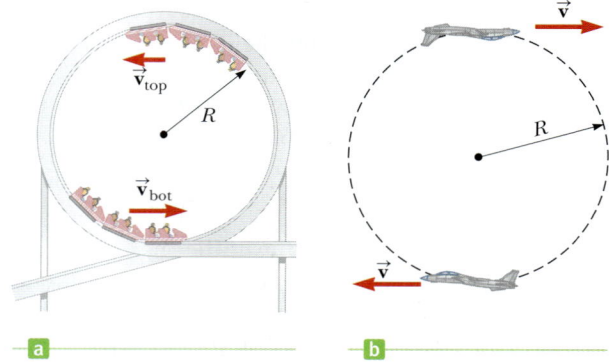

Figure 7.15 (a) (Example 7.8) A roller coaster traveling around a nearly circular track. (b) (Exercise 7.8) A jet executing a vertical loop.

SOLUTION

(a) Find the speed at the top of the loop.

Write Newton's second law for the car:

$$(1) \quad m\vec{a}_c = \vec{n} + m\vec{g}$$

At the top of the loop, set $n = 0$. The force of gravity acts toward the center and provides the centripetal acceleration $a_c = v^2/R$:

$$m\frac{v_{top}^2}{R} = mg$$

Solve the foregoing equation for v_{top}:

$$v_{top} = \boxed{\sqrt{gR}}$$

(b) Find the speed at the bottom of the loop.

Apply conservation of mechanical energy to find the total mechanical energy at the top of the loop:

$$E_{top} = \tfrac{1}{2}mv_{top}^2 + mgh = \tfrac{1}{2}mgR + mg(2R) = 2.5\,mgR$$

Find the total mechanical energy at the bottom of the loop:

$$E_{bot} = \tfrac{1}{2}mv_{bot}^2$$

Energy is conserved, so these two energies may be equated and solved for v_{bot}:

$$\tfrac{1}{2}mv_{bot}^2 = 2.5\,mgR$$

$$v_{bot} = \boxed{\sqrt{5gR}}$$

(c) Find the normal force on a passenger at the bottom. (This is the passenger's perceived weight.)

Use Equation (1). The net centripetal force is $n - mg$:

$$m\frac{v_{bot}^2}{R} = n - mg$$

Solve for n:
$$n = mg + m\frac{v_{bot}^2}{R} = mg + m\frac{5gR}{R} = 6mg$$

REMARKS The final answer for n shows that the rider experiences a force six times normal weight at the bottom of the loop! Astronauts experience a similar force during space launches.

QUESTION 7.8 Suppose the car subsequently goes over a rise with the same radius of curvature and at the same speed as part (a). What is the normal force in this case?

EXERCISE 7.8 A jet traveling at a constant speed of 1.20×10^2 m/s executes a vertical loop with a radius of 5.00×10^2 m. (See Fig. 7.15b.) Find the magnitude of the force of the seat on a 70.0-kg pilot at (a) the top and (b) the bottom of the loop.

ANSWERS (a) 1.33×10^3 N (b) 2.70×10^3 N

Fictitious Forces

Anyone who has ridden a merry-go-round as a child (or as a fun-loving grown-up) has experienced what feels like a "center-fleeing" force. Holding onto the railing and moving toward the center feels like a walk up a steep hill.

Actually, this so-called centrifugal force is *fictitious*. In reality, the rider is exerting a centripetal force on her body with her hand and arm muscles. In addition, a smaller centripetal force is exerted by the static friction between her feet and the platform. If the rider's grip slipped, she wouldn't be flung radially away; rather, she would go off on a straight line, tangent to the point in space where she let go of the railing. The rider lands at a point that is farther away from the center, but not by "fleeing the center" along a radial line. Instead, she travels perpendicular to a radial line, traversing an angular displacement while increasing her radial displacement. (See Fig. 7.16.)

> **Tip 7.3 Centrifugal Force**
> A so-called centrifugal force is very often just the *absence* of an adequate *centripetal force*, arising from measuring phenomena from a noninertial (accelerating) frame of reference such as a merry-go-round.

Figure 7.16 A fun-loving student loses her grip and falls along a line tangent to the rim of the merry-go-round.

7.5 Newtonian Gravitation

LEARNING OBJECTIVES
1. Apply the law of gravitation to calculate gravitational forces and their consequences.
2. Apply the general form of gravitational potential energy to the motion of interacting bodies.

Prior to 1686, a great deal of data had been collected on the motions of the Moon and planets, but no one had a clear understanding of the forces affecting them. In that year, Isaac Newton provided the key that unlocked the secrets of the heavens. He knew from the first law that a net force had to be acting on the Moon. If it were not, the Moon would move in a straight-line path rather than in its almost circular orbit around Earth. Newton reasoned that it was the same kind of force that attracted objects—such as apples—close to the surface of the Earth. He called it the force of gravity.

In 1687 Newton published his work on the law of universal gravitation:

> If two particles with masses m_1 and m_2 are separated by a distance r a gravitational force F acts along a line joining them, with magnitude given by
>
> $$F = G\frac{m_1 m_2}{r^2} \qquad [7.20]$$
>
> where $G = 6.673 \times 10^{-11}$ kg$^{-1} \cdot$ m$^3 \cdot$ s^{-2} is a constant of proportionality called the **constant of universal gravitation**. The gravitational force is always attractive.

This force law is an example of an **inverse-square law,** in that it varies as one over the square of the distance between particles. From Newton's third law, we know that the force exerted by m_1 on m_2, designated $\vec{F}_{12}$ in Figure 7.17, is equal in magnitude

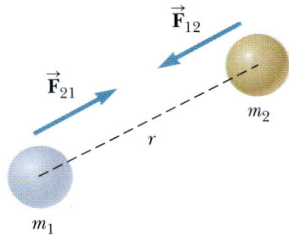

Figure 7.17 The gravitational force between two particles is attractive and acts along the line joining the particles. Note that according to Newton's third law, $\vec{F}_{12} = -\vec{F}_{21}$.

Table 7.1 Free-Fall Acceleration g at Various Altitudes

Altitude (km)[a]	g (m/s^2)
1 000	7.33
2 000	5.68
3 000	4.53
4 000	3.70
5 000	3.08
6 000	2.60
7 000	2.23
8 000	1.93
9 000	1.69
10 000	1.49
50 000	0.13

[a]All figures are distances above Earth's surface.

but opposite in direction to the force $\vec{F}_{21}$ exerted by m_2 on m_1, forming an action–reaction pair.

Another important fact is that **the gravitational force exerted by a uniform sphere on a particle outside the sphere is the same as the force exerted if the entire mass of the sphere were concentrated at its center.** This is called Gauss's law, after the German mathematician and astronomer Karl Friedrich Gauss, and is also true of electric fields, which we will encounter in Chapter 15. Gauss's law is a mathematical result, true because the force falls off as an inverse square of the separation between the particles.

Near the surface of the Earth, the expression $F = mg$ is valid. As shown in Table 7.1, however, the free-fall acceleration g varies considerably with altitude above the Earth.

Quick Quiz

7.8 A ball is falling toward the ground. Which of the following statements are false? (a) The force that the ball exerts on Earth is equal in magnitude to the force that Earth exerts on the ball. (b) The ball undergoes the same acceleration as Earth. (c) The magnitude of the force the Earth exerts on the ball is greater than the magnitude of the force the ball exerts on the Earth.

7.9 A planet has two moons with identical mass. Moon 1 is in a circular orbit of radius r. Moon 2 is in a circular orbit of radius $2r$. The magnitude of the gravitational force exerted by the planet on Moon 2 is (a) four times as large (b) twice as large (c) the same (d) half as large (e) one-fourth as large as the gravitational force exerted by the planet on Moon 1.

Measurement of the Gravitational Constant

The gravitational constant G in Equation 7.20 was first measured in an important experiment by Henry Cavendish in 1798. His apparatus consisted of two small spheres, each of mass m, fixed to the ends of a light horizontal rod suspended by a thin metal wire, as in Figure 7.18. Two large spheres, each of mass M, were placed near the smaller spheres. The attractive force between the smaller and larger spheres caused the rod to rotate in a horizontal plane and the wire to twist. The angle through which the suspended rod rotated was measured with a light beam reflected from a mirror attached to the vertical suspension. (Such a moving spot of light is an effective technique for amplifying motion.) The experiment was carefully repeated with different masses at various separations. In addition to providing a value for G, the results showed that the force is attractive, proportional to the product mM, and inversely proportional to the square of the distance r. Modern forms of such experiments are carried out regularly today in an effort to determine G with greater precision.

Figure 7.18 A schematic diagram of the Cavendish apparatus for measuring G. The smaller spheres of mass m are attracted to the large spheres of mass M, and the rod rotates through a small angle. A light beam reflected from a mirror on the rotating apparatus measures the angle of rotation.

EXAMPLE 7.9 | Billiards, Anyone?

GOAL Use vectors to find the net gravitational force on an object.

PROBLEM (a) Three 0.300-kg billiard balls are placed on a table at the corners of a right triangle, as shown from overhead in Figure 7.19. Find the net gravitational force on the cue ball (designated as m_1) resulting from the forces exerted by the other two balls. (b) Find the components of the gravitational force of m_2 on m_3.

STRATEGY (a) To find the net gravitational force on the cue ball of mass m_1, we first calculate the force $\vec{F}_{21}$ exerted by m_2 on m_1. This force is the y-component of the net force acting on m_1. Then we find the force $\vec{F}_{31}$ exerted by m_3 on m_1, which is the x-component of the net force acting on m_1. With these two components, we can find the magnitude and direction of the net force on the cue ball. (b) In this case, we must use trigonometry to find the components of the force $\vec{F}_{23}$.

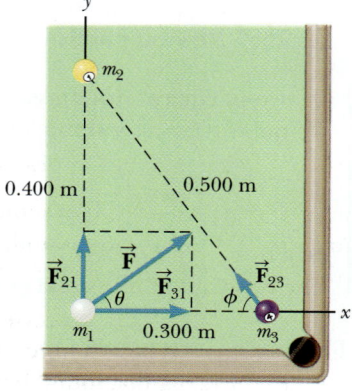

Figure 7.19 (Example 7.9)

SOLUTION

(a) Find the net gravitational force on the cue ball.

Find the magnitude of the force $\vec{F}_{21}$ exerted by m_2 on m_1 using the law of gravitation, Equation 7.20:

$$F_{21} = G\frac{m_2 m_1}{r_{21}^2} = (6.67 \times 10^{-11} \text{ N} \cdot \text{m}^2/\text{kg}^2)\frac{(0.300 \text{ kg})(0.300 \text{ kg})}{(0.400 \text{ m})^2}$$

$$F_{21} = 3.75 \times 10^{-11} \text{ N}$$

Find the magnitude of the force $\vec{F}_{31}$ exerted by m_3 on m_1, again using Newton's law of gravity:

$$F_{31} = G\frac{m_3 m_1}{r_{31}^2} = (6.67 \times 10^{-11} \text{ N} \cdot \text{m}^2/\text{kg}^2)\frac{(0.300 \text{ kg})(0.300 \text{ kg})}{(0.300 \text{ m})^2}$$

$$F_{31} = 6.67 \times 10^{-11} \text{ N}$$

The net force has components $F_x = F_{31}$ and $F_y = F_{21}$. Compute the magnitude of this net force:

$$F = \sqrt{F_x^2 + F_y^2} = \sqrt{(6.67)^2 + (3.75)^2} \times 10^{-11} \text{ N}$$

$$= \boxed{7.65 \times 10^{-11} \text{ N}}$$

Use the inverse tangent to obtain the direction of $\vec{F}$:

$$\theta = \tan^{-1}\left(\frac{F_y}{F_x}\right) = \tan^{-1}\left(\frac{3.75 \times 10^{-11} \text{ N}}{6.67 \times 10^{-11} \text{ N}}\right) = \boxed{29.3°}$$

(b) Find the components of the force of m_2 on m_3.

First, compute the magnitude of $\vec{F}_{23}$:

$$F_{23} = G\frac{m_2 m_1}{r_{23}^2}$$

$$= (6.67 \times 10^{-11} \text{ kg}^{-1}\text{m}^3\text{s}^{-2})\frac{(0.300 \text{ kg})(0.300 \text{ kg})}{(0.500 \text{ m})^2}$$

$$= 2.40 \times 10^{-11} \text{ N}$$

To obtain the x- and y-components of F_{23}, we need $\cos\varphi$ and $\sin\varphi$. Use the sides of the large triangle in Figure 7.19:

$$\cos\varphi = \frac{\text{adj}}{\text{hyp}} = \frac{0.300 \text{ m}}{0.500 \text{ m}} = 0.600$$

$$\sin\varphi = \frac{\text{opp}}{\text{hyp}} = \frac{0.400 \text{ m}}{0.500 \text{ m}} = 0.800$$

Compute the components of $\vec{F}_{23}$. A minus sign must be supplied for the x-component because it's in the negative x-direction.

$$F_{23x} = -F_{23}\cos\varphi = -(2.40 \times 10^{-11} \text{ N})(0.600)$$

$$= \boxed{-1.44 \times 10^{-11} \text{ N}}$$

$$F_{23y} = F_{23}\sin\varphi = (2.40 \times 10^{-11} \text{ N})(0.800) = \boxed{1.92 \times 10^{-11} \text{ N}}$$

REMARKS Notice how small the gravity forces are between everyday objects. Nonetheless, such forces can be measured directly with torsion balances.

QUESTION 7.9 Is the gravity force a significant factor in a game of billiards? Explain.

EXERCISE 7.9 Find magnitude and direction of the force exerted by m_1 and m_3 on m_2.

ANSWERS 5.85×10^{-11} N, $-75.8°$

■ EXAMPLE 7.10 | Ceres

GOAL Relate Newton's universal law of gravity to mg and show how g changes with position.

PROBLEM An astronaut standing on the surface of Ceres, the largest asteroid, drops a rock from a height of 10.0 m. It takes 8.06 s to hit the ground. **(a)** Calculate the acceleration of gravity on Ceres. **(b)** Find the mass of Ceres, given that the radius of Ceres is $R_C = 5.10 \times 10^2$ km. **(c)** Calculate the gravitational acceleration 50.0 km from the surface of Ceres.

STRATEGY Part (a) is a review of one-dimensional kinematics. In part (b) the weight of an object, $w = mg$, is the same as the magnitude of the force given by the universal law of gravity. Solve for the unknown mass of Ceres, after which the answer for (c) can be found by substitution into the universal law of gravity, Equation 7.20.

(Continued)

SOLUTION

(a) Calculate the acceleration of gravity, g_C, on Ceres.

Apply the kinematics equation of displacement to the falling rock:

(1) $\quad \Delta x = \tfrac{1}{2}at^2 + v_0 t$

Substitute $\Delta x = -10.0$ m, $v_0 = 0$, $a = -g_C$, and $t = 8.06$ s, and solve for the gravitational acceleration on Ceres, g_C:

$-10.0 \text{ m} = -\tfrac{1}{2}g_C(8.06 \text{ s})^2 \quad \rightarrow \quad g_C = \boxed{0.308 \text{ m/s}^2}$

(b) Find the mass of Ceres.

Equate the weight of the rock on Ceres to the gravitational force acting on the rock:

$$mg_C = G\frac{M_C m}{R_C^2}$$

Solve for the mass of Ceres, M_C:

$$M_C = \frac{g_C R_C^2}{G} = \boxed{1.20 \times 10^{21} \text{ kg}}$$

(c) Calculate the acceleration of gravity at a height of 50.0 km above the surface of Ceres.

Equate the weight at 50.0 km to the gravitational force:

$$mg'_C = G\frac{mM_C}{r^2}$$

Cancel m, then substitute $r = 5.60 \times 10^5$ m and the mass of Ceres:

$$g'_C = G\frac{M_C}{r^2}$$

$$= (6.67 \times 10^{-11} \text{ kg}^{-1}\text{m}^3\text{s}^{-2})\frac{1.20 \times 10^{21} \text{ kg}}{(5.60 \times 10^5 \text{ m})^2}$$

$$= \boxed{0.255 \text{ m/s}^2}$$

REMARKS This is the standard method of finding the mass of a planetary body: study the motion of a falling (or orbiting) object.

QUESTION 7.10 Give two reasons Equation (1) could not be used for every asteroid as it is used in part (a).

EXERCISE 7.10 An object takes 2.40 s to fall 5.00 m on a certain planet. (a) Find the acceleration due to gravity on the planet. (b) Find the planet's mass if its radius is 5 250 km.

ANSWERS (a) 1.74 m/s² (b) 7.19×10^{23} kg

Gravitational Potential Energy Revisited

In Chapter 5 we introduced the concept of gravitational potential energy and found that the potential energy associated with an object could be calculated from the equation $PE = mgh$, where h is the height of the object above or below some reference level. This equation, however, is valid *only* when the object is near Earth's surface. For objects high above Earth's surface, such as a satellite, an alternative must be used because g varies with distance from the surface, as shown in Table 7.1.

The gravitational potential energy associated with an object of mass m at a distance r from the center of Earth is

$$PE = -G\frac{M_E m}{r} \quad [7.21]$$

where M_E and R_E are the mass and radius of Earth, respectively, with $r > R_E$.

SI units: Joules (J)

Figure 7.20 As a mass m moves radially away from the Earth, the potential energy of the Earth-mass system, which is $PE = -G(M_E m/R_E)$ at Earth's surface, increases toward a limit of zero as the mass m travels away from Earth, as shown in the graph.

As before, gravitational potential energy is a property of a *system*, in this case the object of mass m and Earth. Equation 7.21, illustrated in Figure 7.20, is valid for the special case where the zero level for potential energy is at an infinite distance from the center of Earth. Recall that the gravitational potential energy associated with an object is nothing more than the negative of the work done by the force of gravity in moving the object. If an object falls under the force of gravity from a great distance (effectively infinity), the change in gravitational potential energy is negative, which corresponds to a positive amount of gravitational work done on the system. This positive work is equal to the (also positive) change in kinetic energy, as the next example shows.

EXAMPLE 7.11 | A Near-Earth Asteroid

GOAL Use gravitational potential energy to calculate the work done by gravity on a falling object.

PROBLEM An asteroid with mass $m = 1.00 \times 10^9$ kg comes from deep space, effectively from infinity, and falls toward Earth. **(a)** Find the change in potential energy when it reaches a point 4.00×10^8 m from the center of the Earth (just beyond the orbital radius of the Moon). In addition, find the work done by the force of gravity. **(b)** Calculate the asteroid's speed at that point, assuming it was initially at rest when it was arbitrarily far away. **(c)** How much work would have to be done on the asteroid by some other agent so the asteroid would be traveling at only half the speed found in (b) at the same point?

STRATEGY Part (a) requires simple substitution into the definition of gravitational potential energy. To find the work done by the force of gravity, recall that the work done on an object by a conservative force is just the negative of the change in potential energy. Part (b) can be solved with conservation of energy, and part (c) is an application of the work–energy theorem.

SOLUTION

(a) Find the change in potential energy and the work done by the force of gravity.

Apply Equation 7.21:

$$\Delta PE = PE_f - PE_i = -\frac{GM_E m}{r_f} - \left(-\frac{GM_E m}{r_i}\right)$$

$$= GM_E m \left(-\frac{1}{r_f} + \frac{1}{r_i}\right)$$

Substitute known quantities. The asteroid's initial position is effectively infinity, so $1/r_i$ is zero:

$$\Delta PE = (6.67 \times 10^{-11} \text{ kg}^{-1} \text{ m}^3/\text{s}^2)(5.98 \times 10^{24} \text{ kg})$$

$$\times (1.00 \times 10^9 \text{ kg})\left(-\frac{1}{4.00 \times 10^8 \text{ m}} + 0\right)$$

$$\Delta PE = \boxed{-9.97 \times 10^{14} \text{ J}}$$

Compute the work done by the force of gravity:

$$W_{grav} = \boxed{-\Delta PE = 9.97 \times 10^{14} \text{ J}}$$

(b) Find the speed of the asteroid when it reaches $r_f = 4.00 \times 10^8$ m.

Use conservation of energy:

$$\Delta KE + \Delta PE = 0$$

$$(\tfrac{1}{2}mv^2 - 0) - 9.97 \times 10^{14} \text{ J} = 0$$

$$v = \boxed{1.41 \times 10^3 \text{ m/s}}$$

(c) Find the work needed to reduce the speed to 7.05×10^2 m/s (half the value just found) at this point.

Apply the work–energy theorem:

$$W = \Delta KE + \Delta PE$$

The change in potential energy remains the same as in part (a), but substitute only half the speed in the kinetic-energy term:

$$W = (\tfrac{1}{2}mv^2 - 0) - 9.97 \times 10^{14} \text{ J}$$

$$W = \tfrac{1}{2}(1.00 \times 10^9 \text{ kg})(7.05 \times 10^2 \text{ m/s})^2 - 9.97 \times 10^{14} \text{ J}$$

$$= \boxed{-7.48 \times 10^{14} \text{ J}}$$

REMARKS The amount of work calculated in part (c) is negative because an external agent must exert a force against the direction of motion of the asteroid. It would take a thruster with a megawatt of output about 24 years to slow down

(Continued)

the asteroid to half its original speed. An asteroid endangering Earth need not be slowed that much: A small change in its speed, if applied early enough, will cause it to miss Earth. Timeliness of the applied thrust, however, is important. By the time an astronaut on the asteroid can look over his shoulder and see the Earth, it's already far too late, despite how these scenarios play out in Hollywood. Last-minute rescues won't work!

QUESTION 7.11 As the asteroid approaches Earth, does the gravitational potential energy associated with the asteroid–Earth system (a) increase, (b) decrease, (c) remain the same?

EXERCISE 7.11 Suppose the asteroid starts from rest at a great distance (effectively infinity), falling toward Earth. How much work would have to be done on the asteroid to slow it to 425 m/s by the time it reached a distance of 2.00×10^8 m from Earth?

ANSWER -1.90×10^{15} J

■ APPLYING PHYSICS 7.3 | Why Is the Sun Hot?

EXPLANATION The Sun formed when particles in a cloud of gas coalesced, due to gravitational attraction, into a massive astronomical object. Before this occurred, the particles in the cloud were widely scattered, representing a large amount of gravitational potential energy. As the particles fell closer together, their kinetic energy increased, but the gravitational potential energy of the system decreased, as required by the conservation of energy. With further slow collapse, the cloud became more dense and the average kinetic energy of the particles increased. This kinetic energy is the internal energy of the cloud, which is proportional to the temperature. If enough particles come together, the temperature can rise to a point at which nuclear fusion occurs and the ball of gas becomes a star. Otherwise, the temperature may rise, but not enough to ignite fusion reactions, and the object becomes a brown dwarf (a failed star) or a planet. ■

On inspecting Equation 7.21, some may wonder what happened to mgh, the gravitational potential energy expression introduced in Chapter 5. That expression is still valid when h is small compared with Earth's radius. To see this, we write the change in potential energy as an object is raised from the ground to height h, using the general form for gravitational potential energy (see Fig. 7.21):

$$PE_2 - PE_1 = -G\frac{M_E m}{(R_E + h)} - \left(-G\frac{M_E m}{R_E}\right)$$

$$= -GM_E m\left[\frac{1}{(R_E + h)} - \frac{1}{R_E}\right]$$

After finding a common denominator and applying some algebra, we obtain

$$PE_2 - PE_1 = \frac{GM_E mh}{R_E(R_E + h)}$$

When the height h is very small compared with R_E, h can be dropped from the second factor in the denominator, yielding

$$\frac{1}{R_E(R_E + h)} \cong \frac{1}{R_E^2}$$

Substituting this into the previous expression, we have

$$PE_2 - PE_1 \cong \frac{GM_E}{R_E^2} mh$$

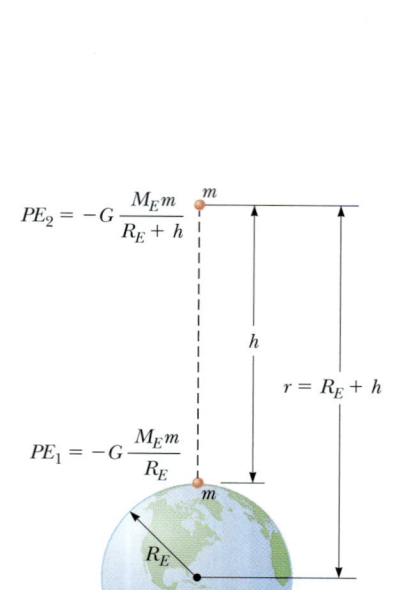

Figure 7.21 Relating the general form of gravitational potential energy to mgh.

Now recall from Chapter 4 that the free-fall acceleration at the surface of Earth is given by $g = GM_E/R_E^2$, giving

$$PE_2 - PE_1 \cong mgh$$

Escape Speed

If an object is projected upward from Earth's surface with a large enough speed, it can soar off into space and never return. This speed is called Earth's **escape speed**. (It is also commonly called the *escape velocity*, but in fact is more properly a speed.)

Earth's escape speed can be found by applying conservation of energy. Suppose an object of mass m is projected vertically upward from Earth's surface with an initial speed v_i. The initial mechanical energy (kinetic plus potential energy) of the object–Earth system is given by

$$KE_i + PE_i = \tfrac{1}{2}mv_i^2 - \frac{GM_E m}{R_E}$$

We neglect air resistance and assume the initial speed is just large enough to allow the object to reach infinity with a speed of zero. This value of v_i is the escape speed v_{esc}. When the object is at an infinite distance from Earth, its kinetic energy is zero because $v_f = 0$, and the gravitational potential energy is also zero because $1/r$ goes to zero as r goes to infinity. Hence the total mechanical energy is zero, and the law of conservation of energy gives

$$\tfrac{1}{2}mv_{esc}^2 - \frac{GM_E m}{R_E} = 0$$

so that

$$v_{esc} = \sqrt{\frac{2GM_E}{R_E}} \qquad [7.22]$$

The escape speed for Earth is about 11.2 km/s, which corresponds to about 25 000 mi/h. (See Example 7.12.) Note that the expression for v_{esc} doesn't depend on the mass of the object projected from Earth, so a spacecraft has the same escape speed as a molecule. Escape speeds for the planets, the Moon, and the Sun are listed in Table 7.2. Escape speed and temperature determine to a large extent whether a world has an atmosphere and, if so, what the constituents of the atmosphere are. Planets with low escape speeds, such as Mercury, generally don't have atmospheres because the average speed of gas molecules is close to the escape speed. Venus has a very thick atmosphere, but it's almost entirely carbon dioxide, a heavy gas. The atmosphere of Earth has very little hydrogen or helium, but has retained the much heavier nitrogen and oxygen molecules.

Table 7.2 Escape Speeds for the Planets and the Moon

Planet	v_{esc} (km/s)
Mercury	4.3
Venus	10.3
Earth	11.2
Moon	2.3
Mars	5.0
Jupiter	60.0
Saturn	36.0
Uranus	22.0
Neptune	24.0
Pluto[a]	1.1

[a]In August 2006, the International Astronomical Union adopted a definition of a planet that separates Pluto from the other eight planets. Pluto is now defined as a "dwarf planet" (like the asteroid Ceres).

■ EXAMPLE 7.12 From the Earth to the Moon

GOAL Apply conservation of energy with the general form of Newton's universal law of gravity.

PROBLEM In Jules Verne's classic novel *From the Earth to the Moon*, a giant cannon dug into the Earth in Florida fired a spacecraft all the way to the Moon. (a) If the spacecraft leaves the cannon at escape speed, at what speed is it moving when 1.50×10^5 km from the center of Earth? Neglect any friction effects. (b) Approximately what constant acceleration is needed to propel the spacecraft to escape speed through a cannon bore 1.00 km long?

STRATEGY For part (a), use conservation of energy and solve for the final speed v_f. Part (b) is an application of the time-independent kinematic equation: solve for the acceleration a.

SOLUTION

(a) Find the speed at $r = 1.50 \times 10^5$ km.

Apply conservation of energy:

$$\tfrac{1}{2}mv_i^2 - \frac{GM_E m}{R_E} = \tfrac{1}{2}mv_f^2 - \frac{GM_E m}{r_f}$$

(Continued)

Multiply by $2/m$ and rearrange, solving for v_f^2. Then substitute known values and take the square root.

$$v_f^2 = v_i^2 + \frac{2GM_E}{r_f} - \frac{2GM_E}{R_E} = v_i^2 + 2GM_E\left(\frac{1}{r_f} - \frac{1}{R_E}\right)$$

$$v_f^2 = (1.12 \times 10^4 \text{ m/s})^2 + 2(6.67 \times 10^{-11} \text{ kg}^{-1}\text{m}^3\text{s}^{-2})$$
$$\times (5.98 \times 10^{24} \text{ kg})\left(\frac{1}{1.50 \times 10^8 \text{ m}} - \frac{1}{6.38 \times 10^6 \text{ m}}\right)$$

$$v_f = \boxed{2.39 \times 10^3 \text{ m/s}}$$

(b) Find the acceleration through the cannon bore, assuming it's constant.

Use the time-independent kinematics equation:

$$v^2 - v_0^2 = 2a\Delta x$$
$$(1.12 \times 10^4 \text{ m/s})^2 - 0 = 2a(1.00 \times 10^3 \text{ m})$$
$$a = \boxed{6.27 \times 10^4 \text{ m/s}^2}$$

REMARKS This result corresponds to an acceleration of over 6 000 times the free-fall acceleration on Earth. Such a huge acceleration is far beyond what the human body can tolerate.

QUESTION 7.12 Suppose the spacecraft managed to go into an elliptical orbit around Earth, with a nearest point (perigee) and farthest point (apogee). At which point is the kinetic energy of the spacecraft higher, and why?

EXERCISE 7.12 Using the data in Table 7.3 (see page 228), find (a) the escape speed from the surface of Mars and (b) the speed of a space vehicle when it is 1.25×10^7 m from the center of Mars if it leaves the surface at the escape speed.

ANSWERS (a) 5.04×10^3 m/s (b) 2.62×10^3 m/s

7.6 Kepler's Laws

LEARNING OBJECTIVES

1. State Kepler's three laws and explain the significance of each.
2. Apply the third law to obtain information about orbiting bodies.

The movements of the planets, stars, and other celestial bodies have been observed for thousands of years. In early history scientists regarded Earth as the center of the Universe. This **geocentric model** was developed extensively by the Greek astronomer Claudius Ptolemy in the second century A.D. and was accepted for the next 1 400 years. In 1543 Polish astronomer Nicolaus Copernicus (1473–1543) showed that Earth and the other planets revolve in circular orbits around the Sun (the **heliocentric model**).

Danish astronomer Tycho Brahe (pronounced Brah or BRAH–huh; 1546–1601) made accurate astronomical measurements over a period of 20 years, providing the data for the currently accepted model of the solar system. Brahe's precise observations of the planets and 777 stars were carried out with nothing more elaborate than a large sextant and compass; the telescope had not yet been invented.

German astronomer Johannes Kepler, who was Brahe's assistant, acquired Brahe's astronomical data and spent about 16 years trying to deduce a mathematical model for the motions of the planets. After many laborious calculations, he found that Brahe's precise data on the motion of Mars about the Sun provided the answer. Kepler's analysis first showed that the concept of circular orbits about the Sun had to be abandoned. He eventually discovered that the orbit of Mars could be accurately described by an ellipse with the Sun at one focus. He then generalized this analysis to include the motions of all planets. The complete analysis is summarized in three statements known as **Kepler's laws**:

1. All planets move in elliptical orbits with the Sun at one of the focal points.
2. A line drawn from the Sun to any planet sweeps out equal areas in equal time intervals.
3. The square of the orbital period of any planet is proportional to the cube of the average distance from the planet to the Sun.

◀ Kepler's Laws

Newton later demonstrated that these laws are consequences of the gravitational force that exists between any two objects. Newton's law of universal gravitation, together with his laws of motion, provides the basis for a full mathematical description of the motions of planets and satellites.

Kepler's First Law

The first law arises as a natural consequence of the inverse-square nature of Newton's law of gravitation. Any object bound to another by a force that varies as $1/r^2$ will move in an elliptical orbit. As shown in Figure 7.22a, an ellipse is a curve drawn so that the sum of the distances from any point on the curve to two internal points called *focal points* or *foci* (singular, *focus*) is always the same. The semimajor axis a is half the length of the line that goes across the ellipse and contains both foci. For the Sun–planet configuration (Fig. 7.22b), the Sun is at one focus and the other focus is empty. Because the orbit is an ellipse, the distance from the Sun to the planet continuously changes.

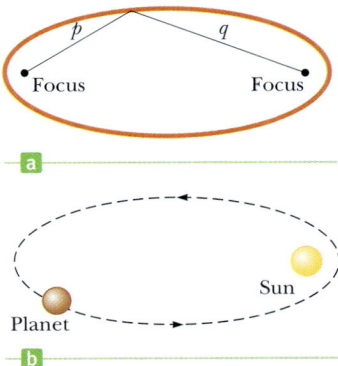

Figure 7.22 (a) The sum $p + q$ is the same for every point on the ellipse. (b) In the Solar System, the Sun is at one focus of the elliptical orbit of each planet and the other focus is empty.

Kepler's Second Law

Kepler's second law states that a line drawn from the Sun to any planet sweeps out equal areas in equal time intervals. Consider a planet in an elliptical orbit about the Sun, as in Figure 7.23. In a given period Δt, the planet moves from point Ⓐ to point Ⓑ. The planet moves more slowly on that side of the orbit because it's farther away from the sun. On the opposite side of its orbit, the planet moves from point Ⓒ to point Ⓓ in the same amount of time, Δt, moving faster because it's closer to the sun. Kepler's second law says that any two wedges formed as in Figure 7.23 will always have the same area. As we will see in Chapter 8, Kepler's second law is related to a physical principle known as conservation of angular momentum.

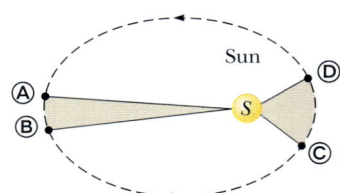

Figure 7.23 The two areas swept out by the planet in its elliptical orbit about the Sun are equal if the time interval between points Ⓐ and Ⓑ is equal to the time interval between points Ⓒ and Ⓓ.

Kepler's Third Law

The derivation of Kepler's third law is simple enough to carry out for the special case of a circular orbit. Consider a planet of mass M_p moving around the Sun, which has a mass of M_S, in a circular orbit. Because the orbit is circular, the planet moves at a constant speed v. Newton's second law, his law of gravitation, and centripetal acceleration then give the following equation:

$$M_p a_c = \frac{M_p v^2}{r} = \frac{G M_S M_p}{r^2}$$

The speed v of the planet in its orbit is equal to the circumference of the orbit divided by the time required for one revolution, T, called the **period** of the planet, so $v = 2\pi r/T$. Substituting, the preceding expression becomes

$$\frac{GM_S}{r^2} = \frac{(2\pi r/T)^2}{r}$$

$$T^2 = \left(\frac{4\pi^2}{GM_S}\right) r^3 = K_S r^3 \qquad [7.23]$$

◀ Kepler's third law

Table 7.3 Useful Planetary Data

Body	Mass (kg)	Mean Radius (m)	Period (s)	Mean Distance from Sun (m)	$\frac{T^2}{r^3} 10^{-19} \left(\frac{s^2}{m^3}\right)$
Mercury	3.18×10^{23}	2.43×10^{6}	7.60×10^{6}	5.79×10^{10}	2.97
Venus	4.88×10^{24}	6.06×10^{6}	1.94×10^{7}	1.08×10^{11}	2.99
Earth	5.98×10^{24}	6.38×10^{6}	3.156×10^{7}	1.496×10^{11}	2.97
Mars	6.42×10^{23}	3.37×10^{6}	5.94×10^{7}	2.28×10^{11}	2.98
Jupiter	1.90×10^{27}	6.99×10^{7}	3.74×10^{8}	7.78×10^{11}	2.97
Saturn	5.68×10^{26}	5.85×10^{7}	9.35×10^{8}	1.43×10^{12}	2.99
Uranus	8.68×10^{25}	2.33×10^{7}	2.64×10^{9}	2.87×10^{12}	2.95
Neptune	1.03×10^{26}	2.21×10^{7}	5.22×10^{9}	4.50×10^{12}	2.99
Pluto[a]	1.27×10^{23}	1.14×10^{6}	7.82×10^{9}	5.91×10^{12}	2.96
Moon	7.36×10^{22}	1.74×10^{6}	—	—	—
Sun	1.991×10^{30}	6.96×10^{8}	—	—	—

[a]In August 2006, the International Astronomical Union adopted a definition of a planet that separates Pluto from the other eight planets. Pluto is now defined as a "dwarf planet" like the asteroid Ceres.

where K_S is a constant given by

$$K_S = \frac{4\pi^2}{GM_S} = 2.97 \times 10^{-19} \text{ s}^2/\text{m}^3$$

Equation 7.23 is Kepler's third law for a circular orbit. The orbits of most of the planets are very nearly circular. Comets and asteroids, however, usually have elliptical orbits. For these orbits, the radius r must be replaced with a, the semimajor axis—half the longest distance across the elliptical orbit. (This is also the average distance of the comet or asteroid from the Sun.) A more detailed calculation shows that K_S actually depends on the sum of both the mass of a given planet and the Sun's mass. The masses of the planets, however, are negligible compared with the Sun's mass; hence can be neglected, meaning Equation 7.23 is valid for any planet in the Sun's family. If we consider the orbit of a satellite such as the Moon around Earth, then the constant has a different value, with the mass of the Sun replaced by the mass of Earth. In that case, K_E equals $4\pi^2/GM_E$.

The mass of the Sun can be determined from Kepler's third law because the constant K_S in Equation 7.23 includes the mass of the Sun and the other variables and constants can be easily measured. The value of this constant can be found by substituting the values of a planet's period and orbital radius and solving for K_S. The mass of the Sun is then

$$M_S = \frac{4\pi^2}{GK_S}$$

This same process can be used to calculate the mass of Earth (by considering the period and orbital radius of the Moon) and the mass of other planets in the solar system that have satellites.

The last column in Table 7.3 confirms that T^2/r^3 is very nearly constant. When time is measured in Earth years and the semimajor axis in astronomical units (1 AU = the distance from Earth to the Sun), Kepler's law takes the following simple form:

$$T^2 = a^3$$

This equation can be easily checked: Earth has a semimajor axis of one astronomical unit (by definition), and it takes one year to circle the Sun. This equation, of course, is valid only for the sun and its planets, asteroids, and comets.

Quick Quiz

7.10 Suppose an asteroid has a semimajor axis of 4 AU. How long does it take the asteroid to go around the Sun? (a) 2 years (b) 4 years (c) 6 years (d) 8 years

EXAMPLE 7.13 | Geosynchronous Orbit and Telecommunications Satellites

GOAL Apply Kepler's third law to an Earth satellite.

PROBLEM From a telecommunications point of view, it's advantageous for satellites to remain at the same location relative to a location on Earth. This can occur only if the satellite's orbital period is the same as the Earth's period of rotation, approximately 24.0 h. **(a)** At what distance from the center of the Earth can this geosynchronous orbit be found? **(b)** What's the orbital speed of the satellite?

STRATEGY This problem can be solved with the same method that was used to derive a special case of Kepler's third law, with Earth's mass replacing the Sun's mass. There's no need to repeat the analysis; just replace the Sun's mass with Earth's mass in Kepler's third law, substitute the period T (converted to seconds), and solve for r. For part (b), find the circumference of the circular orbit and divide by the elapsed time.

SOLUTION

(a) Find the distance r to geosynchronous orbit.

Apply Kepler's third law:
$$T^2 = \left(\frac{4\pi^2}{GM_E}\right)r^3$$

Substitute the period in seconds, $T = 86\,400$ s, the gravity constant $G = 6.67 \times 10^{-11}$ kg^{-1} m^3/s^2, and the mass of the Earth, $M_E = 5.98 \times 10^{24}$ kg. Solve for r:
$$r = 4.23 \times 10^7 \text{ m}$$

(b) Find the orbital speed.

Divide the distance traveled during one orbit by the period:
$$v = \frac{d}{T} = \frac{2\pi r}{T} = \frac{2\pi(4.23 \times 10^7 \text{ m})}{8.64 \times 10^4 \text{ s}} = 3.08 \times 10^3 \text{ m/s}$$

REMARKS Earth's motion around the Sun was neglected; that requires using Earth's "sidereal" period (about four minutes shorter). Notice that Earth's mass could be found by substituting the Moon's distance and period into this form of Kepler's third law.

QUESTION 7.13 If the satellite was placed in an orbit three times as far away, about how long would it take to orbit the Earth once? Answer in days, rounding to one digit.

EXERCISE 7.13 Mars rotates on its axis once every 1.02 days (almost the same as Earth does). (a) Find the distance from the center of Mars at which a satellite would remain in one spot over the Martian surface. (b) Find the speed of the satellite.

ANSWERS (a) 2.03×10^7 m (b) 1.45×10^3 m/s

SUMMARY

7.1 Angular Speed and Angular Acceleration

The **average angular speed** ω_{av} of a rigid object is defined as the ratio of the angular displacement $\Delta\theta$ to the time interval Δt, or

$$\omega_{av} \equiv \frac{\theta_f - \theta_i}{t_f - t_i} = \frac{\Delta\theta}{\Delta t} \quad [7.3]$$

where ω_{av} is in radians per second (rad/s).

The **average angular acceleration** α_{av} of a rotating object is defined as the ratio of the change in angular speed $\Delta\omega$ to the time interval Δt, or

$$\alpha_{av} \equiv \frac{\omega_f - \omega_i}{t_f - t_i} = \frac{\Delta\omega}{\Delta t} \quad [7.5]$$

where α_{av} is in radians per second per second (rad/s^2).

7.2 Rotational Motion Under Constant Angular Acceleration

If an object undergoes rotational motion about a fixed axis under a constant angular acceleration α, its motion can be described with the following set of equations:

$$\omega = \omega_i + \alpha t \quad [7.7]$$

$$\Delta\theta = \omega_i t + \tfrac{1}{2}\alpha t^2 \quad [7.8]$$

$$\omega^2 = \omega_i^2 + 2\alpha\,\Delta\theta \quad [7.9]$$

Problems are solved as in one-dimensional kinematics.

7.3 Relations Between Angular and Linear Quantities

When an object rotates about a fixed axis, the angular speed and angular acceleration are related to the tangential speed and tangential acceleration through the relationships

$$v_t = r\omega \quad [7.10]$$

and

$$a_t = r\alpha \quad [7.11]$$

7.4 Centripetal Acceleration

Any object moving in a circular path has an acceleration directed toward the center of the circular path, called a **centripetal acceleration**. Its magnitude is given by

$$a_c = \frac{v^2}{r} = r\omega^2 \quad [7.13, 7.17]$$

Any object moving in a circular path must have a net force exerted on it that is directed toward the center of the path. Some examples of forces that cause centripetal acceleration are the force of gravity (as in the motion of a satellite) and the force of tension in a string.

7.5 Newtonian Gravitation

Newton's law of universal gravitation states that every particle in the Universe attracts every other particle with a force that is directly proportional to the product of their masses and inversely proportional to the square of the distance r between them:

$$F = G\frac{m_1 m_2}{r^2} \quad [7.20]$$

where $G = 6.673 \times 10^{-11}$ N·m²/kg² is the **constant of universal gravitation**. A general expression for gravitational potential energy is

$$PE = -G\frac{M_E m}{r} \quad [7.21]$$

This expression reduces to $PE = mgh$ close to the surface of Earth and holds for other worlds through replacement of the mass M_E. Problems such as finding the escape velocity from Earth can be solved by using Equation 7.21 in the conservation of energy equation.

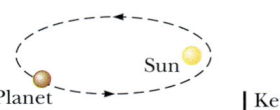

The gravitational force is attractive and acts along the line joining the particles.

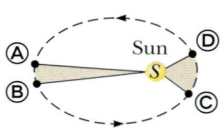

The gravitational potential energy increases towards zero as r increases.

7.6 Kepler's Laws

Kepler derived the following three laws of planetary motion:

1. All planets move in elliptical orbits with the Sun at one of the focal points.

| Kepler's first law. |

2. A line drawn from the Sun to any planet sweeps out equal areas in equal time intervals.

| Kepler's second law. |

3. The square of the orbital period of a planet is proportional to the cube of the average distance from the planet to the Sun:

$$T^2 = \left(\frac{4\pi^2}{GM_S}\right)r^3 \quad [7.23]$$

| Kepler's third law. |

The third law can be applied to any large body and its system of satellites by replacing the Sun's mass with the body's mass. In particular, it can be used to determine the mass of the central body once the average distance to a satellite and its period are known.

■ WARM-UP EXERCISES

The warm-up exercises in this chapter may be assigned online in Enhanced WebAssign.

1. **Math Review** A circular track has a radius of 125 m. (a) Calculate the distance around the track. (b) If a runner jogs 275 m along the track, through what angle has he run?

2. **Math Review** (a) Convert 47.0° to radians, using the appropriate conversion ratio. (b) Convert 2.35 rad to degrees. (c) If a circle has radius 1.70 m, what is the arc length subtended by a 47.0° angle? (See Sections 1.5 and 7.1.)

3. (a) Convert 12.0 rev/min to radians per second. (b) Convert 2.57 rad/s to rev/min. (See Sections 1.5 and 7.1.)

4. A carnival carousel accelerates nonuniformly from rest, moving through an angle of 8.60 rad in 6.00 s. If it's turning at 3.30 rad/s at that time, find (a) its average angular speed, and (b) average angular acceleration during that time interval. (See Section 7.1.)

5. Find the angular speed of a planet that circles its star in 1.00 y, in radians per second. (See Section 7.1.)

6. A grindstone increases in angular speed uniformly from 4.00 rad/s to 12.0 rad/s in 4.00 s. (a) Calculate the grindstone's angular acceleration. (b) Through what angle does it turn during that time? (See Section 7.2.)

7. A bicyclist starting at rest produces a constant angular acceleration of 1.60 rad/s² for wheels that are 38.0 cm in radius. (a) What is the bicyclist's linear acceleration? (b) What is the angular speed of the wheels when the bicyclist reaches 11.0 m/s? (c) How many radians have the wheels turned through in that time? (d) How far has the bicyclist traveled? (See Sections 7.2 and 7.3.)

8. A car of mass 1 230 kg travels along a circular road of radius 60.0 m at 18.0 m/s. (a) Calculate the magnitude of the car's centripetal acceleration. (b) What is the magnitude of the force of static friction acting on the car? (See Section 7.4.)

9. A man whirls a 0.20-kg piece of lead attached to the end of a string of length 0.500 m in a circular path and in a vertical plane. If the man maintains a constant speed of

4.00 m/s, what is the tension in the string when the lead is (a) at the top of the circular path? (b) at the bottom of the circular path? (See Section 7.4.)

10. (a) Find the magnitude of the gravity force between a planet with mass 5.98×10^{24} kg and its moon, with mass 7.36×10^{22} kg, if the average distance between them is 3.84×10^{8} m. (b) What is the acceleration of the moon toward the planet? (c) What is the acceleration of the planet toward the moon? (See Section 7.5.)

11. What is the gravitational acceleration close to the surface of a planet with a mass of $2M_E$ and radius of $2R_E$, where M_E and R_E are the mass and radius of Earth, respectively? Answer as a multiple of g, the magnitude of the gravitational acceleration near Earth's surface. (See Section 7.5.)

12. (a) Find the speed of a satellite in circular orbit 7.20×10^{6} m from the center of a world with mass 9.40×10^{23} kg. (b) How long does it take to orbit the world one time? (See Section 7.5.)

13. Calculate the escape velocity from the surface of a world with mass 9.10×10^{24} kg and radius 6.80×10^{3} km. (See Section 7.5.)

14. A space capsule of mass 645 kg is at rest 1.20×10^{7} m from the center of the Earth. When it has fallen 3.00×10^{6} m closer to the Earth, (a) what is the change in the system's gravitational potential energy? (b) Find the speed of the satellite at that point. (See Section 7.5.)

15. A comet has a period of 76.3 years and moves in an elliptical orbit in which its perihelion (closest approach to the Sun) is 0.610 AU. Find (a) the semi-major axis of the comet and (b) an estimate of the comet's maximum distance from the Sun, both in astronomical units. (See Section 7.6.)

CONCEPTUAL QUESTIONS

The conceptual questions in this chapter may be assigned online in Enhanced WebAssign.

1. In a race like the Indianapolis 500, a driver circles the track counterclockwise and feels his head pulled toward one shoulder. To relieve his neck muscles from having to hold his head erect, the driver fastens a strap to one wall of the car and the other to his helmet. The length of the strap is adjusted to keep his head vertical. (a) Which shoulder does his head tend to lean toward? (b) What force or forces produce the centripetal acceleration when there is no strap? (c) What force or forces do so when there is a strap?

2. If someone told you that astronauts are weightless in Earth orbit because they are beyond the force of gravity, would you accept the statement? Explain.

3. If a car's wheels are replaced with wheels of greater diameter, will the reading of the speedometer change? Explain.

4. At night, you are farther away from the Sun than during the day. What's more, the force exerted by the Sun on you is downward into Earth at night and upward into the sky during the day. If you had a sensitive enough bathroom scale, would you appear to weigh more at night than during the day?

5. A pendulum consists of a small object called a bob hanging from a light cord of fixed length, with the top end of the cord fixed, as represented in Figure CQ7.5. The bob moves without friction, swinging equally high on both sides. It moves from its turning point A through point B and reaches its maximum speed at point C.

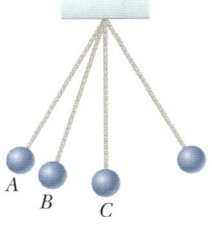

Figure CQ7.5

(a) At what point does the bob have nonzero radial acceleration and zero tangential acceleration? What is the direction of its total acceleration at this point? (b) At what point does the bob have nonzero tangential acceleration and zero radial acceleration? What is the direction of its total acceleration at this point? (c) At what point does the bob have both nonzero tangential and radial acceleration? What is the direction of its total acceleration at this point?

6. Because of Earth's rotation about its axis, you weigh slightly less at the equator than at the poles. Explain.

7. It has been suggested that rotating cylinders about 10 miles long and 5 miles in diameter be placed in space for colonies. The purpose of their rotation is to simulate gravity for the inhabitants. Explain the concept behind this proposal.

8. Describe the path of a moving object in the event that the object's acceleration is constant in magnitude at all times and (a) perpendicular to its velocity; (b) parallel to its velocity.

9. A pail of water can be whirled in a vertical circular path such that no water is spilled. Why does the water remain in the pail, even when the pail is upside down above your head?

10. Use Kepler's second law to convince yourself that Earth must move faster in its orbit during the northern hemisphere winter, when it is closest to the Sun, than during the summer, when it is farthest from the Sun.

11. Is it possible for a car to move in a circular path in such a way that it has a tangential acceleration but no centripetal acceleration?

12. A child is practicing for a BMX race. His speed remains constant as he goes counterclockwise around a level track with two nearly straight sections and two nearly semicircular sections, as shown in the aerial view of Figure CQ7.12. (a) What are the directions of his velocity at points A, B, and C? For each point choose one: north, south, east, west, or nonexistent? (b) What are the directions of his acceleration at points A, B, and C?

Figure CQ7.12

13. An object executes circular motion with constant speed whenever a net force of constant magnitude acts perpendicular to the velocity. What happens to the speed if the force is not perpendicular to the velocity?

PROBLEMS

WebAssign The problems in this chapter may be assigned online in Enhanced WebAssign.

1. denotes straightforward problem; 2. denotes intermediate problem;
3. denotes challenging problem
1. denotes full solution available in *Student Solutions Manual/Study Guide*
1. denotes problems most often assigned in Enhanced WebAssign

BIO denotes biomedical problems
GP denotes guided problems
M denotes Master It tutorial available in Enhanced WebAssign
Q|C denotes asking for quantitative and conceptual reasoning
S denotes symbolic reasoning problem
W denotes Watch It video solution available in Enhanced WebAssign

7.1 Angular Speed and Angular Acceleration

1. **Q|C** (a) Find the angular speed of Earth's rotation about its axis. (b) How does this rotation affect the shape of Earth?

2. A wheel has a radius of 4.1 m. How far (path length) does a point on the circumference travel if the wheel is rotated through angles of (a) 30°, (b) 30 rad, and (c) 30 rev, respectively?

3. The tires on a new compact car have a diameter of 2.0 ft and are warranted for 60 000 miles. (a) Determine the angle (in radians) through which one of these tires will rotate during the warranty period. (b) How many revolutions of the tire are equivalent to your answer in part (a)?

4. **Q|C** A potter's wheel moves uniformly from rest to an angular speed of 1.00 rev/s in 30.0 s. (a) Find its angular acceleration in radians per second per second. (b) Would doubling the angular acceleration during the given period have doubled final angular speed?

7.2 Rotational Motion Under Constant Angular Acceleration

7.3 Relations Between Angular and Linear Quantities

5. A dentist's drill starts from rest. After 3.20 s of constant angular acceleration, it turns at a rate of 2.51×10^4 rev/min. (a) Find the drill's angular acceleration. (b) Determine the angle (in radians) through which the drill rotates during this period.

6. **W** A centrifuge in a medical laboratory rotates at an angular speed of 3 600 rev/min. When switched off, it rotates through 50.0 revolutions before coming to rest. Find the constant angular acceleration (in rad/s²) of the centrifuge.

7. **Q|C** A machine part rotates at an angular speed of 0.06 rad/s; its speed is then increased to 2.2 rad/s at an angular acceleration of 0.70 rad/s². (a) Find the angle through which the part rotates before reaching this final speed. (b) In general, if both the initial and final angular speed are doubled at the same angular acceleration, by what factor is the angular displacement changed? Why? *Hint:* Look at the form of Equation 7.9.

8. **Q|C** A bicycle is turned upside down while its owner repairs a flat tire. A friend spins the other wheel and observes that drops of water fly off tangentially. She measures the heights reached by drops moving vertically (Fig. P7.8). A drop that breaks loose from the tire on one turn rises vertically 54.0 cm above the tangent point. A drop that breaks loose on the next turn rises 51.0 cm above the tangent point. The radius of the wheel is 0.381 m. (a) Why does the first drop rise higher than the second drop? (b) Neglecting air friction and using only the observed heights and the radius of the wheel, find the wheel's angular acceleration (assuming it to be constant).

Figure P7.8 Problems 8 and 69.

9. The diameters of the main rotor and tail rotor of a single-engine helicopter are 7.60 m and 1.02 m, respectively. The respective rotational speeds are 450 rev/min and 4 138 rev/min. Calculate the speeds of the tips of both rotors. Compare these speeds with the speed of sound, 343 m/s.

10. The tub of a washer goes into its spin-dry cycle, starting from rest and reaching an angular speed of 5.0 rev/s in

8.0 s. At this point, the person doing the laundry opens the lid, and a safety switch turns off the washer. The tub slows to rest in 12.0 s. Through how many revolutions does the tub turn during the entire 20-s interval? Assume constant angular acceleration while it is starting and stopping.

11. A car initially traveling at 29.0 m/s undergoes a constant negative acceleration of magnitude 1.75 m/s² after its brakes are applied. (a) How many revolutions does each tire make before the car comes to a stop, assuming the car does not skid and the tires have radii of 0.330 m? (b) What is the angular speed of the wheels when the car has traveled half the total distance?

12. A 45.0-cm diameter disk rotates with a constant angular acceleration of 2.50 rad/s². It starts from rest at $t = 0$, and a line drawn from the center of the disk to a point P on the rim of the disk makes an angle of 57.3° with the positive x-axis at this time. At $t = 2.30$ s, find (a) the angular speed of the wheel, (b) the linear velocity and tangential acceleration of P, and (c) the position of P (in degrees, with respect to the positive x-axis).

13. **M** A rotating wheel requires 3.00 s to rotate 37.0 revolutions. Its angular velocity at the end of the 3.00-s interval is 98.0 rad/s. What is the constant angular acceleration (in rad/s²) of the wheel?

14. An electric motor rotating a workshop grinding wheel at a rate of 1.00×10^2 rev/min is switched off. Assume the wheel has a constant negative angular acceleration of magnitude 2.00 rad/s². (a) How long does it take for the grinding wheel to stop? (b) Through how many radians has the wheel turned during the interval found in part (a)?

7.4 Centripetal Acceleration

15. A car initially traveling eastward turns north by traveling in a circular path at uniform speed as shown in Figure P7.15. The length of the arc ABC is 235 m, and the car completes the turn in 36.0 s. (a) Determine the car's speed. (b) What is the magnitude and direction of the acceleration when the car is at point B?

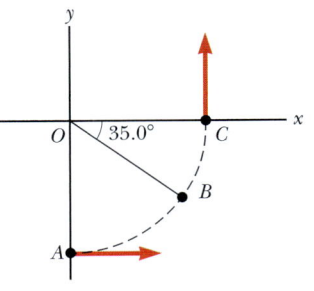

Figure P7.15

16. It has been suggested that rotating cylinders about 10 mi long and 5.0 mi in diameter be placed in space and used as colonies. What angular speed must such a cylinder have so that the centripetal acceleration at its surface equals the free-fall acceleration on Earth?

17. (a) What is the tangential acceleration of a bug on the rim of a 10.0-in.-diameter disk if the disk accelerates uniformly from rest to an angular speed of 78.0 rev/min in 3.00 s? (b) When the disk is at its final speed, what is the tangential velocity of the bug? One second after the bug starts from rest, what are its (c) tangential acceleration, (d) centripetal acceleration, and (e) total acceleration?

18. An adventurous archeologist (m = 85.0 kg) tries to cross a river by swinging from a vine. The vine is 10.0 m long, and his speed at the bottom of the swing is 8.00 m/s. The archeologist doesn't know that the vine has a breaking strength of 1 000 N. Does he make it across the river without falling in?

19. **Q|C** One end of a cord is fixed and a small 0.500-kg object is attached to the other end, where it swings in a section of a vertical circle of radius 2.00 m, as shown in Figure P7.19. When $\theta = 20.0°$, the speed of the object is 8.00 m/s. At this instant, find (a) the tension in the string, (b) the tangential and radial components of acceleration, and (c) the total acceleration. (d) Is your answer changed if the object is swinging down toward its lowest point instead of swinging up? (e) Explain your answer to part (d).

Figure P7.19

20. **Q|C** A coin rests 15.0 cm from the center of a turntable. The coefficient of static friction between the coin and turntable surface is 0.350. The turntable starts from rest at $t = 0$ and rotates with a constant angular acceleration of 0.730 rad/s². (a) Once the turntable starts to rotate, what force causes the centripetal acceleration when the coin is stationary relative to the turntable? Under what condition does the coin begin to move relative to the turntable? (b) After what period of time will the coin start to slip on the turntable?

21. A 55.0-kg ice skater is moving at 4.00 m/s when she grabs the loose end of a rope, the opposite end of which is tied to a pole. She then moves in a circle of radius 0.800 m around the pole. (a) Determine the force exerted by the horizontal rope on her arms. (b) Compare this force with her weight.

22. **M** A 40.0-kg child swings in a swing supported by two chains, each 3.00 m long. The tension in each chain at the lowest point is 350 N. Find (a) the child's speed at the lowest point and (b) the force exerted by the seat on the child at the lowest point. (Ignore the mass of the seat.)

23. A certain light truck can go around a flat curve having a radius of 150 m with a maximum speed of 32.0 m/s. With what maximum speed can it go around a curve having a radius of 75.0 m?

24. **BIO** A sample of blood is placed in a centrifuge of radius 15.0 cm. The mass of a red blood cell is 3.0×10^{-16} kg, and the magnitude of the force acting on it as it settles out of the plasma is 4.0×10^{-11} N. At how many revolutions per second should the centrifuge be operated?

25. A 50.0-kg child stands at the rim of a merry-go-round of radius 2.00 m, rotating with an angular speed of 3.00 rad/s. (a) What is the child's centripetal acceleration? (b) What is the minimum force between her feet and the floor of the carousel that is required to keep her in the circular path? (c) What minimum coefficient of static friction is required? Is the answer you found reasonable? In other words, is she likely to stay on the merry-go-round?

26. GP A space habitat for a long space voyage consists of two cabins each connected by a cable to a central hub as shown in Figure P7.26. The cabins are set spinning around the hub axis, which is connected to the rest of the spacecraft to generate artificial gravity. (a) What forces are acting on an astronaut in one of the cabins? (b) Write Newton's second law for an astronaut lying on the "floor" of one of the habitats, relating the astronaut's mass m, his velocity v, his radial distance from the hub r, and the normal force n. (c) What would n have to equal if the 60.0-kg astronaut is to experience half his normal Earth weight? (d) Calculate the necessary tangential speed of the habitat from Newton's second law. (e) Calculate the angular speed from the tangential speed. (f) Calculate the period of rotation from the angular speed. (g) If the astronaut stands up, will his head be moving faster, slower, or at the same speed as his feet? Why? Calculate the tangential speed at the top of his head if he is 1.80 m tall.

Figure P7.26

27. An air puck of mass $m_1 = 0.25$ kg is tied to a string and allowed to revolve in a circle of radius $R = 1.0$ m on a frictionless horizontal table. The other end of the string passes through a hole in the center of the table, and a mass of $m_2 = 1.0$ kg is tied to it (Fig. P7.27). The suspended mass remains in equilibrium while the puck on the tabletop revolves. (a) What is the tension in the string? (b) What is the horizontal force acting on the puck? (c) What is the speed of the puck?

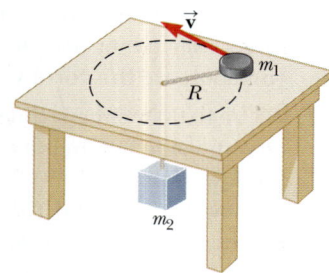

Figure P7.27 Problems 27 and 28.

28. S An air puck of mass m_1 is tied to a string and allowed to revolve in a circle of radius R on a horizontal, frictionless table. The other end of the string passes through a small hole in the center of the table, and an object of mass m_2 is tied to it (Fig. P7.27). The suspended object remains in equilibrium while the puck on the tabletop revolves. (a) Find a symbolic expression for the tension in the string in terms of m_2 and g. (b) Write Newton's second law for the air puck, using the variables m_1, v, R, and T. (c) Eliminate the tension T from the expressions found in parts (a) and (b) and find an expression for the speed of the puck in terms of m_1, m_2, g, and R. (d) Check your answers by substituting the values of Problem 27 and comparing the results with the answers for that problem.

29. Q|C A woman places her briefcase on the backseat of her car. As she drives to work, the car negotiates an unbanked curve in the road that can be regarded as an arc of a circle of radius 62.0 m. While on the curve, the speed of the car is 15.0 m/s at the instant the briefcase starts to slide across the backseat toward the side of the car. (a) What force causes the centripetal acceleration of the briefcase when it is stationary relative to the car? Under what condition does the briefcase begin to move relative to the car? (b) What is the coefficient of static friction between the briefcase and seat surface?

30. Q|C A pail of water is rotated in a vertical circle of radius 1.00 m. (a) What two external forces act on the water in the pail? (b) Which of the two forces is most important in causing the water to move in a circle? (c) What is the pail's minimum speed at the top of the circle if no water is to spill out? (d) If the pail with the speed found in part (c) were to suddenly disappear at the top of the circle, describe the subsequent motion of the water. Would it differ from the motion of a projectile?

31. M A 40.0-kg child takes a ride on a Ferris wheel that rotates four times each minute and has a diameter of 18.0 m. (a) What is the centripetal acceleration of the child? (b) What force (magnitude and direction) does the seat exert on the child at the lowest point of the ride? (c) What force does the seat exert on the child at the highest point of the ride? (d) What force does the seat exert on the child when the child is halfway between the top and bottom?

32. W A roller-coaster vehicle has a mass of 500 kg when fully loaded with passengers (Fig. P7.32). (a) If the

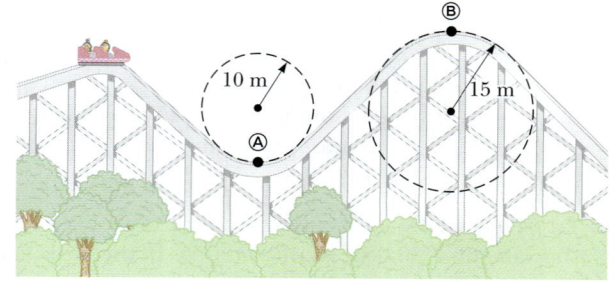

Figure P7.32

vehicle has a speed of 20.0 m/s at point Ⓐ, what is the force of the track on the vehicle at this point? (b) What is the maximum speed the vehicle can have at point Ⓑ for gravity to hold it on the track?

7.5 Newtonian Gravitation

33. The average distance separating Earth and the Moon is 384 000 km. Use the data in Table 7.3 to find the net gravitational force exerted by Earth and the Moon on a 3.00×10^4-kg spaceship located halfway between them.

34. A satellite has a mass of 100 kg and is located at 2.00×10^6 m above the surface of Earth. (a) What is the potential energy associated with the satellite at this location? (b) What is the magnitude of the gravitational force on the satellite?

35. A coordinate system (in meters) is constructed on the surface of a pool table, and three objects are placed on the table as follows: a 2.0-kg object at the origin of the coordinate system, a 3.0-kg object at (0, 2.0), and a 4.0-kg object at (4.0, 0). Find the resultant gravitational force exerted by the other two objects on the object at the origin.

36. After the Sun exhausts its nuclear fuel, its ultimate fate may be to collapse to a *white dwarf* state. In this state, it would have approximately the same mass as it has now, but its radius would be equal to the radius of Earth. Calculate (a) the average density of the white dwarf, (b) the surface free-fall acceleration, and (c) the gravitational potential energy associated with a 1.00-kg object at the surface of the white dwarf.

37. **W** Objects with masses of 200 kg and 500 kg are separated by 0.400 m. (a) Find the net gravitational force exerted by these objects on a 50.0-kg object placed midway between them. (b) At what position (other than infinitely remote ones) can the 50.0-kg object be placed so as to experience a net force of zero?

38. Use the data of Table 7.3 to find the point between Earth and the Sun at which an object can be placed so that the net gravitational force exerted by Earth and the Sun on that object is zero.

39. A projectile is fired straight upward from the Earth's surface at the South Pole with an initial speed equal to one third the escape speed. (a) Ignoring air resistance, determine how far from the center of the Earth the projectile travels before stopping momentarily. (b) What is the altitude of the projectile at this instant?

40. Two objects attract each other with a gravitational force of magnitude 1.00×10^{-8} N when separated by 20.0 cm. If the total mass of the objects is 5.00 kg, what is the mass of each?

7.6 Kepler's Laws

41. **M** A satellite is in a circular orbit around the Earth at an altitude of 2.80×10^6 m. Find (a) the period of the orbit, (b) the speed of the satellite, and (c) the acceleration of the satellite. *Hint:* Modify Equation 7.23 so it is suitable for objects orbiting the Earth rather than the Sun.

42. An artificial satellite circling the Earth completes each orbit in 110 minutes. (a) Find the altitude of the satellite. (b) What is the value of g at the location of this satellite?

43. A satellite of Mars, called Phoebus, has an orbital radius of 9.4×10^6 m and a period of 2.8×10^4 s. Assuming the orbit is circular, determine the mass of Mars.

44. A 600-kg satellite is in a circular orbit about Earth at a height above Earth equal to Earth's mean radius. Find (a) the satellite's orbital speed, (b) the period of its revolution, and (c) the gravitational force acting on it.

45. Two satellites are in circular orbits around the Earth. Satellite A is at an *altitude* equal to the Earth's radius, while satellite B is at an altitude equal to twice the Earth's radius. What is the ratio of their periods, T_B/T_A?

Additional Problems

46. **W** A synchronous satellite, which always remains above the same point on a planet's equator, is put in circular orbit around Jupiter to study that planet's famous red spot. Jupiter rotates once every 9.84 h. Use the data of Table 7.3 to find the altitude of the satellite.

47. (a) One of the moons of Jupiter, named Io, has an orbital radius of 4.22×10^8 m and a period of 1.77 days. Assuming the orbit is circular, calculate the mass of Jupiter. (b) The largest moon of Jupiter, named Ganymede, has an orbital radius of 1.07×10^9 m and a period of 7.16 days. Calculate the mass of Jupiter from this data. (c) Are your results to parts (a) and (b) consistent? Explain.

48. Neutron stars are extremely dense objects that are formed from the remnants of supernova explosions. Many rotate very rapidly. Suppose the mass of a certain spherical neutron star is twice the mass of the Sun and its radius is 10.0 km. Determine the greatest possible angular speed the neutron star can have so that the matter at its surface on the equator is just held in orbit by the gravitational force.

49. **BIO** One method of pitching a softball is called the "windmill" delivery method, in which the pitcher's arm rotates through approximately 360° in a vertical plane before the 198-gram ball is released at the lowest point of the circular motion. An experienced pitcher can throw a ball with a speed of 98.0 mi/h. Assume the angular acceleration is uniform throughout the pitching motion and take the distance between the softball and the shoulder joint to be 74.2 cm. (a) Determine the angular speed of the arm in rev/s at the instant of release. (b) Find the value of the angular acceleration in rev/s² and the radial and tangential acceleration of the ball just before it is released. (c) Determine the force exerted on the ball by the pitcher's hand (both

radial and tangential components) just before it is released.

50. A digital audio compact disc carries data along a continuous spiral track from the inner circumference of the disc to the outside edge. Each bit occupies 0.6 μm of the track. A CD player turns the disc to carry the track counterclockwise above a lens at a constant speed of 1.30 m/s. Find the required angular speed (a) at the beginning of the recording, where the spiral has a radius of 2.30 cm, and (b) at the end of the recording, where the spiral has a radius of 5.80 cm. (c) A full-length recording lasts for 74 min, 33 s. Find the average angular acceleration of the disc. (d) Assuming the acceleration is constant, find the total angular displacement of the disc as it plays. (e) Find the total length of the track.

51. An athlete swings a 5.00-kg ball horizontally on the end of a rope. The ball moves in a circle of radius 0.800 m at an angular speed of 0.500 rev/s. What are (a) the tangential speed of the ball and (b) its centripetal acceleration? (c) If the maximum tension the rope can withstand before breaking is 100 N, what is the maximum tangential speed the ball can have?

52. A car rounds a banked curve where the radius of curvature of the road is R, the banking angle is θ, and the coefficient of static friction is μ. (a) Determine the range of speeds the car can have without slipping up or down the road. (b) What is the range of speeds possible if $R = 100$ m, $\theta = 10°$, and $\mu = 0.10$ (slippery conditions)?

53. The Solar Maximum Mission Satellite was placed in a circular orbit about 150 mi above Earth. Determine (a) the orbital speed of the satellite and (b) the time required for one complete revolution.

54. A 0.400-kg pendulum bob passes through the lowest part of its path at a speed of 3.00 m/s. (a) What is the tension in the pendulum cable at this point if the pendulum is 80.0 cm long? (b) When the pendulum reaches its highest point, what angle does the cable make with the vertical? (c) What is the tension in the pendulum cable when the pendulum reaches its highest point?

55. **Q|C** A car moves at speed v across a bridge made in the shape of a circular arc of radius r. (a) Find an expression for the normal force acting on the car when it is at the top of the arc. (b) At what minimum speed will the normal force become zero (causing the occupants of the car to seem weightless) if $r = 30.0$ m?

56. **S** Show that the escape speed from the surface of a planet of uniform density is directly proportional to the radius of the planet.

57. **Q|C** Because of Earth's rotation about its axis, a point on the equator has a centripetal acceleration of 0.034 0 m/s², whereas a point at the poles has no centripetal acceleration. (a) Show that, at the equator, the gravitational force on an object (the object's true weight) must exceed the object's apparent weight. (b) What are the apparent weights of a 75.0-kg person at the equator and at the poles? (Assume Earth is a uniform sphere and take $g = 9.800$ m/s².)

58. A small block of mass $m = 0.50$ kg is fired with an initial speed of $v_0 = 4.0$ m/s along a horizontal section of frictionless track, as shown in the top portion of Figure P7.58. The block then moves along the frictionless, semicircular, *vertical* tracks of radius $R = 1.5$ m. (a) Determine the force exerted by the track on the block at points Ⓐ and Ⓑ. (b) The bottom of the track consists of a section ($L = 0.40$ m) with friction. Determine the coefficient of kinetic friction between the block and that portion of the bottom track if the block just makes it to point Ⓒ on the first trip. *Hint:* If the block just makes it to point Ⓒ, the force of contact exerted by the track on the block at that point is zero.

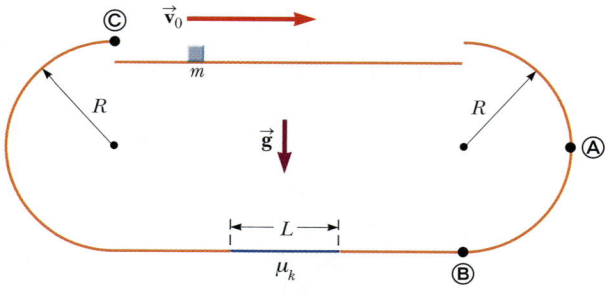

Figure P7.58

59. In Robert Heinlein's *The Moon Is a Harsh Mistress,* the colonial inhabitants of the Moon threaten to launch rocks down onto Earth if they are not given independence (or at least representation). Assuming a gun could launch a rock of mass m at twice the lunar escape speed, calculate the speed of the rock as it enters Earth's atmosphere.

60. **Q|C** A roller coaster travels in a circular path. (a) Identify the forces on a passenger at the top of the circular loop that cause centripetal acceleration. Show the direction of all forces in a sketch. (b) Identify the forces on the passenger at the bottom of the loop that produce centripetal acceleration. Show these in a sketch. (c) Based on your answers to parts (a) and (b), at what point, top or bottom, should the seat exert the greatest force on the passenger? (d) Assume the speed of the roller coaster is 4.00 m/s at the top of the loop of radius 8.00 m. Find the force exerted by the seat on a 70.0-kg passenger at the top of the loop. Then, assume the speed remains the same at the bottom of the loop and find the force exerted by the seat on the passenger at this point. Are your answers consistent with your choice of answers for parts (a) and (b)?

61. In a home laundry dryer, a cylindrical tub containing wet clothes is rotated steadily about a horizontal axis,

as shown in Figure P7.61. So that the clothes will dry uniformly, they are made to tumble. The rate of rotation of the smooth-walled tub is chosen so that a small piece of cloth will lose contact with the tub when the cloth is at an angle of $\theta = 68.0°$ above the horizontal. If the radius of the tub is $r = 0.330$ m, what rate of revolution is needed in revolutions per second?

Figure P7.61

62. **M** A model airplane of mass 0.750 kg flies with a speed of 35.0 m/s in a horizontal circle at the end of a 60.0-m control wire as shown in Figure P7.62a. The forces exerted on the airplane are shown in Figure P7.62b; the tension in the control wire, $\theta = 20.0°$ inward from the vertical. Compute the tension in the wire, assuming the wire makes a constant angle of $\theta = 20.0°$ with the horizontal.

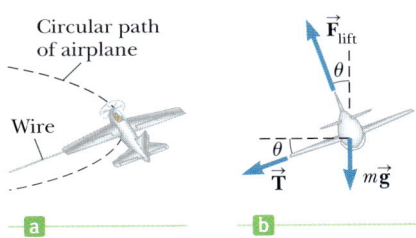

Figure P7.62

63. **S** A skier starts at rest at the top of a large hemispherical hill (Fig. P7.63). Neglecting friction, show that the skier will leave the hill and become airborne at a distance $h = R/3$ below the top of the hill. *Hint:* At this point, the normal force goes to zero.

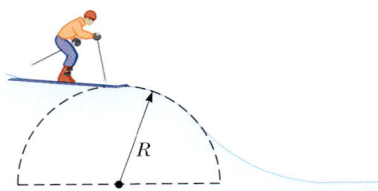

Figure P7.63

64. Casting of molten metal is important in many industrial processes. *Centrifugal casting* is used for manufacturing pipes, bearings, and many other structures. A cylindrical enclosure is rotated rapidly and steadily about a horizontal axis, as in Figure P7.64. Molten metal is poured into the rotating cylinder and then cooled, forming the finished product. Turning the cylinder at a high rotation rate forces the solidifying metal strongly to the outside. Any bubbles are displaced toward the axis so that unwanted voids will not be present in the casting.

Suppose a copper sleeve of inner radius 2.10 cm and outer radius 2.20 cm is to be cast. To eliminate bubbles and give high structural integrity, the centripetal acceleration of each bit of metal should be $100g$. What rate of rotation is required? State the answer in revolutions per minute.

Figure P7.64

65. Suppose a 1 800-kg car passes over a bump in a roadway that follows the arc of a circle of radius 20.4 m, as in Figure P7.65. (a) What force does the road exert on the car as the car passes the highest point of the bump if the car travels at 8.94 m/s? (b) What is the maximum speed the car can have without losing contact with the road as it passes this highest point?

Figure P7.65

66. A stuntman whose mass is 70 kg swings from the end of a 4.0-m-long rope along the arc of a vertical circle. Assuming he starts from rest when the rope is horizontal, find the tensions in the rope that are required to make him follow his circular path (a) at the beginning of his motion, (b) at a height of 1.5 m above the bottom of the circular arc, and (c) at the bottom of the arc.

67. **Q|C** A minimum-energy orbit to an outer planet consists of putting a spacecraft on an elliptical trajectory with the departure planet corresponding to the perihelion of the ellipse, or closest point to the Sun, and the arrival planet corresponding to the aphelion of the ellipse, or farthest point from the Sun. (a) Use Kepler's third law to calculate how long it would take to go from Earth to Mars on such an orbit. (Answer in years.) (b) Can such an orbit be undertaken at any time? Explain.

68. **Q|C** The pilot of an airplane executes a constant-speed loop-the-loop maneuver in a vertical circle as in Figure 7.15b. The speed of the airplane is 2.00×10^2 m/s, and the radius of the circle is 3.20×10^3 m. (a) What is the pilot's apparent weight at the lowest

point of the circle if his true weight is 712 N? (b) What is his apparent weight at the highest point of the circle? (c) Describe how the pilot could experience weightlessness if both the radius and the speed can be varied. *Note:* His apparent weight is equal to the magnitude of the force exerted by the seat on his body. Under what conditions does this occur? (d) What speed would have resulted in the pilot experiencing weightlessness at the top of the loop?

69. **S** A piece of mud is initially at point A on the rim of a bicycle wheel of radius R rotating clockwise about a horizontal axis at a constant angular speed ω (Fig. P7.8). The mud dislodges from point A when the wheel diameter through A is horizontal. The mud then rises vertically and returns to point A. (a) Find a symbolic expression in terms of R, ω, and g for the total time the mud is in the air and returns to point A. (b) If the wheel makes one complete revolution in the time it takes the mud to return to point A, find an expression for the angular speed of the bicycle wheel ω in terms of π, g, and R.

70. **Q|C** A 0.275-kg object is swung in a *vertical* circular path on a string 0.850 m long as in Figure P7.70. (a) What are the forces acting on the ball at any point along this path? (b) Draw free-body diagrams for the ball when it is at the bottom of the circle and when it is at the top. (c) If its speed is 5.20 m/s at the top of the circle, what is the tension in the string there? (d) If the string breaks when its tension exceeds 22.5 N, what is the maximum speed the object can have at the bottom before the string breaks?

Figure P7.70

71. A 4.00-kg object is attached to a vertical rod by two strings as shown in Figure P7.71. The object rotates in a horizontal circle at constant speed 6.00 m/s. Find the tension in (a) the upper string and (b) the lower string.

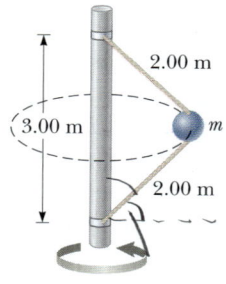

Figure P7.71

72. **BIO** The maximum lift force on a bat is proportional to the square of its flying speed v. For the hoary bat (*Lasiurus cinereus*), the magnitude of the lift force is given by

$$F_L \leq (0.018 \text{ N} \cdot \text{s}^2/\text{m}^2)v^2$$

The bat can fly in a horizontal circle by "banking" its wings at an angle θ, as shown in Figure P7.72. In this situation, the magnitude of the vertical component of the lift force must equal the bat's weight. The horizontal component of the force provides the centripetal acceleration. (a) What is the minimum speed that the bat can have if its mass is 0.031 kg? (b) If the maximum speed of the bat is 10 m/s, what is the maximum banking angle that allows the bat to stay in a horizontal plane? (c) What is the radius of the circle of its flight when the bat flies at its maximum speed? (d) Can the bat turn with a smaller radius by flying more slowly?

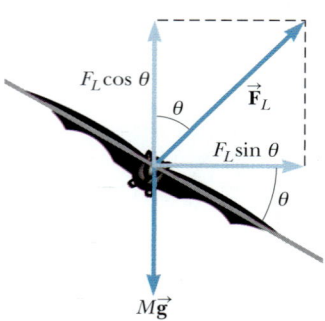

Figure P7.72

73. (a) A luggage carousel at an airport has the form of a section of a large cone, steadily rotating about its vertical axis. Its metallic surface slopes downward toward the outside, making an angle of 20.0° with the horizontal. A 30.0-kg piece of luggage is placed on the carousel, 7.46 m from the axis of rotation. The travel bag goes around once in 38.0 s. Calculate the force of static friction between the bag and the carousel. (b) The drive motor is shifted to turn the carousel at a higher constant rate of rotation, and the piece of luggage is bumped to a position 7.94 m from the axis of rotation. The bag is on the verge of slipping as it goes around once every 34.0 s. Calculate the coefficient of static friction between the bag and the carousel.

74. A 0.50-kg ball that is tied to the end of a 1.5-m light cord is revolved in a horizontal plane, with the cord making a 30° angle with the vertical. (See Fig. P7.74.) (a) Determine the ball's speed. (b) If, instead, the ball is revolved so that its speed is 4.0 m/s, what angle does the cord make with the vertical? (c) If the cord can withstand a maximum tension of 9.8 N, what is the highest speed at which the ball can move?

Figure P7.74

75. In a popular amusement park ride, a rotating cylinder of radius 3.00 m is set in rotation at an angular speed of 5.00 rad/s, as in Figure P7.75. The floor then drops away, leaving the riders suspended against the wall in a vertical position. What minimum coefficient of friction between a rider's clothing and the wall is needed to keep the rider from slipping? *Hint:* Recall that the magnitude of the maximum force

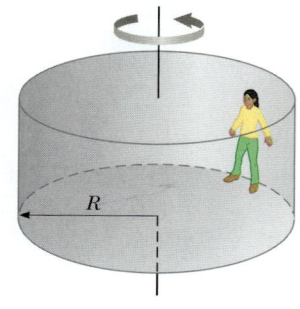

Figure P7.75

of static friction is equal to $\mu_s n$, where n is the normal force—in this case, the force causing the centripetal acceleration.

76. A massless spring of constant $k = 78.4$ N/m is fixed on the left side of a level track. A block of mass $m = 0.50$ kg is pressed against the spring and compresses it a distance d, as in Figure P7.76. The block (initially at rest) is then released and travels toward a circular loop-the-loop of radius $R = 1.5$ m. The entire track and the loop-the-loop are frictionless, except for the section of track between points A and B. Given that the coefficient of kinetic friction between the block and the track along AB is $\mu_k = 0.30$, and that the length of AB is 2.5 m, determine the minimum compression d of the spring that enables the block to just make it through the loop-the-loop at point C. *Hint:* The force exerted by the track on the block will be zero if the block barely makes it through the loop-the-loop.

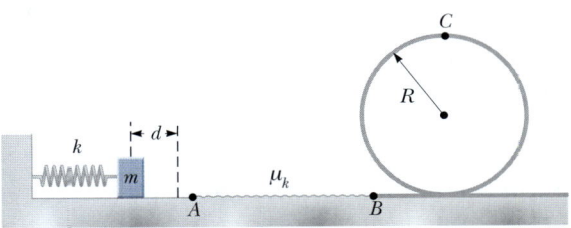

Figure P7.76

Wind exerts forces on the propellers of this wind turbine, producing a torque that causes the turbine to rotate. This process converts the kinetic energy of wind to rotational kinetic energy, which is transformed by electromagnetic induction to electrical energy.

8 Rotational Equilibrium and Rotational Dynamics

8.1 Torque

8.2 Torque and the Two Conditions for Equilibrium

8.3 The Center of Gravity

8.4 Examples of Objects in Equilibrium

8.5 Relationship Between Torque and Angular Acceleration

8.6 Rotational Kinetic Energy

8.7 Angular Momentum

In the study of linear motion, objects were treated as point particles without structure. It didn't matter *where* a force was applied, only *whether* it was applied or not.

The reality is that the point of application of a force *does* matter. In football, for example, if the ball carrier is tackled near his midriff, he might carry the tackler several yards before falling. If tackled well below the waistline, however, his center of mass rotates toward the ground, and he can be brought down immediately. Tennis provides another good example. If a tennis ball is struck with a strong horizontal force acting through its center of mass, it may travel a long distance before hitting the ground, far out of bounds. Instead, the same force applied in an upward, glancing stroke will impart topspin to the ball, which can cause it to land in the opponent's court.

The concepts of rotational equilibrium and rotational dynamics are also important in other disciplines. For example, students of architecture benefit from understanding the forces that act on buildings, and biology students should understand the forces at work in muscles and on bones and joints. These forces create torques, which tell us how the forces affect an object's equilibrium and rate of rotation.

We will find that an object remains in a state of uniform rotational motion unless acted on by a net torque. That principle is the equivalent of Newton's first law. Further, the angular acceleration of an object is proportional to the net torque acting on it, which is the analog of Newton's second law. A net torque acting on an object causes a change in its rotational energy.

Finally, torques applied to an object through a given time interval can change the object's angular momentum. In the absence of external torques, angular momentum is conserved, a property that explains some of the mysterious and formidable properties of pulsars, remnants of supernova explosions that rotate at equatorial speeds approaching that of light.

8.1 Torque

LEARNING OBJECTIVES
1. Define torque and state the rotational analog of the first law.
2. Apply the definition of torque to elementary systems.

Figure 8.1 A bird's-eye view of a door hinged at O, with a force applied perpendicular to the door.

Forces cause accelerations; *torques* cause angular accelerations. There is a definite relationship, however, between the two concepts.

Figure 8.1 depicts an overhead view of a door hinged at point O. From this viewpoint, the door is free to rotate around an axis perpendicular to the page and passing through O. If a force $\vec{F}$ is applied to the door, there are three factors that determine the effectiveness of the force in opening the door: the *magnitude* of the force, the *position* of application of the force, and the *angle* at which it is applied.

For simplicity, we restrict our discussion to position and force vectors lying in a plane. When the applied force $\vec{F}$ is perpendicular to the outer edge of the door, as in Figure 8.1, the door rotates counterclockwise with constant angular acceleration. The same perpendicular force applied at a point nearer the hinge results in a smaller angular acceleration. In general, a larger radial distance r between the applied force and the axis of rotation results in a larger angular acceleration. Similarly, a larger applied force will also result in a larger angular acceleration. These considerations motivate the basic definition of **torque** for the special case of forces perpendicular to the position vector:

◀ Basic definition of torque

Let $\vec{F}$ be a force acting on an object, and let $\vec{r}$ be a position vector from a chosen point O to the point of application of the force, with $\vec{F}$ perpendicular to $\vec{r}$. The magnitude of the torque $\vec{\tau}$ exerted by the force $\vec{F}$ is given by

$$\tau = rF \quad [8.1]$$

where r is the length of the position vector and F is the magnitude of the force.

SI unit: Newton-meter (N · m)

The vectors $\vec{r}$ and $\vec{F}$ lie in a plane. Figure 8.2 illustrates how the point of the force's application affects the magnitude of the torque. As discussed in detail shortly in conjunction with Figure 8.6, the torque $\vec{\tau}$ is then perpendicular to this plane. The point O is usually chosen to coincide with the axis the object is rotating around, such as the hinge of a door or hub of a merry-go-round. (Other choices are possible as well.) In addition, we consider only forces acting in the plane perpendicular to the axis of rotation. This criterion excludes, for example, a force with upward component on a merry-go-round railing, which cannot affect the merry-go-round's rotation.

Under these conditions, an object can rotate around the chosen axis in one of two directions. By convention, counterclockwise is taken to be the positive direction, clockwise the negative direction. When an applied force causes an object to

Figure 8.2 As the force is applied farther out along the wrench, the magnitude of the torque increases.

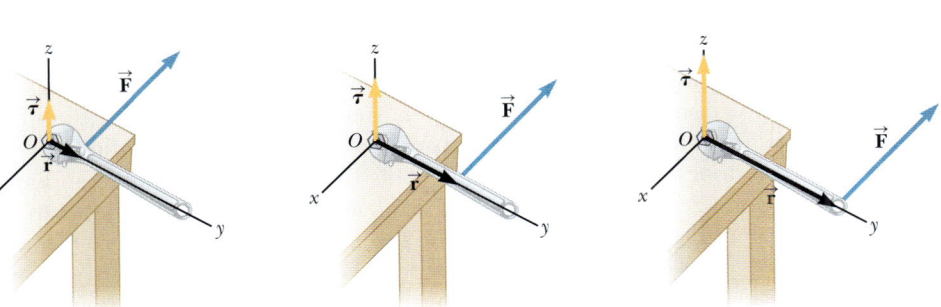

a b c

rotate counterclockwise, the torque on the object is positive. When the force causes the object to rotate clockwise, the torque on the object is negative. When two or more torques act on an object at rest, the torques are added. If the net torque isn't zero, the object starts rotating at an ever-increasing rate. If the net torque is zero, the object's rate of rotation doesn't change. These considerations lead to the rotational analog of the first law: **the rate of rotation of an object doesn't change, unless the object is acted on by a net torque.**

■ EXAMPLE 8.1 | Battle of the Revolving Door

GOAL Apply the basic definition of torque.

PROBLEM Two disgruntled businesspeople are trying to use a revolving door, as in Figure 8.3. The woman on the left exerts a force of 625 N perpendicular to the door and 1.20 m from the hub's center, while the man on the right exerts a force of 8.50×10^2 N perpendicular to the door and 0.800 m from the hub's center. Find the net torque on the revolving door.

STRATEGY Calculate the individual torques on the door using the definition of torque, Equation 8.1, and then sum to find the net torque on the door. The woman exerts a negative torque, the man a positive torque. Their positions of application also differ.

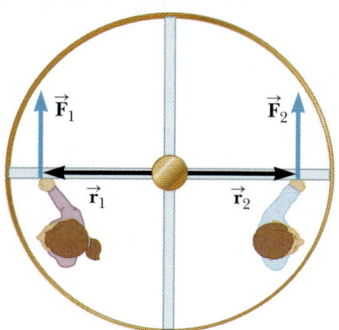

Figure 8.3 (Example 8.1)

SOLUTION

Calculate the torque exerted by the woman. A negative sign must be supplied because $\vec{F}_1$, if unopposed, would cause a clockwise rotation:

$$\tau_1 = -r_1 F_1 = -(1.20 \text{ m})(625 \text{ N}) = -7.50 \times 10^2 \text{ N} \cdot \text{m}$$

Calculate the torque exerted by the man. The torque is positive because $\vec{F}_2$, if unopposed, would cause a counterclockwise rotation:

$$\tau_2 = r_2 F_2 = (0.800 \text{ m})(8.50 \times 10^2 \text{ N}) = 6.80 \times 10^2 \text{ N} \cdot \text{m}$$

Sum the torques to find the net torque on the door:

$$\tau_{\text{net}} = \tau_1 + \tau_2 = \boxed{-7.0 \times 10^1 \text{ N} \cdot \text{m}}$$

REMARKS The negative result here means that the net torque will produce a clockwise rotation.

QUESTION 8.1 What happens if the woman suddenly slides closer to the hub by 0.400 m?

EXERCISE 8.1 A businessman enters the same revolving door on the right, pushing with 576 N of force directed perpendicular to the door and 0.700 m from the hub, while a boy exerts a force of 365 N perpendicular to the door, 1.25 m to the left of the hub. Find (a) the torques exerted by each person and (b) the net torque on the door.

ANSWERS (a) $\tau_{\text{boy}} = -456 \text{ N} \cdot \text{m}$, $\tau_{\text{man}} = 403 \text{ N} \cdot \text{m}$ (b) $\tau_{\text{net}} = -53 \text{ N} \cdot \text{m}$

The applied force isn't always perpendicular to the position vector $\vec{r}$. Suppose the force $\vec{F}$ exerted on a door is directed away from the axis, as in Figure 8.4a, say, by someone's grasping the doorknob and pushing to the right. Exerting the force in this direction couldn't possibly open the door. However, if the applied force acts at an angle to the door as in Figure 8.4b, the component of the force *perpendicular* to the door will cause it to rotate. This figure shows that the component of the force perpendicular to the door is $F \sin \theta$, where θ is the angle between the position vector $\vec{r}$ and the force $\vec{F}$. When the force is directed away from the axis, $\theta = 0°$, $\sin(0°) = 0$, and $F \sin(0°) = 0$. When the force is directed toward the axis, $\theta = 180°$ and $F \sin(180°) = 0$. The maximum absolute value of $F \sin \theta$ is attained only when $\vec{F}$ is perpendicular to $\vec{r}$—that is, when $\theta = 90°$ or $\theta = 270°$. These considerations motivate a more general definition of torque:

Figure 8.5 As the angle between the position vector and force vector increases in parts (a)–(b), the torque exerted by the wrench increases.

Figure 8.4 (a) A force $\vec{F}$ acting at an angle $\theta = 0°$ exerts zero torque about the pivot O. (b) The part of the force perpendicular to the door, $F\sin\theta$, exerts torque $rF\sin\theta$ about O. (c) An alternate interpretation of torque in terms of a *lever arm* $d = r\sin\theta$.

Let $\vec{F}$ be a force acting on an object, and let $\vec{r}$ be a position vector from a chosen point O to the point of application of the force. The magnitude of the torque $\vec{\tau}$ exerted by the force $\vec{F}$ is

$$\tau = rF\sin\theta \qquad [8.2]$$

where r is the length of the position vector, F the magnitude of the force, and θ the angle between $\vec{r}$ and $\vec{F}$.

SI unit: Newton-meter (N · m)

◀ General definition of torque

Again the vectors $\vec{r}$ and $\vec{F}$ lie in a plane, and for our purposes the chosen point O will usually correspond to an axis of rotation perpendicular to the plane. Figure 8.5 illustrates how the magnitude of the torque exerted by a wrench increases as the angle between the position vector and the force vector increases at 90°, where the torque is a maximum.

A second way of understanding the $\sin\theta$ factor is to associate it with the magnitude r of the position vector $\vec{r}$. The quantity $d = r\sin\theta$ is called the **lever arm**, which is the perpendicular distance from the axis of rotation to a line drawn along the direction of the force. This alternate interpretation is illustrated in Figure 8.4c.

It's important to remember that **the value of τ depends on the chosen axis of rotation.** Torques can be computed around any axis, regardless of whether there is some actual, physical rotation axis present. Once the point is chosen, however, it must be used consistently throughout a given problem.

Torque is a vector perpendicular to the plane determined by the position and force vectors, as illustrated in Figure 8.6. The direction can be determined by the *right-hand rule*:

1. Point the fingers of your right hand in the direction of $\vec{r}$.
2. Curl your fingers toward the direction of vector $\vec{F}$.
3. Your thumb then points approximately in the direction of the torque, in this case out of the page.

Notice the two choices of angle in Figure 8.6. The angle θ is the actual angle between the directions of the two vectors. The angle θ' is literally "between"

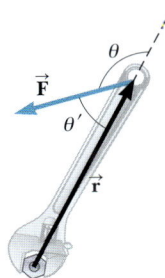

Figure 8.6 The right-hand rule: Point the fingers of your right hand along $\vec{r}$ and curl them in the direction of $\vec{F}$. Your thumb then points in the direction of the torque (out of the page, in this case). Note that either θ or θ' can be used in the definition of torque.

the two vectors. Which angle is correct? Because $\sin \theta = \sin(180° - \theta) = \sin(180°)\cos\theta - \sin\theta\cos(180°) = 0 - \sin\theta \cdot (-1) = \sin\theta$, either angle is correct. Problems used in this book will be confined to objects rotating around an axis perpendicular to the plane containing $\vec{r}$ and $\vec{F}$, so if these vectors are in the plane of the page, the torque will always point either into or out of the page, parallel to the axis of rotation. If your right thumb is pointed in the direction of a torque, your fingers curl naturally in the direction of rotation that the torque would produce on an object at rest.

■ EXAMPLE 8.2 | The Swinging Door

GOAL Apply the more general definition of torque.

PROBLEM (a) A man applies a force of $F = 3.00 \times 10^2$ N at an angle of 60.0° to the door of Figure 8.7a, 2.00 m from well-oiled hinges. Find the torque on the door, choosing the position of the hinges as the axis of rotation. (b) Suppose a wedge is placed 1.50 m from the hinges on the other side of the door. What minimum force must the wedge exert so that the force applied in part (a) won't open the door?

STRATEGY Part (a) can be solved by substitution into the general torque equation. In part (b) the hinges, the wedge, and the applied force all exert torques on the door. The door doesn't open, so the sum of these torques must be zero, a condition that can be used to find the wedge force.

Figure 8.7 (Example 8.2a) (a) Top view of a door being pushed by a 300-N force. (b) The components of the 300-N force.

SOLUTION

(a) Compute the torque due to the applied force exerted at 60.0°.

Substitute into the general torque equation:

$$\tau_F = rF\sin\theta = (2.00 \text{ m})(3.00 \times 10^2 \text{ N})\sin 60.0°$$
$$= (2.00 \text{ m})(2.60 \times 10^2 \text{ N}) = \boxed{5.20 \times 10^2 \text{ N} \cdot \text{m}}$$

(b) Calculate the force exerted by the wedge on the other side of the door.

Set the sum of the torques equal to zero:

$$\tau_{\text{hinge}} + \tau_{\text{wedge}} + \tau_F = 0$$

The hinge force provides no torque because it acts at the axis ($r = 0$). The wedge force acts at an angle of $-90.0°$, opposite the upward 260 N component.

$$0 + F_{\text{wedge}}(1.50 \text{ m})\sin(-90.0°) + 5.20 \times 10^2 \text{ N} \cdot \text{m} = 0$$
$$F_{\text{wedge}} = \boxed{347 \text{ N}}$$

REMARKS Notice that the angle from the position vector to the wedge force is $-90°$. That's because, starting at the position vector, it's necessary to go 90° clockwise (the negative angular direction) to get to the force vector. Measuring the angle that way automatically supplies the correct sign for the torque term and is consistent with the right-hand rule. Alternately, the magnitude of the torque can be found and the correct sign chosen based on physical intuition. Figure 8.7b illustrates the fact that the component of the force perpendicular to the lever arm causes the torque.

QUESTION 8.2 To make the wedge more effective in keeping the door closed, should it be placed closer to the hinge or to the doorknob?

EXERCISE 8.2 A man ties one end of a strong rope 8.00 m long to the bumper of his truck, 0.500 m from the ground, and the other end to a vertical tree trunk at a height of 3.00 m. He uses the truck to create a tension of 8.00×10^2 N in the rope. Compute the magnitude of the torque on the tree due to the tension in the rope, with the base of the tree acting as the reference point.

ANSWER 2.28×10^3 N · m

8.2 Torque and the Two Conditions for Equilibrium

LEARNING OBJECTIVE

1. State the two conditions of mechanical equilibrium and apply them to elementary systems.

An object in mechanical equilibrium must satisfy the following two conditions:

1. **The net external force must be zero:** $\sum \vec{F} = 0$
2. **The net external torque must be zero:** $\sum \vec{\tau} = 0$

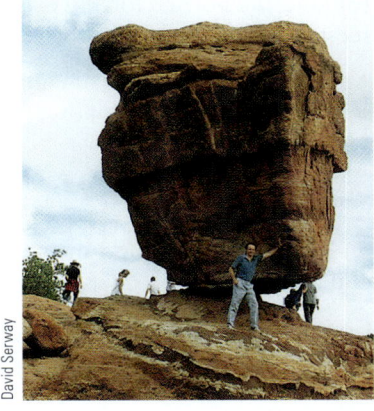

This large balanced rock at the Garden of the Gods in Colorado Springs, Colorado, is in mechanical equilibrium.

The first condition is a statement of translational equilibrium: The sum of all forces acting on the object must be zero, so the object has no translational acceleration, $\vec{a} = 0$. The second condition is a statement of rotational equilibrium: The sum of all torques on the object must be zero, so the object has no angular acceleration, $\vec{\alpha} = 0$. For an object to be in equilibrium, it must move through space at a constant speed and rotate at a constant angular speed.

Because we can choose any location for calculating torques, it's usually best to select an axis that will make at least one torque equal to zero, just to simplify the net torque equation.

■ EXAMPLE 8.3 | Balancing Act

GOAL Apply the conditions of equilibrium and illustrate the use of different axes for calculating the net torque on an object.

PROBLEM A woman of mass $m = 55.0$ kg sits on the left end of a seesaw—a plank of length $L = 4.00$ m, pivoted in the middle as in Figure 8.8. **(a)** First compute the torques on the seesaw about an axis that passes through the pivot point. Where should a man of mass $M = 75.0$ kg sit if the system (seesaw plus man and woman) is to be balanced? **(b)** Find the normal force exerted by the pivot if the plank has a mass of $m_{pl} = 12.0$ kg. **(c)** Repeat part (a), but this time compute the torques about an axis through the left end of the plank.

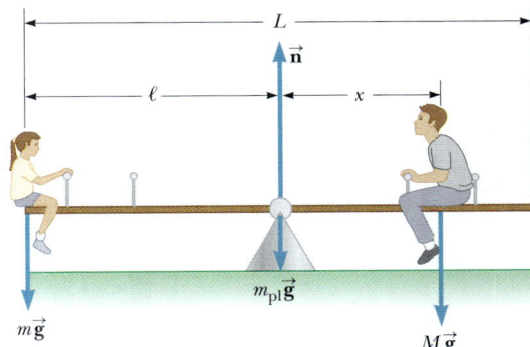

Figure 8.8 (Example 8.3) The system consists of two people and a seesaw. Because the sum of the forces and the sum of the torques acting on the system are both zero, the system is said to be in equilibrium.

STRATEGY In part (a), apply the second condition of equilibrium, $\sum \tau = 0$, computing torques around the pivot point. The mass of the plank forming the seesaw is distributed evenly on either side of the pivot point, so the torque exerted by gravity on the plank, τ_{plank}, can be computed as if all the plank's mass is concentrated at the pivot point. Then τ_{plank} is zero, as is the torque exerted by the pivot, because their lever arms are zero. In part (b) the first condition of equilibrium, $\sum \vec{F} = 0$, must be applied. Part (c) is a repeat of part (a) showing that choice of a different axis yields the same answer.

SOLUTION

(a) Where should the man sit to balance the seesaw?

Apply the second condition of equilibrium to the plank by setting the sum of the torques equal to zero:

$$\tau_{pivot} + \tau_{plank} + \tau_{man} + \tau_{woman} = 0$$

The first two torques are zero. Let x represent the man's distance from the pivot. The woman is at a distance $\ell = L/2$ from the pivot.

$$0 + 0 - Mgx + mg(L/2) = 0$$

Solve this equation for x and evaluate it:

$$x = \frac{m(L/2)}{M} = \frac{(55.0 \text{ kg})(2.00 \text{ m})}{75.0 \text{ kg}} = \boxed{1.47 \text{ m}}$$

(Continued)

(b) Find the normal force n exerted by the pivot on the seesaw.

Apply for first condition of equilibrium to the plank, solving the resulting equation for the unknown normal force, n:

$$-Mg - mg - m_{pl}g + n = 0$$
$$n = (M + m + m_{pl})g$$
$$= (75.0 \text{ kg} + 55.0 \text{ kg} + 12.0 \text{ kg})(9.80 \text{ m/s}^2)$$
$$n = \boxed{1.39 \times 10^3 \text{ N}}$$

(c) Repeat part (a), choosing a new axis through the left end of the plank.

Compute the torques using this axis, and set their sum equal to zero. Now the pivot and gravity forces on the plank result in nonzero torques.

$$\tau_{man} + \tau_{woman} + \tau_{plank} + \tau_{pivot} = 0$$
$$-Mg(L/2 + x) + mg(0) - m_{pl}g(L/2) + n(L/2) = 0$$

Substitute all known quantities:

$$-(75.0 \text{ kg})(9.80 \text{ m/s}^2)(2.00 \text{ m} + x) + 0$$
$$- (12.0 \text{ kg})(9.80 \text{ m/s}^2)(2.00 \text{ m}) + n(2.00 \text{ m}) = 0$$
$$-(1.47 \times 10^3 \text{ N} \cdot \text{m}) - (735 \text{ N})x - (235 \text{ N} \cdot \text{m})$$
$$+ (2.00 \text{ m})n = 0$$

Solve for x, substituting the normal force found in part (b):

$$x = \boxed{1.46 \text{ m}}$$

REMARKS The answers for x in parts (a) and (c) agree except for a small rounding discrepancy. That illustrates how choosing a different axis leads to the same solution.

QUESTION 8.3 What happens if the woman now leans backwards?

EXERCISE 8.3 Suppose a 30.0-kg child sits 1.50 m to the left of center on the same seesaw. A second child sits at the end on the opposite side, and the system is balanced. (a) Find the mass of the second child. (b) Find the normal force acting at the pivot point.

ANSWERS (a) 22.5 kg (b) 632 N

8.3 The Center of Gravity

LEARNING OBJECTIVES

1. Define center of gravity and qualitatively determine it for homogeneous, symmetric bodies.
2. Calculate the center of gravity for individual objects and for systems of objects.

In the example of the seesaw in the previous section, we guessed that the torque due to the force of gravity on the plank was the same as if all the plank's weight were concentrated at its center. That's a general procedure: To compute the torque on a rigid body due to the force of gravity, the body's entire weight can be thought of as concentrated at a single point. The problem then reduces to finding the location of that point. If the body is homogeneous (its mass is distributed evenly) and symmetric, it's usually possible to guess the location of that point, as in Example 8.3. Otherwise, it's necessary to calculate the point's location, as explained in this section.

Consider an object of arbitrary shape lying in the xy-plane, as in Figure 8.9. The object is divided into a large number of very small particles of weight m_1g, m_2g, m_3g, ... having coordinates (x_1, y_1), (x_2, y_2), (x_3, y_3), ... If the object is free to rotate around the origin, each particle contributes a torque about the origin that is equal to its weight multiplied by its lever arm. For example, the torque due to the weight m_1g is m_1gx_1, and so forth.

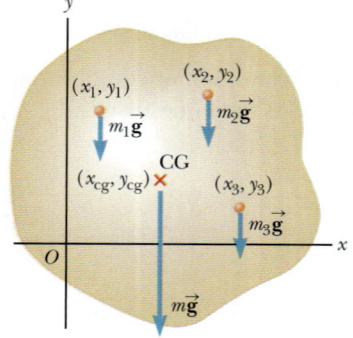

Figure 8.9 The net gravitational torque on an object is zero if computed around the center of gravity. The object will balance if supported at that point (or at any point along a vertical line above or below that point).

8.3 The Center of Gravity

We wish to locate the point of application of the single force of magnitude $w = F_g = Mg$ (the total weight of the object), where the effect on the rotation of the object is the same as that of the individual particles. That point is called the object's **center of gravity**. Equating the torque exerted by w at the center of gravity to the sum of the torques acting on the individual particles gives

$$(m_1 g + m_2 g + m_3 g + \cdots)x_{cg} = m_1 g x_1 + m_2 g x_2 + m_3 g x_3 + \cdots$$

We assume that g is the same everywhere in the object (which is true for all objects we will encounter). Then the g factors in the preceding equation cancel, resulting in

$$x_{cg} = \frac{m_1 x_1 + m_2 x_2 + m_3 x_3 + \cdots}{m_1 + m_2 + m_3 + \cdots} = \frac{\sum m_i x_i}{\sum m_i} \quad [8.3a]$$

where x_{cg} is the x-coordinate of the center of gravity. Similarly, the y-coordinate and z-coordinate of the center of gravity of the system can be found from

$$y_{cg} = \frac{\sum m_i y_i}{\sum m_i} \quad [8.3b]$$

and

$$z_{cg} = \frac{\sum m_i z_i}{\sum m_i} \quad [8.3c]$$

Tip 8.1 Specify Your Axis
Choose the axis of rotation and use that axis exclusively throughout a given problem. The axis need not correspond to a physical axle or pivot point. Any convenient point will do.

These three equations are identical to the equations for a similar concept called **center of mass**. The center of mass and center of gravity of an object are exactly the same when g doesn't vary significantly over the object.

It's often possible to guess the location of the center of gravity. **The center of gravity of a homogeneous, symmetric body must lie on the axis of symmetry.** For example, the center of gravity of a homogeneous rod lies midway between the ends of the rod, and the center of gravity of a homogeneous sphere or a homogeneous cube lies at the geometric center of the object. The center of gravity of an irregularly shaped object, such as a wrench, can be determined experimentally by suspending the wrench from two different arbitrary points (Fig. 8.10). The wrench is first hung from point A, and a vertical line AB (which can be established with a plumb bob) is drawn when the wrench is in equilibrium. The wrench is then hung from point C, and a second vertical line CD is drawn. The center of gravity coincides with the intersection of these two lines. In fact, if the wrench is hung freely from any point, the center of gravity always lies straight below the point of support, so the vertical line through that point must pass through the center of gravity.

Several examples in Section 8.4 involve homogeneous, symmetric objects where the centers of gravity coincide with their geometric centers. A rigid object in a uniform gravitational field can be balanced by a single force equal in magnitude to the weight of the object, as long as the force is directed upward through the object's center of gravity.

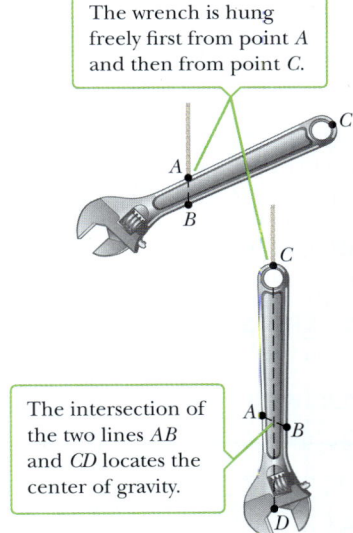

The wrench is hung freely first from point A and then from point C.

The intersection of the two lines AB and CD locates the center of gravity.

Figure 8.10 An experimental technique for determining the center of gravity of a wrench.

■ EXAMPLE 8.4 Where Is the Center of Gravity?

GOAL Find the center of gravity of a system of objects.

PROBLEM (a) Three objects are located in a coordinate system as shown in Figure 8.11a. Find the center of gravity. (b) How does the answer change if the object on the left is displaced upward by 1.00 m and the object on the right is displaced downward by 0.500 m (Figure 8.11b)? Treat the objects as point particles.

STRATEGY The y-coordinate and z-coordinate of the center of gravity in part (a) are both zero because all the objects are on the x-axis. We can find the x-coordinate of the center of gravity using Equation 8.3a. Part (b) requires Equation 8.3b.

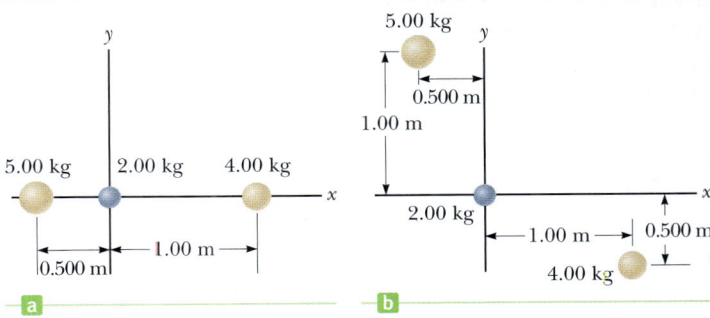

Figure 8.11 (Example 8.4) Locating the center of gravity of a system of three particles.

(Continued)

SOLUTION

(a) Find the center of gravity of the system in Figure 8.11a.

Apply Equation 8.3a to the system of three objects:

$$(1) \quad x_{cg} = \frac{\sum m_i x_i}{\sum m_i} = \frac{m_1 x_1 + m_2 x_2 + m_3 x_3}{m_1 + m_2 + m_3}$$

Compute the numerator of Equation (1):

$$\sum m_i x_i = m_1 x_1 + m_2 x_2 + m_3 x_3$$
$$= (5.00 \text{ kg})(-0.500 \text{ m}) + (2.00 \text{ kg})(0 \text{ m}) + (4.00 \text{ kg})(1.00 \text{ m})$$
$$= 1.50 \text{ kg} \cdot \text{m}$$

Substitute the denominator, $\sum m_i = 11.0$ kg, and the numerator into Equation (1).

$$x_{cg} = \frac{1.50 \text{ kg} \cdot \text{m}}{11.0 \text{ kg}} = \boxed{0.136 \text{ m}}$$

(b) How does the answer change if the positions of the objects are changed as in Figure 8.11b?

Because the x-coordinates have not been changed, the x-coordinate of the center of gravity is also unchanged:

$$x_{cg} = \boxed{0.136 \text{ m}}$$

Write Equation 8.3b:

$$y_{cg} = \frac{\sum m_i y_i}{\sum m_i} = \frac{m_1 y_1 + m_2 y_2 + m_3 y_3}{m_1 + m_2 + m_3}$$

Substitute values:

$$y_{cg} = \frac{(5.00 \text{ kg})(1.00 \text{ m}) + (2.00 \text{ kg})(0 \text{ m}) + (4.00 \text{ kg})(-0.500 \text{ m})}{5.00 \text{ kg} + 2.00 \text{ kg} + 4.00 \text{ kg}}$$

$$y_{cg} = \boxed{0.273 \text{ m}}$$

REMARKS Notice that translating objects in the y-direction doesn't change the x-coordinate of the center of gravity. The three components of the center of gravity are each independent of the other two coordinates.

QUESTION 8.4 If 1.00 kg is added to the masses on the left and right in Figure 8.11a, does the center of mass (a) move to the left, (b) move to the right, or (c) remain in the same position?

EXERCISE 8.4 If a fourth particle of mass 2.00 kg is placed at (0, 0.25 m) in Figure 8.11a, find the x- and y-coordinates of the center of gravity for this system of four particles.

ANSWER $x_{cg} = 0.115$ m; $y_{cg} = 0.0385$ m

EXAMPLE 8.5 Locating Your Lab Partner's Center of Gravity BIO

GOAL Use torque to find a center of gravity.

PROBLEM In this example we show how to find the location of a person's center of gravity. Suppose your lab partner has a height L of 173 cm (5 ft, 8 in.) and a weight w of 715 N (160 lb). You can determine the position of his center of gravity by having him stretch out on a uniform board supported at one end by a scale, as shown in Figure 8.12. If the board's weight w_b is 49 N and the scale reading F is 3.50×10^2 N, find the distance of your lab partner's center of gravity from the left end of the board.

Figure 8.12 (Example 8.5) Determining your lab partner's center of gravity.

STRATEGY To find the position x_{cg} of the center of gravity, compute the torques using an axis through O. There is no torque due to the normal force $\vec{n}$ because its moment arm is zero about an axis through O. Set the sum of the torques equal to zero and solve for x_{cg}.

SOLUTION

Apply the second condition of equilibrium:

$$\sum \tau_i = \tau_n + \tau_w + \tau_{w_b} + \tau_F = 0$$

Substitute expressions for the torques:

$$0 - wx_{cg} - w_b(L/2) + FL = 0$$

Solve for x_{cg} and substitute known values:

$$x_{cg} = \frac{FL - w_b(L/2)}{w}$$

$$= \frac{(350 \text{ N})(173 \text{ cm}) - (49 \text{ N})(86.5 \text{ cm})}{715 \text{ N}} = \boxed{79 \text{ cm}}$$

REMARKS The given information is sufficient only to determine the x-coordinate of the center of gravity. The other two coordinates can be estimated, based on the body's symmetry.

QUESTION 8.5 What would happen if a support is placed exactly at $x = 79$ cm followed by the removal of the supports at the subject's head and feet?

EXERCISE 8.5 Suppose a 416-kg alligator of length 3.5 m is stretched out on a board of the same length weighing 65 N. If the board is supported on the ends as in Figure 8.12, and the scale reads 1 880 N, find the x-component of the alligator's center of gravity.

ANSWER 1.59 m

8.4 Examples of Objects in Equilibrium

LEARNING OBJECTIVE
1. Apply the conditions of mechanical equilibrium to rigid bodies.

Recall from Chapter 4 that when an object is treated as a geometric point, equilibrium requires only that the net force on the object is zero. In this chapter we have shown that for extended objects a second condition for equilibrium must also be satisfied: The net torque on the object must be zero. The following general procedure is recommended for solving problems that involve objects in equilibrium.

> **Tip 8.2 Rotary Motion Under Zero Torque**
> If a net torque of zero is exerted on an object, it will continue to rotate at a constant angular speed—which need not be zero. However, zero torque *does* imply that the angular acceleration is zero.

■ PROBLEM-SOLVING STRATEGY

Objects in Equilibrium

1. **Diagram the system.** Include coordinates and choose a convenient rotation axis for computing the net torque on the object.
2. **Draw a force diagram** of the object of interest, showing all external forces acting on it. For systems with more than one object, draw a *separate* diagram for each object. (Most problems will have a single object of interest.)
3. **Apply $\Sigma \tau_i = 0$, the second condition of equilibrium.** This condition yields a single equation for each object of interest. If the axis of rotation has been carefully chosen, the equation often has only one unknotwn and can be solved immediately.
4. **Apply $\Sigma F_x = 0$ and $\Sigma F_y = 0$, the first condition of equilibrium.** This yields two more equations per object of interest.
5. **Solve the system of equations.** For each object, the two conditions of equilibrium yield three equations, usually with three unknowns. Solve by substitution.

■ EXAMPLE 8.6 A Weighted Forearm BIO

GOAL Apply the equilibrium conditions to the human body.

PROBLEM A 50.0-N (11-lb) bowling ball is held in a person's hand with the forearm horizontal, as in Figure 8.13a. The biceps muscle is attached 0.030 0 m from the joint, and the ball is 0.350 m from the joint. Find the upward

(Continued)

force $\vec{F}$ exerted by the biceps on the forearm (the ulna) and the downward force $\vec{R}$ exerted by the humerus on the forearm, acting at the joint. Neglect the weight of the forearm and slight deviation from the vertical of the biceps.

STRATEGY The forces acting on the forearm are equivalent to those acting on a bar of length 0.350 m, as shown in Figure 8.13b. Choose the usual x- and y-coordinates as shown and the axis at O on the left end. (This completes Steps 1 and 2.) Use the conditions of equilibrium to generate equations for the unknowns, and solve.

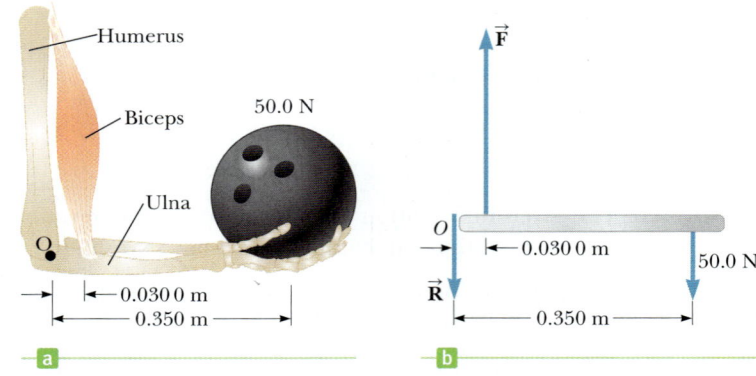

Figure 8.13 (Example 8.6) (a) A weight held with the forearm horizontal. (b) The mechanical model for the system.

SOLUTION

Apply the second condition for equilibrium (Step 3) and solve for the upward force F:

$$\sum \tau_i = \tau_R + \tau_F + \tau_{BB} = 0$$
$$R(0) + F(0.030\ 0\ \text{m}) - (50.0\ \text{N})(0.350\ \text{m}) = 0$$
$$F = \boxed{583\ \text{N (131 lb)}}$$

Apply the first condition for equilibrium (Step 4) and solve (Step 5) for the downward force R:

$$\sum F_y = F - R - 50.0\ \text{N} = 0$$
$$R = F - 50.0\ \text{N} = 583\ \text{N} - 50\ \text{N} = \boxed{533\ \text{N (120 lb)}}$$

REMARKS The magnitude of the force supplied by the biceps must be about ten times as large as the bowling ball it is supporting!

QUESTION 8.6 Suppose the biceps were surgically reattached three centimeters farther toward the person's hand. If the same bowling ball were again held in the person's hand, how would the force required of the biceps be affected? Explain.

EXERCISE 8.6 Suppose you wanted to limit the force acting on your joint to a maximum value of 8.00×10^2 N. (a) Under these circumstances, what maximum weight would you attempt to lift? (b) What force would your biceps apply while lifting this weight?

ANSWERS (a) 75.0 N   (b) 875 N

■ EXAMPLE 8.7   Don't Climb the Ladder

GOAL Apply the two conditions of equilibrium.

PROBLEM A uniform ladder 10.0 m long and weighing 50.0 N rests against a frictionless vertical wall as in Figure 8.14a. If the ladder is just on the verge of slipping when it makes a 50.0° angle with the ground, find the coefficient of static friction between the ladder and ground.

STRATEGY Figure 8.14b is the force diagram for the ladder. The first condition of equilibrium, $\Sigma \vec{F}_i = 0$, gives two equations for three unknowns: the magnitudes of the static friction force f and the normal force n, both acting on the base of the ladder, and the magnitude of the force of the wall, P, acting on the top of the ladder. The second condition of equilibrium, $\Sigma \tau_i = 0$, gives a third equation (for P), so all three quantities can be found. The definition of static friction then allows computation of the coefficient of static friction.

Figure 8.14 (Example 8.7) (a) A ladder leaning against a frictionless wall. (b) A force diagram of the ladder. (c) Lever arms for the force of gravity and $\vec{P}$.

SOLUTION

Apply the first condition of equilibrium to the ladder:

(1) $\quad \Sigma F_x = f - P = 0 \quad \rightarrow \quad f = P$

(2) $\quad \Sigma F_y = n - 50.0\ \text{N} = 0 \quad \rightarrow \quad n = 50.0\ \text{N}$

Apply the second condition of equilibrium, computing torques around the base of the ladder, with τ_{grav} standing for the torque due to the ladder's 50.0-N weight:

$$\Sigma \tau_i = \tau_f + \tau_n + \tau_{grav} + \tau_P = 0$$

The torques due to friction and the normal force are zero about O because their moment arms are zero. (Moment arms can be found from Fig. 8.14c.)

$$0 + 0 - (50.0 \text{ N})(5.00 \text{ m}) \sin 40.0° + P(10.0 \text{ m}) \sin 50.0° = 0$$

$$P = 21.0 \text{ N}$$

From Equation (1), we now have $f = P = 21.0$ N. The ladder is on the verge of slipping, so write an expression for the maximum force of static friction and solve for μ_s:

$$21.0 \text{ N} = f = f_{s,\text{max}} = \mu_s n = \mu_s (50.0 \text{ N})$$

$$\mu_s = \frac{21.0 \text{ N}}{50.0 \text{ N}} = \boxed{0.420}$$

REMARKS Note that torques were computed around an axis through the bottom of the ladder so that only $\vec{P}$ and the force of gravity contributed nonzero torques. This choice of axis reduces the complexity of the torque equation, often resulting in an equation with only one unknown.

QUESTION 8.7 If a 50.0 N monkey hangs from the middle rung, would the coefficient of static friction be (a) doubled, (b) halved, or (c) unchanged?

EXERCISE 8.7 If the coefficient of static friction is 0.360, and the same ladder makes a 60.0° angle with respect to the horizontal, how far along the length of the ladder can a 70.0-kg painter climb before the ladder begins to slip?

ANSWER 6.33 m

■ EXAMPLE 8.8 | Walking a Horizontal Beam

GOAL Apply the two conditions of equilibrium.

PROBLEM A uniform horizontal beam 5.00 m long and weighing 3.00×10^2 N is attached to a wall by a pin connection that allows the beam to rotate. Its far end is supported by a cable that makes an angle of 53.0° with the horizontal (Fig. 8.15a). If a person weighing 6.00×10^2 N stands 1.50 m from the wall, find the magnitude of the tension $\vec{T}$ in the cable and the components of the force $\vec{R}$ exerted by the wall on the beam.

STRATEGY See Figure 8.15a–c (Steps 1 and 2). The second condition of equilibrium, $\Sigma \tau_i = 0$, with torques computed around the pin, can be solved for the tension T in the cable. The first condition of equilibrium, $\Sigma \vec{F}_i = 0$, gives two equations and two unknowns for the two components of the force exerted by the wall, R_x and R_y.

Figure 8.15 (Example 8.8) (a) A uniform beam attached to a wall and supported by a cable. (b) A force diagram for the beam. (c) The component form of the force diagram. (d) (Exercise 8.8)

(Continued)

SOLUTION

From Figure 8.15, the forces causing torques are the wall force $\vec{R}$, the gravity forces on the beam and the man, w_B and w_M, and the tension force $\vec{T}$. Apply the condition of rotational equilibrium (Step 3):

$$\sum \tau_i = \tau_R + \tau_B + \tau_M + \tau_T = 0$$

Compute torques around the pin at O, so $\tau_R = 0$ (zero moment arm). The torque due to the beam's weight acts at the beam's center of gravity.

$$\sum \tau_i = 0 - w_B(L/2) - w_M(1.50 \text{ m}) + TL \sin(53°) = 0$$

Substitute $L = 5.00$ m and the weights, solving for T:

$$-(3.00 \times 10^2 \text{ N})(2.50 \text{ m}) - (6.00 \times 10^2 \text{ N})(1.50 \text{ m}) + (T \sin 53.0°)(5.00 \text{ m}) = 0$$

$$T = \boxed{413 \text{ N}}$$

Now apply the first condition of equilibrium to the beam (Step 4):

(1) $\sum F_x = R_x - T \cos 53.0° = 0$
(2) $\sum F_y = R_y - w_B - w_M + T \sin 53.0° = 0$

Substituting the value of T found in the previous step and the weights, obtain the components of $\vec{R}$ (Step 5):

$R_x = \boxed{249 \text{ N}}$ $R_y = \boxed{5.70 \times 10^2 \text{ N}}$

REMARKS Even if we selected some other axis for the torque equation, the solution would be the same. For example, if the axis were to pass through the center of gravity of the beam, the torque equation would involve both T and R_y. Together with Equations (1) and (2), however, the unknowns could still be found—a good exercise. In both Example 8.6 and Example 8.8, notice the steps of the Problem-Solving Strategy could be carried out in the explicit recommended order.

QUESTION 8.8 What happens to the tension in the cable if the man in Figure 8.15a moves farther away from the wall?

EXERCISE 8.8 A person with mass 55.0 kg stands 2.00 m away from the wall on a uniform 6.00-m beam, as shown in Figure 8.15d. The mass of the beam is 40.0 kg. Find the hinge force components and the tension in the wire.

ANSWERS $T = 751$ N, $R_x = -6.50 \times 10^2$ N, $R_y = 556$ N

8.5 Relationship Between Torque and Angular Acceleration

LEARNING OBJECTIVES

1. Define moment of inertia and state the rotational analog of Newton's second law.
2. Calculate the moment of inertia for a variety of different objects.
3. Apply the rotational second law to physical systems.

Figure 8.16 An object of mass m attached to a light rod of length r moves in a circular path on a frictionless horizontal surface while a tangential force $\vec{F}_t$ acts on it.

When a rigid object is subject to a net torque, it undergoes an angular acceleration that is directly proportional to the net torque. This result, which is analogous to Newton's second law, is derived as follows.

The system shown in Figure 8.16 consists of an object of mass m connected to a very light rod of length r. The rod is pivoted at the point O, and its movement is confined to rotation on a frictionless *horizontal* table. Assume that a force F_t acts perpendicular to the rod and hence is tangent to the circular path of the object. Because there is no force to oppose this tangential force, the object undergoes a tangential acceleration a_t in accordance with Newton's second law:

$$F_t = ma_t$$

Multiply both sides of this equation by r:

$$F_t r = mra_t$$

Substituting the equation $a_t = r\alpha$ relating tangential and angular acceleration into the above expression gives

$$F_t r = mr^2 \alpha \qquad [8.4]$$

The left side of Equation 8.4 is the torque acting on the object about its axis of rotation, so we can rewrite it as

$$\tau = mr^2 \alpha \qquad [8.5]$$

Equation 8.5 shows that the torque on the object is proportional to the angular acceleration of the object, where the constant of proportionality mr^2 is called the **moment of inertia** of the object of mass m. (Because the rod is very light, its moment of inertia can be neglected.)

Quick Quiz

8.1 Using a screwdriver, you try to remove a screw from a piece of furniture, but can't get it to turn. To increase the chances of success, you should use a screwdriver that (a) is longer, (b) is shorter, (c) has a narrower handle, or (d) has a wider handle.

Torque on a Rotating Object

Consider a solid disk rotating about its axis as in Figure 8.17a. The disk consists of many particles at various distances from the axis of rotation. (See Fig. 8.17b.) The torque on each one of these particles is given by Equation 8.5. The *net* torque on the disk is given by the sum of the individual torques on all the particles:

$$\sum \tau = \left(\sum mr^2 \right) \alpha \qquad [8.6]$$

Because the disk is rigid, all of its particles have the *same* angular acceleration, so α is not involved in the sum. If the masses and distances of the particles are labeled with subscripts as in Figure 8.17b, then

$$\sum mr^2 = m_1 r_1^2 + m_2 r_2^2 + m_3 r_3^2 + \cdots$$

This quantity is the moment of inertia, I, of the whole body:

$$I \equiv \sum mr^2 \qquad [8.7] \qquad \blacktriangleleft \text{ Moment of inertia}$$

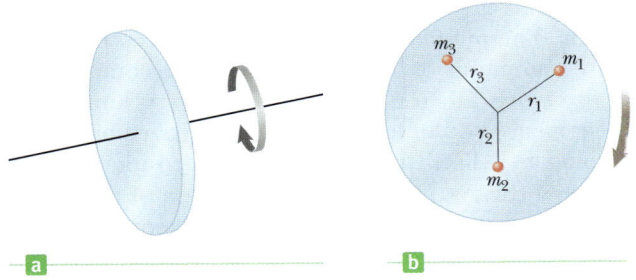

Figure 8.17 (a) A solid disk rotating about its axis. (b) The disk consists of many particles, all with the same angular acceleration.

254 CHAPTER 8 | Rotational Equilibrium and Rotational Dynamics

The moment of inertia has the SI units kg · m². Using this result in Equation 8.6, we see that the net torque on a rigid body rotating about a fixed axis is given by

$$\sum \tau = I\alpha \qquad [8.8]$$

Equation 8.8 says that **the angular acceleration of an extended rigid object is proportional to the net torque acting on it.** This equation is the rotational analog of Newton's second law of motion, with torque replacing force, moment of inertia replacing mass, and angular acceleration replacing linear acceleration. Although the moment of inertia of an object is related to its mass, there is an important difference between them. The mass m depends only on the quantity of matter in an object, whereas the moment of inertia, I, depends on both the quantity of matter and its distribution (through the r^2 term in $I = \Sigma mr^2$) in the rigid object.

◀ Rotational analog of Newton's second law

A

B

■ Quick Quiz

8.2 A constant net torque is applied to an object. Which one of the following will *not* be constant? (a) angular acceleration, (b) angular velocity, (c) moment of inertia, or (d) center of gravity.

8.3 The two rigid objects shown in Figure 8.18 have the same mass, radius, and angular speed, each spinning around an axis through the center of its circular shape. If the same braking torque is applied to each, which takes longer to stop? (a) A (b) B (c) more information is needed

Figure 8.18 (Quick Quiz 8.3)

The gear system on a bicycle provides an easily visible example of the relationship between torque and angular acceleration. Consider first a five-speed gear system in which the drive chain can be adjusted to wrap around any of five gears attached to the back wheel (Fig. 8.19). The gears, with different radii, are concentric with the wheel hub. When the cyclist begins pedaling from rest, the chain is attached to the largest gear. Because it has the largest radius, this gear provides the largest torque to the drive wheel. A large torque is required initially, because the bicycle starts from rest. As the bicycle rolls faster, the tangential speed of the chain increases, eventually becoming too fast for the cyclist to maintain by pushing the pedals. The chain is then moved to a gear with a smaller radius, so the chain has a smaller tangential speed that the cyclist can more easily maintain. This gear doesn't provide as much torque as the first, but the cyclist needs to accelerate only to a somewhat higher speed. This process continues as the bicycle moves faster and faster and the cyclist shifts through all five gears. The fifth gear supplies the lowest torque, but now the main function of that torque is to counter the frictional torque from the rolling tires, which tends to reduce the speed of the bicycle. The small radius of the fifth gear allows the cyclist to keep up with the chain's movement by pushing the pedals.

APPLICATION
Bicycle Gears

Figure 8.19 The drive wheel and gears of a bicycle.

A 15-speed bicycle has the same gear structure on the drive wheel, but has three gears on the sprocket connected to the pedals. By combining different positions of the chain on the rear gears and the sprocket gears, 15 different torques are available.

More on the Moment of Inertia

As we have seen, a small object (or a particle) has a moment of inertia equal to mr^2 about some axis. The moment of inertia of a *composite* object about some axis is just the sum of the moments of inertia of the object's components. For example, suppose a majorette twirls a baton as in Figure 8.20. Assume that the baton can be modeled as a very light rod of length 2ℓ with a heavy object at each end. (The rod of a real baton has a significant mass relative to its ends.) Because we are neglecting the mass of the rod, the moment of inertia of the baton about an axis through its center and perpendicular to its length is given by Equation 8.7:

$$I = \Sigma mr^2$$

Figure 8.20 A baton of length 2ℓ and mass $2m$. (The mass of the connecting rod is neglected.) The moment of inertia about the axis through the baton's center and perpendicular to its length is $2m\ell^2$.

Because this system consists of two objects with equal masses equidistant from the axis of rotation, $r = \ell$ for each object, and the sum is

$$I = \Sigma mr^2 = m\ell^2 + m\ell^2 = 2m\ell^2$$

If the mass of the rod were not neglected, we would have to include its moment of inertia to find the total moment of inertia of the baton.

We pointed out earlier that I is the rotational counterpart of m. However, there are some important distinctions between the two. For example, mass is an intrinsic property of an object that doesn't change, whereas **the moment of inertia of a system depends on how the mass is distributed and on the location of the axis of rotation.** Example 8.9 illustrates this point.

EXAMPLE 8.9 | The Baton Twirler

GOAL Calculate a moment of inertia.

PROBLEM In an effort to be the star of the halftime show, a majorette twirls an unusual baton made up of four balls fastened to the ends of very light rods (Fig. 8.21). Each rod is 1.0 m long. **(a)** Find the moment of inertia of the baton about an axis perpendicular to the page and passing through the point where the rods cross. **(b)** The majorette tries spinning her strange baton about the axis OO', as shown in Figure 8.22 on page 256. Calculate the moment of inertia of the baton about this axis.

Figure 8.21 (Example 8.9a) Four balls connected to light rods rotating in the plane of the page.

STRATEGY In Figure 8.21, all four balls contribute to the moment of inertia, whereas in Figure 8.22, with the new axis, only the two balls on the left and right contribute. Technically, the balls on the top and bottom in Figure 8.22 still make a small contribution because they're not really point particles. However, their contributions can be neglected because the distance from the axis of rotation of the balls on the horizontal rod is much greater than the radii of the balls on the vertical rod.

SOLUTION

(a) Calculate the moment of inertia of the baton when oriented as in Figure 8.21.

Apply Equation 8.7, neglecting the mass of the connecting rods:

$$I = \Sigma mr^2 = m_1 r_1^2 + m_2 r_2^2 + m_3 r_3^2 + m_4 r_4^2$$
$$= (0.20 \text{ kg})(0.50 \text{ m})^2 + (0.30 \text{ kg})(0.50 \text{ m})^2$$
$$+ (0.20 \text{ kg})(0.50 \text{ m})^2 + (0.30 \text{ kg})(0.50 \text{ m})^2$$
$$I = \boxed{0.25 \text{ kg} \cdot \text{m}^2}$$

(b) Calculate the moment of inertia of the baton when oriented as in Figure 8.22.

Apply Equation 8.7 again, neglecting the radii of the 0.20-kg balls.

$$I = \Sigma mr^2 = m_1 r_1^2 + m_2 r_2^2 + m_3 r_3^2 + m_4 r_4^2$$
$$= (0.20 \text{ kg})(0)^2 + (0.30 \text{ kg})(0.50 \text{ m})^2 + (0.20 \text{ kg})(0)^2$$
$$+ (0.30 \text{ kg})(0.50 \text{ m})^2$$
$$I = \boxed{0.15 \text{ kg} \cdot \text{m}^2}$$

REMARKS The moment of inertia is smaller in part (b) because in this configuration the 0.20-kg balls are essentially located on the axis of rotation.

(Continued)

QUESTION 8.9 If one of the rods is lengthened, which one would cause the larger change in the moment of inertia, the rod connecting balls one and three or the rod connecting balls two and four?

EXERCISE 8.9 Yet another bizarre baton is created by taking four identical balls, each with mass 0.300 kg, and fixing them as before, except that one of the rods has a length of 1.00 m and the other has a length of 1.50 m. Calculate the moment of inertia of this baton (a) when oriented as in Figure 8.21; (b) when oriented as in Figure 8.22, with the shorter rod vertical; and (c) when oriented as in Figure 8.22, but with longer rod vertical.

ANSWERS (a) $0.488 \text{ kg} \cdot \text{m}^2$ (b) $0.338 \text{ kg} \cdot \text{m}^2$ (c) $0.150 \text{ kg} \cdot \text{m}^2$

Figure 8.22 (Example 8.9b) A double baton rotating about the axis OO'.

Calculation of Moments of Inertia for Extended Objects

The method used for calculating moments of inertia in Example 8.9 is simple when only a few small objects rotate about an axis. When the object is an extended one, such as a sphere, a cylinder, or a cone, techniques of calculus are often required, unless some simplifying symmetry is present. One such extended object amenable to a simple solution is a hoop rotating about an axis perpendicular to its plane and passing through its center, as shown in Figure 8.23. (A bicycle tire, for example, would approximately fit into this category.)

To evaluate the moment of inertia of the hoop, we can still use the equation $I = \Sigma mr^2$ and imagine that the mass of the hoop M is divided into n small segments having masses $m_1, m_2, m_3, \cdots, m_n$, as in Figure 8.23, with $M = m_1 + m_2 + m_3 + \cdots + m_n$. This approach is just an extension of the baton problem described in the preceding examples, except that now we have a large number of small masses in rotation instead of only four.

We can express the sum for I as

$$I = \Sigma mr^2 = m_1 r_1^2 + m_2 r_2^2 + m_3 r_3^2 + \cdots + m_n r_n^2$$

All of the segments around the hoop are at the *same distance R* from the axis of rotation, so we can drop the subscripts on the distances and factor out R^2 to obtain

$$I = (m_1 + m_2 + m_3 + \cdots + m_n)R^2 = MR^2 \quad [8.9]$$

This expression can be used for the moment of inertia of any ring-shaped object rotating about an axis through its center and perpendicular to its plane. Note that the result is strictly valid only if the thickness of the ring is small relative to its inner radius.

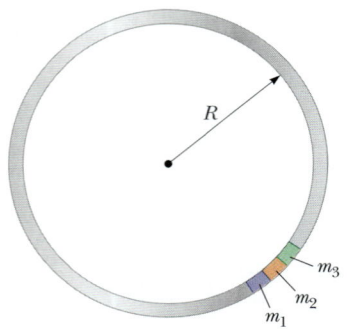

Figure 8.23 A uniform hoop can be divided into a large number of small segments that are equidistant from the center of the hoop.

The hoop we selected as an example is unique in that we were able to find an expression for its moment of inertia by using only simple algebra. Unfortunately, for most extended objects the calculation is much more difficult because the mass elements are not all located at the same distance from the axis, so the methods of integral calculus are required. The moments of inertia for some other common shapes are given without proof in Table 8.1. You can use this table as needed to determine the moment of inertia of a body having any one of the listed shapes.

If mass elements in an object are redistributed parallel to the axis of rotation, the moment of inertia of the object doesn't change. Consequently, the expression $I = MR^2$ can be used equally well to find the axial moment of inertia of an embroidery hoop or of a long sewer pipe. Likewise, a door turning on its hinges is described by the same moment-of-inertia expression as that tabulated for a long, thin rod rotating about an axis through its end.

> **Tip 8.3 No Single Moment of Inertia**
> Moment of inertia is analogous to mass, but there are major differences. Mass is an inherent property of an object. The moment of inertia of an object depends on the shape of the object, its mass, and the choice of rotation axis.

8.5 | Relationship Between Torque and Angular Acceleration

Table 8.1 Moments of Inertia for Various Rigid Objects of Uniform Composition

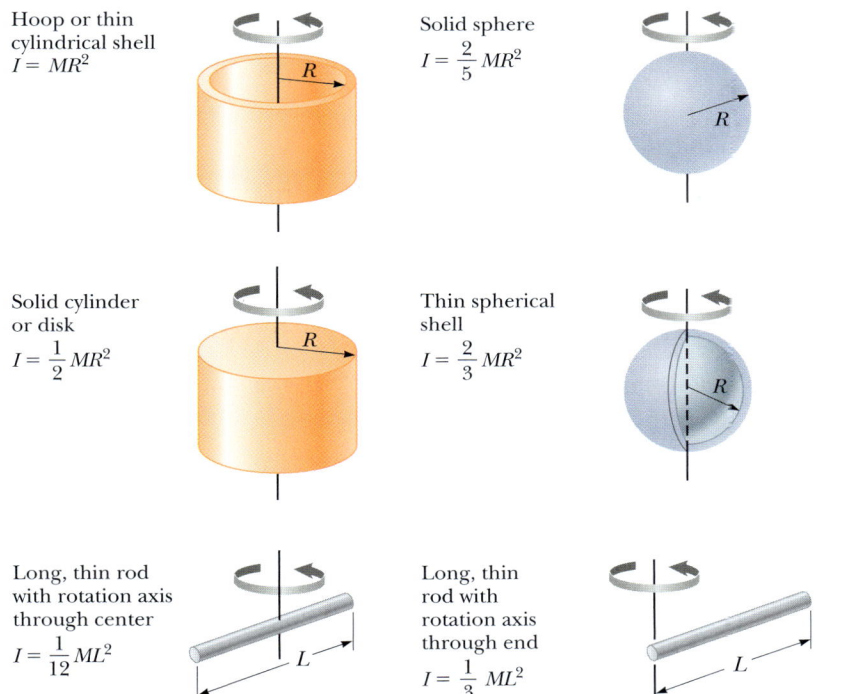

Hoop or thin cylindrical shell
$I = MR^2$

Solid sphere
$I = \frac{2}{5}MR^2$

Solid cylinder or disk
$I = \frac{1}{2}MR^2$

Thin spherical shell
$I = \frac{2}{3}MR^2$

Long, thin rod with rotation axis through center
$I = \frac{1}{12}ML^2$

Long, thin rod with rotation axis through end
$I = \frac{1}{3}ML^2$

■ EXAMPLE 8.10 | Warming Up

GOAL Find a moment of inertia and apply the rotational analog of Newton's second law.

PROBLEM A baseball player loosening up his arm before a game tosses a 0.150-kg baseball, using only the rotation of his forearm to accelerate the ball (Fig. 8.24). The forearm has a mass of 1.50 kg and the length from the elbow to the ball's center is 0.350 m. The ball starts at rest and is released with a speed of 30.0 m/s in 0.300 s. **(a)** Find the constant angular acceleration of the arm and ball. **(b)** Calculate the moment of inertia of the system consisting of the forearm and ball. **(c)** Find the torque exerted on the system that results in the angular acceleration found in part (a).

Figure 8.24 (Example 8.10) A ball being tossed by a pitcher. The forearm is used to accelerate the ball.

STRATEGY The angular acceleration can be found with rotational kinematic equations, while the moment of inertia of the system can be obtained by summing the separate moments of inertia of the ball and forearm. The ball is treated as a point particle. Multiplying these two results together gives the torque.

SOLUTION

(a) Find the angular acceleration of the ball.

The angular acceleration is constant, so use the angular velocity kinematic equation with $\omega_i = 0$:

$$\omega = \omega_i + \alpha t \quad \rightarrow \quad \alpha = \frac{\omega}{t}$$

The ball accelerates along a circular arc with radius given by the length of the forearm. Solve $v = r\omega$ for ω and substitute:

$$\alpha = \frac{\omega}{t} = \frac{v}{rt} = \frac{30.0 \text{ m/s}}{(0.350 \text{ m})(0.300 \text{ s})} = \boxed{286 \text{ rad/s}^2}$$

(Continued)

(b) Find the moment of inertia of the system (forearm plus ball).

Find the moment of inertia of the ball about an axis that passes through the elbow, perpendicular to the arm:

$$I_{ball} = mr^2 = (0.150 \text{ kg})(0.350 \text{ m})^2 = 1.84 \times 10^{-2} \text{ kg} \cdot \text{m}^2$$

Obtain the moment of inertia of the forearm, modeled as a rod rotating about an axis through one end, by consulting Table 8.1:

$$I_{forearm} = \tfrac{1}{3} ML^2 = \tfrac{1}{3}(1.50 \text{ kg})(0.350 \text{ m})^2$$
$$= 6.13 \times 10^{-2} \text{ kg} \cdot \text{m}^2$$

Sum the individual moments of inertia to obtain the moment of inertia of the system (ball plus forearm):

$$I_{system} = I_{ball} + I_{forearm} = \boxed{7.97 \times 10^{-2} \text{ kg} \cdot \text{m}^2}$$

(c) Find the torque exerted on the system.

Apply Equation 8.8, using the results of parts (a) and (b):

$$\tau = I_{system}\alpha = (7.97 \times 10^{-2} \text{ kg} \cdot \text{m}^2)(286 \text{ rad/s}^2)$$
$$= \boxed{22.8 \text{ N} \cdot \text{m}}$$

REMARKS Notice that having a long forearm can greatly increase the torque and hence the acceleration of the ball. This is one reason it's advantageous for a pitcher to be tall: the pitching arm is proportionately longer. A similar advantage holds in tennis, where taller players can usually deliver faster serves.

QUESTION 8.10 Why do pitchers step forward when delivering the pitch? Why is the timing important?

EXERCISE 8.10 A catapult with a radial arm 4.00 m long accelerates a ball of mass 20.0 kg through a quarter circle. The ball leaves the apparatus at 45.0 m/s. If the mass of the arm is 25.0 kg and the acceleration is constant, find (a) the angular acceleration, (b) the moment of inertia of the arm and ball, and (c) the net torque exerted on the ball and arm.

Hint: Use the time-independent rotational kinematics equation to find the angular acceleration, rather than the angular velocity equation.

ANSWERS (a) 40.3 rad/s² (b) 453 kg · m² (c) 1.83×10^4 N · m

■ EXAMPLE 8.11 | The Falling Bucket

GOAL Combine Newton's second law with its rotational analog.

PROBLEM A solid, uniform, frictionless cylindrical reel of mass $M = 3.00$ kg and radius $R = 0.400$ m is used to draw water from a well (Fig. 8.25a). A bucket of mass $m = 2.00$ kg is attached to a cord that is wrapped around the cylinder. **(a)** Find the tension T in the cord and acceleration a of the bucket. **(b)** If the bucket starts from rest at the top of the well and falls for 3.00 s before hitting the water, how far does it fall?

Figure 8.25 (Example 8.11) (a) A water bucket attached to a rope passing over a frictionless reel. (b) A force diagram for the bucket. (c) The tension produces a torque on the cylinder about its axis of rotation. (d) A falling cylinder (Exercise 8.11).

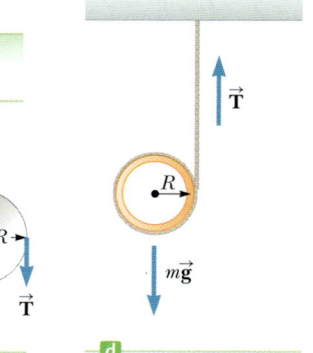

STRATEGY This problem involves three equations and three unknowns. The three equations are Newton's second law applied to the bucket, $ma = \Sigma F_i$; the rotational version of the second law applied to the cylinder, $I\alpha = \Sigma \tau_i$; and the relationship between linear and angular acceleration, $a = r\alpha$, which connects the dynamics of the bucket and cylinder. The three unknowns are the acceleration a of the bucket, the angular acceleration α of the cylinder, and the tension T in the rope. Assemble the terms of the three equations and solve for the three unknowns by substitution. Part (b) is a review of kinematics.

SOLUTION

(a) Find the tension in the cord and the acceleration of the bucket.

Apply Newton's second law to the bucket in Figure 8.25b. There are two forces: the tension $\vec{T}$ acting upward and gravity $m\vec{g}$ acting downward.

(1) $\quad ma = -mg + T$

Apply $\tau = I\alpha$ to the cylinder in Figure 8.25c:

$\sum \tau = I\alpha = \tfrac{1}{2}MR^2\alpha \quad$ (solid cylinder)

Notice the angular acceleration is clockwise, so the torque is negative. The normal and gravity forces have zero moment arm and don't contribute any torque.

(2) $\quad -TR = \tfrac{1}{2}MR^2\alpha$

Solve for T and substitute $\alpha = a/R$ (notice that both α and a are negative):

(3) $\quad T = -\tfrac{1}{2}MR\alpha = -\tfrac{1}{2}Ma$

Substitute the expression for T in Equation (3) into Equation (1), and solve for the acceleration:

$ma = -mg - \tfrac{1}{2}Ma \quad \rightarrow \quad a = -\dfrac{mg}{m + \tfrac{1}{2}M}$

Substitute the values for m, M, and g, getting a, then substitute a into Equation (3) to get T:

$a = \boxed{-5.60 \text{ m/s}^2} \qquad T = \boxed{8.40 \text{ N}}$

(b) Find the distance the bucket falls in 3.00 s.

Apply the displacement kinematic equation for constant acceleration, with $t = 3.00$ s and $v_0 = 0$:

$\Delta y = v_0 t + \tfrac{1}{2}at^2 = -\tfrac{1}{2}(5.60 \text{ m/s}^2)(3.00 \text{ s})^2 = \boxed{-25.2 \text{ m}}$

REMARKS Proper handling of signs is very important in these problems. All such signs should be chosen initially and checked mathematically and physically. In this problem, for example, both the angular acceleration α and the acceleration a are negative, so $\alpha = a/R$ applies. If the rope had been wound the other way on the cylinder, causing counterclockwise rotation, the torque would have been positive, and the relationship would have been $\alpha = -a/R$, with the double negative making the right-hand side positive, just like the left-hand side.

QUESTION 8.11 How would the acceleration and tension change if most of the reel's mass were at its rim?

EXERCISE 8.11 A hollow cylinder of mass 0.100 kg and radius 4.00 cm has a string wrapped several times around it, as in Figure 8.25d. If the string is attached to a rigid support and the cylinder allowed to drop from rest, find (a) the acceleration of the cylinder and (b) the speed of the cylinder when a meter of string has unwound off of it.

ANSWERS (a) -4.90 m/s^2 (b) 3.13 m/s

8.6 Rotational Kinetic Energy

LEARNING OBJECTIVES

1. Define the kinetic energy of a rotating object and extend the work-energy theorem to include it.
2. Apply the extended work-energy theorem to systems involving rotation.

In Chapter 5 we defined the kinetic energy of a particle moving through space with a speed v as the quantity $\tfrac{1}{2}mv^2$. Analogously, **an object rotating about some axis with an angular speed ω has rotational kinetic energy given by $\tfrac{1}{2}I\omega^2$.** To prove this, consider an object in the shape of a thin, rigid plate rotating around some axis perpendicular to its plane, as in Figure 8.26. The plate consists of many small particles, each of mass m. All these particles rotate in circular paths around the axis. If r is the distance of one of the particles from the

260 CHAPTER 8 | Rotational Equilibrium and Rotational Dynamics

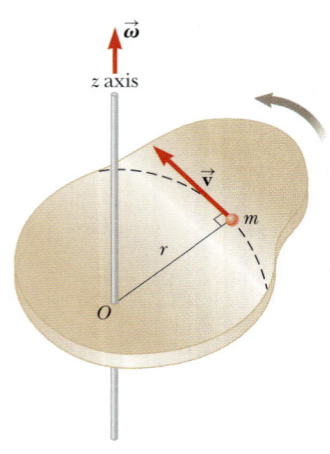

Figure 8.26 A rigid plate rotates about the z-axis with angular speed v. The kinetic energy of a particle of mass m is $\frac{1}{2}mv^2$. The total kinetic energy of the plate is $\frac{1}{2}I\omega^2$.

axis of rotation, the speed of that particle is $v = r\omega$. Because the *total* kinetic energy of the plate's rotation is the sum of all the kinetic energies associated with its particles, we have

$$KE_r = \sum(\tfrac{1}{2}mv^2) = \sum(\tfrac{1}{2}mr^2\omega^2) = \tfrac{1}{2}(\sum mr^2)\omega^2$$

In the last step, the ω^2 term is factored out because it's the same for every particle. Now, the quantity in parentheses on the right is the moment of inertia of the plate in the limit as the particles become vanishingly small, so

$$KE_r = \tfrac{1}{2}I\omega^2 \qquad [8.10]$$

where $I = \sum mr^2$ is the moment of inertia of the plate.

A system such as a bowling ball rolling down a ramp is described by three types of energy: **gravitational potential energy PE_g**, **translational kinetic energy KE_t**, and **rotational kinetic energy KE_r**. All these forms of energy, plus the potential energies of any other conservative forces, must be included in our equation for the conservation of mechanical energy of an isolated system:

◀ Conservation of mechanical energy

$$(KE_t + KE_r + PE)_i = (KE_t + KE_r + PE)_f \qquad [8.11]$$

where i and f refer to initial and final values, respectively, and PE includes the potential energies of all conservative forces in a given problem. This relation is true *only* if we ignore dissipative forces such as friction. Otherwise, it's necessary to resort to a generalization of the work–energy theorem:

◀ Work–energy theorem including rotational energy

$$W_{nc} = \Delta KE_t + \Delta KE_r + \Delta PE \qquad [8.12]$$

■ PROBLEM-SOLVING STRATEGY

Energy Methods and Rotation

1. **Choose two points of interest,** one where all necessary information is known, and the other where information is desired.
2. **Identify** the conservative and nonconservative forces acting on the system being analyzed.
3. **Write the general work–energy theorem,** Equation 8.12, or Equation 8.11 if all forces are conservative.
4. **Substitute general expressions** for the terms in the equation.
5. **Use $v = r\omega$** to eliminate either ω or v from the equation.
6. **Solve** for the unknown.

■ EXAMPLE 8.12 | A Ball Rolling Down an Incline

GOAL Combine gravitational, translational, and rotational energy.

PROBLEM A ball of mass M and radius R starts from rest at a height of $h = 2.00$ m and rolls down a $\theta = 30.0°$ slope, as in Figure 8.27. What is the linear speed of the ball when it leaves the incline? Assume that the ball rolls without slipping.

STRATEGY The two points of interest are the top and bottom of the incline, with the bottom acting as the zero point of gravitational potential energy. As the ball rolls down the ramp, gravitational potential energy is converted into both translational and rotational kinetic energy without dissipation, so conservation of mechanical energy can be applied with the use of Equation 8.11.

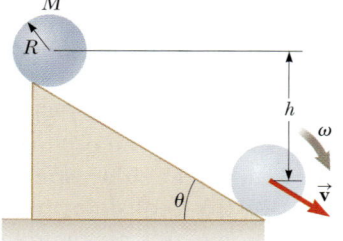

Figure 8.27 (Example 8.12) A ball starts from rest at the top of an incline and rolls to the bottom without slipping.

SOLUTION

Apply conservation of energy with $PE = PE_g$, the potential energy associated with gravity:

$$(KE_t + KE_r + PE_g)_i = (KE_t + KE_r + PE_g)_f$$

Substitute the appropriate general expressions, noting that $(KE_t)_i = (KE_r)_i = 0$ and $(PE_g)_f = 0$ (obtain the moment of inertia of a ball from Table 8.1):

$$0 + 0 + Mgh = \tfrac{1}{2}Mv^2 + \tfrac{1}{2}(\tfrac{2}{5}MR^2)\omega^2 + 0$$

The ball rolls without slipping, so $R\omega = v$, the "no-slip condition," can be applied:

$$Mgh = \tfrac{1}{2}Mv^2 + \tfrac{1}{5}Mv^2 = \tfrac{7}{10}Mv^2$$

Solve for v, noting that M cancels.

$$v = \sqrt{\frac{10gh}{7}} = \sqrt{\frac{10(9.80 \text{ m/s}^2)(2.00 \text{ m})}{7}} = \boxed{5.29 \text{ m/s}}$$

REMARKS Notice the translational speed is less than that of a block sliding down a frictionless slope, $v = \sqrt{2gh}$. That's because some of the original potential energy must go to increasing the rotational kinetic energy.

QUESTION 8.12 Rank from fastest to slowest: (a) a solid ball rolling down a ramp without slipping, (b) a cylinder rolling down the same ramp without slipping, (c) a block sliding down a frictionless ramp with the same height and slope.

EXERCISE 8.12 Repeat this example for a solid cylinder of the same mass and radius as the ball and released from the same height. In a race between the two objects on the incline, which one would win?

ANSWER $v = \sqrt{4gh/3} = 5.11$ m/s; the ball would win.

■ Quick Quiz

8.4 Two spheres, one hollow and one solid, are rotating with the same angular speed around an axis through their centers. Both spheres have the same mass and radius. Which sphere, if either, has the higher rotational kinetic energy? (a) The hollow sphere. (b) The solid sphere. (c) They have the same kinetic energy.

■ EXAMPLE 8.13 | Blocks and Pulley

GOAL Solve a system requiring rotation concepts and the work–energy theorem.

PROBLEM Two blocks with masses $m_1 = 5.00$ kg and $m_2 = 7.00$ kg are attached by a string as in Figure 8.28a, over a pulley with mass $M = 2.00$ kg. The pulley, which turns on a frictionless axle, is a hollow cylinder with radius 0.050 0 m over

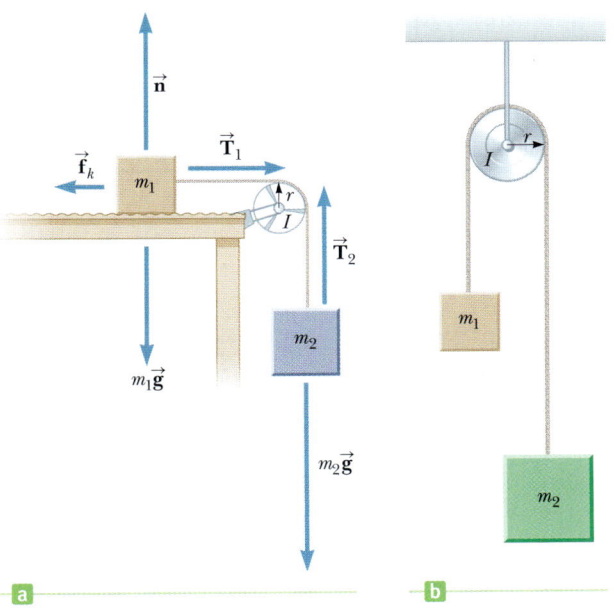

Figure 8.28 (a) (Example 8.13) $\vec{T}_1$ and $\vec{T}_2$ exert torques on the pulley. (b) (Exercise 8.13)

(*Continued*)

which the string moves without slipping. The horizontal surface has coefficient of kinetic friction 0.350. Find the speed of the system when the block of mass m_2 has dropped 2.00 m.

STRATEGY This problem can be solved with the extension of the work–energy theorem, Equation 8.12. If the block of mass m_2 falls from height h to 0, then the block of mass m_1 moves the same distance, $\Delta x = h$. Apply the work-energy theorem, solve for v, and substitute. Kinetic friction is the sole nonconservative force.

SOLUTION

Apply the work–energy theorem, with $PE = PE_g$, the potential energy associated with gravity:

$$W_{nc} = \Delta KE_t + \Delta KE_r + \Delta PE_g$$

Substitute the frictional work for W_{nc}, kinetic energy changes for the two blocks, the rotational kinetic energy change for the pulley, and the potential energy change for the second block:

$$-\mu_k n \Delta x = -\mu_k(m_1 g)\Delta x = (\tfrac{1}{2}m_1 v^2 - 0) + (\tfrac{1}{2}m_2 v^2 - 0) + (\tfrac{1}{2}I\omega^2 - 0) + (0 - m_2 g h)$$

Substitute $\Delta x = h$, and write I as $(I/r^2)r^2$:

$$-\mu_k(m_1 g)h = \tfrac{1}{2}m_1 v^2 + \tfrac{1}{2}m_2 v^2 + \tfrac{1}{2}\left(\frac{I}{r^2}\right)r^2\omega^2 - m_2 g h$$

For a hoop, $I = Mr^2$ so $(I/r^2) = M$. Substitute this quantity and $v = r\omega$:

$$-\mu_k(m_1 g)h = \tfrac{1}{2}m_1 v^2 + \tfrac{1}{2}m_2 v^2 + \tfrac{1}{2}Mv^2 - m_2 g h$$

Solve for v:

$$m_2 g h - \mu_k(m_1 g)h = \tfrac{1}{2}m_1 v^2 + \tfrac{1}{2}m_2 v^2 + \tfrac{1}{2}Mv^2$$
$$= \tfrac{1}{2}(m_1 + m_2 + M)v^2$$
$$v = \sqrt{\frac{2gh(m_2 - \mu_k m_1)}{m_1 + m_2 + M}}$$

Substitute $m_1 = 5.00$ kg, $m_2 = 7.00$ kg, $M = 2.00$ kg, $g = 9.80$ m/s², $h = 2.00$ m, and $\mu_k = 0.350$:

$$v = \boxed{3.83 \text{ m/s}}$$

REMARKS In the expression for the speed v, the mass m_1 of the first block and the mass M of the pulley all appear in the denominator, reducing the speed, as they should. In the numerator, m_2 is positive while the friction term is negative. Both assertions are reasonable because the force of gravity on m_2 increases the speed of the system while the force of friction on m_1 slows it down. This problem can also be solved with Newton's second law together with $\tau = I\alpha$, a good exercise.

QUESTION 8.13 How would increasing the radius of the pulley affect the final answer? Assume the angles of the cables are unchanged and the mass is the same as before.

EXERCISE 8.13 Two blocks with masses $m_1 = 2.00$ kg and $m_2 = 9.00$ kg are attached over a pulley with mass $M = 3.00$ kg, hanging straight down as in Atwood's machine (Fig. 8.28b). The pulley is a solid cylinder with radius 0.050 0 m, and there is some friction in the axle. The system is released from rest, and the string moves without slipping over the pulley. If the larger mass is traveling at a speed of 2.50 m/s when it has dropped 1.00 m, how much mechanical energy was lost due to friction in the pulley's axle?

ANSWER 29.5 J

8.7 Angular Momentum

LEARNING OBJECTIVES

1. Define angular momentum and state the rotational second law in terms of it.
2. State and apply the principle of the conservation of angular momentum.

In Figure 8.29, an object of mass m rotates in a circular path of radius r, acted on by a net force, $\vec{F}_{net}$. The resulting net torque on the object increases its angular speed from the value ω_0 to the value ω in a time interval Δt. Therefore, we can write

$$\sum \tau = I\alpha = I\frac{\Delta \omega}{\Delta t} = I\left(\frac{\omega - \omega_0}{\Delta t}\right) = \frac{I\omega - I\omega_0}{\Delta t}$$

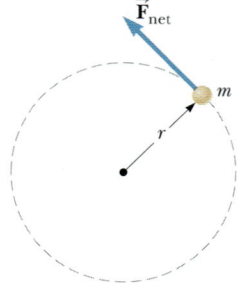

Figure 8.29 An object of mass m rotating in a circular path under the action of a constant torque.

If we define the product

$$L \equiv I\omega \quad [8.13]$$

as the **angular momentum** of the object, then we can write

$$\sum \tau = \frac{\text{change in angular momentum}}{\text{time interval}} = \frac{\Delta L}{\Delta t} \quad [8.14]$$

Equation 8.14 is the rotational analog of Newton's second law, which can be written in the form $F = \Delta p/\Delta t$ and states that **the net torque acting on an object is equal to the time rate of change of the object's angular momentum.** Recall that this equation also parallels the impulse–momentum theorem.

When the net external torque ($\Sigma \tau$) acting on a system is zero, Equation 8.14 gives $\Delta L/\Delta t = 0$, which says that the time rate of change of the system's angular momentum is zero. We then have the following important result:

Let L_i and L_f be the angular momenta of a system at two different times, and suppose there is no net external torque, so $\Sigma \tau = 0$. Then

$$L_i = L_f \quad [8.15]$$

and angular momentum is said to be *conserved*.

◀ Conservation of angular momentum

Equation 8.15 gives us a third conservation law to add to our list: **conservation of angular momentum.** We can now state that **the mechanical energy, linear momentum, and angular momentum of an isolated system all remain constant.**

If the moment of inertia of an isolated rotating system changes, the system's angular speed will change. Conservation of angular momentum then requires that

$$I_i\omega_i = I_f\omega_f \quad \text{if} \quad \sum \tau = 0 \quad [8.16]$$

Note that conservation of angular momentum applies to macroscopic objects such as planets and people, as well as to atoms and molecules. There are many examples of conservation of angular momentum; one of the most dramatic is that of a figure skater spinning in the finale of his act. In Figure 8.30a, the skater has pulled his arms and legs close to his body, reducing their distance from his axis of rotation and hence also reducing his moment of inertia. By conservation of angular momentum, a reduction in his moment of inertia must increase his angular

APPLICATION
Figure Skating

By pulling in his arms and legs, he reduces his moment of inertia and increases his angular speed (rate of spin).

Upon landing, extending his arms and legs increases his moment of inertia and helps slow his spin.

Figure 8.30 Evgeni Plushenko varies his moment of inertia to change his angular speed.

Tightly curling her body, a diver decreases her moment of inertia, increasing her angular speed.

speed. Coming out of the spin in Figure 8.30b, he needs to reduce his angular speed, so he extends his arms and legs again, increasing his moment of inertia and thereby slowing his rotation.

APPLICATION
Aerial Somersaults

Similarly, when a diver or an acrobat wishes to make several somersaults, she pulls her hands and feet close to the trunk of her body in order to rotate at a greater angular speed. In this case, the external force due to gravity acts through her center of gravity and hence exerts no torque about her axis of rotation, so the angular momentum about her center of gravity is conserved. For example, when a diver wishes to double her angular speed, she must reduce her moment of inertia to half its initial value.

An interesting astrophysical example of conservation of angular momentum occurs when a massive star, at the end of its lifetime, uses up all its fuel and collapses under the influence of gravitational forces, causing a gigantic outburst of energy called a supernova. The best-studied example of a remnant of a supernova explosion is the Crab Nebula, a chaotic, expanding mass of gas (Fig. 8.31). In a supernova, part of the star's mass is ejected into space, where it eventually condenses into new stars and planets. Most of what is left behind typically collapses into a **neutron star**—an extremely dense sphere of matter with a diameter of about 10 km, greatly reduced from the 10^6-km diameter of the original star and containing a large fraction of the star's original mass. In a neutron star, pressures become so great that atomic electrons combine with protons, becoming neutrons. As the moment of inertia of the system decreases during the collapse, the star's rotational speed increases. More than 700 rapidly rotating neutron stars have been identified since their first discovery in 1967, with periods of rotation ranging from a millisecond to several seconds. The neutron star is an amazing system—an object with a mass greater than the Sun, fitting comfortably within the space of a small county and rotating so fast that the tangential speed of the surface approaches a sizable fraction of the speed of light!

APPLICATION
Rotating Neutron Stars

Quick Quiz

8.5 A horizontal disk with moment of inertia I_1 rotates with angular speed ω_1 about a vertical frictionless axle. A second horizontal disk having moment of inertia I_2 drops onto the first, initially not rotating but sharing the same axis as the first disk. Because their surfaces are rough, the two disks eventually reach the same angular speed ω. The ratio ω/ω_1 is equal to (a) I_1/I_2 (b) I_2/I_1 (c) $I_1/(I_1 + I_2)$ (d) $I_2/(I_1 + I_2)$

8.6 If global warming continues, it's likely that some ice from the polar ice caps of the Earth will melt and the water will be distributed closer to the equator. If this occurs, would the length of the day (one rotation) (a) increase, (b) decrease, or (c) remain the same?

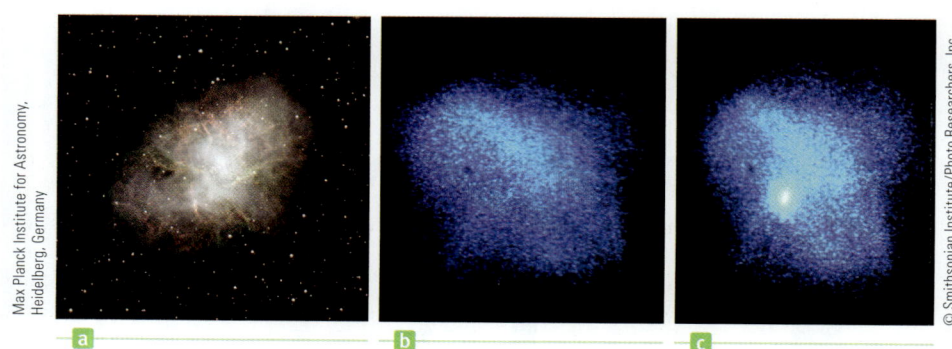

Figure 8.31 (a) The Crab Nebula in the constellation Taurus. This nebula is the remnant of a supernova seen on Earth in A.D. 1054. It is located some 6 300 light-years away and is approximately 6 light-years in diameter, still expanding outward. A pulsar deep inside the nebula flashes 30 times every second. (b) Pulsar off. (c) Pulsar on.

EXAMPLE 8.14 The Spinning Stool

GOAL Apply conservation of angular momentum to a simple system.

PROBLEM A student sits on a pivoted stool while holding a pair of weights. (See Fig. 8.32.) The stool is free to rotate about a vertical axis with negligible friction. The moment of inertia of student, weights, and stool is $2.25 \text{ kg} \cdot \text{m}^2$. The student is set in rotation with arms outstretched, making one complete turn every 1.26 s, arms outstretched. **(a)** What is the initial angular speed of the system? **(b)** As he rotates, he pulls the weights inward so that the new moment of inertia of the system (student, objects, and stool) becomes $1.80 \text{ kg} \cdot \text{m}^2$. What is the new angular speed of the system? **(c)** Find the work done by the student on the system while pulling in the weights. (Ignore energy lost through dissipation in his muscles.)

Figure 8.32 (Example 8.14) (a) The student is given an initial angular speed while holding two weights out. (b) The angular speed increases as the student draws the weights inwards.

STRATEGY (a) The angular speed can be obtained from the frequency, which is the inverse of the period. (b) There are no external torques acting on the system, so the new angular speed can be found with the principle of conservation of angular momentum. (c) The work done on the system during this process is the same as the system's change in rotational kinetic energy.

SOLUTION

(a) Find the initial angular speed of the system.

Invert the period to get the frequency, and multiply by 2π: $\quad \omega_i = 2\pi f = 2\pi/T =$ **4.99 rad/s**

(b) After he pulls the weights in, what's the system's new angular speed?

Equate the initial and final angular momenta of the system:

$$(1) \quad L_i = L_f \quad \rightarrow \quad I_i \omega_i = I_f \omega_f$$

Substitute and solve for the final angular speed ω_f:

$$(2) \quad (2.25 \text{ kg} \cdot \text{m}^2)(4.99 \text{ rad/s}) = (1.80 \text{ kg} \cdot \text{m}^2)\omega_f$$

$$\omega_f = \boxed{6.24 \text{ rad/s}}$$

(c) Find the work the student does on the system.

Apply the work–energy theorem:

$$W_{\text{student}} = \Delta K_r = \tfrac{1}{2} I_f \omega_f^2 - \tfrac{1}{2} I_i \omega_i^2$$
$$= \tfrac{1}{2}(1.80 \text{ kg} \cdot \text{m}^2)(6.24 \text{ rad/s})^2$$
$$- \tfrac{1}{2}(2.25 \text{ kg} \cdot \text{m}^2)(4.99 \text{ rad/s})^2$$
$$W_{\text{student}} = \boxed{7.03 \text{ J}}$$

REMARKS Although the angular momentum of the system is conserved, mechanical energy is not conserved because the student does work on the system.

QUESTION 8.14 If the student suddenly releases the weights, will his angular speed increase, decrease, or remain the same?

EXERCISE 8.14 A star with an initial radius of 1.0×10^8 m and period of 30.0 days collapses suddenly to a radius of 1.0×10^4 m. (a) Find the period of rotation after collapse. (b) Find the work done by gravity during the collapse if the mass of the star is 2.0×10^{30} kg. (c) What is the speed of an indestructible person standing on the equator of the collapsed star? (Neglect any relativistic or thermal effects, and assume the star is spherical before and after it collapses.)

ANSWERS (a) 2.6×10^{-2} s (b) 2.3×10^{42} J (c) 2.4×10^6 m/s

EXAMPLE 8.15 The Merry-Go-Round

GOAL Apply conservation of angular momentum while combining two moments of inertia.

PROBLEM A merry-go-round modeled as a disk of mass $M = 1.00 \times 10^2$ kg and radius $R = 2.00$ m is rotating in a horizontal plane about a frictionless vertical axle (Fig. 8.33 is an overhead view of the system). **(a)** After a student with mass $m = 60.0$ kg

(*Continued*)

jumps on the rim of the merry-go-round, the system's angular speed decreases to 2.00 rad/s. If the student walks slowly from the edge toward the center, find the angular speed of the system when she reaches a point 0.500 m from the center. **(b)** Find the change in the system's rotational kinetic energy caused by her movement to $r = 0.500$ m. **(c)** Find the work done on the student as she walks to $r = 0.500$ m.

Figure 8.33 (Example 8.15) As the student walks toward the center of the merry-go-round, the moment of inertia I of the system becomes smaller. Because angular momentum is conserved and $L = I\omega$, the angular speed must increase.

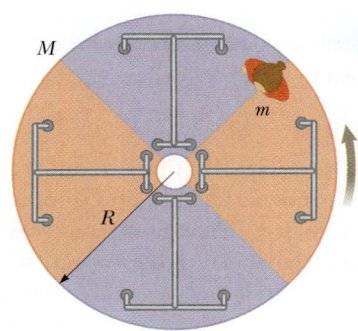

STRATEGY This problem can be solved with conservation of angular momentum by equating the system's initial angular momentum when the student stands at the rim to the angular momentum when the student has reached $r = 0.500$ m. The key is to find the different moments of inertia.

SOLUTION

(a) Find the angular speed when the student reaches a point 0.500 m from the center.

Calculate the moment of inertia of the disk, I_D:

$$I_D = \tfrac{1}{2}MR^2 = \tfrac{1}{2}(1.00 \times 10^2 \text{ kg})(2.00 \text{ m})^2$$
$$= 2.00 \times 10^2 \text{ kg} \cdot \text{m}^2$$

Calculate the initial moment of inertia of the student. This is the same as the moment of inertia of a mass a distance R from the axis:

$$I_{Si} = mR^2 = (60.0 \text{ kg})(2.00 \text{ m})^2 = 2.40 \times 10^2 \text{ kg} \cdot \text{m}^2$$

Sum the two moments of inertia and multiply by the initial angular speed to find L_i, the initial angular momentum of the system:

$$L_i = (I_D + I_{Si})\omega_i$$
$$= (2.00 \times 10^2 \text{ kg} \cdot \text{m}^2 + 2.40 \times 10^2 \text{ kg} \cdot \text{m}^2)(2.00 \text{ rad/s})$$
$$= 8.80 \times 10^2 \text{ kg} \cdot \text{m}^2/\text{s}$$

Calculate the student's final moment of inertia, I_{Sf}, when she is 0.500 m from the center:

$$I_{Sf} = mr_f^2 = (60.0 \text{ kg})(0.50 \text{ m})^2 = 15.0 \text{ kg} \cdot \text{m}^2$$

The moment of inertia of the platform is unchanged. Add it to the student's final moment of inertia, and multiply by the unknown final angular speed to find L_f:

$$L_f = (I_D + I_{Sf})\omega_f = (2.00 \times 10^2 \text{ kg} \cdot \text{m}^2 + 15.0 \text{ kg} \cdot \text{m}^2)\omega_f$$
$$= (2.15 \times 10^2 \text{ kg} \cdot \text{m}^2)\omega_f$$

Equate the initial and final angular momenta and solve for the final angular speed of the system:

$$L_i = L_f$$
$$(8.80 \times 10^2 \text{ kg} \cdot \text{m}^2/\text{s}) = (2.15 \times 10^2 \text{ kg} \cdot \text{m}^2)\omega_f$$
$$\omega_f = \boxed{4.09 \text{ rad/s}}$$

(b) Find the change in the rotational kinetic energy of the system.

Calculate the initial kinetic energy of the system:

$$KE_i = \tfrac{1}{2}I_i\omega_i^2 = \tfrac{1}{2}(4.40 \times 10^2 \text{ kg} \cdot \text{m}^2)(2.00 \text{ rad/s})^2$$
$$= 8.80 \times 10^2 \text{ J}$$

Calculate the final kinetic energy of the system:

$$KE_f = \tfrac{1}{2}I_f\omega_f^2 = \tfrac{1}{2}(215 \text{ kg} \cdot \text{m}^2)(4.09 \text{ rad/s})^2 = 1.80 \times 10^3 \text{ J}$$

Calculate the change in kinetic energy of the system:

$$KE_f - KE_i = \boxed{920 \text{ J}}$$

(c) Find the work done on the student.

The student undergoes a change in kinetic energy that equals the work done on her. Apply the work–energy theorem:

$$W = \Delta KE_{\text{student}} = \tfrac{1}{2}I_{Sf}\omega_f^2 - \tfrac{1}{2}I_{Si}\omega_i^2$$
$$= \tfrac{1}{2}(15.0 \text{ kg} \cdot \text{m}^2)(4.09 \text{ rad/s})^2$$
$$- \tfrac{1}{2}(2.40 \times 10^2 \text{ kg} \cdot \text{m}^2)(2.00 \text{ rad/s})^2$$
$$W = \boxed{-355 \text{ J}}$$

REMARKS The angular momentum is unchanged by internal forces; however, the kinetic energy increases because the student must perform positive work in order to walk toward the center of the platform.

QUESTION 8.15 Is energy conservation violated in this example? Explain why there is a positive net change in mechanical energy. What is the origin of this energy?

EXERCISE 8.15 (a) Find the angular speed of the merry-go-round before the student jumped on, assuming the student didn't transfer any momentum or energy as she jumped on the merry-go-round. (b) By how much did the kinetic energy of the system change when the student jumped on? Notice that energy is lost in this process, as should be expected, since it is essentially a perfectly inelastic collision.

ANSWERS (a) 4.40 rad/s (b) $KE_f - KE_i = -1.06 \times 10^3$ J.

SUMMARY

8.1 Torque

Let $\vec{F}$ be a force acting on an object, and let $\vec{r}$ be a position vector from a chosen point O to the point of application of the force. Then the magnitude of the torque $\vec{\tau}$ of the force $\vec{F}$ is given by

$$\tau = rF\sin\theta \quad [8.2]$$

where r is the length of the position vector, F the magnitude of the force, and θ the angle between $\vec{F}$ and $\vec{r}$.

The torque at O depends on the distance to the point of application of the force $\vec{F}$ and the force's magnitude and direction.

An alternate interpretation of torque involves the concept of a lever arm $d = r\sin\theta$ that is perpendicular to the force.

The quantity $d = r\sin\theta$ is called the *lever arm* of the force.

8.2 Torque and the Two Conditions for Equilibrium

An object in mechanical equilibrium must satisfy the following two conditions:

1. The net external force must be zero: $\sum \vec{F} = 0$.
2. The net external torque must be zero: $\sum \vec{\tau} = 0$.

These two conditions, used in solving problems involving rotation in a plane—result in three equations and three unknowns—two from the first condition (corresponding to the x- and y-components of the force) and one from the second condition, on torques. These equations must be solved simultaneously.

8.5 Relationship Between Torque and Angular Acceleration

The **moment of inertia** of a group of particles is

$$I \equiv \sum mr^2 \quad [8.7]$$

If a rigid object free to rotate about a fixed axis has a net external torque $\Sigma\tau$ acting on it, then the object undergoes an angular acceleration a, where

$$\Sigma\tau = I\alpha \quad [8.8]$$

This equation is the rotational equivalent of the second law of motion.

Problems are solved by using Equation 8.8 together with Newton's second law and solving the resulting equations simultaneously. The relation $a = r\alpha$ is often key in relating the translational equations to the rotational equations.

8.6 Rotational Kinetic Energy

If a rigid object rotates about a fixed axis with angular speed ω, its **rotational kinetic energy** is

$$KE_r = \tfrac{1}{2}I\omega^2 \quad [8.10]$$

where I is the moment of inertia of the object around the axis of rotation.

A system involving rotation is described by three types of energy: potential energy PE, translational kinetic energy KE_t, and rotational kinetic energy KE_r. All these forms of energy must be included in the equation for conservation of mechanical energy for an isolated system:

$$(KE_t + KE_r + PE)_i = (KE_t + KE_r + PE)_f \quad [8.11]$$

where i and f refer to initial and final values, respectively. When non-conservative forces are present, it's necessary to use a generalization of the work–energy theorem:

$$W_{nc} = \Delta KE_t + \Delta KE_r + \Delta PE \quad [8.12]$$

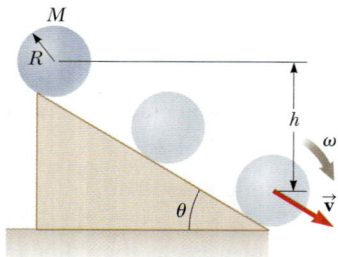

A ball rolling down an incline converts potential energy to translational and rotational kinetic energy.

8.7 Angular Momentum

The **angular momentum** of a rotating object is given by

$$L \equiv I\omega \qquad [8.13]$$

Angular momentum is related to torque in the following equation:

$$\sum \tau = \frac{\text{change in angular momentum}}{\text{time interval}} = \frac{\Delta L}{\Delta t} \qquad [8.14]$$

If the net external torque acting on a system is zero, the total angular momentum of the system is constant,

$$L_i = L_f \qquad [8.15]$$

and is said to be conserved. Solving problems usually involves substituting into the expression

$$I_i \omega_i = I_f \omega_f \qquad [8.16]$$

and solving for the unknown.

■ WARM-UP EXERCISES

The warm-up exercises in this chapter may be assigned online in Enhanced WebAssign.

1. **Math Review** The two conditions for equilibrium (see Sections 8.2 and 8.4) often result in a system of equations such as $F_1 + F_2 = 60.0$ N and $2.00F_1 - 3.00F_2 = 0$. Find (a) F_1 and (b) F_2.

2. **Math Review** Solve the equations $\frac{1}{2}mv^2 + \frac{1}{2}I\omega^2 = mgh$ and $v = r\omega$ for the speed v using substitution, given that $I = mr^2$ and $h = 3.00$ m. (See Section 8.6. Note that mass m and radius r will both cancel, so their numerical values aren't required.)

3. **Physics Review** A spinning wheel steadily slows from an initial angular velocity of 2.00 rev/s to 0.500 rev/s in 10.0 s. (a) Calculate the wheel's angular acceleration in radians per second squared. (b) What angle does it go through during that time? (See Sections 7.1 and 7.2.)

4. **Physics Review** A construction crane's cable lifts a 50.0-kg box upward with an acceleration of 1.50 m/s². Find the tension in the rope. (See Section 4.5.)

5. A man opens a 1.00-m wide door by pushing on it with a force of 50.0 N directed perpendicular to its surface. What magnitude torque does he apply about an axis through the hinges if the force is applied (a) at the center of the door? (b) at the edge farthest from the hinges? (See Section 8.1.)

6. A worker applies a torque to a nut with a wrench 0.500 m long. Because of the cramped space, she must exert a force upward at an angle of 60.0° with respect to a line from the nut through the end of the wrench. If the force she exerts has magnitude 80.0 N, what magnitude torque does she apply to the nut? (See Section 8.1.)

7. A mass of 1.00 kg is at (−2.00 m, 0) and a 2.00-kg mass at (3.00 m, 3.00 m). Find the center of mass of the system. (See Section 8.3.)

8. A horizontal plank 4.00 m long and having mass 20.0 kg rests on two pivots, one at the left end and a second 1.00 m from the right end. Find the magnitude of the force exerted on the plank by the second pivot. (See Section 8.4.)

9. A student rides his bicycle at a constant speed of 3.00 m/s along a straight, level road. If the bike's tires each have a radius of 0.350 m, (a) what is the tires' angular speed? (See Section 7.3.) (b) What is the net torque on each tire? (See Section 8.5.)

10. What is the magnitude of the angular acceleration of a 25.0-kg disk of radius 0.800 m when a torque of magnitude 40.0 N · m is applied to it? (See Section 8.5.)

11. A bicycle tire has a mass of 2.70 kg and a radius of 0.350 m. (a) Treating the tire as a hoop, what is its moment of inertia about an axis passing through the hub at its center? (b) What torque is required to produce an angular acceleration of 0.750 rad/s²? (c) What friction force applied tangentially to the edge of the tire will create a torque of that magnitude? (See Section 8.5.)

12. A bowling ball of mass 7.00 kg is rolling at 3.00 m/s along a level surface. Calculate (a) the ball's translational kinetic energy, (b) the ball's rotational kinetic energy, and (c) the ball's total kinetic energy. (d) How much work would have to be done on the ball to bring it to rest? (See Section 8.6.)

13. A basketball player entertains the crowd by spinning a basketball on his nose. The basketball has a mass of 0.600 kg and a radius of 0.121 m. If the basketball is spinning at a rate of 3.00 revolutions per second, (a) what is its rotational kinetic energy? (See Section 8.6.)

(b) What is the magnitude of its angular momentum? Treat the ball as a thin, spherical shell. (See Section 8.7.)

14. A disk of mass m is spinning freely at 6.00 rad/s when a second disk of identical mass, initially not spinning, is dropped on it so that their axes coincide. In a short time the two disks are corotating. (a) What is the angular speed of the new system? (b) If a third such disk is dropped on the first two, find the final angular speed of the system. (See Section 8.7.)

CONCEPTUAL QUESTIONS

The conceptual questions in this chapter may be assigned online in Enhanced WebAssign.

1. Why can't you put your heels firmly against a wall and then bend over without falling?

2. Explain why changing the axis of rotation of an object changes its moment of inertia.

3. If you see an object rotating, is there necessarily a net torque acting on it?

4. (a) Is it possible to calculate the torque acting on a rigid object without specifying an origin? (b) Is the torque independent of the location of the origin?

5. Why does a long pole help a tightrope walker stay balanced?

6. In the movie *Jurassic Park*, there is a scene in which some members of the visiting group are trapped in the kitchen with dinosaurs outside. The paleontologist is pressing against the center of the door, trying to keep out the dinosaurs on the other side. The botanist throws herself against the door at the edge near the hinge. A pivotal point in the film is that she cannot reach a gun on the floor because she is trying to hold the door closed. If the paleontologist is pressing at the center of the door, and the botanist is pressing at the edge about 8 cm from the hinge, estimate how far the paleontologist would have to relocate in order to have a greater effect on keeping the door closed than both of them pushing together have in their original positions. (Question 6 is courtesy of Edward F. Redish. For more questions of this type, see www.physics.umd.edu/perg/.)

7. In some motorcycle races, the riders drive over small hills and the motorcycle becomes airborne for a short time. If the motorcycle racer keeps the throttle open while leaving the hill and going into the air, the motorcycle's nose tends to rise upwards. Why does this happen?

8. If you toss a textbook into the air, rotating it each time about one of the three axes perpendicular to it, you will find that it will not rotate smoothly about one of those axes. (Try placing a strong rubber band around the book before the toss so that it will stay closed.) The book's rotation is stable about those axes having the largest and smallest moments of inertia, but unstable about the axis of intermediate moment. Try this on your own to find the axis that has this intermediate moment of inertia.

9. Stars originate as large bodies of slowly rotating gas. Because of gravity, these clumps of gas slowly decrease in size. What happens to the angular speed of a star as it shrinks? Explain.

10. If a high jumper positions his body correctly when going over the bar, the center of gravity of the athlete may actually pass under the bar. (See Fig. CQ8.10.) Explain how this is possible.

Figure CQ8.10

11. In a tape recorder, the tape is pulled past the read–write heads at a constant speed by the drive mechanism. Consider the reel from which the tape is pulled: As the tape is pulled off, the radius of the roll of remaining tape decreases. (a) How does the torque on the reel change with time? (b) If the tape mechanism is suddenly turned on so that the tape is quickly pulled with a large force, is the tape more likely to break when pulled from a nearly full reel or from a nearly empty reel?

12. (a) Give an example in which the net force acting on an object is zero, yet the net torque is nonzero. (b) Give an example in which the net torque acting on an object is zero, yet the net force is nonzero.

13. A ladder rests inclined against a wall. Would you feel safer climbing up the ladder if you were told that the floor was frictionless, but the wall was rough, or that the wall was frictionless, but the floor was rough? Justify your answer.

14. A cat usually lands on its feet regardless of the position from which it is dropped. A slow-motion film of a cat falling shows that the upper half of its body twists in one direction while the lower half twists in the opposite direction. (See Fig. CQ8.14.) Why does this type of rotation occur?

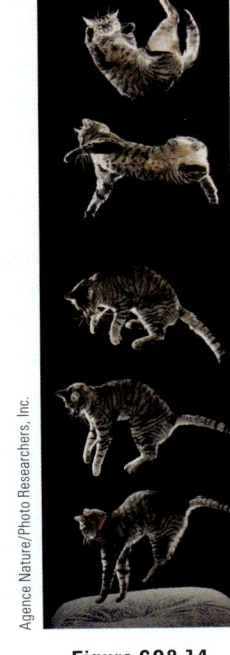

Figure CQ8.14

270 CHAPTER 8 | Rotational Equilibrium and Rotational Dynamics

15. A solid disk and a hoop are simultaneously released from rest at the top of an incline and roll down without slipping. Which object reaches the bottom first? (a) The one that has the largest mass arrives first. (b) The one that has the largest radius arrives first. (c) The hoop arrives first. (d) The disk arrives first. (e) The hoop and the disk arrive at the same time.

16. A mouse is initially at rest on a horizontal turntable mounted on a frictionless, vertical axle. As the mouse begins to walk clockwise around the perimeter, which of the following statements *must* be true of the turntable? (a) It also turns clockwise. (b) It turns counterclockwise with the same angular velocity as the mouse. (c) It remains stationary. (d) It turns counterclockwise because angular momentum is conserved. (e) It turns clockwise because mechanical energy is conserved.

17. The cars in a soapbox derby have no engines; they simply coast downhill. Which of the following design criteria is best from a competitive point of view? The car's wheels should (a) have large moments of inertia, (b) be massive, (c) be hoop-like wheels rather than solid disks, (d) be large wheels rather than small wheels, or (e) have small moments of inertia.

PROBLEMS

WebAssign The problems in this chapter may be assigned online in Enhanced WebAssign.

1. denotes straightforward problem; 2. denotes intermediate problem;
3. denotes challenging problem
1. denotes full solution available in *Student Solutions Manual/Study Guide*
1. denotes problems most often assigned in Enhanced WebAssign

BIO denotes biomedical problems
GP denotes guided problems
M denotes Master It tutorial available in Enhanced WebAssign
QC denotes asking for quantitative and conceptual reasoning
S denotes symbolic reasoning problem
W denotes Watch It video solution available in Enhanced WebAssign

8.1 Torque

1. The fishing pole in Figure P8.1 makes an angle of 20.0° with the horizontal. What is the magnitude of the torque exerted by the fish about an axis perpendicular to the page and passing through the angler's hand if the fish pulls on the fishing line with a force $\vec{F} = 100$ N at an angle 37.0° below the horizontal? The force is applied at a point 2.00 m from the angler's hands.

Figure P8.1

2. **M** Find the net torque on the wheel in Figure P8.2 about the axle through O perpendicular to the page, taking $a = 10.0$ cm and $b = 25.0$ cm.

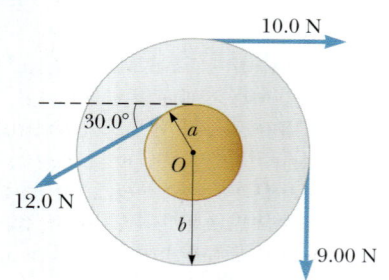

Figure P8.2

3. Calculate the net torque (magnitude and direction) on the beam in Figure P8.3 about (a) an axis through O perpendicular to the page and (b) an axis through C perpendicular to the page.

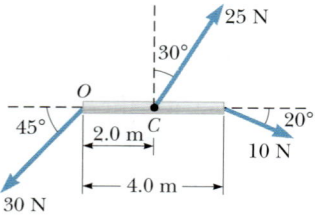

Figure P8.3

4. **BIO** A dental bracket exerts a horizontal force of 80.0 N on a tooth at point B in Figure P8.4. What is the torque on the root of the tooth about point A?

Figure P8.4

5. **QC** A simple pendulum consists of a small object of mass 3.0 kg hanging at the end of a 2.0-m-long light string that is connected to a pivot point. (a) Calculate the magnitude of the torque (due to the force of gravity) about this pivot point when the string makes a 5.0° angle with the vertical. (b) Does the torque increase or decrease as the angle increases? Explain.

6. **S** Write the necessary equations of equilibrium of the object shown in Figure P8.6. Take the origin of the torque equation about an axis perpendicular to the page through the point O.

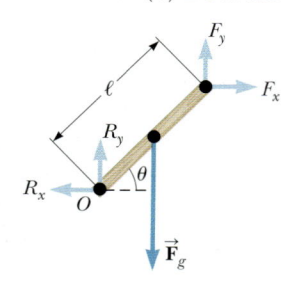

Figure P8.6

8.2 Torque and the Two Conditions for Equilibrium
8.3 The Center of Gravity
8.4 Examples of Objects in Equilibrium

7. **BIO** The arm in Figure P8.7 weighs 41.5 N. The force of gravity acting on the arm acts through point A. Determine the magnitudes of the tension force $\vec{F}_t$ in the deltoid muscle and the force $\vec{F}_s$ exerted by the shoulder on the humerus (upper-arm bone) to hold the arm in the position shown.

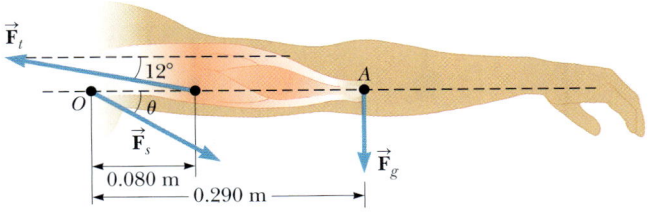

Figure P8.7

8. A uniform 35.0-kg beam of length $\ell = 5.00$ m is supported by a vertical rope located $d = 1.20$ m from its left end as in Figure P8.8. The right end of the beam is supported by a vertical column. Find (a) the tension in the rope and (b) the force that the column exerts on the right end of the beam.

9. **BIO** A cook holds a 2.00-kg carton of milk at arm's length (Fig. P8.9). What force $\vec{F}_B$ must be exerted by the biceps muscle? (Ignore the weight of the forearm.)

Figure P8.8

Figure P8.9

10. A meter stick is found to balance at the 49.7-cm mark when placed on a fulcrum. When a 50.0-gram mass is attached at the 10.0-cm mark, the fulcrum must be moved to the 39.2-cm mark for balance. What is the mass of the meter stick?

11. Find the x- and y-coordinates of the center of gravity of a 4.00-ft by 8.00-ft uniform sheet of plywood with the upper right quadrant removed as shown in Figure P8.11. *Hint:* The mass of any segment of the plywood sheet is proportional to the area of that segment.

Figure P8.11

12. **GP** A beam resting on two pivots has a length of $L = 6.00$ m and mass $M = 90.0$ kg. The pivot under the left end exerts a normal force n_1 on the beam, and the second pivot placed a distance $\ell = 4.00$ m from the left end exerts a normal force n_2. A woman of mass $m = 55.0$ kg steps onto the left end of the beam and begins walking to the right as in Figure P8.12. The goal is to find the woman's position when the beam begins to tip. (a) Sketch a free-body diagram, labeling the gravitational and normal forces acting on the beam and placing the woman x meters to the right of the first pivot, which is the origin. (b) Where is the woman when the normal force n_1 is the greatest? (c) What is n_1 when the beam is about to tip? (d) Use the force equation of equilibrium to find the value of n_2 when the beam is about to tip. (e) Using the result of part (c) and the torque equilibrium equation, with torques computed around the second pivot point, find the woman's position when the beam is about to tip. (f) Check the answer to part (e) by computing torques around the first pivot point. Except for possible slight differences due to rounding, is the answer the same?

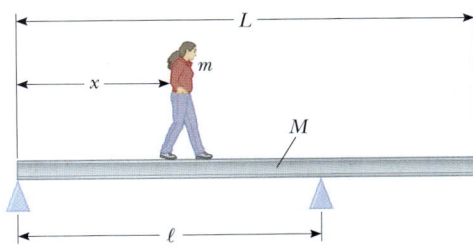

Figure P8.12 Problems 12 and 14.

13. Consider the following mass distribution, where x- and y-coordinates are given in meters: 5.0 kg at (0.0, 0.0) m, 3.0 kg at (0.0, 4.0) m, and 4.0 kg at (3.0, 0.0) m. Where should a fourth object of 8.0 kg be placed so that the center of gravity of the four-object arrangement will be at (0.0, 0.0) m?

14. **S** A beam of length L and mass M rests on two pivots. The first pivot is at the left end, taken as the origin, and the second pivot is at a distance ℓ from the left end. A woman of mass m starts at the left end and walks toward the right end as in Figure P8.12. When the beam is on the verge of tipping, find symbolic expressions for (a) the normal force exerted by the second pivot in terms of M, m, and g and (b) the woman's position in terms of M, m, L, and ℓ. (c) Find the minimum value of ℓ that will allow the woman to reach the end of the beam without it tipping.

15. **BIO** Many of the elements in horizontal-bar exercises can be modeled by representing the gymnast by four segments consisting of arms, torso (including the head), thighs, and lower legs, as shown in Figure P8.15a. Inertial parameters for a particular gymnast are as follows:

Segment	Mass (kg)	Length (m)	r_{cg} (m)	I (kg · m²)
Arms	6.87	0.548	0.239	0.205
Torso	33.57	0.601	0.337	1.610
Thighs	14.07	0.374	0.151	0.173
Legs	7.54	—	0.227	0.164

Note that in Figure P8.15a r_{cg} is the distance to the center of gravity measured from the joint closest to the bar and the masses for the arms, thighs, and legs include both appendages. I is the moment of inertia of each segment about its center of gravity. Determine the distance from the bar to the center of gravity of the gymnast for the two positions shown in Figures P8.15b and P8.15c.

Figure P8.15

16. **BIO** Using the data given in Problem 15 and the coordinate system shown in Figure P8.16b, calculate the position of the center of gravity of the gymnast shown in Figure P8.16a. Pay close attention to the definition of r_{cg} in the table.

Figure P8.16

17. **BIO** A person bending forward to lift a load "with his back" (Fig. P8.17a) rather than "with his knees" can be injured by large forces exerted on the muscles and vertebrae. The spine pivots mainly at the fifth lumbar vertebra, with the principal supporting force provided by the erector spinalis muscle in the back. To see the magnitude of the forces involved, and to understand why back problems are common among humans, consider the model shown in Figure P8.17b of a person bending forward to lift a 200-N object. The spine and upper body are represented as a uniform horizontal rod of weight 350 N, pivoted at the base of the spine. The erector spinalis muscle, attached at a point two-thirds of the way up the spine, maintains the position of the back. The angle between the spine and this muscle is 12.0°. Find (a) the tension in the back muscle and (b) the compressional force in the spine.

Figure P8.17

18. **BIO** When a person stands on tiptoe (a strenuous position), the position of the foot is as shown in Figure P8.18a. The total gravitational force on the body, $\vec{F}_g$, is supported by the force $\vec{n}$ exerted by the floor on the toes of one foot. A mechanical model of the situation is shown in Figure P8.18b, where $\vec{T}$ is the force exerted by the Achilles tendon on the foot and $\vec{R}$ is the force exerted by the tibia on the foot. Find the values of T, R, and θ when $F_g = n = 700$ N.

Figure P8.18

19. **M** A 500-N uniform rectangular sign 4.00 m wide and 3.00 m high is suspended from a horizontal, 6.00-m-long, uniform, 100-N rod as indicated in Figure P8.19. The left end of the rod is supported by a hinge, and the right end is supported by a thin cable making a 30.0° angle with the vertical. (a) Find the tension T in the cable. (b) Find the horizontal and vertical components of force exerted on the left end of the rod by the hinge.

Figure P8.19

20. A window washer is standing on a scaffold supported by a vertical rope at each end. The scaffold weighs 200 N and is 3.00 m long. What is the tension in each

rope when the 700-N worker stands 1.00 m from one end?

21. **W** A uniform plank of length 2.00 m and mass 30.0 kg is supported by three ropes, as indicated by the blue vectors in Figure P8.21. Find the tension in each rope when a 700-N person is $d = 0.500$ m from the left end.

Figure P8.21

22. A hungry bear weighing 700 N walks out on a beam in an attempt to retrieve a basket of goodies hanging at the end of the beam (Fig. P8.22). The beam is uniform, weighs 200 N, and is 6.00 m long, and it is supported by a wire at an angle of $\theta = 60.0°$. The basket weighs 80.0 N. (a) Draw a force diagram for the beam. (b) When the bear is at $x = 1.00$ m, find the tension in the wire supporting the beam and the components of the force exerted by the wall on the left end of the beam. (c) If the wire can withstand a maximum tension of 900 N, what is the maximum distance the bear can walk before the wire breaks?

Figure P8.22

23. **S** Figure P8.23 shows a uniform beam of mass m pivoted at its lower end, with a horizontal spring attached between its top end and a vertical wall. The beam makes an angle θ with the horizontal. Find expressions for (a) the distance d the spring is stretched from equilibrium and (b) the components of the force exerted by the pivot on the beam.

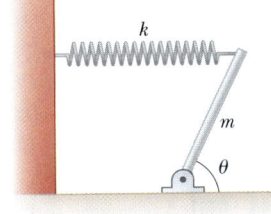

Figure P8.23

24. **GP** A strut of length $L = 3.00$ m and mass $m = 16.0$ kg is held by a cable at an angle of $\theta = 30.0°$ with respect to the horizontal as shown in Figure P8.24. (a) Sketch a force diagram, indicating all the forces and their placement on the strut. (b) Why is the hinge a good place to use for calculating torques? (c) Write the condition for rotational equilibrium symbolically, calculating the torques around the hinge. (d) Use the torque equation to calculate the tension in the cable. (e) Write the x- and y-components of Newton's second law for equilibrium. (f) Use the force equation to find the x- and y-components of the force on the hinge. (g) Assuming the strut position is to remain the same, would it be advantageous to attach the cable higher up on the wall? Explain the benefit in terms of the force on the hinge and cable tension.

Figure P8.24

25. **S** A refrigerator of width w and height h rests on a rough incline as in Figure P8.25. Find an expression for the maximum value θ can have before the refrigerator tips over. Note, the contact point between the refrigerator and incline shifts as θ increases and treat the refrigerator as a uniform box.

Figure P8.25

26. **QC S** A uniform beam of length L and mass m shown in Figure P8.26 is inclined at an angle θ to the horizontal. Its upper end is connected to a wall by a rope, and its lower end rests on a rough horizontal surface. The coefficient of static friction between the beam and surface is μ_s. Assume the angle is such that the static friction force is at its maximum value. (a) Draw a force diagram for the beam. (b) Using the condition of rotational equilibrium, find an expression for the tension T in the rope in terms of m, g, and θ. (c) Using Newton's second law for equilibrium, find a second expression for T in terms of μ_s, m, and g. (d) Using the foregoing results, obtain a relationship involving only μ_s and the angle θ. (e) What happens if the angle gets smaller? Is this equation valid for all values of θ? Explain.

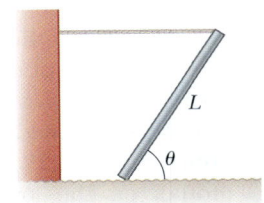

Figure P8.26

27. **BIO** The chewing muscle, the masseter, is one of the strongest in the human body. It is attached to the mandible (lower jawbone) as shown in Figure P8.27a. The jawbone is pivoted about a socket just in front of the auditory canal. The forces acting on the jawbone are equivalent to those acting on the curved bar in

Figure P8.27b. $\vec{F}_C$ is the force exerted by the food being chewed against the jawbone, $\vec{T}$ is the force of tension in the masseter, and $\vec{R}$ is the force exerted by the socket on the mandible. Find $\vec{T}$ and $\vec{R}$ for a person who bites down on a piece of steak with a force of 50.0 N.

Figure P8.27

28. A 1 200-N uniform boom at $\phi = 65°$ to the horizontal is supported by a cable at an angle $\theta = 25.0°$ to the horizontal as shown in Figure P8.28. The boom is pivoted at the bottom, and an object of weight $w = 2\,000$ N hangs from its top. Find (a) the tension in the support cable and (b) the components of the reaction force exerted by the pivot on the boom.

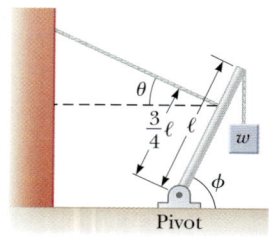

Figure P8.28

29. **BIO** The large quadriceps muscle in the upper leg terminates at its lower end in a tendon attached to the upper end of the tibia (Fig. P8.29a). The forces on the lower leg when the leg is extended are modeled as in Figure P8.29b, where $\vec{T}$ is the force of tension in the tendon, $\vec{w}$ is the force of gravity acting on the lower leg, and $\vec{F}$ is the force of gravity acting on the foot. Find $\vec{T}$ when the tendon is at an angle of 25.0° with the tibia, assuming that $w = 30.0$ N, $F = 12.5$ N, and the leg is extended at an angle θ of 40.0° with the vertical. Assume that the center of gravity of the lower leg is at its center and that the tendon attaches to the lower leg at a point one-fifth of the way down the leg.

Figure P8.29

30. One end of a uniform 4.0-m-long rod of weight w is supported by a cable at an angle of $\theta = 37°$ with the rod. The other end rests against a wall, where it is held by friction. (See Fig. P8.30.) The coefficient of static friction between the wall and the rod is $\mu_s = 0.50$. Determine the minimum distance x from point A at which an additional weight w (the same as the weight of the rod) can be hung without causing the rod to slip at point A.

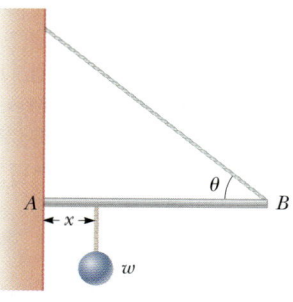

Figure P8.30

8.5 Relationship Between Torque and Angular Acceleration

31. Four objects are held in position at the corners of a rectangle by light rods as shown in Figure P8.31. Find the moment of inertia of the system about (a) the x-axis, (b) the y-axis, and (c) an axis through O and perpendicular to the page.

32. If the system shown in Figure P8.31 is set in rotation about each of the axes mentioned in Problem 31, find the torque that will produce an angular acceleration of 1.50 rad/s² in each case.

Figure P8.31 Problems 31 and 32.

33. A large grinding wheel in the shape of a solid cylinder of radius 0.330 m is free to rotate on a frictionless, vertical axle. A constant tangential force of 250 N applied to its edge causes the wheel to have an angular acceleration of 0.940 rad/s². (a) What is the moment of inertia of the wheel? (b) What is the mass of the wheel? (c) If the wheel starts from rest, what is its angular velocity after 5.00 s have elapsed, assuming the force is acting during that time?

34. **GP** An oversized yo-yo is made from two identical solid disks each of mass $M = 2.00$ kg and radius $R = 10.0$ cm. The two disks are joined by a solid cylinder of radius $r = 4.00$ cm and mass $m = 1.00$ kg as in Figure P8.34. Take the center of the cylinder as the axis of the system, with positive torques directed to the left along this axis. All torques and angular variables are to be calculated around this axis. Light string is wrapped around the cylinder, and the system is then allowed to drop from rest. (a) What is the moment of inertia of the system? Give a symbolic answer. (b) What torque does gravity exert on the system with respect to the given axis? (c) Take downward as the negative coordinate direction. As depicted in Figure P8.34, is the torque exerted by the tension positive or negative? Is the angular acceleration positive or negative? What about

the translational acceleration? (d) Write an equation for the angular acceleration α in terms of the translational acceleration a and radius r. (Watch the sign!) (e) Write Newton's second law for the system in terms of m, M, a, T, and g. (f) Write Newton's second law for rotation in terms of I, α, T, and r. (g) Eliminate α from the rotational second law with the expression found in part (d) and find a symbolic expression for the acceleration a in terms of m, M, g, r and R. (h) What is the numeric value for the system's acceleration? (i) What is the tension in the string? (j) How long does it take the system to drop 1.00 m from rest?

Figure P8.34

35. **QC** A rope of negligible mass is wrapped around a 225-kg solid cylinder of radius 0.400 m. The cylinder is suspended several meters off the ground with its axis oriented horizontally, and turns on that axis without friction. (a) If a 75.0-kg man takes hold of the free end of the rope and falls under the force of gravity, what is his acceleration? (b) What is the angular acceleration of the cylinder? (c) If the mass of the rope were not neglected, what would happen to the angular acceleration of the cylinder as the man falls?

36. **W** A potter's wheel having a radius of 0.50 m and a moment of inertia of 12 kg · m² is rotating freely at 50 rev/min. The potter can stop the wheel in 6.0 s by pressing a wet rag against the rim and exerting a radially inward force of 70 N. Find the effective coefficient of kinetic friction between the wheel and the wet rag.

37. A model airplane with mass 0.750 kg is tethered by a wire so that it flies in a circle 30.0 m in radius. The airplane engine provides a net thrust of 0.800 N perpendicular to the tethering wire. (a) Find the torque the net thrust produces about the center of the circle. (b) Find the angular acceleration of the airplane when it is in level flight. (c) Find the linear acceleration of the airplane tangent to its flight path.

38. A bicycle wheel has a diameter of 64.0 cm and a mass of 1.80 kg. Assume that the wheel is a hoop with all the mass concentrated on the outside radius. The bicycle is placed on a stationary stand, and a resistive force of 120 N is applied tangent to the rim of the tire. (a) What force must be applied by a chain passing over a 9.00-cm-diameter sprocket in order to give the wheel an acceleration of 4.50 rad/s²? (b) What force is required if you shift to a 5.60-cm-diameter sprocket?

39. **M** A 150-kg merry-go-round in the shape of a uniform, solid, horizontal disk of radius 1.50 m is set in motion by wrapping a rope about the rim of the disk and pulling on the rope. What constant force must be exerted on the rope to bring the merry-go-round from rest to an angular speed of 0.500 rev/s in 2.00 s?

40. **QC S** An Atwood's machine consists of blocks of masses m_1 = 10.0 kg and m_2 = 20.0 kg attached by a cord running over a pulley as in Figure P8.40. The pulley is a solid cylinder with mass M = 8.00 kg and radius r = 0.200 m. The block of mass m_2 is allowed to drop, and the cord turns the pulley without slipping. (a) Why must the tension T_2 be greater than the tension T_1? (b) What is the acceleration of the system, assuming the pulley axis is frictionless? (c) Find the tensions T_1 and T_2.

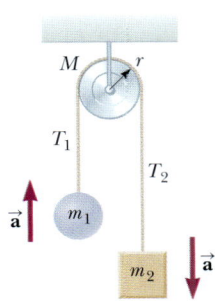

Figure P8.40

41. An airliner lands with a speed of 50.0 m/s. Each wheel of the plane has a radius of 1.25 m and a moment of inertia of 110 kg · m². At touchdown, the wheels begin to spin under the action of friction. Each wheel supports a weight of 1.40×10^4 N, and the wheels attain their angular speed in 0.480 s while rolling without slipping. What is the coefficient of kinetic friction between the wheels and the runway? Assume that the speed of the plane is constant.

8.6 Rotational Kinetic Energy

42. A car is designed to get its energy from a rotating flywheel with a radius of 2.00 m and a mass of 500 kg. Before a trip, the flywheel is attached to an electric motor, which brings the flywheel's rotational speed up to 5 000 rev/min. (a) Find the kinetic energy stored in the flywheel. (b) If the flywheel is to supply energy to the car as a 10.0-hp motor would, find the length of time the car could run before the flywheel would have to be brought back up to speed.

43. A horizontal 800-N merry-go-round of radius 1.50 m is started from rest by a constant horizontal force of 50.0 N applied tangentially to the merry-go-round. Find the kinetic energy of the merry-go-round after 3.00 s. (Assume it is a solid cylinder.)

44. **QC** Four objects—a hoop, a solid cylinder, a solid sphere, and a thin, spherical shell—each have a mass of 4.80 kg and a radius of 0.230 m. (a) Find the moment of inertia for each object as it rotates about the axes shown in Table 8.1. (b) Suppose each object is rolled down a ramp. Rank the translational speed of each object from highest to lowest. (c) Rank the objects' rotational kinetic energies from highest to lowest as the objects roll down the ramp.

45. A light rod of length $\ell = 1.00$ m rotates about an axis perpendicular to its length and passing through its center as in Figure P8.45. Two particles of masses $m_1 = 4.00$ kg and $m_2 = 3.00$ kg are connected to the ends of the rod. (a) Neglecting the mass of the rod, what is the system's kinetic energy when its angular speed is 2.50 rad/s? (b) Repeat the problem, assuming the mass of the rod is taken to be 2.00 kg.

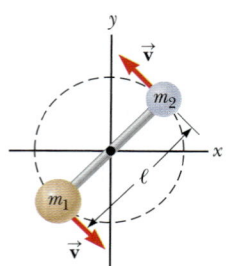

Figure P8.45 Problems 45 and 57.

46. A 240-N sphere 0.20 m in radius rolls without slipping 6.0 m down a ramp that is inclined at 37° with the horizontal. What is the angular speed of the sphere at the bottom of the slope if it starts from rest?

47. A solid, uniform disk of radius 0.250 m and mass 55.0 kg rolls down a ramp of length 4.50 m that makes an angle of 15.0° with the horizontal. The disk starts from rest from the top of the ramp. Find (a) the speed of the disk's center of mass when it reaches the bottom of the ramp and (b) the angular speed of the disk at the bottom of the ramp.

48. Q|C S A solid uniform sphere of mass m and radius R rolls without slipping down an incline of height h. (a) What forms of mechanical energy are associated with the sphere at any point along the incline when its angular speed is ω? Answer in words and symbolically in terms of the quantities m, g, y, I, ω, and v. (b) What force acting on the sphere causes it to roll rather than slip down the incline? (c) Determine the ratio of the sphere's rotational kinetic energy to its total kinetic energy at any instant.

49. The top in Figure P8.49 has a moment of inertia of 4.00×10^{-4} kg·m² and is initially at rest. It is free to rotate about a stationary axis AA'. A string wrapped around a peg along the axis of the top is pulled in such a manner as to maintain a constant tension of 5.57 N in the string. If the string does not slip while wound around the peg, what is the angular speed of the top after 80.0 cm of string has been pulled off the peg? *Hint:* Consider the work that is done.

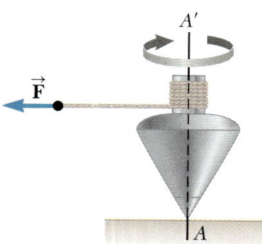

Figure P8.49

50. A constant torque of 25.0 N·m is applied to a grindstone whose moment of inertia is 0.130 kg·m². Using energy principles and neglecting friction, find the angular speed after the grindstone has made 15.0 revolutions. *Hint:* The angular equivalent of $W_{net} = F\Delta x = \frac{1}{2}mv_f^2 - \frac{1}{2}mv_i^2$ is $W_{net} = \tau\Delta\theta = \frac{1}{2}I\omega_f^2 - \frac{1}{2}I\omega_i^2$. You should convince yourself that this last relationship is correct.

51. A 10.0-kg cylinder rolls without slipping on a rough surface. At an instant when its center of gravity has a speed of 10.0 m/s, determine (a) the translational kinetic energy of its center of gravity, (b) the rotational kinetic energy about its center of gravity, and (c) its total kinetic energy.

52. W Use conservation of energy to determine the angular speed of the spool shown in Figure P8.52 after the 3.00-kg bucket has fallen 4.00 m, starting from rest. The light string attached to the bucket is wrapped around the spool and does not slip as it unwinds.

Figure P8.52

53. Q|C A giant swing at an amusement park consists of a 365-kg uniform arm 10.0 m long, with two seats of negligible mass connected at the lower end of the arm (Fig. P8.53). (a) How far from the upper end is the center of mass of the arm? (b) The gravitational potential energy of the arm is the same as if all its mass were concentrated at the center of mass. If the arm is raised through a 45.0° angle, find the gravitational potential energy, where the zero level is taken to be 10.0 m below the axis. (c) The arm drops from rest from the position described in part (b). Find the gravitational potential energy of the system when it reaches the vertical orientation. (d) Find the speed of the seats at the bottom of the swing.

Figure P8.53

8.7 Angular Momentum

54. Each of the following objects has a radius of 0.180 m and a mass of 2.40 kg, and each rotates about an axis through its center (as in Table 8.1) with an angular speed of 35.0 rad/s. Find the magnitude of the angular momentum of each object. (a) a hoop (b) a solid cylinder (c) a solid sphere (d) a hollow spherical shell

55. (a) Calculate the angular momentum of Earth that arises from its spinning motion on its axis, treating Earth as a uniform solid sphere. (b) Calculate the angular momentum of Earth that arises from its orbital motion about the Sun, treating Earth as a point particle.

56. **Q|C** A 0.005 00-kg bullet traveling horizontally with a speed of 1.00×10^3 m/s enters an 18.0-kg door, embedding itself 10.0 cm from the side opposite the hinges as in Figure P8.56. The 1.00-m-wide door is free to swing on its hinges. (a) Before it hits the door, does the bullet have angular momentum relative the door's axis of rotation? Explain. (b) Is mechanical energy conserved in this collision? Answer without doing a calculation. (c) At what angular speed does the door swing open immediately after the collision? (The door has the same moment of inertia as a rod with axis at one end.) (d) Calculate the energy of the door–bullet system and determine whether it is less than or equal to the kinetic energy of the bullet before the collision.

Figure P8.56. An overhead view of a bullet striking a door.

57. A light rigid rod of length $\ell = 1.00$ m rotates about an axis perpendicular to its length and through its center, as shown in Figure P8.45. Two particles of masses $m_1 = 4.00$ kg and $m_2 = 3.00$ kg are connected to the ends of the rod. What is the angular momentum of the system if the speed of each particle is 5.00 m/s? (Neglect the rod's mass.)

58. Halley's comet moves about the Sun in an elliptical orbit, with its closest approach to the Sun being 0.59 A.U. and its greatest distance being 35 A.U. (1 A.U. is the Earth–Sun distance). If the comet's speed at closest approach is 54 km/s, what is its speed when it is farthest from the Sun? You may neglect any change in the comet's mass and assume that its angular momentum about the Sun is conserved.

59. **S** A rigid, massless rod has three particles with equal masses attached to it as shown in Figure P8.59. The rod is free to rotate in a vertical plane about a frictionless axle perpendicular to the rod through the point P and is released from rest in the horizontal position at $t = 0$. Assuming m and d are known, find (a) the moment of inertia of the system (rod plus particles) about the pivot, (b) the torque acting on the system at $t = 0$, (c) the angular acceleration of the system at $t = 0$, (d) the linear acceleration of the particle labeled 3 at $t = 0$, (e) the maximum kinetic energy of the system, (f) the maximum angular speed reached by the rod, (g) the maximum angular momentum of the system, and (h) the maximum translational speed reached by the particle labeled 2.

Figure P8.59

60. **M** A 60.0-kg woman stands at the rim of a horizontal turntable having a moment of inertia of 500 kg · m² and a radius of 2.00 m. The turntable is initially at rest and is free to rotate about a frictionless, vertical axle through its center. The woman then starts walking around the rim clockwise (as viewed from above the system) at a constant speed of 1.50 m/s relative to Earth. (a) In what direction and with what angular speed does the turntable rotate? (b) How much work does the woman do to set herself and the turntable into motion?

61. A solid, horizontal cylinder of mass 10.0 kg and radius 1.00 m rotates with an angular speed of 7.00 rad/s about a fixed vertical axis through its center. A 0.250-kg piece of putty is dropped vertically onto the cylinder at a point 0.900 m from the center of rotation and sticks to the cylinder. Determine the final angular speed of the system.

62. A student sits on a rotating stool holding two 3.0-kg objects. When his arms are extended horizontally, the objects are 1.0 m from the axis of rotation and he rotates with an angular speed of 0.75 rad/s. The moment of inertia of the student plus stool is 3.0 kg · m² and is assumed to be constant. The student then pulls in the objects horizontally to 0.30 m from the rotation axis. (a) Find the new angular speed of the student. (b) Find the kinetic energy of the student before and after the objects are pulled in.

63. **W** The puck in Figure P8.63 has a mass of 0.120 kg. Its original distance from the center of rotation is 40.0 cm, and it moves with a speed of 80.0 cm/s. The string is pulled downward 15.0 cm through the hole in the frictionless table. Determine the work done on the puck. *Hint:* Consider the change in kinetic energy of the puck.

Figure P8.63

64. A space station shaped like a giant wheel has a radius of 100 m and a moment of inertia of 5.00×10^8 kg · m². A crew of 150 lives on the rim, and the station is rotating so that the crew experiences an apparent acceleration of $1g$ (Fig. P8.64). When 100 people move to the center of the station for a union meeting, the angular speed changes. What apparent acceleration is experienced by the managers remaining at the rim? Assume the average mass of a crew member is 65.0 kg.

Figure P8.64

278 CHAPTER 8 | Rotational Equilibrium and Rotational Dynamics

65. **QC** **S** A cylinder with moment of inertia I_1 rotates with angular velocity ω_0 about a frictionless vertical axle. A second cylinder, with moment of inertia I_2, initially not rotating, drops onto the first cylinder (Fig. P8.65). Because the surfaces are rough, the two cylinders eventually reach the same angular speed ω. (a) Calculate ω. (b) Show that kinetic energy is lost in this situation, and calculate the ratio of the final to the initial kinetic energy.

Figure P8.65

66. **M** A particle of mass 0.400 kg is attached to the 100-cm mark of a meter stick of mass 0.100 kg. The meter stick rotates on a horizontal, frictionless table with an angular speed of 4.00 rad/s. Calculate the angular momentum of the system when the stick is pivoted about an axis (a) perpendicular to the table through the 50.0-cm mark and (b) perpendicular to the table through the 0-cm mark.

Additional Problems

67. **GP** A typical propeller of a turbine used to generate electricity from the wind consists of three blades as in Figure P8.67. Each blade has a length of $L = 35$ m and a mass of $m = 420$ kg. The propeller rotates at the rate of 25 rev/min. (a) Convert the angular speed of the propeller to units of rad/s. Find (b) the moment of inertia of the propeller about the axis of rotation and (c) the total kinetic energy of the propeller.

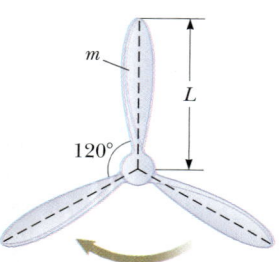

Figure P8.67

68. Figure P8.68 shows a clawhammer as it is being used to pull a nail out of a horizontal board. If a force of magnitude 150 N is exerted horizontally as shown, find (a) the force exerted by the hammer claws on the nail and (b) the force exerted by the surface at the point of contact with the hammer head. Assume that the force the hammer exerts on the nail is parallel to the nail and perpendicular to the position vector from the point of contact.

Figure P8.68

69. **QC** A 40.0-kg child stands at one end of a 70.0-kg boat that is 4.00 m long (Fig. P8.69). The boat is initially 3.00 m from the pier. The child notices a turtle on a rock beyond the far end of the boat and proceeds to walk to that end to catch the turtle. (a) Neglecting friction between the boat and water, describe the motion of the system (child plus boat). (b) Where will the child be relative to the pier when he reaches the far end of the boat? (c) Will he catch the turtle? (Assume that he can reach out 1.00 m from the end of the boat.)

Figure P8.69

70. An object of mass $M = 12.0$ kg is attached to a cord that is wrapped around a wheel of radius $r = 10.0$ cm (Fig. P8.70). The acceleration of the object down the frictionless incline is measured to be $a = 2.00$ m/s^2 and the incline makes an angle $\theta = 37.0°$ with the horizontal. Assuming the axle of the wheel to be frictionless, determine (a) the tension in the rope, (b) the moment of inertia of the wheel, and (c) the angular speed of the wheel 2.00 s after it begins rotating, starting from rest.

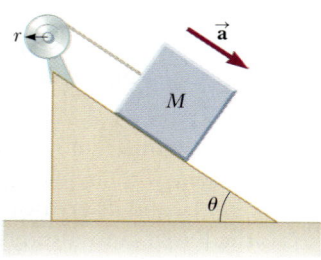

Figure P8.70

71. A uniform ladder of length L and weight w is leaning against a vertical wall. The coefficient of static friction between the ladder and the floor is the same as that between the ladder and the wall. If this coefficient of static friction is $\mu_s = 0.500$, determine the smallest angle the ladder can make with the floor without slipping.

72. Two astronauts (Fig. P8.72), each having a mass of 75.0 kg, are connected by a 10.0-m rope of negligible mass. They are isolated in space, moving in circles around the point halfway between them at a speed of 5.00 m/s. Treating the astronauts as particles, calculate (a) the magnitude of the angular momentum and (b) the rotational energy of the system. By pulling on the rope, the astronauts shorten the distance between them to 5.00 m. (c) What is the new angular momentum of the system? (d) What are their new speeds? (e) What is the new rotational energy of the system? (f) How much work is done by the astronauts in shortening the rope?

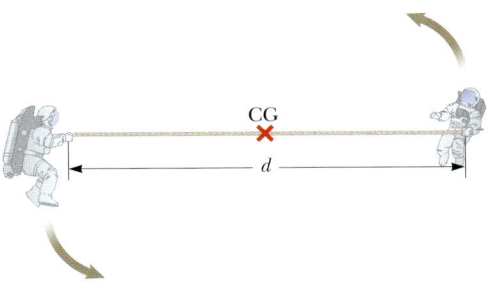

Figure P8.72 Problems 72 and 73.

73. **S** *This is a symbolic version of problem 72.* Two astronauts (Fig. P8.72), each having a mass M, are connected by a rope of length d having negligible mass. They are isolated in space, moving in circles around the point halfway between them at a speed v. (a) Calculate the magnitude of the angular momentum of the system by treating the astronauts as particles. (b) Calculate the rotational energy of the system. By pulling on the rope, the astronauts shorten the distance between them to $d/2$. (c) What is the new angular momentum of the system? (d) What are their new speeds? (e) What is the new rotational energy of the system? (f) How much work is done by the astronauts in shortening the rope?

74. Two window washers, Bob and Joe, are on a 3.00-m-long, 345-N scaffold supported by two cables attached to its ends. Bob weighs 750 N and stands 1.00 m from the left end, as shown in Figure P8.74. Two meters from the left end is the 500-N washing equipment. Joe is 0.500 m from the right end and weighs 1 000 N. Given that the scaffold is in rotational and translational equilibrium, what are the forces on each cable?

Figure P8.74

75. A 2.35-kg uniform bar of length $\ell = 1.30$ m is held in a horizontal position by three vertical springs as in Figure P8.75. The two lower springs are compressed and exert upward forces on the bar of magnitude $F_1 = 6.80$ N and $F_2 = 9.50$ N,

Figure P8.75

respectively. Find (a) the force F_s exerted by the top spring on the bar, and (b) the location x of the upper spring that will keep the bar in equilibrium.

76. **Q|C** **S** A light rod of length $2L$ is free to rotate in a *vertical* plane about a frictionless pivot through its center. A particle of mass m_1 is attached at one end of the rod, and a mass m_2 is at the opposite end, where $m_1 > m_2$. The system is released from rest in the vertical position shown in Figure P8.76a, and at some later time the system is rotating in the position shown in Figure P8.76b. Take the reference point of the gravitational potential energy to be at the pivot. (a) Find an expression for the system's total mechanical energy in the vertical position. (b) Find an expression for the total mechanical energy in the rotated position shown in Figure P8.76b. (c) Using the fact that the mechanical energy of the system is conserved, how would you determine the angular speed ω of the system in the rotated position? (d) Find the magnitude of the torque on the system in the vertical position and in the rotated position. Is the torque constant? Explain what these results imply regarding the angular momentum of the system. (e) Find an expression for the magnitude of the angular acceleration of the system in the rotated position. Does your result make sense when the rod is horizontal? When it is vertical? Explain.

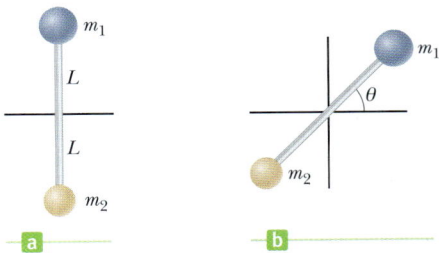

Figure P8.76

77. **Q|C** A light rope passes over a light, frictionless pulley. One end is fastened to a bunch of bananas of mass M, and a monkey of mass M clings to the other end (Fig. P8.77). The monkey climbs the rope in an attempt to reach the bananas. (a) Treating the system as consisting of the monkey, bananas, rope, and pulley, find the net torque of the system about the pulley axis. (b) Using the result of part (a), determine the total angular momentum about the pulley axis and describe the motion of the system. (c) Will the monkey reach the bananas before they get stuck in the pulley?

Figure P8.77

78. An electric motor turns a flywheel through a drive belt that joins a pulley on the motor and a pulley that is rigidly attached to the flywheel, as shown in Figure P8.78. The flywheel is a uniform disk with a mass of 80.0 kg and a radius of $R = 0.625$ m. It turns on a frictionless

axle. Its pulley has much smaller mass and a radius of $r = 0.230$ m. The tension T_u in the upper (taut) segment of the belt is 135 N and the flywheel has a clockwise angular acceleration of 1.67 rad/s². Find the tension in the lower (slack) segment of the belt.

Figure P8.78

79. BIO In exercise physiology studies, it is sometimes important to determine the location of a person's center of gravity. This can be done with the arrangement shown in Figure P8.79. A light plank rests on two scales that read $F_{g1} = 380$ N and $F_{g2} = 320$ N. The scales are separated by a distance of 2.00 m. How far from the woman's feet is her center of gravity?

Figure P8.79

80. S A uniform thin rod of length L and mass M is free to rotate on a frictionless pin passing through one end (Fig. P8.80). The rod is released from rest in the horizontal position. (a) What is the speed of its center of gravity when the rod reaches its lowest position? (b) What is the tangential speed of the lowest point on the rod when it is in the vertical position?

Figure P8.80

81. S A uniform solid cylinder of mass M and radius R rotates on a frictionless horizontal axle (Fig. P8.81). Two objects with equal masses m hang from light cords wrapped around the cylinder. If the system is released from rest, find (a) the tension in each cord and (b) the acceleration of each object after the objects have descended a distance h.

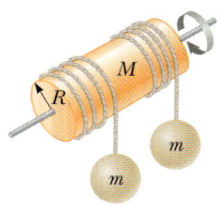

Figure P8.81

82. Q|C S A painter climbs a ladder leaning against a smooth wall. At a certain height, the ladder is on the verge of slipping. (a) Explain why the force exerted by the vertical wall on the ladder is horizontal. (b) If the ladder of length L leans at an angle θ with the horizontal, what is the lever arm for this horizontal force with the axis of rotation taken at the base of the ladder? (c) If the ladder is uniform, what is the lever arm for the force of gravity acting on the ladder? (d) Let the mass of the painter be 80 kg, $L = 4.0$ m, the ladder's mass be 30 kg, $\theta = 53°$, and the coefficient of friction between ground and ladder be 0.45. Find the maximum distance the painter can climb up the ladder.

83. A *war-wolf*, or *trebuchet*, is a device used during the Middle Ages to throw rocks at castles and now sometimes used to fling pumpkins and pianos. A simple trebuchet is shown in Figure P8.83. Model it as a stiff rod of negligible mass 3.00 m long and joining particles of mass $m_1 = 0.120$ kg and $m_2 = 60.0$ kg at its ends. It can turn on a frictionless horizontal axle perpendicular to the rod and 14.0 cm from the particle of larger mass. The rod is released from rest in a horizontal orientation. Find the maximum speed that the object of smaller mass attains.

Figure P8.83

84. A string is wrapped around a uniform cylinder of mass M and radius R. The cylinder is released from rest with the string vertical and its top end tied to a fixed bar (Fig. P8.84). Show that (a) the tension in the string is one-third the weight of the cylinder, (b) the magnitude of the acceleration of the center of gravity is $2g/3$, and (c) the speed of the center of gravity is $(4gh/3)^{1/2}$ after the cylinder has descended through distance h. Verify your answer to part (c) with the energy approach.

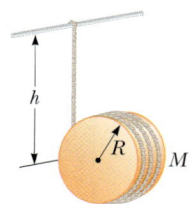

Figure P8.84

85. BIO **The Iron Cross** When a gymnast weighing 750 N executes the iron cross as in Figure P8.85a, the primary muscles involved in supporting this position are the latissimus dorsi ("lats") and the pectoralis major ("pecs"). The rings exert an upward force on the arms and support the weight of the gymnast. The force exerted by the shoulder joint on the arm is labeled $\vec{F}_s$ while the two muscles exert a total force $\vec{F}_m$ on the arm. Estimate the magnitude of the force $\vec{F}_m$. Note that one ring supports half the weight of the gymnast, which is 375 N as indicated in Figure P8.85b. Assume that the force $\vec{F}_m$ acts at an angle of 45° below the horizontal at a distance of 4.0 cm from the shoulder joint. In your estimate, take the distance from the shoulder joint to the hand to be $L = 70$ cm and ignore the weight of the arm.

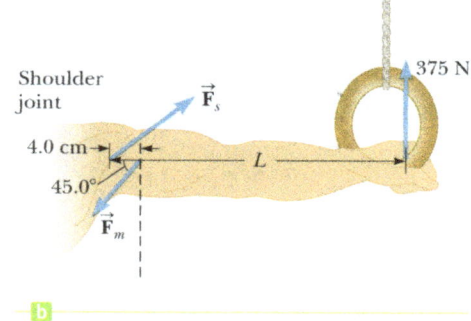

Figure P8.85

86. **BIO** In an emergency situation, a person with a broken forearm ties a strap from his hand to clip on his shoulder as in Figure P8.86. His 1.60-kg forearm remains in a horizontal position and the strap makes an angle of $\theta = 50.0°$ with the horizontal. Assume the forearm is uniform, has a length of $\ell = 0.320$ m, assume the biceps muscle is relaxed, and ignore the mass and length of the hand. Find (a) the tension in the strap and (b) the components of the reaction force exerted by the humerus on the forearm.

Figure P8.86

87. An object of mass $m_1 = 4.00$ kg is connected by a light cord to an object of mass $m_2 = 3.00$ kg on a frictionless surface (Fig. P8.87). The pulley rotates about a frictionless axle and has a moment of inertia of 0.500 kg $\cdot$ m^2 and a radius of 0.300 m. Assuming that the cord does not slip on the pulley, find (a) the acceleration of the two masses and (b) the tensions T_1 and T_2.

Figure P8.87

88. **Q|C** A 10.0-kg monkey climbs a uniform ladder with weight $w = 1.20 \times 10^2$ N and length $L = 3.00$ m as shown in Figure P8.88. The ladder rests against the wall at an angle of $\theta = 60.0°$. The upper and lower ends of the ladder rest on frictionless surfaces, with the lower end fastened to the wall by a horizontal rope that is frayed and that can support a maximum tension of only 80.0 N. (a) Draw a force diagram for the ladder. (b) Find the normal force exerted by the bottom of the ladder. (c) Find the tension in the rope when the monkey is two-thirds of the way up the ladder. (d) Find the maximum distance d that the monkey can climb up the ladder before the rope breaks. (e) If the horizontal surface were rough and the rope were removed, how would your analysis of the problem be changed and what other information would you need to answer parts (c) and (d)?

Figure P8.88

89. A 3.2-kg sphere is suspended by a cord that passes over a 1.8-kg pulley of radius 3.8 cm. The cord is attached to a spring whose force constant is $k = 86$ N/m as in Figure P8.89. Assume the pulley is a solid disk. (a) If the sphere is released from rest with the spring unstretched, what distance does the sphere fall through before stopping? (b) Find the speed of the sphere after it has fallen 25 cm.

Figure P8.89

Ocean waves combine properties of both transverse and longitudinal waves. With proper balance and timing, a surfer can capture some of the wave's energy and take it for a ride.

Vibrations and Waves 13

Periodic motion, from masses on springs to vibrations of atoms, is one of the most important kinds of physical behavior. In this chapter we take a more detailed look at Hooke's law, where the force is proportional to the displacement, tending to restore objects to some equilibrium position. A large number of physical systems can be successfully modeled with this simple idea, including the vibrations of strings, the swinging of a pendulum, and the propagation of waves of all kinds. All these physical phenomena involve periodic motion.

Periodic vibrations can cause disturbances that move through a medium in the form of waves. Many kinds of waves occur in nature, such as sound waves, water waves, seismic waves, and electromagnetic waves. These very different physical phenomena are described by common terms and concepts introduced here.

- 13.1 Hooke's Law
- 13.2 Elastic Potential Energy
- 13.3 Comparing Simple Harmonic Motion with Uniform Circular Motion
- 13.4 Position, Velocity, and Acceleration as a Function of Time
- 13.5 Motion of a Pendulum
- 13.6 Damped Oscillations
- 13.7 Waves
- 13.8 Frequency, Amplitude, and Wavelength
- 13.9 The Speed of Waves on Strings
- 13.10 Interference of Waves
- 13.11 Reflection of Waves

13.1 Hooke's Law

LEARNING OBJECTIVES

1. Define Hooke's force law for springs and describe the elements of simple harmonic motion that arise from it.
2. Apply Hooke's law and the second law of motion to spring systems.

One of the simplest types of vibrational motion is that of an object attached to a spring, previously discussed in the context of energy in Chapter 5. We assume the object moves on a frictionless horizontal surface. If the spring is stretched or compressed a small distance x from its unstretched or equilibrium position and then released, it exerts a force on the object as shown in Figure 13.1. From experiment, the spring force F_s is found to obey the equation

Hooke's law ▶

$$F_s = -kx \quad [13.1]$$

where x is the displacement of the object from its equilibrium position ($x = 0$) and k is a positive constant called the **spring constant.** This force law for springs was discovered by Robert Hooke in 1678 and is known as **Hooke's law.** The value of k is a measure of the stiffness of the spring. Stiff springs have large k values, and soft springs have small k values.

The negative sign in Equation 13.1 means that the force exerted by the spring is always directed *opposite* the displacement of the object. When the object is to the right of the equilibrium position, as in Figure 13.1a, x is positive and F_s is negative. This means that the force is in the negative direction, to the left. When the object is to the left of the equilibrium position, as in Figure 13.1c, x is negative and F_s is positive, indicating that the direction of the force is to the right. Of course, when $x = 0$, as in Figure 13.1b, the spring is unstretched and $F_s = 0$. Because the spring force always acts toward the equilibrium position, it is sometimes called a restoring force. **A restoring force always pushes or pulls the object toward the equilibrium position.**

Suppose the object is initially pulled a distance A to the right and released from rest. The force exerted by the spring on the object pulls it back toward the equilibrium position. As the object moves toward $x = 0$, the magnitude of the force decreases (because x decreases) and reaches zero at $x = 0$. The object gains speed as it moves toward the equilibrium position, however, reaching its maximum speed when $x = 0$. The momentum gained by the object causes it to overshoot the equilibrium position and compress the spring. As the object moves to the left of the equilibrium position (negative x-values), the spring force acts on it to the right, steadily increasing in strength, and the speed of the object decreases. The object finally comes briefly to rest at $x = -A$ before accelerating back towards $x = 0$ and ultimately returning to the original position at $x = A$. The process is then repeated, and the object continues to oscillate back and forth over the same path. This type of motion is called **simple harmonic motion. Simple harmonic motion occurs when the net force along the direction of motion obeys Hooke's law—when the net force is proportional to the displacement from the equilibrium point and is always directed toward the equilibrium point.**

Not all periodic motions over the same path can be classified as simple harmonic motion. A ball being tossed back and forth between a parent and a child moves repetitively, but the motion isn't simple harmonic motion because the force acting on the ball doesn't take the form of Hooke's law, Equation 13.1.

The motion of an object suspended from a vertical spring is also simple harmonic. In this case the force of gravity acting on the attached object stretches the spring until equilibrium is reached and the object is suspended at rest. By definition, the equilibrium position of the object is $x = 0$. When the object is moved away from equilibrium by a distance x and released, a net force acts toward the equilibrium position. Because the net force is proportional to x, the motion is simple harmonic.

The following three concepts are important in discussing any kind of periodic motion:

- The **amplitude** A is the maximum distance of the object from its equilibrium position. In the absence of friction, an object in simple harmonic motion oscillates between the positions $x = -A$ and $x = +A$.
- The **period** T is the time it takes the object to move through one complete cycle of motion, from $x = A$ to $x = -A$ and back to $x = A$.
- The **frequency** f is the number of complete cycles or vibrations per unit of time, and is the reciprocal of the period ($f = 1/T$).

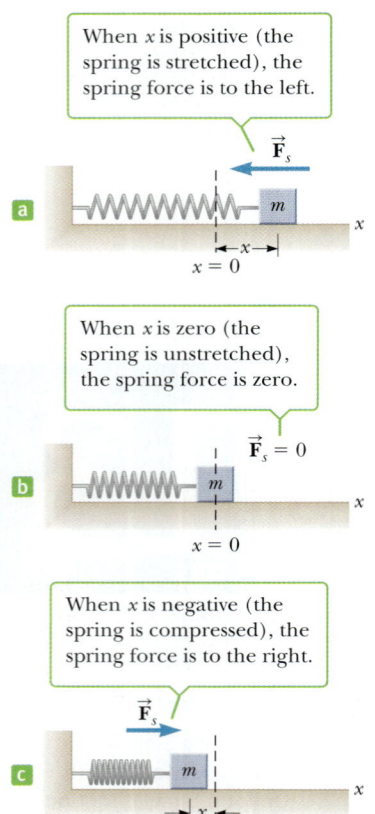

Figure 13.1 The force exerted by a spring on an object varies with the displacement of the object from the equilibrium position, $x = 0$.

The acceleration of an object moving with simple harmonic motion can be found by using Hooke's law in the equation for Newton's second law, $F = ma$. This gives

$$ma = F = -kx$$

$$a = -\frac{k}{m}x \quad [13.2]$$

◀ Acceleration in simple harmonic motion

Equation 13.2, an example of a *harmonic oscillator equation*, gives the acceleration as a function of position. Because the maximum value of x is defined to be the amplitude A, the acceleration ranges over the values $-kA/m$ to $+kA/m$. In the next section we will find equations for velocity as a function of position and for position as a function of time. Springs satisfying Hooke's law are also called *ideal* springs. In real springs, spring mass, internal friction, and varying elasticity affect the force law and motion.

Tip 13.1 Constant-Acceleration Equations Don't Apply

The acceleration a of a particle in simple harmonic motion is *not* constant; it changes, varying with x, so we can't apply the constant acceleration kinematic equations of Chapter 2.

■ Quick Quiz

13.1 A block on the end of a horizontal spring is pulled from equilibrium at $x = 0$ to $x = A$ and released. Through what total distance does it travel in one full cycle of its motion? (a) $A/2$ (b) A (c) $2A$ (d) $4A$

13.2 For a simple harmonic oscillator, which of the following pairs of vector quantities can't both point in the same direction? (The position vector is the displacement from equilibrium.) (a) position and velocity (b) velocity and acceleration (c) position and acceleration

■ EXAMPLE 13.1 | Simple Harmonic Motion on a Frictionless Surface

GOAL Calculate forces and accelerations for a horizontal spring system.

PROBLEM A 0.350-kg object attached to a spring of force constant 1.30×10^2 N/m is free to move on a frictionless horizontal surface, as in Figure 13.1. If the object is released from rest at $x = 0.100$ m, find the force on it and its acceleration at $x = 0.100$ m, $x = 0.050\,0$ m, $x = 0$ m, $x = -0.050\,0$ m, and $x = -0.100$ m.

STRATEGY Substitute given quantities into Hooke's law to find the forces, then calculate the accelerations with Newton's second law. The amplitude A is the same as the point of release from rest, $x = 0.100$ m.

SOLUTION

Write Hooke's force law:

$$F_s = -kx$$

Substitute the value for k and take $x = A = 0.100$ m, finding the spring force at that point:

$$F_{max} = -kA = -(1.30 \times 10^2 \text{ N/m})(0.100 \text{ m})$$
$$= \boxed{-13.0 \text{ N}}$$

Solve Newton's second law for a and substitute to find the acceleration at $x = A$:

$$ma = F_{max}$$
$$a = \frac{F_{max}}{m} = \frac{-13.0 \text{ N}}{0.350 \text{ kg}} = \boxed{-37.1 \text{ m/s}^2}$$

Repeat the same process for the other four points, assembling a table:

Position (m)	Force (N)	Acceleration (m/s²)
0.100	−13.0	−37.1
0.050	−6.50	−18.6
0	0	0
−0.050	+6.50	+18.6
−0.100	+13.0	+37.1

REMARKS The table above shows that when the initial position is halved, the force and acceleration are also halved. Further, positive values of x give negative values of the force and acceleration, whereas negative values of x give positive values of the force and acceleration. As the object moves to the left and passes the equilibrium point, the spring force becomes positive (for negative values of x), slowing the object down.

(Continued)

QUESTION 13.1 Will doubling a given displacement always result in doubling the magnitude of the spring force? Explain.

EXERCISE 13.1 For the same spring and mass system, find the force exerted by the spring and the position x when the object's acceleration is $+9.00$ m/s^2.

ANSWERS 3.15 N, -2.42 cm

■ EXAMPLE 13.2 | Mass on a Vertical Spring

GOAL Apply Newton's second law together with the force of gravity and Hooke's law.

PROBLEM A spring is hung vertically (Fig. 13.2a), and an object of mass m attached to the lower end is then slowly lowered a distance d to the equilibrium point (Fig. 13.2b). **(a)** Find the value of the spring constant if the magnitude of the displacement d is 2.0 cm and the mass is 0.55 kg. **(b)** If a second identical spring is attached to the object in parallel with the first spring (Fig. 13.2d), where is the new equilibrium point of the system? **(c)** What is the effective spring constant of the two springs acting as one?

Figure 13.2 (Example 13.2) (a)–(c) Determining the spring constant. Because the upward spring force balances the weight when the system is in equilibrium, it follows that $k = mg/d$. (d) A system involving two springs in parallel.

STRATEGY This example is an application of Newton's second law. The spring force is upward, balancing the downward force of gravity mg when the system is in equilibrium. (See Fig. 13.2c.) Because the suspended object is in equilibrium, the forces on the object sum to zero, and it's possible to solve for the spring constant k. Part (b) is solved the same way, but has two spring forces balancing the force of gravity. The spring constants are known, so the second law for equilibrium can be solved for the displacement of the spring. Part (c) involves using the displacement found in part (b). Treating the two springs as one equivalent spring, the second law then leads to the effective spring constant of the two-spring system.

SOLUTION

(a) Find the value of the spring constant if the spring is displaced by 2.0 cm and the mass of the object is 0.55 kg.

Apply Newton's second law to the object (with $a = 0$) and solve for the spring constant k:

$$\sum F = F_g + F_s = -mg + kd = 0$$

$$k = \frac{mg}{d} = \frac{(0.55 \text{ kg})(9.80 \text{ m/s}^2)}{2.0 \times 10^{-2} \text{ m}} = \boxed{2.7 \times 10^2 \text{ N/m}}$$

(b) If a second identical spring is attached to the object in parallel with the first spring (Fig. 13.2d), find the new equilibrium point of the system.

Apply Newton's second law, but with two springs acting on the object:

$$\sum F = F_g + F_{s1} + F_{s2} = -mg + kd_2 + kd_2 = 0$$

Solve for d_2:

$$d_2 = \frac{mg}{2k} = \frac{(0.55 \text{ kg})(9.80 \text{ m/s}^2)}{2(2.7 \times 10^2 \text{ N/m})} = \boxed{1.0 \times 10^{-2} \text{ m}}$$

(c) What is the effective spring constant of the two springs acting as one?

Write the second law for the system, with an effective spring constant k_{eff}:

$$\sum F = F_g + F_s = -mg + k_{\text{eff}} d_2 = 0$$

Solve for k_{eff}:

$$k_{\text{eff}} = \frac{mg}{d_2} = \frac{(0.55 \text{ kg})(9.80 \text{ m/s}^2)}{1.0 \times 10^{-2} \text{ m}} = \boxed{5.4 \times 10^2 \text{ N/m}}$$

REMARKS In this example, the spring force is positive because it's directed upward. If the object were displaced from the equilibrium position and released, it would oscillate around that point, just like a horizontal spring. Notice that attaching an extra identical spring in parallel is equivalent to having a single spring with twice the force constant. With springs attached end to end in series, however, the exercise illustrates that, all other things being equal, longer springs have smaller force constants than shorter springs.

QUESTION 13.2 Generalize: When two springs with force constants k_1 and k_2 act in parallel on an object, what is the spring constant k_{eff} of the single spring that would be equivalent to the two springs, in terms of k_1 and k_2?

EXERCISE 13.2 When a 75.0-kg man slowly adds his weight to a vertical spring attached to the ceiling, he reaches equilibrium when the spring is stretched by 6.50 cm. (a) Find the spring constant. (b) If a second, identical spring is hung on the first in series, and the man again adds his weight to the system, by how much does the system of springs stretch? (c) What would be the spring constant of a single, equivalent spring?

ANSWERS (a) 1.13×10^4 N/m (b) 13.0 cm (c) 5.65×10^3 N/m

13.2 Elastic Potential Energy

LEARNING OBJECTIVE

1. Review the application of the work-energy theorem to systems involving spring potential energy.

In this section we review the material covered in Section 4 of Chapter 5.

A system of interacting objects has potential energy associated with the configuration of the system. A compressed spring has potential energy that, when allowed to expand, can do work on an object, transforming spring potential energy into the object's kinetic energy. As an example, Figure 13.3 shows a ball being projected from a spring-loaded toy gun, where the spring is compressed a distance x. As the gun is fired, the compressed spring does work on the ball and imparts kinetic energy to it.

Recall that the energy stored in a stretched or compressed spring or some other elastic material is called elastic potential energy, PE_s, given by

$$PE_s \equiv \tfrac{1}{2}kx^2 \qquad [13.3]$$

◀ Elastic potential energy

Recall also that the law of conservation of energy, including both gravitational and spring potential energy, is given by

$$(KE + PE_g + PE_s)_i = (KE + PE_g + PE_s)_f \qquad [13.4]$$

If nonconservative forces such as friction are present, then the change in mechanical energy must equal the work done by the nonconservative forces:

$$W_{nc} = (KE + PE_g + PE_s)_f - (KE + PE_g + PE_s)_i \qquad [13.5]$$

Rotational kinetic energy must be included in both Equation 13.4 and Equation 13.5 for systems involving torques.

Figure 13.3 A ball projected from a spring-loaded gun. The elastic potential energy stored in the spring is transformed into the kinetic energy of the ball.

Figure 13.4 A block sliding on a frictionless horizontal surface collides with a light spring. In the absence of friction, the mechanical energy in this process remains constant.

Initially, the mechanical energy is entirely the block's kinetic energy.
$E = \frac{1}{2}mv_i^2$

Here the mechanical energy is the sum of the block's kinetic energy and the elastic potential energy stored in the compressed spring.
$E = \frac{1}{2}mv^2 + \frac{1}{2}kx^2$

When the block comes to rest, the mechanical energy is entirely elastic potential energy.
$E = \frac{1}{2}kx_m^2$

When the block leaves the spring, the mechanical energy is again solely the block's kinetic energy.
$E = \frac{1}{2}mv_i^2$

APPLICATION
Archery

As an example of the energy conversions that take place when a spring is included in a system, consider Figure 13.4. A block of mass m slides on a frictionless horizontal surface with constant velocity $\vec{v}_i$ and collides with a coiled spring. The description that follows is greatly simplified by assuming the spring is very light (an ideal spring) and therefore has negligible kinetic energy. As the spring is compressed, it exerts a force to the left on the block. At maximum compression, the block comes to rest for just an instant (Fig. 13.4c). The initial total energy in the system (block plus spring) before the collision is the kinetic energy of the block. After the block collides with the spring and the spring is partially compressed, as in Figure 13.4b, the block has kinetic energy $\frac{1}{2}mv^2$ (where $v < v_i$) and the spring has potential energy $\frac{1}{2}kx^2$. When the block stops for an instant at the point of maximum compression, the kinetic energy is zero. Because the spring force is conservative and because there are no external forces that can do work on the system, **the total mechanical energy of the system consisting of the block and spring remains constant.** Energy is transformed from the kinetic energy of the block to the potential energy stored in the spring. As the spring expands, the block moves in the opposite direction and regains all its initial kinetic energy, as in Figure 13.4d.

When an archer pulls back on a bowstring, elastic potential energy is stored in both the bent bow and stretched bowstring (Fig. 13.5). When the arrow is released, the potential energy stored in the system is transformed into the kinetic energy of the arrow. Devices such as crossbows and slingshots work the same way.

Figure 13.5 Elastic potential energy is stored in this drawn bow.

■ **Quick Quiz**

13.3 When an object moving in simple harmonic motion is at its maximum displacement from equilibrium, which of the following is at a maximum? (a) velocity, (b) acceleration, or (c) kinetic energy

■ **EXAMPLE 13.3** | **Stop That Car!**

GOAL Apply conservation of energy and the work–energy theorem with spring and gravitational potential energy.

PROBLEM A 13 000-N car starts at rest and rolls down a hill from a height of 10.0 m (Fig. 13.6). It then moves across a level surface and collides with a light spring-loaded guardrail. **(a)** Neglecting any losses due to friction, and ignoring the rotational kinetic energy of the wheels, find the maximum distance the spring is compressed. Assume a spring constant of 1.0×10^6 N/m. **(b)** Calculate the magnitude of the car's maximum acceleration after contact with the spring, assuming no frictional losses. **(c)** If the spring is compressed by only 0.30 m, find the change in the mechanical energy due to friction.

STRATEGY Because friction losses are neglected, use conservation of energy in the form of Equation 13.4 to solve for the spring displacement in part (a). The initial and final values of the car's kinetic energy are zero, so the initial potential energy of the car–spring–Earth system is completely converted to elastic potential energy in the spring at the end of the ride. In part (b) apply Newton's second law, substituting the answer to part (a) for x because the maximum compression will give the maximum acceleration. In part (c) friction is no longer neglected, so use the work–energy theorem, Equation 13.5. The change in mechanical energy must equal the mechanical energy lost due to friction.

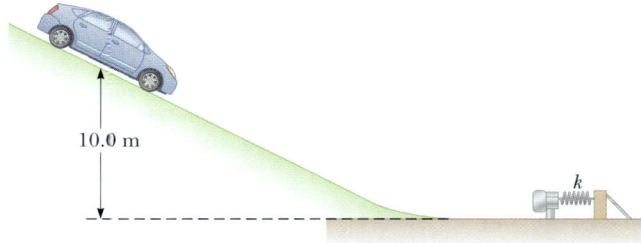

Figure 13.6 (Example 13.3) A car starts from rest on a hill at the position shown. When the car reaches the bottom of the hill, it collides with a spring-loaded guardrail.

SOLUTION

(a) Find the maximum spring compression, assuming no energy losses due to friction.

Apply conservation of mechanical energy. Initially, there is only gravitational potential energy, and at maximum compression of the guardrail, there is only spring potential energy.

$$(KE + PE_g + PE_s)_i = (KE + PE_g + PE_s)_f$$
$$0 + mgh + 0 = 0 + 0 + \tfrac{1}{2}kx^2$$

Solve for x:

$$x = \sqrt{\frac{2mgh}{k}} = \sqrt{\frac{2(13\,000\text{ N})(10.0\text{ m})}{1.0 \times 10^6 \text{ N/m}}} = \boxed{0.51 \text{ m}}$$

(b) Calculate the magnitude of the car's maximum acceleration by the spring, neglecting friction.

Apply Newton's second law to the car:

$$ma = -kx \quad \rightarrow \quad a = -\frac{kx}{m} = -\frac{kxg}{mg} = -\frac{kxg}{w}$$

Substitute values:

$$a = -\frac{(1.0 \times 10^6 \text{ N/m})(0.51 \text{ m})(9.8 \text{ m/s}^2)}{13\,000 \text{ N}}$$
$$= -380 \text{ m/s}^2 \quad \rightarrow \quad |a| = \boxed{380 \text{ m/s}^2}$$

(c) If the compression of the guardrail is only 0.30 m, find the change in the mechanical energy due to friction.

Use the work–energy theorem:

$$W_{nc} = (KE + PE_g + PE_s)_f - (KE + PE_g + PE_s)_i$$
$$= (0 + 0 + \tfrac{1}{2}kx^2) - (0 + mgh + 0)$$
$$= \tfrac{1}{2}(1.0 \times 10^6 \text{ N/m})(0.30)^2 - (13\,000 \text{ N})(10.0 \text{ m})$$
$$W_{nc} = \boxed{-8.5 \times 10^4 \text{ J}}$$

REMARKS The answer to part (b) is about 40 times greater than the acceleration of gravity, so we'd better be wearing our seat belts. Note that the solution didn't require calculation of the velocity of the car.

QUESTION 13.3 True or False: In the absence of energy losses due to friction, doubling the height of the hill doubles the maximum acceleration delivered by the spring.

EXERCISE 13.3 A spring-loaded gun fires a 0.100-kg puck along a tabletop. The puck slides up a curved ramp and flies straight up into the air. If the spring is displaced 12.0 cm from equilibrium and the spring constant is 875 N/m, how high does the puck rise, neglecting friction? (b) If instead it only rises to a height of 5.00 m because of friction, what is the change in mechanical energy?

ANSWERS (a) 6.43 m (b) −1.40 J

Figure 13.7 (a) An object attached to a spring on a frictionless surface is released from rest with the spring extended a distance A. Just before the object is released, the total energy is the elastic potential energy $\frac{1}{2}kA^2$. (b) When the object reaches position x, it has kinetic energy $\frac{1}{2}mv^2$ and the elastic potential energy has decreased to $\frac{1}{2}kx^2$.

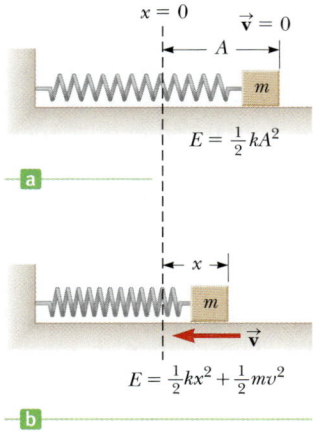

In addition to studying the preceding example, it's a good idea to review those given in Section 5.4.

Velocity as a Function of Position

Conservation of energy provides a simple method of deriving an expression for the velocity of an object undergoing periodic motion as a function of position. The object in question is initially at its maximum extension A (Fig. 13.7a) and is then released from rest. The initial energy of the system is entirely elastic potential energy stored in the spring, $\frac{1}{2}kA^2$. As the object moves toward the origin to some new position x (Fig. 13.7b), part of this energy is transformed into kinetic energy, and the potential energy stored in the spring is reduced to $\frac{1}{2}kx^2$. Because the total energy of the system is equal to $\frac{1}{2}kA^2$ (the initial energy stored in the spring), we can equate this quantity to the sum of the kinetic and potential energies at the position x:

$$\tfrac{1}{2}kA^2 = \tfrac{1}{2}mv^2 + \tfrac{1}{2}kx^2$$

Solving for v, we get

$$v = \pm\sqrt{\frac{k}{m}(A^2 - x^2)} \qquad [13.6]$$

This expression shows that the object's speed is a maximum at $x = 0$ and is zero at the extreme positions $x = \pm A$.

The right side of Equation 13.6 is preceded by the $\pm$ sign because the square root of a number can be either positive or negative. If the object in Figure 13.7 is moving to the right, v is positive; if the object is moving to the left, v is negative.

■ EXAMPLE 13.4 | The Object–Spring System Revisited

GOAL Apply the time-independent velocity expression, Equation 13.6, to an object–spring system.

PROBLEM A 0.500-kg object connected to a light spring with a spring constant of 20.0 N/m oscillates on a frictionless horizontal surface. (a) Calculate the total energy of the system and the maximum speed of the object if the amplitude of the motion is 3.00 cm. (b) What is the velocity of the object when the displacement is 2.00 cm? (c) Compute the kinetic and potential energies of the system when the displacement is 2.00 cm.

STRATEGY The total energy of the system can be found most easily at $x = A$, where the kinetic energy is zero. There, the potential energy alone is equal to the total energy. Conservation of energy then yields the speed at $x = 0$. For part (b), obtain the velocity by substituting the given value of x into the time-independent velocity equation. Using this result, the kinetic energy asked for in part (c) can be found by substitution, and the potential energy can be found by substitution into Equation 13.3.

SOLUTION

(a) Calculate the total energy and maximum speed if the amplitude is 3.00 cm.

Substitute $x = A = 3.00$ cm and $k = 20.0$ N/m into the equation for the total mechanical energy E:

$$E = KE + PE_g + PE_s$$
$$= 0 + 0 + \tfrac{1}{2}kA^2 = \tfrac{1}{2}(20.0 \text{ N/m})(3.00 \times 10^{-2} \text{ m})^2$$
$$= \boxed{9.00 \times 10^{-3} \text{ J}}$$

Use conservation of energy with $x_i = A$ and $x_f = 0$ to compute the speed of the object at the origin:

$$(KE + PE_g + PE_s)_i = (KE + PE_g + PE_s)_f$$
$$0 + 0 + \tfrac{1}{2}kA^2 = \tfrac{1}{2}mv_{max}^2 + 0 + 0$$
$$\tfrac{1}{2}mv_{max}^2 = 9.00 \times 10^{-3} \text{ J}$$
$$v_{max} = \sqrt{\frac{18.0 \times 10^{-3} \text{ J}}{0.500 \text{ kg}}} = \boxed{0.190 \text{ m/s}}$$

(b) Compute the velocity of the object when the displacement is 2.00 cm.

Substitute known values directly into Equation 13.6:

$$v = \pm\sqrt{\frac{k}{m}(A^2 - x^2)}$$
$$= \pm\sqrt{\frac{20.0 \text{ N/m}}{0.500 \text{ kg}}[(0.030\,0 \text{ m})^2 - (0.020\,0 \text{ m})^2]}$$
$$= \boxed{\pm 0.141 \text{ m/s}}$$

(c) Compute the kinetic and potential energies when the displacement is 2.00 cm.

Substitute into the equation for kinetic energy:

$$KE = \tfrac{1}{2}mv^2 = \tfrac{1}{2}(0.500 \text{ kg})(0.141 \text{ m/s})^2 = \boxed{4.97 \times 10^{-3} \text{ J}}$$

Substitute into the equation for spring potential energy:

$$PE_s = \tfrac{1}{2}kx^2 = \tfrac{1}{2}(20.0 \text{ N/m})(2.00 \times 10^{-2} \text{ m})^2$$
$$= \boxed{4.00 \times 10^{-3} \text{ J}}$$

REMARKS With the given information, it is impossible to choose between the positive and negative solutions in part (b). Notice that the sum $KE + PE_s$ in part (c) equals the total energy E found in part (a), as it should (except for a small discrepancy due to rounding).

QUESTION 13.4 True or False: Doubling the initial displacement doubles the speed of the object at the equilibrium point.

EXERCISE 13.4 For what values of x is the speed of the object 0.10 m/s?

ANSWER ± 2.55 cm

13.3 Comparing Simple Harmonic Motion with Uniform Circular Motion

LEARNING OBJECTIVES

1. Describe the relationship between simple harmonic motion and uniform circular motion.
2. Define and apply the related concepts of the period, frequency, and angular frequency of a spring harmonic oscillator.

We can better understand and visualize many aspects of simple harmonic motion along a straight line by looking at its relationship to uniform circular motion. Figure 13.8 is a top view of an experimental arrangement that is useful for this purpose. A ball is attached to the rim of a turntable of radius A, illuminated from

Figure 13.8 An experimental setup for demonstrating the connection between simple harmonic motion and uniform circular motion.

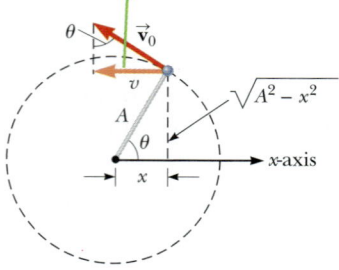

Figure 13.9 The ball rotates with constant speed v_0.

APPLICATION
Pistons and Drive Wheels

Figure 13.10 The drive wheel mechanism of an old locomotive.

the side by a lamp. We find that **as the turntable rotates with constant angular speed, the shadow of the ball moves back and forth with simple harmonic motion.**

This fact can be understood from Equation 13.6, which says that the velocity of an object moving with simple harmonic motion is related to the displacement by

$$v = C\sqrt{A^2 - x^2}$$

where C is a constant. To see that the shadow also obeys this relation, consider Figure 13.9, which shows the ball moving with a constant speed v_0 in a direction tangent to the circular path. At this instant, the velocity of the ball in the x-direction is given by $v = v_0 \sin \theta$, or

$$\sin \theta = \frac{v}{v_0}$$

From the larger triangle in Figure 13.9 we can obtain a second expression for $\sin \theta$:

$$\sin \theta = \frac{\sqrt{A^2 - x^2}}{A}$$

Equating the right-hand sides of the two expressions for $\sin \theta$, we find the following relationship between the velocity v and the displacement x:

$$\frac{v}{v_0} = \frac{\sqrt{A^2 - x^2}}{A}$$

or

$$v = \frac{v_0}{A}\sqrt{A^2 - x^2} = C\sqrt{A^2 - x^2}$$

The velocity of the ball in the x-direction is related to the displacement x in exactly the same way as the velocity of an object undergoing simple harmonic motion. The shadow therefore moves with simple harmonic motion.

A valuable example of the relationship between simple harmonic motion and circular motion can be seen in vehicles and machines that use the back-and-forth motion of a piston to create rotational motion in a wheel. Consider the drive wheel of a locomotive. In Figure 13.10, the rods are connected to a piston that moves back and forth in simple harmonic motion. The rods transform the back-and-forth motion of the piston into rotational motion of the wheels. A similar mechanism in an automobile engine transforms the back-and-forth motion of the pistons to rotational motion of the crankshaft.

Period and Frequency

The period T of the shadow in Figure 13.8, which represents the time required for one complete trip back and forth, is also the time it takes the ball to make one complete circular trip on the turntable. Because the ball moves through the distance $2\pi A$ (the circumference of the circle) in the time T, the speed v_0 of the ball around the circular path is

$$v_0 = \frac{2\pi A}{T}$$

and the period is

$$T = \frac{2\pi A}{v_0} \qquad [13.7]$$

Imagine that the ball moves from P to Q, a quarter of a revolution, in Figure 13.8. The motion of the shadow is equivalent to the horizontal motion of an object on the end of a spring. For this reason, the radius A of the circular motion is the same as the amplitude A of the simple harmonic motion of the shadow. During the quarter of a cycle shown, the shadow moves from a point where the energy of the system (ball and spring) is solely elastic potential energy to a point where the energy is solely kinetic energy. By conservation of energy, we have

$$\tfrac{1}{2}kA^2 = \tfrac{1}{2}mv_0^2$$

which can be solved for A/v_0:

$$\frac{A}{v_0} = \sqrt{\frac{m}{k}}$$

Substituting this expression for A/v_0 in Equation 13.7, we find that the period is

$$T = 2\pi\sqrt{\frac{m}{k}} \qquad [13.8]$$

◀ The period of an object–spring system moving with simple harmonic motion

Equation 13.8 represents the time required for an object of mass m attached to a spring with spring constant k to complete one cycle of its motion. The square root of the mass is in the numerator, so a large mass will mean a large period, in agreement with intuition. The square root of the spring constant k is in the denominator, so a large spring constant will yield a small period, again agreeing with intuition. It's also interesting that the period doesn't depend on the amplitude A.

The inverse of the period is the frequency of the motion:

$$f = \frac{1}{T} \qquad [13.9]$$

Therefore, the frequency of the periodic motion of a mass on a spring is

$$f = \frac{1}{2\pi}\sqrt{\frac{k}{m}} \qquad [13.10]$$

◀ Frequency of an object–spring system

The units of frequency are cycles per second (s^{-1}), or **hertz** (Hz). The **angular frequency** ω is

$$\omega = 2\pi f = \sqrt{\frac{k}{m}} \qquad [13.11]$$

◀ Angular frequency of an object–spring system

The frequency and angular frequency are actually closely related concepts. The unit of frequency is cycles per second, where a cycle may be thought of as a unit of angular measure corresponding to 2π radians, or 360°. Viewed in this way, angular frequency is just a unit conversion of frequency. Radian measure is used for angles mainly because it provides a convenient and natural link between linear and angular quantities.

Although an ideal mass–spring system has a period proportional to the square root of the object's mass m, experiments show that a graph of T^2 versus m doesn't pass through the origin. This is because the spring itself has a mass. The coils of the spring oscillate just like the object, except the amplitudes are smaller for all coils but the last. For a cylindrical spring, energy arguments can be used to show that the *effective* additional mass of a light spring is one-third the mass of the spring. The square of the period is proportional to the total oscillating mass, so a graph of T^2 versus total mass (the mass hung on the spring plus the effective oscillating mass of the spring) would pass through the origin.

Tip 13.2 Twin Frequencies
The *frequency* gives the number of cycles per second, whereas the *angular frequency* gives the number of radians per second. These two physical concepts are nearly identical and are linked by the conversion factor 2π rad/cycle.

■ Quick Quiz

13.4 An object of mass m is attached to a horizontal spring, stretched to a displacement A from equilibrium and released, undergoing harmonic oscillations on a frictionless surface with period T_0. The experiment is then repeated with a mass of $4m$. What's the new period of oscillation? (a) $2T_0$ (b) T_0 (c) $T_0/2$ (d) $T_0/4$

13.5 Consider the situation in Quick Quiz 13.4. Is the subsequent total mechanical energy of the object with mass $4m$ (a) greater than, (b) less than, or (c) equal to the original total mechanical energy?

APPLYING PHYSICS 13.1 | Bungee Jumping

A bungee cord can be roughly modeled as a spring. If you go bungee jumping, you will bounce up and down at the end of the elastic cord after your dive off a bridge (Fig. 13.11). Suppose you perform a dive and measure the frequency of your bouncing. You then move to another bridge, but find that the bungee cord is too long for dives off this bridge. What possible solutions might be applied? In terms of the original frequency, what is the frequency of vibration associated with the solution?

EXPLANATION There are two possible solutions: Make the bungee cord smaller or fold it in half. The latter would be the safer of the two choices, as we'll see. The force exerted by the bungee cord, modeled as a spring, is proportional to the separation of the coils as the spring is extended. First, we extend the spring by a given distance and measure the distance between coils. We then cut the spring in half. If one of the half-springs is now extended by the same distance, the coils will be twice as far apart as they were for the complete spring. Therefore, it takes twice as much force to stretch the half-spring through the same displacement, so the half-spring has a spring constant twice that of the complete spring. The folded bungee cord can then be modeled as two half-springs in parallel. Each half has a spring constant that is twice the original spring constant of the bungee cord. In addition, an object hanging on the folded bungee cord will experience two forces, one from each half-spring. As a result, the required force for a given extension will be four times as much as for the original bungee cord. The effective spring constant of the folded bungee cord is therefore four times as large as the original spring constant. Because the frequency of oscillation is proportional to the square root of the spring constant, your bouncing frequency on the folded cord will be twice what it was on the original cord.

This discussion neglects the fact that the coils of a spring have an initial separation. It's also important to remember that a shorter coil may lose elasticity more readily, possibly even going beyond the elastic limit for the material, with disastrous results. Bungee jumping is dangerous; discretion is advised! ■

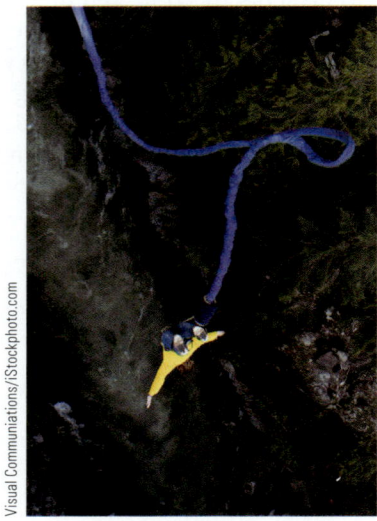

Figure 13.11 (Applying Physics 13.1) A bungee jumper relies on elastic forces to pull him up short of a deadly impact.

EXAMPLE 13.5 | That Car Needs Shock Absorbers!

GOAL Understand the relationships between period, frequency, and angular frequency.

PROBLEM A 1.30×10^3-kg car is constructed on a frame supported by four springs. Each spring has a spring constant of 2.00×10^4 N/m. If two people riding in the car have a combined mass of 1.60×10^2 kg, find the frequency of vibration of the car when it is driven over a pothole in the road. Find also the period and the angular frequency. Assume the weight is evenly distributed.

STRATEGY Because the weight is evenly distributed, each spring supports one-fourth of the mass. Substitute this value and the spring constant into Equation 13.10 to get the frequency. The reciprocal is the period, and multiplying the frequency by 2π gives the angular frequency.

SOLUTION

Compute one-quarter of the total mass:

$$m = \tfrac{1}{4}(m_{\text{car}} + m_{\text{pass}}) = \tfrac{1}{4}(1.30 \times 10^3 \text{ kg} + 1.60 \times 10^2 \text{ kg})$$
$$= 365 \text{ kg}$$

Substitute into Equation 13.10 to find the frequency:

$$f = \frac{1}{2\pi}\sqrt{\frac{k}{m}} = \frac{1}{2\pi}\sqrt{\frac{2.00 \times 10^4 \text{ N/m}}{365 \text{ kg}}} = \boxed{1.18 \text{ Hz}}$$

Invert the frequency to get the period:

$$T = \frac{1}{f} = \frac{1}{1.18 \text{ Hz}} = \boxed{0.847 \text{ s}}$$

Multiply the frequency by 2π to get the angular frequency:

$$\omega = 2\pi f = 2\pi(1.18 \text{ Hz}) = \boxed{7.41 \text{ rad/s}}$$

REMARKS Solving this problem didn't require any knowledge of the size of the pothole because the frequency doesn't depend on the amplitude of the motion.

QUESTION 13.5 True or False: The frequency of vibration of a heavy vehicle is greater than that of a lighter vehicle, assuming the two vehicles are supported by the same set of springs.

EXERCISE 13.5 A 45.0-kg boy jumps on a 5.00-kg pogo stick with spring constant 3 650 N/m. Find (a) the angular frequency, (b) the frequency, and (c) the period of the boy's motion.

ANSWERS (a) 8.54 rad/s (b) 1.36 Hz (c) 0.735 s

13.4 Position, Velocity, and Acceleration as a Function of Time

LEARNING OBJECTIVE

1. Describe and apply the position, velocity, and acceleration of simple harmonic oscillators as functions of time.

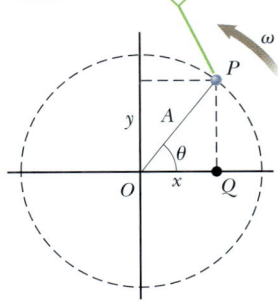

As the ball at P rotates in a circle with uniform angular speed, its projection Q along the x-axis moves with simple harmonic motion.

Figure 13.12 A reference circle.

We can obtain an expression for the position of an object moving with simple harmonic motion as a function of time by returning to the relationship between simple harmonic motion and uniform circular motion. Again, consider a ball on the rim of a rotating turntable of radius A, as in Figure 13.12. We refer to the circle made by the ball as the *reference circle* for the motion. We assume the turntable revolves at a *constant* angular speed ω. As the ball rotates on the reference circle, the angle θ made by the line OP with the x-axis changes with time. Meanwhile, the projection of P on the x-axis, labeled point Q, moves back and forth along the axis with simple harmonic motion.

From the right triangle OPQ, we see that $\cos\theta = x/A$. Therefore, the x-coordinate of the ball is

$$x = A\cos\theta$$

Because the ball rotates with constant angular speed, it follows that $\theta = \omega t$ (see Chapter 7), so we have

$$x = A\cos(\omega t) \quad [13.12]$$

In one complete revolution, the ball rotates through an angle of 2π rad in a time equal to the period T. In other words, the motion repeats itself every T seconds. Therefore,

$$\omega = \frac{\Delta\theta}{\Delta t} = \frac{2\pi}{T} = 2\pi f \quad [13.13]$$

where f is the frequency of the motion. The angular speed of the ball as it moves around the reference circle is the same as the angular frequency of the projected simple harmonic motion. Consequently, Equation 13.12 can be written

$$x = A\cos(2\pi f t) \quad [13.14a]$$

This cosine function represents the position of an object moving with simple harmonic motion as a function of time, and is graphed in Figure 13.13a (page 458). Because the cosine function varies between 1 and -1, x varies between A and $-A$. The shape of the graph is called *sinusoidal*.

Figures 13.13b and 13.13c represent curves for velocity and acceleration as a function of time. To find the equation for the velocity, use Equations 13.6 and 13.14a (page 458) together with the identity $\cos^2\theta + \sin^2\theta = 1$, obtaining

$$v = -A\omega\sin(2\pi f t) \quad [13.14b]$$

Figure 13.13 (a) Displacement, (b) velocity, and (c) acceleration versus time for an object moving with simple harmonic motion under the initial conditions $x_0 = A$ and $v_0 = 0$ at $t = 0$.

Figure 13.14 An experimental apparatus for demonstrating simple harmonic motion. A pen attached to the oscillating object traces out a sinusoidal wave on the moving chart paper.

where we have used the fact that $\omega = \sqrt{k/m}$. The $\pm$ sign is no longer needed, because sine can take both positive and negative values. Deriving an expression for the acceleration involves substituting Equation 13.14a into Equation 13.2, Newton's second law for springs:

$$a = -A\omega^2 \cos(2\pi f t) \qquad [13.14c]$$

The detailed steps of these derivations are left as an exercise for the student. Notice that when the displacement x is at a maximum, at $x = A$ or $x = -A$, the velocity is zero, and when x is zero, the magnitude of the velocity is a maximum. Further, when $x = +A$, its most positive value, the acceleration is a maximum but in the negative x-direction, and when x is at its most negative position, $x = -A$, the acceleration has its maximum value in the positive x-direction. These facts are consistent with our earlier discussion of the points at which v and a reach their maximum, minimum, and zero values.

The maximum values of the position, velocity, and acceleration are always equal to the magnitude of the expression in front of the trigonometric function in each equation because the largest value of either cosine or sine is 1.

Figure 13.14 illustrates one experimental arrangement that demonstrates the sinusoidal nature of simple harmonic motion. An object connected to a spring has a marking pen attached to it. While the object vibrates vertically, a sheet of paper is moved horizontally with constant speed. The pen traces out a sinusoidal pattern.

■ Quick Quiz

13.6 If the amplitude of a system moving in simple harmonic motion is doubled, which of the following quantities *doesn't* change? (a) total energy (b) maximum speed (c) maximum acceleration (d) period

■ **EXAMPLE 13.6** | **The Vibrating Object–Spring System**

GOAL Identify the physical parameters of a harmonic oscillator from its mathematical description.

PROBLEM (a) Find the amplitude, frequency, and period of motion for an object vibrating at the end of a horizontal spring if the equation for its position as a function of time is

$$x = (0.250 \text{ m}) \cos\left(\frac{\pi}{8.00}t\right)$$

(b) Find the maximum magnitude of the velocity and acceleration. (c) What are the position, velocity, and acceleration of the object after 1.00 s has elapsed?

STRATEGY In part (a) the amplitude and frequency can be found by comparing the given equation with the standard form in Equation 13.14a, matching up the numerical values with the corresponding terms in the standard form. In part (b) the maximum speed will occur when the sine function in Equation 13.14b equals 1 or −1, the extreme values of the sine function (and similarly for the acceleration and the cosine function). In each case, find the magnitude of the expression in front of the trigonometric function. Part (c) is just a matter of substituting values into Equations 13.14a–13.14c.

SOLUTION

(a) Find the amplitude, frequency, and period.

Write the standard form given by Equation 13.14a and underneath it write the given equation:

(1) $\quad x = A \cos(2\pi f t)$

(2) $\quad x = (0.250 \text{ m}) \cos\left(\dfrac{\pi}{8.00} t\right)$

Equate the factors in front of the cosine functions to find the amplitude:

$A =$ $\boxed{0.250 \text{ m}}$

The angular frequency ω is the factor in front of t in Equations (1) and (2). Equate these factors:

$\omega = 2\pi f = \dfrac{\pi}{8.00}$ rad/s $= 0.393$ rad/s

Divide ω by 2π to get the frequency f:

$f = \dfrac{\omega}{2\pi} =$ $\boxed{0.062\ 5 \text{ Hz}}$

The period T is the reciprocal of the frequency:

$T = \dfrac{1}{f} =$ $\boxed{16.0 \text{ s}}$

(b) Find the maximum magnitudes of the velocity and the acceleration.

Calculate the maximum speed from the factor in front of the sine function in Equation 13.14b:

$v_{\text{max}} = A\omega = (0.250 \text{ m})(0.393 \text{ rad/s}) =$ $\boxed{0.098\ 3 \text{ m/s}}$

Calculate the maximum acceleration from the factor in front of the cosine function in Equation 13.14c:

$a_{\text{max}} = A\omega^2 = (0.250 \text{ m})(0.393 \text{ rad/s})^2 =$ $\boxed{0.038\ 6 \text{ m/s}^2}$

(c) Find the position, velocity, and acceleration of the object after 1.00 s.

Substitute $t = 1.00$ s in the given equation:

$x = (0.250 \text{ m}) \cos(0.393 \text{ rad}) =$ $\boxed{0.231 \text{ m}}$

Substitute values into the velocity equation:

$v = -A\omega \sin(\omega t)$
$ = -(0.250 \text{ m})(0.393 \text{ rad/s}) \sin(0.393 \text{ rad/s} \cdot 1.00 \text{ s})$
$v =$ $\boxed{-0.027\ 6 \text{ m/s}}$

Substitute values into the acceleration equation:

$a = -A\omega^2 \cos(\omega t)$
$ = -(0.250 \text{ m})(0.393 \text{ rad/s}^2)^2 \cos(0.393 \text{ rad/s} \cdot 1.00 \text{ s})$
$a =$ $\boxed{-0.035\ 7 \text{ m/s}^2}$

REMARKS In evaluating the sine or cosine function, the angle is in radians, so you should either set your calculator to evaluate trigonometric functions based on radian measure or convert from radians to degrees.

QUESTION 13.6 If the mass is doubled, is the magnitude of the acceleration of the system at any position (a) doubled, (b) halved, or (c) unchanged?

EXERCISE 13.6 If the object–spring system is described by $x = (0.330 \text{ m}) \cos(1.50 t)$, find (a) the amplitude, the angular frequency, the frequency, and the period, (b) the maximum magnitudes of the velocity and acceleration, and (c) the position, velocity, and acceleration when $t = 0.250$ s.

ANSWERS (a) $A = 0.330$ m, $\omega = 1.50$ rad/s, $f = 0.239$ Hz, $T = 4.18$ s (b) $v_{\text{max}} = 0.495$ m/s, $a_{\text{max}} = 0.743$ m/s^2 (c) $x = 0.307$ m, $v = -0.181$ m/s, $a = -0.691$ m/s^2

13.5 Motion of a Pendulum

LEARNING OBJECTIVES

1. Define the simple pendulum and the condition under which it executes simple harmonic motion.
2. Define and apply the angular frequency, frequency, and period of a simple pendulum.
3. Generalize the concept of a pendulum to pendulums of arbitrary shape, called physical pendulums.

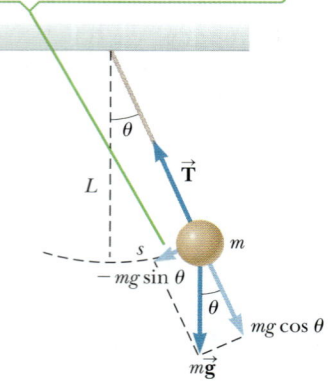

The restoring force causing the pendulum to oscillate harmonically is the tangential component of the gravity force $-mg \sin \theta$.

Figure 13.15
A simple pendulum consists of a bob of mass m suspended by a light string of length L. (L is the distance from the pivot to the center of mass of the bob.)

A simple pendulum is another mechanical system that exhibits periodic motion. It consists of a small bob of mass m suspended by a light string of length L fixed at its upper end, as in Figure 13.15. (By a light string, we mean that the string's mass is assumed to be very small compared with the mass of the bob and hence can be ignored.) When released, the bob swings to and fro over the same path, but is its motion simple harmonic?

Answering this question requires examining the restoring force—the force of gravity—that acts on the pendulum. The pendulum bob moves along a circular arc, rather than back and forth in a straight line. When the oscillations are small, however, the motion of the bob is nearly straight, so Hooke's law may apply approximately.

In Figure 13.15, s is the displacement of the bob from equilibrium along the arc. Hooke's law is $F = -kx$, so we are looking for a similar expression involving s, $F_t = -ks$, where F_t is the force acting in a direction tangent to the circular arc. From the figure, the restoring force is

$$F_t = -mg \sin \theta$$

Since $s = L\theta$, the equation for F_t can be written as

$$F_t = -mg \sin\left(\frac{s}{L}\right)$$

This expression isn't of the form $F_t = -ks$, so in general, the motion of a pendulum is *not* simple harmonic. For small angles less than about 15 degrees, however, the angle θ measured in radians and the sine of the angle are approximately equal. For example, $\theta = 10.0° = 0.175$ rad, and $\sin(10.0°) = 0.174$. Therefore, if we restrict the motion to *small* angles, the approximation $\sin \theta \approx \theta$ is valid, and the restoring force can be written

$$F_t = -mg \sin \theta \approx -mg\theta$$

Substituting $\theta = s/L$, we obtain

$$F_t = -\left(\frac{mg}{L}\right)s$$

This equation follows the general form of Hooke's force law $F_t = -ks$, with $k = mg/L$. We are justified in saying that a pendulum undergoes simple harmonic motion only when it swings back and forth at small amplitudes (or, in this case, small values of θ, so that $\sin \theta \cong \theta$).

Recall that for the object–spring system, the angular frequency is given by Equation 13.11:

$$\omega = 2\pi f = \sqrt{\frac{k}{m}}$$

Substituting the expression of k for a pendulum, we obtain

$$\omega = \sqrt{\frac{mg/L}{m}} = \sqrt{\frac{g}{L}}$$

Tip 13.3 Pendulum Motion Is Not Harmonic
Remember that the pendulum *does not* exhibit true simple harmonic motion for *any* angle. If the angle is less than about 15°, the motion can be *modeled* as approximately simple harmonic.

This angular frequency can be substituted into Equation 13.12, which then mathematically describes the motion of a pendulum. The frequency is just the angular frequency divided by 2π, while the period is the reciprocal of the frequency, or

$$T = 2\pi\sqrt{\frac{L}{g}} \qquad [13.15]$$

◀ The period of a simple pendulum depends only on L and g

This equation reveals the somewhat surprising result that the period of a simple pendulum doesn't depend on the mass, but only on the pendulum's length and on the free-fall acceleration. Further, the amplitude of the motion isn't a factor as long as it's relatively small. The analogy between the motion of a simple pendulum and the object–spring system is illustrated in Figure 13.16.

Galileo first noted that the period of a pendulum was independent of its amplitude. He supposedly observed this while attending church services at the cathedral in Pisa. The pendulum he studied was a swinging chandelier that was set in motion when someone bumped it while lighting candles. Galileo was able to measure its period by timing the swings with his pulse.

The dependence of the period of a pendulum on its length and on the free-fall acceleration allows us to use a pendulum as a timekeeper for a clock. A

APPLICATION
Pendulum Clocks

Figure 13.16
Simple harmonic motion for an object–spring system, and its analogy, the motion of a simple pendulum.

number of clock designs employ a pendulum, with the length adjusted so that its period serves as the basis for the rate at which the clock's hands turn. Of course, these clocks are used at different locations on the Earth, so there will be some variation of the free-fall acceleration. To compensate for this variation, the pendulum of a clock should have some movable mass so that the effective length can be adjusted.

APPLICATION
Use of Pendulum in Prospecting

Geologists often make use of the simple pendulum and Equation 13.15 when prospecting for oil or minerals. Deposits beneath the Earth's surface can produce irregularities in the free-fall acceleration over the region being studied. A specially designed pendulum of known length is used to measure the period, which in turn is used to calculate g. Although such a measurement in itself is inconclusive, it's an important tool for geological surveys.

■ Quick Quiz

13.7 A simple pendulum is suspended from the ceiling of a stationary elevator, and the period is measured. If the elevator moves with constant velocity, does the period (a) increase, (b) decrease, or (c) remain the same? If the elevator accelerates upward, does the period (a) increase, (b) decrease, or (c) remain the same?

13.8 A pendulum clock depends on the period of a pendulum to keep correct time. Suppose a pendulum clock is keeping correct time and then Dennis the Menace slides the bob of the pendulum downward on the oscillating rod. Does the clock run (a) slow, (b) fast, or (c) correctly?

13.9 The period of a simple pendulum is measured to be T on the Earth. If the same pendulum were set in motion on the Moon, would its period be (a) less than T, (b) greater than T, or (c) equal to T?

■ **EXAMPLE 13.7** | Measuring the Value of g

GOAL Determine g from pendulum motion.

PROBLEM Using a small pendulum of length 0.171 m, a geophysicist counts 72.0 complete swings in a time of 60.0 s. What is the value of g in this location?

STRATEGY First calculate the period of the pendulum by dividing the total time by the number of complete swings. Solve Equation 13.15 for g and substitute values.

SOLUTION

Calculate the period by dividing the total elapsed time by the number of complete oscillations:

$$T = \frac{\text{time}}{\text{\# of oscillations}} = \frac{60.0 \text{ s}}{72.0} = 0.833 \text{ s}$$

Solve Equation 13.15 for g and substitute values:

$$T = 2\pi\sqrt{\frac{L}{g}} \quad \rightarrow \quad T^2 = 4\pi^2 \frac{L}{g}$$

$$g = \frac{4\pi^2 L}{T^2} = \frac{(39.5)(0.171 \text{ m})}{(0.833 \text{ s})^2} = \boxed{9.73 \text{ m/s}^2}$$

REMARKS Measuring such a vibration is a good way of determining the local value of the acceleration of gravity.

QUESTION 13.7 True or False: A simple pendulum of length 0.50 m has a larger frequency of vibration than a simple pendulum of length 1.0 m.

EXERCISE 13.7 What would be the period of the 0.171-m pendulum on the Moon, where the acceleration of gravity is 1.62 m/s²?

ANSWER 2.04 s

The Physical Pendulum

The simple pendulum discussed thus far consists of a mass attached to a string. A pendulum, however, can be made from an object of any shape. The general case is called the *physical pendulum*.

In Figure 13.17 a rigid object is pivoted at point O, which is a distance L from the object's center of mass. The center of mass oscillates along a circular arc, just like the simple pendulum. The period of a physical pendulum is given by

$$T = 2\pi \sqrt{\frac{I}{mgL}} \quad [13.16]$$

where I is the object's moment of inertia and m is the object's mass. As a check, notice that in the special case of a simple pendulum with an arm of length L and negligible mass, the moment of inertia is $I = mL^2$. Substituting into Equation 13.16 results in

$$T = 2\pi \sqrt{\frac{mL^2}{mgL}} = 2\pi \sqrt{\frac{L}{g}}$$

which is the correct period for a simple pendulum.

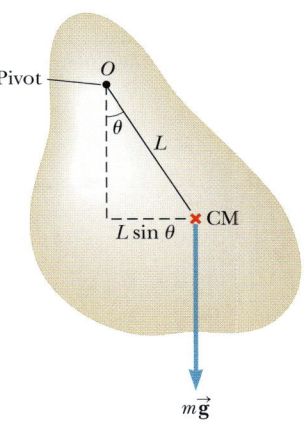

Figure 13.17 A physical pendulum pivoted at O.

13.6 Damped Oscillations

LEARNING OBJECTIVE

1. Describe and contrast the three classes of damped oscillations.

The vibrating motions we have discussed so far have taken place in ideal systems that *oscillate indefinitely* under the action of a linear restoring force. In all real mechanical systems, forces of friction retard the motion, so the systems don't oscillate indefinitely. The friction reduces the mechanical energy of the system as time passes, and the motion is said to be **damped.**

Shock absorbers in automobiles (Fig. 13.18) are one practical application of damped motion. A shock absorber consists of a piston moving through a liquid such as oil. The upper part of the shock absorber is firmly attached to the body of the car. When the car travels over a bump in the road, holes in the piston allow it to move up and down through the fluid in a damped fashion.

Damped motion varies with the fluid used. For example, if the fluid has a relatively low viscosity, the vibrating motion is preserved but the amplitude of vibration decreases in time and the motion ultimately ceases. This process is known as *underdamped* oscillation. The position vs. time curve for an object undergoing such

APPLICATION
Shock Absorbers

Figure 13.18 (a) A shock absorber consists of a piston oscillating in a chamber filled with oil. As the piston oscillates, the oil is squeezed through holes between the piston and the chamber, causing a damping of the piston's oscillations. (b) One type of automotive suspension system, in which a shock absorber is placed inside a coil spring at each wheel.

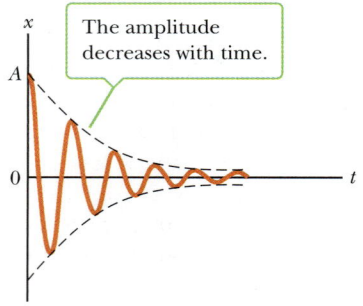

Figure 13.19 A graph of displacement versus time for an underdamped oscillator.

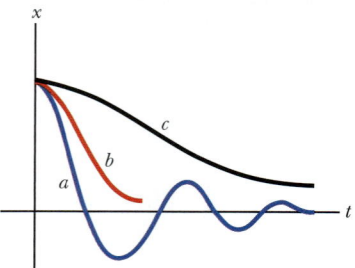

Figure 13.20 Plots of displacement versus time for (a) an underdamped oscillator, (b) a critically damped oscillator, and (c) an overdamped oscillator.

oscillation appears in Figure 13.19. Figure 13.20 compares three types of damped motion, with curve (a) representing underdamped oscillation. If the fluid viscosity is increased, the object returns rapidly to equilibrium after it's released and doesn't oscillate. In this case the system is said to be *critically damped*, and is shown as curve (b) in Figure 13.20. The piston returns to the equilibrium position in the shortest time possible without once overshooting the equilibrium position. If the viscosity is made greater still, the system is said to be *overdamped*. In this case the piston returns to equilibrium without ever passing through the equilibrium point, but the time required to reach equilibrium is greater than in critical damping, as illustrated by curve (c) in Figure 13.20.

To make automobiles more comfortable to ride in, shock absorbers are designed to be slightly underdamped. This can be demonstrated by a sharp downward push on the hood of a car. After the applied force is removed, the body of the car oscillates a few times about the equilibrium position before returning to its fixed position.

13.7 Waves

LEARNING OBJECTIVES

1. Describe the concept of a wave and discuss physical examples.
2. Contrast transverse and longitudinal waves.

The world is full of waves: sound waves, waves on a string, seismic waves, and electromagnetic waves, such as visible light, radio waves, television signals, and x-rays. All these waves have as their source a vibrating object, so we can apply the concepts of simple harmonic motion in describing them.

In the case of sound waves, the vibrations that produce waves arise from sources such as a person's vocal chords or a plucked guitar string. The vibrations of electrons in an antenna produce radio or television waves, and the simple up-and-down motion of a hand can produce a wave on a string. Certain concepts are common to all waves, regardless of their nature. In the remainder of this chapter, we focus our attention on the general properties of waves. In later chapters we will study specific types of waves, such as sound waves and electromagnetic waves.

What Is a Wave?

When you drop a pebble into a pool of water, the disturbance produces water waves, which move away from the point where the pebble entered the water. A leaf floating near the disturbance moves up and down and back and forth about its original position, but doesn't undergo any net displacement attributable to the disturbance. This means that the water wave (or disturbance) moves from one place to another, *but the water isn't carried with it.*

When we observe a water wave, we see a rearrangement of the water's surface. Without the water, there wouldn't be a wave. Similarly, a wave traveling on a string wouldn't exist without the string. Sound waves travel through air as a result of pressure variations from point to point. Therefore, we can consider a wave to be *the motion of a disturbance*. In Chapter 21 we discuss electromagnetic waves, which don't require a medium.

The mechanical waves discussed in this chapter require (1) some source of disturbance, (2) a medium that can be disturbed, and (3) some physical connection or mechanism through which adjacent portions of the medium can influence each other. All waves carry energy and momentum. The amount of energy transmitted through a medium and the mechanism responsible for the transport of energy differ from case to case. The energy carried by ocean waves during a storm, for example, is much greater than the energy carried by a sound wave generated by a single human voice.

APPLYING PHYSICS 11.2 | Burying Bond

At one point in *On Her Majesty's Secret Service,* a James Bond film from the 1960s, Bond was escaping on skis. He had a good lead and was a hard-to-hit moving target. There was no point in wasting bullets shooting at him, so why did the bad guys open fire?

EXPLANATION These misguided gentlemen had a good understanding of the physics of waves. An impulsive sound, like a gunshot, can cause an acoustical disturbance that propagates through the air. If it impacts a ledge of snow that is ready to break free, an avalanche can result. Such a disaster occurred in 1916 during World War I when Austrian soldiers in the Alps were smothered by an avalanche caused by cannon fire. So the bad guys, who have never been able to hit Bond with a bullet, decided to use the sound of gunfire to start an avalanche. ∎

Types of Waves

One of the simplest ways to demonstrate wave motion is to flip one end of a long string that is under tension and has its opposite end fixed, as in Figure 13.21. The bump (called a pulse) travels to the right with a definite speed. A disturbance of this type is called a **traveling wave**. The figure shows the shape of the string at three closely spaced times.

As such a wave pulse travels along the string, **each segment of the string that is disturbed moves in a direction perpendicular to the wave motion.** Figure 13.22 illustrates this point for a particular tiny segment *P*. The string never moves in the direction of the wave. A traveling wave in which the particles of the disturbed medium move in a direction perpendicular to the wave velocity is called a **transverse wave**. Figure 13.23a illustrates the formation of transverse waves on a long spring.

In another class of waves, called **longitudinal waves, the elements of the medium undergo displacements parallel to the direction of wave motion.** Sound waves in air are longitudinal. Their disturbance corresponds to a series of high- and low-pressure regions that may travel through air or through any material medium with a certain speed. A longitudinal pulse can easily be produced in a stretched spring, as in Figure 13.23b. The free end is pumped back and forth along the length of the spring. This action produces compressed and

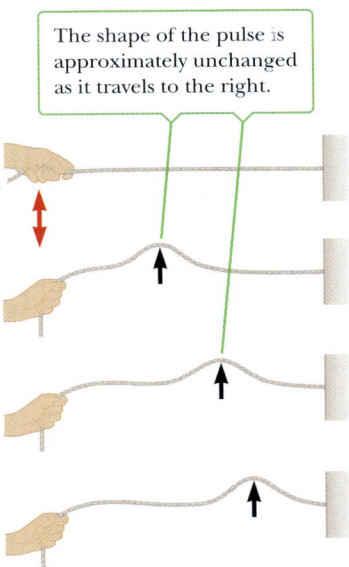

Figure 13.21 A hand moves the end of a stretched string up and down once (red arrow), causing a pulse to travel along the string.

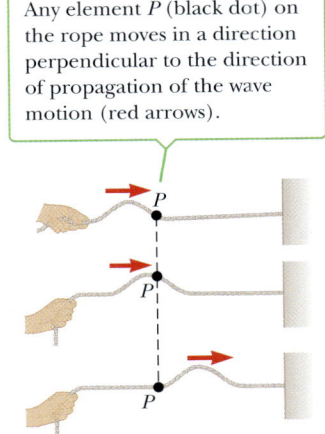

Figure 13.22 A pulse traveling on a stretched string is a transverse wave.

Figure 13.23 (a) A transverse wave is set up in a spring by moving one end of the spring perpendicular to its length. (b) A longitudinal wave along a stretched spring.

stretched regions of the coil that travel along the spring, parallel to the wave motion.

Waves need not be purely transverse or purely longitudinal: ocean waves exhibit a superposition of both types. When an ocean wave encounters a cork, the cork executes a circular motion, going up and down while going forward and back.

Another type of wave, called a **soliton,** consists of a solitary wave front that propagates in isolation. Ordinary water waves generally spread out and dissipate, but solitons tend to maintain their form. The study of solitons began in 1849, when Scottish engineer John Scott Russell noticed a solitary wave leaving the turbulence in front of a barge and propagating forward all on its own. The wave maintained its shape and traveled down a canal at about 10 mi/h. Russell chased the wave two miles on horseback before losing it. Only in the 1960s did scientists take solitons seriously; they are now widely used to model physical phenomena, from elementary particles to the Giant Red Spot of Jupiter.

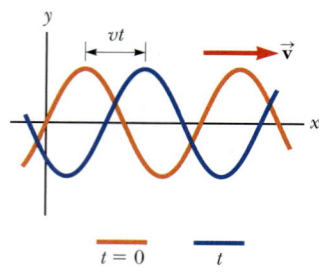

Figure 13.24 A one-dimensional sinusoidal wave traveling to the right with a speed v. The brown curve is a snapshot of the wave at $t = 0$, and the blue curve is another snapshot at some later time t.

Picture of a Wave

Figure 13.24 shows the curved shape of a vibrating string. This pattern is a sinusoidal curve, the same as in simple harmonic motion. The brown curve can be thought of as a snapshot of a traveling wave taken at some instant of time, say, $t = 0$; the blue curve is a snapshot of the same traveling wave at a later time. This picture can also be used to represent a wave on water. In such a case, a high point would correspond to the *crest* of the wave and a low point to the *trough* of the wave.

The same waveform can be used to describe a longitudinal wave, even though no up-and-down motion is taking place. Consider a longitudinal wave traveling on a spring. Figure 13.25a is a snapshot of this wave at some instant, and Figure 13.25b shows the sinusoidal curve that represents the wave. Points where the coils of the spring are compressed correspond to the crests of the waveform, and stretched regions correspond to troughs.

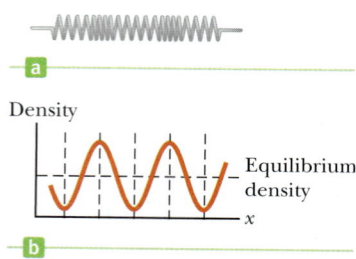

Figure 13.25 (a) A longitudinal wave on a spring. (b) The crests of the waveform correspond to compressed regions of the spring, and the troughs correspond to stretched regions of the spring.

The type of wave represented by the curve in Figure 13.25b is often called a *density wave* or *pressure wave*, because the crests, where the spring coils are compressed, are regions of high density, and the troughs, where the coils are stretched, are regions of low density. Sound waves are longitudinal waves, propagating as a series of high- and low-density regions.

13.8 Frequency, Amplitude, and Wavelength

LEARNING OBJECTIVES
1. Discuss the physical meaning of the term wavelength.
2. Relate the wave speed to its frequency and wavelength

Figure 13.26 illustrates a method of producing a continuous wave or a steady stream of pulses on a very long string. One end of the string is connected to a blade that is set vibrating. As the blade oscillates vertically with simple harmonic motion, a traveling wave moving to the right is set up in the string. Figure 13.26 shows the wave at intervals of one-quarter of a period. Note that **each small segment of the string, such as *P*, oscillates vertically in the *y*-direction with simple harmonic motion.** That must be the case because each segment follows the simple harmonic motion of the blade. Every segment of the string can therefore be treated as a simple harmonic oscillator vibrating with the same frequency as the blade that drives the string.

13.8 | Frequency, Amplitude, and Wavelength

The frequencies of the waves studied in this course will range from rather low values for waves on strings and waves on water, to values for sound waves between 20 Hz and 20 000 Hz (recall that 1 Hz = 1 s^{-1}), to much higher frequencies for electromagnetic waves. These waves have different physical sources, but can be described with the same concepts.

The horizontal dashed line in Figure 13.26 represents the position of the string when no wave is present. The maximum distance the string moves above or below this equilibrium value is called the **amplitude** A of the wave. For the waves we work with, the amplitudes at the crest and the trough will be identical.

Figure 13.26a illustrates another characteristic of a wave. The horizontal arrows show the distance between two successive points that behave identically. This distance is called the **wavelength** λ (the Greek letter lambda).

We can use these definitions to derive an expression for the speed of a wave. We start with the defining equation for the **wave speed** v:

$$v = \frac{\Delta x}{\Delta t}$$

The wave speed is the speed at which a particular part of the wave—say, a crest—moves through the medium.

A wave advances a distance of one wavelength in a time interval equal to one period of the vibration. Taking $\Delta x = \lambda$ and $\Delta t = T$, we see that

$$v = \frac{\lambda}{T}$$

Because the frequency is the reciprocal of the period, we have

$$v = f\lambda \qquad [13.17]$$ ◀ Wave speed

Figure 13.26 One method for producing traveling waves on a continuous string. The left end of the string is connected to a blade that is set vibrating. Every part of the string, such as point P, oscillates vertically with simple harmonic motion.

This important general equation applies to many different types of waves, such as sound waves and electromagnetic waves.

■ EXAMPLE 13.8 | A Traveling Wave

GOAL Obtain information about a wave directly from its graph.

PROBLEM A wave traveling in the positive x-direction is pictured in Figure 13.27a. Find the amplitude, wavelength, speed, and period of the wave if it has a frequency of 8.00 Hz. In Figure 13.27a, $\Delta x = 40.0$ cm and $\Delta y = 15.0$ cm.

STRATEGY The amplitude and wavelength can be read directly from the figure: The maximum vertical displacement is the amplitude, and the distance from one crest to the next is the wavelength. Multiplying the wavelength by the frequency gives the speed, whereas the period is the reciprocal of the frequency.

Figure 13.27 (a) (Example 13.8) (b) (Exercise 13.8)

SOLUTION

The maximum wave displacement is the amplitude A: $A = \Delta y = 15.0$ cm = 0.150 m

The distance from crest to crest is the wavelength: $\lambda = \Delta x = 40.0$ cm = 0.400 m

Multiply the wavelength by the frequency to get the speed: $v = f\lambda = (8.00 \text{ Hz})(0.400 \text{ m}) = 3.20$ m/s

Take the reciprocal of the frequency to get the period: $T = \dfrac{1}{f} = \dfrac{1}{8.00 \text{ Hz}} = 0.125$ s

(Continued)

REMARKS It's important not to confuse the wave with the medium it travels in. A wave is energy transmitted through a medium; some waves, such as light waves, don't require a medium.

QUESTION 13.8 Is the frequency of a wave affected by the wave's amplitude?

EXERCISE 13.8 A wave traveling in the positive x-direction is pictured in Figure 13.27b. Find the amplitude, wavelength, speed, and period of the wave if it has a frequency of 15.0 Hz. In the figure, $\Delta x = 72.0$ cm and $\Delta y = 25.0$ cm.

ANSWERS $A = 0.250$ m, $\lambda = 0.720$ m, $v = 10.8$ m/s, $T = 0.006\ 7$ s

---

■ EXAMPLE 13.9 | Sound and Light

GOAL Perform elementary calculations using speed, wavelength, and frequency.

PROBLEM A wave has a wavelength of 3.00 m. Calculate the frequency of the wave if it is **(a)** a sound wave and **(b)** a light wave. Take the speed of sound as 343 m/s and the speed of light as 3.00×10^8 m/s.

SOLUTION

(a) Find the frequency of a sound wave with $\lambda = 3.00$ m.

Solve Equation 3.17 for the frequency and substitute:

$$(1) \quad f = \frac{v}{\lambda} = \frac{343 \text{ m/s}}{3.00 \text{ m}} = \boxed{114 \text{ Hz}}$$

(b) Find the frequency of a light wave with $\lambda = 3.00$ m.

Substitute into Equation (1), using the speed of light for c:

$$f = \frac{c}{\lambda} = \frac{3.00 \times 10^8 \text{ m/s}}{3.00 \text{ m}} = \boxed{1.00 \times 10^8 \text{ Hz}}$$

REMARKS The same equation can be used to find the frequency in each case, despite the great difference between the physical phenomena. Notice how much larger frequencies of light waves are than frequencies of sound waves.

QUESTION 13.9 A wave in one medium encounters a new medium and enters it. Which of the following wave properties will be affected in this process? (a) wavelength (b) frequency (c) speed

EXERCISE 13.9 (a) Find the wavelength of an electromagnetic wave with frequency 9.00 GHz = 9.00×10^9 Hz (G = giga = 10^9), which is in the microwave range. (b) Find the speed of a sound wave in an unknown fluid medium if a frequency of 567 Hz has a wavelength of 2.50 m.

ANSWERS (a) 0.033 3 m (b) 1.42×10^3 m/s

13.9 The Speed of Waves on Strings

LEARNING OBJECTIVES

1. Discuss the dependence of the speed of waves on a string on the string tension and linear mass density.
2. Calculate the speed of waves on strings.

In this section we focus our attention on the speed of a transverse wave on a stretched string.

For a vibrating string, there are two speeds to consider. One is the speed of the physical string that vibrates up and down, transverse to the string, in the y-direction. The other is the *wave speed*, which is the rate at which the disturbance propagates along the length of the string in the x-direction. We wish to find an expression for the wave speed.

If a horizontal string under tension is pulled vertically and released, it starts at its maximum displacement, $y = A$, and takes a certain amount of time to go to $y = -A$

13.9 | The Speed of Waves on Strings

and back to A again. This amount of time is the period of the wave, and is the same as the time needed for the wave to advance *horizontally* by one wavelength. Dividing the wavelength by the period of one transverse oscillation gives the wave speed.

For a fixed wavelength, a string under greater tension F has a greater wave speed because the period of vibration is shorter, and the wave advances one wavelength during one period. It also makes sense that a string with greater mass per unit length, μ, vibrates more slowly, leading to a longer period and a slower wave speed. The wave speed is given by

$$v = \sqrt{\frac{F}{\mu}} \qquad [13.18]$$

where F is the tension in the string and μ is the mass of the string per unit length, called the *linear density*. From Equation 13.18, it's clear that a larger tension F results in a larger wave speed, whereas a larger linear density μ gives a slower wave speed, as expected.

According to Equation 13.18, the propagation speed of a mechanical wave, such as a wave on a string, depends only on the properties of the string through which the disturbance travels. It doesn't depend on the amplitude of the vibration. This turns out to be generally true of waves in various media.

A proof of Equation 13.18 requires calculus, but dimensional analysis can easily verify that the expression is dimensionally correct. Note that the dimensions of F are ML/T^2, and the dimensions of μ are M/L. The dimensions of F/μ are therefore L^2/T^2, so those of $\sqrt{F/\mu}$ are L/T, giving the dimensions of speed. No other combination of F and μ is dimensionally correct, so in the case in which the tension and mass density are the only relevant physical factors, we have verified Equation 13.18 up to an overall constant.

According to Equation 13.18, we can increase the speed of a wave on a stretched string by increasing the tension in the string. Increasing the mass per unit length, on the other hand, decreases the wave speed. These physical facts lie behind the metallic windings on the bass strings of pianos and guitars. The windings increase the mass per unit length, μ, decreasing the wave speed and hence the frequency, resulting in a lower tone. Tuning a string to a desired frequency is a simple matter of changing the tension in the string.

APPLICATION
Bass Guitar Strings

■ EXAMPLE 13.10 | A Pulse Traveling on a String

GOAL Calculate the speed of a wave on a string.

PROBLEM A uniform string has a mass M of 0.030 0 kg and a length L of 6.00 m. Tension is maintained in the string by suspending a block of mass $m = 2.00$ kg from one end (Fig. 13.28). (a) Find the speed of a transverse wave pulse on this string. (b) Find the time it takes the pulse to travel from the wall to the pulley. Neglect the mass of the hanging part of the string.

Figure 13.28 (Example 13.10) The tension F in the string is maintained by the suspended block. The wave speed is given by the expression $v = \sqrt{F/\mu}$.

STRATEGY The tension F can be obtained from Newton's second law for equilibrium applied to the block, and the mass per unit length of the string is $\mu = M/L$. With these quantities, the speed of the transverse pulse can be found by substitution into Equation 13.18. Part (b) requires the formula $d = vt$.

SOLUTION

(a) Find the speed of the wave pulse.

Apply the second law to the block: the tension F is equal and opposite to the force of gravity.

$\Sigma F = F - mg = 0 \quad \rightarrow \quad F = mg$

(Continued)

Substitute expressions for F and μ into Equation 13.18:

$$v = \sqrt{\frac{F}{\mu}} = \sqrt{\frac{mg}{M/L}}$$

$$= \sqrt{\frac{(2.00 \text{ kg})(9.80 \text{ m/s}^2)}{(0.030\ 0 \text{ kg})/(6.00 \text{ m})}} = \sqrt{\frac{19.6 \text{ N}}{0.005\ 00 \text{ kg/m}}}$$

$$= 62.6 \text{ m/s}$$

(b) Find the time it takes the pulse to travel from the wall to the pulley.

Solve the distance formula for time:

$$t = \frac{d}{v} = \frac{5.00 \text{ m}}{62.6 \text{ m/s}} = 0.079\ 9 \text{ s}$$

REMARKS Don't confuse the speed of the wave on the string with the speed of the sound wave produced by the vibrating string. (See Chapter 14.)

QUESTION 13.10 If the mass of the block is quadrupled, what happens to the speed of the wave?

EXERCISE 13.10 To what tension must a string with mass 0.010 0 kg and length 2.50 m be tightened so that waves will travel on it at a speed of 125 m/s?

ANSWER 62.5 N

13.10 Interference of Waves

LEARNING OBJECTIVE

1. Discuss the superposition principle and use it to explain the phenomena of both constructive and destructive interference of two waves passing through each other.

Many interesting wave phenomena in nature require two or more waves passing through the same region of space at the same time. **Two traveling waves can meet and pass through each other without being destroyed or even altered.** For instance, when two pebbles are thrown into a pond, the expanding circular waves don't destroy each other. In fact, the ripples pass through each other. Likewise, when sound waves from two sources move through air, they pass through each other. In the region of overlap, the resultant wave is found by adding the displacements of the individual waves. For such analyses, the **superposition principle** applies:

> When two or more traveling waves encounter each other while moving through a medium, the resultant wave is found by adding together the displacements of the individual waves point by point.

Experiments show that the superposition principle is valid only when the individual waves have small amplitudes of displacement, which is an assumption we make in all our examples.

Figures 13.29a and 13.29b show two waves of the same amplitude and frequency. If at some instant of time these two waves were traveling through the same region of space, the resultant wave at that instant would have a shape like that shown in Figure 13.29c. For example, suppose the waves are water waves of amplitude 1 m. At the instant they overlap so that crest meets crest and trough meets trough, the resultant wave has an amplitude of 2 m. Waves coming together like that are said to be *in phase* and to exhibit **constructive interference.**

Figures 13.30a and 13.30b show two similar waves. In this case, however, the crest of one coincides with the trough of the other, so one wave is *inverted* relative

Combining the two waves in parts (a) and (b) results in a wave with twice the amplitude.

Figure 13.29 Constructive interference. If two waves having the same frequency and amplitude are in phase, as in (a) and (b), the resultant wave when they combine (c) has the same frequency as the individual waves, but twice their amplitude.

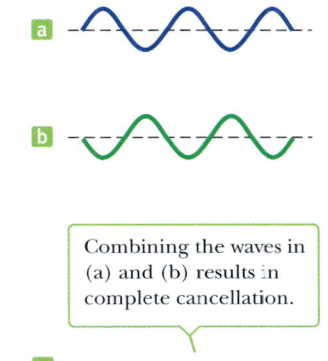

Figure 13.32 Two wave pulses traveling on a stretched string in opposite directions pass through each other.

Figure 13.30 Destructive interference. The two waves in (a) and (b) have the same frequency and amplitude but are 180° out of phase.

to the other. The resultant wave, shown in Figure 13.30c, is seen to be a state of complete cancellation. If these were water waves coming together, one of the waves would exert an upward force on an individual drop of water at the same instant the other wave was exerting a downward force. The result would be no motion of the water at all. In such a situation the two waves are said to be 180° out of phase and to exhibit **destructive interference.** Figure 13.31 illustrates the interference of water waves produced by drops of water falling into a pond.

Figure 13.32 shows constructive interference in two pulses moving toward each other along a stretched string; Figure 13.33 shows destructive interference in two pulses. Notice in both figures that when the two pulses separate, their shapes are unchanged, as if they had never met!

Figure 13.31 Interference patterns produced by outward-spreading waves from many drops of liquid falling into a body of water.

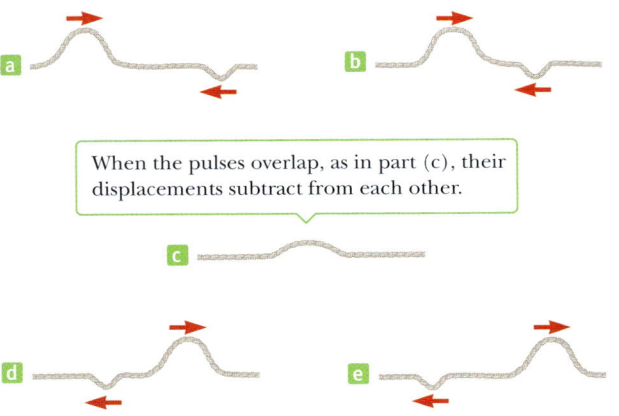

Figure 13.33 Two wave pulses traveling in opposite directions with displacements that are inverted relative to each other.

13.11 Reflection of Waves

LEARNING OBJECTIVE

1. Describe qualitatively the reflection of waves on strings from fixed and free ends.

In our discussion so far, we have assumed waves could travel indefinitely without striking anything. Such conditions are not often realized in practice. Whenever a traveling wave reaches a boundary, part or all of the wave is reflected. For example,

Figure 13.34 The reflection of a traveling wave at the fixed end of a stretched string. Note that the reflected pulse is inverted, but its shape remains the same.

consider a pulse traveling on a string that is fixed at one end (Fig. 13.34). When the pulse reaches the wall, it is reflected.

Note that the reflected pulse is inverted. This can be explained as follows: When the pulse meets the wall, the string exerts an upward force on the wall. According to Newton's third law, the wall must exert an equal and opposite (downward) reaction force on the string. This downward force causes the pulse to invert on reflection.

Now suppose the pulse arrives at the string's end, and the end is attached to a ring of negligible mass that is free to slide along the post without friction (Fig. 13.35). Again the pulse is reflected, but this time it is not inverted. On reaching the post, the pulse exerts a force on the ring, causing it to accelerate upward. The ring is then returned to its original position by the downward component of the tension in the string.

An alternate method of showing that a pulse is reflected without inversion when it strikes a free end of a string is to send the pulse down a string hanging vertically. When the pulse hits the free end, it's reflected without inversion, like the pulse in Figure 13.35.

Finally, when a pulse reaches a boundary, it's partly reflected and partly transmitted past the boundary into the new medium. This effect is easy to observe in the case of two ropes of different density joined at some boundary.

Figure 13.35 The reflection of a traveling wave at the free end of a stretched string. In this case the reflected pulse is not inverted.

SUMMARY

13.1 Hooke's Law

Simple harmonic motion occurs when the net force on an object along the direction of motion is proportional to the object's displacement and in the opposite direction:

$$F_s = -kx \quad [13.1]$$

This equation is called Hooke's law. The time required for one complete vibration is called the **period** of the motion. The reciprocal of the period is the **frequency** of the motion, which is the number of oscillations per second.

When an object moves with simple harmonic motion, its **acceleration** as a function of position is

$$a = -\frac{k}{m}x \quad [13.2]$$

13.2 Elastic Potential Energy

The energy stored in a stretched or compressed spring or in some other elastic material is called **elastic potential energy**:

$$PE_s \equiv \tfrac{1}{2}kx^2 \quad [13.3]$$

The **velocity** of an object as a function of position, when the object is moving with simple harmonic motion, is

$$v = \pm\sqrt{\frac{k}{m}(A^2 - x^2)} \quad [13.6]$$

13.3 Comparing Simple Harmonic Motion with Uniform Circular Motion

The **period** of an object of mass m moving with simple harmonic motion while attached to a spring of spring constant k is

$$T = 2\pi\sqrt{\frac{m}{k}} \quad [13.8]$$

where T is independent of the amplitude A.

The **frequency** of an object–spring system is $f = 1/T$. The **angular frequency** ω of the system in rad/s is

$$\omega = 2\pi f = \sqrt{\frac{k}{m}} \quad [13.11]$$

13.4 Position, Velocity, and Acceleration as a Function of Time

When an object is moving with simple harmonic motion, the **position, velocity,** and **acceleration** of the object as a function of time are given by

$$x = A \cos(2\pi f t) \quad [13.14a]$$
$$v = -A\omega \sin(2\pi f t) \quad [13.14b]$$
$$a = -A\omega^2 \cos(2\pi f t) \quad [13.14c]$$

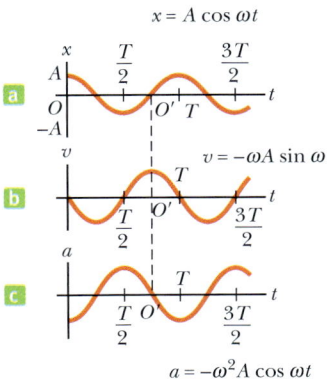

(a) Displacement, (b) velocity, and (c) acceleration versus time for an object moving with simple harmonic motion under the initial conditions $x_0 = A$ and $v_0 = 0$ at $t = 0$.

13.5 Motion of a Pendulum

A **simple pendulum** of length L moves with simple harmonic motion for small angular displacements from the vertical, with a period of

$$T = 2\pi \sqrt{\frac{L}{g}} \quad [13.15]$$

13.7 Waves

In a **transverse wave** the elements of the medium move in a direction perpendicular to the direction of the wave. An example is a wave on a stretched string.

In a **longitudinal wave** the elements of the medium move parallel to the direction of the wave velocity. An example is a sound wave.

13.8 Frequency, Amplitude, and Wavelength

The relationship of the speed, wavelength, and frequency of a wave is

$$v = f\lambda \quad [13.17]$$

This relationship holds for a wide variety of different waves.

13.9 The Speed of Waves on Strings

The speed of a wave traveling on a stretched string of mass per unit length μ and under tension F is

$$v = \sqrt{\frac{F}{\mu}} \quad [13.18]$$

13.10 Interference of Waves

The **superposition principle** states that if two or more traveling waves are moving through a medium, the resultant wave is found by adding the individual waves together point by point. When waves meet crest to crest and trough to trough, they undergo **constructive interference.** When crest meets trough, the waves undergo **destructive interference.**

13.11 Reflection of Waves

When a wave pulse reflects from a rigid boundary, the pulse is inverted. When the boundary is free, the reflected pulse is not inverted.

■ WARM-UP EXERCISES

WebAssign The warm-up exercises in this chapter may be assigned online in Enhanced WebAssign.

1. **Math Review** Suppose a function is given by $y(t) = (5.00 \text{ m}) \sin(8.00\pi t)$. Determine (a) the maximum value of $y(t)$, (b) the minimum value of $y(t)$, and (c) the period of the function.

2. A horizontal spring system consists of a block of mass 4.00 kg on a frictionless surface attached to a horizontal spring with constant 624 N/m. (a) Calculate the magnitude of the spring force acting on the block when displaced from equilibrium by 0.250 m. (b) If the block is released, what is the magnitude of its initial acceleration? (See Section 13.1.)

3. A light spring with force constant 575 N/m is hung vertically from the ceiling. A 7.20-kg bowling ball is attached to the end of the spring and the ball is slowly lowered. How far does the spring stretch from its equilibrium position when the gravity and spring forces balance? (See Section 13.1.)

4. A 5.00-kg mass attached to a horizontal spring oscillates back and forth in simple harmonic motion with an amplitude of 0.200 m. If the spring has a force constant of 75.0 N/m, find (a) the potential energy of the system at its maximum amplitude, and (b) the speed of the object as it passes through its equilibrium point. (See Section 13.2.)

5. A 4.00-kg block is sliding on a frictionless surface at 7.00 m/s toward a horizontal spring of constant 1 830 N/m that is attached to the wall. (a) Calculate the kinetic energy of the block. (b) By how much will the block compress the spring after striking it? (See Section 13.2.)

6. A horizontal spring system has a spring with constant 628 N/m and a block of mass 8.00 kg, lying on a frictionless surface, attached to the free end. The block is pulled a short distance from equilibrium and released. Calculate (a) the angular frequency, (b) the frequency, and (c) the period of the oscillating spring system. (See Section 13.3.)

7. The position of a 5.00-kg object moving with simple harmonic motion is given by $x = (4.00 \text{ m}) \cos(6.00\pi t)$, where x is in meters and t is in seconds. Find (a) the angular frequency, (b) the frequency, (c) the period, and (d) the spring constant. (See Section 13.4.)

8. A mass of 0.400 kg, hanging from a spring with a spring constant of 90.0 N/m, is displaced from equilibrium by 0.200 m and released from rest. Find (a) the amplitude of the oscillations, (b) the angular frequency of the oscillations, (c) the maximum speed of the mass, and (d) the maximum acceleration of the mass. (See Sections 13.3 and 13.4.)

9. A simple pendulum has a period of 2.50 s. Find (a) the frequency, (b) the angular frequency, and (c) the length of the pendulum. (See Section 13.5.)

10. A man on a dock observes that the distance between two successive crests of a water wave is 4.00 m. Just as a crest passes he starts a stopwatch, and finds it takes 1.40 s for the next crest to reach him. Find (a) the wavelength of the wave, (b) the frequency of the wave, and (c) the speed of the wave. (See Section 13.8.)

11. A uniform string has mass 0.005 00 kg and length 0.800 m. (a) Calculate the linear density, or mass divided by length, of the string. (b) If the string is under a tension of 275 N, find the velocity of waves on the string. (See Section 13.9.)

CONCEPTUAL QUESTIONS

The conceptual questions in this chapter may be assigned online in Enhanced WebAssign.

1. An object–spring system undergoes simple harmonic motion with an amplitude A. Does the total energy change if the mass is doubled but the amplitude isn't changed? Are the kinetic and potential energies at a given point in its motion affected by the change in mass? Explain.

2. If an object–spring system is hung vertically and set into oscillation, why does the motion eventually stop?

3. An object is hung on a spring, and the frequency of oscillation of the system, f, is measured. The object, a second identical object, and the spring are carried to space in the space shuttle. The two objects are attached to the ends of the spring, and the system is taken out into space on a space walk. The spring is extended, and the system is released to oscillate while floating in space. The coils of the spring don't bump into one another. What is the frequency of oscillation for this system in terms of f?

4. If a spring is cut in half, what happens to its spring constant?

5. A pendulum bob is made from a sphere filled with water. What would happen to the frequency of vibration of this pendulum if the sphere had a hole in it that allowed the water to leak out slowly?

6. If a pendulum clock keeps perfect time at the base of a mountain, will it also keep perfect time when it is moved to the top of the mountain? Explain.

7. (a) Is a bouncing ball an example of simple harmonic motion? (b) Is the daily movement of a student from home to school and back simple harmonic motion?

8. If a grandfather clock were running slow, how could we adjust the length of the pendulum to correct the time?

9. What happens to the speed of a wave on a string when the frequency is doubled? Assume the tension in the string remains the same.

10. If you stretch a rubber hose and pluck it, you can observe a pulse traveling up and down the hose. What happens to the speed of the pulse if you stretch the hose more tightly? What happens to the speed if you fill the hose with water?

11. Explain why the kinetic and potential energies of an object–spring system can never be negative.

12. A grandfather clock depends on the period of a pendulum to keep correct time. Suppose such a clock is calibrated correctly and then the temperature of the room in which it resides increases. Does the clock run slow, fast, or correctly? *Hint:* A material expands when its temperature increases.

PROBLEMS

The problems in this chapter may be assigned online in Enhanced WebAssign.

1. denotes straightforward problem; 2. denotes intermediate problem;
3. denotes challenging problem
1. denotes full solution available in *Student Solutions Manual/Study Guide*
1. denotes problems most often assigned in Enhanced WebAssign

BIO denotes biomedical problems
GP denotes guided problems
M denotes Master It tutorial available in Enhanced WebAssign
QC denotes asking for quantitative and conceptual reasoning
S denotes symbolic reasoning problem
W denotes Watch It video solution available in Enhanced WebAssign

13.1 Hooke's Law

1. A block of mass $m = 0.60$ kg attached to a spring with force constant 130 N/m is free to move on a frictionless, horizontal surface as in Figure P13.1. The block is released from rest after the spring is stretched a distance $A = 0.13$ m. At that instant, find (a) the force on the block and (b) its acceleration.

Figure P13.1

2. A spring oriented vertically is attached to a hard horizontal surface as in Figure P13.2. The spring has a force constant of 1.46 kN/m. How much is the spring compressed when a object of mass $m = 2.30$ kg is placed on top of the spring and the system is at rest?

Figure P13.2

3. The force constant of a spring is 137 N/m. Find the magnitude of the force required to (a) compress the spring by 4.80 cm from its unstretched length and (b) stretch the spring by 7.36 cm from its unstretched length.

4. A load of 50 N attached to a spring hanging vertically stretches the spring 5.0 cm. The spring is now placed horizontally on a table and stretched 11 cm. (a) What force is required to stretch the spring by that amount? (b) Plot a graph of force (on the y-axis) versus spring displacement from the equilibrium position along the x-axis.

5. A spring is hung from a ceiling, and an object attached to its lower end stretches the spring by a distance $d = 5.00$ cm from its unstretched position when the system is in equilibrium as in Figure P13.5. If the spring constant is 47.5 N/m, determine the mass of the object.

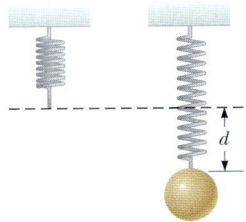

Figure P13.5

6. An archer must exert a force of 375 N on the bowstring shown in Figure P13.6a such that the string makes an angle of $\theta = 35.0°$ with the vertical. (a) Determine the tension in the bowstring. (b) If the applied force is replaced by a stretched spring as in Figure P13.6b and the spring is stretched 30.0 cm from its unstretched length, what is the spring constant?

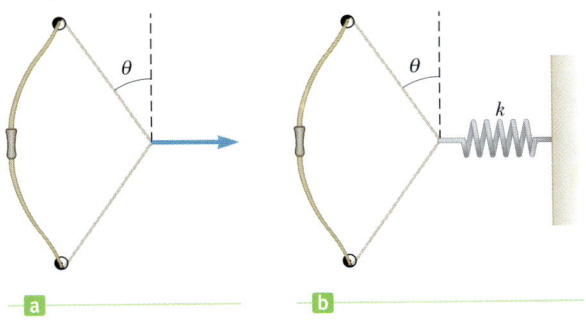

Figure P13.6

7. **Q|C** A spring 1.50 m long with force constant 475 N/m is hung from the ceiling of an elevator, and a block of mass 10.0 kg is attached to the bottom of the spring. (a) By how much is the spring stretched when the block is slowly lowered to its equilibrium point? (b) If the elevator subsequently accelerates upward at 2.00 m/s², what is the position of the block, taking the equilibrium position found in part (a) as $y = 0$ and upwards as the positive y-direction. (c) If the elevator cable snaps during the acceleration, describe the subsequent motion of the block relative to the freely falling elevator. What is the amplitude of its motion?

13.2 Elastic Potential Energy

8. A block of mass $m = 2.00$ kg is attached to a spring of force constant $k = 5.00 \times 10^2$ N/m that lies on a horizontal frictionless surface as shown in Figure P13.8. The block is pulled to a position $x_i = 5.00$ cm to the right of equilibrium and released from rest. Find (a) the work required to stretch the spring and (b) the speed the block has as it passes through equilibrium.

Figure P13.8

9. A slingshot consists of a light leather cup containing a stone. The cup is pulled back against two parallel rubber bands. It takes a force of 15 N to stretch either one of these bands 1.0 cm. (a) What is the potential energy stored in the two bands together when a 50-g stone is placed in the cup and pulled back 0.20 m from the equilibrium position? (b) With what speed does the stone leave the slingshot?

10. An archer pulls her bowstring back 0.400 m by exerting a force that increases uniformly from zero to 230 N. (a) What is the equivalent spring constant of the bow? (b) How much work is done in pulling the bow?

11. A child's toy consists of a piece of plastic attached to a spring (Fig. P13.11). The spring is compressed against the floor a distance of 2.00 cm, and the toy is released. If the toy has a mass of 100 g and rises to a maximum height of 60.0 cm, estimate the force constant of the spring.

Figure P13.11

12. An automobile having a mass of 1 000 kg is driven into a brick wall in a safety test. The bumper behaves like a spring with constant 5.00×10^6 N/m and is compressed 3.16 cm as the car is brought to rest. What was the speed of the car before impact, assuming no energy is lost in the collision with the wall?

13. A 10.0-g bullet is fired into, and embeds itself in, a 2.00-kg block attached to a spring with a force constant of 19.6 N/m and whose mass is negligible. How far is the spring compressed if the bullet has a speed of 300 m/s just before it strikes the block and the block

slides on a frictionless surface? *Note:* You must use conservation of momentum in this problem because of the inelastic collision between the bullet and block.

14. **S** An object–spring system moving with simple harmonic motion has an amplitude A. (a) What is the total energy of the system in terms of k and A only? (b) Suppose at a certain instant the kinetic energy is twice the elastic potential energy. Write an equation describing this situation, using only the variables for the mass m, velocity v, spring constant k, and position x. (c) Using the results of parts (a) and (b) and the conservation of energy equation, find the positions x of the object when its kinetic energy equals twice the potential energy stored in the spring. (The answer should in terms of A only.)

15. **GP** A horizontal block–spring system with the block on a frictionless surface has total mechanical energy $E = 47.0$ J and a maximum displacement from equilibrium of 0.240 m. (a) What is the spring constant? (b) What is the kinetic energy of the system at the equilibrium point? (c) If the maximum speed of the block is 3.45 m/s, what is its mass? (d) What is the speed of the block when its displacement is 0.160 m? (e) Find the kinetic energy of the block at $x = 0.160$ m. (f) Find the potential energy stored in the spring when $x = 0.160$ m. (g) Suppose the same system is released from rest at $x = 0.240$ m on a rough surface so that it loses 14.0 J by the time it reaches its first turning point (after passing equilibrium at $x = 0$). What is its position at that instant?

16. **Q|C** A 0.250-kg block resting on a frictionless, horizontal surface is attached to a spring having force constant 83.8 N/m as in Figure P13.16. A horizontal force $\vec{F}$ causes the spring to stretch a distance of 5.46 cm from its equilibrium position. (a) Find the value of F. (b) What is the total energy stored in the system when the spring is stretched? (c) Find the magnitude of the acceleration of the block immediately after the applied force is removed. (d) Find the speed of the block when it first reaches the equilibrium position. (e) If the surface is not frictionless but the block still reaches the equilibrium position, how would your answer to part (d) change? (f) What other information would you need to know to find the actual answer to part (d) in this case?

Figure P13.16

13.3 Comparing Simple Harmonic Motion with Uniform Circular Motion

13.4 Position, Velocity, and Acceleration as a Function of Time

17. A 0.40-kg object connected to a light spring with a force constant of 19.6 N/m oscillates on a frictionless horizontal surface. If the spring is compressed 4.0 cm and released from rest, determine (a) the maximum speed of the object, (b) the speed of the object when the spring is compressed 1.5 cm, and (c) the speed of the object as it passes the point 1.5 cm from the equilibrium position. (d) For what value of x does the speed equal one-half the maximum speed?

18. A horizontal object-spring system oscillates with an amplitude of 3.5 cm on a frictionless surface. If the spring constant is 250 N/m and the object has a mass of 0.50 kg, determine (a) the mechanical energy of the system, (b) the maximum speed of the object, and (c) the maximum acceleration of the object.

19. **M** At an outdoor market, a bunch of bananas attached to the bottom of a vertical spring of force constant 16.0 N/m is set into oscillatory motion with an amplitude of 20.0 cm. It is observed that the maximum speed of the bunch of bananas is 40.0 cm/s. What is the weight of the bananas in newtons?

20. A 50.0-g object is attached to a horizontal spring with a force constant of 10.0 N/m and released from rest with an amplitude of 25.0 cm. What is the velocity of the object when it is halfway to the equilibrium position if the surface is frictionless?

21. A horizontal spring attached to a wall has a force constant of $k = 850$ N/m. A block of mass $m = 1.00$ kg is attached to the spring and rests on a frictionless, horizontal surface as in Figure P13.21. (a) The block is pulled to a position $x_i = 6.00$ cm from equilibrium and released. Find the potential energy stored in the spring when the block is 6.00 cm from equilibrium. (b) Find the speed of the block as it passes through the equilibrium position. (c) What is the speed of the block when it is at a position $x_i/2 = 3.00$ cm?

Figure P13.21

22. An object moves uniformly around a circular path of radius 20.0 cm, making one complete revolution every 2.00 s. What are (a) the translational speed of the object, (b) the frequency of motion in hertz, and (c) the angular speed of the object?

23. Consider the simplified single-piston engine in Figure P13.23. If the wheel rotates at a constant angular speed ω, explain why the piston oscillates in simple harmonic motion.

Figure P13.23

24. The period of motion of an object–spring system is $T = 0.528$ s when an object of mass $m = 238$ g is attached to the spring. Find (a) the frequency of motion in hertz and (b) the force constant of the spring. (c) If the total energy of the oscillating motion is 0.234 J, find the amplitude of the oscillations.

25. **M** A vertical spring stretches 3.9 cm when a 10-g object is hung from it. The object is replaced with a block of mass 25 g that oscillates up and down in simple harmonic motion. Calculate the period of motion.

26. When four people with a combined mass of 320 kg sit down in a 2.0×10^3-kg car, they find that their weight compresses the springs an additional 0.80 cm. (a) What is the effective force constant of the springs? (b) The four people get out of the car and bounce it up and down. What is the frequency of the car's vibration?

27. A cart of mass 250 g is placed on a frictionless horizontal air track. A spring having a spring constant of 9.5 N/m is attached between the cart and the left end of the track. If the cart is displaced 4.5 cm from its equilibrium position, find (a) the period at which it oscillates, (b) its maximum speed, and (c) its speed when it is located 2.0 cm from its equilibrium position.

28. The position of an object connected to a spring varies with time according to the expression $x = (5.2 \text{ cm}) \sin(8.0\pi t)$. Find (a) the period of this motion, (b) the frequency of the motion, (c) the amplitude of the motion, and (d) the first time after $t = 0$ that the object reaches the position $x = 2.6$ cm.

29. **W** A 326-g object is attached to a spring and executes simple harmonic motion with a period of 0.250 s. If the total energy of the system is 5.83 J, find (a) the maximum speed of the object, (b) the force constant of the spring, and (c) the amplitude of the motion.

30. **S** An object executes simple harmonic motion with an amplitude A. (a) At what values of its position does its speed equal half its maximum speed? (b) At what values of its position does its potential energy equal half the total energy?

31. A 2.00-kg object on a frictionless horizontal track is attached to the end of a horizontal spring whose force constant is 5.00 N/m. The object is displaced 3.00 m to the right from its equilibrium position and then released, initiating simple harmonic motion. (a) What is the force (magnitude and direction) acting on the object 3.50 s after it is released? (b) How many times does the object oscillate in 3.50 s?

32. **GP** A spring of negligible mass stretches 3.00 cm from its relaxed length when a force of 7.50 N is applied. A 0.500-kg particle rests on a frictionless horizontal surface and is attached to the free end of the spring. The particle is displaced from the origin to $x = 5.00$ cm and released from rest at $t = 0$. (a) What is the force constant of the spring? (b) What are the angular frequency ω, the frequency, and the period of the motion? (c) What is the total energy of the system? (d) What is the amplitude of the motion? (e) What are the maximum velocity and the maximum acceleration of the particle? (f) Determine the displacement x of the particle from the equilibrium position at $t = 0.500$ s. (g) Determine the velocity and acceleration of the particle when $t = 0.500$ s.

33. Given that $x = A \cos(\omega t)$ is a sinusoidal function of time, show that v (velocity) and a (acceleration) are also sinusoidal functions of time. *Hint:* Use Equations 13.6 and 13.2.

13.5 Motion of a Pendulum

34. **W** A man enters a tall tower, needing to know its height. He notes that a long pendulum extends from the ceiling almost to the floor and that its period is 15.5 s. (a) How tall is the tower? (b) If this pendulum is taken to the Moon, where the free-fall acceleration is 1.67 m/s^2, what is the period there?

35. A simple pendulum has a length of 52.0 cm and makes 82.0 complete oscillations in 2.00 min. Find (a) the period of the pendulum and (b) the value of g at the location of the pendulum.

36. A "seconds" pendulum is one that moves through its equilibrium position once each second. (The period of the pendulum is 2.000 s.) The length of a seconds pendulum is 0.992 7 m at Tokyo and 0.994 2 m at Cambridge, England. What is the ratio of the free-fall accelerations at these two locations?

37. A pendulum clock that works perfectly on the Earth is taken to the Moon. (a) Does it run fast or slow there? (b) If the clock is started at 12:00 midnight, what will it read after one Earth day (24.0 h)? Assume the free-fall acceleration on the Moon is 1.63 m/s^2.

38. A coat hanger of mass $m = 0.238$ kg oscillates on a peg as a physical pendulum as shown in Figure P13.38. The distance from the pivot to the center of mass of the coat hanger is $d = 18.0$ cm and the period of the motion is $T = 1.25$ s. Find the moment of inertia of the coat hanger about the pivot.

Figure P13.38

39. **M** The free-fall acceleration on Mars is 3.7 m/s^2. (a) What length of pendulum has a period of 1 s on Earth? (b) What length of pendulum would have a 1-s period on Mars? An object is suspended from a spring with force constant 10 N/m. Find the mass suspended from this spring that would result in a period of 1 s (c) on Earth and (d) on Mars.

40. A simple pendulum is 5.00 m long. (a) What is the period of simple harmonic motion for this pendulum if it is located in an elevator accelerating upward at 5.00 m/s^2? (b) What is its period if the elevator is accelerating downward at 5.00 m/s^2? (c) What is the period of simple harmonic motion for the pendulum if it is placed in a truck that is accelerating horizontally at 5.00 m/s^2?

13.6 Damped Oscillations
13.7 Waves
13.8 Frequency, Amplitude, and Wavelength

41. The sinusoidal wave shown in Figure P13.41 is traveling in the positive x-direction and has a frequency of 18.0 Hz. Find the (a) amplitude, (b) wavelength, (c) period, and (d) speed of the wave.

Figure P13.41

42. An object attached to a spring vibrates with simple harmonic motion as described by Figure P13.42. For this motion, find (a) the amplitude, (b) the period, (c) the angular frequency, (d) the maximum speed, (e) the maximum acceleration, and (f) an equation for its position x in terms of a sine function.

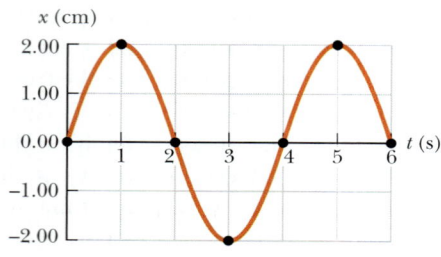

Figure P13.42

43. Light waves are electromagnetic waves that travel at 3.00×10^8 m/s. The eye is most sensitive to light having a wavelength of 5.50×10^{-7} m. Find (a) the frequency of this light wave and (b) its period.

44. The distance between two successive minima of a transverse wave is 2.76 m. Five crests of the wave pass a given point along the direction of travel every 14.0 s. Find (a) the frequency of the wave and (b) the wave speed.

45. **W** A harmonic wave is traveling along a rope. It is observed that the oscillator that generates the wave completes 40.0 vibrations in 30.0 s. Also, a given maximum travels 425 cm along the rope in 10.0 s. What is the wavelength?

46. **BIO** A bat can detect small objects, such as an insect, whose size is approximately equal to one wavelength of the sound the bat makes. If bats emit a chirp at a frequency of 60.0×10^3 Hz and the speed of sound in air is 343 m/s, what is the smallest insect a bat can detect?

47. A cork on the surface of a pond bobs up and down two times per second on ripples having a wavelength of 8.50 cm. If the cork is 10.0 m from shore, how long does it take a ripple passing the cork to reach the shore?

48. Ocean waves are traveling to the east at 4.0 m/s with a distance of 20 m between crests. With what frequency do the waves hit the front of a boat (a) when the boat is at anchor and (b) when the boat is moving westward at 1.0 m/s?

13.9 The Speed of Waves on Strings

49. An ethernet cable is 4.00 m long and has a mass of 0.200 kg. A transverse wave pulse is produced by plucking one end of the taut cable. The pulse makes four trips down and back along the cable in 0.800 s. What is the tension in the cable?

50. A circus performer stretches a tightrope between two towers. He strikes one end of the rope and sends a wave along it toward the other tower. He notes that it takes the wave 0.800 s to reach the opposite tower, 20.0 m away. If a 1.00-m length of the rope has a mass of 0.350 kg, find the tension in the tightrope.

51. A piano string of mass per unit length 5.00×10^{-3} kg/m is under a tension of 1 350 N. Find the speed with which a wave travels on this string.

52. **Q|C S** A student taking a quiz finds on a reference sheet the two equations

$$f = \frac{1}{T} \quad \text{and} \quad v = \sqrt{\frac{T}{\mu}}$$

She has forgotten what T represents in each equation. (a) Use dimensional analysis to determine the units required for T in each equation. (b) Explain how you can identify the physical quantity each T represents from the units.

53. Transverse waves with a speed of 50.0 m/s are to be produced on a stretched string. A 5.00-m length of string with a total mass of 0.060 0 kg is used. (a) What is the required tension in the string? (b) Calculate the wave speed in the string if the tension is 8.00 N.

54. **W** An astronaut on the Moon wishes to measure the local value of g by timing pulses traveling down a wire that has a large object suspended from it. Assume a wire of mass 4.00 g is 1.60 m long and has a 3.00-kg object suspended from it. A pulse requires 36.1 ms to traverse the length of the wire. Calculate g_{Moon} from these data. (You may neglect the mass of the wire when calculating the tension in it.)

55. A simple pendulum consists of a ball of mass 5.00 kg hanging from a uniform string of mass 0.060 0 kg and length L. If the period of oscillation of the pendulum is 2.00 s, determine the speed of a transverse wave in the string when the pendulum hangs vertically.

56. A string is 50.0 cm long and has a mass of 3.00 g. A wave travels at 5.00 m/s along this string. A second string has the same length, but half the mass of the first. If the two strings are under the same tension, what is the speed of a wave along the second string?

57. **M** Tension is maintained in a string as in Figure P13.57. The observed wave speed is $v = 24.0$ m/s when the suspended mass is $m = 3.00$ kg. (a) What is the mass per unit length of the string? (b) What is the wave speed when the suspended mass is $m = 2.00$ kg?

Figure P13.57

58. The elastic limit of a piece of steel wire is 2.70×10^9 Pa. What is the maximum speed at which transverse wave pulses can propagate along the wire without exceeding its elastic limit? (The density of steel is 7.86×10^3 kg/m^3.)

59. **Q|C** A 2.65-kg power line running between two towers has a length of 38.0 m and is under a tension of 12.5 N. (a) What is the speed of a transverse pulse set up on the line? (b) If the tension in the line was unknown, describe a procedure a worker on the ground might use to estimate the tension.

60. **S** A taut clothesline has length L and a mass M. A transverse pulse is produced by plucking one end of the clothesline. If the pulse makes n round trips along the clothesline in t seconds, find expressions for (a) the speed of the pulse in terms of n, L, and t and (b) the tension F in the clothesline in terms of the same variables and mass M.

13.10 Interference of Waves

13.11 Reflection of Waves

61. A wave of amplitude 0.30 m interferes with a second wave of amplitude 0.20 m traveling in the same direction. What are (a) the largest and (b) the smallest resultant amplitudes that can occur, and under what conditions will these maxima and minima arise?

Additional Problems

62. The position of a 0.30-kg object attached to a spring is described by

$$x = (0.25 \text{ m}) \cos (0.4\pi t)$$

Find (a) the amplitude of the motion, (b) the spring constant, (c) the position of the object at $t = 0.30$ s, and (d) the object's speed at $t = 0.30$ s.

63. An object of mass 2.00 kg is oscillating freely on a vertical spring with a period of 0.600 s. Another object of unknown mass on the same spring oscillates with a period of 1.05 s. Find (a) the spring constant k and (b) the unknown mass.

64. A certain tuning fork vibrates at a frequency of 196 Hz while each tip of its two prongs has an amplitude of 0.850 mm. (a) What is the period of this motion? (b) Find the wavelength of the sound produced by the vibrating fork, taking the speed of sound in air to be 343 m/s.

65. A simple pendulum has mass 1.20 kg and length 0.700 m. (a) What is the period of the pendulum near the surface of Earth? (b) If the same mass is attached to a spring, what spring constant would result in the period of motion found in part (a)?

66. A 500-g block is released from rest and slides down a frictionless track that begins 2.00 m above the horizontal, as shown in Figure P13.66. At the bottom of the track, where the surface is horizontal, the block strikes and sticks to a light spring with a spring constant of 20.0 N/m. Find the maximum distance the spring is compressed.

Figure P13.66

67. A 3.00-kg object is fastened to a light spring, with the intervening cord passing over a pulley (Fig. P13.67). The pulley is frictionless, and its inertia may be neglected. The object is released from rest when the spring is unstretched. If the object drops 10.0 cm before stopping, find (a) the spring constant of the spring and (b) the speed of the object when it is 5.00 cm below its starting point.

Figure P13.67

68. A 5.00-g bullet moving with an initial speed of 400 m/s is fired into and passes through a 1.00-kg block, as in Figure P13.68. The block, initially at rest on a frictionless horizontal surface, is connected to a spring with a spring constant of 900 N/m. If the block moves 5.00 cm to the right after impact, find (a) the speed at which the bullet emerges from the block and (b) the mechanical energy lost in the collision.

Figure P13.68

69. **M** A large block P executes horizontal simple harmonic motion as it slides across a frictionless surface with a frequency $f = 1.50$ Hz. Block B rests on it, as shown in Figure P13.69 (page 480), and the coefficient of static

friction between the two is $\mu_s = 0.600$. What maximum amplitude of oscillation can the system have if block B is not to slip?

Figure P13.69

70. A spring in a toy gun has a spring constant of 9.80 N/m and can be compressed 20.0 cm beyond the equilibrium position. A 1.00-g pellet resting against the spring is propelled forward when the spring is released. (a) Find the muzzle speed of the pellet. (b) If the pellet is fired horizontally from a height of 1.00 m above the floor, what is its range?

71. **QC** A light balloon filled with helium of density 0.179 kg/m^3 is tied to a light string of length $L = 3.00$ m. The string is tied to the ground, forming an "inverted" simple pendulum (Fig. P13.71a). If the balloon is displaced slightly from equilibrium, as in Figure P13.71b, (a) show that the motion is simple harmonic and (b) determine the period of the motion. Take the density of air to be 1.29 kg/m^3. *Hint:* Use an analogy with the simple pendulum discussed in the text, and see Chapter 9.

Figure P13.71

72. **S** An object of mass m is connected to two rubber bands of length L, each under tension F, as in Figure P13.72. The object is displaced vertically by a small distance y. Assuming the tension does not change, show that (a) the restoring force is $-(2F/L)y$ and (b) the system exhibits simple harmonic motion with an angular frequency $\omega = \sqrt{2F/mL}$.

Figure P13.72

73. Assume a hole is drilled through the center of the Earth. It can be shown that an object of mass m at a distance r from the center of the Earth is pulled toward the center only by the material in the shaded portion of Figure P13.73. Assume Earth has a uniform density ρ. Write down Newton's law of gravitation for an object at a distance r from the center of the Earth and show that the force on it is of the form of Hooke's law, $F = -kr$, with an effective force constant of $k = (\frac{4}{3})\pi\rho Gm$, where G is the gravitational constant.

Figure P13.73

74. **BIO** Figure P13.74 shows a crude model of an insect wing. The mass m represents the entire mass of the wing, which pivots about the fulcrum F. The spring represents the surrounding connective tissue. Motion of the wing corresponds to vibration of the spring. Suppose the mass of the wing is 0.30 g and the effective spring constant of the tissue is 4.7×10^{-4} N/m. If the mass m moves up and down a distance of 2.0 mm from its position of equilibrium, what is the maximum speed of the outer tip of the wing?

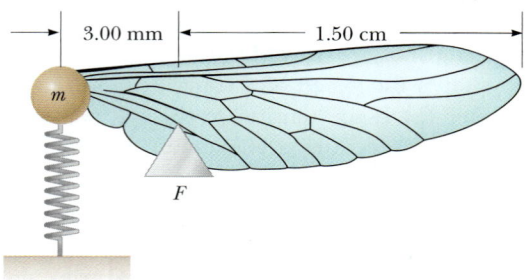

Figure P13.74

75. A 2.00-kg block hangs without vibrating at the end of a spring ($k = 500$ N/m) that is attached to the ceiling of an elevator car. The car is rising with an upward acceleration of $g/3$ when the acceleration suddenly ceases (at $t = 0$). (a) What is the angular frequency of oscillation of the block after the acceleration ceases? (b) By what amount is the spring stretched during the time that the elevator car is accelerating?

76. **QC** **S** A system consists of a vertical spring with force constant $k = 1\,250$ N/m, length $L = 1.50$ m, and object of mass $m = 5.00$ kg attached to the end (Fig. P13.76). The object is placed at the level of the point of attachment with the spring unstretched, at position $y_i = L$, and then it is released so that it swings like a pendulum. (a) Write Newton's second law symbolically for the system as the object passes through its lowest point. (Note that at the lowest point, $r = L - y_f$.) (b) Write the conservation of energy equation symbolically, equating the total mechanical energies at the initial point and lowest point. (c) Find the coordinate position of the lowest point. (d) Will this pendulum's period be greater or less than the period of a simple pendulum with the same mass m and length L? Explain.

Figure P13.76

View of lightning over a city at night. During a thunderstorm, a high concentration of electrical charge in a thundercloud creates a higher-than-normal electric field between the thundercloud and the negatively charged Earth's surface. This strong electric field creates an electric discharge—an enormous spark—between the charged cloud and the ground. Other discharges observed in the sky include cloud-to-cloud discharges and the more frequent intracloud discharges.

Electric Forces and Electric Fields

Electricity is the lifeblood of technological civilization and modern society. Without it, we revert to the mid-nineteenth century: no telephones, no television, none of the household appliances that we take for granted. Modern medicine would be a fantasy, and due to the lack of sophisticated experimental equipment and fast computers—and especially the slow dissemination of information—science and technology would grow at a glacial pace.

Instead, with the discovery and harnessing of electric forces and fields, we can view arrangements of atoms, probe the inner workings of the cell, and send spacecraft beyond the limits of the solar system. All this has become possible in just the last few generations of human life, a blink of the eye compared to the million years our kind spent foraging the savannahs of Africa.

Around 700 B.C. the ancient Greeks conducted the earliest known study of electricity. It all began when someone noticed that a fossil material called amber would attract small objects after being rubbed with wool. Since then we have learned that this phenomenon is not restricted to amber and wool, but occurs (to some degree) when almost any two nonconducting substances are rubbed together.

In this chapter we use the effect of charging by friction to begin an investigation of electric forces. We then discuss Coulomb's law, which is the fundamental law of force between any two stationary charged particles. The concept of an electric field associated with charges is introduced and its effects on other charged particles described. We end with discussions of the Van de Graaff generator and Gauss's law.

- 15.1 Properties of Electric Charges
- 15.2 Insulators and Conductors
- 15.3 Coulomb's Law
- 15.4 The Electric Field
- 15.5 Electric Field Lines
- 15.6 Conductors in Electrostatic Equilibrium
- 15.7 The Millikan Oil-Drop Experiment
- 15.8 The Van de Graaff Generator
- 15.9 Electric Flux and Gauss's Law

524 CHAPTER 15 | Electric Forces and Electric Fields

15.1 Properties of Electric Charges

LEARNING OBJECTIVES

1. Define the SI unit of electric charge and identify the basic carriers of positive and negative charge.
2. Discuss the concept of charge conservation and the role of experiments in early studies of electricity.

Benjamin Franklin
(1706–1790)
Franklin was a printer, author, physical scientist, inventor, diplomat, and a founding father of the United States. His work on electricity in the late 1740s changed a jumbled, unrelated set of observations into a coherent science.

After running a plastic comb through your hair, you will find that the comb attracts bits of paper. The attractive force is often strong enough to suspend the paper from the comb, defying the gravitational pull of the entire Earth. The same effect occurs with other rubbed materials, such as glass and hard rubber.

Another simple experiment is to rub an inflated balloon against wool (or across your hair). On a dry day, the rubbed balloon will then stick to the wall of a room, often for hours. These materials have become electrically charged. You can give your body an electric charge by vigorously rubbing your shoes on a wool rug or by sliding across a car seat. You can then surprise and annoy a friend or coworker with a light touch on the arm, delivering a slight shock to both yourself and your victim. (If the coworker is your boss, don't expect a promotion!) These experiments work best on a dry day because excessive moisture can facilitate a leaking away of the charge.

Experiments also demonstrate that there are two kinds of electric charge, which Benjamin Franklin (1706–1790) named **positive** and **negative.** Figure 15.1 illustrates the interaction of the two charges. A hard rubber (or plastic) rod that has been rubbed with fur is suspended by a piece of string. When a glass rod that has been rubbed with silk is brought near the rubber rod, the rubber rod is attracted toward the glass rod (Fig. 15.1a). If two charged rubber rods (or two charged glass rods) are brought near each other, as in Figure 15.1b, the force between them is repulsive. These observations may be explained by assuming the rubber and glass rods have acquired different kinds of excess charge. We use the convention suggested by Franklin, where the excess electric charge on the glass rod is called positive and that on the rubber rod is called negative. On the basis of such observations, we conclude that **like charges repel one another and unlike charges attract one another.** Objects usually contain equal amounts of positive and negative charge; electrical forces between objects arise when those objects have net negative or positive charges.

▶ Like charges repel; unlike charges attract.

Nature's basic carriers of positive charge are protons, which, along with neutrons, are located in the nuclei of atoms. The nucleus, about 10^{-15} m in radius, is surrounded by a cloud of negatively charged electrons with a radius about ten

Figure 15.1 An experimental setup for observing the electrical force between two charged objects.

thousand times larger. An electron has the same magnitude charge as a proton, but the opposite sign. In a gram of matter there are approximately 10^{23} positively charged protons and just as many negatively charged electrons, so the net charge is zero. Because the nucleus of an atom is held firmly in place inside a solid, protons never move from one material to another. Electrons are far lighter than protons and hence more easily accelerated by forces. Further, they occupy the outer regions of the atom. Consequently, objects become charged by gaining or losing electrons.

Charge transfers readily from one type of material to another. Rubbing the two materials together serves to increase the area of contact, facilitating the transfer process.

An important characteristic of charge is that **electric charge is always conserved.** Charge isn't *created* when two neutral objects are rubbed together; rather, the objects become charged because **negative charge is transferred from one object to the other.** One object gains a negative charge while the other loses an equal amount of negative charge and hence is left with a net positive charge. When a glass rod is rubbed with silk, as in Figure 15.2, electrons are transferred from the rod to the silk. As a result, the glass rod carries a net positive charge, the silk a net negative charge. Likewise, when rubber is rubbed with fur, electrons are transferred from the fur to the rubber.

In 1909 Robert Millikan (1886–1953) discovered that if an object is charged, its charge is always a multiple of a fundamental unit of charge, designated by the symbol e. In modern terms, the charge is said to be **quantized,** meaning that charge occurs in discrete chunks that can't be further subdivided. An object may have a charge of $\pm e$, $\pm 2e$, $\pm 3e$, and so on, but never[1] a fractional charge of $\pm 0.5e$ or $\pm 0.22e$. Other experiments in Millikan's time showed that the electron has a charge of $-e$ and the proton has an equal and opposite charge of $+e$. Some particles, such as a neutron, have no net charge. A neutral atom (an atom with no net charge) contains as many protons as electrons. The value of e is now known to be 1.60219×10^{-19} C. (The SI unit of electric charge is the **coulomb,** or C.)

◀ Charge is conserved

Each (negatively-charged) electron transferred from the rod to the silk leaves an equal positive charge on the rod.

Figure 15.2 When a glass rod is rubbed with silk, electrons are transferred from the glass to the silk. Because the charges are transferred in discrete bundles, the charges on the two objects are $\pm e$, $\pm 2e$, $\pm 3e$, and so on.

15.2 Insulators and Conductors

LEARNING OBJECTIVES

1. Describe conductors, insulators, and semi-conductors on the basis of their relative abilities to conduct electric charge.
2. Describe the physical processes of polarization and charging by conduction or induction.

Substances can be classified in terms of their ability to conduct electric charge.

In **conductors,** electric charges move freely in response to an electric force. All other materials are called **insulators.**

Glass and rubber are insulators. When such materials are charged by rubbing, only the rubbed area becomes charged, and there is no tendency for the charge to move into other regions of the material. In contrast, materials such as copper, aluminum, and silver are good conductors. When such materials are charged in some small region, the charge readily distributes itself over the entire surface of the material. If you hold a copper rod in your bare hand and rub the rod with wool or fur, it will not attract a piece of paper. This might suggest that a metal can't be charged. However, if you hold the copper rod with an insulator and then rub it with wool or fur, the rod remains charged and attracts the paper. In the first case,

---

[1]There is strong evidence for the existence of fundamental particles called **quarks** that have charges of $\pm e/3$ or $\pm 2e/3$. The charge is *still* quantized, but in units of $\pm e/3$ rather than $\pm e$. A more complete discussion of quarks and their properties is presented in Chapter 30.

a) Before contact, the negative rod repels the sphere's electrons, inducing a local positive charge.

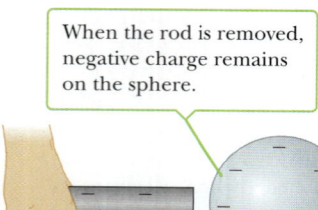

b) After contact, electrons from the rod flow onto the sphere, neutralizing the local positive charges.

c) When the rod is removed, negative charge remains on the sphere.

Figure 15.3 Charging a metallic object by conduction.

the electric charges produced by rubbing readily move from the copper through your body and finally to ground. In the second case, the insulating handle prevents the flow of charge to ground.

Semiconductors are a third class of materials, and their electrical properties are somewhere between those of insulators and those of conductors. Silicon and germanium are well-known semiconductors that are widely used in the fabrication of a variety of electronic devices.

Charging by Conduction

Consider a negatively charged rubber rod brought into contact with an insulated neutral conducting sphere. The excess electrons on the rod repel electrons on the sphere, creating local positive charges on the neutral sphere. On contact, some electrons on the rod are now able to move onto the sphere, as in Figure 15.3, neutralizing the positive charges. When the rod is removed, the sphere is left with a net negative charge. This process is referred to as charging by **conduction.** The object being charged in such a process (the sphere) is always left with a charge having the same sign as the object doing the charging (the rubber rod).

Charging by Induction

An object connected to a conducting wire or copper pipe buried in the Earth is said to be **grounded.** The Earth can be considered an infinite reservoir for electrons; in effect, it can accept or supply an unlimited number of electrons. With this idea in mind, we can understand the charging of a conductor by a process known as **induction.**

Consider a negatively charged rubber rod brought near a neutral (uncharged) conducting sphere that is insulated, so there is no conducting path to ground (Fig. 15.4). Initially the sphere is electrically neutral (Fig. 15.4a). When the negatively charged rod is brought close to the sphere, the repulsive force between the electrons in the rod and those in the sphere causes some electrons to move to the side of the sphere farthest away from the rod (Fig. 15.4b). The region of the sphere nearest the negatively charged rod has an excess of positive charge because of the migration of electrons away from that location. If a grounded conducting wire is then connected to the sphere, as in Figure 15.4c, some of the electrons leave the sphere and travel to ground. If the wire to ground is then removed (Fig. 15.4d), the conducting sphere is left with an excess of induced positive charge. Finally, when the rubber rod is removed from the vicinity of the sphere (Fig. 15.4e), the induced positive charge remains on the ungrounded sphere. Even though the positively charged atomic nuclei remain fixed, this excess positive charge becomes uniformly distributed over the surface of the ungrounded sphere because of the repulsive forces among the like charges and the high mobility of electrons in a metal.

In the process of inducing a charge on the sphere, the charged rubber rod doesn't lose any of its negative charge because it never comes in contact with the sphere. Further, the sphere is left with a charge opposite that of the rubber rod. **Charging an object by induction requires no contact with the object inducing the charge.**

A process similar to charging by induction in conductors also takes place in insulators. In most neutral atoms or molecules, the center of positive charge coincides with the center of negative charge. In the presence of a charged object, however, these centers may separate slightly, resulting in more positive charge on one side of the molecule than on the other. This effect is known as **polarization.** The realignment of charge within individual molecules produces an induced charge on the surface of the insulator, as shown in Figure 15.5a. This property explains why a balloon charged through rubbing will stick to an electrically neutral wall or why the comb you just used on your hair attracts tiny bits of neutral paper.

■ Quick Quiz

15.1 A suspended object A is attracted to a neutral wall. It's also attracted to a positively charged object B. Which of the following is true about object A? (a) It is uncharged. (b) It has a negative charge. (c) It has a positive charge. (d) It may be either charged or uncharged.

15.3 Coulomb's Law

LEARNING OBJECTIVES

1. State Coulomb's law and summarize the properties of the electric force between two charged particles.
2. Apply Coulomb's law and the superposition principle to systems of charged particles.

In 1785 Charles Coulomb (1736–1806) experimentally established the fundamental law of electric force between two stationary charged particles.

An **electric force** has the following properties:

1. It is directed along a line joining the two particles and is inversely proportional to the square of the separation distance r, between them.
2. It is proportional to the product of the magnitudes of the charges, $|q_1|$ and $|q_2|$, of the two particles.
3. It is attractive if the charges are of opposite sign and repulsive if the charges have the same sign.

From these observations, Coulomb proposed the following mathematical form for the electric force between two charges:

The magnitude of the electric force F between charges q_1 and q_2 separated by a distance r is given by

$$F = k_e \frac{|q_1||q_2|}{r^2} \quad [15.1]$$

where k_e is a constant called the *Coulomb constant*.

Figure 15.5 (a) A charged balloon is brought near an insulating wall. (b) A charged rod is brought close to bits of paper.

The neutral sphere has equal numbers of positive and negative charges.

Electrons redistribute when a charged rod is brought close.

Some electrons leave the grounded sphere through the ground wire.

The excess positive charge is nonuniformly distributed.

The remaining electrons redistribute uniformly, and there is a net uniform distribution of positive charge on the sphere's surface.

Figure 15.4 Charging a metallic object by induction. (a) A neutral metallic sphere. (b) A charged rubber rod is placed near the sphere. (c) The sphere is grounded. (d) The ground connection is removed. (e) The rod is removed.

528 CHAPTER 15 | Electric Forces and Electric Fields

Charles Coulomb
(1736–1806)
Coulomb's major contribution to science was in the field of electrostatics and magnetism. During his lifetime, he also investigated the strengths of materials and identified the forces that affect objects on beams, thereby contributing to the field of structural mechanics.

Table 15.1 Charge and Mass of the Electron, Proton, and Neutron

Particle	Charge (C)	Mass (kg)
Electron	-1.60×10^{-19}	9.11×10^{-31}
Proton	$+1.60 \times 10^{-19}$	1.67×10^{-27}
Neutron	0	1.67×10^{-27}

Equation 15.1, known as **Coulomb's law,** applies exactly only to point charges and to spherical distributions of charges, in which case r is the distance between the two centers of charge. Electric forces between unmoving charges are called *electrostatic* forces. Moving charges, in addition, create magnetic forces, studied in Chapter 19.

The value of the Coulomb constant in Equation 15.1 depends on the choice of units. The SI unit of charge is the **coulomb** (C). From experiment, we know that the **Coulomb constant** in SI units has the value

$$k_e = 8.9875 \times 10^9 \text{ N} \cdot \text{m}^2/\text{C}^2 \quad [15.2]$$

This number can be rounded, depending on the accuracy of other quantities in a given problem. We'll use either two or three significant digits, as usual.

The charge on the proton has a magnitude of $e = 1.6 \times 10^{-19}$ C. Therefore, it would take $1/e = 6.3 \times 10^{18}$ protons to create a total charge of $+1.0$ C. Likewise, 6.3×10^{18} electrons would have a total charge of -1.0 C. Compare this charge with the number of free electrons in 1 cm^3 of copper, which is on the order of 10^{23}. Even so, 1.0 C is a very large amount of charge. In typical electrostatic experiments in which a rubber or glass rod is charged by friction, there is a net charge on the order of 10^{-6} C (= 1 μC). Only a very small fraction of the total available charge is transferred between the rod and the rubbing material. Table 15.1 lists the charges and masses of the electron, proton, and neutron.

When using Coulomb's force law, remember that force is a vector quantity and must be treated accordingly. Figure 15.6a shows the electric force of repulsion between two positively charged particles. Like other forces, electric forces obey Newton's third law; hence, the forces $\vec{F}_{12}$ and $\vec{F}_{21}$ are equal in magnitude but opposite in direction. (The notation $\vec{F}_{12}$ denotes the force exerted by particle 1 on particle 2; likewise, $\vec{F}_{21}$ is the force exerted by particle 2 on particle 1.) From Newton's third law, F_{12} and F_{21} are always equal regardless of whether q_1 and q_2 have the same magnitude.

■ **Quick Quiz**

15.2 Object A has a charge of $+2$ μC, and object B has a charge of $+6$ μC. Which statement is true?
(a) $\vec{F}_{AB} = -3\vec{F}_{BA}$ (b) $\vec{F}_{AB} = -\vec{F}_{BA}$ (c) $3\vec{F}_{AB} = -\vec{F}_{BA}$

Figure 15.6 Two point charges separated by a distance r exert a force on each other given by Coulomb's law. The force on q_1 is equal in magnitude and opposite in direction to the force on q_2.

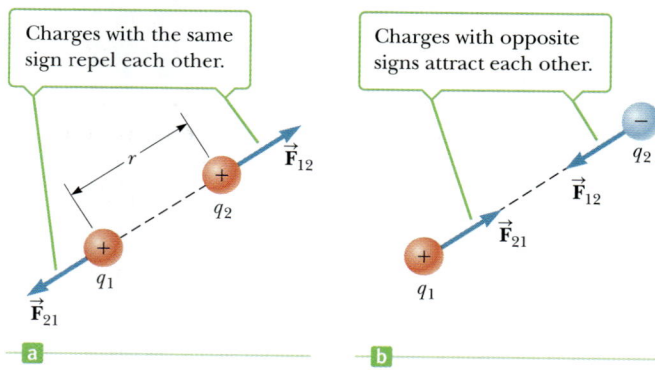

15.3 | Coulomb's Law

The Coulomb force is similar to the gravitational force. Both act at a distance without direct contact. Both are inversely proportional to the distance squared, with the force directed along a line connecting the two bodies. The mathematical form is the same, with the masses m_1 and m_2 in Newton's law replaced by q_1 and q_2 in Coulomb's law and with Newton's constant G replaced by Coulomb's constant k_e. There are two important differences: (1) electric forces can be either attractive or repulsive, but gravitational forces are always attractive, and (2) the electric force between charged elementary particles is far stronger than the gravitational force between the same particles, as the next example shows.

■ EXAMPLE 15.1 | Forces in a Hydrogen Atom

GOAL Contrast the magnitudes of an electric force and a gravitational force.

PROBLEM The electron and proton of a hydrogen atom are separated (on the average) by a distance of about 5.3×10^{-11} m. **(a)** Find the magnitudes of the electric force and the gravitational force that each particle exerts on the other, and the ratio of the electric force F_e to the gravitational force F_g. **(b)** Compute the acceleration caused by the electric force of the proton on the electron. Repeat for the gravitational acceleration.

STRATEGY Solving this problem is just a matter of substituting known quantities into the two force laws and then finding the ratio.

SOLUTION

(a) Compute the magnitudes of the electric and gravitational forces, and find the ration F_e/F_g.

Substitute $|q_1| = |q_2| = e$ and the distance into Coulomb's law to find the electric force:

$$F_e = k_e \frac{|e|^2}{r^2} = \left(8.99 \times 10^9 \, \frac{\text{N} \cdot \text{m}^2}{\text{C}^2}\right) \frac{(1.6 \times 10^{-19} \, \text{C})^2}{(5.3 \times 10^{-11} \, \text{m})^2}$$

$$= \boxed{8.2 \times 10^{-8} \, \text{N}}$$

Substitute the masses and distance into Newton's law of gravity to find the gravitational force:

$$F_g = G \frac{m_e m_p}{r^2}$$

$$= \left(6.67 \times 10^{-11} \, \frac{\text{N} \cdot \text{m}^2}{\text{kg}^2}\right) \frac{(9.11 \times 10^{-31} \, \text{kg})(1.67 \times 10^{-27} \, \text{kg})}{(5.3 \times 10^{-11} \, \text{m})^2}$$

$$= \boxed{3.6 \times 10^{-47} \, \text{N}}$$

Find the ratio of the two forces:

$$\frac{F_e}{F_g} = 2.3 \times 10^{39}$$

(b) Compute the acceleration of the electron caused by the electric force. Repeat for the gravitational acceleration.

Use Newton's second law and the electric force found in part (a):

$$m_e a_e = F_e \quad \rightarrow \quad a_e = \frac{F_e}{m_e} = \frac{8.2 \times 10^{-8} \, \text{N}}{9.11 \times 10^{-31} \, \text{kg}} = \boxed{9.0 \times 10^{22} \, \text{m/s}^2}$$

Use Newton's second law and the gravitational force found in part (a):

$$m_e a_g = F_g \quad \rightarrow \quad a_g = \frac{F_g}{m_e} = \frac{3.6 \times 10^{-47} \, \text{N}}{9.11 \times 10^{-31} \, \text{kg}} = \boxed{4.0 \times 10^{-17} \, \text{m/s}^2}$$

REMARKS The gravitational force between the charged constituents of the atom is negligible compared with the electric force between them. The electric force is so strong, however, that any net charge on an object quickly attracts nearby opposite charges, neutralizing the object. As a result, gravity plays a greater role in the mechanics of moving objects in everyday life.

QUESTION 15.1 If the distance between two charges is doubled, by what factor is the magnitude of the electric force changed?

(Continued)

EXERCISE 15.1 (a) Find the magnitude of the electric force between two protons separated by 1 femtometer (10^{-15} m), approximately the distance between two protons in the nucleus of a helium atom. (b) If the protons were not held together by the strong nuclear force, what would be their initial acceleration due to the electric force between them?

ANSWERS (a) 2×10^2 N (b) 1×10^{29} m/s^2

The Superposition Principle

When a number of separate charges act on the charge of interest, each exerts an electric force. These electric forces can all be computed separately, one at a time, then added as vectors. This is another example of the **superposition principle**. The following example illustrates this procedure in one dimension.

■ EXAMPLE 15.2 Finding Electrostatic Equilibrium

GOAL Apply Coulomb's law in one dimension.

PROBLEM Three charges lie along the x-axis as in Figure 15.7. The positive charge $q_1 = 15$ μC is at $x = 2.0$ m, and the positive charge $q_2 = 6.0$ μC is at the origin. Where must a *negative* charge q_3 be placed on the x-axis so that the resultant electric force on it is zero?

STRATEGY If q_3 is to the right or left of the other two charges, the net force on q_3 can't be zero because then $\vec{F}_{13}$ and $\vec{F}_{23}$ act in the same direction. Consequently, q_3 must lie between the two other charges. Write $\vec{F}_{13}$ and $\vec{F}_{23}$ in terms of the unknown coordinate position x, then sum them and set them equal to zero, solving for the unknown. The solution can be obtained with the quadratic formula.

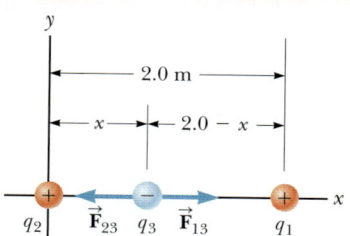

Figure 15.7 (Example 15.2) Three point charges are placed along the x-axis. The charge q_3 is negative, whereas q_1 and q_2 are positive. If the resultant force on q_3 is zero, the force $\vec{F}_{13}$ exerted by q_1 on q_3 must be equal in magnitude and opposite the force $\vec{F}_{23}$ exerted by q_2 on q_3.

SOLUTION

Write the x-component of $\vec{F}_{13}$:

$$F_{13x} = +k_e \frac{(15 \times 10^{-6} \text{ C})|q_3|}{(2.0 \text{ m} - x)^2}$$

Write the x-component of $\vec{F}_{23}$:

$$F_{23x} = -k_e \frac{(6.0 \times 10^{-6} \text{ C})|q_3|}{x^2}$$

Set the sum equal to zero:

$$k_e \frac{(15 \times 10^{-6} \text{ C})|q_3|}{(2.0 \text{ m} - x)^2} - k_e \frac{(6.0 \times 10^{-6} \text{ C})|q_3|}{x^2} = 0$$

Cancel k_e, 10^{-6}, and q_3 from the equation and rearrange terms (explicit significant figures and units are temporarily suspended for clarity):

(1) $\quad 6(2-x)^2 = 15x^2$

Put this equation into standard quadratic form, $ax^2 + bx + c = 0$:

$$6(4 - 4x + x^2) = 15x^2 \;\rightarrow\; 2(4 - 4x + x^2) = 5x^2$$
$$3x^2 + 8x - 8 = 0$$

Apply the quadratic formula:

$$x = \frac{-8 \pm \sqrt{64 - (4)(3)(-8)}}{2 \cdot 3} = \frac{-4 \pm 2\sqrt{10}}{3}$$

Only the positive root makes sense:

$$x = \boxed{0.77 \text{ m}}$$

REMARKS Notice that physical reasoning was required to choose between the two possible answers for x, which is nearly always the case when quadratic equations are involved. Use of the quadratic formula could have been avoided by taking the square root of both sides of Equation (1); however, this shortcut is often unavailable.

QUESTION 15.2 If q_1 has the same magnitude as before but is negative, in what region along the x-axis would it be possible for the net electric force on q_3 to be zero? (a) $x < 0$ (b) $0 < x < 2$ m (c) 2 m $< x$

EXERCISE 15.2 Three charges lie along the x-axis. A positive charge $q_1 = 10.0$ μC is at $x = 1.00$ m, and a negative charge $q_2 = -2.00$ μC is at the origin. Where must a positive charge q_3 be placed on the x-axis so that the resultant force on it is zero?

ANSWER $x = -0.809$ m

EXAMPLE 15.3 | A Charge Triangle

GOAL Apply Coulomb's law in two dimensions.

PROBLEM Consider three point charges at the corners of a triangle, as shown in Figure 15.8, where $q_1 = 6.00 \times 10^{-9}$ C, $q_2 = -2.00 \times 10^{-9}$ C, and $q_3 = 5.00 \times 10^{-9}$ C. **(a)** Find the components of the force $\vec{F}_{23}$ exerted by q_2 on q_3. **(b)** Find the components of the force $\vec{F}_{13}$ exerted by q_1 on q_3. **(c)** Find the resultant force on q_3, in terms of components and also in terms of magnitude and direction.

STRATEGY Coulomb's law gives the magnitude of each force, which can be split with right-triangle trigonometry into x- and y-components. Sum the vectors componentwise and then find the magnitude and direction of the resultant vector.

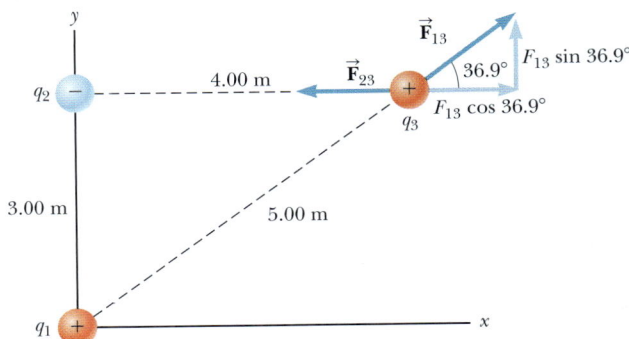

Figure 15.8 (Example 15.3) The force exerted by q_1 on q_3 is $\vec{F}_{13}$. The force exerted by q_2 on q_3 is $\vec{F}_{23}$. The *resultant force* $\vec{F}_3$ exerted on q_3 is the *vector* sum $\vec{F}_{13} + \vec{F}_{23}$.

SOLUTION
(a) Find the components of the force exerted by q_2 on q_3.
Find the magnitude of $\vec{F}_{23}$ with Coulomb's law:

$$F_{23} = k_e \frac{|q_2||q_3|}{r^2}$$

$$= (8.99 \times 10^9 \text{ N} \cdot \text{m}^2/\text{C}^2) \frac{(2.00 \times 10^{-9} \text{ C})(5.00 \times 10^{-9} \text{ C})}{(4.00 \text{ m})^2}$$

$$F_{23} = 5.62 \times 10^{-9} \text{ N}$$

Because $\vec{F}_{23}$ is horizontal and points in the negative x-direction, the negative of the magnitude gives the x-component, and the y-component is zero:

$F_{23x} = -5.62 \times 10^{-9}$ N
$F_{23y} = 0$

(b) Find the components of the force exerted by q_1 on q_3.
Find the magnitude of $\vec{F}_{13}$:

$$F_{13} = k_e \frac{|q_1||q_3|}{r^2}$$

$$= (8.99 \times 10^9 \text{ N} \cdot \text{m}^2/\text{C}^2) \frac{(6.00 \times 10^{-9} \text{ C})(5.00 \times 10^{-9} \text{ C})}{(5.00 \text{ m})^2}$$

$$F_{13} = 1.08 \times 10^{-8} \text{ N}$$

Use the given triangle to find the components of $\vec{F}_{13}$:

$F_{13x} = F_{13} \cos \theta = (1.08 \times 10^{-8} \text{ N}) \cos (36.9°)$
$= 8.64 \times 10^{-9}$ N

$F_{13y} = F_{13} \sin \theta = (1.08 \times 10^{-8} \text{ N}) \sin (36.9°)$
$= 6.48 \times 10^{-9}$ N

(c) Find the components of the resultant vector.
Sum the x-components to find the resultant F_x:

$F_x = -5.62 \times 10^{-9}$ N $+ 8.64 \times 10^{-9}$ N
$= 3.02 \times 10^{-9}$ N

(Continued)

Sum the y-components to find the resultant F_y:

$$F_y = 0 + 6.48 \times 10^{-9} \text{ N} = \boxed{6.48 \times 10^{-9} \text{ N}}$$

Find the magnitude of the resultant force on the charge q_3, using the Pythagorean theorem:

$$|\vec{F}| = \sqrt{F_x^2 + F_y^2}$$
$$= \sqrt{(3.02 \times 10^{-9} \text{ N})^2 + (6.48 \times 10^{-9} \text{ N})^2}$$
$$= \boxed{7.15 \times 10^{-9} \text{ N}}$$

Find the angle the resultant force makes with respect to the positive x-axis:

$$\theta = \tan^{-1}\left(\frac{F_y}{F_x}\right) = \tan^{-1}\left(\frac{6.48 \times 10^{-9} \text{ N}}{3.02 \times 10^{-9} \text{ N}}\right) = \boxed{65.0°}$$

REMARKS The methods used here are just like those used with Newton's law of gravity in two dimensions.

QUESTION 15.3 Without actually calculating the electric force on q_2, determine the quadrant into which the electric force vector points.

EXERCISE 15.3 Using the same triangle, find the vector components of the electric force on q_1 and the vector's magnitude and direction.

ANSWERS $F_x = -8.64 \times 10^{-9}$ N, $F_y = 5.52 \times 10^{-9}$ N, $F = 1.03 \times 10^{-8}$ N, $\theta = 147°$

15.4 The Electric Field

LEARNING OBJECTIVE

1. Define the electric field and apply it to systems of charged particles.

The gravitational force and the electrostatic force are both capable of acting through space, producing an effect even when there isn't any physical contact between the objects involved. Field forces can be discussed in a variety of ways, but an approach developed by Michael Faraday (1791–1867) is the most practical. In this approach an **electric field** is said to exist in the region of space around a charged object. The electric field exerts an electric force on any other charged object within the field. This differs from the Coulomb's law concept of a force exerted at a distance in that the force is now exerted by something—the field—that is in the same location as the charged object.

Figure 15.9 shows an object with a small positive charge q_0 placed near a second object with a much larger positive charge Q.

The electric field $\vec{E}$ produced by a charge Q at the location of a small "test" charge q_0 is defined as the electric force $\vec{F}$ exerted by Q on q_0 divided by the test charge q_0:

$$\vec{E} \equiv \frac{\vec{F}}{q_0} \qquad [15.3]$$

SI unit: newton per coulomb (N/C)

Conceptually and experimentally, the test charge q_0 is required to be very small (arbitrarily small, in fact), so it doesn't cause any significant rearrangement of the charge creating the electric field $\vec{E}$. Mathematically, however, the size of the test charge makes no difference: the calculation comes out the same, regardless. In view of this, using $q_0 = 1$ C in Equation 15.3 can be convenient if not rigorous.

When a positive test charge is used, the electric field always has the same direction as the electric force on the test charge, which follows from Equation 15.3. Hence, in Figure 15.9, the direction of the electric field is horizontal and to the right. The electric field at point A in Figure 15.10a is vertical and downward because at that point a positive test charge would be attracted toward the negatively charged sphere.

Source charge / Test charge

Figure 15.9 A small object with a positive charge q_0 placed near an object with a larger positive charge Q is subject to an electric field $\vec{E}$ directed as shown. The magnitude of the electric field at the location of q_0 is defined as the electric force on q_0 divided by the charge q_0.

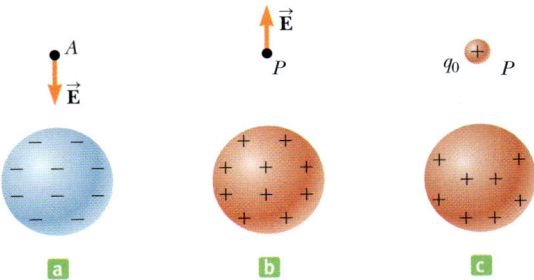

Figure 15.10 (a) The electric field at A due to the negatively charged sphere is downward, toward the negative charge. (b) The electric field at P due to the positively charged conducting sphere is upward, away from the positive charge. (c) A test charge q_0 placed at P will cause a rearrangement of charge on the sphere unless q_0 is negligibly small compared with the charge on the sphere.

Once the electric field due to a given arrangement of charges is known at some point, the force on *any* particle with charge q placed at that point can be calculated from a rearrangement of Equation 15.3:

$$\vec{F} = q\vec{E} \quad [15.4]$$

Here q_0 has been replaced by q, which need not be a mere test charge.

As shown in Figure 15.11, the direction of $\vec{E}$ is the direction of the force that acts on a positive test charge q_0 placed in the field. We say that **an electric field exists at a point if a test charge at that point is subject to an electric force.**

Consider a point charge q located a distance r from a test charge q_0. According to Coulomb's law, the *magnitude* of the electric force of the charge q on the test charge is

$$F = k_e \frac{|q||q_0|}{r^2} \quad [15.5]$$

Because the magnitude of the electric field at the position of the test charge is defined as $E = F/q_0$, we see that the *magnitude* of the electric field due to the charge q at the position of q_0 is

$$E = k_e \frac{|q|}{r^2} \quad [15.6]$$

Equation 15.6 points out an important property of electric fields that makes them useful quantities for describing electrical phenomena. As the equation indicates, an electric field at a given point depends only on the charge q on the object setting up the field and the distance r from that object to a specific point in space. As a result, we can say that an electric field exists at point P in Figure 15.11 whether or not there is a test charge at P.

The principle of superposition holds when the electric field due to a group of point charges is calculated. We first use Equation 15.6 to calculate the electric field produced by each charge individually at a point and then add the electric fields together as vectors.

It's also important to exploit any symmetry of the charge distribution. For example, if equal charges are placed at $x = a$ and at $x = -a$, the electric field is zero at the origin, by symmetry. Similarly, if the x-axis has a uniform distribution of positive charge, it can be guessed by symmetry that the electric field points away from the x axis and is zero parallel to that axis.

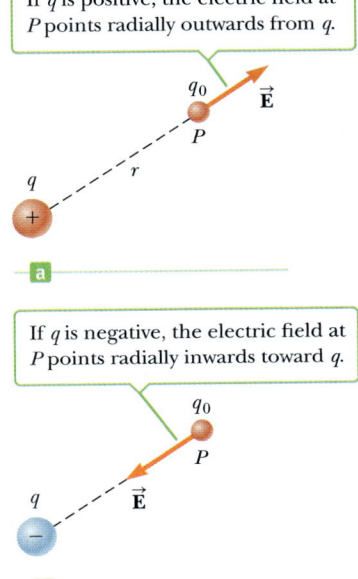

Figure 15.11 A test charge q_0 at P is a distance r from a point charge q.

Quick Quiz

15.3 A test charge of $+3\ \mu C$ is at a point P where the electric field due to other charges is directed to the right and has a magnitude of 4×10^6 N/C. If the test charge is replaced with a charge of $-3\ \mu C$, the electric field at P (a) has the same magnitude as before, but changes direction, (b) increases in magnitude and changes direction, (c) remains the same, or (d) decreases in magnitude and changes direction.

15.4 A circular ring of charge of radius b has a total charge q uniformly distributed around it. Find the magnitude of the electric field in the center of the ring.
(a) 0   (b) $k_e q/b^2$ (c) $k_e q^2/b^2$ (d) $k_e q^2/b$ (e) None of these answers is correct.

15.5 A "free" electron and a "free" proton are placed in an identical electric field. Which of the following statements are true? (a) Each particle is acted upon by the same electric force and has the same acceleration. (b) The electric force on the proton is greater in magnitude than the electric force on the electron, but in the opposite direction. (c) The electric force on the proton is equal in magnitude to the electric force on the electron, but in the opposite direction. (d) The magnitude of the acceleration of the electron is greater than that of the proton. (e) Both particles have the same acceleration.

EXAMPLE 15.4 Electrified Oil

GOAL Use electric forces and fields together with Newton's second law in a one-dimensional problem.

PROBLEM Tiny droplets of oil acquire a small negative charge while dropping through a vacuum (pressure = 0) in an experiment. An electric field of magnitude 5.92×10^4 N/C points straight down. **(a)** One particular droplet is observed to remain suspended against gravity. If the mass of the droplet is 2.93×10^{-15} kg, find the charge carried by the droplet. **(b)** Another droplet of the same mass falls 10.3 cm from rest in 0.250 s, again moving through a vacuum. Find the charge carried by the droplet.

STRATEGY We use Newton's second law with both gravitational and electric forces. In both parts the electric field $\vec{E}$ is pointing down, taken as the negative direction, as usual. In part (a) the acceleration is equal to zero. In part (b) the acceleration is uniform, so the kinematic equations yield the acceleration. Newton's law can then be solved for q.

SOLUTION

(a) Find the charge on the suspended droplet.

Apply Newton's second law to the droplet in the vertical direction:

$$(1) \quad ma = \sum F = -mg + Eq$$

E points downward, hence is negative. Set $a = 0$ in Equation (1) and solve for q:

$$q = \frac{mg}{E} = \frac{(2.93 \times 10^{-15} \text{ kg})(9.80 \text{ m/s}^2)}{-5.92 \times 10^4 \text{ N/C}}$$

$$= \boxed{-4.85 \times 10^{-19} \text{ C}}$$

(b) Find the charge on the falling droplet.

Use the kinematic displacement equation to find the acceleration:

$$\Delta y = \tfrac{1}{2} a t^2 + v_0 t$$

Substitute $\Delta y = -0.103$ m, $t = 0.250$ s, and $v_0 = 0$:

$$-0.103 \text{ m} = \tfrac{1}{2} a (0.250 \text{ s})^2 \quad \rightarrow \quad a = -3.30 \text{ m/s}^2$$

Solve Equation (1) for q and substitute:

$$q = \frac{m(a + g)}{E}$$

$$= \frac{(2.93 \times 10^{-15} \text{ kg})(-3.30 \text{ m/s}^2 + 9.80 \text{ m/s}^2)}{-5.92 \times 10^4 \text{ N/C}}$$

$$= \boxed{-3.22 \times 10^{-19} \text{ C}}$$

REMARKS This example exhibits features similar to the Millikan Oil-Drop experiment discussed in Section 15.7, which determined the value of the fundamental electric charge e. Notice that in both parts of the example, the charge is very nearly a multiple of e.

QUESTION 15.4 What would be the acceleration of the oil droplet in part (a) if the electric field suddenly reversed direction without changing in magnitude?

EXERCISE 15.4 Suppose a droplet of unknown mass remains suspended against gravity when $E = -2.70 \times 10^5$ N/C. What is the minimum mass of the droplet?

ANSWER 4.41×10^{-15} kg

■ PROBLEM-SOLVING STRATEGY

Calculating Electric Forces and Fields

The following procedure is used to calculate electric forces. The same procedure can be used to calculate an electric field, a simple matter of replacing the charge of interest, q, with a convenient test charge and dividing by the test charge at the end:

1. **Draw** a diagram of the charges in the problem.
2. **Identify** the charge of interest, q, and circle it.
3. **Convert all units** to SI, with charges in coulombs and distances in meters, so as to be consistent with the SI value of the Coulomb constant k_e.
4. **Apply Coulomb's law.** For each charge Q, find the electric force on the charge of interest, q. The magnitude of the force can be found using Coulomb's law. The vector direction of the electric force is along the line of the two charges, directed away from Q if the charges have the same sign, toward Q if the charges have the opposite sign. Find the angle θ this vector makes with the positive x-axis. The x-component of the electric force exerted by Q on q will be $F \cos \theta$, and the y-component will be $F \sin \theta$.
5. **Sum all the x-components,** getting the x-component of the resultant electric force.
6. **Sum all the y-components,** getting the y-component of the resultant electric force.
7. **Use the Pythagorean theorem and trigonometry** to find the magnitude and direction of the resultant force if desired.

■ EXAMPLE 15.5 | Electric Field Due to Two Point Charges

GOAL Use the superposition principle to calculate the electric field due to two point charges.

PROBLEM Charge $q_1 = 7.00 \ \mu C$ is at the origin, and charge $q_2 = -5.00 \ \mu C$ is on the x-axis, 0.300 m from the origin (Fig. 15.12). **(a)** Find the magnitude and direction of the electric field at point P, which has coordinates (0, 0.400) m. **(b)** Find the force on a charge of 2.00×10^{-8} C placed at P.

STRATEGY Follow the problem-solving strategy, finding the electric field at point P due to each individual charge in terms of x- and y-components, then adding the components of each type to get the x- and y-components of the resultant electric field at P. The magnitude of the force in part (b) can be found by simply multiplying the magnitude of the electric field by the charge.

Figure 15.12 (Example 15.5) The resultant electric field $\vec{E}$ at P equals the vector sum $\vec{E}_1 + \vec{E}_2$, where $\vec{E}_1$ is the field due to the positive charge q_1 and $\vec{E}_2$ is the field due to the negative charge q_2.

SOLUTION

(a) Calculate the electric field at P.

Find the magnitude of $\vec{E}_1$ with Equation 15.6:

$$E_1 = k_e \frac{|q_1|}{r_1^2} = (8.99 \times 10^9 \ \text{N} \cdot \text{m}^2/\text{C}^2) \frac{(7.00 \times 10^{-6} \ \text{C})}{(0.400 \ \text{m})^2}$$

$$= 3.93 \times 10^5 \ \text{N/C}$$

The vector $\vec{E}_1$ is vertical, making an angle of 90° with respect to the positive x-axis. Use this fact to find its components:

$$E_{1x} = E_1 \cos(90°) = 0$$
$$E_{1y} = E_1 \sin(90°) = 3.93 \times 10^5 \ \text{N/C}$$

Next, find the magnitude of $\vec{E}_2$, again with Equation 15.6:

$$E_2 = k_e \frac{|q_2|}{r_2^2} = (8.99 \times 10^9 \ \text{N} \cdot \text{m}^2/\text{C}^2) \frac{(5.00 \times 10^{-6} \ \text{C})}{(0.500 \ \text{m})^2}$$

$$= 1.80 \times 10^5 \ \text{N/C}$$

(Continued)

Obtain the x-component of $\vec{E}_2$, using the triangle in Figure 15.12 to find $\cos\theta$:	$\cos\theta = \dfrac{\text{adj}}{\text{hyp}} = \dfrac{0.300}{0.500} = 0.600$ $E_{2x} = E_2 \cos\theta = (1.80 \times 10^5 \text{ N/C})(0.600)$ $\quad = 1.08 \times 10^5 \text{ N/C}$
Obtain the y-component in the same way, but a minus sign has to be provided for $\sin\theta$ because this component is directed downwards:	$\sin\theta = \dfrac{\text{opp}}{\text{hyp}} = \dfrac{0.400}{0.500} = 0.800$ $E_{2y} = E_2 \sin\theta = (1.80 \times 10^5 \text{ N/C})(-0.800)$ $\quad = -1.44 \times 10^5 \text{ N/C}$
Sum the x-components to get the x-component of the resultant vector:	$E_x = E_{1x} + E_{2x} = 0 + 1.08 \times 10^5 \text{ N/C} = 1.08 \times 10^5 \text{ N/C}$
Sum the y-components to get the y-component of the resultant vector:	$E_y = E_{1y} + E_{2y} = 3.93 \times 10^5 \text{ N/C} - 1.44 \times 10^5 \text{ N/C}$ $E_y = 2.49 \times 10^5 \text{ N/C}$
Use the Pythagorean theorem to find the magnitude of the resultant vector:	$E = \sqrt{E_x^2 + E_y^2} = \boxed{2.71 \times 10^5 \text{ N/C}}$
The inverse tangent function yields the direction of the resultant vector:	$\phi = \tan^{-1}\left(\dfrac{E_y}{E_x}\right) = \tan^{-1}\left(\dfrac{2.49 \times 10^5 \text{ N/C}}{1.08 \times 10^5 \text{ N/C}}\right) = \boxed{66.6°}$
(b) Find the force on a charge of 2.00×10^{-8} C placed at P.	
Calculate the magnitude of the force (the direction is the same as that of $\vec{E}$ because the charge is positive):	$F = Eq = (2.71 \times 10^5 \text{ N/C})(2.00 \times 10^{-8} \text{ C})$ $\quad = \boxed{5.42 \times 10^{-3} \text{ N}}$

REMARKS There were numerous steps to this problem, but each was very short. When attacking such problems, it's important to focus on one small step at a time. The solution comes not from a leap of genius, but from the assembly of a number of relatively easy parts.

QUESTION 15.5 Suppose q_2 were moved slowly to the right. What would happen to the angle ϕ?

EXERCISE 15.5 (a) Place a charge of -7.00 μC at point P and find the magnitude and direction of the electric field at the location of q_2 due to q_1 and the charge at P. (b) Find the magnitude and direction of the force on q_2.

ANSWERS (a) 5.84×10^5 N/C, $\phi = 20.2°$ (b) $F = 2.92$ N, $\phi = 200.°$

15.5 Electric Field Lines

LEARNING OBJECTIVES
1. Relate the electric field to electric field lines.
2. State the rules for drawing electric field lines and sketch field lines for simple charge configurations.

A convenient aid for visualizing electric field patterns is to draw lines pointing in the direction of the electric field vector at any point. These lines, introduced by Michael Faraday and called **electric field lines,** are related to the electric field in any region of space in the following way:

1. The electric field vector $\vec{E}$ is tangent to the electric field lines at each point.
2. The number of lines per unit area through a surface perpendicular to the lines is proportional to the strength of the electric field in a given region.

> **Tip 15.1 Electric Field Lines Aren't Paths of Particles**
> Electric field lines are *not* material objects. They are used only as a pictorial representation of the electric field at various locations. Except in special cases, they *do not* represent the path of a charged particle released in an electric field.

15.5 | Electric Field Lines

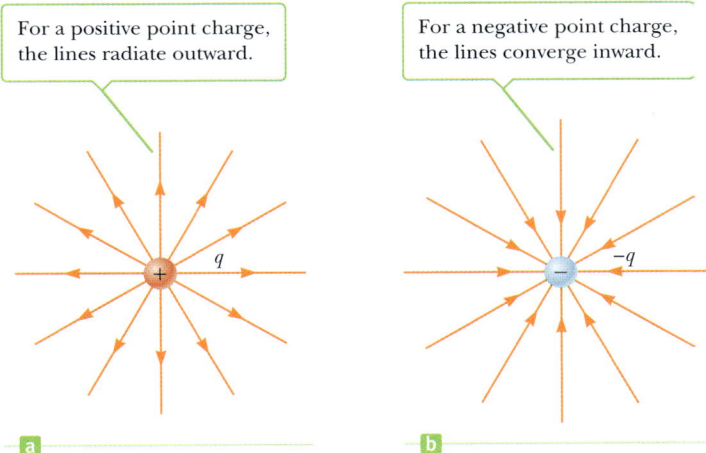

Figure 15.13 (a), (b) The electric field lines for a point charge. Notice that the figures show only those field lines that lie in the plate of the page.

Note that $\vec{\mathbf{E}}$ is large when the field lines are close together and small when the lines are far apart.

Figure 15.13a shows some representative electric field lines for a single positive point charge. This two-dimensional drawing contains only the field lines that lie in the plane containing the point charge. The lines are actually directed radially outward from the charge in *all* directions, somewhat like the quills of an angry porcupine. Because a positive test charge placed in this field would be repelled by the charge q, the lines are directed radially away from the positive charge. The electric field lines for a single negative point charge are directed toward the charge (Fig. 15.13b) because a positive test charge is attracted by a negative charge. In either case the lines are radial and extend all the way to infinity. Note that the lines are closer together as they get near the charge, indicating that the strength of the field is increasing. Equation 15.6 verifies that this is indeed the case.

The rules for drawing electric field lines for any charge distribution follow directly from the relationship between electric field lines and electric field vectors:

1. The lines for a group of point charges must begin on positive charges and end on negative charges. In the case of an excess of charge, some lines will begin or end infinitely far away.
2. The number of lines drawn leaving a positive charge or ending on a negative charge is proportional to the magnitude of the charge.
3. No two field lines can cross each other.

Figure 15.14 shows the beautifully symmetric electric field lines for two point charges of equal magnitude but opposite sign. This charge configuration is called an **electric dipole**. Note that the number of lines that begin at the positive charge must equal the number that terminate at the negative charge. At points very near either charge, the lines are nearly radial. The high density of lines between the charges indicates a strong electric field in this region.

Figure 15.15 shows the electric field lines in the vicinity of two equal positive point charges. Again, close to either charge the lines are nearly radial. The same number of lines emerges from each charge because the charges are equal in magnitude. At great distances from the charges, the field is approximately equal to that of a single point charge of magnitude $2q$. The bulging out of the electric field lines between the charges reflects the repulsive nature of the electric force between like charges. Also, the low density of field lines between the charges indicates a weak field in this region, unlike the dipole.

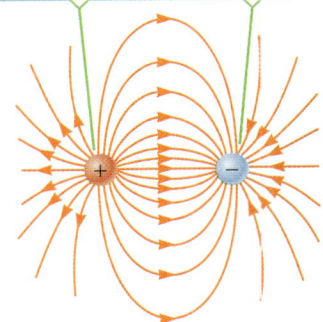

Figure 15.14 The electric field lines for two equal and opposite point charges (an electric dipole).

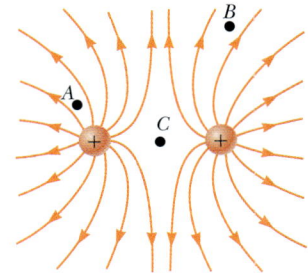

Figure 15.15 The electric field lines for two positive point charges. The points A, B, and C are discussed in Quick Quiz 15.6.

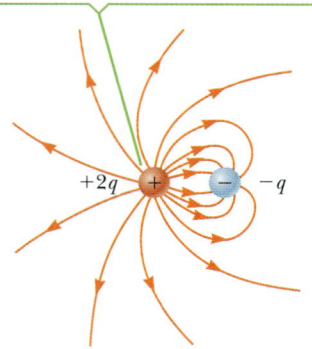

Two field lines leave $+2q$ for every one that terminates on $-q$.

Figure 15.16 The electric field lines for a point charge of $+2q$ and a second point charge of $-q$.

Finally, Figure 15.16 is a sketch of the electric field lines associated with the positive charge $+2q$ and the negative charge $-q$. In this case the number of lines leaving charge $+2q$ is twice the number terminating on charge $-q$. Hence, only half of the lines that leave the positive charge end at the negative charge. The remaining half terminate on negative charges that we assume to be located at infinity. At great distances from the charges (great compared with the charge separation), the electric field lines are equivalent to those of a single charge $+q$.

■ Quick Quiz

15.6 Rank the magnitudes of the electric field at points A, B, and C in Figure 15.15, with the largest magnitude first.
(a) A, B, C (b) A, C, B (c) C, A, B (d) The answer can't be determined by visual inspection.

■ APPLYING PHYSICS 15.1 | Measuring Atmospheric Electric Fields

The electric field near the surface of the Earth in fair weather is about 100 N/C downward. Under a thundercloud, the electric field can be very large, on the order of 20 000 N/C. How are these electric fields measured?

EXPLANATION A device for measuring these fields is called the *field mill*. Figure 15.17 shows the fundamental components of a field mill: two metal plates parallel to the ground. Each plate is connected to ground with a wire, with an ammeter (a low-resistance device for measuring the flow of charge, to be discussed in Section 17.3) in one path. Consider first just the lower plate. Because it's connected to ground and the ground carries a negative charge, the plate is negatively charged. The electric field lines are therefore directed downward, ending on the plate as in Figure 15.17a. Now imagine that the upper plate is suddenly moved over the lower plate, as in Figure 15.17b. This plate is also connected to ground and is also negatively charged, so the field lines now end on the upper plate. The negative charges in the lower plate are repelled by those on the upper plate and must pass through the ammeter, registering a flow of charge. The amount of charge that was on the lower plate is related to the strength of the electric field. In this way, the flow of charge through the ammeter can be calibrated to measure the electric field. The plates are normally designed like the blades of a fan, with the upper plate rotating so that the lower plate is alternately covered and uncovered. As a result, charges flow back and forth continually through the ammeter, and the reading can be related to the electric field strength. ■

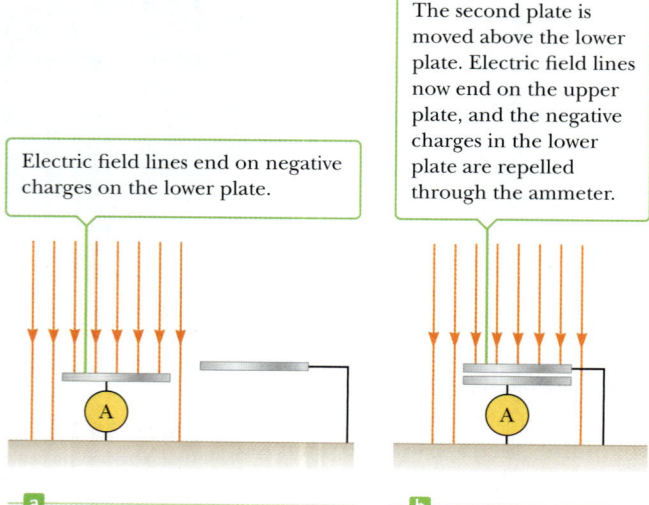

Electric field lines end on negative charges on the lower plate.

The second plate is moved above the lower plate. Electric field lines now end on the upper plate, and the negative charges in the lower plate are repelled through the ammeter.

Figure 15.17 Experimental setup for Applying Physics 15.1.

15.6 Conductors in Electrostatic Equilibrium

LEARNING OBJECTIVE

1. Discuss the properties of an isolated conductor in electrostatic equilibrium.

A good electric conductor like copper, although electrically neutral, contains charges (electrons) that aren't bound to any atom and are free to move about within the material. When no net motion of charge occurs within a conductor, the conductor is said to be in **electrostatic equilibrium.** An isolated conductor (one that is insulated from ground) has the following properties:

1. The electric field is zero everywhere inside the conducting material.
2. Any excess charge on an isolated conductor resides entirely on its surface.
3. The electric field just outside a charged conductor is perpendicular to the conductor's surface.
4. On an irregularly shaped conductor, the charge accumulates at sharp points, where the radius of curvature of the surface is smallest.

◀ Properties of an isolated conductor

The first property can be understood by examining what would happen if it were *not* true. If there were an electric field inside a conductor, the free charge there would move and a flow of charge, or current, would be created. If there were a net movement of charge, however, the conductor would no longer be in electrostatic equilibrium.

Property 2 is a direct result of the $1/r^2$ repulsion between like charges described by Coulomb's law. If by some means an excess of charge is placed inside a conductor, the repulsive forces between the like charges push them as far apart as possible, causing them to quickly migrate to the surface. (We won't prove it here, but the excess charge resides on the surface because Coulomb's law is an inverse-square law. With any other power law, an excess of charge would exist on the surface, but there would be a distribution of charge, of either the same or opposite sign, inside the conductor.)

Property 3 can be understood by again considering what would happen if it were not true. If the electric field in Figure 15.18 were not perpendicular to the surface, it would have a component along the surface, which would cause the free charges of the conductor to move (to the left in the figure). If the charges moved, however, a current would be created and the conductor would no longer be in electrostatic equilibrium. Therefore, $\vec{E}$ must be perpendicular to the surface.

To see why property 4 must be true, consider Figure 15.19a (page 540), which shows a conductor that is fairly flat at one end and relatively pointed at the other. Any excess charge placed on the object moves to its surface. Figure 15.19b shows the forces between two such charges at the flatter end of the object. These forces

Figure 15.18 This situation is *impossible* if the conductor is in electrostatic equilibrium. If the electric field $\vec{E}$ had a component parallel to the surface, an electric force would be exerted on the charges along the surface and they would move to the left.

Figure 15.19 (a) A conductor with a flatter end *A* and a relatively sharp end *B*. Excess charge placed on this conductor resides entirely at its surface and is distributed so that (b) there is less charge per unit area on the flatter end and (c) there is a large charge per unit area on the sharper end.

Figure 15.20 An experiment showing that any charge transferred to a conductor resides on its surface in electrostatic equilibrium. The hollow conductor is insulated from ground, and the small metal ball is supported by an insulating thread.

are predominantly directed parallel to the surface, so the charges move apart until repulsive forces from other nearby charges establish an equilibrium. At the sharp end, however, the forces of repulsion between two charges are directed predominantly away from the surface, as in Figure 15.19c. As a result, there is less tendency for the charges to move apart along the surface here, and the amount of charge per unit area is greater than at the flat end. The cumulative effect of many such outward forces from nearby charges at the sharp end produces a large resultant force directed away from the surface that can be great enough to cause charges to leap from the surface into the surrounding air.

Many experiments have shown that the net charge on a conductor resides on its surface. One such experiment was first performed by Michael Faraday and is referred to as *Faraday's ice pail experiment*. Faraday lowered a negatively-charged metal ball at the end of a silk thread (an insulator) into an uncharged hollow conductor insulated from ground, a metal ice pail as in Figure 15.20a. As the ball entered the pail, the needle on an electrometer attached to the outer surface of the pail was observed to deflect. (An electrometer is a device used to measure charge.) The needle deflected because the charged ball induced a positive charge on the inner wall of the pail, which left an equal negative charge on the outer wall (Fig. 15.20b).

Faraday next touched the inner surface of the pail with the ball and noted that the deflection of the needle did not change, either when the ball touched the inner surface of the pail (Fig. 15.20c) or when it was removed (Fig. 15.20d). Further, he found that the ball was now uncharged because when it touched the inside of the pail, the excess negative charge on the ball had been drawn off, neutralizing the induced positive charge on the inner surface of the pail. In this way Faraday discovered the useful result that *all* the excess charge on an object can be transferred to an already charged metal shell if the object is touched to the *inside* of the shell. As we will see, this result is the principle of operation of the Van de Graaff generator.

Faraday concluded that because the deflection of the needle in the electrometer didn't change when the charged ball touched the inside of the pail, the positive charge induced on the inside surface of the pail was just enough to neutralize the negative charge on the ball. As a result of his investigations, he concluded that a charged object suspended inside a metal container rearranged the charge on the container so that the sign of the charge on its inside surface was *opposite* the sign of the charge on the suspended object. This produced a charge on the outside surface of the container of the same sign as that on the suspended object.

Faraday also found that if the electrometer was connected to the inside surface of the pail after the experiment had been run, the needle showed no deflection. Thus, the *excess* charge acquired by the pail when contact was made between ball and pail appeared on the outer surface of the pail.

If a metal rod having sharp points is attached to a house, most of the charge on the house passes through these points, eliminating the induced charge on the house produced by storm clouds. In addition, a lightning discharge striking the house passes through the metal rod and is safely carried to the ground through wires leading from the rod to the Earth. Lightning rods using this principle were first developed by Benjamin Franklin. Some European countries couldn't accept the fact that such a worthwhile idea could have originated in the New World, so they "improved" the design by eliminating the sharp points!

APPLICATION
Lightning Rods

■ APPLYING PHYSICS 15.2 | Conductors and Field Lines

Suppose a point charge $+Q$ is in empty space. Wearing rubber gloves, you proceed to surround the charge with a concentric spherical conducting shell. What effect does that have on the field lines from the charge?

EXPLANATION When the spherical shell is placed around the charge, the charges in the shell rearrange to satisfy the rules for a conductor in equilibrium. A net charge of $-Q$ moves to the interior surface of the conductor, so the electric field inside the conductor becomes zero. This means the field lines originating on the $+Q$ charge now terminate on the negative charges. The movement of the negative charges to the inner surface of the sphere leaves a net charge of $+Q$ on the outer surface of the sphere. Then the field lines outside the sphere look just as before: the only change, overall, is the absence of field lines within the conductor. ■

■ APPLYING PHYSICS 15.3 | Driver Safety During Electrical Storms

Why is it safe to stay inside an automobile during a lightning storm?

EXPLANATION Many people believe that staying inside the car is safe because of the insulating characteristics of the rubber tires, but in fact that isn't true. Lightning can travel through several kilometers of air, so it can certainly penetrate a centimeter of rubber. The safety of remaining in the car is due to the fact that charges on the metal shell of the car will reside on the outer surface of the car, as noted in property 2 discussed earlier. As a result, an occupant in the automobile touching the inner surfaces is not in danger. ■

15.7 The Millikan Oil-Drop Experiment

LEARNING OBJECTIVE

1. Describe Robert Millikan's oil-drop experiment and how it measured the electron's charge.

From 1909 to 1913, Robert Andrews Millikan (1868–1953) performed a brilliant set of experiments at the University of Chicago in which he measured the elementary charge e of the electron and demonstrated the quantized nature of the electronic charge. The apparatus he used, diagrammed in Figure 15.21, contains two parallel metal plates. Oil droplets that have been charged by friction in an atomizer are allowed to pass through a small hole in the upper plate. A horizontal light beam is used to illuminate the droplets, which are viewed by a telescope with axis at right angles to the beam. The droplets then appear as shining stars against a dark background, and the rate of fall of individual drops can be determined.

We assume a single drop having a mass of m and carrying a charge of q is being viewed and its charge is negative. If no electric field is present between the plates,

Figure 15.21 A schematic view of Millikan's oil-drop apparatus.

Figure 15.22 The forces on a negatively charged oil droplet in Millikan's experiment.

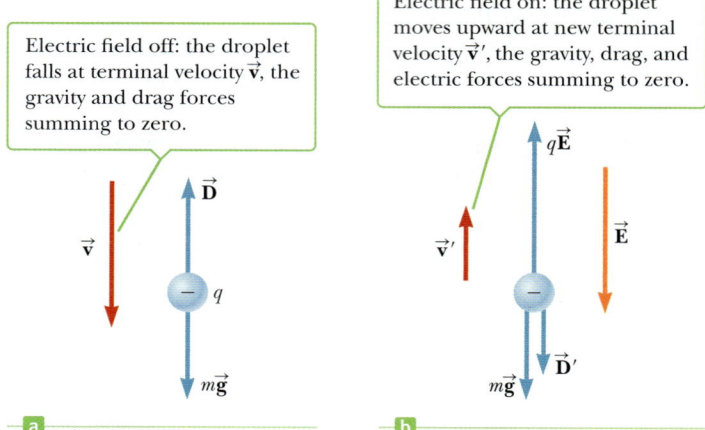

the two forces acting on the charge are the force of gravity, $m\vec{g}$, acting downward, and an upward viscous drag force $\vec{D}$ (Fig. 15.22a). The drag force is proportional to the speed of the drop. When the drop reaches its terminal speed, v, the two forces balance each other ($mg = D$).

Now suppose an electric field is set up between the plates by a battery connected so that the upper plate is positively charged. In this case a third force, $q\vec{E}$, acts on the charged drop. Because q is negative and $\vec{E}$ is downward, the electric force is *upward* as in Figure 15.22b. If this force is great enough, the drop moves upward and the drag force $\vec{D}'$ acts downward. When the upward electric force, $q\vec{E}$, balances the sum of the force of gravity and the drag force, both acting downward, the drop reaches a new terminal speed v'.

With the field turned on, a drop moves slowly upward, typically at a rate of *hundredths* of a centimeter per second. The rate of fall in the absence of a field is comparable. Hence, a single droplet with constant mass and radius can be followed for hours as it alternately rises and falls, simply by turning the electric field on and off.

After making measurements on thousands of droplets, Millikan and his coworkers found that, to within about 1% precision, every drop had a charge equal to some positive or negative integer multiple of the elementary charge e,

$$q = ne \quad n = 0, \pm 1, \pm 2, \pm 3, \ldots \quad [15.7]$$

where $e = 1.60 \times 10^{-19}$ C. It was later established that positive integer multiples of e would arise when an oil droplet had lost one or more electrons. Likewise, negative integer multiples of e would arise when a drop had gained one or more electrons. Gains or losses in integral numbers provide conclusive evidence that charge is quantized. In 1923 Millikan was awarded the Nobel Prize in Physics for this work.

15.8 The Van de Graaff Generator

LEARNING OBJECTIVE

1. Describe the operating principles of Robert Van de Graaff's electrostatic generator.

In 1929 Robert J. Van de Graaff (1901–1967) designed and built an electrostatic generator that has been used extensively in nuclear physics research. The principles of its operation can be understood with knowledge of the properties of electric fields and charges already presented in this chapter. Figure 15.23 shows the basic construction of this device. A motor-driven pulley P moves a belt past positively charged comb-like metallic needles positioned at A. Negative charges are attracted to these needles from the belt, leaving the left side of the belt with a net

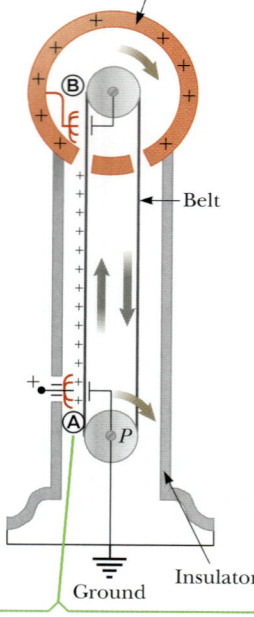

The charge is deposited on the belt at point Ⓐ and transferred to the hollow conductor at point Ⓑ.

Figure 15.23 A schematic diagram of a Van de Graaff generator. Charge is transferred to the dome by means of a rotating belt.

positive charge. The positive charges attract electrons onto the belt as it moves past a second comb of needles at B, increasing the excess positive charge on the dome. Because the electric field inside the metal dome is negligible, the positive charge on it can easily be increased regardless of how much charge is already present. The result is that the dome is left with a large amount of positive charge.

This accumulation of charge on the dome can't continue indefinitely. As more and more charge appears on the surface of the dome, the magnitude of the electric field at that surface also increases. Finally, the strength of the field becomes great enough to partially ionize the air near the surface, increasing the conductivity of the air. Charges on the dome now have a pathway to leak off into the air, producing some spectacular "lightning bolts" as the discharge occurs. As noted earlier, charges find it easier to leap off a surface at points where the curvature is great. As a result, one way to inhibit the electric discharge, and to increase the amount of charge that can be stored on the dome, is to increase its radius. Another method for inhibiting discharge is to place the entire system in a container filled with a high-pressure gas, which is significantly more difficult to ionize than air at atmospheric pressure.

If protons (or other charged particles) are introduced into a tube attached to the dome, the large electric field of the dome exerts a repulsive force on the protons, causing them to accelerate to energies high enough to initiate nuclear reactions between the protons and various target nuclei.

15.9 Electric Flux and Gauss's Law

LEARNING OBJECTIVES

1. Define electric flux and calculate it in elementary contexts.
2. State Gauss's law relating electric flux through a closed surface to the charge inside the surface.
3. Apply Gauss's law to distributions of charge.

Gauss's law is essentially a technique for calculating the average electric field on a closed surface, developed by Karl Friedrich Gauss (1777–1855). When the electric field, because of its symmetry, is constant everywhere on that surface and perpendicular to it, the exact electric field can be found. In such special cases, Gauss's law is far easier to apply than Coulomb's law.

Gauss's law relates the electric flux through a closed surface and the total charge inside that surface. A *closed surface* has an inside and an outside: an example is a sphere. *Electric flux* is a measure of how much the electric field vectors penetrate through a given surface. If the electric field vectors are tangent to the surface at all points, for example, they don't penetrate the surface and the electric flux through the surface is zero. These concepts will be discussed more fully in the next two subsections. As we'll see, Gauss's law states that the electric flux through a closed surface is proportional to the charge contained *inside* the surface.

Electric Flux

Consider an electric field that is uniform in both magnitude and direction, as in Figure 15.24. The electric field lines penetrate a surface of area A, which is perpendicular to the field. The technique used for drawing a figure such as Figure 15.24 is that the number of lines per unit area, N/A, is proportional to the magnitude of the electric field, or $E \propto N/A$. We can rewrite this proportion as $N \propto EA$, which means that the number of field lines is proportional to the *product* of E and A, called the **electric flux** and represented by the symbol Φ_E:

$$\Phi_E = EA \qquad [15.8]$$

Note that Φ_E has SI units of $N \cdot m^2/C$ and is proportional to the number of field lines that pass through some area A oriented perpendicular to the field. (It's called

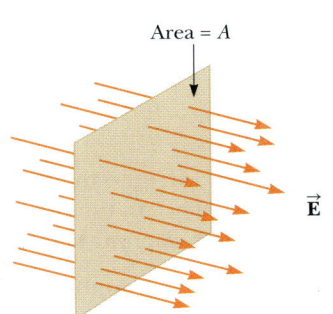

Figure 15.24 Field lines of a uniform electric field penetrating a plane of area A perpendicular to the field. The electric flux Φ_E through this area is equal to EA.

flux by analogy with the term *flux* in fluid flow, which is the volume of liquid flowing through a perpendicular area per second.) If the surface under consideration is not perpendicular to the field, as in Figure 15.25, the expression for the electric flux is

◀ Electric flux

$$\Phi_E = EA \cos \theta \quad [15.9]$$

where a vector perpendicular to the area A is at an angle θ with respect to the field. This vector is often said to be *normal* to the surface, and we will refer to it as "the normal vector to the surface." The number of lines that cross this area is equal to the number that cross the projected area A', which is perpendicular to the field. We see that the two areas are related by $A' = A \cos \theta$. From Equation 15.9, we see that the flux through a surface of fixed area has the maximum value EA when the surface is perpendicular to the field (when $\theta = 0°$) and that the flux is zero when the surface is parallel to the field (when $\theta = 90°$). **By convention, for a closed surface, the flux lines passing into the interior of the volume are negative and those passing out of the interior of the volume are positive.** This convention is equivalent to requiring the normal vector of the surface to point outward when computing the flux through a closed surface.

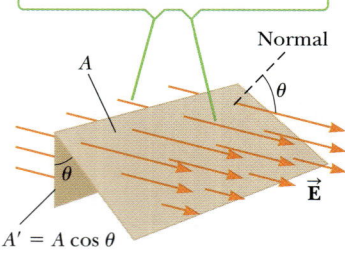

Figure 15.25 Field lines for a uniform electric field through an area A that is at an angle of $(90° - \theta)$ to the field.

■ Quick Quiz

15.7 Calculate the magnitude of the flux of a constant electric field of 5.00 N/C in the z-direction through a rectangle with area 4.00 m² in the xy-plane. (a) 0 (b) 10.0 N · m²/C (c) 20.0 N · m²/C (d) More information is needed

15.8 Suppose the electric field of Quick Quiz 15.7 is tilted 60° away from the positive z-direction. Calculate the magnitude of the flux through the same area. (a) 0 (b) 10.0 N · m²/C (c) 20.0 N · m²/C (d) More information is needed

■ **EXAMPLE 15.6** | Flux Through a Cube

GOAL Calculate the electric flux through a closed surface.

PROBLEM Consider a uniform electric field oriented in the x-direction. Find the electric flux through each surface of a cube with edges L oriented as shown in Figure 15.26, and the net flux.

STRATEGY This problem involves substituting into the definition of electric flux given by Equation 15.9. In each case E and $A = L^2$ are the same; the only difference is the angle θ that the electric field makes with respect to a vector perpendicular to a given surface and pointing outward (the normal vector to the surface). The angles can be determined by inspection. The flux through a surface parallel to the xy-plane will be labeled Φ_{xy} and further designated by position (front, back); others will be labeled similarly: Φ_{xz} top or bottom, and Φ_{yz} left or right.

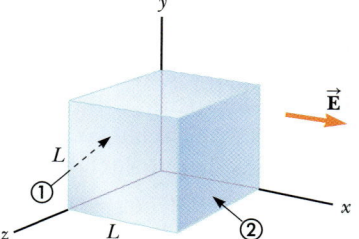

Figure 15.26 (Example 15.6) A hypothetical surface in the shape of a cube in a uniform electric field parallel to the x-axis. The net flux through the surface is zero when the net charge inside the cube is zero.

SOLUTION

The normal vector to the xy-plane points in the negative z-direction. This, in turn, is perpendicular to $\vec{E}$, so $\theta = 90°$. (The opposite side works similarly.)

$$\Phi_{xy} = EA \cos(90°) = \boxed{0} \text{ (back and front surfaces)}$$

The normal vector to the xz-plane points in the negative y-direction. This, in turn, is perpendicular to $\vec{E}$, so again $\theta = 90°$. (The opposite side works similarly.)

$$\Phi_{xz} = EA \cos(90°) = \boxed{0} \text{ (top and bottom surfaces)}$$

The normal vector to surface ① (the yz-plane) points in the negative x-direction. This is antiparallel to $\vec{E}$, so $\theta = 180°$.

$$\Phi_{yz} = EA \cos(180°) = \boxed{-EL^2} \text{ (surface ①)}$$

Surface ② has normal vector pointing in the positive x-direction, so $\theta = 0°$.

$\Phi_{yz} = EA \cos(0°) = EL^2$ (surface ②)

We calculate the net flux by summing:

$\Phi_{net} = 0 + 0 + 0 + 0 - EL^2 + EL^2 = 0$

REMARKS In doing this calculation, it is necessary to remember that the angle in the definition of flux is measured from the normal vector to the surface and that this vector must point outwards for a closed surface. As a result, the normal vector for the yz-plane on the left points in the negative x-direction, and the normal vector to the plane parallel to the yz-plane on the right points in the positive x-direction. Notice that there aren't any charges in the box. The net electric flux is always zero for closed surfaces that contain a net charge of zero.

QUESTION 15.6 If the surface in Figure 15.26 were spherical, would the answer be (a) greater than, (b) less than, or (c) the same as the net electric flux found for the cubical surface?

EXERCISE 15.6 Suppose the constant electric field in Example 15.6 points in the positive y-direction instead. Calculate the flux through the xz-plane and the surface parallel to it. What's the net electric flux through the surface of the cube?

ANSWERS $\Phi_{xz} = -EL^2$ (bottom surface), $\Phi_{xz} = +EL^2$ (top surface). The net flux is still zero.

Gauss's Law

Consider a point charge q surrounded by a spherical surface of radius r centered on the charge, as in Figure 15.27a. The magnitude of the electric field everywhere on the surface of the sphere is

$$E = k_e \frac{q}{r^2}$$

Note that the electric field is perpendicular to the spherical surface at all points on the surface. The electric flux through the surface is therefore EA, where $A = 4\pi r^2$ is the surface area of the sphere:

$$\Phi_E = EA = k_e \frac{q}{r^2}(4\pi r^2) = 4\pi k_e q$$

It's sometimes convenient to express k_e in terms of another constant, ϵ_0, as $k_e = 1/(4\pi\epsilon_0)$. The constant ϵ_0 is called the **permittivity of free space** and has the value

$$\epsilon_0 = \frac{1}{4\pi k_e} = 8.85 \times 10^{-12} \text{ C}^2/\text{N} \cdot \text{m}^2 \qquad [15.10]$$

The use of k_e or ϵ_0 is strictly a matter of taste. The electric flux through the closed spherical surface that surrounds the charge q can now be expressed as

$$\Phi_E = 4\pi k_e q = \frac{q}{\epsilon_0}$$

This result says that the electric flux through a sphere that surrounds a charge q is equal to the charge divided by the constant ϵ_0. Using calculus, this result can be proven for *any* closed surface that surrounds the charge q. For example, if the surface surrounding q is irregular, as in Figure 15.27b, the flux through that surface is also q/ϵ_0. This leads to the following general result, known as Gauss's law:

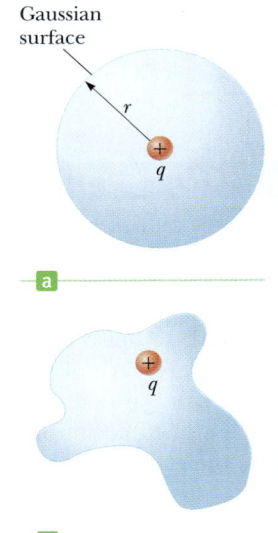

Figure 15.27 (a) The flux through a spherical surface of radius r surrounding a point charge q is $\Phi_E = q/\epsilon_0$. (b) The flux through any arbitrary surface surrounding the charge is also equal to q/ϵ_0.

> The electric flux Φ_E through any closed surface is equal to the net charge inside the surface, Q_{inside}, divided by ϵ_0:
>
> $$\Phi_E = \frac{Q_{inside}}{\epsilon_0} \qquad [15.11]$$

◄ Gauss's Law

Although it's not obvious, Gauss's law describes how charges create electric fields. In principle it can always be used to calculate the electric field of a system of charges or a continuous distribution of charge. In practice, the technique is

Tip 15.2 Gaussian Surfaces Aren't Real

A Gaussian surface is an imaginary surface, created solely to facilitate a mathematical calculation. It doesn't necessarily coincide with the surface of a physical object.

useful only in a limited number of cases in which there is a high degree of symmetry, such as spheres, cylinders, or planes. With the symmetry of these special shapes, the charges can be surrounded by an imaginary surface, called a Gaussian surface. This imaginary surface is used strictly for mathematical calculation, and need not be an actual, physical surface. If the imaginary surface is chosen so that the electric field is constant everywhere on it, the electric field can be computed with

$$EA = \Phi_E = \frac{Q_{inside}}{\epsilon_0} \qquad [15.12]$$

as will be seen in the examples. Although Gauss's law in this form can be used to obtain the electric field only for problems with a lot of symmetry, it can *always* be used to obtain the *average* electric field on *any* surface.

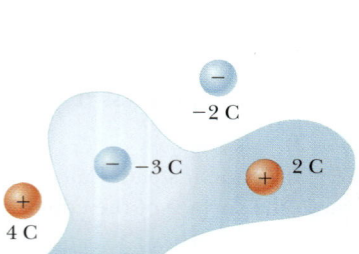

Figure 15.28 (Quick Quiz 15.9)

■ **Quick Quiz**

15.9 Find the electric flux through the surface in Figure 15.28. Assume all charges in the shaded area are inside the surface. (a) $-(3\text{ C})/\epsilon_0$ (b) $(3\text{ C})/\epsilon_0$ (c) 0 (d) $-(6\text{ C})/\epsilon_0$

15.10 For a closed surface through which the net flux is zero, each of the following four statements *could* be true. Which of the statements *must* be true? (There may be more than one.) (a) There are no charges inside the surface. (b) The net charge inside the surface is zero. (c) The electric field is zero everywhere on the surface. (d) The number of electric field lines entering the surface equals the number leaving the surface.

■ **EXAMPLE 15.7** | **The Electric Field of a Charged Spherical Shell**

GOAL Use Gauss's law to determine electric fields when the symmetry is spherical.

PROBLEM A spherical conducting shell of inner radius a and outer radius b carries a total charge $+Q$ distributed on the surface of a conducting shell (Fig. 15.29a). The quantity Q is taken to be positive. **(a)** Find the electric field in the interior of the conducting shell, for $r < a$, and **(b)** the electric field outside the shell, for $r > b$. **(c)** If an additional charge of $-2Q$ is placed at the center, find the electric field for $r > b$. **(d)** What is the distribution of charge on the sphere in part (c)?

STRATEGY For each part, draw a spherical Gaussian surface in the region of interest. Add up the charge inside the Gaussian surface, substitute it and the area into Gauss's law, and solve for the electric field. To find the distribution of charge in part (c), use Gauss's law in reverse: the charge distribution must be such that the electrostatic field is zero inside a conductor.

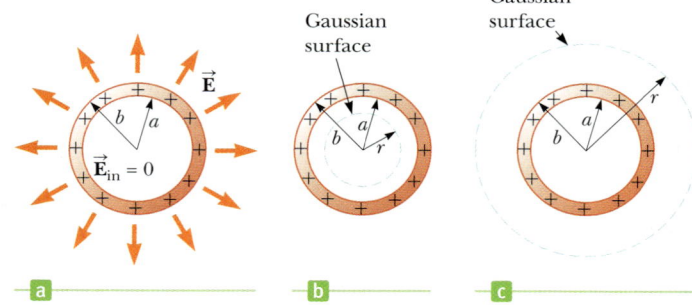

Figure 15.29 (Example 15.7) (a) The electric field inside a uniformly charged spherical shell is *zero*. It is also zero for the conducting material in the region $a < r < b$. The field outside is the same as that of a point charge having a total charge Q located at the center of the shell. (b) The construction of a Gaussian surface for calculating the electric field *inside* a spherical shell. (c) The construction of a Gaussian surface for calculating the electric field *outside* a spherical shell.

SOLUTION

(a) Find the electric field for $r < a$.

Apply Gauss's law, Equation 15.12, to the Gaussian surface illustrated in Figure 15.29b (note that there isn't any charge inside this surface):

$$EA = E(4\pi r^2) = \frac{Q_{inside}}{\epsilon_0} = 0 \quad \rightarrow \quad E = 0$$

(b) Find the electric field for $r > b$.

Apply Gauss's law, Equation 15.12, to the Gaussian surface illustrated in Figure 15.29c:

$$EA = E(4\pi r^2) = \frac{Q_{inside}}{\epsilon_0} = \frac{Q}{\epsilon_0}$$

Divide by the area:

$$E = \frac{Q}{4\pi\epsilon_0 r^2}$$

(c) Now an additional charge of $-2Q$ is placed at the center of the sphere. Compute the new electric field outside the sphere, for $r > b$.

Apply Gauss's law as in part (b), including the new charge in Q_{inside}:

$$EA = E(4\pi r^2) = \frac{Q_{inside}}{\epsilon_0} = \frac{+Q - 2Q}{\epsilon_0}$$

Solve for the electric field:

$$E = -\frac{Q}{4\pi\epsilon_0 r^2}$$

(d) Find the charge distribution on the sphere for part (c).

Write Gauss's law for the interior of the shell:

$$EA = \frac{Q_{inside}}{\epsilon_0} = \frac{Q_{center} + Q_{inner\ surface}}{\epsilon_0}$$

Find the charge on the inner surface of the shell, noting that the electric field in the conductor is zero:

$$Q_{center} + Q_{inner\ surface} = 0$$
$$Q_{inner\ surface} = -Q_{center} = \boxed{+2Q}$$

Find the charge on the outer surface, noting that the inner and outer surface charges must sum to $+Q$:

$$Q_{outer\ surface} + Q_{inner\ surface} = Q$$
$$Q_{outer\ surface} = -Q_{inner\ surface} + Q = \boxed{-Q}$$

REMARKS The important thing to notice is that in each case, the charge is spread out over a region with spherical symmetry or is located at the exact center. That is what allows the computation of a value for the electric field.

QUESTION 15.7 If the charge at the center of the sphere is made positive, how is the charge on the inner surface of the sphere affected?

EXERCISE 15.7 Suppose the charge at the center is now increased to $+2Q$, while the surface of the conductor still retains a charge of $+Q$. (a) Find the electric field exterior to the sphere, for $r > b$. (b) What's the electric field inside the conductor, for $a < r < b$? (c) Find the charge distribution on the conductor.

ANSWERS (a) $E = 3Q/4\pi\epsilon_0 r^2$ (b) $E = 0$, which is always the case when charges are not moving in a conductor. (c) Inner surface: $-2Q$; outer surface: $+3Q$

Problems like Example 15.7 sometimes involve "thin, nonconducting shells" carrying a uniformly distributed charge. In these cases no distinction need be made between the outer surface and inner surface of the shell. The next example makes that implicit assumption.

EXAMPLE 15.8 A Nonconducting Plane Sheet of Charge

GOAL Apply Gauss's law to a problem with plane symmetry.

PROBLEM Find the electric field above and below a nonconducting infinite plane sheet of charge with uniform positive charge per unit area σ (Fig. 15.30a, page 548).

STRATEGY By symmetry, the electric field must be perpendicular to the plane and directed away from it on either side, as shown in Figure 15.30b. For the Gaussian surface, choose a small cylinder with axis perpendicular to the plane, each end having area A_0. No electric field lines pass through the curved surface of the cylinder, only through the two ends, which have total area $2A_0$. Apply Gauss's law, using Figure 15.30b.

(Continued)

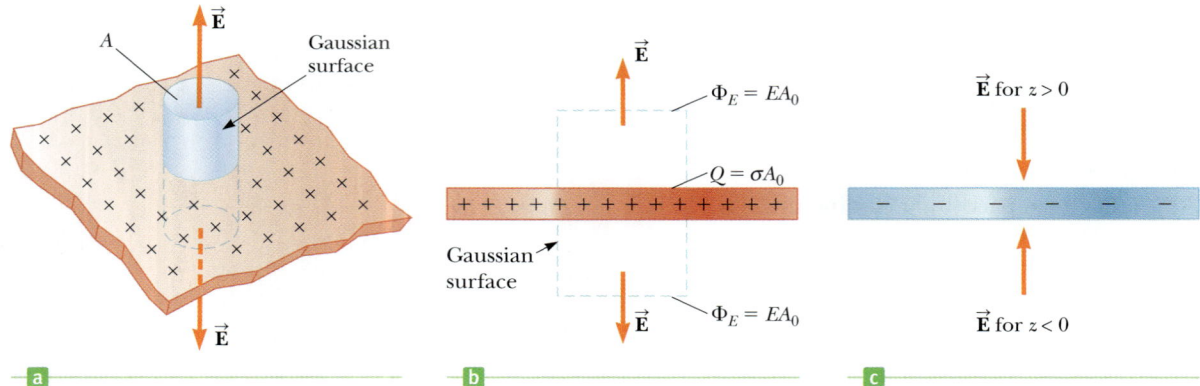

Figure 15.30 (Example 15.8) (a) A cylindrical Gaussian surface penetrating an infinite sheet of charge. (b) A cross section of the same Gaussian cylinder. The flux through each end of the Gaussian surface is EA_0. There is no flux through the cylindrical surface. (c) (Exercise 15.8).

SOLUTION

Find the electric field above and below a plane of uniform charge.

Apply Gauss's law, Equation 15.12:

$$EA = \frac{Q_{inside}}{\epsilon_0}$$

The total charge inside the Gaussian cylinder is the charge density times the cross-sectional area:

$$Q_{inside} = \sigma A_0$$

The electric flux comes entirely from the two ends, each having area A_0. Substitute $A = 2A_0$ and Q_{inside} and solve for E.

$$E = \frac{\sigma A_0}{(2A_0)\epsilon_0} = \boxed{\frac{\sigma}{2\epsilon_0}}$$

This is the *magnitude* of the electric field. Find the z-component of the field above and below the plane. The electric field points away from the plane, so it's positive above the plane and negative below the plane.

$$\boxed{E_z = \frac{\sigma}{2\epsilon_0} \quad z > 0}$$

$$\boxed{E_z = -\frac{\sigma}{2\epsilon_0} \quad z < 0}$$

REMARKS Notice here that the plate was taken to be a thin, nonconducting shell. If it's made of metal, of course, the electric field inside it is zero, with half the charge on the upper surface and half on the lower surface.

QUESTION 15.8 In reality, the sheet carrying charge would likely be metallic and have a small but nonzero thickness. If it carries the same charge per unit area, what is the electric field inside the sheet between the two surfaces?

EXERCISE 15.8 Suppose an infinite nonconducting plane of charge as in Example 15.8 has a uniform negative charge density of $-\sigma$. Find the electric field above and below the plate. Sketch the field.

ANSWERS $E_z = \frac{-\sigma}{2\epsilon_0}$, $z > 0$; $E_z = \frac{\sigma}{2\epsilon_0}$, $z < 0$. See Figure 15.30c for the sketch.

Figure 15.31 Cross section of an idealized parallel-plate capacitor. Electric field vector contributions sum together in between the plates, but cancel outside.

An important circuit element that will be studied extensively in the next chapter is the parallel-plate capacitor. The device consists of a plate of positive charge, as in Example 15.8, with the negative plate of Exercise 15.8 placed above it. The sum of these two fields is illustrated in Figure 15.31. The result is an electric field with double the magnitude in between the two plates:

$$E = \frac{\sigma}{\epsilon_0} \qquad [15.13]$$

Outside the plates, the electric fields cancel.

SUMMARY

15.1 Properties of Electric Charges

Electric charges have the following properties:

1. Unlike charges attract one another and like charges repel one another.
2. Electric charge is always conserved.
3. Charge comes in discrete packets that are integral multiples of the basic electric charge $e = 1.6 \times 10^{-19}$ C.
4. The force between two charged particles is proportional to the inverse square of the distance between them.

15.2 Insulators and Conductors

Conductors are materials in which charges move freely in response to an electric field. All other materials are called **insulators**.

15.3 Coulomb's Law

Coulomb's law states that the electric force between two stationary charged particles separated by a distance r has the magnitude

$$F = k_e \frac{|q_1||q_2|}{r^2} \quad [15.1]$$

where $|q_1|$ and $|q_2|$ are the magnitudes of the charges on the particles in coulombs and

$$k_e \approx 8.99 \times 10^9 \text{ N} \cdot \text{m}^2/\text{C}^2 \quad [15.2]$$

is the **Coulomb constant**.

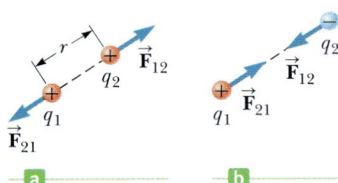

(a) The electric force between two charges with the same sign is repulsive, and (b) it is attractive when the charges have opposite signs.

15.4 The Electric Field

An electric field $\vec{E}$ exists at some point in space if a small test charge q_0 placed at that point is acted upon by an electric force $\vec{F}$. The electric field is defined as

$$\vec{E} \equiv \frac{\vec{F}}{q_0} \quad [15.3]$$

The **direction** of the electric field at a point in space is defined to be the direction of the electric force that would be exerted on a small positive charge placed at that point.

Source charge   Test charge

The electric force of charge Q on a test charge q_0 divided by q_0 gives the electric field $\vec{E}$ of Q at that point.

The magnitude of the electric field due to a *point charge* q at a distance r from the point charge is

$$E = k_e \frac{|q|}{r^2} \quad [15.6]$$

15.5 Electric Field Lines

Electric field lines are useful for visualizing the electric field in any region of space. The electric field vector $\vec{E}$ is tangent to the electric field lines at every point. Further, the number of electric field lines per unit area through a surface perpendicular to the lines is proportional to the strength of the electric field at that surface.

15.6 Conductors in Electrostatic Equilibrium

A **conductor in electrostatic equilibrium** has the following properties:

1. The electric field is zero everywhere inside the conducting material.
2. Any excess charge on an isolated conductor must reside entirely on its surface.
3. The electric field just outside a charged conductor is perpendicular to the conductor's surface.
4. On an irregularly shaped conductor, charge accumulates where the radius of curvature of the surface is smallest, at sharp points.

15.9 Electric Flux and Gauss's Law

Gauss's law states that the electric flux through any closed surface is equal to the net charge Q inside the surface divided by the permittivity of free space, ϵ_0:

$$EA = \Phi_E = \frac{Q_{\text{inside}}}{\epsilon_0} \quad [15.12]$$

For highly symmetric distributions of charge, Gauss's law can be used to calculate electric fields.

The electric flux Φ through any arbitrary surface surrounding a charge q is q/ϵ_0.

WARM-UP EXERCISES

WebAssign The warm-up exercises in this chapter may be assigned online in Enhanced WebAssign.

1. **Math Review** A force vector has components given by $F_x = -7.00$ N and $F_y = 4.50$ N. Find (a) the magnitude and (b) the direction of the force, measured counterclockwise from the positive x-axis.

2. **Math Review** The force acting on a particle has a magnitude of 125 N and is directed 30.0° above the positive x-axis. Determine (a) the x-component and (b) y-component of the force.

3. **Math Review** Two electric force vectors act on a particle. Their x-components are 15.0 N and −7.50 N and their y-components are −11.5 N and −4.50 N, respectively. For the resultant electric force, find (a) the x-component, (b) the y-component, and the (c) magnitude and (d) direction of the resultant electric force, measured counterclockwise from the positive x-axis. (See also Section 15.4.)

4. **Physics Review** A large firecracker explodes under a soda can, sending it straight upward with an initial speed of 7.27 m/s. Neglecting air friction, find (a) the can's maximum height, (b) the first time at which it reaches a height of 1.00 m, and (c) the velocity at that time. (See Section 2.5.)

5. **Physics Review** A simple pendulum with mass 0.250 kg hangs from string 1.40 m long. (See Fig. WU15.5.) A leaf blower blows air at it, exerting a constant horizontal drag force $\vec{F}_{drag}$ of magnitude 1.67 N. Find (a) the tension in the string and (b) the angle the pendulum makes with respect to the vertical. (See Section 4.5.)

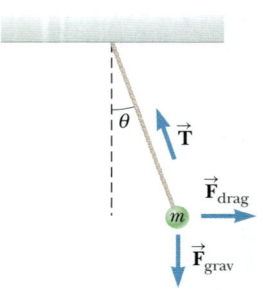

Figure WU15.5

6. **Physics Review** The Falcon 9 rocket, manufactured by the commercial SpaceX Corporation, has a takeoff mass of 4.80×10^5 kg and an initial thrust of 5.88×10^6 N. At takeoff, determine (a) the net upward force acting on the rocket, and (b) the upward acceleration of the rocket. (See Section 4.5.)

7. **Physics Review** A horizontal spring with force constant $k = 57.0$ N/m is used to accelerate a 0.500-kg mass on a frictionless surface. If the spring is compressed by 5.00×10^{-2} m and released, find (a) the initial magnitude of the spring force when the mass is released and (b) the initial acceleration of the mass. (See Section 5.4.)

8. Two protons are separated by a distance of 0.100 m. Given the proton charge of 1.60×10^{-19} C, determine the magnitude of the electric force that one proton exerts on the other. (See Section 15.3.)

9. A charge of 16.0 nC is located at the origin. Find (a) the magnitude and (b) the direction of the displacement vector pointing from the origin to (3.00, 4.00) m. Calculate (c) the magnitude, and (d) the x- and (e) y-components of the electric field at that point. (See Section 15.4.)

10. Consider a system of two charges where the first charge is on the positive x-axis and the second charge is on the positive y-axis. If these two charges produce electric fields at the origin given by (2.50, 0) N/C and (0, −2.50) N/C, respectively, determine (a) the magnitude and (b) the direction of the resultant electric field at the origin, measured counterclockwise from the positive x-axis. (c) What would be the magnitude of the force on a 1.41×10^{-9} C charge placed at the origin? (See Sections 15.3 and 15.4.)

11. A tiny droplet of oil of mass 1.50×10^{-15} kg having a charge of -2.56×10^{-18} C accelerates straight down inside a vacuum chamber. If the electric field is −5 625 N/C, find (a) the force of gravity on the droplet, (b) the electric force, and (c) the droplet's acceleration. (Give the correct signs, with down the negative direction.)

12. A constant electric field of magnitude 5.00 N/C and pointing in the positive z-direction passes through a square with sides of length 0.500 m, located in the xy-plane. (a) Calculate the magnitude of the electric flux through the square. (b) Repeat the calculation if the electric field is oriented at an angle of 30.0° with respect to the positive z-direction. (See Section 15.9.)

13. Three charges are placed inside a basketball. If the charges are 2.50 nC, −1.25 nC and 0.500 nC, respectively, what is the electric flux through the basketball? (See Section 15.9.)

14. A thin, hollow sphere has a charge $q = -2.00$ nC evenly distributed over its surface. The radius of the charged sphere is 1.00 m and the radii of two spherical Gaussian surfaces are $a = 0.500$ m and $b = 1.50$ m, respectively. (See Figure WU15.14.) A charge of $Q = 5.00$ nC is placed at the center of the charged sphere. (a) How much charge is inside the inner Gaussian surface? (b) Find the electric flux through the inner Gaussian surface, and (c) the electric field there, by dividing by the sphere's area. (d) How much charge is inside the outer dashed sphere? Find (e) the electric flux through the outer Gaussian surface and (f) the electric field strength on that surface. (See Section 15.9.)

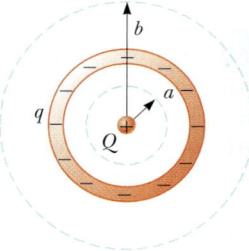

Figure WU15.14

■ CONCEPTUAL QUESTIONS

WebAssign The conceptual questions in this chapter may be assigned online in Enhanced WebAssign.

1. A glass object receives a positive charge of +3 nC by rubbing it with a silk cloth. In the rubbing process, have protons been added to the object or have electrons been removed from it?

2. Explain from an atomic viewpoint why charge is usually transferred by electrons.

3. A person is placed in a large, hollow metallic sphere that is insulated from ground. If a large charge is placed on the sphere, will the person be harmed upon touching the inside of the sphere?

4. Why must hospital personnel wear special conducting shoes while working around oxygen in an operating

room? What might happen if the personnel wore shoes with rubber soles?

5. (a) Would life be different if the electron were positively charged and the proton were negatively charged? (b) Does the choice of signs have any bearing on physical and chemical interactions? Explain your answers.

6. If a suspended object A is attracted to a charged object B, can we conclude that A is charged? Explain.

7. Explain how a positively charged object can be used to leave another metallic object with a net negative charge. Discuss the motion of charges during the process.

8. Consider point A in Figure CQ15.8 located an arbitrary distance from two point charges in otherwise empty space. (a) Is it possible for an electric field to exist at point A in empty space? (b) Does charge exist at this point? (c) Does a force exist at this point?

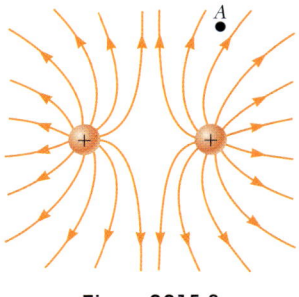

Figure CQ15.8

9. A student stands on a thick piece of insulating material, places her hand on top of a Van de Graaff generator, and then turns on the generator. Does she receive a shock?

10. In fair weather, there is an electric field at the surface of the Earth, pointing down into the ground. What is the sign of the electric charge on the ground in this situation?

11. A charged comb often attracts small bits of dry paper that then fly away when they touch the comb. Explain why that occurs.

12. Why should a ground wire be connected to the metal support rod for a television antenna?

13. There are great similarities between electric and gravitational fields. A room can be electrically shielded so that there are no electric fields in the room by surrounding it with a conductor. Can a room be gravitationally shielded? Explain.

14. A spherical surface surrounds a point charge q. Describe what happens to the total flux through the surface if (a) the charge is tripled, (b) the volume of the sphere is doubled, (c) the surface is changed to a cube, (d) the charge is moved to another location inside the surface, and (e) the charge is moved outside the surface.

15. If more electric field lines leave a Gaussian surface than enter it, what can you conclude about the net charge enclosed by that surface?

16. A student who grew up in a tropical country and is studying in the United States may have no experience with static electricity sparks and shocks until his or her first American winter. Explain.

17. What happens when a charged insulator is placed near an uncharged metallic object? (a) They repel each other. (b) They attract each other. (c) They may attract or repel each other, depending on whether the charge on the insulator is positive or negative. (d) They exert no electrostatic force on each other. (e) The charged insulator always spontaneously discharges.

■ PROBLEMS

WebAssign The problems in this chapter may be assigned online in Enhanced WebAssign.

1. denotes straightforward problem; 2. denotes intermediate problem;
3. denotes challenging problem

1. denotes full solution available in *Student Solutions Manual/Study Guide*

1. denotes problems most often assigned in Enhanced WebAssign

BIO denotes biomedical problems
GP denotes guided problems
M denotes Master It tutorial available in Enhanced WebAssign
QC denotes asking for quantitative and conceptual reasoning
S denotes symbolic reasoning problem
W denotes Watch It video solution available in Enhanced WebAssign

15.3 Coulomb's Law

1. **QC** A 7.50-nC charge is located 1.80 m from a 4.20-nC charge. (a) Find the magnitude of the electrostatic force that one particle exerts on the other. (b) Is the force attractive or repulsive?

2. A charged particle A exerts a force of 2.62 N to the right on charged particle B when the particles are 13.7 mm apart. Particle B moves straight away from A to make the distance between them 17.7 mm. What vector force does particle B then exert on A?

3. **QC** Two metal balls A and B of negligible radius are floating at rest on Space Station Freedom between two metal bulkheads, connected by a taut nonconducting thread of length 2.00 m. Ball A carries charge q, and ball B carries charge $2q$. Each ball is 1.00 m away from a bulkhead. (a) If the tension in the string is 2.50 N, what is the magnitude of q? (b) What happens to the system as time passes? Explain.

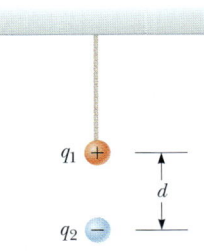

Figure P15.4

4. A small sphere of mass $m = 7.50$ g and charge $q_1 = 32.0$ nC is attached to the end of a string and hangs vertically as in Figure P15.4. A second charge of equal mass and charge $q_2 = -58.0$ nC is located below the first charge a

distance $d = 2.00$ cm below the first charge as in Figure P15.4. (a) Find the tension in the string. (b) If the string can withstand a maximum tension of 0.180 N, what is the smallest value d can have before the string breaks?

5. The nucleus of ^{8}Be, which consists of 4 protons and 4 neutrons, is very unstable and spontaneously breaks into two alpha particles (helium nuclei, each consisting of 2 protons and 2 neutrons). (a) What is the force between the two alpha particles when they are 5.00×10^{-15} m apart, and (b) what is the initial magnitude of the acceleration of the alpha particles due to this force? Note that the mass of an alpha particle is 4.002 6 u.

6. **BIO** A molecule of DNA (deoxyribonucleic acid) is 2.17 μm long. The ends of the molecule become singly ionized: negative on one end, positive on the other. The helical molecule acts like a spring and compresses 1.00% upon becoming charged. Determine the effective spring constant of the molecule.

7. A small sphere of charge 0.800 μC hangs from the end of a spring as in Figure P15.7a. When another small sphere of charge -0.600 μC is held beneath the first sphere as in Figure P15.7b, the spring stretches by $d = 3.50$ cm from its original length and reaches a new equilibrium position with a separation between the charges of $r = 5.00$ cm. What is the force constant of the spring?

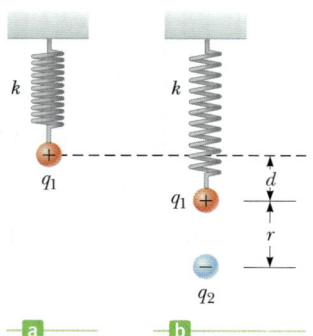

Figure P15.7

8. **S** Four point charges are at the corners of a square of side a as shown in Figure P15.8. Determine the magnitude and direction of the resultant electric force on q, with k_e, q, and a left in symbolic form.

Figure P15.8

9. Two small identical conducting spheres are placed with their centers 0.30 m apart. One is given a charge of 12×10^{-9} C, the other a charge of -18×10^{-9} C. (a) Find the electrostatic force exerted on one sphere by the other. (b) The spheres are connected by a conducting wire. Find the electrostatic force between the two after equilibrium is reached, where both spheres have the same charge.

10. Calculate the magnitude and direction of the Coulomb force on each of the three charges shown in Figure P15.10.

Figure P15.10 Problems 10 and 18.

11. **M** Three charges are arranged as shown in Figure P15.11. Find the magnitude and direction of the electrostatic force on the charge at the origin.

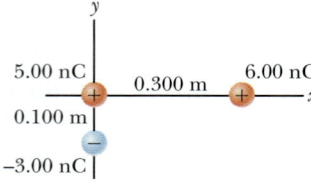

Figure P15.11

12. A positive charge $q_1 = 2.70$ μC on a frictionless horizontal surface is attached to a spring of force constant k as in Figure P15.12. When a charge of $q_2 = -8.60$ μC is placed 9.50 cm away from the positive charge, the spring stretches by 5.00 mm, reducing the distance between charges to $d = 9.00$ cm. Find the value of k.

Figure P15.12

13. Three point charges are located at the corners of an equilateral triangle as in Figure P15.13. Find the magnitude and direction of the net electric force on the 2.00 μC charge.

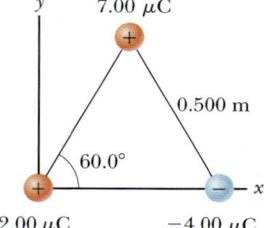

Figure P15.13 Problems 13 and 24.

14. Two identical metal blocks resting on a frictionless horizontal surface are connected by a light metal spring having constant $k = 100$ N/m and unstretched length $L_i = 0.400$ m as in Figure P15.14a. A charge Q is slowly placed on each block causing the spring to stretch to an equilibrium length $L = 0.500$ m as in Figure P15.14b. Determine the value of Q, modeling the blocks as charged particles.

Figure P15.14

15. **W** Two small metallic spheres, each of mass $m = 0.20$ g, are suspended as pendulums by light strings from a common point as shown in Figure P15.15. The spheres are given the same electric charge, and it is found that they come to equilibrium when each string is at an angle of $\theta = 5.0°$ with the vertical. If each string has length $L = 30.0$ cm, what is the magnitude of the charge on each sphere?

Figure P15.15

16. **GP** Particle A of charge 3.00×10^{-4} C is at the origin, particle B of charge -6.00×10^{-4} C is at (4.00 m, 0) and particle C of charge 1.00×10^{-4} C is at (0, 3.00 m). (a) What is the x-component of the electric force exerted by A on C? (b) What is the y-component of the force exerted by A on C? (c) Find the magnitude of the force exerted by B on C. (d) Calculate the x-component of the force exerted by B on C. (e) Calculate the y-component of the force exerted by B on C. (f) Sum the two x-components to obtain the resultant x-component of the electric force acting on C. (g) Repeat part (f) for the y-component. (h) Find the magnitude and direction of the resultant electric force acting on C.

15.4 The Electric Field

17. A small object of mass 3.80 g and charge -18 μC is suspended motionless above the ground when immersed in a uniform electric field perpendicular to the ground. What is the magnitude and direction of the electric field?

18. (a) Determine the electric field strength at a point 1.00 cm to the left of the middle charge shown in Figure P15.10. (b) If a charge of -2.00 μC is placed at this point, what are the magnitude and direction of the force on it?

19. An electric field of magnitude 5.25×10^5 N/C points due south at a certain location. Find the magnitude and direction of the force on a -6.00 μC charge at this location.

20. **W** An electron is accelerated by a constant electric field of magnitude 300 N/C. (a) Find the acceleration of the electron. (b) Use the equations of motion with constant acceleration to find the electron's speed after 1.00×10^{-8} s, assuming it starts from rest.

21. A small block of mass m and charge Q is placed on an insulated, frictionless, inclined plane of angle θ as in Figure P15.21. An electric field is applied parallel to the incline. (a) Find an expression for the magnitude of the electric field that enables the block to remain at rest. (b) If $m = 5.40$ g, $Q = -7.00$ μC, and

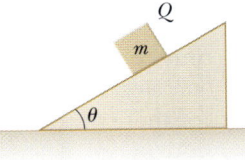

Figure P15.21

$\theta = 25.0°$, determine the magnitude and direction of the electric field that enables the block to remain at rest on the incline.

22. A small sphere of charge $q = +68$ μC and mass $m = 5.8$ g is attached to a light string and placed in a uniform electric field $\vec{E}$ that makes an angle $\theta = 37°$ with the horizontal. The opposite end of the string is attached to a wall and the sphere is in static equilibrium when the string is horizontal as in Figure P15.22. (a) Construct a free body diagram for the sphere. Find (b) the magnitude of the electric field and (c) the tension in the string.

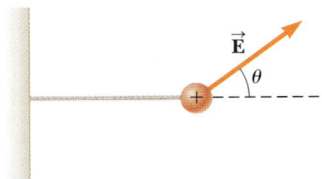

Figure P15.22

23. **M** A proton accelerates from rest in a uniform electric field of 640 N/C. At some later time, its speed is 1.20×10^6 m/s. (a) Find the magnitude of the acceleration of the proton. (b) How long does it take the proton to reach this speed? (c) How far has it moved in that interval? (d) What is its kinetic energy at the later time?

24. **Q|C** (a) Find the magnitude and direction of the electric field at the position of the 2.00 μC charge in Figure P15.13. (b) How would the electric field at that point be affected if the charge there were doubled? Would the magnitude of the electric force be affected?

25. **S** Two equal positive charges are at opposite corners of a trapezoid as in Figure P15.25. Find symbolic expressions for the components of the electric field at the point P.

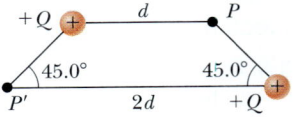

Figure P15.25

26. Three point charges are located on a circular arc as shown in Figure P15.26. (a) What is the total electric field at P, the center of the arc? (b) Find the electric force that would be exerted on a -5.00-nC charge placed at P.

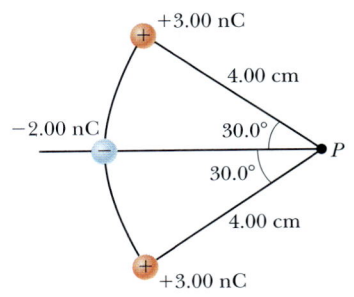

Figure P15.26

27. In Figure P15.27 determine the point (other than infinity) at which the total electric field is zero.

Figure P15.27

28. Three charges are at the corners of an equilateral triangle, as shown in Figure P15.28. Calculate the electric field at a point midway between the two charges on the x-axis.

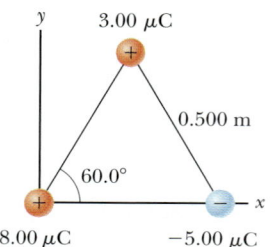

Figure P15.28

29. Three identical charges ($q = -5.0\ \mu C$) lie along a circle of radius 2.0 m at angles of 30°, 150°, and 270°, as shown in Figure P15.29. What is the resultant electric field at the center of the circle?

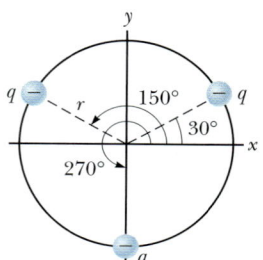

Figure P15.29

15.5 Electric Field Lines
15.6 Conductors in Electrostatic Equilibrium

30. Figure P15.30 shows the electric field lines for two point charges separated by a small distance. (a) Determine the ratio q_1/q_2. (b) What are the signs of q_1 and q_2?

31. (a) Sketch the electric field lines around an isolated point charge $q > 0$. (b) Sketch the electric field pattern around an isolated negative point charge of magnitude $-2q$.

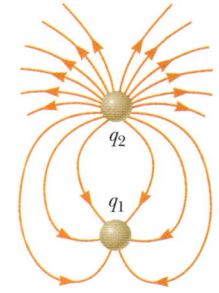

Figure P15.30

32. (a) Sketch the electric field pattern around two positive point charges of magnitude $1\ \mu C$ placed close together. (b) Sketch the electric field pattern around two negative point charges of $-2\ \mu C$, placed close together. (c) Sketch the pattern around two point charges of $+1\ \mu C$ and $-2\ \mu C$, placed close together.

33. Two point charges are a small distance apart. (a) Sketch the electric field lines for the two if one has a charge four times that of the other and both charges are positive. (b) Repeat for the case in which both charges are negative.

34. S Three equal positive charges are at the corners of an equilateral triangle of side a as in Figure P15.34. Assume the three charges together create an electric field. (a) Sketch the electric field lines in the plane of the charges. (b) Find the location of one point (other than ∞) where the electric field is zero. What are (c) the magnitude and (d) the direction of the electric field at P due to the two charges at the base?

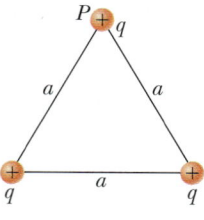

Figure P15.34

35. Refer to Figure 15.20. The charge lowered into the center of the hollow conductor has a magnitude of $5\ \mu C$. Find the magnitude and sign of the charge on the inside and outside of the hollow conductor when the charge is as shown in (a) Figure 15.20a, (b) Figure 15.20b, (c) Figure 15.20c, and (d) Figure 15.20d.

15.8 The Van de Graaff Generator

36. The dome of a Van de Graaff generator receives a charge of 2.0×10^{-4} C. Find the strength of the electric field (a) inside the dome, (b) at the surface of the dome, assuming it has a radius of 1.0 m, and (c) 4.0 m from the center of the dome. *Hint:* See Section 15.6 to review properties of conductors in electrostatic equilibrium. Also, note that the points on the surface are outside a spherically symmetric charge distribution; the total charge may be considered to be located at the center of the sphere.

37. If the electric field strength in air exceeds 3.0×10^6 N/C, the air becomes a conductor. Using this fact, determine the maximum amount of charge that can be carried by a metal sphere 2.0 m in radius. (See the hint in Problem 36.)

38. In the Millikan oil-drop experiment illustrated in Figure 15.21, an atomizer (a sprayer with a fine nozzle) is used to introduce many tiny droplets of oil between two oppositely charged parallel metal plates. Some of the droplets pick up one or more excess electrons. The charge on the plates is adjusted so that the electric force on the excess electrons exactly balances the weight of the droplet. The idea is to look for a droplet that has the smallest electric force and assume it has only one excess electron. This strategy lets the observer measure the charge on the electron. Suppose we are using an electric field of 3×10^4 N/C. The charge on one electron is about 1.6×10^{-19} C. Estimate the radius of an oil drop of density 858 kg/m³ for which its weight could be balanced by the electric force of this field on one electron. (Problem 38 is courtesy of E. F. Redish. For more problems of this type, visit www.physics.umd.edu/perg/.)

39. W A Van de Graaff generator is charged so that a proton at its surface accelerates radially outward at 1.52×10^{12} m/s². Find (a) the magnitude of the electric force on the proton at that instant and (b) the magnitude and direction of the electric field at the surface of the generator.

15.9 Electric Flux and Gauss's Law

40. A uniform electric field of magnitude $E = 435$ N/C makes an angle of $\theta = 65.0°$ with a plane surface of area $A = 3.50$ m^2 as in Figure P15.40. Find the electric flux through this surface.

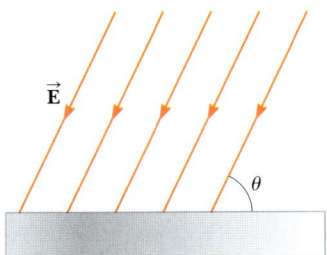

Figure P15.40

41. **M** An electric field of intensity 3.50 kN/C is applied along the x-axis. Calculate the electric flux through a rectangular plane 0.350 m wide and 0.700 m long if (a) the plane is parallel to the yz-plane, (b) the plane is parallel to the xy-plane, and (c) the plane contains the y-axis and its normal makes an angle of 40.0° with the x-axis.

42. **QC** The electric field everywhere on the surface of a charged sphere of radius 0.230 m has a magnitude of 575 N/C and points radially outward from the center of the sphere. (a) What is the net charge on the sphere? (b) What can you conclude about the nature and distribution of charge inside the sphere?

43. **S** Four closed surfaces, S_1 through S_4, together with the charges $-2Q$, Q, and $-Q$, are sketched in Figure P15.43. (The colored lines are the intersections of the surfaces with the page.) Find the electric flux through each surface.

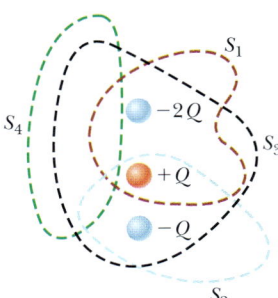

Figure P15.43

44. A charge $q = +5.80$ μC is located at the center of a regular tetrahedron (a four-sided surface) as in Figure P15.44. Find (a) the total electric flux through the tetrahedron and (b) the electric flux through one face of the tetrahedron.

Figure P15.44

45. A point charge q is located at the center of a spherical shell of radius a that has a charge $-q$ uniformly distributed on its surface. Find the electric field (a) for all points outside the spherical shell and (b) for a point inside the shell a distance r from the center.

46. **QC** A charge of 1.70×10^2 μC is at the center of a cube of edge 80.0 cm. No other charges are nearby. (a) Find the flux through the whole surface of the cube. (b) Find the flux through each face of the cube. (c) Would your answers to parts (a) or (b) change if the charge were not at the center? Explain.

47. Suppose the conducting spherical shell of Figure 15.29 carries a charge of 3.00 nC and that a charge of -2.00 nC is at the center of the sphere. If $a = 2.00$ m and $b = 2.40$ m, find the electric field at (a) $r = 1.50$ m, (b) $r = 2.20$ m, and (c) $r = 2.50$ m. (d) What is the charge distribution on the sphere?

48. **S** A very large nonconducting plate lying in the xy-plane carries a charge per unit area of σ. A second such plate located at $z = 2.00$ cm and oriented parallel to the xy-plane carries a charge per unit area of -2σ. Find the electric field (a) for $z < 0$, (b) $0 < z < 2.00$ cm, and (c) $z > 2.00$ cm.

Additional Problems

49. In deep space two spheres each of radius 5.00 m are connected by a 3.00×10^2 m nonconducting cord. If a uniformly distributed charge of 35.0 mC resides on the surface of each sphere, calculate the tension in the cord.

50. **QC** A nonconducting, thin plane sheet of charge carries a uniform charge per unit area of 5.20 μC/m^2 as in Figure 15.30. (a) Find the electric field at a distance of 8.70 cm from the plate. (b) Explain whether your result changes as the distance from the sheet is varied.

51. **W** Three point charges are aligned along the x-axis as shown in Figure P15.51. Find the electric field at the position $x = +2.0$ m, $y = 0$.

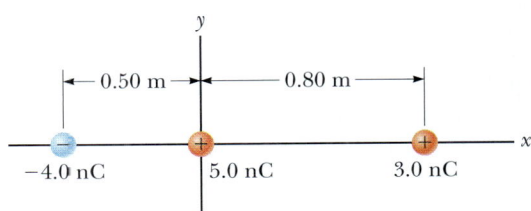

Figure P15.51

52. **M** A small plastic ball of mass $m = 2.00$ g is suspended by a string of length $L = 20.0$ cm in a uniform electric field, as shown in Figure P15.52. If the ball is in

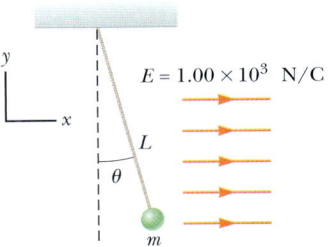

Figure P15.52

equilibrium when the string makes a $\theta = 15.0°$ angle with the vertical as indicated, what is the net charge on the ball?

53. (a) Two identical point charges $+q$ are located on the y-axis at $y = +a$ and $y = -a$. What is the electric field along the x-axis at $x = b$? (b) A circular ring of charge of radius a has a total positive charge Q distributed uniformly around it. The ring is in the $x = 0$ plane with its center at the origin. What is the electric field along the x-axis at $x = b$ due to the ring of charge? *Hint:* Consider the charge Q to consist of many pairs of identical point charges positioned at the ends of diameters of the ring.

54. **S** The electrons in a particle beam each have a kinetic energy K. Find the magnitude of the electric field that will stop these electrons in a distance d, expressing the answer symbolically in terms of K, e, and d. Should the electric field point in the direction of the motion of the electron, or should it point in the opposite direction?

55. **S** A point charge $+2Q$ is at the origin and a point charge $-Q$ is located along the x-axis at $x = d$ as in Figure P15.55. Find symbolic expressions for the components of the net force on a third point charge $+Q$ located along the y-axis at $y = d$.

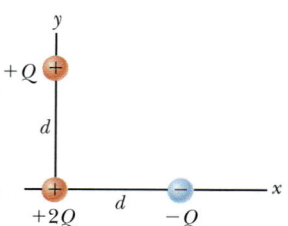

Figure P15.55

56. **QC** A 1.00-g cork ball having a positive charge of 2.00 mC is suspended vertically on a 0.500-m-long light string in the presence of a uniform downward-directed electric field of magnitude $E = 1.00 \times 10^5$ N/C as in Figure P15.56. If the ball is displaced slightly from the vertical, it oscillates like a simple pendulum. (a) Determine the period of the ball's oscillation. (b) Should gravity be included in the calculation for part (a)? Explain.

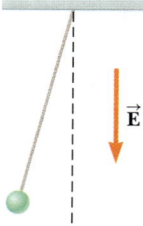

Figure P15.56

57. Two 2.0-g spheres are suspended by 10.0-cm-long light strings (Fig. P15.57). A uniform electric field is applied in the x-direction. If the spheres have charges of -5.0×10^{-8} C and $+5.0 \times 10^{-8}$ C, determine the electric field intensity that enables the spheres to be in equilibrium at $\theta = 10°$.

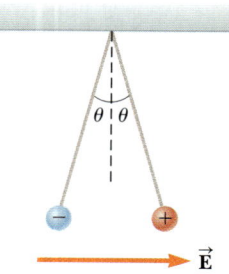

Figure P15.57

58. **QC** A point charge of magnitude 5.00 µC is at the origin of a coordinate system, and a charge of -4.00 µC is

at the point $x = 1.00$ m. There is a point on the x-axis, at x less than infinity, where the electric field goes to zero. (a) Show by conceptual arguments that this point cannot be located between the charges. (b) Show by conceptual arguments that the point cannot be at any location between $x = 0$ and negative infinity. (c) Show by conceptual arguments that the point must be between $x = 1.00$ m and $x =$ positive infinity. (d) Use the values given to find the point and show that it is consistent with your conceptual argument.

59. Two hard rubber spheres, each of mass $m = 15.0$ g, are rubbed with fur on a dry day and are then suspended with two insulating strings of length $L = 5.00$ cm whose support points are a distance $d = 3.00$ cm from each other as shown in Figure P15.59. During the rubbing process, one sphere receives exactly twice the charge of the other. They are observed to hang at equilibrium,

Figure P15.59

each at an angle of $\theta = 10.0°$ with the vertical. Find the amount of charge on each sphere.

60. Two small beads having positive charges $q_1 = 3q$ and $q_2 = q$ are fixed at the opposite ends of a horizontal insulating rod of length $d = 1.50$ m. The bead with charge q_1 is at the origin. As shown in Figure P15.60, a

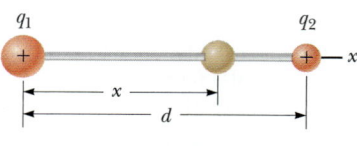

Figure P15.60

third small charged bead is free to slide on the rod. At what position x is the third bead in equilibrium?

61. **M** A solid conducting sphere of radius 2.00 cm has a charge of 8.00 µC. A conducting spherical shell of inner radius 4.00 cm and outer radius 5.00 cm is concentric with the solid sphere and has a charge of -4.00 µC. Find the electric field at (a) $r = 1.00$ cm, (b) $r = 3.00$ cm, (c) $r = 4.50$ cm, and (d) $r = 7.00$ cm from the center of this charge configuration.

62. Three identical point charges, each of mass $m = 0.100$ kg, hang from three strings, as shown in Figure P15.62. If the lengths of the left and right strings are each $L = 30.0$ cm and if the angle θ is 45.0°, determine the value of q.

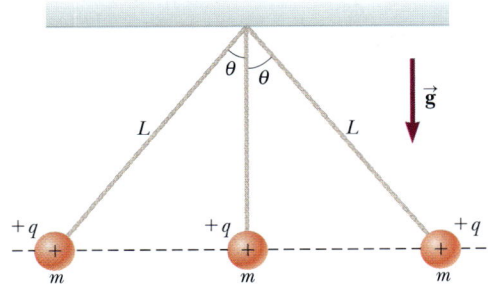

Figure P15.62

63. Each of the electrons in a particle beam has a kinetic energy of 1.60×10^{-17} J. (a) What is the magnitude of the uniform electric field (pointing in the direction of the electrons' movement) that will stop these electrons in a distance of 10.0 cm? (b) How long will it take to stop the electrons? (c) After the electrons stop, what will they do? Explain.

64. Protons are projected with an initial speed $v_0 = 9\,550$ m/s into a region where a uniform electric field of magnitude $E = 720$ N/C is present (Fig. P15.64). The protons are to hit a target that lies a horizontal distance of 1.27 mm from the point where the protons are launched. Find (a) the two projection angles θ that will result in a hit and (b) the total duration of flight for each of the two trajectories.

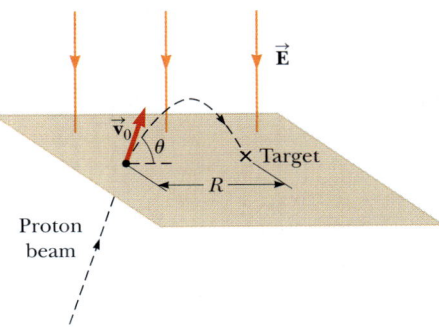

Figure P15.64

The only effective treatment for a patient with a heart in ventricular fibrillation, a spastic quivering of the heart muscle that is fatal in minutes, is an electrical shock delivered by a defibrillator. A capacitor in the defibrillator stores a large charge at high voltage and delivers it rapidly, shocking the heart and restoring a normal heart beat.

16 Electrical Energy and Capacitance

16.1 Electric Potential Energy and Electric Potential

16.2 Electric Potential and Potential Energy Due to Point Charges

16.3 Potentials and Charged Conductors

16.4 Equipotential Surfaces

16.5 Applications

16.6 Capacitance

16.7 The Parallel-Plate Capacitor

16.8 Combinations of Capacitors

16.9 Energy Stored in a Charged Capacitor

16.10 Capacitors with Dielectrics

The concept of potential energy was first introduced in Chapter 5 in connection with the conservative forces of gravity and springs. By using the principle of conservation of energy, we were often able to avoid working directly with forces when solving problems. Here we learn that the potential energy concept is also useful in the study of electricity. Because the Coulomb force is conservative, we can define an electric potential energy corresponding to that force. In addition, we define an electric potential—the potential energy per unit charge—corresponding to the electric field.

With the concept of electric potential in hand, we can begin to understand electric circuits, starting with an investigation of common circuit elements called capacitors. These simple devices store electrical energy and have found uses virtually everywhere, from etched circuits on a microchip to the creation of enormous bursts of power in fusion experiments.

16.1 Electric Potential Energy and Electric Potential

LEARNING OBJECTIVES

1. Define electric potential energy difference for a constant electric field in terms of the work done by the field.
2. Contrast the concept of electric potential energy with that of electric potential.
3. Apply the work-energy theorem to systems involving electric potential energy and electric potential.

Electric potential energy and electric potential are closely related concepts. The electric potential turns out to be just the electric potential energy per unit charge. This relationship is similar to that between electric force and the electric field, which is the electric force per unit charge.

Work and Electric Potential Energy

Recall from Chapter 5 that the work done by a conservative force $\vec{F}$ on an object depends only on the initial and final positions of the object and not on the path taken between those two points. This, in turn, means that a potential energy function PE exists. As we have seen, potential energy is a scalar quantity with the change in potential energy equal by definition to the negative of the work done by the conservative force: $\Delta PE = PE_f - PE_i = -W_F$.

Both the Coulomb force law and the universal law of gravity are proportional to $1/r^2$. Because they have the same mathematical form and because the gravity force is conservative, it follows that **the Coulomb force is also conservative.** As with gravity, an electrical potential energy function can be associated with this force.

To make these ideas more quantitative, imagine a small positive charge placed at point A in a *uniform* electric field $\vec{E}$, as in Figure 16.1. For simplicity, we first consider only constant electric fields and charges that move parallel to that field in one dimension (taken to be the x-axis). The electric field between equally and oppositely charged parallel plates is an example of a field that is approximately constant. (See Chapter 15.) As the charge moves from point A to point B under the influence of the electric field $\vec{E}$, the work done on the charge by the electric field is equal to the part of the electric force $q\vec{E}$ acting parallel to the displacement times the displacement $\Delta x = x_f - x_i$:

$$W_{AB} = F_x \Delta x = qE_x(x_f - x_i)$$

In this expression q is the charge and E_x is the vector component of $\vec{E}$ in the x-direction (*not* the magnitude of $\vec{E}$). Unlike the magnitude of $\vec{E}$, the component E_x can be positive or negative, depending on the direction of $\vec{E}$, although in Figure 16.1 E_x is positive. Finally, note that the displacement, like q and E_x, can also be either positive or negative, depending on the direction of the displacement.

The preceding expression for the work done by an electric field on a charge moving in one dimension is valid for both positive and negative charges and for constant electric fields pointing in *any* direction. When numbers are substituted with correct signs, the overall correct sign automatically results. In some books the expression $W = qEd$ is used, instead, where E is the magnitude of the electric field and d is the distance the particle travels. The weakness of this formulation is that it doesn't allow, mathematically, for negative electric work on positive charges, nor for positive electric work on negative charges! Nonetheless, the expression is easy to remember and useful for finding magnitudes: the magnitude of the work done by a constant electric field on a charge moving parallel to the field is always given by $|W| = |q|Ed$.

We can substitute our definition of electric work into the work–energy theorem (assume other forces are absent):

$$W = qE_x \Delta x = \Delta KE$$

The electric force is conservative, so the electric work depends only on the endpoints of the path, A and B, not on the path taken. Therefore, as the charge accelerates to the right in Figure 16.1, it gains kinetic energy and loses an equal amount of potential energy. Recall from Chapter 5 that **the work done by a conservative force can be reinterpreted as the negative of the change in a potential energy**

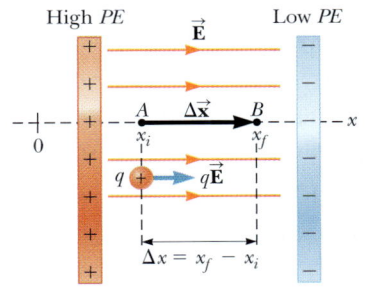

Figure 16.1 When a charge q moves in a uniform electric field $\vec{E}$ from point A to point B, the work done on the charge by the electric force is $qE_x \Delta x$.

associated with that force. This interpretation motivates the definition of the change in electric potential energy:

▶ Change in electric potential energy

The change in the electric potential energy, ΔPE, of a system consisting of an object of charge q moving through a displacement Δx in a constant electric field $\vec{E}$ is given by

$$\Delta PE = -W_{AB} = -qE_x \Delta x \quad [16.1]$$

where E_x is the x-component of the electric field and $\Delta x = x_f - x_i$ is the displacement of the charge along the x-axis.

SI unit: joule (J)

Although potential energy can be defined for any electric field, **Equation 16.1 is valid only for the case of a uniform (i.e., constant) electric field, for a particle that undergoes a displacement along a given axis (here called the x-axis).** Because the electric field is conservative, the change in potential energy doesn't depend on the path. Consequently, it's unimportant whether or not the charge remains on the axis at all times during the displacement: the change in potential energy will be the same. In subsequent sections we will examine situations in which the electric field is not uniform.

Electric and gravitational potential energy can be compared in Figure 16.2. In this figure the electric and gravitational fields are both directed downwards. We see that positive charge in an electric field acts very much like mass in a gravity field: a positive charge at point A falls in the direction of the electric field, just as a positive mass falls in the direction of the gravity field. Let point B be the zero point for potential energy in both Figures 16.2a and 16.2b. From conservation of energy, in falling from point A to point B the positive charge gains kinetic energy equal in magnitude to the loss of electric potential energy:

$$\Delta KE + \Delta PE_{el} = \Delta KE + (0 - |q|Ed) = 0 \quad \rightarrow \quad \Delta KE = |q|Ed$$

The absolute-value signs on q are there only to make explicit that the charge is positive in this case. Similarly, the object in Figure 16.2b gains kinetic energy equal in magnitude to the loss of gravitational potential energy:

$$\Delta KE + \Delta PE_g = \Delta KE + (0 - mgd) = 0 \quad \rightarrow \quad \Delta KE = mgd$$

So for positive charges, electric potential energy works very much like gravitational potential energy. In both cases moving an object opposite the direction of the field results in a gain of potential energy, and upon release, the potential energy is converted to the object's kinetic energy.

Electric potential energy differs significantly from gravitational potential energy, however, in that there are two kinds of electrical charge—positive and negative—whereas gravity has only positive "gravitational charge" (i.e., mass). A negatively charged particle at rest at point A in Figure 16.2a would have to be *pushed* down to point B. To see why, apply the work–energy theorem to a negative charge at rest at point A and assumed to have some speed v on arriving at point B:

$$W = \Delta KE + \Delta PE_{el} = (\tfrac{1}{2}mv^2 - 0) + [0 - (-|q|Ed)]$$
$$W = \tfrac{1}{2}mv^2 + |q|Ed$$

Notice that the negative charge, $-|q|$, unlike the positive charge, had a positive change in electric potential energy in moving from point A to point B. If the negative charge has any speed at point B, the kinetic energy corresponding to that speed is also positive. Because both terms on the right-hand side of the work–energy equation are positive, there is no way of getting the negative charge from point A to point B without doing positive work W on it. In fact, if the negative charge is simply released at point A, it will "fall" upwards against the direction of the field!

When a positive test charge moves from A to B, the electric potential energy decreases.

When an object with mass moves from A to B, the gravitational potential energy decreases.

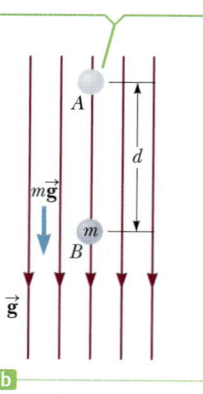

Figure 16.2 (a) When the electric field $\vec{E}$ is directed downward, point B is at a lower electric potential than point A. (b) An object of mass m moves in the direction of the gravitational field $\vec{g}$.

Quick Quiz

16.1 If an electron is released from rest in a uniform electric field, does the electric potential energy of the charge–field system (a) increase, (b) decrease, or (c) remain the same?

EXAMPLE 16.1 Potential Energy Differences in an Electric Field

GOAL Illustrate the concept of electric potential energy.

PROBLEM A proton is released from rest at $x = -2.00$ cm in a constant electric field with magnitude 1.50×10^3 N/C, pointing in the positive x-direction. **(a)** Calculate the change in the electric potential energy associated with the proton when it reaches $x = 5.00$ cm. **(b)** An electron is now fired in the same direction from the same position. What is the change in electric potential energy associated with the electron if it reaches $x = 12.0$ cm? **(c)** If the direction of the electric field is reversed and an electron is released from rest at $x = 3.00$ cm, by how much has the electric potential energy changed when the electron reaches $x = 7.00$ cm?

STRATEGY This problem requires a straightforward substitution of given values into the definition of electric potential energy, Equation 16.1.

SOLUTION

(a) Calculate the change in the electric potential energy associated with the proton.

Apply Equation 16.1:
$$\Delta PE = -qE_x \Delta x = -qE_x(x_f - x_i)$$
$$= -(1.60 \times 10^{-19} \text{ C})(1.50 \times 10^3 \text{ N/C})$$
$$\times [0.050\,0 \text{ m} - (-0.020\,0 \text{ m})]$$
$$= \boxed{-1.68 \times 10^{-17} \text{ J}}$$

(b) Find the change in electric potential energy associated with an electron fired from $x = -0.020\,0$ m and reaching $x = 0.120$ m.

Apply Equation 16.1, but in this case note that the electric charge q is negative:
$$\Delta PE = -qE_x \Delta x = -qE_x(x_f - x_i)$$
$$= -(-1.60 \times 10^{-19} \text{ C})(1.50 \times 10^3 \text{ N/C})$$
$$\times [(0.120 \text{ m} - (-0.020\,0 \text{ m})]$$
$$= \boxed{+3.36 \times 10^{-17} \text{ J}}$$

(c) Find the change in potential energy associated with an electron traveling from $x = 3.00$ cm to $x = 7.00$ cm if the direction of the electric field is reversed.

Substitute, but now the electric field points in the negative x-direction, hence carries a minus sign:
$$\Delta PE = -qE_x \Delta x = -qE_x(x_f - x_i)$$
$$= -(-1.60 \times 10^{-19} \text{ C})(-1.50 \times 10^3 \text{ N/C})$$
$$\times (0.070 \text{ m} - 0.030 \text{ m})$$
$$= \boxed{-9.60 \times 10^{-18} \text{ J}}$$

REMARKS Notice that the proton (actually the proton–field system) lost potential energy when it moved in the positive x-direction, whereas the electron gained potential energy when it moved in the same direction. Finding changes in potential energy with the field reversed was only a matter of supplying a minus sign, bringing the total number in this case to three! It's important not to drop any of the signs.

QUESTION 16.1 True or False: When an electron is released from rest in a constant electric field, the change in the electric potential energy associated with the electron becomes more negative with time.

EXERCISE 16.1 Find the change in electric potential energy associated with the electron in part (b) as it goes on from $x = 0.120$ m to $x = -0.180$ m. (Note that the electron must turn around and go back at some point. The location of the turning point is unimportant because changes in potential energy depend only on the endpoints of the path.)

ANSWER -7.20×10^{-17} J

EXAMPLE 16.2 Dynamics of Charged Particles

GOAL Use electric potential energy in conservation of energy problems.

PROBLEM (a) Find the speed of the proton at $x = 0.050\ 0$ m in part (a) of Example 16.1. (b) Find the initial speed of the electron (at $x = -2.00$ cm) in part (b) of Example 16.1 given that its speed has fallen by half when it reaches $x = 0.120$ m.

STRATEGY Apply conservation of energy, solving for the unknown speeds. Part (b) involves two equations: the conservation of energy equation and the condition $v_f = \frac{1}{2}v_i$ for the unknown initial and final speeds. The changes in electric potential energy have already been calculated in Example 16.1.

SOLUTION

(a) Calculate the proton's speed at $x = 0.050$ m.

Use conservation of energy, with an initial speed of zero:

$$\Delta KE + \Delta PE = 0 \quad \rightarrow \quad (\tfrac{1}{2}m_p v^2 - 0) + \Delta PE = 0$$

Solve for v and substitute the change in potential energy found in Example 16.1a:

$$v^2 = -\frac{2}{m_p}\Delta PE$$

$$v = \sqrt{-\frac{2}{m_p}\Delta PE}$$

$$= \sqrt{-\frac{2}{(1.67 \times 10^{-27}\ \text{kg})}(-1.68 \times 10^{-17}\ \text{J})}$$

$$= \boxed{1.42 \times 10^5\ \text{m/s}}$$

(b) Find the electron's initial speed (at $x = -2.00$ cm) given that its speed has fallen by half at $x = 0.120$ m.

Apply conservation of energy once again, substituting expressions for the initial and final kinetic energies:

$$\Delta KE + \Delta PE = 0$$
$$(\tfrac{1}{2}m_e v_f^2 - \tfrac{1}{2}m_e v_i^2) + \Delta PE = 0$$

Substitute the condition $v_f = \frac{1}{2}v_i$ and subtract the change in potential energy from both sides:

$$\tfrac{1}{2}m_e(\tfrac{1}{2}v_i)^2 - \tfrac{1}{2}m_e v_i^2 = -\Delta PE$$

Combine terms and solve for v_i, the initial speed, and substitute the change in potential energy found in Example 16.1b:

$$-\tfrac{3}{8}m_e v_i^2 = -\Delta PE$$

$$v_i = \sqrt{\frac{8\Delta PE}{3m_e}} = \sqrt{\frac{8(3.36 \times 10^{-17}\ \text{J})}{3(9.11 \times 10^{-31}\ \text{kg})}}$$

$$= \boxed{9.92 \times 10^6\ \text{m/s}}$$

REMARKS Although the changes in potential energy associated with the proton and electron were similar in magnitude, the effect on their speeds differed dramatically. The change in potential energy had a far greater effect on the much lighter electron than on the proton.

QUESTION 16.2 True or False: If a proton and electron both move through the same displacement in an electric field, the change in potential energy associated with the proton must be equal in magnitude and opposite in sign to the change in potential energy associated with the electron.

EXERCISE 16.2 Refer to Exercise 16.1. Find the electron's speed at $x = -0.180$ m. *Note*: Use the initial velocity from part (b) of Example 16.2.

ANSWER 1.35×10^7 m/s   The answer is 4.5% of the speed of light.

Electric Potential

In Chapter 15 it was convenient to define an electric field $\vec{E}$ related to the electric force $\vec{F} = q\vec{E}$. In this way the properties of fixed collections of charges could be easily studied, and the force on any particle in the electric field could be obtained simply by multiplying by the particle's charge q. For the same reasons, it's useful to define an *electric potential difference* ΔV related to the potential energy by $\Delta PE = q\Delta V$:

> The electric potential difference ΔV between points A and B is the change in electric potential energy as a charge q moves from A to B divided by the charge q:
>
> $$\Delta V = V_B - V_A = \frac{\Delta PE}{q} \qquad [16.2]$$
>
> **SI unit: joule per coulomb, or volt (J/C, or V)**

◀ Potential difference between two points

This definition is completely general, although in many cases calculus would be required to compute the change in potential energy of the system. Because electric potential energy is a scalar quantity, **electric potential is also a scalar quantity**. From Equation 16.2, we see that electric potential difference is a measure of the change in electric potential energy per unit charge. Alternately, the electric potential difference is the work per unit charge that would have to be done by some force to move a charge from point A to point B in the electric field. The SI unit of electric potential is the joule per coulomb, called the volt (V). From the definition of that unit, 1 J of work must be done to move a 1-C charge between two points that are at a potential difference of 1 V. In the process of moving through a potential difference of 1 V, the 1-C charge gains 1 J of energy.

For the special case of a uniform electric field such as that between charged parallel plates, dividing Equation 16.1 by q gives

$$\frac{\Delta PE}{q} = -E_x \Delta x$$

Comparing this equation with Equation 16.2, we find that

$$\Delta V = -E_x \Delta x \qquad [16.3]$$

Equation 16.3 shows that potential difference also has units of electric field times distance. It then follows that the SI unit of the electric field, the newton per coulomb, can also be expressed as volts per meter:

$$1 \text{ N/C} = 1 \text{ V/m}$$

Because Equation 16.3 is directly related to Equation 16.1, remember that it's valid only for the system consisting of a uniform electric field and a charge moving in one dimension.

Released from rest, positive charges accelerate spontaneously from regions of high potential to low potential. If a positive charge is given some initial velocity in the direction of high potential, it can move in that direction, but will slow and finally turn around, just like a ball tossed upwards in a gravity field. Negative charges do exactly the opposite: released from rest, they accelerate from regions of low potential toward regions of high potential. Work must be done on negative charges to make them go in the direction of lower electric potential.

Tip 16.1 Potential and Potential Energy

Electric potential is characteristic of the field only, independent of a test charge that may be placed in that field. On the other hand, potential energy is a characteristic of the charge-field system due to an interaction between the field and a charge placed in the field.

■ **Quick Quiz**

16.2 If a negatively charged particle is placed at rest in an electric potential field that increases in the positive x-direction, will the particle (a) accelerate in the positive x-direction, (b) accelerate in the negative x-direction, or (c) remain at rest?

16.3 Figure 16.3 is a graph of an electric potential as a function of position. If a positively charged particle is placed at point A, what will its subsequent motion be? Will it (a) go to the right, (b) go to the left, (c) remain at point A, or (d) oscillate around point B?

(*Continued*)

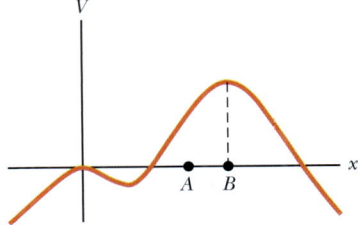

Figure 16.3 (Quick Quizzes 16.3 and 16.4)

16.4 If a negatively charged particle is placed at point B in Figure 16.3 and given a very small kick to the right, what will its subsequent motion be? Will it (a) go to the right and not return, (b) go to the left, (c) remain at point B, or (d) oscillate around point B?

APPLICATION
Automobile Batteries

An application of potential difference is the 12-V battery found in an automobile. Such a battery maintains a potential difference across its terminals, with the positive terminal 12 V higher in potential than the negative terminal. In practice the negative terminal is usually connected to the metal body of the car, which can be considered to be at a potential of zero volts. The battery provides the electrical current necessary to operate headlights, a radio, power windows, motors, and so forth. Now consider a charge of +1 C, to be moved around a circuit that contains the battery and some of these external devices. As the charge is moved inside the battery from the negative terminal (at 0 V) to the positive terminal (at 12 V), the work done on the charge by the battery is 12 J. Every coulomb of positive charge that leaves the positive terminal of the battery carries an energy of 12 J. As the charge moves through the external circuit toward the negative terminal, it gives up its 12 J of electrical energy to the external devices. When the charge reaches the negative terminal, its electrical energy is zero again. At this point, the battery takes over and restores 12 J of energy to the charge as it is moved from the negative to the positive terminal, enabling it to make another transit of the circuit. The actual amount of charge that leaves the battery each second and traverses the circuit depends on the properties of the external devices, as seen in the next chapter.

EXAMPLE 16.3 | TV Tubes and Atom Smashers

GOAL Relate electric potential to an electric field and conservation of energy.

PROBLEM In atom smashers (also known as cyclotrons and linear accelerators) charged particles are accelerated in much the same way they are accelerated in TV tubes: through potential differences. Suppose a proton is injected at a speed of 1.00×10^6 m/s between two plates 5.00 cm apart, as shown in Figure 16.4. The proton subsequently accelerates across the gap and exits through the opening. **(a)** What must the electric potential difference be if the exit speed is to be 3.00×10^6 m/s? **(b)** What is the electric field between the plates, assuming it's constant? The positive x-direction is to the right.

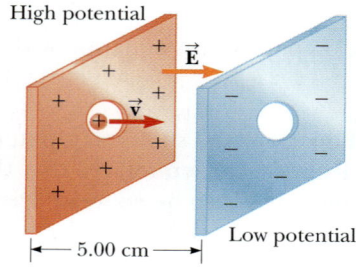

Figure 16.4 (Example 16.3) A proton enters a cavity and accelerates from one charged plate toward the other in an electric field $\vec{E}$.

STRATEGY Use conservation of energy, writing the change in potential energy in terms of the change in electric potential, ΔV, and solve for ΔV. For part (b), solve Equation 16.3 for the electric field.

SOLUTION

(a) Find the electric potential yielding the desired exit speed of the proton.

Apply conservation of energy, writing the potential energy in terms of the electric potential:

$$\Delta KE + \Delta PE = \Delta KE + q\Delta V = 0$$

Solve the energy equation for the change in potential:

$$\Delta V = -\frac{\Delta KE}{q} = -\frac{\frac{1}{2}m_p v_f^2 - \frac{1}{2}m_p v_i^2}{q} = -\frac{m_p}{2q}(v_f^2 - v_i^2)$$

Substitute the given values, obtaining the necessary potential difference:

$$\Delta V = -\frac{(1.67 \times 10^{-27} \text{ kg})}{2(1.60 \times 10^{-19} \text{ C})}[(3.00 \times 10^6 \text{ m/s})^2 - (1.00 \times 10^6 \text{ m/s})^2]$$

$$\Delta V = \boxed{-4.18 \times 10^4 \text{ V}}$$

(b) What electric field must exist between the plates?

Solve Equation 16.3 for the electric field and substitute:

$$E = -\frac{\Delta V}{\Delta x} = \frac{4.18 \times 10^4 \text{ V}}{0.050\ 0 \text{ m}} = \boxed{8.36 \times 10^5 \text{ N/C}}$$

REMARKS Systems of such cavities, consisting of alternating positive and negative plates, are used to accelerate charged particles to high speed before smashing them into targets. To prevent a slowing of, say, a positively charged particle after it passes through the negative plate of one cavity and enters the next, the charges on the plates are reversed. Otherwise, the particle would be traveling from the negative plate to a positive plate in the second cavity, and the kinetic energy gained in the previous cavity would be lost in the second.

QUESTION 16.3 True or False: A more massive particle gains less energy in traversing a given potential difference than does a lighter particle having the same charge.

EXERCISE 16.3 Suppose electrons in a TV tube are accelerated through a potential difference of 2.00×10^4 V from the heated cathode (negative electrode), where they are produced, toward the screen, which also serves as the anode (positive electrode), 25.0 cm away. (a) At what speed would the electrons impact the phosphors on the screen? Assume they accelerate from rest and ignore relativistic effects (Chapter 26). (b) What's the magnitude of the electric field, if it is assumed constant?

ANSWERS (a) 8.38×10^7 m/s (b) 8.00×10^4 V/m

16.2 Electric Potential and Potential Energy Due to Point Charges

LEARNING OBJECTIVES
1. Define the electric potential of a point charge and the potential energy of a pair of point charges.
2. Apply electric potential and electric potential energy to systems of charged particles.

In electric circuits a point of zero electric potential is often defined by grounding (connecting to the Earth) some point in the circuit. For example, if the negative terminal of a 12-V battery were connected to ground, it would be considered to have a potential of zero, whereas the positive terminal would have a potential of +12 V. The potential difference created by the battery, however, is only locally defined. In this section we describe the electric potential of a point charge, which is defined throughout space.

The electric field of a point charge extends throughout space, so its electric potential does, also. The zero point of electric potential could be taken anywhere, but is usually taken to be an infinite distance from the charge, far from its influence and the influence of any other charges. With this choice, the methods of calculus can be used to show that the electric potential created by a point charge q at any distance r from the charge is given by

Figure 16.5 Electric field and electric potential versus distance from a point charge of 1.11×10^{-10} C. Note that V is proportional to $1/r$, whereas E is proportional to $1/r^2$.

$$V = k_e \frac{q}{r} \quad [16.4]$$

◀ Electric potential created by a point charge

Equation 16.4 shows that the electric potential, or work per unit charge, required to move a positive test charge in from infinity to a distance r from a positive point charge q increases as the test charge moves closer to q. A plot of Equation 16.4 in Figure 16.5 shows that the potential associated with a point charge decreases as $1/r$ with increasing r, in contrast to the magnitude of the charge's electric field, which decreases as $1/r^2$.

The electric potential of two or more charges is obtained by applying the **superposition principle: the total electric potential at some point P due to several point charges is the algebraic sum of the electric potentials due to the individual charges.** This method is similar to the one used in Chapter 15 to find the resultant electric field at a point in space. Unlike electric field superposition, which involves a sum of vectors, the superposition of electric potentials requires evaluating a sum

◀ Superposition principle

Figure 16.6 The electric potential (in arbitrary units) in the plane containing an electric dipole. Potential is plotted in the vertical dimension.

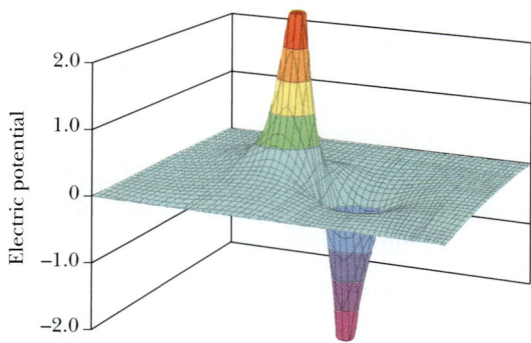

of scalars. As a result, it's much easier to evaluate the electric potential at some point due to several charges than to evaluate the electric field, which is a vector quantity.

Figure 16.6 is a computer-generated plot of the electric potential associated with an electric dipole, which consists of two charges of equal magnitude but opposite in sign. The charges lie in a horizontal plane at the center of the potential spikes. The value of the potential is plotted in the vertical dimension. The computer program has added the potential of each charge to arrive at total values of the potential.

Just as in the case of constant electric fields, there is a relationship between electric potential and electric potential energy. If V_1 is the electric potential due to charge q_1 at a point P (Fig. 16.7a) the work required to bring charge q_2 from infinity to P without acceleration is $q_2 V_1$. By definition, this work equals the potential energy PE of the two-particle system when the particles are separated by a distance r (Fig. 16.7b).

We can therefore express the electrical potential energy of the *pair* of charges as

▶ Potential energy of a pair of charges

$$PE = q_2 V_1 = k_e \frac{q_1 q_2}{r} \quad [16.5]$$

If the charges are of the *same* sign, PE is positive. Because like charges repel, positive work must be done on the system by an external agent to force the two charges near each other. Conversely, if the charges are of *opposite* sign, the force is attractive and PE is negative. This means that negative work must be done to prevent unlike charges from accelerating toward each other as they are brought close together.

Quick Quiz

16.5 Consider a collection of charges in a given region and suppose all other charges are distant and have a negligible effect. Further, the electric potential is taken to be zero at infinity. If the electric potential at a given point in the region is zero, which of the following statements must be true? (a) The electric field is zero at that point. (b) The electric potential energy is a minimum at that point. (c) There is no net charge in the region. (d) Some charges in the region are positive, and some are negative. (e) The charges have the same sign and are symmetrically arranged around the given point.

16.6 A spherical balloon contains a positively charged particle at its center. As the balloon is inflated to a larger volume while the charged particle remains at the center, which of the following are true? (a) The electric potential at the surface of the balloon increases. (b) The magnitude of the electric field at the surface of the balloon increases. (c) The electric flux through the balloon remains the same. (d) None of these.

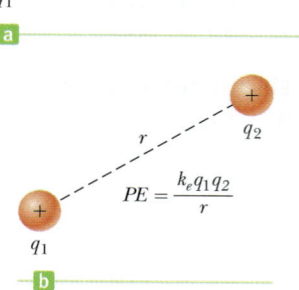

Figure 16.7 (a) The electric potential V_1 at P due to the point charge q_1 is $V_1 = k_e q_1/r$. (b) If a second charge, q_2, is brought from infinity to P, the potential energy of the pair is $PE = k_e q_1 q_2 / r$.

16.2 Electric Potential and Potential Energy Due to Point Charges

■ PROBLEM-SOLVING STRATEGY

Electric Potential

1. Draw a diagram of all charges and circle the point of interest.
2. Calculate the distance from each charge to the point of interest, labeling it on the diagram.
3. For each charge q, calculate the scalar quantity $V = \dfrac{k_e q}{r}$. The sign of each charge must be included in your calculations!
4. Sum all the numbers found in the previous step, obtaining the electric potential at the point of interest.

■ EXAMPLE 16.4 Finding the Electric Potential

GOAL Calculate the electric potential due to a collection of point charges.

PROBLEM A 5.00-μC point charge is at the origin, and a point charge $q_2 = -2.00$ μC is on the x-axis at (3.00, 0) m, as in Figure 16.8. **(a)** If the electric potential is taken to be zero at infinity, find the electric potential due to these charges at point P with coordinates (0, 4.00) m. **(b)** How much work is required to bring a third point charge of 4.00 μC from infinity to P?

STRATEGY For part (a), the electric potential at P due to each charge can be calculated from $V = k_e q/r$. The electric potential at P is the sum of these two quantities. For part (b), use the work–energy theorem, together with Equation 16.5, recalling that the potential at infinity is taken to be zero.

Figure 16.8 (Example 16.4) The electric potential at point P due to the point charges q_1 and q_2 is the algebraic sum of the potentials due to the individual charges.

SOLUTION

(a) Find the electric potential at point P.

Calculate the electric potential at P due to the 5.00-μC charge:

$$V_1 = k_e \dfrac{q_1}{r_1} = \left(8.99 \times 10^9 \dfrac{\text{N} \cdot \text{m}^2}{\text{C}^2}\right)\left(\dfrac{5.00 \times 10^{-6} \text{ C}}{4.00 \text{ m}}\right)$$
$$= 1.12 \times 10^4 \text{ V}$$

Find the electric potential at P due to the -2.00-μC charge:

$$V_2 = k_e \dfrac{q_2}{r_2} = \left(8.99 \times 10^9 \dfrac{\text{N} \cdot \text{m}^2}{\text{C}^2}\right)\left(\dfrac{-2.00 \times 10^{-6} \text{ C}}{5.00 \text{ m}}\right)$$
$$= -0.360 \times 10^4 \text{ V}$$

Sum the two quantities to find the total electric potential at P:

$$V_P = V_1 + V_2 = 1.12 \times 10^4 \text{ V} + (-0.360 \times 10^4 \text{ V})$$
$$= \boxed{7.6 \times 10^3 \text{ V}}$$

(b) Find the work needed to bring the 4.00-μC charge from infinity to P.

Apply the work–energy theorem, with Equation 16.5:

$$W = \Delta PE = q_3 \Delta V = q_3(V_P - V_\infty)$$
$$= (4.00 \times 10^{-6} \text{ C})(7.6 \times 10^3 \text{ V} - 0)$$
$$W = \boxed{3.0 \times 10^{-2} \text{ J}}$$

REMARKS Unlike the electric field, where vector addition is required, the electric potential due to more than one charge can be found with ordinary addition of scalars. Further, notice that the work required to move the charge is equal to the change in electric potential energy. The sum of the work done moving the particle plus the work done by the electric field is zero ($W_{\text{other}} + W_{\text{electric}} = 0$) because the particle starts and ends at rest. Therefore, $W_{\text{other}} = -W_{\text{electric}} = \Delta U_{\text{electric}} = q \Delta V$.

(Continued)

QUESTION 16.4 If q_2 were moved to the right, what would happen to the electric potential V_p at point P? (a) It would increase. (b) It would decrease. (c) It would remain the same.

EXERCISE 16.4 Suppose a charge of -2.00 μC is at the origin and a charge of 3.00 μC is at the point $(0, 3.00)$ m. (a) Find the electric potential at $(4.00, 0)$ m, assuming the electric potential is zero at infinity, and (b) find the work necessary to bring a 4.00 μC charge from infinity to the point $(4.00, 0)$ m.

ANSWERS (a) 8.99×10^2 V (b) 3.60×10^{-3} J

■ EXAMPLE 16.5 | Electric Potential Energy and Dynamics

GOAL Apply conservation of energy and electrical potential energy to a configuration of charges.

Figure 16.9 (Example 16.5)

PROBLEM Suppose three protons lie on the x-axis, at rest relative to one another at a given instant of time, as in Figure 16.9. If proton q_3 on the right is released while the others are held fixed in place, find a symbolic expression for the proton's speed at infinity and evaluate this speed when $r_0 = 2.00$ fm. (*Note*: 1 fm = 10^{-15} m.)

STRATEGY First calculate the initial electric potential energy associated with the system of three particles. There will be three terms, one for each interacting pair. Then calculate the final electric potential energy associated with the system when the proton on the right is arbitrarily far away. Because the electric potential energy falls off as $1/r$, two of the terms will vanish. Using conservation of energy then yields the speed of the particle in question.

SOLUTION

Calculate the electric potential energy associated with the initial configuration of charges:

$$PE_i = \frac{k_e q_1 q_2}{r_{12}} + \frac{k_e q_1 q_3}{r_{13}} + \frac{k_e q_2 q_3}{r_{23}} = \frac{k_e e^2}{r_0} + \frac{k_e e^2}{2r_0} + \frac{k_e e^2}{r_0}$$

Calculate the electric potential energy associated with the final configuration of charges:

$$PE_f = \frac{k_e q_1 q_2}{r_{12}} = \frac{k_e e^2}{r_0}$$

Write the conservation of energy equation:

$$\Delta KE + \Delta PE = KE_f - KE_i + PE_f - PE_i = 0$$

Substitute appropriate terms:

$$\tfrac{1}{2} m_3 v_3^2 - 0 + \frac{k_e e^2}{r_0} - \left(\frac{k_e e^2}{r_0} + \frac{k_e e^2}{2r_0} + \frac{k_e e^2}{r_0} \right) = 0$$

$$\tfrac{1}{2} m_3 v_3^2 - \left(\frac{k_e e^2}{2r_0} + \frac{k_e e^2}{r_0} \right) = 0$$

Solve for v_3 after combining the two remaining potential energy terms:

$$v_3 = \sqrt{\frac{3 k_e e^2}{m_3 r_0}}$$

Evaluate taking $r_0 = 2.00$ fm:

$$v_3 = \sqrt{\frac{3(8.99 \times 10^9 \text{ N} \cdot \text{m}^2/\text{C}^2)(1.60 \times 10^{-19} \text{ C})^2}{(1.67 \times 10^{-27} \text{ kg})(2.00 \times 10^{-15} \text{ m})}} = 1.44 \times 10^7 \text{ m/s}$$

REMARKS The difference in the initial and final potential energies yields the energy available for motion. This calculation is somewhat contrived because it would be difficult, although not impossible, to arrange such a configuration of protons; it could conceivably occur by chance inside a star.

QUESTION 16.5 If a fourth proton were placed to the right of q_3, how many additional potential energy terms would have to be calculated in the initial configuration?

EXERCISE 16.5 Starting from the initial configuration of three protons, suppose the end two particles are released simultaneously and the middle particle is fixed. Obtain a numerical answer for the speed of the two particles at infinity. (Note that their speeds, by symmetry, must be the same.)

ANSWER 1.31×10^7 m/s

16.3 Potentials and Charged Conductors

LEARNING OBJECTIVES
1. Discuss the electric potential of a perfect conductor.
2. Define the electron volt unit of energy.

The electric potential at all points on a charged conductor can be determined by combining Equations 16.1 and 16.2. From Equation 16.1, we see that the work done on a charge by electric forces is related to the change in electrical potential energy of the charge by

$$W = -\Delta PE$$

From Equation 16.2, we see that the change in electric potential energy between two points A and B is related to the potential difference between those points by

$$\Delta PE = q(V_B - V_A)$$

Combining these two equations, we find that

$$W = -q(V_B - V_A) \quad [16.6]$$

Using this equation, we obtain the following general result: **No net work is required to move a charge between two points that are at the same electric potential.** In mathematical terms this result says that $W = 0$ whenever $V_B = V_A$.

In Chapter 15 we found that when a conductor is in electrostatic equilibrium, a net charge placed on it resides entirely on its surface. Further, we showed that the electric field just outside the surface of a charged conductor in electrostatic equilibrium is perpendicular to the surface and that the field inside the conductor is zero. We now show that **all points on the surface of a charged conductor in electrostatic equilibrium are at the same potential**.

Consider a surface path connecting any points A and B on a charged conductor, as in Figure 16.10. The charges on the conductor are assumed to be in equilibrium with each other, so none are moving. In this case the electric field $\vec{E}$ is always perpendicular to the displacement along this path. This must be so, for otherwise the part of the electric field tangent to the surface would move the charges. Because $\vec{E}$ is perpendicular to the path, no work is done by the electric field if a charge is moved between the given two points. From Equation 16.6 we see that if the work done is zero, the difference in electric potential, $V_B - V_A$, is also zero. It follows that **the electric potential is a constant everywhere on the surface of a charged conductor in equilibrium**. Further, because the electric field inside a conductor is zero, no work is required to move a charge between two points inside the conductor. Again, Equation 16.6 shows that if the work done is zero, the difference in electric potential between any two points inside a conductor must also be zero. We conclude that the electric potential is constant everywhere inside a conductor.

Finally, because one of the points inside the conductor could be arbitrarily close to the surface of the conductor, we conclude that **the electric potential is constant everywhere inside a conductor and equal to that same value at the surface**. As a consequence, no work is required to move a charge from the interior of a charged conductor to its surface. (It's important to realize that the potential inside a conductor is not necessarily zero, even though the interior electric field is zero.)

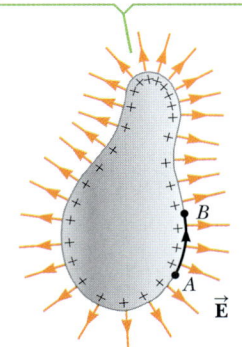

Notice from the spacing of the positive signs that the surface charge density is nonuniform.

Figure 16.10 An arbitrarily shaped conductor with an excess positive charge. When the conductor is in electrostatic equilibrium, all the charge resides at the surface, $\vec{E} = 0$, inside the conductor, and the electric field just outside the conductor is perpendicular to the surface. The potential is constant inside the conductor and is equal to the potential at the surface.

The Electron Volt

An appropriately sized unit of energy commonly used in atomic and nuclear physics is the electron volt (eV). For example, electrons in normal atoms typically have energies of tens of eV's, excited electrons in atoms emitting x-rays have energies of thousands of eV's, and high-energy gamma rays (electromagnetic waves) emitted by the nucleus have energies of millions of eV's.

▶ Definition of the electron volt

The **electron volt** is defined as the kinetic energy that an electron gains when accelerated through a potential difference of 1 V.

Because 1 V = 1 J/C and because the magnitude of the charge on the electron is 1.60×10^{-19} C, we see that the electron volt is related to the joule by

$$1 \text{ eV} = 1.60 \times 10^{-19} \text{ C} \cdot \text{V} = 1.60 \times 10^{-19} \text{ J} \quad [16.7]$$

■ Quick Quiz

16.7 An electron initially at rest accelerates through a potential difference of 1 V, gaining kinetic energy KE_e, whereas a proton, also initially at rest, accelerates through a potential difference of -1 V, gaining kinetic energy KE_p. Which of the following relationships holds? (a) $KE_e = KE_p$ (b) $KE_e < KE_p$ (c) $KE_e > KE_p$ (d) The answer can't be determined from the given information.

16.4 Equipotential Surfaces

LEARNING OBJECTIVE

1. Define an equipotential surface and discuss its electrical properties.

A surface on which all points are at the same potential is called an **equipotential surface**. The potential difference between any two points on an equipotential surface is zero. Hence, **no work is required to move a charge at constant speed on an equipotential surface**.

Equipotential surfaces have a simple relationship to the electric field: **The electric field at every point of an equipotential surface is perpendicular to the surface**. If the electric field $\vec{E}$ had a component parallel to the surface, that component would produce an electric force on a charge placed on the surface. This force would do work on the charge as it moved from one point to another, in contradiction to the definition of an equipotential surface.

Equipotential surfaces can be represented on a diagram by drawing equipotential contours, which are two-dimensional views of the intersections of the equipotential surfaces with the plane of the drawing. These equipotential contours are generally referred to simply as **equipotentials**. Figure 16.11a shows the equipotentials (in blue) associated with a positive point charge. Note that the equipotentials are perpendicular to the electric field lines (in orange) at all points. Recall that the electric potential created by a point charge q is given by $V = k_e q/r$. This relation shows that, for a single point charge, the potential is constant on any surface on which r is constant. It follows that the equipotentials of a point charge are a family

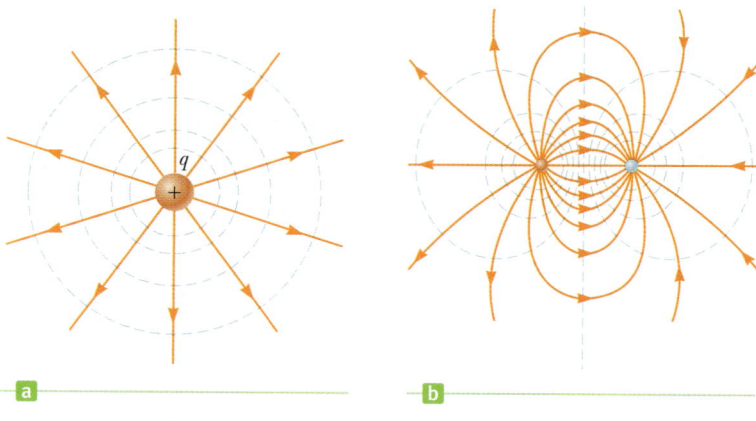

Figure 16.11 Equipotentials (dashed blue lines) and electric field lines (orange lines) for (a) a positive point charge and (b) two point charges of equal magnitude and opposite sign. In all cases the equipotentials are *perpendicular* to the electric field lines at every point.

of spheres centered on the point charge. Figure 16.11b shows the equipotentials associated with two charges of equal magnitude but opposite sign.

16.5 Applications

LEARNING OBJECTIVE

1. Describe several important applications of electrically charged systems.

The Electrostatic Precipitator

One important application of electric discharge in gases is a device called an *electrostatic precipitator*. This device removes particulate matter from combustion gases, thereby reducing air pollution. It's especially useful in coal-burning power plants and in industrial operations that generate large quantities of smoke. Systems currently in use can eliminate approximately 90% by mass of the ash and dust from smoke. Unfortunately, a very high percentage of the lighter particles still escape, and they contribute significantly to smog and haze.

Figure 16.12 illustrates the basic idea of the electrostatic precipitator. A high voltage (typically 40 kV to 100 kV) is maintained between a wire running down the center of a duct and the outer wall, which is grounded. The wire is maintained at a negative electric potential with respect to the wall, so the electric field is directed toward the wire. The electric field near the wire reaches a high enough value to cause a discharge around the wire and the formation of positive ions, electrons, and negative ions, such as O_2^-. As the electrons and negative ions are accelerated toward the outer wall by the nonuniform electric field, the dirt particles in the streaming gas become charged by collisions and ion capture. Because most of the charged dirt particles are negative, they are also drawn to the outer wall by the electric field. When the duct is shaken, the particles fall loose and are collected at the bottom.

In addition to reducing the amounts of harmful gases and particulate matter in the atmosphere, the electrostatic precipitator recovers valuable metal oxides from the stack.

APPLICATION
The Electrostatic Precipitator

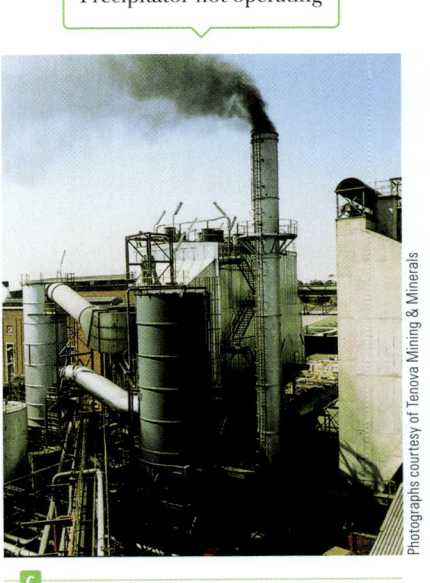

Figure 16.12 (a) A schematic diagram of an electrostatic precipitator. Compare the air pollution when the precipitator is (b) operating and (c) turned off.

APPLICATION
The Electrostatic Air Cleaner

A similar device called an *electrostatic air cleaner* is used in homes to relieve the discomfort of allergy sufferers. Air laden with dust and pollen is drawn into the device across a positively charged mesh screen. The airborne particles become positively charged when they make contact with the screen, and then they pass through a second, negatively charged mesh screen. The electrostatic force of attraction between the positively charged particles in the air and the negatively charged screen causes the particles to precipitate out on the surface of the screen, removing a very high percentage of contaminants from the air stream.

APPLICATION
Xerographic Copiers

Xerography and Laser Printers

Xerography is widely used to make photocopies of printed materials. The basic idea behind the process was developed by Chester Carlson, who was granted a patent for his invention in 1940. In 1947 the Xerox Corporation launched a full-scale program to develop automated duplicating machines using Carlson's process. The huge success of that development is evident: today, practically all offices and libraries have one or more duplicating machines, and the capabilities of these machines continue to evolve.

Some features of the xerographic process involve simple concepts from electrostatics and optics. The one idea that makes the process unique, however, is the use of photoconductive material to form an image. A photoconductor is a material that is a poor conductor of electricity in the dark, but a reasonably good conductor when exposed to light.

Figure 16.13 illustrates the steps in the xerographic process. First, the surface of a plate or drum is coated with a thin film of the photoconductive material (usually selenium or some compound of selenium), and the photoconductive surface is given a positive electrostatic charge in the dark (Fig. 16.13a). The page to be copied is then projected onto the charged surface (Fig. 16.13b). The photoconducting surface becomes conducting only in areas where light strikes; there the light produces charge carriers in the photoconductor that neutralize the positively charged surface. The charges remain on those areas of the photoconductor not exposed to light, however, leaving a hidden image of the object in the form of a positive distribution of surface charge.

Next, a negatively charged powder called a *toner* is dusted onto the photoconducting surface (Fig. 16.13c). The charged powder adheres only to the areas that contain the positively charged image. At this point, the image becomes visible. It is then transferred to the surface of a sheet of positively charged paper. Finally,

APPLICATION
Laser Printers

a Charging the drum
b Imaging the document
c Applying the toner
d Transferring the toner to the paper
e Laser printer drum

Figure 16.13 The xerographic process. (a) The photoconductive surface is positively charged. (b) Through the use of a light source and a lens, a hidden image is formed on the charged surface in the form of positive charges. (c) The surface containing the image is covered with a negatively charged powder, which adheres only to the image area. (d) A piece of paper is placed over the surface and given a charge. This transfers the image to the paper, which is then heated to "fix" the powder to the paper. (e) The image on the drum of a laser printer is produced by turning a laser beam on and off as it sweeps across the selenium-coated drum.

the toner is "fixed" to the surface of the paper by heat (Fig. 16.13d), resulting in a permanent copy of the original.

The steps for producing a document on a laser printer are similar to those used in a photocopy machine in that parts (a), (c), and (d) of Figure 16.13 remain essentially the same. The difference between the two techniques lies in the way the image is formed on the selenium-coated drum. In a laser printer the command to print the letter O, for instance, is sent to a laser from the memory of a computer. A rotating mirror inside the printer causes the beam of the laser to sweep across the selenium-coated drum in an interlaced pattern (Fig. 16.13e). Electrical signals generated by the printer turn the laser beam on and off in a pattern that traces out the letter O in the form of positive charges on the selenium. Toner is then applied to the drum, and the transfer to paper is accomplished as in a photocopy machine.

16.6 Capacitance

LEARNING OBJECTIVE

1. Describe a capacitor and define capacitance.

A **capacitor** is a device used in a variety of electric circuits, such as to tune the frequency of radio receivers, eliminate sparking in automobile ignition systems, or store short-term energy for rapid release in electronic flash units. Figure 16.14 shows a typical design for a capacitor. It consists of two parallel metal plates separated by a distance d. Used in an electric circuit, the plates are connected to the positive and negative terminals of a battery or some other voltage source. When this connection is made, electrons are pulled off one of the plates, leaving it with a charge of $+Q$, and are transferred through the battery to the other plate, leaving it with a charge of $-Q$, as shown in the figure. The transfer of charge stops when the potential difference across the plates equals the potential difference of the battery. A charged capacitor is a device that stores energy that can be reclaimed when needed for a specific application.

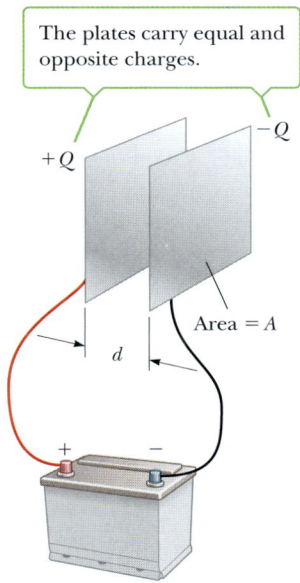

Figure 16.14 A parallel-plate capacitor consists of two parallel plates, each of area A, separated by a distance d.

The capacitance C of a capacitor is the ratio of the magnitude of the charge on either conductor (plate) to the magnitude of the potential difference between the conductors (plates):

$$C \equiv \frac{Q}{\Delta V} \quad [16.8]$$

◄ Capacitance of a pair of conductors

SI unit: farad (F) = coulomb per volt (C/V)

The quantities Q and ΔV are always taken to be positive when used in Equation 16.8. For example, if a 3.0-μF capacitor is connected to a 12-V battery, the magnitude of the charge on each plate of the capacitor is

$$Q = C\Delta V = (3.0 \times 10^{-6} \text{ F})(12 \text{ V}) = 36 \ \mu\text{C}$$

From Equation 16.8, we see that a large capacitance is needed to store a large amount of charge for a given applied voltage. The farad is a very large unit of capacitance. In practice, most typical capacitors have capacitances ranging from microfarads (1 μF = 1 × 10^{-6} F) to picofarads (1 pF = 1 × 10^{-12} F).

16.7 The Parallel-Plate Capacitor

LEARNING OBJECTIVES

1. Derive an expression for the capacitance of a parallel-plate capacitor.
2. Calculate the fundamental physical properties of a parallel-plate capacitor.

Figure 16.15 The electric field between the plates of a parallel-plate capacitor is uniform near the center, but nonuniform near the edges.

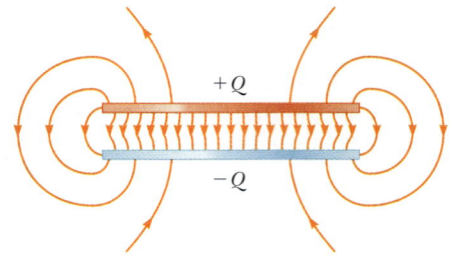

> **Tip 16.2** Potential Difference Is ΔV, Not V
>
> Use the symbol ΔV for the potential difference across a circuit element or a device (many other books use simply V for potential difference). The dual use of V to represent potential in one place and a potential difference in another can lead to unnecessary confusion.

The capacitance of a device depends on the geometric arrangement of the conductors. The capacitance of a parallel-plate capacitor with plates separated by air (see Fig. 16.14) can be easily calculated from three facts. First, recall from Chapter 15 that the magnitude of the electric field between two plates is given by $E = \sigma/\epsilon_0$, where σ is the magnitude of the charge per unit area on each plate. Second, we found earlier in this chapter that the potential difference between two plates is $\Delta V = Ed$, where d is the distance between the plates. Third, the charge on one plate is given by $Q = \sigma A$, where A is the area of the plate. Substituting these three facts into the definition of capacitance gives the desired result:

$$C = \frac{Q}{\Delta V} = \frac{\sigma A}{Ed} = \frac{\sigma A}{(\sigma/\epsilon_0)d}$$

Canceling the charge per unit area, σ, yields

▶ Capacitance of a parallel-plate capacitor

$$C = \epsilon_0 \frac{A}{d} \quad [16.9]$$

where A is the area of one of the plates, d is the distance between the plates, and ϵ_0 is the permittivity of free space.

From Equation 16.9, we see that plates with larger area can store more charge. The same is true for a small plate separation d because then the positive charges on one plate exert a stronger force on the negative charges on the other plate, allowing more charge to be held on the plates.

Figure 16.15 shows the electric field lines of a more realistic parallel-plate capacitor. The electric field is very nearly constant in the center between the plates, but becomes less so when approaching the edges. For most purposes, however, the field may be taken as constant throughout the region between the plates.

APPLICATION
Camera Flash Attachments

One practical device that uses a capacitor is the flash attachment on a camera. A battery is used to charge the capacitor, and the stored charge is then released when the shutter-release button is pressed to take a picture. The stored charge is delivered to a flash tube very quickly, illuminating the subject at the instant more light is needed.

APPLICATION
Computer Keyboards

Computers make use of capacitors in many ways. For example, one type of computer keyboard has capacitors at the bases of its keys, as in Figure 16.16. Each key is connected to a movable plate, which represents one side of the capacitor; the fixed plate on the bottom of the keyboard represents the other side of the capacitor. When a key is pressed, the capacitor spacing decreases, causing an increase in capacitance. External electronic circuits recognize each key by the *change* in its capacitance when it is pressed.

Figure 16.16 When the key of one type of keyboard is pressed, the capacitance of a parallel-plate capacitor increases as the plate spacing decreases. The substance labeled "dielectric" is an insulating material, as described in Section 16.10.

Capacitors are useful for storing a large amount of charge that needs to be delivered quickly. A good example on the forefront of fusion research is electrostatic confinement. In this role capacitors discharge their electrons through a grid. The negatively charged electrons in the grid draw positively charged particles to them and therefore to each other, causing some particles to fuse and release energy in the process.

APPLICATION
Electrostatic Confinement

EXAMPLE 16.6 A Parallel-Plate Capacitor

GOAL Calculate fundamental physical properties of a parallel-plate capacitor.

PROBLEM A parallel-plate capacitor has an area $A = 2.00 \times 10^{-4}$ m² and a plate separation $d = 1.00 \times 10^{-3}$ m. **(a)** Find its capacitance. **(b)** How much charge is on the positive plate if the capacitor is connected to a 3.00-V battery? Calculate **(c)** the charge density on the positive plate, assuming the density is uniform, and **(d)** the magnitude of the electric field between the plates.

STRATEGY Parts (a) and (b) can be solved by substituting into the basic equations for capacitance. In part (c) use the definition of charge density, and in part (d) use the fact that the voltage difference equals the electric field times the distance.

SOLUTION

(a) Find the capacitance.

Substitute into Equation 16.9:

$$C = \epsilon_0 \frac{A}{d} = (8.85 \times 10^{-12} \, \text{C}^2/\text{N} \cdot \text{m}^2)\left(\frac{2.00 \times 10^{-4} \, \text{m}^2}{1.00 \times 10^{-3} \, \text{m}}\right)$$

$$C = \boxed{1.77 \times 10^{-12} \, \text{F} = 1.77 \, \text{pF}}$$

(b) Find the charge on the positive plate after the capacitor is connected to a 3.00-V battery.

Substitute into Equation 16.8:

$$C = \frac{Q}{\Delta V} \quad \rightarrow \quad Q = C \, \Delta V = (1.77 \times 10^{-12} \, \text{F})(3.00 \, \text{V})$$

$$= \boxed{5.31 \times 10^{-12} \, \text{C}}$$

(c) Calculate the charge density on the positive plate.

Charge density is charge divided by area:

$$\sigma = \frac{Q}{A} = \frac{5.31 \times 10^{-12} \, \text{C}}{2.00 \times 10^{-4} \, \text{m}^2} = \boxed{2.66 \times 10^{-8} \, \text{C/m}^2}$$

(d) Calculate the magnitude of the electric field between the plates.

Apply $\Delta V = Ed$:

$$E = \frac{\Delta V}{d} = \frac{3.00 \, \text{V}}{1.00 \times 10^{-3} \, \text{m}} = \boxed{3.00 \times 10^3 \, \text{V/m}}$$

REMARKS The answer to part (d) could also have been obtained from the electric field derived for a parallel plate capacitor, Equation 15.13, $E = \sigma/\epsilon_0$.

QUESTION 16.6 How do the answers change if the distance between the plates is doubled?

EXERCISE 16.6 Two plates, each of area 3.00×10^{-4} m², are used to construct a parallel-plate capacitor with capacitance 1.00 pF. (a) Find the necessary separation distance. (b) If the positive plate is to hold a charge of 5.00×10^{-12} C, find the charge density. (c) Find the electric field between the plates. (d) What voltage battery should be attached to the plate to obtain the preceding results?

ANSWERS (a) 2.66×10^{-3} m (b) 1.67×10^{-8} C/m² (c) 1.89×10^3 N/C (d) 5.00 V

Symbols for Circuit Elements and Circuits

The symbol that is commonly used to represent a capacitor in a circuit is —| |— or sometimes —| (—. Don't confuse either of these symbols with the circuit symbol, —||—, which is used to designate a battery (or any other source of

Figure 16.17 (a) A real circuit and (b) its equivalent circuit diagram.

direct current). The positive terminal of the battery is at the higher potential and is represented by the longer vertical line in the battery symbol. In the next chapter we discuss another circuit element, called a resistor, represented by the symbol —⋁⋁⋁—. When wires in a circuit don't have appreciable resistance compared with the resistance of other elements in the circuit, the wires are represented by straight lines.

It's important to realize that a circuit is a collection of real objects, usually containing a source of electrical energy (such as a battery) connected to elements that convert electrical energy to other forms (light, heat, sound) or store the energy in electric or magnetic fields for later retrieval. A real circuit and its schematic diagram are sketched in Figure 16.17. The circuit symbol for a lightbulb shown in Figure 16.17b is —⊗—.

If you are not familiar with circuit diagrams, trace the path of the real circuit with your finger to see that it is equivalent to the geometrically regular schematic diagram.

16.8 Combinations of Capacitors

LEARNING OBJECTIVES

1. Derive the equivalent capacitance of capacitors in parallel.
2. Analyze a circuit with capacitors in parallel.
3. Derive the equivalent capacitance of capacitors in series.
4. Analyze a circuit with a series of capacitors.
5. Analyze a circuit with combinations of both parallel and series capacitors.

Two or more capacitors can be combined in circuits in several ways, but most reduce to two simple configurations, called *parallel* and *series*. The idea, then, is to find the single equivalent capacitance due to a combination of several different capacitors that are in parallel or in series with each other. Capacitors are manufactured with a number of different standard capacitances, and by combining them in different ways, any desired value of capacitance can be obtained.

Capacitors in Parallel

Two capacitors connected as shown in Figure 16.18a are said to be in *parallel*. The left plate of each capacitor is connected to the positive terminal of the battery by a conducting wire, so the left plates are at the same potential. In the same way, the right plates, both connected to the negative terminal of the battery, are also at the same potential. This means that **capacitors in parallel both have the**

Figure 16.18 (a) A parallel connection of two capacitors. (b) The circuit diagram for the parallel combination. (c) The potential differences across the capacitors are the same, and the equivalent capacitance is $C_{eq} = C_1 + C_2$.

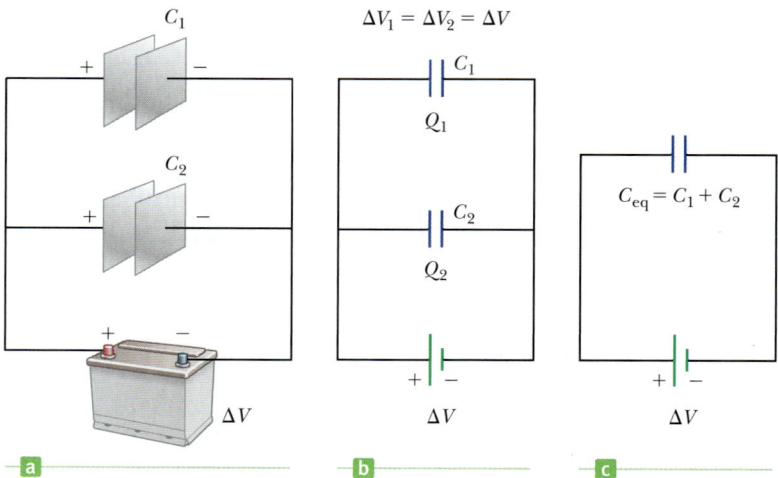

same potential difference ΔV across them. Capacitors in parallel are illustrated in Figure 16.18b.

When the capacitors are first connected in the circuit, electrons are transferred from the left plates through the battery to the right plates, leaving the left plates positively charged and the right plates negatively charged. The energy source for this transfer of charge is the internal chemical energy stored in the battery, which is converted to electrical energy. The flow of charge stops when the voltage across the capacitors equals the voltage of the battery, at which time the capacitors have their maximum charges. If the maximum charges on the two capacitors are Q_1 and Q_2, respectively, the *total charge*, Q, stored by the two capacitors is

$$Q = Q_1 + Q_2 \quad [16.10]$$

We can replace these two capacitors with one equivalent capacitor having a capacitance of C_{eq}. This equivalent capacitor must have exactly the same external effect on the circuit as the original two, so it must store Q units of charge and have the same potential difference across it. The respective charges on each capacitor are

$$Q_1 = C_1 \Delta V \quad \text{and} \quad Q_2 = C_2 \Delta V$$

The charge on the equivalent capacitor is

$$Q = C_{eq} \Delta V$$

Substituting these relationships into Equation 16.10 gives

$$C_{eq} \Delta V = C_1 \Delta V + C_2 \Delta V$$

or

$$C_{eq} = C_1 + C_2 \quad \begin{pmatrix} \text{parallel} \\ \text{combination} \end{pmatrix} \quad [16.11]$$

If we extend this treatment to three or more capacitors connected in parallel, the equivalent capacitance is found to be

$$C_{eq} = C_1 + C_2 + C_3 + \cdots \quad \begin{pmatrix} \text{parallel} \\ \text{combination} \end{pmatrix} \quad [16.12]$$

We see that **the equivalent capacitance of a parallel combination of capacitors is larger than any of the individual capacitances**.

> **Tip 16.3** Voltage Is the Same as Potential Difference
>
> A voltage *across* a device, such as a capacitor, has the same meaning as the potential difference across the device. For example, if we say that the voltage across a capacitor is 12 V, we mean that the potential difference between its plates is 12 V.

■ EXAMPLE 16.7 | Four Capacitors Connected in Parallel

GOAL Analyze a circuit with several capacitors in parallel.

PROBLEM (a) Determine the capacitance of the single capacitor that is equivalent to the parallel combination of capacitors shown in Figure 16.19. Find (b) the charge on the 12.0-μF capacitor and (c) the total charge contained in the configuration. (d) Derive a symbolic expression for the fraction of the total charge contained on one of the capacitors.

STRATEGY For part (a), add the individual capacitances. For part (b), apply the formula $C = Q/\Delta V$ to the 12.0-μF capacitor. The voltage difference is the same as the difference across the battery. To find the total charge contained in all four capacitors, use the equivalent capacitance in the same formula.

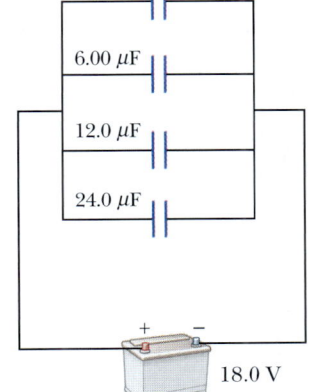

Figure 16.19 (Example 16.7) Four capacitors connected in parallel.

(Continued)

SOLUTION

(a) Find the equivalent capacitance.

Apply Equation 16.12:
$$C_{eq} = C_1 + C_2 + C_3 + C_4$$
$$= 3.00\ \mu F + 6.00\ \mu F + 12.0\ \mu F + 24.0\ \mu F$$
$$= \boxed{45.0\ \mu F}$$

(b) Find the charge on the 12-μF capacitor (designated C_3).

Solve the capacitance equation for Q and substitute:
$$Q = C_3\ \Delta V = (12.0 \times 10^{-6}\ F)(18.0\ V) = 216 \times 10^{-6}\ C$$
$$= \boxed{216\ \mu C}$$

(c) Find the total charge contained in the configuration.

Use the equivalent capacitance:
$$C_{eq} = \frac{Q}{\Delta V} \quad \rightarrow \quad Q = C_{eq}\Delta V = (45.0\ \mu F)(18.0\ V) = \boxed{8.10 \times 10^2\ \mu C}$$

(d) Derive a symbolic expression for the fraction of the total charge contained in one of the capacitors.

Write a symbolic expression for the fractional charge in the ith capacitor and use the capacitor definition:
$$\frac{Q_i}{Q_{tot}} = \frac{C_i \Delta V}{C_{eq} \Delta V} = \frac{C_i}{C_{eq}}$$

REMARKS The charge on any one of the parallel capacitors can be found as in part (b) because the potential difference is the same. Notice that finding the total charge does not require finding the charge on each individual capacitor and adding. It's easier to use the equivalent capacitance in the capacitance definition.

QUESTION 16.7 If all four capacitors had the same capacitance, what fraction of the total charge would be held by each?

EXERCISE 16.7 Find the charge on the 24.0-μF capacitor.

ANSWER 432 μC

Capacitors in Series

> Q is the same for all capacitors connected in series

Now consider two capacitors connected in *series*, as illustrated in Figure 16.20a. **For a series combination of capacitors, the magnitude of the charge must be the same on all the plates.** To understand this principle, consider the charge transfer process in some detail. When a battery is connected to the circuit, electrons with total charge $-Q$ are transferred from the left plate of C_1 to the right plate of C_2 through the battery, leaving the left plate of C_1 with a charge of $+Q$. As a consequence, the magnitudes of the charges on the left plate of C_1 and the right plate of C_2 must be the same. Now consider the right plate of C_1 and the left plate of C_2, in the middle. These plates are not connected to the battery (because of the gap across the plates) and, taken together, are electrically neutral. The charge of $+Q$ on the left plate of C_1, however, attracts negative charges to the right plate of C_1. These charges will continue to accumulate until the left and right plates of C_1, taken together, become electrically neutral, which means that the charge on the right plate of C_1 is $-Q$. This negative charge could only have come from the left plate of C_2, so C_2 has a charge of $+Q$.

Therefore, regardless of how many capacitors are in series or what their capacitances are, **all the right plates gain charges of $-Q$ and all the left plates have charges of $+Q$** (a consequence of the conservation of charge).

16.8 | **Combinations of Capacitors** 579

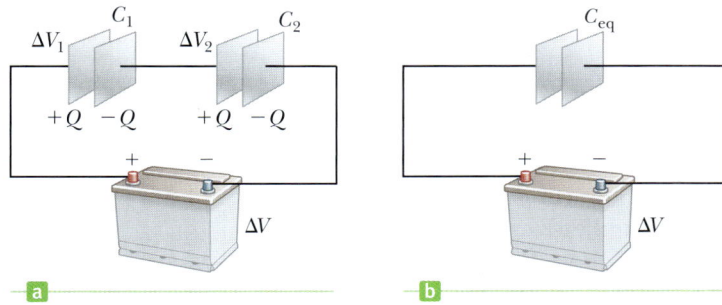

Figure 16.20 A series combination of two capacitors. The charges on the capacitors are the same, and the equivalent capacitance can be calculated from the reciprocal relationship $1/C_{eq} = (1/C_1) + (1/C_2)$.

After an equivalent capacitor for a series of capacitors is fully charged, **the equivalent capacitor must end up with a charge of $-Q$ on its right plate and a charge of $+Q$ on its left plate.** Applying the definition of capacitance to the circuit in Figure 16.20b, we have

$$\Delta V = \frac{Q}{C_{eq}}$$

where ΔV is the potential difference between the terminals of the battery and C_{eq} is the equivalent capacitance. Because $Q = C\Delta V$ can be applied to each capacitor, the potential differences across them are given by

$$\Delta V_1 = \frac{Q}{C_1} \qquad \Delta V_2 = \frac{Q}{C_2}$$

From Figure 16.20a, we see that

$$\Delta V = \Delta V_1 + \Delta V_2 \quad [16.13]$$

where ΔV_1 and ΔV_2 are the potential differences across capacitors C_1 and C_2 (a consequence of the conservation of energy).

The potential difference across any number of capacitors (or other circuit elements) in series equals the sum of the potential differences across the individual capacitors. Substituting these expressions into Equation 16.13 and noting that $\Delta V = Q/C_{eq}$, we have

$$\frac{Q}{C_{eq}} = \frac{Q}{C_1} + \frac{Q}{C_2}$$

Canceling Q, we arrive at the following relationship:

$$\frac{1}{C_{eq}} = \frac{1}{C_1} + \frac{1}{C_2} \quad \binom{\text{series}}{\text{combination}} \quad [16.14]$$

If this analysis is applied to three or more capacitors connected in series, the equivalent capacitance is found to be

$$\frac{1}{C_{eq}} = \frac{1}{C_1} + \frac{1}{C_2} + \frac{1}{C_3} + \cdots \quad \binom{\text{series}}{\text{combination}} \quad [16.15]$$

As we will show in Example 16.8, Equation 16.15 implies that **the equivalent capacitance of a series combination is always smaller than any individual capacitance in the combination**.

■ Quick Quiz

16.8 A capacitor is designed so that one plate is large and the other is small. If the plates are connected to a battery, (a) the large plate has a greater charge than the small plate, (b) the large plate has less charge than the small plate, or (c) the plates have equal, but opposite, charge.

EXAMPLE 16.8 Four Capacitors Connected in Series

GOAL Find an equivalent capacitance of capacitors in series, and the charge and voltage on each capacitor.

PROBLEM Four capacitors are connected in series with a battery, as in Figure 16.21. (a) Calculate the capacitance of the equivalent capacitor. (b) Compute the charge on the 12-μF capacitor. (c) Find the voltage drop across the 12-μF capacitor.

STRATEGY Combine all the capacitors into a single, equivalent capacitor using Equation 16.15. Find the charge on this equivalent capacitor using $C = Q/\Delta V$. This charge is the same as on the individual capacitors. Use this same equation again to find the voltage drop across the 12-μF capacitor.

Figure 16.21 (Example 16.8) Four capacitors connected in series.

SOLUTION

(a) Calculate the equivalent capacitance of the series.

Apply Equation 16.15:

$$\frac{1}{C_{eq}} = \frac{1}{3.0\ \mu F} + \frac{1}{6.0\ \mu F} + \frac{1}{12\ \mu F} + \frac{1}{24\ \mu F}$$

$$C_{eq} = \boxed{1.6\ \mu F}$$

(b) Compute the charge on the 12-μF capacitor.

The desired charge equals the charge on the equivalent capacitor:

$$Q = C_{eq}\Delta V = (1.6 \times 10^{-6}\ F)(18\ V) = \boxed{29\ \mu C}$$

(c) Find the voltage drop across the 12-μF capacitor.

Apply the basic capacitance equation:

$$C = \frac{Q}{\Delta V} \rightarrow \Delta V = \frac{Q}{C} = \frac{29\ \mu C}{12\ \mu F} = \boxed{2.4\ V}$$

REMARKS Notice that the equivalent capacitance is less than that of any of the individual capacitors. The relationship $C = Q/\Delta V$ can be used to find the voltage drops on the other capacitors, just as in part (c).

QUESTION 16.8 Over which capacitor is the voltage drop the smallest? The largest?

EXERCISE 16.8 The 24-μF capacitor is removed from the circuit, leaving only three capacitors in series. Find (a) the equivalent capacitance, (b) the charge on the 6-μF capacitor, and (c) the voltage drop across the 6-μF capacitor.

ANSWERS (a) 1.7 μF (b) 31 μC (c) 5.2 V

PROBLEM-SOLVING STRATEGY

Complex Capacitor Combinations

1. **Combine** capacitors that are in series or in parallel, following the derived formulas.
2. **Redraw** the circuit after every combination.
3. **Repeat** the first two steps until there is only a single equivalent capacitor.
4. **Find the charge** on the single equivalent capacitor, using $C = Q/\Delta V$.
5. **Work backwards** through the diagrams to the original one, finding the charge and voltage drop across each capacitor along the way. To do this, use the following collection of facts:
    A. The capacitor equation: $C = Q/\Delta V$
    B. Capacitors in parallel: $C_{eq} = C_1 + C_2$
    C. Capacitors in parallel all have the same voltage difference, ΔV, as does their equivalent capacitor.
    D. Capacitors in series: $\dfrac{1}{C_{eq}} = \dfrac{1}{C_1} + \dfrac{1}{C_2}$
    E. Capacitors in series all have the same charge, Q, as does their equivalent capacitor.

EXAMPLE 16.9 | Equivalent Capacitance

GOAL Solve a complex combination of series and parallel capacitors.

PROBLEM (a) Calculate the equivalent capacitance between a and b for the combination of capacitors shown in Figure 16.22a. All capacitances are in microfarads. (b) If a 12-V battery is connected across the system between points a and b, find the charge on the 4.0-μF capacitor in the first diagram and the voltage drop across it.

STRATEGY For part (a), use Equations 16.12 and 16.15 to reduce the combination step by step, as indicated in the figure. For part (b), to find the charge on the 4.0-μF capacitor, start with Figure 16.22c, finding the charge on the 2.0-μF capacitor. This same charge is on each of the 4.0-μF capacitors in the second diagram, by fact 5E of the Problem-Solving Strategy. One of these 4.0-μF capacitors in the second diagram is simply the original 4.0-μF capacitor in the first diagram.

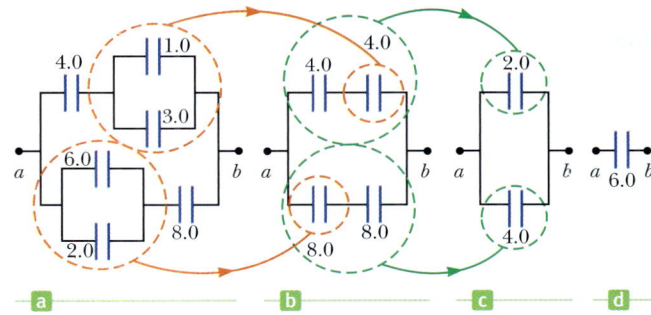

Figure 16.22 (Example 16.9) To find the equivalent capacitance of the circuit in (a), use the series and parallel rules described in the text to successively reduce the circuit as indicated in (b), (c), and (d). All capacitances are in microfarads.

SOLUTION

(a) Calculate the equivalent capacitance.

Find the equivalent capacitance of the parallel 1.0-μF and 3.0-μF capacitors in Figure 16.22a:

$$C_{eq} = C_1 + C_2 = 1.0\ \mu F + 3.0\ \mu F = 4.0\ \mu F$$

Find the equivalent capacitance of the parallel 2.0-μF and 6.0-μF capacitors in Figure 16.22a:

$$C_{eq} = C_1 + C_2 = 2.0\ \mu F + 6.0\ \mu F = 8.0\ \mu F$$

Combine the two series 4.0-μF capacitors in Figure 16.22b:

$$\frac{1}{C_{eq}} = \frac{1}{C_1} + \frac{1}{C_2} = \frac{1}{4.0\ \mu F} + \frac{1}{4.0\ \mu F}$$

$$= \frac{1}{2.0\ \mu F} \rightarrow C_{eq} = 2.0\ \mu F$$

Combine the two series 8.0-μF capacitors in Figure 16.22b:

$$\frac{1}{C_{eq}} = \frac{1}{C_1} + \frac{1}{C_2} = \frac{1}{8.0\ \mu F} + \frac{1}{8.0\ \mu F}$$

$$= \frac{1}{4.0\ \mu F} \rightarrow C_{eq} = 4.0\ \mu F$$

Finally, combine the two parallel capacitors in Figure 16.22c to find the equivalent capacitance between a and b:

$$C_{eq} = C_1 + C_2 = 2.0\ \mu F + 4.0\ \mu F = \boxed{6.0\ \mu F}$$

(b) Find the charge on the 4.0-μF capacitor and the voltage drop across it.

Compute the charge on the 2.0-μF capacitor in Figure 16.22c, which is the same as the charge on the 4.0-μF capacitor in Figure 16.22a:

$$C = \frac{Q}{\Delta V} \rightarrow Q = C\Delta V = (2.0\ \mu F)(12\ V) = \boxed{24\ \mu C}$$

Use the basic capacitance equation to find the voltage drop across the 4.0-μF capacitor in Figure 16.22a:

$$C = \frac{Q}{\Delta V} \rightarrow \Delta V = \frac{Q}{C} = \frac{24\ \mu C}{4.0\ \mu F} = \boxed{6.0\ V}$$

REMARKS To find the rest of the charges and voltage drops, it's just a matter of using $C = Q/\Delta V$ repeatedly, together with facts 5C and 5E in the Problem-Solving Strategy. The voltage drop across the 4.0-μF capacitor could also have been found by noticing, in Figure 16.22b, that both capacitors had the same value and so by symmetry would split the total drop of 12 volts between them.

QUESTION 16.9 Which capacitor holds more charge, the 1.0-μF capacitor or the 3.0-μF capacitor?

(Continued)

EXERCISE 16.9 (a) In Example 16.9 find the charge on the 8.0-μF capacitor in Figure 16.22a and the voltage drop across it. (b) Do the same for the 6.0-μF capacitor in Figure 16.22a.

ANSWERS (a) 48 μC, 6.0 V (b) 36 μC, 6.0 V

16.9 Energy Stored in a Charged Capacitor

LEARNING OBJECTIVES

1. Obtain an expression for the energy stored by a capacitor.
2. Apply energy and power concepts to a capacitor.

Almost everyone who works with electronic equipment has at some time verified that a capacitor can store energy. If the plates of a charged capacitor are connected by a conductor such as a wire, charge transfers from one plate to the other until the two are uncharged. The discharge can often be observed as a visible spark. If you accidentally touched the opposite plates of a charged capacitor, your fingers would act as a pathway by which the capacitor could discharge, inflicting an electric shock. The degree of shock would depend on the capacitance and voltage applied to the capacitor. *Where high voltages and large quantities of charge are present, as in the power supply of a television set, such a shock can be fatal.*

Capacitors store electrical energy, and that energy is the same as the work required to move charge onto the plates. If a capacitor is initially uncharged (both plates are neutral) so that the plates are at the same potential, very little work is required to transfer a small amount of charge ΔQ from one plate to the other. Once this charge has been transferred, however, a small potential difference $\Delta V = \Delta Q/C$ appears between the plates, so work must be done to transfer additional charge against this potential difference. From Equation 16.6, if the potential difference at any instant during the charging process is ΔV, the work ΔW required to move more charge ΔQ through this potential difference is given by

$$\Delta W = \Delta V \Delta Q$$

We know that $\Delta V = Q/C$ for a capacitor that has a total charge of Q. Therefore, a plot of voltage versus total charge gives a straight line with a slope of $1/C$, as shown in Figure 16.23. The work ΔW, for a particular ΔV, is the area of the blue rectangle. Adding up all the rectangles gives an approximation of the total work needed to fill the capacitor. In the limit as ΔQ is taken to be infinitesimally small, the total work needed to charge the capacitor to a final charge Q and voltage ΔV is the area under the line. This is just the area of a triangle, one-half the base times the height, so it follows that

$$W = \tfrac{1}{2} Q \Delta V \qquad [16.16]$$

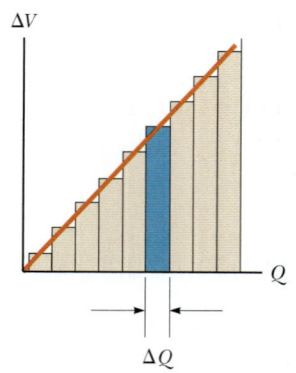

Figure 16.23 A plot of voltage vs. charge for a capacitor is a straight line with slope $1/C$. The work required to move a charge of ΔQ through a potential difference of ΔV across the capacitor plates is $\Delta W = \Delta V \Delta Q$, which equals the area of the blue rectangle. The *total work* required to charge the capacitor to a final charge of Q is the area under the straight line, which equals $Q\Delta V/2$.

As previously stated, W is also the energy stored in the capacitor. From the definition of capacitance, we have $Q = C \Delta V$; hence, we can express the energy stored three different ways:

$$\text{Energy stored} = \tfrac{1}{2}Q\Delta V = \tfrac{1}{2}C(\Delta V)^2 = \frac{Q^2}{2C} \qquad [16.17]$$

For example, the amount of energy stored in a 5.0-μF capacitor when it is connected across a 120-V battery is

$$\text{Energy stored} = \tfrac{1}{2}C(\Delta V)^2 = \tfrac{1}{2}(5.0 \times 10^{-6}\text{ F})(120\text{ V})^2 = 3.6 \times 10^{-2}\text{ J}$$

In practice, there is a limit to the maximum energy (or charge) that can be stored in a capacitor. At some point, the Coulomb forces between the charges on the plates become so strong that electrons jump across the gap, discharging the capacitor. For this reason, capacitors are usually labeled with a maximum operating voltage. (This physical fact can actually be exploited to yield a circuit with a regularly blinking light.)

Large capacitors can store enough electrical energy to cause severe burns or even death if they are discharged so that the flow of charge can pass through the heart. Under the proper conditions, however, they can be used to *sustain* life by stopping cardiac fibrillation in heart attack victims. When fibrillation occurs, the heart produces a rapid, irregular pattern of beats. A fast discharge of electrical energy through the heart can return the organ to its normal beat pattern. Emergency medical teams use portable defibrillators that contain batteries capable of charging a capacitor to a high voltage. (The circuitry actually permits the capacitor to be charged to a much higher voltage than the battery.) In this case and others (camera flash units and lasers used for fusion experiments), capacitors serve as energy reservoirs that can be slowly charged and then quickly discharged to provide large amounts of energy in a short pulse. The stored electrical energy is released through the heart by conducting electrodes, called paddles, placed on both sides of the victim's chest. The paramedics must wait between applications of electrical energy because of the time it takes the capacitors to become fully charged. The high voltage on the capacitor can be obtained from a low-voltage battery in a portable machine through the phenomenon of *electromagnetic induction*, to be studied in Chapter 20.

APPLICATION BIO
Defibrillators

EXAMPLE 16.10 | Typical Voltage, Energy, and Discharge Time for a Defibrillator

GOAL Apply energy and power concepts to a capacitor.

PROBLEM A fully charged defibrillator contains 1.20 kJ of energy stored in a 1.10×10^{-4} F capacitor. In a discharge through a patient, 6.00×10^2 J of electrical energy is delivered in 2.50 ms. (a) Find the voltage needed to store 1.20 kJ in the unit. (b) What average power is delivered to the patient?

STRATEGY Because we know the energy stored and the capacitance, we can use Equation 16.17 to find the required voltage in part (a). For part (b), dividing the energy delivered by the time gives the average power.

SOLUTION

(a) Find the voltage needed to store 1.20 kJ in the unit.

Solve Equation 16.17 for ΔV:

$$\text{Energy stored} = \tfrac{1}{2} C \Delta V^2$$

$$\Delta V = \sqrt{\frac{2 \times (\text{energy stored})}{C}}$$

$$= \sqrt{\frac{2(1.20 \times 10^3 \text{ J})}{1.10 \times 10^{-4} \text{ F}}}$$

$$= 4.67 \times 10^3 \text{ V}$$

(b) What average power is delivered to the patient?

Divide the energy delivered by the time:

$$P_{av} = \frac{\text{energy delivered}}{\Delta t} = \frac{6.00 \times 10^2 \text{ J}}{2.50 \times 10^{-3} \text{ s}}$$

$$= 2.40 \times 10^5 \text{ W}$$

REMARKS The power delivered by a draining capacitor isn't constant, as we'll find in the study of *RC* circuits in Chapter 18. For that reason, we were able to find only an average power. Capacitors are necessary in defibrillators because they can deliver energy far more quickly than batteries. Batteries provide current through relatively slow chemical reactions, whereas capacitors can quickly release charge that has already been produced and stored.

QUESTION 16.10 If the voltage across the capacitor were doubled, would the energy stored be (a) halved, (b) doubled, or (c) quadrupled?

(Continued)

EXERCISE 16.10 (a) Find the energy contained in a 2.50×10^{-5} F parallel-plate capacitor if it holds 1.75×10^{-3} C of charge. (b) What's the voltage between the plates? (c) What new voltage will result in a doubling of the stored energy?

ANSWERS (a) 6.13×10^{-2} J (b) 70.0 V (c) 99.0 V

■ APPLYING PHYSICS 16.1 | Maximum Energy Design

How should three capacitors and two batteries be connected so that the capacitors will store the maximum possible energy?

EXPLANATION The energy stored in the capacitor is proportional to the capacitance and the square of the potential difference, so we would like to maximize each of these quantities. If the three capacitors are connected in parallel, their capacitances add, and if the batteries are in series, their potential differences, similarly, also add together. ■

■ Quick Quiz

16.9 A parallel-plate capacitor is disconnected from a battery, and the plates are pulled a small distance farther apart. Do the following quantities increase, decrease, or stay the same?
(a) C (b) Q (c) E between the plates   (d) ΔV (e) energy stored in the capacitor

16.10 Capacitors with Dielectrics

LEARNING OBJECTIVES

1. Discuss the physical origins and practical applications of using dielectric materials in capacitors.
2. Evaluate the capacitance for capacitors with dielectrics.
3. Describe the physical origins of the polarization of molecules.

A **dielectric** is an insulating material, such as rubber, plastic, or waxed paper. When a dielectric is inserted between the plates of a capacitor, the capacitance increases. If the dielectric completely fills the space between the plates, the capacitance is multiplied by the factor κ, called the **dielectric constant**.

The following experiment illustrates the effect of a dielectric in a capacitor. Consider a parallel-plate capacitor of charge Q_0 and capacitance C_0 in the absence of a dielectric. The potential difference across the capacitor plates can be measured, and is given by $\Delta V_0 = Q_0/C_0$ (Fig. 16.24a). Because the capacitor is not connected to an external circuit, there is no pathway for charge to leave or be added to the plates. If a dielectric is now inserted between the plates as in Figure 16.24b, the voltage across the plates is *reduced* by the factor κ to the value

$$\Delta V = \frac{\Delta V_0}{\kappa}$$

Because $\kappa > 1$, ΔV is less than ΔV_0. Because the charge Q_0 on the capacitor doesn't change, we conclude that the capacitance in the presence of the dielectric must change to the value

$$C = \frac{Q_0}{\Delta V} = \frac{Q_0}{\Delta V_0/\kappa} = \frac{\kappa Q_0}{\Delta V_0}$$

or

$$C = \kappa C_0 \qquad [16.18]$$

Figure 16.24 When a dielectric with dielectric constant κ is inserted in a charged capacitor that is *not* connected to a battery, the potential difference is reduced to $\Delta V = \Delta V_0/\kappa$ and the capacitance increases to $C = \kappa C_0$.

According to this result, the capacitance is *multiplied* by the factor κ when the dielectric fills the region between the plates. For a parallel-plate capacitor, where the capacitance in the absence of a dielectric is $C_0 = \epsilon_0 A/d$, we can express the capacitance in the presence of a dielectric as

$$C = \kappa \epsilon_0 \frac{A}{d} \qquad [16.19]$$

From this result, it appears that the capacitance could be made very large by decreasing d, the separation between the plates. In practice the lowest value of d is limited by the electric discharge that can occur through the dielectric material separating the plates. For any given plate separation, there is a maximum electric field that can be produced in the dielectric before it breaks down and begins to conduct. This maximum electric field is called the **dielectric strength**, and for air its value is about 3×10^6 V/m. Most insulating materials have dielectric strengths greater than that of air, as indicated by the values listed in Table 16.1. Figure 16.25 shows an instance of dielectric breakdown in air.

Commercial capacitors are often made by using metal foil interlaced with thin sheets of paraffin-impregnated paper or Mylar®, which serves as the dielectric material. These alternate layers of metal foil and dielectric are rolled into a small cylinder (Fig. 16.26a on page 586). One type of a high-voltage capacitor consists of

Figure 16.25 Dielectric breakdown in air. Sparks are produced when a large alternating voltage is applied across the wires by a high-voltage induction coil power supply.

Table 16.1 Dielectric Constants and Dielectric Strengths of Various Materials at Room Temperature

Material	Dielectric Constant κ	Dielectric Strength (V/m)
Air	1.000 59	3×10^6
Bakelite®	4.9	24×10^6
Fused quartz	3.78	8×10^6
Neoprene rubber	6.7	12×10^6
Nylon	3.4	14×10^6
Paper	3.7	16×10^6
Polystyrene	2.56	24×10^6
Pyrex® glass	5.6	14×10^6
Silicone oil	2.5	15×10^6
Strontium titanate	233	8×10^6
Teflon®	2.1	60×10^6
Vacuum	1.000 00	—
Water	80	—

Figure 16.26 Three commercial capacitor designs.

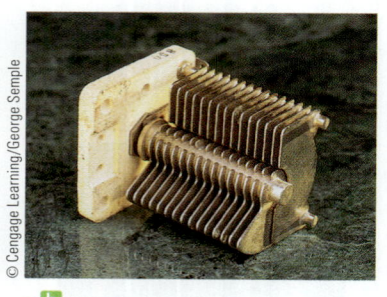

Figure 16.27 (a) A collection of capacitors used in a variety of applications. (b) A variable capacitor. When one set of metal plates is rotated so as to lie between a fixed set of plates, the capacitance of the device changes.

a number of interwoven metal plates immersed in silicone oil (Fig. 16.26b). Small capacitors are often constructed from ceramic materials. Variable capacitors (typically 10 pF to 500 pF) usually consist of two interwoven sets of metal plates, one fixed and the other movable, with air as the dielectric.

An electrolytic capacitor (Fig. 16.26c) is often used to store large amounts of charge at relatively low voltages. It consists of a metal foil in contact with an electrolyte—a solution that conducts charge by virtue of the motion of the ions contained in it. When a voltage is applied between the foil and the electrolyte, a thin layer of metal oxide (an insulator) is formed on the foil, and this layer serves as the dielectric. Enormous capacitances can be attained because the dielectric layer is very thin.

Figure 16.27 shows a variety of commercially available capacitors. Variable capacitors are used in radios to adjust the frequency.

When electrolytic capacitors are used in circuits, the polarity (the plus and minus signs on the device) must be observed. If the polarity of the applied voltage is opposite that intended, the oxide layer will be removed and the capacitor will conduct rather than store charge. Further, reversing the polarity can result in such a large current that the capacitor may either burn or produce steam and explode.

APPLYING PHYSICS 16.2 | Stud Finders

If you have ever tried to hang a picture on a wall securely, you know that it can be difficult to locate a wooden stud in which to anchor your nail or screw. The principles discussed in this section can be used to detect a stud electronically. The primary element of an electronic stud finder is a capacitor with its plates arranged side by side instead of facing one another, as in Figure 16.28. How does this device work?

EXPLANATION As the detector is moved along a wall, its capacitance changes when it passes across a stud because the dielectric constant of the material "between" the plates changes. The change in capacitance can be used to cause a light to come on, signaling the presence of the stud. ■

Figure 16.28 (Applying Physics 16.2) A stud finder produces an electric field that is affected by the dielectric constant of the materials placed in its field. When the device moves across a stud, the change in dielectric constant activates a signal light.

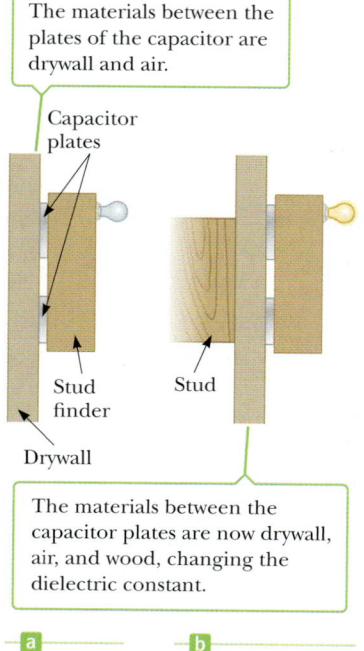

Quick Quiz

16.10 A fully charged parallel-plate capacitor remains connected to a battery while a dielectric is slid between the plates. Do the following quantities increase, decrease, or stay the same? (a) C (b) Q (c) E between the plates (d) ΔV (e) energy stored in the capacitor

EXAMPLE 16.11 | A Paper-Filled Capacitor

GOAL Calculate fundamental physical properties of a parallel-plate capacitor with a dielectric.

PROBLEM A parallel-plate capacitor has plates 2.0 cm by 3.0 cm. The plates are separated by a 1.0-mm thickness of paper. Find **(a)** the capacitance of this device and **(b)** the maximum charge that can be placed on the capacitor. **(c)** After the fully charged capacitor is disconnected from the battery, the dielectric is subsequently removed. Find the new electric field across the capacitor. Does the capacitor discharge?

STRATEGY For part (a), obtain the dielectric constant for paper from Table 16.1 and substitute, with other given quantities, into Equation 16.19. For part (b), note that Table 16.1 also gives the dielectric strength of paper, which is the maximum electric field that can be applied before electrical breakdown occurs. Use Equation 16.3, $\Delta V = Ed$, to obtain the maximum voltage and substitute into the basic capacitance equation. For part (c), remember that disconnecting the battery traps the extra charge on the plates, which must remain even after the dielectric is removed. Find the charge density on the plates and use Gauss's law to find the new electric field between the plates.

SOLUTION

(a) Find the capacitance of this device.

Substitute into Equation 16.19:

$$C = \kappa \epsilon_0 \frac{A}{d}$$
$$= 3.7 \left(8.85 \times 10^{-12} \frac{C^2}{N \cdot m^2} \right) \left(\frac{6.0 \times 10^{-4} \, m^2}{1.0 \times 10^{-3} \, m} \right)$$
$$= \boxed{2.0 \times 10^{-11} \, F}$$

(b) Find the maximum charge that can be placed on the capacitor.

Calculate the maximum applied voltage, using the dielectric strength of paper, E_{max}:

$$\Delta V_{max} = E_{max} d = (16 \times 10^6 \, V/m)(1.0 \times 10^{-3} \, m)$$
$$= 1.6 \times 10^4 \, V$$

Solve the basic capacitance equation for Q_{max} and substitute ΔV_{max} and C:

$$Q_{max} = C \Delta V_{max} = (2.0 \times 10^{-11} \, F)(1.6 \times 10^4 \, V)$$
$$= \boxed{0.32 \, \mu C}$$

(c) Suppose the fully charged capacitor is disconnected from the battery and the dielectric is subsequently removed. Find the new electric field between the plates of the capacitor. Does the capacitor discharge?

Compute the charge density on the plates:

$$\sigma = \frac{Q_{max}}{A} = \frac{3.2 \times 10^{-7} \, C}{6.0 \times 10^{-4} \, m^2} = 5.3 \times 10^{-4} \, C/m^2$$

Calculate the electric field from the charge density:

$$E = \frac{\sigma}{\epsilon_0} = \frac{5.3 \times 10^{-4} \, C/m^2}{8.85 \times 10^{-12} \, C^2/m^2 \cdot N} = \boxed{6.0 \times 10^7 \, N/C}$$

Because the electric field without the dielectric exceeds the value of the dielectric strength of air, the capacitor discharges across the gap.

REMARKS Dielectrics allow κ times as much charge to be stored on a capacitor for a given voltage. They also allow an increase in the applied voltage by increasing the threshold of electrical breakdown.

QUESTION 16.11 Without the paper dielectric, is the maximum charge that can be stored on this capacitor (a) larger than, (b) smaller than, or (c) the same as found in part (b)?

(Continued)

EXERCISE 16.11 A parallel-plate capacitor has plate area of 2.50×10^{-3} m^2 and distance between the plates of 2.00 mm. (a) Find the maximum charge that can be placed on the capacitor if air is between the plates. (b) Find the maximum charge if the air is replaced by polystyrene.

ANSWERS (a) 7×10^{-8} C (b) 1.4×10^{-6} C

EXAMPLE 16.12 | Capacitors with Two Dielectrics

GOAL Derive a symbolic expression for a parallel-plate capacitor with two dielectrics.

PROBLEM A parallel-plate capacitor has dielectrics with constants κ_1 and κ_2 between the two plates, as shown in Figure 16.29. Each dielectric fills exactly half the volume between the plates. Derive expressions for **(a)** the potential difference between the two plates and **(b)** the resulting capacitance of the system.

STRATEGY The magnitude of the potential difference between the two plates of a capacitor is equal to the electric field multiplied by the plate separation. The electric field in a region is reduced by a factor of $1/\kappa$ when a dielectric is introduced, so $E = \sigma/\epsilon = \sigma/\kappa\epsilon_0$. Add the potential difference across each dielectric to find the total potential difference ΔV between the plates. The voltage difference across each dielectric is given by $\Delta V = Ed$, where E is the electric field and d the displacement. Obtain the capacitance from the relationship $C = Q/\Delta V$.

Figure 16.29 (Exercise 16.12)

SOLUTION

(a) Derive an expression for the potential difference between the two plates.

Write a general expression for the potential difference across both slabs:

$$\Delta V = \Delta V_1 + \Delta V_2 = E_1 d_1 + E_2 d_2$$

Substitute expressions for the electric fields and dielectric thicknesses, $d_1 = d_2 = d/2$:

$$\Delta V = \frac{\sigma}{\kappa_1 \epsilon_0} \frac{d}{2} + \frac{\sigma}{\kappa_2 \epsilon_0} \frac{d}{2} = \boxed{\frac{\sigma d}{2\epsilon_0}\left(\frac{1}{\kappa_1} + \frac{1}{\kappa_2}\right)}$$

(b) Derive an expression for the resulting capacitance of the system.

Write the general expression for capacitance:

$$C = \frac{Q}{\Delta V}$$

Substitute $Q = \sigma A$ and the expression for the potential difference from part (a):

$$C = \frac{\sigma A}{\dfrac{\sigma d}{2\epsilon_0}\left(\dfrac{1}{\kappa_1} + \dfrac{1}{\kappa_2}\right)} = \boxed{\frac{2\epsilon_0 A}{d}\frac{\kappa_1 \kappa_2}{\kappa_1 + \kappa_2}}$$

REMARKS The answer is the same as if there had been two capacitors in series with the respective dielectrics. When a capacitor consists of two dielectrics as shown in Figure 16.30, however, it's equivalent to two different capacitors in parallel.

QUESTION 16.12 What answer is obtained when the two dielectrics are removed so there is vacuum between the plates?

EXERCISE 16.12 Suppose a capacitor has two dielectrics arranged as shown in Figure 16.30, each dielectric filling exactly half of the volume between the two plates. Derive an expression for the capacitance if each dielectric fills exactly half the volume between the plates.

ANSWER $C = \dfrac{\kappa_1 + \kappa_2}{2} \dfrac{\epsilon_0 A}{d}$

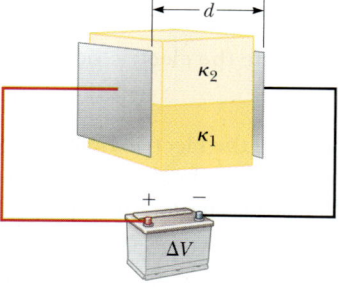

Figure 16.30 (Exercise 16.12)

An Atomic Description of Dielectrics

The explanation of why a dielectric increases the capacitance of a capacitor is based on an atomic description of the material, which in turn involves a property of some molecules called **polarization**. A molecule is said to be polarized when there is a separation between the average positions of its negative charge and its positive charge. In some molecules, such as water, this condition is always present. To see why, consider the geometry of a water molecule (Fig. 16.31).

The molecule is arranged so that the negative oxygen atom is bonded to the positively charged hydrogen atoms with a 105° angle between the two bonds. The center of negative charge is at the oxygen atom, and the center of positive charge lies at a point midway along the line joining the hydrogen atoms (point x in the diagram). Materials composed of molecules that are permanently polarized in this way have large dielectric constants, and indeed, Table 16.1 shows that the dielectric constant of water is large ($\kappa = 80$) compared with other common substances.

A symmetric molecule (Fig. 16.32a) can have no permanent polarization, but a polarization can be induced in it by an external electric field. A field directed to the left, as in Figure 16.32b, would cause the center of positive charge to shift to the left from its initial position and the center of negative charge to shift to the right. This *induced polarization* is the effect that predominates in most materials used as dielectrics in capacitors.

To understand why the polarization of a dielectric can affect capacitance, consider the slab of dielectric shown in Figure 16.33. Before placing the slab between the plates of the capacitor, the polar molecules are randomly oriented (Fig. 16.33a). The polar molecules are dipoles, and each creates a dipole electric field, but because of their random orientation, this field averages to zero.

After insertion of the dielectric slab into the electric field $\vec{E}_0$ between the plates (Fig. 16.33b), the positive plate attracts the negative ends of the dipoles and the negative plate attracts the positive ends of the dipoles. These forces exert a torque on the molecules making up the dielectric, reorienting them so that on average the negative pole is more inclined toward the positive plate and the positive pole is more aligned toward the negative plate. The positive and negative charges in the middle still cancel each other, but there is a net accumulation of negative charge in the dielectric next to the positive plate and a net accumulation of positive charge next to the negative plate. This configuration can be modeled as an additional pair of charged plates, as in Figure 16.33c, creating an induced electric field $\vec{E}_{ind}$ that partly cancels the original electric field $\vec{E}_0$. If the battery is not connected when the dielectric is inserted, the potential difference ΔV_0 across the plates is reduced to $\Delta V_0/\kappa$.

If the capacitor is still connected to the battery, however, the negative poles push more electrons off the positive plate, making it more positive. Meanwhile, the

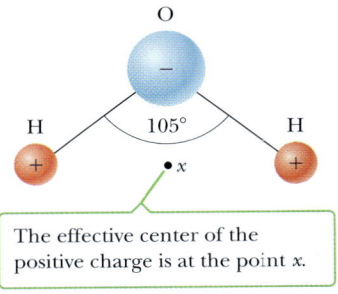

The effective center of the positive charge is at the point x.

Figure 16.31 The water molecule, H_2O, has a permanent polarization resulting from its bent geometry.

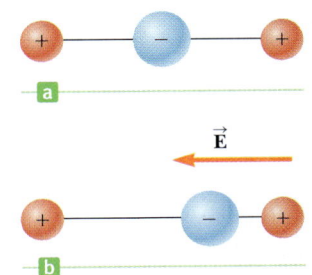

Figure 16.32 (a) A symmetric molecule has no permanent polarization. (b) An external electric field induces a polarization in the molecule.

Polar molecules are randomly oriented in the absence of an external electric field.

When an external electric field is applied, the molecules partially align with the field.

The charged edges create an induced electric field $\vec{E}_{ind}$ pointing in the direction opposite $\vec{E}_0$.

Figure 16.33 (a) Polar molecules are randomly oriented in a dielectric. (b) An electric field is applied to the dielectric. (c) The charged edges of the dielectric act like an additional pair of parallel plates, reducing the overall field between the actual plates. The interior of the dielectric is still neutral.

positive poles attract more electrons onto the negative plate. This situation continues until the potential difference across the battery reaches its original magnitude, equal to the potential gain across the battery. The net effect is an increase in the amount of charge stored on the capacitor. Because the plates can store more charge for a given voltage, it follows from $C = Q\Delta V$ that the capacitance must increase.

■ Quick Quiz

16.11 Consider a parallel-plate capacitor with a dielectric material between the plates. If the temperature of the dielectric increases, does the capacitance (a) decrease, (b) increase, or (c) remain the same?

■ SUMMARY

16.1 Potential Difference and Electric Potential

The change in the electric potential energy of a system consisting of an object of charge q moving through a displacement Δx in a constant electric field $\vec{E}$ is given by

$$\Delta PE = -W_{AB} = -qE_x\,\Delta x \qquad [16.1]$$

When a charge q moves in a uniform electric field $\vec{E}$ from point A to point B, the work done on the charge by the electric force is $qE_x\,\Delta x$.

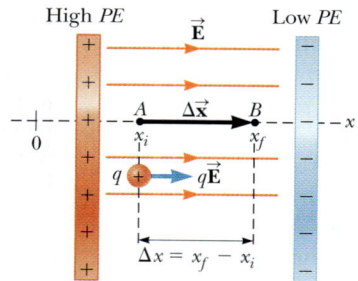

where E_x is the component of the electric field in the x-direction and $\Delta x = x_f - x_i$. The **difference in electric potential** between two points A and B is

$$\Delta V = V_B - V_A = \frac{\Delta PE}{q} \qquad [16.2]$$

where ΔPE is the *change* in electrical potential energy as a charge q moves between A and B. The units of potential difference are joules per coulomb, or **volts**; $1\,\text{J/C} = 1\,\text{V}$.

The **electric potential difference** between two points A and B in a *uniform* electric field $\vec{E}$ is

$$\Delta V = -E_x\,\Delta x \qquad [16.3]$$

where $\Delta x = x_f - x_i$ is the displacement between A and B and E_x is the x-component of the electric field in that region.

16.2 Electric Potential and Potential Energy Due to Point Charges

The **electric potential** due to a point charge q at distance r from the point charge is

$$V = k_e \frac{q}{r} \qquad [16.4]$$

Electric field and electric potential versus distance from a point charge of 1.11×10^{-10} C. Note that V is proportional to $1/r$, whereas E is proportional to $1/r^2$.

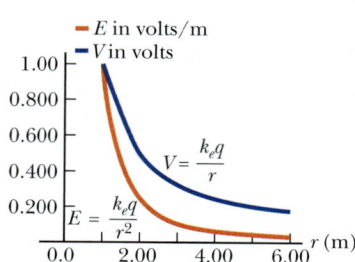

The **electric potential energy** of a pair of point charges separated by distance r is

$$PE = k_e \frac{q_1 q_2}{r} \qquad [16.5]$$

These equations can be used in the solution of conservation of energy problems and in the work–energy theorem.

16.3 Potentials and Charged Conductors
16.4 Equipotential Surfaces

Every point on the surface of a charged conductor in electrostatic equilibrium is at the same potential. Further, the potential is constant everywhere inside the conductor and equals its value on the surface.

The **electron volt** is defined as the energy that an electron (or proton) gains when accelerated through a potential difference of 1 V. The conversion between electron volts and joules is

$$1\,\text{eV} = 1.60 \times 10^{-19}\,\text{C}\cdot\text{V} = 1.60 \times 10^{-19}\,\text{J} \qquad [16.7]$$

Any surface on which the potential is the same at every point is called an equipotential surface. The electric field is always oriented perpendicular to an equipotential surface.

16.6 Capacitance

A capacitor consists of two metal plates with charges that are equal in magnitude but opposite in sign. The capacitance C of any capacitor is the ratio of the magnitude of the charge Q on either plate to the magnitude of potential difference ΔV between them:

$$C \equiv \frac{Q}{\Delta V} \qquad [16.8]$$

Capacitance has the units coulombs per volt, or farads; 1 C/V = 1 F.

16.7 The Parallel-Plate Capacitor

The capacitance of two parallel metal plates of area A separated by distance d is

$$C = \epsilon_0 \frac{A}{d} \quad [16.9]$$

where $\epsilon_0 = 8.85 \times 10^{-12}$ $C^2/N \cdot m^2$ is a constant called the **permittivity of free space.**

A parallel-plate capacitor consists of two parallel plates, each of area A, separated by a distance d.

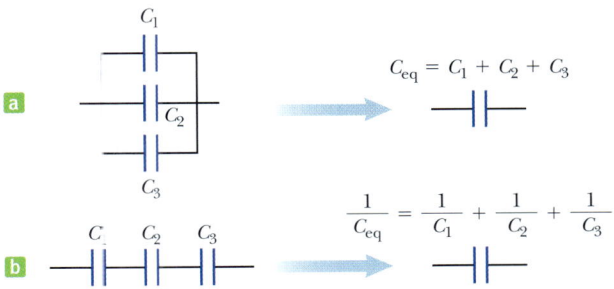

Capacitors in (a) parallel or in (b) series can be written as a single equivalent capacitor.

Problems involving a combination of capacitors can be solved by applying Equations 16.12 and 16.15 repeatedly to a circuit diagram, simplifying it as much as possible. This step is followed by working backwards to the original diagram, applying $C = Q/\Delta V$, that parallel capacitors have the same voltage drop, and that series capacitors have the same charge.

16.9 Energy Stored in a Charged Capacitor

Three equivalent expressions for calculating the **energy stored** in a charged capacitor are

$$\text{Energy stored} = \tfrac{1}{2}Q\,\Delta V = \tfrac{1}{2}C(\Delta V)^2 = \frac{Q^2}{2C} \quad [16.17]$$

16.8 Combinations of Capacitors

The **equivalent capacitance of a parallel combination** of capacitors is

$$C_{eq} = C_1 + C_2 + C_3 + \cdots \quad [16.12]$$

If two or more capacitors are connected in series, the **equivalent capacitance of the series combination** is

$$\frac{1}{C_{eq}} = \frac{1}{C_1} + \frac{1}{C_2} + \frac{1}{C_3} + \cdots \quad [16.15]$$

16.10 Capacitors with Dielectrics

When a nonconducting material, called a **dielectric,** is placed between the plates of a capacitor, the capacitance is multiplied by the factor κ, which is called the **dielectric constant,** a property of the dielectric material. The capacitance of a parallel-plate capacitor filled with a dielectric is

$$C = \kappa \epsilon_0 \frac{A}{d} \quad [16.19]$$

■ WARM-UP EXERCISES

WebAssign The warm-up exercises in this chapter may be assigned online in Enhanced WebAssign.

1. **Math Review** Determine the value of C_{eq} in the following equation if $C_1 = 5.00$ Ω, $C_2 = 6.00$ Ω, and $C_3 = 7.00$ Ω: $\frac{1}{C_{eq}} = \frac{1}{C_1} + \frac{1}{C_2} + \frac{1}{C_3}$. (See also Section 16.8.)

2. **Physics Review** A 2.50-kg object initially at rest has a gravitational potential energy of 72.0 J. Determine the speed of the object when it has moved under the influence of gravity to a location where its gravitational potential energy is 45.0 J. (See Section 5.3.)

3. A uniform electric field of magnitude 3.00 N/C is directed along the $+x$-axis. If a 2.00 μC charge moves from (1.00, 0) m to (2.50, 0) m in this field, determine (a) the work done by the electric force, (b) the change in the electric potential energy of the particle, and (c) the electric potential difference between the particle's initial and final points. (See Section 16.1.)

4. A +4.00 μC charge is located at the origin. Determine (a) the electric potential at a distance of 2.00 m from the charge, and (b) the electric potential energy of a -2.03 μC charge located at (0, 2.00) m. (See Section 16.2.)

5. Two protons are located at (1.00, 0) m and (0,1.50) m, respectively. Determine (a) the electric potential at the origin, and (b) the electric potential energy of a third proton located at the origin. (See Section 16.2.)

6. (a) A hydrogen atom can be ionized by a photon having an energy of 13.6 eV. Convert that quantity to joules. (b) It requires 4.186 J of thermal energy to warm a gram of water by 1.00 K. Convert that energy to electron volts. (See Section 16.3.)

7. A capacitor with capacitance 3.00 μF is connected to a 9.00-V battery. (a) Find the charge on the capacitor in coulombs. (b) What voltage battery would be required to store 7.20×10^{-5} C on the capacitor? (See Section 16.6.)

8. A parallel-plate capacitor with no dielectric is made from two plates separated by 1.15×10^{-3} m. (a) Determine the device's capacitance if each plate has an area of 0.250 m². (See Section 16.6.) (b) What magnitude of charge is stored on each plate when the capacitor is connected to a 12.0-V battery? (See Section 16.6.) (c) If a material of dielectric constant 12.5 is inserted between the plates, what are the new values of the capacitance and (d) the magnitude of charge stored on each plate? (See Section 16.10.)

9. Two capacitors have capacitance 2.00 μF and 3.00 μF, respectively. Calculate the equivalent capacitance in microfarads if the capacitors are put (a) in parallel with each other, and (b) in series with each other. (See Section 16.8.)

10. Two capacitors are connected in parallel across a 12.0-V battery. If their capacitances are 12.0 μF and 25.0 μF, determine (a) the voltage across each capacitor, (b) the magnitude of charge stored on each plate of the 12.0 μF capacitor, (c) the magnitude of charge stored on each plate of the 25.0 μF capacitor, and (d) the equivalent capacitance of the system. (See Section 16.8.)

11. Two capacitors are connected in series between the terminals of a 12.0-V battery. If their capacitances are 12.0 μF and 25.0 μF, determine (a) the equivalent capacitance of the system, (b) the magnitude of charge stored on each plate of either capacitor, (c) the voltage across the 12.0 μF capacitor, and (d) the voltage across the 25.0 μF capacitor. (See Section 16.8.)

12. The plates of a parallel-plate capacitor are charged to a potential difference of 12.0 V. If the capacitance is 15.0 μF, calculate (a) the energy stored in the capacitor and (b) the magnitude of charge stored on each plate of the capacitor. (See Section 16.9.)

CONCEPTUAL QUESTIONS

The conceptual questions in this chapter may be assigned online in Enhanced WebAssign.

1. (a) Describe the motion of a proton after it is released from rest in a uniform electric field. (b) Describe the changes (if any) in its kinetic energy and the electric potential energy associated with the proton.

2. Rank the potential energies of the four systems of particles shown in Figure CQ16.2 from largest to smallest. Include equalities if appropriate.

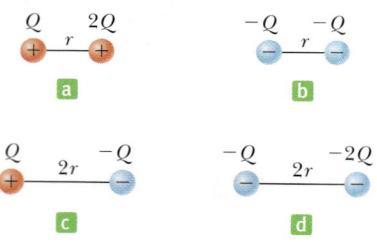

Figure CQ16.2

3. A parallel-plate capacitor is charged by a battery, and the battery is then disconnected from the capacitor. Because the charges on the capacitor plates are opposite in sign, they attract each other. Hence, it takes positive work to increase the plate separation. Show that the external work done when the plate separation is increased leads to an increase in the energy stored in the capacitor.

4. When charged particles are separated by an infinite distance, the electric potential energy of the pair is zero. When the particles are brought close, the electric potential energy of a pair with the same sign is positive, whereas the electric potential energy of a pair with opposite signs is negative. Explain.

5. Suppose you are sitting in a car and a 20-kV power line drops across the car. Should you stay in the car or get out? The power line potential is 20 kV compared to the potential of the ground.

6. Why is it important to avoid sharp edges or points on conductors used in high-voltage equipment?

7. Explain why, under static conditions, all points in a conductor must be at the same electric potential.

8. If you are given three different capacitors C_1, C_2, and C_3, how many different combinations of capacitance can you produce, using all capacitors in your circuits?

9. (a) Why is it dangerous to touch the terminals of a high-voltage capacitor even after the voltage source that charged the battery is disconnected from the capacitor? (b) What can be done to make the capacitor safe to handle after the voltage source has been removed?

10. The plates of a capacitor are connected to a battery. (a) What happens to the charge on the plates if the connecting wires are removed from the battery? (b) What happens to the charge if the wires are removed from the battery and connected to each other?

11. Rank the electric potentials at the four points shown in Figure CQ16.11 from largest to smallest.

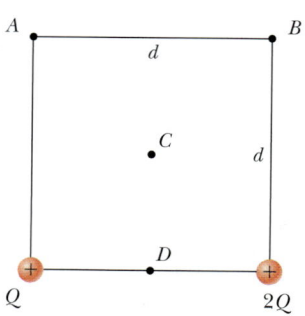

Figure CQ16.11

12. If you were asked to design a capacitor in which small size and large capacitance were required, what would be the two most important factors in your design?

13. Is it always possible to reduce a combination of capacitors to one equivalent capacitor with the rules developed in this chapter? Explain.

14. Explain why a dielectric increases the maximum operating voltage of a capacitor even though the physical size of the capacitor doesn't change.

PROBLEMS

WebAssign The problems in this chapter may be assigned online in Enhanced WebAssign.
1. denotes straightforward problem; 2. denotes intermediate problem;
3. denotes challenging problem
[1.] denotes full solution available in *Student Solutions Manual/Study Guide*
1. denotes problems most often assigned in Enhanced WebAssign

BIO denotes biomedical problems
GP denotes guided problems
M denotes Master It tutorial available in Enhanced WebAssign
QC denotes asking for quantitative and conceptual reasoning
S denotes symbolic reasoning problem
W denotes Watch It video solution available in Enhanced WebAssign

16.1 Potential Difference and Electric Potential

1. A uniform electric field of magnitude 375 N/C pointing in the positive *x*-direction acts on an electron, which is initially at rest. After the electron has moved 3.20 cm, what is (a) the work done by the field on the electron, (b) the change in potential energy associated with the electron, and (c) the velocity of the electron?

2. A proton is released from rest in a uniform electric field of magnitude 385 N/C. Find (a) the electric force on the proton, (b) the acceleration of the proton, and (c) the distance it travels in 2.00 μs.

3. **BIO W** A potential difference of 90 mV exists between the inner and outer surfaces of a cell membrane. The inner surface is negative relative to the outer surface. How much work is required to eject a positive sodium ion (Na$^+$) from the interior of the cell?

4. A metal sphere of radius 5.00 cm is initially uncharged. How many electrons would have to be placed on the sphere to produce an electric field of magnitude 1.50×10^5 N/C at a point 8.00 cm from the center of the sphere?

5. The potential difference between the accelerating plates of a TV set is about 25 kV. If the distance between the plates is 1.5 cm, find the magnitude of the uniform electric field in the region between the plates.

6. A point charge $q = +40.0$ μC moves from *A* to *B* separated by a distance $d = 0.180$ m in the presence of an external electric field $\vec{E}$ of magnitude 275 N/C directed toward the right as in Figure P16.6. Find (a) the electric force exerted on the charge, (b) the work done by the electric force, (c) the change in the electric potential energy of the charge, and (d) the potential difference between *A* and *B*.

7. **M** Oppositely charged parallel plates are separated by 5.33 mm. A potential difference of 600 V exists between the plates. (a) What is the magnitude of the electric field between the plates? (b) What is the magnitude of the force on an electron between the plates? (c) How much work must be done on the electron to move it to the negative plate if it is initially positioned 2.90 mm from the positive plate?

8. **QC S** (a) Find the potential difference ΔV_e required to stop an electron (called a "stopping potential") moving with an initial speed of 2.85×10^7 m/s. (b) Would a proton traveling at the same speed require a greater or lesser magnitude potential difference? Explain. (c) Find a symbolic expression for the ratio of the proton stopping potential and the electron stopping potential, $\Delta V_p/\Delta V_e$. The answer should be in terms of the proton mass m_p and electron mass m_e.

9. **QC** A 74.0-g block carrying a charge $Q = 35.0$ μC is connected to a spring for which $k = 78.0$ N/m. The block lies on a frictionless, horizontal surface and is immersed in a uniform electric field of magnitude $E = 4.86 \times 10^4$ N/C directed as shown in Figure P16.9. If the block is released from rest when the spring is unstretched ($x = 0$), (a) by what maximum distance does the block move from its initial position? (b) Find the subsequent equilibrium position of the block and the amplitude of its motion. (c) Using conservation of energy, find a symbolic relationship giving the potential difference between its initial position and the point of maximum extension in terms of the spring constant *k*, the amplitude *A*, and the charge *Q*.

Figure P16.6

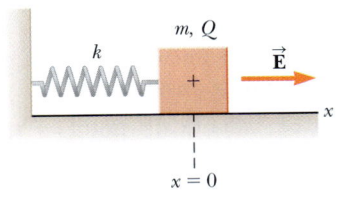

Figure P16.9

10. On planet Tehar, the free-fall acceleration is the same as that on the Earth, but there is also a strong downward electric field that is uniform close to the planet's surface. A 2.00-kg ball having a charge of 5.00 μC is thrown upward at a speed of 20.1 m/s. It hits the ground after an interval of 4.10 s. What is the potential difference between the starting point and the top point of the trajectory?

16.2 Electric Potential and Potential Energy Due to Point Charges

16.3 Potentials and Charged Conductors

16.4 Equipotential Surfaces

11. **Q|C** An electron is at the origin. (a) Calculate the electric potential V_A at point A, $x = 0.250$ cm. (b) Calculate the electric potential V_B at point B, $x = 0.750$ cm. What is the potential difference $V_B - V_A$? (c) Would a negatively charged particle placed at point A necessarily go through this same potential difference upon reaching point B? Explain.

12. The two charges in Figure P16.12 are separated by $d = 2.00$ cm. Find the electric potential at (a) point A and (b) point B, which is halfway between the charges.

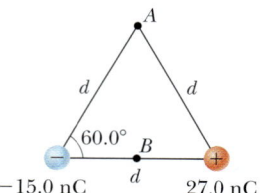

Figure P16.12

13. (a) Find the electric potential, taking zero at infinity, at the upper right corner (the corner without a charge) of the rectangle in Figure P16.13. (b) Repeat if the 2.00-μC charge is replaced with a charge of -2.00 μC.

Figure P16.13 Problems 13 and 14.

14. Three charges are situated at corners of a rectangle as in Figure P16.13. How much work must an external agent do to move the 8.00-μC charge to infinity?

15. **M Q|C** Two point charges $Q_1 = +5.00$ nC and $Q_2 = -3.00$ nC are separated by 35.0 cm. (a) What is the electric potential at a point midway between the charges? (b) What is the potential energy of the pair of charges? What is the significance of the algebraic sign of your answer?

16. **Q|C S** Three identical point charges each of charge q are located at the vertices of an equilateral triangle as in Figure P16.16.

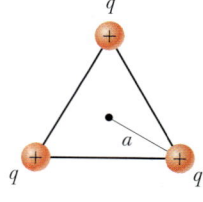

Figure P16.16

The distance from the center of the triangle to each vertex is a. (a) Show that the electric field at the center of the triangle is zero. (b) Find a symbolic expression for the electric potential at the center of the triangle. (c) Give a physical explanation of the fact that the electric potential is not zero, yet the electric field is zero at the center.

17. The three charges in Figure P16.17 are at the vertices of an isosceles triangle. Let $q = 7.00$ nC and calculate the electric potential at the midpoint of the base.

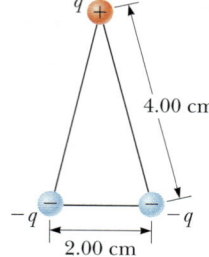

Figure P16.17

18. **S** A positive point charge $q = +2.50$ nC is located at $x = 1.20$ m and a negative charge of $-2q = -5.00$ nC is located at the origin as in Figure P16.18. (a) Sketch the electric potential versus x for points along the x-axis in the range -1.50 m $< x <$ 1.50 m. (b) Find a symbolic expression for the potential on the x-axis at an arbitrary point P between the two charges. (c) Find the electric potential at $x = 0.600$ m. (d) Find the point along the x-axis between the two charges where the electric potential is zero.

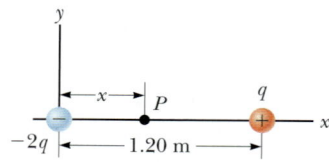

Figure P16.18

19. **GP** A proton is located at the origin, and a second proton is located on the x-axis at $x = 6.00$ fm (1 fm = 10^{-15} m). (a) Calculate the electric potential energy associated with this configuration. (b) An alpha particle (charge = $2e$, mass = 6.64×10^{-27} kg) is now placed at $(x, y) = (3.00, 3.00)$ fm. Calculate the electric potential energy associated with this configuration. (c) Starting with the three-particle system, find the change in electric potential energy if the alpha particle is allowed to escape to infinity while the two protons remain fixed in place. (Throughout, neglect any radiation effects.) (d) Use conservation of energy to calculate the speed of the alpha particle at infinity. (e) If the two protons are released from rest and the alpha particle remains fixed, calculate the speed of the protons at infinity.

20. **Q|C** A proton and an alpha particle (charge = $2e$, mass = 6.64×10^{-27} kg) are initially at rest, separated by 4.00×10^{-15} m. (a) If they are both released simultaneously, explain why you can't find their velocities at infinity using only conservation of energy. (b) What other conservation law can be applied in this case? (c) Find the speeds of the proton and alpha particle, respectively, at infinity.

21. A tiny sphere of mass 8.00 μg and charge −2.80 nC is initially at a distance of 1.60 μm from a fixed charge of +8.50 nC. If the 8.00-mg sphere is released from rest, find (a) its kinetic energy when it is 0.500 μm from the fixed charge and (b) its speed when it is 0.500 μm from the fixed charge.

22. The metal sphere of a small Van de Graaff generator illustrated in Figure 15.23 has a radius of 18 cm. When the electric field at the surface of the sphere reaches 3.0×10^6 V/m, the air breaks down, and the generator discharges. What is the maximum potential the sphere can have before breakdown occurs?

23. **W** In Rutherford's famous scattering experiments that led to the planetary model of the atom, alpha particles (having charges of $+2e$ and masses of 6.64×10^{-27} kg) were fired toward a gold nucleus with charge $+79e$. An alpha particle, initially very far from the gold nucleus, is fired at 2.00×10^7 m/s directly toward the nucleus, as in Figure P16.23. How close does the alpha particle get to the gold nucleus before turning around? Assume the gold nucleus remains stationary.

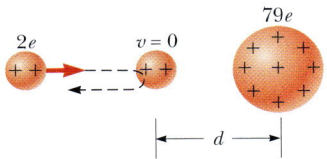

Figure P16.23

24. **S** Four point charges each having charge Q are located at the corners of a square having sides of length a. Find symbolic expressions for (a) the total electric potential at the center of the square due to the four charges and (b) the work required to bring a fifth charge q from infinity to the center of the square.

16.6 Capacitance

16.7 The Parallel-Plate Capacitor

25. **W** Consider the Earth and a cloud layer 800 m above the planet to be the plates of a parallel-plate capacitor. (a) If the cloud layer has an area of 1.0 km² = 1.0×10^6 m², what is the capacitance? (b) If an electric field strength greater than 3.0×10^6 N/C causes the air to break down and conduct charge (lightning), what is the maximum charge the cloud can hold?

26. (a) When a 9.00-V battery is connected to the plates of a capacitor, it stores a charge of 27.0 μC. What is the value of the capacitance? (b) If the same capacitor is connected to a 12.0-V battery, what charge is stored?

27. An air-filled parallel-plate capacitor has plates of area 2.30 cm² separated by 1.50 mm. The capacitor is connected to a 12.0-V battery. (a) Find the value of its capacitance. (b) What is the charge on the capacitor? (c) What is the magnitude of the uniform electric field between the plates?

28. Two conductors having net charges of +10.0 μC and −10.0 μC have a potential difference of 10.0 V between them. (a) Determine the capacitance of the system. (b) What is the potential difference between the two conductors if the charges on each are increased to +100 μC and −100 μC?

29. **M** An air-filled capacitor consists of two parallel plates, each with an area of 7.60 cm² and separated by a distance of 1.80 mm. If a 20.0-V potential difference is applied to these plates, calculate (a) the electric field between the plates, (b) the capacitance, and (c) the charge on each plate.

30. A 1-megabit computer memory chip contains many 60.0×10^{-15}-F capacitors. Each capacitor has a plate area of 21.0×10^{-12} m². Determine the plate separation of such a capacitor. (Assume a parallel-plate configuration.) The diameter of an atom is on the order of 10^{-10} m = 1 Å. Express the plate separation in angstroms.

31. **Q|C** A parallel-plate capacitor with area 0.200 m² and plate separation of 3.00 mm is connected to a 6.00-V battery. (a) What is the capacitance? (b) How much charge is stored on the plates? (c) What is the electric field between the plates? (d) Find the magnitude of the charge density on each plate. (e) Without disconnecting the battery, the plates are moved farther apart. Qualitatively, what happens to each of the previous answers?

32. A small object with a mass of 350 μg carries a charge of 30.0 nC and is suspended by a thread between the vertical plates of a parallel-plate capacitor. The plates are separated by 4.00 cm. If the thread makes an angle of 15.0° with the vertical, what is the potential difference between the plates?

16.8 Combinations of Capacitors

33. Given a 2.50-μF capacitor, a 6.25-μF capacitor, and a 6.00-V battery, find the charge on each capacitor if you connect them (a) in series across the battery and (b) in parallel across the battery.

34. Two capacitors, $C_1 = 5.00$ μF and $C_2 = 12.0$ μF, are connected in parallel, and the resulting combination is connected to a 9.00-V battery. Find (a) the equivalent capacitance of the combination, (b) the potential difference across each capacitor, and (c) the charge stored on each capacitor.

35. Find (a) the equivalent capacitance of the capacitors in Figure P16.35, (b) the charge on each capacitor, and (c) the potential difference across each capacitor.

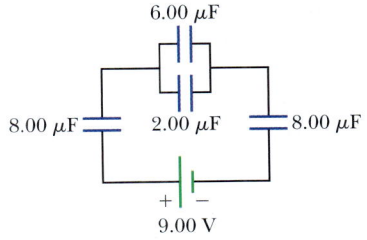

Figure P16.35

36. W Two capacitors give an equivalent capacitance of 9.00 pF when connected in parallel and an equivalent capacitance of 2.00 pF when connected in series. What is the capacitance of each capacitor?

37. For the system of capacitors shown in Figure P16.37, find (a) the equivalent capacitance of the system, (b) the charge on each capacitor, and (c) the potential difference across each capacitor.

Figure P16.37 Problems 37 and 56.

38. GP Consider the combination of capacitors in Figure P16.38. (a) Find the equivalent single capacitance of the two capacitors in series and redraw the diagram (called diagram 1) with this equivalent capacitance. (b) In diagram 1 find the equivalent capacitance of the three capacitors in parallel and redraw the diagram as a single battery and single capacitor in a loop. (c) Compute the charge on the single equivalent capacitor. (d) Returning to diagram 1, compute the charge on each individual capacitor. Does the sum agree with the value found in part (c)? (e) What is the charge on the 24.0-μF capacitor and on the 8.00-μF capacitor? Compute the voltage drop across (f) the 24.0-μF capacitor and (g) the 8.00-μF capacitor.

Figure P16.38

39. Find the charge on each of the capacitors in Figure P16.39.

Figure P16.39

40. QC S Three capacitors are connected to a battery as shown in Figure P16.40. Their capacitances are $C_1 = 3C$, $C_2 = C$, and $C_3 = 5C$. (a) What is the equivalent capacitance of this set of capacitors? (b) State the ranking of the capacitors according to the charge they store from largest to smallest. (c) Rank the capacitors

Figure P16.40

according to the potential differences across them from largest to smallest. (d) Assume C_3 is increased. Explain what happens to the charge stored by each capacitor.

41. A 25.0-μF capacitor and a 40.0-μF capacitor are charged by being connected across separate 50.0-V batteries. (a) Determine the resulting charge on each capacitor. (b) The capacitors are then disconnected from their batteries and connected to each other, with each negative plate connected to the other positive plate. What is the final charge of each capacitor? (c) What is the final potential difference across the 40.0-μF capacitor?

42. (a) Find the equivalent capacitance between points a and b for the group of capacitors connected as shown in Figure P16.42 if $C_1 = 5.00\ \mu$F, $C_2 = 10.00\ \mu$F, and $C_3 = 2.00\ \mu$F. (b) If the potential between points a and b is 60.0 V, what charge is stored on C_3?

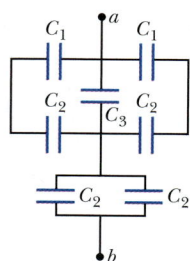

Figure P16.42

43. A 1.00-μF capacitor is charged by being connected across a 10.0-V battery. It is then disconnected from the battery and connected across an uncharged 2.00-μF capacitor. Determine the resulting charge on each capacitor.

44. M Four capacitors are connected as shown in Figure P16.44. (a) Find the equivalent capacitance between points a and b. (b) Calculate the charge on each capacitor, taking $\Delta V_{ab} = 15.0$ V.

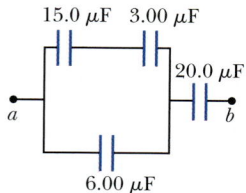

Figure P16.44

16.9 Energy Stored in a Charged Capacitor

45. A 12.0-V battery is connected to a 4.50-μF capacitor. How much energy is stored in the capacitor?

46. QC Two capacitors, $C_1 = 18.0\ \mu$F and $C_2 = 36.0\ \mu$F, are connected in series, and a 12.0-V battery is connected across them. (a) Find the equivalent capacitance, and the energy contained in this equivalent capacitor. (b) Find the energy stored in each individual capacitor. Show that the sum of these two energies is the same as the energy found in part (a). Will this equality always be true, or does it depend on the number of capacitors and their capacitances? (c) If the same capacitors were connected in parallel, what potential difference would be required across them so that the combination stores the same energy as in part (a)? Which capacitor stores more energy in this situation, C_1 or C_2?

47. A parallel-plate capacitor has capacitance 3.00 μF. (a) How much energy is stored in the capacitor if it is connected to a 6.00-V battery? (b) If the battery is disconnected and the distance between the charged plates doubled, what is the energy stored? (c) The battery is subsequently reattached to the capacitor, but the plate separation remains as in part (b). How much energy is stored? (Answer each part in microjoules.)

48. A certain storm cloud has a potential difference of 1.00×10^8 V relative to a tree. If, during a lightning storm, 50.0 C of charge is transferred through this potential difference and 1.00% of the energy is absorbed by the tree, how much water (sap in the tree) initially at 30.0°C can be boiled away? Water has a specific heat of 4 186 J/kg·°C, a boiling point of 100°C, and a heat of vaporization of 2.26×10^6 J/kg.

16.10 Capacitors with Dielectrics

49. Q|C The voltage across an air-filled parallel-plate capacitor is measured to be 85.0 V. When a dielectric is inserted and completely fills the space between the plates as in Figure P16.49, the voltage drops to 25.0 V. (a) What is the dielectric constant of the inserted material? Can you identify the dielectric? (b) If the dielectric doesn't completely fill the space between the plates, what could you conclude about the voltage across the plates?

Figure P16.49

50. (a) How much charge can be placed on a capacitor with air between the plates before it breaks down if the area of each plate is 5.00 cm²? (b) Find the maximum charge if polystyrene is used between the plates instead of air. Assume the dielectric strength of air is 3.00×10^6 V/m and that of polystyrene is 24.0×10^6 V/m.

51. Determine (a) the capacitance and (b) the maximum voltage that can be applied to a Teflon-filled parallel-plate capacitor having a plate area of 175 cm² and an insulation thickness of 0.040 0 mm.

52. Consider a plane parallel-plate capacitor made of two strips of aluminum foil separated by a layer of paraffin-coated paper. Each strip of foil and paper is 7.00 cm wide. The foil is 0.004 00 mm thick, and the paper is 0.025 0 mm thick and has a dielectric constant of 3.70. What length should the strips be if a capacitance of 9.50×10^{-8} F is desired? (If, after this plane capacitor is formed, a second paper strip can be added below the foil-paper-foil stack and the resulting assembly rolled into a cylindrical form—similar to that shown in Figure 16.26—the capacitance can be doubled because both surfaces of each foil strip would then store charge. Without the second strip of paper, however, rolling the layers would result in a short circuit.)

53. BIO A model of a red blood cell portrays the cell as a spherical capacitor, a positively charged liquid sphere of surface area A separated from the surrounding negatively charged fluid by a membrane of thickness t. Tiny electrodes introduced into the interior of the cell show a potential difference of 100 mV across the membrane. The membrane's thickness is estimated to be 100 nm and has a dielectric constant of 5.00. (a) If an average red blood cell has a mass of 1.00×10^{-12} kg, estimate the volume of the cell and thus find its surface area. The density of blood is 1 100 kg/m³. (b) Estimate the capacitance of the cell by assuming the membrane surfaces act as parallel plates. (c) Calculate the charge on the surface of the membrane. How many electronic charges does the surface charge represent?

Additional Problems

54. When a potential difference of 150 V is applied to the plates of an air-filled parallel-plate capacitor, the plates carry a surface charge density of 3.00×10^{-10} C/cm². What is the spacing between the plates?

55. S Three parallel-plate capacitors are constructed, each having the same plate area A and with C_1 having plate spacing d_1, C_2 having plate spacing d_2, and C_3 having plate spacing d_3. Show that the total capacitance C of the three capacitors connected in series is the same as a capacitor of plate area A and with plate spacing $d = d_1 + d_2 + d_3$.

56. Q|C For the system of four capacitors shown in Figure P16.37, find (a) the total energy stored in the system and (b) the energy stored by each capacitor. (c) Compare the sum of the answers in part (b) with your result to part (a) and explain your observation.

57. S A parallel-plate capacitor with a plate separation d has a capacitance C_0 in the absence of a dielectric. A slab of dielectric material of dielectric constant κ and thickness $d/3$ is then inserted between the plates as in Figure P16.57a. Show that the capacitance of this partially filled capacitor is given by

$$C = \left(\frac{3\kappa}{2\kappa + 1}\right)C_0$$

Hint: Treat the system as two capacitors connected in series as in Figure P16.57b, one with dielectric in it and the other one empty.

Figure P16.57

58. **S** Two capacitors give an equivalent capacitance of C_p when connected in parallel and an equivalent capacitance of C_s when connected in series. What is the capacitance of each capacitor?

59. **M** A parallel-plate capacitor is constructed using a dielectric material whose dielectric constant is 3.00 and whose dielectric strength is 2.00×10^8 V/m. The desired capacitance is 0.250 μF, and the capacitor must withstand a maximum potential difference of 4.00 kV. Find the minimum area of the capacitor plates.

60. Two charges of 1.0 μC and -2.0 μC are 0.50 m apart at two vertices of an equilateral triangle as in Figure P16.60. (a) What is the electric potential due to the 1.0-μC charge at the third vertex, point P? (b) What is the electric potential due to the -2.0-μC charge at P? (c) Find the total electric potential at P. (d) What is the work required to move a 3.0-μC charge from infinity to P?

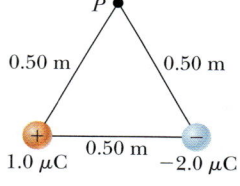

Figure P16.60

61. Find the equivalent capacitance of the group of capacitors shown in Figure P16.61.

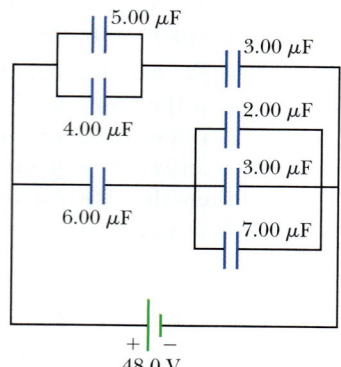

Figure P16.61

62. A spherical capacitor consists of a spherical conducting shell of radius b and charge $-Q$ concentric with a smaller conducting sphere of radius a and charge Q. (a) Find the capacitance of this device. (b) Show that as the radius b of the outer sphere approaches infinity, the capacitance approaches the value $a/k_e = 4\pi\epsilon_0 a$.

63. **BIO** The immediate cause of many deaths is ventricular fibrillation, an uncoordinated quivering of the heart, as opposed to proper beating. An electric shock to the chest can cause momentary paralysis of the heart muscle, after which the heart will sometimes start organized beating again. A *defibrillator* is a device that applies a strong electric shock to the chest over a time of a few milliseconds. The device contains a capacitor of a few microfarads, charged to several thousand volts. Electrodes called paddles, about 8 cm across and coated with conducting paste, are held against the chest on both sides of the heart. Their handles are insulated to prevent injury to the operator, who calls "Clear!" and pushes a button on one paddle to discharge the capacitor through the patient's chest. Assume an energy of 300 W · s is to be delivered from a 30.0-μF capacitor. To what potential difference must it be charged?

64. When a certain air-filled parallel-plate capacitor is connected across a battery, it acquires a charge of 150 μC on each plate. While the battery connection is maintained, a dielectric slab is inserted into, and fills, the region between the plates. This results in the accumulation of an additional charge of 200 μC on each plate. What is the dielectric constant of the slab?

65. Capacitors $C_1 = 6.0$ μF and $C_2 = 2.0$ μF are charged as a parallel combination across a 250-V battery. The capacitors are disconnected from the battery and from each other. They are then connected positive plate to negative plate and negative plate to positive plate. Calculate the resulting charge on each capacitor.

66. **S** Two positive charges each of charge q are fixed on the y-axis, one at $y = d$ and the other at $y = -d$ as in Figure P16.66. A third positive charge $2q$ located on the x-axis at $x = 2d$ is released from rest. Find symbolic expressions for (a) the total electric potential due to the first two charges at the location of the charge $2q$, (b) the electric potential energy of the charge $2q$, (c) the kinetic energy of the charge $2q$ after it has moved infinitely far from the other charges, and (d) the speed of the charge $2q$ after it has moved infinitely far from the other charges if its mass is m.

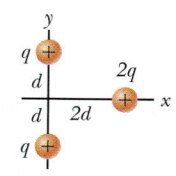

Figure P16.66

67. Metal sphere A of radius 12.0 cm carries 6.00 μC of charge, and metal sphere B of radius 18.0 cm carries -4.00 μC of charge. If the two spheres are attached by

a very long conducting thread, what is the final distribution of charge on the two spheres?

68. An electron is fired at a speed $v_0 = 5.6 \times 10^6$ m/s and at an angle $\theta_0 = -45°$ between two parallel conducting plates that are $D = 2.0$ mm apart, as in Figure P16.68. If the voltage difference between the plates is $\Delta V = 100$ V, determine (a) how close, d, the electron will get to the bottom plate and (b) where the electron will strike the top plate.

Figure P16.68

The blue glow comes from positively-charged xenon atoms that are electrostatically accelerated, then expelled from an ion engine prototype. The current of ions produces ninety millinewtons of thrust continuously for months at a time. Electrons must be fed back into the exhaust to prevent a buildup of negative charge. Such engines are highly efficient and suitable for extended deep space missions.

17 Current and Resistance

17.1 Electric Current

17.2 A Microscopic View: Current and Drift Speed

17.3 Current and Voltage Measurements in Circuits

17.4 Resistance, Resistivity, and Ohm's Law

17.5 Temperature Variation of Resistance

17.6 Electrical Energy and Power

17.7 Superconductors

17.8 Electrical Activity in the Heart

Many practical applications and devices are based on the principles of static electricity, but electricity was destined to become an inseparable part of our daily lives when scientists learned how to produce a continuous flow of charge for relatively long periods of time using batteries. The battery or voltaic cell was invented in 1800 by Italian physicist Alessandro Volta. Batteries supplied a continuous flow of charge at low potential, in contrast to earlier electrostatic devices that produced a tiny flow of charge at high potential for brief periods. This steady source of electric current allowed scientists to perform experiments to learn how to control the flow of electric charges in circuits. Today, electric currents power our lights, radios, television sets, air conditioners, computers, and refrigerators. They ignite the gasoline in automobile engines, travel through miniature components making up the chips of microcomputers, and provide the power for countless other invaluable tasks.

In this chapter we define current and discuss some of the factors that contribute to the resistance to the flow of charge in conductors. We also discuss energy transformations in electric circuits. These topics will be the foundation for additional work with circuits in later chapters.

17.1 Electric Current

LEARNING OBJECTIVES

1. Define both average and instantaneous electrical current and discuss their physical meaning.
2. Apply the concept of current to simple electrical systems.

Tip 17.1 *Current Flow Is Redundant*

The phrases *flow of current* and *current flow* are commonly used, but here the word *flow* is redundant because current is already defined as a flow (of charge). Avoid this construction!

In Figure 17.1 charges move in a direction perpendicular to a surface of area A. (That area could be the cross-sectional area of a wire, for example.) **The current is the rate at which charge flows through this surface**.

Suppose ΔQ is the amount of charge that flows through an area A in a time interval Δt and that the direction of flow is perpendicular to the area. Then the **average current** I_{av} is equal to the amount of charge divided by the time interval:

$$I_{av} \equiv \frac{\Delta Q}{\Delta t} \quad [17.1a]$$

SI unit: coulomb/second (C/s), or the ampere (A)

Current is composed of individual moving charges, so for an extremely low current, it is conceivable that a single charge could pass through area A in one instant and no charge in the next instant. All currents, then, are essentially averages over time. Given the very large number of charges usually involved, however, it makes sense to define an instantaneous current.

The **instantaneous current** I is the limit of the average current as the time interval goes to zero:

$$I = \lim_{\Delta t \to 0} I_{av} = \lim_{\Delta t \to 0} \frac{\Delta Q}{\Delta t} \quad [17.1b]$$

SI unit: coulomb/second (C/s), or the ampere (A)

When the current is steady, the average and instantaneous currents are the same. Note that one ampere of current is equivalent to one coulomb of charge passing through a given area A in a time interval of 1 s.

When charges flow through a surface as in Figure 17.1, they can be positive, negative, or both. **The direction of conventional current used in this book is the direction positive charges flow.** (This historical convention originated about 200 years ago, when the ideas of positive and negative charges were introduced.) In a common conductor such as copper, the current is due to the motion of negatively charged electrons, so the direction of the current is opposite the direction of motion of the electrons. On the other hand, for a beam of positively charged protons in an accelerator, the current is in the same direction as the motion of the protons. In some cases—gases and electrolytes, for example—the current is the result of the flows of both positive and negative charges. Moving charges, whether positive or negative, are referred to as *charge carriers*. In a metal, for example, the charge carriers are electrons.

In electrostatics, where charges are stationary, the electric potential is the same everywhere in a conductor. That is no longer true for conductors carrying current: as charges move along a wire, the electric potential is continually decreasing (except in the special case of superconductors). The decreasing electric potential means that the moving charges lose energy according to the relationship $\Delta U_{charges} = q\Delta V$, while

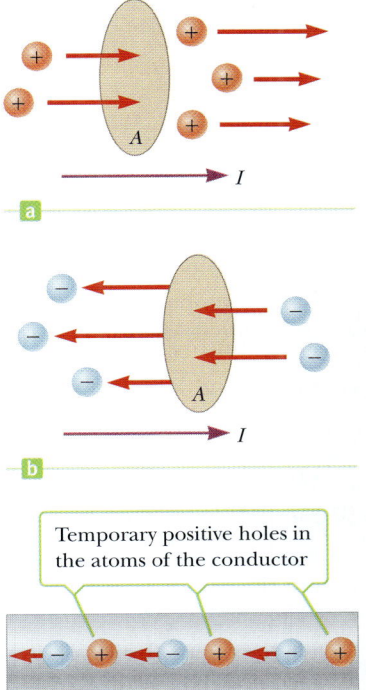

Figure 17.1 The time rate of flow of charge through area A is the current I. (a) The direction of current is the same as the flow of positive charge. (b) Negative charge flowing to the left is equivalent to an equal amount of positive charge flowing to the right. (c) In a conductor, positive holes open in the lattice of the conductor's atoms as electrons move in response to a potential. Negative electrons moving actively to the left are equivalent to positive holes migrating to the right.

an energy $\Delta U_{wire} = -q\Delta V$ is deposited in the current-carrying wire. (Those expressions derive from Equation 16.2.) If q is taken to be positive, corresponding to the convention of positive current, then $\Delta V = V_f - V_i$ is negative because, in a circuit, positive charges move from regions of high potential to regions of low potential. That in turn means $\Delta U_{charges} = q\Delta V$ is negative, as it should be, because the moving charges lose energy. Often only the magnitude is desired, however, in which case absolute values are substituted into q and ΔV. If the current is constant, then dividing the energy by the elapsed time yields the power delivered to the circuit element, such as a lightbulb filament.

EXAMPLE 17.1 Turn On the Light

GOAL Apply the concept of current.

PROBLEM The amount of charge that passes through the filament of a certain lightbulb in 2.00 s is 1.67 C. Find **(a)** the average current in the lightbulb and **(b)** the number of electrons that pass through the filament in 5.00 s. **(c)** If the current is supplied by a 12.0-V battery, what total energy is delivered to the lightbulb filament during 2.00 s? What is the average power?

STRATEGY Substitute into Equation 17.1a for part (a), then multiply the answer by the time given in part (b) to get the total charge that passes in that time. The total charge equals the number N of electrons going through the circuit times the charge per electron. To obtain the energy delivered to the filament, multiply the potential difference, ΔV, by the total charge. Dividing the energy by time yields the average power.

SOLUTION

(a) Compute the average current in the lightbulb.

Substitute the charge and time into Equation 17.1a:

$$I_{av} = \frac{\Delta Q}{\Delta t} = \frac{1.67 \text{ C}}{2.00 \text{ s}} = \boxed{0.835 \text{ A}}$$

(b) Find the number of electrons passing through the filament in 5.00 s.

The total number N of electrons times the charge per electron equals the total charge, $I_{av} \Delta t$:

(1) $\quad Nq = I_{av} \Delta t$

Substitute and solve for N:

$N(1.60 \times 10^{-19} \text{ C/electron}) = (0.835 \text{ A})(5.00 \text{ s})$

$N = \boxed{2.61 \times 10^{19} \text{ electrons}}$

(c) What total energy is delivered to the lightbulb filament? What is the average power?

Multiply the potential difference by the total charge to obtain the energy transferred to the filament:

(2) $\quad \Delta U = q\Delta V = (1.67 \text{ C})(12.0 \text{ V}) = \boxed{20.0 \text{ J}}$

Divide the energy by the elapsed time to calculate the average power:

$$P_{av} = \frac{\Delta U}{\Delta t} = \frac{20.0 \text{ J}}{2.00 \text{ s}} = \boxed{10.0 \text{ W}}$$

REMARKS It's important to use units to ensure the correctness of equations such as Equation (1). Notice the enormous number of electrons that pass through a given point in a typical circuit. Magnitudes were used in calculating the energies in Equation (2). Technically, the charge carriers are electrons with negative charge moving from a lower potential to a higher potential, so the change in their energy is $\Delta U_{charge} = q\Delta V = (-1.67 \text{ C})(+12.0 \text{ V}) = -20.0 \text{ J}$, a loss of energy that is delivered to the filament, $\Delta U_{fil} = -\Delta U_{charge} = +20.0 \text{ J}$. The energy and power, calculated here using the definitions of Chapter 16, will be further addressed in Section 17.6.

QUESTION 17.1 Is it possible to have an instantaneous current of $e/2$ per second? Explain. Can the average current take this value?

EXERCISE 17.1 A 9.00-V battery delivers a current of 1.34 A to the lightbulb filament of a pocket flashlight. (a) How much charge passes through the filament in 2.00 min? (b) How many electrons pass through the filament? Calculate (c) the energy delivered to the filament during that time and (d) the power delivered by the battery.

ANSWERS (a) 161 C (b) 1.01×10^{21} electrons (c) 1.45×10^3 J (d) 12.1 W

17.2 | A Microscopic View: Current and Drift Speed

Figure 17.2 (Quick Quiz 17.1)

■ Quick Quiz

17.1 Consider positive and negative charges all moving horizontally with the same speed through the four regions in Figure 17.2. Rank the magnitudes of the currents in these four regions from lowest to highest. (I_a is the current in Figure 17.2a, I_b the current in Figure 17.2b, etc.) (a) I_d, I_a, I_c, I_b (b) I_a, I_c, I_b, I_d (c) I_c, I_a, I_d, I_b (d) I_d, I_b, I_c, I_a (e) I_a, I_b, I_c, I_d (f) None of these

17.2 A Microscopic View: Current and Drift Speed

LEARNING OBJECTIVES

1. Relate electrical current to the drift speed of charge carriers.
2. Evaluate the drift speed in typical electrical conductors.

Macroscopic currents can be related to the motion of the microscopic charge carriers making up the current. It turns out that current depends on the average speed of the charge carriers in the direction of the current, the number of charge carriers per unit volume, and the of the charge carried by each charge carrier.

Consider identically charged particles moving in a conductor of cross-sectional area A (Fig. 17.3). The volume of an element of length Δx of the conductor is $A\Delta x$. If n represents the number of mobile charge carriers per unit volume, the number of carriers in the volume element is $nA\Delta x$. The mobile charge ΔQ in this element is therefore

$$\Delta Q = \text{number of carriers} \times \text{charge per carrier} = (nA\Delta x)q$$

where q is the charge on each carrier. If the carriers move with a constant average speed called the **drift speed** v_d, the distance they move in the time interval Δt is $\Delta x = v_d \Delta t$. We can therefore write

$$\Delta Q = (nAv_d \Delta t)q$$

If we divide both sides of this equation by Δt and take the limit as Δt goes to zero, we see that the current in the conductor is

$$I = \lim_{\Delta t \to 0} \frac{\Delta Q}{\Delta t} = nqv_d A \quad [17.2]$$

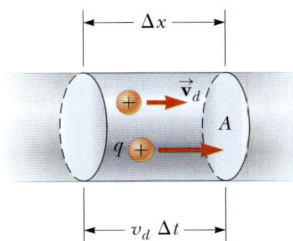

Figure 17.3 A section of a uniform conductor of cross-sectional area A. The charge carriers move with a speed v_d, and the distance they travel in time Δt is given by $\Delta x = v_d \Delta t$. The number of mobile charge carriers in the section of length Δx is given by $nAv_d \Delta t$, where n is the number of mobile carriers per unit volume.

To understand the meaning of drift speed, consider a conductor in which the charge carriers are free electrons. If the conductor is isolated, these electrons undergo random motion similar to the motion of the molecules in a gas. The drift speed is normally much smaller than the free electrons' average speed between collisions with the fixed atoms of the conductor. When a potential difference is applied between the ends of the conductor (say, with a battery), an electric field is set up in the conductor, creating an electric force on the electrons and hence a current. In reality, the electrons don't simply move in straight lines along the conductor. Instead, they undergo repeated collisions with the atoms of the metal, and the result is a complicated zigzag motion with only a small average drift speed along the wire (Fig. 17.4). The energy transferred from the electrons to the metal atoms during a collision increases the vibrational energy of the atoms and causes

Although electrons move with average velocity $\vec{v}_d$, collisions with atoms cause sharp, momentary changes of direction.

Figure 17.4 A schematic representation of the zigzag motion of a charge carrier in a conductor. Notice that the drift velocity $\vec{v}_d$ is opposite the direction of the electric field.

a corresponding increase in the temperature of the conductor. Despite the collisions, however, the electrons move slowly along the conductor in a direction opposite $\vec{\mathbf{E}}$ with the drift velocity $\vec{\mathbf{v}}_d$.

EXAMPLE 17.2 | Drift Speed of Electrons

GOAL Calculate a drift speed and compare it with the rms speed of an electron gas.

PROBLEM A copper wire of cross-sectional area 3.00×10^{-6} m² carries a current of 10.0 A. **(a)** Assuming each copper atom contributes one free electron to the metal, find the drift speed of the electrons in this wire. **(b)** Use the ideal gas model to compare the drift speed with the random rms speed an electron would have at 20.0°C. The density of copper is 8.92 g/cm³, and its atomic mass is 63.5 u.

STRATEGY All the variables in Equation 17.2 are known except for n, the number of free charge carriers per unit volume. We can find n by recalling that one mole of copper contains an Avogadro's number (6.02×10^{23}) of atoms and each atom contributes one charge carrier to the metal. The volume of one mole can be found from copper's known density and atomic mass. The atomic mass is the same, numerically, as the number of grams in a mole of the substance.

SOLUTION

(a) Find the drift speed of the electrons.

Calculate the volume of one mole of copper from its density and its atomic mass:

$$V = \frac{m}{\rho} = \frac{63.5 \text{ g/mol}}{8.92 \text{ g/cm}^3} = 7.12 \text{ cm}^3/\text{mol}$$

Convert the volume from cm³ to m³:

$$7.12 \text{ cm}^3/\text{mol} \left(\frac{1 \text{ m}}{10^2 \text{ cm}}\right)^3 = 7.12 \times 10^{-6} \text{ m}$$

Divide Avogadro's number (the number of electrons in one mole) by the volume per mole to obtain the number density:

$$n = \frac{6.02 \times 10^{23} \text{ electrons/mole}}{7.12 \times 10^{-6} \text{ m}^3/\text{mole}}$$
$$= 8.46 \times 10^{28} \text{ electrons/m}^3$$

Solve Equation 17.2 for the drift speed and substitute:

$$v_d = \frac{I}{nqA}$$
$$= \frac{10.0 \text{ C/s}}{(8.46 \times 10^{28} \text{ electrons/m}^3)(1.60 \times 10^{-19} \text{ C})(3.00 \times 10^{-6} \text{ m}^2)}$$
$$v_d = \boxed{2.46 \times 10^{-4} \text{ m/s}}$$

(b) Find the rms speed of a gas of electrons at 20.0°C.

Apply Equation 10.18:

$$v_{\text{rms}} = \sqrt{\frac{3k_B T}{m_e}}$$

Convert the temperature to the Kelvin scale and substitute values:

$$v_{\text{rms}} = \sqrt{\frac{3(1.38 \times 10^{-23} \text{ J/K})(293 \text{ K})}{9.11 \times 10^{-31} \text{ kg}}}$$
$$= \boxed{1.15 \times 10^5 \text{ m/s}}$$

REMARKS The drift speed of an electron in a wire is very small, only about one-billionth of its random thermal speed.

QUESTION 17.2 True or False: The drift velocity in a wire of a given composition is inversely proportional to the number density of charge carriers.

EXERCISE 17.2 What current in a copper wire with a cross-sectional area of 7.50×10^{-7} m² would result in a drift speed equal to 5.00×10^{-4} m/s?

ANSWER 5.08 A

Example 17.2 shows that drift speeds are typically very small. In fact, the drift speed is much smaller than the average speed between collisions. Electrons traveling at 2.46×10^{-4} m/s, as in the example, would take about 68 min to travel 1 m! In view of this low speed, why does a lightbulb turn on almost instantaneously when a switch is thrown? Think of the flow of water through a pipe. If a drop of water is forced into one end of a pipe that is already filled with water, a drop must be pushed out the other end of the pipe. Although it may take an individual drop a long time to make it through the pipe, a flow initiated at one end produces a similar flow at the other end very quickly. Another familiar analogy is the motion of a bicycle chain. When the sprocket moves one link, the other links all move more or less immediately, even though it takes a given link some time to make a complete rotation. In a conductor, the change in the electric field that drives the free electrons travels at a speed close to that of light, so when you flip a light switch, the message for the electrons to start moving through the wire (the electric field) reaches them at a speed on the order of 10^8 m/s!

> **Tip 17.2 Electrons Are Everywhere in the Circuit**
>
> Electrons don't have to travel from the light switch to the lightbulb for the lightbulb to operate. Electrons already in the filament of the lightbulb move in response to the electric field set up by the battery. Also, the battery does *not* provide electrons to the circuit; it provides *energy* to the existing electrons.

■ Quick Quiz

17.2 Suppose a current-carrying wire has a cross-sectional area that gradually becomes smaller along the wire so that the wire has the shape of a very long, truncated cone. How does the drift speed vary along the wire? (a) It slows down as the cross section becomes smaller. (b) It speeds up as the cross section becomes smaller. (c) It doesn't change. (d) More information is needed.

17.3 Current and Voltage Measurements In Circuits

LEARNING OBJECTIVES

1. Discuss the concept of an electrical circuit.
2. Discuss ammeters and voltmeters, instruments used to measure currents and potential differences in circuits.

To study electric current in circuits, we need to understand how to measure currents and voltages.

The circuit shown in Figure 17.5a is a drawing of the actual circuit necessary for measuring the current in Example 17.1. Figure 17.5b shows a stylized figure called a circuit diagram that represents the actual circuit of Figure 17.5a. This circuit consists of only a battery and a lightbulb. The word *circuit* means "a closed loop of some sort around which current circulates." The battery pumps charge through the bulb and around the loop. No charge would flow without a complete conducting path from the positive terminal of the battery into one side of the bulb, out the other side, and

Figure 17.5 (a) A sketch of an actual circuit used to measure the current in a flashlight bulb and the potential difference across it. (b) A schematic diagram of the circuit shown in (a). (c) A digital multimeter can be used to measure both current and potential difference.

through the copper conducting wires back to the negative terminal of the battery. The most important quantities that characterize how the bulb works in different situations are the current I in the bulb and the potential difference ΔV across the bulb. To measure the current in the bulb, we place an ammeter, the device for measuring current, in line with the bulb so there is no path for the current to bypass the meter; all the charge passing through the bulb must also pass through the ammeter. The voltmeter measures the potential difference, or voltage, between the two ends of the bulb's filament. If we use two meters simultaneously as in Figure 17.5a, we can remove the voltmeter and see if its presence affects the current reading. Figure 17.5c shows a digital multimeter, a convenient device, with a digital readout, that can be used to measure voltage, current, or resistance. An advantage of using a digital multimeter as a voltmeter is that it will usually not affect the current because a digital meter has enormous resistance to the flow of charge in the voltmeter mode.

At this point, you can measure the current as a function of voltage (an I–ΔV curve) of various devices in the lab. All you need is a variable voltage supply (an adjustable battery) capable of supplying potential differences from about -5 V to $+5$ V, a bulb, a resistor, some wires and alligator clips, and a couple of multimeters. Be sure to always start your measurements using the highest multimeter scales (say, 10 A and 1 000 V), and increase the sensitivity one scale at a time to obtain the highest accuracy without overloading the meters. (Increasing the sensitivity means lowering the maximum current or voltage that the scale reads.) Note that the meters must be connected with the proper polarity with respect to the voltage supply, as shown in Figure 17.5b. Finally, follow your instructor's directions carefully to avoid damaging the meters and incurring a soaring lab fee.

Quick Quiz

17.3 Look at the four "circuits" shown in Figure 17.6 and select those that will light the bulb.

Figure 17.6 (Quick Quiz 17.3)

17.4 Resistance, Resistivity, and Ohm's Law

LEARNING OBJECTIVES

1. Define electrical resistance and discuss its physical origins.
2. Relate resistance to resistivity.
3. Apply the concepts of resistivity and resistance to electrical systems.

Resistance and Ohm's Law

When a voltage (potential difference) ΔV is applied across the ends of a metallic conductor as in Figure 17.7, the current in the conductor is found to be proportional to the applied voltage; $I \propto \Delta V$. If the proportionality holds, we can write $\Delta V = IR$, where the proportionality constant R is called the *resistance* of the conductor. In fact, we define the **resistance** as the ratio of the voltage across the conductor to the current it carries:

$$R \equiv \frac{\Delta V}{I} \qquad [17.3]$$

◀ Resistance

Resistance has SI units of volts per ampere, called **ohms** (Ω). If a potential difference of 1 V across a conductor produces a current of 1 A, the resistance of the conductor is 1 Ω. For example, if an electrical appliance connected to a 120-V source carries a current of 6 A, its resistance is 20 Ω.

The concepts of electric current, voltage, and resistance can be compared to the flow of water in a river. As water flows downhill in a river of constant width and depth, the flow rate (water current) depends on the steepness of descent of the river and the effects of rocks, the riverbank, and other obstructions. The voltage difference is analogous to the steepness, and the resistance to the obstructions. Based on this analogy, it seems reasonable that increasing the voltage applied to a circuit should increase the current in the circuit, just as increasing the steepness of descent increases the water current. Also, increasing the obstructions in the river's path will reduce the water current, just as increasing the resistance in a circuit will lower the electric current. Resistance in a circuit arises due to collisions between the electrons carrying the current with fixed atoms inside the conductor. These collisions inhibit the movement of charges in much the same way as would a force of friction. For many materials, including most metals, experiments show that **the resistance remains constant over a wide range of applied voltages or currents**. This statement is known as **Ohm's law**, after Georg Simon Ohm (1789–1854), who was the first to conduct a systematic study of electrical resistance.

Ohm's law is given by

$$\Delta V = IR \qquad [17.4]$$

where R is understood to be independent of ΔV, the potential drop across the resistor, and I, the current in the resistor. We will continue to use this traditional form of Ohm's law when discussing electrical circuits. A **resistor** is a conductor that provides a specified resistance in an electric circuit. The symbol for a resistor in circuit diagrams is a zigzag line: —⋀⋀⋀—.

Ohm's law is an empirical relationship valid only for certain materials. Materials that obey Ohm's law, and hence have a constant resistance over a wide range of voltages, are said to be **ohmic**. Materials having resistance that changes with voltage or current are **nonohmic**. Ohmic materials have a linear current–voltage relationship over a large range of applied voltages (Fig. 17.8a page 608). Nonohmic materials have a nonlinear current–voltage relationship (Fig. 17.8b). One common semiconducting device that is nonohmic is the *diode*, a circuit element that acts like a one-way valve for current. Its resistance is small for currents in one direction (positive ΔV) and large for currents in the reverse direction (negative ΔV). Most modern electronic devices, such as transistors, have nonlinear current–voltage relationships; their operation depends on the particular ways in which they violate Ohm's law.

Figure 17.7 A uniform conductor of length ℓ and cross-sectional area A. The current I is proportional to the potential difference or, equivalently, to the electric field and length.

The potential difference $\Delta V = V_b - V_a$ creates the electric field $\vec{E}$ that produces the current I.

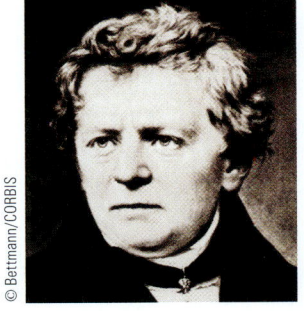

Georg Simon Ohm
(1787–1854)
A high school teacher in Cologne and later a professor at Munich, Ohm formulated the concept of resistance and discovered the proportionalities expressed in Equation 17.5.

■ Quick Quiz

17.4 In Figure 17.8b does the resistance of the diode (a) increase or (b) decrease as the positive voltage ΔV increases?

17.5 All electric devices are required to have identifying plates that specify their electrical characteristics. The plate on a certain steam iron states that the iron carries a current of 6.00 A when connected to a source of 1.20×10^2 V. What is the resistance of the steam iron? (a) 0.050 0 Ω (b) 20.0 Ω (c) 36.0 Ω

An assortment of resistors used for a variety of applications in electronic circuits.

Resistivity

Electrons don't move in straight-line paths through a conductor. Instead, they undergo repeated collisions with the metal atoms. Consider a conductor with a voltage applied across its ends. An electron gains speed as the electric force associated with the internal electric field accelerates it, giving it a velocity in the direction

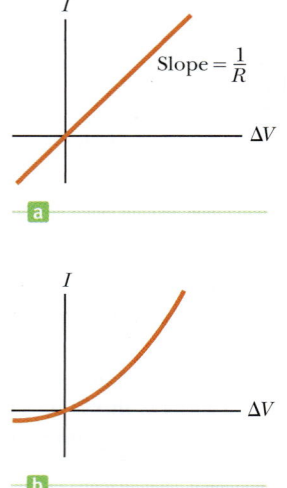

Figure 17.8 (a) The current–voltage curve for an ohmic material. The curve is linear, and the slope gives the resistance of the conductor. (b) A nonlinear current–voltage curve for a semiconducting diode. This device doesn't obey Ohm's law.

opposite that of the electric field. A collision with an atom randomizes the electron's velocity, reducing it in the direction opposite the field. The process then repeats itself. Together, these collisions affect the electron somewhat as a force of internal friction would. This step is the origin of a material's resistance.

The resistance of an ohmic conductor increases with length, which makes sense because the electrons going through it must undergo more collisions in a longer conductor. A smaller cross-sectional area also increases the resistance of a conductor, just as a smaller pipe slows the fluid moving through it. The resistance, then, is proportional to the conductor's length ℓ and inversely proportional to its cross-sectional area A,

$$R = \rho \frac{\ell}{A} \qquad [17.5]$$

where the constant of proportionality, ρ, is called the **resistivity** of the material. Every material has a characteristic resistivity that depends on its electronic structure and on temperature. Good electric conductors have very low resistivities, and good insulators have very high resistivities. Table 17.1 lists the resistivities of various materials at 20°C. Because resistance values are in ohms, resistivity values must be in ohm-meters ($\Omega \cdot m$).

Table 17.1 Resistivities and Temperature Coefficients of Resistivity for Various Materials (at 20°C)

Material	Resistivity ($\Omega \cdot m$)	Temperature Coefficient of Resistivity [(°C)$^{-1}$]
Silver	1.59×10^{-8}	3.8×10^{-3}
Copper	1.7×10^{-8}	3.9×10^{-3}
Gold	2.44×10^{-8}	3.4×10^{-3}
Aluminum	2.82×10^{-8}	3.9×10^{-3}
Tungsten	5.6×10^{-8}	4.5×10^{-3}
Iron	10.0×10^{-8}	5.0×10^{-3}
Platinum	11×10^{-8}	3.92×10^{-3}
Lead	22×10^{-8}	3.9×10^{-3}
Nichrome[a]	150×10^{-8}	0.4×10^{-3}
Carbon	3.5×10^{-5}	-0.5×10^{-3}
Germanium	0.46	-48×10^{-3}
Silicon	640	-75×10^{-3}
Glass	10^{10}–10^{14}	
Hard rubber	$\approx 10^{13}$	
Sulfur	10^{15}	
Quartz (fused)	75×10^{16}	

[a]A nickel-chromium alloy commonly used in heating elements.

APPLYING PHYSICS 17.1 | Dimming of Aging Lightbulbs

As a lightbulb ages, why does it gives off less light than when new?

EXPLANATION There are two reasons for the lightbulb's behavior, one electrical and one optical, but both are related to the same phenomenon occurring within the bulb. The filament of an old lightbulb is made of a tungsten wire that has been kept at a high temperature for many hours. High temperatures evaporate tungsten from the filament, decreasing its radius. From $R = \rho \ell / A$, we see that a decreased cross-sectional area leads to an increase in the resistance of the filament. This increasing resistance with age means that the filament will carry less current for the same applied voltage. With less current in the filament, there is less light output, and the filament glows more dimly.

At the high operating temperature of the filament, tungsten atoms leave its surface, much as water molecules evaporate from a puddle of water. The atoms are carried away by convection currents in the gas in the bulb and are deposited on the inner surface of the glass. In time, the glass becomes less transparent because of the tungsten coating, which decreases the amount of light that passes through the glass. ∎

EXAMPLE 17.3 The Resistance of Nichrome Wire

GOAL Combine the concept of resistivity with Ohm's law.

PROBLEM (a) Calculate the resistance per unit length of a 22-gauge Nichrome wire of radius 0.321 mm. (b) If a potential difference of 10.0 V is maintained across a 1.00-m length of the Nichrome wire, what is the current in the wire? (c) The wire is melted down and recast with twice its original length. Find the new resistance R_N as a multiple of the old resistance R_O.

STRATEGY Part (a) requires substitution into Equation 17.5, after calculating the cross-sectional area, whereas part (b) is a matter of substitution into Ohm's law. Part (c) requires some algebra. The idea is to take the expression for the new resistance and substitute expressions for ℓ_N and A_N, the new length and cross-sectional area, in terms of the old length and cross section. For the area substitution, remember that the volumes of the old and new wires are the same.

SOLUTION

(a) Calculate the resistance per unit length.

Find the cross-sectional area of the wire:

$$A = \pi r^2 = \pi (0.321 \times 10^{-3} \text{ m})^2 = 3.24 \times 10^{-7} \text{ m}^2$$

Obtain the resistivity of Nichrome from Table 17.1, solve Equation 17.5 for R/ℓ, and substitute:

$$\frac{R}{\ell} = \frac{\rho}{A} = \frac{1.5 \times 10^{-6} \, \Omega \cdot \text{m}}{3.24 \times 10^{-7} \text{ m}^2} = \boxed{4.6 \, \Omega/\text{m}}$$

(b) Find the current in a 1.00-m segment of the wire if the potential difference across it is 10.0 V.

Substitute given values into Ohm's law:

$$I = \frac{\Delta V}{R} = \frac{10.0 \text{ V}}{4.6 \, \Omega} = \boxed{2.2 \text{ A}}$$

(c) If the wire is melted down and recast with twice its original length, find the new resistance as a multiple of the old.

Find the new area A_N in terms of the old area A_O, using the fact the volume doesn't change and $\ell_N = 2\ell_O$:

$$V_N = V_O \;\rightarrow\; A_N \ell_N = A_O \ell_O \;\rightarrow\; A_N = A_O(\ell_O/\ell_N)$$
$$A_N = A_O(\ell_O/2\ell_O) = A_O/2$$

Substitute into Equation 17.5:

$$R_N = \frac{\rho \ell_N}{A_N} = \frac{\rho(2\ell_O)}{(A_O/2)} = 4\frac{\rho \ell_O}{A_O} = \boxed{4R_O}$$

REMARKS From Table 17.1, the resistivity of Nichrome is about 100 times that of copper, a typical good conductor. Therefore, a copper wire of the same radius would have a resistance per unit length of only 0.052 Ω/m, and a 1.00-m length of copper wire of the same radius would carry the same current (2.2 A) with an applied voltage of only 0.115 V.

Because of its resistance to oxidation, Nichrome is often used for heating elements in toasters, irons, and electric heaters.

QUESTION 17.3 Would replacing the Nichrome with copper result in a higher current or lower current?

EXERCISE 17.3 What is the resistance of a 6.0-m length of Nichrome wire that has a radius 0.321 mm? How much current does it carry when connected to a 120-V source?

ANSWERS 28 Ω; 4.3 A

■ Quick Quiz

17.6 Suppose an electrical wire is replaced with one having every linear dimension doubled (i.e., the length and radius have twice their original values). Does the wire now have (a) more resistance than before, (b) less resistance, or (c) the same resistance?

17.5 Temperature Variation of Resistance

LEARNING OBJECTIVES

1. Explain the physical reasons for the increase in resistivity and resistance with increasing temperature.
2. State the dependence of resistivity on temperature for limited temperature ranges, and extend it to the temperature dependence of resistance.
3. Apply the temperature dependence of resistance to electrical systems.

In an old-fashioned carbon filament incandescent lamp, the electrical resistance is typically 10 Ω, but changes with temperature.

The resistivity ρ, and hence the resistance, of a conductor depends on a number of factors. One of the most important is the temperature of the metal. For most metals, resistivity increases with increasing temperature. This correlation can be understood as follows: as the temperature of the material increases, its constituent atoms vibrate with greater amplitudes. As a result, the electrons find it more difficult to get by those atoms, just as it is more difficult to weave through a crowded room when the people are in motion than when they are standing still. The increased electron scattering with increasing temperature results in increased resistivity. Technically, thermal expansion also affects resistance; however, this is a very small effect.

Over a limited temperature range, the resistivity of most metals increases linearly with increasing temperature according to the expression

$$\rho = \rho_0[1 + \alpha(T - T_0)] \qquad [17.6]$$

where ρ is the resistivity at some temperature T (in Celsius degrees), ρ_0 is the resistivity at some reference temperature T_0 (usually taken to be 20°C), and α is a parameter called the **temperature coefficient of resistivity**. Temperature coefficients for various materials are provided in Table 17.1. The interesting negative values of α for semiconductors arise because these materials possess weakly bound charge carriers that become free to move and contribute to the current as the temperature rises.

Because the resistance of a conductor with a uniform cross section is proportional to the resistivity according to Equation 17.5 ($R = \rho l/A$), the temperature variation of resistance can be written

$$R = R_0[1 + \alpha(T - T_0)] \qquad [17.7]$$

Precise temperature measurements are often made using this property, as shown by the following example.

EXAMPLE 17.4 | A Platinum Resistance Thermometer

GOAL Apply the temperature dependence of resistance.

PROBLEM A resistance thermometer, which measures temperature by measuring the change in resistance of a conductor, is made of platinum and has a resistance of 50.0 Ω at 20.0°C. **(a)** When the device is immersed in a vessel containing indium at its melting point, its resistance increases to 76.8 Ω. From this information, find the melting point of indium. **(b)** The indium is heated further until it reaches a temperature of 235°C. Estimate the ratio of the new current in the platinum to the current I_{mp} at the melting point, assuming the coefficient of resistivity for platinum doesn't change significantly with temperature.

STRATEGY For part (a), solve Equation 17.7 for $T - T_0$ and get α for platinum from Table 17.1, substituting known quantities. For part (b), use Ohm's law in Equation 17.7.

SOLUTION

(a) Find the melting point of indium.

Solve Equation 17.7 for $T - T_0$:

$$T - T_0 = \frac{R - R_0}{\alpha R_0} = \frac{76.8 \, \Omega - 50.0 \, \Omega}{[3.92 \times 10^{-3} \, (°C)^{-1}][50.0 \, \Omega]}$$

$$= 137°C$$

Substitute $T_0 = 20.0°C$ and obtain the melting point of indium:

$$T = 157°C$$

(b) Estimate the ratio of the new current to the old when the temperature rises from 157°C to 235°C.

Write Equation 17.7, with R_0 and T_0 replaced by R_{mp} and T_{mp}, the resistance and temperature at the melting point.

$$R = R_{mp}[1 + \alpha(T - T_{mp})]$$

According to Ohm's law, $R = \Delta V/I$ and $R_{mp} = \Delta V/I_{mp}$. Substitute these expressions into Equation 17.7:

$$\frac{\Delta V}{I} = \frac{\Delta V}{I_{mp}}[1 + \alpha(T - T_{mp})]$$

Cancel the voltage differences, invert the two expressions, and then divide both sides by I_{mp}:

$$\frac{I}{I_{mp}} = \frac{1}{1 + \alpha(T - T_{mp})}$$

Substitute $T = 235°C$, $T_{mp} = 157°C$, and the value for α, obtaining the desired ratio:

$$\frac{I}{I_{mp}} = 0.766$$

REMARKS The answer to part (b) is only an estimate because the temperature coefficient is, in fact, temperature dependent. As the temperature rises, both the rms speed of the electrons in the metal and the resistance increase.

QUESTION 17.4 What happens to the drift speed of the electrons as the temperature rises? (a) It becomes larger. (b) It becomes smaller. (c) It remains unchanged.

EXERCISE 17.4 Suppose a wire made of an unknown alloy and having a temperature of 20.0°C carries a current of 0.450 A. At 52.0°C the current is 0.370 A for the same potential difference. Find the temperature coefficient of resistivity of the alloy.

ANSWER $6.76 \times 10^{-3} \, (°C)^{-1}$

17.6 Electrical Energy and Power

LEARNING OBJECTIVES

1. Derive an expression for the electrical power delivered to a resistor and discuss the transfer of energy in a simple circuit.
2. Apply the concept of electrical power to practical systems.

If a battery is used to establish an electric current in a conductor, chemical energy stored in the battery is continuously transformed into kinetic energy of the charge carriers. This kinetic energy is quickly lost as a result of collisions between the charge carriers and fixed atoms in the conductor, causing an increase in the temperature of the conductor. In this way the chemical energy stored in the battery is continuously transformed into thermal energy.

To understand the process of energy transfer in a simple circuit, consider a battery with terminals connected to a resistor (Fig. 17.9; remember that the positive terminal of the battery is always at the higher potential). Now imagine following a quantity of positive charge ΔQ around the circuit from point A, through the battery and resistor, and back to A. Point A is a reference point that

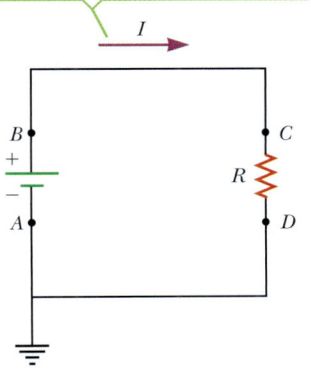

Figure 17.9 A circuit consisting of a battery and a resistance R. Point A is grounded.

is grounded (the ground symbol is ⏚), and its potential is taken to be zero. As the charge ΔQ moves from A to B through the battery, the electrical potential energy of the system increases by the amount $\Delta Q \Delta V$ and the chemical potential energy in the battery decreases by the same amount. (Recall from Chapter 16 that $\Delta PE = q \Delta V$.) As the charge moves from C to D through the resistor, however, it loses this electrical potential energy during collisions with atoms in the resistor. In the process the energy is transformed to internal energy corresponding to increased vibrational motion of those atoms. Because we can ignore the very small resistance of the interconnecting wires, no energy transformation occurs for paths BC and DA. When the charge returns to point A, the net result is that some of the chemical energy in the battery has been delivered to the resistor and has caused its temperature to rise.

The charge ΔQ loses energy $\Delta Q \Delta V$ as it passes through the resistor. If Δt is the time it takes the charge to pass through the resistor, the instantaneous rate at which it loses electric potential energy is

$$\lim_{\Delta t \to 0} \frac{\Delta Q}{\Delta t} \Delta V = I \Delta V$$

where I is the current in the resistor and ΔV is the potential difference across it. Of course, the charge regains this energy when it passes through the battery, at the expense of chemical energy in the battery. The rate at which the system loses potential energy as the charge passes through the resistor is equal to the rate at which the system gains internal energy in the resistor. Therefore, the power P, representing the rate at which energy is delivered to the resistor, is

Power ▶

$$P = I \Delta V \qquad [17.8]$$

Although this result was developed by considering a battery delivering energy to a resistor, Equation 17.8 can be used to determine the power transferred from a voltage source to *any* device carrying a current I and having a potential difference ΔV between its terminals.

Using Equation 17.8 and the fact that $\Delta V = IR$ for a resistor, we can express the power delivered to the resistor in the alternate forms

Power delivered to a resistor ▶

$$P = I^2 R = \frac{\Delta V^2}{R} \qquad [17.9]$$

Tip 17.3 Misconception About Current

Current is *not* "used up" in a resistor. Rather, some of the energy the charges have received from the voltage source is delivered to the resistor, making it hot and causing it to radiate. Also, the current doesn't slow down when going through the resistor: it's the same throughout the circuit.

When I is in amperes, ΔV in volts, and R in ohms, the SI unit of power is the watt (introduced in Chapter 5). The power delivered to a conductor of resistance R is often referred to as an I^2R *loss*. Note that Equation 17.9 applies only to resistors and not to nonohmic devices such as lightbulbs and diodes.

Regardless of the ways in which you use electrical energy in your home, you ultimately must pay for it or risk having your power turned off. The unit of energy used by electric companies to calculate consumption, the **kilowatt-hour**, is defined in terms of the unit of power and the amount of time it's supplied. One kilowatt-hour (kWh) is the energy converted or consumed in 1 h at the constant rate of 1 kW. It has the numerical value

$$1 \text{ kWh} = (10^3 \text{ W})(3\,600 \text{ s}) = 3.60 \times 10^6 \text{ J} \qquad [17.10]$$

On an electric bill, the amount of electricity used in a given period is usually stated in multiples of kilowatt-hours.

APPLYING PHYSICS 17.2 | Lightbulb Failures

Why do lightbulbs fail so often immediately after they're turned on?

EXPLANATION Once the switch is closed, the line voltage is applied across the bulb. As the voltage is applied across the cold filament when the bulb is first turned on, the resistance of the filament is low, the current is high, and a relatively large amount of power is delivered to the bulb. This current spike at the beginning of operation is the reason lightbulbs often fail immediately after they are turned on. As the filament warms, its resistance rises and the current decreases. As a result, the power delivered to the bulb decreases and the bulb is less likely to burn out. ∎

Quick Quiz

17.7 A voltage ΔV is applied across the ends of a Nichrome heater wire having a cross-sectional area A and length L. The same voltage is applied across the ends of a second Nichrome heater wire having a cross-sectional area A and length $2L$. Which wire gets hotter? (a) The shorter wire does. (b) The longer wire does. (c) More information is needed.

17.8 For the two resistors shown in Figure 17.10, rank the currents at points a through f from largest to smallest. (a) $I_a = I_b > I_e = I_f > I_c = I_d$ (b) $I_a = I_b > I_c = I_d > I_e = I_f$ (c) $I_e = I_f > I_c = I_d > I_a = I_b$

17.9 Two resistors, A and B, are connected in a series circuit with a battery. The resistance of A is twice that of B. Which resistor dissipates more power? (a) Resistor A does. (b) Resistor B does. (c) More information is needed.

17.10 The diameter of wire A is greater than the diameter of wire B, but their lengths and resistivities are identical. For a given voltage difference across the ends, what is the relationship between P_A and P_B, the dissipated power for wires A and B, respectively? (a) $P_A = P_B$ (b) $P_A < P_B$ (c) $P_A > P_B$

Figure 17.10 (Quick Quiz 17.8)

EXAMPLE 17.5 | The Cost of Lighting Up Your Life

GOAL Apply the electric power concept and calculate the cost of power usage using kilowatt-hours.

PROBLEM A circuit provides a maximum current of 20.0 A at an operating voltage of 1.20×10^2 V. **(a)** How many 75 W bulbs can operate with this voltage source? **(b)** At $0.120 per kilowatt-hour, how much does it cost to operate these bulbs for 8.00 h?

STRATEGY Find the necessary power with $P = I \Delta V$ then divide by 75.0 W per bulb to get the total number of bulbs. To find the cost, convert power to kilowatts and multiply by the number of hours, then multiply by the cost per-kilowatt-hour.

SOLUTION

(a) Find the number of bulbs that can be lighted.

Substitute into Equation 17.8 to get the total power:

$$P_{total} = I \Delta V = (20.0 \text{ A})(1.20 \times 10^2 \text{ V}) = 2.40 \times 10^3 \text{ W}$$

Divide the total power by the power per bulb to get the number of bulbs:

$$\text{Number of bulbs} = \frac{P_{total}}{P_{bulb}} = \frac{2.40 \times 10^3 \text{ W}}{75.0 \text{ W}} = \boxed{32.0}$$

(b) Calculate the cost of this electricity for an 8.00-h day.

Find the energy in kilowatt-hours:

$$\text{Energy} = Pt = (2.40 \times 10^3 \text{ W})\left(\frac{1.00 \text{ kW}}{1.00 \times 10^3 \text{ W}}\right)(8.00 \text{ h})$$
$$= 19.2 \text{ kWh}$$

Multiply the energy by the cost per kilowatt-hour:

$$\text{Cost} = (19.2 \text{ kWh})(\$0.12/\text{kWh}) = \boxed{\$2.30}$$

(Continued)

REMARKS This amount of energy might correspond to what a small office uses in a working day, taking into account all power requirements (not just lighting). In general, resistive devices can have variable power output, depending on how the circuit is wired. Here, power outputs were specified, so such considerations were unnecessary.

QUESTION 17.5 Considering how hot the parts of an incandescent light bulb get during operation, guess what fraction of the energy emitted by an incandescent lightbulb is in the form of visible light. (a) 10% (b) 50% (c) 80%

EXERCISE 17.5 (a) How many Christmas tree lights drawing 5.00 W of power each could be run on a circuit operating at 1.20×10^2 V and providing 15.0 A of current? (b) Find the cost to operate one such string 24.0 h per day for the Christmas season (two weeks), using the rate \$0.12/kWh.

ANSWERS (a) 3.60×10^2 bulbs (b) \$72.60

■ EXAMPLE 17.6 | The Power Converted by an Electric Heater

GOAL Calculate an electrical power output and link to its effect on the environment through the first law of thermodynamics.

PROBLEM An electric heater is operated by applying a potential difference of 50.0 V to a Nichrome wire of total resistance 8.00 Ω. **(a)** Find the current carried by the wire and the power rating of the heater. **(b)** Using this heater, how long would it take to heat 2.50×10^3 moles of diatomic gas (e.g., a mixture of oxygen and nitrogen, or air) from a chilly 10.0°C to 25.0°C? Take the molar specific heat at constant volume of air to be $\frac{5}{2}R$. **(c)** How many kilowatt-hours of electricity are used during the time calculated in part (b) and at what cost, at \$0.12 per kilowatt-hour?

STRATEGY For part (a), find the current with Ohm's law and substitute into the expression for power. Part (b) is an isovolumetric process, so the thermal energy provided by the heater all goes into the change in internal energy, ΔU. Calculate this quantity using the first law of thermodynamics and divide by the power to get the time. Finding the number of kilowatt-hours used requires a simple unit conversion technique. Multiplying by the cost per kilowatt-hour yields the total cost of operating the heater for the given time.

SOLUTION

(a) Compute the current and power output.

Apply Ohm's law to get the current:

$$I = \frac{\Delta V}{R} = \frac{50.0 \text{ V}}{8.00 \text{ Ω}} = \boxed{6.25 \text{ A}}$$

Substitute into Equation 17.9 to find the power:

$$P = I^2 R = (6.25 \text{ A})^2 (8.00 \text{ Ω}) = \boxed{313 \text{ W}}$$

(b) How long does it take to heat the gas?

Calculate the thermal energy transfer from the first law. Note that $W = 0$ because the volume doesn't change.

$$Q = \Delta U = n C_v \Delta T$$
$$= (2.50 \times 10^3 \text{ mol})(\tfrac{5}{2} \cdot 8.31 \text{ J/mol} \cdot \text{K})(298 \text{ K} - 283 \text{ K})$$
$$= 7.79 \times 10^5 \text{ J}$$

Divide the thermal energy by the power to get the time:

$$t = \frac{Q}{P} = \frac{7.79 \times 10^5 \text{ J}}{313 \text{ W}} = \boxed{2.49 \times 10^3 \text{ s}}$$

(c) Calculate the kilowatt-hours of electricity used and the cost.

Convert the energy to kilowatt-hours, noting that $1 \text{ J} = 1 \text{ W} \cdot \text{s}$:

$$U = (7.79 \times 10^5 \text{ W} \cdot \text{s})\left(\frac{1.00 \text{ kW}}{1.00 \times 10^3 \text{ W}}\right)\left(\frac{1.00 \text{ h}}{3.60 \times 10^3 \text{ s}}\right) = \boxed{0.216 \text{ kWh}}$$

Multiply by \$0.12/kWh to obtain the total cost of operation:

$$\text{Cost} = (0.216 \text{ kWh})(\$0.12/\text{kWh}) = \boxed{\$0.026}$$

REMARKS The number of moles of gas given here is approximately what would be found in a bedroom. Warming the air with this space heater requires only about 40 minutes. The calculation, however, doesn't take into account conduction losses. Recall that a 20-cm-thick concrete wall, as calculated in Chapter 11, permitted the loss of more than 2 megajoules an hour by conduction!

QUESTION 17.6 If the heater wire is replaced by a wire with lower resistance, is the time required to heat the gas (a) unchanged, (b) increased, or (c) decreased?

EXERCISE 17.6 A hot-water heater is rated at 4.50×10^3 W and operates at 2.40×10^2 V. (a) Find the resistance in the heating element and the current. (b) How long does it take to heat 125 L of water from 20.0°C to 50.0°C, neglecting conduction and other losses? (c) How much does it cost at \$0.12/kWh?

ANSWERS (a) 12.8 Ω, 18.8 A (b) 3.49×10^3 s (c) \$0.52

17.7 Superconductors

LEARNING OBJECTIVE
1. Discuss superconductivity and describe important applications.

There is a class of metals and compounds with resistances that fall virtually to zero below a certain temperature T_c called the *critical temperature*. These materials are known as **superconductors**. The resistance vs. temperature graph for a superconductor follows that of a normal metal at temperatures above T_c (Fig. 17.11). When the temperature is at or below T_c, however, the resistance suddenly drops to zero. This phenomenon was discovered in 1911 by Dutch physicist H. Kamerlingh Onnes as he and a graduate student worked with mercury, which is a superconductor below 4.1 K. Recent measurements have shown that the resistivities of superconductors below T_c are less than 4×10^{-25} Ω·m, around 10^{17} times smaller than the resistivity of copper and in practice considered to be zero.

Today, thousands of superconductors are known, including such common metals as aluminum, tin, lead, zinc, and indium. Table 17.2 lists the critical temperatures of several superconductors. The value of T_c is sensitive to chemical composition, pressure, and crystalline structure. Interestingly, copper, silver, and gold, which are excellent conductors, don't exhibit superconductivity.

A truly remarkable feature of superconductors is that once a current is set up in them, it persists *without any applied voltage* (because $R = 0$). In fact, steady currents in superconducting loops have been observed to persist for years with no apparent decay!

An important development in physics that created much excitement in the scientific community was the discovery of high-temperature copper-oxide-based superconductors. The excitement began with a 1986 publication by J. Georg Bednorz and K. Alex Müller, scientists at the IBM Zurich Research Laboratory in Switzerland, in which they reported evidence for superconductivity at a temperature near 30 K in an oxide of barium, lanthanum, and copper. Bednorz and Müller were awarded the Nobel Prize in Physics in 1987 for their important discovery. The discovery was remarkable because the critical temperature was significantly higher than that of any previously known superconductor. Shortly thereafter a new family of compounds was investigated, and research activity in the field of superconductivity proceeded vigorously. In early 1987 groups at the University of Alabama at Huntsville and the University of Houston announced the discovery of superconductivity at about 92 K in an oxide of yttrium, barium, and copper ($YBa_2Cu_3O_7$). Late in 1987, teams of scientists from Japan and the United States reported superconductivity at 105 K in an oxide of bismuth, strontium, calcium, and copper. More recently, scientists have reported superconductivity at temperatures as high as 150 K in an oxide containing mercury. The search for novel superconducting materials continues, with the hope of someday obtaining a room-temperature superconducting material. This research is

Figure 17.11 Resistance versus temperature for a sample of mercury (Hg). The graph follows that of a normal metal above the critical temperature T_c. The resistance drops to zero at the critical temperature, which is 4.2 K for mercury, and remains at zero for lower temperatures.

Table 17.2 Critical Temperatures for Various Superconductors

Material	T_c(K)
Zn	0.88
Al	1.19
Sn	3.72
Hg	4.15
Pb	7.18
Nb	9.46
Nb_3Sn	18.05
Nb_3Ge	23.2
$YBa_2Cu_3O_7$	92
Bi–Sr–Ca–Cu–O	105
Tl–Ba–Ca–Cu–O	125
$HgBa_2Ca_2Cu_3O_8$	134

important both for scientific reasons and for practical applications. A superconducting ceramic levitates a permanent magnet in Figure 17.12.

An important and useful application is the construction of superconducting magnets in which the magnetic field intensities are about ten times greater than those of the best normal electromagnets. Such magnets are being considered as a means of storing energy. The idea of using superconducting power lines to transmit power efficiently is also receiving serious consideration. Modern superconducting electronic devices consisting of two thin-film superconductors separated by a thin insulator have been constructed. Among these devices are magnetometers (magnetic-field measuring devices) and various microwave devices.

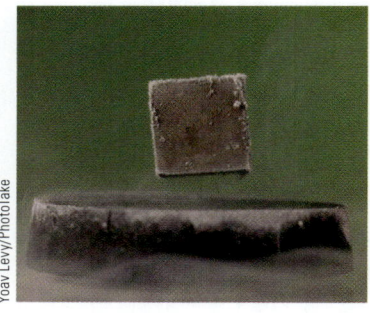

Figure 17.12 A small permanent magnet floats freely above a nitrogen-cooled ceramic superconductor. The superconductor has zero resistance and expels any magnetic field from its interior by creating a mirror image of the magnetic poles of the permanent magnet. This "Meissner effect" results in magnetic levitation of the permanent magnet.

17.8 Electrical Activity in the Heart BIO

LEARNING OBJECTIVES

1. Discuss the critical role of electrical activity in the human body.
2. Discuss medical instruments used to monitor and regulate the body's electrical activity.

Electrocardiograms

APPLICATION
Electrocardiograms

Every action involving the body's muscles is initiated by electrical activity. The voltages produced by muscular action in the heart are particularly important to physicians. Voltage pulses cause the heart to beat, and the waves of electrical excitation that sweep across the heart associated with the heartbeat are conducted through the body via the body fluids. These voltage pulses are large enough to be detected by suitable monitoring equipment attached to the skin. A sensitive voltmeter making good electrical contact with the skin by means of contacts attached with conducting paste can be used to measure heart pulses, which are typically of the order of 1 mV at the surface of the body. The voltage pulses can be recorded on an instrument called an **electrocardiograph**, and the pattern recorded by this instrument is called an **electrocardiogram** (EKG). To understand the information contained in an EKG pattern, it is necessary first to describe the underlying principles concerning electrical activity in the heart.

The right atrium of the heart contains a specialized set of muscle fibers called the SA (sinoatrial) node that initiates the heartbeat (Fig. 17.13). Electric impulses that originate in these fibers gradually spread from cell to cell throughout the right and left atrial muscles, causing them to contract. The pulse that passes through the muscle cells is often called a *depolarization wave* because of its effect on individual cells. If an individual muscle cell were examined in its resting state, a double-layer electric charge distribution would be found on its surface, as shown in Figure 17.14a. The impulse generated by the SA node momentarily and locally allows positive charge on the outside of the

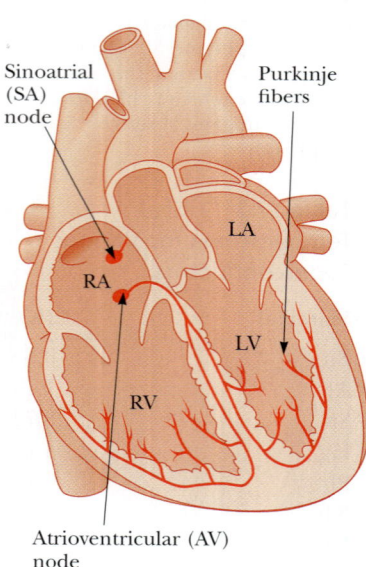

Figure 17.13 The electrical conduction system of the human heart. (RA: right atrium; LA: left atrium; RV: right ventricle; LV: left ventricle.)

Figure 17.14 (a) Charge distribution of a muscle cell in the atrium before a depolarization wave has passed through the cell. (b) Charge distribution as the wave passes.

cell to flow in and neutralize the negative charge on the inside layer. This effect changes the cell's charge distribution to that shown in Figure 17.14b. Once the depolarization wave has passed through an individual heart muscle cell, the cell recovers the resting-state charge distribution (positive out, negative in) shown in Figure 17.14a in about 250 ms. When the impulse reaches the atrioventricular (AV) node (Fig. 17.13), the muscles of the atria begin to relax, and the pulse is directed to the ventricular muscles by the AV node. The muscles of the ventricles contract as the depolarization wave spreads through the ventricles along a group of fibers called the *Purkinje fibers*. The ventricles then relax after the pulse has passed through. At this point, the SA node is triggered again and the cycle is repeated.

A sketch of the electrical activity registered on an EKG for one beat of a normal heart is shown in Figure 17.15. The pulse indicated by *P* occurs just before the atria begin to contract. The *QRS* pulse occurs in the ventricles just before they contract, and the *T* pulse occurs when the cells in the ventricles begin to recover. EKGs for an abnormal heart are shown in Figure 17.16. The *QRS* portion of the pattern shown in Figure 17.16a is wider than normal, indicating that the patient may have an enlarged heart. (Why?) Figure 17.16b indicates that there is no constant relationship between the *P* pulse and the *QRS* pulse. This suggests a blockage in the electrical conduction path between the SA and AV nodes that results in the atria and ventricles beating independently and inefficient heart pumping. Finally, Figure 17.16c shows a situation in which there is no *P* pulse and an irregular spacing between the *QRS* pulses. This is symptomatic of irregular atrial contraction, which is called *fibrillation*. In this condition, the atrial and ventricular contractions are irregular.

As noted previously, the sinoatrial node directs the heart to beat at the appropriate rate, usually about 72 beats per minute. Disease or the aging process, however, can damage the heart and slow its beating, and a medical assist may be necessary in the form of a *cardiac pacemaker* attached to the heart. This matchbox-sized electrical device implanted under the skin has a lead that is connected to the wall of the right ventricle. Pulses from this lead stimulate the heart to maintain its proper rhythm. In general, a pacemaker is designed to produce pulses at a rate of about 60 per minute, slightly slower than the normal number of beats per minute, but sufficient to maintain life. The circuitry

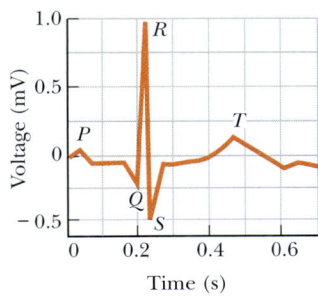

Figure 17.15 An EKG response for a normal heart.

BIO APPLICATION
Cardiac Pacemakers

Figure 17.16 Abnormal EKGs.

Figure 17.17 (a) A dual-chamber ICD with leads in the heart. One lead monitors and stimulates the right atrium, and the other monitors and stimulates the right ventricle. (b) Medtronic Dual Chamber ICD.

consists of a capacitor charging up to a certain voltage from a lithium battery and then discharging. The design of the circuit is such that if the heart is beating normally, the capacitor is not allowed to charge completely and send pulses to the heart.

An Emergency Room in Your Chest

BIO APPLICATION
Implanted Cardioverter Defibrillators

In June 2001 an operation on then–Vice President Dick Cheney focused attention on the progress in treating heart problems with tiny implanted electrical devices. Aptly called "an emergency room in your chest" by Cheney's attending physician, devices called *I*mplanted *C*ardioverter *D*efibrillators (**ICDs**) can monitor, record, and logically process heart signals and then supply different corrective signals to hearts beating too slowly, too rapidly, or irregularly. ICDs can even monitor and send signals to the atria and ventricles independently! Figure 17.17a shows a sketch of an ICD with conducting leads that are implanted in the heart. Figure 17.17b shows an actual titanium-encapsulated dual-chamber ICD.

The latest ICDs are sophisticated devices capable of a number of functions:

1. monitoring both atrial and ventricular chambers to differentiate between atrial and potentially fatal ventricular arrhythmias, which require prompt regulation
2. storing about a half hour of heart signals that can easily be read out by a physician
3. being easily reprogrammed with an external magnetic wand
4. performing complicated signal analysis and comparison
5. supplying 0.25- to 10-V repetitive pacing signals to speed up or slow down a malfunctioning heart, or a high-voltage pulse of about 800 V to halt the potentially fatal condition of ventricular fibrillation, in which the heart quivers rapidly rather than beats (people who have experienced such a high-voltage jolt say that it feels like a kick or a bomb going off in the chest)
6. automatically adjusting the number of pacing pulses per minute to match the patient's activity

ICDs are powered by lithium batteries and have implanted lifetimes of 4 to 6 years. Some basic properties of these adjustable ICDs are given in Table 17.3. In the table *tachycardia* means "rapid heartbeat" and *bradycardia* means "slow heartbeat." A key factor in developing tiny electrical implants that serve as defibrillators is the development of capacitors with relatively large capacitance (125 μF) and small physical size.

Table 17.3 Properties of Implanted Cardioverter Defibrillators

Physical Specifications	
Mass (g)	85
Size (cm)	7.3 × 6.2 × 1.3 (about five stacked silver dollars)
Antitachycardia Pacing	ICD delivers a burst of critically timed low-energy pulses
Number of bursts	1–15
Burst cycle length (ms)	200–552
Number of pulses per burst	2–20
Pulse amplitude (V)	7.5 or 10
Pulse width (ms)	1.0 or 1.9
High-Voltage Defibrillation	
Pulse energy (J)	37 stored/33 delivered
Pulse amplitude (V)	801
Bradycardia Pacing	A dual-chamber ICD can steadily deliver repetitive pulses to both the atrium and the ventricle
Base frequency (beats/minute)	40–100
Pulse amplitude (V)	0.25–7.5
Pulse width (ms)	0.05, 0.1–1.5, 1.9

Note: For more information, go to **www.photonicd.com/specs.html**.

SUMMARY

17.1 Electric Current

The **average electric current** I in a conductor is defined as

$$I_{av} \equiv \frac{\Delta Q}{\Delta t} \quad [17.1a]$$

Current is the time rate of flow of charge through a surface.

where ΔQ is the charge that passes through a cross section of the conductor in time Δt. The SI unit of current is the **ampere** (A); 1 A = 1 C/s. By convention, the direction of current is the direction of flow of positive charge.

The **instantaneous current** I is the limit of the average current as the time interval goes to zero:

$$I = \lim_{\Delta t \to 0} I_{av} = \lim_{\Delta t \to 0} \frac{\Delta Q}{\Delta t} \quad [17.1b]$$

17.2 A Microscopic View: Current and Drift Speed

The current in a conductor is related to the motion of the charge carriers by

$$I = nqv_d A \quad [17.2]$$

where n is the number of mobile charge carriers per unit volume, q is the charge on each carrier, v_d is the drift speed of the charges, and A is the cross-sectional area of the conductor.

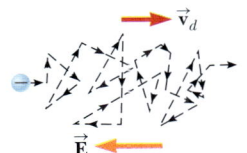

The current I in a conductor is related to the number density n of charge carriers, the charge q per carrier, the drift speed $\vec{v}_d$, and the cross-sectional area of the conductor, A.

17.4 Resistance, Resistivity, and Ohm's Law

The **resistance** R of a conductor is defined as the ratio of the potential difference across the conductor to the current in it:

$$R \equiv \frac{\Delta V}{I} \quad [17.3]$$

The potential difference ΔV is proportional to the current, I.

The SI units of resistance are volts per ampere, or **ohms** (Ω); 1 Ω = 1 V/A.

Ohm's law describes many conductors for which the applied voltage is directly proportional to the current it causes. The proportionality constant is the resistance:

$$\Delta V = IR \quad [17.4]$$

If a conductor has length ℓ and cross-sectional area A, its **resistance** is

$$R = \rho \frac{\ell}{A} \quad [17.5]$$

where ρ is an intrinsic property of the conductor called the **electrical resistivity**. The SI unit of resistivity is the **ohm-meter** ($\Omega \cdot$ m).

17.5 Temperature Variation of Resistance

Over a limited temperature range, the resistivity of a conductor varies with temperature according to the expression

$$\rho = \rho_0[1 + \alpha(T - T_0)] \quad [17.6]$$

where α is the **temperature coefficient of resistivity** and ρ_0 is the resistivity at some reference temperature T_0 (usually taken to be 20°C).

The resistance of a conductor varies with temperature according to the expression

$$R = R_0[1 + \alpha(T - T_0)] \quad [17.7]$$

17.6 Electrical Energy and Power

If a potential difference ΔV is maintained across an electrical device, the **power**, or rate at which energy is supplied to the device, is

$$P = I\Delta V \quad [17.8]$$

Because the potential difference across a resistor is $\Delta V = IR$, the **power delivered to a resistor** can be expressed as

$$P = I^2R = \frac{\Delta V^2}{R} \quad [17.9]$$

A **kilowatt-hour** is the amount of energy converted or consumed in one hour by a device supplied with power at the rate of 1 kW. It is equivalent to

$$1 \text{ kWh} = 3.60 \times 10^6 \text{ J} \quad [17.10]$$

WARM-UP EXERCISES

The warm-up exercises in this chapter may be assigned online in Enhanced WebAssign.

1. **Physics Review** A 5.00-kg box slides across a rough horizontal floor, initially at 2.50 m/s. If friction brings the box to rest after 1.50 s, determine the magnitude of the average rate in watts at which friction dissipates the block's mechanical energy. (See Sections 5.5 and 5.6.)

2. **Physics Review** Suppose an isolated conductor is given −5.00 nC of electrical charge. Determine the number of excess electrons on the conductor. (See Section 15.1.)

3. **Physics Review** A proton, initially at a location where the electric potential is 2.00 V, moves past a point where the potential is −4.00 V. (a) Calculate the change in the proton's speed at that point if it is initially at rest. (The mass of a proton is 1.673×10^{-27} kg.) (See Section 16.1.) (b) Repeat the calculation if the proton is initially traveling at 1.20×10^4 m/s.

4. A wire carries a current of 1.60 A. How many electrons per second pass a given point in the wire? (See Section 17.1.)

5. Copper contains approximately 8.46×10^{28} electrons/m^3 available to carry current. Suppose a copper wire has a diameter of 2.05 mm. (a) Calculate the cross-sectional area of the wire. (b) If the wire carries a current of 10.0 A, determine the drift speed of electrons through the wire. (See Section 17.2.)

6. A simple circuit consists of a light bulb connected to the terminals of a battery. A voltmeter shows a potential difference of 12.0 V across the light bulb, whereas an ammeter registers a current of 0.500 A in the circuit. Determine the resistance of the bulb. (See Section 17.4.)

7. Determine the length of copper wire having a resistance of 2.00 Ω at 20°C if the wire's cross-sectional area is 3.31×10^{-6} m^2. (See Section 17.4.)

8. Suppose a spool of copper wire is measured to have a resistance of 0.750 Ω at room temperature (21.0°C). Find its resistance on a cold winter day when the temperature is −25.0°C. (See Section 17.5.)

9. A simple circuit is constructed so that a 12.0-V source supplies current to a 1.00×10^3 Ω resistor. (a) Find the current in the resistor. (b) Calculate the power delivered to the resistor. (c) Determine the energy delivered to the resistor in 1.00 h. (See Section 17.6.)

10. A typical hair dryer consumes 1.20×10^3 W of electrical power. If the hair dryer runs for 10.0 minutes, determine (a) the amount of energy consumed, (b) the amount of energy consumed in kWh, (c) the cost of this energy, assuming the electrical power company charges $0.120 per kWh. (See Section 17.6.)

CONCEPTUAL QUESTIONS

The conceptual questions in this chapter may be assigned online in Enhanced WebAssign.

1. We have seen that an electric field must exist inside a conductor that carries a current. How is that possible in view of the fact that in electrostatics we concluded that the electric field must be zero inside a conductor?

2. Use the atomic theory of matter to explain why the resistance of a material should increase as its temperature increases.

3. If charges flow very slowly through a metal, why does it not require several hours for a light to come on when you throw a switch?

4. In an analogy between traffic flow and electrical current, (a) what would correspond to the charge Q? (b) What would correspond to the current I?

5. When the voltage across a certain conductor is doubled, the current is observed to triple. What can you conclude about the conductor?

6. Two lightbulbs are each connected to a voltage of 120 V. One has a power of 25 W, the other 100 W. (a) Which lightbulb has the higher resistance? (b) Which lightbulb carries more current?

7. Newspaper articles often have statements such as "10 000 volts of electricity surged *through* the victim's body." What is wrong with this statement?

8. There is an old admonition given to experimenters to "keep one hand in the pocket" when working around high voltages. Why is this warning a good idea?

9. What could happen to the drift velocity of the electrons in a wire and to the current in the wire if the electrons could move through it freely without resistance?

10. Some homes have light dimmers that are operated by rotating a knob. What is being changed in the electric circuit when the knob is rotated?

11. When is more power delivered to a lightbulb, immediately after it is turned on and the glow of the filament is increasing or after it has been on for a few seconds and the glow is steady?

PROBLEMS

WebAssign The problems in this chapter may be assigned online in Enhanced WebAssign.

1. denotes straightforward problem; 2. denotes intermediate problem;
3. denotes challenging problem
[1.] denotes full solution available in *Student Solutions Manual/Study Guide*
1. denotes problems most often assigned in Enhanced WebAssign

BIO denotes biomedical problems
GP denotes guided problems
M denotes Master It tutorial available in Enhanced WebAssign
Q|C denotes asking for quantitative and conceptual reasoning
S denotes symbolic reasoning problem
W denotes Watch It video solution available in Enhanced WebAssign

17.1 Electric Current
17.2 A Microscopic View: Current and Drift Speed

1. **W** If a current of 80.0 mA exists in a metal wire, (a) how many electrons flow past a given cross section of the wire in 10.0 min? (b) In what direction do the electrons travel with respect to the current?

2. **Q|C** A copper wire has a circular cross section with a radius of 1.25 mm. (a) If the wire carries a current of 3.70 A, find the drift speed of the electrons in the wire. (See Example 17.2 for relevant data on copper.) (b) All other things being equal, what happens to the drift speed in wires made of metal having a larger number of conduction electrons per atom than copper? Explain.

3. In the Bohr model of the hydrogen atom, an electron in the lowest energy state moves at a speed equal to 2.19×10^6 m/s in a circular path having a radius of 5.29×10^{-11} m. What is the effective current associated with this orbiting electron?

4. The current supplied by a battery in a portable device is typically 0.15 A. Find the number of electrons passing through the device in one hour.

5. A certain laboratory experiment requires an aluminum wire of length of 32.0 m and a resistance of 2.50 Ω at 20.0°C. What diameter wire must be used?

6. If 3.25×10^{-3} kg of gold is deposited on the negative electrode of an electrolytic cell in a period of 2.78 h, what is the current in the cell during that period? Assume the gold ions carry one elementary unit of positive charge.

7. **M** A 200-km-long high-voltage transmission line 2.0 cm in diameter carries a steady current of 1 000 A. If the conductor is copper with a free charge density of 8.5×10^{28} electrons per cubic meter, how many years does it take one electron to travel the full length of the cable?

8. An aluminum wire having a cross-sectional area of 4.00×10^{-6} m^2 carries a current of 5.00 A. The density of aluminum is 2.70 g/cm^3. Assume each aluminum atom supplies one conduction electron per atom. Find the drift speed of the electrons in the wire.

9. **GP** An iron wire has a cross-sectional area of 5.00×10^{-6} m^2. Carry out steps (a) through (e) to compute the drift speed of the conduction electrons in the wire. (a) How many kilograms are there in 1 mole of iron? (b) Starting with the density of iron and the result of part (a), compute the molar density of iron (the number of moles of iron per cubic meter). (c) Calculate the number density of iron atoms using Avogadro's number. (d) Obtain the number density of conduction electrons given that there are two conduction electrons per

iron atom. (e) If the wire carries a current of 30.0 A, calculate the drift speed of conduction electrons.

17.4 Resistance, Resistivity, and Ohm's Law

10. An electric heater carries a current of 13.5 A when operating at a voltage of 1.20×10^2 V. What is the resistance of the heater?

11. **BIO** A person notices a mild shock if the current along a path through the thumb and index finger exceeds 80 μA. Compare the maximum possible voltage without shock across the thumb and index finger with a dry-skin resistance of 4.0×10^5 Ω and a wet-skin resistance of 2 000 Ω.

12. **W** Suppose you wish to fabricate a uniform wire out of 1.00 g of copper. If the wire is to have a resistance $R = 0.500$ Ω, and if all the copper is to be used, what will be (a) the length and (b) the diameter of the wire?

13. Nichrome wire of cross-sectional radius 0.791 mm is to be used in winding a heating coil. If the coil must carry a current of 9.25 A when a voltage of 1.20×10^2 V is applied across its ends, find (a) the required resistance of the coil and (b) the length of wire you must use to wind the coil.

14. A wire of diameter 0.800 mm and length 25.0 m has a measured resistance of 1.60 Ω. What is the resistivity of the wire?

15. A potential difference of 12 V is found to produce a current of 0.40 A in a 3.2-m length of wire with a uniform radius of 0.40 cm. What is (a) the resistance of the wire? (b) The resistivity of the wire?

16. **S** Aluminum and copper wires of equal length are found to have the same resistance. What is the ratio of their radii?

17. **M** A wire 50.0 m long and 2.00 mm in diameter is connected to a source with a potential difference of 9.11 V, and the current is found to be 36.0 A. Assume a temperature of 20°C and, using Table 17.1, identify the metal out of which the wire is made.

18. A rectangular block of copper has sides of length 10 cm, 20 cm, and 40 cm. If the block is connected to a 6.0-V source across two of its opposite faces, what are (a) the maximum current and (b) the minimum current the block can carry?

19. A wire of initial length L_0 and radius r_0 has a measured resistance of 1.0 Ω. The wire is drawn under tensile stress to a new uniform radius of $r = 0.25 r_0$. What is the new resistance of the wire?

20. **BIO** **QC** The human body can exhibit a wide range of resistances to current depending on the path of the current, contact area, and sweatiness of the skin. Suppose the resistance across the chest from the left hand to the right hand is 1.0×10^6 Ω. (a) How much voltage is required to cause possible heart fibrillation in a man, which corresponds to 500 mA of direct current? (b) Why should rubber-soled shoes and rubber gloves be worn when working around electricity?

21. **S** Starting from Ohm's law, show that $E = J\rho$, where E is the magnitude of the electric field (assumed constant) and $J = I/A$ is called the current density. The result is in fact true in general.

17.5 Temperature Variation of Resistance

22. If a certain silver wire has a resistance of 6.00 Ω at 20.0°C, what resistance will it have at 34.0°C?

23. While taking photographs in Death Valley on a day when the temperature is 58.0°C, Bill Hiker finds that a certain voltage applied to a copper wire produces a current of 1.000 A. Bill then travels to Antarctica and applies the same voltage to the same wire. What current does he register there if the temperature is −88.0°C? Assume no change occurs in the wire's shape and size.

24. **QC** A length of aluminum wire has a resistance of 30.0 Ω at 20.0°C. When the wire is warmed in an oven and reaches thermal equilibrium, the resistance of the wire increases to 46.2 Ω. (a) Neglecting thermal expansion, find the temperature of the oven. (b) Qualitatively, how would thermal expansion be expected to affect the answer?

25. Gold is the most ductile of all metals. For example, one gram of gold can be drawn into a wire 2.40 km long. The density of gold is 19.3×10^3 kg/m^3, and its resistivity is 2.44×10^{-8} $\Omega \cdot$ m. What is the resistance of such a wire at 20.0°C?

26. At what temperature will aluminum have a resistivity that is three times the resistivity of copper at room temperature?

27. At 20°C, the carbon resistor in an electric circuit connected to a 5.0-V battery has a resistance of 2.0×10^2 Ω. What is the current in the circuit when the temperature of the carbon rises to 80°C?

28. A wire 3.00 m long and 0.450 mm^2 in cross-sectional area has a resistance of 41.0 Ω at 20°C. If its resistance increases to 41.4 Ω at 29.0°C, what is the temperature coefficient of resistivity?

29. **M** (a) A 34.5-m length of copper wire at 20.0°C has a radius of 0.25 mm. If a potential difference of 9.0 V is applied across the length of the wire, determine the current in the wire. (b) If the wire is heated to 30.0°C while the 9.0-V potential difference is maintained, what is the resulting current in the wire?

30. An engineer needs a resistor with a zero overall temperature coefficient of resistance at 20.0°C. She designs a pair of circular cylinders, one of carbon and one of Nichrome as shown in Figure P17.30. The device must have an overall resistance of $R_1 + R_2 = 10.0$ Ω independent of temperature and a uniform radius of $r = 1.50$ mm. Ignore thermal expansion of the cylinders and assume both are always at the same temperature. (a) Can she meet the design goal with this method? (b) If so, state what you can determine about the lengths L_1 and L_2 of each segment. If not, explain.

Figure P17.30

31. **BIO** In one form of plethysmograph (a device for measuring volume), a rubber capillary tube with an inside diameter of 1.00 mm is filled with mercury at 20°C. The resistance of the mercury is measured with the aid of electrodes sealed into the ends of the tube. If 100.00 cm of the tube is wound in a spiral around a patient's upper arm, the blood flow during a heartbeat causes the arm to expand, stretching the tube to a length of 100.04 cm. From this observation, and assuming cylindrical symmetry, you can find the change in volume of the arm, which gives an indication of blood flow. (a) Calculate the resistance of the mercury. (b) Calculate the fractional change in resistance during the heartbeat. Take $\rho_{Hg} = 9.4 \times 10^{-7}\ \Omega \cdot m$. *Hint:* Because the cylindrical volume is constant, $V = A_i L_i = A_f L_f$ and $A_f = A_i(L_i/L_f)$.

32. A platinum resistance thermometer has resistances of 200.0 Ω when placed in a 0°C ice bath and 253.8 Ω when immersed in a crucible containing melting potassium. What is the melting point of potassium? *Hint:* First determine the resistance of the platinum resistance thermometer at room temperature, 20°C.

17.6 Electrical Energy and Power

33. Suppose your waffle iron is rated at 1.00 kW when connected to a 1.20×10^2 V source. (a) What current does the waffle iron carry? (b) What is its resistance?

34. If electrical energy costs $0.12 per kilowatt-hour, how much does it cost to (a) burn a 100-W lightbulb for 24 h? (b) Operate an electric oven for 5.0 h if it carries a current of 20.0 A at 220 V?

35. **Q|C** Residential building codes typically require the use of 12-gauge copper wire (diameter 0.205 cm) for wiring receptacles. Such circuits carry currents as large as 20.0 A. If a wire of smaller diameter (with a higher gauge number) carried that much current, the wire could rise to a high temperature and cause a fire. (a) Calculate the rate at which internal energy is produced in 1.00 m of 12-gauge copper wire carrying 20.0 A. (b) Repeat the calculation for a 12-gauge aluminum wire. (c) Explain whether a 12-gauge aluminum wire would be as safe as a copper wire.

36. A high-voltage transmission line with a resistance of 0.31 Ω/km carries a current of 1 000 A. The line is at a potential of 700 kV at the power station and carries the current to a city located 160 km from the station. (a) What is the power loss due to resistance in the line? (b) What fraction of the transmitted power does this loss represent?

37. **M** The heating element of a coffeemaker operates at 120 V and carries a current of 2.00 A. Assuming the water absorbs all the energy converted by the resistor, calculate how long it takes to heat 0.500 kg of water from room temperature (23.0°C) to the boiling point.

38. **W** The power supplied to a typical black-and-white television set is 90 W when the set is connected to 120 V. (a) How much electrical energy does this set consume in 1 hour? (b) A color television set draws about 2.5 A when connected to 120 V. How much time is required for it to consume the same energy as the black-and-white model consumes in 1 hour?

39. **Q|C** Lightbulb A is marked "25.0 W 120 V," and lightbulb B is marked "100 W 120 V." These labels mean that each lightbulb has its respective power delivered to it when it is connected to a constant 120-V source. (a) Find the resistance of each lightbulb. (b) During what time interval does 1.00 C pass into lightbulb A? (c) Is this charge different upon its exit versus its entry into the lightbulb? Explain. (d) In what time interval does 1.00 J pass into lightbulb A? (e) By what mechanisms does this energy enter and exit the lightbulb? Explain. (f) Find the cost of running lightbulb A continuously for 30.0 days, assuming the electric company sells its product at $0.110 per kWh.

40. A certain toaster has a heating element made of Nichrome resistance wire. When the toaster is first connected to a 120-V source of potential difference (and the wire is at a temperature of 20.0°C), the initial current is 1.80 A but the current begins to decrease as the resistive element warms up. When the toaster reaches its final operating temperature, the current has dropped to 1.53 A. (a) Find the power the toaster converts when it is at its operating temperature. (b) What is the final temperature of the heating element?

41. A copper cable is designed to carry a current of 300 A with a power loss of 2.00 W/m. What is the required radius of this cable?

42. Batteries are rated in terms of ampere-hours (A · h). For example, a battery that can deliver a current of 3.0 A for 5.0 h is rated at 15 A · h. (a) What is the total energy, in kilowatt-hours, stored in a 12-V battery rated at 55 A · h? (b) At $0.12 per kilowatt-hour, what is the value of the electricity that can be produced by this battery?

43. **BIO** The potential difference across a resting neuron in the human body is about 75.0 mV and carries a current of about 0.200 mA. How much power does the neuron release?

44. The cost of electricity varies widely throughout the United States; $0.120/kWh is a typical value. At this unit price, calculate the cost of (a) leaving a 40.0-W porch light on for 2 weeks while you are on vacation, (b) making a piece of dark toast in 3.00 min with a 970-W toaster, and (c) drying a load of clothes in 40.0 min in a 5 200-W dryer.

45. An electric utility company supplies a customer's house from the main power lines (120 V) with two copper wires, each of which is 50.0 m long and has a resistance of 0.108 Ω per 300 m. (a) Find the potential difference at the customer's house for a load current of 110 A. For this load current, find (b) the power delivered to the customer.

46. **Q|C** An office worker uses an immersion heater to warm 250 g of water in a light, covered, insulated cup from 20°C to 100°C in 4.00 minutes. The heater is a Nichrome resistance wire connected to a 120-V power supply. Assume the wire is at 100°C throughout the 4.00-min time interval. (a) Calculate the average power required to warm the water to 100°C in 4.00 min. (b) Calculate the required resistance in the heating element at 100°C. (c) Calculate the resistance of the heating element at 20.0°C. (d) Derive a relationship between the diameter of the wire, the resistivity at 20.0°C, ρ_0, the resistance at 20.0°C, R_0, and the length L. (e) If $L = 3.00$ m, what is the diameter of the wire?

47. **S** Two wires A and B made of the same material and having the same lengths are connected across the same voltage source. If the power supplied to wire A is three times the power supplied to wire B, what is the ratio of their diameters?

48. **Q|C** A tungsten wire in a vacuum has length 15.0 cm and radius 1.00 mm. A potential difference is applied across it. (a) What is the resistance of the wire at 293 K? (b) Suppose the wire reaches an equilibrium temperature such that it emits 75.0 W in the form of radiation. Neglecting absorption of any radiation from its environment, what is the temperature of the wire? (*Note:* $e = 0.320$ for tungsten.) (c) What is the resistance of the wire at the temperature found in part (b)? Assume the temperature changes linearly over this temperature range. (d) What voltage drop is required across the wire? (e) Why are tungsten lightbulbs energetically inefficient as light sources?

Additional Problems

49. If a battery is rated at 60.0 A · h, how much total charge can it deliver before it goes "dead"?

50. A car owner forgets to turn off the headlights of his car while it is parked in his garage. If the 12.0-V battery in his car is rated at 90.0 A · h and each headlight requires 36.0 W of power, how long will it take the battery to completely discharge?

51. Consider an aluminum wire of diameter 0.600 mm and length 15.0 m. The resistivity of aluminum at 20.0°C is 2.82×10^{-8} Ω · m. (a) Find the resistance of this wire at 20.0°C. (b) If a 9.00 V battery is connected across the ends of the wire, find the current in the wire.

52. A given copper wire has a resistance of 5.00 Ω at 20.0°C while a tungsten wire of the same diameter has a resistance of 4.75 Ω at 20.0°C. At what temperature will the two wires have the same resistance?

53. **M** A particular wire has a resistivity of 3.0×10^{-8} Ω · m and a cross-sectional area of 4.0×10^{-6} m^2. A length of this wire is to be used as a resistor that will develop 48 W of power when connected across a 20-V battery. What length of wire is required?

54. Birds resting on high-voltage power lines are a common sight. The copper wire on which a bird stands is 2.2 cm in diameter and carries a current of 50 A. If the bird's feet are 4.0 cm apart, calculate the potential difference between its feet.

55. An experiment is conducted to measure the electrical resistivity of Nichrome in the form of wires with different lengths and cross-sectional areas. For one set of measurements, a student uses 30-gauge wire, which has a cross-sectional area of 7.30×10^{-8} m^2. The student measures the potential difference across the wire and the current in the wire with a voltmeter and an ammeter, respectively. For each of the measurements given in the following table taken on wires of three different lengths, calculate the resistance of the wires and the corresponding value of the resistivity.

L (m)	ΔV(V)	I(A)	R (Ω)	ρ (Ω · m)
0.540		5.22		0.500
1.028		5.82		0.276
1.543		5.94		0.187

What is the average value of the resistivity, and how does this value compare with the value given in Table 17.1?

56. A 50.0-g sample of a conducting material is all that is available. The resistivity of the material is measured to be 11×10^{-8} Ω · m, and the density is 7.86 g/cm^3. The material is to be shaped into a solid cylindrical wire that has a total resistance of 1.5 Ω. (a) What length of wire is required? (b) What must be the diameter of the wire?

57. You are cooking breakfast for yourself and a friend using a 1 200-W waffle iron and a 500-W coffeepot. Usually, you operate these appliances from a 110-V outlet for 0.500 h each day. (a) At 12 cents per kWh, how much do you spend to cook breakfast during a 30.0-day period? (b) You find yourself addicted to waffles and would like to upgrade to a 2 400-W waffle iron that will enable you to cook twice as many waffles during a half-hour period, but you know that the circuit breaker in your kitchen is a 20-A breaker. Can you do the upgrade?

58. The current in a conductor varies in time as shown in Figure P17.58. (a) How many coulombs of charge pass through a cross section of the conductor in the interval from $t = 0$ to $t = 5.0$ s? (b) What constant current would transport the same total charge during the 5.0-s interval as does the actual current?

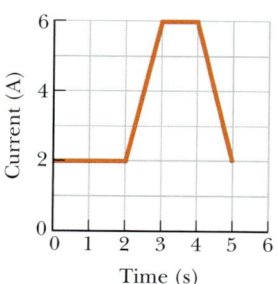

Figure P17.58

59. A 120-V motor has mechanical power output of 2.50 hp. It is 90.0% efficient in converting power that it takes in by electrical transmission into mechanical power. (a) Find the current in the motor. (b) Find the energy delivered to the motor by electrical transmission in 3.00 h of operation. (c) If the electric company charges $0.110/kWh, what does it cost to run the motor for 3.00 h?

60. **QC** A Nichrome heating element in an oven has a resistance of 8.0 Ω at 20°C. (a) What is its resistance at 350°C? (b) What assumption did you have to make to obtain your answer to part (a)?

61. A length of metal wire has a radius of 5.00×10^{-3} m and a resistance of 0.100 Ω. When the potential difference across the wire is 15.0 V, the electron drift speed is found to be 3.17×10^{-4} m/s. On the basis of these data, calculate the density of free electrons in the wire.

62. In a certain stereo system, each speaker has a resistance of 4.00 Ω. The system is rated at 60.0 W in each channel. Each speaker circuit includes a fuse rated at a maximum current of 4.00 A. Is this system adequately protected against overload?

63. A resistor is constructed by forming a material of resistivity 3.5×10^5 $\Omega \cdot$ m into the shape of a hollow cylinder of length 4.0 cm and inner and outer radii 0.50 cm and 1.2 cm, respectively. In use, a potential difference is applied between the ends of the cylinder, producing a current parallel to the length of the cylinder. Find the resistance of the cylinder.

64. **S** When a straight wire is heated, its resistance changes according to the equation

$$R = R_0[1 + \alpha(T - T_0)]$$

(Eq. 17.7), where α is the temperature coefficient of resistivity. (a) Show that a more precise result, which includes the length and area of a wire change when it is heated, is

$$R = \frac{R_0[1 + \alpha(T - T_0)][1 + \alpha'(T - T_0)]}{[1 + 2\alpha'(T - T_0)]}$$

where α' is the coefficient of linear expansion. (See Chapter 10.) (b) Compare the two results for a 2.00-m-long copper wire of radius 0.100 mm, starting at 20.0°C and heated to 100.0°C.

65. **BIO** An x-ray tube used for cancer therapy operates at 4.0 MV, with a beam current of 25 mA striking the metal target. Nearly all the power in the beam is transferred to a stream of water flowing through holes drilled in the target. What rate of flow, in kilograms per second, is needed if the rise in temperature (ΔT) of the water is not to exceed 50°C?

66. A man wishes to vacuum his car with a canister vacuum cleaner marked 535 W at 120 V. The car is parked far from the building, so he uses an extension cord 15.0 m long to plug the cleaner into a 120-V source. Assume the cleaner has constant resistance. (a) If the resistance of each of the two conductors of the extension cord is 0.900 Ω, what is the actual power delivered to the cleaner? (b) If, instead, the power is to be at least 525 W, what must be the diameter of each of two identical copper conductors in the cord the young man buys? (c) Repeat part (b) if the power is to be at least 532 W. *Suggestion:* A symbolic solution can simplify the calculations.

This vision-impaired patient is wearing an artificial vision device. A camera on the lens sends video to a portable computer. The computer then converts the video to signals that stimulate implants in the patient's visual cortex, giving him enough sight to move around on his own.

18 Direct-Current Circuits

18.1 Sources of emf
18.2 Resistors in Series
18.3 Resistors in Parallel
18.4 Kirchhoff's Rules and Complex DC Circuits
18.5 *RC* Circuits
18.6 Household Circuits
18.7 Electrical Safety
18.8 Conduction of Electrical Signals by Neurons

Batteries, resistors, and capacitors can be used in various combinations to construct electric circuits, which direct and control the flow of electricity and the energy it conveys. Such circuits make possible all the modern conveniences in a home: electric lights, electric stove tops and ovens, washing machines, and a host of other appliances and tools. Electric circuits are also found in our cars, in tractors that increase farming productivity, and in all types of medical equipment that saves so many lives every day.

In this chapter we study and analyze a number of simple direct-current circuits. The analysis is simplified by the use of two rules known as Kirchhoff's rules, which follow from the principle of conservation of energy and the law of conservation of charge. Most of the circuits are assumed to be in *steady state,* which means that the currents are constant in magnitude and direction. We close the chapter with a discussion of circuits containing resistors and capacitors, in which current varies with time.

18.1 Sources of emf

LEARNING OBJECTIVES

1. Describe in physical terms the concepts of emf, terminal voltage, internal resistance, and load resistance.
2. Evaluate the current and total power output of an emf source.

A current is maintained in a closed circuit by a source of emf.[1] Among such sources are any devices (for example, batteries and generators) that increase the potential

---
[1]The term was originally an abbreviation for *electromotive force,* but emf is not really a force, so the long form is discouraged.

energy of the circulating charges. A source of emf can be thought of as a "charge pump" that forces electrons to move in a direction opposite the electrostatic field inside the source. The emf $\mathcal{E}$ of a source is the work done per unit charge; hence the SI unit of emf is the volt.

Consider the circuit in Figure 18.1a consisting of a battery connected to a resistor. We assume the connecting wires have no resistance. If we neglect the internal resistance of the battery, the potential drop across the battery (the terminal voltage) equals the emf of the battery. Because a real battery always has some internal resistance r, however, the terminal voltage is not equal to the emf. The circuit of Figure 18.1a can be described schematically by the diagram in Figure 18.1b. The battery, represented by the dashed rectangle, consists of a source of emf $\mathcal{E}$ in series with an internal resistance r. Now imagine a positive charge moving through the battery from a to b in the figure. As the charge passes from the negative to the positive terminal of the battery, the potential of the charge increases by $\mathcal{E}$. As the charge moves through the resistance r, however, its potential decreases by the amount Ir, where I is the current in the circuit. The terminal voltage of the battery, $\Delta V = V_b - V_a$, is therefore given by

$$\Delta V = \mathcal{E} - Ir \qquad [18.1]$$

From this expression, we see that **$\mathcal{E}$ is equal to the terminal voltage when the current is zero**, called the **open-circuit voltage**. By inspecting Figure 18.1b, we find that the terminal voltage ΔV must also equal the potential difference across the external resistance R, often called the **load resistance**; that is, $\Delta V = IR$. Combining this relationship with Equation 18.1, we arrive at

$$\mathcal{E} = IR + Ir \qquad [18.2]$$

Solving for the current gives

$$I = \frac{\mathcal{E}}{R + r}$$

The preceding equation shows that the current in this simple circuit depends on both the resistance external to the battery and the internal resistance of the battery. If R is much greater than r, we can neglect r in our analysis (an option we usually select).

If we multiply Equation 18.2 by the current I, we get

$$I\mathcal{E} = I^2R + I^2r$$

This equation tells us that the total power output $I\mathcal{E}$ of the source of emf is converted at the rate I^2R at which energy is delivered to the load resistance, plus the rate I^2r at which energy is delivered to the internal resistance. Again, if $r \ll R$, most of the power delivered by the battery is transferred to the load resistance.

Unless otherwise stated, in our examples and end-of-chapter problems we will assume the internal resistance of a battery in a circuit is negligible.

Figure 18.1 (a) A circuit consisting of a resistor connected to the terminals of a battery. (b) A circuit diagram of a source of emf $\mathcal{E}$ having internal resistance r connected to an external resistor R.

An assortment of batteries.

Quick Quiz

18.1 True or False: While discharging, the terminal voltage of a battery can never be greater than the emf of the battery.

18.2 Why does a battery get warm while in use?

Tip 18.1 What's Constant in a Battery?

Equation 18.2 shows that the current in a circuit depends on the resistance of the battery, so a battery can't be considered a source of constant current. Even the terminal voltage of a battery given by Equation 18.1 can't be considered constant because the internal resistance can change (due to warming, for example, during the operation of the battery). A battery is, however, a source of constant emf.

18.2 Resistors in Series

LEARNING OBJECTIVES

1. Derive an expression for the equivalent resistance of a series of resistors.
2. Analyze circuits having resistors in series.

When two or more resistors are connected end to end as in Figure 18.2, they are said to be in *series*. The resistors could be simple devices, such as lightbulbs or heating elements. When two resistors R_1 and R_2 are connected to a battery as in Figure 18.2, **the current is the same in the two resistors because any charge that flows through R_1 must also flow through R_2**. This is analogous to water flowing through a pipe with two constrictions, corresponding to R_1 and R_2. Whatever volume of water flows in one end in a given time interval must exit the opposite end.

Because the potential difference between a and b in Figure 18.2b equals IR_1 and the potential difference between b and c equals IR_2, the potential difference between a and c is

$$\Delta V = IR_1 + IR_2 = I(R_1 + R_2)$$

Regardless of how many resistors we have in series, the sum of the potential differences across the resistors is equal to the total potential difference across the combination. As we will show later, this result is a consequence of the conservation of energy. Figure 18.2c shows an equivalent resistor R_{eq} that can replace the two resistors of the original circuit. The equivalent resistor has the same effect on the circuit because it results in the same current in the circuit as the two resistors. Applying Ohm's law to this equivalent resistor, we have

$$\Delta V = IR_{eq}$$

Equating the preceding two expressions, we have

$$IR_{eq} = I(R_1 + R_2)$$

or

$$R_{eq} = R_1 + R_2 \quad \text{(series combination)} \qquad [18.3]$$

An extension of the preceding analysis shows that the equivalent resistance of three or more resistors connected in series is

▶ Equivalent resistance of a series combination of resistors

$$R_{eq} = R_1 + R_2 + R_3 + \cdots \qquad [18.4]$$

Figure 18.2 Two resistors, R_1 and R_2, in the form of incandescent light bulbs, in series with a battery. The currents in the resistors are the same, and the equivalent resistance of the combination is $R_{eq} = R_1 + R_2$.

Therefore, **the equivalent resistance of a series combination of resistors is the algebraic sum of the individual resistances and is always greater than any individual resistance.**

Note that if the filament of one lightbulb in Figure 18.2 were to fail, the circuit would no longer be complete (an open-circuit condition would exist) and the second bulb would also go out.

APPLYING PHYSICS 18.1 | Christmas Lights in Series

A new design for Christmas lights allows them to be connected in series. A failed bulb in such a string would result in an open circuit, and all the bulbs would go out. How can the bulbs be redesigned to prevent that from happening?

EXPLANATION If the string of lights contained the usual kind of bulbs, a failed bulb would be hard to locate. Each bulb would have to be replaced with a good bulb, one by one, until the failed bulb was found. If there happened to be two or more failed bulbs in the string of lights, finding them would be a lengthy and annoying task.

Christmas lights use special bulbs that have an insulated loop of wire (a jumper) across the conducting supports to the bulb filaments (Fig. 18.3). If the filament breaks and the bulb fails, the bulb's resistance increases dramatically. As a result, most of the applied voltage appears across the loop of wire. This voltage causes the insulation around the loop of wire to burn, causing the metal wire to make electrical contact with the supports. This produces a conducting path through the bulb, so the other bulbs remain lit. ∎

Figure 18.3 (Applying Physics 18.1) (a) Schematic diagram of a modern "miniature" holiday light bulb, with a jumper connection to provide a current path if the filament breaks. (b) A holiday lightbulb with a broken filament.

Quick Quiz

18.3 In Figure 18.4 the current is measured with the ammeter at the bottom of the circuit. When the switch is opened, does the reading on the ammeter (a) increase, (b) decrease, or (c) not change?

18.4 The circuit in Figure 18.4 consists of two resistors, a switch, an ammeter, and a battery. When the switch is closed, power P_c is delivered to resistor R_1. When the switch is opened, which of the following statements is true about the power P_o delivered to R_1? (a) $P_o < P_c$ (b) $P_o = P_c$ (c) $P_o > P_c$

Figure 18.4 (Quick Quizzes 18.3 and 18.4)

EXAMPLE 18.1 | Four Resistors in Series

GOAL Analyze several resistors connected in series.

PROBLEM Four resistors are arranged as shown in Figure 18.5a. Find **(a)** the equivalent resistance of the circuit and **(b)** the current in the circuit if the closed-circuit terminal voltage of the battery is 6.0 V. **(c)** Calculate the electric potential at point A if the potential at the positive terminal is 6.0 V. **(d)** Suppose the open circuit voltage, or emf $\mathcal{E}$, is 6.2 V. Calculate the battery's internal resistance. **(e)** What fraction f of the battery's power is delivered to the load resistors?

(Continued)

630 CHAPTER 18 | Direct-Current Circuits

STRATEGY Because the resistors are connected in series, summing their resistances gives the equivalent resistance. Ohm's law can then be used to find the current. To find the electric potential at point A, calculate the voltage drop ΔV across the 2.0-Ω resistor and subtract the result from 6.0 V. In part (d) use Equation 18.1 to find the internal resistance of the battery. The fraction of the power delivered to the load resistance is just the power delivered to load, $I\Delta V$, divided by the total power, $I\mathcal{E}$.

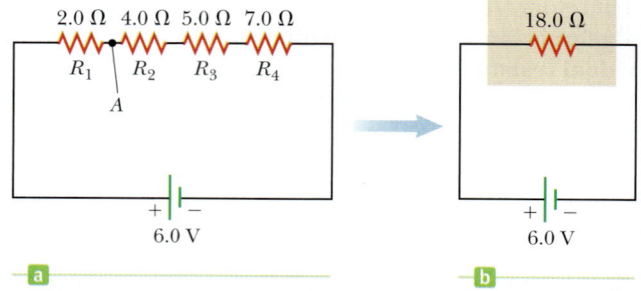

Figure 18.5 (Example 18.1) (a) Four resistors connected in series. (b) The equivalent resistance of the circuit in (a).

SOLUTION

(a) Find the equivalent resistance of the circuit.

Apply Equation 18.4, summing the resistances:

$R_{eq} = R_1 + R_2 + R_3 + R_4 = 2.0\,\Omega + 4.0\,\Omega + 5.0\,\Omega + 7.0\,\Omega$
$= \boxed{18.0\,\Omega}$

(b) Find the current in the circuit.

Apply Ohm's law to the equivalent resistor in Figure 18.5b, solving for the current:

$I = \dfrac{\Delta V}{R_{eq}} = \dfrac{6.0\text{ V}}{18.0\,\Omega} = \boxed{0.33\text{ A}}$

(c) Calculate the electric potential at point A.

Apply Ohm's law to the 2.0-Ω resistor to find the voltage drop across it:

$\Delta V = IR = (0.33\text{ A})(2.0\,\Omega) = 0.66\text{ V}$

To find the potential at A, subtract the voltage drop from the potential at the positive terminal:

$V_A = 6.0\text{ V} - 0.66\text{ V} = \boxed{5.3\text{ V}}$

(d) Calculate the battery's internal resistance if the battery's emf is 6.2 V.

Write Equation 18.1:

$\Delta V = \mathcal{E} - Ir$

Solve for the internal resistance r and substitute values:

$r = \dfrac{\mathcal{E} - \Delta V}{I} = \dfrac{6.2\text{ V} - 6.0\text{ V}}{0.33\text{ A}} = \boxed{0.6\,\Omega}$

(e) What fraction f of the battery's power is delivered to the load resistors?

Divide the power delivered to the load by the total power output:

$f = \dfrac{I\Delta V}{I\mathcal{E}} = \dfrac{\Delta V}{\mathcal{E}} = \dfrac{6.0\text{ V}}{6.2\text{ V}} = \boxed{0.97}$

REMARKS A common misconception is that the current is "used up" and steadily declines as it progresses through a series of resistors. That would be a violation of the conservation of charge. What is actually used up is the electric potential energy of the charge carriers, some of which is delivered to each resistor.

QUESTION 18.1 Explain why the current in a real circuit very slowly decreases with time compared to its initial value.

EXERCISE 18.1 A closed circuit consists of a battery with a terminal voltage of 12.0 V, and 3.0-Ω, 6.0-Ω, 8.0-Ω, and 9.0-Ω resistors connected in series, oriented as in Figure 18.5a, with the battery in the bottom of the loop, positive terminal on the left, and resistors in increasing order, left to right, in the top of the loop. Calculate (a) the equivalent resistance of the circuit, (b) the current, and (c) the total power dissipated by the load resistors. (d) What is the electric potential at a point between the 6.0-Ω and 8.0-Ω resistors, if the electric potential at the positive terminal is 12.0 V? (e) If the battery has an emf of 12.1 V, find the battery's internal resistance.

ANSWERS (a) 26.0 Ω (b) 0.462 A (c) 5.54 W (d) 7.84 V (e) 0.2 Ω

18.3 Resistors in Parallel

LEARNING OBJECTIVES

1. Derive an expression for the equivalent resistance of resistors in parallel.
2. Analyze circuits having resistors in parallel.
3. Simplify circuits having combinations of parallel and series resistors.

Now consider two resistors connected in parallel, as in Figure 18.6. In this case **the potential differences across the resistors are the same because each is connected directly across the battery terminals**. The currents are generally not the same. When charges reach point a (called a junction) in Figure 18.6b, the current splits into two parts: I_1, flowing through R_1; and I_2, flowing through R_2. If R_1 is greater than R_2, then I_1 is less than I_2. In general, more charge travels through the path with less resistance. **Because charge is conserved, the current I that enters point a must equal the total current $I_1 + I_2$ leaving that point.** Mathematically, this is written

$$I = I_1 + I_2$$

The potential drop must be the same for the two resistors and must also equal the potential drop across the battery. Ohm's law applied to each resistor yields

$$I_1 = \frac{\Delta V}{R_1} \qquad I_2 = \frac{\Delta V}{R_2}$$

Ohm's law applied to the equivalent resistor in Figure 18.6c gives

$$I = \frac{\Delta V}{R_{eq}}$$

When these expressions for the currents are substituted into the equation $I = I_1 + I_2$ and the ΔV's are canceled, we obtain

$$\frac{1}{R_{eq}} = \frac{1}{R_1} + \frac{1}{R_2} \quad \text{(parallel combination)} \qquad [18.5]$$

Figure 18.6 Two resistors, R_1 and R_2, in the form of incandescent light bulbs, in parallel with a battery. The potential differences across R_1 and R_2 are the same. Currents in the resistors are the same, and the equivalent resistance of the combination is $1/R_{eq} = 1/R_1 + 1/R_2$.

An extension of this analysis to three or more resistors in parallel produces the following general expression for the equivalent resistance:

◀ Equivalent resistance of a parallel combination of resistors

$$\frac{1}{R_{eq}} = \frac{1}{R_1} + \frac{1}{R_2} + \frac{1}{R_3} + \cdots \quad [18.6]$$

From this expression, we see that **the inverse of the equivalent resistance of two or more resistors connected in parallel is the sum of the inverses of the individual resistances and is always less than the smallest resistance in the group.**

Figure 18.7 (Quick Quizzes 18.5 and 18.6)

Quick Quiz

18.5 In Figure 18.7 the current is measured with the ammeter on the right side of the circuit diagram. When the switch is closed, does the reading on the ammeter (a) increase, (b) decrease, or (c) remain the same?

18.6 When the switch is open in Figure 18.7, power P_o is delivered to the resistor R_1. When the switch is closed, which of the following is true about the power P_c delivered to R_1? (Neglect the internal resistance of the battery.) (a) $P_c < P_o$ (b) $P_c = P_o$ (c) $P_c > P_o$

■ EXAMPLE 18.2 | Three Resistors in Parallel

GOAL Analyze a circuit that contains resistors connected in parallel.

PROBLEM Three resistors are connected in parallel as in Figure 18.8. A potential difference of 18 V is maintained between points a and b. (a) Find the current in each resistor. (b) Calculate the power delivered to each resistor and the total power. (c) Find the equivalent resistance of the circuit. (d) Find the total power delivered to the equivalent resistance.

Figure 18.8 (Example 18.2) Three resistors connected in parallel. The voltage across each resistor is 18 V.

STRATEGY To get the current in each resistor we can use Ohm's law and the fact that the voltage drops across parallel resistors are all the same. The rest of the problem just requires substitution into the equation for power delivered to a resistor, $P = I^2R$, and the reciprocal-sum law for parallel resistors.

SOLUTION
(a) Find the current in each resistor.

Apply Ohm's law, solved for the current I delivered by the battery to find the current in each resistor:

$$I_1 = \frac{\Delta V}{R_1} = \frac{18\text{ V}}{3.0\text{ }\Omega} = \boxed{6.0\text{ A}}$$

$$I_2 = \frac{\Delta V}{R_2} = \frac{18\text{ V}}{6.0\text{ }\Omega} = \boxed{3.0\text{ A}}$$

$$I_3 = \frac{\Delta V}{R_3} = \frac{18\text{ V}}{9.0\text{ }\Omega} = \boxed{2.0\text{ A}}$$

(b) Calculate the power delivered to each resistor and the total power.

Apply $P = I^2R$ to each resistor, substituting the results from part (a):

3 Ω: $P_1 = I_1^2 R_1 = (6.0\text{ A})^2(3.0\text{ }\Omega) = \boxed{110\text{ W}}$

6 Ω: $P_2 = I_2^2 R_2 = (3.0\text{ A})^2(6.0\text{ }\Omega) = \boxed{54\text{ W}}$

9 Ω: $P_3 = I_3^2 R_3 = (2.0\text{ A})^2(9.0\text{ }\Omega) = \boxed{36\text{ W}}$

Sum to get the total power:

$P_{tot} = 110\text{ W} + 54\text{ W} + 36\text{ W} = \boxed{2.0 \times 10^2\text{ W}}$

(c) Find the equivalent resistance of the circuit.
Apply the reciprocal-sum rule, Equation 18.6:

$$\frac{1}{R_{eq}} = \frac{1}{R_1} + \frac{1}{R_2} + \frac{1}{R_3}$$

$$\frac{1}{R_{eq}} = \frac{1}{3.0\ \Omega} + \frac{1}{6.0\ \Omega} + \frac{1}{9.0\ \Omega} = \frac{11}{18\ \Omega}$$

$$R_{eq} = \frac{18}{11}\ \Omega = \boxed{1.6\ \Omega}$$

(d) Compute the power dissipated by the equivalent resistance.
Use the alternate power equation:

$$P = \frac{(\Delta V)^2}{R_{eq}} = \frac{(18\ V)^2}{1.6\ \Omega} = \boxed{2.0 \times 10^2\ W}$$

REMARKS There's something important to notice in part (a): the smallest 3.0 Ω resistor carries the largest current, whereas the other, larger resistors of 6.0 Ω and 9.0 Ω carry smaller currents. The largest current is always found in the path of least resistance. In part (b) the power could also be found with $P = (\Delta V)^2/R$. Note that $P_1 = $ 108 W, but is rounded to 110 W because there are only two significant figures. Finally, notice that the total power dissipated in the equivalent resistor is the same as the sum of the power dissipated in the individual resistors, as it should be.

Tip 18.2 Don't Forget to Flip It!
The most common mistake in calculating the equivalent resistance for resistors in parallel is to forget to invert the answer after summing the reciprocals. Don't forget to flip it!

QUESTION 18.2 If a fourth resistor were added in parallel to the other three, how would the equivalent resistance change? (a) It would be larger. (b) It would be smaller. (c) More information is required to determine the effect.

EXERCISE 18.2 Suppose the resistances in the example are 1.0 Ω, 2.0 Ω, and 3.0 Ω, respectively, and a new voltage source is provided. If the current measured in the 3.0-Ω resistor is 2.0 A, find (a) the potential difference provided by the new battery and the currents in each of the remaining resistors, (b) the power delivered to each resistor and the total power, (c) the equivalent resistance, and (d) the total current and the power dissipated by the equivalent resistor.

ANSWERS (a) $\mathcal{E} = 6.0$ V, $I_1 = 6.0$ A, $I_2 = 3.0$ A (b) $P_1 = 36$ W, $P_2 = 18$ W, $P_3 = 12$ W, $P_{tot} = 66$ W (c) $\frac{6}{11}$ Ω (d) $I = 11$ A, $P_{eq} = 66$ W

Quick Quiz

18.7 Suppose you have three identical lightbulbs, some wire, and a battery. You connect one lightbulb to the battery and take note of its brightness. You add a second lightbulb, connecting it in parallel with the previous lightbulbs, and again take note of the brightness. Repeat the process with the third lightbulb, connecting it in parallel with the other two. As the lightbulbs are added, what happens to (a) the brightness of the lightbulbs? (b) The individual currents in the lightbulbs? (c) The power delivered by the battery? (d) The lifetime of the battery? (Neglect the battery's internal resistance.)

18.8 If the lightbulbs in Quick Quiz 18.7 are connected one by one in series instead of in parallel, what happens to (a) the brightness of the lightbulbs? (b) The individual currents in the lightbulbs? (c) The power delivered by the battery? (d) The lifetime of the battery? (Again, neglect the battery's internal resistance.)

Household circuits are always wired so that the electrical devices are connected in parallel, as in Figure 18.6a. In this way each device operates independently of the others so that if one is switched off, the others remain on. For example, if one of the lightbulbs in Figure 18.6 were removed from its socket, the other would continue to operate. Equally important is that each device operates at the same voltage. If the devices were connected in series, the voltage across each one would depend on how many there were in the combination and on their individual resistances.

APPLICATION
Circuit Breakers

In many household circuits, circuit breakers are used in series with other circuit elements for safety purposes. A circuit breaker is designed to switch off and open the circuit at some maximum value of current (typically 15 A or 20 A) that depends on the nature of the circuit. If a circuit breaker were not used, excessive currents caused by operating several devices simultaneously could result in excessive wire temperatures, perhaps causing a fire. In older homes, fuses were used in place of circuit breakers. When the current in a circuit exceeded some value, the conductor in a fuse melted and opened the circuit. The disadvantage of fuses is that they are destroyed in the process of opening the circuit, whereas circuit breakers can be reset.

APPLYING PHYSICS 18.2 | Lightbulb Combinations

Compare the brightness of the four identical lightbulbs shown in Figure 18.9. What happens if bulb A fails and so cannot conduct current? What if C fails? What if D fails?

EXPLANATION Bulbs A and B are connected in series across the emf of the battery, whereas bulb C is connected by itself across the battery. This means the voltage drop across C has the same magnitude as the battery emf, whereas this same emf is split between bulbs A and B. As a result, bulb C will glow more brightly than either of bulbs A and B, which will glow equally brightly. Bulb D has a wire connected across it—a short circuit—so the potential difference across bulb D is zero and it doesn't glow. If bulb A fails, B goes out, but C stays lit. If C fails, there is no effect on the other bulbs. If D fails, the event is undetectable because D was not glowing initially. ∎

Figure 18.9 (Applying Physics 18.2)

APPLYING PHYSICS 18.3 | Three-Way Lightbulbs

Figure 18.10 illustrates how a three-way lightbulb is constructed to provide three levels of light intensity. The socket of the lamp is equipped with a three-way switch for selecting different light intensities. The bulb contains two filaments. Why are the filaments connected in parallel? Explain how the two filaments are used to provide the three different light intensities.

EXPLANATION If the filaments were connected in series and one of them were to fail, there would be no current in the bulb and the bulb would not glow, regardless of the position of the switch. When the filaments are connected in parallel and one of them (say, the 75-W filament) fails, however, the bulb will still operate in one of the switch positions because there is current in the other (100-W) filament. The three light intensities are made possible by selecting one of three values of filament resistance, using a single value of 120 V for the applied voltage. The 75-W filament offers one value of resistance, the 100-W filament offers a second value, and the third resistance is obtained by combining the two filaments in parallel. When switch S_1 is closed and switch S_2 is opened, only the 75-W filament carries current. When switch S_1 is opened and switch S_2 is closed, only the 100-W filament carries current. When both switches are closed, both filaments carry current and a total illumination corresponding to 175 W is obtained. ∎

Figure 18.10 (Applying Physics 18.3)

■ PROBLEM-SOLVING STRATEGY

Simplifying Circuits with Resistors

1. **Combine all resistors in series** by summing the individual resistances and draw the new, simplified circuit diagram.
   Useful facts: $R_{eq} = R_1 + R_2 + R_3 + \cdots$
   The current in each resistor is the same.

2. **Combine all resistors in parallel** by summing the reciprocals of the resistances and then taking the reciprocal of the result. Draw the new, simplified circuit diagram.
   Useful facts: $\dfrac{1}{R_{eq}} = \dfrac{1}{R_1} + \dfrac{1}{R_2} + \dfrac{1}{R_3} + \cdots$
   The potential difference across each resistor is the same.

3. **Repeat** the first two steps as necessary, until no further combinations can be made. If there is only a single battery in the circuit, the result will usually be a single equivalent resistor in series with the battery.

4. **Use Ohm's law**, $\Delta V = IR$, to determine the current in the equivalent resistor. Then work backwards through the diagrams, applying the useful facts listed in step 1 or step 2 to find the currents in the other resistors. (In more complex circuits, Kirchhoff's rules will be needed, as described in the next section.)

■ EXAMPLE 18.3 | Equivalent Resistance

GOAL Solve a problem involving both series and parallel resistors.

PROBLEM Four resistors are connected as shown in Figure 18.11a. **(a)** Find the equivalent resistance between points a and c. **(b)** What is the current in each resistor if a 42-V battery is connected between a and c?

STRATEGY Reduce the circuit in steps, as shown in Figures 18.11b and 18.11c, using the sum rule for resistors in series and the reciprocal-sum rule for resistors in parallel. Finding the currents is a matter of applying Ohm's law while working backwards through the diagrams.

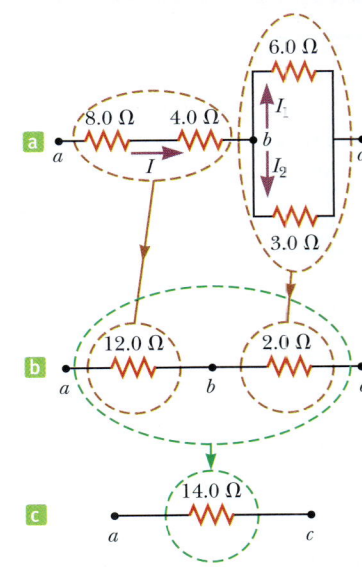

Figure 18.11 (Example 18.3) The four resistors shown in (a) can be reduced in steps to an equivalent 14-Ω resistor.

SOLUTION

(a) Find the equivalent resistance of the circuit.

The 8.0-Ω and 4.0-Ω resistors are in series, so use the sum rule to find the equivalent resistance between a and b:

$$R_{eq} = R_1 + R_2 = 8.0 \text{ Ω} + 4.0 \text{ Ω} = 12.0 \text{ Ω}$$

The 6.0-Ω and 3.0-Ω resistors are in parallel, so use the reciprocal-sum rule to find the equivalent resistance between b and c (don't forget to invert!):

$$\dfrac{1}{R_{eq}} = \dfrac{1}{R_1} + \dfrac{1}{R_2} = \dfrac{1}{6.0 \text{ Ω}} + \dfrac{1}{3.0 \text{ Ω}} = \dfrac{1}{2.0 \text{ Ω}}$$

$$R_{eq} = 2.0 \text{ Ω}$$

(Continued)

In the new diagram, 18.11b, there are now two resistors in series. Combine them with the sum rule to find the equivalent resistance of the circuit:

$R_{eq} = R_1 + R_2 = 12.0\ \Omega + 2.0\ \Omega = 14.0\ \Omega$

(b) Find the current in each resistor if a 42-V battery is connected between points a and c.

Find the current in the equivalent resistor in Figure 18.11c, which is the total current. Resistors in series all carry the same current, so this value is the current in the 12-Ω resistor in Figure 18.11b and also in the 8.0-Ω and 4.0-Ω resistors in Figure 18.11a.

$I = \dfrac{\Delta V_{ac}}{R_{eq}} = \dfrac{42\ \text{V}}{14\ \Omega} = 3.0\ \text{A}$

Calculate the voltage drop ΔV_{para} across the parallel circuit, which has an equivalent resistance of 2.0 Ω:

$\Delta V_{\text{para}} = IR = (3.0\ \text{A})(2.0\ \Omega) = 6.0\ \text{V}$

Apply Ohm's law again to find the currents in each resistor of the parallel circuit:

$I_1 = \dfrac{\Delta V_{\text{para}}}{R_{6.0\ \Omega}} = \dfrac{6.0\ \text{V}}{6.0\ \Omega} = 1.0\ \text{A}$

$I_2 = \dfrac{\Delta V_{\text{para}}}{R_{3.0\ \Omega}} = \dfrac{6.0\ \text{V}}{3.0\ \Omega} = 2.0\ \text{A}$

REMARKS As a final check, note that $\Delta V_{bc} = (6.0\ \Omega)I_1 = (3.0\ \Omega)I_2 = 6.0\ \text{V}$ and $\Delta V_{ab} = (12\ \Omega)I_1 = 36\ \text{V}$; therefore, $\Delta V_{ac} = \Delta V_{ab} + \Delta V_{bc} = 42\ \text{V}$, as expected.

QUESTION 18.3 Which of the original resistors dissipates energy at the greatest rate?

EXERCISE 18.3 Suppose the series resistors in Example 18.3 are now 6.00 Ω and 3.00 Ω while the parallel resistors are 8.00 Ω (top) and 4.00 Ω (bottom), and the battery provides an emf of 27.0 V. Find (a) the equivalent resistance and (b) the currents I, I_1, and I_2.

ANSWERS (a) 11.7 Ω (b) $I = 2.31$ A, $I_1 = 0.770$ A, $I_2 = 1.54$ A

The current I_1 entering the junction must equal the sum of the currents I_2 and I_3 leaving the junction.

The net volume flow rate in must equal the net volume flow rate out.

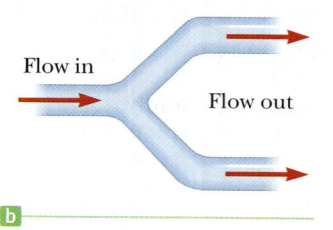

Flow in

Flow out

Figure 18.12 (a) Kirchoff's junction rule. (b) A mechanical analog of the junction rule.

18.4 Kirchhoff's Rules and Complex DC Circuits

LEARNING OBJECTIVES

1. State Kirchoff's rules and discuss their physical origins.
2. Apply Kirchoff's rules to DC circuits.

As demonstrated in the preceding section, we can analyze simple circuits using Ohm's law and the rules for series and parallel combinations of resistors. There are, however, many ways in which resistors can be connected so that the circuits formed can't be reduced to a single equivalent resistor. The procedure for analyzing more complex circuits can be facilitated by the use of two simple rules called **Kirchhoff's rules**:

1. The sum of the currents entering any junction must equal the sum of the currents leaving that junction. (This rule is often referred to as the **junction rule**.)
2. The sum of the potential differences across all the elements around any closed circuit loop must be zero. (This rule is usually called the **loop rule**.)

The junction rule is a statement of *conservation of charge*. Whatever current enters a given point in a circuit must leave that point because charge can't build up or disappear at a point. If we apply this rule to the junction in Figure 18.12a, we get

$I_1 = I_2 + I_3$

18.4 | Kirchhoff's Rules and Complex DC Circuits

Figure 18.12b represents a mechanical analog of the circuit shown in Figure 18.12a. In this analog, water flows through a branched pipe with no leaks. The flow rate into the pipe equals the total flow rate out of the two branches.

The loop rule is equivalent to the principle of *conservation of energy*. Any charge that moves around any closed loop in a circuit (starting and ending at the same point) must gain as much energy as it loses. It gains energy as it is pumped through a source of emf. Its energy may decrease in the form of a potential drop $-IR$ across a resistor or as a result of flowing backward through a source of emf, from the positive to the negative terminal inside the battery. In the latter case, electrical energy is converted to chemical energy as the battery is charged.

When applying Kirchhoff's rules, you must make two decisions at the beginning of the problem:

1. Assign symbols and directions to the currents in all branches of the circuit. Don't worry about guessing the direction of a current incorrectly; the resulting answer will be negative, but *its magnitude will be correct*. (Because the equations are *linear* in the currents, all currents are to the first power.)
2. When applying the loop rule, you must choose a direction for traversing the loop and be consistent in going either clockwise or counterclockwise. As you traverse the loop, record voltage drops and rises according to the following rules (summarized in Fig. 18.13, where it is assumed that movement is from point a toward point b):
    (a) If a resistor is traversed in the direction of the current, the change in electric potential across the resistor is $-IR$ (Fig. 18.13a).
    (b) If a resistor is traversed in the direction opposite the current, the change in electric potential across the resistor is $+IR$ (Fig. 18.13b).
    (c) If a source of emf is traversed in the direction of the emf (from $-$ to $+$ on the terminals), the change in electric potential is $+\mathcal{E}$ (Fig. 18.13c).
    (d) If a source of emf is traversed in the direction opposite the emf (from $+$ to $-$ on the terminals), the change in electric potential is $-\mathcal{E}$ (Fig. 18.13d).

In each diagram, $\Delta V = V_b - V_a$ and the circuit element is traversed from a to b, left to right.

Figure 18.13 Rules for determining the potential differences across a resistor and a battery, assuming the battery has no internal resistance.

There are limits to the number of times the junction rule and the loop rule can be used. You can use the junction rule as often as needed as long as you include a current in each new junction equation that has not been used in a previous junction-rule equation. (If this procedure isn't followed, the new equation will just be a combination of two other equations that you already have.) In general, the number of times the junction rule can be used is one fewer than the number of junction points in the circuit. The loop rule can also be used as often as needed, so long as a new circuit element (resistor or battery) or a new current appears in each new equation. **To solve a particular circuit problem, you need as many independent equations as you have unknowns.**

■ PROBLEM-SOLVING STRATEGY

Applying Kirchhoff's Rules to a Circuit

1. **Assign labels and symbols** to all the known and unknown quantities.
2. **Assign *directions* to the currents** in each part of the circuit. Although the assignment of current directions is arbitrary, you must stick with your original choices throughout the problem as you apply Kirchhoff's rules.

(Continued)

Gustav Kirchhoff
German Physicist (1824–1887)
Together with German chemist Robert Bunsen, Kirchhoff, a professor at Heidelberg, invented the spectroscopy that we study in Chapter 28. He also formulated another rule that states, "A cool substance will absorb light of the same wavelengths that it emits when hot."

3. **Apply the junction rule** to any junction in the circuit. The rule may be applied as many times as a new current (one not used in a previously found equation) appears in the resulting equation.
4. **Apply Kirchhoff's loop rule** to as many loops in the circuit as are needed to solve for the unknowns. To apply this rule, you must correctly identify the change in electric potential as you cross each element in traversing the closed loop. Watch out for signs!
5. **Solve the equations** simultaneously for the unknown quantities, using substitution or any other method familiar to the student.
6. **Check your answers** by substituting them into the original equations.

EXAMPLE 18.4 | Applying Kirchhoff's Rules

GOAL Use Kirchhoff's rules to find currents in a circuit with three currents and one battery.

PROBLEM Find the currents in the circuit shown in Figure 18.14 by using Kirchhoff's rules.

STRATEGY There are three unknown currents in this circuit, so we must obtain three independent equations, which then can be solved by substitution. We can find the equations with one application of the junction rule and two applications of the loop rule. We choose junction c. (Junction d gives the same equation.) For the loops, we choose the bottom loop and the top loop, both shown by blue arrows, which indicate the direction we are going to traverse the circuit mathematically (not necessarily the direction of the current). The third loop gives an equation that can be obtained by a linear combination of the other two, so it provides no additional information and isn't used.

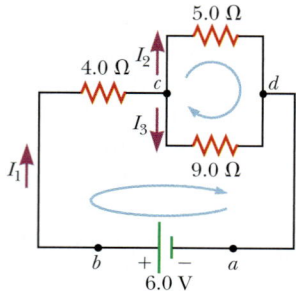

Figure 18.14 (Example 18.4)

SOLUTION

Apply the junction rule to point c. I_1 is directed into the junction, I_2 and I_3 are directed out of the junction.

$$I_1 = I_2 + I_3$$

Select the bottom loop and traverse it clockwise starting at point a, generating an equation with the loop rule:

$$\Sigma \Delta V = \Delta V_{bat} + \Delta V_{4.0\,\Omega} + \Delta V_{9.0\,\Omega} = 0$$
$$6.0\text{ V} - (4.0\text{ }\Omega)I_1 - (9.0\text{ }\Omega)I_3 = 0$$

Select the top loop and traverse it clockwise from point c. Notice the gain across the 9.0-Ω resistor because it is traversed *against* the direction of the current!

$$\Sigma \Delta V = \Delta V_{5.0\,\Omega} + \Delta V_{9.0\,\Omega} = 0$$
$$-(5.0\text{ }\Omega)I_2 + (9.0\text{ }\Omega)I_3 = 0$$

Rewrite the three equations, rearranging terms and dropping units for the moment, for convenience:

(1) $I_1 = I_2 + I_3$
(2) $4.0I_1 + 9.0I_3 = 6.0$
(3) $-5.0I_2 + 9.0I_3 = 0$

Solve Equation (3) for I_2 and substitute into Equation (1):

$$I_2 = 1.8I_3$$
$$I_1 = I_2 + I_3 = 1.8I_3 + I_3 = 2.8I_3$$

Substitute the latter expression into Equation (2) and solve for I_3:

$$4.0(2.8I_3) + 9.0I_3 = 6.0 \quad \rightarrow \quad I_3 = \boxed{0.30\text{ A}}$$

Substitute I_3 back into Equation (3) to get I_2:

$$-5.0I_2 + 9.0(0.30\text{ A}) = 0 \quad \rightarrow \quad I_2 = \boxed{0.54\text{ A}}$$

Substitute I_3 into Equation (2) to get I_1:

$$4.0I_1 + 9.0(0.30\text{ A}) = 6.0 \quad \rightarrow \quad I_1 = \boxed{0.83\text{ A}}$$

REMARKS Substituting these values back into the original equations verifies that they are correct, with any small discrepancies due to rounding. The problem can also be solved by first combining resistors.

QUESTION 18.4 How would the answers change if the indicated directions of the currents in Figure 18.14 were all reversed?

EXERCISE 18.4 Suppose the 6.0-V battery is replaced by a battery of unknown emf and an ammeter measures $I_1 = 1.5$ A. Find the other two currents and the emf of the battery.

ANSWERS $I_2 = 0.96$ A, $I_3 = 0.54$ A, $\mathcal{E} = 11$ V

> **Tip 18.3** More Current Goes in the Path of Less Resistance
>
> You may have heard the statement "Current takes the path of least resistance." For a parallel combination of resistors, this statement is inaccurate because current actually follows all paths. The most current, however, travels in the path of least resistance.

EXAMPLE 18.5 | Another Application of Kirchhoff's Rules

GOAL Find the currents in a circuit with three currents and two batteries when some current directions are chosen inaccurately.

PROBLEM Find I_1, I_2, and I_3 in Figure 18.15a.

STRATEGY Use Kirchhoff's two rules, the junction rule once and the loop rule twice, to develop three equations for the three unknown currents. Solve the equations simultaneously.

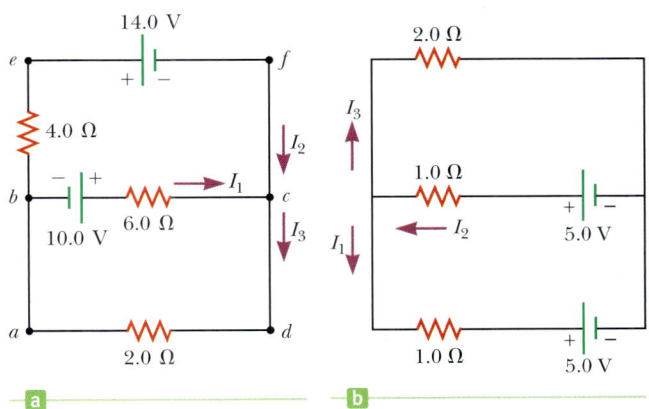

Figure 18.15
(a) (Example 18.5)
(b) (Exercise 18.5)

SOLUTION

Apply Kirchhoff's junction rule to junction c. Because of the chosen current directions, I_1 and I_2 are directed into the junction and I_3 is directed out of the junction.

(1) $I_3 = I_1 + I_2$

Apply Kirchhoff's loop rule to the loops $abcda$ and $befcb$. (Loop $aefda$ gives no new information.) In loop $befcb$, a positive sign is obtained when the 6.0-Ω resistor is traversed because the direction of the path is opposite the direction of the current I_1.

(2) Loop $abcda$: 10 V $-$ $(6.0\ \Omega)I_1 - (2.0\ \Omega)I_3 = 0$
(3) Loop $befcb$: $-(4.0\ \Omega)I_2 - 14$ V $+ (6.0\ \Omega)I_1 - 10$ V $= 0$

Using Equation (1), eliminate I_3 from Equation (2) (ignore units for the moment):

$10 - 6.0I_1 - 2.0(I_1 + I_2) = 0$
(4) $10 = 8.0I_1 + 2.0I_2$

Divide each term in Equation (3) by 2 and rearrange the equation so that the currents are on the right side:

(5) $-12 = -3.0I_1 + 2.0I_2$

Subtracting Equation (5) from Equation (4) eliminates I_2 and gives I_1:

$22 = 11I_1 \rightarrow I_1 = \boxed{2.0\ \text{A}}$

Substituting this value of I_1 into Equation (5) gives I_2:

$2.0I_2 = 3.0I_1 - 12 = 3.0(2.0) - 12 = -6.0$ A
$I_2 = \boxed{-3.0\ \text{A}}$

Finally, substitute the values found for I_1 and I_2 into Equation (1) to obtain I_3:

$I_3 = I_1 + I_2 = 2.0$ A $-$ 3.0 A $= \boxed{-1.0\ \text{A}}$

(Continued)

REMARKS The fact that I_2 and I_3 are both negative indicates that the wrong directions were chosen for these currents. Nonetheless, the magnitudes are correct. Choosing the right directions of the currents at the outset is unimportant because the equations are linear, and wrong choices result only in a minus sign in the answer.

QUESTION 18.5 Is it possible for the current in a battery to be directed from the positive terminal toward the negative terminal?

EXERCISE 18.5 Find the three currents in Figure 18.15b. (Note that the direction of one current was deliberately chosen wrongly!)

ANSWERS $I_1 = -1.0$ A, $I_2 = 1.0$ A, $I_3 = 2.0$ A

18.5 RC Circuits

LEARNING OBJECTIVES

1. Describe the time dependence of charge on a capacitor in an *RC* circuit.
2. Define the time constant and discuss its physical significance.
3. Evaluate elementary properties of *RC* circuits.

So far, we have been concerned with circuits with constant currents. We now consider direct-current circuits containing capacitors, in which the currents vary with time. Consider the series circuit in Figure 18.16. We assume the capacitor is initially uncharged with the switch opened. After the switch is closed, the battery begins to charge the plates of the capacitor and the charge passes through the resistor. As the capacitor is being charged, the circuit carries a changing current. The charging process continues until the capacitor is charged to its maximum equilibrium value, $Q = C\mathcal{E}$, where $\mathcal{E}$ is the maximum voltage across the capacitor. Once the capacitor is fully charged, the current in the circuit is zero. If we assume the capacitor is uncharged before the switch is closed, and if the switch is closed at $t = 0$, we find that the charge on the capacitor varies with time according to the equation

$$q = Q(1 - e^{-t/RC}) \qquad [18.7]$$

where $e = 2.718\ldots$ is Euler's constant, the base of the natural logarithm. Figure 18.16b is a graph of this equation. The charge is zero at $t = 0$ and

Figure 18.16 (a) A capacitor in series with a resistor, a battery, and a switch. (b) A plot of the charge on the capacitor versus time after the switch on the circuit is closed.

approaches its maximum value, Q, as t approaches infinity. The voltage ΔV across the capacitor at any time is obtained by dividing the charge by the capacitance: $\Delta V = q/C$.

As you can see from Equation 18.7, it would take an infinite amount of time, in this model, for the capacitor to become fully charged. The reason is mathematical: in obtaining that equation, charges are assumed to be infinitely small, whereas in reality the smallest charge is that of an electron, with a magnitude equal to 1.60×10^{-19} C. For all practical purposes, the capacitor is fully charged after a finite amount of time. The term RC that appears in Equation 18.7 is called the **time constant** τ (Greek letter tau), so

$$\tau = RC \qquad [18.8]$$

The time constant represents the time required for the charge to increase from zero to 63.2% of its maximum equilibrium value. This means that in a period of time equal to one time constant, the charge on the capacitor increases from zero to $0.632Q$. This can be seen by substituting $t = \tau = RC$ into Equation 18.7 and solving for q. (Note that $1 - e^{-1} = 0.632$.) It's important to note that a capacitor charges very slowly in a circuit with a long time constant, whereas it charges very rapidly in a circuit with a short time constant. After a time equal to ten time constants, the capacitor is more than 99.99% charged.

Now consider the circuit in Figure 18.17a, consisting of a capacitor with an initial charge Q, a resistor, and a switch. Before the switch is closed, the potential difference across the charged capacitor is Q/C. Once the switch is closed, the charge begins to flow through the resistor from one capacitor plate to the other until the capacitor is fully discharged. If the switch is closed at $t = 0$, it can be shown that the charge q on the capacitor varies with time according to the equation

$$q = Qe^{-t/RC} \qquad [18.9]$$

The charge decreases exponentially with time, as shown in Figure 18.17b. In the interval $t = \tau = RC$, the charge decreases from its initial value Q to $0.368Q$. In other words, in a time equal to one time constant, the capacitor loses 63.2% of its initial charge. Because $\Delta V = q/C$, the voltage across the capacitor also decreases exponentially with time according to the equation $\Delta V = \mathcal{E}e^{-t/RC}$, where $\mathcal{E}$ (which equals Q/C) is the initial voltage across the fully charged capacitor.

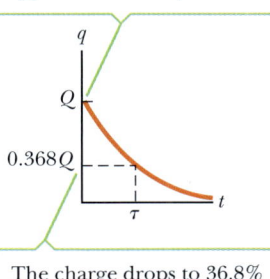

Figure 18.17 (a) A charged capacitor connected to a resistor and a switch. (b) A graph of the charge on the capacitor versus time after the switch is closed.

The charge has its maximum value Q at $t = 0$ and decays to zero exponentially as t approaches infinity.

The charge drops to 36.8% of its initial value when one time constant has elapsed.

■ **APPLYING PHYSICS 18.4** | **Timed Windshield Wipers**

Many automobiles are equipped with windshield wipers that can be used intermittently during a light rainfall. How does the operation of this feature depend on the charging and discharging of a capacitor?

EXPLANATION The wipers are part of an RC circuit with time constant that can be varied by selecting different values of R through a multiposition switch. The brief time that the wipers remain on and the time they are off are determined by the value of the time constant of the circuit. ■

■ **APPLYING PHYSICS 18.5** | **Bacterial Growth**

In biological research concerning population growth, an equation is used that is similar to the exponential equations encountered in the analysis of RC circuits. Applied to a number of bacteria, this equation is

$$N_f = N_i 2^n$$

where N_f is the number of bacteria present after n doubling times, N_i is the number present initially, and n is the number of growth cycles or doubling times. Doubling times vary according to the organism. The doubling time is about 30 days for the bacteria responsible for leprosy, and about 20 minutes for the salmonella bacteria responsible for food poisoning. Suppose only 10 salmonella bacteria find their way onto a turkey leg after your Thanksgiving meal. Four hours later you come back for a midnight snack. How many bacteria are present now?

EXPLANATION The number of doubling times is 240 min/20 min = 12. Thus,

$$N_f = N_i 2^n = (10 \text{ bacteria})(2^{12}) = 40\,960 \text{ bacteria}$$

So your system will have to deal with an invading host of about 41 000 bacteria, which are going to continue to double in a very promising environment. ■

642 CHAPTER 18 | Direct-Current Circuits

APPLYING PHYSICS 18.6 | Roadway Flashers

Many roadway construction sites have flashing yellow lights to warn motorists of possible dangers. What causes the lights to flash?

EXPLANATION A typical circuit for such a flasher is shown in Figure 18.18. The lamp L is a gas-filled lamp that acts as an open circuit until a large potential difference causes a discharge, which gives off a bright light. During this discharge, charge flows through the gas between the electrodes of the lamp. When the switch is closed, the battery charges the capacitor. At the beginning, the current is high and the charge on the capacitor is low, so most of the potential difference appears across the resistance R. As the capacitor charges, more potential difference appears across it, reflecting the lower current and lower potential difference across the resistor. Eventually, the potential difference across the capacitor reaches a value at which the lamp will conduct, causing a flash. This flash discharges the capacitor through the lamp, and the process of charging begins again. The period between flashes can be adjusted by changing the time constant of the RC circuit. ∎

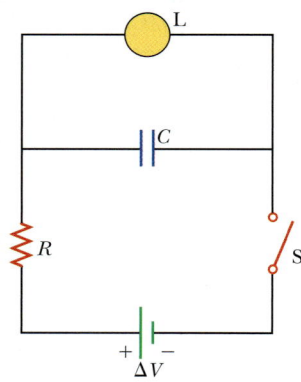

Figure 18.18 (Applying Physics 18.6)

Quick Quiz

18.9 The switch is closed in Figure 18.19. After a long time compared with the time constant of the circuit, what will the current be in the 2-Ω resistor? (a) 4 A (b) 3 A (c) 2 A (d) 1 A (e) More information is needed.

Figure 18.19 (Quick Quiz 18.9)

EXAMPLE 18.6 | Charging a Capacitor in an RC Circuit

GOAL Calculate elementary properties of a simple RC circuit.

PROBLEM An uncharged capacitor and a resistor are connected in series to a battery, as in Figure 18.16a. If $\mathcal{E}$ = 12.0 V, C = 5.00 μF, and R = 8.00 × 10⁵ Ω, find (a) the time constant of the circuit, (b) the maximum charge on the capacitor, (c) the charge on the capacitor after 6.00 s, (d) the potential difference across the resistor after 6.00 s, and (e) the current in the resistor at that time.

STRATEGY Finding the time constant in part (a) requires substitution into Equation 18.8. For part (b), the maximum charge occurs after a long time, when the current has dropped to zero. By Ohm's law, $\Delta V = IR$, the potential difference across the resistor is also zero at that time, and Kirchhoff's loop rule then gives the maximum charge. Finding the charge at some particular time, as in part (c), is a matter of substituting into Equation 18.7. Kirchhoff's loop rule and the capacitance equation can be used to indirectly find the potential drop across the resistor in part (d), and then Ohm's law yields the current.

SOLUTION

(a) Find the time constant of the circuit.

Use the definition of the time constant, Equation 18.8:

$$\tau = RC = (8.00 \times 10^5 \, \Omega)(5.00 \times 10^{-6} \, \text{F}) = 4.00 \text{ s}$$

(b) Calculate the maximum charge on the capacitor.

Apply Kirchhoff's loop rule to the RC circuit, going clockwise, which means that the voltage difference across the battery is positive and the differences across the capacitor and resistor are negative:

(1) $\Delta V_{\text{bat}} + \Delta V_C + \Delta V_R = 0$

From the definition of capacitance (Eq. 16.8) and Ohm's law, we have $\Delta V_C = -q/C$ and $\Delta V_R = -IR$. These are voltage drops, so they're negative. Also, $\Delta V_{\text{bat}} = +\mathcal{E}$.

(2) $\mathcal{E} - \dfrac{q}{C} - IR = 0$

When the maximum charge $q = Q$ is reached, $I = 0$. Solve Equation (2) for the maximum charge:

$$\mathcal{E} - \frac{Q}{C} = 0 \quad \rightarrow \quad Q = C\mathcal{E}$$

Substitute to find the maximum charge:

$$Q = (5.00 \times 10^{-6}\text{ F})(12.0\text{ V}) = \boxed{60.0\ \mu\text{C}}$$

(c) Find the charge on the capacitor after 6.00 s.

Substitute into Equation 18.7:

$$q = Q(1 - e^{-t/\tau}) = (60.0\ \mu\text{C})(1 - e^{-6.00\text{ s}/4.00\text{ s}})$$
$$= \boxed{46.6\ \mu\text{C}}$$

(d) Compute the potential difference across the resistor after 6.00 s.

Compute the voltage drop ΔV_C across the capacitor at that time:

$$\Delta V_C = -\frac{q}{C} = -\frac{46.6\ \mu\text{C}}{5.00\ \mu\text{F}} = -9.32\text{ V}$$

Solve Equation (1) for ΔV_R and substitute:

$$\Delta V_R = -\Delta V_{\text{bat}} - \Delta V_C = -12.0\text{ V} - (-9.32\text{ V})$$
$$= \boxed{-2.68\text{ V}}$$

(e) Find the current in the resistor after 6.00 s.

Apply Ohm's law, using the results of part (d) (remember that $\Delta V_R = -IR$ here):

$$I = \frac{-\Delta V_R}{R} = \frac{-(-2.68\text{ V})}{(8.00 \times 10^5\ \Omega)}$$
$$= \boxed{3.35 \times 10^{-6}\text{ A}}$$

REMARKS In solving this problem, we paid scrupulous attention to signs. These signs must always be chosen when applying Kirchhoff's loop rule and must remain consistent throughout the problem. Alternately, magnitudes can be used and the signs chosen by physical intuition. For example, the magnitude of the potential difference across the resistor must equal the magnitude of the potential difference across the battery minus the magnitude of the potential difference across the capacitor.

QUESTION 18.6 In an RC circuit as depicted in Figure 18.16a, what happens to the time required for the capacitor to be charged to half its maximum value if either the resistance or capacitance is increased? (a) It increases. (b) It decreases. (c) It remains the same.

EXERCISE 18.6 Find (a) the charge on the capacitor after 2.00 s have elapsed, (b) the magnitude of the potential difference across the capacitor after 2.00 s, and (c) the magnitude of the potential difference across the resistor at that same time.

ANSWERS (a) 23.6 μC (b) 4.72 V (c) 7.28 V

■ EXAMPLE 18.7 Discharging a Capacitor in an RC Circuit

GOAL Calculate some elementary properties of a discharging capacitor in an RC circuit.

PROBLEM Consider a capacitor C being discharged through a resistor R as in Figure 18.17a (page 641). **(a)** How long does it take the charge on the capacitor to drop to one-fourth its initial value? Answer as a multiple of τ. **(b)** Compute the initial charge and time constant, and **(c)** the time it takes to discharge all but the last quantum of charge, 1.60×10^{-19} C, if the initial potential difference across the capacitor is 12.0 V, the capacitance is equal to 3.50×10^{-6} F, and the resistance is 2.00 Ω. (Assume an exponential decrease during the entire discharge process.)

STRATEGY This problem requires substituting given values into various equations, as well as a few algebraic manipulations involving the natural logarithm. In part (a) set $q = \frac{1}{4}Q$ in Equation 18.9 for a discharging capacitor, where Q is the initial charge, and solve for time t. In part (b) substitute into Equations 16.8 and 18.8 to find the initial capacitor charge and time constant, respectively. In part (c) substitute the results of part (b) and the final charge $q = 1.60 \times 10^{-19}$ C into the discharging-capacitor equation, again solving for time.

SOLUTION

(a) How long does it take the charge on the capacitor to reduce to one-fourth its initial value?

Apply Equation 18.9:

$$q(t) = Qe^{-t/RC}$$

(Continued)

Substitute $q(t) = Q/4$ into the preceding equation and cancel Q:

$$\tfrac{1}{4}Q = Qe^{-t/RC} \quad \rightarrow \quad \tfrac{1}{4} = e^{-t/RC}$$

Take natural logarithms of both sides and solve for the time t:

$$\ln(\tfrac{1}{4}) = -t/RC$$

$$t = -RC\ln(\tfrac{1}{4}) = 1.39RC = \boxed{1.39\tau}$$

(b) Compute the initial charge and time constant from the given data.

Use the capacitance equation to find the initial charge:

$$C = \frac{Q}{\Delta V} \quad \rightarrow \quad Q = C\,\Delta V = (3.50 \times 10^{-6}\text{ F})(12.0\text{ V})$$

$$Q = \boxed{4.20 \times 10^{-5}\text{ C}}$$

Now calculate the time constant:

$$\tau = RC = (2.00\text{ }\Omega)(3.50 \times 10^{-6}\text{ F}) = \boxed{7.00 \times 10^{-6}\text{ s}}$$

(c) How long does it take to drain all but the last quantum of charge?

Apply Equation 18.9, divide by Q, and take natural logarithms of both sides:

$$q(t) = Qe^{-t/\tau} \quad \rightarrow \quad e^{-t/\tau} = \frac{q}{Q}$$

Take the natural logs of both sides:

$$-t/\tau = \ln\left(\frac{q}{Q}\right) \quad \rightarrow \quad t = -\tau\ln\left(\frac{q}{Q}\right)$$

Substitute $q = 1.60 \times 10^{-19}$ C and the values for Q and τ found in part (b):

$$t = -(7.00 \times 10^{-6}\text{ s})\ln\left(\frac{1.60 \times 10^{-19}\text{ C}}{4.20 \times 10^{-5}\text{ C}}\right)$$

$$= \boxed{2.32 \times 10^{-4}\text{ s}}$$

REMARKS Part (a) shows how useful information can often be obtained even when no details concerning capacitances, resistances, or voltages are known. Part (c) demonstrates that capacitors can be rapidly discharged (or conversely, charged), despite the mathematical form of Equations 18.7 and 18.9, which indicate an infinite time would be required.

QUESTION 18.7 Suppose the initial voltage used to charge the capacitor were doubled. Would the time required for discharging all but the last quantum of charge (a) increase, (b) decrease, (c) remain the same?

EXERCISE 18.7 Suppose the same type of series circuit has $R = 8.00 \times 10^4$ Ω, $C = 5.00$ μF, and an initial voltage across the capacitor of 6.00 V. (a) How long does it take the capacitor to lose half its initial charge? (b) How long does it take to lose all but the last 10 electrons on the negative plate?

ANSWERS (a) 0.277 s (b) 12.2 s

18.6 Household Circuits

LEARNING OBJECTIVE
1. Describe fundamental properties of household circuits.

Household circuits are a practical application of some of the ideas presented in this chapter. In a typical installation the utility company distributes electric power to individual houses with a pair of wires, or power lines. Electrical devices in a house are then connected in parallel to these lines, as shown in Figure 18.20. The potential difference between the two wires is about 120 V. (These currents and voltages are actually alternating currents and voltages, but for the present discussion we will assume they are direct currents and voltages.) One of the wires is connected to ground, and the other wire, sometimes called the "hot" wire, is at a potential of 120 V. A meter and a circuit breaker (or a fuse) are connected in series with the wire entering the house, as indicated in the figure.

In modern homes, circuit breakers are used in place of fuses. When the current in a circuit exceeds some value (typically 15 A or 20 A), the circuit breaker acts as a switch and opens the circuit. Figure 18.21 shows one design for a circuit breaker. Current passes through a bimetallic strip, the top of which bends to the left when excessive current heats it. If the strip bends far enough to the left, it settles into a groove in the spring-loaded metal bar. When this settling occurs, the bar drops enough to open the circuit at the contact point. The bar also flips a switch that indicates that the circuit breaker is not operational. (After the overload is removed, the switch can be flipped back on.) Circuit breakers based on this design have the disadvantage that some time is required for the heating of the strip, so the circuit may not be opened rapidly enough when it is overloaded. Therefore, many circuit breakers are now designed to use electromagnets (discussed in Chapter 19).

The wire and circuit breaker are carefully selected to meet the current demands of a circuit. If the circuit is to carry currents as large as 30 A, a heavy-duty wire and an appropriate circuit breaker must be used. Household circuits that are normally used to power lamps and small appliances often require only 20 A. Each circuit has its own circuit breaker to accommodate its maximum safe load.

As an example, consider a circuit that powers a toaster, a microwave oven, and a heater (represented by R_1, R_2, and R_3 in Fig. 18.20). Using the equation $P = I\Delta V$, we can calculate the current carried by each appliance. The toaster, rated at 1 000 W, draws a current of $1\,000/120 = 8.33$ A. The microwave oven, rated at 800 W, draws a current of 6.67 A, and the heater, rated at 1 300 W, draws a current of 10.8 A. If the three appliances are operated simultaneously, they draw a total current of 25.8 A. Therefore, the breaker should be able to handle at least this much current, or else it will be tripped. As an alternative, the toaster and microwave oven could operate on one 20-A circuit with the heater on a separate 20-A circuit.

Many heavy-duty appliances, such as electric ranges and clothes dryers, require 240 V to operate. The power company supplies this voltage by providing, in addition to a live wire that is 120 V above ground potential, another wire, also considered live, that is 120 V below ground potential (Fig. 18.22, page 646). Therefore, the potential drop across the two live wires is 240 V. An appliance operating from a 240-V line requires half the current of one operating from a 120-V line; consequently, smaller wires can be used in the higher-voltage circuit without becoming overheated.

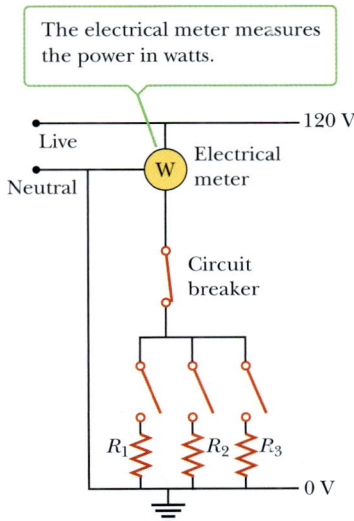

Figure 18.20 A wiring diagram for a household circuit. The resistances R_1, R_2, and R_3 represent appliances or other electrical devices that operate at an applied voltage of 120 V.

APPLICATION
Fuses and Circuit Breakers

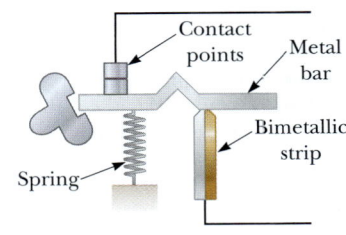

Figure 18.21 A circuit breaker that uses a bimetallic strip for its operation.

18.7 Electrical Safety

LEARNING OBJECTIVE

1. Describe several important aspects of electrical safety.

A person can be electrocuted by touching a live wire while in contact with ground. Such a hazard is often due to frayed insulation that exposes the conducting wire. The ground contact might be made by touching a water pipe (which is normally at ground potential) or by standing on the ground with wet feet because impure water is a good conductor. Obviously, such situations should be avoided at all costs.

Electric shock can result in fatal burns or cause the muscles of vital organs, such as the heart, to malfunction. The degree of damage to the body depends on the magnitude of the current, the length of time it acts, and the part of the body through which it passes. Currents of 5 mA or less can cause a sensation of shock, but ordinarily do little or no damage. If the current is larger than about 10 mA, the hand muscles contract and the person may be unable to let go of the live wire. If a current of about 100 mA passes through the body for just a few seconds, it

APPLICATION
Third Wire on Consumer Appliances

Figure 18.22 (a) The connections for each of the openings in a 240-V outlet. (b) An outlet for connection to a 240-V supply.

Figure 18.23 The "hot" (or "live") wire, at 120 V, always includes a circuit breaker for safety. (a) When the drill is operated with two wires, the normal current path is from the "hot" wire, through the motor connections, and back to ground through the "neutral" wire. (b) Shock can be prevented by a third wire running from the drill case to the ground. The wire colors represent electrical standards in the United States: the "hot" wire is black, the ground wire is green, and the neutral wire is white (shown as gray in the figure).

can be fatal. Such large currents paralyze the respiratory muscles. In some cases, currents of about 1 A through the body produce serious (and sometimes fatal) burns.

As an additional safety feature for consumers, electrical equipment manufacturers now use electrical cords that have a third wire, called a *case ground*. To understand how this works, consider the drill being used in Figure 18.23. A two-wire device has one wire, called the "hot" wire, connected to the high-potential (120-V) side of the input power line, while the second wire is connected to ground (0 V). If the high-voltage wire comes in contact with the case of the drill (Fig. 18.23a), a short circuit occurs. In this undesirable circumstance, the pathway for the current is from the high-voltage wire through the person holding the drill and to Earth, a pathway that can be fatal. Protection is provided by a third wire, connected to the case of the drill (Fig. 18.23b). In this case, if a short occurs, the path of least resistance for the current is from the high-voltage wire through the case and back to ground through the third wire. The resulting high current produced will blow a fuse or trip a circuit breaker before the consumer is injured.

Special power outlets called ground-fault interrupters (GFIs) are now being used in kitchens, bathrooms, basements, and other hazardous areas of new homes. They are designed to protect people from electrical shock by sensing small currents—approximately 5 mA and greater—leaking to ground. When current above this level is detected, the device shuts off (interrupts) the current in less than a millisecond. (Ground-fault interrupters will be discussed in Chapter 20.)

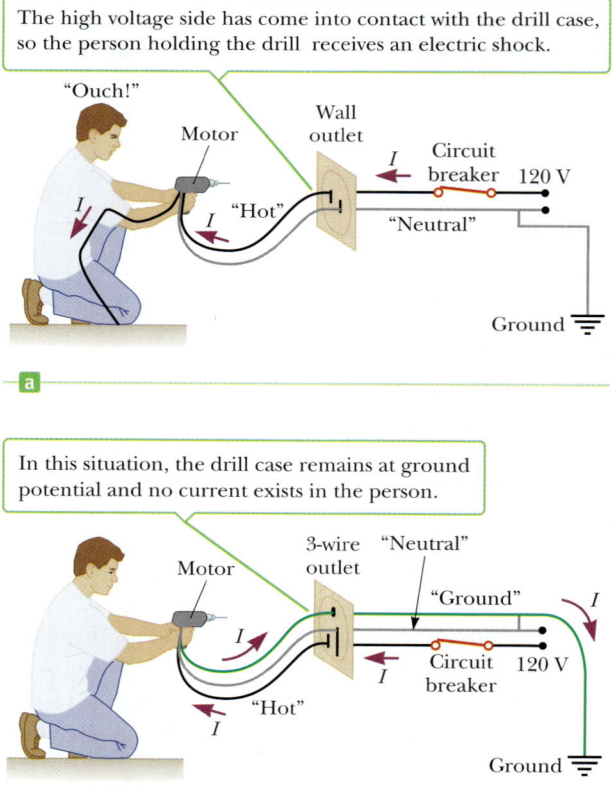

18.8 Conduction of Electrical Signals by Neurons[2] BIO

LEARNING OBJECTIVES

1. Describe the conduction of electrical signals by neurons.
2. Describe the structure and operational principles of neurons.

The most remarkable use of electrical phenomena in living organisms is found in the nervous system of animals. Specialized cells in the body called **neurons** form a complex network that receives, processes, and transmits information from one part of the body to another. The center of this network is located in the brain, which has the ability to store and analyze information. On the basis of this information, the nervous system controls parts of the body.

The nervous system is highly complex and consists of about 10^{10} interconnected neurons. Some aspects of the nervous system are well known. Over the past several decades the method of signal propagation through the nervous system has been established. The messages transmitted by neurons are voltage pulses called *action potentials*. When a neuron receives a strong enough stimulus, it produces identical voltage pulses that are actively propagated along its structure. The strength of the stimulus is conveyed by the number of pulses produced. When the pulses reach the end of the neuron, they activate either muscle cells or other neurons. There is a "firing threshold" for neurons: action potentials propagate along a neuron only if the stimulus is sufficiently strong.

Neurons can be divided into three classes: sensory neurons, motor neurons, and interneurons. The sensory neurons receive stimuli from sensory organs that monitor the external and internal environment of the body. Depending on their specialized functions, the sensory neurons convey messages about factors such as light, temperature, pressure, muscle tension, and odor to higher centers in the nervous system. The motor neurons carry messages that control the muscle cells. Those messages are based on the information provided by the sensory neurons and by the brain. The interneurons transmit information from one neuron to another.

Each neuron consists of a cell body to which are attached input ends called **dendrites** and a long tail called the **axon**, which transmits signals away from the cell (Fig. 18.24). The far end of the axon branches into nerve endings that transmit signals across small gaps to other neurons or to muscle cells. A simple sensorimotor neuron circuit is shown in Figure 18.25. A stimulus from a muscle produces nerve impulses that travel to the spine. Here the signal is transmitted to a motor neuron, which in turn sends impulses to control the muscle. Figure 18.26 shows an electron microscope image of neurons in the brain.

The axon, which is an extension of the neuron cell, conducts electric impulses away from the cell body. Some axons are extremely long. In humans, for example, the axons connecting the spine with the fingers and toes are more than 1 m long. The neuron can transmit messages because of the special active electrical characteristics of the axon. (The axon acts as an **active** source of energy like a battery, rather than like a **passive** stretch of resistive wire.) Much of the information about the electrical and chemical properties of the axon is obtained by inserting small needlelike probes into it. Figure 18.27 (page 648) shows an experimental setup.

Note that the outside of the axon is grounded, so all measured voltages are with respect to a zero potential on the outside. With these probes it is possible to inject current into the axon, measure the resulting action potential as a function of time at a fixed point, and sample the cell's chemical composition. Such experiments are usually difficult to run because the diameter of most axons is very small. Even the largest axons in the human nervous system have a diameter of only about 20×10^{-4} cm. The

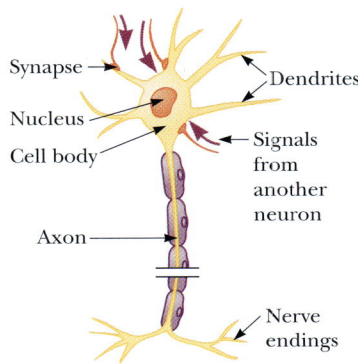

Figure 18.24 Diagram of a neuron.

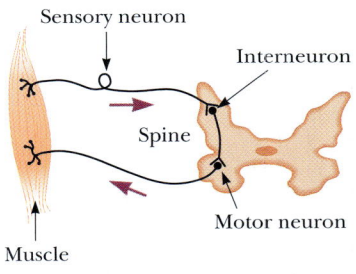

Figure 18.25 A simple neural circuit.

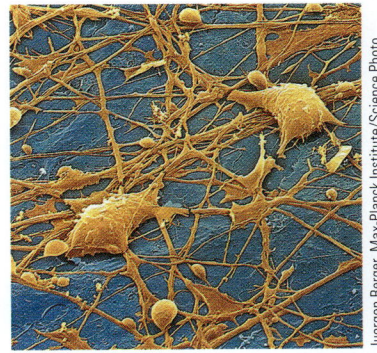

Figure 18.26 Stellate neuron from human cortex.

---

[2]This section is based on an essay by Paul Davidovits of Boston College.

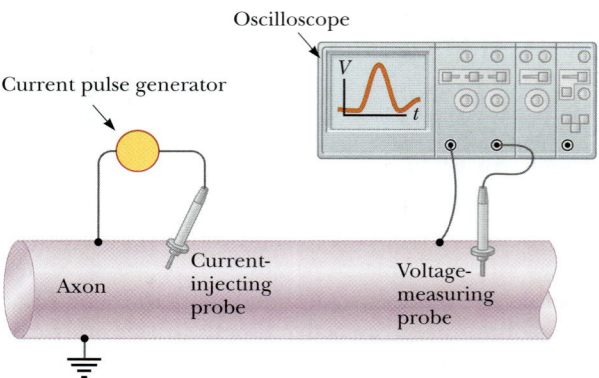

Figure 18.27 An axon stimulated electrically. The left probe injects a short pulse of current, and the right probe measures the resulting action potential as a function of time.

giant squid, however, has an axon with a diameter of about 0.5 mm, which is large enough for the convenient insertion of probes. Much of the information about signal transmission in the nervous system has come from experiments with the squid axon.

In the aqueous environment of the body, salts and other molecules dissociate into positive and negative ions. As a result, body fluids are relatively good conductors of electricity. The inside of the axon is filled with an ionic fluid that is separated from the surrounding body fluid by a thin membrane that is only about 5 nm to 10 nm thick. The resistivities of the internal and external fluids are about the same, but their chemical compositions are substantially different. The external fluid is similar to seawater: its ionic solutes are mostly positive sodium ions and negative chloride ions. Inside the axon, the positive ions are mostly potassium ions and the negative ions are mostly large organic ions.

Ordinarily, the concentrations of sodium and potassium ions inside and outside the axon would be equalized by diffusion. The axon, however, is a living cell with an energy supply and can change the permeability of its membranes on a time scale of milliseconds.

When the axon is not conducting an electric pulse, the axon membrane is highly permeable to potassium ions, slightly permeable to sodium ions, and impermeable to large organic ions. Consequently, although sodium ions cannot easily enter the axon, potassium ions can leave it. As the potassium ions leave the axon, however, they leave behind large negative organic ions, which cannot follow them through the membrane. As a result, a negative potential builds up inside the axon with respect to the outside. The final negative potential reached, which has been measured at about -70 mV, holds back the outflow of potassium ions so that at equilibrium the concentration of ions remains as stated above.

The mechanism for the production of an electric signal by the neuron is conceptually simple, but was experimentally difficult to sort out. When a neuron changes its resting potential because of an appropriate stimulus, the properties of its membrane change locally. As a result, there is a sudden flow of sodium ions into the cell that lasts for about 2 ms. This flow produces the $+30$ mV peak in the action potential shown in Figure 18.28a. Immediately afterward, there is an increase in potassium ion flow out of the cell that restores the resting action potential of -70 mV in an additional 3 ms. Both the Na^+ and K^+ ion flows have been measured by using radioactive Na and K tracers. The nerve signal has been measured to travel along the axon at speeds from 50 m/s to about 150 m/s. This flow of charged particles (or signal transmission) in a nerve axon is *unlike* signal transmission in a metal wire. In an axon, charges move perpendicular to the direction of travel of the nerve signal, and the signal moves much more slowly than a voltage pulse traveling along a metallic wire.

Although the axon is a highly complex structure and much of how Na^+ and K^+ ion channels open and close is not understood, standard electric circuit concepts of current and capacitance can be used to analyze axons. It is left as a problem (Problem 41) to show that the axon, having equal and opposite charges separated by a thin dielectric membrane, acts like a capacitor.

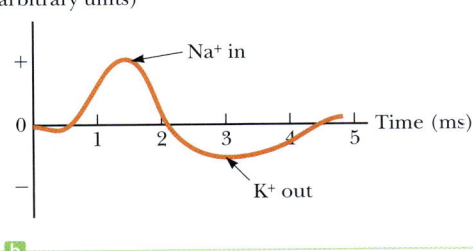

Figure 18.28 (a) Typical action potential as a function of time. (b) Current in the axon membrane wall as a function of time.

SUMMARY

18.1 Sources of emf
Any device, such as a battery or generator, that increases the electric potential energy of charges in an electric circuit is called a **source of emf**. Batteries convert chemical energy into electrical potential energy, and generators convert mechanical energy into electrical potential energy.

The terminal voltage ΔV of a battery is given by
$$\Delta V = \mathcal{E} - Ir \quad [18.1]$$

where ε is the emf of the battery, I is the current, and r is the internal resistance of the battery. Generally, the internal resistance is small enough to be neglected.

18.2 Resistors in Series
The **equivalent resistance** of a set of resistors connected in series is
$$R_{eq} = R_1 + R_2 + R_3 + \cdots \quad [18.4]$$

Several resistors in series can be replaced with a single equivalent resistor.

The current remains at a constant value as it passes through a series of resistors. The potential difference across any two resistors in series is different, unless the resistors have the same resistance.

18.3 Resistors in Parallel
The **equivalent resistance** of a set of resistors connected in parallel is
$$\frac{1}{R_{eq}} = \frac{1}{R_1} + \frac{1}{R_2} + \frac{1}{R_3} + \cdots \quad [18.6]$$

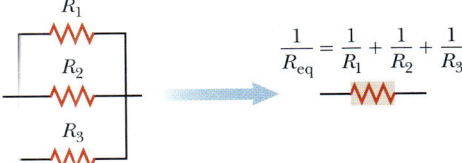

Several resistors in parallel can be replaced with a single equivalent resistor.

The potential difference across any two parallel resistors is the same; the current in each resistor, however, will be different unless the two resistances are equal.

18.4 Kirchhoff's Rules and Complex DC Circuits
Complex circuits can be analyzed using **Kirchhoff's rules**:

1. The sum of the currents entering any junction must equal the sum of the currents leaving that junction.
2. The sum of the potential differences across all the elements around any closed circuit loop must be zero.

The first rule, called the junction rule, is a statement of **conservation of charge**. The second rule, called the loop rule, is a statement of **conservation of energy**. Solving problems involves using these rules to generate as many equations as there are unknown currents. The equations can then be solved simultaneously.

18.5 RC Circuits

In a simple *RC* circuit with a battery, a resistor, and a capacitor in series, the charge on the capacitor increases according to the equation

$$q = Q(1 - e^{-t/RC}) \quad [18.7]$$

An *RC* circuit with a battery and a resistor charges a capacitor when the switch is closed.

A simple *RC* circuit consisting of a charged capacitor in series with a resistor discharges according to the expression

$$q = Qe^{-t/RC} \quad [18.9]$$

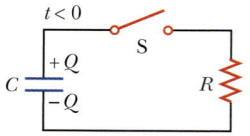

An *RC* circuit with charged capacitor discharges across a resistor when the switch is closed.

The term *RC* in Equation 18.7 is called the **time constant** τ (Greek letter tau), so

$$\tau = RC \quad [18.8]$$

The time constant represents the time required for the charge to increase from zero to 63.2% of its maximum equilibrium value.

Problems can be solved by substituting values for q, Q, t or τ into these equations. The voltage ΔV across the capacitor at any time is obtained by dividing the charge by the capacitance: $\Delta V = q/C$. Using Kirchhoff's loop rule yields the potential difference across the resistor. Ohm's law applied to the resistor then gives the current.

■ WARM-UP EXERCISES

The warm-up exercises in this chapter may be assigned online in Enhanced WebAssign.

1. **Math Review** Determine the values of (a) I_1 and (b) I_2 that satisfy the following two equations: $9.00 + 3.00\,I_1 - 2.00\,I_2 = 0$ and $-2.00\,I_1 + I_2 = 0$. (See also Section 18.4.)

2. **Math Review** Determine the values of (a) I_1, (b) I_2, and (c) I_3 that satisfy the following three equations: $-I_1 + I_2 + I_3 = 5.00$, $-3.00\,I_1 + I_2 = 0$, and $I_1 + I_3 = 0$.

3. **Math Review** Suppose the charge q on a capacitor is given by $q = (3.00\text{ C})\,e^{-t/(0.200\text{ s})}$. Find (a) the value of q when $t = 0.040\ 0$ s, and (b) the value of t when $q = 1.20$ C. (See also Section 18.5.)

4. **Physics Review** Find the charge on a 25.0 μF capacitor when connected to a 12.0-V battery. (See Section 16.6.)

5. **Physics Review** Two capacitors have capacitances of 2.00 μF and 0.500 μF. Determine the equivalent capacitance if they are connected (a) in parallel, and (b) in series. (See Section 16.8.)

6. **Physics Review** A resistor with resistance 30.0 Ω is connected to the terminals of a voltage source. Calculate (a) the current through the resistor, and (b) the power dissipated by the resistor when the voltage is set at 6.00 V. (c) If the voltage is doubled, recalculate the power. (See Sections 17.4 and 17.6.)

7. A battery has an open-circuit voltage (emf) of 9.20 V and an internal resistance of 0.400 Ω. Determine (a) the current when the terminals are connected across a 3.00-Ω resistor, (b) the terminal voltage ΔV, and (c) the total power supplied by the battery. (See Section 18.1.)

8. Two resistors have resistances of 2.00 Ω and 0.500 Ω. Determine the equivalent resistance if they are connected (a) in series, (b) in parallel. (See Sections 18.2 and 18.3.)

9. Three resistors have resistances of 2.00 Ω, 10.0 Ω and 25.0 Ω. Determine the equivalent resistance if they are connected (a) in parallel, and (b) in series. (See Sections 18.2 and 18.3.)

10. A simple circuit is constructed from a 12.0-V battery, a 3.00-Ω resistor and a 5.00-Ω resistor. Ignoring the internal resistance of the battery, determine the current supplied by the battery if the resistors are connected across the battery terminals (a) in series, and (b) in parallel. (See Sections 18.2 and 18.3.)

11. Consider the circuit shown in Figure WU18.11. Determine (a) the equivalent resistance of the three resistors in parallel, (b) the equivalent resistance of all four resistors in the circuit, (c) the current supplied by the battery, and (d) the magnitude of the potential difference across the 2.00-Ω resistor, (e) the magnitude of the potential difference across the 10.0-Ω resistor, and (f) the current through the 10.0-Ω resistor. (See Sections 18.2, 18.3 and 18.4)

Figure WU18.11

12. A flashing light blub is driven by a simple series RC circuit with a time constant of 1.30 s. If the circuit's equivalent resistance is $5.00 \times 10^3 \, \Omega$, determine its equivalent capacitance. (See Section 18.5.)
13. A 275-Ω resistor is in series with a 36.0 μF capacitor and a 12.0-V voltage source with the circuit switch open and the capacitor uncharged. (a) What is the time constant of this RC circuit? Calculate (b) the maximum charge the capacitor can accumulate, and (c) the charge on the capacitor 5.00 ms after the switch is closed. (See Section 18.5.)

CONCEPTUAL QUESTIONS

The conceptual questions in this chapter may be assigned online in Enhanced WebAssign.

1. Is the direction of current in a battery always from the negative terminal to the positive one? Explain.
2. Given three lightbulbs and a battery, sketch as many different circuits as you can.
3. Suppose the energy transferred to a dead battery during charging is W. The recharged battery is then used until fully discharged again. Is the total energy transferred out of the battery during use also W?
4. A short circuit is a circuit containing a path of very low resistance in parallel with some other part of the circuit. Discuss the effect of a short circuit on the portion of the circuit it parallels. Use a lamp with a frayed line cord as an example.
5. Connecting batteries in series increases the emf applied to a circuit. What advantage might there be to connecting them in parallel?
6. If electrical power is transmitted over long distances, the resistance of the wires becomes significant. Why? Which mode of transmission would result in less energy loss, high current and low voltage or low current and high voltage? Discuss.
7. If you have your headlights on while you start your car, why do they dim while the car is starting?
8. Two sets of Christmas lights are available. For set A, when one bulb is removed, the remaining bulbs remain illuminated. For set B, when one bulb is removed, the remaining bulbs do not operate. Explain the difference in wiring for the two sets.
9. Why is it possible for a bird to sit on a high-voltage wire without being electrocuted? (See Fig. CQ18.9.)

Figure CQ18.9

10. (a) Two resistors are connected in series across a battery. Is the power delivered to each resistor (i) the same or (ii) not necessarily the same? (b) Two resistors are connected in parallel across a battery. Is the power delivered to each resistor (i) the same or (ii) not necessarily the same?
11. Suppose a parachutist lands on a high-voltage wire and grabs the wire as she prepares to be rescued. Will she be electrocuted? If the wire then breaks, should she continue to hold onto the wire as she falls to the ground?
12. A ski resort consists of a few chairlifts and several interconnected downhill runs on the side of a mountain, with a lodge at the bottom. The lifts are analogous to batteries, and the runs are analogous to resistors. Describe how two runs can be in series. Describe how three runs can be in parallel. Sketch a junction of one lift and two runs. One of the skiers is carrying an altimeter. State Kirchhoff's junction rule and Kirchhoff's loop rule for ski resorts.
13. Embodied in Kirchhoff's rules are two conservation laws. What are they?
14. Why is it dangerous to turn on a light when you are in a bathtub?

PROBLEMS

The problems in this chapter may be assigned online in Enhanced WebAssign.

1. denotes straightforward problem; 2. denotes intermediate problem;
3. denotes challenging problem
[1.] denotes full solution available in *Student Solutions Manual/Study Guide*
1. denotes problems most often assigned in Enhanced WebAssign

BIO denotes biomedical problems
GP denotes guided problems
M denotes Master It tutorial available in Enhanced WebAssign
Q|C denotes asking for quantitative and conceptual reasoning
S denotes symbolic reasoning problem
W denotes Watch It video solution available in Enhanced WebAssign

18.1 Sources of emf
18.2 Resistors in Series
18.3 Resistors in Parallel

1. A battery having an emf of 9.00 V delivers 117 mA when connected to a 72.0-Ω load. Determine the internal resistance of the battery.

2. Three 9.0-Ω resistors are connected in series with a 12-V battery. Find (a) the equivalent resistance of the circuit and (b) the current in each resistor. (c) Repeat for the case in which all three resistors are connected in parallel across the battery.

3. A lightbulb marked "75 W [at] 120 V" is screwed into a socket at one end of a long extension cord in which each of the two conductors has a resistance of 0.800 Ω. The other end of the extension cord is plugged into a 120-V outlet. (a) Draw a circuit diagram, and (b) find the actual power of the bulb in the circuit described.

4. (a) Find the current in an 8.00-Ω resistor connected to a battery that has an internal resistance of 0.15 Ω if the voltage across the battery (the terminal voltage) is 9.00 V. (b) What is the emf of the battery?

5. M (a) Find the equivalent resistance between points a and b in Figure P18.5. (b) Calculate the current in each resistor if a potential difference of 34.0 V is applied between points a and b.

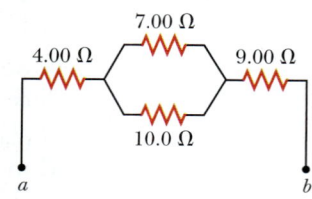

Figure P18.5

6. Consider the combination of resistors shown in Figure P18.6. (a) Find the equivalent resistance between point a and b. (b) If a voltage of 35.0 V is applied between points a and b, find the current in each resistor.

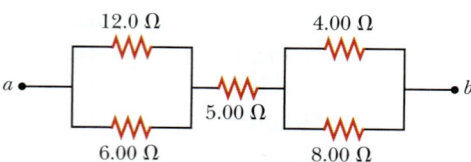

Figure P18.6

7. Two resistors connected in series have an equivalent resistance of 690 Ω. When they are connected in parallel, their equivalent resistance is 150 Ω. Find the resistance of each resistor.

8. GP Consider the circuit shown in Figure P18.8. (a) Calculate the equivalent resistance of the 10.0-Ω and 5.00-Ω resistors connected in parallel. (b) Using the result of part (a), calculate the combined resistance of the 10.0-Ω, 5.00-Ω, and 4.00-Ω resistors. (c) Calculate the equivalent resistance of the combined resistance found in part (b) and the parallel 3.00-Ω resistor. (d) Combine the equivalent resistance found in part (c) with the 2.00-Ω resistor. (e) Calculate the total current in the circuit. (f) What is the voltage drop across the 2.00-Ω resistor? (g) Subtracting the result of part (f) from the battery voltage, find the voltage across the 3.00-Ω resistor. (h) Calculate the current in the 3.00-Ω resistor.

Figure P18.8

9. Consider the circuit shown in Figure P18.9. Find (a) the potential difference between points a and b and (b) the current in the 20.0-Ω resistor.

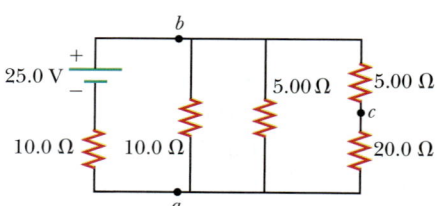

Figure P18.9

10. S Four resistors are connected to a battery as shown in Figure P18.10. (a) Determine the potential difference across each resistor in terms of $\mathcal{E}$. (b) Determine the current in each resistor in terms of I.

Figure P18.10

11. The resistance between terminals a and b in Figure P18.11 is 75 Ω. If the resistors labeled R have the same value, determine R.

Figure P18.11

12. A battery with $\mathcal{E} = 6.00$ V and no internal resistance supplies current to the circuit shown in Figure P18.12. When the double-throw switch S is open as shown in the figure, the current in the battery is 1.00 mA. When the switch is closed in position a, the current in the battery is 1.20 mA. When the switch is closed in position b, the current in the battery is 2.00 mA. Find the resistances (a) R_1, (b) R_2, and (c) R_3.

Figure P18.12

13. Find the current in the 12-Ω resistor in Figure P18.13.

Figure P18.13

14. **Q|C** (a) Is it possible to reduce the circuit shown in Figure P18.14 to a single equivalent resistor connected across the battery? Explain. (b) Find the current in the 2.00-Ω resistor. (c) Calculate the power delivered by the battery to the circuit.

Figure P18.14

15. **W** (a) You need a 45-Ω resistor, but the stockroom has only 20-Ω and 50-Ω resistors. How can the desired resistance be achieved under these circumstances? (b) What can you do if you need a 35-Ω resistor?

18.4 Kirchhoff's Rules and Complex DC Circuits

Note: For some circuits, the currents are not necessarily in the direction shown.

16. **S** (a) Find the current in each resistor of Figure P18.16 by using the rules for resistors in series and parallel. (b) Write three independent equations for the three currents using Kirchhoff's laws: one with the node rule; a second using the loop rule through the battery, the 6.0-Ω resistor, and the 24.0-Ω resistor; and the third using the loop rule through the 12.0-Ω and 24.0-Ω resistors. Solve to check the answers found in part (a).

Figure P18.16

17. The ammeter shown in Figure P18.17 reads 2.00 A. Find I_1, I_2, and $\mathcal{E}$.

18. **W** For the circuit shown in Figure P18.18, calculate (a) the current in the 2.00-Ω resistor and (b) the potential difference between points a and b, $\Delta V = V_b - V_a$.

19. Taking $R = 1.00$ kΩ and $\mathcal{E} = 250$ V in Figure P18.19, determine the direction and magnitude of the current in the horizontal wire between a and e.

Figure P18.18

Figure P18.17

Figure P18.19

20. **W** In the circuit of Figure P18.20, the current I_1 is 3.0 A and the values of $\mathcal{E}$ and R are unknown. What are the currents I_2 and I_3?

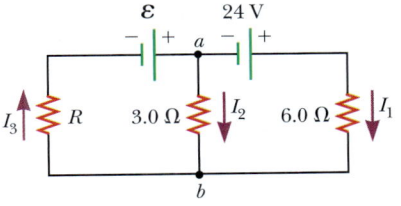

Figure P18.20

21. In the circuit of Figure P18.21, determine (a) the current in each resistor and (b) the potential difference across the 200-Ω resistor.

Figure P18.21

22. QC Four resistors are connected to a battery with a terminal voltage of 12 V, as shown in Figure P18.22. (a) How would you reduce the circuit to an equivalent single resistor connected to the battery? Use this procedure to find the equivalent resistance of the circuit. (b) Find the current delivered by the battery to this equivalent resistance. (c) Determine the power delivered by the battery. (d) Determine the power delivered to the 50.0-Ω resistor.

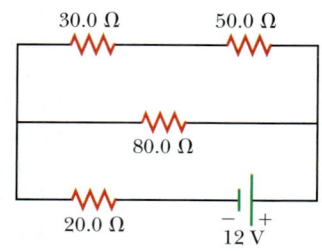

Figure P18.22

23. M Using Kirchhoff's rules, (a) find the current in each resistor shown in Figure P18.23 and (b) find the potential difference between points c and f.

Figure P18.23

24. Two 1.50-V batteries—with their positive terminals in the same direction—are inserted in series into the barrel of a flashlight. One battery has an internal resistance of 0.255 Ω, the other an internal resistance of 0.153 Ω. When the switch is closed, a current of 0.600 A passes through the lamp. (a) What is the lamp's resistance? (b) What fraction of the power dissipated is dissipated in the batteries?

25. QC (a) Can the circuit shown in Figure P18.25 be reduced to a single resistor connected to the batteries? Explain. (b) Calculate each of the unknown currents I_1, I_2, and I_3 for the circuit.

Figure P18.25

26. A dead battery is charged by connecting it to the live battery of another car with jumper cables (Fig. P18.26). Determine the current in (a) the starter and in (b) the dead battery.

Figure P18.26

27. QC (a) Can the circuit shown in Figure P18.27 be reduced to a single resistor connected to the batteries? Explain. (b) Find the magnitude of the current and its direction in each resistor.

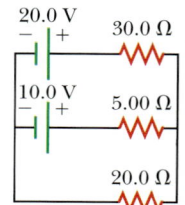

Figure P18.27

28. GP S For the circuit shown in Figure P18.28, use Kirchhoff's rules to obtain equations for (a) the upper loop, (b) the lower loop, and (c) the node on the left side. In each case suppress units for clarity and simplify, combining like terms. (d) Solve the node equation for I_{36}. (e) Using the equation found in (d), eliminate I_{36} from the equation found in part (b). (f) Solve the equations found in part (a) and part (e) simultaneously for the two

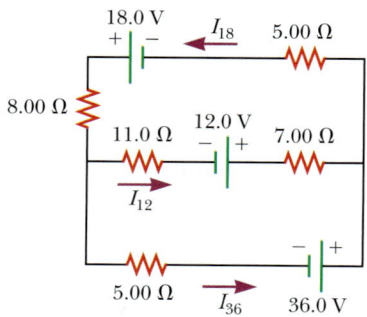

Figure P18.28

unknowns for I_{18} and I_{12}, respectively. (g) Substitute the answers found in part (f) into the node equation found in part (d), solving for I_{36}. (h) What is the significance of the negative answer for I_{12}?

29. **M** Find the potential difference across each resistor in Figure P18.29.

Figure P18.29

18.5 RC Circuits

30. Show that $\tau = RC$ has units of time.

31. Consider the series RC-circuit shown in Figure 18.16 for which $R = 75.0$ kΩ, $C = 25.0$ μF, and $\mathcal{E} = 12.0$ V. Find (a) the time constant of the circuit and (b) the charge on the capacitor one time constant after the switch is closed.

32. An uncharged capacitor and a resistor are connected in series to a source of emf. If $\mathcal{E} = 9.00$ V, $C = 20.0$ μF, and $R = 100$ Ω, find (a) the time constant of the circuit, (b) the maximum charge on the capacitor, and (c) the charge on the capacitor after one time constant.

33. Consider a series RC circuit as in Figure P18.33 for which $R = 1.00$ MΩ, $C = 5.00$ μF, and $\mathcal{E} = 30.0$ V. Find (a) the time constant of the circuit and (b) the maximum charge on the capacitor after the switch is thrown closed. (c) Find the current in the resistor 10.0 s after the switch is closed.

Figure P18.33
Problems 33 and 34.

34. **W** Consider a series RC circuit for which $R = 1.0$ MΩ, $C = 5.0$ μF, and $\mathcal{E} = 30$ V as in Figure P18.33. Find the charge on the capacitor 10 s after the switch is closed.

35. A charged capacitor is connected to a resistor and a switch as in Figure P18.35. The circuit has a time constant of 1.50 s. Soon after the switch is closed, the charge on the capacitor is 75.0% of its initial charge. (a) Find the time interval required for the capacitor to reach this charge. (b) If $R = 250$ kΩ, what is the value of C?

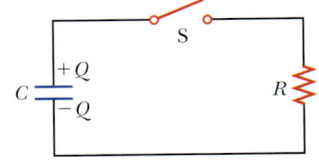

Figure P18.35

36. A 10.0-μF capacitor is charged by a 10.0-V battery through a resistance R. The capacitor reaches a potential difference of 4.00 V in a time interval of 3.00 s after charging begins. Find R.

18.6 Household Circuits

37. What minimum number of 75-W lightbulbs must be connected in parallel to a single 120-V household circuit to trip a 30.0-A circuit breaker?

38. A lamp ($R = 150$ Ω), an electric heater ($R = 25$ Ω), and a fan ($R = 50$ Ω) are connected in parallel across a 120-V line. (a) What total current is supplied to the circuit? (b) What is the voltage across the fan? (c) What is the current in the lamp? (d) What power is expended in the heater?

39. **M** A heating element in a stove is designed to dissipate 3 000 W when connected to 240 V. (a) Assuming the resistance is constant, calculate the current in the heating element if it is connected to 120 V. (b) Calculate the power it dissipates at that voltage.

40. **Q|C** A coffee maker is rated at 1 200 W, a toaster at 1 100 W, and a waffle maker at 1 400 W. The three appliances are connected in parallel to a common 120-V household circuit. (a) What is the current in each appliance when operating independently? (b) What total current is delivered to the appliances when all are operating simultaneously? (c) Is a 15-A circuit breaker sufficient in this situation? Explain.

18.8 Conduction of Electrical Signals by Neurons

41. **BIO** Assume a length of axon membrane of about 0.10 m is excited by an action potential (length excited = nerve speed $\times$ pulse duration = 50.0 m/s $\times$ 2.0×10^{-3} s = 0.10 m). In the resting state, the outer surface of the axon wall is charged positively with K^+ ions and the inner wall has an equal and opposite charge of negative organic ions, as shown in Figure P18.41. Model the axon as a parallel-plate capacitor and take $C = \kappa\epsilon_0 A/d$ and $Q = C\Delta V$ to investigate the charge as follows. Use typical values for a cylindrical axon of cell wall thickness $d = 1.0 \times 10^{-8}$ m, axon radius $r = 1.0 \times 10^1$ μm, and cell-wall dielectric constant

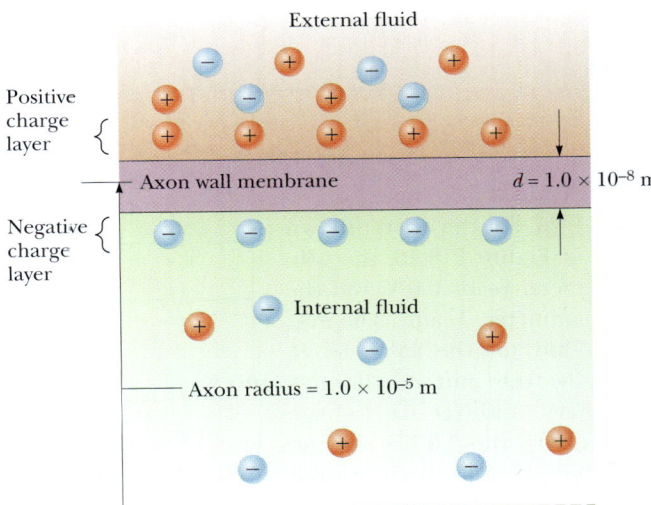

Figure P18.41 Problems 41 and 42.

$\kappa = 3.0$. (a) Calculate the positive charge on the outside of a 0.10-m piece of axon when it is not conducting an electric pulse. How many K^+ ions are on the outside of the axon assuming an initial potential difference of 7.0×10^{-2} V? Is this a large charge per unit area? *Hint:* Calculate the charge per unit area in terms of electronic charge e per squared (Å^2). An atom has a cross section of about 1 Å^2 (1 Å = 10^{-10} m). (b) How much positive charge must flow through the cell membrane to reach the excited state of $+3.0 \times 10^{-2}$ V from the resting state of -7.0×10^{-2} V? How many sodium ions (Na^+) is this? (c) If it takes 2.0 ms for the Na^+ ions to enter the axon, what is the average current in the axon wall in this process? (d) How much energy does it take to raise the potential of the inner axon wall to $+3.0 \times 10^{-2}$ V, starting from the resting potential of -7.0×10^{-2} V?

42. **BIO** Consider the model of the axon as a capacitor from Problem 41 and Figure P18.41. (a) How much energy does it take to restore the inner wall of the axon to -7.0×10^{-2} V, starting from $+3.0 \times 10^{-2}$ V? (b) Find the average current in the axon wall during this process.

43. **BIO** Using Figure 18.28b and the results of Problems 18.41d and 18.42a, find the power supplied by the axon per action potential.

Additional Problems

44. How many different resistance values can be constructed from a 2.0-Ω, a 4.0-Ω, and a 6.0-Ω resistor? Show how you would get each resistance value either individually or by combining them.

45. (a) Calculate the potential difference between points a and b in Figure P18.45 and (b) identify which point is at the higher potential.

Figure P18.45

46. **Q|C** For the circuit shown in Figure P18.46, the voltmeter reads 6.0 V and the ammeter reads 3.0 mA. Find (a) the value of R, (b) the emf of the battery, and (c) the voltage across the 3.0-kΩ resistor. (d) What assumptions did you have to make to solve this problem?

Figure P18.46

47. Find (a) the equivalent resistance of the circuit in Figure P18.47, (b) each current in the circuit, (c) the potential difference across each resistor, and (d) the power dissipated by each resistor.

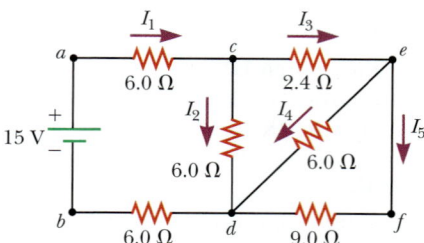

Figure P18.47

48. Three 60.0-W, 120-V lightbulbs are connected across a 120-V power source, as shown in Figure P18.48. Find (a) the total power delivered to the three bulbs and (b) the potential difference across each. Assume the resistance of each bulb is constant (even though, in reality, the resistance increases markedly with current).

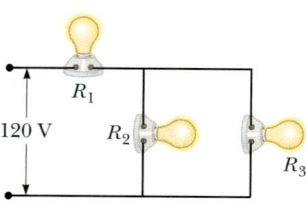

Figure P18.48

49. When two unknown resistors are connected in series with a battery, the battery delivers 225 W and carries a total current of 5.00 A. For the same total current, 50.0 W is delivered when the resistors are connected in parallel. Determine the value of each resistor.

50. **Q|C** The circuit in Figure P18.50a consists of three resistors and one battery with no internal resistance. (a) Find the current in the 5.00-Ω resistor. (b) Find the power delivered to the 5.00-Ω resistor. (c) In each of the circuits in Figures P18.50b, P18.50c, and P18.50d, an additional 15.0-V battery has been inserted into the circuit. Which diagram or diagrams represent a circuit that requires the use of Kirchhoff's rules to find the

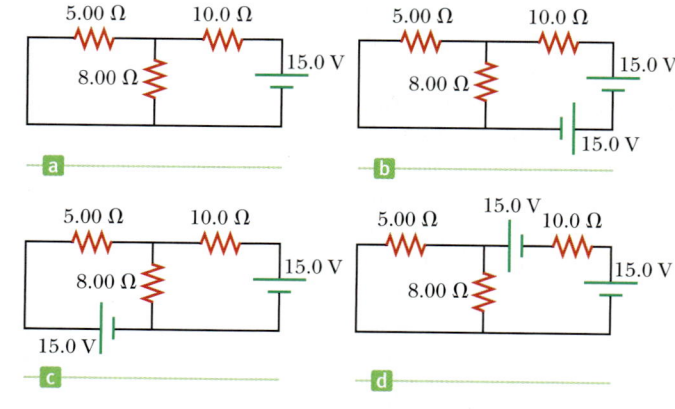

Figure P18.50

currents? Explain why. (d) In which of these three new circuits is the smallest amount of power delivered to the 10.0-Ω resistor? (You need not calculate the power in each circuit if you explain your answer.)

51. **QC S** A circuit consists of three identical lamps, each of resistance R, connected to a battery as in Figure P18.51. (a) Calculate an expression for the equivalent resistance of the circuit when the switch is open. Repeat the calculation when the switch is closed. (b) Write an expression for the power supplied by the battery when the switch is open. Repeat the calculation when the switch is closed. (c) Using the results already obtained, explain what happens to the brightness of the lamps when the switch is closed.

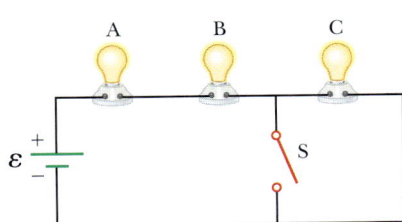

Figure P18.51

52. The resistance between points a and b in Figure P18.52 drops to one-half its original value when switch S is closed. Determine the value of R.

Figure P18.52

53. The circuit in Figure P18.53 has been connected for several seconds. Find the current (a) in the 4.00-V battery, (b) in the 3.00-Ω resistor, (c) in the 8.00-V battery, and (d) in the 3.00-V battery. (e) Find the charge on the capacitor.

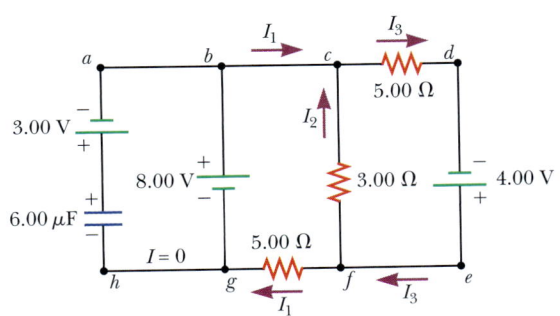

Figure P18.53

54. An emf of 10 V is connected to a series RC circuit consisting of a resistor of 2.0×10^6 Ω and an initially uncharged capacitor of 3.0 μF. Find the time required for the charge on the capacitor to reach 90% of its final value.

55. The student engineer of a campus radio station wishes to verify the effectiveness of the lightning rod on the antenna mast (Fig. P18.55). The unknown resistance R_x is between points C and E. Point E is a "true ground," but is inaccessible for direct measurement because the stratum in which it is located is several meters below Earth's surface. Two identical rods are driven into the ground at A and B, introducing an unknown resistance R_y. The procedure for finding the unknown resistance R_x is as follows. Measure resistance R_1 between points A and B. Then connect A and B with a heavy conducting wire and measure resistance R_2 between points A and C. (a) Derive a formula for R_x in terms of the observable resistances R_1 and R_2. (b) A satisfactory ground resistance would be $R_x < 2.0$ Ω. Is the grounding of the station adequate if measurements give $R_1 = 13$ Ω and $R_2 = 6.0$ Ω?

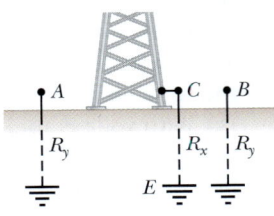

Figure P18.55

56. The resistor R in Figure P18.56 dissipates 20 W of power. Determine the value of R.

Figure P18.56

57. A voltage ΔV is applied to a series configuration of n resistors, each of resistance R. The circuit components are reconnected in a parallel configuration, and voltage ΔV is again applied. Show that the power consumed by the series configuration is $1/n^2$ times the power consumed by the parallel configuration.

58. For the network in Figure P18.58, show that the resistance between points a and b is $R_{ab} = \frac{27}{17}$ Ω. (*Hint:* Connect a battery with emf $\mathcal{E}$ across points a and b and determine $\mathcal{E}/I$, where I is the current in the battery.)

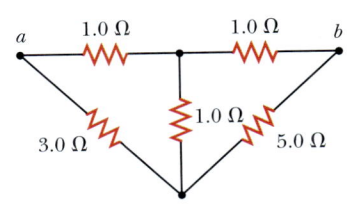

Figure P18.58

59. A battery with an internal resistance of 10.0 Ω produces an open circuit voltage of 12.0 V. A variable load resistance with a range from 0 to 30.0 Ω is connected across the battery. (*Note:* A battery has a resistance that depends on the condition of its chemicals and that increases as the battery ages. This internal resistance can be represented in a simple circuit diagram as a resistor in series with the battery.) (a) Graph the power dissipated in the load resistor as a function of the load resistance. (b) With your graph, demonstrate the following important theorem: *The power delivered to a load is a maximum if the load resistance equals the internal resistance of the source.*

60. The circuit in Figure P18.60 contains two resistors, $R_1 = 2.0$ kΩ and $R_2 = 3.0$ kΩ, and two capacitors, $C_1 = 2.0$ μF and $C_2 = 3.0$ μF, connected to a battery with emf $\mathcal{E} = 120$ V. If there are no charges on the capacitors before switch S is closed, determine the charges q_1 and q_2 on capacitors C_1 and C_2, respectively, as functions of time, after the switch is closed. *Hint:* First reconstruct the circuit so that it becomes a simple *RC* circuit containing a single resistor and single capacitor in series, connected to the battery, and then determine the total charge q stored in the circuit.

Figure P18.60

61. **BIO** An electric eel generates electric currents through its highly specialized Hunter's organ, in which thousands of disk-shaped cells called electrocytes are lined up in series, very much in the same way batteries are lined up inside a flashlight. When activated, each electrocyte can maintain a potential difference of about 150 mV at a current of 1 A for about 2.0 ms. Suppose a grown electric eel has 4.0×10^3 electrocytes and can deliver up to 300 shocks in rapid series over about 1 s. (a) What maximum electrical power can an electric eel generate? (b) Approximately how much energy does it release in one shock? (c) How high would a mass of 1 kg have to be lifted so that its gravitational potential energy equals the energy released in 300 such shocks?

62. In Figure P18.62, $R_1 = 0.100$ Ω, $R_2 = 1.00$ Ω, and $R_3 = 10.0$ Ω. Find the equivalent resistance of the circuit and the current in each resistor when a 5.00-V power supply is connected between (a) points *A* and *B*, (b) points *A* and *C*, and (c) points *A* and *D*.

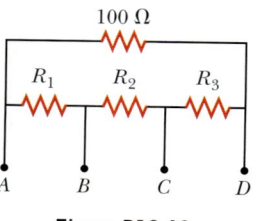

Figure P18.62

63. **M** What are the expected readings of the ammeter and voltmeter for the circuit in Figure P18.63?

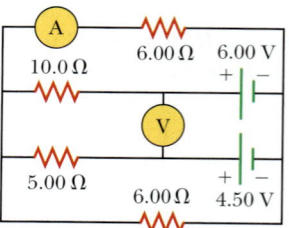

Figure P18.63

64. Consider the two arrangements of batteries and bulbs shown in Figure P18.64. The two bulbs are identical and have resistance *R*, and the two batteries are identical with output voltage Δ*V*. (a) In case 1, with the two bulbs in series, compare the brightness of each bulb, the current in each bulb, and the power delivered to each bulb. (b) In case 2, with the two bulbs in parallel, compare the brightness of each bulb, the current in each bulb, and the power supplied to each bulb. (c) Which bulbs are brighter, those in case 1 or those in case 2? (d) In each case, if one bulb fails, will the other go out as well? If the other bulb doesn't fail, will it get brighter or stay the same? (Problem 64 is courtesy of E. F. Redish. For other problems of this type, visit http://www.physics.umd.edu/perg/.)

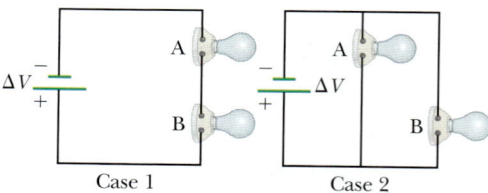

Figure P18.64

65. The given pair of capacitors in Figure P18.65 is fully charged by a 12.0-V battery. The battery is disconnected and the circuit closed. After 1.00 ms, how much charge remains on (a) the 3.00-μF capacitor? (b) The 2.00-μF capacitor? (c) What is the current in the resistor?

Figure P18.65

66. A 2.00-nF capacitor with an initial charge of 5.10 μC is discharged through a 1.30-kΩ resistor. (a) Calculate the magnitude of the current in the resistor 9.00 μs after the resistor is connected across the terminals of the capacitor. (b) What charge remains on the capacitor after 8.00 μs? (c) What is the maximum current in the resistor?

Aurora borealis, the northern lights. Displays such as this one are caused by cosmic ray particles trapped in Earth's magnetic field. When the particles collide with atoms in the atmosphere, they cause the atoms to emit visible light.

Magnetism 19

In terms of applications, magnetism is one of the most important fields in physics. Large electromagnets are used to pick up heavy loads. Magnets are used in such devices as meters, motors, and loudspeakers. Magnetic tapes and disks are used routinely in sound- and video-recording equipment and to store computer data. Intense magnetic fields are used in magnetic resonance imaging (MRI) devices to explore the human body with better resolution and greater safety than x-rays can provide. Giant superconducting magnets are used in the cyclotrons that guide particles into targets at nearly the speed of light. Rail guns (Fig. 19.1) use magnetic forces to fire high-speed projectiles, and magnetic bottles hold antimatter, a possible key to future space propulsion systems.

Magnetism is closely linked with electricity. Magnetic fields affect moving charges, and moving charges produce magnetic fields. Changing magnetic fields can even create electric fields. These phenomena signify an underlying unity of electricity and magnetism, which James Clerk Maxwell first described in the 19th century. The ultimate source of any magnetic field is electric current.

- 19.1 Magnets
- 19.2 Earth's Magnetic Field
- 19.3 Magnetic Fields
- 19.4 Magnetic Force on a Current-Carrying Conductor
- 19.5 Torque on a Current Loop and Electric Motors
- 19.6 Motion of a Charged Particle in a Magnetic Field
- 19.7 Magnetic Field of a Long, Straight Wire and Ampère's Law
- 19.8 Magnetic Force Between Two Parallel Conductors
- 19.9 Magnetic Fields of Current Loops and Solenoids
- 19.10 Magnetic Domains

19.1 Magnets

LEARNING OBJECTIVES

1. Discuss the basic properties of magnets and magnetic fields, contrasting them with electric charges and electric fields.
2. Characterize magnetic materials as hard or soft by the extent to which they retain their magnetism. Give examples of each kind.
3. Identify the origin of magnetic fields in moving electric charges. State some applications of magnets and their fields.

Figure 19.1 Rail guns launch projectiles at high speed using magnetic force. Larger versions could send payloads into space or be used as thrusters to move metal-rich asteroids from deep space to Earth orbit. In this photo a rail gun at Sandia National Laboratories in Albuquerque, New Mexico, fires a projectile at over three kilometers per second. (For more information on this, see Applying Physics 20.2 on pages 709–710.)

Most people have had experience with some form of magnet. You are most likely familiar with the common iron horseshoe magnet that can pick up iron-containing objects such as paper clips and nails. In the discussion that follows, we assume the magnet has the shape of a bar. Iron objects are most strongly attracted to either end of such a bar magnet, called its **poles**. One end is called the **north pole** and the other the **south pole**. The names come from the behavior of a magnet in the presence of Earth's magnetic field. If a bar magnet is suspended from its midpoint by a piece of string so that it can swing freely in a horizontal plane, it will rotate until its north pole points to the north and its south pole points to the south. The same idea is used to construct a simple compass. Magnetic poles also exert attractive or repulsive forces on each other similar to the electrical forces between charged objects. In fact, simple experiments with two bar magnets show that **like poles repel each other and unlike poles attract each other**.

Although the force between opposite magnetic poles is similar to the force between positive and negative electric charges, there is an important difference: positive and negative electric charges can exist in isolation of each other; north and south poles don't. No matter how many times a permanent magnet is cut, each piece always has a north pole and a south pole. There is some theoretical basis, however, for the speculation that magnetic monopoles (isolated north or south poles) exist in nature, and the attempt to detect them is currently an active experimental field of investigation.

An unmagnetized piece of iron can be magnetized by stroking it with a magnet. Magnetism can also be induced in iron (and other materials) by other means. For example, if a piece of unmagnetized iron is placed near a strong permanent magnet, the piece of iron eventually becomes magnetized. The process can be accelerated by heating and then cooling the iron.

Naturally occurring magnetic materials such as magnetite are magnetized in this way because they have been subjected to Earth's magnetic field for long periods of time. The extent to which a piece of material retains its magnetism depends on whether it is classified as magnetically hard or soft. **Soft** magnetic materials, such as iron, are easily magnetized but tend to lose their magnetization easily. These materials are used in the cores of transformers, generators, and motors. Iron is the most common choice because it's inexpensive. Other magnetically soft materials include nickel, nickel-iron alloys, and ferrites. Ferrites are combinations of a divalent metal oxide of nickel or magnesium with ferric oxide. Ferrites are used in high-frequency applications, such as radar.

Hard magnetic materials are used in permanent magnets. Such magnets provide magnetic fields without the use of electricity. Permanent magnets are used in many devices, including loudspeakers, permanent-magnet motors, and the read/write heads of computer hard drives. There are a large number of different materials used in permanent magnets. Alnico is a generic name for various alloys of iron, cobalt, and nickel, together with smaller amounts of aluminum, copper, or other elements. Rare earths such as samarium and neodymium are also used in conjunction with other elements to make strong permanent magnets.

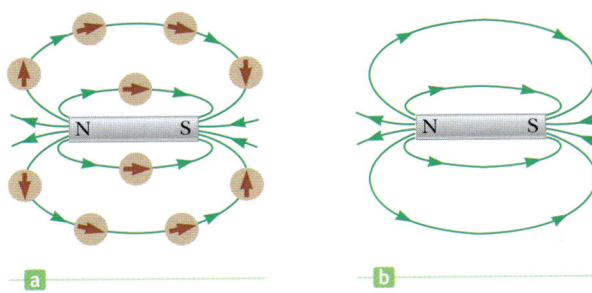

Figure 19.2 (a) Tracing the magnetic field of a bar magnet with compasses. (b) Several magnetic field lines of a bar magnet.

In earlier chapters we described the interaction between charged objects in terms of electric fields. Recall that an electric field surrounds any stationary electric charge. The region of space surrounding a *moving* charge includes a magnetic field as well. A magnetic field also surrounds a properly magnetized magnetic material.

To describe any type of vector field, we must define its magnitude, or strength, and its direction. The direction of a magnetic field $\vec{B}$ at any location is the direction in which the north pole of a compass needle points at that location. Figure 19.2a shows how the magnetic field of a bar magnet can be traced with the aid of a compass, defining a **magnetic field line**. Several magnetic field lines of a bar magnet traced out in this way appear in the two-dimensional representation in Figure 19.2b. Magnetic field patterns can be displayed by placing small iron filings in the vicinity of a magnet, as in Figure 19.3.

Forensic scientists use a technique similar to that shown in Figure 19.3 to find fingerprints at a crime scene. One way to find latent, or invisible, prints is by sprinkling a powder of iron dust on a surface. The iron adheres to any perspiration or body oils that are present and can be spread around on the surface with a magnetic brush that never comes into contact with the powder or the surface.

APPLICATION
Dusting for Fingerprints

19.2 Earth's Magnetic Field

LEARNING OBJECTIVE

1. Describe the Earth's magnetic field geographically, and discuss its possible origins and the ability of some life to sense and react to the field.

A small bar magnet is said to have north and south poles, but it's more accurate to say it has a "north-seeking" pole and a "south-seeking" pole. By these expressions, we mean that if such a magnet is used as a compass, one end will "seek," or point to, the geographic North Pole of Earth and the other end will "seek," or point to, the geographic South Pole of Earth. We conclude that **the geographic North Pole of Earth corresponds to a magnetic south pole, and the geographic**

Tip 19.1 The Geographic North Pole Is the Magnetic South Pole

The north pole of a magnet in a compass points north because it's attracted to Earth's *magnetic* south pole, located near Earth's *geographic* north pole.

Figure 19.3 (a) The magnetic field pattern of a bar magnet, as displayed with iron filings on a sheet of paper. (b) The magnetic field pattern between *unlike* poles of two bar magnets, as displayed with iron filings. (c) The magnetic field pattern between two *like* poles.

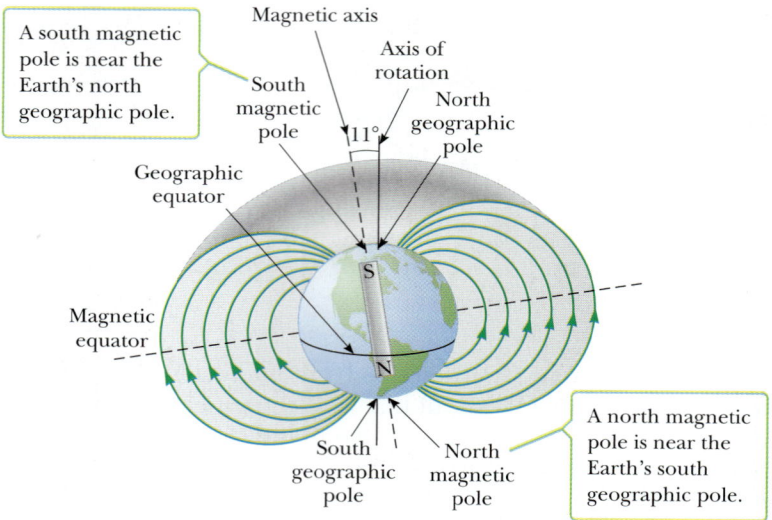

Figure 19.4 Earth's magnetic field lines. The lines leading away from the immediate vicinity of the north magnetic pole and entering the vicinity of the south magnetic pole have been left out for clarity.

Figure 19.5 A map of the continental United States showing the declination of a compass from true north.

South Pole of Earth corresponds to a magnetic north pole. In fact, the configuration of Earth's magnetic field, pictured in Figure 19.4, very much resembles what would be observed if a huge bar magnet were buried deep in the Earth's interior.

If a compass needle is suspended in bearings that allow it to rotate in the vertical plane as well as in the horizontal plane, the needle is horizontal with respect to Earth's surface only near the equator. As the device is moved northward, the needle rotates so that it points more and more toward the surface of Earth. The angle between the direction of the magnetic field and the horizontal is called the **dip angle**. Finally, at a point just north of Hudson Bay in Canada, the north pole of the needle points directly downward, with a dip angle of 90°. That site, first found in 1832, is considered to be the location of the south magnetic pole of Earth. It is approximately 1 300 miles from Earth's geographic North Pole and varies with time. Similarly, Earth's magnetic north pole is about 1 200 miles from its geographic South Pole. This means that compass needles point only approximately north. The difference between true north, defined as the geographic North Pole, and north indicated by a compass varies from point to point on Earth, a difference referred to as *magnetic declination*. For example, along a line through South Carolina and the Great Lakes a compass indicates true north, whereas in Washington state it aligns 25° east of true north (Fig. 19.5).

Although the magnetic field pattern of Earth is similar to the pattern that would be set up by a bar magnet placed at its center, the source of Earth's field can't consist of large masses of permanently magnetized material. Earth does have large deposits of iron ore deep beneath its surface, but the high temperatures in the core prevent the iron from retaining any permanent magnetization. It's considered more likely that the true source of Earth's magnetic field is electric current in the liquid part of its core. This current, which is not well understood, may be driven by an interaction between the planet's rotation and convection in the hot liquid core. There is some evidence that the strength of a planet's magnetic field is related to the planet's rate of rotation. For example, Jupiter rotates faster than Earth, and recent space probes indicate that Jupiter's magnetic field is stronger than Earth's, even though Jupiter lacks an iron core. Venus, on the other hand, rotates more slowly than Earth, and its magnetic field is weaker. Investigation into the cause of Earth's magnetism continues.

An interesting fact concerning Earth's magnetic field is that its direction reverses every few million years. Evidence for this phenomenon is provided by basalt (an iron-containing rock) that is sometimes spewed forth by volcanic activity on the ocean floor. As the lava cools, it solidifies and retains a picture of the direction of Earth's magnetic field. When the basalt deposits are dated, they provide evidence for periodic reversals of the magnetic field. The cause of these field reversals is still not understood.

It has long been speculated that some animals, such as birds, use the magnetic field of Earth to guide their migrations. Studies have shown that a type of anaerobic bacterium that lives in swamps has a magnetized chain of magnetite as part of its internal structure. (The term *anaerobic* means that these bacteria live and grow without oxygen; in fact, oxygen is toxic to them.) The magnetized chain acts as a compass needle that enables the bacteria to align with Earth's magnetic field. When they find themselves out of the mud on the bottom of the swamp, they return to their oxygen-free environment by following the magnetic field lines of Earth. Further evidence for their magnetic sensing ability is that bacteria found in the northern hemisphere have internal magnetite chains that are opposite in polarity to those of similar bacteria in the southern hemisphere. Similarly, in the northern hemisphere, Earth's field has a downward component, whereas in the southern hemisphere it has an upward component. Recently, a meteorite originating on Mars has been found to contain a chain of magnetite. NASA scientists believe it may be a fossil of ancient Martian bacterial life.

BIO APPLICATION
Magnetic Bacteria

The magnetic field of Earth is used to label runways at airports according to their direction. A large number is painted on the end of the runway so that it can be read by the pilot of an incoming airplane. This number describes the direction in which the airplane is traveling, expressed as the magnetic heading, in degrees measured clockwise from magnetic north divided by 10. A runway marked 9 would be directed toward the east (90° divided by 10), whereas a runway marked 18 would be directed toward magnetic south.

APPLICATION
Labeling Airport Runways

APPLYING PHYSICS 19.1 | Compasses Down Under

On a business trip to Australia, you take along your American-made compass that you may have used on a camping trip. Does this compass work correctly in Australia?

EXPLANATION There's no problem with using the compass in Australia. The north pole of the magnet in the compass will be attracted to the south magnetic pole near the geographic North Pole, just as it was in the United States. The only difference in the magnetic field lines is that they have an upward component in Australia, whereas they have a downward component in the United States. Held in a horizontal plane, your compass can't detect this difference; it only displays the direction of the horizontal component of the magnetic field. ■

19.3 Magnetic Fields

LEARNING OBJECTIVES

1. Define the magnetic force on a charged particle in terms of experimental observations.
2. Master the right hand rule that gives the direction of the magnetic force on a charged particle.
3. Apply the magnetic force in physical contexts.

Experiments show that a stationary charged particle doesn't interact with a static magnetic field. **When a charged particle is *moving* through a magnetic field, however, a magnetic force acts on it**. This force has its maximum value when the charge moves in a direction perpendicular to the magnetic field lines, decreases in value at other angles, and becomes zero when the particle moves along the field lines. This is quite different from the electric force, which exerts a force on a charged particle whether it's moving or at rest. Further, the electric force is directed parallel to the electric field whereas the magnetic force on a moving charge is directed perpendicular to the magnetic field.

In our discussion of electricity, the electric field at some point in space was defined as the electric force per unit charge acting on some test charge placed at that point. In a similar way, we can describe the properties of the magnetic field $\vec{B}$ at some point in terms of the magnetic force exerted on a test charge at that point. Our test object is a charge q moving with velocity $\vec{v}$. It is found experimentally that the strength of the magnetic force on the particle is proportional to the magnitude of the charge q, the magnitude of the velocity $\vec{v}$, the strength of the external magnetic field $\vec{B}$, and the sine of the angle θ between the direction of $\vec{v}$ and the direction of $\vec{B}$. These observations can be summarized by writing the magnitude of the magnetic force as

$$F = qvB \sin \theta \qquad [19.1]$$

This expression is used to define the magnitude of the magnetic field as

$$B \equiv \frac{F}{qv \sin \theta} \qquad [19.2]$$

If F is in newtons, q in coulombs, and v in meters per second, the SI unit of magnetic field is the **tesla** (T), also called the **weber** (Wb) **per square meter** (1 T = 1 Wb/m^2). If a 1-C charge moves in a direction perpendicular to a magnetic field of magnitude 1 T with a speed of 1 m/s, the magnetic force exerted on the charge is 1 N. We can express the units of $\vec{B}$ as

$$[B] = T = \frac{Wb}{m^2} = \frac{N}{C \cdot m/s} = \frac{N}{A \cdot m} \qquad [19.3]$$

In practice, the cgs unit for magnetic field, the **gauss** (G), is often used. The gauss is related to the tesla through the conversion

$$1 \text{ T} = 10^4 \text{ G}$$

Conventional laboratory magnets can produce magnetic fields as large as about 25 000 G, or 2.5 T. Superconducting magnets that can generate magnetic fields as great as 3×10^5 G, or 30 T, have been constructed. These values can be compared with the small value of Earth's magnetic field near its surface, which is only about 0.5 G, or 0.5×10^{-4} T.

From Equation 19.1 we see that the force on a charged particle moving in a magnetic field has its maximum value when the particle's motion is *perpendicular* to the magnetic field, corresponding to $\theta = 90°$, so that $\sin \theta = 1$. The magnitude of this maximum force has the value

$$F_{max} = qvB \qquad [19.4]$$

Also from Equation 19.1, F is zero when $\vec{v}$ is parallel to $\vec{B}$ (corresponding to $\theta = 0°$ or $180°$), so no magnetic force is exerted on a charged particle when it moves in the direction of the magnetic field or opposite the field.

Experiments show that the direction of the magnetic force is always perpendicular to both $\vec{v}$ and $\vec{B}$, as shown in Figure 19.6 for a positively charged particle. To determine the direction of the force, we employ **right-hand rule number 1**:

1. Point the fingers of your right hand in the direction of the velocity $\vec{v}$.
2. Curl the fingers in the direction of the magnetic field $\vec{B}$, moving through the smallest angle (as in Fig. 19.7).
3. Your thumb is now pointing in the direction of the magnetic force $\vec{F}$ exerted on a positive charge.

19.3 | Magnetic Fields

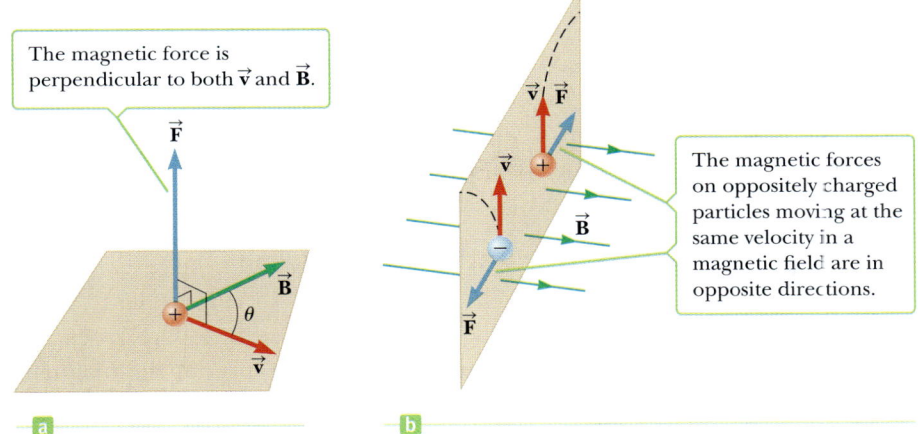

Figure 19.6 (a) The direction of the magnetic force $\vec{F}$ acting on a charged particle moving with a velocity $\vec{v}$ in the presence of a magnetic field $\vec{B}$. (b) Magnetic forces on positive and negative charges. The dashed lines show the paths of the particles, which are investigated in Section 19.6.

If the charge is negative rather than positive, the force $\vec{F}$ is directed *opposite* to what's shown in Figures 19.6a and 19.7. So if q is negative, simply use the right-hand rule to find the direction for positive q and then reverse that direction for the negative charge. Figure 19.6b illustrates the effect of a magnetic field on charged particles with opposite signs.

■ Quick Quiz

19.1 A charged particle moves in a straight line through a region of space. Which of the following answers *must* be true? (Assume any other fields are negligible.) The magnetic field (a) has a magnitude of zero (b) has a zero component perpendicular to the particle's velocity (c) has a zero component parallel to the particle's velocity in that region.

19.2 The north-pole end of a bar magnet is held near a stationary positively charged piece of plastic. Is the plastic (a) attracted, (b) repelled, or (c) unaffected by the magnet?

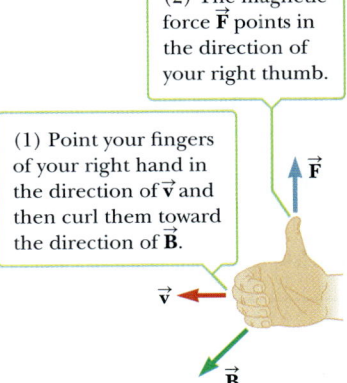

Figure 19.7 Right-hand rule number 1 for determining the direction of the magnetic force on a positive charge moving with a velocity $\vec{v}$ in a magnetic field $\vec{B}$.

■ EXAMPLE 19.1 | A Proton Traveling in Earth's Magnetic Field

GOAL Calculate the magnitude and direction of a magnetic force.

PROBLEM A proton moves with a speed of 1.00×10^5 m/s through Earth's magnetic field, which has a value of 55.0 μT at a particular location. When the proton moves eastward, the magnetic force acting on it is directed straight upward, and when it moves northward, no magnetic force acts on it. (a) What is the direction of the magnetic field, and (b) what is the strength of the magnetic force when the proton moves eastward? (c) Calculate the gravitational force on the proton and compare it with the magnetic force. Compare it also with the electric force if there were an electric field with a magnitude equal to $E = 1.50 \times 10^2$ N/C at that location, a common value at Earth's surface. Note that the mass of the proton is 1.67×10^{-27} kg.

STRATEGY The direction of the magnetic field can be found from an application of the right-hand rule, together with the fact that no force is exerted on the proton when it's traveling north. Substituting into Equation 19.1 yields the magnitude of the magnetic field.

SOLUTION

(a) Find the direction of the magnetic field.

No magnetic force acts on the proton when it's going north, so the angle such a proton makes with the magnetic field direction must be either 0° or 180°. Therefore, the magnetic field $\vec{B}$ must point either north or south. Now apply the right-hand rule. When the particle travels east, the magnetic force is directed upward. Point your thumb in the direction of the force and your fingers in the direction of the velocity eastward. When you curl your fingers, they point north, which must therefore be the direction of the magnetic field.

(Continued)

(b) Find the magnitude of the magnetic force.

Substitute the given values and the charge of a proton into Equation 19.1. From part (a), the angle between the velocity $\vec{v}$ of the proton and the magnetic field $\vec{B}$ is 90.0°.

$$F = qvB \sin \theta$$
$$= (1.60 \times 10^{-19} \text{ C})(1.00 \times 10^5 \text{ m/s}) \times (55.0 \times 10^{-6} \text{ T}) \sin (90.0°)$$
$$= 8.80 \times 10^{-19} \text{ N}$$

(c) Calculate the gravitational force on the proton and compare it with the magnetic force and also with the electric force if $E = 1.50 \times 10^2$ N/C.

$$F_{\text{grav}} = mg = (1.67 \times 10^{-27} \text{ kg})(9.80 \text{ m/s}^2)$$
$$= 1.64 \times 10^{-26} \text{ N}$$
$$F_{\text{elec}} = qE = (1.60 \times 10^{-19} \text{ C})(1.50 \times 10^2 \text{ N/C})$$
$$= 2.40 \times 10^{-17} \text{ N}$$

REMARKS The information regarding a proton moving north was necessary to fix the direction of the magnetic field. Otherwise, an upward magnetic force on an eastward-moving proton could be caused by a magnetic field pointing anywhere northeast or northwest. Notice in part (c) the relative strengths of the forces, with the electric force larger than the magnetic force and both much larger than the gravitational force, all for typical field values found in nature.

QUESTION 19.1 An electron and proton moving with the same velocity enter a uniform magnetic field. In what two ways does the magnetic field affect the electron differently from the proton?

EXERCISE 19.1 Suppose an electron is moving due west in the same magnetic field as in Example 19.1 at a speed of 2.50×10^5 m/s. Find the magnitude and direction of the magnetic force on the electron.

ANSWER 2.20×10^{-18} N, straight up. (Don't forget, the electron is negatively charged!)

EXAMPLE 19.2 | A Proton Moving in a Magnetic Field

GOAL Calculate the magnetic force and acceleration when a particle moves at an angle other than 90° to the field.

PROBLEM A proton moves at 8.00×10^6 m/s along the x-axis. It enters a region in which there is a magnetic field of magnitude 2.50 T, directed at an angle of 60.0° with the x-axis and lying in the xy-plane (Fig. 19.8). **(a)** Find the initial magnitude and direction of the magnetic force on the proton. **(b)** Calculate the proton's initial acceleration.

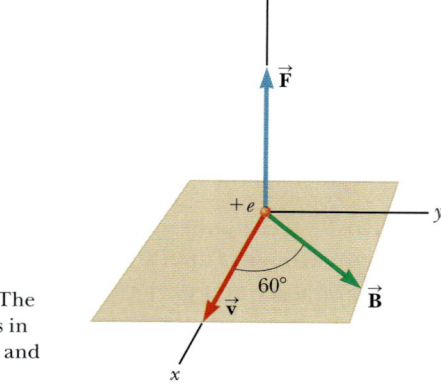

Figure 19.8 (Example 19.2) The magnetic force $\vec{F}$ on a proton is in the positive z-direction when $\vec{v}$ and $\vec{B}$ lie in the xy-plane.

STRATEGY Finding the magnitude and direction of the magnetic force requires substituting values into the equation for magnetic force, Equation 19.1, and using the right-hand rule. Applying Newton's second law solves part (b).

SOLUTION

(a) Find the magnitude and direction of the magnetic force on the proton.

Substitute $v = 8.00 \times 10^6$ m/s, the magnetic field strength $B = 2.50$ T, the angle, and the charge of a proton into Equation 19.1:

$$F = qvB \sin \theta$$
$$= (1.60 \times 10^{-19} \text{ C})(8.00 \times 10^6 \text{ m/s})(2.50 \text{ T})(\sin 60°)$$
$$F = 2.77 \times 10^{-12} \text{ N}$$

Apply right-hand rule number 1 to find the initial direction of the magnetic force:

Point the fingers of the right hand in the x-direction (the direction of $\vec{v}$) and then curl them toward $\vec{B}$. The thumb points upward, in the positive z-direction.

(b) Calculate the proton's initial acceleration.

Substitute the force and the mass of a proton into Newton's second law:

$$ma = F \quad \rightarrow \quad (1.67 \times 10^{-27} \text{ kg})a = 2.77 \times 10^{-12} \text{ N}$$

$$a = \boxed{1.66 \times 10^{15} \text{ m/s}^2}$$

REMARKS The initial acceleration is also in the positive z-direction. Because the direction of $\vec{v}$ changes, however, the subsequent direction of the magnetic force also changes. In applying right-hand rule number 1 to find the direction, it was important to take into consideration the charge. A negatively charged particle accelerates in the opposite direction.

QUESTION 19.2 Can a constant magnetic field change the speed of a charged particle? Explain.

EXERCISE 19.2 Calculate the acceleration of an electron that moves through the same magnetic field as in Example 19.2, at the same velocity as the proton. The mass of an electron is 9.11×10^{-31} kg.

ANSWER 3.04×10^{18} m/s² in the negative z-direction

19.4 Magnetic Force on a Current-Carrying Conductor

LEARNING OBJECTIVES
1. Derive the magnetic force exerted by a magnetic field on a long, current-carrying wire.
2. Calculate magnetic forces on current-carrying wires.

If a magnetic field exerts a force on a single charged particle when it moves through a magnetic field, it should be no surprise that magnetic forces are exerted on a current-carrying wire as well (see Fig. 19.9). Because the current is a collection of many charged particles in motion, the resultant force on the wire is due to the sum of the individual forces on the charged particles. The force on the particles is transmitted to the "bulk" of the wire through collisions with the atoms making up the wire.

Some explanation is in order concerning notation in many of the figures. To indicate the direction of $\vec{B}$, we use the following conventions:

> If $\vec{B}$ is directed into the page, as in Figure 19.10, we use a series of green crosses, representing the tails of arrows. If $\vec{B}$ is directed out of the page, we use a series of green dots, representing the tips of arrows. If $\vec{B}$ lies in the plane of the page, we use a series of green field lines with arrowheads.

Figure 19.9 This apparatus demonstrates the force on a current-carrying conductor in an external magnetic field. Why does the bar swing *toward* the magnet after the switch is closed?

The force on a current-carrying conductor can be demonstrated by hanging a wire between the poles of a magnet, as in Figure 19.10 (page 668). In this figure, the magnetic field is directed into the page and covers the region within the shaded area. The wire deflects to the right or left when it carries a current.

We can quantify this discussion by considering a straight segment of wire of length ℓ and cross-sectional area A carrying current I in a uniform external magnetic field $\vec{B}$, as in Figure 19.11 (page 668). We assume that the magnetic field is perpendicular to the wire and is directed into the page. A force of magnitude $F_{max} = qv_dB$ is exerted on each charge carrier in the wire, where v_d is the drift velocity of the charge. To find the total force on the wire, we multiply the force on one charge carrier by the number of carriers in the segment. Because the volume of the segment is $A\ell$, the number of carriers is $nA\ell$, where n is the number of carriers per unit volume. Hence, the magnitude of the total magnetic force on the wire of length ℓ is as follows:

Total force = force on each charge carrier × total number of carriers

$$F_{max} = (qv_dB)(nA\ell)$$

Tip 19.2 The Origin of the Magnetic Force on a Wire
When a magnetic field is applied at some angle to a wire carrying a current, a magnetic force is exerted on each moving charge in the wire. The total magnetic force on the wire is the sum of all the magnetic forces on the individual charges producing the current.

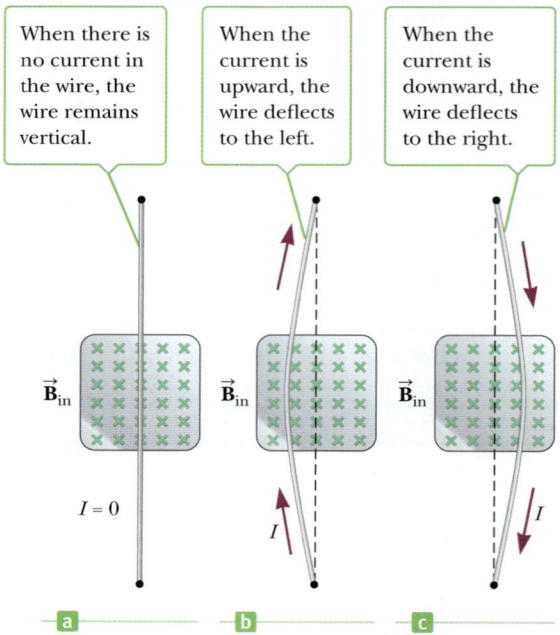

Figure 19.10 A segment of a flexible vertical wire partially stretched between the poles of a magnet, with the field (green crosses) directed into the page.

- When there is no current in the wire, the wire remains vertical.
- When the current is upward, the wire deflects to the left.
- When the current is downward, the wire deflects to the right.

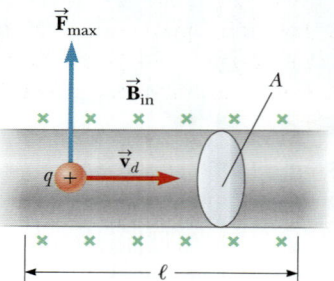

Figure 19.11 A section of a wire containing moving charges in an external magnetic field $\vec{B}$.

From Chapter 17, however, we know that the current in the wire is given by the expression $I = nqv_d A$, so

$$F_{max} = BI\ell \quad [19.5]$$

This equation can be used only when the current and the magnetic field are at right angles to each other.

If the wire is not perpendicular to the field but is at some arbitrary angle, as in Figure 19.12, the magnitude of the magnetic force on the wire is

$$F = BI\ell \sin\theta \quad [19.6]$$

where θ is the angle between $\vec{B}$ and the direction of the current. The direction of this force can be obtained by the use of right-hand rule number 1. In this case, however, you must place your fingers in the direction of the positive current I, rather than in the direction of $\vec{v}$, before curling them in the direction of $\vec{B}$. The thumb then points in the direction of the force, as before. The current, naturally, is made up of charges moving at some velocity, so this really isn't a separate rule. In Figure 19.12 the direction of the magnetic force on the wire is out of the page.

Finally, when the current is either in the direction of the field or opposite the direction of the field, the magnetic force on the wire is zero.

The way a magnetic force acts on a current-carrying wire in a magnetic field is the operating principle of most speakers in sound systems. One speaker design, shown in Figure 19.13, consists of a coil of wire called the voice coil, a flexible paper cone, and a permanent magnet. The coil of wire surrounding the north

APPLICATION
Loudspeaker Operation

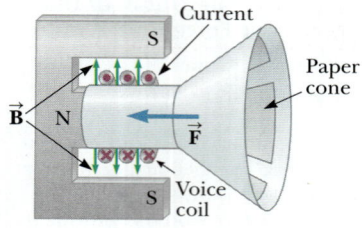

Figure 19.13 A diagram of a loudspeaker.

Figure 19.12 A wire carrying a current I in the presence of an external magnetic field $\vec{B}$ that makes an angle θ with the wire. The magnetic force vector comes out of the page.

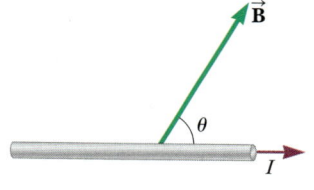

19.4 | Magnetic Force on a Current-Carrying Conductor

pole of the magnet is shaped so that the magnetic field lines are directed radially outward from the coil's axis. When an electrical signal is sent to the coil, producing a current in the coil as in Figure 19.13, a magnetic force to the left acts on the coil. (This can be seen by applying right-hand rule number 1 to each turn of wire.) When the current reverses direction, as it would for a current that varied sinusoidally, the magnetic force on the coil also reverses direction, and the cone, which is attached to the coil, accelerates to the right. An alternating current through the coil causes an alternating force on the coil, which results in vibrations of the cone. The vibrating cone creates sound waves as it pushes and pulls on the air in front of it. In this way, a 1-kHz electrical signal is converted to a 1-kHz sound wave.

An application of the force on a current-carrying conductor is illustrated by the electromagnetic pump shown in Figure 19.14. Artificial hearts require a pump to keep the blood flowing, and kidney dialysis machines also require a pump to assist the heart in pumping blood that is to be cleansed. Ordinary mechanical pumps create problems because they damage the blood cells as they move through the pump. The mechanism shown in the figure has demonstrated some promise in such applications. A magnetic field is established across a segment of the tube containing the blood, flowing in the direction of the velocity $\vec{v}$. An electric current passing through the fluid in the direction shown has a magnetic force acting on it in the direction of $\vec{v}$, as applying the right-hand rule shows. This force helps to keep the blood in motion.

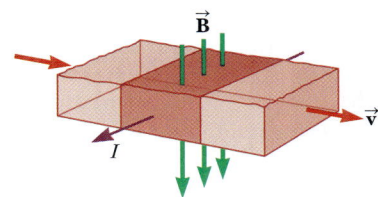

Figure 19.14 A simple electromagnetic pump has no moving parts that might damage a conducting fluid, such as blood passing through it. Application of right-hand rule number 1 (right fingers in the direction of the current I, curl them in the direction of $\vec{B}$, thumb points in the direction of the force) shows that the force on the current-carrying segment of the fluid is in the direction of the velocity.

BIO APPLICATION
Electromagnetic Pumps for Artificial Hearts and Kidneys

■ APPLYING PHYSICS 19.2 | Lightning Strikes

In a lightning strike there is a rapid movement of negative charge from a cloud to the ground. In what direction is a lightning strike deflected by Earth's magnetic field?

EXPLANATION The downward flow of negative charge in a lightning strike is equivalent to a current moving upward. Consequently, we have an upward-moving current in a northward-directed magnetic field. According to right-hand rule number 1, the lightning strike would be deflected toward the west. ■

■ EXAMPLE 19.3 | A Current-Carrying Wire in Earth's Magnetic Field

GOAL Compare the magnetic force on a current-carrying wire with the gravitational force exerted on the wire.

PROBLEM A wire carries a current of 22.0 A from west to east. Assume the magnetic field of Earth at this location is horizontal and directed from south to north and it has a magnitude of 0.500×10^{-4} T. **(a)** Find the magnitude and direction of the magnetic force on a 36.0-m length of wire. **(b)** Calculate the gravitational force on the same length of wire if it's made of copper and has a cross-sectional area of 2.50×10^{-6} m².

SOLUTION

(a) Calculate the magnetic force on the wire.

Substitute into Equation 19.6, using the fact that the magnetic field and the current are at right angles to each other:

$F = BI\ell \sin\theta = (0.500 \times 10^{-4} \text{ T})(22.0 \text{ A})(36.0 \text{ m}) \sin 90.0°$
$= 3.96 \times 10^{-2}$ N

Apply right-hand rule number 1 to find the direction of the magnetic force:

With the fingers of your right hand pointing west to east in the direction of the current, curl them north in the direction of the magnetic field. Your thumb points upward.

(b) Calculate the gravitational force on the wire segment.

First, obtain the mass of the wire from the density of copper, the length, and cross-sectional area of the wire:

$m = \rho V = \rho(A\ell)$
$= (8.92 \times 10^3 \text{ kg/m}^3)(2.50 \times 10^{-6} \text{ m}^2 \cdot 36.0 \text{ m})$
$= 0.803$ kg

(Continued)

To get the gravitational force, multiply the mass by the acceleration of gravity:

$$F_{grav} = mg = \boxed{7.87 \text{ N}}$$

REMARKS This calculation demonstrates that under normal circumstances, the gravitational force on a current-carrying conductor is much greater than the magnetic force due to Earth's magnetic field.

QUESTION 19.3 What magnetic force is exerted on a wire carrying current parallel to the direction of the magnetic field?

EXERCISE 19.3 What current would make the magnetic force in the example equal in magnitude to the gravitational force?

ANSWER 4.37×10^3 A, a large current that would very rapidly heat and melt the wire.

19.5 Torque on a Current Loop and Electric Motors

LEARNING OBJECTIVES

1. Derive the magnetic torque on a rectangular loop of current-carrying wire.
2. Define the magnetic moment of a current loop in a plane.
3. Calculate the magnetic torque on simple current loops.
4. Discuss the application of magnetic torque to electric motors.

In the preceding section we showed how a magnetic force is exerted on a current-carrying conductor when the conductor is placed in an external magnetic field. With this starting point, we now show that a torque is exerted on a current loop placed in a magnetic field. The results of this analysis will be of great practical value when we discuss generators and motors in Chapter 20.

Consider a rectangular loop carrying current I in the presence of an external uniform magnetic field in the plane of the loop, as shown in Figure 19.15a.

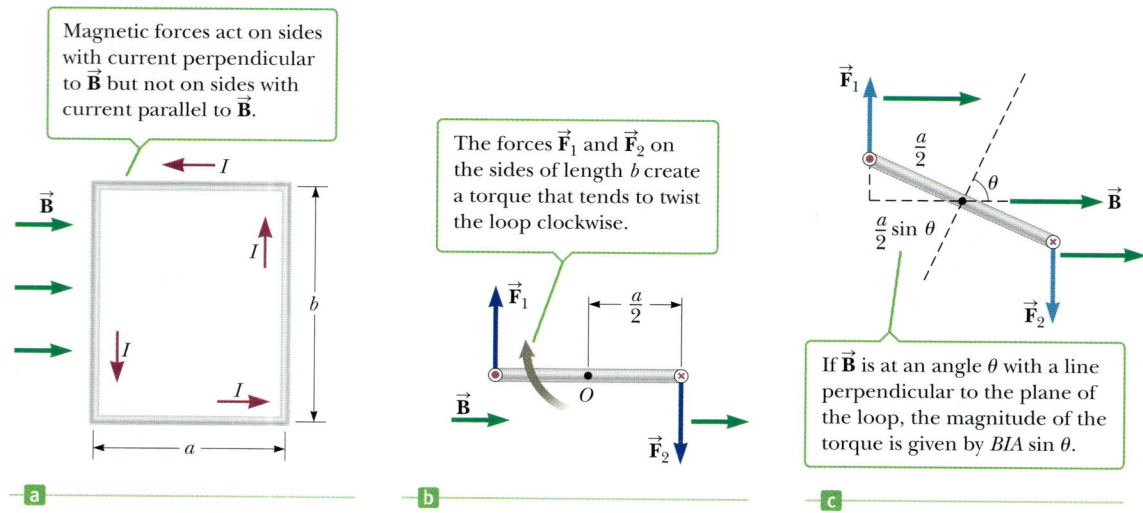

Figure 19.15 (a) Top view of a rectangular loop in a uniform magnetic field. (b) An edge view of the rectangular loop in part (a). (c) An edge view of the loop in part (a) with the normal to the loop at angle θ with respect to the magnetic field.

The forces on the sides of length a are zero because these wires are parallel to the field. The magnitudes of the magnetic forces on the sides of length b, however, are

$$F_1 = F_2 = BIb$$

The direction of $\vec{F}_1$, the force on the left side of the loop, is out of the page and that of $\vec{F}_2$, the force on the right side of the loop, is into the page. If we view the loop from the side, as in Figure 19.15b, the forces are directed as shown. If we assume the loop is pivoted so that it can rotate about point O, we see that these two forces produce a torque about O that rotates the loop clockwise. The magnitude of this torque, τ_{max}, is

$$\tau_{max} = F_1 \frac{a}{2} + F_2 \frac{a}{2} = (BIb)\frac{a}{2} + (BIb)\frac{a}{2} = BIab$$

where the moment arm about O is $a/2$ for both forces. Because the area of the loop is $A = ab$, the maximum torque can be expressed as

$$\tau_{max} = BIA \quad \text{[19.7]}$$

This result is valid only when the magnetic field is *parallel* to the plane of the loop, as in Figure 19.15b. If the field makes an angle θ with a line perpendicular to the plane of the loop, as in Figure 19.15c, the moment arm for each force is given by $(a/2) \sin \theta$. An analysis similar to the previous one gives, for the magnitude of the torque,

$$\tau = BIA \sin \theta \quad \text{[19.8]}$$

This result shows that the torque has the *maximum* value BIA when the field is parallel to the plane of the loop ($\theta = 90°$) and is *zero* when the field is perpendicular to the plane of the loop ($\theta = 0$). As seen in Figure 19.15c, the loop tends to rotate to smaller values of θ (so that the normal to the plane of the loop rotates toward the direction of the magnetic field).

Although the foregoing analysis was for a rectangular loop, a more general derivation shows that Equation 19.8 applies regardless of the shape of the loop. Further, the torque on a coil with N turns is

$$\tau = BIAN \sin \theta \quad \text{[19.9a]}$$

The quantity $\mu = IAN$ is defined as the magnitude of a vector $\vec{\mu}$ called the *magnetic moment* of the coil. The vector $\vec{\mu}$ always points perpendicular to the plane of the loop(s) and is such that if the thumb of the right hand points in the direction of $\vec{\mu}$, the fingers of the right hand point in the direction of the current. The angle θ in Equations 19.8 and 19.9 lies between the directions of the magnetic moment $\vec{\mu}$ and the magnetic field $\vec{B}$. The magnetic torque can then be written

$$\tau = \mu B \sin \theta \quad \text{[19.9b]}$$

Note that the torque $\vec{\tau}$ is always perpendicular to both the magnetic moment $\vec{\mu}$ and the magnetic field $\vec{B}$.

Quick Quiz

19.3 A square and a circular loop with the same area lie in the xy-plane, where there is a uniform magnetic field $\vec{B}$ pointing at some angle θ with respect to the positive z-direction. Each loop carries the same current, in the same direction. Which magnetic torque is larger? (a) the torque on the square loop (b) the torque on the circular loop (c) the torques are the same (d) more information is needed

EXAMPLE 19.4 | The Torque on a Circular Loop in a Magetic Field

GOAL Calculate a magnetic torque on a loop of current.

PROBLEM A circular wire loop of radius 1.00 m is placed in a magnetic field of magnitude 0.500 T. The normal to the plane of the loop makes an angle of 30.0° with the magnetic field (Fig. 19.16a). The current in the loop is 2.00 A in the direction shown. (a) Find the magnetic moment of the loop and the magnitude of the torque at this instant. (b) The same current is carried by the rectangular 2.00-m by 3.00-m coil with three loops shown in Figure 19.16b. Find the magnetic moment of the coil and the magnitude of the torque acting on the coil at that instant.

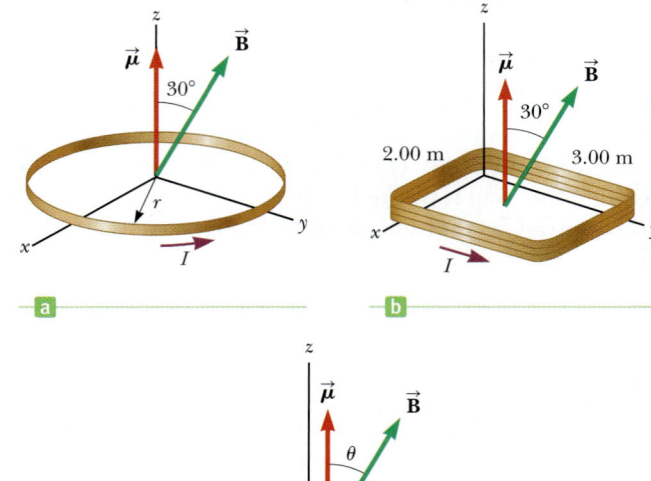

Figure 19.16 (Example 19.4) (a) A circular current loop lying in the xy-plane in an external magnetic field $\vec{B}$. (b) A rectangular coil lying in the xy-plane in the same field. (c) (Exercise 19.4)

STRATEGY For each part, we just have to calculate the area, use it in the calculation of the magnetic moment, and multiply the result by $B \sin \theta$. Altogether, this process amounts to substituting values into Equation 19.9b.

SOLUTION

(a) Find the magnetic moment of the circular loop and the magnetic torque exerted on it.

First, calculate the enclosed area of the circular loop:

$$A = \pi r^2 = \pi(1.00 \text{ m})^2 = 3.14 \text{ m}^2$$

Calculate the magnetic moment of the loop:

$$\mu = IAN = (2.00 \text{ A})(3.14 \text{ m}^2)(1) = \boxed{6.28 \text{ A} \cdot \text{m}^2}$$

Now substitute values for the magnetic moment, magnetic field, and θ into Equation 19.9b:

$$\tau = \mu B \sin \theta = (6.28 \text{ A} \cdot \text{m}^2)(0.500 \text{ T})(\sin 30.0°)$$
$$= \boxed{1.57 \text{ N} \cdot \text{m}}$$

(b) Find the magnetic moment of the rectangular coil and the magnetic torque exerted on it.

Calculate the area of the coil:

$$A = L \times H = (2.00 \text{ m})(3.00 \text{ m}) = 6.00 \text{ m}^2$$

Calculate the magnetic moment of the coil:

$$\mu = IAN = (2.00 \text{ A})(6.00 \text{ m}^2)(3) = \boxed{36.0 \text{ A} \cdot \text{m}^2)}$$

Substitute values into Equation 19.9b:

$$\tau = \mu B \sin \theta = (0.500 \text{ T})(36.0 \text{ A} \cdot \text{m}^2)(\sin 30.0°)$$
$$= \boxed{9.00 \text{ N} \cdot \text{m}}$$

REMARKS In calculating a magnetic torque, it's not strictly necessary to calculate the magnetic moment. Instead, Equation 19.9a can be used directly.

QUESTION 19.4 What happens to the magnitude of the torque if the angle increases toward 90°? Goes beyond 90°?

EXERCISE 19.4 Suppose a right triangular coil with base of 2.00 m and height 3.00 m having two loops carries a current of 2.00 A as shown in Figure 19.16c. Find the magnetic moment and the torque on the coil. The magnetic field is again 0.500 T and makes an angle of 30.0° with respect to the normal direction.

ANSWERS $\mu = 12.0 \text{ A} \cdot \text{m}^2$, $\tau = 3.00 \text{ N} \cdot \text{m}$

Electric Motors

It's hard to imagine life in the 21st century without electric motors. Some appliances that contain motors include computer disk drives, CD players, DVD players, food processors and blenders, car starters, furnaces, and air conditioners. The motors convert electrical energy to kinetic energy of rotation and consist of a rigid current-carrying loop that rotates when placed in a magnetic field.

As we have just seen (Fig. 19.15), the torque on such a loop rotates the loop to smaller values of θ until the torque becomes zero, where the magnetic field is perpendicular to the plane of the loop and θ = 0. If the loop turns past this angle and the current remains in the direction shown in the figure, the torque reverses direction and turns the loop in the opposite direction, that is, counterclockwise. To overcome this difficulty and provide continuous rotation in one direction, the current in the loop must periodically reverse direction. In alternating-current (AC) motors, such a reversal occurs naturally 120 times each second. In direct-current (DC) motors, the reversal is accomplished mechanically with split-ring contacts (commutators) and brushes, as shown in Figure 19.17.

Although actual motors contain many current loops and commutators, for simplicity Figure 19.17 shows only a single loop and a single set of split-ring contacts rigidly attached to and rotating with the loop. Electrical stationary contacts called *brushes* are maintained in electrical contact with the rotating split ring. These brushes are usually made of graphite because it is a good electrical conductor as well as a good lubricant. Just as the loop becomes perpendicular to the magnetic field and the torque becomes zero, inertia carries the loop forward in the clockwise direction and the brushes cross the gaps in the ring, causing the loop current to reverse its direction. This reversal provides another pulse of torque in the clockwise direction for another 180°, the current reverses, and the process repeats itself. Figure 19.18 shows a modern motor used to power a hybrid gas–electric car.

APPLICATION
Electric Motors

Figure 19.17 Simplified sketch of a DC electric motor.

19.6 Motion of a Charged Particle in a Magnetic Field

LEARNING OBJECTIVES
1. Describe the motion of a charged particle in a uniform magnetic field.
2. Apply the second law in polar coordinates to particles moving in a uniform magnetic field.

Consider the case of a positively charged particle moving in a uniform magnetic field so that the direction of the particle's velocity is *perpendicular to the field*, as in Figure 19.19 (page 674). The label $\vec{B}_{in}$ and the crosses indicate that $\vec{B}$ is directed into the page. Application of the right-hand rule to the particle at the bottom of the circle shows that the direction of the magnetic force $\vec{F}$ at that location is upward. The force causes the particle to alter its direction of travel and to follow a curved path. Application of the right-hand rule to the particle at other points on the circle shows that **the magnetic force is always directed toward the center of the circular path**; therefore, the magnetic force causes a centripetal acceleration, which changes only the direction of $\vec{v}$ and not its magnitude. Because $\vec{F}$ produces the centripetal acceleration, we can equate its magnitude, qvB in this case, to the mass of the particle multiplied by the centripetal acceleration v^2/r. From Newton's second law, we find that

$$F = qvB = \frac{mv^2}{r}$$

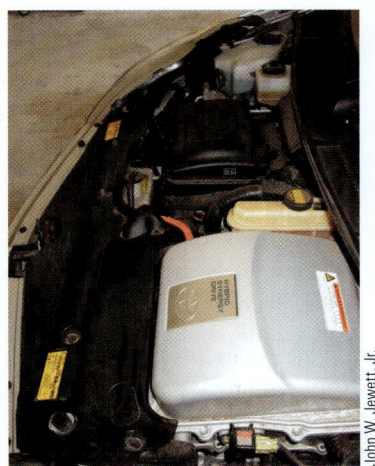

Figure 19.18 The engine compartment of the Toyota Prius, a hybrid vehicle.

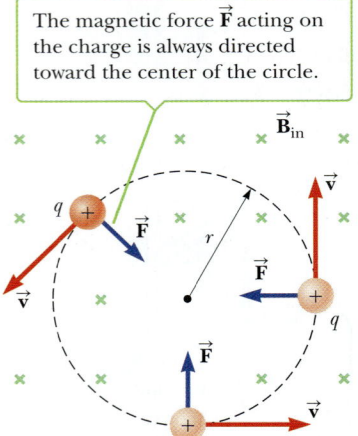

Figure 19.19 When the velocity of a charged particle is perpendicular to a uniform magnetic field, the particle moves in a circle in a plane perpendicular to $\vec{B}$, which is directed into the page. (The crosses represent the tails of the magnetic field vectors.)

Figure 19.20 A charged particle having a velocity directed at an angle with a uniform magnetic field moves in a helical path.

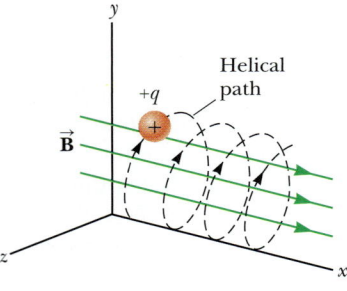

which gives

$$r = \frac{mv}{qB} \quad [19.10]$$

This equation says that the radius of the path is proportional to the momentum mv of the particle and is inversely proportional to the charge and the magnetic field. Equation 19.10 is often called the *cyclotron equation* because it's used in the design of these instruments (popularly known as atom smashers).

If the initial direction of the velocity of the charged particle is not perpendicular to the magnetic field, as shown in Figure 19.20, the path followed by the particle is a spiral (called a helix) along the magnetic field lines.

■ APPLYING PHYSICS 19.3 | Trapping Charges

Storing charged particles is important for a variety of applications. Suppose a uniform magnetic field exists in a finite region of space. Can a charged particle be injected into this region from the outside and remain trapped in the region by magnetic force alone?

EXPLANATION It's best to consider separately the components of the particle velocity parallel and perpendicular to the field lines in the region. There is no magnetic force on the particle associated with the velocity component parallel to the field lines, so that velocity component remains unchanged.

Now consider the component of velocity perpendicular to the field lines. This component will result in a magnetic force that is perpendicular to both the field lines and the velocity component itself. The path of a particle for which the force is always perpendicular to the velocity is a circle. The particle therefore follows a circular arc and exits the field on the other side of the circle, as shown in Figure 19.21 for a particle with constant kinetic energy. On the other hand, a particle can become trapped if it loses some kinetic energy in a collision after entering the field, so that it turns in a smaller circle and stays within the field.

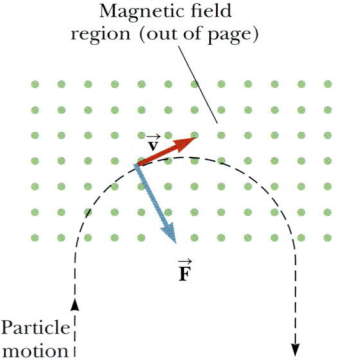

Figure 19.21 (Applying Physics 19.3)

Particles *can* be injected and contained if, in addition to the magnetic field, electrostatic fields are involved. These fields are used in the *Penning trap*. With these devices, it's possible to store charged particles for extended periods. Such traps are useful, for example, in the storage of antimatter, which disintegrates completely on contact with ordinary matter. ■

■ Quick Quiz

19.4 As a charged particle moves freely in a circular path in the presence of a constant magnetic field applied perpendicular to the particle's velocity, the particle's kinetic energy (a) remains constant, (b) increases, or (c) decreases.

19.6 | Motion of a Charged Particle in a Magnetic Field

EXAMPLE 19.5 | The Mass Spectrometer: Identifying Particles

GOAL Use the cyclotron equation to identify a particle.

PROBLEM A charged particle enters the magnetic field of a mass spectrometer at a speed of 1.79×10^6 m/s. It subsequently moves in a circular orbit with a radius of 16.0 cm in a uniform magnetic field of magnitude 0.350 T having a direction perpendicular to the particle's velocity. Find the particle's mass-to-charge ratio and identify it based on the table below.

STRATEGY After finding the mass-to-charge ratio with Equation 19.10, compare it with the values in the table, identifying the particle.

SOLUTION

Write the cyclotron equation:

$$r = \frac{mv}{qB}$$

Solve this equation for the mass divided by the charge, m/q, and substitute values:

$$\frac{m}{q} = \frac{rB}{v} = \frac{(0.160 \text{ m})(0.350 \text{ T})}{1.79 \times 10^6 \text{ m/s}} = 3.13 \times 10^{-8} \frac{\text{kg}}{\text{C}}$$

Identify the particle from the table. All particles are completely ionized, bare nuclei.

Nucleus	m (kg)	q (C)	m/q (kg/C)
Hydrogen	1.67×10^{-27}	1.60×10^{-19}	1.04×10^{-8}
Deuterium	3.35×10^{-27}	1.60×10^{-19}	2.09×10^{-8}
Tritium	5.01×10^{-27}	1.60×10^{-19}	3.13×10^{-8}
Helium-3	5.01×10^{-27}	3.20×10^{-19}	1.57×10^{-8}

The particle is a tritium nucleus.

REMARKS The mass spectrometer is an important tool in both chemistry and physics. Unknown chemicals can be heated and ionized, and the resulting particles passed through the mass spectrometer and subsequently identified.

QUESTION 19.5 What happens to the momentum of a charged particle in a uniform magnetic field?

EXERCISE 19.5 Suppose a second charged particle enters the mass spectrometer at the same speed as the particle in Example 19.5. If it travels in a circle with radius 10.7 cm, find the mass-to-charge ratio and identify the particle from the table above.

ANSWERS 2.09×10^{-8} kg/C; deuterium nucleus

EXAMPLE 19.6 | The Mass Spectrometer: Separating Isotopes

GOAL Apply the cyclotron equation to the process of separating isotopes.

APPLICATION Mass Spectrometers

PROBLEM Two singly ionized atoms move out of a slit at point S in Figure 19.22 and into a magnetic field of magnitude 0.100 T pointing into the page. Each has a speed of 1.00×10^6 m/s. The nucleus of the first atom contains one proton and has a mass of 1.67×10^{-27} kg, whereas the nucleus of the second atom contains a proton and a neutron and has a mass of 3.34×10^{-27} kg. Atoms with the same number of protons in the nucleus but different masses are called isotopes. The two isotopes here are hydrogen and deuterium. Find their distance of separation when they strike a photographic plate at P.

Figure 19.22 (Example 19.6) Two isotopes leave the slit at point S and travel in different circular paths before striking a photographic plate at P.

STRATEGY Apply the cyclotron equation to each atom, finding the radius of the path of each. Double the radii to find the path diameters and then find their difference.

(Continued)

SOLUTION

Use Equation 19.10 to find the radius of the circular path followed by the lighter isotope, hydrogen:

$$r_1 = \frac{m_1 v}{qB} = \frac{(1.67 \times 10^{-27} \text{ kg})(1.00 \times 10^6 \text{ m/s})}{(1.60 \times 10^{-19} \text{ C})(0.100 \text{ T})}$$
$$= 0.104 \text{ m}$$

Use the same equation to calculate the radius of the path of deuterium, the heavier isotope:

$$r_2 = \frac{m_2 v}{qB} = \frac{(3.34 \times 10^{-27} \text{ kg})(1.00 \times 10^6 \text{ m/s})}{(1.60 \times 10^{-19} \text{ C})(0.100 \text{ T})}$$
$$= 0.209 \text{ m}$$

Multiply the radii by 2 to find the diameters and take the difference, getting the separation x between the isotopes:

$$x = 2r_2 - 2r_1 = \boxed{0.210 \text{ m}}$$

REMARKS During World War II, mass spectrometers were used to separate the radioactive uranium isotope U-235 from its far more common isotope, U-238.

QUESTION 19.6 Estimate the radius of the circle traced out by a singly ionized lead atom moving at the same speed.

EXERCISE 19.6 Use the same mass spectrometer as in Example 19.6 to find the separation between two isotopes of helium: normal helium-4, which has a nucleus consisting of two protons and two neutrons, and helium-3, which has two protons and a single neutron. Assume both nuclei, doubly ionized (having a charge of $2e = 3.20 \times 10^{-19}$ C), enter the field at 1.00×10^6 m/s. The helium-4 nucleus has a mass of 6.64×10^{-27} kg, and the helium-3 nucleus has a mass equal to 5.01×10^{-27} kg.

ANSWER 0.102 m

19.7 Magnetic Field of a Long, Straight Wire and Ampère's Law

LEARNING OBJECTIVES

1. State the equation giving the magnitude of the magnetic field of a long straight wire and generalize it to Ampère's Law.
2. Apply Ampère's Law to determine the magnetic fields of wires and cylinders carrying current.

During a lecture demonstration in 1819, Danish scientist Hans Oersted (1777–1851) found that an electric current in a wire deflected a nearby compass needle. This momentous discovery, linking a magnetic field with an electric current for the first time, was the beginning of our understanding of the origin of magnetism.

A simple experiment first carried out by Oersted in 1820 clearly demonstrates that a current-carrying conductor produces a magnetic field. In this experiment, several compass needles are placed in a horizontal plane near a long vertical wire, as in Figure 19.23a. When there is no current in the wire, all needles point in the

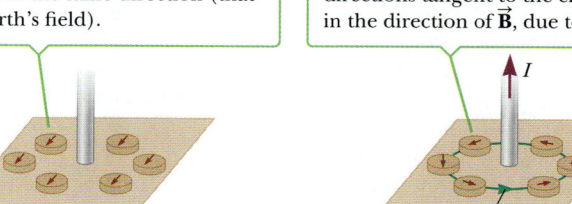

Figure 19.23 (a), (b) Compasses shown the effects of the current in a nearby wire.

When no current is present in the vertical wire, all compass needles point in the same direction (that of Earth's field).

When the wire carries a strong current, the compass needles deflect in directions tangent to the circle, pointing in the direction of $\vec{B}$, due to the current.

19.7 | Magnetic Field of a Long, Straight Wire and Ampère's Law

Figure 19.24 (a) Right-hand rule number 2 for determining the direction of the magnetic field due to a long, straight wire carrying a current. Note that the magnetic field lines form circles around the wire. (b) Circular magnetic field lines surrounding a current-carrying wire, displayed by iron filings.

same direction (that of Earth's field), as one would expect. When the wire carries a strong, steady current, however, the needles all deflect in directions tangent to the circle, as in Figure 19.23b. These observations show that the direction of $\vec{B}$ is consistent with the following convenient rule, **right-hand rule number 2**:

> Point the thumb of your right hand along a wire in the direction of positive current, as in Figure 19.24a. Your fingers then naturally curl in the direction of the magnetic field $\vec{B}$.

Tip 19.3 Raise Your Right Hand!
We have introduced two right-hand rules in this chapter. Be sure to use *only* your right hand when applying these rules.

When the current is reversed, the filings in Figure 19.24b also reverse.

Because the filings point in the direction of $\vec{B}$, we conclude that the lines of $\vec{B}$ form circles about the wire. By symmetry, the magnitude of $\vec{B}$ is the same everywhere on a circular path centered on the wire and lying in a plane perpendicular to the wire. By varying the current and distance from the wire, it can be experimentally determined that $\vec{B}$ is proportional to the current and inversely proportional to the distance from the wire. These observations lead to a mathematical expression for the strength of the magnetic field due to the current I in a long, straight wire:

$$B = \frac{\mu_0 I}{2\pi r} \quad [19.11]$$

◀ Magnetic field due to a long, straight wire

The proportionality constant μ_0, called the **permeability of free space**, has the value

$$\mu_0 \equiv 4\pi \times 10^{-7} \, \text{T} \cdot \text{m/A} \quad [19.12]$$

Ampère's Law and a Long, Straight Wire

Equation 19.11 enables us to calculate the magnetic field due to a long, straight wire carrying a current. A general procedure for deriving such equations was proposed by French scientist André-Marie Ampère (1775–1836); it provides a relation between the current in an arbitrarily shaped wire and the magnetic field produced by the wire.

Consider an arbitrary closed path surrounding a current as in Figure 19.25 (page 678). The path consists of many short segments, each of length $\Delta \ell$. Multiply one of these lengths by the component of the magnetic field parallel to that segment, where the product is labeled $B_{\parallel} \Delta \ell$. According to Ampère, the sum of all such products over the closed path is equal to μ_0 times the net current I that passes through the surface bounded by the closed path. This statement, known as **Ampère's circuital law**, can be written

$$\sum B_{\parallel} \Delta \ell = \mu_0 I \quad [19.13]$$

Hans Christian Oersted
Danish Physicist and Chemist (1777–1851)
Oersted is best known for observing that a compass needle deflects when placed near a wire carrying a current. This important discovery was the first evidence of the connection between electric and magnetic phenomena. Oersted was also the first to prepare pure aluminum.

678 CHAPTER 19 | Magnetism

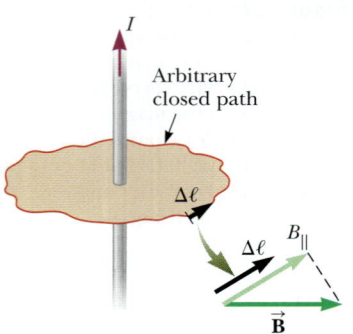

Figure 19.25 An arbitrary closed path around a current is used to calculate the magnetic field of the current by the use of Ampère's rule.

Figure 19.26 A closed, circular path of radius r around a long, straight, current-carrying wire is used to calculate the magnetic field set up by the wire.

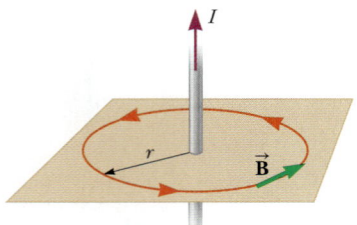

where $B_\parallel$ is the component of $\vec{B}$ parallel to the segment of length $\Delta\ell$ and $\Sigma B_\parallel \Delta\ell$ means that we take the sum over all the products $B_\parallel \Delta\ell$ around the closed path. Ampère's law is the fundamental law describing how electric currents create magnetic fields in the surrounding empty space.

We can use Ampère's circuital law to derive the magnetic field due to a long, straight wire carrying a current I. As discussed earlier, each of the magnetic field lines of this configuration forms a circle with the wire at its center. The magnetic field is tangent to this circle at every point, and its magnitude has the same value B over the entire circumference of a circle of radius r, so $B_\parallel = B$, as shown in Figure 19.26. In calculating the sum $\sum B_\parallel \Delta\ell$ over the circular path, notice that $B_\parallel$ can be removed from the sum (because it has the same value B for each element on the circle). Equation 19.13 then gives

$$\sum B_\parallel \Delta\ell = B_\parallel \sum \Delta\ell = B(2\pi r) = \mu_0 I$$

Dividing both sides by the circumference $2\pi r$, we obtain

$$B = \frac{\mu_0 I}{2\pi r}$$

This result is identical to Equation 19.11, which is the magnetic field due to the current I in a long, straight wire.

Ampère's circuital law provides an elegant and simple method for calculating the magnetic fields of highly symmetric current configurations, but it can't easily be used to calculate magnetic fields for complex current configurations that lack symmetry. In addition, Ampère's circuital law in this form is valid only when the currents and fields don't change with time.

André-Marie Ampère (1775–1836)
Ampère, a Frenchman, is credited with the discovery of electromagnetism, the relationship between electric currents and magnetic fields.

■ EXAMPLE 19.7 | The Magnetic Field of a Coaxial Cable

GOAL Use Ampère's law to calculate the magnetic field produced by current-carrying wires and cylinders.

PROBLEM A coaxial cable consists of an insulated wire carrying current $I_1 = 3.00$ A surrounded by a cylindrical conductor carrying current $I_2 = 1.00$ A in the opposite direction, as in Figure 19.27. **(a)** Calculate the magnetic field inside the cylindrical conductor at $r_{int} = 0.500$ cm. **(b)** Calculate the magnetic field outside the cylindrical conductor at $r_{ext} = 1.50$ cm.

STRATEGY Construct a circular path around the interior wire, as in Figure 19.27. Only the current inside that circle contributes to the magnetic field B_{int} at points on the circle. To compute the magnetic field B_{ext} exterior to the cylinder, construct a circular path outside the cylinder. Now both currents must be included in the calculation, but the current going down the page must be subtracted from the current in the wire.

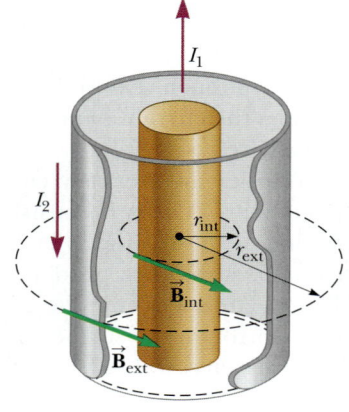

Figure 19.27 (Example 19.7)

SOLUTION

(a) Calculate the magnetic field B_{int} inside the cylindrical conductor at $r_{int} = 0.500$ cm.

Write Ampère's law:

$$\sum B_{\parallel} \Delta \ell = \mu_0 I$$

The magnetic field is constant on the given path and the total path length is $2\pi r_{int}$:

$$B_{int}(2\pi r_{int}) = \mu_0 I_1$$

Solve for B_{int} and substitute values:

$$B_{int} = \frac{\mu_0 I_1}{2\pi r_{int}} = \frac{(4\pi \times 10^{-7} \text{ T} \cdot \text{m/A})(3.00 \text{ A})}{2\pi(0.005 \text{ m})}$$

$$= 1.20 \times 10^{-4} \text{ T}$$

(b) Calculate the magnetic field B_{ext} outside the cylindrical conductor at $r_{ext} = 1.50$ cm.

Write Ampère's law:

$$\sum B_{\parallel} \Delta \ell = \mu_0 I$$

The magnetic field is again constant on the given path and the total path length is $2\pi r_{ext}$:

$$B_{ext}(2\pi r_{ext}) = \mu_0(I_1 - I_2)$$

Solve for B_{ext} and substitute values:

$$B_{ext} = \frac{\mu_0(I_1 - I_2)}{2\pi r_{ext}} = \frac{(4\pi \times 10^{-7} \text{ T} \cdot \text{m/A})(3.00 \text{ A} - 1.00 \text{ A})}{2\pi(0.015 \text{ m})}$$

$$= 2.67 \times 10^{-5} \text{ T}$$

REMARKS The direction of the field both inside and outside the cylinder is given by the right-hand rule, or counterclockwise in the perspective of Figure 19.27. Coaxial cables can be used to minimize the magnetic effects of current, provided that the currents inside the wire and the cylinder are equal in magnitude and opposite in direction.

QUESTION 19.7 What direction is the magnetic force on a proton traveling up the page (a) to the right of the cable? (b) On the left side?

EXERCISE 19.7 Suppose the current in the wire is 4.00 A downward and the current in the cylindrical conductor is 5.00 A upward. Find the magnitudes of the magnetic field (a) inside the cable at $r_{int} = 0.25$ cm and (b) outside the cable at $r_{ext} = 1.25$ cm.

ANSWERS (a) 3.20×10^{-4} T (b) 1.60×10^{-5} T

19.8 Magnetic Force Between Two Parallel Conductors

LEARNING OBJECTIVE

1. Calculate the magnetic force between two parallel current-carrying conductors.

As we have seen, a magnetic force acts on a current-carrying conductor when the conductor is placed in an external magnetic field. Because a conductor carrying a current creates a magnetic field around itself, it is easy to understand that two current-carrying wires placed close together exert magnetic forces on each other. Consider two long, straight, parallel wires separated by the distance d and carrying currents I_1 and I_2 in the same direction, as shown in Figure 19.28. Wire 1 is directly above wire 2. What's the magnetic force on one wire due to a magnetic field set up by the other wire?

In this calculation we are finding the force on wire 1 due to the magnetic field of wire 2. The current I_2 sets up magnetic field $\vec{B}_2$ at wire 1. The direction of $\vec{B}_2$ is

Figure 19.28 Two parallel wires, oriented vertically, carry steady currents and exert forces on each other. The force is attractive if the currents have the same direction, as shown, and repulsive if the two currents have opposite directions.

The field $\vec{B}_2$ at wire 1 due to wire 2 produces a force on wire 1 given by $F_1 = B_2 \ell I_1$.

perpendicular to the wire, as shown in the figure. Using Equation 19.11, we find that the magnitude of this magnetic field is

$$B_2 = \frac{\mu_0 I_2}{2\pi d}$$

According to Equation 19.5, the magnitude of the magnetic force on wire 1 in the presence of field $\vec{B}_2$ due to I_2 is

$$F_1 = B_2 I_1 \ell = \left(\frac{\mu_0 I_2}{2\pi d}\right) I_1 \ell = \frac{\mu_0 I_1 I_2 \ell}{2\pi d}$$

We can rewrite this relationship in terms of the force per unit length:

$$\frac{F_1}{\ell} = \frac{\mu_0 I_1 I_2}{2\pi d} \quad [19.14]$$

The direction of $\vec{F}_1$ is downward, toward wire 2, as indicated by right-hand rule number 1. This calculation is completely symmetric, which means that the force $\vec{F}_2$ on wire 2 is equal to and opposite $\vec{F}_1$, as expected from Newton's third law of action–reaction.

We have shown that parallel conductors carrying currents in the same direction *attract* each other. You should use the approach indicated by Figure 19.28 and the steps leading to Equation 19.14 to show that parallel conductors carrying currents in opposite directions *repel* each other.

The force between two parallel wires carrying a current is used to define the SI unit of current, the **ampere** (A), as follows:

▶ Definition of the ampere

If two long, parallel wires 1 m apart carry the same current and the magnetic force per unit length on each wire is 2×10^{-7} N/m, the current is defined to be 1 A.

The SI unit of charge, the **coulomb** (C), can now be defined in terms of the ampere as follows:

▶ Definition of the coulomb

If a conductor carries a steady current of 1 A, the quantity of charge that flows through any cross section in 1 s is 1 C.

■ Quick Quiz

19.5 Which of the following actions would double the magnitude of the magnetic force per unit length between two parallel current-carrying wires? Choose all correct answers. (a) Double one of the currents. (b) Double the distance between them. (c) Reduce the distance between them by half. (d) Double both currents.

19.6 If, in Figure 19.28, $I_1 = 2$ A and $I_2 = 6$ A, which of the following is true? (Note that F_2 represents the magnitude of the force on wire 2.) (a) $F_1 = 3F_2$ (b) $F_1 = F_2$ (c) $F_1 = F_2/3$

■ **EXAMPLE 19.8** | Levitating a Wire

GOAL Calculate the magnetic force of one current-carrying wire on a parallel current-carrying wire.

PROBLEM Two wires, each having a weight per unit length of 1.00×10^{-4} N/m, are parallel with one directly above the other. Assume the wires carry currents that are equal in magnitude and opposite in direction. The wires are 0.10 m apart, and the sum of the magnetic force and gravitational force on the upper wire is zero. Find the current in the wires. (Neglect Earth's magnetic field.)

STRATEGY The upper wire must be in equilibrium under the forces of magnetic repulsion and gravity. Set the sum of the forces equal to zero and solve for the unknown current, I.

SOLUTION

Set the sum of the forces equal to zero and substitute the appropriate expressions. Notice that the magnetic force between the wires is repulsive.

$$\vec{F}_{grav} + \vec{F}_{mag} = 0$$

$$-mg + \frac{\mu_0 I_1 I_2}{2\pi d} \ell = 0$$

The currents are equal, so $I_1 = I_2 = I$. Make these substitutions and solve for I^2:

$$\frac{\mu_0 I^2}{2\pi d} \ell = mg \quad \rightarrow \quad I^2 = \frac{(2\pi d)(mg/\ell)}{\mu_0}$$

Substitute given values, finding I^2, then take the square root. Notice that the weight per unit length, mg/ℓ, is given.

$$I^2 = \frac{(2\pi \cdot 0.100 \text{ m})(1.00 \times 10^{-4} \text{ N/m})}{(4\pi \times 10^{-7} \text{ T} \cdot \text{m})} = 50.0 \text{ A}^2$$

$$I = \boxed{7.07 \text{ A}}$$

REMARKS Exercise 19.3 showed that using Earth's magnetic field to levitate a wire required extremely large currents. Currents in wires can create much stronger magnetic fields than Earth's magnetic field in regions near the wire.

QUESTION 19.8 Why can't cars be constructed that can magnetically levitate in Earth's magnetic field?

EXERCISE 19.8 If the current in each wire is doubled, how far apart should the wires be placed if the magnitudes of the gravitational and magnetic forces on the upper wire are to be equal?

ANSWER 0.400 m

19.9 Magnetic Fields of Current Loops and Solenoids

LEARNING OBJECTIVES

1. Generalize the form of a magnetic field inside a loop to that inside a solenoid.
2. Calculate the magnetic field inside a solenoid.
3. Use Ampere's Law to derive the magnetic field inside a solenoid.

The strength of the magnetic field set up by a piece of wire carrying a current can be enhanced at a specific location if the wire is formed into a loop. You can understand this by considering the effect of several small segments of the current loop, as in Figure 19.29. The small segment at the bottom of the loop, labeled Δx_1, produces a magnetic field of magnitude B_1 at the loop's center, directed upward. The direction of $\vec{B}$ can be verified using right-hand rule number 2 for a long, straight wire. Imagine holding the wire with your right hand, with your thumb pointing in the direction of the current. Your fingers then curl around in the direction of $\vec{B}$.

A segment of length Δx_2 at the top of the loop also contributes to the field at the center, increasing its strength. The field produced at the center by the segment Δx_2 has the same magnitude as B_1 and is also directed upward. Similarly, all other such segments of the current loop contribute to the field. The net effect is a magnetic field for the current loop as pictured in Figure 19.30a.

Notice in Figure 19.30a that the magnetic field lines enter at the bottom of the current loop and exit at the top. Compare this figure with Figure 19.30b, illustrating the field of a bar magnet. The two fields are similar. One side of the loop acts as though it were the north pole of a magnet, and the other acts as a south pole. The similarity of these two fields will be used to discuss magnetism in matter in an upcoming section.

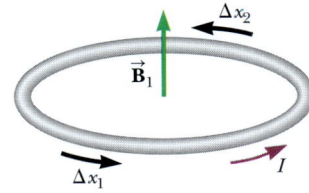

Figure 19.29 All segments of the current loop produce a magnetic field at the center of the loop, directed *upward*.

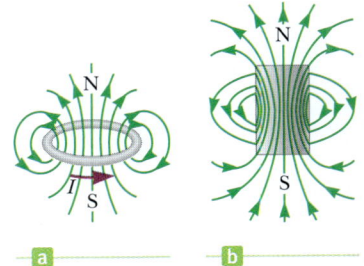

Figure 19.30 (a) Magnetic field lines for a current loop. Note that the lines resemble those of a bar magnet. (b) The magnetic field of a bar magnet is similar to that of a current loop.

APPLYING PHYSICS 19.4 | Twisted Wires

In electrical circuits it is often the case that insulated wires carrying currents in opposite directions are twisted together. What is the advantage of doing this?

EXPLANATION If the wires are not twisted together, the combination of the two wires forms a current loop, which produces a relatively strong magnetic field. This magnetic field generated by the loop could be strong enough to affect adjacent circuits or components. When the wires are twisted together, their magnetic fields tend to cancel. ■

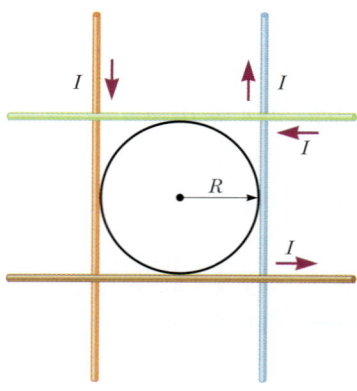

Figure 19.31 The field of a circular loop carrying current I can be approximated by the field due to four straight wires, each carrying current I.

The magnitude of the magnetic field at the center of a circular loop carrying current I is given by

$$B = \frac{\mu_0 I}{2R}$$

This equation must be derived with calculus. It can be shown, however, to be reasonable by calculating the field at the center of four long wires, each carrying current I and forming a square, as in Figure 19.31, with a circle of radius R inscribed within it. Intuitively, this arrangement should give a magnetic field at the center that is similar in magnitude to the field produced by the circular loop. The current in the circular wire is closer to the center, so that wire would have a magnetic field somewhat stronger than just the four legs of the rectangle, but the lengths of the straight wires beyond the rectangle compensate for it. Each wire contributes the same magnetic field at the exact center, so the total field is given by

$$B = 4 \times \frac{\mu_0 I}{2\pi R} = \frac{4}{\pi}\left(\frac{\mu_0 I}{2R}\right) = (1.27)\left(\frac{\mu_0 I}{2R}\right)$$

This result is *approximately* the same as the field produced by the circular loop of current.

When the coil has N loops, each carrying current I, the magnetic field at the center is given by

$$B = N\frac{\mu_0 I}{2R} \qquad [19.15]$$

Magnetic Field of a Solenoid

If a long, straight wire is bent into a coil of several closely spaced loops, the resulting device is a **solenoid**, often called an **electromagnet**. This device is important in many applications because it acts as a magnet only when it carries a current. The magnetic field inside a solenoid increases with the current and is proportional to the number of coils per unit length.

Figure 19.32 shows the magnetic field lines of a loosely wound solenoid of length ℓ and total number of turns N. Notice that the field lines inside the solenoid are nearly parallel, uniformly spaced, and close together. As a result, the field inside the solenoid is strong and approximately uniform. The exterior field at the sides of the solenoid is nonuniform, much weaker than the interior field, and *opposite in direction* to the field inside the solenoid.

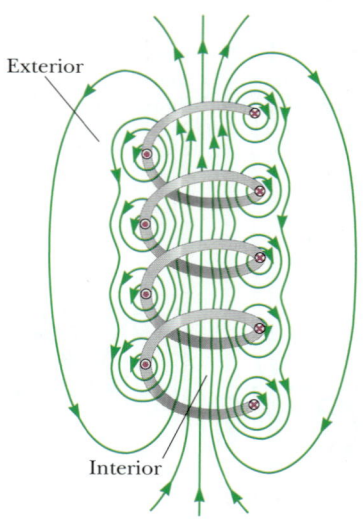

Figure 19.32 The magnetic field lines for a loosely wound solenoid.

If the turns are closely spaced, the field lines are as shown in Figure 19.33a, entering at one end of the solenoid and emerging at the other. One end of the solenoid acts as a north pole and the other end acts as a south pole. If the length of the solenoid is much greater than its radius, the lines that leave the north end of the solenoid spread out over a wide region before returning to enter the south end. The more widely separated the field lines are, the weaker the field. This is in contrast to a much stronger field *inside* the solenoid, where the lines are close together. Also, the field inside the solenoid has a constant magnitude at all points

19.9 | Magnetic Fields of Current Loops and Solenoids

Figure 19.33 (a) Magnetic field lines for a tightly wound solenoid of finite length carrying a steady current. The field inside the solenoid is nearly uniform and strong. (b) The magnetic field pattern of a bar magnet, displayed by small iron filings on a sheet of paper.

The magnetic field lines resemble those of a bar magnet, meaning that the solenoid effectively has north and south poles.

far from its ends. As will be shown subsequently, these considerations allow the application of Ampère's law to the solenoid, giving a result of

$$B = \mu_0 n I \quad [19.16]$$

◀ The magnetic field inside a solenoid

for the field inside the solenoid, where $n = N/\ell$ is the number of turns per unit length of the solenoid.

Numerous devices create beams of charged particles for various purposes, and those particles are usually controlled and directed by electromagnetic fields. Old-style cathode ray TV sets use steering magnets that rapidly and accurately direct an electron beam across a screen of phosphors in a scanning motion, creating an illusion of a moving picture out of a series of bright dots. (See Fig. 19.34.) Electron microscopes (see Fig. 27.17b, page 937) use a similar gun and both electrostatic and electromagnetic lenses to focus the beam. Particle accelerators require very large electromagnets to turn particles moving at nearly the speed of light. Tokamaks, experimental devices used in fusion power research, use magnetic fields to contain hot plasmas. Figure 19.35 is a photograph of one such device.

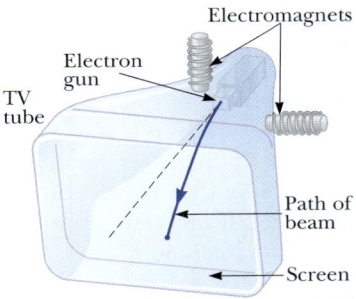

Figure 19.34 Electromagnets are used to deflect electrons to desired positions on the screen of a television tube.

Figure 19.35 Interior view of the closed Tokamak Fusion Test Reactor (TFTR) vacuum vessel at the Princeton Plasma Physics Laboratory.

■ EXAMPLE 19.9 | The Magnetic Field Inside a Solenoid

GOAL Calculate the magnetic field of a solenoid from given data and the momentum of a charged particle in this field.

PROBLEM A certain solenoid consists of 100 turns of wire and has a length of 10.0 cm. **(a)** Find the magnitude of the magnetic field inside the solenoid when it carries a current of 0.500 A. **(b)** What is the momentum of a proton orbiting inside the solenoid in a circle with a radius of 0.020 m? The axis of the solenoid is perpendicular to the plane of the orbit. **(c)** Approximately how much wire would be needed to build this solenoid? Assume the solenoid's radius is 5.00 cm.

STRATEGY In part (a) calculate the number of turns per meter and substitute that and given information into Equation 19.16, getting the magnitude of the magnetic field. Part (b) is an application of Newton's second law.

(Continued)

SOLUTION

(a) Find the magnitude of the magnetic field inside the solenoid when it carries a current of 0.500 A.

Calculate the number of turns per unit length:

$$n = \frac{N}{\ell} = \frac{100 \text{ turns}}{0.100 \text{ m}} = 1.00 \times 10^3 \text{ turns/m}$$

Substitute n and I into Equation 19.16 to find the magnitude of the magnetic field:

$$B = \mu_0 n I$$
$$= (4\pi \times 10^{-7} \text{ T} \cdot \text{m/A})(1.00 \times 10^3 \text{ turns/m})(0.500 \text{ A})$$
$$= \boxed{6.28 \times 10^{-4} \text{ T}}$$

(b) Find the momentum of a proton orbiting in a circle of radius 0.020 m near the center of the solenoid.

Write Newton's second law for the proton:

$$ma = F = qvB$$

Substitute the centripetal acceleration $a = v^2/r$:

$$m\frac{v^2}{r} = qvB$$

Cancel one factor of v on both sides and multiply by r, getting the momentum mv:

$$mv = rqB = (0.020 \text{ m})(1.60 \times 10^{-19} \text{ C})(6.28 \times 10^{-4} \text{ T})$$
$$p = mv = \boxed{2.01 \times 10^{-24} \text{ kg} \cdot \text{m/s}}$$

(c) Approximately how much wire would be needed to build this solenoid?

Multiply the number of turns by the circumference of one loop:

$$\text{Length of wire} \approx (\text{number of turns})(2\pi r)$$
$$= (1.00 \times 10^2 \text{ turns})(2\pi \cdot 0.050 \, 0 \text{ m})$$
$$= \boxed{31.4 \text{ m}}$$

REMARKS An electron in part (b) would have the same momentum as the proton, but a much higher speed. It would also orbit in the opposite direction. The length of wire in part (c) is only an estimate because the wire has a certain thickness, slightly increasing the size of each loop. In addition, the wire loops aren't perfect circles because they wind slowly up along the solenoid.

QUESTION 19.9 What would happen to the orbiting proton if the solenoid were oriented vertically?

EXERCISE 19.9 Suppose you have a 32.0-m length of copper wire. If the wire is wrapped into a solenoid 0.240 m long and having a radius of 0.040 0 m, how strong is the resulting magnetic field in its center when the current is 12.0 A?

ANSWER 8.00×10^{-3} T

Ampère's Law Applied to a Solenoid

We can use Ampère's law to obtain the expression for the magnetic field inside a solenoid carrying a current I. A cross section taken along the length of part of our solenoid is shown in Figure 19.36. $\vec{B}$ inside the solenoid is uniform and parallel to the axis, and $\vec{B}$ outside is approximately zero. Consider a rectangular path of length L and width w, as shown in the figure. We can apply Ampère's law to this path by evaluating the sum of $B_\parallel \Delta\ell$ over each side of the rectangle. The contribution along side 3 is clearly zero because $\vec{B} = 0$ in this region. The contributions from sides 2 and 4 are both zero because $\vec{B}$ is perpendicular to $\Delta\vec{\ell}$ along these paths. Side 1 of length L gives a contribution BL to the sum because $\vec{B}$ is uniform along this path and parallel to $\Delta\vec{\ell}$. Therefore, the sum over the closed rectangular path has the value

$$\sum B_\parallel \Delta\ell = BL$$

The right side of Ampère's law involves the total current that passes through the area bounded by the path chosen. In this case, the total current through the rectangular path equals the current through each turn of the solenoid, multiplied by the number of turns. If N is the number of turns in the length L, then the total current through the rectangular path equals NI. Ampère's law applied to this path therefore gives

$$\sum B_\parallel \Delta \ell = BL = \mu_0 NI$$

or

$$B = \mu_0 \frac{N}{L} I = \mu_0 n I$$

where $n = N/L$ is the number of turns per unit length.

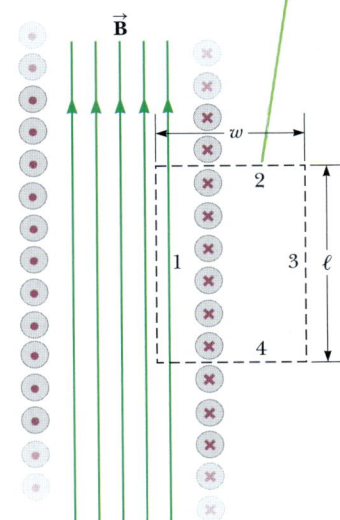

Ampère's law applied to the rectangular dashed path can be used to calculate the field inside the solenoid.

Figure 19.36 A cross-sectional view of a tightly wound solenoid. If the solenoid is long relative to its radius, we can assume the magnetic field inside is uniform and the field outside is zero.

19.10 Magnetic Domains

LEARNING OBJECTIVES

1. Discuss the creation of magnet fields at the atomic level by orbiting electrons and electron spin.
2. Define ferromagnetic materials and magnetic domains.
3. Contrast the magnetic properties of ferromagnetic, paramagnetic, and diamagnetic materials.

The magnetic field produced by a current in a coil of wire gives us a hint as to what might cause certain materials to exhibit strong magnetic properties. A single coil like that in Figure 19.30a has a north pole and a south pole, but if that is true for a coil of wire, it should also be true for any current confined to a circular path. In particular, *an individual atom should act as a magnet because of the motion of the electrons about the nucleus.* Each electron, with its charge of 1.6×10^{-19} C, circles the atom once in about 10^{-16} s. If we divide the electric charge by this time interval, we see that the orbiting electron is equivalent to a current of 1.6×10^{-3} A. Such a current produces a magnetic field on the order of 20 T at the center of the circular path. From this we see that a very strong magnetic field would be produced if several of these atomic magnets could be aligned inside a material. This doesn't occur, however, because the simple model we have described is not the complete story. A thorough analysis of atomic structure shows that the magnetic field produced by one electron in an atom is often canceled by an oppositely revolving electron in the same atom. The net result is that **the magnetic effect produced by the electrons orbiting the nucleus is either zero or very small for most materials**.

The magnetic properties of many materials can be explained by the fact that an electron not only circles in an orbit, but also spins on its axis like a top, with spin magnetic moment as shown (Fig. 19.37). (This classical description should not be taken too literally. The property of electron *spin* can be understood only in the context of quantum mechanics, which we will not discuss here.) The spinning electron represents a charge in motion that produces a magnetic field. The field due to the spinning is generally stronger than the field due to the orbital motion. In atoms containing many electrons, the electrons usually pair up with their spins opposite each other so that their fields cancel. That is why most substances are not magnets. In certain strongly magnetic materials, such as iron, cobalt, and nickel, however, the magnetic fields produced by the electron spins don't cancel completely. Such materials are said to be **ferromagnetic**. In ferromagnetic materials strong coupling occurs between neighboring atoms,

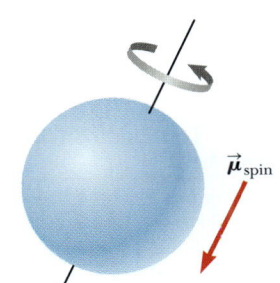

Figure 19.37 Classical model of a spinning electron.

686 CHAPTER 19 | Magnetism

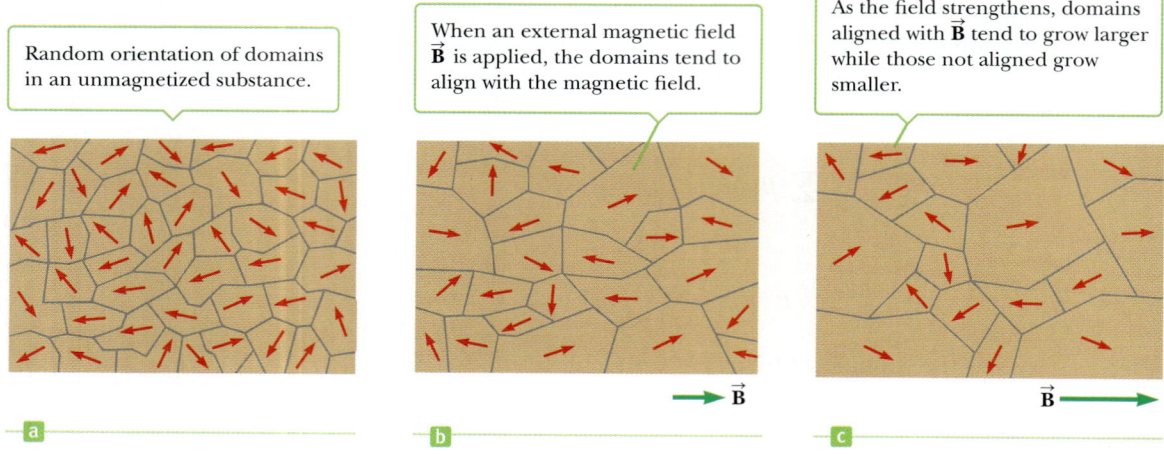

Random orientation of domains in an unmagnetized substance.

When an external magnetic field $\vec{B}$ is applied, the domains tend to align with the magnetic field.

As the field strengthens, domains aligned with $\vec{B}$ tend to grow larger while those not aligned grow smaller.

Figure 19.38 Orientation of magnetic dipoles before and after a magnetic field is applied to a ferromagnetic substance.

Figure 19.39 The permanent magnet (red) temporarily magnetizes some paper clips, which then cling to each other through magnetic forces.

Tip 19.4 The Electron Spins, but Doesn't!
Even though we use the word *spin*, the electron, unlike a child's top, isn't physically spinning in this sense. The electron has an intrinsic angular momentum that causes it to act *as if it were spinning*, but the concept of spin angular momentum is actually a relativistic quantum effect.

forming large groups of atoms with spins that are aligned. Called **domains**, the sizes of these groups typically range from about 10^{-4} cm to 0.1 cm. In an unmagnetized substance the domains are randomly oriented, as shown in Figure 19.38a. When an external field is applied, as in Figures 19.38b and 19.38c, the magnetic field of each domain tends to come nearer to alignment with the external field, resulting in magnetization.

In what are called hard magnetic materials, domains remain aligned even after the external field is removed; the result is a **permanent magnet**. In soft magnetic materials, such as iron, once the external field is removed, thermal agitation produces motion of the domains and the material quickly returns to an unmagnetized state.

The alignment of domains explains why the strength of an electromagnet is increased dramatically by the insertion of an iron core into the magnet's center. The magnetic field produced by the current in the loops causes the domains to align, thus producing a large net external field. The use of iron as a core is also advantageous because it is a soft magnetic material that loses its magnetism almost instantaneously after the current in the coils is turned off.

The formation of domains in ferromagnetic substances also explains why such substances are attracted to permanent magnets. The magnetic field of a permanent magnet realigns domains in a ferromagnetic object so that the object becomes temporarily magnetized. The object's poles are then attracted to the corresponding opposite poles of the permanent magnet. The object can similarly attract other ferromagnetic objects, as illustrated in Figure 19.39.

Types of Magnetic Materials

Magnetic materials can be classified according to how they react to the application of a magnetic field. In **ferromagnetic** materials the atoms have permanent magnetic moments that align readily with an externally applied magnetic field. Examples of ferromagnetic materials are iron, cobalt, and nickel. Such substances can retain some of their magnetization even after the applied magnetic field is removed.

Paramagnetic materials also have magnetic moments that tend to align with an externally applied magnetic field, but the response is extremely weak compared with that of ferromagnetic materials. Examples of paramagnetic substances are aluminum, calcium, and platinum. A ferromagnetic material can

Summary

Figure 19.40 Diamagnetism. A frog is levitated in a 16-T magnetic field at the Nijmegen High Field Magnet Laboratory in the Netherlands. The levitation force is exerted on the diamagnetic water molecules in the frog's body. The frog suffered no ill effects from the levitation experience.

become paramagnetic when warmed to a certain critical temperature, the Curie temperature, that depends on the material.

In **diamagnetic** materials, an externally applied magnetic field induces a very weak magnetization that is opposite the applied field. Ordinarily diamagnetism isn't observed because paramagnetic and ferromagnetic effects are far stronger. In Figure 19.40, however, a very high magnetic field exerts a levitating force on the diamagnetic water molecules in a frog.

■ SUMMARY

19.3 Magnetic Fields

The **magnetic force** that acts on a charge q moving with velocity $\vec{v}$ in a magnetic field $\vec{B}$ has magnitude

$$F = qvB \sin \theta \quad [19.1]$$

where θ is the angle between $\vec{v}$ and $\vec{B}$.

(a) The direction of the magnetic force $\vec{F}$ acting on a charged particle moving with a velocity $\vec{v}$ in the presence of a magnetic field $\vec{B}$. The magnetic force is perpendicular to both $\vec{v}$ and $\vec{B}$. (b) Magnetic forces on positive and negative charges. The dashed lines show the paths of the particles, which are investigated in Section 19.6. The magnetic forces on oppositely charged particles moving at the same velocity in a magnetic field are in opposite directions.

To find the direction of this force, use **right-hand rule number 1**: point the fingers of your open right hand in the direction of $\vec{v}$ and then curl them in the direction of $\vec{B}$. Your thumb then points in the direction of the magnetic force $\vec{F}$.

If the charge is *negative* rather than positive, the force is directed opposite the force given by the right-hand rule.

The SI unit of the magnetic field is the **tesla** (T), or weber per square meter (Wb/m²). An additional commonly used unit for the magnetic field is the **gauss** (G); $1 \text{ T} = 10^4 \text{ G}$.

19.4 Magnetic Force on a Current-Carrying Conductor

If a straight conductor of length ℓ carries current I, the magnetic force on that conductor when it is placed in a uniform external magnetic field $\vec{B}$ is

$$F = BI\ell \sin \theta \quad [19.6]$$

where θ is the angle between the direction of the current and the direction of the magnetic field.

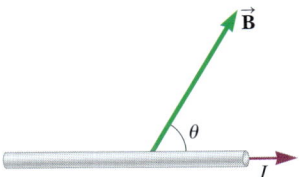

The magnetic force on this current-carrying conductor is directed straight up out of the page.

Right-hand rule number 1 also gives the direction of the magnetic force on the conductor. In this case, however, you must point your fingers in the direction of the current rather than in the direction of $\vec{v}$.

19.5 Torque on a Current Loop and Electric Motors

The torque τ on a current-carrying loop of wire in a magnetic field $\vec{B}$ has magnitude

$$\tau = BIA \sin \theta \qquad [19.8]$$

where I is the current in the loop and A is its cross-sectional area. The magnitude of the magnetic moment of a current-carrying coil is defined by $\mu = IAN$, where N is the number of loops. The magnetic moment is considered a vector, $\vec{\mu}$, that is perpendicular to the plane of the loop. The angle between $\vec{B}$ and $\vec{\mu}$ is θ.

19.6 Motion of a Charged Particle in a Magnetic Field

If a charged particle moves in a uniform magnetic field so that its initial velocity is perpendicular to the field, it will move in a circular path in a plane perpendicular to the magnetic field. The radius r of the circular path can be found from Newton's second law and centripetal acceleration, and is given by

$$r = \frac{mv}{qB} \qquad [19.10]$$

where m is the mass of the particle and q is its charge.

19.7 Magnetic Field of a Long, Straight Wire and Ampère's Law

The magnetic field at distance r from a **long, straight wire** carrying current I has the magnitude

$$B = \frac{\mu_0 I}{2\pi r} \qquad [19.11]$$

where $\mu_0 = 4\pi \times 10^{-7}$ T·m/A is the **permeability of free space**. The magnetic field lines around a long, straight wire are circles concentric with the wire.

Ampère's law can be used to find the magnetic field around certain simple current-carrying conductors. It can be written

$$\sum B_\parallel \Delta \ell = \mu_0 I \qquad [19.13]$$

where $B_\parallel$ is the component of $\vec{B}$ tangent to a small current element of length $\Delta \ell$ that is part of a closed path and I is the total current that penetrates the closed path.

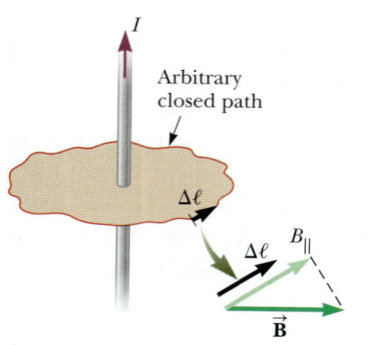

An arbitrary closed path around a current is used to calculate the magnetic field of the current by the use of Ampère's rule.

19.8 Magnetic Force Between Two Parallel Conductors

The force per unit length on each of two parallel wires separated by the distance d and carrying currents I_1 and I_2 has the magnitude

$$\frac{F}{\ell} = \frac{\mu_0 I_1 I_2}{2\pi d} \qquad [19.14]$$

The forces are attractive if the currents are in the same direction and repulsive if they are in opposite directions.

19.9 Magnetic Field of Current Loops and Solenoids

The magnetic field at the center of a coil of N circular loops of radius R, each carrying current I, is given by

$$B = N \frac{\mu_0 I}{2R} \qquad [19.15]$$

(a) Magnetic field lines for a current loop. Note that the lines resemble those of a bar magnet.
(b) The magnetic field of a bar magnet is similar to that of a current loop.

The magnetic field inside a solenoid has the magnitude

$$B = \mu_0 n I \qquad [19.16]$$

where $n = N/\ell$ is the number of turns of wire per unit length.

Magnetic field lines for a tightly wound solenoid of finite length carrying a steady current. The field inside the solenoid is nearly uniform and strong. Note that the field lines resemble those of a bar magnet, so the solenoid effectively has north and south poles.

WARM-UP EXERCISES

The warm-up exercises in this chapter may be assigned online in Enhanced WebAssign.

1. **Physics Review** A man of mass 70.0 kg sits on the end of a seesaw of length 4.00 m, pivoted in the middle. His end of the seesaw is on the ground. A standing woman grasps and exerts a force straight down on the raised end, which is initially 2.00 m off the ground. (a) What torque must she exceed if she is to begin lifting the man off the ground? (b) What minimum torque must she exert just as the seesaw becomes level? (See Section 8.1.)

2. **Physics Review** Suppose a futuristic tractor beam on a large space station exerts a constant force on a satellite of mass 2 420 kg, keeping it in a circular orbit of radius 785 m. If the satellite is traveling at 1 250 m/s, what magnitude force must the tractor beam exert on it? (See Section 7.4.)

3. A wire of length 0.500 m carries a current of 0.100 A in the positive *x*-direction, parallel to the ground. If the wire has a weight of 1.00×10^{-2} N, what is the minimum magnitude magnetic field that exerts a magnetic force on the wire equal to its weight? (See Section 19.4.)

4. An electron moves across Earth's equator at a speed of 2.53×10^6 m/s and in a direction 35.0° N of E. At this point, Earth's magnetic field has a direction due north, is parallel to the surface, and has a magnitude of 0.100×10^{-4} T. (a) What is the magnitude of the force acting on the electron due to its interaction with Earth's magnetic field? (b) Is the force toward, away, or parallel to the Earth's surface? (See Section 19.3.)

5. A rectangular coil of wire consisting of 10.0 loops, each with length 0.200 m and width 0.300 m, lies in the *xy*-plane. If the coil carries a current of 2.00 A, what is the magnitude of the torque exerted by a magnetic field of magnitude 0.010 0 T directed at an angle of 30.0° with respect to the positive *z*-axis? (See Section 19.5.)

6. A long wire carries a current of 0.500 A. Find the magnitude of the magnetic field 0.700 m away from the wire. (See Section 19.7.)

7. A proton enters a constant magnetic field of magnitude 0.050 0 T and traverses a semicircle of radius 1.00 mm before leaving the field. What is the proton's speed? (See Section 19.6.)

8. Calculate the magnitude of the magnetic force per unit length between a pair of parallel wires separated by 2.00 m if they each carry a current of 3.00 A. (See Section 19.8.)

9. What is the magnitude of the magnetic field at the core of a 120-turn solenoid of length 0.50 m carrying a current of 2.0 A? (See Section 19.9.)

CONCEPTUAL QUESTIONS

The conceptual questions in this chapter may be assigned online in Enhanced WebAssign.

1. In older television sets, a beam of electrons moves from the back of the picture tube to the screen, where it strikes a fluorescent dot that glows with a particular color when hit. Earth's magnetic field at the location of the television set is horizontal and toward the north. In which direction(s) should the set be oriented so that the beam undergoes the largest deflection?

2. Which way would a compass point if you were at Earth's north magnetic pole?

3. How can the motion of a charged particle be used to distinguish between a magnetic field and an electric field in a certain region? Give a specific example to justify your answer.

4. Can a constant magnetic field set a proton at rest into motion? Explain your answer.

5. Explain why two parallel wires carrying currents in opposite directions repel each other.

6. Will a nail be attracted to either pole of a magnet? Explain what is happening inside the nail when it is placed near the magnet.

7. A Hindu ruler once suggested that he be entombed in a magnetic coffin with the polarity arranged so that he could be forever suspended between heaven and Earth. Is such magnetic levitation possible? Discuss.

8. A magnet attracts a piece of iron. The iron can then attract another piece of iron. On the basis of domain alignment, explain what happens in each piece of iron.

9. Can you use a compass to detect the currents in wires in the walls near light switches in your home?

10. Is the magnetic field created by a current loop uniform? Explain.

11. Suppose you move along a wire at the same speed as the drift speed of the electrons in the wire. Do you now measure a magnetic field of zero?

12. Why do charged particles from outer space, called cosmic rays, strike Earth more frequently at the poles than at the equator?

13. A hanging Slinky® toy is attached to a powerful battery and a switch. When the switch is closed so that the toy now carries current, does the Slinky compress or expand?

690 CHAPTER 19 | Magnetism

14. How can a current loop be used to determine the presence of a magnetic field in a given region of space?

15. Parallel wires exert magnetic forces on each other. What about perpendicular wires? Imagine two wires oriented perpendicular to each other and almost touching. Each wire carries a current. Is there a force between the wires?

16. Figure CQ19.16 shows four permanent magnets, each having a hole through its center. Notice that the blue and yellow magnets are levitated above the red ones. (a) How does this levitation occur? (b) What purpose do the rods serve? (c) What can you say about the poles of the magnets from this observation? (d) If the upper magnet were inverted, what do you suppose would happen?

Figure CQ19.16

17. Two charged particles are projected in the same direction into a magnetic field perpendicular to their velocities. If the two particles are deflected in opposite directions, what can you say about them?

18. Two long, straight wires cross each other at right angles, and each carries the same current as in Figure CQ19.18. Which of the following statements are true regarding the total magnetic field at the various points due to the two wires? (There may be more than one correct statement.) (a) The field is strongest at points B and D. (b) The field is strongest at points A and C. (c) The field is out of the page at point B and into the page at point D. (d) The field is out of the page at point C and out of the page at point D. (e) The field has the same magnitude at all four points.

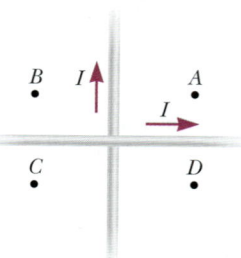

Figure CQ19.18

19. A magnetic field exerts a torque on each of the current-carrying single loops of wire shown in Figure CQ19.19. The loops lie in the xy-plane, each carrying the same magnitude current, and the uniform magnetic field points in the positive x-direction. Rank the coils by the magnitude of the torque exerted on them by the field, from largest to smallest. (a) A, B, C (b) A, C, B (c) B, A, C (d) B, C, A (e) C, A, B

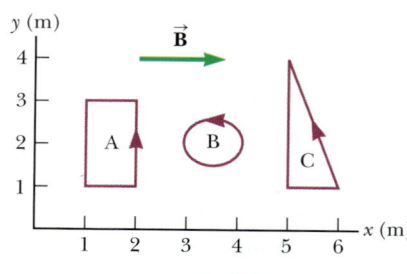

Figure CQ19.19

PROBLEMS

WebAssign The problems in this chapter may be assigned online in Enhanced WebAssign.

1. denotes straightforward problem; 2. denotes intermediate problem; 3. denotes challenging problem

1. denotes full solution available in *Student Solutions Manual/Study Guide*

1. denotes problems most often assigned in Enhanced WebAssign

BIO denotes biomedical problems
GP denotes guided problems
M denotes Master It tutorial available in Enhanced WebAssign
QC denotes asking for quantitative and conceptual reasoning
S denotes symbolic reasoning problem
W denotes Watch It video solution available in Enhanced WebAssign

19.3 Magnetic Fields

1. Consider an electron near the Earth's equator. In which direction does it tend to deflect if its velocity is (a) directed downward? (b) Directed northward? (c) Directed westward? (d) Directed southeastward?

2. (a) Find the direction of the force on a proton (a positively charged particle) moving through the magnetic fields in Figure P19.2, as shown. (b) Repeat part (a), assuming the moving particle is an electron.

Problems 691

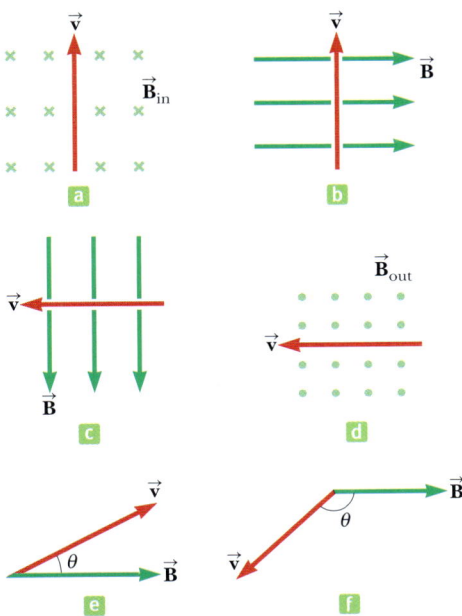

Figure P19.2 Problems 2 and 12.

3. Find the direction of the magnetic field acting on the positively charged particle moving in the various situations shown in Figure P19.3 if the direction of the magnetic force acting on it is as indicated.

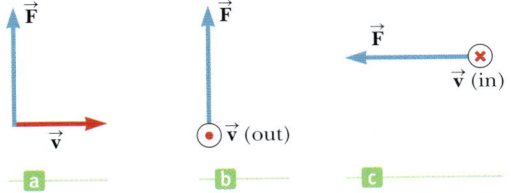

Figure P19.3 (Problems 3 and 15) For Problem 15, replace the velocity vector with a current in that direction.

4. Determine the initial direction of the deflection of charged particles as they enter the magnetic fields, as shown in Figure P19.4.

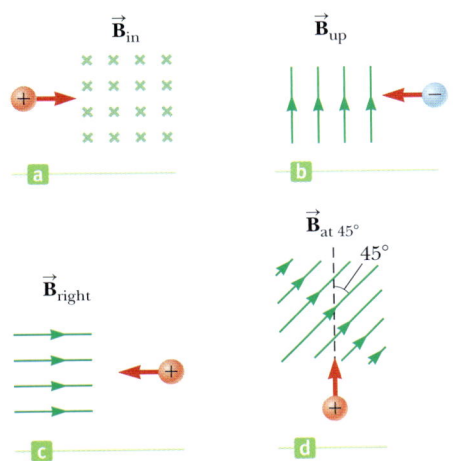

Figure P19.4

5. **Q|C** A laboratory electromagnet produces a magnetic field of magnitude 1.50 T. A proton moves through this field with a speed of 6.00×10^6 m/s. (a) Find the magnitude of the maximum magnetic force that could be exerted on the proton. (b) What is the magnitude of the maximum acceleration of the proton? (c) Would the field exert the same magnetic force on an electron moving through the field with the same speed? (d) Would the electron undergo the same acceleration? Explain.

6. **M** A proton moves perpendicular to a uniform magnetic field $\vec{B}$ at a speed of 1.00×10^7 m/s and experiences an acceleration of 2.00×10^{13} m/s² in the positive x-direction when its velocity is in the positive z-direction. Determine the magnitude and direction of the field.

7. What velocity would a proton need to circle Earth 1 000 km above the magnetic equator, where Earth's magnetic field is directed horizontally north and has a magnitude of 4.00×10^{-8} T?

8. **W** An electron is accelerated through 2 400 V from rest and then enters a region where there is a uniform 1.70-T magnetic field. What are (a) the maximum and (b) the minimum magnitudes of the magnetic force acting on this electron?

9. **M** A proton moving at 4.00×10^6 m/s through a magnetic field of magnitude 1.70 T experiences a magnetic force of magnitude 8.20×10^{-13} N. What is the angle between the proton's velocity and the field?

10. **BIO** Sodium ions (Na⁺) move at 0.851 m/s through a bloodstream in the arm of a person standing near a large magnet. The magnetic field has a strength of 0.254 T and makes an angle of 51.0° with the motion of the sodium ions. The arm contains 100 cm³ of blood with a concentration of 3.00×10^{20} Na⁺ ions per cubic centimeter. If no other ions were present in the arm, what would be the magnetic force on the arm?

11. At the equator, near the surface of Earth, the magnetic field is approximately 50.0 μT northward, and the electric field is about 100 N/C downward in fair weather. Find the gravitational, electric, and magnetic forces on an electron with an instantaneous velocity of 6.00×10^6 m/s directed to the east in this environment.

19.4 Magnetic Force on a Current-Carrying Conductor

12. In Figure P19.2 assume in each case the velocity vector shown is replaced with a wire carrying a current in the direction of the velocity vector. For each case, find the direction of the magnetic force acting on the wire.

13. **W** A current $I = 15$ A is directed along the positive x-axis and perpendicular to a magnetic field. A magnetic force per unit length of 0.12 N/m acts on the conductor in the negative y-direction. Calculate the magnitude and direction of the magnetic field in the region through which the current passes.

14. **Q|C** A straight wire carrying a 3.0-A current is placed in a uniform magnetic field of magnitude 0.28 T directed perpendicular to the wire. (a) Find the magnitude of the magnetic force on a section of the wire having a length of 14 cm. (b) Explain why you can't determine the direction of the magnetic force from the information given in the problem.

15. In Figure P19.3 assume in each case the velocity vector shown is replaced with a wire carrying a current in the direction of the velocity vector. For each case, find the direction of the magnetic field that will produce the magnetic force shown.

16. A wire having a mass per unit length of 0.500 g/cm carries a 2.00-A current horizontally to the south. What are the direction and magnitude of the minimum magnetic field needed to lift this wire vertically upward?

17. A wire carries a current of 10.0 A in a direction that makes an angle of 30.0° with the direction of a magnetic field of strength 0.300 T. Find the magnetic force on a 5.00-m length of the wire.

18. At a certain location, Earth has a magnetic field of 0.60×10^{-4} T, pointing 75° below the horizontal in a north–south plane. A 10.0-m-long straight wire carries a 15-A current. (a) If the current is directed horizontally toward the east, what are the magnitude and direction of the magnetic force on the wire? (b) What are the magnitude and direction of the force if the current is directed vertically upward?

19. A wire with a mass of 1.00 g/cm is placed on a horizontal surface with a coefficient of friction of 0.200. The wire carries a current of 1.50 A eastward and moves horizontally to the north. What are the magnitude and the direction of the *smallest* vertical magnetic field that enables the wire to move in this fashion?

20. A conductor suspended by two flexible wires as shown in Figure P19.20 has a mass per unit length of 0.040 0 kg/m. (a) What current must exist in the conductor for the tension in the supporting wires to be zero when the magnetic field is 3.60 T into the page? (b) What is the required direction for the current?

Figure P19.20

21. **Q|C** Consider the system pictured in Figure P19.21. A 15-cm length of conductor of mass 15 g, free to move vertically, is placed between two thin, vertical conductors, and a uniform magnetic field acts perpendicular to the page. When a 5.0-A current is directed as shown in the figure, the horizontal wire moves upward at constant velocity in the presence of gravity. (a) What forces act on the horizontal wire, and under what condition is the wire able to move upward at constant velocity? (b) Find the magnitude and direction of the minimum magnetic field required to move the wire at constant speed. (c) What happens if the magnetic field exceeds this minimum value? (The wire slides without friction on the two vertical conductors.)

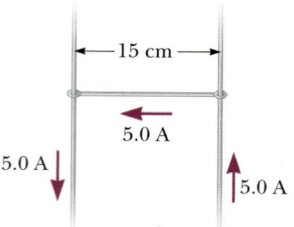

Figure P19.21

22. **S** A metal rod of mass m carrying a current I glides on two horizontal rails a distance d apart. If the coefficient of kinetic friction between the rod and rails is μ_k, what vertical magnetic field is required to keep the rod moving at a constant speed?

23. In Figure P19.23 the cube is 40.0 cm on each edge. Four straight segments of wire—ab, bc, cd, and da—form a closed loop that carries a current $I = 5.00$ A in the direction shown. A uniform magnetic field of magnitude $B = 0.020\ 0$ T is in the positive y-direction. Determine the magnitude and direction of the magnetic force on each segment.

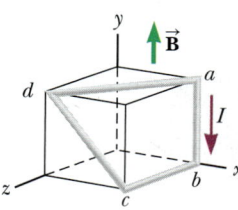

Figure P19.23

24. A horizontal power line of length 58 m carries a current of 2.2 kA as shown in Figure P19.24. Earth's magnetic field at this location has a magnitude equal to 5.0×10^{-5} T and makes an angle of 65° with the power line. Find the magnitude and direction of the magnetic force on the power line.

Figure P19.24

19.5 Torque on a Current Loop and Electric Motors

25. A wire is formed into a circle having a diameter of 10.0 cm and is placed in a uniform magnetic field of 3.00 mT. The wire carries a current of 5.00 A. Find the maximum torque on the wire.

26. A current of 17.0 mA is maintained in a single circular loop with a circumference of 2.00 m. A magnetic field of 0.800 T is directed parallel to the plane of the loop. What is the magnitude of the torque exerted by the magnetic field on the loop?

27. **M** An eight-turn coil encloses an elliptical area having a major axis of 40.0 cm and a minor axis of 30.0 cm (Fig. P19.27). The coil lies in the plane of the page and carries a clockwise current of 6.00 A. If the coil is in a uniform magnetic field of 2.00×10^{-4} T directed toward the left of the page, what is the magnitude of the torque on the coil? *Hint:* The area of an ellipse is $A = \pi ab$, where a and b are, respectively, the semimajor and semiminor axes of the ellipse.

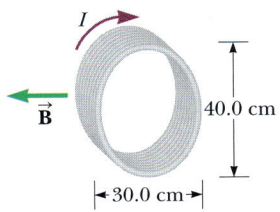

Figure P19.27

28. **W** A rectangular loop consists of 100 closely wrapped turns and has dimensions 0.400 m by 0.300 m. The loop is hinged along the y-axis, and the plane of the coil makes an angle of 30.0° with the x-axis (Fig. P19.28). What is the magnitude of the torque exerted on the loop by a uniform magnetic field of 0.800 T directed along the x-axis when the current in the windings has a value of 1.20 A in the direction shown? What is the expected direction of rotation of the loop?

Figure P19.28

29. A 200-turn rectangular coil having dimensions of 3.0 cm by 5.0 cm is placed in a uniform magnetic field of magnitude 0.90 T. (a) Find the current in the coil if the maximum torque exerted on it by the magnetic field is 0.15 N · m. (b) Find the magnitude of the torque on the coil when the magnetic field makes an angle of 25° with the normal to the plane of the coil.

30. A copper wire is 8.00 m long and has a cross-sectional area of 1.00×10^{-4} m². The wire forms a one-turn loop in the shape of a square and is then connected to a battery that applies a potential difference of 0.100 V. If the loop is placed in a uniform magnetic field of magnitude 0.400 T, what is the maximum torque that can act on it? The resistivity of copper is 1.70×10^{-8} Ω · m.

31. A long piece of wire with a mass of 0.100 kg and a total length of 4.00 m is used to make a square coil with a side of 0.100 m. The coil is hinged along a horizontal side, carries a 3.40-A current, and is placed in a vertical magnetic field with a magnitude of 0.010 0 T. (a) Determine the angle that the plane of the coil makes with the vertical when the coil is in equilibrium. (b) Find the torque acting on the coil due to the magnetic force at equilibrium.

32. **GP** A rectangular loop has dimensions 0.500 m by 0.300 m. The loop is hinged along the x-axis and lies in the xy-plane (Fig. P19.32). A uniform magnetic field of 1.50 T is directed at an angle of 40.0° with respect to the positive y-axis and lies parallel everywhere to the yz-plane. The loop carries a current of 0.900 A in the direction shown. (Ignore gravitation.) (a) In what direction is magnetic force exerted on wire segment ab? What is the direction of the magnetic torque associated with this force, as computed with respect to the x-axis? (b) What is the direction of the magnetic force exerted on segment cd? What is the direction of the magnetic torque associated with this force, again computed with respect to the x-axis? (c) Can the forces examined in parts (a) and (b) combine to cause the loop to rotate around the x-axis? Can they affect the motion of the loop in any way? Explain. (d) What is the direction (in the yz-plane) of the magnetic force exerted on segment bc? Measuring torques with respect to the x-axis, what is the direction of the torque exerted by the force on segment bc? (e) Looking toward the origin along the positive x-axis, will the loop rotate clockwise or counterclockwise? (f) Compute the magnitude of the magnetic moment of the loop. (g) What is the angle between the magnetic moment vector and the magnetic field? (h) Compute the torque on the loop using the values found for the magnetic moment and magnetic field.

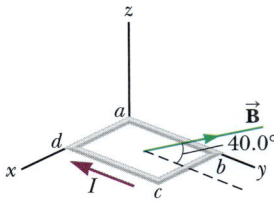

Figure P19.32

19.6 Motion of a Charged Particle in a Magnetic Field

33. An electron moves in a circular path perpendicular to a magnetic field of magnitude 0.235 T. If the kinetic energy of the electron is 3.30×10^{-19} J, find (a) the speed of the electron and (b) the radius of the circular path.

34. A proton travels with a speed of 5.02×10^6 m/s at an angle of 60° with the direction of a magnetic field of magnitude 0.180 T in the positive x-direction. What are (a) the magnitude of the magnetic force on the proton and (b) the proton's acceleration?

35. Figure P19.35a is a diagram of a device called a velocity selector, in which particles of a specific velocity pass through undeflected while those with greater or lesser velocities are deflected either upwards or downwards. An electric field is directed perpendicular to a magnetic field, producing an electric force and a magnetic force on the charged particle that can be equal in magnitude and opposite in direction (Fig. P19.35b) and hence cancel. Show that particles with a speed of $v = E/B$ will pass through the velocity selector undeflected.

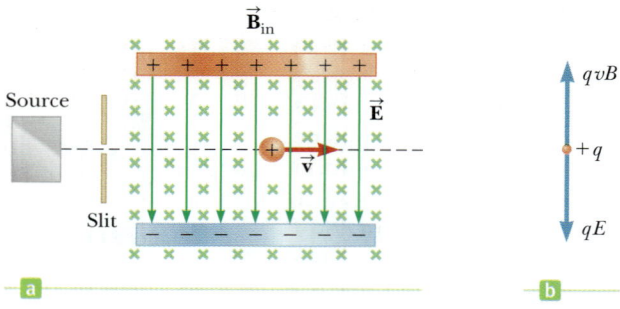

Figure P19.35

36. **W** Consider the mass spectrometer shown schematically in Figure P19.36. The electric field between the plates of the velocity selector is 950 V/m, and the magnetic fields in both the velocity selector and the deflection chamber have magnitudes of 0.930 T. Calculate the radius of the path in the system for a singly charged ion with mass $m = 2.18 \times 10^{-26}$ kg. *Hint:* See Problem 35.

Figure P19.36 (Problems 36 and 41) A mass spectrometer. Charged particles are first sent through a velocity selector. They then enter a region where a magnetic field $\vec{B}_0$ (directed inward) causes positive ions to move in a semicircular path and strike a photographic film at P.

37. A singly charged positive ion has a mass of 2.50×10^{-26} kg. After being accelerated through a potential difference of 250 V, the ion enters a magnetic field of 0.500 T, in a direction perpendicular to the field. Calculate the radius of the path of the ion in the field.

38. A mass spectrometer is used to examine the isotopes of uranium. Ions in the beam emerge from the velocity selector at a speed of 3.00×10^5 m/s and enter a uniform magnetic field of 0.600 T directed perpendicularly to the velocity of the ions. What is the distance between the impact points formed on the photographic plate by singly charged ions of ^{235}U and ^{238}U?

39. A proton is at rest at the plane vertical boundary of a region containing a uniform vertical magnetic field B (Fig. P19.39). An alpha particle moving horizontally makes a head-on elastic collision with the proton. Immediately after the collision, both particles enter the magnetic field, moving perpendicular to the direction of the field. The radius of the proton's trajectory is R. The mass of the alpha particle is four times that of the proton, and its charge is twice that of the proton. Find the radius of the alpha particle's trajectory.

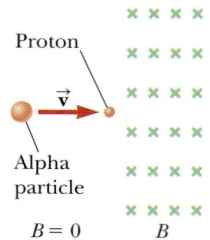

Figure P19.39

40. **S** A particle with charge q and kinetic energy KE travels in a uniform magnetic field of magnitude B. If the particle moves in a circular path of radius R, find expressions for (a) its speed and (b) its mass.

41. A particle passes through a mass spectrometer as illustrated in Figure P19.36. The electric field between the plates of the velocity selector has a magnitude of 8 250 V/m, and the magnetic fields in both the velocity selector and the deflection chamber have magnitudes of 0.093 1 T. In the deflection chamber the particle strikes a photographic plate 39.6 cm removed from its exit point after traveling in a semicircle. (a) What is the mass-to-charge ratio of the particle? (b) What is the mass of the particle if it is doubly ionized? (c) What is its identity, assuming it's an element?

42. **S** A proton (charge $+e$, mass m_p), a deuteron (charge $+e$, mass $2m_p$), and an alpha particle (charge $+2e$, mass $4m_p$) are accelerated from rest through a common potential difference ΔV. Each of the particles enters a uniform magnetic field $\vec{B}$, with its velocity in a direction perpendicular to $\vec{B}$. The proton moves in a circular path of radius r_p. In terms of r_p, determine (a) the radius r_d of the circular orbit for the deuteron and (b) the radius r_α for the alpha particle.

19.7 Magnetic Field of a Long, Straight Wire and Ampère's Law

43. A lightning bolt may carry a current of 1.00×10^4 A for a short time. What is the resulting magnetic field 100 m from the bolt? Suppose the bolt extends far above and below the point of observation.

44. In each of parts (a), (b), and (c) of Figure P19.44, find the direction of the current in the wire that would produce a magnetic field directed as shown.

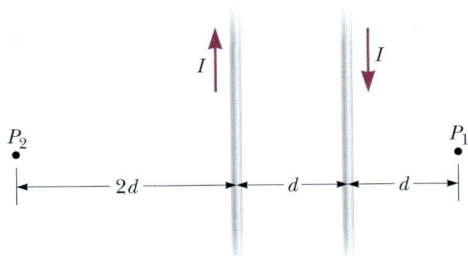

Figure P19.44

directions. Find the magnitude and direction of the net magnetic field (a) at a point midway between the wires; (b) at point P_1, 10.0 cm to the right of the wire on the right; and (c) at point P_2, $2d = 20.0$ cm to the left of the wire on the left.

49. Four long, parallel conductors carry equal currents of $I = 5.00$ A. Figure P19.49 is an end view of the conductors. The direction of the current is into the page at points A and B (indicated by the crosses) and out of the page at C and D (indicated by the dots). Calculate the magnitude and direction of the magnetic field at point P, located at the center of the square with edge of length 0.200 m.

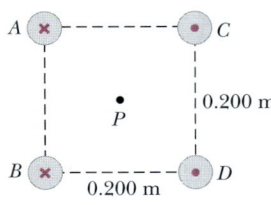

Figure P19.49

50. Two long, parallel wires carry currents of $I_1 = 3.00$ A and $I_2 = 5.00$ A in the direction indicated in Figure P19.50. (a) Find the magnitude and direction of the magnetic field at a point midway between the wires ($d = 20.0$ cm). (b) Find the magnitude and direction of the magnetic field at point P, located $d = 20.0$ cm above the wire carrying the 5.00-A current.

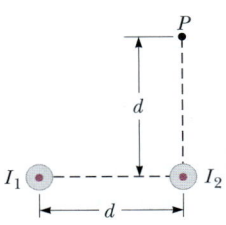

Figure P19.50

51. **M** A wire carries a 7.00-A current along the x-axis, and another wire carries a 6.00-A current along the y-axis, as shown in Figure P19.51. What is the magnetic field at point P, located at $x = 4.00$ m, $y = 3.00$ m?

45. **BIO** Neurons in our bodies carry weak currents that produce detectable magnetic fields. A technique called *magnetoencephalography*, or MEG, is used to study electrical activity in the brain using this concept. This technique is capable of detecting magnetic fields as weak as 1.0×10^{-15} T. Model the neuron as a long wire carrying a current and find the current it must carry to produce a field of this magnitude at a distance of 4.0 cm from the neuron.

46. In 1962 measurements of the magnetic field of a large tornado were made at the Geophysical Observatory in Tulsa, Oklahoma. If the magnitude of the tornado's field was $B = 1.50 \times 10^{-8}$ T pointing north when the tornado was 9.00 km east of the observatory, what current was carried up or down the funnel of the tornado? Model the vortex as a long, straight wire carrying a current.

47. **BIO** A cardiac pacemaker can be affected by a static magnetic field as small as 1.7 mT. How close can a pacemaker wearer come to a long, straight wire carrying 20 A?

48. The two wires shown in Figure P19.48 are separated by $d = 10.0$ cm and carry currents of $I = 5.00$ A in opposite

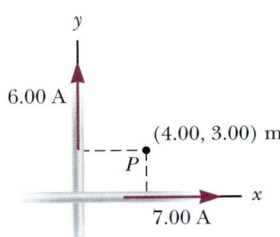

Figure P19.51

52. **GP** A long, straight wire lies on a horizontal table in the xy-plane and carries a current of 1.20 μA in the positive x-direction along the x-axis. A proton is traveling in the negative x-direction at speed 2.30×10^4 m/s a distance d above the wire (i.e., $z = d$). (a) What is the direction of the magnetic field of the wire at the position of the proton? (b) What is the direction of the magnetic force acting on the proton? (c) Explain why the direction of the proton's motion doesn't change. (d) Using Newton's second law, find a symbolic expression for d in terms of the acceleration of gravity g, the proton mass m, its speed v, charge q, and the current I.

Figure P19.48

(e) Find the numeric answer for the distance *d* using the results of part (d).

53. The magnetic field 40.0 cm away from a long, straight wire carrying current 2.00 A is 1.00 μT. (a) At what distance is it 0.100 μT? (b) At one instant, the two conductors in a long household extension cord carry equal 2.00-A currents in opposite directions. The two wires are 3.00 mm apart. Find the magnetic field 40.0 cm away from the middle of the straight cord, in the plane of the two wires. (c) At what distance is it one-tenth as large? (d) The center wire in a coaxial cable carries current 2.00 A in one direction, and the sheath around it carries current 2.00 A in the opposite direction. What magnetic field does the cable create at points outside?

54. Q|C S Two long, parallel wires separated by a distance $2d$ carry equal currents in the same direction. An end view of the two wires is shown in Figure P19.54, where the currents are out of the page. (a) What is the direction of the magnetic field at *P* on the *x*-axis set up by the two wires? (b) Find an expression for the magnitude of the field at *P*. (c) From your result to part (b), determine the field at a point midway between the two wires. Does your result meet with your expectation? Explain.

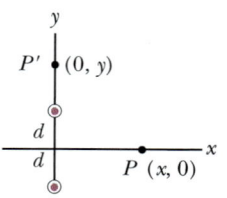

Figure P19.54

19.8 Magnetic Force Between Two Parallel Conductors

55. Two long, parallel wires separated by 2.50 cm carry currents in opposite directions. The current in one wire is 1.25 A, and the current in the other is 3.50 A. (a) Find the magnitude of the force per unit length that one wire exerts on the other. (b) Is the force attractive or repulsive?

56. Q|C Two parallel wires separated by 4.0 cm repel each other with a force per unit length of 2.0×10^{-4} N/m. The current in one wire is 5.0 A. (a) Find the current in the other wire. (b) Are the currents in the same direction or in opposite directions? (c) What would happen if the direction of one current were reversed and doubled?

57. A wire with a weight per unit length of 0.080 N/m is suspended directly above a second wire. The top wire carries a current of 30.0 A, and the bottom wire carries a current of 60.0 A. Find the distance of separation between the wires so that the top wire will be held in place by magnetic repulsion.

58. In Figure P19.58 the current in the long, straight wire is $I_1 = 5.00$ A, and the wire lies in the plane of the rectangular loop, which carries 10.0 A. The dimensions shown are $c = 0.100$ m, $a = 0.150$ m, and $\ell = 0.450$ m. Find the magnitude and direction of the net force exerted by the magnetic field due to the straight wire on the loop.

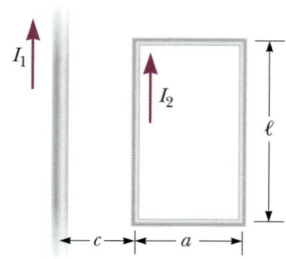

Figure P19.58

19.9 Magnetic Fields of Current Loops and Solenoids

59. M A long solenoid that has 1 000 turns uniformly distributed over a length of 0.400 m produces a magnetic field of magnitude 1.00×10^{-4} T at its center. What current is required in the windings for that to occur?

60. A certain superconducting magnet in the form of a solenoid of length 0.50 m can generate a magnetic field of 9.0 T in its core when its coils carry a current of 75 A. The windings, made of a niobium–titanium alloy, must be cooled to 4.2 K. Find the number of turns in the solenoid.

61. It is desired to construct a solenoid that will have a resistance of 5.00 Ω (at 20°C) and produce a magnetic field of 4.00×10^{-2} T at its center when it carries a current of 4.00 A. The solenoid is to be constructed from copper wire having a diameter of 0.500 mm. If the radius of the solenoid is to be 1.00 cm, determine (a) the number of turns of wire needed and (b) the length the solenoid should have.

62. A solenoid 10.0 cm in diameter and 75.0 cm long is made from copper wire of diameter 0.100 cm, with very thin insulation. The wire is wound onto a cardboard tube in a single layer, with adjacent turns touching each other. What power must be delivered to the solenoid if it is to produce a field of 8.00 mT at its center?

63. An electron is moving at a speed of 1.0×10^4 m/s in a circular path of radius 2.0 cm inside a solenoid. The magnetic field of the solenoid is perpendicular to the plane of the electron's path. Find (a) the strength of the magnetic field inside the solenoid and (b) the current in the solenoid if it has 25 turns per centimeter.

Additional Problems

64. Q|C Figure P19.64 is a setup that can be used to measure magnetic fields. A rectangular coil of wire contains *N* turns has a width *w*. The coil is attached to one arm of a balance and is suspended between the poles of a magnet. The field is uniform and perpendicular to the plane of the coil. The system is first balanced when the current in the coil is zero. When the switch is closed and the coil carries a current *I*, a mass *m* must be added to the right side to balance the system. (a) Find an expression for the magnitude of the magnetic field and determine its direction. (b) Why is the result independent of the vertical dimension of the coil? (c) Suppose the coil has 50 turns and width of 5.0 cm. When the switch is closed, the coil carries a current of 0.30 A, and a mass of 20.0 g must be added to the right side to balance the system. What is the magnitude of the magnetic field?

Figure P19.64

65. Two coplanar and concentric circular loops of wire carry currents of $I_1 = 5.00$ A and $I_2 = 3.00$ A in opposite directions as in Figure P19.65. (a) If $r_1 = 12.0$ cm and $r_2 = 9.00$ cm, what are (a) the magnitude and (b) the direction of the net magnetic field at the center of the two loops? (c) Let r_1 remain fixed at 12.0 cm and let r_2 be a variable. Determine the value of r_2 such that the net field at the center of the loop is zero.

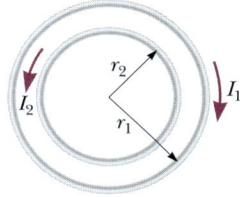

Figure P19.65

66. An electron moves in a circular path perpendicular to a constant magnetic field of magnitude 1.00 mT. The angular momentum of the electron about the center of the circle is 4.00×10^{-25} kg · m²/s. Determine (a) the radius of the circular path and (b) the speed of the electron.

67. Two long, straight wires cross each other at right angles, as shown in Figure P19.67. (a) Find the direction and magnitude of the magnetic field at point P, which is in the same plane as the two wires. (b) Find the magnetic field at a point 30.0 cm above the point of intersection (30.0 cm out of the page, toward you).

Figure P19.67

68. A 0.200-kg metal rod carrying a current of 10.0 A glides on two horizontal rails 0.500 m apart. What vertical magnetic field is required to keep the rod moving at a constant speed if the coefficient of kinetic friction between the rod and rails is 0.100?

69. BIO Using an electromagnetic flowmeter (Fig. P19.69), a heart surgeon monitors the flow rate of blood through an artery. Electrodes A and B make contact with the outer surface of the blood vessel, which has interior diameter 3.00 mm. (a) For a magnetic field magnitude of 0.040 0 T, a potential difference of 160 μV appears between the electrodes. Calculate the speed of the blood. (b) Verify that electrode A is positive, as shown. Does the sign of the emf depend on whether the mobile ions in the blood are predominantly positively or negatively charged? Explain.

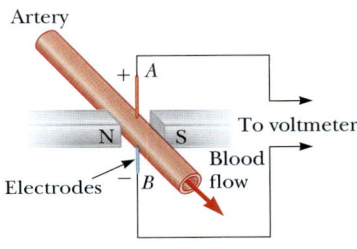

Figure P19.69

70. A uniform horizontal wire with a linear mass density of 0.50 g/m carries a 2.0-A current. It is placed in a constant magnetic field with a strength of 4.0×10^{-3} T. The field is horizontal and perpendicular to the wire. As the wire moves upward starting from rest, (a) what is its acceleration and (b) how long does it take to rise 50 cm? Neglect the magnetic field of Earth.

71. Three long, parallel conductors carry currents of $I = 2.0$ A. Figure P19.71 is an end view of the conductors, with each current coming out of the page. Given that $a = 1.0$ cm, determine the magnitude and direction of the magnetic field at points A, B, and C.

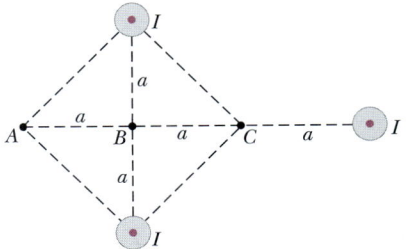

Figure P19.71

72. Two long, parallel wires, each with a mass per unit length of 40 g/m, are supported in a horizontal plane by 6.0-cm-long strings, as shown in Figure P19.72. Each wire carries the same current I, causing the wires to

repel each other so that the angle θ between the supporting strings is 16°. (a) Are the currents in the same or opposite directions? (b) Determine the magnitude of each current.

Figure P19.72

73. Protons having a kinetic energy of 5.00 MeV are moving in the positive x-direction and enter a magnetic field of 0.050 0 T in the z-direction, out of the plane of the page, and extending from $x = 0$ to $x = 1.00$ m as in Figure P19.73. (a) Calculate the y-component of the protons' momentum as they leave the magnetic field. (b) Find the angle α between the initial velocity vector of the proton beam and the velocity vector after the beam emerges from the field. *Hint:* Neglect relativistic effects and note that 1 eV = 1.60×10^{-19} J.

Figure P19.73

74. A straight wire of mass 10.0 g and length 5.0 cm is suspended from two identical springs that, in turn, form a closed circuit (Fig. P19.74). The springs stretch a distance of 0.50 cm under the weight of the wire. The circuit has a total resistance of 12 Ω. When a magnetic field directed out of the page (indicated by the dots in the figure) is turned on, the springs are observed to stretch an additional 0.30 cm. What is the strength of the magnetic field? (The upper portion of the circuit is fixed.)

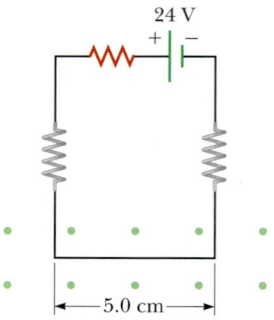

Figure P19.74

75. A 1.00-kg ball having net charge $Q = 5.00$ μC is thrown out of a window horizontally at a speed $v = 20.0$ m/s. The window is at a height $h = 20.0$ m above the ground. A uniform horizontal magnetic field of magnitude $B = 0.010\ 0$ T is perpendicular to the plane of the ball's trajectory. Find the magnitude of the magnetic force acting on the ball just before it hits the ground. *Hint:* Ignore magnetic forces in finding the ball's final velocity.

76. Two long, parallel conductors separated by 10.0 cm carry currents in the same direction. The first wire carries a current $I_1 = 5.00$ A, and the second carries $I_2 = 8.00$ A. (a) What is the magnitude of the magnetic field created by I_1 at the location of I_2? (b) What is the force per unit length exerted by I_1 on I_2? (c) What is the magnitude of the magnetic field created by I_2 at the location of I_1? (d) What is the force per length exerted by I_2 on I_1?

A forest of fluorescent lights, not wired to any power source, are lighted by electromagnetic induction. Changing currents in the power wires overhead create time-dependent magnetic flux in the vicinity of the tubes, inducing a voltage across them.

Induced Voltages and Inductance 20

In 1819 Hans Christian Oersted discovered that an electric current exerted a force on a magnetic compass. Although there had long been speculation that such a relationship existed, Oersted's finding was the first evidence of a link between electricity and magnetism. Because nature is often symmetric, the discovery that electric currents produce magnetic fields led scientists to suspect that magnetic fields could produce electric currents. Indeed, experiments conducted by Michael Faraday in England and independently by Joseph Henry in the United States in 1831 showed that a changing magnetic field could induce an electric current in a circuit. The results of these experiments led to a basic and important law known as Faraday's law. In this chapter we discuss Faraday's law and several practical applications, one of which is the production of electrical energy in power plants throughout the world.

20.1 Induced emf and Magnetic Flux
20.2 Faraday's Law of Induction and Lenz's Law
20.3 Motional emf
20.4 Generators
20.5 Self-Inductance
20.6 RL Circuits
20.7 Energy Stored in a Magnetic Field

20.1 Induced emf and Magnetic Flux

LEARNING OBJECTIVES

1. Define magnetic flux and discuss its role in producing an induced emf.
2. Evaluate the magnitude and change in the magnetic flux through a given area.

An experiment first conducted by Faraday demonstrated that a current can be produced by a changing magnetic field. The apparatus shown in Figure 20.1 consists of a coil connected to a switch and a battery. We call this coil the *primary coil* and the corresponding circuit the primary circuit. The coil is wrapped around an iron ring to intensify the magnetic field produced by the current in the coil. A

Figure 20.1 Faraday's experiment.

The emf in the secondary circuit is induced by the changing magnetic field through the coil in that circuit.

When the switch in the primary circuit is closed, the ammeter in the secondary circuit at the right measures a momentary current.

Michael Faraday
British physicist and chemist (1791–1867)
Faraday is often regarded as the greatest experimental scientist of the 1800s. His many contributions to the study of electricity include the inventions of the electric motor, electric generator, and transformer, as well as the discovery of electromagnetic induction and the laws of electrolysis. Greatly influenced by religion, he refused to work on military poison gas for the British government.

secondary coil, at the right, is wrapped around the iron ring and is connected to an ammeter. The corresponding circuit is called the secondary circuit. It's important to notice that **there is no battery in the secondary circuit**.

At first glance, you might guess that no current would ever be detected in the secondary circuit. When the switch in the primary circuit in Figure 20.1 is suddenly closed, however, something amazing happens: the ammeter measures a current in the secondary circuit and then returns to zero! When the switch is opened again, the ammeter reads a current in the opposite direction and again returns to zero. Finally, whenever there is a steady current in the primary circuit, the ammeter reads zero.

From such observations, Faraday concluded that an electric current could be produced by a *changing* magnetic field. (A steady magnetic field doesn't produce a current unless the coil is moving, as explained below.) The current produced in the secondary circuit occurs for only an instant while the magnetic field through the secondary coil is changing. In effect, the secondary circuit behaves as though a source of emf were connected to it for a short time. It's customary to say that **an induced emf is produced in the secondary circuit by the changing magnetic field**.

Magnetic Flux

To evaluate induced emfs quantitatively, we need to understand what factors affect the phenomenon. Although changing magnetic fields always induce electric fields, in other situations the magnetic field remains constant, yet an induced electric field is still produced. The best example of this is an electric generator: A loop of conductor rotating in a constant magnetic field creates an electric current.

The physical quantity associated with magnetism that creates an electric field is a **changing magnetic flux**. Magnetic flux is defined in the same way as electric flux (Section 15.9) and is proportional to both the strength of the magnetic field passing through the plane of a loop of wire and the area of the loop.

▶ Magnetic flux

The **magnetic flux** Φ_B through a loop of wire with area A is defined by

$$\Phi_B \equiv B_\perp A = BA \cos\theta \qquad [20.1]$$

where $B_\perp$ is the component of a uniform magnetic field $\vec{B}$ perpendicular to the plane of the loop, as in Figure 20.2a, and θ is the angle between $\vec{B}$ and the normal (perpendicular) to the plane of the loop.

SI unit: weber (Wb)

20.1 | Induced emf and Magnetic Flux

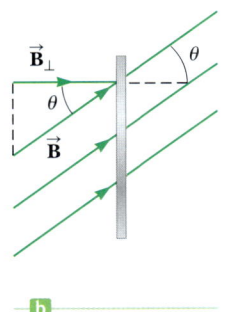

Figure 20.2 (a) A uniform magnetic field $\vec{B}$ making an angle θ with a direction normal to the plane of a wire loop of area A. (b) An edge view of the loop.

Note that there are always two directions normal to a given plane surface. In Figure 20.2, for example, that direction could be chosen to be to the right, resulting in positive flux. The normal direction could also be chosen to point to the left, which would result in a negative flux of the same magnitude. The choice of normal direction is called the orientation of the surface. Once chosen in a given problem, the normal direction remains fixed. A good default is to choose the normal direction so that the initial angle between the magnetic field and the normal direction is less than 90°.

From Equation 20.1, it follows that $B_\perp = B \cos \theta$. The magnetic flux, in other words, is the magnitude of the part of $\vec{B}$ that is perpendicular to the plane of the loop times the area of the loop. Figure 20.2b is an edge view of the loop and the penetrating magnetic field lines. When the field is perpendicular to the plane of the loop as in Figure 20.3a, $\theta = 0$ and Φ_B has a maximum value, $\Phi_{B,\text{max}} = BA$. When the plane of the loop is parallel to $\vec{B}$ as in Figure 20.3b, $\theta = 90°$ and $\Phi_B = 0$. The flux can also be negative. For example, when $\theta = 180°$, the flux is equal to $-BA$. Because the SI unit of B is the tesla, or weber per square meter, the unit of flux is $T \cdot m^2$, or weber (Wb).

We can emphasize the qualitative meaning of Equation 20.1 by first drawing magnetic field lines, as in Figure 20.3. The number of lines per unit area increases as the field strength increases. **The value of the magnetic flux is proportional to the total number of lines passing through the loop.** We see that the most lines pass through the loop when its plane is perpendicular to the field, as in Figure 20.3a, so the flux has its maximum value at that time. As Figure 20.3b shows, no lines pass through the loop when its plane is parallel to the field, so in that case $\Phi_B = 0$.

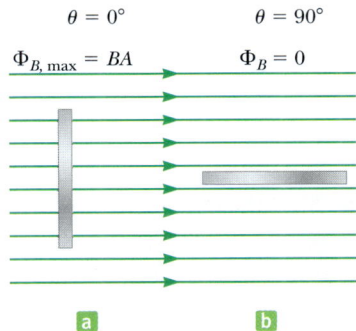

Figure 20.3 An edge view of a loop in a uniform magnetic field. (a) When the field lines are perpendicular to the plane of the loop, the magnetic flux through the loop is a maximum and equal to $\Phi_B = BA$. (b) When the field lines are parallel to the plane of the loop, the magnetic flux through the loop is zero.

■ APPLYING PHYSICS 20.1 | Flux Compared

Argentina has more land area (2.8×10^6 km^2) than Greenland (2.2×10^6 km^2). Why is the magnetic flux of Earth's magnetic field larger through Greenland than through Argentina?

EXPLANATION Greenland (latitude 60° north to 80° north) is closer to a magnetic pole than Argentina (latitude 20° south to 50° south), so the magnetic field is stronger there. That in itself isn't sufficient to conclude that the magnetic flux is greater, but Greenland's proximity to a pole also means that the angle the magnetic field lines make with the vertical is smaller than in Argentina. As a result, more field lines penetrate the surface in Greenland, despite Argentina's slightly larger area. ■

■ EXAMPLE 20.1 | Magnetic Flux

GOAL Calculate magnetic flux and a change in flux.

PROBLEM A conducting circular loop of radius 0.250 m is placed in the xy-plane in a uniform magnetic field of 0.360 T that points in the positive z-direction, the same direction as the normal to the plane. **(a)** Calculate the magnetic flux through the loop. **(b)** Suppose the loop is rotated clockwise around the x-axis, so the normal direction now points at a

(Continued)

45.0° angle with respect to the z-axis. Recalculate the magnetic flux through the loop. **(c)** What is the change in flux due to the rotation of the loop?

STRATEGY After finding the area, substitute values into the equation for magnetic flux for each part. Because the normal direction was chosen to be the same direction as the magnetic field, the angle between the magnetic field and the normal is initially 0°. After the rotation, that angle becomes 45°.

SOLUTION

(a) Calculate the initial magnetic flux through the loop.

First, calculate the area of the loop:

$$A = \pi r^2 = \pi(0.250 \text{ m})^2 = 0.196 \text{ m}^2$$

Substitute A, B, and $\theta = 0°$ into Equation 20.1 to find the initial magnetic flux:

$$\Phi_B = AB \cos \theta = (0.196 \text{ m}^2)(0.360 \text{ T}) \cos(0°)$$
$$= 0.070 \text{ 6 T} \cdot \text{m}^2 = \boxed{0.070 \text{ 6 Wb}}$$

(b) Calculate the magnetic flux through the loop after it has rotated 45.0° around the x-axis.

Make the same substitutions as in part (a), except the angle between $\vec{B}$ and the normal is now $\theta = 45.0°$:

$$\Phi_B = AB \cos \theta = (0.196 \text{ m}^2)(0.360 \text{ T}) \cos(45.0°)$$
$$= 0.049 \text{ 9 T} \cdot \text{m}^2 = \boxed{0.049 \text{ 9 Wb}}$$

(c) Find the change in the magnetic flux due to the rotation of the loop.

Subtract the result of part (a) from the result of part (b):

$$\Delta \Phi_B = 0.049 \text{ 9 Wb} - 0.070 \text{ 6 Wb} = \boxed{-0.020 \text{ 7 Wb}}$$

REMARKS Notice that the rotation of the loop, not any change in the magnetic field, is responsible for the change in flux. This changing magnetic flux is essential in the functioning of electric motors and generators.

QUESTION 20.1 True or False: If the loop is rotated in the opposite direction by the same amount, the change in magnetic flux has the same magnitude but opposite sign.

EXERCISE 20.1 The loop, having rotated by 45°, rotates clockwise another 30°, so the normal to the plane points at an angle of 75° with respect to the direction of the magnetic field. Find (a) the magnetic flux through the loop when $\theta = 75°$ and (b) the change in magnetic flux during the rotation from 45° to 75°.

ANSWERS (a) 0.018 3 Wb (b) −0.031 6 Wb

20.2 Faraday's Law of Induction and Lenz's Law

LEARNING OBJECTIVES

1. State Faraday's and Lenz's laws and describe applications based on them.
2. Apply Faraday's and Lenz's laws to systems with changing magnetic flux.

The usefulness of the concept of magnetic flux can be made obvious by another simple experiment that demonstrates the basic idea of electromagnetic induction. Consider a wire loop connected to an ammeter as in Figure 20.4. If a magnet is moved toward the loop, the ammeter reads a current in one direction, as in Figure 20.4a. When the magnet is held stationary, as in Figure 20.4b, the ammeter reads zero current. If the magnet is moved away from the loop, the ammeter reads a current in the opposite direction, as in Figure 20.4c. If the magnet is held stationary and the loop is moved either toward or away from the magnet, the ammeter also reads a current. From these observations, it can be concluded that **a current is established in the circuit as long as there is relative motion between the magnet and the loop**. The same experimental results are found whether the loop moves or the magnet moves. We call such a current an **induced current** because it is produced by an **induced emf**.

Tip 20.1 Induced Current Requires a Change in Magnetic Flux

The existence of magnetic flux through an area is not sufficient to create an induced emf. A *change* in the magnetic flux over some time interval Δt must occur for an emf to be induced.

20.2 Faraday's Law of Induction and Lenz's Law

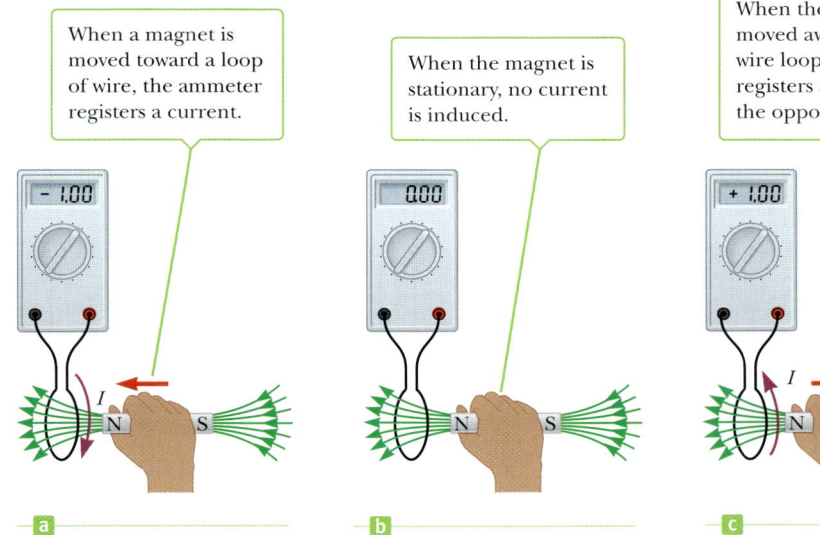

Figure 20.4 A simple experiment showing that a current is induced in a loop when a magnet is moved toward or away from the loop.

> **Tip 20.2 There Are Two Magnetic Fields to Consider**
>
> When applying Lenz's law, there are *two* magnetic fields to consider. The first is the external changing magnetic field that induces the current in a conducting loop. The second is the magnetic field produced by the induced current in the loop.

This experiment is similar to the Faraday experiment discussed in Section 20.1. In each case, an emf is induced in a circuit when the magnetic flux through the circuit changes with time. It turns out that the instantaneous emf induced in a circuit equals the negative of the rate of change of magnetic flux with respect to time through the circuit. This is **Faraday's law of magnetic induction**.

If a circuit contains N tightly wound loops and the magnetic flux through each loop changes by the amount $\Delta\Phi_B$ during the interval Δt, the average emf induced in the circuit during time Δt is

$$\mathcal{E} = -N\frac{\Delta\Phi_B}{\Delta t} \quad [20.2]$$

◀ Faraday's law

Because $\Phi_B = BA\cos\theta$, a change of any of the factors B, A, or θ with time produces an emf. We explore the effect of a change in each of these factors in the following sections. The minus sign in Equation 20.2 is included to indicate the polarity of the induced emf. This polarity determines the direction of the current in the loop, and is given by **Lenz's law**:

> The current caused by the induced emf travels in the direction that creates a magnetic field with flux opposing the change in the original flux through the circuit.

Lenz's law says that if the magnetic flux through a loop is becoming more positive, say, then the induced emf creates a current and associated magnetic field that produces negative magnetic flux. Some mistakenly think this "counter magnetic field" created by the induced current, called $\vec{B}_{ind}$ ("ind" for induced), will always point in a direction opposite the applied magnetic field $\vec{B}$, but that is only true half the time! Figure 20.5a shows a field penetrating a loop. The graph in Figure 20.5b shows that the magnitude of the magnetic field $\vec{B}$ shrinks with time, which means that the flux of $\vec{B}$ is shrinking with time, so the induced field $\vec{B}_{ind}$ will actually be in the same direction as $\vec{B}$. In effect, $\vec{B}_{ind}$ "shores up" the field $\vec{B}$, slowing the loss of flux through the loop.

The direction of the current in Figure 20.5a can be determined by right-hand rule number 2: Point your right thumb in the direction that will cause the fingers

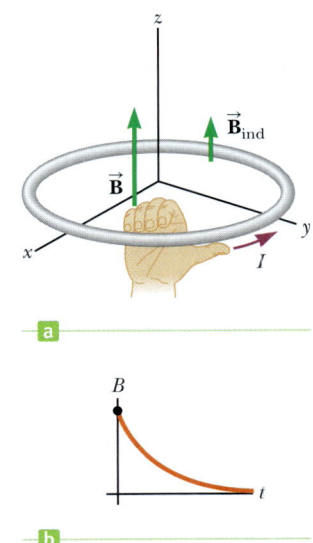

Figure 20.5 (a) The magnetic field $\vec{B}$ becomes smaller with time, reducing the flux, so current is induced in a direction that creates an induced magnetic field $\vec{B}_{ind}$ opposing the change in magnetic flux. (b) Graph of the magnitude of the magnetic field as a function of time.

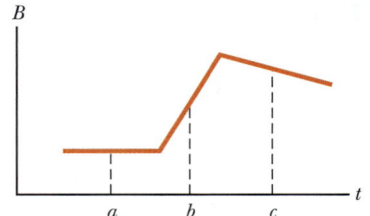

Figure 20.6 (Quick Quiz 20.1)

on your right hand to curl in the direction of the induced field $\vec{B}_{ind}$. In this case, that direction is counterclockwise: with the right thumb pointed in the direction of the current, your fingers curl down outside the loop and around and *up through the inside of the loop*. Remember, inside the loop is where it's important for the induced magnetic field to be pointing up.

■ **Quick Quiz**

20.1 Figure 20.6 is a graph of the magnitude B versus time for a magnetic field that passes through a fixed loop and is oriented perpendicular to the plane of the loop. Rank the magnitudes of the emf generated in the loop from largest to smallest at the three instants indicated.

■ **EXAMPLE 20.2** Faraday and Lenz to the Rescue

GOAL Calculate an induced emf and current with Faraday's law and apply Lenz's law when the magnetic field changes with time.

PROBLEM A coil with 25 turns of wire is wrapped on a frame with a square cross section 1.80 cm on a side. Each turn has the same area, equal to that of the frame, and the total resistance of the coil is 0.350 Ω. An applied uniform magnetic field is perpendicular to the plane of the coil, as in Figure 20.7. **(a)** If the field changes uniformly from 0.00 T to 0.500 T in 0.800 s, what is the induced emf in the coil while the field is changing? Find **(b)** the magnitude and **(c)** the direction of the induced current in the coil while the field is changing.

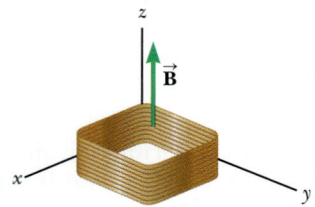

Figure 20.7 (Example 20.2)

STRATEGY Part (a) requires substituting into Faraday's law, Equation 20.2. The necessary information is given, except for $\Delta \Phi_B$, the change in the magnetic flux during the elapsed time. Using the normal direction to coincide with the positive z-axis, compute the initial and final magnetic fluxes with Equation 20.1, find the difference, and assemble all terms in Faraday's law. The current can then be found with Ohm's law, and its direction with Lenz's law.

SOLUTION

(a) Find the induced emf in the coil.

To compute the flux, the area of the coil is needed:

$$A = L^2 = (0.018\ 0\ \text{m})^2 = 3.24 \times 10^{-4}\ \text{m}^2$$

The magnetic flux $\Phi_{B,i}$ through the coil at $t = 0$ is zero because $B = 0$. Calculate the flux at $t = 0.800$ s:

$$\Phi_{B,f} = BA \cos\theta = (0.500\ \text{T})(3.24 \times 10^{-4}\ \text{m}^2) \cos (0°)$$
$$= 1.62 \times 10^{-4}\ \text{Wb}$$

Compute the change in the magnetic flux through the cross section of the coil over the 0.800-s interval:

$$\Delta \Phi_B = \Phi_{B,f} - \Phi_{B,i} = 1.62 \times 10^{-4}\ \text{Wb}$$

Substitute into Faraday's law of induction to find the induced emf in the coil:

$$\mathcal{E} = -N \frac{\Delta \Phi_B}{\Delta t} = -(25\ \text{turns})\left(\frac{1.62 \times 10^{-4}\ \text{Wb}}{0.800\ \text{s}}\right)$$
$$= \boxed{-5.06 \times 10^{-3}\ \text{V}}$$

(b) Find the magnitude of the induced current in the coil.

Substitute the voltage difference and the resistance into Ohm's law, where $\Delta V = \mathcal{E}$:

$$I = \frac{\Delta V}{R} = \frac{5.06 \times 10^{-3}\ \text{V}}{0.350\ \Omega} = \boxed{1.45 \times 10^{-2}\ \text{A}}$$

(c) Find the direction of the induced current in the coil.

The magnetic field is increasing up through the loop, in the same direction as the normal to the plane; hence, the flux is positive and is also increasing. A downward-pointing induced magnetic field will create negative flux, opposing the change. If you point your right thumb in the clockwise direction along the loop as viewed from above, your fingers curl down through the loop, which is the correct direction for the counter magnetic field. Hence the current must proceed in a clockwise direction as viewed from above the coil.

REMARKS Lenz's law can best be handled by first sketching a diagram.

QUESTION 20.2 What average emf is induced in the loop if, instead, the magnetic field changes uniformly from 0.500 T to 0 in 0.800 s? How would that affect the induced current?

EXERCISE 20.2 Suppose the magnetic field changes uniformly from 0.500 T to 0.200 T in the next 0.600 s. Compute (a) the induced emf in the coil and (b) the magnitude and direction of the induced current.

ANSWERS (a) 4.05×10^{-3} V (b) 1.16×10^{-2} A (counterclockwise as viewed from above the coil)

Finding the Direction of the Induced Current

Finding the direction of the induced current can be tricky. The following three examples illustrate how the direction is found using Lenz's law.

Lenz's Law Example 1 The current in the wire of Figure 20.8 is steadily increasing in the direction indicated. Let's choose the normal direction to be out of the page so that magnetic field vectors coming out of the page will produce positive magnetic flux. The magnetic field created by the current I circulates around the wire, going into the page on the right side of the long wire in the region of the rectangular coil and coming out of the page on the left side of the long straight wire. Therefore, the magnetic flux through the rectangular coil due to the current I is negative. Because the current is increasing up the page, the magnetic field is becoming stronger, increasing the magnitude of the negative flux through the rectangular coil. By Lenz's law, the induced current in the coil must produce positive flux, countering the increasing negative flux. That requires an induced magnetic field pointing out of the page through the coil. Mentally curl the fingers of the right hand around the right branch of the rectangular coil, and note that the fingers come straight up out of the page through the coil, as required. The right thumb, meanwhile, points up the page, indicating the current direction in that part of the coil. Therefore the induced current in the coil is counterclockwise.

Lenz's Law Example 2 In Figure 20.9a, the north pole of the magnet moves toward the coil. If the normal direction is chosen to the right, then the magnetic flux through the coil due to the magnet is positive and increases with time. A negative flux to the left must therefore be created by the induced current in the coil, so the induced magnetic field must also point to the left as indicated in Figure 20.9b. Imagine curling the fingers of the right hand around the coil so they point through the coil to the left. The right thumb then points upward, indicating the induced current is counterclockwise as viewed from the left side of the loop.

Lenz's Law Example 3 Consider a coil of wire placed near a solenoid in Figure 20.10a (page 706). The wire is wrapped in such a way as to create a south magnetic pole at the right end when the switch is closed in Figure 20.10b. Choose left as the normal direction. When the switch is closed, the current in the solenoid begins to increase, and the magnetic flux through the coil is positive and increasing with time. Therefore the induced current in the coil must create negative magnetic flux to counteract the increasing positive flux created by the current in the

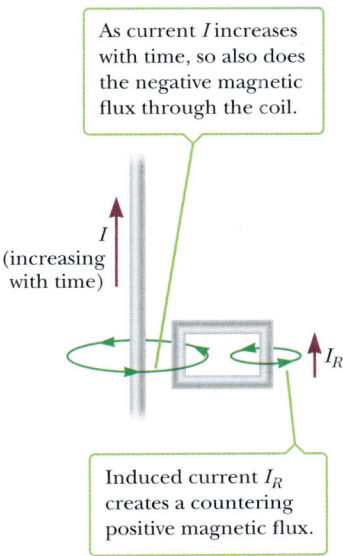

Figure 20.8 (Lenz's Law Example 1) Current I increases in magnitude with time, strengthening the magnetic field that circulates around the wire.

Figure 20.9 (Lenz's Law Example 2) (a) The north pole of the magnet approaches the coil from the left, with the normal direction taken to the right. (b) A current is induced in the coil.

Figure 20.10 (Lenz's Law Example 3) (a) The turns of the solenoid create a magnetic field with north pole pointing left, which is also taken as the normal direction. (b) When the switch is closed, positive flux begins increasing through the coil as field lines converge on the solenoid's south pole. (c) Opening the switch causes the solenoid's field to rapidly decrease.

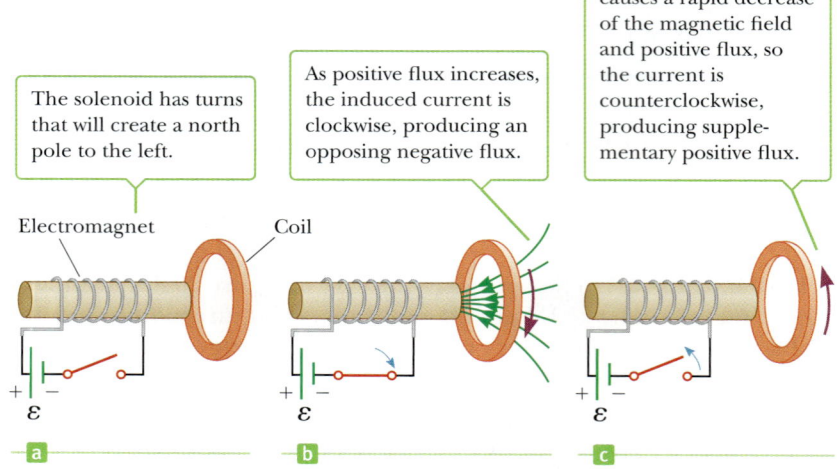

solenoid. That requires an induced magnetic field directed to the right through the coil. Turning the right hand so the thumb is pointed downward, the right fingers can curl through the coil and to the right. The induced current in the coil follows the direction of the right thumb, which is clockwise as viewed from the left end of the coil. When the switch is opened again in Figure 20.10c the current in the solenoid changes direction because the magnetic field and positive flux begin to decrease. Moving counterclockwise, the induced current creates positive flux through the coil, opposing the decrease in positive flux.

In all three of these examples, the critical idea is that a changing flux causes an induced current, and the associated induced magnetic field produces flux opposing the change in flux in accordance with Lenz's law. When the flux stops changing, the induced current stops. Although in each case the magnetic flux changed because of a changing magnetic field, induced currents can result even when the magnetic field is constant provided the flux through the loop changes. That fact will become clear when discussing motional emf in Section 20.3 and generators in Section 20.4.

■ **Quick Quiz**

20.2 A bar magnet is falling toward the center of a loop of wire, with the north pole oriented downward. Viewed from the same side of the loop as the magnet, as the north pole approaches the loop, what is the direction of the induced current? (a) clockwise (b) zero (c) counterclockwise (d) along the length of the magnet

20.3 Two circular loops are side by side and lie in the xy-plane. A switch is closed, starting a counterclockwise current in the left-hand loop, as viewed from a point on the positive z-axis passing through the center of the loop. Which of the following statements is true of the right-hand loop? (a) The current remains zero. (b) An induced current moves counterclockwise. (c) An induced current moves clockwise.

The ground fault interrupter (GFI) is an interesting safety device that protects people against electric shock when they touch appliances and power tools. Its operation makes use of Faraday's law. Figure 20.11 shows the essential parts of a ground fault interrupter. Wire 1 leads from the wall outlet to the appliance to be protected, and wire 2 leads from the appliance back to the wall outlet. An iron ring surrounds the two wires to confine the magnetic field set up by each wire. A sensing coil, which can activate a circuit breaker when changes in magnetic flux occur, is wrapped around part of the iron ring. Because the currents in the wires are in opposite directions, the net magnetic field through the sensing coil due to

Figure 20.11 Essential components of a ground fault interrupter (contents of the gray box in Fig. 20.12a). In newer homes such devices are built directly into wall outlets. The purpose of the sensing coil and circuit breaker is to cut off the current before damage is done.

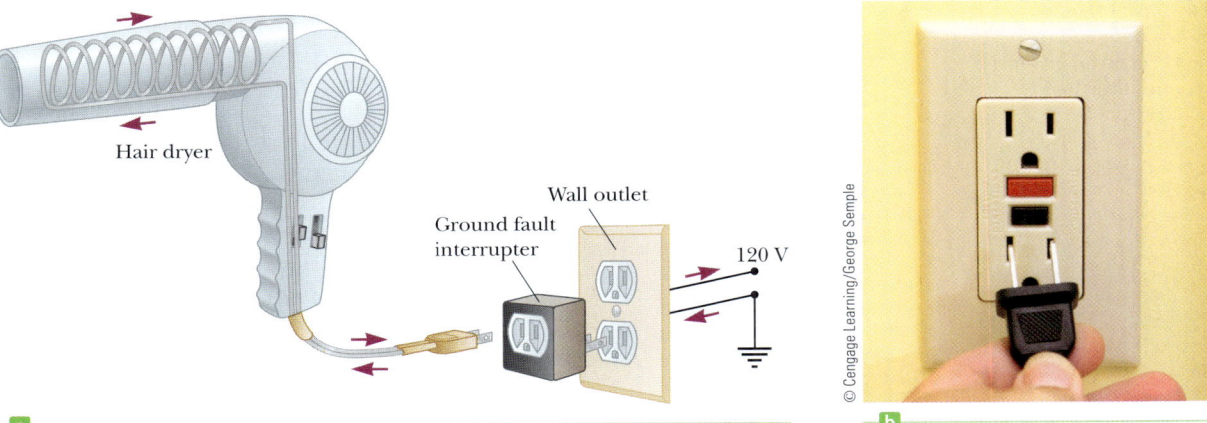

Figure 20.12 (a) This hair dryer has been plugged into a ground fault interrupter that is in turn plugged into an unprotected wall outlet. (b) You likely have seen this kind of ground fault interrupter in a hotel bathroom, where hair dryers and electric shavers are often used by people just out of the shower or who might touch a water pipe, providing a ready path to ground in the event of a short circuit.

the currents is zero. If a short circuit occurs in the appliance so that there is no returning current, however, the net magnetic field through the sensing coil is no longer zero. A short circuit can happen if, for example, one of the wires loses its insulation, providing a path through you to ground if you happen to be touching the appliance and are grounded as in Figure 18.23a. Because the current is alternating, the magnetic flux through the sensing coil changes with time, producing an induced voltage in the coil. This induced voltage is used to trigger a circuit breaker, stopping the current quickly (in about 1 ms) before it reaches a level that might be harmful to the person using the appliance. A ground fault interrupter provides faster and more complete protection than even the case-ground-and-circuit-breaker combination shown in Figure 18.23b. For this reason, ground fault interrupters are commonly found in bathrooms, where electricity poses a hazard to people. (See Fig. 20.12.)

Another interesting application of Faraday's law is the production of sound in an electric guitar. A vibrating string induces an emf in a coil (Fig. 20.13). The pickup coil is placed near the vibrating guitar string, which is made of a metal that can be magnetized. The permanent magnet inside the coil magnetizes the portion of the string nearest the coil. When the guitar string vibrates at some frequency, its magnetized segment produces a changing magnetic flux through the pickup coil. The changing flux induces a voltage in the coil, which is fed to an amplifier. The output of the amplifier is sent to the loudspeakers, producing the sound waves that we hear.

Sudden infant death syndrome, or SIDS, is a devastating affliction in which a baby suddenly stops breathing during sleep without an apparent cause. One type of monitoring device, called an apnea monitor, is sometimes used to alert caregivers

APPLICATION
Ground fault interrupters

APPLICATION
Electric guitar pickups

Figure 20.13 (a) In an electric guitar a vibrating string induces a voltage in the pickup coil. (b) Several pickups allow the vibration to be detected from different portions of the string.

Figure 20.14 This infant is wearing a monitor designed to alert care-givers if breathing stops. Notice the two wires attached to opposite sides of the chest.

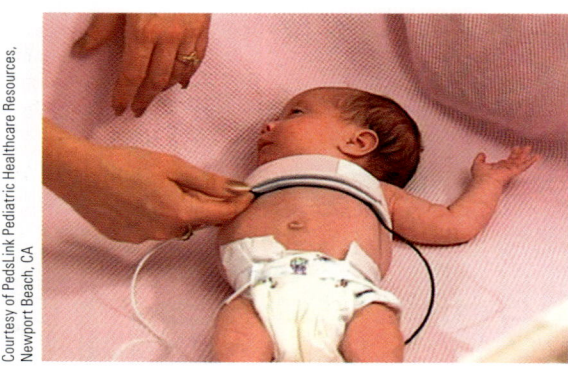

BIO APPLICATION
Apnea monitors

of the cessation of breathing. The device uses induced currents, as shown in Figure 20.14. A coil of wire attached to one side of the chest carries an alternating current. The varying magnetic flux produced by this current passes through a pickup coil attached to the opposite side of the chest. Expansion and contraction of the chest caused by breathing or movement change the strength of the voltage induced in the pickup coil. If breathing stops, however, the pattern of the induced voltage stabilizes, and external circuits monitoring the voltage sound an alarm to the caregivers after a momentary pause to ensure that a problem actually does exist.

20.3 Motional emf

LEARNING OBJECTIVES

1. Define motional emf and use Faraday's law to discuss the potential difference across a moving conductor.
2. Apply Faraday's law to systems involving a conductor moving through a magnetic field.

In Section 20.2 we considered emfs induced in a circuit when the magnetic field changes with time. In this section we describe a particular application of Faraday's law in which a so-called **motional emf** is produced. It is the emf induced in a conductor moving through a magnetic field.

First consider a straight conductor of length ℓ moving with constant velocity through a uniform magnetic field directed into the paper, as in Figure 20.15. For simplicity, we assume the conductor moves in a direction perpendicular to the field. A magnetic force of magnitude $F_m = qvB$, directed downward, acts on the electrons in the conductor. Because of this magnetic force, the free electrons move to the lower end of the conductor and accumulate there, leaving a net positive charge at the upper end. As a result of this charge separation, an electric field is produced in the conductor. The charge at the ends builds up until the downward magnetic force qvB is balanced by the upward electric force qE. At this point, charge stops flowing and the condition for equilibrium requires that

$$qE = qvB \quad \text{or} \quad E = vB$$

Figure 20.15 A straight conductor of length ℓ moving with velocity $\vec{v}$ through a uniform magnetic field $\vec{B}$ directed perpendicular to $\vec{v}$. The vector $\vec{F}_m$ is the magnetic force on an electron in the conductor. An emf of $B\ell v$ is induced between the ends of the bar.

Because the electric field is uniform, the field produced in the conductor is related to the potential difference across the ends by $\Delta V = E\ell$, giving

$$\Delta V = E\ell = B\ell v \qquad [20.3]$$

Because there is an excess of positive charge at the upper end of the conductor and an excess of negative charge at the lower end, the upper end is at a higher potential than the lower end. There is a potential difference across a conductor as long as it moves through a field. If the motion is reversed, the polarity of the potential difference is also reversed.

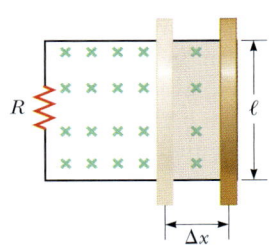

Figure 20.17 As the bar moves to the right, the area of the loop increases by the amount $\ell \Delta x$ and the magnetic flux through the loop increases by $B\ell \Delta x$.

A more interesting situation occurs if the moving conductor is part of a closed conducting path. This situation is particularly useful for illustrating how a changing loop area induces a current in a closed circuit described by Faraday's law. Consider a circuit consisting of a conducting bar of length ℓ, sliding along two fixed, parallel conducting rails, as in Figure 20.16a. For simplicity, assume the moving bar has zero resistance and the stationary part of the circuit has constant resistance R. Take the normal direction to coincide with the z-axis, out of the page. A uniform and constant magnetic field $\vec{B}$ is applied perpendicular to the plane of the circuit. As the bar is pulled to the right in the positive x-direction with velocity $\vec{v}$ under the influence of an applied force $\vec{F}_{app}$, a magnetic force along the length of the bar acts on the free charges in the bar. This force in turn sets up an induced current because the charges are free to move in a closed conducting path. In this case, the changing magnetic flux through the loop and the corresponding induced emf across the moving bar arise from the *change in area of the loop* as the bar moves through the magnetic field. Because the flux into the page increases, by Lenz's law the induced current circulates counterclockwise, producing flux out of the page that opposes the change.

Assume the bar moves a distance Δx in time Δt, as shown in Figure 20.17. The increase in flux $\Delta \Phi_B$ through the loop in that time is the amount of flux that now passes through the portion of the circuit that has area $\ell \Delta x$:

$$\Delta \Phi_B = BA = B\ell \Delta x$$

Using Faraday's law and noting that there is one loop ($N = 1$), we find that the magnitude of the induced emf is

$$|\mathcal{E}| = \frac{\Delta \Phi_B}{\Delta t} = B\ell \frac{\Delta x}{\Delta t} = B\ell v \quad [20.4]$$

This induced emf is often called a **motional emf** because it arises from the motion of a conductor through a magnetic field.

Further, if the resistance of the circuit is R, the magnitude of the induced current in the circuit is

$$I = \frac{|\mathcal{E}|}{R} = \frac{B\ell v}{R} \quad [20.5]$$

Figure 20.16b shows the equivalent circuit diagram for this example.

Figure 20.16 (a) A conducting bar sliding with velocity $\vec{v}$ along two conducting rails under the action of an applied force $\vec{F}_{app}$. (b) The equivalent circuit of that in (a).

APPLYING PHYSICS 20.2 | Space Catapult

Applying a force on the bar will result in an induced emf in the circuit shown in Figure 20.16. Suppose we remove the external magnetic field in the diagram and replace the resistor with a high-voltage source and a switch, as in Figure 20.18. What will happen when the switch is closed? Will the bar move, and does it matter which way we connect the high-voltage source?

EXPLANATION Suppose the source is capable of establishing high current. Then the two horizontal conducting rods will create a strong magnetic field in the area between

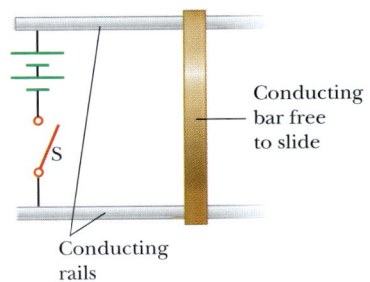

Figure 20.18 (Applying Physics 20.2)

(*Continued*)

them, directed into the page. (The movable bar also creates a magnetic field, but this field can't exert force on the bar itself.) Because the moving bar carries a downward current, a magnetic force is exerted on the bar, directed to the right. Hence, the bar accelerates along the rails away from the power supply. If the polarity of the power were reversed, the magnetic field would be out of the page, the current in the bar would be upward, and the force on the bar would still be directed to the right. The $BI\ell$ force exerted by a magnetic field according to Equation 19.6 causes the bar to accelerate away from the voltage source. Studies have shown that it's possible to launch payloads into space with this technology. (This is the working principle of a rail gun.) Very large accelerations can be obtained with currently available technology, with payloads being accelerated to a speed of several kilometers per second in a fraction of a second. This acceleration is larger than humans can tolerate.

Rail guns have been proposed as propulsion systems for moving asteroids into more useful orbits. The material of an asteroid could be mined and launched off the surface by a rail gun, which would act like a rocket engine, modifying the velocity and hence the orbit of the asteroid. Some asteroids contain trillions of dollars worth of valuable metals. ■

■ Quick Quiz

20.4 A horizontal metal bar oriented east-west drops straight down in a location where Earth's magnetic field is due north. As a result, an emf develops between the ends. Which end is positively charged? (a) the east end (b) the west end (c) neither end carries a charge

20.5 You intend to move a rectangular loop of wire into a region of uniform magnetic field at a given speed so as to induce an emf in the loop. The plane of the loop must remain perpendicular to the magnetic field lines. In which orientation should you hold the loop while you move it into the region with the magnetic field to generate the largest emf? (a) with the long dimension of the loop parallel to the velocity vector (b) with the short dimension of the loop parallel to the velocity vector (c) either way because the emf is the same regardless of orientation

■ EXAMPLE 20.3 | A Potential Difference Induced Across Airplane Wings

GOAL Find the emf induced by motion through a magnetic field.

PROBLEM An airplane with a wingspan of 30.0 m flies due north at a location where the downward component of Earth's magnetic field is 0.600×10^{-4} T. There is also a component pointing due north that has a magnitude of 0.470×10^{-4} T. **(a)** Find the difference in potential between the wingtips when the speed of the plane is 2.50×10^2 m/s. **(b)** Which wingtip is positive?

STRATEGY Because the plane is flying north, the northern component of magnetic field won't have any effect on the induced emf. The induced emf across the wing is caused solely by the downward component of the Earth's magnetic field. Substitute the given quantities into Equation 20.4. Use right-hand rule number 1 to find the direction positive charges would be propelled by the magnetic force.

SOLUTION

(a) Calculate the difference in potential across the wingtips.

Write the motional emf equation and substitute the given quantities:

$\mathcal{E} = B\ell v = (0.600 \times 10^{-4}\text{ T})(30.0\text{ m})(2.50 \times 10^2\text{ m/s})$
$= \boxed{0.450 \text{ V}}$

(b) Which wingtip is positive?

Apply right-hand rule number 1:

Point the fingers of your right hand north, in the direction of the velocity, and curl them down, in the direction of the magnetic field. Your thumb points $\boxed{\text{west.}}$ Therefore the west wingtip is positive.

REMARKS An induced emf such as this one can cause problems on an aircraft.

QUESTION 20.3 In what directions are magnetic forces exerted on electrons in the metal aircraft if it is flying due west? (a) north (b) south (c) east (d) west (e) up (f) down

20.3 | Motional emf

EXERCISE 20.3 Suppose a space station is in orbit where the magnetic field is parallel to Earth's surface, points north, and has magnitude 1.80×10^{-4} T. A metal cable attached to the space station stretches radially outwards 2.50 km. (a) Estimate the potential difference that develops between the ends of the cable if it's traveling eastward around Earth at a speed of 7.70×10^3 m/s. (b) Which end of the cable is positive, the lower end or the upper end?

ANSWERS (a) 3.47×10^3 V (b) The upper end is positive.

■ EXAMPLE 20.4   Where Is the Energy Source?

GOAL Use motional emf to find an induced emf and a current.

PROBLEM (a) The sliding bar in Figure 20.16a has a length of 0.500 m and moves at 2.00 m/s in a magnetic field of magnitude 0.250 T. Using the concept of motional emf, find the induced voltage in the moving rod. (b) If the resistance in the circuit is 0.500 Ω, find the current in the circuit and the power delivered to the resistor. (*Note:* The current in this case goes counterclockwise around the loop.) (c) Calculate the magnetic force on the bar. (d) Use the concepts of work and power to calculate the applied force.

STRATEGY For part (a), substitute into Equation 20.4 for the motional emf. Once the emf is found, substitution into Ohm's law gives the current. In part (c), use Equation 19.6 for the magnetic force on a current-carrying conductor. In part (d), use the fact that the power dissipated by the resistor multiplied by the elapsed time must equal the work done by the applied force.

SOLUTION

(a) Find the induced emf with the concept of motional emf.

Substitute into Equation 20.4 to find the induced emf:

$\mathcal{E} = B\ell v = (0.250 \text{ T})(0.500 \text{ m})(2.00 \text{ m/s}) = $ **0.250 V**

(b) Find the induced current in the circuit and the power dissipated by the resistor.

Substitute the emf and the resistance into Ohm's law to find the induced current:

$I = \dfrac{\mathcal{E}}{R} = \dfrac{0.250 \text{ V}}{0.500 \text{ Ω}} = $ **0.500 A**

Substitute $I = 0.500$ A and $\mathcal{E} = 0.250$ V into Equation 17.8 to find the power dissipated by the 0.500-Ω resistor:

$P = I\Delta V = (0.500 \text{ A})(0.250 \text{ V}) = $ **0.125 W**

(c) Calculate the magnitude and direction of the magnetic force on the bar.

Substitute values for I, B, and ℓ into Equation 19.6, with $\sin\theta = \sin(90°) = 1$, to find the magnitude of the force:

$F_m = IB\ell = (0.500 \text{ A})(0.250 \text{ T})(0.500 \text{ m}) = $ **6.25×10^{-2} N**

Apply right-hand rule number 1 to find the direction of the force:

Point the fingers of your right hand in the direction of the positive current, then curl them in the direction of the magnetic field. Your thumb points in the **negative x-direction**.

(d) Find the value of F_{app}, the applied force.

Set the work done by the applied force equal to the dissipated power times the elapsed time:

$W_{\text{app}} = F_{\text{app}} d = P\Delta t$

Solve for F_{app} and substitute $d = v\Delta t$:

$F_{\text{app}} = \dfrac{P\Delta t}{d} = \dfrac{P\Delta t}{v\Delta t} = \dfrac{P}{v} = \dfrac{0.125 \text{ W}}{2.00 \text{ m/s}} = $ **6.25×10^{-2} N**

REMARKS Part (d) could be solved by using Newton's second law for an object in equilibrium: Two forces act horizontally on the bar and the acceleration of the bar is zero, so the forces must be equal in magnitude and opposite in direction. Notice the agreement between the answers for F_m and F_{app}, despite the very different concepts used.

QUESTION 20.4 Suppose the applied force and magnetic field in Figure 20.16a are removed, but a battery creates a current in the same direction as indicated. What happens to the bar?

(*Continued*)

EXERCISE 20.4 Suppose the current suddenly increases to 1.25 A in the same direction as before due to an increase in speed of the bar. Find (a) the emf induced in the rod and (b) the new speed of the rod.

ANSWERS (a) 0.625 V (b) 5.00 m/s

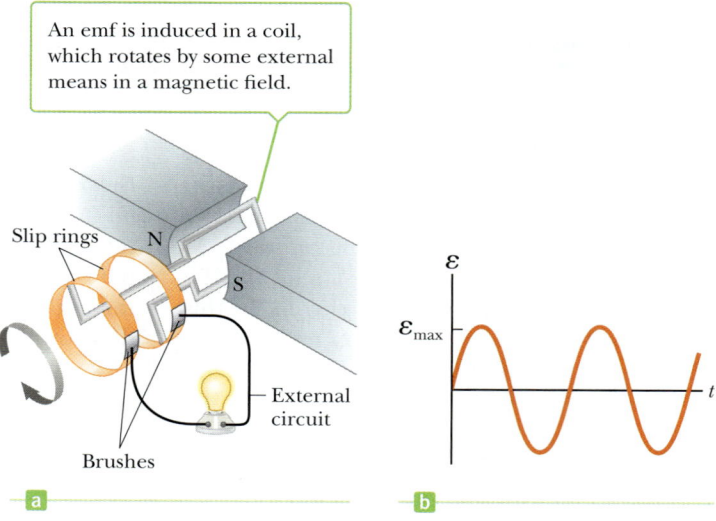

Figure 20.19 (a) A schematic diagram of an AC generator. An emf is induced in a coil, which rotates by some external means in a magnetic field. (b) A plot of the alternating emf induced in the loop versus time.

20.4 Generators

LEARNING OBJECTIVES

1. Contrast the operating principles of AC and DC generators.
2. Describe the operating principles of motors and the phenomenon of back emf.
3. Apply Faraday's law to generators and motors.

APPLICATION
Alternating-current generators

Generators and motors are important practical devices that operate on the principle of electromagnetic induction. First, consider the **alternating-current** (AC) **generator**, a device that converts mechanical energy to electrical energy. In its simplest form, the AC generator consists of a wire loop rotated in a magnetic field by some external means (Fig. 20.19a). In commercial power plants, the energy required to rotate the loop can be derived from a variety of sources. In a hydroelectric plant, for example, falling water directed against the blades of a turbine produces the rotary motion; in a coal-fired plant, heat produced by burning coal is used to convert water to steam, and this steam is directed against the turbine blades. As the loop rotates, the magnetic flux through it changes with time, inducing an emf and a current in an external circuit. The ends of the loop are connected to slip rings that rotate with the loop. Connections to the external circuit are made by stationary brushes in contact with the slip rings.

We can derive an expression for the emf generated in the rotating loop by making use of the equation for motional emf, $\mathcal{E} = B\ell v$. Figure 20.20a shows a loop of wire rotating clockwise in a uniform magnetic field directed to the right. The magnetic force (qvB) on the charges in wires AB and CD is not along the lengths of the wires. (The force on the electrons in these wires is perpendicular to the wires.) Hence, an emf is generated only in wires BC and AD. At any instant, wire BC has velocity $\vec{v}$ at an angle θ with the magnetic field, as shown in Figure 20.20b. (Note that the component of velocity parallel to the field has no effect on the charges in the wire, whereas the component of velocity perpendicular to the field produces a magnetic force on the charges that moves electrons from C to B.) The emf generated in wire BC equals $B\ell v_\perp$,

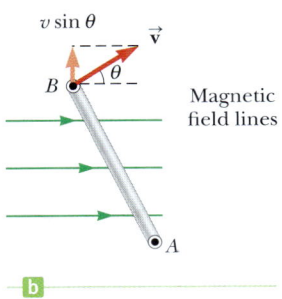

Figure 20.20 (a) A loop rotating at a constant angular velocity in an external magnetic field. The emf induced in the loop varies sinusoidally with time. (b) An edge view of the rotating loop.

where ℓ is the length of the wire and $v_\perp$ is the component of velocity perpendicular to the field. An emf of $B\ell v_\perp$ is also generated in wire DA, and the sense of this emf is the same as that in wire BC. Because $v_\perp = v \sin \theta$, the total induced emf is

$$\mathcal{E} = 2B\ell v_\perp = 2B\ell v \sin \theta \qquad [20.6]$$

If the loop rotates with a constant angular speed ω, we can use the relation $\theta = \omega t$ in Equation 20.6. Furthermore, because every point on the wires BC and DA rotates in a circle about the axis of rotation with the same angular speed ω, we have $v = r\omega = (a/2)\omega$, where a is the length of sides AB and CD. Equation 20.6 therefore reduces to

$$\mathcal{E} = 2B\ell \left(\frac{a}{2}\right) \omega \sin \omega t = B\ell a \omega \sin \omega t$$

If a coil has N turns, the emf is N times as large because each loop has the same emf induced in it. Further, because the area of the loop is $A = \ell a$, the total emf is

$$\mathcal{E} = NBA\omega \sin \omega t \qquad [20.7]$$

This result shows that the emf varies sinusoidally with time, as plotted in Figure 20.19b. Note that the maximum emf has the value

$$\mathcal{E}_{max} = NBA\omega \qquad [20.8]$$

which occurs when $\omega t = 90°$ or $270°$. In other words, $\mathcal{E} = \mathcal{E}_{max}$ when the plane of the loop is parallel to the magnetic field. Further, the emf is zero when $\omega t = 0$ or $180°$, which happens whenever the magnetic field is perpendicular to the plane of the loop. In the United States and Canada the frequency of rotation for commercial generators is 60 Hz, whereas in some European countries 50 Hz is used. (Recall that $\omega = 2\pi f$, where f is the frequency in hertz.)

The **direct-current** (DC) **generator** is illustrated in Figure 20.21a. The components are essentially the same as those of the AC generator except that the contacts to the rotating loop are made by a split ring, or commutator. In this design the output voltage always has the same polarity and the current is a pulsating direct current, as in Figure 20.21b. Note that the contacts to the split ring reverse their

Turbines turn electric generators at a hydroelectric power plant.

APPLICATION
Direct-current generators

Figure 20.21 (a) A schematic diagram of a DC generator. (b) The emf fluctuates in magnitude, but always has the same polarity.

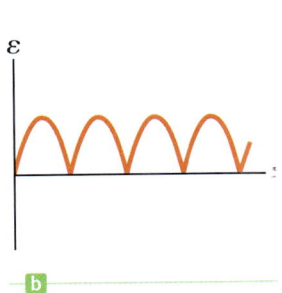

CHAPTER 20 | Induced Voltages and Inductance

roles every half cycle. At the same time, the polarity of the induced emf reverses. Hence, the polarity of the split ring remains the same.

A pulsating DC current is not suitable for most applications. To produce a steady DC current, commercial DC generators use many loops and commutators distributed around the axis of rotation so that the sinusoidal pulses from the loops overlap in phase. When these pulses are superimposed, the DC output is almost free of fluctuations.

EXAMPLE 20.5 | emf Induced in an AC Generator

GOAL Understand physical aspects of an AC generator.

PROBLEM An AC generator consists of eight turns of wire, each having area $A = 0.090\ 0\ \text{m}^2$, with a total resistance of $12.0\ \Omega$. The coil rotates in a magnetic field of $0.500\ \text{T}$ at a constant frequency of $60.0\ \text{Hz}$, with axis of rotation perpendicular to the direction of the magnetic field. **(a)** Find the maximum induced emf. **(b)** What is the maximum induced current? **(c)** Determine the induced emf and current as functions of time. **(d)** What maximum torque must be applied to keep the coil turning?

STRATEGY From the given frequency, calculate the angular frequency ω and substitute it, together with given quantities, into Equation 20.8. As functions of time, the emf and current have the form $A \sin \omega t$, where A is the maximum emf or current, respectively. For part (d), calculate the magnetic torque on the coil when the current is at a maximum. (See Chapter 19.) The applied torque must do work against this magnetic torque to keep the coil turning.

SOLUTION

(a) Find the maximum induced emf.

First, calculate the angular frequency of the rotational motion:

$$\omega = 2\pi f = 2\pi(60.0\ \text{Hz}) = 377\ \text{rad/s}$$

Substitute the values for N, A, B, and ω into Equation 20.8, obtaining the maximum induced emf:

$$\mathcal{E}_{max} = NAB\omega = 8(0.090\ 0\ \text{m}^2)(0.500\ \text{T})(377\ \text{rad/s})$$
$$= \boxed{136\ \text{V}}$$

(b) What is the maximum induced current?

Substitute the maximum induced emf $\mathcal{E}_{max}$ and the resistance R into Ohm's law to find the maximum induced current:

$$I_{max} = \frac{\mathcal{E}_{max}}{R} = \frac{136\ \text{V}}{12.0\ \Omega} = \boxed{11.3\ \text{A}}$$

(c) Determine the induced emf and the current as functions of time.

Substitute $\mathcal{E}_{max}$ and ω into Equation 20.7 to obtain the variation of $\mathcal{E}$ with time t in seconds:

$$\mathcal{E} = \mathcal{E}_{max} \sin \omega t = \boxed{(136\ \text{V}) \sin 377t}$$

The time variation of the current looks just like this expression, except with the maximum current out in front:

$$I = \boxed{(11.3\ \text{A}) \sin 377t}$$

(d) Calculate the maximum applied torque necessary to keep the coil turning.

Write the equation for magnetic torque:

$$\tau = \mu B \sin \theta$$

Calculate the maximum magnetic moment of the coil, μ:

$$\mu = I_{max} A N = (11.3\ \text{A})(0.090\ \text{m}^2)(8) = 8.14\ \text{A} \cdot \text{m}^2$$

Substitute into the magnetic torque equation, with $\theta = 90°$ to find the maximum applied torque:

$$\tau_{max} = (8.14\ \text{A} \cdot \text{m}^2)(0.500\ \text{T}) \sin 90° = \boxed{4.07\ \text{N} \cdot \text{m}}$$

REMARKS The number of loops, N, can't be arbitrary because there must be a force strong enough to turn the coil.

QUESTION 20.5 What effect does doubling the frequency have on the maximum induced emf?

EXERCISE 20.5 An AC generator is to have a maximum output of 301 V. Each circular turn of wire has an area of 0.100 m² and a resistance of 0.80 Ω. The coil rotates in a magnetic field of 0.600 T with a frequency of 40.0 Hz, with the axis of rotation perpendicular to the direction of the magnetic field. (a) How many turns of wire should the coil have to produce the desired emf? (b) Find the maximum current induced in the coil. (c) Determine the induced emf as a function of time.

ANSWERS (a) 20 turns (b) 18.8 A (c) $\mathcal{E} = (301 \text{ V}) \sin 251t$

Motors and Back emf

Motors are devices that convert electrical energy to mechanical energy. Essentially, **a motor is a generator run in reverse**: instead of a current being generated by a rotating loop, a current is supplied to the loop by a source of emf, and the magnetic torque on the current-carrying loop causes it to rotate.

A motor can perform useful mechanical work when a shaft connected to its rotating coil is attached to some external device. As the coil in the motor rotates, however, the changing magnetic flux through it induces an emf that acts to reduce the current in the coil. If it *increased* the current, Lenz's law would be violated. The phrase **back emf** is used for an emf that tends to reduce the applied current. The back emf increases in magnitude as the rotational speed of the coil increases. We can picture this state of affairs as the equivalent circuit in Figure 20.22. For illustrative purposes, assume the external power source supplying current in the coil of the motor has a voltage of 120 V, the coil has a resistance of 10 Ω, and the back emf induced in the coil at this instant is 70 V. The voltage available to supply current equals the difference between the applied voltage and the back emf, or 50 V in this case. The current is always reduced by the back emf.

When a motor is turned on, there is no back emf initially and the current is very large because it's limited only by the resistance of the coil. As the coil begins to rotate, the induced back emf opposes the applied voltage and the current in the coil is reduced. If the mechanical load increases, the motor slows down, which decreases the back emf. This reduction in the back emf increases the current in the coil and therefore also increases the power needed from the external voltage source. As a result, the power requirements for starting a motor and for running it under heavy loads are greater than those for running the motor under average loads. If the motor is allowed to run under no mechanical load, the back emf reduces the current to a value just large enough to balance energy losses by heat and friction.

APPLICATION
Motors

Figure 20.22 A motor can be represented as a resistance plus a back emf.

EXAMPLE 20.6 | Induced Current in a Motor

GOAL Apply the concept of a back emf in calculating the induced current in a motor.

PROBLEM A motor has coils with a resistance of 10.0 Ω and is supplied by a voltage of $\Delta V = 1.20 \times 10^2$ V. When the motor is running at its maximum speed, the back emf is 70.0 V. Find the current in the coils **(a)** when the motor is first turned on and **(b)** when the motor has reached its maximum rotation rate.

STRATEGY For each part, find the net voltage, which is the applied voltage minus the induced emf. Divide the net voltage by the resistance to get the current.

SOLUTION

(a) Find the initial current, when the motor is first turned on.

If the coil isn't rotating, the back emf is zero and the current has its maximum value. Calculate the difference between the emf and the initial back emf and divide by the resistance R, obtaining the initial current:

$$I = \frac{\mathcal{E} - \mathcal{E}_{back}}{R} = \frac{1.20 \times 10^2 \text{ V} - 0}{10.0 \text{ Ω}} = \boxed{12.0 \text{ A}}$$

(Continued)

(b) Find the current when the motor is rotating at its maximum rate.

Repeat the calculation, using the maximum value of the back emf:

$$I = \frac{\mathcal{E} - \mathcal{E}_{back}}{R} = \frac{1.20 \times 10^2 \text{ V} - 70.0 \text{ V}}{10.0 \text{ }\Omega} = \frac{50.0 \text{ V}}{10.0 \text{ }\Omega}$$

$$= 5.00 \text{ A}$$

REMARKS The phenomenon of back emf is one way in which the rotation rate of electric motors is limited.

QUESTION 20.6 As a motor speeds up, what happens to the magnitude of the magnetic torque? (a) It increases. (b) It decreases. (c) It remains constant.

EXERCISE 20.6 If the current in the motor is 8.00 A at some instant, what is the back emf at that time?

ANSWER 40.0 V

20.5 Self-Inductance

LEARNING OBJECTIVES

1. Describe the concepts of self-induction and self-induced emf.
2. Evaluate the self-inductance and self-induced emf in simple electrical systems.

Consider a circuit consisting of a switch, a resistor, and a source of emf, as in Figure 20.23. When the switch is closed, the current doesn't immediately change from zero to its maximum value, $\mathcal{E}/R$. The law of electromagnetic induction, Faraday's law, prevents this change. What happens instead is the following: as the current increases with time, the magnetic flux through the loop due to this current also increases. The increasing flux induces an emf in the circuit that opposes the change in magnetic flux. By Lenz's law, the induced emf is in the direction indicated by the dashed battery in the figure. The net potential difference across the resistor is the emf of the battery minus the opposing induced emf. As the magnitude of the current increases, the *rate* of increase lessens and hence the induced emf decreases. This opposing emf results in a gradual increase in the current. For the same reason, when the switch is opened, the current doesn't immediately fall to zero. This effect is called **self-induction** because the changing flux through the circuit arises from the circuit itself. The emf that is set up in the circuit is called a **self-induced emf**.

As a second example of self-inductance, consider Figure 20.24, which shows a coil wound on a cylindrical iron core. (A practical device would have several hundred turns.) Assume the current changes with time. When the current is in the direction shown, a magnetic field is set up inside the coil, directed from right to left. As a result, some lines of magnetic flux pass through the cross-sectional area of the coil. As the current changes with time, the flux through the coil changes and induces an emf in the coil. Lenz's law shows that this induced emf has a direction that opposes the change in the current. If the current is increasing, the induced emf is as pictured in Figure 20.24b, and if the current is decreasing, the induced emf is as shown in Figure 20.24c.

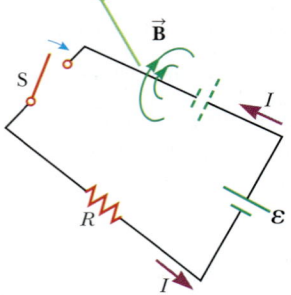

Figure 20.23 After the switch in the circuit is closed, the current produces its own magnetic flux through the loop. An opposing emf is therefore induced, meaning the current relatively slowly increases toward its maximum value rather than jumping to that value right away. The battery with the dashed lines is a symbol for the self-induced emf.

As the current increases toward its maximum value it creates changing magnetic flux, inducing an opposing emf in the loop.

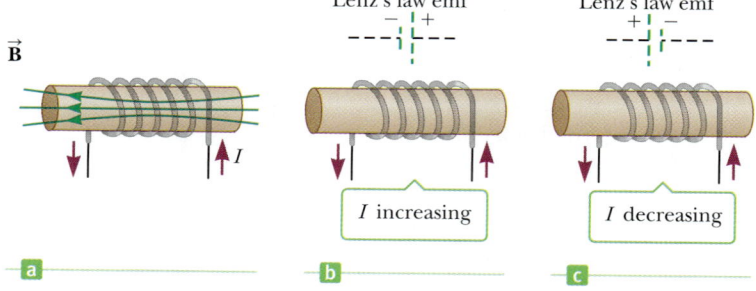

Figure 20.24 (a) A current in the coil produces a magnetic field directed to the left. (b) If the current increases, the coil acts as a source of emf directed as shown by the dashed battery. (c) The induced emf in the coil changes its polarity if the current decreases. The battery symbols drawn with dashed lines represent the included emf in the coil.

To evaluate self-inductance quantitatively, first note that, according to Faraday's law, the induced emf is given by Equation 20.2:

$$\mathcal{E} = -N\frac{\Delta \Phi_B}{\Delta t}$$

The magnetic flux is proportional to the magnetic field, which is proportional to the current in the coil. Therefore, **the self-induced emf must be proportional to the rate of change of the current with time,** or

$$\mathcal{E} \equiv -L\frac{\Delta I}{\Delta t} \qquad [20.9]$$

where L is a proportionality constant called the **inductance** of the device. The negative sign indicates that a changing current induces an emf in opposition to the change. In other words, if the current is increasing (ΔI positive), the induced emf is negative, indicating opposition to the increase in current. Likewise, if the current is decreasing (ΔI negative), the sign of the induced emf is positive, indicating that the emf is acting to oppose the decrease.

The inductance of a coil depends on the cross-sectional area of the coil and other quantities, which can all be grouped under the general heading of geometric factors. The SI unit of inductance is the **henry** (H), which, from Equation 20.9, is equal to 1 volt-second per ampere:

$$1 \text{ H} = 1 \text{ V} \cdot \text{s/A}$$

In the process of calculating self-inductance, it is often convenient to equate Equations 20.2 and 20.9 to find an expression for L:

$$N\frac{\Delta \Phi_B}{\Delta t} = L\frac{\Delta I}{\Delta t}$$

$$L = N\frac{\Delta \Phi_B}{\Delta I} = \frac{N\Phi_B}{I} \qquad [20.10]$$

◀ Inductance

Joseph Henry
American physicist (1797–1878)
Henry became the first director of the Smithsonian Institution and first president of the Academy of Natural Science. He was the first to produce an electric current with a magnetic field, but he failed to publish his results as early as Faraday because of his heavy teaching duties at the Albany Academy in New York State. He improved the design of the electromagnet and constructed one of the first motors. He also discovered the phenomenon of self-induction. The unit of inductance, the henry, is named in his honor.

APPLYING PHYSICS 20.3 | Making Sparks Fly

In some circuits a spark occurs between the poles of a switch when the switch is opened. Why isn't there a spark when the switch for this circuit is closed?

EXPLANATION According to Lenz's law, the direction of induced emfs is such that the induced magnetic field opposes change in the original magnetic flux. When the switch is opened, the sudden drop in the magnetic field in the circuit induces an emf in a direction that opposes change in the original current. This induced emf can cause a spark as the current bridges the air gap between the poles of the switch. The spark doesn't occur when the switch is closed, because the original current is zero and the induced emf opposes any change in that current. ■

In general, determining the inductance of a given current element can be challenging. Finding an expression for the inductance of a common solenoid, however, is straightforward. Let the solenoid have N turns and length ℓ. Assume ℓ is large compared with the radius and the core of the solenoid is air. We take the interior magnetic field to be uniform and given by Equation 19.16,

$$B = \mu_0 nI = \mu_0 \frac{N}{\ell} I$$

where $n = N/\ell$ is the number of turns per unit length. The magnetic flux through each turn is therefore

$$\Phi_B = BA = \mu_0 \frac{N}{\ell} AI$$

where A is the cross-sectional area of the solenoid. From this expression and Equation 20.10, we find that

$$L = \frac{N\Phi_B}{I} = \frac{\mu_0 N^2 A}{\ell} \quad \text{[20.11a]}$$

This equation shows that L depends on the geometric factors ℓ and A and on μ_0 and is proportional to the square of the number of turns. Because $N = n\ell$, we can also express the result in the form

$$L = \mu_0 \frac{(n\ell)^2}{\ell} A = \mu_0 n^2 A \ell = \mu_0 n^2 V \quad \text{[20.11b]}$$

where $V = A\ell$ is the volume of the solenoid.

■ EXAMPLE 20.7 | Inductance, Self-Induced emf, and Solenoids

GOAL Calculate the inductance and self-induced emf of a solenoid.

PROBLEM (a) Calculate the inductance of a solenoid containing 300 turns if the length of the solenoid is 25.0 cm and its cross-sectional area is 4.00×10^{-4} m². (b) Calculate the self-induced emf in the solenoid described in part (a) if the current in the solenoid decreases at the rate of 50.0 A/s.

STRATEGY Substituting given quantities into Equation 20.11a gives the inductance L. For part (b), substitute the result of part (a) and $\Delta I/\Delta t = -50.0$ A/s into Equation 20.9 to get the self-induced emf.

SOLUTION

(a) Calculate the inductance of the solenoid.

Substitute the number N of turns, the area A, and the length ℓ into Equation 20.11a to find the inductance:

$$L = \frac{\mu_0 N^2 A}{\ell}$$

$$= (4\pi \times 10^{-7} \, \text{T} \cdot \text{m/A}) \frac{(300)^2 (4.00 \times 10^{-4} \, \text{m}^2)}{25.0 \times 10^{-2} \, \text{m}}$$

$$= 1.81 \times 10^{-4} \, \text{T} \cdot \text{m}^2/\text{A} = \boxed{0.181 \text{ mH}}$$

(b) Calculate the self-induced emf in the solenoid.

Substitute L and $\Delta I/\Delta t = -50.0$ A/s into Equation 20.9, finding the self-induced emf:

$$\mathcal{E} = -L\frac{\Delta I}{\Delta t} = -(1.81 \times 10^{-4} \, \text{H})(-50.0 \, \text{A/s})$$

$$= \boxed{9.05 \text{ mV}}$$

REMARKS Notice that $\Delta I/\Delta t$ is negative because the current is decreasing with time. The expression for the inductance in part (a) relies on the assumption that the radius of the solenoid is small compared to its length.

QUESTION 20.7 If the solenoid were wrapped into a circle so as to become a toroidal solenoid, what would be true of its self-inductance? (a) It would be the same. (b) It would be larger. (c) It would be smaller.

EXERCISE 20.7 A solenoid is to have an inductance of 0.285 mH, a cross-sectional area of 6.00×10^{-4} m², and a length of 36.0 cm. (a) How many turns per unit length should it have? (b) If the self-induced emf is -12.5 mV at a given time, at what rate is the current changing at that instant?

ANSWERS (a) 1.02×10^3 turns/m (b) 43.9 A/s

20.6 RL Circuits

LEARNING OBJECTIVES

1. Describe the physical effect of introducing inductance into a circuit with changing current.
2. State the time constant for an RL circuit.
3. Apply the RL circuit time constant to circuits containing resistors and inductors.

A circuit element that has a large inductance, such as a closely wrapped coil of many turns, is called an **inductor**. The circuit symbol for an inductor is ⎯⎯⎯⎯. We will always assume the self-inductance of the remainder of the circuit is negligible compared with that of the inductor in the circuit.

To gain some insight into the effect of an inductor in a circuit, consider the two circuits in Figure 20.25. Figure 20.25a shows a resistor connected to the terminals of a battery. For this circuit, Kirchhoff's loop rule is $\mathcal{E} - IR = 0$. The voltage drop across the resistor is

$$\Delta V_R = -IR \qquad [20.12]$$

In this case, **we interpret resistance as a measure of opposition to the current**. Now consider the circuit in Figure 20.25b, consisting of an inductor connected to the terminals of a battery. At the instant the switch in this circuit is closed, because $IR = 0$, the emf of the battery equals the back emf generated in the coil. Hence, we have

$$\mathcal{E}_L = -L \frac{\Delta I}{\Delta t} \qquad [20.13]$$

From this expression, **we can interpret L as a measure of opposition to the rate of change of current**.

Figure 20.26 shows a circuit consisting of a resistor, an inductor, and a battery. Suppose the switch is closed at $t = 0$. The current begins to increase, but the inductor produces an emf that opposes the increasing current. As a result, the current can't change from zero to its maximum value of $\mathcal{E}/R$ instantaneously. Equation 20.13 shows that the induced emf is a maximum when the current is changing most rapidly, which occurs when the switch is first closed. As the current approaches its steady-state value, the back emf of the coil falls off because the current is changing more slowly. Finally, when the current reaches its steady-state value, the rate of change is zero and the back emf is also zero. Figure 20.27 plots current in the circuit as a function of time. This plot is similar to that of the charge on a capacitor as a function of time, discussed in Chapter 18, section 5, in connection with RC circuits. In that case, we found it convenient to introduce a quantity called the *time constant of the circuit*, which told us something about the time required for the capacitor to approach its steady-state charge. In the same way, time constants are defined for circuits containing resistors and inductors. The **time constant τ** for an RL circuit is the time required for the current in the circuit to reach 63.2% of its final value $\mathcal{E}/R$; the time constant of an RL circuit is given by

$$\tau = \frac{L}{R} \qquad [20.14]$$

◄ Time constant for an RL circuit

Using methods of calculus, it can be shown that the current in such a circuit is given by

$$I = \frac{\mathcal{E}}{R}(1 - e^{-t/\tau}) \qquad [20.15]$$

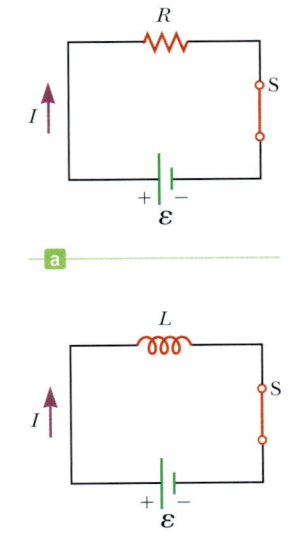

Figure 20.25 A comparison of the effect of a resistor with that of an inductor in a simple circuit.

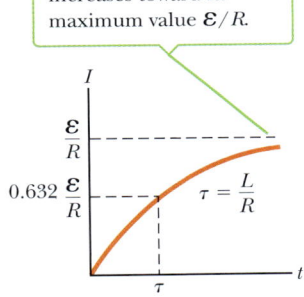

After the switch is closed at $t = 0$, the current increases toward its maximum value $\mathcal{E}/R$.

Figure 20.27 A plot of current versus time for the RL circuit shown in Figure 20.26. The switch is closed at $t = 0$, and the current increases toward its maximum value $\mathcal{E}/R$. The time constant τ is the time it takes the current to reach 63.2% of its maximum value.

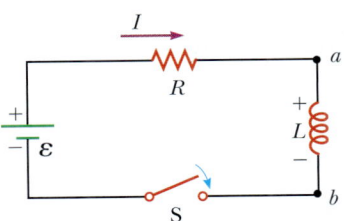

Figure 20.26 A series RL circuit. As the current increases toward its maximum value, the inductor produces an emf that opposes the increasing current.

This equation is consistent with our intuition: when the switch is closed at $t = 0$, the current is initially zero, rising with time to some maximum value. Notice the mathematical similarity between Equation 20.15 and Equation 18.7, which features a capacitor instead of an inductor. As in the case of a capacitor, the equation's form suggests an infinite amount of time is required for the current in the inductor to reach its maximum value. This is an artifact of assuming current is composed of moving charges that are infinitesimal, as will be demonstrated in Example 20.9.

Figure 20.28 (Quick Quiz 20.6)

Quick Quiz

20.6 The switch in the circuit shown in Figure 20.28 is closed, and the lightbulb glows steadily. The inductor is a simple air-core solenoid. An iron rod is inserted into the interior of the solenoid, increasing the magnitude of the magnetic field in the solenoid. As the rod is inserted, the brightness of the lightbulb (a) increases, (b) decreases, or (c) remains the same.

EXAMPLE 20.8 | An *RL* Circuit

GOAL Calculate a time constant and relate it to current in an *RL* circuit.

PROBLEM A 12.6-V battery is in a circuit with a 30.0-mH inductor and a 0.150-Ω resistor, as in Figure 20.26. The switch is closed at $t = 0$. **(a)** Find the time constant of the circuit. **(b)** Find the current after one time constant has elapsed. **(c)** Find the voltage drops across the resistor when $t = 0$ and $t =$ one time constant, τ. **(d)** What's the rate of change of the current after one time constant?

STRATEGY Part (a) requires only substitution into the definition of time constant. With this value and Ohm's law, the current after one time constant can be found, and multiplying this current by the resistance yields the voltage drop across the resistor after one time constant. With the voltage drop and Kirchhoff's loop law, the voltage across the inductor can be found. This value can be substituted into Equation 20.13 to obtain the rate of change of the current.

SOLUTION

(a) What's the time constant of the circuit?

Substitute the inductance L and resistance R into Equation 20.14, finding the time constant:

$$\tau = \frac{L}{R} = \frac{30.0 \times 10^{-3}\,\text{H}}{0.150\,\Omega} = 0.200\,\text{s}$$

(b) Find the current after one time constant has elapsed.

First, use Ohm's law to compute the final value of the current after many time constants have elapsed:

$$I_{\text{max}} = \frac{\mathcal{E}}{R} = \frac{12.6\,\text{V}}{0.150\,\Omega} = 84.0\,\text{A}$$

After one time constant, the current rises to 63.2% of its final value:

$$I_{1\tau} = (0.632)I_{\text{max}} = (0.632)(84.0\,\text{A}) = 53.1\,\text{A}$$

(c) Find the voltage drops across the resistance when $t = 0$ and $t =$ one time constant.

Initially, the current in the circuit is zero, so, from Ohm's law, the voltage across the resistor is zero:

$$\Delta V_R = IR$$
$$\Delta V_R\,(t = 0\,\text{s}) = (0\,\text{A})(0.150\,\Omega) = 0$$

Next, using Ohm's law, find the magnitude of the voltage drop across the resistor after one time constant:

$$\Delta V_R\,(t = 0.200\,\text{s}) = (53.1\,\text{A})(0.150\,\Omega) = 7.97\,\text{V}$$

(d) What's the rate of change of the current after one time constant?

Using Kirchhoff's voltage rule, calculate the voltage drop across the inductor at that time:

$$\mathcal{E} + \Delta V_R + \Delta V_L = 0$$

Solve for ΔV_L:

$$\Delta V_L = -\mathcal{E} - \Delta V_R = -12.6\,\text{V} - (-7.97\,\text{V}) = -4.6\,\text{V}$$

Now solve Equation 20.13 for $\Delta I/\Delta t$ and substitute:

$$\Delta V_L = -L\frac{\Delta I}{\Delta t}$$

$$\frac{\Delta I}{\Delta t} = -\frac{\Delta V_L}{L} = -\frac{-4.6 \text{ V}}{30.0 \times 10^{-3} \text{ H}} = \boxed{150 \text{ A/s}}$$

REMARKS The values used in this problem were taken from actual components salvaged from the starter system of a car. Because the current in such an RL circuit is initially zero, inductors are sometimes referred to as "chokes" because they temporarily choke off the current. In solving part (d), we traversed the circuit in the direction of positive current, so the voltage difference across the battery was positive and the differences across the resistor and inductor were negative.

QUESTION 20.8 Find the current in the circuit after two time constants.

EXERCISE 20.8 A 12.6-V battery is in series with a resistance of 0.350 Ω and an inductor. (a) After a long time, what is the current in the circuit? (b) What is the current after one time constant? (c) What's the voltage drop across the inductor at this time? (d) Find the inductance if the time constant is 0.130 s.

ANSWERS (a) 36.0 A (b) 22.8 A (c) 4.62 V (d) 4.55×10^{-2} H

■ EXAMPLE 20.9 | Formation of a Magnetic Field

GOAL Understand the role of time in setting up an inductor's magnetic field.

PROBLEM Given the RL circuit of Example 20.8, find the time required for the current to reach 99.9% of its maximum value after the switch is closed.

STRATEGY The solution requires solving Equation 20.15 for time followed by substitution. Notice that the maximum current is $I_{max} = \mathcal{E}/R$.

SOLUTION

Write Equation 20.15, with I_f substituted for the current:

$$I_f = \frac{\mathcal{E}}{R}(1 - e^{-t/\tau}) = I_{max}(1 - e^{-t/\tau})$$

Divide both sides by I_{max}:

$$\frac{I_f}{I_{max}} = 1 - e^{-t/\tau}$$

Subtract 1 from both sides and then multiply both sides by -1:

$$1 - \frac{I_f}{I_{max}} = e^{-t/\tau}$$

Take the natural log of both sides:

$$\ln\left(1 - \frac{I_f}{I_{max}}\right) = \ln(e^{-t/\tau}) = -t/\tau$$

Solve for t and substitute the expression for τ from Equation 20.14:

$$t = -\frac{L}{R}\ln\left(1 - \frac{I_f}{I_{max}}\right)$$

Substitute values, obtaining the desired time:

$$t = -\frac{30.0 \times 10^{-3} \text{ H}}{0.150 \text{ Ω}}\ln(1 - 0.999) = \boxed{1.38 \text{ s}}$$

REMARKS From this calculation, it's found that forming the magnetic field in an inductor and approaching the maximum current occurs relatively rapidly. Contrary to what might be expected from the mathematical form of Equation 20.15, an infinite amount of time is not actually required.

QUESTION 20.9 If the inductance were doubled, by what factor would the length of time found be changed? (a) 1 (i.e., no change) (b) 2 (c) $\frac{1}{2}$

EXERCISE 20.9 Suppose a series RL circuit is composed of a 2.00-Ω resistor, a 15.0 H inductor, and a 6.00 V battery. (a) What is the time constant for this circuit? (b) Once the switch is closed, how long does it take the current to reach half its maximum value?

ANSWERS (a) 7.50 s (b) 5.20 s

20.7 Energy Stored in a Magnetic Field

LEARNING OBJECTIVES

1. Define the energy stored in an inductor and contrast it with the energy stored in a capacitor.
2. Apply the energy stored by an inductor to *RL* circuits.

The emf induced by an inductor prevents a battery from establishing an instantaneous current in a circuit. The battery has to do work to produce a current. We can think of this needed work as energy stored in the inductor in its magnetic field. In a manner similar to that used in Section 16.9 to find the energy stored in a capacitor, we find that the energy stored by an inductor is

▶ Energy stored in an inductor

$$PE_L = \tfrac{1}{2}LI^2 \qquad [20.16]$$

Notice that the result is similar in form to the expression for the energy stored in a charged capacitor (Eq. 16.18):

▶ Energy stored in a capacitor

$$PE_C = \tfrac{1}{2}C(\Delta V)^2$$

■ EXAMPLE 20.10 | Magnetic Energy

GOAL Relate the storage of magnetic energy to currents in an *RL* circuit.

PROBLEM A 12.0-V battery is connected in series to a 25.0-Ω resistor and a 5.00-H inductor. **(a)** Find the maximum current in the circuit. **(b)** Find the energy stored in the inductor at this time. **(c)** How much energy is stored in the inductor when the current is changing at a rate of 1.50 A/s?

STRATEGY In part (a) Ohm's law and Kirchhoff's voltage rule yield the maximum current because the voltage across the inductor is zero when the current is maximal. Substituting the current into Equation 20.16 gives the energy stored in the inductor. In part (c) the given rate of change of the current can be used to calculate the voltage drop across the inductor at the specified time. Kirchhoff's voltage rule and Ohm's law then give the current *I* at that time, which can be used to find the energy stored in the inductor.

SOLUTION

(a) Find the maximum current in the circuit.

Apply Kirchhoff's voltage rule to the circuit:

$$\Delta V_{\text{batt}} + \Delta V_R + \Delta V_L = 0$$

$$\mathcal{E} - IR - L\frac{\Delta I}{\Delta t} = 0$$

When the maximum current is reached, $\Delta I/\Delta t$ is zero, so the voltage drop across the inductor is zero. Solve for the maximum current $I_{\max}$:

$$I_{\max} = \frac{\mathcal{E}}{R} = \frac{12.0\text{ V}}{25.0\text{ Ω}} = \boxed{0.480\text{ A}}$$

(b) Find the energy stored in the inductor at this time.

Substitute known values into Equation 20.16:

$$PE_L = \tfrac{1}{2}LI_{\max}^2 = \tfrac{1}{2}(5.00\text{ H})(0.480\text{ A})^2 = \boxed{0.576\text{ J}}$$

(c) Find the energy in the inductor when the current changes at a rate of 1.50 A/s.

Apply Kirchhoff's voltage rule to the circuit once again:

$$\mathcal{E} - IR - L\frac{\Delta I}{\Delta t} = 0$$

Solve this equation for the current *I* and substitute:

$$I = \frac{1}{R}\left(\mathcal{E} - L\frac{\Delta I}{\Delta t}\right)$$

$$= \frac{1}{25.0\text{ Ω}}[12.0\text{ V} - (5.00\text{ H})(1.50\text{ A/s})] = 0.180\text{ A}$$

Finally, substitute the value for the current into Equation 20.15, finding the energy stored in the inductor:

$$PE_L = \tfrac{1}{2}Li^2 = \tfrac{1}{2}(5.00 \text{ H})(0.180 \text{ A})^2 = \boxed{0.081\ 0 \text{ J}}$$

REMARKS Notice how important it is to combine concepts from previous chapters. Here, Ohm's law and Kirchhoff's loop rule were essential to the solution of the problem.

QUESTION 20.10 True or False: The larger the value of the inductance in such an RL circuit, the larger the maximum current.

EXERCISE 20.10 For the same circuit, find the energy stored in the inductor when the rate of change of the current is 1.00 A/s.

ANSWER 0.196 J

SUMMARY

20.1 Induced emf and Magnetic Flux

The magnetic flux Φ_B through a closed loop is defined as

$$\Phi_B \equiv BA\cos\theta \qquad [20.1]$$

where B is the strength of the uniform magnetic field, A is the cross-sectional area of the loop, and θ is the angle between $\vec{B}$ and a direction perpendicular to the plane of the loop.

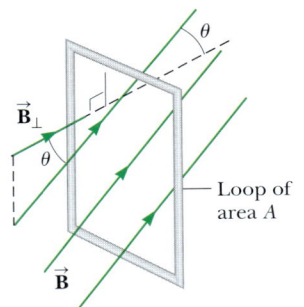

In this view of a loop of area A, the component of the magnetic field $\vec{B}$ perpendicular to surface multiplied by the area gives the magnetic flux through the surface.

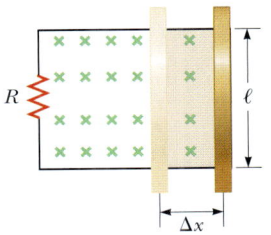

As the bar moves to the right, the area of the loop increases by the amount $\ell\Delta x$ and the magnetic flux through the loop increases by $B\ell\Delta x$.

20.2 Faraday's Law of Induction and Lenz's Law

Faraday's law of induction states that the instantaneous emf induced in a circuit equals the negative of the rate of change of magnetic flux through the circuit,

$$\mathcal{E} = -N\frac{\Delta\Phi_B}{\Delta t} \qquad [20.2]$$

where N is the number of loops in the circuit. The magnetic flux Φ_B can change with time whenever the magnetic field $\vec{B}$, the area A, or the angle θ changes with time.

Lenz's law states that the current from the induced emf creates a magnetic field with flux opposing the *change* in magnetic flux through a circuit.

20.3 Motional emf

If a conducting bar of length ℓ moves through a magnetic field with a speed v so that $\vec{B}$ is perpendicular to the bar, the emf induced in the bar, often called a **motional emf**, is

$$|\mathcal{E}| = B\ell v \qquad [20.4]$$

20.4 Generators

When a coil of wire with N turns, each of area A, rotates with constant angular speed ω in a uniform magnetic field $\vec{B}$, the emf induced in the coil is

$$\mathcal{E} = NBA\omega\sin\omega t \qquad [20.7]$$

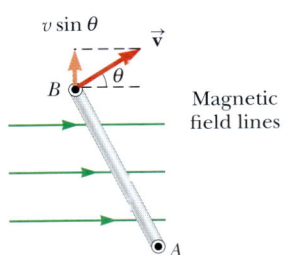

In this edge view of a rotating loop, the magnetic flux changes continuously, generating an alternating current in the loop.

Such generators naturally produce alternating current (AC), which changes direction with frequency $\omega/2\pi$. The AC current can be transformed to direct current.

20.5 Self-Inductance

20.6 RL Circuits

When the current in a coil changes with time, an emf is induced in the coil according to Faraday's law. This **self-induced emf** is defined by the expression

$$\mathcal{E} \equiv -L\frac{\Delta I}{\Delta t} \qquad [20.9]$$

where L is the inductance of the coil. The SI unit for inductance is the henry (H); 1 H = 1 V·s/A.

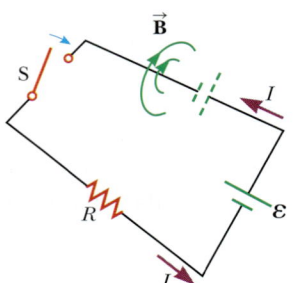

When the switch is closed, a magnetic field begins to develop as shown. The changing magnetic flux creates a self-induced emf in the opposite direction, represented by the dashed lines.

The **inductance** of a coil can be found from the expression

$$L = \frac{N\Phi_B}{I} \qquad [20.10]$$

where N is the number of turns on the coil, I is the current in the coil, and Φ_B is the magnetic flux through the coil produced by that current. For a solenoid, the inductance is given by

$$L = \frac{\mu_0 N^2 A}{\ell} \qquad [20.11a]$$

If a resistor and inductor are connected in series to a battery and a switch is closed at $t = 0$, the current in the circuit doesn't rise instantly to its maximum value. After one **time constant** $\tau = L/R$, the current in the circuit is 63.2% of its final value $\mathcal{E}/R$. As the current approaches its final, maximum value, the voltage drop across the inductor approaches zero. The current I in such a circuit any time t after the circuit is completed is

$$I = \frac{\mathcal{E}}{R}(1 - e^{-t/\tau}) \qquad [20.15]$$

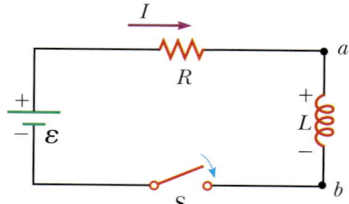

A series RL circuit. As the current increases toward its maximum value, the inductor produces an emf that opposes the increasing current.

20.7 Energy Stored in a Magnetic Field

The **energy stored** in the magnetic field of an inductor carrying current I is

$$PE_L = \tfrac{1}{2}LI^2 \qquad [20.16]$$

As the current in an RL circuit approaches its maximum value, the stored energy also approaches a maximum value.

■ WARM-UP EXERCISES

 The warm-up exercises in this chapter may be assigned online in Enhanced WebAssign.

1. **Physics Review** A simple pendulum undergoes one complete oscillation in 4.30 s. Find (a) the frequency and (b) the angular frequency of the oscillation. (See Section 13.5.)

2. **Physics Review** An electric field of magnitude 2.75 V/m is oriented at an angle of 30.0° with respect to the positive z-direction. Determine the magnitude of the electric flux through a rectangular area of 2.00 m² in the xy-plane. (See Section 15.9.)

3. **Physics Review** Determine the time constant for an RC circuit having an effective resistance of 1.25×10^4 Ω and an effective capacitance of 2.22×10^{-6} F. (See Section 18.5.)

4. A magnetic field of magnitude 5.00×10^{-3} T is oriented at an angle of 30.0° with respect to the positive z-direction. Determine the magnitude of the magnetic flux through a rectangular surface in the xy-plane with an area of 3.00 m². (See Section 20.1.)

5. A coil of wire has 6.00 turns, each turn enclosing an area of 2.00×10^{-3} m², while the total resistance of the coil is 1.20×10^{-3} Ω. If a magnet is brought closer to the coil from a distance so that the magnetic field through each loop increases steadily from zero to 4.00×10^{-2} T over the course of 0.750 s, determine the magnitude of (a) the induced emf in the coil and (b) the current in the coil while the flux is increasing. (See Section 20.2.)

6. In one of NASA's space tether experiments, a 20.0-km long conducting wire was deployed by the space shuttle as it orbited at 7.86×10^3 m/s around earth and across earth's magnetic field lines. The resulting motional emf was used as a power source. If the component of earth's magnetic field perpendicular to the tether was 1.50×10^{-5} T, determine the maximum possible potential difference between the two ends of the tether. (See Section 20.3.)

7. Wind turbines use a generator to convert the wind's kinetic energy into electrical energy. Determine (a) the maximum emf of a generator with a 100-turn, 0.500 m² coil that is rotated by the wind at an angular frequency of 6.00 rad/s through a 0.450-T magnetic field and (b) the maximum induced current if the coil has a total resistance of 10.0 Ω. (See Section 20.4.)

8. Suppose insulated wire of length 5.28 m is coiled into a 0.015 0-m radius solenoid of length 0.060 0 m. Determine (a) the number of loops in the coil, (b) the inductance of the solenoid, and (c) the magnitude of the induced emf if the current is changing at a rate of 6.00×10^2 A/s. (See Section 20.5.)

9. Determine the time constant for an RL circuit having an effective resistance of 1.25×10^4 Ω and an inductance of 2.00×10^{-2} H. (See Section 20.6.)

10. Determine the energy stored in a 2.50-mH inductor carrying a current of 10.0 A. (See Section 20.7.)

CONCEPTUAL QUESTIONS

The conceptual questions in this chapter may be assigned online in Enhanced WebAssign.

1. A spacecraft orbiting Earth has a coil of wire in it. An astronaut measures a small current in the coil, although there is no battery connected to it and there are no magnets in the spacecraft. What is causing the current?

2. Does dropping a magnet down a copper tube produce a current in the tube? Explain.

3. A circular loop is located in a uniform and constant magnetic field. Describe how an emf can be induced in the loop in this situation.

4. A loop of wire is placed in a uniform magnetic field. (a) For what orientation of the loop is the magnetic flux a maximum? (b) For what orientation is the flux zero?

5. As the conducting bar in Figure CQ20.5 moves to the right, an electric field directed downward is set up. If the bar were moving to the left, explain why the electric field would be upward.

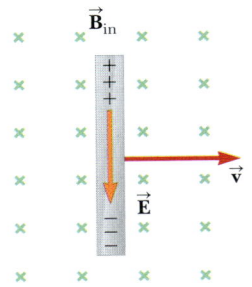

Figure CQ20.5 Conceptual Questions 5 and 8.

6. How is electrical energy produced in dams? (That is, how is the energy of motion of the water converted to AC electricity?)

7. Wearing a metal bracelet in a region of strong magnetic field could be hazardous. Discuss this statement.

8. As the bar in Figure CQ20.5 moves perpendicular to the field, is an external force required to keep it moving with constant speed?

9. Eddy currents are induced currents set up in a piece of metal when it moves through a nonuniform magnetic field. For example, consider the flat metal plate swinging at the end of a bar as a pendulum, as shown in Figure CQ20.9. (a) At position 1, the pendulum is moving from a region where there is no magnetic field into a region where the field $\vec{B}_{in}$ is directed into the paper. Show that at position 1 the direction of the eddy current is counterclockwise. (b) At position 2, the pendulum is moving out of the field into a region of zero field. Show that the direction of the eddy current is clockwise in this case. (c) Use right-hand rule number 2 to show that these eddy currents lead to a magnetic force on the plate directed as shown in the figure. Because the induced eddy current always produces a retarding force when the plate enters or leaves the field, the swinging plate quickly comes to rest.

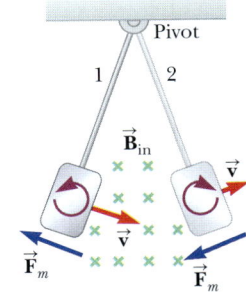

Figure CQ20.9

10. A bar magnet is dropped toward a conducting ring lying on the floor. As the magnet falls toward the ring, does it move as a freely falling object? Explain.

11. A piece of aluminum is dropped vertically downward between the poles of an electromagnet. Does the magnetic field affect the velocity of the aluminum? *Hint:* See Conceptual Question 9.

12. When the switch in Figure CQ20.12a is closed, a current is set up in the coil and the metal ring springs upward (Fig. CQ20.12b). Explain this behavior.

Figure CQ20.12 Conceptual Questions 12 and 13.

13. Assume the battery in Figure CQ20.12a is replaced by an AC source and the switch is held closed. If held down, the metal ring on top of the solenoid becomes hot. Why?

14. A magneto is used to cause the spark in a spark plug in many lawn mowers today. A magneto consists of a permanent magnet mounted on a flywheel so that it spins past a fixed coil. Explain how this arrangement generates a large enough potential difference to cause the spark.

PROBLEMS

20.1 Induced EMF and Magnetic Flux

1. A uniform magnetic field of magnitude 0.50 T is directed perpendicular to the plane of a rectangular loop having dimensions 8.0 cm by 12 cm. Find the magnetic flux through the loop.

2. Find the flux of Earth's magnetic field of magnitude 5.00×10^{-5} T through a square loop of area 20.0 cm^2 (a) when the field is perpendicular to the plane of the loop, (b) when the field makes a 30.0° angle with the normal to the plane of the loop, and (c) when the field makes a 90.0° angle with the normal to the plane.

3. A circular loop of radius 12.0 cm is placed in a uniform magnetic field. (a) If the field is directed perpendicular to the plane of the loop and the magnetic flux through the loop is 8.00×10^{-3} T · m^2, what is the strength of the magnetic field? (b) If the magnetic field is directed parallel to the plane of the loop, what is the magnetic flux through the loop?

4. A long, straight wire carrying a current of 2.00 A is placed along the axis of a cylinder of radius 0.500 m and a length of 3.00 m. Determine the total magnetic flux through the cylinder.

5. A long, straight wire lies in the plane of a circular coil with a radius of 0.010 m. The wire carries a current of 2.0 A and is placed along a diameter of the coil. (a) What is the net flux through the coil? (b) If the wire passes through the center of the coil and is perpendicular to the plane of the coil, what is the net flux through the coil?

6. A 400-turn solenoid of length 36.0 cm and radius 3.00 cm carries a current of 5.00 A. Find (a) the magnetic field strength inside the coil at its midpoint and (b) the magnetic flux through a circular cross-sectional area of the solenoid at its midpoint.

7. **M** A cube of edge length ℓ = 2.5 cm is positioned as shown in Figure P20.7. There is a uniform magnetic field throughout the region with components B_x = +5.0 T, B_y = +4.0 T, and B_z = +3.0 T. (a) Calculate the flux through the shaded face of the cube. (b) What is the total flux emerging from the volume enclosed by the cube (i.e., the total flux through all six faces)?

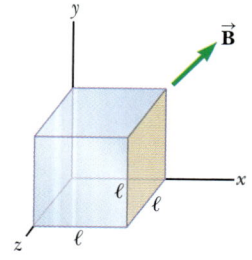

Figure P20.7

20.2 Faraday's Law of Induction and Lenz's Law

8. **BIO** Transcranial magnetic stimulation (TMS) is a noninvasive technique used to stimulate regions of the human brain. A small coil is placed on the scalp, and a brief burst of current in the coil produces a rapidly changing magnetic field inside the brain. The induced emf can be sufficient to stimulate neuronal activity. One such device generates a magnetic field within the brain that rises from zero to 1.5 T in 120 ms. Determine the induced emf within a circle of tissue of radius 1.6 mm and that is perpendicular to the direction of the field.

9. Three loops of wire move near a long straight wire carrying a current as in Figure P20.9. What is the direction of the induced current, if any, in (a) loop A, (b) loop B, and (c) loop C.

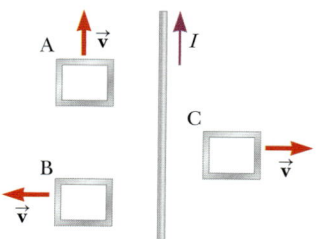

Figure P20.9

10. **W** The flexible loop in Figure P20.10 has a radius of 12 cm and is in a magnetic field of strength 0.15 T. The loop is grasped at points A and B and stretched until its area is nearly zero. If it takes 0.20 s to close the loop, what is the magnitude of the average induced emf in it during this time?

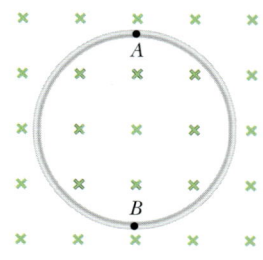

Figure P20.10
Problems 10, 12, and 22.

11. A wire loop of radius 0.30 m lies so that an external magnetic field of magnitude 0.30 T is perpendicular to the loop. The field reverses its direction, and its magnitude changes to 0.20 T in 1.5 s. Find the magnitude of the average induced emf in the loop during this time.

12. A circular loop of wire of radius 12.0 cm is placed in a magnetic field directed perpendicular to the plane of the loop, as shown in Figure P20.10. If the field decreases at the rate of 0.050 0 T/s in some time interval, what is the magnitude of the emf induced in the loop during this interval?

13. A technician wearing a circular metal band on his wrist moves his hand into a uniform magnetic field of magnitude 2.5 T in a time of 0.18 s. If the diameter of the band is 6.5 cm and the field is at an angle of 45° with the plane of the metal band while the hand is in the field, find the magnitude of the average emf induced in the band.

14. In Figure P20.14 what is the direction of the current induced in the resistor at the instant the switch is closed?

Figure P20.14

15. A bar magnet is positioned near a coil of wire, as shown in Figure P20.15. What is the direction of the current in the resistor when the magnet is moved (a) to the left and (b) to the right?

Figure P20.15

16. Find the direction of the current in the resistor shown in Figure P20.16 (a) at the instant the switch is closed, (b) after the switch has been closed for several minutes, and (c) at the instant the switch is opened.

Figure P20.16

17. A circular loop of wire lies below a long wire carrying a current that is increasing as in Figure P20.17a. (a) What is the direction of the induced current in the loop, if any? (b) Now suppose the loop is next to the same wire as in P20.17b. What is the direction of the induced current in the loop, if any? Explain your answers.

Figure P20.17

18. A square, single-turn wire loop $\ell = 1.00$ cm on a side is placed inside a solenoid that has a circular cross section of radius $r = 3.00$ cm, as shown in the end view of Figure P20.18. The solenoid is 20.0 cm long and wound with 100 turns of wire. (a) If the current in the solenoid is 3.00 A, what is the flux through the square loop? (b) If the current in the solenoid is reduced to zero in 3.00 s, what is the magnitude of the average induced emf in the square loop?

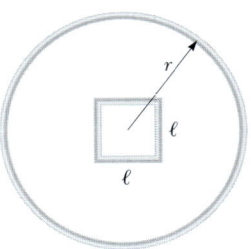

Figure P20.18

19. **GP** A 300-turn solenoid with a length of 20.0 cm and a radius of 1.50 cm carries a current of 2.00 A. A second coil of four turns is wrapped tightly around this solenoid, so it can be considered to have the same radius as the solenoid. The current in the 300-turn solenoid increases steadily to 5.00 A in 0.900 s. (a) Use Ampere's law to calculate the initial magnetic field in the middle of the 300-turn solenoid. (b) Calculate the magnetic field of the 300-turn solenoid after 0.900 s. (c) Calculate the area of the 4-turn coil. (d) Calculate the change in the magnetic flux through the 4-turn coil during the same period. (e) Calculate the average induced emf in the 4-turn coil. Is it equal to the instantaneous induced emf? Explain. (f) Why could contributions to the magnetic field by the current in the 4-turn coil be neglected in this calculation?

20. **M** A circular coil enclosing an area of 100 cm² is made of 200 turns of copper wire. The wire making up the coil has resistance of 5.0 Ω, and the ends of the wire are connected to form a closed circuit. Initially, a 1.1-T uniform magnetic field points perpendicularly upward through the plane of the coil. The direction of the field then reverses so that the final magnetic field has a magnitude of 1.1 T and points downward through the coil. If the time required for the field to reverse directions is 0.10 s, what is the average current in the coil during that time?

21. **BIO** To monitor the breathing of a hospital patient, a thin belt is girded around the patient's chest as in Figure P20.21. The belt is a 200-turn coil. When the patient inhales, the area encircled by the coil increases

Figure P20.21

by 39.0 cm². The magnitude of Earth's magnetic field is 50.0 μT and makes an angle of 28.0° with the plane of the coil. Assuming a patient takes 1.80 s to inhale, find the magnitude of the average induced emf in the coil during that time.

22. **Q|C S** An N-turn circular wire coil of radius r lies in the xy-plane (the plane of the page), as in Figure P20.10. A uniform magnetic field is turned on, increasing steadily from 0 to B_0 in the positive z-direction in t seconds. (a) Find a symbolic expression for the emf, $\mathcal{E}$, induced in the coil in terms of the variables given. (b) Looking down on at the xy-plane from the positive z-axis, is the direction of the induced current clockwise or counterclockwise? (c) If each loop has resistance R, find an expression for the magnitude of the induced current, I.

20.3 Motional emf

23. A truck is carrying a steel beam of length 15.0 m on a freeway. An accident causes the beam to be dumped off the truck and slide horizontally along the ground at a speed of 25.0 m/s. The velocity of the center of mass of the beam is northward while the length of the beam maintains an east–west orientation. The vertical component of the Earth's magnetic field at this location has a magnitude of 35.0 μT. What is the magnitude of the induced emf between the ends of the beam?

24. A 2.00-m length of wire is held in an east–west direction and moves horizontally to the north with a speed of 15.0 m/s. The vertical component of Earth's magnetic field in this region is 40.0 μT directed downward. Calculate the induced emf between the ends of the wire and determine which end is positive.

25. A pickup truck has a width of 79.8 in. If it is traveling north at 37 m/s through a magnetic field with vertical component of 35 μT, what magnitude emf is induced between the driver and passenger sides of the truck?

26. **Q|C** A bar magnet is held stationary while a circular loop of wire is moved toward the magnet at constant velocity at position A as in Figure P20.26. The loop passes over the magnet at position B and moves away from the magnet at position C. Find the direction of the induced current in the loop using Lenz's law (a) at position A and (b) at position C. (c) What is the induced current in the loop at position B? Explain.

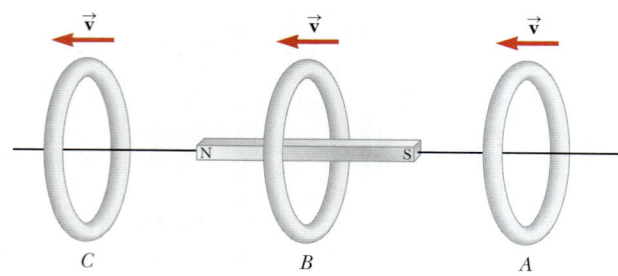

Figure P20.26 Problems 26 and 43.

27. **M** An automobile has a vertical radio antenna 1.20 m long. The automobile travels at 65.0 km/h on a horizontal road where Earth's magnetic field is 50.0 μT, directed toward the north and downward at an angle of 65.0° below the horizontal. (a) Specify the direction the automobile should move so as to generate the maximum motional emf in the antenna, with the top of the antenna positive relative to the bottom. (b) Calculate the magnitude of this induced emf.

28. **Q|C** An astronaut is connected to her spacecraft by a 25-m-long tether cord as she and the spacecraft orbit Earth in a circular path at a speed of 3.0×10^3 m/s. At one instant, the voltage measured between the ends of a wire embedded in the cord is measured to be 0.45 V. Assume the long dimension of the cord is perpendicular to the vertical component of Earth's magnetic field at that instant. (a) What is the magnitude of the vertical component of Earth's field at this location? (b) Does the measured voltage change as the system moves from one location to another? Explain.

29. Figure P20.29 shows a bar of mass $m = 0.200$ kg that can slide without friction on a pair of rails separated by a distance $\ell = 1.20$ m and located on an inclined plane that makes an angle $\theta = 25.0°$ with respect to the ground. The resistance of the resistor is $R = 1.00$ Ω, and a uniform magnetic field of magnitude $B = 0.500$ T is directed downward, perpendicular to the ground, over the entire region through which the bar moves. With what constant speed v does the bar slide along the rails?

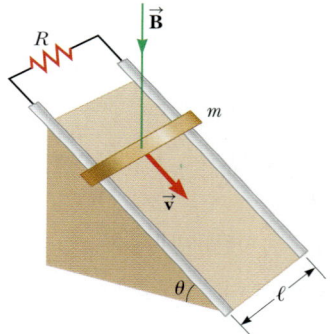

Figure P20.29

30. **W** Consider the arrangement shown in Figure P20.30. Assume $R = 6.00$ Ω, $\ell = 1.20$ m, and a uniform 2.50-T magnetic field is directed *into* the page. At what speed should the bar be moved to produce a current of 0.500 A in the resistor?

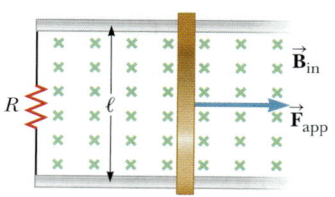

Figure P20.30 Problems 30, 56, and 59.

20.4 Generators

31. A square coil of wire of side 2.80 cm is placed in a uniform magnetic field of magnitude 1.25 T directed into the page as in Figure P20.31. The coil has 28.0 turns and a resistance of 0.780 Ω. If the coil is rotated through an angle of 90.0° about the horizontal axis shown in 0.335 s, find (a) the magnitude of the average emf induced in the coil during this rotation and (b) the average current induced in the coil during this rotation.

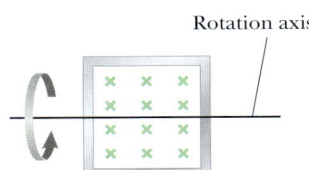

Figure P20.31

32. W A 100-turn square wire coil of area 0.040 m² rotates about a vertical axis at 1 500 rev/min, as indicated in Figure P20.32. The horizontal component of Earth's magnetic field at the location of the loop is 2.0×10^{-5} T. Calculate the maximum emf induced in the coil by Earth's field.

Figure P20.32

33. BIO Considerable scientific work is currently under way to determine whether weak oscillating magnetic fields such as those found near outdoor electric power lines can affect human health. One study indicated that a magnetic field of magnitude 1.0×10^{-3} T, oscillating at 60 Hz, might stimulate red blood cells to become cancerous. If the diameter of a red blood cell is 8.0 μm, determine the maximum emf that can be generated around the perimeter of the cell.

34. A flat coil enclosing an area of 0.10 m² is rotating at 60 rev/s, with its axis of rotation perpendicular to a 0.20-T magnetic field. (a) If there are 1 000 turns on the coil, what is the maximum voltage induced in the coil? (b) When the maximum induced voltage occurs, what is the orientation of the coil with respect to the magnetic field?

35. In a model AC generator, a 500-turn rectangular coil 8.0 cm by 20 cm rotates at 120 rev/min in a uniform magnetic field of 0.60 T. (a) What is the maximum emf induced in the coil? (b) What is the instantaneous value of the emf in the coil at $t = (\pi/32)$ s? Assume the emf is zero at $t = 0$. (c) What is the smallest value of t for which the emf will have its maximum value?

36. A motor has coils with a resistance of 30 Ω and operates from a voltage of 240 V. When the motor is operating at its maximum speed, the back emf is 145 V. Find the current in the coils (a) when the motor is first turned on and (b) when the motor has reached maximum speed. (c) If the current in the motor is 6.0 A at some instant, what is the back emf at that time?

37. A coil of 10.0 turns is in the shape of an ellipse having a major axis of 10.0 cm and a minor axis of 4.00 cm. The coil rotates at 100 rpm in a region in which the magnitude of Earth's magnetic field is 55 μT. What is the maximum voltage induced in the coil if the axis of rotation of the coil is along its major axis and is aligned (a) perpendicular to Earth's magnetic field and (b) parallel to Earth's magnetic field? *Note:* The area of an ellipse is given by $A = \pi ab$, where a is the length of the *semi*major axis and b is the length of the *semi*minor axis.

20.5 Self-Inductance

38. A technician wraps wire around a tube of length 36 cm having a diameter of 8.0 cm. When the windings are evenly spread over the full length of the tube, the result is a solenoid containing 580 turns of wire. (a) Find the self-inductance of this solenoid. (b) If the current in this solenoid increases at the rate of 4.0 A/s, what is the self-induced emf in the solenoid?

39. The current in a coil drops from 3.5 A to 2.0 A in 0.50 s. If the average emf induced in the coil is 12 mV, what is the self-inductance of the coil?

40. S Show that the two expressions for inductance given by

$$L = \frac{N\Phi_B}{I} \quad \text{and} \quad L = \frac{-\mathcal{E}}{\Delta I/\Delta t}$$

have the same units.

41. M A solenoid of radius 2.5 cm has 400 turns and a length of 20 cm. Find (a) its inductance and (b) the rate at which current must change through it to produce an emf of 75 mV.

42. An emf of 24.0 mV is induced in a 500-turn coil when the current is changing at a rate of 10.0 A/s. What is the magnetic flux through each turn of the coil at an instant when the current is 4.00 A?

20.6 RL Circuits

43. An electromagnet can be modeled as an inductor in series with a resistor. Consider a large electromagnet of inductance $L = 12.0$ H and resistance $R = 4.50 \Omega$ connected to a 24.0 V battery and switch as in Figure P20.43. After the switch is closed, find (a) the maximum current carried by the electromagnet, (b) the time constant of the circuit, and (c) the time it takes the current to reach 95.0% of its maximum value.

Figure P20.43
Problems 43, 45, 46, and 48

44. An RL circuit with $L = 3.00$ H and an RC circuit with $C = 3.00$ μF have the same time constant. If the two circuits have the same resistance R, (a) what is the value of R and (b) what is this common time constant?

45. A battery is connected in series with a 0.30-Ω resistor and an inductor, as in Figure P20.43. The switch is closed at $t = 0$. The time constant of the circuit is 0.25 s, and the maximum current in the circuit is 8.0 A. Find (a) the emf of the battery, (b) the inductance of the circuit, (c) the current in the circuit after one time constant has elapsed, (d) the voltage across the resistor after one time constant has elapsed, and (e) the voltage across the inductor after one time constant has elapsed.

46. A 25-mH inductor, an 8.0-Ω resistor, and a 6.0-V battery are connected in series as in Figure P20.43. The switch is closed at $t = 0$. Find the voltage drop across the resistor (a) at $t = 0$ and (b) after one time constant has passed. Also, find the voltage drop across the inductor (c) at $t = 0$ and (d) after one time constant has elapsed.

47. Calculate the resistance in an RL circuit in which $L = 2.50$ H and the current increases to 90.0% of its final value in 3.00 s.

48. W Consider the circuit shown in Figure P20.43. Take $\mathcal{E} = 6.00$ V, $L = 8.00$ mH, and $R = 4.00$ Ω. (a) What is the inductive time constant of the circuit? (b) Calculate the current in the circuit 250 μs after the switch is closed. (c) What is the value of the final steady-state current? (d) How long does it take the current to reach 80.0% of its maximum value?

20.7 Energy Stored in a Magnetic Field

49. (a) If an inductor carrying a 1.70-A current stores an energy of 0.300 mJ, what is its inductance? (b) How much energy does the same inductor store if it carries a 3.0-A current?

50. A 300-turn solenoid has a radius of 5.00 cm and a length of 20.0 cm. Find (a) the inductance of the solenoid and (b) the energy stored in the solenoid when the current in its windings is 0.500 A.

51. M A 24-V battery is connected in series with a resistor and an inductor, with $R = 8.0$ Ω and $L = 4.0$ H, respectively. Find the energy stored in the inductor (a) when the current reaches its maximum value and (b) one time constant after the switch is closed.

52. GP A 60.0-m length of insulated copper wire is wound to form a solenoid of radius 2.0 cm. The copper wire has a radius of 0.50 mm. (a) What is the resistance of the wire? (b) Treating each turn of the solenoid as a circle, how many turns can be made with the wire? (c) How long is the resulting solenoid? (d) What is the self-inductance of the solenoid? (e) If the solenoid is attached to a battery with an emf of 6.0 V and internal resistance of 350 mΩ, compute the time constant of the circuit. (f) What is the maximum current attained?

(g) How long would it take to reach 99.9% of its maximum current? (h) What maximum energy is stored in the inductor?

Additional Problems

53. Two circular loops of wire surround an insulating rod as in Figure P20.53. Loop 1 carries a current I in the clockwise direction when viewed from the left end. If loop 1 moves toward loop 2, which remains stationary, what is the direction of the induced current in loop 2 when viewed from the left end?

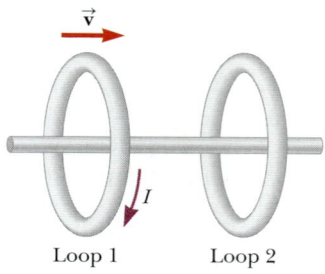

Figure P20.53

54. A circular loop of wire of resistance $R = 0.500$ Ω and radius $r = 8.00$ cm is in a uniform magnetic field directed out of the page as in Figure P20.54. If a clockwise current of $I = 2.50$ mA is induced in the loop, (a) is the magnetic field increasing or decreasing in time? (b) Find the rate at which the field is changing with time.

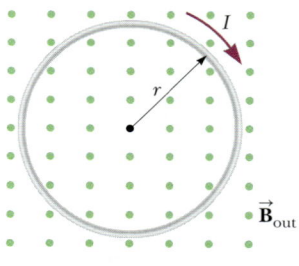

Figure P20.54

55. A rectangular coil with resistance R has N turns, each of length ℓ and width w, as shown in Figure P20.55. The coil moves into a uniform magnetic field $\vec{B}_{in}$ with constant velocity $\vec{v}$. What are the magnitude and direction of the total magnetic force on the coil (a) as it enters the magnetic field, (b) as it moves within the field, and (c) as it leaves the field?

Figure P20.55

56. **Q|C** A conducting bar of length ℓ moves to the right on two frictionless rails, as shown in Figure P20.30. A uniform magnetic field directed into the page has a magnitude of 0.30 T. Assume $\ell = 35$ cm and $R = 9.0\ \Omega$. (a) At what constant speed should the bar move to produce an 8.5-mA current in the resistor? What is the direction of this induced current? (b) At what rate is energy delivered to the resistor? (c) Explain the origin of the energy being delivered to the resistor.

57. An 820-turn wire coil of resistance 24.0 Ω is placed on top of a 12 500-turn, 7.00-cm-long solenoid, as in Figure P20.57. Both coil and solenoid have cross-sectional areas of $1.00 \times 10^{-4}\ m^2$. (a) How long does it take the solenoid current to reach 0.632 times its maximum value? (b) Determine the average back emf caused by the self-inductance of the solenoid during this interval. The magnetic field produced by the solenoid at the location of the coil is one-half as strong as the field at the center of the solenoid. (c) Determine the average rate of change in magnetic flux through each turn of the coil during the stated interval. (d) Find the magnitude of the average induced current in the coil.

Figure P20.57

58. **Q|C** A spacecraft is in a circular orbit of radius equal to 3.0×10^4 km around a 2.0×10^{30} kg pulsar. The magnetic field of the pulsar at that radial distance is 1.0×10^2 T directed perpendicular to the velocity of the spacecraft. The spacecraft is 0.20 km long with a radius of 0.040 km and moves counterclockwise in the xy-plane around the pulsar. (a) What is the speed of the spacecraft? (b) If the magnetic field points in the positive z-direction, is the emf induced from the back to the front of the spacecraft or from side to side? (c) Compute the induced emf. (d) Describe the hazards for astronauts inside any spacecraft moving in the vicinity of a pulsar.

59. **Q|C** A conducting rod of length ℓ moves on two horizontal frictionless rails, as in Figure P20.30. A constant force of magnitude 1.00 N moves the bar at a uniform speed of 2.00 m/s through a magnetic field $\vec{B}$ that is directed into the page. (a) What is the current in an 8.00-Ω resistor R? (b) What is the rate of energy dissipation in the resistor? (c) What is the mechanical power delivered by the constant force?

60. A long solenoid of radius $r = 2.00$ cm is wound with 3.50×10^3 turns/m and carries a current that changes at the rate of 28.5 A/s as in Figure P20.60. What is the magnitude of the emf induced in the square conducting loop surrounding the center of the solenoid?

Figure P20.60

61. The bolt of lightning depicted in Figure P20.61 passes 200 m from a 100-turn coil oriented as shown. If the current in the lightning bolt falls from 6.02×10^6 A to zero in 10.5 μs, what is the average voltage induced in the coil? Assume the distance to the center of the coil determines the average magnetic field at the coil's position. Treat the lightning bolt as a long, vertical wire.

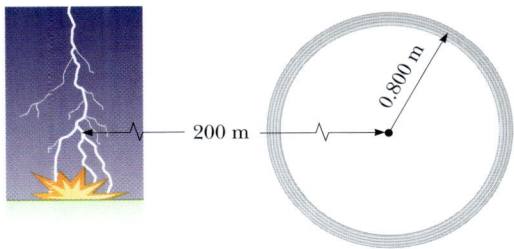

Figure P20.61

62. The square loop in Figure P20.62 is made of wires with a total series resistance of 10.0 Ω. It is placed in a uniform 0.100-T magnetic field directed perpendicular into the plane of the paper. The loop, which is hinged at each corner, is pulled as shown until the separation between points A and B is 3.00 m. If this process takes 0.100 s, what is the average current generated in the loop? What is the direction of the current?

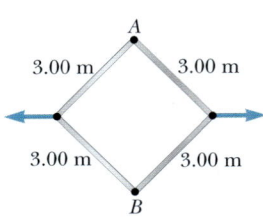

Figure P20.62

63. The magnetic field shown in Figure P20.63 (page 732) has a uniform magnitude of 25.0 mT directed into the paper. The initial diameter of the kink is 2.00 cm. (a) The wire is quickly pulled taut, and the kink shrinks to a diameter of zero in 50.0 ms. Determine the average voltage induced between endpoints A and B. Include the polarity. (b) Suppose the kink is undisturbed, but the magnetic field increases to 100 mT in 4.00×10^{-3} s. Determine the average voltage across terminals A and B, including polarity, during this period.

Figure P20.63

64. An aluminum ring of radius 5.00 cm and resistance 3.00×10^{-4} Ω is placed around the top of a long air-core solenoid with 1 000 turns per meter and a smaller radius of 3.00 cm, as in Figure P20.64. If the current in the solenoid is increasing at a constant rate of 270 A/s, what is the induced current in the ring? Assume the magnetic field produced by the solenoid over the area at the end of the solenoid is one-half as strong as the field at the center of the solenoid. Assume also the solenoid produces a negligible field outside its cross-sectional area.

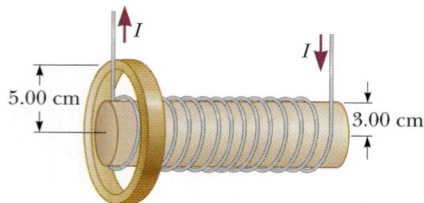

Figure P20.64

65. In Figure P20.65 the rolling axle of length 1.50 m is pushed along horizontal rails at a constant speed $v = 3.00$ m/s. A resistor $R = 0.400$ Ω is connected to the rails at points a and b, directly opposite each other. (The wheels make good electrical contact with the rails, so the axle, rails, and R form a closed-loop circuit. The only significant resistance in the circuit is R.) A uniform magnetic field $B = 0.800$ T is directed vertically downward. (a) Find the induced current I in the resistor. (b) What horizontal force $\vec{F}$ is required to keep the axle rolling at constant speed? (c) Which end of the resistor, a or b, is at the higher electric potential? (d) After the axle rolls past the resistor, does the current in R reverse direction? Explain your answer.

66. **QC S** An N-turn square coil with side ℓ and resistance R is pulled to the right at constant speed v in the positive x-direction in the presence of a uniform magnetic field B acting perpendicular to the coil, as shown in Figure P20.66. At $t = 0$, the right side of the coil is at the edge of the field. After a time t has elapsed, the entire coil is in the region where $B = 0$. In terms of the quantities N, B, ℓ, v, and R, find symbolic expressions for (a) the magnitude of the induced emf in the loop during the time interval t, (b) the magnitude of the induced current in the coil, (c) the power delivered to the coil, and (d) the force required to remove the coil from the field. (e) What is the direction of the induced current in the loop? (f) What is the direction of the magnetic force on the loop while it is being pulled out of the field?

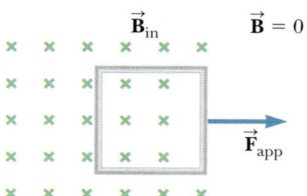

Figure P20.66

67. **S** A conducting rectangular loop of mass M, resistance R, and dimensions w by ℓ falls from rest into a magnetic field $\vec{B}$, as shown in Figure P20.67. During the time interval before the top edge of the loop reaches the field, the loop approaches a terminal speed v_T. (a) Show that

$$v_T = \frac{MgR}{B^2 w^2}$$

(b) Why is v_T proportional to R? (c) Why is it inversely proportional to B^2?

Figure P20.67

Figure P20.65

Arecibo, a large radio telescope in Puerto Rico, gathers electromagnetic radiation in the form of radio waves. These long wavelengths pass through obscuring dust clouds, allowing astronomers to create images of the core region of the Milky Way galaxy, which can't be observed in the visible spectrum.

Alternating-Current Circuits and Electromagnetic Waves

21

Every time we turn on a television set, a stereo system, or any other electric appliances, we call on alternating currents (AC) to provide the power to operate them. We begin our study of AC circuits by examining the characteristics of a circuit containing a source of emf and one other circuit element: a resistor, a capacitor, or an inductor. Then we examine what happens when these elements are connected in combination with each other. Our discussion is limited to simple series configurations of the three kinds of elements.

We conclude this chapter with a discussion of **electromagnetic waves,** which are composed of fluctuating electric and magnetic fields. Electromagnetic waves in the form of visible light enable us to view the world around us; infrared waves warm our environment; radio-frequency waves carry our television and radio programs, as well as information about processes in the core of our galaxy; and X-rays allow us to perceive structures hidden inside our bodies and study properties of distant, collapsed stars. Light is key to our understanding of the universe.

- **21.1** Resistors in an AC Circuit
- **21.2** Capacitors in an AC Circuit
- **21.3** Inductors in an AC Circuit
- **21.4** The *RLC* Series Circuit
- **21.5** Power in an AC Circuit
- **21.6** Resonance in a Series *RLC* Circuit
- **21.7** The Transformer
- **21.8** Maxwell's Predictions
- **21.9** Hertz's Confirmation of Maxwell's Predictions
- **21.10** Production of Electromagnetic Waves by an Antenna
- **21.11** Properties of Electromagnetic Waves
- **21.12** The Spectrum of Electromagnetic Waves
- **21.13** The Doppler Effect for Electromagnetic Waves

21.1 Resistors in an AC Circuit

LEARNING OBJECTIVES

1. Define an AC circuit.
2. Define rms current, rms voltage differences, and power in circuits with alternating current.
3. Discuss the phase relationship between voltage difference and current in a resistive AC circuit.
4. Apply the concepts of rms voltage and current to resistive AC circuits.

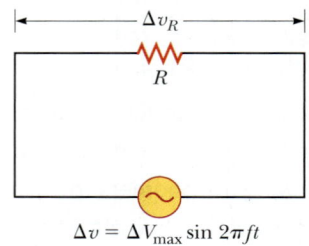

Figure 21.1 A series circuit consisting of a resistor R connected to an AC generator, designated by the symbol —⊙—.

An AC circuit consists of combinations of circuit elements and an AC generator or an AC source, which provides the alternating current. We have seen that the output of an AC generator is sinusoidal and varies with time according to

$$\Delta v = \Delta V_{max} \sin 2\pi f t \quad [21.1]$$

where Δv is the instantaneous voltage, ΔV_{max} is the maximum voltage of the AC generator, and f is the frequency at which the voltage changes, measured in hertz (Hz). (Compare Equations 20.7 and 20.8 with Equation 21.1, and recall that $\omega = 2\pi f$.) We first consider a simple circuit consisting of a resistor and an AC source (designated by the symbol —⊙—), as in Figure 21.1. The current and the voltage versus time across the resistor are shown in Figure 21.2.

To explain the concept of alternating current, we begin by discussing the current versus time curve in Figure 21.2. At point a on the curve, the current has a maximum value in one direction, arbitrarily called the positive direction. Between points a and b, the current is decreasing in magnitude but is still in the positive direction. At point b, the current is momentarily zero; it then begins to increase in the opposite (negative) direction between points b and c. At point c, the current has reached its maximum value in the negative direction.

The current and voltage are in step with each other because they vary identically with time. **Because the current and the voltage reach their maximum values at the same time, they are said to be in phase.** Notice that **the average value of the current over one cycle is zero** because the current is maintained in one direction (the positive direction) for the same amount of time and at the same magnitude as it is in the opposite direction (the negative direction). The direction of the current, however, has no effect on the behavior of the resistor in the circuit: the collisions between electrons and the fixed atoms of the resistor result in an increase in the resistor's temperature regardless of the direction of the current.

We can quantify this discussion by recalling that the rate at which electrical energy is dissipated in a resistor, the power P, is

$$P = i^2 R$$

where i is the *instantaneous* current in the resistor. Because the heating effect of a current is proportional to the *square* of the current, it makes no difference whether the sign associated with the current is positive or negative. The heating effect produced by an alternating current with a maximum value of I_{max} is not the same as that produced by a direct current of the same value, however. The reason is that the alternating current has this maximum value for only an instant of time during a cycle. The important quantity in an AC circuit is a special kind of average value of current, called the **rms current**: the direct current that dissipates the

Figure 21.2 A plot of current and voltage difference in a resistor versus time.

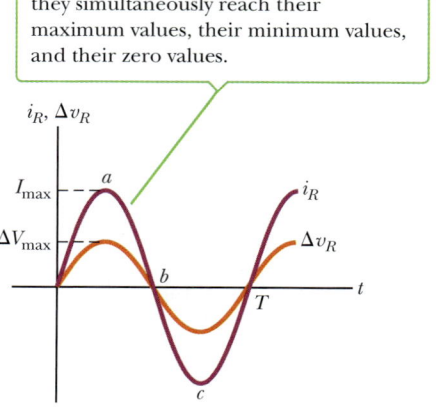

The current and the voltage are in phase: they simultaneously reach their maximum values, their minimum values, and their zero values.

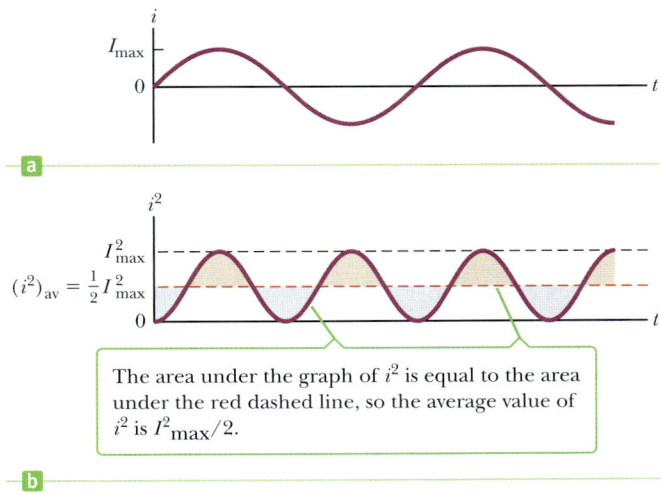

Figure 21.3 (a) Plot of the current in a resistor as a function of time. (b) Plot of the square of the current in a resistor as a function of time. The area under the red-dashed line omits the tan-shaded regions, but that is made up by the inclusion of the gray-shaded regions with the same area. That shows the area under the curve of i^2 is the same as the area under the red dashed line, so $(i^2)_{av} = I_{max}^2/2$.

same amount of energy in a resistor as the actual alternating current. Finding the average of the alternating current i, depicted in Figure 21.3a, would not be useful because that average is zero, whereas the rms current is always positive. To find the rms current, we first square the current, then find its average value, and finally take the square root of this average value. Hence, the rms current is the square *root* of the average (*mean*) of the *square* of the current. Because i^2 varies as $\sin^2 2\pi ft$, the average value of i^2 is $\frac{1}{2}I_{max}^2$ (Fig. 21.3b).[1] Therefore, the rms current I_{rms} is related to the maximum value of the alternating current I_{max} by

$$I_{rms} = \frac{I_{max}}{\sqrt{2}} = 0.707 I_{max} \quad [21.2]$$

This equation says that an alternating current with a maximum value of 3 A produces the same heating effect in a resistor as a direct current of $(3/\sqrt{2})$ A. We can therefore say that the average power dissipated in a resistor that carries alternating current I is

$$P_{av} = I_{rms}^2 R$$

Alternating voltages are also best discussed in terms of rms voltages, with a relationship identical to the preceding one,

$$\Delta V_{rms} = \frac{\Delta V_{max}}{\sqrt{2}} = 0.707\, \Delta V_{max} \quad [21.3] \quad \blacktriangleleft \text{ rms voltage}$$

where ΔV_{rms} is the rms voltage and ΔV_{max} is the maximum value of the alternating voltage.

---

[1] We can show that $(i^2)_{av} = I_{max}^2/2$ as follows: The current in the circuit varies with time according to the expression $i = I_{max} \sin 2\pi ft$, so $i^2 = I_{max}^2 \sin^2 2\pi ft$. Therefore, we can find the average value of i^2 by calculating the average value of $\sin^2 2\pi ft$. Note that a graph of $\cos^2 2\pi ft$ versus time is identical to a graph of $\sin^2 2\pi ft$ versus time, except that the points are shifted on the time axis. Thus, the time average of $\sin^2 2\pi ft$ is equal to the time average of $\cos^2 2\pi ft$, taken over one or more cycles. That is,

$$(\sin^2 2\pi ft)_{av} = (\cos^2 2\pi ft)_{av}$$

With this fact and the trigonometric identity $\sin^2 \theta + \cos^2 \theta = 1$, we get

$$(\sin^2 2\pi ft)_{av} + (\cos^2 2\pi ft)_{av} = 2(\sin^2 2\pi ft)_{av} = 1$$

$$(\sin^2 2\pi ft)_{av} = \tfrac{1}{2}$$

When this result is substituted into the expression $i^2 = I_{max}^2 \sin^2 2\pi ft$, we get $(i^2)_{av} = I_{rms}^2 = I_{max}^2/2$ or $I_{rms} = I_{max}/\sqrt{2}$, where I_{rms} is the rms current.

Table 21.1 Notation Used in This Chapter

	Voltage	Current
Instantaneous value	Δv	i
Maximum value	ΔV_{max}	I_{max}
rms value	ΔV_{rms}	I_{rms}

When we speak of measuring an AC voltage of 120 V from an electric outlet, we actually mean an rms voltage of 120 V. A quick calculation using Equation 21.3 shows that such an AC voltage actually has a peak value of about 170 V. In this chapter we use rms values when discussing alternating currents and voltages. One reason is that AC ammeters and voltmeters are designed to read rms values. Further, if we use rms values, many of the equations for alternating current will have the same form as those used in the study of direct-current (DC) circuits. Table 21.1 summarizes the notations used throughout this chapter.

Consider the series circuit in Figure 21.1 (page 734), consisting of a resistor connected to an AC generator. A resistor impedes the current in an AC circuit, just as it does in a DC circuit. Ohm's law is therefore valid for an AC circuit, and we have

$$\Delta V_{R,rms} = I_{rms} R \qquad [21.4a]$$

The rms voltage across a resistor is equal to the rms current in the circuit times the resistance. This equation is also true if maximum values of current and voltage are used:

$$\Delta V_{R,max} = I_{max} R \qquad [21.4b]$$

Quick Quiz

21.1 Which of the following statements can be true for a resistor connected in a simple series circuit to an operating AC generator? (a) $P_{av} = 0$ and $i_{av} = 0$ (b) $P_{av} = 0$ and $i_{av} > 0$ (c) $P_{av} > 0$ and $i_{av} = 0$ (d) $P_{av} > 0$ and $i_{av} > 0$

■ EXAMPLE 21.1 What Is the rms Current?

GOAL Perform basic AC circuit calculations for a purely resistive circuit.

PROBLEM An AC voltage source has an output of $\Delta v = (2.00 \times 10^2 \text{ V}) \sin 2\pi f t$. This source is connected to a $1.00 \times 10^2 \, \Omega$ resistor as in Figure 21.1. Find the rms voltage and rms current in the resistor.

STRATEGY Compare the expression for the voltage output just given with the general form, $\Delta v = \Delta V_{max} \sin 2\pi f t$, finding the maximum voltage. Substitute this result into the expression for the rms voltage.

SOLUTION

Obtain the maximum voltage by comparison of the given expression for the output with the general expression:

$$\Delta v = (2.00 \times 10^2 \text{ V}) \sin 2\pi f t \qquad \Delta v = \Delta V_{max} \sin 2\pi f t$$
$$\rightarrow \Delta V_{max} = 2.00 \times 10^2 \text{ V}$$

Next, substitute into Equation 21.3 to find the rms voltage of the source:

$$\Delta V_{rms} = \frac{\Delta V_{max}}{\sqrt{2}} = \frac{2.00 \times 10^2 \text{ V}}{\sqrt{2}} = \boxed{141 \text{ V}}$$

Substitute this result into Ohm's law to find the rms current:

$$I_{rms} = \frac{\Delta V_{rms}}{R} = \frac{141 \text{ V}}{1.00 \times 10^2 \, \Omega} = \boxed{1.41 \text{ A}}$$

REMARKS Notice how the concept of rms values allows the handling of an AC circuit quantitatively in much the same way as a DC circuit.

QUESTION 21.1 True or False: The rms current in an AC circuit oscillates sinusoidally with time.

EXERCISE 21.1 Find the maximum current in the circuit and the average power delivered to the circuit.

ANSWER 2.00 A; 2.00×10^2 W

APPLYING PHYSICS 21.1 | Electric Fields and Cancer Treatment BIO

Cancer cells multiply far more frequently than most normal cells, spreading throughout the body, using its resources and interfering with normal functioning. Most therapies damage both cancerous and healthy cells, so finding methods that target cancer cells is important in developing better treatments for the disease.

Because cancer cells multiply so rapidly, it's natural to consider treatments that prevent or disrupt cell division. Treatments such as chemotherapy interfere with the cell division cycle, but can also damage healthy cells. It has recently been found that alternating electric fields produced by alternating currents in the range of 100 kHz can disrupt the cell division cycle, either by slowing the division or by causing a dividing cell to disintegrate. Healthy cells that divide at only a very slow rate are less vulnerable than the rapidly dividing cancer cells, so such therapy holds out promise for certain types of cancer.

The alternating electric fields are thought to affect the process of mitosis, which is the dividing of the cell nucleus into two sets of identical chromosomes. Near the end of the first phase of mitosis, called the prophase, the mitotic spindle forms, a structure of fine filaments that guides the two replicated sets of chromosomes into separate daughter cells. The mitotic spindle is made up of a polymerization of dimers of tubulin, a protein with a large electric dipole moment. The alternating electric field exerts forces on these dipoles, disrupting their proper functioning.

Electric field therapy is especially promising for the treatment of brain tumors because healthy brain cells don't divide and therefore would be unharmed by the alternating electric fields. Research on such therapies is ongoing. ■

21.2 Capacitors in an AC Circuit

LEARNING OBJECTIVES

1. Define capacitive reactance.
2. Discuss the phase relationship between current and voltage difference in an AC circuit capacitor.
3. Calculate the capacitive reactance and rms current of capacitive AC circuits.

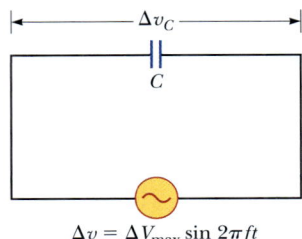

Figure 21.4 A series circuit consisting of a capacitor C connected to an AC generator.

To understand the effect of a capacitor on the behavior of a circuit containing an AC voltage source, we first review what happens when a capacitor is placed in a circuit containing a DC source, such as a battery. When the switch is closed in a series circuit containing a battery, a resistor, and a capacitor, the initial charge on the plates of the capacitor is zero. The motion of charge through the circuit is therefore relatively free, and there is a large current in the circuit. As more charge accumulates on the capacitor, the voltage across it increases, opposing the current. After some time interval, which depends on the time constant RC, the current approaches zero. Consequently, a capacitor in a DC circuit limits or impedes the current so that it approaches zero after a brief time.

Now consider the simple series circuit in Figure 21.4, consisting of a capacitor connected to an AC generator. We sketch curves of current versus time and voltage versus time, and then attempt to make the graphs seem reasonable. The curves are shown in Figure 21.5. First, notice that the segment of the current curve from a to b indicates that the current starts out at a rather large value. This large value can be understood by recognizing that there is no charge on the capacitor at $t = 0$; as a consequence, there is nothing in the circuit except the resistance of the wires to hinder the flow of charge at this instant. The current decreases, however, as the voltage across the capacitor increases from c to d on the voltage curve. When the voltage is at point d, the current reverses and begins to increase in the opposite direction (from b to e on the current curve). During this time, the voltage across the capacitor decreases from d to f because the plates are now losing the charge they accumulated earlier. The remainder of the cycle for both voltage and current is a repeat of what happened during the first half of the cycle. The current reaches a maximum value in the opposite direction at point e on the current curve and then decreases as the voltage across the capacitor builds up.

In a purely resistive circuit, the current and voltage are always in step with each other. That isn't the case when a capacitor is in the circuit. In Figure 21.5, when an

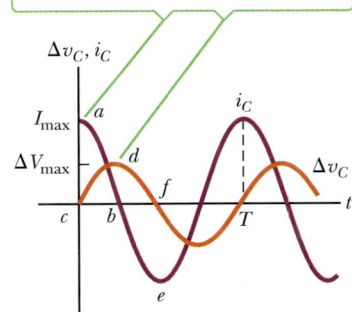

Figure 21.5 Plots of current and voltage across a capacitor versus time in an AC circuit.

The voltage across a capacitor lags the current by 90° ▶ alternating voltage is applied across a capacitor, the voltage reaches its maximum value one-quarter of a cycle after the current reaches its maximum value. We say that **the voltage across a capacitor always lags the current by 90°.**

Capacitive reactance ▶ The impeding effect of a capacitor on the current in an AC circuit is expressed in terms of a factor called the **capacitive reactance** X_C, defined as

$$X_C \equiv \frac{1}{2\pi f C} \quad [21.5]$$

When C is in farads and f is in hertz, the unit of X_C is the ohm. Notice that $2\pi f = \omega$, the angular frequency.

From Equation 21.5, as the frequency f of the voltage source increases, the capacitive reactance X_C (the impeding effect of the capacitor) decreases, so the current increases. At high frequency, there is less time available to charge the capacitor, so less charge and voltage accumulate on the capacitor, which translates into less opposition to the flow of charge and, consequently, a higher current. The analogy between capacitive reactance and resistance means that we can write an equation of the same form as Ohm's law to describe AC circuits containing capacitors. This equation relates the rms voltage and rms current in the circuit to the capacitive reactance:

$$\Delta V_{C,\text{rms}} = I_{\text{rms}} X_C \quad [21.6]$$

■ EXAMPLE 21.2 | A Purely Capacitive AC Circuit

GOAL Perform basic AC circuit calculations for a capacitive circuit.

PROBLEM An 8.00-μF capacitor is connected to the terminals of an AC generator with an rms voltage of 1.50×10^2 V and a frequency of 60.0 Hz. Find the capacitive reactance and the rms current in the circuit.

STRATEGY Substitute values into Equations 21.5 and 21.6.

SOLUTION

Substitute the values of f and C into Equation 21.5:

$$X_C = \frac{1}{2\pi f C} = \frac{1}{2\pi(60.0 \text{ Hz})(8.00 \times 10^{-6} \text{ F})} = 332 \text{ }\Omega$$

Solve Equation 21.6 for the current and substitute the values for X_C and the rms voltage to find the rms current:

$$I_{\text{rms}} = \frac{\Delta V_{C,\text{rms}}}{X_C} = \frac{1.50 \times 10^2 \text{ V}}{332 \text{ }\Omega} = 0.452 \text{ A}$$

REMARKS Again, notice how similar the technique is to that of analyzing a DC circuit with a resistor.

QUESTION 21.2 True or False: The larger the capacitance of a capacitor, the larger the capacitive reactance.

EXERCISE 21.2 If the frequency is doubled, what happens to the capacitive reactance and the rms current?

ANSWER X_C is halved, and I_{rms} is doubled.

21.3 Inductors in an AC Circuit

LEARNING OBJECTIVES

1. Define inductive reactance.
2. Discuss the phase relationship between current and voltage difference in an AC circuit inductor.
3. Calculate the inductive reactance and rms current in inductive AC circuits.

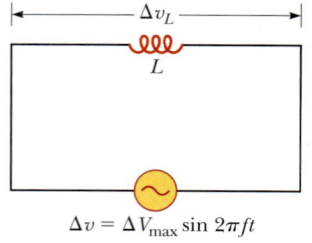

Figure 21.6 A series circuit consisting of an inductor L connected to an AC generator.

Now consider an AC circuit consisting only of an inductor connected to the terminals of an AC source, as in Figure 21.6. (In any real circuit there is some resistance in the wire forming the inductive coil, but we ignore this consideration for now.) The changing current output of the generator produces a back emf that impedes the current in the circuit. The magnitude of this back emf is

$$\Delta v_L = L \frac{\Delta I}{\Delta t} \quad [21.7]$$

The effective resistance of the coil in an AC circuit is measured by a quantity called the **inductive reactance**, X_L:

$$X_L \equiv 2\pi f L \quad [21.8]$$

When f is in hertz and L is in henries, the unit of X_L is the ohm. The inductive reactance *increases* with increasing frequency and increasing inductance. Contrast this with capacitors, where increasing frequency or capacitance *decreases* the capacitive reactance.

To understand the meaning of inductive reactance, compare Equation 21.8 with Equation 21.7. First, note from Equation 21.8 that the inductive reactance depends on the inductance L, which is reasonable because the back emf (Eq. 21.7) is large for large values of L. Second, note that the inductive reactance depends on the frequency f. This dependence, too, is reasonable because the back emf depends on $\Delta I/\Delta t$, a quantity that is large when the current changes rapidly, as it would for high frequencies.

With inductive reactance defined in this way, we can write an equation of the same form as Ohm's law for the voltage across the coil or inductor:

$$\Delta V_{L,\mathrm{rms}} = I_{\mathrm{rms}} X_L \quad [21.9]$$

where $\Delta V_{L,\mathrm{rms}}$ is the rms voltage across the coil and I_{rms} is the rms current in the coil.

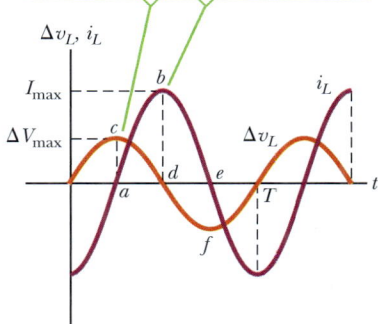

Figure 21.7 Plots of current and voltage across an inductor versus time in an AC circuit.

Figure 21.7 shows the instantaneous voltage and instantaneous current across the coil as functions of time. When a sinusoidal voltage is applied across an inductor, the voltage reaches its maximum value one-quarter of an oscillation period before the current reaches its maximum value. In this situation we say that **the voltage across an inductor always leads the current by 90°.**

To see why there is a phase relationship between voltage and current, we examine a few points on the curves of Figure 21.7. At point a on the current curve, the current is beginning to increase in the positive direction. At this instant the rate of change of current, $\Delta I/\Delta t$ (the slope of the current curve), is at a maximum, and we see from Equation 21.7 that the voltage across the inductor is consequently also at a maximum. As the current rises between points a and b on the curve, $\Delta I/\Delta t$ gradually decreases until it reaches zero at point b. As a result, the voltage across the inductor is decreasing during this same time interval, as the segment between c and d on the voltage curve indicates. Immediately after point b, the current begins to decrease, although it still has the same direction it had during the previous quarter cycle. As the current decreases to zero (from b to e on the curve), a voltage is again induced in the coil (from d to f), but the polarity of this voltage is opposite the polarity of the voltage induced between c and d. This occurs because back emfs always oppose the change in the current.

We could continue to examine other segments of the curves, but no new information would be gained because the current and voltage variations are repetitive.

■ EXAMPLE 21.3 | A Purely Inductive AC Circuit

GOAL Perform basic AC circuit calculations for an inductive circuit.

PROBLEM In a purely inductive AC circuit (see Fig. 21.6), $L = 25.0$ mH and the rms voltage is 1.50×10^2 V. Find the inductive reactance and rms current in the circuit if the frequency is 60.0 Hz.

(Continued)

SOLUTION

Substitute L and f into Equation 21.8 to get the inductive reactance:

$$X_L = 2\pi f L = 2\pi(60.0\text{ s}^{-1})(25.0 \times 10^{-3}\text{ H}) = 9.42\ \Omega$$

Solve Equation 21.9 for the rms current and substitute:

$$I_{rms} = \frac{\Delta V_{L,rms}}{X_L} = \frac{1.50 \times 10^2\text{ V}}{9.42\ \Omega} = 15.9\text{ A}$$

REMARKS The analogy with DC circuits is even closer than in the capacitive case because in the inductive equivalent of Ohm's law, the voltage across an inductor is *proportional* to the inductance L, just as the voltage across a resistor is proportional to R in Ohm's law.

QUESTION 21.3 True or False: A larger inductance or frequency results in a larger inductive reactance.

EXERCISE 21.3 Calculate the inductive reactance and rms current in a similar circuit if the frequency is again 60.0 Hz, but the rms voltage is 85.0 V and the inductance is 47.0 mH.

ANSWER $X_L = 17.7\ \Omega$, $I_{rms} = 4.80\text{ A}$

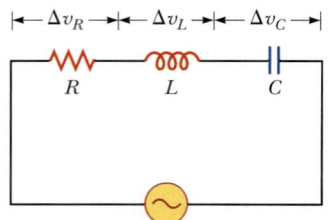

Figure 21.8 A series circuit consisting of a resistor, an inductor, and a capacitor connected to an AC generator.

21.4 The *RLC* Series Circuit

LEARNING OBJECTIVES

1. Define impedance in an AC circuit.
2. Discuss the AC relationship between current and voltage differences for elements in a series *RLC* circuit.
3. Apply the concepts of impedance, phasors and phase diagrams to series *RLC* circuits.

In the foregoing sections we examined the effects of an inductor, a capacitor, and a resistor when they are connected separately across an AC voltage source. We now consider what happens when these elements are combined.

Figure 21.8 shows a circuit containing a resistor, an inductor, and a capacitor connected in series across an AC source that supplies a total voltage Δv at some instant. The current in the circuit is the same at all points in the circuit at any instant and varies sinusoidally with time, as indicated in Figure 21.9a. This fact can be expressed mathematically as

$$i = I_{max}\sin 2\pi ft$$

Earlier, we learned that the voltage across each element may or may not be in phase with the current. The instantaneous voltages across the three elements, shown in Figure 21.9, have the following phase relations to the instantaneous current:

1. The instantaneous voltage Δv_R across the resistor is *in phase* with the instantaneous current. (See Fig. 21.9b.)
2. The instantaneous voltage Δv_L across the inductor *leads* the current by 90°. (See Fig. 21.9c.)
3. The instantaneous voltage Δv_C across the capacitor *lags* the current by 90°. (See Fig. 21.9d.)

The net instantaneous voltage Δv supplied by the AC source equals the sum of the instantaneous voltages across the separate elements: $\Delta v = \Delta v_R + \Delta v_C + \Delta v_L$. This doesn't mean, however, that the voltages measured with an AC voltmeter across R, C, and L sum to the measured source voltage! In fact, the measured voltages *don't* sum to the measured source voltage because the voltages across R, C, and L all have different phases.

To account for the different phases of the voltage drops, we use a technique involving vectors. We represent the voltage across each element with a rotating vector, as in

Figure 21.9 Phase relations in the series *RLC* circuit shown in Figure 21.8.

Figure 21.10. The rotating vectors are referred to as **phasors**, and the diagram is called a **phasor diagram**. This particular diagram represents the circuit voltage given by the expression $\Delta v = \Delta V_{max} \sin(2\pi f t + \phi)$, where ΔV_{max} is the maximum voltage (the magnitude or length of the rotating vector or phasor) and ϕ is the angle between the phasor and the positive x-axis when $t = 0$. The phasor can be viewed as a vector of magnitude ΔV_{max} rotating at a constant frequency f so that its projection along the y-axis is the instantaneous voltage in the circuit. Because ϕ is the phase angle between the voltage and current in the circuit, the phasor for the current (not shown in Fig. 21.10) lies along the positive x-axis when $t = 0$ and is expressed by the relation $i = I_{max} \sin(2\pi f t)$.

The phasor diagrams in Figure 21.11 are useful for analyzing the *series RLC* circuit. Voltages in phase with the current are represented by vectors along the positive x-axis, and voltages out of phase with the current lie along other directions. ΔV_R is horizontal and to the right because it's in phase with the current. Likewise, ΔV_L is represented by a phasor along the positive y-axis because it leads the current by 90°. Finally, ΔV_C is along the negative y-axis because it lags the current[2] by 90°. If the phasors are added as vector quantities so as to account for the different phases of the voltages across R, L, and C, Figure 21.11a shows that the only x-component for the voltages is ΔV_R and the net y-component is $\Delta V_L - \Delta V_C$. We now add the phasors vectorially to find the phasor ΔV_{max} (Fig. 21.11b), which represents the maximum voltage. The right triangle in Figure 21.11b gives the following equations for the maximum voltage and the phase angle ϕ between the maximum voltage and the current:

$$\Delta V_{max} = \sqrt{\Delta V_R^2 + (\Delta V_L - \Delta V_C)^2} \quad [21.10]$$

$$\tan \phi = \frac{\Delta V_L - \Delta V_C}{\Delta V_R} \quad [21.11]$$

In these equations, all voltages are maximum values. Although we choose to use maximum voltages in our analysis, the preceding equations apply equally well to rms voltages because the two quantities are related to each other by the same factor for all circuit elements. The result for the maximum voltage ΔV_{max} given by Equation 21.10 reinforces the fact that **the voltages across the resistor, capacitor, and inductor are not in phase, so one cannot simply add them to get the voltage across the combination of element or to get the source voltage.**

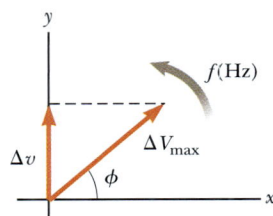

Figure 21.10 A phasor diagram for the voltage in an AC circuit, where ϕ is the phase angle between the voltage and the current and Δv is the instantaneous voltage.

Quick Quiz

21.2 For the circuit in Figure 21.8, is the instantaneous voltage of the source equal to (a) the sum of the maximum voltages across the elements, (b) the sum of the instantaneous voltages across the elements, or (c) the sum of the rms voltages across the elements?

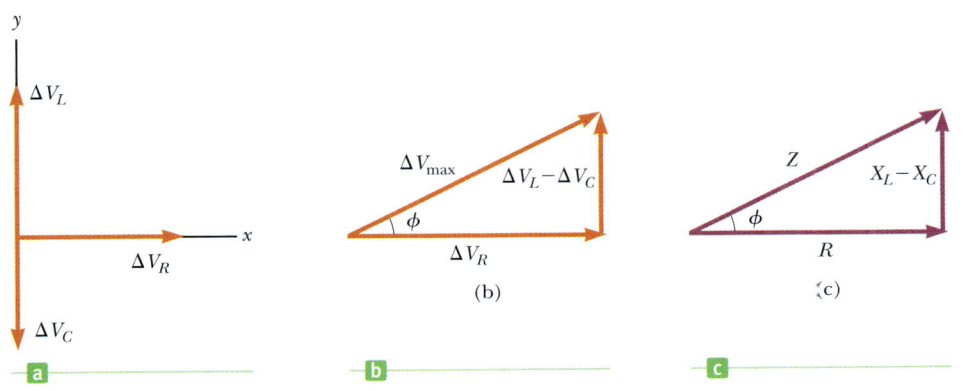

Figure 21.11 (a) A phasor diagram for the RLC circuit. (b) Addition of the phasors as vectors gives $\Delta V_{max} = \sqrt{\Delta V_R^2 + (\Delta V_L - \Delta V_C)^2}$. (c) The reactance triangle that gives the impedance relation $Z = \sqrt{R^2 + (X_L - X_C)^2}$.

---

[2] A mnemonic to help you remember the phase relationships in RLC circuits is "*ELI* the *ICE* man." $\mathcal{E}$ represents the voltage $\mathcal{E}$, I the current, L the inductance, and C the capacitance. Thus, the name *ELI* means that in an inductive circuit, the voltage $\mathcal{E}$ leads the current I. In a capacitive circuit *ICE* means that the current leads the voltage.

We can write Equation 21.10 in the form of Ohm's law, using the relations $\Delta V_R = I_{max}R$, $\Delta V_L = I_{max}X_L$, and $\Delta V_C = I_{max}X_C$, where I_{max} is the maximum current in the circuit:

$$\Delta V_{max} = I_{max}\sqrt{R^2 + (X_L - X_C)^2} \quad [21.12]$$

It's convenient to define a parameter called the **impedance** Z of the circuit as

▶ Impedance

$$Z \equiv \sqrt{R^2 + (X_L - X_C)^2} \quad [21.13]$$

so that Equation 21.12 becomes

$$\Delta V_{max} = I_{max}Z \quad [21.14]$$

Equation 21.14 is in the form of Ohm's law, $\Delta V = IR$, with R replaced by the impedance in ohms. Indeed, Equation 21.14 can be regarded as a generalized form of Ohm's law applied to a series AC circuit. Both the impedance and therefore the current in an AC circuit depend on the resistance, the inductance, the capacitance, *and* the frequency (because the reactances are frequency dependent).

It's useful to represent the impedance Z with a vector diagram such as the one depicted in Figure 21.11c. A right triangle is constructed with right side $X_L - X_C$, base R, and hypotenuse Z. Applying the Pythagorean theorem to this triangle, we see that

$$Z = \sqrt{R^2 + (X_L - X_C)^2}$$

which is Equation 21.13. Furthermore, we see from the vector diagram in Figure 21.11c that the phase angle ϕ between the current and the voltage obeys the relationship

▶ Phase angle ϕ

$$\tan \phi = \frac{X_L - X_C}{R} \quad [21.15]$$

The physical significance of the phase angle will become apparent in Section 21.5. Table 21.2 provides impedance values and phase angles for some series circuits containing different combinations of circuit elements.

Nikola Tesla
(1856–1943)
Tesla was born in Croatia, but spent most of his professional life as an inventor in the United States. He was a key figure in the development of alternating-current electricity, high-voltage transformers, and the transport of electrical power via AC transmission lines. Tesla's viewpoint was at odds with the ideas of Edison, who committed himself to the use of direct current in power transmission. Tesla's AC approach won out.

Table 21.2 Impedance Values and Phase Angles for Various Combinations of Circuit Elements

Circuit Elements	Impedance Z	Phase Angle ϕ
R	R	$0°$
C	X_C	$-90°$
L	X_L	$+90°$
R, C	$\sqrt{R^2 + X_C^2}$	Negative, between $-90°$ and $0°$
R, L	$\sqrt{R^2 + X_L^2}$	Positive, between $0°$ and $90°$
R, L, C	$\sqrt{R^2 + (X_L - X_C)^2}$	Negative if $X_C > X_L$ Positive if $X_C < X_L$

Note: In each case an AC voltage (not shown) is applied across the combination of elements (that is, across the dots).

Parallel alternating current circuits are also useful in everyday applications. We won't discuss them here, however, because their analysis is beyond the scope of this book.

■ Quick Quiz

21.3 If switch A is closed in Figure 21.12, what happens to the impedance of the circuit? (a) It increases. (b) It decreases. (c) It doesn't change.

21.4 Suppose $X_L > X_C$ in Figure 21.12. If switch A is closed, what happens to the phase angle? (a) It increases. (b) It decreases. (c) It doesn't change.

21.5 Suppose $X_L > X_C$ in Figure 21.12. If switch A is left open and switch B is closed, what happens to the phase angle? (a) It increases. (b) It decreases. (c) It doesn't change.

21.6 Suppose $X_L > X_C$ in Figure 21.12 and, with both switches open, a piece of iron is slipped into the inductor. During this process, what happens to the brightness of the bulb? (a) It increases. (b) It decreases. (c) It doesn't change.

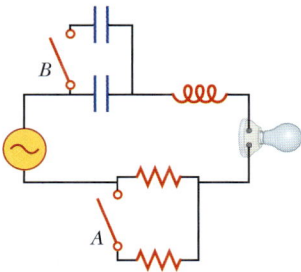

Figure 21.12 (Quick Quizzes 21.3–21.6)

■ PROBLEM-SOLVING STRATEGY

RLC Circuits

The following procedure is recommended for solving series RLC circuit problems:

1. Calculate the inductive and capacitive reactances, X_L and X_C.
2. Use X_L and X_C together with the resistance R to calculate the impedance Z of the circuit.
3. Find the maximum current or maximum voltage drop with the equivalent of Ohm's law, $\Delta V_{max} = I_{max} Z$.
4. Calculate the voltage drops across the individual elements with the appropriate variations of Ohm's law: $\Delta V_{R,max} = I_{max} R$, $\Delta V_{L,max} = I_{max} X_L$, and $\Delta V_{C,max} = I_{max} X_C$.
5. Obtain the phase angle using $\tan \phi = (X_L - X_C)/R$.

■ EXAMPLE 21.4 | An *RLC* Circuit

GOAL Analyze a series *RLC* AC circuit and find the phase angle.

PROBLEM A series *RLC* AC circuit has resistance $R = 2.50 \times 10^2 \, \Omega$, inductance $L = 0.600$ H, capacitance $C = 3.50 \, \mu$F, frequency $f = 60.0$ Hz, and maximum voltage $\Delta V_{max} = 1.50 \times 10^2$ V. Find **(a)** the impedance of the circuit, **(b)** the maximum current in the circuit, **(c)** the phase angle, and **(d)** the maximum voltages across the elements.

STRATEGY Calculate the inductive and capacitive reactances, which can be used with the resistance to calculate the impedance and phase angle. The impedance and Ohm's law yield the maximum current.

SOLUTION

(a) Find the impedance of the circuit.

First, calculate the inductive and capacitive reactances:

$$X_L = 2\pi f L = 226 \, \Omega \qquad X_C = 1/2\pi f C = 758 \, \Omega$$

Substitute these results and the resistance R into Equation 21.13 to obtain the impedance of the circuit:

$$Z = \sqrt{R^2 + (X_L - X_C)^2}$$
$$= \sqrt{(2.50 \times 10^2 \, \Omega)^2 + (226 \, \Omega - 758 \, \Omega)^2} = \boxed{588 \, \Omega}$$

(b) Find the maximum current in the circuit.

Use Equation 21.12, the equivalent of Ohm's law, to find the maximum current:

$$I_{max} = \frac{\Delta V_{max}}{Z} = \frac{1.50 \times 10^2 \, \text{V}}{588 \, \Omega} = \boxed{0.255 \, \text{A}}$$

(Continued)

(c) Find the phase angle.

Calculate the phase angle between the current and the voltage with Equation 21.15:

$$\phi = \tan^{-1}\frac{X_L - X_C}{R} = \tan^{-1}\left(\frac{226\ \Omega - 758\ \Omega}{2.50 \times 10^2\ \Omega}\right) = -64.8°$$

(d) Find the maximum voltages across the elements.

Use the "Ohm's law" expressions for each individual type of current element:

$$\Delta V_{R,\max} = I_{\max}R = (0.255\ \text{A})(2.50 \times 10^2\ \Omega) = 63.8\ \text{V}$$

$$\Delta V_{L,\max} = I_{\max}X_L = (0.255\ \text{A})(2.26 \times 10^2\ \Omega) = 57.6\ \text{V}$$

$$\Delta V_{C,\max} = I_{\max}X_C = (0.255\ \text{A})(7.58 \times 10^2\ \Omega) = 193\ \text{V}$$

REMARKS Because the circuit is more capacitive than inductive ($X_C > X_L$), ϕ is negative. A negative phase angle means that the current leads the applied voltage. Notice also that the sum of the maximum voltages across the elements is $\Delta V_R + \Delta V_L + \Delta V_C = 314$ V, which is much greater than the maximum voltage of the generator, 150 V. As we saw in Quick Quiz 21.2, the sum of the maximum voltages is a meaningless quantity because when alternating voltages are added, *both their amplitudes and their phases* must be taken into account. We know that the maximum voltages across the various elements occur at different times, so it doesn't make sense to add all the maximum values. The correct way to "add" the voltages is through Equation 21.10.

QUESTION 21.4 True or False: In an *RLC* circuit, the impedance must always be greater than or equal to the resistance.

EXERCISE 21.4 Analyze a series *RLC* AC circuit for which $R = 175\ \Omega$, $L = 0.500$ H, $C = 22.5\ \mu\text{F}$, $f = 60.0$ Hz, and $\Delta V_{\max} = 325$ V. Find (a) the impedance, (b) the maximum current, (c) the phase angle, and (d) the maximum voltages across the elements.

ANSWERS (a) 189 Ω (b) 1.72 A (c) 22.0° (d) $\Delta V_{R,\max} = 301$ V, $\Delta V_{L,\max} = 324$ V, $\Delta V_{C,\max} = 203$ V

21.5 Power in an AC Circuit

LEARNING OBJECTIVES

1. Discuss the conservation of energy in purely capacitive and purely inductive AC circuits.
2. Discuss the energy dissipated by resistive *RLC* circuits.
3. Evaluate the average power dissipated by an *RLC* circuit.

No power losses are associated with pure capacitors and pure inductors in an AC circuit. A pure capacitor, by definition, has no resistance or inductance, whereas a pure inductor has no resistance or capacitance. (These definitions are idealizations: in a real capacitor, for example, inductive effects could become important at high frequencies.) We begin by analyzing the power dissipated in an AC circuit that contains only a generator and a capacitor.

When the current increases in one direction in an AC circuit, charge accumulates on the capacitor and a voltage drop appears across it. When the voltage reaches its maximum value, the energy stored in the capacitor is

$$PE_C = \tfrac{1}{2}C(\Delta V_{\max})^2$$

This energy storage is only momentary, however: When the current reverses direction, the charge leaves the capacitor plates and returns to the voltage source. During one-half of each cycle the capacitor is being charged, and during the other half the charge is being returned to the voltage source. Therefore, the average power supplied by the source is zero. In other words, **no power losses occur in a capacitor in an AC circuit**.

Similarly, the source must do work against the back emf of an inductor that is carrying a current. When the current reaches its maximum value, the energy stored in the inductor is a maximum and is given by

$$PE_L = \tfrac{1}{2}LI_{\max}^2$$

When the current begins to decrease in the circuit, this stored energy is returned to the source as the inductor attempts to maintain the current in the circuit. The average power delivered to a resistor in an *RLC* circuit is

$$P_{av} = I_{rms}^2 R \qquad [21.16]$$

The average power delivered by the generator is converted to internal energy in the resistor. No power loss occurs in an ideal capacitor or inductor.

An alternate equation for the average power loss in an AC circuit can be found by substituting (from Ohm's law) $R = \Delta V_{R,rms}/I_{rms}$ into Equation 21.16:

$$P_{av} = I_{rms} \Delta V_{R,rms}$$

It's convenient to refer to a voltage triangle that shows the relationship among ΔV_{rms}, $\Delta V_{R,rms}$, and $\Delta V_{L,rms} - \Delta V_{C,rms}$, such as Figure 21.11b (page 741). (Remember that Fig. 21.11 applies to *both* maximum and rms voltages.) From this figure, we see that the voltage drop across a resistor can be written in terms of the voltage of the source, ΔV_{rms}:

$$\Delta V_{R,rms} = \Delta V_{rms} \cos \phi$$

Hence, the average power delivered by a generator in an AC circuit is

$$P_{av} = I_{rms} \Delta V_{rms} \cos \phi \qquad [21.17] \quad \blacktriangleleft \text{ Average power}$$

where the quantity $\cos \phi$ is called the **power factor**.

Equation 21.17 shows that the power delivered by an AC source to any circuit depends on the phase difference between the source voltage and the resulting current. This fact has many interesting applications. For example, factories often use devices such as large motors in machines, generators, and transformers that have a large inductive load due to all the windings. To deliver greater power to such devices without using excessively high voltages, factory technicians introduce capacitance in the circuits to shift the phase.

APPLICATION
Shifting phase to deliver more power

■ EXAMPLE 21.5 | Average Power in an *RLC* Series Circuit

GOAL Understand power in *RLC* series circuits.

PROBLEM Calculate the average power delivered to the series *RLC* circuit described in Example 21.4.

STRATEGY After finding the rms current and rms voltage with Equations 21.2 and 21.3, substitute into Equation 21.17, using the phase angle found in Example 21.4.

SOLUTION

First, use Equations 21.2 and 21.3 to calculate the rms current and rms voltage:

$$I_{rms} = \frac{I_{max}}{\sqrt{2}} = \frac{0.255 \text{ A}}{\sqrt{2}} = 0.180 \text{ A}$$

$$\Delta V_{rms} = \frac{\Delta V_{max}}{\sqrt{2}} = \frac{1.50 \times 10^2 \text{ V}}{\sqrt{2}} = 106 \text{ V}$$

Substitute these results and the phase angle $\phi = -64.8°$ into Equation 21.17 to find the average power:

$$P_{av} = I_{rms} \Delta V_{rms} \cos \phi = (0.180 \text{ A})(106 \text{ V}) \cos(-64.8°)$$
$$= 8.12 \text{ W}$$

REMARKS The same result can be obtained from Equation 21.16, $P_{av} = I_{rms}^2 R$.

QUESTION 21.5 Under what circumstance can the average power of an *RLC* circuit be zero?

EXERCISE 21.5 Repeat this problem, using the system described in Exercise 21.4.

ANSWER 259 W

21.6 Resonance in a Series RLC Circuit

LEARNING OBJECTIVES

1. Discuss the concept of resonance in a series *RLC* circuit.
2. Evaluate and apply the resonance frequency of a series RLC circuit.

In general, the rms current in a series *RLC* circuit can be written

$$I_{rms} = \frac{\Delta V_{rms}}{Z} = \frac{\Delta V_{rms}}{\sqrt{R^2 + (X_L - X_C)^2}} \quad [21.18]$$

From this equation, we see that if the frequency is varied, the current has its *maximum* value when the impedance has its *minimum* value, which occurs when $X_L = X_C$. In such a circumstance, the impedance of the circuit reduces to $Z = R$. The frequency f_0 at which this happens is called the **resonance frequency** of the circuit. To find f_0, we set $X_L = X_C$, which gives, from Equations 21.5 and 21.8,

$$2\pi f_0 L = \frac{1}{2\pi f_0 C}$$

$$f_0 = \frac{1}{2\pi \sqrt{LC}} \quad [21.19]$$

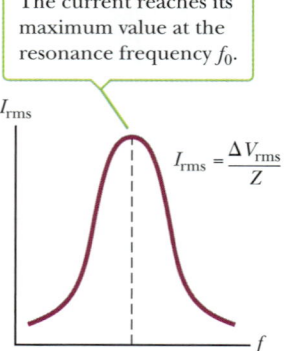

Figure 21.13 A plot of current amplitude in a series *RLC* circuit versus frequency of the generator voltage.

Figure 21.13 is a plot of current as a function of frequency for a circuit containing a fixed value for both the capacitance and the inductance. From Equation 21.18, it must be concluded that the current would become infinite at resonance when $R = 0$. Although Equation 21.18 predicts this result, real circuits always have some resistance, which limits the value of the current.

APPLICATION
Tuning Your Radio

The tuning circuit of a radio is an important application of a series resonance circuit. The radio is tuned to a particular station (which transmits a specific radio-frequency signal) by varying a capacitor, which changes the resonance frequency of the tuning circuit. When this resonance frequency matches that of the incoming radio wave, the current in the tuning circuit increases.

■ APPLYING PHYSICS 21.2 | Metal Detectors at the Courthouse

When you walk through the doorway of a courthouse metal detector, as the person in Figure 21.14 is doing, you are really walking through a coil of many turns. How might the metal detector work?

EXPLANATION The metal detector is essentially a resonant circuit. The portal you step through is an inductor (a large loop of conducting wire) that is part of the circuit. The frequency of the circuit is tuned to the circuit's resonant frequency of the circuit when there is no metal in the inductor. When you walk through with metal in your pocket, you change the effective inductance of the resonance circuit, resulting in a change in the circuit's current. This change in current is detected, and an electronic circuit causes a sound to be emitted as an alarm. ■

Figure 21.14 (Applying Physics 21.2) A courthouse metal detector.

■ EXAMPLE 21.6 | A Circuit in Resonance

GOAL Understand resonance frequency and its relation to inductance, capacitance, and the rms current.

PROBLEM Consider a series *RLC* circuit for which $R = 1.50 \times 10^2 \, \Omega$, $L = 20.0$ mH, $\Delta V_{rms} = 20.0$ V, and $f = 796 \, s^{-1}$. **(a)** Determine the value of the capacitance for which the rms current is a maximum. **(b)** Find the maximum rms current in the circuit.

STRATEGY The current is a maximum at the resonance frequency f_0, which should be set equal to the driving frequency, 796 s^{-1}. The resulting equation can be solved for C. For part (b), substitute into Equation 21.18 to get the maximum rms current.

SOLUTION

(a) Find the capacitance giving the maximum current in the circuit (the resonance condition).

Solve the resonance frequency for the capacitance:

$$f_0 = \frac{1}{2\pi\sqrt{LC}} \rightarrow \sqrt{LC} = \frac{1}{2\pi f_0} \rightarrow LC = \frac{1}{4\pi^2 f_0^2}$$

$$C = \frac{1}{4\pi^2 f_0^2 L}$$

Insert the given values, substituting the source frequency for the resonance frequency, f_0:

$$C = \frac{1}{4\pi^2 (796 \text{ Hz})^2 (20.0 \times 10^{-3} \text{ H})} = 2.00 \times 10^{-6} \text{ F}$$

(b) Find the maximum rms current in the circuit.

The capacitive and inductive reactances are equal, so $Z = R = 1.50 \times 10^2$ Ω. Substitute into Equation 21.18 to find the rms current:

$$I_{rms} = \frac{\Delta V_{rms}}{Z} = \frac{20.0 \text{ V}}{1.50 \times 10^2 \text{ Ω}} = 0.133 \text{ A}$$

REMARKS Because the impedance Z is in the denominator of Equation 21.18, the maximum current will always occur when $X_L = X_C$ because that yields the minimum value of Z.

QUESTION 21.6 True or False: The magnitude of the current in an RLC circuit is never larger than the rms current.

EXERCISE 21.6 Consider a series RLC circuit for which $R = 1.20 \times 10^2$ Ω, $C = 3.10 \times 10^{-5}$ F, $\Delta V_{rms} = 35.0$ V, and $f = 60.0$ s^{-1}. (a) Determine the value of the inductance for which the rms current is a maximum. (b) Find the maximum rms current in the circuit.

ANSWERS (a) 0.227 H (b) 0.292 A

21.7 The Transformer

LEARNING OBJECTIVES

1. Discuss the concept and operation of an AC transformer.
2. Apply the current and voltage values on an AC transformer's primary coil to those on the secondary coil.

It's often necessary to change a small AC voltage to a larger one or vice versa. Such changes are effected with a device called a transformer.

In its simplest form the **AC transformer** consists of two coils of wire wound around a core of soft iron, as shown in Figure 21.15. The coil on the left, which is connected to the input AC voltage source and has N_1 turns, is called the primary winding, or the *primary*. The coil on the right, which is connected to a resistor R and consists of N_2 turns, is the *secondary*. The common iron core is used to increase the magnetic flux and to provide a medium in which nearly all the flux through one coil passes through the other.

When an input AC voltage ΔV_1 is applied to the primary, the induced voltage across it is given by

$$\Delta V_1 = -N_1 \frac{\Delta \Phi_B}{\Delta t} \quad [21.20]$$

where Φ_B is the magnetic flux through each turn. If we assume that no flux leaks from the iron core, then the flux through each turn of the primary equals the

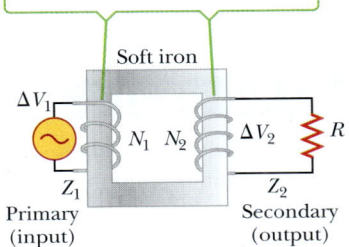

An AC voltage ΔV_1 is applied to the primary coil, and the output voltage ΔV_2 is observed across the load resistance R.

Figure 21.15 An ideal transformer consists of two coils wound on the same soft iron core. An AC voltage ΔV_1 is applied to the primary coil, and the output voltage ΔV_2 is observed across the load resistance R after the switch is closed.

flux through each turn of the secondary. Hence, the voltage across the secondary coil is

$$\Delta V_2 = -N_2 \frac{\Delta \Phi_B}{\Delta t} \qquad [21.21]$$

The term $\Delta \Phi_B/\Delta t$ is common to Equations 21.20 and 21.21 and can be algebraically eliminated, giving

$$\Delta V_2 = \frac{N_2}{N_1} \Delta V_1 \qquad [21.22]$$

When N_2 is greater than N_1, ΔV_2 exceeds ΔV_1 and the transformer is referred to as a *step-up transformer*. When N_2 is less than N_1, making ΔV_2 less than ΔV_1, we have a *step-down transformer*.

By Faraday's law, a voltage is generated across the secondary only when there is a *change* in the number of flux lines passing through the secondary. The input current in the primary must therefore change with time, which is what happens when an alternating current is used. When the input at the primary is a direct current, however, a voltage output occurs at the secondary only at the instant a switch in the primary circuit is opened or closed. Once the current in the primary reaches a steady value, the output voltage at the secondary is zero.

It may seem that a transformer is a device in which it is possible to get something for nothing. For example, a step-up transformer can change an input voltage from, say, 10 V to 100 V. This means that each coulomb of charge leaving the secondary has 100 J of energy, whereas each coulomb of charge entering the primary has only 10 J of energy. That is not the case, however, because **the power input to the primary equals the power output at the secondary**:

▶ In an ideal transformer, the input power equals the output power

$$I_1 \Delta V_1 = I_2 \Delta V_2 \qquad [21.23]$$

Although the *voltage* at the secondary may be, say, ten times greater than the voltage at the primary, the *current* in the secondary will be smaller than the primary's current by a factor of ten. Equation 21.23 assumes an **ideal transformer** in which there are no power losses between the primary and the secondary. Real transformers typically have power efficiencies ranging from 90% to 99%. Power losses occur because of such factors as eddy currents induced in the iron core of the transformer, which dissipate energy in the form of I^2R losses.

APPLICATION
Long-Distance Electric Power Transmission

When electric power is transmitted over large distances, it's economical to use a high voltage and a low current because the power lost via resistive heating in the transmission lines varies as I^2R. If a utility company can reduce the current by a factor of ten, for example, the power loss is reduced by a factor of one hundred. In practice, the voltage is stepped up to around 230 000 V at the generating station, then stepped down to around 20 000 V at a distribution station, and finally stepped down to 120 V at the customer's utility pole.

■ EXAMPLE 21.7 Distributing Power to a City

GOAL Understand transformers and their role in reducing power loss.

PROBLEM A generator at a utility company produces 1.00×10^2 A of current at 4.00×10^3 V. The voltage is stepped up to 2.40×10^5 V by a transformer before being sent on a high-voltage transmission line across a rural area to a city. Assume the effective resistance of the power line is 30.0 Ω and that the transformers are ideal. **(a)** Determine the percentage of power lost in the transmission line. **(b)** What percentage of the original power would be lost in the transmission line if the voltage were not stepped up?

STRATEGY Solving this problem is just a matter of substitution into the equation for transformers and the equation for power loss. To obtain the fraction of power lost, it's also necessary to compute the power output of the generator: the current times the potential difference created by the generator.

SOLUTION

(a) Determine the percentage of power lost in the line.

Substitute into Equation 21.23 to find the current in the transmission line:

$$I_2 = \frac{I_1 \Delta V_1}{\Delta V_2} = \frac{(1.00 \times 10^2 \text{ A})(4.00 \times 10^3 \text{ V})}{2.40 \times 10^5 \text{ V}} = 1.67 \text{ A}$$

Now use Equation 21.16 to find the power lost in the transmission line:

$$(1) \quad P_{\text{lost}} = I_2^2 R = (1.67 \text{ A})^2 (30.0 \text{ }\Omega) = 83.7 \text{ W}$$

Calculate the power output of the generator:

$$P = I_1 \Delta V_1 = (1.00 \times 10^2 \text{ A})(4.00 \times 10^3 \text{ V}) = 4.00 \times 10^5 \text{ W}$$

Finally, divide P_{lost} by the power output and multiply by 100 to find the percentage of power lost:

$$\% \text{ power lost} = \left(\frac{83.7 \text{ W}}{4.00 \times 10^5 \text{ W}}\right) \times 100 = \boxed{0.020\ 9\%}$$

(b) What percentage of the original power would be lost in the transmission line if the voltage were not stepped up?

Replace the stepped-up current in Equation (1) by the original current of 1.00×10^2 A:

$$P_{\text{lost}} = I^2 R = (1.00 \times 10^2 \text{ A})^2 (30.0 \text{ }\Omega) = 3.00 \times 10^5 \text{ W}$$

Calculate the percentage loss, as before:

$$\% \text{ power lost} = \left(\frac{3.00 \times 10^5 \text{ W}}{4.00 \times 10^5 \text{ W}}\right) \times 100 = \boxed{75\%}$$

REMARKS This example illustrates the advantage of high-voltage transmission lines. In the city, a transformer at a substation steps the voltage back down to about 4 000 V, and this voltage is maintained across utility lines throughout the city. When the power is to be used at a home or business, a transformer on a utility pole near the establishment reduces the voltage to 240 V or 120 V.

QUESTION 21.7 If the voltage is stepped up to double the amount in this problem, by what factor is the power loss changed? (a) 2 (b) no change (c) $\frac{1}{2}$ (d) $\frac{1}{4}$

This cylindrical step-down transformer drops the voltage from 4 000 V to 220 V for delivery to a group of residences.

EXERCISE 21.7 Suppose the same generator has the voltage stepped up to only 7.50×10^4 V and the resistance of the line is 85.0 Ω. Find the percentage of power lost in this case.

ANSWER 0.604%

21.8 Maxwell's Predictions

LEARNING OBJECTIVES

1. State and discuss the four fundamental ideas underlying Maxwell's unifying theory of electricity and magnetism.
2. Discuss the concept of electromagnetic waves.

During the early stages of their study and development, electric and magnetic phenomena were thought to be unrelated. In 1865, however, James Clerk Maxwell (1831–1879) provided a mathematical theory that showed a close relationship between all electric and magnetic phenomena. In addition to unifying the formerly separate fields of electricity and magnetism, his brilliant theory predicted

James Clerk Maxwell
Scottish Theoretical Physicist
(1831–1879)
Maxwell developed the electromagnetic theory of light and the kinetic theory of gases, and he explained the nature of Saturn's rings and color vision. Maxwell's successful interpretation of the electromagnetic field resulted in the equations that bear his name. Formidable mathematical ability combined with great insight enabled him to lead the way in the study of electromagnetism and kinetic theory.

that electric and magnetic fields can move through space as waves. The theory he developed is based on the following four pieces of information:

1. Electric field lines originate on positive charges and terminate on negative charges.
2. Magnetic field lines always form closed loops; they don't begin or end anywhere.
3. A varying magnetic field induces an emf and hence an electric field. This fact is a statement of Faraday's law (Chapter 20).
4. Magnetic fields are generated by moving charges (or currents), as summarized in Ampère's law (Chapter 19).

The first statement is a consequence of the nature of the electrostatic force between charged particles, given by Coulomb's law. It embodies the fact that **free charges (electric monopoles) exist in nature**.

The second statement—that magnetic fields form continuous loops—is exemplified by the magnetic field lines around a long, straight wire, which are closed circles, and the magnetic field lines of a bar magnet, which form closed loops. It says, in contrast to the first statement, that **free magnetic charges (magnetic monopoles) don't exist in nature**.

The third statement is equivalent to Faraday's law of induction, and the fourth is equivalent to Ampère's law.

In one of the greatest theoretical developments of the 19th century, Maxwell used these four statements within a corresponding mathematical framework to prove that electric and magnetic fields play symmetric roles in nature. It was already known from experiments that a changing magnetic field produced an electric field according to Faraday's law. Maxwell believed that nature was symmetric, and he therefore hypothesized that a changing electric field should produce a magnetic field. This hypothesis could not be proven experimentally at the time it was developed because the magnetic fields generated by changing electric fields are generally very weak and therefore difficult to detect.

To justify his hypothesis, Maxwell searched for other phenomena that might be explained by it. He turned his attention to the motion of rapidly oscillating (accelerating) charges, such as those in a conducting rod connected to an alternating voltage. Such charges are accelerated and, according to Maxwell's predictions, generate changing electric and magnetic fields. The changing fields cause electromagnetic disturbances that travel through space as waves, similar to the spreading water waves created by a pebble thrown into a pool. The waves sent out by the oscillating charges are fluctuating electric and magnetic fields, so they are called *electromagnetic waves*. From Faraday's law and from Maxwell's own generalization of Ampère's law, Maxwell calculated the speed of the waves to be equal to the speed of light, $c = 3 \times 10^8$ m/s. He concluded that visible light and other electromagnetic waves consist of fluctuating electric and magnetic fields traveling through empty space, with each varying field inducing the other! His was truly one of the greatest discoveries of science, on a par with Newton's discovery of the laws of motion. Like Newton's laws, it had a profound influence on later scientific developments.

21.9 Hertz's Confirmation of Maxwell's Predictions

LEARNING OBJECTIVE

1. Discuss Heinrich Hertz's experimental confirmation of Maxwell's predictions about electromagnetic waves.

In 1887, after Maxwell's death, Heinrich Hertz (1857–1894) was the first to generate and detect electromagnetic waves in a laboratory setting, using LC circuits. In such a circuit a charged capacitor is connected to an inductor, as in Figure 21.16.

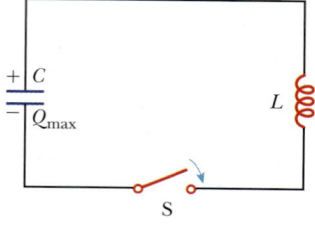

Figure 21.16 A simple LC circuit. The capacitor has an initial charge of Q_{max}, and the switch is closed at $t = 0$.

When the switch is closed, oscillations occur in the current in the circuit and in the charge on the capacitor. If the resistance of the circuit is neglected, no energy is dissipated and the oscillations continue.

In the following analysis, we neglect the resistance in the circuit. We assume the capacitor has an initial charge of Q_{max} and the switch is closed at $t = 0$. When the capacitor is fully charged, the total energy in the circuit is stored in the electric field of the capacitor and is equal to $Q_{max}^2/2C$. At this time, the current is zero, so no energy is stored in the inductor. As the capacitor begins to discharge, the energy stored in its electric field decreases. At the same time, the current increases and energy equal to $LI^2/2$ is now stored in the magnetic field of the inductor. Thus, energy is transferred from the electric field of the capacitor to the magnetic field of the inductor. When the capacitor is fully discharged, it stores no energy. At this time, the current reaches its maximum value and all the energy is stored in the inductor. The process then repeats in the reverse direction. The energy continues to transfer between the inductor and the capacitor, corresponding to oscillations in the current and charge.

As we saw in Section 21.6, the frequency of oscillation of an LC circuit is called the *resonance frequency* of the circuit and is given by

$$f_0 = \frac{1}{2\pi\sqrt{LC}}$$

The circuit Hertz used in his investigations of electromagnetic waves is similar to that just discussed and is shown schematically in Figure 21.17. An induction coil (a large coil of wire) is connected to two metal spheres with a narrow gap between them to form a capacitor. Oscillations are initiated in the circuit by short voltage pulses sent via the coil to the spheres, charging one positive, the other negative. Because L and C are quite small in this circuit, the frequency of oscillation is quite high, $f \approx 100$ MHz. This circuit is called a transmitter because it produces electromagnetic waves.

Several meters from the transmitter circuit, Hertz placed a second circuit, the receiver, which consisted of a single loop of wire connected to two spheres. It had its own effective inductance, capacitance, and natural frequency of oscillation. Hertz found that energy was being sent from the transmitter to the receiver when the resonance frequency of the receiver was adjusted to match that of the transmitter. The energy transfer was detected when the voltage across the spheres in the receiver circuit became high enough to produce ionization in the air, which caused sparks to appear in the air gap separating the spheres. Hertz's experiment is analogous to the mechanical phenomenon in which a tuning fork picks up the vibrations from another, identical tuning fork.

Hertz hypothesized that the energy transferred from the transmitter to the receiver is carried in the form of waves, now recognized as electromagnetic waves. In a series of experiments, he also showed that the radiation generated by the transmitter exhibits wave properties: interference, diffraction, reflection, refraction, and polarization. As you will see shortly, all these properties are exhibited by light. It became evident that Hertz's electromagnetic waves had the same known properties of light waves and differed only in frequency and wavelength. Hertz effectively confirmed Maxwell's theory by showing that Maxwell's mysterious electromagnetic waves existed and had all the properties of light waves.

Perhaps the most convincing experiment Hertz performed was the measurement of the speed of waves from the transmitter, accomplished as follows: waves of known frequency from the transmitter were reflected from a metal sheet so that an interference pattern was set up, much like the standing-wave pattern on a stretched string. As we learned in our discussion of standing waves, the distance between nodes is $\lambda/2$, so Hertz was able to determine the wavelength λ. Using the relationship $v = \lambda f$, he found that v was close to 3×10^8 m/s, the known speed of visible light. Hertz's experiments thus provided the first evidence in support of Maxwell's theory.

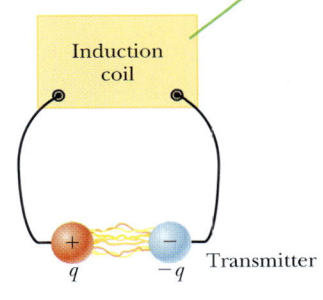

The transmitter consists of two spherical electrodes connected to an induction coil, which provides short voltage surges to the spheres, setting up oscillations in the discharge.

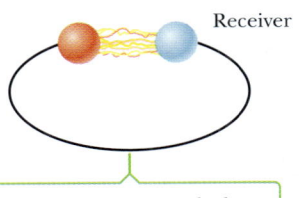

The receiver is a nearby loop of wire containing a second spark gap.

Figure 21.17 A schematic diagram of Hertz's apparatus for generating and detecting electromagnetic waves.

Heinrich Rudolf Hertz
German Physicist (1857–1894)
Hertz made his most important discovery of radio waves in 1887. After finding that the speed of a radio wave was the same as that of light, Hertz showed that radio waves, like light waves, could be reflected, refracted, and diffracted. Hertz died of blood poisoning at the age of 36. During his short life, he made many contributions to science. The hertz, equal to one complete vibration or cycle per second, is named after him.

21.10 Production of Electromagnetic Waves by an Antenna

LEARNING OBJECTIVES

1. Discuss the radiation of energy by the accelerated charges in an AC circuit.
2. Discuss the transverse nature of electromagnetic waves.

In the previous section we found that the energy stored in an *LC* circuit is continually transferred between the electric field of the capacitor and the magnetic field of the inductor. This energy transfer, however, continues for prolonged periods of time only when the changes occur slowly. If the current alternates rapidly, the circuit loses some of its energy in the form of electromagnetic waves. In fact, electromagnetic waves are radiated by *any* circuit carrying an alternating current. The fundamental mechanism responsible for this radiation is the acceleration of a charged particle. **Whenever a charged particle accelerates, it radiates energy.**

APPLICATION
Radio-Wave Transmission

An alternating voltage applied to the wires of an antenna forces electric charges in the antenna to oscillate. This common technique for accelerating charged particles is the source of the radio waves emitted by the broadcast antenna of a radio station.

Figure 21.18 illustrates the production of an electromagnetic wave by oscillating electric charges in an antenna. Two metal rods are connected to an AC source, which causes charges to oscillate between the rods. The output voltage of the generator is sinusoidal. At $t = 0$, the upper rod is given a maximum positive charge and the bottom rod an equal negative charge, as in Figure 21.18a. The electric field near the antenna at this instant is also shown in the figure. As the charges oscillate, the rods become less charged, the field near the rods decreases in strength, and the downward-directed maximum electric field produced at $t = 0$ moves away from the rod. When the charges are neutralized, as in Figure 21.18b, the electric field has dropped to zero, after an interval equal to one-quarter of the period of oscillation. Continuing in this fashion, the upper rod soon obtains a maximum negative charge and the lower rod becomes positive, as in Figure 21.18c, resulting in an electric field directed upward. This occurs after an interval equal to one-half the period of oscillation. The oscillations continue as indicated in Figure 21.18d. Note that the electric field near the antenna oscillates in phase with the charge distribution: the field points down when the upper rod is positive and up when the upper rod is negative. Further, the magnitude of the field at any instant depends on the amount of charge on the rods at that instant.

As the charges continue to oscillate (and accelerate) between the rods, the electric field set up by the charges moves away from the antenna in all directions at the speed of light. Figure 21.18 shows the electric field pattern on one side of the antenna at certain times during the oscillation cycle. As you can see, one cycle of charge oscillation produces one full wavelength in the electric field pattern.

Tip 21.1 Accelerated Charges Produce Electromagnetic Waves

Stationary charges produce only electric fields, whereas charges in uniform motion (i.e., constant velocity) produce electric and magnetic fields, but no electromagnetic waves. In contrast, accelerated charges produce electromagnetic waves as well as electric and magnetic fields. An accelerating charge also radiates energy.

Figure 21.18 An electric field set up by oscillating charges in an antenna. The field moves away from the antenna at the speed of light.

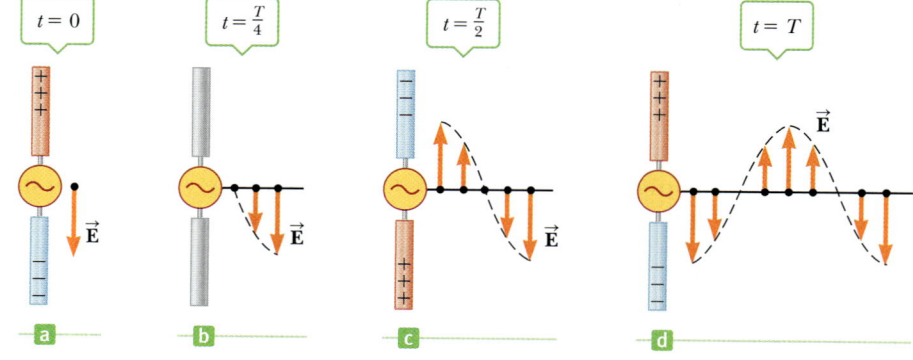

Because the oscillating charges create a current in the rods, a magnetic field is also generated when the current in the rods is upward, as shown in Figure 21.19. The magnetic field lines circle the antenna (recall right-hand rule number 2) and are perpendicular to the electric field at all points. As the current changes with time, the magnetic field lines spread out from the antenna. At great distances from the antenna, the strengths of the electric and magnetic fields become very weak. At these distances, however, it is necessary to take into account the facts that (1) a changing magnetic field produces an electric field and (2) a changing electric field produces a magnetic field, as predicted by Maxwell. These induced electric and magnetic fields are in phase: at any point, the two fields reach their maximum values at the same instant. This synchrony is illustrated at one instant of time in Figure 21.20. Note that (1) the $\vec{E}$ and $\vec{B}$ fields are perpendicular to each other and (2) both fields are perpendicular to the direction of motion of the wave. This second property is characteristic of transverse waves. Hence, we see that **an electromagnetic wave is a transverse wave**.

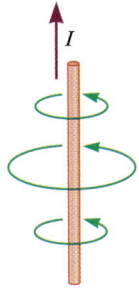

Figure 21.19 Magnetic field lines around an antenna carrying a changing current.

21.11 Properties of Electromagnetic Waves

> **LEARNING OBJECTIVES**
> 1. Discuss the concept of a plane wave.
> 2. Relate the speed of light to the permeability and permittivity of the propagation medium.
> 3. Relate the speed of light to the ratio of the electric and magnetic field magnitudes in an electromagnetic wave.
> 4. Discuss the fundamental properties of electromagnetic waves.
> 5. Evaluate the average intensity of an electromagnetic wave and the momentum delivered by the wave to perfectly absorbing or reflecting surfaces.
> 6. Calculate thermal and mechanical effects due to the interaction of electromagnetic waves and matter.

We have seen that Maxwell's detailed analysis predicted the existence and properties of electromagnetic waves. In this section we summarize what we know about electromagnetic waves thus far and consider some additional properties. In our discussion here and in future sections, we will often make reference to a type of wave called a **plane wave**. A plane electromagnetic wave is a wave traveling from a very distant source. Figure 21.20 pictures such a wave at a given instant of time. In this case the oscillations of the electric and magnetic fields take place in planes perpendicular

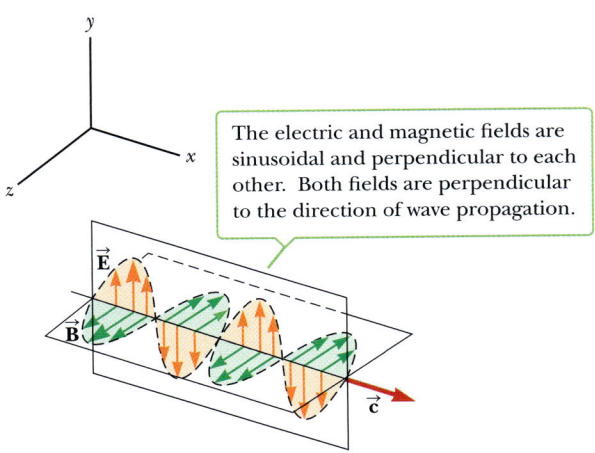

Figure 21.20 An electromagnetic wave sent out by oscillating charges in an antenna, represented at one instant of time and far from the antenna, moving in the positive x-direction with speed c.

to the x-axis and are therefore perpendicular to the direction of travel of the wave. Because of the latter property, electromagnetic waves are transverse waves. In the figure the electric field $\vec{E}$ is in the y-direction and the magnetic field $\vec{B}$ is in the z-direction. Light propagates in a direction perpendicular to these two fields. That direction is determined by yet another right-hand rule: (1) point the fingers of your right hand in the direction of $\vec{E}$, (2) curl them in the direction of $\vec{B}$, and (3) the right thumb then points in the direction of propagation of the wave.

Electromagnetic waves travel with the speed of light. In fact, it can be shown that the speed of an electromagnetic wave is related to the permeability and permittivity of the medium through which it travels. Maxwell found this relationship for free space to be

◂ Speed of light

$$c = \frac{1}{\sqrt{\mu_0 \epsilon_0}} \qquad [21.24]$$

where c is the speed of light, $\mu_0 = 4\pi \times 10^{-7}$ N·s²/C² is the permeability constant of vacuum, and $\epsilon_0 = 8.854\ 19 \times 10^{-12}$ C²/N·m² is the permittivity of free space. Substituting these values into Equation 21.24, we find that

$$c = 2.997\ 92 \times 10^8 \text{ m/s} \qquad [21.25]$$

Because electromagnetic waves travel at the same speed as light in vacuum, scientists concluded (correctly) that **light is an electromagnetic wave**.

Maxwell also proved the following relationship for electromagnetic waves:

> **Tip 21.2 E Stronger than B?**
> The relationship $E = Bc$ makes it appear that the electric fields associated with light are much larger than the magnetic fields. That is not the case: The units are different, so the quantities can't be directly compared. The two fields contribute equally to the energy of a light wave.

$$\frac{E}{B} = c \qquad [21.26]$$

which states that the ratio of the magnitude of the electric field to the magnitude of the magnetic field equals the speed of light.

Electromagnetic waves carry energy as they travel through space, and this energy can be transferred to objects placed in their paths. The average rate at which energy passes through an area perpendicular to the direction of travel of a wave, or the average power per unit area, is called the **intensity** I of the wave and is given by

$$I = \frac{E_{max} B_{max}}{2\mu_0} \qquad [21.27]$$

where E_{max} and B_{max} are the *maximum* values of E and B. The quantity I is analogous to the intensity of sound waves introduced in Chapter 14. From Equation 21.26, we see that $E_{max} = cB_{max} = B_{max}/\sqrt{\mu_0 \epsilon_0}$. Equation 21.27 can therefore also be expressed as

$$I = \frac{E_{max}^2}{2\mu_0 c} = \frac{c}{2\mu_0} B_{max}^2 \qquad [21.28]$$

Note that in these expressions we use the *average* power per unit area. A detailed analysis would show that the energy carried by an electromagnetic wave is shared equally by the electric and magnetic fields.

◂ Light is an electromagnetic wave and transports energy and momentum

Electromagnetic waves have an average intensity given by Equation 21.28. When the waves strike an area A of an object's surface for a given time Δt, energy $U = IA\Delta t$ is transferred to the surface. Momentum is transferred, as well. Hence, pressure is exerted on a surface when an electromagnetic wave impinges on it. In what follows, we assume the electromagnetic wave transports a total energy U to a surface in a time Δt. If the surface absorbs all the incident energy U in this time, Maxwell showed that the total momentum $\vec{p}$ delivered to this surface has a magnitude

$$p = \frac{U}{c} \qquad \text{(complete absorption)} \qquad [21.29]$$

If the surface is a perfect reflector, then the momentum transferred in a time Δt for normal incidence is twice that given by Equation 21.29. This is analogous to a molecule of gas bouncing off the wall of a container in a perfectly elastic collision. If the molecule is initially traveling in the positive x-direction at velocity v and after the collision is traveling in the negative x-direction at velocity $-v$, its change in momentum is given by $\Delta p = mv - (-mv) = 2mv$. Light bouncing off a perfect reflector is a similar process, so for complete reflection,

$$p = \frac{2U}{c} \quad \text{(complete reflection)} \quad [21.30]$$

Figure 21.21 An apparatus for measuring the radiation pressure of light. In practice, the system is contained in a high vacuum.

Although radiation pressures are very small (about 5×10^{-6} N/m² for direct sunlight), they have been measured with a device such as the one shown in Figure 21.21. Light is allowed to strike a mirror and a black disk that are connected to each other by a horizontal bar suspended from a fine fiber. Light striking the black disk is completely absorbed, so *all* the momentum of the light is transferred to the disk. Light striking the mirror head-on is totally reflected; hence, the momentum transfer to the mirror is twice that transmitted to the disk. As a result, the horizontal bar supporting the disks twists counterclockwise as seen from above. The bar comes to equilibrium at some angle under the action of the torques caused by radiation pressure and the twisting of the fiber. The radiation pressure can be determined by measuring the angle at which equilibrium occurs. The apparatus must be placed in a high vacuum to eliminate the effects of air currents. It's interesting that similar experiments demonstrate that electromagnetic waves carry angular momentum, as well.

In summary, electromagnetic waves traveling through free space have the following properties:

1. Electromagnetic waves travel at the speed of light.
2. Electromagnetic waves are transverse waves because the electric and magnetic fields are perpendicular to the direction of propagation of the wave and to each other.
3. The ratio of the electric field to the magnetic field in an electromagnetic wave equals the speed of light.
4. Electromagnetic waves carry both energy and momentum, which can be delivered to a surface.

◀ Some properties of electromagnetic waves

APPLYING PHYSICS 21.3 | Solar System Dust

In the interplanetary space in the solar system, there is a large amount of dust. Although interplanetary dust can in theory have a variety of sizes—from molecular size upward—why are there very few dust particles smaller than about 0.2 µm in the solar system? *Hint:* The solar system originally contained dust particles of all sizes.

EXPLANATION Dust particles in the solar system are subject to two forces: the gravitational force toward the Sun and the force from radiation pressure, which is directed away from the Sun. The gravitational force is proportional to the cube of the radius of a spherical dust particle because it is proportional to the mass (ρV) of the particle. The radiation pressure is proportional to the square of the radius because it depends on the cross-sectional area of the particle. For large particles, the gravitational force is larger than the force of radiation pressure, and the weak attraction to the Sun causes such particles to move slowly toward it. For small particles, less than about 0.2 µm, the larger force from radiation pressure sweeps them out of the solar system. ■

■ Quick Quiz

21.7 In an apparatus such as the one in Figure 21.21, suppose the black disk is replaced by one with half the radius. Which of the following are different after the disk is replaced? (a) radiation pressure on the disk (b) radiation force on the disk (c) radiation momentum delivered to the disk in a given time interval

EXAMPLE 21.8 | A Hot Tin Roof (Solar-Powered Homes)

GOAL Calculate some basic properties of light and relate them to thermal radiation.

PROBLEM Assume the Sun delivers an average power per unit area of about 1.00×10^3 W/m^2 to Earth's surface. **(a)** Calculate the total power incident on a flat tin roof 8.00 m by 20.0 m. Assume the radiation is incident *normal* (perpendicular) to the roof. **(b)** The tin roof reflects some light, and convection, conduction, and radiation transport the rest of the thermal energy away until some equilibrium temperature is established. If the roof is a perfect blackbody and rids itself of one-half of the incident radiation through thermal radiation, what's its equilibrium temperature? Assume the ambient temperature is 298 K.

(Example 21.8) A solar home.

SOLUTION

(a) Calculate the power delivered to the roof.

Multiply the intensity by the area to get the power:

$$P = IA = (1.00 \times 10^3 \text{ W/m}^2)(8.00 \text{ m} \times 20.0 \text{ m})$$
$$= \boxed{1.60 \times 10^5 \text{ W}}$$

(b) Find the equilibrium temperature of the roof.

Substitute into Stefan's law. Only one-half the incident power should be substituted, and twice the area of the roof (both the top and the underside of the roof count).

$$P = \sigma e A (T^4 - T_0^4)$$

$$T^4 = T_0^4 + \frac{P}{\sigma e A}$$

$$= (298 \text{ K})^4 + \frac{(0.500)(1.60 \times 10^5 \text{ W/m}^2)}{(5.67 \times 10^{-8} \text{ W/m}^2 \cdot \text{K}^4)(1)(3.20 \times 10^2 \text{ m}^2)}$$

$$T = 333 \text{ K} = \boxed{6.0 \times 10^1 \text{ °C}}$$

REMARKS If the incident power could *all* be converted to electric power, it would be more than enough for the average home. Unfortunately, solar energy isn't easily harnessed, and the prospects for large-scale conversion are not as bright as they may appear from this simple calculation. For example, the conversion efficiency from solar to electrical energy is far less than 100%; 10–20% is typical for photovoltaic cells. Roof systems for using solar energy to raise the temperature of water with efficiencies of around 50% have been built. Other practical problems must be considered, however, such as overcast days, geographic location, and energy storage.

QUESTION 21.8 Does the angle the roof makes with respect to the horizontal affect the amount of power absorbed by the roof? Explain.

EXERCISE 21.8 A spherical satellite orbiting Earth is lighted on one side by the Sun, with intensity 1 340 W/m^2. If the radius of the satellite is 1.00 m, what power is incident upon it? *Note:* The satellite effectively intercepts radiation only over a cross section, an area equal to that of a disk, πr^2.

ANSWERS 4.21×10^3 W

EXAMPLE 21.9 | Clipper Ships of Space

GOAL Relate the intensity of light to its mechanical effect on matter.

PROBLEM Aluminized Mylar film is a highly reflective, lightweight material that could be used to make sails for spacecraft driven by the light of the Sun. Suppose a sail with area 1.00 km^2 is orbiting the Sun at a distance of 1.50×10^{11} m. The sail has a mass of 5.00×10^3 kg and is tethered to a payload of mass 2.00×10^4 kg. **(a)** If the intensity of sunlight is 1.34×10^3 W/m^2 and the sail is oriented perpendicular to the incident light, what radial force is exerted on the sail? **(b)** About how long would it take to change the radial speed of the sail by 1.00 km/s? Assume the sail is perfectly reflecting. **(c)** Suppose the light were supplied by a large, powerful laser beam instead of the Sun. (Such systems have been proposed.) Calculate the peak electric and magnetic fields of the laser light.

STRATEGY Equation 21.30 gives the momentum imparted when light strikes an object and is totally reflected. The change in this momentum with time is a force.

For part (b), use Newton's second law to obtain the acceleration. The velocity kinematics equation then yields the necessary time to achieve the desired change in speed. Part (c) follows from Equation 21.27 and $E = Bc$.

SOLUTION

(a) Find the force exerted on the sail.

Write Equation 21.30 and substitute $U = P\Delta t = IA\,\Delta t$ for the energy delivered to the sail:

$$\Delta p = \frac{2U}{c} = \frac{2P\Delta t}{c} = \frac{2IA\,\Delta t}{c}$$

Divide both sides by Δt, obtaining the force $\Delta p/\Delta t$ exerted by the light on the sail:

$$F = \frac{\Delta p}{\Delta t} = \frac{2IA}{c} = \frac{2(1340\text{ W/m}^2)(1.00 \times 10^6\text{ m}^2)}{3.00 \times 10^8\text{ m/s}}$$

$$= \boxed{8.93\text{ N}}$$

(b) Find the time it takes to change the radial speed by 1.00 km/s.

Substitute the force into Newton's second law and solve for the acceleration of the sail:

$$a = \frac{F}{m} = \frac{8.93\text{ N}}{2.50 \times 10^4\text{ kg}} = 3.57 \times 10^{-4}\text{ m/s}^2$$

Apply the kinematics velocity equation:

$$v = at + v_0$$

Solve for t:

$$t = \frac{v - v_0}{a} = \frac{1.00 \times 10^3\text{ m/s}}{3.57 \times 10^{-4}\text{ m/s}^2} = \boxed{2.80 \times 10^6\text{ s}}$$

(c) Calculate the peak electric and magnetic fields if the light is supplied by a laser.

Solve Equation 21.28 for E_{max}:

$$I = \frac{E_{max}^2}{2\mu_0 c} \quad \rightarrow \quad E_{max} = \sqrt{2\mu_0 cI}$$

$$E_{max} = \sqrt{2(4\pi \times 10^{-7}\text{ N}\cdot\text{s}^2/\text{C}^2)(3.00 \times 10^8\text{ m/s})(1.34 \times 10^3\text{ W/m}^2)}$$

$$= \boxed{1.01 \times 10^3\text{ N/C}}$$

Obtain B_{max} using $E_{max} = B_{max}c$:

$$B_{max} = \frac{E_{max}}{c} = \frac{1.01 \times 10^3\text{ N/C}}{3.00 \times 10^8\text{ m/s}} = \boxed{3.37 \times 10^{-6}\text{ T}}$$

REMARKS The answer to part (b) is a little over a month. While the acceleration is very low, there are no fuel costs, and within a few months the velocity can change sufficiently to allow the spacecraft to reach any planet in the solar system. Such spacecraft may be useful for certain purposes and are highly economical, but require a considerable amount of patience.

QUESTION 21.9 By what factor will the force exerted by the Sun's light be changed when the spacecraft is twice as far from the Sun? (a) no change (b) $\frac{1}{2}$ (c) $\frac{1}{4}$ (d) $\frac{1}{8}$

EXERCISE 21.9 A laser has a power of 22.0 W and a beam radius of 0.500 mm. (a) Find the intensity of the laser. (b) Suppose you were floating in space and pointed the laser beam away from you. What would your acceleration be? Assume your total mass, including equipment, is 72.0 kg and the force is directed through your center of mass. *Hint:* The change in momentum is the same as in the nonreflective case. (c) Calculate your acceleration if it were due to the gravity of a space station with mass 1.00×10^6 kg and center of mass 100.0 m away. (d) Calculate the peak electric and magnetic fields of the laser.

ANSWERS (a) 2.80×10^7 W/m^2 (b) 1.02×10^{-9} m/s^2 (c) 6.67×10^{-9} m/s^2 (d) 1.45×10^5 N/C, 4.84×10^{-4} T.
Remark If you were planning to use your laser welding torch as a thruster to get you back to the station, don't bother, because the force of gravity is stronger. Better yet, get somebody to toss you a line.

21.12 The Spectrum of Electromagnetic Waves

LEARNING OBJECTIVE

1. Describe and discuss the electromagnetic spectrum.

All electromagnetic waves travel in a vacuum with the speed of light, c. These waves transport energy and momentum from some source to a receiver. In 1887 Hertz successfully generated and detected the radio-frequency electromagnetic

758 CHAPTER 21 | Alternating-Current Circuits and Electromagnetic Waves

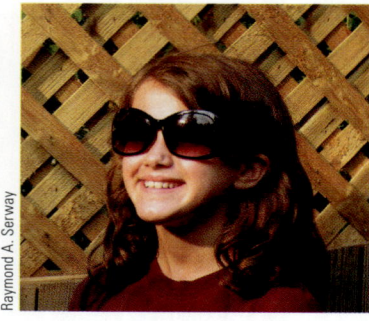

Wearing sunglasses lacking ultraviolet (UV) protection is worse for your eyes than wearing no sunglasses at all. Sunglasses without protection absorb some visible light, causing the pupils to dilate. This allows more UV light to enter the eye, increasing the damage to the lens of the eye over time. Without the sunglasses, the pupils constrict, reducing both visible and dangerous UV radiation. Be cool: wear sunglasses with UV protection.

waves predicted by Maxwell. Maxwell himself had recognized as electromagnetic waves both visible light and the infrared radiation discovered in 1800 by William Herschel. It is now known that other forms of electromagnetic waves exist that are distinguished by their frequencies and wavelengths.

Because all electromagnetic waves travel through free space with a speed c, their frequency f and wavelength λ are related by the important expression

$$c = f\lambda \qquad [21.31]$$

The various types of electromagnetic waves are presented in Figure 21.22. Notice the wide and overlapping range of frequencies and wavelengths. For instance, an AM radio wave with a frequency of 1.50 MHz (a typical value) has a wavelength of

$$\lambda = \frac{c}{f} = \frac{3.00 \times 10^8 \text{ m/s}}{1.50 \times 10^6 \text{ s}^{-1}} = 2.00 \times 10^2 \text{ m}$$

The following abbreviations are often used to designate short wavelengths and distances:

$$1 \text{ micrometer } (\mu\text{m}) = 10^{-6} \text{ m}$$
$$1 \text{ nanometer } (\text{nm}) = 10^{-9} \text{ m}$$
$$1 \text{ angstrom } (\text{Å}) = 10^{-10} \text{ m}$$

The wavelengths of visible light, for example, range from 0.4 μm to 0.7 μm, or 400 nm to 700 nm, or 4 000 Å to 7 000 Å.

■ Quick Quiz

21.8 Which of the following statements are true about light waves? (a) The higher the frequency, the longer the wavelength. (b) The lower the frequency, the longer the wavelength. (c) Higher-frequency light travels faster than lower-frequency light. (d) The shorter the wavelength, the higher the frequency. (e) The lower the frequency, the shorter the wavelength.

Figure 21.22 The electromagnetic spectrum.

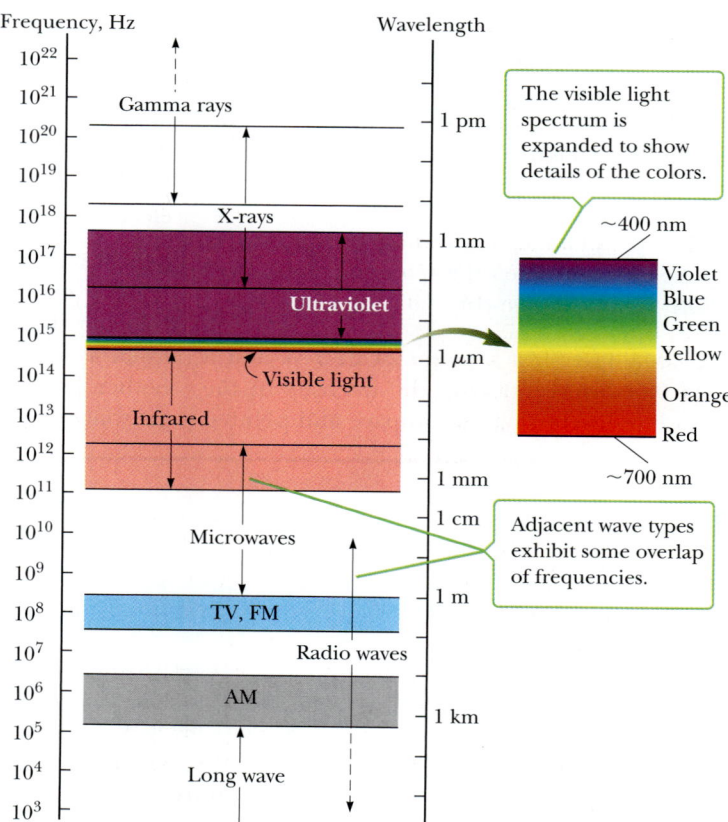

Brief descriptions of the wave types follow, in order of decreasing wavelength. There is no sharp division between one kind of electromagnetic wave and the next. All forms of electromagnetic radiation are produced by accelerating charges.

Radio waves, which were discussed in Section 21.10, are the result of charges accelerating through conducting wires. They are, of course, used in radio and television communication systems.

Microwaves (short-wavelength radio waves) have wavelengths ranging between about 1 mm and 30 cm and are generated by electronic devices. Their short wavelengths make them well suited for the radar systems used in aircraft navigation and for the study of atomic and molecular properties of matter. Microwave ovens are an interesting domestic application of these waves. It has been suggested that solar energy might be harnessed by beaming microwaves to Earth from a solar collector in space.

Infrared waves (sometimes incorrectly called "heat waves"), produced by hot objects and molecules, have wavelengths ranging from about 1 mm to the longest wavelength of visible light, 7×10^{-7} m. They are readily absorbed by most materials. The infrared energy absorbed by a substance causes it to get warmer because the energy agitates the atoms of the object, increasing their vibrational or translational motion. The result is a rise in temperature. Infrared radiation has many practical and scientific applications, including physical therapy, infrared photography, and the study of the vibrations of atoms.

Visible light, the most familiar form of electromagnetic waves, may be defined as the part of the spectrum that is detected by the human eye. Light is produced by the rearrangement of electrons in atoms and molecules. The wavelengths of visible light are classified as colors ranging from violet ($\lambda \approx 4 \times 10^{-7}$ m) to red ($\lambda \approx 7 \times 10^{-7}$ m). The eye's sensitivity is a function of wavelength and is greatest at a wavelength of about 5.6×10^{-7} m (yellow green).

Ultraviolet (UV) light covers wavelengths ranging from about 4×10^{-7} m (400 nm) down to 6×10^{-10} m (0.6 nm). The Sun is an important source of ultraviolet light (which is the main cause of suntans). Most of the ultraviolet light from the Sun is absorbed by atoms in the upper atmosphere, or stratosphere, which is fortunate, because UV light in large quantities has harmful effects on humans. One important constituent of the stratosphere is ozone (O_3), produced from reactions of oxygen with ultraviolet radiation. The resulting ozone shield causes lethal high-energy ultraviolet radiation to warm the stratosphere.

X-rays are electromagnetic waves with wavelengths from about 10^{-8} m (10 nm) down to 10^{-13} m (10^{-4} nm). The most common source of x-rays is the acceleration of high-energy electrons bombarding a metal target. X-rays are used as a diagnostic tool in medicine and as a treatment for certain forms of cancer. Because x-rays easily penetrate and damage or destroy living tissues and organisms, care must be taken to avoid unnecessary exposure and overexposure.

Gamma rays—electromagnetic waves emitted by radioactive nuclei—have wavelengths ranging from about 10^{-10} m to less than 10^{-14} m. They are highly penetrating and cause serious damage when absorbed by living tissues. Accordingly, those working near such radiation must be protected by garments containing heavily absorbing materials, such as layers of lead.

When astronomers observe the same celestial object using detectors sensitive to different regions of the electromagnetic spectrum, striking variations in the object's features can be seen. Figure 21.23 (page 760) shows images of the Crab Nebula made in three different wavelength ranges. The Crab Nebula is the remnant of a supernova explosion that was seen on Earth in 1054 A.D. (Compare with Fig. 8.31.)

Figure 21.23 Observations in different parts of the electromagnetic spectrum show different features of the Crab Nebula. (a) X-ray image. (b) Optical image. (c) Infrared. (d) Radio image.

APPLYING PHYSICS 21.4 — Light and Wound Treatment BIO

An important issue in human health is wound management. Chronic wounds affect five million to seven million people in the United States at an annual cost of more than twenty billion dollars. Low-level laser therapy has been shown to facilitate the healing and closure of wounds.

Infrared light increases the generation of adenosine triphosphate (ATP) in mitochondria and may stimulate the activation of genes and enzymes associated with cellular respiration. (ATP molecules provide energy for a variety of important cell functions.) Infrared light may also increase the concentration of reactive oxygen molecules, which could increase communication between the nucleus, cytosol, and the mitochondria. This mechanism may enhance and accelerate the healing process.

Green laser light can also be used to stimulate the body's repair mechanisms via a different process. A pink dye, called "rose bengal," is applied to the tissue, which is then exposed to the laser light for a few minutes. When the dye absorbs the light, it causes cross-linkages between collagen molecules in the tissue. The cross-linked molecules promote the closing of the tissue while reducing or eliminating the formation of scar tissue. Figure 21.24 shows the contrast between tissue receiving the normal treatment and tissue irradiated by lasers. The technique is also being studied for application to damaged peripheral nerves, blood vessels, and other tissues, such as incisions made in the cornea during eye surgery. ■

Figure 21.24 After bringing the sides of this wound together with deep sutures, closure on the left side was achieved with light-activated technology, whereas on the right side closure was carried out with sutures. This photo, taken at the end of two weeks, shows that healing was enhanced with light activation.

APPLYING PHYSICS 21.5 — The Sun and the Evolution of the Eye BIO

The center of sensitivity of our eyes coincides with the center of the wavelength distribution of the Sun. Is this an amazing coincidence?

EXPLANATION This fact is not a coincidence; rather, it's the result of biological evolution. Humans have evolved with vision most sensitive to wavelengths that are strongest from the Sun. If aliens from another planet ever arrived at Earth, their eyes would have the center of sensitivity at wavelengths different from ours. If their sun were a red dwarf, for example, the alien's eyes would be most sensitive to red light. ■

21.13 The Doppler Effect for Electromagnetic Waves

LEARNING OBJECTIVE

1. Describe the Doppler effect for electromagnetic waves and discuss its role in scientific discoveries.

As we saw in Section 14.6, sound waves exhibit the Doppler effect when the observer, the source, or both are moving relative to the medium of propagation.

Recall that in the Doppler effect, the observed frequency of the wave is larger or smaller than the frequency emitted by the source of the wave.

A Doppler effect also occurs for electromagnetic waves, but it differs from the Doppler effect for sound waves in two ways. First, in the Doppler effect for sound waves, motion relative to the medium is most important because sound waves require a medium in which to propagate. In contrast, the medium of propagation plays no role in the Doppler effect for electromagnetic waves because the waves require no medium in which to propagate. Second, the speed of sound that appears in the equation for the Doppler effect for sound depends on the reference frame in which it is measured. In contrast, as we'll see in Chapter 26, the speed of electromagnetic waves has the same value in all coordinate systems that are either at rest or moving at constant velocity with respect to one another.

The single equation that describes the Doppler effect for electromagnetic waves is given by the approximate expression

$$f_O \approx f_S\left(1 \pm \frac{u}{c}\right) \quad \text{if } u \ll c \quad [21.32]$$

where f_O is the observed frequency, f_S is the frequency emitted by the source, u is the *relative* speed of the observer and source, and c is the speed of light in a vacuum. Note that Equation 21.32 is valid only if u is much smaller than c. Further, it can also be used for sound as long as the relative velocity of the source and observer is much less than the velocity of sound. The positive sign in the equation must be used when the source and observer are moving toward each other, whereas the negative sign must be used when they are moving away from each other. Thus, we anticipate an increase in the observed frequency if the source and observer are approaching each other and a decrease if the source and observer recede from each other.

Astronomers have made important discoveries using Doppler observations on light reaching Earth from distant galaxies. Such measurements have shown that the more distant a galaxy is from Earth, the more its light is shifted toward the red end of the spectrum. This *cosmological red shift* is evidence that the Universe is expanding. The stretching and expanding of space, like a rubber sheet being pulled in all directions, is consistent with Einstein's theory of general relativity. A given star or galaxy, however, can have a peculiar motion toward or away from Earth. For example, Doppler effect measurements made with the Hubble Space Telescope have shown that a galaxy labeled M87 is rotating, with one edge moving toward us and the other moving away. Its measured speed of rotation was used to identify a supermassive black hole located at its center.

SUMMARY

21.1 Resistors in an AC Circuit

If an AC circuit consists of a generator and a resistor, the current in the circuit is in phase with the voltage, which means that the current and voltage reach their maximum values at the same time.

In discussions of voltages and currents in AC circuits, **rms values** of voltages are usually used. One reason is that AC ammeters and voltmeters are designed to read rms values. The rms values of currents and voltages (I_{rms} and ΔV_{rms}) are related to the maximum values of these quantities (I_{max} and ΔV_{max}) as follows:

$$I_{rms} = \frac{I_{max}}{\sqrt{2}} \quad \text{and} \quad \Delta V_{rms} = \frac{\Delta V_{max}}{\sqrt{2}} \quad [21.2, 21.3]$$

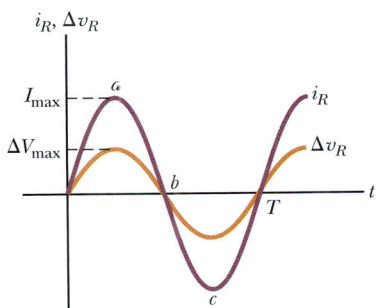

The voltage across a resistor and the current are in phase: they simultaneously reach their maximum values, their minimum values, and their zero values.

The rms voltage across a resistor is related to the rms current in the resistor by **Ohm's law**:

$$\Delta V_{R,rms} = I_{rms}R \quad [21.4a]$$

21.2 Capacitors in an AC Circuit

If an AC circuit consists of a generator and a capacitor, the voltage lags behind the current by 90°. This means that the voltage reaches its maximum value one-quarter of a period after the current reaches its maximum value.

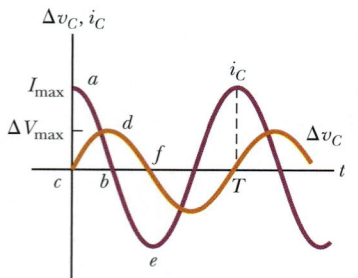

The voltage across a capacitor reaches its maximum value 90° after the current reaches its maximum value, so the voltage "lags" the current.

The impeding effect of a capacitor on current in an AC circuit is given by the **capacitive reactance** X_C, defined as

$$X_C \equiv \frac{1}{2\pi f C} \quad [21.5]$$

where f is the frequency of the AC voltage source.

The rms voltage across and the rms current in a capacitor are related by

$$\Delta V_{C,rms} = I_{rms} X_C \quad [21.6]$$

21.3 Inductors in an AC Circuit

If an AC circuit consists of a generator and an inductor, the voltage leads the current by 90°. This means the voltage reaches its maximum value one-quarter of a period before the current reaches its maximum value.

The effective impedance of a coil in an AC circuit is measured by a quantity called the **inductive reactance** X_L, defined as

$$X_L \equiv 2\pi f L \quad [21.8]$$

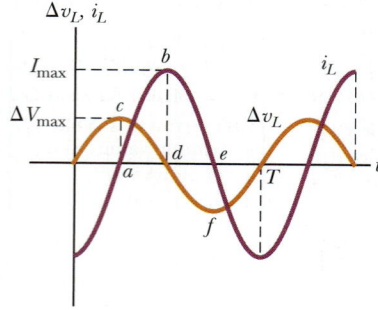

The voltage across an inductor reaches its maximum value 90° before the current reaches its maximum value, so the voltage "leads" the current.

The rms voltage across a coil is related to the rms current in the coil by

$$\Delta V_{L,rms} = I_{rms} X_L \quad [21.9]$$

21.4 The RLC Series Circuit

In an RLC series AC circuit, the maximum applied voltage ΔV is related to the maximum voltages across the resistor (ΔV_R), capacitor (ΔV_C), and inductor (ΔV_L) by

$$\Delta V_{max} = \sqrt{\Delta V_R^2 + (\Delta V_L - \Delta V_C)^2} \quad [21.10]$$

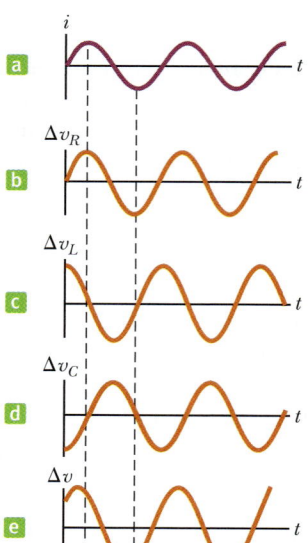

The time dependence of voltage differences of different circuit elements of an RLC series circuit are shown in these graphs. Notice that Δv_R is in phase with the current, Δv_L leads the current, and Δv_C lags the current.

If an AC circuit contains a resistor, an inductor, and a capacitor connected in series, the limit they place on the current is given by the **impedance Z** of the circuit, defined as

$$Z \equiv \sqrt{R^2 + (X_L - X_C)^2} \quad [21.13]$$

The relationship between the maximum voltage supplied to an RLC series AC circuit and the maximum current in the circuit, which is the same in every element, is

$$\Delta V_{max} = I_{max} Z \quad [21.14]$$

In an RLC series AC circuit, the applied rms voltage and current are out of phase. The **phase angle** ϕ between the current and voltage is given by

$$\tan \phi = \frac{X_L - X_C}{R} \quad [21.15]$$

21.5 Power in an AC Circuit

The **average power** delivered by the voltage source in an RLC series AC circuit is

$$P_{av} = I_{rms} \Delta V_{rms} \cos \phi \quad [21.17]$$

where the constant $\cos \phi$ is called the **power factor**.

21.6 Resonance in a Series RLC Circuit

In general, the rms current in a series RLC circuit can be written

$$I_{rms} = \frac{\Delta V_{rms}}{Z} = \frac{\Delta V_{rms}}{\sqrt{R^2 + (X_L - X_C)^2}} \quad [21.18]$$

The current has its *maximum* value when the impedance has its *minimum* value, corresponding to $X_L = X_C$ and $Z = R$. The frequency f_0 at which this happens is called the **resonance frequency** of the circuit, given by

$$f_0 = \frac{1}{2\pi\sqrt{LC}} \quad [21.19]$$

21.7 The Transformer

If the primary winding of a transformer has N_1 turns and the secondary winding consists of N_2 turns and then an input AC voltage ΔV_1 is applied to the primary, the induced voltage in the secondary winding is given by

$$\Delta V_2 = \frac{N_2}{N_1} \Delta V_1 \quad [21.22]$$

An AC voltage ΔV_1 is applied to the primary coil, and the output voltage ΔV_2 is observed across the load resistance R.

When N_2 is greater than N_1, ΔV_2 exceeds ΔV_1 and the transformer is referred to as a *step-up transformer*. When N_2 is less than N_1, making ΔV_2 less than ΔV_1, we have a *step-down transformer*. In an ideal transformer, the power output equals the power input.

$$I_1 \Delta V_1 = I_2 \Delta V_2 \quad [21.23]$$

21.8–21.13 Electromagnetic Waves and Their Properties

Electromagnetic waves were predicted by James Clerk Maxwell and experimentally confirmed by Heinrich Hertz. These waves are created by accelerating electric charges and have the following properties:

1. Electromagnetic waves are transverse waves because the electric and magnetic fields are perpendicular to the direction of propagation of the waves.
2. Electromagnetic waves travel at the speed of light.
3. The ratio of the electric field to the magnetic field at a given point in an electromagnetic wave equals the speed of light:

$$\frac{E}{B} = c \quad [21.26]$$

4. Electromagnetic waves carry energy as they travel through space. The average power per unit area is the intensity I, given by

$$I = \frac{E_{max}B_{max}}{2\mu_0} = \frac{E_{max}^2}{2\mu_0 c} = \frac{c}{2\mu_0} B_{max}^2 \quad [21.27, 21.28]$$

where E_{max} and B_{max} are the maximum values of the electric and magnetic fields.

5. Electromagnetic waves transport linear and angular momentum as well as energy. The momentum p delivered in time Δt at normal incidence to an object that completely absorbs light energy U is given by

$$p = \frac{U}{c} \quad \text{(complete absorption)} \quad [21.29]$$

If the surface is a perfect reflector, the momentum delivered in time Δt at normal incidence is twice that given by Equation 21.29:

$$p = \frac{2U}{c} \quad \text{(complete reflection)} \quad [21.30]$$

6. The speed c, frequency f, and wavelength λ of an electromagnetic wave are related by

$$c = f\lambda \quad [21.31]$$

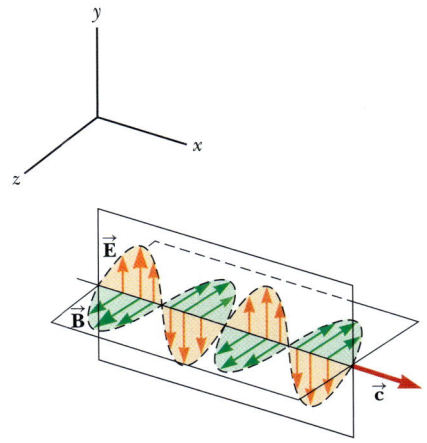

An electromagnetic wave sent out by oscillating charges in an antenna, represented at one instant of time and far from the antenna, moving in the positive x-direction with speed c.

The **electromagnetic spectrum** includes waves covering a broad range of frequencies and wavelengths. These waves have a variety of applications and characteristics, depending on their frequencies or wavelengths. The frequency of a given wave can be shifted by the relative velocity of observer and source, with the observed frequency f_O given by

$$f_O \approx f_S\left(1 \pm \frac{u}{c}\right) \quad \text{if } u \ll c \quad [21.32]$$

where f_S is the frequency of the source, u is the *relative* speed of the observer and source, and c is the speed of light in a vacuum. The positive sign is used when the source and observer approach each other, the negative sign when they recede from each other.

WARM-UP EXERCISES

The warm-up exercises in this chapter may be assigned online in Enhanced WebAssign.

1. **Math Review** The Cartesian components of a vector are (1.50 m, 2.50 m). Find (a) the magnitude and (b) the polar coordinate angle of the vector. (See Sections 1.7 and 1.8.)

2. **Physics Review** A potential difference of 12.0 V is measured across a 1.10×10^3 Ω resistor. Determine (a) the current through the resistor and (b) the power dissipated by the resistor. (See Sections 17.4 and 17.6.)

3. **Physics Review** Determine the frequency of a sound wave if it has a speed is 343 m/s and a wavelength of 1.50 m.

4. An rms potential difference of 1.20×10^2 V is measured across an AC circuit element where the rms current is 8.50 A. Determine the maximum (a) current through and (b) potential difference across the element. (See Section 21.1.)

5. An AC power source has an rms voltage of 1.20×10^2 V and operates at a frequency of 60.0 Hz. If a purely capacitive circuit is made from the power source and a 0.470 μF capacitor, determine (a) the capacitive reactance and (b) the rms current through the capacitor. (See Section 21.2.)

6. An AC power source has an rms voltage of 1.20×10^2 V and operates at a frequency of 60.0 Hz. If a purely inductive circuit is made from the power source and a 47.0 H inductor, determine (a) the inductive reactance and (b) the rms current through the inductor. (See Section 21.3.)

7. An AC power source has an rms voltage of 1.20×10^2 V and operates at a frequency of 60.0 Hz. If a series *RLC* circuit is made from the power source, a 0.850 μF capacitor, a 13.0 H inductor and a 1.50×10^3 Ω resistor, determine (a) the capacitive reactance, (b) the inductive reactance, (c) the circuit impedance, and (d) the maximum current in the circuit. (See Section 21.4.)

8. A series *RLC* circuit has an inductive reactance of 5.50×10^2 Ω, a capacitive reactance of 2.75×10^2 Ω, and an equivalent resistance of 1.70×10^2 Ω. Determine (a) the circuit impedance, (b) the phase angle between the current and the voltage, and (c) the power factor of the circuit. (See Sections 21.4 and 21.5.)

9. An analog receiver tunes in a radio station when the circuit's *RLC* resonance frequency matches the frequency transmitted by the station. For a receiver tuned to 6.90×10^2 kHz with an equivalent inductance of 10.0 μH, determine the equivalent capacitance. (See Section 21.6.)

10. An AC transformer powering a neon sign has potential differences of 1.20×10^2 V and 9.00×10^3 V across the primary coil and high-voltage secondary coils, respectively. If the current through the secondary coil is 30.0 mA, determine the current through the primary coil. (See Section 21.7.)

11. Suppose the Sun delivers an average power of 1.00×10^3 W/m^2 to Earth's surface. Determine the average pressure on a perfectly reflecting surface pointed directly at the Sun. (See Section 21.11.)

CONCEPTUAL QUESTIONS

The conceptual questions in this chapter may be assigned online in Enhanced WebAssign.

1. Despite the advent of digital television, some viewers still use "rabbit ears" atop their sets (Fig. CQ21.1) instead of purchasing cable television service or satellite dishes. Certain orientations of the receiving antenna on a television set give better reception than others. Furthermore, the best orientation varies from station to station. Explain.

Figure CQ21.1

2. (a) Does the phase angle in an *RLC* series circuit depend on frequency? (b) What is the phase angle for the circuit when the inductive reactance equals the capacitive reactance?

3. If the fundamental source of a sound wave is a vibrating object, what is the fundamental source of an electromagnetic wave?

4. Receiving radio antennas can be in the form of conducting lines or loops. What should the orientation of each of these antennas be relative to a broadcasting antenna that is vertical?

5. In radio transmission a radio wave serves as a carrier wave, and the sound signal is superimposed on the carrier wave. In amplitude modulation (AM) radio, the amplitude of the carrier wave varies according to the sound wave. The U.S. Navy sometimes uses flashing lights to send Morse code between neighboring ships, a process that has similarities to radio broadcasting.

(a) Is this process AM or FM? (b) What is the carrier frequency? (c) What is the signal frequency? (d) What is the broadcasting antenna? (e) What is the receiving antenna?

6. When light (or other electromagnetic radiation) travels across a given region, (a) what is it that oscillates? (b) What is it that is transported?

7. In space sailing, which is a proposed alternative for transport to the planets, a spacecraft carries a very large sail. Sunlight striking the sail exerts a force, accelerating the spacecraft. Should the sail be absorptive or reflective to be most effective?

8. What does a radio wave do to the charges in the receiving antenna to provide a signal for your car radio?

9. Does a wire connected to a battery emit an electromagnetic wave?

10. Suppose a creature from another planet had eyes that were sensitive to infrared radiation. Describe what it would see if it looked around the room that you are now in. That is, what would be bright and what would be dim?

11. Why should an infrared photograph of a person look different from a photograph taken using visible light?

12. If a high-frequency current is passed through a solenoid containing a metallic core, the core becomes warm due to induction. Explain why the temperature of the material rises in this situation.

13. What is the advantage of transmitting power at high voltages?

14. Why is the sum of the maximum voltages across each of the elements in a series RLC circuit usually greater than the maximum applied voltage? Doesn't this violate Kirchhoff's loop rule?

15. If the resistance in an RLC circuit remains the same, but the capacitance and inductance are each doubled, how will the resonance frequency change?

16. An inductor and a resistor are connected in series across an AC generator, as shown in Figure CQ21.16. Immediately after the switch is closed, which of the following statements is true? (a) The current is $\Delta V/R$. (b) The voltage across the inductor is zero. (c) The current in the circuit is zero. (d) The voltage across the resistor is ΔV. (e) The voltage across the inductor is half its maximum value.

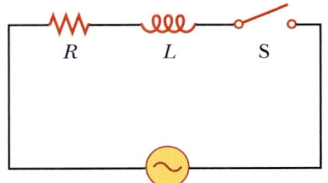

Figure CQ21.16

17. A capacitor and a resistor are connected in series across an AC generator, as shown in Figure CQ21.17. After the switch is closed, which of the following statements is true? (a) The voltage across the capacitor lags the current by 90°. (b) The voltage across the resistor is out of phase with the current. (c) The voltage across the capacitor leads the current by 90°. (d) The current decreases as the frequency of the generator is increased, but its peak voltage remains the same. (e) None of these

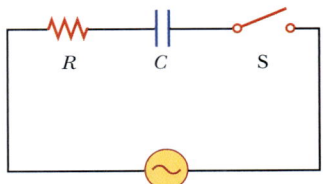

Figure CQ21.17

18. What is the impedance of a series RLC circuit at resonance? (a) X_L (b) X_C (c) R (d) $X_L - X_C$ (e) 0

19. Which of the following statements are true regarding electromagnetic waves traveling through a vacuum? More than one statement may be correct. (a) All waves have the same wavelength. (b) All waves have the same frequency. (c) All waves travel at 3.00×10^8 m/s. (d) The electric and magnetic fields associated with the waves are perpendicular to each other and to the direction of wave propagation. (e) The speed of the waves depends on their frequency.

PROBLEMS

WebAssign The problems in this chapter may be assigned online in Enhanced WebAssign.

1. denotes straightforward problem; 2. denotes intermediate problem;
3. denotes challenging problem

1. denotes full solution available in *Student Solutions Manual/Study Guide*

1. denotes problems most often assigned in Enhanced WebAssign

BIO denotes biomedical problems
GP denotes guided problems
M denotes Master It tutorial available in Enhanced WebAssign
QC denotes asking for quantitative and conceptual reasoning
S denotes symbolic reasoning problem
W denotes Watch It video solution available in Enhanced WebAssign

21.1 Resistors in an AC Circuit

1. (a) What is the resistance of a lightbulb that uses an average power of 75.0 W when connected to a 60.0-Hz power source having a maximum voltage of 170 V? (b) What is the resistance of a 100-W lightbulb?

2. **QC** A certain lightbulb is rated at 60.0 W when operating at an rms voltage of 120 V. (a) What is the peak voltage applied across the bulb? (b) What is the resistance of the bulb? (c) Does a 100-W bulb have greater or less resistance than a 60.0-W bulb? Explain.

3. The current in the circuit shown in Figure P21.3 equals 60.0% of the peak current at $t = 7.00$ ms. What is the lowest source frequency that gives this current?

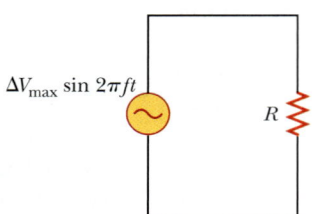

Figure P21.3

4. Figure P21.4 shows three lamps connected to a 120-V AC (rms) household supply voltage. Lamps 1 and 2 have 150-W bulbs; lamp 3 has a 100-W bulb. For each bulb, find (a) the rms current and (b) the resistance.

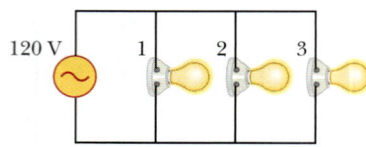

Figure P21.4

5. An audio amplifier, represented by the AC source and the resistor R in Figure P21.5, delivers alternating voltages at audio frequencies to the speaker. If the source puts out an alternating voltage of 15.0 V (rms), the resistance R is 8.20 Ω, and the speaker is equivalent to a resistance of 10.4 Ω, what is the time-averaged power delivered to the speaker?

Figure P21.5

6. **GP** The output voltage of an AC generator is given by $\Delta v = (170 \text{ V}) \sin(60\pi t)$. The generator is connected across a 20.0-Ω resistor. By inspection, what are the (a) maximum voltage and (b) frequency? Find the (c) rms voltage across the resistor, (d) rms current in the resistor, (e) maximum current in the resistor, (f) power delivered to the resistor, and (g) current when $t = 0.005\ 0$ s. (h) Should the argument of the sine function be in degrees or radians?

21.2 Capacitors in an AC Circuit

7. (a) For what frequencies does a 22.0-μF capacitor have a reactance below 175 Ω? (b) What is the reactance of a 44.0-μF capacitor over this same frequency range?

8. What is the maximum current delivered to a circuit containing a 2.20-μF capacitor when it is connected across (a) a North American outlet having $\Delta V_{rms} = 120$ V and $f = 60.0$ Hz and (b) a European outlet having $\Delta V_{rms} = 240$ V and $f = 50.0$ Hz?

9. **W** When a 4.0-μF capacitor is connected to a generator whose rms output is 30 V, the current in the circuit is observed to be 0.30 A. What is the frequency of the source?

10. **QC** An AC generator with an output rms voltage of 36.0 V at a frequency of 60.0 Hz is connected across a 12.0-μF capacitor. Find the (a) capacitive reactance, (b) rms current, and (c) maximum current in the circuit. (d) Does the capacitor have its maximum charge when the current takes its maximum value? Explain.

11. **M** What maximum current is delivered by an AC source with $\Delta V_{max} = 48.0$ V and $f = 90.0$ Hz when connected across a 3.70-μF capacitor?

12. A generator delivers an AC voltage of the form $\Delta v = (98.0 \text{ V}) \sin(80\pi t)$ to a capacitor. The maximum current in the circuit is 0.500 A. Find the (a) rms voltage of the generator, (b) frequency of the generator, (c) rms current, (d) reactance, and (e) value of the capacitance.

21.3 Inductors in an AC Circuit

13. An inductor has a 54.0-Ω reactance when connected to a 60.0-Hz source. The inductor is removed and then connected to a 50.0-Hz source that produces a 100-V rms voltage. What is the maximum current in the inductor?

14. An AC generator has an output rms voltage of 78.0 V at a frequency of 80.0 Hz. If the generator is connected across a 25.0-mH inductor, find the (a) inductive reactance, (b) rms current, and (c) maximum current in the circuit.

15. **M** In a purely inductive AC circuit as shown in Figure P21.15, $\Delta V_{max} = 100$ V. (a) The maximum current is 7.50 A at 50.0 Hz. Calculate the inductance L. (b) At what angular frequency ω is the maximum current 2.50 A?

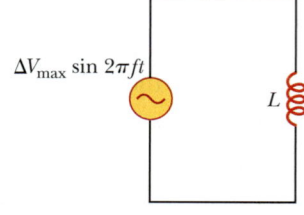

Figure P21.15

16. **GP** The output voltage of an AC generator is given by $\Delta v = (1.20 \times 10^2 \text{ V}) \sin(30\pi t)$. The generator is connected across a 0.500-H inductor. Find the (a) frequency of the generator, (b) rms voltage across the inductor, (c) inductive reactance, (d) rms current in the inductor, (e) maximum current in the inductor, and (f) average power delivered to the inductor. (g) Find an expression for the instantaneous current. (h) At what time after $t = 0$ does the instantaneous current first reach 1.00 A? (Use the inverse sine function.)

17. Determine the maximum magnetic flux through an inductor connected to a standard outlet ($\Delta V_{rms} = 120$ V, $f = 60.0$ Hz).

21.4 The RLC Series Circuit

18. A sinusoidal voltage $\Delta v = (80.0 \text{ V}) \sin (150t)$ is applied to a series RLC circuit with $L = 80.0$ mH, $C = 125.0$ μF, and $R = 40.0$ Ω. (a) What is the impedance of the circuit? (b) What is the maximum current in the circuit?

19. A 40.0-μF capacitor is connected to a 50.0-Ω resistor and a generator whose rms output is 30.0 V at 60.0 Hz. Find (a) the rms current in the circuit, (b) the rms voltage drop across the resistor, (c) the rms voltage drop across the capacitor, and (d) the phase angle for the circuit.

20. **M** An inductor ($L = 400$ mH), a capacitor ($C = 4.43$ μF), and a resistor ($R = 500$ Ω) are connected in series. A 50.0-Hz AC generator connected in series to these elements produces a maximum current of 250 mA in the circuit. (a) Calculate the required maximum voltage ΔV_{max}. (b) Determine the phase angle by which the current leads or lags the applied voltage.

21. A resistor ($R = 9.00 \times 10^2$ Ω), a capacitor ($C = 0.250$ μF), and an inductor ($L = 2.50$ H) are connected in series across a 2.40×10^2-Hz AC source for which $\Delta V_{max} = 1.40 \times 10^2$ V. Calculate (a) the impedance of the circuit, (b) the maximum current delivered by the source, and (c) the phase angle between the current and voltage. (d) Is the current leading or lagging the voltage?

22. A 50.0-Ω resistor, a 0.100-H inductor, and a 10.0-μF capacitor are connected in series to a 60.0-Hz source. The rms current in the circuit is 2.75 A. Find the rms voltages across (a) the resistor, (b) the inductor, (c) the capacitor, and (d) the RLC combination. (e) Sketch the phasor diagram for this circuit.

23. An RLC circuit consists of a 150-Ω resistor, a 21.0-μF capacitor, and a 460-mH inductor connected in series with a 120-V, 60.0-Hz power supply. (a) What is the phase angle between the current and the applied voltage? (b) Which reaches its maximum earlier, the current or the voltage?

24. An AC source operating at 60 Hz with a maximum voltage of 170 V is connected in series with a resistor ($R = 1.2$ kΩ) and an inductor ($L = 2.8$ H). (a) What is the maximum value of the current in the circuit? (b) What are the maximum values of the potential difference across the resistor and the inductor? (c) When the current is at a maximum, what are the magnitudes of the potential differences across the resistor, the inductor, and the AC source? (d) When the current is zero, what are the magnitudes of the potential difference across the resistor, the inductor, and the AC source?

25. A person is working near the secondary of a transformer, as shown in Figure P21.25. The primary voltage is 120 V (rms) at 60.0 Hz. The capacitance C_s, which is the stray capacitance between the hand and the secondary winding, is 20.0 pF. Assuming the person has a body resistance to ground of $R_b = 50.0$ kΩ, determine

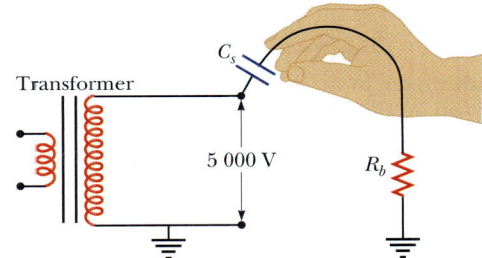

Figure P21.25

the rms voltage across the body. *Hint:* Redraw the circuit with the secondary of the transformer as a simple AC source.

26. **Q|C** A 60.0-Ω resistor is connected in series with a 30.0-μF capacitor and a generator having a maximum voltage of 1.20×10^2 V and operating at 60.0 Hz. Find the (a) capacitive reactance of the circuit, (b) impedance of the circuit, and (c) maximum current in the circuit. (d) Does the voltage lead or lag the current? (e) How will putting an inductor in series with the existing capacitor and resistor affect the current? Explain.

27. **GP** A series AC circuit contains a resistor, an inductor of 150 mH, a capacitor of 5.00 μF, and a generator with $\Delta V_{max} = 240$ V operating at 50.0 Hz. The maximum current in the circuit is 100 mA. Calculate (a) the inductive reactance, (b) the capacitive reactance, (c) the impedance, (d) the resistance in the circuit, and (e) the phase angle between the current and the generator voltage.

28. At what frequency does the inductive reactance of a 57.0-μH inductor equal the capacitive reactance of a 57.0-μF capacitor?

29. **W** An AC source with a maximum voltage of 150 V and $f = 50.0$ Hz is connected between points a and d in Figure P21.29. Calculate the rms voltages between points (a) a and b, (b) b and c, (c) c and d, and (d) b and d.

Figure P21.29

21.5 Power in an AC Circuit

30. An AC source operating at 60 Hz with a maximum voltage of 170 V is connected in series with a resistor ($R = 1.2$ kΩ) and a capacitor ($C = 2.5$ μF). (a) What is the maximum value of the current in the circuit? (b) What are the maximum values of the potential difference across the resistor and the capacitor? (c) When the current is zero, what are the magnitudes of the potential difference across the resistor, the capacitor, and the AC source? How much charge is on the capacitor at this instant? (d) When the current is at a maximum, what are the magnitudes of the potential differences across the

resistor, the capacitor, and the AC source? How much charge is on the capacitor at this instant?

31. A multimeter in an RL circuit records an rms current of 0.500 A and a 60.0-Hz rms generator voltage of 104 V. A wattmeter shows that the average power delivered to the resistor is 10.0 W. Determine (a) the impedance in the circuit, (b) the resistance R, and (c) the inductance L.

32. An AC voltage of the form $\Delta v = (90.0 \text{ V}) \sin (350t)$ is applied to a series RLC circuit. If $R = 50.0 \text{ }\Omega$, $C = 25.0 \text{ }\mu\text{F}$, and $L = 0.200 \text{ H}$, find the (a) impedance of the circuit, (b) rms current in the circuit, and (c) average power delivered to the circuit.

33. **W** An AC voltage of the form $\Delta v = 100 \sin (1\,000t)$, where Δv is in volts and t is in seconds, is applied to a series RLC circuit. Assume the resistance is 400 Ω, the capacitance is 5.00 μF, and the inductance is 0.500 H. Find the average power delivered to the circuit.

34. A series RLC circuit has a resistance of 22.0 Ω and an impedance of 80.0 Ω. If the rms voltage applied to the circuit is 160 V, what average power is delivered to the circuit?

35. An inductor and a resistor are connected in series. When connected to a 60-Hz, 90-V (rms) source, the voltage drop across the resistor is found to be 50 V (rms) and the power delivered to the circuit is 14 W. Find (a) the value of the resistance and (b) the value of the inductance.

36. **Q|C** Consider a series RLC circuit with $R = 25 \text{ }\Omega$, $L = 6.0$ mH, and $C = 25 \text{ }\mu\text{F}$. The circuit is connected to a 10-V (rms), 600-Hz AC source. (a) Is the sum of the voltage drops across R, L, and C equal to 10 V (rms)? (b) Which is greatest, the power delivered to the resistor, to the capacitor, or to the inductor? (c) Find the average power delivered to the circuit.

21.6 Resonance in a Series RLC Circuit

37. **M** An RLC circuit is used in a radio to tune into an FM station broadcasting at $f = 99.7$ MHz. The resistance in the circuit is $R = 12.0 \text{ }\Omega$, and the inductance is $L = 1.40 \text{ }\mu\text{H}$. What capacitance should be used?

38. The resonant frequency of a certain series RLC circuit is 2.84 kHz, and the value of its capacitance is 6.50 μF. What is the value of the resonant frequency when the capacitance of the circuit is 9.80 μF?

39. The AM band extends from approximately 500 kHz to 1 600 kHz. If a 2.0-μH inductor is used in a tuning circuit for a radio, what are the extremes that a capacitor must reach to cover the complete band of frequencies?

40. Consider a series RLC circuit with $R = 15 \text{ }\Omega$, $L = 200$ mH, $C = 75 \text{ }\mu\text{F}$, and a maximum voltage of 150 V.
(a) What is the impedance of the circuit at resonance?
(b) What is the resonance frequency of the circuit?
(c) When will the current be greatest: at resonance, at 10% below the resonant frequency, or at 10% above the resonant frequency? (d) What is the rms current in the circuit at a frequency of 60 Hz?

41. Two electrical oscillators are used in a heterodyne metal detector to detect buried metal objects (see Fig. P21.41). The detector uses two identical electrical oscillators in the form of LC circuits having resonant frequencies of 725 kHz. When the signals from the two oscillating circuits are combined, the beat frequency is zero because each has the same resonant frequency. However, when the coil of one circuit encounters a buried metal object, the inductance of this circuit increases by 1.000%, while that of the second is unchanged. Determine the beat frequency that would be detected in this situation.

Figure P21.41

42. A series circuit contains a 3.00-H inductor, a 3.00-μF capacitor, and a 30.0-Ω resistor connected to a 120-V (rms) source of variable frequency. Find the power delivered to the circuit when the frequency of the source is (a) the resonance frequency, (b) one-half the resonance frequency, (c) one-fourth the resonance frequency, (d) two times the resonance frequency, and (e) four times the resonance frequency. From your calculations, can you draw a conclusion about the frequency at which the maximum power is delivered to the circuit?

21.7 The Transformer

43. The primary coil of a transformer has $N_1 = 250$ turns, and its secondary coil has $N_2 = 1\,500$ turns. If the input voltage across the primary coil is $\Delta v = (170 \text{ V}) \sin \omega t$, what rms voltage is developed across the secondary coil?

44. A step-down transformer is used for recharging the batteries of portable devices. The turns ratio N_2/N_1 for a particular transformer used in a CD player is 1:13. When used with 120-V (rms) household service, the transformer draws an rms current of 250 mA. Find the (a) rms output voltage of the transformer and (b) power delivered to the CD player.

45. An AC power generator produces 50 A (rms) at 3 600 V. The voltage is stepped up to 100 000 V by an ideal transformer, and the energy is transmitted through a long-distance power line that has a resistance of 100 Ω. What percentage of the power delivered by the generator is dissipated as heat in the power line?

46. **W** A transformer is to be used to provide power for a computer disk drive that needs 6.0 V (rms) instead of the 120 V (rms) from the wall outlet. The number of turns in the primary is 400, and it delivers 500 mA

(the secondary current) at an output voltage of 6.0 V (rms). (a) Should the transformer have more turns in the secondary compared with the primary, or fewer turns? (b) Find the current in the primary. (c) Find the number of turns in the secondary.

47. A transformer on a pole near a factory steps the voltage down from 3 600 V (rms) to 120 V (rms). The transformer is to deliver 1 000 kW to the factory at 90% efficiency. Find (a) the power delivered to the primary, (b) the current in the primary, and (c) the current in the secondary.

48. A transmission line that has a resistance per unit length of 4.50×10^{-4} Ω/m is to be used to transmit 5.00 MW over 400 miles (6.44×10^5 m). The output voltage of the generator is 4.50 kV (rms). (a) What is the line loss if a transformer is used to step up the voltage to 500 kV (rms)? (b) What fraction of the input power is lost to the line under these circumstances? (c) What difficulties would be encountered on attempting to transmit the 5.00 MW at the generator voltage of 4.50 kV (rms)?

21.10 Production of Electromagnetic Waves by an Antenna

21.11 Properties of Electromagnetic Waves

49. The U.S. Navy has long proposed the construction of extremely low frequency (ELF waves) communications systems; such waves could penetrate the oceans to reach distant submarines. Calculate the length of a quarter-wavelength antenna for a transmitter generating ELF waves of frequency 75 Hz. How practical is this antenna?

50. (a) The distance to Polaris, the North Star, is approximately 6.44×10^{18} m. If Polaris were to burn out today, how many years would it take to see it disappear? (b) How long does it take sunlight to reach Earth? (c) How long does it take a microwave signal to travel from Earth to the Moon and back? (The distance from Earth to the Moon is 3.84×10^5 km.)

51. **Q|C** The Earth reflects approximately 38.0% of the incident sunlight from its clouds and surface. (a) Given that the intensity of solar radiation at the top of the atmosphere is 1 370 W/m², find the radiation pressure on the Earth, in pascals, at the location where the Sun is straight overhead. (b) State how this quantity compares with normal atmospheric pressure at the Earth's surface, which is 101 kPa.

52. Experimenters at the National Institute of Standards and Technology have made precise measurements of the speed of light using the fact that, in vacuum, the speed of electromagnetic waves is $c = 1/\sqrt{\mu_0 \epsilon_0}$, where the constants $\mu_0 = 4\pi \times 10^{-7}$ N·s²/C² and $\epsilon_0 = 8.854 \times 10^{-12}$ C²/N·m². What value (to four significant figures) does this formula give for the speed of light in vacuum?

53. **BIO** Oxygenated hemoglobin absorbs weakly in the red (hence its red color) and strongly in the near infrared, whereas deoxygenated hemoglobin has the opposite absorption. This fact is used in a "pulse oximeter" to measure oxygen saturation in arterial blood. The device clips onto the end of a person's finger and has two light-emitting diodes—a red (660 nm) and an infrared (940 nm)—and a photocell that detects the amount of light transmitted through the finger at each wavelength. (a) Determine the frequency of each of these light sources. (b) If 67% of the energy of the red source is absorbed in the blood, by what factor does the amplitude of the electromagnetic wave change? *Hint:* The intensity of the wave is equal to the average power per unit area as given by Equation 21.28.

54. **BIO Operation of the pulse oximeter (see previous problem).** The transmission of light energy as it passes through a solution of light-absorbing molecules is described by the Beer–Lambert law

$$I = I_0 10^{-\epsilon CL} \quad \text{or} \quad \log_{10}\left(\frac{I}{I_0}\right) = -\epsilon CL$$

which gives the decrease in intensity I in terms of the distance L the light has traveled through a fluid with a concentration C of the light-absorbing molecule. The quantity ϵ is called the extinction coefficient, and its value depends on the frequency of the light. (It has units of m²/mol.) Assume the extinction coefficient for 660-nm light passing through a solution of oxygenated hemoglobin is identical to the coefficient for 940-nm light passing through deoxygenated hemoglobin. Also assume 940-nm light has zero absorption ($\epsilon = 0$) in oxygenated hemoglobin and 660-nm light has zero absorption in deoxygenated hemoglobin. If 33% of the energy of the red source and 76% of the infrared energy is transmitted through the blood, what is the fraction of hemoglobin that is oxygenated?

55. The Sun delivers an average power of 1 370 W/m² to the top of Earth's atmosphere. Find the magnitudes of $\vec{E}_{max}$ and $\vec{B}_{max}$ for the electromagnetic waves at the top of the atmosphere.

56. **Q|C** A laser beam is used to levitate a metal disk against the force of Earth's gravity. (a) Derive an equation giving the required intensity of light, I, in terms of the mass m of the disk, the gravitational acceleration g, the speed of light c, and the cross-sectional area of the disk A. Assume the disk is perfectly reflecting and the beam is directed perpendicular to the disk. (b) If the disk has mass 5.00 g and radius 4.00 cm, find the necessary light intensity. (c) Give two reasons why using light pressure as propulsion near Earth's surface is impractical.

57. A microwave oven is powered by an electron tube called a magnetron that generates electromagnetic waves of frequency 2.45 GHz. The microwaves enter the oven and are reflected by the walls. The standing-wave pattern

produced in the oven can cook food unevenly, with hot spots in the food at antinodes and cool spots at nodes, so a turntable is often used to rotate the food and distribute the energy. If a microwave oven is used with a cooking dish in a fixed position, the antinodes can appear as burn marks on foods such as carrot strips or cheese. The separation distance between the burns is measured to be 6.00 cm. Calculate the speed of the microwaves from these data.

58. Consider a bright star in our night sky. Assume its distance from the Earth is 20.0 light-years (ly) and its power output is 4.00×10^{28} W, about 100 times that of the Sun. (a) Find the intensity of the starlight at the Earth. (b) Find the power of the starlight the Earth intercepts. One light-year is the distance traveled by light through a vacuum in one year.

21.12 The Spectrum of Electromagnetic Waves

59. What are the wavelengths of electromagnetic waves in free space that have frequencies of (a) 5.00×10^{19} Hz and (b) 4.00×10^9 Hz?

60. **BIO** A diathermy machine, used in physiotherapy, generates electromagnetic radiation that gives the effect of "deep heat" when absorbed in tissue. One assigned frequency for diathermy is 27.33 MHz. What is the wavelength of this radiation?

61. What are the wavelength ranges in (a) the AM radio band (540–1 600 kHz) and (b) the FM radio band (88–108 MHz)?

62. An important news announcement is transmitted by radio waves to people who are 100 km away, sitting next to their radios, and by sound waves to people sitting across the newsroom, 3.0 m from the newscaster. Who receives the news first? Explain. Take the speed of sound in air to be 343 m/s.

63. Infrared spectra are used by chemists to help identify an unknown substance. Atoms in a molecule that are bound together by a particular bond vibrate at a predictable frequency, and light at that frequency is absorbed strongly by the atom. In the case of the C=O double bond, for example, the oxygen atom is bound to the carbon by a bond that has an effective spring constant of 2 800 N/m. If we assume the carbon atom remains stationary (it is attached to other atoms in the molecule), determine the resonant frequency of this bond and the wavelength of light that matches that frequency. Verify that this wavelength lies in the infrared region of the spectrum. (The mass of an oxygen atom is 2.66×10^{-26} kg.)

21.13 The Doppler Effect for Electromagnetic Waves

64. A spaceship is approaching a space station at a speed of 1.8×10^5 m/s. The space station has a beacon that emits green light with a frequency of 6.0×10^{14} Hz. (a) What is the frequency of the beacon observed on the spaceship? (b) What is the change in frequency? (Carry five digits in these calculations.)

65. While driving at a constant speed of 80 km/h, you are passed by a car traveling at 120 km/h. If the frequency of light emitted by the taillights of the car that passes you is 4.3×10^{14} Hz, what frequency will you observe? What is the change in frequency?

66. A speeder tries to explain to the police that the yellow warning lights she was approaching on the side of the road looked green to her because of the Doppler shift. How fast would she have been traveling if yellow light of wavelength 580 nm had been shifted to green with a wavelength of 560 nm? *Note:* For speeds less than 0.03c, Equation 21.32 will lead to a value for the observed frequency accurate to approximately two significant digits.

Additional Problems

67. A 25.0-mW laser beam of diameter 2.00 mm is reflected at normal incidence by a perfectly reflecting mirror. Calculate the radiation pressure on the mirror.

68. The intensity of solar radiation at the top of Earth's atmosphere is 1 370 W/m^3. Assuming 60% of the incoming solar energy reaches Earth's surface and assuming you absorb 50% of the incident energy, make an order-of-magnitude estimate of the amount of solar energy you absorb in a 60-minute sunbath.

69. A 200-Ω resistor is connected in series with a 5.0-μF capacitor and a 60-Hz, 120-V rms line. If electrical energy costs $0.080/kWh, how much does it cost to leave this circuit connected for 24 h?

70. **S** In an *RLC* series circuit that includes a source of alternating current operating at fixed frequency and voltage, the resistance R is equal to the inductive reactance. If the plate separation of the parallel-plate capacitor is reduced to one-half its original value, the current in the circuit doubles. Find the initial capacitive reactance in terms of R.

71. As a way of determining the inductance of a coil used in a research project, a student first connects the coil to a 12.0-V battery and measures a current of 0.630 A. The student then connects the coil to a 24.0-V (rms), 60.0-Hz generator and measures an rms current of 0.570 A. What is the inductance?

72. (a) What capacitance will resonate with a one-turn loop of inductance 400 pH to give a radar wave of wavelength 3.0 cm? (b) If the capacitor has square parallel plates separated by 1.0 mm of air, what should the edge length of the plates be? (c) What is the common reactance of the loop and capacitor at resonance?

73. **M** A dish antenna with a diameter of 20.0 m receives (at normal incidence) a radio signal from a distant source, as shown in Figure P21.73. The radio signal is a continuous sinusoidal wave with amplitude E_{max} = 0.20 μV/m. Assume the antenna absorbs all the radiation that falls

on the dish. (a) What is the amplitude of the magnetic field in this wave? (b) What is the intensity of the radiation received by the antenna? (c) What is the power received by the antenna?

Figure P21.73

74. A particular inductor has appreciable resistance. When the inductor is connected to a 12-V battery, the current in the inductor is 3.0 A. When it is connected to an AC source with an rms output of 12 V and a frequency of 60 Hz, the current drops to 2.0 A. What are (a) the impedance at 60 Hz and (b) the inductance of the inductor?

75. One possible means of achieving space flight is to place a perfectly reflecting aluminized sheet into Earth's orbit and to use the light from the Sun to push this solar sail. Suppose such a sail, of area 6.00×10^4 m^2 and mass 6 000 kg, is placed in orbit facing the Sun. (a) What force is exerted on the sail? (b) What is the sail's acceleration? (c) How long does it take this sail to reach the Moon, 3.84×10^8 m away? Ignore all gravitational effects and assume a solar intensity of 1 340 W/m^2. *Hint:* The radiation pressure by a reflected wave is given by 2 (average power per unit area)/c.

76. **BIO** **Q|C** The U.S. Food and Drug Administration limits the radiation leakage of microwave ovens to no more than 5.0 mW/cm^2 at a distance of 2.0 in. A typical cell phone, which also transmits microwaves, has a peak output power of about 2.0 W. (a) Approximating the cell phone as a point source, calculate the radiation intensity of a cell phone at a distance of 2.0 in. How does the answer compare with the maximum allowable microwave oven leakage? (b) The distance from your ear to your brain is about 2 in. What would the radiation intensity in your brain be if you used a Bluetooth headset, keeping the phone in your pocket, 1.0 m away from your brain? Most headsets are so-called Class 2 devices with a maximum output power of 2.5 mW.

Light is bent (refracted) as it passes through water, with different wavelengths bending by different amounts, a phenomenon called dispersion. Together with reflection, these physical phenomena lead to the creation of a rainbow when light passes through small, suspended droplets of water.

Dave Walker/Shutterstock.com

22 Reflection and Refraction of Light

22.1 The Nature of Light
22.2 Reflection and Refraction
22.3 The Law of Refraction
22.4 Dispersion and Prisms
22.5 The Rainbow
22.6 Huygens' Principle
22.7 Total Internal Reflection

Light has a dual nature. In some experiments it acts like a particle, while in others it acts like a wave. In this and the next two chapters, we concentrate on the aspects of light that are best understood through the wave model. First we discuss the reflection of light at the boundary between two media and the refraction (bending) of light as it travels from one medium into another. We use these ideas to study the refraction of light as it passes through lenses and the reflection of light from mirrored surfaces. In Chapter 25, we describe how lenses and mirrors can be used to view objects with telescopes and microscopes and how lenses are used in photography. The ability to manipulate light has greatly enhanced our capacity to investigate and understand the nature of the Universe.

22.1 The Nature of Light

LEARNING OBJECTIVES

1. Discuss the dual nature of light as both a wave and as particles called photons.
2. State the equation yielding the energy of a photon.

Until the beginning of the 19th century, light was modeled as a stream of particles emitted by a source that stimulated the sense of sight on entering the eye. The chief architect of the particle theory of light was Newton. With this theory, he provided simple explanations of some known experimental facts concerning the nature of light, namely, the laws of reflection and refraction.

Most scientists accepted Newton's particle theory of light. During Newton's lifetime, however, another theory was proposed. In 1678 Dutch physicist and astronomer Christian Huygens (1629–1695) showed that a wave theory of light could also explain the laws of reflection and refraction.

The wave theory didn't receive immediate acceptance, for several reasons. First, all the waves known at the time (sound, water, and so on) traveled through some sort of medium, but light from the Sun could travel to Earth through empty space. Further, it was argued that if light were some form of wave, it would bend around obstacles; hence, we should be able to see around corners. It is now known that light does indeed bend around the edges of objects. This phenomenon, known as *diffraction*, is difficult to observe because light waves have such short wavelengths. Even though experimental evidence for the diffraction of light was discovered by Francesco Grimaldi (1618–1663) around 1660, for more than a century most scientists rejected the wave theory and adhered to Newton's particle theory, probably due to Newton's great reputation as a scientist.

The first clear demonstration of the wave nature of light was provided in 1801 by Thomas Young (1773–1829), who showed that under appropriate conditions, light exhibits interference behavior. Light waves emitted by a single source and traveling along two different paths can arrive at some point and combine and cancel each other by destructive interference. Such behavior couldn't be explained at that time by a particle model because scientists couldn't imagine how two or more particles could come together and cancel one another.

The most important development in the theory of light was the work of Maxwell, who predicted in 1865 that light was a form of high-frequency electromagnetic wave (Chapter 21). His theory also predicted that these waves should have a speed of 3×10^8 m/s, in agreement with the measured value.

Although the classical theory of electricity and magnetism explained most known properties of light, some subsequent experiments couldn't be explained by the assumption that light was a wave. The most striking experiment was the *photoelectric effect* (which we examine more closely in Chapter 27), discovered by Hertz. Hertz found that clean metal surfaces emit charges when exposed to ultraviolet light.

In 1905, Einstein published a paper that formulated the theory of light quanta ("particles") and explained the photoelectric effect. He reached the conclusion that light was composed of corpuscles, or discontinuous quanta of energy. These corpuscles or quanta are now called *photons* to emphasize their particle-like nature. According to Einstein's theory, the energy of a photon is proportional to the frequency of the electromagnetic wave associated with it, or

$$E = hf \quad [22.1]$$

◄ Energy of a photon

Christian Huygens
(1629–1695), Dutch Physicist and Astronomer
Huygens is best known for his contributions to the fields of optics and dynamics. To Huygens, light was a vibratory motion in the ether, spreading out and producing the sensation of light when impinging on the eye. On the basis of this theory, he deduced the laws of reflection and refraction and explained the phenomenon of double refraction.

where $h = 6.63 \times 10^{-34}$ J · s is *Planck's constant*. This theory retains some features of both the wave and particle theories of light. As we discuss later, the photoelectric effect is the result of energy transfer from a single photon to an electron in the metal. This means the electron interacts with one photon of light as if the electron had been struck by a particle. Yet the photon has wave-like characteristics, as implied by the fact that a frequency is used in its definition.

In view of these developments, light must be regarded as having a *dual nature*: **In some experiments light acts as a wave and in others it acts as a particle.** Classical electromagnetic wave theory provides adequate explanations of light propagation and of the effects of interference, whereas the photoelectric effect and other experiments involving the interaction of light with matter are best explained by assuming light is a particle.

So in the final analysis, is light a wave or a particle? The answer is neither and both: light has a number of physical properties, some associated with waves and others with particles.

22.2 Reflection and Refraction

LEARNING OBJECTIVES
1. State and apply the ray approximation.
2. Apply the law of reflection of light.
3. Discuss the physical meaning of the refraction of light.

When light traveling in one medium encounters a boundary leading into a second medium, the processes of reflection and refraction can occur. In **reflection** part of the light encountering the second medium bounces off that medium. In **refraction** the light passing into the second medium bends through an angle with respect to the normal to the boundary. Often, both processes occur at the same time, with part of the light being reflected and part refracted. To study reflection and refraction we need a way of thinking about beams of light, and this is given by the ray approximation.

The Ray Approximation in Geometric Optics

An important property of light that can be understood based on common experience is the following: **light travels in a straight-line path in a homogeneous medium, until it encounters a boundary between two different materials.** When light strikes a boundary, it is reflected from that boundary, passes into the material on the other side of the boundary, or partially does both.

The preceding observation leads us to use what is called the **ray approximation** to represent beams of light. As shown in Figure 22.1, a ray of light is an imaginary line drawn along the direction of travel of the light beam. For example, a beam of sunlight passing through a darkened room traces out the path of a light ray. We also make use of the concept of wave fronts of light. A **wave front** is a surface passing through the points of a wave that have the same phase and amplitude. For instance, the wave fronts in Figure 22.1 could be surfaces passing through the crests of waves. The rays, corresponding to the direction of wave motion, are straight lines perpendicular to the wave fronts. When light rays travel in parallel paths, the wave fronts are planes perpendicular to the rays.

The rays are straight lines perpendicular to the wave fronts and pointing in the direction of the wave motion.

Figure 22.1 A plane wave traveling to the right.

Reflection of Light

When a light ray traveling in a transparent medium encounters a boundary leading into a second medium, part of the incident ray is reflected back into the first medium. Figure 22.2a shows several rays of a beam of light incident on a smooth, mirror-like reflecting surface. The reflected rays are parallel to one another, as indicated in the figure. The reflection of light from such a smooth surface is called **specular reflection**. On the other hand, if the reflecting surface is rough, as in Figure 22.2b, the surface reflects the rays in a variety of directions. Reflection from any rough surface is known as **diffuse reflection**. A surface behaves as a smooth surface as long as its variations are small compared with the wavelength of the incident light. Figures 22.2c and 22.2d are photographs of specular and diffuse reflection of laser light, respectively.

As an example, consider the two types of reflection from a road surface that someone might observe while driving at night. When the road is dry, light from oncoming vehicles is scattered off the road in different directions (diffuse reflection) and the road is clearly visible. On a rainy night when the road is wet, the road's irregularities are filled with water. Because the wet surface is smooth, the light undergoes specular reflection. This means that the light is reflected straight ahead, and the driver of a car sees only what is directly in front of him. Light from the side never reaches the driver's eye. In this book we concern ourselves only with specular reflection, and we use the term *reflection* to mean specular reflection.

APPLICATION
Seeing the Road on a Rainy Night

22.2 | Reflection and Refraction   775

Figure 22.2 A schematic representation of (a) specular reflection, where the reflected rays are all parallel to one another, and (b) diffuse reflection, where the reflected rays travel in random directions. (c, d) Photographs of specular and diffuse reflection, made with laser light.

■ Quick Quiz

22.1 Which part of Figure 22.3, (a) or (b), better shows specular reflection of light from the roadway?

Figure 22.3 (Quick Quiz 22.1)

Consider a light ray traveling in air and incident at some angle on a flat, smooth surface, as in Figure 22.4. The incident and reflected rays make angles θ_1 and θ_1', respectively, **with a line perpendicular to the surface** at the point where the incident ray strikes the surface. We call this line the *normal* to the surface. Experiments show that **the angle of reflection equals the angle of incidence**:

$$\theta_1' = \theta_1 \qquad [22.2]$$

You may have noticed a common occurrence in photographs of individuals: their eyes appear to be glowing red. "Red-eye" occurs when a photographic flash is used and the flash unit is close to the camera lens. Light from the flash unit enters the eye and is reflected back along its original path from the retina. This type of reflection back along the original direction is called *retroreflection*. If the flash unit and lens are close together, retroreflected light can enter the lens. Most of the light reflected from the retina is red due to the blood vessels at the back of the eye, giving the red-eye effect in the photograph.

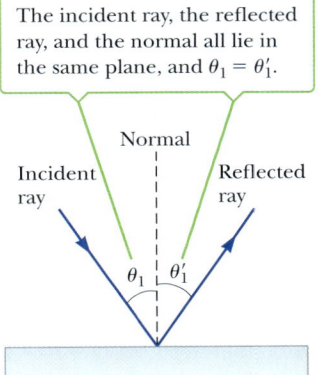

The incident ray, the reflected ray, and the normal all lie in the same plane, and $\theta_1 = \theta_1'$.

Figure 22.4 The wave under reflection model.

BIO APPLICATION
Red Eyes in Flash Photographs

CHAPTER 22 | Reflection and Refraction of Light

APPLYING PHYSICS 22.1 | The Colors of Water Ripples at Sunset

An observer on the west-facing beach of a large lake is watching the beginning of a sunset. The water is very smooth except for some areas with small ripples. The observer notices that some areas of the water are blue and some are pink. Why does the water appear to be different colors in different areas?

EXPLANATION The different colors arise from specular and diffuse reflection. The smooth areas of the water will specularly reflect the light from the west, which is the pink light from the sunset. The areas with small ripples will reflect the light diffusely, so light from all parts of the sky will be reflected into the observer's eyes. Because most of the sky is still blue at the beginning of the sunset, these areas will appear to be blue. ∎

APPLYING PHYSICS 22.2 | Double Images

When standing outside in the Sun close to a single-pane window looking to the darker interior of a building, why can you often see two images of yourself, one superposed on the other?

EXPLANATION Reflection occurs whenever there is an interface between two different media. For the glass in the window, there are two such surfaces, the window surface facing outdoors and the window surface facing indoors. Each of these interfaces results in an image. You will notice that one image is slightly smaller than the other, because the reflecting surface is farther away. ∎

EXAMPLE 22.1 | The Double-Reflecting Light Ray

GOAL Calculate a resultant angle from two reflections.

PROBLEM Two mirrors make an angle of 120° with each other, as in Figure 22.5. A ray is incident on mirror M_1 at an angle of 65° to the normal. Find the angle the ray makes with the normal to M_2 after it is reflected from both mirrors.

STRATEGY Apply the law of reflection twice. Given the incident ray at angle θ_{inc}, find the final resultant angle, β_{ref}.

Figure 22.5 (Example 22.1) Mirrors M_1 and M_2 make an angle of 120° with each other.

SOLUTION

Apply the law of reflection to M_1 to find the angle of reflection, θ_{ref}:

$$\theta_{ref} = \theta_{inc} = 65°$$

Find the angle ϕ that is the complement of the angle θ_{ref}:

$$\phi = 90° - \theta_{ref} = 90° - 65° = 25°$$

Find the unknown angle α in the triangle of M_1, M_2, and the ray traveling from M_1 to M_2, using the fact that the three angles sum to 180°:

$$180° = 25° + 120° + \alpha \rightarrow \alpha = 35°$$

The angle α is complementary to the angle of incidence, β_{inc}, for M_2:

$$\alpha + \beta_{inc} = 90° \rightarrow \beta_{inc} = 90° - 35° = 55°$$

Apply the law of reflection a second time, obtaining β_{ref}:

$$\beta_{ref} = \beta_{inc} = 55°$$

REMARKS Notice the heavy reliance on elementary geometry and trigonometry in these reflection problems.

QUESTION 22.1 In general, what is the relationship between the incident angle θ_{inc} and the final reflected angle β_{ref} when the angle between the mirrors is 90.0°? (a) $\theta_{inc} + \beta_{ref} = 90.0°$ (b) $\theta_{inc} - \beta_{ref} = 90.0°$ (c) $\theta_{inc} + \beta_{ref} = 180°$

EXERCISE 22.1 Repeat the problem if the angle of incidence is 55° and the second mirror makes an angle of 100° with the first mirror.

ANSWER 45°

Refraction of Light

When a ray of light traveling through a transparent medium encounters a boundary leading into another transparent medium, as in Figure 22.6a, part of the ray is reflected and part enters the second medium. The ray that enters the second medium is bent at the boundary and is said to be *refracted*. The incident ray, the reflected ray, the refracted ray, and the normal at the point of incidence all lie in the same plane. The **angle of refraction**, θ_2, in Figure 22.6a depends on the properties of the two media and on the angle of incidence, through the relationship

$$\frac{\sin \theta_2}{\sin \theta_1} = \frac{v_2}{v_1} = \text{constant} \quad [22.3]$$

where v_1 is the speed of light in medium 1 and v_2 is the speed of light in medium 2. Note that the angle of refraction is also measured with respect to the normal. In Section 22.7 we derive the laws of reflection and refraction using Huygens' principle.

Experiment shows that **the path of a light ray through a refracting surface is reversible**. For example, the ray in Figure 22.6a travels from point A to point B. If the ray originated at B, it would follow the same path to reach point A, but the reflected ray would be in the glass.

■ Quick Quiz

22.2 If beam 1 is the incoming beam in Figure 22.6b, which of the other four beams are due to reflection? Which are due to refraction?

When light moves from a material in which its speed is high to a material in which its speed is lower, the angle of refraction θ_2 is less than the angle of incidence. The refracted ray therefore bends toward the normal, as shown in Figure 22.7a (page 778). If the ray moves from a material in which it travels slowly to a material in which it travels more rapidly, θ_2 is greater than θ_1, so the ray bends away from the normal, as shown in Figure 22.7b.

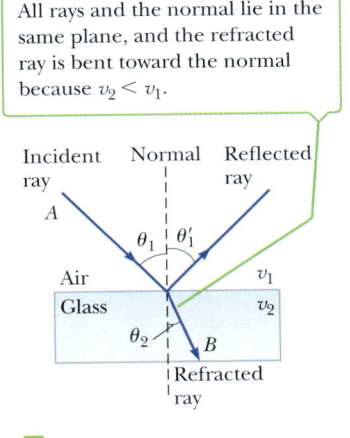

All rays and the normal lie in the same plane, and the refracted ray is bent toward the normal because $v_2 < v_1$.

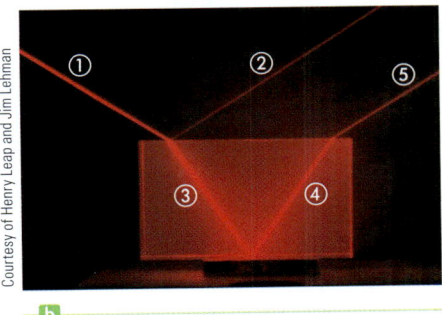

Figure 22.6 (a) The wave under refraction model. (b) Light incident on the Lucite block refracts both when it enters the block and when it leaves the block.

Figure 22.7 The refraction of light as it (a) moves from air into glass and (b) moves from glass into air.

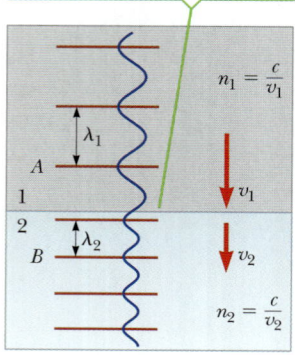

Figure 22.8 A wave travels from medium 1 to medium 2, in which it moves with lower speed.

22.3 The Law of Refraction

LEARNING OBJECTIVES

1. Define the index of refraction.
2. Derive the wavelength of light in different media.
3. Apply Snell's Law.

When light passes from one transparent medium to another, it's refracted because the speed of light is different in the two media.[1] The **index of refraction**, n, of a medium is defined as the ratio c/v;

$$n \equiv \frac{\text{speed of light in vacuum}}{\text{speed of light in a medium}} = \frac{c}{v} \qquad [22.4]$$

From this definition, we see that the index of refraction is a dimensionless number that is greater than or equal to 1 because v is always less than c. Further, n is equal to one for vacuum. Table 22.1 lists the indices of refraction for various substances.

As light travels from one medium to another, its frequency doesn't change. To see why, consider Figure 22.8. Wave fronts pass an observer at point A in medium 1 with a certain frequency and are incident on the boundary between

Table 22.1 Indices of Refraction for Various Substances, Measured with Light of Vacuum Wavelength $\lambda_0 = 589$ mn

Substance	Index of Refraction	Substance	Index of Refraction
Solids at 20°C		**Liquids at 20°C**	
Diamond (C)	2.419	Benzene	1.501
Fluorite (CaF_2)	1.434	Carbon disulfide	1.628
Fused quartz (SiO_2)	1.458	Carbon tetrachloride	1.461
Glass, crown	1.52	Ethyl alcohol	1.361
Glass, flint	1.66	Glycerine	1.473
Ice (H_2O) (at 0°C)	1.309	Water	1.333
Polystyrene	1.49		
Sodium chloride (NaCl)	1.544	**Gases at 0°C, 1 atm**	
Zircon	1.923	Air	1.000 293
		Carbon dioxide	1.000 45

[1] The speed of light varies between media because the time lags caused by the absorption and reemission of light as it travels from atom to atom depend on the particular electronic structure of the atoms constituting each material.

medium 1 and medium 2. The frequency at which the wave fronts pass an observer at point B in medium 2 must equal the frequency at which they arrive at point A. If not, the wave fronts would either pile up at the boundary or be destroyed or created at the boundary. Because neither of these events occurs, the frequency must remain the same as a light ray passes from one medium into another.

Therefore, because the relation $v = f\lambda$ must be valid in both media and because $f_1 = f_2 = f$, we see that

$$v_1 = f\lambda_1 \text{ and } v_2 = f\lambda_2$$

Because $v_1 \neq v_2$, it follows that $\lambda_1 \neq \lambda_2$. A relationship between the index of refraction and the wavelength can be obtained by dividing these two equations and making use of the definition of the index of refraction given by Equation 22.4:

$$\frac{\lambda_1}{\lambda_2} = \frac{v_1}{v_2} = \frac{c/n_1}{c/n_2} = \frac{n_2}{n_1} \quad [22.5]$$

which gives

$$\lambda_1 n_1 = \lambda_2 n_2 \quad [22.6]$$

Let medium 1 be the vacuum so that $n_1 = 1$. It follows from Equation 22.6 that the index of refraction of any medium can be expressed as the ratio

$$n = \frac{\lambda_0}{\lambda_n} \quad [22.7]$$

where λ_0 is the wavelength of light in vacuum and λ_n is the wavelength in the medium having index of refraction n. Figure 22.9 is a schematic representation of this reduction in wavelength when light passes from a vacuum into a transparent medium.

We are now in a position to express Equation 22.3 in an alternate form. If we substitute Equation 22.5 into Equation 22.3, we get

$$n_1 \sin \theta_1 = n_2 \sin \theta_2 \quad [22.8]$$

◀ Snell's law of refraction

The experimental discovery of this relationship is usually credited to Willebrørd Snell (1591–1626) and is therefore known as **Snell's law of refraction**.

> **Tip 22.1 An Inverse Relationship**
> The index of refraction is *inversely* proportional to the wave speed. Therefore, as the wave speed v decreases, the index of refraction, n, increases.

> **Tip 22.2 The Frequency Remains the Same**
> The *frequency* of a wave does *not* change as the wave passes from one medium to another. Both the wave speed and the wavelength *do* change, but the frequency remains the same.

■ **Quick Quiz**

22.3 A material has an index of refraction that increases continuously from top to bottom. Of the three paths shown in Figure 22.10, which path will a light ray follow as it passes through the material?

22.4 As light travels from a vacuum ($n = 1$) to a medium such as glass ($n > 1$), which of the following properties remains the same, the (a) wavelength, (b) wave speed, or (c) frequency?

Figure 22.9 A schematic diagram of the *reduction* in wavelength when light travels from a medium with a low index of refraction to one with a higher index of refraction.

Figure 22.10 (Quick Quiz 22.3)

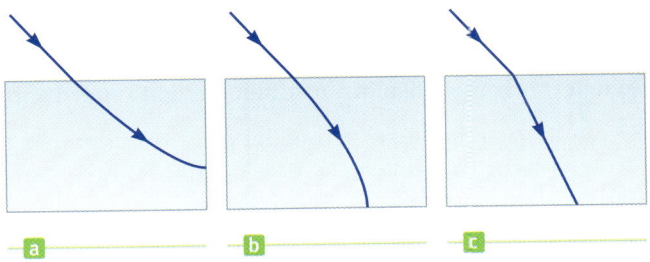

EXAMPLE 22.2 | Angle of Refraction for Glass

GOAL Apply Snell's law.

PROBLEM A light ray of wavelength 589 nm (produced by a sodium lamp) traveling through air is incident on a smooth, flat slab of crown glass at an angle θ_1 of 30.0° to the normal, as sketched in Figure 22.11. **(a)** Find the angle of refraction, θ_2. **(b)** At what angle θ_3 does the ray leave the glass as it re-enters the air? **(c)** How does the answer for θ_3 change if the ray enters water below the slab instead of the air?

STRATEGY Substitute quantities into Snell's law and solve for the unknown angles of refraction.

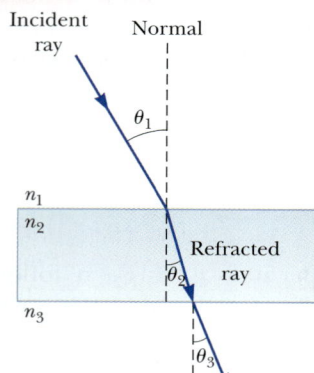

Figure 22.11 (Example 22.2) Refraction of light by glass.

SOLUTION

(a) Find the angle of refraction, θ_2.

Solve Snell's law (Eq. 22.8) for $\sin \theta_2$:

$$(1) \quad \sin \theta_2 = \frac{n_1}{n_2} \sin \theta_1$$

From Table 22.1, find $n_1 = 1.00$ for air and $n_2 = 1.52$ for crown glass. Substitute these values into Equation (1) and take the inverse sine of both sides:

$$\sin \theta_2 = \left(\frac{1.00}{1.52}\right)(\sin 30.0°) = 0.329$$

$$\theta_2 = \sin^{-1}(0.329) = \boxed{19.2°}$$

(b) At what angle θ_3 does the ray leave the glass as it re-enters the air?

Write Equation (1), replacing θ_3 with θ_2 and θ_1 with θ_2:

$$(2) \quad \sin \theta_3 = \frac{n_2}{n_3} \sin \theta_2 = \frac{1.52}{1.00} \sin(19.2°) = 0.500$$

Take the inverse sine of both sides to find θ_3:

$$\theta_3 = \sin^{-1}(0.500) = \boxed{30.0°}$$

(c) How does the answer for θ_3 change if the ray enters water below the slab instead of air?

Write Equation (2) and substitute a different value for n_3:

$$\sin \theta_3 = \frac{n_2}{n_3} \sin \theta_2 = \frac{1.52}{1.333} \sin(19.2°) = 0.375$$

$$\theta_3 = \sin^{-1}(0.375) = \boxed{22.0°}$$

REMARKS Notice that the light ray bends toward the normal when it enters a material of a higher index of refraction, and away from the normal when entering a material with a lower index of refraction. In passing through a slab of material with parallel surfaces, for example, from air to glass and back to air, the final direction of the ray is parallel to the direction of the incident ray. The only effect in that case is a lateral displacement of the light ray.

QUESTION 22.2 If the glass is replaced by a transparent material with smaller index of refraction, will the refraction angle θ_2 be (a) smaller, (b) larger, or (c) unchanged?

EXERCISE 22.2 Suppose a light ray in air ($n = 1.00$) enters a cube of material ($n = 2.50$) at a 45.0° angle with respect to the normal and then exits the bottom of the cube into water ($n = 1.333$). At what angle to the normal does the ray leave the slab?

ANSWER 32.0°

EXAMPLE 22.3 | Light in Fused Quartz

GOAL Use the index of refraction to determine the effect of a medium on light's speed and wavelength.

PROBLEM Light of wavelength 589 nm in vacuum passes through a piece of fused quartz of index of refraction $n = 1.458$. **(a)** Find the speed of light in fused quartz. **(b)** What is the wavelength of this light in fused quartz? **(c)** What is the frequency of the light in fused quartz?

STRATEGY Substitute values into Equations 22.4 and 22.7.

SOLUTION

(a) Find the speed of light in fused quartz.

Obtain the speed from Equation 22.4:

$$v = \frac{c}{n} = \frac{3.00 \times 10^8 \text{ m/s}}{1.458} = 2.06 \times 10^8 \text{ m/s}$$

(b) What is the wavelength of this light in fused quartz?

Use Equation 22.7 to calculate the wavelength:

$$\lambda_n = \frac{\lambda_0}{n} = \frac{589 \text{ nm}}{1.458} = 404 \text{ nm}$$

(c) What is the frequency of the light in fused quartz?

The frequency in quartz is the same as in vacuum. Solve $c = f\lambda$ for the frequency:

$$f = \frac{c}{\lambda} = \frac{3.00 \times 10^8 \text{ m/s}}{589 \times 10^{-9} \text{ m}} = 5.09 \times 10^{14} \text{ Hz}$$

REMARKS It's interesting to note that the speed of light in vacuum, 3.00×10^8 m/s, is an upper limit for the speed of material objects. In our treatment of relativity in Chapter 26, we will find that this upper limit is consistent with experimental observations. However, it's possible for a particle moving in a medium to have a speed that exceeds the speed of light in that medium. For example, it's theoretically possible for a particle to travel through fused quartz at a speed greater than 2.06×10^8 m/s, but it must still have a speed less than 3.00×10^8 m/s.

QUESTION 22.3 True or False: If light with wavelength λ in glass passes into water with index n_w, the new wavelength of the light is λ/n_w.

EXERCISE 22.3 Light with wavelength 589 nm passes through crystalline sodium chloride. In this medium, find (a) the speed of light, (b) the wavelength, and (c) the frequency of the light.

ANSWER (a) 1.94×10^8 m/s (b) 381 nm (c) 5.09×10^{14} Hz

■ EXAMPLE 22.4 | Refraction of Laser Light in a Digital Videodisc (DVD)

GOAL Apply Snell's law together with geometric constraints.

PROBLEM A DVD is a video recording consisting of a spiral track about 1.0 μm wide with digital information. (See Fig. 22.12a.) The digital information consists of a series of pits that are "read" by a laser beam sharply focused on a track in the information layer. If the width a of the beam at the information layer must equal 1.0 μm to distinguish individual tracks and the width w of the beam as it enters the plastic is 0.700 0 mm, find the angle θ_1 at which the conical beam should enter the plastic. (See Fig. 22.12b.) Assume the plastic has a thickness $t = 1.20$ mm and an index of refraction $n = 1.55$. Note that this system is relatively immune to small dust particles degrading the video quality because particles would have to be as large as 0.700 mm to obscure the beam at the point where it enters the plastic.

STRATEGY Use right-triangle trigonometry to determine the angle θ_2 and then apply Snell's law to obtain the angle θ_1.

Figure 22.12 (Example 22.4) A micrograph of a DVD surface showing tracks and pits along each track. (b) Cross section of a cone-shaped laser beam used to read a DVD.

(Continued)

SOLUTION

From the top and bottom of Figure 22.12b, obtain an equation relating w, b, and a:

$$w = 2b + a$$

Solve this equation for b and substitute given values:

$$b = \frac{w - a}{2} = \frac{700.0 \times 10^{-6}\,\text{m} - 1.0 \times 10^{-6}\,\text{m}}{2} = 349.5\,\mu\text{m}$$

Now use the tangent function to find θ_2:

$$\tan \theta_2 = \frac{b}{t} = \frac{349.5\,\mu\text{m}}{1.20 \times 10^3\,\mu\text{m}} \rightarrow \theta_2 = 16.2°$$

Finally, use Snell's law to find θ_1:

$$n_1 \sin \theta_1 = n_2 \sin \theta_2$$

$$\sin \theta_1 = \frac{n_2 \sin \theta_2}{n_1} = \frac{1.55 \sin 16.2°}{1.00} = 0.432$$

$$\theta_1 = \sin^{-1}(0.432) = \boxed{25.6°}$$

REMARKS Despite its apparent complexity, the problem isn't that different from Example 22.2.

QUESTION 22.4 Suppose the plastic were replaced by a material with a higher index of refraction. How would the width of the beam at the information layer be affected? (a) It would remain the same. (b) It would decrease. (c) It would increase.

EXERCISE 22.4 Suppose you wish to redesign the system to decrease the initial width of the beam from 0.700 0 mm to 0.600 0 mm but leave the incident angle θ_1 and all other parameters the same as before, except the index of refraction for the plastic material (n_2) and the angle θ_2. What index of refraction should the plastic have?

ANSWER 1.79

22.4 Dispersion and Prisms

LEARNING OBJECTIVES
1. Define dispersion.
2. Describe how prisms work.

In Table 22.1 we presented values for the index of refraction of various materials. If we make careful measurements, however, we find that the index of refraction in anything but vacuum depends on the wavelength of light. The dependence of the index of refraction on wavelength is called **dispersion**. Figure 22.13 is a graphical representation of this variation in the index of refraction with wavelength. Because n is a function of wavelength, Snell's law indicates that **the angle of refraction made when light enters a material depends on the wavelength of the light**. As seen in the figure, the index of refraction for a material usually decreases with increasing wavelength. This means that violet light ($\lambda \cong 400$ nm) refracts more than red light ($\lambda \cong 650$ nm) when passing from air into a material.

To understand the effects of dispersion on light, consider what happens when light strikes a prism, as in Figure 22.14a. A ray of light of a single wavelength that is incident on the prism from the left emerges bent away from its original direction of travel by an angle δ, called the **angle of deviation**. Now suppose a beam of white light (a combination of all visible wavelengths) is incident on a prism. Because of dispersion, the different colors refract through different angles of deviation, as illustrated in Figure 22.14b. The rays that emerge from the second face of the prism spread out in a series of colors known as a visible **spectrum**, as shown in Figure 22.15. These colors, in order of decreasing wavelength, are red, orange, yellow, green, blue, and violet. Violet light deviates the most, red light the least, and the remaining colors in the visible spectrum fall between these extremes.

Figure 22.13 Variations of index of refraction in the visible spectrum with respect to vacuum wavelength for three materials.

Tip 22.3 Dispersion
Light of shorter wavelength, such as violet light, refracts more than light of longer wavelengths, such as red light.

22.4 | Dispersion and Prisms

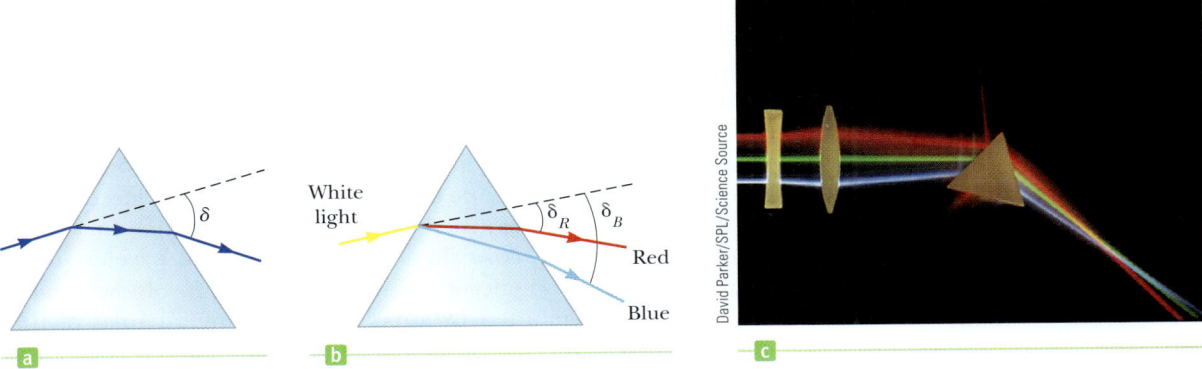

Figure 22.14 (a) A prism refracts a light ray and deviates the light through the angle δ. (b) When light is incident on a prism, the blue light is bent more than the red light. (c) Light of different colors passes through a prism and two lenses. Note that as the light passes through the prism, different wavelengths are refracted at different angles.

Figure 22.15 (a) Dispersion of white light by a prism. (b) White light enters a glass prism at the upper left.

Prisms are often used in an instrument known as a **prism spectrometer**, the essential elements of which are shown in Figure 22.16a. This instrument is commonly used to study the wavelengths emitted by a light source, such as a sodium vapor lamp. Light from the source is sent through a narrow, adjustable slit and lens to produce a parallel, or collimated, beam. The light then passes through the prism and is dispersed into a spectrum. The refracted light is observed through a telescope. The experimenter sees different colored images of the slit through the

Figure 22.16 (a) A diagram of a prism spectroscope. The colors in the spectrum are viewed through a telescope. (b) A prism spectrometer with interchangeable components. A spectrometer is a spectroscope with a scale or detector that measures wavelengths.

eyepiece of the telescope. The telescope can be moved or the prism can be rotated to view the various wavelengths, which have different angles of deviation. Figure 22.16b shows one type of prism spectrometer used in undergraduate laboratories.

APPLICATION
Identifying Gases with a Spectrometer

All hot, low-pressure gases emit their own characteristic spectra, so one use of a prism spectrometer is to identify gases. For example, sodium emits only two wavelengths in the visible spectrum: two closely spaced yellow lines. (The bright line-like images of the slit seen in a spectroscope are called *spectral lines*.) A gas emitting these, and only these, colors can be identified as sodium. Likewise, mercury vapor has its own characteristic spectrum, consisting of four prominent wavelengths—orange, green, blue, and violet lines—along with some wavelengths of lower intensity. The particular wavelengths emitted by a gas serve as "fingerprints" of that gas. Spectral analysis, which is the measurement of the wavelengths emitted or absorbed by a substance, is a powerful general tool in many scientific areas. As examples, chemists and biologists use infrared spectroscopy to identify molecules, astronomers use visible-light spectroscopy to identify elements on distant stars, and geologists use spectral analysis to identify minerals.

■ APPLYING PHYSICS 22.3 | Dispersion

When a beam of light enters a glass prism, which has nonparallel sides, the rainbow of color exiting the prism is a testimonial to the dispersion occurring in the glass. Suppose a beam of light enters a slab of material with parallel sides. When the beam exits the other side, traveling in the same direction as the original beam, is there any evidence of dispersion?

EXPLANATION Due to dispersion, light at the violet end of the spectrum exhibits a larger angle of refraction on entering the glass than light at the red end. All colors of light return to their original direction of propagation as they refract back out into the air. As a result, the outgoing beam is white. The net shift in the position of the violet light along the edge of the slab is larger than the shift of the red light, however, so one edge of the outgoing beam has a bluish tinge to it (it appears blue rather than violet because the eye is not very sensitive to violet light), whereas the other edge has a reddish tinge. This effect is indicated in Figure 22.17. The colored edges of the outgoing beam of white light are evidence of dispersion. ■

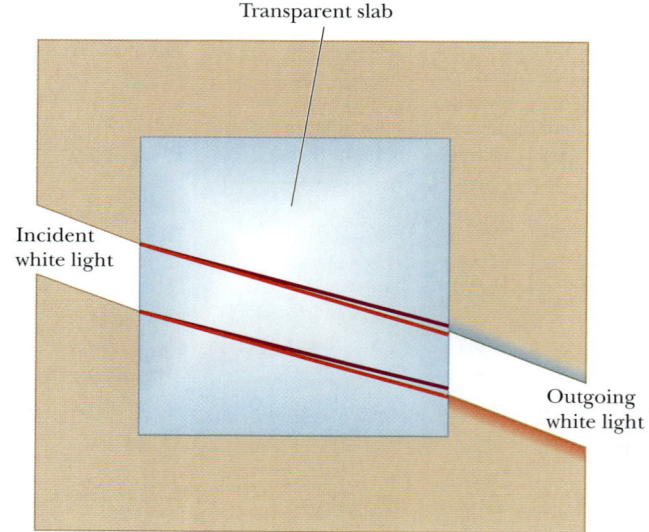

Figure 22.17 (Applying Physics 22.3)

■ EXAMPLE 22.5 | Light Through a Prism

GOAL Calculate the consequences of dispersion.

PROBLEM A beam of light is incident on a prism of a certain glass at an angle of $\theta_1 = 30.0°$, as shown in Figure 22.18. If the index of refraction of the glass for violet light is 1.80, find **(a)** θ_2, the angle of refraction at the air–glass interface, **(b)** ϕ_2, the angle of incidence at the glass–air interface, and **(c)** ϕ_1, the angle of refraction when the violet light exits the prism. **(d)** What is the value of Δy, the amount by which the violet light is displaced vertically?

STRATEGY This problem requires Snell's law to find the refraction angles and some elementary geometry and trigonometry based on Figure 22.18.

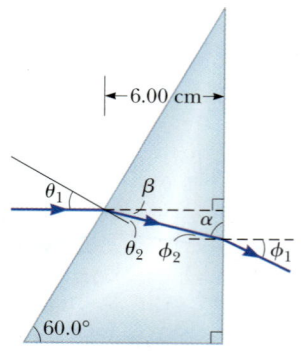

Figure 22.18 (Example 22.5)

SOLUTION

(a) Find θ_2, the angle of refraction at the air–glass interface.

Use Snell's law to find the first angle of refraction:

$$n_1 \sin\theta_1 = \sin\theta_2 \rightarrow (1.00)\sin 30.0 = (1.80)\sin\theta_2$$

$$\theta_2 = \sin^{-1}\left(\frac{0.500}{1.80}\right) = \boxed{16.1°}$$

(b) Find ϕ_2, the angle of incidence at the glass–air interface.

Compute the angle β:

$$\beta = 30.0° - \theta_2 = 30.0° - 16.1° = 13.9°$$

Compute the angle α using the fact that the sum of the interior angles of a triangle equals 180°:

$$180° = 13.9° + 90° + \alpha \rightarrow \alpha = 76.1°$$

The incident angle ϕ_2 at the glass–air interface is complementary to α:

$$\phi_2 = 90° - \alpha = 90° - 76.1° = \boxed{13.9°}$$

(c) Find ϕ_1, the angle of refraction when the violet light exits the prism.

Apply Snell's law:

$$\phi_1 = \left(\frac{1}{n_1}\right)\sin^{-1}(n_2\sin\phi_2)$$

$$= \left(\frac{1}{1.00}\right)\sin^{-1}[(1.80)\sin 13.9°] = \boxed{25.6°}$$

(d) What is the value of Δy, the amount by which the violet light is displaced vertically?

Use the tangent function to find the vertical displacement:

$$\tan\beta = \frac{\Delta y}{\Delta x} \rightarrow \Delta y = \Delta x \tan\beta$$

$$\Delta y = (6.00 \text{ cm})\tan(13.9°) = \boxed{1.48 \text{ cm}}$$

REMARKS The same calculation for red light is left as an exercise. The violet light is bent more and displaced farther down the face of the prism. Notice that a theorem in geometry about parallel lines and the angles created by a transverse line give $\phi_2 = \beta$ immediately, which would have saved some calculation. In general, however, this tactic might not be available.

QUESTION 22.5 On passing through the prism, will yellow light bend through a larger angle or smaller angle than the violet light? (a) Yellow light bends through a larger angle. (b) Yellow light bends through a smaller angle. (c) The angles are the same.

EXERCISE 22.5 Repeat parts (a) through (d) of the example for red light passing through the prism, given that the index of refraction for red light is 1.72.

ANSWERS (a) 16.9° (b) 13.1° (c) 22.9° (d) 1.40 cm

22.5 The Rainbow

LEARNING OBJECTIVE

1. Describe how the dispersion of light forms rainbows.

The dispersion of light into a spectrum is demonstrated most vividly in nature through the formation of a rainbow, often seen by an observer positioned between the Sun and a rain shower. To understand how a rainbow is formed, consider Figure 22.19 (page 786). A ray of light passing overhead strikes a drop of water in the atmosphere and is refracted and reflected as follows: It is first refracted at the front surface of the drop, with the violet light deviating the most and the red light

786 CHAPTER 22 | Reflection and Refraction of Light

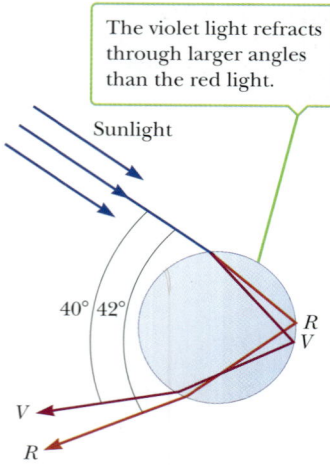

Figure 22.19 Refraction of sunlight by a spherical raindrop.

Figure 22.20 The formation of a rainbow seen by an observer standing with the Sun behind his back. (b) This photograph of a rainbow shows a distinct secondary rainbow with the colors reversed.

▶ Huygens' principle

the least. At the back surface of the drop, the light is reflected and returns to the front surface, where it again undergoes refraction as it moves from water into air. The rays leave the drop so that the angle between the incident white light and the returning violet ray is 40° and the angle between the white light and the returning red ray is 42°. This small angular difference between the returning rays causes us to see the bow as explained in the next paragraph.

Now consider an observer viewing a rainbow, as in Figure 22.20a. If a raindrop high in the sky is being observed, the red light returning from the drop can reach the observer because it is deviated the most, but the violet light passes over the observer because it is deviated the least. Hence, the observer sees this drop as being red. Similarly, a drop lower in the sky would direct violet light toward the observer and appear to be violet. (The red light from this drop would strike the ground and not be seen.) The remaining colors of the spectrum would reach the observer from raindrops lying between these two extreme positions. Figure 22.20b shows a beautiful rainbow and a secondary rainbow with its colors reversed.

22.6 Huygens' Principle

LEARNING OBJECTIVE

1. State Huygens' principle, and apply it to the understanding of reflection and refraction.

The laws of reflection and refraction can be deduced using a geometric method proposed by Huygens in 1678. Huygens assumed light is a form of wave motion rather than a stream of particles. He had no knowledge of the nature of light or of its electromagnetic character. Nevertheless, his simplified wave model is adequate for understanding many practical aspects of the propagation of light.

Huygens' principle is a geometric construction for determining at some instant the position of a new wave front from knowledge of the wave front that preceded it. (A wave front is a surface passing through those points of a wave which have the same phase and amplitude. For instance, a wave front could be a surface passing through the crests of waves.) In Huygens' construction, **all points on a given wave front are taken as point sources for the production of spherical secondary waves, called wavelets, that propagate in the forward direction with speeds characteristic of waves in that medium. After some time has elapsed, the new position of the wave front is the surface tangent to the wavelets.**

Figure 22.21 illustrates two simple examples of Huygens' construction. First, consider a plane wave moving through free space, as in Figure 22.21a. At $t = 0$, the wave front is indicated by the plane labeled AA'. In Huygens' construction, each point on this wave front is considered a point source. For clarity, only a few points

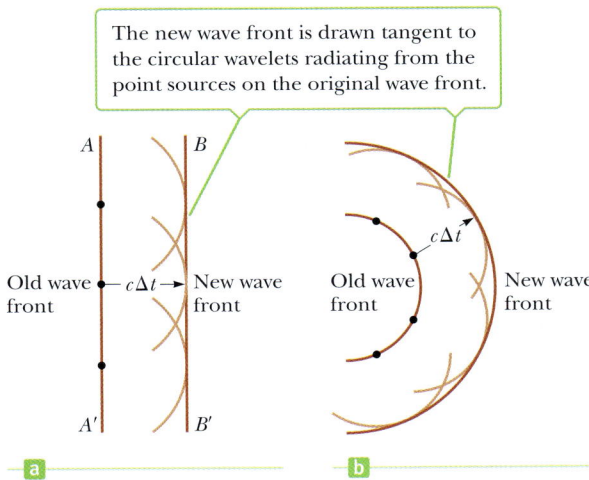

Figure 22.21 Huygens' constructions for (a) a plane wave propagating to the right and (b) a spherical wave.

on AA' are shown. With these points as sources for the wavelets, we draw circles of radius $c\,\Delta t$, where c is the speed of light in vacuum and Δt is the period of propagation from one wave front to the next. The surface drawn tangent to these wavelets is the plane BB', which is parallel to AA'. In a similar manner, Figure 22.21b shows Huygens' construction for an outgoing spherical wave.

Huygens' Principle Applied to Reflection and Refraction

The laws of reflection and refraction were stated earlier in the chapter without proof. We now derive these laws using Huygens' principle. Figure 22.22a illustrates the law of reflection. The line AA' represents a wave front of the incident light. As ray 3 travels from A' to C, ray 1 reflects from A and produces a spherical wavelet of radius AD. (Recall that the radius of a Huygens wavelet is $v\,\Delta t$.) Because the two wavelets having radii $A'C$ and AD are in the same medium, they have the same speed v, so $AD = A'C$. Meanwhile, the spherical wavelet centered at B has spread only half as far as the one centered at A because ray 2 strikes the surface later than ray 1.

From Huygens' principle, we find that the reflected wave front is CD, a line tangent to all the outgoing spherical wavelets. The remainder of our analysis depends on geometry, as summarized in Figure 22.22b. Note that the right triangles ADC and $AA'C$ are congruent because they have the same hypotenuse, AC, and because $AD = A'C$. From the figure, we have

$$\sin \theta_1 = \frac{A'C}{AC} \quad \text{and} \quad \sin \theta_1' = \frac{AD}{AC}$$

The right-hand sides are equal, so $\sin \theta_1 = \sin \theta_1'$, and it follows that $\theta_1 = \theta_1'$, which is the law of reflection.

Huygens' principle and Figure 22.23a (page 788) can be used to derive Snell's law of refraction. In the time interval Δt, ray 1 moves from A to B and ray 2 moves from A' to C. The radius of the outgoing spherical wavelet centered at A is equal

 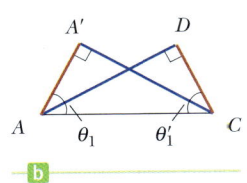

Figure 22.22 (a) Huygens' construction for proving the law of reflection. (b) Triangle ADC is congruent to triangle $AA'C$.

Figure 22.23 (a) Huygens' construction for proving the law of refraction. (b) Overhead view of a barrel rolling from concrete onto grass.

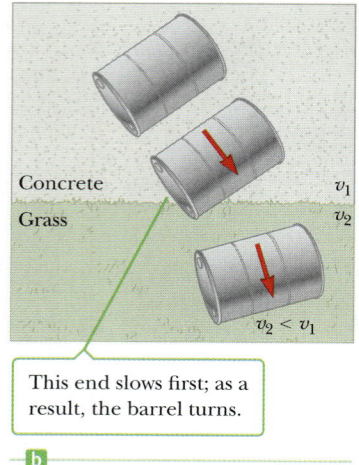

to $v_2 \Delta t$. The distance $A'C$ is equal to $v_1 \Delta t$. Geometric considerations show that angle $A'AC$ equals θ_1 and angle ACB equals θ_2. From triangles $AA'C$ and ACB, we find that

$$\sin \theta_1 = \frac{v_1 \Delta t}{AC} \quad \text{and} \quad \sin \theta_2 = \frac{v_2 \Delta t}{AC}$$

If we divide the first equation by the second, we get

$$\frac{\sin \theta_1}{\sin \theta_2} = \frac{v_1}{v_2}$$

From Equation 22.4, though, we know that $v_1 = c/n_1$ and $v_2 = c/n_2$. Therefore,

$$\frac{\sin \theta_1}{\sin \theta_2} = \frac{c/n_1}{c/n_2} = \frac{n_2}{n_1}$$

and it follows that

$$n_1 \sin \theta_1 = n_2 \sin \theta_2$$

which is the law of refraction.

A mechanical analog of refraction is shown in Figure 22.23b. When the left end of the rolling barrel reaches the grass, it slows down, while the right end remains on the concrete and moves at its original speed. This difference in speeds causes the barrel to pivot, changing the direction of its motion.

22.7 Total Internal Reflection

LEARNING OBJECTIVE

1. Define total internal reflection and apply it in basic physical contexts.

An interesting effect called *total internal reflection* can occur when light encounters the boundary between a medium with a *higher* index of refraction and one with a *lower* index of refraction. Consider a light beam traveling in medium 1 and meeting the boundary between medium 1 and medium 2, where n_1 is greater than n_2 (Fig. 22.24). Possible directions of the beam are indicated by rays 1 through 5. Note that the refracted rays are bent away from the normal because n_1 is greater than n_2. At some particular angle of incidence θ_c, called the **critical angle**, the refracted light ray moves parallel to the boundary so that $\theta_2 = 90°$ (Fig. 22.24b). *For angles of incidence greater than θ_c*, the beam is entirely reflected at the boundary, as is ray 5 in Figure 22.24a. This ray is reflected as though it had struck a perfectly reflecting surface. It and all rays like it obey the law of reflection: the angle of incidence equals the angle of reflection.

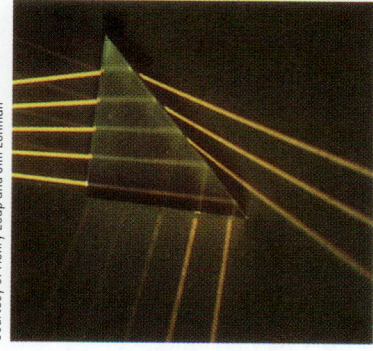

This photograph shows nonparallel light rays entering a glass prism. The bottom two rays undergo total internal reflection at the longest side of the prism. The top three rays are refracted at the longest side as they leave the prism.

22.7 | Total Internal Reflection

Figure 22.24 (a) Rays from a medium with index of refraction n_1 travel to a medium with index of refraction n_2, where $n_1 > n_2$. (b) Ray 4 is singled out.

As the angle of incidence θ_1 increases, the angle of refraction θ_2 increases until θ_2 is 90° (ray 4). The dashed line indicates that no energy actually propagates in this direction.

The angle of incidence producing an angle of refraction equal to 90° is the *critical angle* θ_c. At this angle of incidence, all the energy of the incident light is reflected.

For even larger angles of incidence, total internal reflection occurs (ray 5).

We can use Snell's law to find the critical angle. When $\theta_1 = \theta_c$ and $\theta_2 = 90°$, Snell's law (Eq. 22.8) gives

$$n_1 \sin \theta_c = n_2 \sin 90° = n_2$$

$$\sin \theta_c = \frac{n_2}{n_1} \qquad \text{for } n_1 > n_2 \qquad [22.9]$$

Equation 22.9 can be used only when n_1 is greater than n_2 because **total internal reflection occurs only when light is incident on the boundary of a medium having a lower index of refraction than the medium in which it's traveling.** If n_1 were less than n_2, Equation 22.9 would give $\sin \theta_c > 1$, which is an absurd result because the sine of an angle can never be greater than 1.

When medium 2 is air, the critical angle is small for substances with large indices of refraction, such as diamond, where $n = 2.42$ and $\theta_c = 24.0°$. By comparison, for crown glass, $n = 1.52$ and $\theta_c = 41.0°$. This property, combined with proper faceting, causes a diamond to sparkle brilliantly.

A prism and the phenomenon of total internal reflection can alter the direction of travel of a light beam. Figure 22.25 illustrates two such possibilities. In one case the light beam is deflected by 90° (Fig. 22.25a), and in the second case the path of the beam is reversed (Fig. 22.25b). A common application of total internal reflection is a submarine periscope. In this device two prisms are arranged as in Figure 22.25c so that an incident beam of light follows the path shown and the user can "see around corners."

APPLICATION
Submarine Periscopes

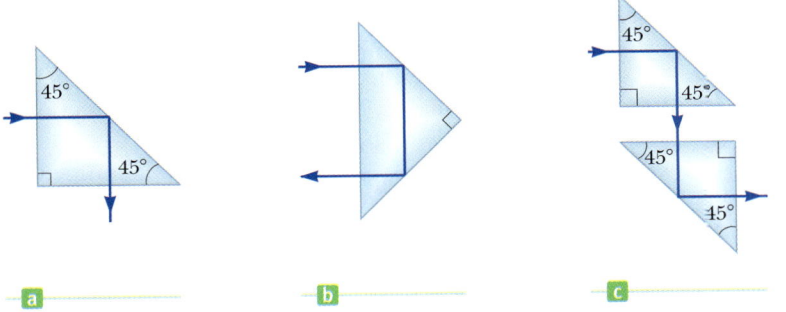

Figure 22.25 Internal reflection in a prism. (a) The ray is deviated by 90°. (b) The direction of the ray is reversed. (c) Two prisms used as a periscope.

APPLYING PHYSICS 22.4 | Total Internal Reflection and Dispersion

A beam of white light is incident on the curved edge of a semicircular piece of glass, as shown in Figure 22.26. The light enters the curved surface along the normal, so it shows no refraction. It encounters the straight side of the glass at the center of curvature of the curved side and refracts into the air. The incoming beam is moved clockwise (so that the angle θ increases) such that the beam always enters along the normal to the curved side and encounters the straight side at the center of curvature of the curved side. Why does the refracted beam become redder as it approaches a direction parallel to the straight side?

EXPLANATION When the outgoing beam approaches the direction parallel to the straight side, the incident angle is approaching the critical angle for total internal reflection. Dispersion occurs as the light passes out of the glass. The index of refraction for light at the violet end of the visible spectrum is larger than at the red end. As a result, as the outgoing beam approaches the straight side, the violet light undergoes total internal reflection, followed by the other colors. The red light is the last to undergo total internal reflection, so just before the outgoing light disappears, it's composed of light from the red end of the visible spectrum. ■

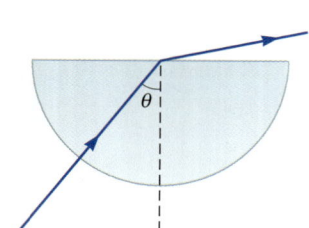

Figure 22.26 (Applying Physics 22.4)

■ EXAMPLE 22.6 | A View from the Fish's Eye

GOAL Apply the concept of total internal reflection.

PROBLEM (a) Find the critical angle for a water–air boundary. (b) Use the result of part (a) to predict what a fish will see (Fig. 22.27) if it looks up toward the water surface at angles of 40.0°, 48.6°, and 60.0°.

STRATEGY After finding the critical angle by substitution, use the fact that the path of a light ray is reversible: at a given angle, wherever a light beam can go is also where a light beam can come from, along the same path.

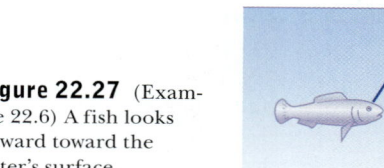

Figure 22.27 (Example 22.6) A fish looks upward toward the water's surface.

SOLUTION

(a) Find the critical angle for a water–air boundary.

Substitute into Equation 22.9 to find the critical angle:

$$\sin \theta_c = \frac{n_2}{n_1} = \frac{1.00}{1.333} = 0.750$$

$$\theta_c = \sin^{-1}(0.750) = \boxed{48.6°}$$

(b) Predict what a fish will see if it looks up toward the water surface at angles of 40.0°, 48.6°, and 60.0°.

At an angle of 40.0°, a beam of light from underwater will be refracted at the surface and enter the air above. Because the path of a light ray is reversible (Snell's law works both going and coming), light from above can follow the same path and be perceived by the fish. At an angle of 48.6°, the critical angle for water, light from underwater is bent so that it travels along the surface. So light following the same path in reverse can reach the fish only by skimming along the water surface before being refracted toward the fish's eye. At angles greater than the critical angle of 48.6°, a beam of light shot toward the surface will be completely reflected down toward the bottom of the pool. Reversing the path, the fish sees a reflection of some object on the bottom.

QUESTION 22.6 If the water is replaced by a transparent fluid with a higher index of refraction, is the critical angle of the fluid–air boundary (a) larger, (b) smaller, or (c) the same as for water?

EXERCISE 22.6 Suppose a layer of oil with $n = 1.50$ coats the surface of the water. What is the critical angle for total internal reflection for light traveling in the oil layer and encountering the oil–water boundary?

ANSWER 62.7°

Fiber Optics

Another interesting application of total internal reflection is the use of solid glass or transparent plastic rods to "pipe" light from one place to another. As indicated in Figure 22.28, light is confined to traveling within the rods, even around gentle curves, as a result of successive internal reflections. Such a light pipe can be quite flexible if thin fibers are used rather than thick rods. If a bundle of parallel fibers is used to construct an optical transmission line, images can be transferred from one point to another.

Very little light intensity is lost in these fibers as a result of reflections on the sides. Any loss of intensity is due essentially to reflections from the two ends and absorption by the fiber material. Fiber-optic devices are particularly useful for viewing images produced at inaccessible locations. Physicians often use fiber-optic cables to aid in the diagnosis and correction of certain medical problems without the intrusion of major surgery. For example, a fiber-optic cable can be threaded through the esophagus and into the stomach to look for ulcers. In this application the cable consists of two fiber-optic lines: one to transmit a beam of light into the stomach for illumination and the other to allow the light to be transmitted out of the stomach. The resulting image can, in some cases, be viewed directly by the physician, but more often is displayed on a television monitor or saved in digital form. In a similar way, fiber-optic cables can be used to examine the colon or to help physicians perform surgery without the need for large incisions.

The field of fiber optics has revolutionized the entire communications industry. Billions of kilometers of optical fiber have been installed in the United States to carry high-speed Internet traffic, radio and television signals, and telephone calls. The fibers can carry much higher volumes of telephone calls and other forms of communication than electrical wires because of the higher frequency of the infrared light used to carry the information on optical fibers. Optical fibers are also preferable to copper wires because they are insulators and don't pick up stray electric and magnetic fields or electronic "noise."

Figure 22.28 Light travels in a curved transparent rod by multiple internal reflections.

BIO APPLICATION
Fiber Optics in Medical Diagnosis and Surgery

APPLICATION
Fiber Optics in Telecommunications

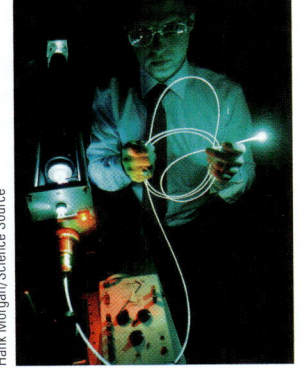

(*Left*) Strands of glass optical fibers are used to carry voice, video, and data signals in telecommunication networks. (*Right*) A bundle of optical fibers is illuminated by a laser.

APPLYING PHYSICS 22.5 | Design of an Optical Fiber

An optical fiber consists of a transparent core surrounded by cladding, which is a material with a lower index of refraction than the core (Fig. 22.29). A cone of angles, called the acceptance cone, is at the entrance to the fiber. Incoming light at angles within this cone will be transmitted through the fiber, whereas light entering the core from angles outside the cone will not be transmitted. The figure shows a light ray entering the fiber just within the acceptance cone and undergoing total internal reflection at the interface between the core and the cladding. If it is

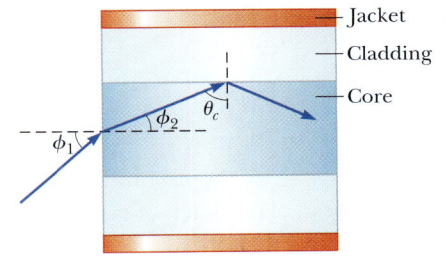

Figure 22.29 (Applying Physics 22.5)

(*Continued*)

technologically difficult to produce light so that it enters the fiber from a small range of angles, how could you adjust the indices of refraction of the core and cladding to increase the size of the acceptance cone? Would you design the indices to be farther apart or closer together?

EXPLANATION The acceptance cone would become larger if the critical angle (θ_c in the figure) could be made smaller. This requires making the index of refraction of the cladding material smaller so that the indices of refraction of the core and cladding material are farther apart. ■

SUMMARY

22.1 The Nature of Light
Light has a dual nature. In some experiments it acts like a wave, in others like a particle, called a photon by Einstein. The energy of a photon is proportional to its frequency,

$$E = hf \quad [22.1]$$

where $h = 6.63 \times 10^{-34}$ J · s is *Planck's constant*.

22.2 Reflection and Refraction
In the reflection of light off a flat, smooth surface, the angle of incidence, θ_1, with respect to a line perpendicular to the surface is equal to the angle of reflection, θ_1':

$$\theta_1' = \theta_1 \quad [22.2]$$

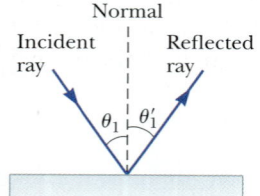

The wave under reflection model.

Light that passes into a transparent medium is bent at the boundary and is said to be *refracted*. The angle of refraction is the angle the ray makes with respect to a line perpendicular to the surface after it has entered the new medium.

22.3 The Law of Refraction
The **index of refraction** of a material, n, is defined as

$$n = \frac{c}{v} \quad [22.4]$$

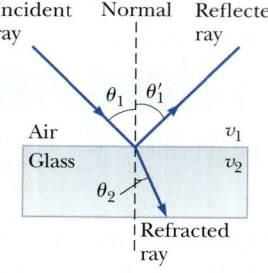

The wave under refraction model.

where c is the speed of light in a vacuum and v is the speed of light in the material. The index of refraction of a material is also

$$n = \frac{\lambda_0}{\lambda_n} \quad [22.7]$$

where λ_0 is the wavelength of the light in vacuum and λ_n is its wavelength in the material.

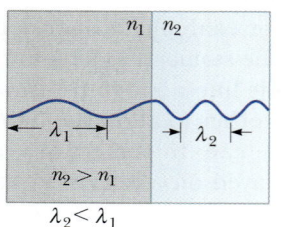

When entering a material with higher index of refraction, the wavelength of light is reduced.

The **law of refraction**, or **Snell's law**, states that

$$n_1 \sin \theta_1 = n_2 \sin \theta_2 \quad [22.8]$$

where n_1 and n_2 are the indices of refraction in the two media. The incident ray, the reflected ray, the refracted ray, and the normal to the surface all lie in the same plane.

22.4 Dispersion and Prisms
22.5 The Rainbow
The index of refraction of a material depends on the wavelength of the incident light, an effect called *dispersion*. Light at the violet end of the spectrum exhibits a larger angle of refraction on entering glass than light at the red end. Rainbows are a consequence of dispersion.

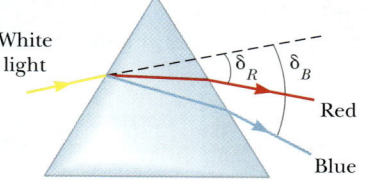

Due to dispersion, blue light is bent more than red light.

22.6 Huygens' Principle
Huygens' principle states that all points on a wave front are point sources for the production of spherical secondary waves called wavelets. These wavelets propagate forward at a speed characteristic of waves in a particular medium. After some time has elapsed, the new position of the wave front is the surface tangent to the wavelets. This principle can be used to deduce the laws of reflection and refraction.

22.7 Total Internal Reflection

When light propagating in a medium with index of refraction n_1 is incident on the boundary of a region with index of refraction n_2, and $n_1 > n_2$, then total internal reflection can occur if the angle of incidence equals or exceeds a **critical angle** θ_c given by

$$\sin \theta_c = \frac{n_2}{n_1} \quad (n_1 > n_2) \quad [22.9]$$

Total internal reflection is used in the optical fibers that carry data at high speed around the world.

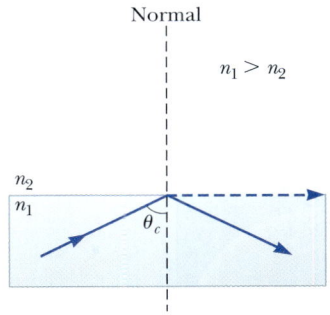

A ray undergoes total internal reflection at angles equal to or greater than the critical angle.

■ WARM-UP EXERCISES

WebAssign The warm-up exercises in this chapter may be assigned online in Enhanced WebAssign.

1. How many 800-nm photons does it take to have the same total energy as four 200-nm photons? (See Section 22.1.)

2. Two mirrors are set up at an angle of 90.0°. If light is incident on the first mirror at an angle of 30.0° from the normal, at what angle is it incident on the second mirror? (See Section 22.2.)

3. A ray of light in air is incident at an angle of 30.0° on a glass slide with index of refraction 1.73. (a) At what angle is the ray refracted? (b) If the wavelength of the light in vacuum is 564 nm, find its wavelength in the glass. (See Section 22.3.)

4. A source emits monochromatic light of wavelength 495 nm in air. When the light passes through a liquid, its wavelength reduces to 434 nm. (a) What is the liquid's index of refraction? (b) Find the speed of light in the liquid. (See Section 22.3.)

5. Carbon disulfide ($n = 1.63$) is poured into a container made of crown glass ($n = 1.52$). What is the critical angle for total internal reflection of a light ray in the liquid when it is incident on the liquid-to-glass surface? (See Section 22.7.)

■ CONCEPTUAL QUESTIONS

WebAssign The conceptual questions in this chapter may be assigned online in Enhanced WebAssign.

1. Why does the arc of a rainbow appear with red on top and violet on the bottom?

2. A ray of light is moving from a material having a high index of refraction into a material with a lower index of refraction. (a) Is the ray bent toward the normal or away from it? (b) If the wavelength is 600 nm in the material with the high index of refraction, is it greater, smaller, or the same in the material with the lower index of refraction? (c) How does the frequency change as the light moves between the two materials? Does it increase, decrease, or remain the same?

3. A light ray travels through three parallel slabs having different indices of refraction as in Figure CQ22.3. The rays shown are only the refracted rays. Rank the materials according to the size of their indices of refraction, from largest to smallest.

Figure CQ22.3

4. Under what conditions is a mirage formed? On a hot day, what are we seeing when we observe a mirage of a water puddle on the road?

5. Explain why a diamond loses most of its sparkle when submerged in carbon disulfide.

6. A type of mirage called a *pingo* is often observed in Alaska. Pingos occur when the light from a small hill passes to an observer by a path that takes the light over a body of water warmer than the air. What is seen is the hill and an inverted image directly below it. Explain how these mirages are formed.

7. In dispersive materials, the angle of refraction for a light ray depends on the wavelength of the light. Does the angle of reflection from the surface of the material depend on the wavelength? Why or why not?

8. The level of water in a clear, colorless glass can easily be observed with the naked eye. The level of liquid helium in a clear glass vessel is extremely difficult to see with the naked eye. Explain. *Hint:* The index of refraction of liquid helium is close to that of air.

9. Suppose you are told that only two colors of light (X and Y) are sent through a glass prism and that X is bent more than Y. Which color travels more slowly in the prism?

10. Is it possible to have total internal reflection for light incident from air on water? Explain.

11. Figure CQ22.11 shows a pencil partially immersed in a cup of water. Why does the pencil appear to be bent?

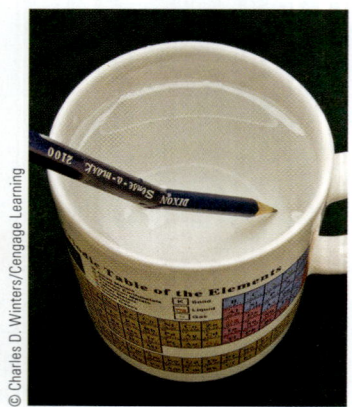

Figure CQ22.11

12. Try this simple experiment on your own. Take two opaque cups, place a coin at the bottom of each cup near the edge, and fill one cup with water. Next, view the cups at some angle from the side so that the coin in water is just visible as shown on the left in Figure CQ22.12. Notice that the coin in air is not visible as shown on the right in Figure CQ22.12. Explain this observation.

Figure CQ22.12

13. Why do astronomers looking at distant galaxies talk about looking backward in time?

14. Light can travel from air into water. Some possible paths for the light ray in the water are shown in Figure CQ22.14. Which path will the light most likely follow? (a) A (b) B (c) C (d) D (e) E

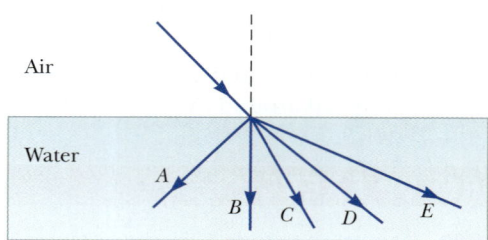

Figure CQ22.14

15. A light ray containing both blue and red wavelengths is incident at an angle on a slab of glass. Which of the sketches in Figure CQ22.15 represents the most likely outcome? (a) A (b) B (c) C (d) D (e) none of these

Figure CQ22.15

■ PROBLEMS

WebAssign The problems in this chapter may be assigned online in Enhanced WebAssign.

1. denotes straightforward problem; 2. denotes intermediate problem;
3. denotes challenging problem
1. denotes full solution available in *Student Solutions Manual/Study Guide*
1. denotes problems most often assigned in Enhanced WebAssign

BIO denotes biomedical problems
GP denotes guided problems
M denotes Master It tutorial available in Enhanced WebAssign
Q|C denotes asking for quantitative and conceptual reasoning
S denotes symbolic reasoning problem
W denotes Watch It video solution available in Enhanced WebAssign

22.1 The Nature of Light

1. During the Apollo XI Moon landing, a retroreflecting panel was erected on the Moon's surface. The speed of light can be found by measuring the time it takes a laser beam to travel from Earth, reflect from the panel, and return to Earth. If this interval is found to be 2.51 s, what is the measured speed of light? Take the center-to-center distance from Earth to the Moon to be 3.84×10^8 m. Assume the Moon is directly overhead and do not neglect the sizes of Earth and the Moon.

2. **Q|C** (a) What is the energy in joules of an x-ray photon with wavelength 1.00×10^{-10} m? (b) Convert the

energy to electron volts. (c) If more penetrating x-rays are desired, should the wavelength be increased or decreased? (d) Should the frequency be increased or decreased?

3. **M** Find the energy of (a) a photon having a frequency of 5.00×10^{17} Hz and (b) a photon having a wavelength of 3.00×10^2 nm. Express your answers in units of electron volts, noting that $1 \text{ eV} = 1.60 \times 10^{-19}$ J.

4. **Q|C** (a) Calculate the wavelength of light in vacuum that has a frequency of 5.45×10^{14} Hz. (b) What is its wavelength in benzene? (c) Calculate the energy of one photon of such light in vacuum. Express the answer in electron volts. (d) Does the energy of the photon change when it enters the benzene? Explain.

5. Find the speed of light in (a) water, (b) crown glass, and (c) diamond.

6. **S** (a) Find a symbolic expression for the wavelength λ of a photon in terms of its energy E, Planck's constant h, and the speed of light c. (b) What does the equation say about the wavelengths of higher-energy photons?

7. A ray of light travels from air into another medium, making an angle of $\theta_1 = 45.0°$ with the normal as in Figure P22.7. Find the angle of refraction θ_2 if the second medium is (a) fused quartz, (b) carbon disulfide, and (c) water.

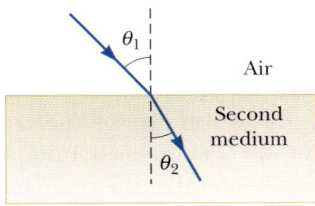

Figure P22.7

22.2 Reflection and Refraction
22.3 The Law of Refraction

8. **W** The two mirrors in Figure P22.8 meet at a right angle. The beam of light in the vertical plane P strikes mirror 1 as shown. (a) Determine the distance the reflected light beam travels before striking mirror 2. (b) In what direction does the light beam travel after being reflected from mirror 2?

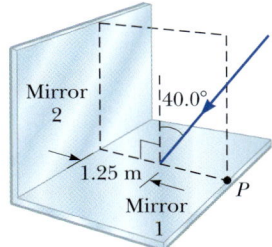

Figure P22.8

9. An underwater scuba diver sees the Sun at an apparent angle of 45.0° from the vertical. What is the actual direction of the Sun?

10. Two plane mirrors are at right angles to each other as shown by the side view in Figure P22.10. A light ray is incident on mirror 1 at an angle θ with the vertical. Using the law of reflection and geometry, show that after the ray is reflected off of both mirrors, the outgoing reflected ray is parallel to the incident ray.

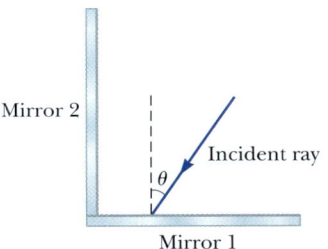

Figure P22.10

11. A laser beam is incident at an angle of 30.0° to the vertical onto a solution of corn syrup in water. If the beam is refracted to 19.24° to the vertical, (a) what is the index of refraction of the syrup solution? Suppose the light is red, with wavelength 632.8 nm in a vacuum. Find its (b) wavelength, (c) frequency, and (d) speed in the solution.

12. Light containing wavelengths of 400 nm, 500 nm, and 650 nm is incident from air on a block of crown glass at an angle of 25.0°. (a) Are all colors refracted alike, or is one color bent more than the others? (b) Calculate the angle of refraction in each case to verify your answer.

13. A ray of light is incident on the surface of a block of clear ice at an angle of 40.0° with the normal. Part of the light is reflected, and part is refracted. Find the angle between the reflected and refracted light.

14. Two plane mirrors are at an angle of $\theta_1 = 50.0°$ with each other as in the side view shown in Figure P22.14. If a horizontal ray is incident on mirror 1, at what angle θ_2 does the outgoing reflected ray make with the surface of mirror 2?

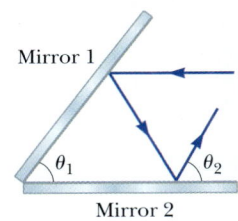

Figure P22.14

15. The light emitted by a helium–neon laser has a wavelength of 632.8 nm in air. As the light travels from air into zircon, find its (a) speed, (b) wavelength, and (c) frequency, all in the zircon.

16. Figure P22.16 shows a light ray traveling in a slab of crown glass surrounded by air. The ray is incident on the right surface at an angle of 55° with the normal and then reflects from points A, B, and C. (a) At which of these points does part of the ray enter the air? (b) If the glass slab is surrounded by carbon disulfide, at which point does part of the ray enter the carbon disulfide?

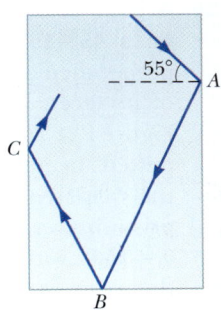

Figure P22.16

17. How many times will the incident beam shown in Figure P22.17 be reflected by each of the parallel mirrors?

Figure P22.17

18. QC A ray of light strikes a flat, 2.00-cm-thick block of glass ($n = 1.50$) at an angle of 30.0° with respect to the normal (Fig. P22.18). (a) Find the angle of refraction at the top surface. (b) Find the angle of incidence at the bottom surface and the refracted angle. (c) Find the lateral distance d by which the light beam is shifted. (d) Calculate the speed of light in the glass and (e) the time required for the light to pass through the glass block. (f) Is the travel time through the block affected by the angle of incidence? Explain.

Figure P22.18

19. M The light beam shown in Figure P22.19 makes an angle of 20.0° with the normal line NN' in the linseed oil. Determine the angles θ and θ'. (The refractive index for linseed oil is 1.48.)

Figure P22.19

20. A laser beam is incident on a 45°–45°–90° prism perpendicular to one of its faces, as shown in Figure P22.20. The transmitted beam that exits the hypotenuse of the prism makes an angle of $\theta = 15.0°$ with the direction of the incident beam. Find the index of refraction of the prism.

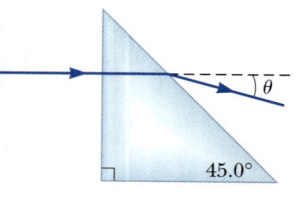

Figure P22.20

21. A block of crown glass is immersed in water as in Figure P22.21. A light ray is incident on the top face at an angle of $\theta_1 = 42.0°$ with the normal and exits the block at point P. (a) Find the vertical distance y from the top of the block to P. (b) Find the angle of refraction θ_2 of the light ray leaving the block at P.

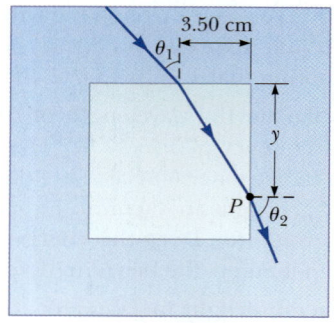

Figure P22.21

22. W BIO A narrow beam of ultrasonic waves reflects off the liver tumor in Figure P22.22. If the speed of the wave is 10.0% less in the liver than in the surrounding medium, determine the depth of the tumor.

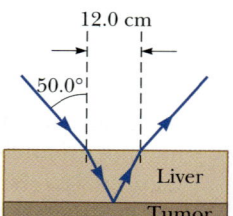

Figure P22.22

23. S A person looking into an empty container is able to see the far edge of the container's bottom, as shown in Figure P22.23a. The height of the container is h, and its width is d. When the container is completely filled with a fluid of index of refraction n and viewed from the same angle, the person can see the center of a coin at the middle of the container's bottom, as shown in Figure P22.23b. (a) Show that the ratio h/d is given by

$$\frac{h}{d} = \sqrt{\frac{n^2 - 1}{4 - n^2}}$$

(b) Assuming the container has a width of 8.00 cm and is filled with water, use the expression above to find the height of the container.

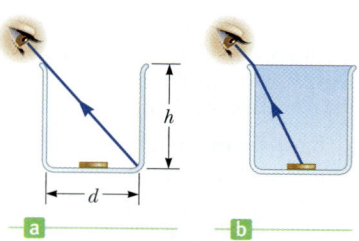

Figure P22.23

24. GP A submarine is 3.00×10^2 m horizontally from shore and 1.00×10^2 m beneath the surface of the water. A laser beam is sent from the submarine so that the beam strikes the surface of the water 2.10×10^2 m from the shore. A building stands on the shore, and the laser beam hits a target at the top of the building. The goal is to find the height of the target above sea level. (a) Draw a diagram of the situation, identifying the two triangles that are important to finding the solution. (b) Find the angle of incidence of the beam striking the water–air interface. (c) Find the angle of refraction. (d) What angle does the refracted beam make with respect to the horizontal? (e) Find the height of the target above sea level.

25. **S** A beam of light both reflects and refracts at the surface between air and glass, as shown in Figure P22.25. If the index of refraction of the glass is n_g, find the angle of incidence, θ_1, in the air that would result in the reflected ray and the refracted ray being perpendicular to each other. *Hint:* Remember the identity $\sin(90° - \theta) = \cos\theta$.

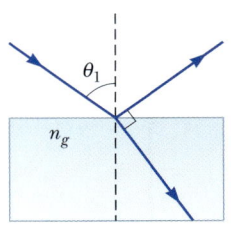

Figure P22.25

26. Figure P22.26 shows a light ray incident on a series of slabs having different refractive indices, where $n_1 < n_2 < n_3 < n_4$. Notice that the path of the ray steadily bends toward the normal. If the variation in n were continuous, the path would form a smooth curve. Use this idea and a ray diagram to explain why you can see the Sun at sunset after it has fallen below the horizon.

Figure P22.26

27. **M** An opaque cylindrical tank with an open top has a diameter of 3.00 m and is completely filled with water. When the afternoon Sun reaches an angle of 28.0° above the horizon, sunlight ceases to illuminate the bottom of the tank. How deep is the tank?

22.4 Dispersion and Prisms

28. **W** A certain kind of glass has an index of refraction of 1.650 for blue light of wavelength 430 nm and an index of 1.615 for red light of wavelength 680 nm. If a beam containing these two colors is incident at an angle of 30.00° on a piece of this glass, what is the angle between the two beams inside the glass?

29. The index of refraction for red light in water is 1.331 and that for blue light is 1.340. If a ray of white light enters the water at an angle of incidence of 83.00°, what are the underwater angles of refraction for the (a) blue and (b) red components of the light?

30. The index of refraction for crown glass is 1.512 at a wavelength of 660 nm (red), whereas its index of refraction is 1.530 at a wavelength of 410 nm (violet). If both wavelengths are incident on a slab of crown glass at the same angle of incidence, 60.0°, what is the angle of refraction for each wavelength?

31. A light beam containing red and violet wavelengths is incident on a slab of quartz at an angle of incidence of 50.00°. The index of refraction of quartz is 1.455 at 660 nm (red light), and its index of refraction is 1.468 at 410 nm (violet light). Find the dispersion of the slab, which is defined as the difference in the angles of refraction for the two wavelengths.

32. The index of refraction for violet light in silica flint glass is 1.66 and that for red light is 1.62. What is the angular dispersion of visible light passing through an equilateral prism of apex angle 60.0° if the angle of incidence is 50.0°? (See Fig. P22.32.)

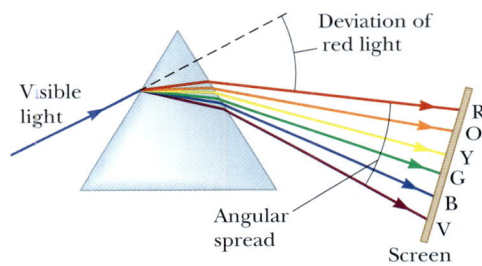

Figure P22.32

33. **M** A ray of light strikes the midpoint of one face of an equiangular (60°–60°–60°) glass prism ($n = 1.5$) at an angle of incidence of 30°. (a) Trace the path of the light ray through the glass and find the angles of incidence and refraction at each surface. (b) If a small fraction of light is also reflected at each surface, what are the angles of reflection at the surfaces?

22.7 Total Internal Reflection

34. For light of wavelength 589 nm, calculate the critical angles for the following substances when surrounded by air: (a) fused quartz, (b) polystyrene, and (c) sodium chloride.

35. Repeat Problem 34, but this time assume the quartz, polystyrene, and sodium chloride are surrounded by water.

36. **M** A beam of light is incident from air on the surface of a liquid. If the angle of incidence is 30.0° and the angle of refraction is 22.0°, find the critical angle for the liquid when surrounded by air.

37. A plastic light pipe has an index of refraction of 1.53. For total internal reflection, what is the minimum angle of incidence if the pipe is in (a) air and (b) water?

38. Determine the maximum angle θ for which the light rays incident on the end of the light pipe in Figure P22.38 are subject to total internal reflection along the walls of the pipe. Assume the light pipe has an index of refraction of 1.36 and the outside medium is air.

Figure P22.38

39. A light ray is incident normally to the long face (the hypotenuse) of a 45°–45°–90° prism surrounded by air, as shown in Figure 22.26b. Calculate the minimum index of refraction of the prism for which the ray will totally internally reflect at each of the two sides making the right angle.

40. Q|C A beam of laser light with wavelength 612 nm is directed through a slab of glass having index of refraction 1.78. (a) For what minimum incident angle would a ray of light undergo total internal reflection? (b) If a layer of water is placed over the glass, what is the minimum angle of incidence on the glass–water interface that will result in total internal reflection at the water–air interface? (c) Does the thickness of the water layer or glass affect the result? (d) Does the index of refraction of the intervening layer affect the result?

41. A room contains air in which the speed of sound is 343 m/s. The walls of the room are made of concrete, in which the speed of sound is 1 850 m/s. (a) Find the critical angle for total internal reflection of sound at the concrete–air boundary. (b) In which medium must the sound be traveling in order to undergo total internal reflection? (c) "A bare concrete wall is a highly efficient mirror for sound." Give evidence for or against this statement.

42. GP Q|C Consider a light ray traveling between air and a diamond cut in the shape shown in Figure P22.42. (a) Find the critical angle for total internal reflection for light in the diamond incident on the interface between the diamond and the outside air. (b) Consider the light ray incident normally on the top surface of the diamond as shown in Figure P22.42. Show that the light traveling toward point P in the diamond is totally reflected. (c) If the diamond is immersed in water, find the critical angle at the diamond–water interface. (d) When the diamond is immersed in water, does the light ray entering the top surface in Figure P22.42 undergo total internal reflection at P? Explain. (e) If the light ray entering the diamond remains vertical as shown in Figure P22.42, which way should the diamond in the water be rotated about an axis perpendicular to the page through O so that light will exit the diamond at P? (f) At what angle of rotation in part (e) will light first exit the diamond at point P?

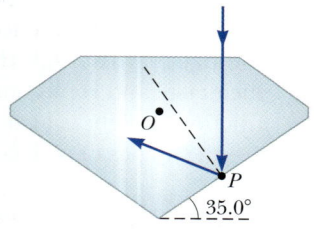

Figure P22.42

43. The light beam in Figure P22.43 strikes surface 2 at the critical angle. Determine the angle of incidence, θ_1.

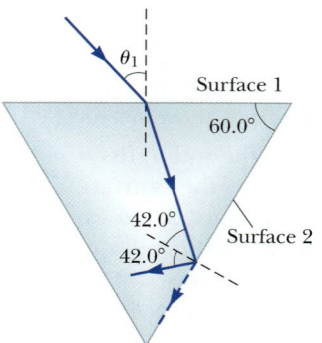

Figure P22.43

44. W A jewel thief hides a diamond by placing it on the bottom of a public swimming pool. He places a circular raft on the surface of the water directly above and centered over the diamond, as shown in Figure P22.44. If the surface of the water is calm and the pool is 2.00 m deep, find the minimum diameter of the raft that would prevent the diamond from being seen.

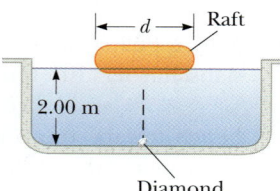

Figure P22.44

Additional Problems

45. A layer of ice having parallel sides floats on water. If light is incident on the upper surface of the ice at an angle of incidence of 30.0°, what is the angle of refraction in the water?

46. Q|C A ray of light is incident at an angle 30.0° on a plane slab of flint glass surrounded by water. (a) Find the refraction angle. (b) Suppose the index of refraction of the surrounding medium can be adjusted, but the incident angle of the light remains the same. As the index of refraction of the medium approaches that of the glass, what happens to the refraction angle? (c) What happens to the refraction angle when the medium's index of refraction exceeds that of the glass?

47. When a man stands near the edge of an empty drainage ditch of depth 2.80 m, he can barely see the boundary between the opposite wall and bottom of the ditch as in Figure P22.47a. The distance from his eyes to the ground is 1.85 m. (a) What is the horizontal distance d from the man to the edge of the drainage ditch? (b) After the drainage ditch is filled with water as in Figure P22.47b, what is the maximum distance x the man can stand from the edge and still see the same boundary?

Figure P22.47

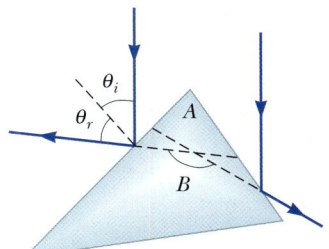

Figure P22.51

48. A light ray of wavelength 589 nm is incident at an angle θ on the top surface of a block of polystyrene surrounded by air, as shown in Figure P22.48. (a) Find the maximum value of θ for which the refracted ray will undergo total internal reflection at the left vertical face of the block. (b) Repeat the calculation for the case in which the polystyrene block is immersed in water. (c) What happens if the block is immersed in carbon disulfide?

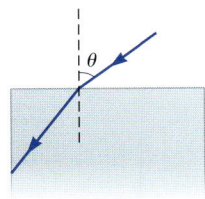

Figure P22.48

49. As shown in Figure P22.49, a light ray is incident normal to one face of a 30°–60°–90° block of flint glass (a prism) that is immersed in water. (a) Determine the exit angle θ_3 of the ray. (b) A substance is dissolved in the water to increase the index of refraction n_2. At what value of n_2 does total internal reflection cease at point P?

Figure P22.49

50. A narrow beam of light is incident from air onto a glass surface with index of refraction 1.56. Find the angle of incidence for which the corresponding angle of refraction is one-half the angle of incidence. *Hint:* You might want to use the trigonometric identity $\sin 2\theta = 2 \sin \theta \cos \theta$.

51. One technique for measuring the angle of a prism is shown in Figure P22.51. A parallel beam of light is directed onto the apex of the prism so that the beam reflects from opposite faces of the prism. Show that the angular separation of the two reflected beams is given by $B = 2A$.

52. An optical fiber with index of refraction n and diameter d is surrounded by air. Light is sent into the fiber along its axis, as shown in Figure P22.52. (a) Find the smallest outside radius R permitted for a bend in the fiber if no light is to escape. (b) Does the result for part (a) predict reasonable behavior as d approaches zero? As n increases? As n approaches unity? (c) Evaluate R, assuming the diameter of the fiber is 100 μm and its index of refraction is 1.40.

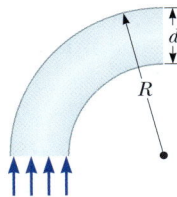

Figure P22.52

53. A piece of wire is bent through an angle θ. The bent wire is partially submerged in benzene (index of refraction = 1.50) so that, to a person looking along the dry part, the wire appears to be straight and makes an angle of 30.0° with the horizontal. Determine the value of θ.

54. A light ray traveling in air is incident on one face of a right-angle prism with index of refraction $n = 1.50$, as shown in Figure P22.54, and the ray follows the path shown in the figure. Assuming $\theta = 60.0°$ and the base of the prism is mirrored, determine the angle ϕ made by the outgoing ray with the normal to the right face of the prism.

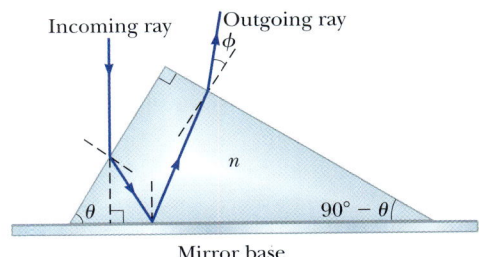

Figure P22.54

55. A transparent cylinder of radius $R = 2.00$ m has a mirrored surface on its right half, as shown in Figure P22.55 (page 800). A light ray traveling in air is incident on the left side of the cylinder. The incident light ray and the exiting light ray are parallel, and $d = 2.00$ m. Determine the index of refraction of the material.

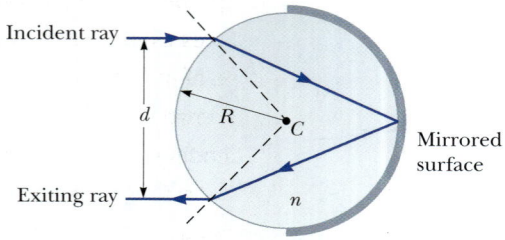

Figure P22.55

56. A laser beam strikes one end of a slab of material, as in Figure P22.56. The index of refraction of the slab is 1.48. Determine the number of internal reflections of the beam before it emerges from the opposite end of the slab.

Figure P22.56

57. A light ray enters a rectangular block of plastic at an angle $\theta_1 = 45.0°$ and emerges at an angle $\theta_2 = 76.0°$, as shown in Figure P22.57. (a) Determine the index of refraction of the plastic. (b) If the light ray enters the plastic at a point $L = 50.0$ cm from the bottom edge, how long does it take the light ray to travel through the plastic?

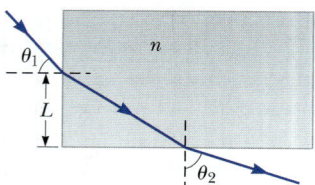

Figure P22.57

58. Students allow a narrow beam of laser light to strike a water surface. They arrange to measure the angle of refraction for selected angles of incidence and record the data shown in the following table:

Angle of Incidence (degrees)	Angle of Refraction (degrees)
10.0	7.5
20.0	15.1
30.0	22.3
40.0	28.7
50.0	35.2
60.0	40.3
70.0	45.3
80.0	47.7

Use the data to verify Snell's law of refraction by plotting the sine of the angle of incidence versus the sine of the angle of refraction. From the resulting plot, deduce the index of refraction of water.

59. Figure P22.59 shows the path of a beam of light through several layers with different indices of refraction. (a) If $\theta_1 = 30.0°$, what is the angle θ_2 of the emerging beam? (b) What must the incident angle θ_1 be to have total internal reflection at the surface between the medium with $n = 1.20$ and the medium with $n = 1.00$?

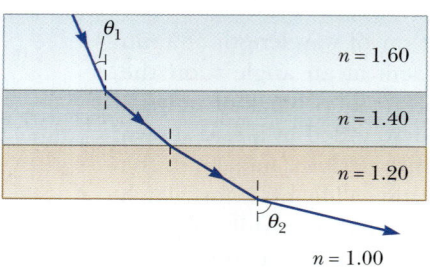

Figure P22.59

60. Three sheets of plastic have unknown indices of refraction. Sheet 1 is placed on top of sheet 2, and a laser beam is directed onto the sheets from above so that it strikes the interface at an angle of 26.5° with the normal. The refracted beam in sheet 2 makes an angle of 31.7° with the normal. The experiment is repeated with sheet 3 on top of sheet 2, and with the same angle of incidence, the refracted beam makes an angle of 36.7° with the normal. If the experiment is repeated again with sheet 1 on top of sheet 3, what is the expected angle of refraction in sheet 3? Assume the same angle of incidence.

61. A thick piece of Lucite ($n = 1.50$) is has the shape of a quarter circle of radius $R = 12.0$ cm as shown in the side view of Figure P22.61. A light ray traveling in air parallel to the base of the Lucite is incident at a distance $h = 6.00$ cm above the base and emerges out of the Lucite at an angle θ with the horizontal. Determine the value of θ.

Figure P22.61

Funhouse mirrors distort images because the curved surfaces essentially change the angle of incidence of incoming rays, change that differs depending on the mirror's shape in a given location. In every case, however, the angle of reflection equals the angle of incidence.

Mirrors and Lenses 23

The development of the technology of mirrors and lenses led to a revolution in the progress of science. These devices, relatively simple to construct from cheap materials, led to microscopes and telescopes, extending human sight and opening up new pathways to knowledge, from microbes to distant planets.

This chapter covers the formation of images when plane and spherical light waves fall on plane and spherical surfaces. Images can be formed by reflection from mirrors or by refraction through lenses. In our study of mirrors and lenses, we continue to assume light travels in straight lines (the ray approximation), ignoring diffraction.

- 23.1 Flat Mirrors
- 23.2 Images Formed by Concave Mirrors
- 23.3 Convex Mirrors and Sign Conventions
- 23.4 Images Formed by Refraction
- 23.5 Atmospheric Refraction
- 23.6 Thin Lenses
- 23.7 Lens and Mirror Aberrations

23.1 Flat Mirrors

LEARNING OBJECTIVE

1. Discuss and apply the properties of flat mirrors.

We begin by examining the flat mirror. Consider a point source of light placed at O in Figure 23.1, a distance p in front of a flat mirror. The distance p is called the **object distance**. Light rays leave the source and are reflected from the mirror. After reflection, the rays diverge (spread apart), but they appear to the viewer to come from a point I behind the mirror. Point I is called the **image** of the object at O. Regardless of the system under study, **images are formed at the point where rays of light actually intersect or where they appear to originate**. Because the rays in the figure appear to originate at I, which is a distance q behind the mirror, that is the location of the image. The distance q is called the **image distance**.

802 CHAPTER 23 | Mirrors and Lenses

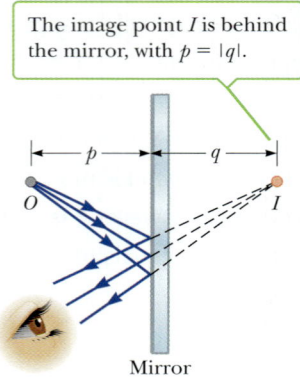

Figure 23.1 An image formed by reflection from a flat mirror. The image at point I is virtual. In Section 23.3, it will be shown that q must be taken as negative for virtual images: the object distance p, therefore, equals the absolute value of the image distance q.

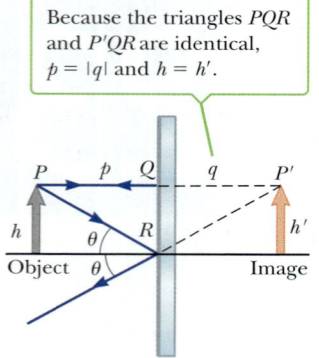

Figure 23.2 A geometric construction to locate the image of an object placed in front of a flat mirror.

Images are classified as real or virtual. In the formation of a *real image*, light actually passes through the image point. For a *virtual image*, light doesn't pass through the image point, but appears to come (diverge) from there. The image formed by the flat mirror in Figure 23.1 is a virtual image. In fact, the images seen in flat mirrors are always virtual (for real objects). Real images can be displayed on a screen (as at a movie), but virtual images cannot.

We examine some of the properties of the images formed by flat mirrors by using the simple geometric techniques. To find out where an image is formed, it's necessary to follow at least two rays of light as they reflect from the mirror as in Figure 23.2. One of those rays starts at P, follows the horizontal path PQ to the mirror, and reflects back on itself. The second ray follows the oblique path PR and reflects as shown. An observer to the left of the mirror would trace the two reflected rays back to the point from which they appear to have originated: point P'. A continuation of this process for points other than P on the object would result in a virtual image (drawn as a yellow arrow) to the right of the mirror. Because triangles PQR and $P'QR$ are identical, $PQ = P'Q$. Hence, we conclude that **the image formed by an object placed in front of a flat mirror is as far behind the mirror as the object is in front of the mirror**. Geometry also shows that the object height h equals the image height h'. The **lateral magnification** M is defined as

$$M \equiv \frac{\text{image height}}{\text{object height}} = \frac{h'}{h} \qquad [23.1]$$

Equation 23.1 is a general definition of the lateral magnification of any type of mirror. For a flat mirror, $M = 1$ because $h' = h$.

In summary, the image formed by a flat mirror has the following properties:

1. The image is as far behind the mirror as the object is in front.
2. The image is unmagnified, virtual, and upright. (By *upright*, we mean that if the object arrow points upward, as in Figure 23.2, so does the image arrow. The opposite of an upright image is an inverted image.)

Finally, note that a flat mirror produces an image having an *apparent* left–right reversal. You can see this reversal standing in front of a mirror and raising your right hand. Your image in the mirror raises the left hand. Likewise, your hair appears to be parted on the opposite side, and a mole on your right cheek appears to be on your image's left cheek.

Quick Quiz

23.1 In the overhead view of Figure 23.3, the image of the stone seen by observer 1 is at C. Where does observer 2 see the image: at A, at B, at C, at D, at E, or not at all?

Figure 23.3 (Quick Quiz 23.1)

Tip 23.1 Magnification ≠ Enlargement

Note that the word *magnification*, as used in optics, doesn't always mean *enlargement* because the image could be smaller than the object.

EXAMPLE 23.1 "Mirror, Mirror, on the Wall"

GOAL Apply the properties of a flat mirror.

PROBLEM A man 1.80 m tall stands in front of a mirror and sees his full height, no more and no less. If his eyes are 0.14 m from the top of his head, what is the minimum height of the mirror?

STRATEGY Figure 23.4 shows two rays of light, one from the man's feet and the other from the top of his head, reflecting off the mirror and entering his eye. The ray from his feet just strikes the bottom of the mirror, so if the mirror were longer, it would be too long, and if shorter, the ray would not be reflected. The angle of incidence and the angle of reflection are equal, labeled θ. This means the two triangles, ABD and DBC, are identical because they are right triangles with a common side (DB) and two identical angles θ. Use this key fact and the small isosceles triangle FEC to solve the problem.

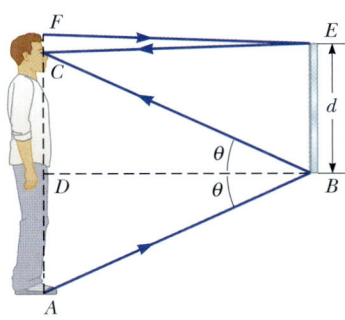

Figure 23.4 (Example 23.1)

SOLUTION

We need to find BE, which equals d. Relate this length to lengths on the man's body:

(1) $\quad BE = DC + \frac{1}{2}CF$

We need the lengths DC and CF. Set the sum of sides opposite the identical angles θ equal to AC:

(2) $\quad AD + DC = AC = (1.80 - 0.14) = 1.66 \text{ m}$

$AD = DC$, so substitute into Equation (2) and solve for DC:

$AD + DC = 2DC = 1.66 \text{ m} \quad \rightarrow \quad DC = 0.83 \text{ m}$

CF is given as 0.14 m. Substitute this value and DC into Equation (1):

$BE = d = DC + \frac{1}{2}CF = 0.83 \text{ m} + \frac{1}{2}(0.14 \text{ m}) = \boxed{0.90 \text{ m}}$

REMARKS The mirror must be exactly equal to half the height of the man for him to see only his full height and nothing more or less. Notice that the answer doesn't depend on his distance from the mirror.

QUESTION 23.1 Would a taller man be able to see his full height in the same mirror?

EXERCISE 23.1 How large should the mirror be if he wants to see only the upper third of his full height?

ANSWER 0.30 m

Most rearview mirrors in cars have a day setting and a night setting. The night setting greatly diminishes the intensity of the image so that lights from trailing cars will not blind the driver. To understand how such a mirror works, consider Figure 23.5. The mirror is a wedge of glass with a reflecting metallic coating on the back side. When the mirror is in the day setting, as in Figure 23.5a, light from an object behind the car strikes the mirror at point 1. Most of the light enters the

APPLICATION
Day and Night Settings for Rearview Mirrors

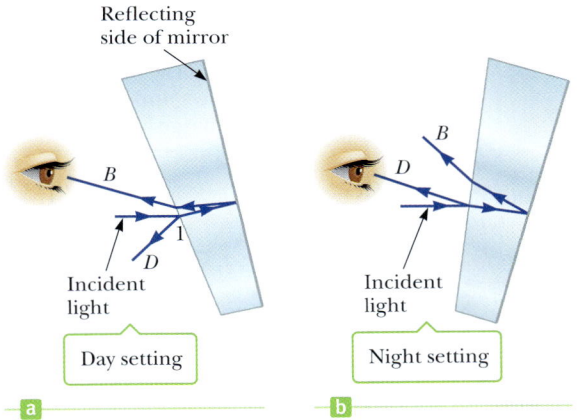

Figure 23.5 Cross-sectional views of a rearview mirror. (a) With the day setting, the silvered back surface of the mirror reflects a bright ray B into the driver's eyes. (b) With the night setting, the glass of the unsilvered front surface of the mirror reflects a dim ray D into the driver's eyes.

wedge, is refracted, and reflects from the back of the mirror to return to the front surface, where it is refracted again as it reenters the air as ray *B* (for *bright*). In addition, a small portion of the light is reflected at the front surface, as indicated by ray *D* (for *dim*). This dim reflected light is responsible for the image observed when the mirror is in the night setting, as in Figure 23.5b. Now the wedge is rotated so that the path followed by the bright light (ray *B*) doesn't lead to the eye. Instead, the dim light reflected from the front surface travels to the eye, and the brightness of trailing headlights doesn't become a hazard.

APPLYING PHYSICS 23.1 | Illusionist's Trick

The professor in the box shown in Figure 23.6 appears to be balancing himself on a few fingers with both of his feet elevated from the floor. He can maintain this position for a long time, and appears to defy gravity. How do you suppose this illusion was created?

EXPLANATION This trick is an example of an optical illusion, used by magicians, that makes use of a mirror. The box the professor is standing in is a cubical open frame that contains a flat, vertical mirror through a diagonal plane. The professor straddles the mirror so that one leg is in front of the mirror and the other leg is behind it, out of view. When he raises his front leg, that leg's reflection rises also, making it appear both his feet are off the ground, creating the illusion that he's floating in the air. In fact, he supports himself with the leg behind the mirror, which remains in contact with the ground. ∎

Figure 23.6 (Applying Physics 23.1)

23.2 Images Formed by Concave Mirrors

LEARNING OBJECTIVES

1. Use ray diagrams to understand relationships between the radius, focal length, object position, and image position for concave mirrors.
2. Define the focal length.
3. State the mirror equation.

A **spherical mirror**, as its name implies, has the shape of a segment of a sphere. Figure 23.7 shows a spherical mirror with a silvered inner, concave surface; this type of mirror is called a **concave mirror**. The mirror has radius of curvature R,

Figure 23.7 (a) A concave mirror of radius R. The center of curvature, C, is located on the principal axis. (b) A point object placed at O in front of a concave spherical mirror of radius R, where O is any point on the principal axis farther than R from the surface of the mirror, forms a real image at I.

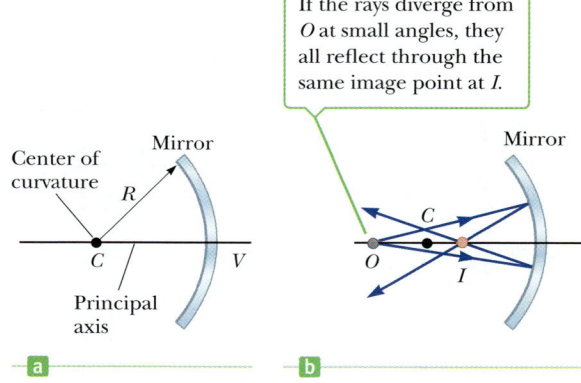

and its center of curvature is at point C. Point V is the center of the spherical segment, and a line drawn from C to V is called the **principal axis** of the mirror.

Now consider a point source of light placed at point O in Figure 23.7b, on the principal axis and outside point C. Several diverging rays originating at O are shown. After reflecting from the mirror, these rays converge to meet at I, called the **image point**. The rays then continue and diverge from I as if there were an object there. As a result, a real image is formed. **Whenever reflected light actually passes through a point, the image formed there is real.**

We often assume all rays that diverge from the object make small angles with the principal axis. All such rays reflect through the image point, as in Figure 23.7b. Rays that make a large angle with the principal axis, as in Figure 23.8, converge to other points on the principal axis, producing a blurred image. This effect, called **spherical aberration**, is present to some extent with any spherical mirror and will be discussed in Section 23.7.

We can use the geometry shown in Figure 23.9 to calculate the image distance q from the object distance p and radius of curvature R. By convention, these distances are measured from point V. The figure shows two rays of light leaving the tip of the object. One ray passes through the center of curvature, C, of the mirror, hitting the mirror head-on (perpendicular to the mirror surface) and reflecting back on itself. The second ray strikes the mirror at point V and reflects as shown, obeying the law of reflection. The image of the tip of the arrow is at the point where the two rays intersect. From the largest triangle in Figure 23.9, we see that $\tan \theta = h/p$; the light-blue triangle gives $\tan \theta = -h'/q$. The negative sign has been introduced to satisfy our convention that h' is negative when the image is inverted with respect to the object, as it is here. From Equation 23.1 and these results, we find that the magnification of the mirror is

$$M = \frac{h'}{h} = -\frac{q}{p} \qquad [23.2]$$

From two other triangles in the figure, we get

$$\tan \alpha = \frac{h}{p - R} \quad \text{and} \quad \tan \alpha = -\frac{h'}{R - q}$$

from which we find that

$$\frac{h'}{h} = -\frac{R - q}{p - R} \qquad [23.3]$$

If we compare Equation 23.2 with Equation 23.3, we see that

$$\frac{R - q}{p - R} = \frac{q}{p}$$

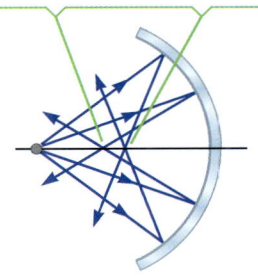

Figure 23.8 A spherical concave mirror exhibits *spherical aberration* when light rays make large angles with the principal axis.

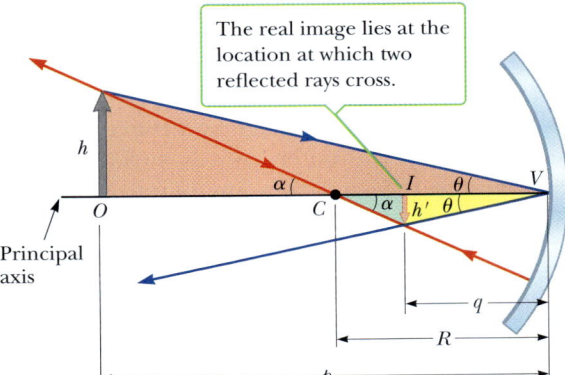

Figure 23.9 The image formed by a spherical concave mirror, where the object at O lies outside the center of curvature, C.

Figure 23.10 (a) Light rays from a distant object ($p = \infty$) reflect from a concave mirror through the focal point F. (b) A photograph of the reflection of parallel rays from a concave mirror.

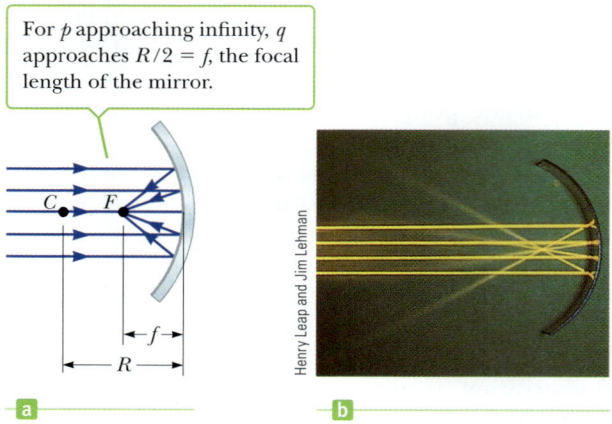

For p approaching infinity, q approaches $R/2 = f$, the focal length of the mirror.

Simple algebra reduces this equation to

Mirror equation ▶

$$\frac{1}{p} + \frac{1}{q} = \frac{2}{R}$$ [23.4]

This expression is called the **mirror equation**.

If the object is very far from the mirror—if the object distance p is great enough compared with R that p can be said to approach infinity—then $1/p \approx 0$, and we see from Equation 23.4 that $q \approx R/2$. In other words, when the object is very far from the mirror, **the image point is halfway between the center of curvature and the center of the mirror**, as in Figure 23.10a. The incoming rays are essentially parallel in that figure because the source is assumed to be very far from the mirror. In this special case we call the image point the **focal point** F and the image distance the **focal length** f, where

Focal length ▶

$$f = \frac{R}{2}$$ [23.5]

The mirror equation can therefore be expressed in terms of the focal length:

$$\frac{1}{p} + \frac{1}{q} = \frac{1}{f}$$ [23.6]

Tip 23.2 Focal Point ≠ Focus Point
The focal point is *not* the point at which light rays focus to form an image. The focal point of a mirror is determined *solely* by its curvature; it doesn't depend on the location of any object.

Note that rays from objects at infinity are always focused at the focal point.

23.3 Convex Mirrors and Sign Conventions

LEARNING OBJECTIVES

1. Understand the optical properties of convex mirrors.
2. Master the sign conventions for concave and convex mirrors.
3. Analyze concave and convex mirrors, finding image positions, magnifications, and stating whether images are upright or inverted, real or virtual.

Figure 23.11 shows the formation of an image by a **convex mirror**, which is silvered so that light is reflected from the outer, convex surface. It is sometimes called a **diverging mirror** because the rays from any point on the object diverge after reflection, as though they were coming from some point behind the mirror. The image in Figure 23.11 is virtual rather than real because it lies behind the mirror at the point the reflected rays appear to originate. In general, the image formed by a convex mirror is upright, virtual, and smaller than the object.

We won't derive any equations for convex spherical mirrors. If we did, we would find that the equations developed for concave mirrors can be used with convex

23.3 | Convex Mirrors and Sign Conventions

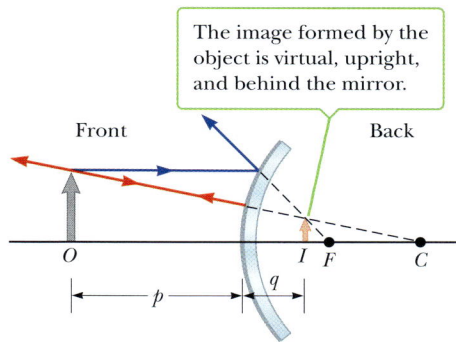

Figure 23.11 Formation of an image by a spherical, convex mirror.

mirrors if particular sign conventions are used. We call the region in which light rays move the *front side* of the mirror, and the other side, where virtual images are formed, the *back side*. For example, in Figures 23.9 and 23.11, the side to the left of the mirror is the front side and the side to the right is the back side. Figure 23.12 is helpful for understanding the rules for object and image distances, and Table 23.1 summarizes the sign conventions for all the necessary quantities. Notice that when the quantities p, q, and f (and R) are located where the light is—in front of the mirror—they are positive, whereas when they are located behind the mirror (where the light isn't), they are negative.

> **Tip 23.3 Positive Is Where the Light Is**
>
> The quantities p, q, and f are all positive when they are located where the light is—in front of the mirror as indicated in Figure 23.12.

Ray Diagrams for Mirrors

We can conveniently determine the positions and sizes of images formed by mirrors by constructing *ray diagrams* similar to the ones we have been using. This kind of graphical construction tells us the overall nature of the image and can be used to check parameters calculated from the mirror and magnification equations. Making a ray diagram requires knowing the position of the object and the location of the center of curvature. To locate the image, three rays are constructed (rather than only the two we have been constructing so far), as shown by the examples in Figure 23.13 (page 808). All three rays start from the same object point; for these examples, the tip of the arrow was chosen. For the concave mirrors in Figures 23.13a and 23.13b, the rays are drawn as follows:

1. Ray 1 is drawn parallel to the principal axis and is reflected back through the focal point F.
2. Ray 2 is drawn through the focal point and is reflected parallel to the principal axis.
3. Ray 3 is drawn through the center of curvature, C, and is reflected back on itself.

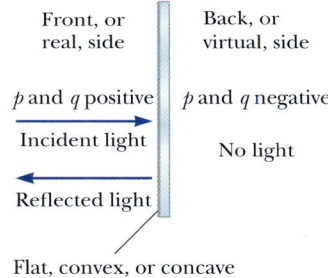

Figure 23.12 A diagram describing the signs of p and q for convex and concave mirrors.

Note that rays actually go in all directions from the object; we choose to follow those moving in a direction that simplifies our drawing.

The intersection of any *two* of these rays at a point locates the image. The third ray serves as a check of our construction. The image point obtained in

Table 23.1 Sign Conventions for Mirrors

Quantity	Symbol	In Front	In Back	Upright Image	Inverted Image
Object location	p	+	−		
Image location	q	+	−		
Focal length	f	+	−		
Image height	h'			+	−
Magnification	M			+	−

808 **CHAPTER 23** | Mirrors and Lenses

Figure 23.13 Ray diagrams for spherical mirrors and corresponding photographs of the images of bottles.

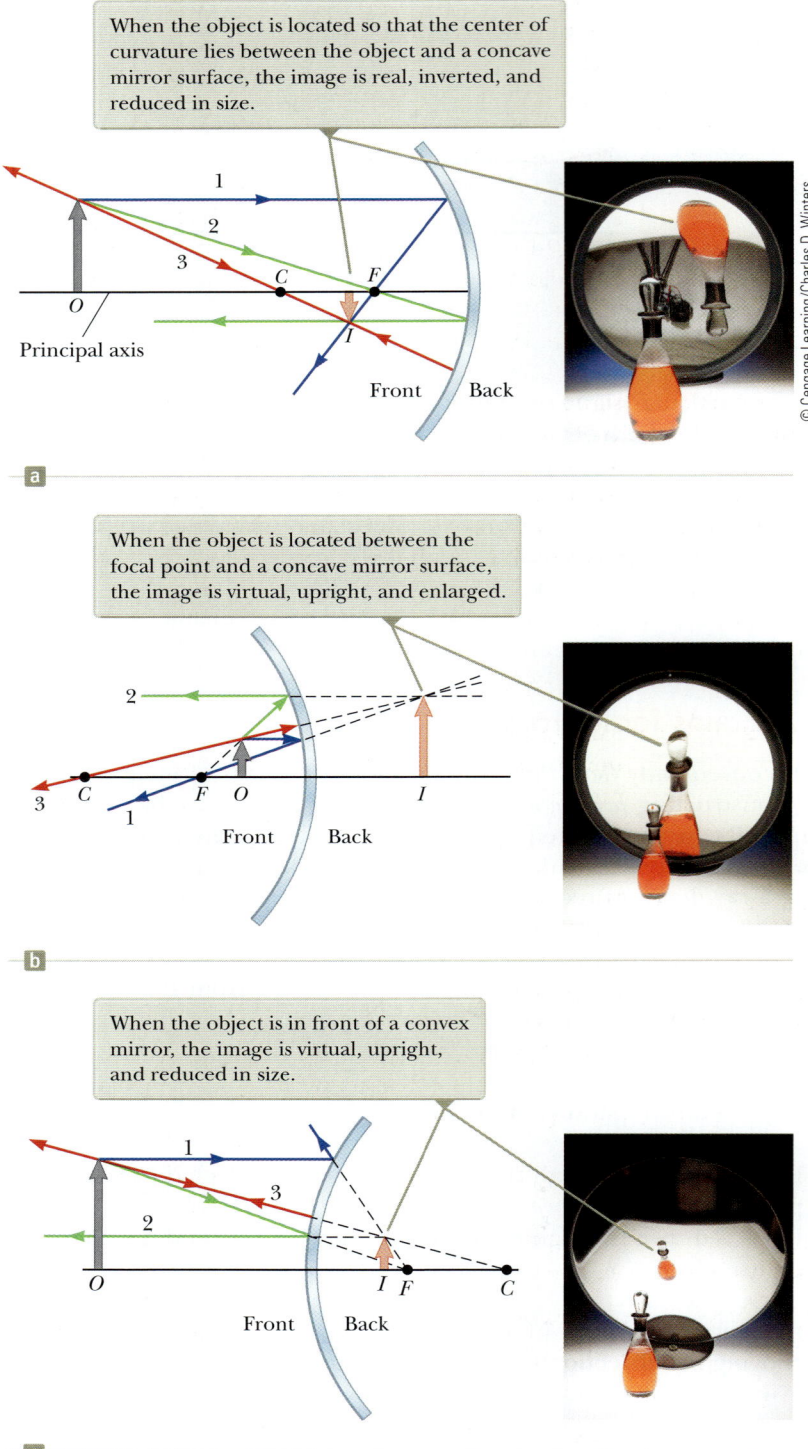

this fashion must always agree with the value of q calculated from the mirror formula.

In the case of a concave mirror, note what happens as the object is moved closer to the mirror. The real, inverted image in Figure 23.13a moves to the left as the object approaches the focal point. When the object is at the focal point, the image is infinitely far to the left. When the object lies between the focal point and the mirror surface, as in Figure 23.13b, however, the image is virtual and upright.

With the convex mirror shown in Figure 23.13c, the image of a real object is always virtual and upright. As the object distance increases, the virtual image shrinks and approaches the focal point as p approaches infinity. You should construct a ray diagram to verify these statements.

The image-forming characteristics of curved mirrors obviously determine their uses. For example, suppose you want to design a mirror that will help people shave or apply cosmetics. For this, you need a concave mirror that puts the user inside the focal point, such as the mirror in Figure 23.13b. With that mirror, the image is upright and greatly enlarged. In contrast, suppose the primary purpose of a mirror is to observe a large field of view. In that case you need a convex mirror such as the one in Figure 23.13c. The diminished size of the image means that a fairly large field of view is seen in the mirror. Mirrors like this one are often placed in stores to help employees watch for shoplifters. A second use of such a mirror is as a side-view mirror on a car (Fig. 23.14). This kind of mirror is usually placed on the passenger side of the car and carries the warning "Objects are closer than they appear." Without such warning, a driver might think she is looking into a flat mirror, which doesn't alter the size of the image. She could be fooled into believing that a truck is far away because it looks small, when it's actually a large semi very close behind her, but diminished in size because of the image formation characteristics of the convex mirror.

Figure 23.14 A convex side-view mirror on a vehicle produces an upright image that is smaller than the object. The smaller image means that the object is closer than its apparent distance as observed in the mirror.

APPLYING PHYSICS 23.2 | Concave Versus Convex

A virtual image can be anywhere behind a concave mirror. Why is there a maximum distance at which the image can exist behind a *convex* mirror?

EXPLANATION Consider the concave mirror first and imagine two different light rays leaving a tiny object and striking the mirror. If the object is at the focal point, the light rays reflecting from the mirror will be parallel to the mirror axis. They can be interpreted as forming a virtual image infinitely far away behind the mirror. As the object is brought closer to the mirror, the reflected rays will diverge through larger and larger angles, resulting in their extensions converging closer and closer to the back of the mirror. When the object is brought right up to the mirror, the image is right behind the mirror. When the object is much closer to the mirror than the focal length, the mirrors acts like a flat mirror and the image is just as far behind the mirror as the object is in front of it. The image can therefore be anywhere from infinitely far away to right at the surface of the mirror. For the convex mirror, an object at infinity produces a virtual image at the focal point. As the object is brought closer, the reflected rays diverge more sharply and the image moves closer to the mirror. As a result, the virtual image is restricted to the region between the mirror and the focal point. ■

APPLYING PHYSICS 23.3 | Reversible Waves

Large trucks often have a sign on the back saying, "If you can't see my mirror, I can't see you." Explain this sign.

EXPLANATION The trucking companies are making use of the principle of the reversibility of light rays. For an image of you to be formed in the driver's mirror, there must be a pathway for rays of light to reach the mirror, allowing the driver to see your image. If you can't see the mirror, this pathway doesn't exist. ■

EXAMPLE 23.2 | Images Formed by a Concave Mirror

GOAL Calculate properties of a concave mirror.

PROBLEM Assume a certain concave, spherical mirror has a focal length of 10.0 cm. **(a)** Locate the image and find the magnification for an object distance of 25.0 cm. Determine whether the image is real or virtual, inverted or upright, and larger or smaller. Do the same for object distances of **(b)** 10.0 cm and **(c)** 5.00 cm.

STRATEGY For each part, substitute into the mirror and magnification equations. Part (b) involves a limiting process because the answers are infinite. Notice that when the magnification M is positive, the image is upright, and when M is negative, the image is inverted. Similarly, when q is positive the image is real, and when q is negative the image is virtual.

(Continued)

SOLUTION

(a) Find the image position for an object distance of 25.0 cm. Calculate the magnification and describe the image.

Use the mirror equation to find the image distance:

$$\frac{1}{p} + \frac{1}{q} = \frac{1}{f}$$

Substitute and solve for q. According to Table 23.1, p and f are positive.

$$\frac{1}{25.0 \text{ cm}} + \frac{1}{q} = \frac{1}{10.0 \text{ cm}}$$

$$q = \boxed{16.7 \text{ cm}}$$

Because q is positive, the image is in front of the mirror and is real. The magnification is given by substituting into Equation 23.2:

$$M = -\frac{q}{p} = -\frac{16.7 \text{ cm}}{25.0 \text{ cm}} = \boxed{-0.668}$$

The image is smaller than the object because $|M| < 1$, and it is inverted because M is negative. (See Fig. 23.13a.)

(b) Locate the image when the object distance is 10.0 cm. Calculate the magnification and describe the image.

The object is at the focal point. Substitute $p = 10.0$ cm and $f = 10.0$ cm into the mirror equation:

$$\frac{1}{10.0 \text{ cm}} + \frac{1}{q} = \frac{1}{10.0 \text{ cm}}$$

$$\frac{1}{q} = 0 \quad \rightarrow \quad \boxed{q = \infty}$$

Because $M = -q/p$, the magnification is also infinite.

(c) Locate the image when the object distance is 5.00 cm. Calculate the magnification and describe the image.

Once again, substitute into the mirror equation:

$$\frac{1}{5.00 \text{ cm}} + \frac{1}{q} = \frac{1}{10.0 \text{ cm}}$$

$$\frac{1}{q} = \frac{1}{10.0 \text{ cm}} - \frac{1}{5.00 \text{ cm}} = -\frac{1}{10.0 \text{ cm}}$$

$$q = \boxed{-10.0 \text{ cm}}$$

The image is virtual (behind the mirror) because q is negative. Use Equation 23.2 to calculate the magnification:

$$M = -\frac{q}{p} = -\left(\frac{-10.0 \text{ cm}}{5.00 \text{ cm}}\right) = \boxed{2.00}$$

The image is larger (magnified by a factor of 2) because $|M| > 1$, and upright because M is positive. (See Fig. 23.13b.)

REMARKS Note the characteristics of an image formed by a concave, spherical mirror. When the object is outside the focal point, the image is inverted and real; at the focal point, the image is formed at infinity; inside the focal point, the image is upright and virtual.

QUESTION 23.2 What location does the image approach as the object gets arbitrarily far away from the mirror? (a) infinity (b) the focal point (c) the radius of curvature of the mirror (d) the mirror itself

EXERCISE 23.2 If the object distance is 20.0 cm, find the image distance and the magnification of the mirror.

ANSWER $q = 20.0$ cm, $M = -1.00$

EXAMPLE 23.3 Images Formed by a Convex Mirror

GOAL Calculate properties of a convex mirror.

PROBLEM An object 3.00 cm high is placed 20.0 cm from a convex mirror with a focal length of magnitude 8.00 cm. Find **(a)** the position of the image, **(b)** the magnification of the mirror, and **(c)** the height of the image.

STRATEGY This problem again requires only substitution into the mirror and magnification equations. Multiplying the object height by the magnification gives the image height.

SOLUTION

(a) Find the position of the image.

Because the mirror is convex, its focal length is negative. Substitute into the mirror equation:

$$\frac{1}{p} - \frac{1}{q} = \frac{1}{f}$$

$$\frac{1}{20.0 \text{ cm}} - \frac{1}{q} = \frac{1}{-8.00 \text{ cm}}$$

Solve for q:

$$q = \boxed{-5.71 \text{ cm}}$$

(b) Find the magnification of the mirror.

Substitute into Equation 23.2:

$$M = -\frac{q}{p} = -\left(\frac{-5.71 \text{ cm}}{20.0 \text{ cm}}\right) = \boxed{0.286}$$

(c) Find the height of the image.

Multiply the object height by the magnification:

$$h' = hM = (3.00 \text{ cm})(0.286) = \boxed{0.858 \text{ cm}}$$

REMARKS The negative value of q indicates the image is virtual, or behind the mirror, as in Figure 23.13c. The image is upright because M is positive.

QUESTION 23.3 True or False: A convex mirror can produce only virtual images.

EXERCISE 23.3 Suppose the object is moved so that it is 4.00 cm from the same mirror. Repeat parts (a) through (c).

ANSWERS (a) -2.67 cm (b) 0.668 (c) 2.00 cm; the image is upright and virtual.

EXAMPLE 23.4 The Face in the Mirror

GOAL Find a focal length from a magnification and an object distance.

PROBLEM When a woman stands with her face 40.0 cm from a cosmetic mirror, the upright image is twice as tall as her face. What is the focal length of the mirror?

STRATEGY To find f in this example, we must first find q, the image distance. Because the problem states that the image is upright, the magnification must be positive (in this case, $M = +2$), and because $M = -q/p$, we can determine q.

SOLUTION

Obtain q from the magnification equation:

$$M = -\frac{q}{p} = 2$$

$$q = -2p = -2(40.0 \text{ cm}) = -80.0 \text{ cm}$$

Because q is negative, the image is on the opposite side of the mirror and hence is virtual. Substitute q and p into the mirror equation and solve for f:

$$\frac{1}{40.0 \text{ cm}} - \frac{1}{80.0 \text{ cm}} = \frac{1}{f}$$

$$f = \boxed{80.0 \text{ cm}}$$

REMARKS The positive sign for the focal length tells us that the mirror is concave, a fact we already knew because the mirror magnified the object. (A convex mirror would have produced a smaller image.)

(Continued)

QUESTION 23.4 If she moves the mirror closer to her face, what happens to the image? (a) It becomes inverted and smaller. (b) It remains upright and becomes smaller. (c) It becomes inverted and larger. (d) It remains upright and becomes larger.

EXERCISE 23.4 Suppose a fun-house spherical mirror makes you appear to be one-third your normal height. If you are 1.20 m away from the mirror, find its focal length. Is the mirror concave or convex?

ANSWERS -0.600 m, convex

23.4 Images Formed by Refraction

LEARNING OBJECTIVES

1. State the generalized equations relating object distance, image distance, and radius of curvature for images formed by transparent spherical surfaces.
2. Master the sign conventions for refracting surfaces.
3. Analyze the images formed by refracting surfaces.

In this section we describe how images are formed by refraction at a spherical surface. Consider two transparent media with indices of refraction n_1 and n_2, where the boundary between the two media is a spherical surface of radius R (Fig. 23.15). We assume the medium to the right has a higher index of refraction than the one to the left: $n_2 > n_1$. That would be the case for light entering a curved piece of glass from air or for light entering the water in a fishbowl from air. The rays originating at the object location O are refracted at the spherical surface and then converge to the image point I. We can begin with Snell's law of refraction and use simple geometric techniques to show that the object distance, image distance, and radius of curvature are related by the equation

$$\frac{n_1}{p} + \frac{n_2}{q} = \frac{n_2 - n_1}{R} \qquad [23.7]$$

Further, the magnification of a refracting surface is

$$M = \frac{h'}{h} = -\frac{n_1 q}{n_2 p} \qquad [23.8]$$

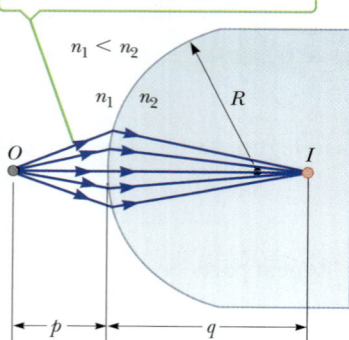

Rays making small angles with the principal axis diverge from a point object at O and pass through the image point I.

Figure 23.15 An image formed by refraction at a spherical surface.

As with mirrors, certain sign conventions hold, depending on circumstances. First note that real images are formed by refraction on the side of the surface *opposite* the side from which the light comes, in contrast to mirrors, where real images are formed on the *same* side of the reflecting surface. This makes sense because light reflects off mirrors, so any real images must form on the same side the light comes from. With a transparent medium, the rays pass through and naturally form real images on the opposite side. We define the side of the surface where light rays originate as the front side. The other side is called the back side. Because of the difference in location of real images, the refraction sign conventions for q and R are the opposite of those for reflection. For example, p, q, and R are all positive in Figure 23.15. The sign conventions for spherical refracting surfaces are summarized in Table 23.2.

Table 23.2 Sign Conventions for Refracting Surfaces

Quantity	Symbol	In Front	In Back	Upright Image	Inverted Image
Object location	p	+	−		
Image location	q	−	+		
Radius	R	−	+		
Image height	h'			+	−

■ APPLYING PHYSICS 23.4 | Underwater Vision BIO

Why does a person with normal vision see a blurry image if the eyes are opened underwater with no goggles or diving mask in use?

EXPLANATION The eye presents a spherical refraction surface. The eye normally functions so that light entering from the air is refracted to form an image in the retina located at the back of the eyeball. The difference in the index of refraction between water and the eye is smaller than the difference in the index of refraction between air and the eye. Consequently, light entering the eye from the water doesn't undergo as much refraction as does light entering from the air, and the image is formed behind the retina. A diving mask or swimming goggles have no optical action of their own; they are simply flat pieces of glass or plastic in a rubber mount. They do, however, provide a region of air adjacent to the eyes so that the correct refraction relationship is established and images will be in focus. ■

Flat Refracting Surfaces

If the refracting surface is flat, then R approaches infinity and Equation 23.7 reduces to

$$\frac{n_1}{p} = -\frac{n_2}{q}$$

$$q = -\frac{n_2}{n_1} p \qquad [23.9]$$

From Equation 23.9, we see that the sign of q is opposite that of p. Consequently, **the image formed by a flat refracting surface is on the same side of the surface as the object.** This statement is illustrated in Figure 23.16 for the situation in which n_1 is greater than n_2, where a virtual image is formed between the object and the surface. Note that the refracted ray bends *away* from the normal in this case because $n_1 > n_2$.

Figure 23.16 The image formed by a flat refracting surface.

■ Quick Quiz

23.2 A person spearfishing from a boat sees a fish located 3 m from the boat at an apparent depth of 1 m. To spear the fish, should the person aim (a) at, (b) above, or (c) below the image of the fish?

23.3 True or False: (a) The image of an object placed in front of a concave mirror is always upright. (b) The height of the image of an object placed in front of a concave mirror must be smaller than or equal to the height of the object. (c) The image of an object placed in front of a convex mirror is always upright and smaller than the object.

■ EXAMPLE 23.5 | Gaze into the Crystal Ball

GOAL Calculate the properties of an image created by a spherical lens.

PROBLEM A coin 2.00 cm in diameter is embedded in a solid glass ball of radius 30.0 cm (Fig. 23.17). The index of refraction of the ball is 1.50, and the coin is 20.0 cm from the surface. Find the position of the image of the coin and the height of the coin's image.

Figure 23.17 (Example 23.5) A coin embedded in a glass ball forms a virtual image between the coin and the surface of the glass.

STRATEGY Because the rays are moving from a medium of high index of refraction (the glass ball) to a medium of lower index of refraction (air), the rays originating at the coin are refracted away from the normal at the surface and diverge outward. The image is formed in the glass and is virtual. Substitute into Equations 23.7 and 23.8 for the image position and magnification, respectively.

(Continued)

SOLUTION

Apply Equation 23.7 and take $n_1 = 1.50$, $n_2 = 1.00$, $p = 20.0$ cm, and $R = -30.0$ cm:

$$\frac{n_1}{p} + \frac{n_2}{q} = \frac{n_2 - n_1}{R}$$

$$\frac{1.50}{20.0 \text{ cm}} + \frac{1.00}{q} = \frac{1.00 - 1.50}{-30.0 \text{ cm}}$$

Solve for q:

$$q = \boxed{-17.1 \text{ cm}}$$

To find the image height, use Equation 23.8 for the magnification:

$$M = -\frac{n_1 q}{n_2 p} = -\frac{1.50(-17.1 \text{ cm})}{1.00(20.0 \text{ cm})} = \frac{h'}{h}$$

$$h' = 1.28h = (1.28)(2.00 \text{ cm}) = \boxed{2.56 \text{ cm}}$$

REMARKS The negative sign on q indicates that the image is in the same medium as the object (the side of incident light), in agreement with our ray diagram, and therefore must be virtual. The positive value for M means that the image is upright.

QUESTION 23.5 How would the final answer be affected if the ball and observer were immersed in water? (a) It would be smaller. (b) It would be larger. (c) There would be no change.

EXERCISE 23.5 A coin is embedded 20.0 cm from the surface of a similar ball of transparent substance having radius 30.0 cm and unknown composition. If the coin's image is virtual and located 15.0 cm from the surface, find the (a) index of refraction of the substance and (b) magnification.

ANSWERS (a) 2.00 (b) 1.50

■ EXAMPLE 23.6 | The One That Got Away

GOAL Calculate the properties of an image created by a flat refractive surface.

PROBLEM A small fish is swimming at a depth d below the surface of a pond (Fig. 23.18). **(a)** What is the *apparent depth* of the fish as viewed from directly overhead? **(b)** If the fish is 12 cm long, how long is its image?

STRATEGY In this example the refracting surface is flat, so R is infinite. Hence, we can use Equation 23.9 to determine the location of the image, which is the apparent location of the fish.

Figure 23.18 (Example 23.6) The apparent depth q of the fish is less than the true depth d.

SOLUTION

(a) Find the apparent depth of the fish.

Substitute $n_1 = 1.33$ for water and $p = d$ into Equation 23.9:

$$q = -\frac{n_2}{n_1}p = -\frac{1}{1.33}d = \boxed{-0.752d}$$

(b) What is the size of the fish's image?

Use Equation 23.9 to eliminate q from Equation 23.8, the magnification equation:

$$M = \frac{h'}{h} = -\frac{n_1 q}{n_2 p} = -\frac{n_1\left(-\frac{n_2}{n_1}p\right)}{n_2 p} = 1$$

$$h' = h = \boxed{12 \text{ cm}}$$

REMARKS Again, because q is negative, the image is virtual, as indicated in Figure 23.18. The apparent depth is approximately three-fourths the actual depth. For instance, if $d = 4.0$ m, then $q = -3.0$ m.

QUESTION 23.6 Suppose a similar experiment is carried out with an object immersed in oil ($n = 1.5$) the same distance below the surface. How does the apparent depth of the object compare with its apparent depth when immersed in water? (a) The apparent depth is unchanged. (b) The apparent depth is larger. (c) The apparent depth is smaller.

EXERCISE 23.6 A spear fisherman estimates that a trout is 1.5 m below the water's surface. What is the actual depth of the fish?

ANSWER 2.0 m

23.5 Atmospheric Refraction

LEARNING OBJECTIVE

1. Discuss images formed by atmospheric refraction, such as mirages.

Images formed by refraction in our atmosphere lead to some interesting phenomena. One such phenomenon that occurs daily is the visibility of the Sun at dusk even though it has passed below the horizon. Figure 23.19 shows why it occurs. Rays of light from the Sun strike Earth's atmosphere (represented by the shaded area around the planet) and are bent as they pass into a medium that has an index of refraction different from that of the almost empty space in which they have been traveling. The bending in this situation differs somewhat from the bending we have considered previously in that it is gradual and continuous as the light moves through the atmosphere toward an observer at point O. This is because the light moves through layers of air that have a continuously changing index of refraction. When the rays reach the observer, the eye follows them back along the direction from which they appear to have come (indicated by the dashed path in the figure). The end result is that the Sun appears to be above the horizon even after it has fallen below it.

Figure 23.19 Because light is refracted by Earth's atmosphere, an observer at O sees the Sun even though it has fallen below the horizon.

The **mirage** is another phenomenon of nature produced by refraction in the atmosphere. A mirage can be observed when the ground is so hot that the air directly above it is warmer than the air at higher elevations. The desert is a region in which such circumstances prevail, but mirages are also seen on heated roadways during the summer. The layers of air at different heights above Earth have different densities and different refractive indices. The effect these differences can have is pictured in Figure 23.20a. The observer sees the sky and a cactus in two different ways. One group of light rays reaches the observer by the straight-line path A, and the eye traces these rays back to see the cactus in the normal fashion. In addition, a second group of rays travels along the curved path B. These rays are directed toward the ground and are then bent as a result of refraction. As a consequence,

Figure 23.20 (a) A mirage is produced by the bending of light rays in the atmosphere when there are large temperature differences between the ground and the air. (b) Notice the reflection of the cars in this photograph of a mirage. The road looks like it's flooded with water, but it is actually dry.

816 CHAPTER 23 | Mirrors and Lenses

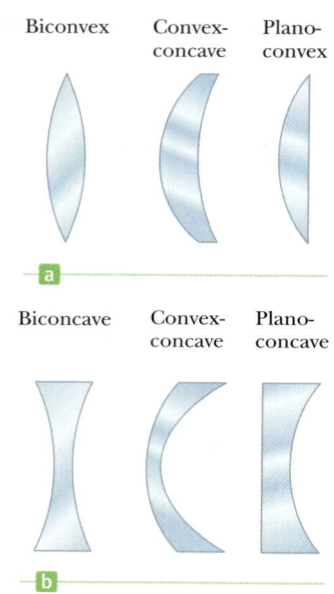

Figure 23.21 Various lens shapes. (a) Converging lenses have positive focal lengths and are thickest at the middle. (b) Diverging lenses have negative focal lengths and are thickest at the edges.

the observer also sees an inverted image of the cactus and the background of the sky as he traces the rays back to the point at which they appear to have originated. Because both an upright image and an inverted image are seen when the image of a cactus or other object is observed in a reflecting pool of water, the observer unconsciously calls on this past experience and concludes that the sky is reflected by a pool of water in front of the cactus.

23.6 Thin Lenses

LEARNING OBJECTIVES

1. Discuss converging and diverging thin lenses and their properties.
2. Master the sign conventions for thin lenses.
3. Apply the lens-maker's equation to calculating the physical properties of thin lenses.
4. Analyze systems involving more than one thin lens or having a lens and a mirror.

A typical **thin lens** consists of a piece of glass or plastic, ground so that each of its two refracting surfaces is a segment of either a sphere or a plane. Lenses are commonly used to form images by refraction in optical instruments, such as cameras, telescopes, and microscopes. The equation that relates object and image distances for a lens is virtually identical to the mirror equation derived earlier, and the method used to derive it is also similar.

Figure 23.21 shows some representative shapes of lenses. Notice that we have placed these lenses in two groups. Those in Figure 23.21a are thicker at the center than at the rim, and those in Figure 23.21b are thinner at the center than at the rim. The lenses in the first group are examples of **converging lenses**, and those in the second group are **diverging lenses**. The reason for these names will become apparent shortly.

As we did for mirrors, it is convenient to define a point called the **focal point** for a lens. For example, in Figure 23.22a, a group of rays parallel to the axis passes

Figure 23.22 (*Left*) Photographs of the effects of converging and diverging lenses on parallel rays. (*Right*) The focal points of the (a) biconvex lens and (b) biconcave lens.

23.6 | Thin Lenses

through the focal point F after being converged by the lens. The distance from the focal point to the lens is called the **focal length** f. **The focal length is the image distance that corresponds to an infinite object distance.** Recall that we are considering the lens to be very thin. As a result, it makes no difference whether we take the focal length to be the distance from the focal point to the surface of the lens or the distance from the focal point to the center of the lens because the difference between these two lengths is negligible. A thin lens has *two* focal points, as illustrated in Figure 23.22, one on each side of the lens. One focal point corresponds to parallel rays traveling from the left and the other corresponds to parallel rays traveling from the right.

Rays parallel to the axis diverge after passing through a lens of biconcave shape, shown in Figure 23.22b. In this case the focal point is defined to be the point where the diverged rays appear to originate, labeled F in the figure. Figures 23.22a and 23.22b indicate why the names *converging* and *diverging* are applied to these lenses.

Now consider a ray of light passing through the center of a lens. Such a ray is labeled ray 1 in Figure 23.23. For a thin lens, a ray passing through the center is undeflected. Ray 2 in the same figure is parallel to the principal axis of the lens (the horizontal axis passing through O), and as a result it passes through the focal point F after refraction. Rays 1 and 2 intersect at the point that is the tip of the image arrow.

We first note that the tangent of the angle α can be found by using the blue and gold shaded triangles in Figure 23.23:

$$\tan \alpha = \frac{h}{p} \quad \text{or} \quad \tan \alpha = -\frac{h'}{q}$$

From this result, we find that

$$M = \frac{h'}{h} = -\frac{q}{p} \quad [23.10]$$

The equation for magnification by a lens is the same as the equation for magnification by a mirror. We also note from Figure 23.23 that

$$\tan \theta = \frac{PQ}{f} \quad \text{or} \quad \tan \theta = -\frac{h'}{q - f}$$

The height PQ used in the first of these equations, however, is the same as h, the height of the object. Therefore,

$$\frac{h}{f} = -\frac{h'}{q - f}$$

$$\frac{h'}{h} = -\frac{q - f}{f}$$

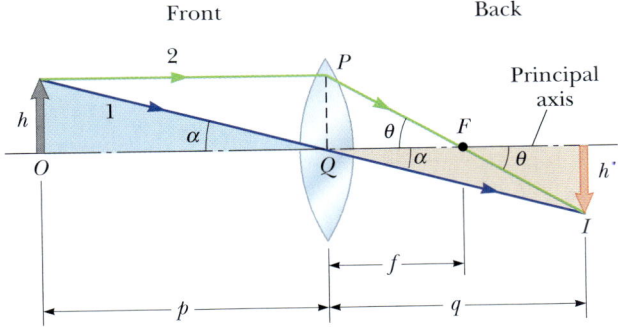

Figure 23.23 A geometric construction for developing the thin-lens equation.

Using the latter equation in combination with Equation 23.10 gives

$$\frac{q}{p} = \frac{q-f}{f}$$

which reduces to

Thin-lens equation ▶

$$\frac{1}{p} + \frac{1}{q} = \frac{1}{f} \qquad [23.11]$$

This equation, called the **thin-lens equation**, can be used with both converging and diverging lenses if we adhere to a set of sign conventions. Figure 23.24 is useful for obtaining the signs of p and q, and Table 23.3 gives the complete sign conventions for lenses. Note that **a converging lens has a positive focal length** under this convention and **a diverging lens has a negative focal length**. Hence, the names *positive* and *negative* are often given to these lenses.

The focal length for a lens in air is related to the curvatures of its front and back surfaces and to the index of refraction n of the lens material by

Lens-maker's equation ▶

$$\frac{1}{f} = (n-1)\left(\frac{1}{R_1} - \frac{1}{R_2}\right) \qquad [23.12]$$

where R_1 is the radius of curvature of the front surface of the lens and R_2 is the radius of curvature of the back surface. (As with mirrors, we arbitrarily call the side from which the light approaches the *front* of the lens.) Table 23.3 gives the sign conventions for R_1 and R_2. Equation 23.12, called the **lens-maker's equation**, enables us to calculate the focal length from the known properties of the lens.

Ray Diagrams for Thin Lenses

Ray diagrams are essential for understanding the overall image formation by a thin lens or a system of lenses. They should also help clarify the sign conventions already discussed. Figure 23.25 illustrates this method for three single-lens situations. To locate the image formed by a converging lens (Figs. 23.25a and 23.25b), the following three rays are drawn from the top of the object:

1. The first ray is drawn parallel to the principal axis. After being refracted by the lens, this ray passes through (or appears to come from) one of the focal points.
2. The second ray is drawn through the center of the lens. This ray continues in a straight line.
3. The third ray is drawn through the other focal point and emerges from the lens parallel to the principal axis.

A similar construction is used to locate the image formed by a diverging lens, as shown in Figure 23.25c. The point of intersection of *any two* of the rays in these diagrams can be used to locate the image. The third ray serves as a check on construction.

For the converging lens in Figure 23.25a, where the object is *outside* the front focal point ($p > f$), the ray diagram shows that the image is real and inverted.

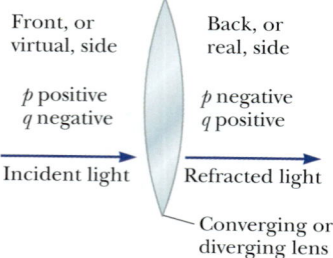

Figure 23.24 A diagram for obtaining the signs of p and q for a thin lens or a refracting surface.

Front, or virtual, side — p positive, q negative — Incident light

Back, or real, side — p negative, q positive — Refracted light

Converging or diverging lens

Tip 23.4 Positive Is Again Where the Light Is

For lenses, p and q are positive where the light is, where the object or image is real. For real objects, the light originates with the object in front of the lens, so p is positive there as indicated in Figure 23.24. If the image forms in back of the lens, q is positive there, as well.

Tip 23.5 We *Choose* Only a Few Rays

Although our ray diagrams in Figure 23.25 only show three rays leaving an object, an infinite number of rays can be drawn between the object and its image.

Table 23.3 Sign Conventions for Thin Lenses

Quantity	Symbol	In Front	In Back	Convergent	Divergent
Object location	p	+	−		
Image location	q	−	+		
Lens radii	R_1, R_2	−	+		
Focal length	f			+	−

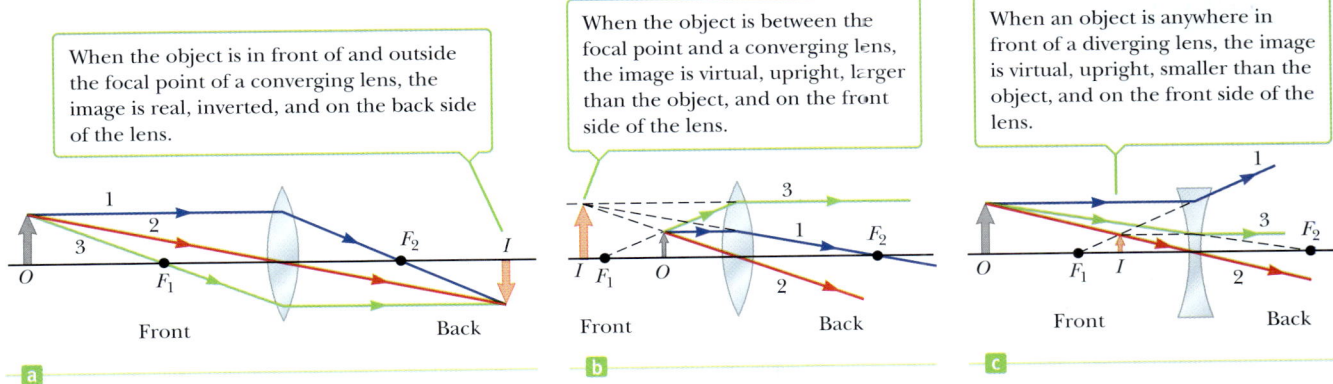

Figure 23.25 Ray diagrams for locating the image formed by a thin lens.

When the real object is *inside* the front focal point ($p < f$), as in Figure 23.25b, the image is virtual and upright. For the diverging lens of Figure 23.25c, the image is virtual and upright.

■ Quick Quiz

23.4 A clear plastic sandwich bag filled with water can act as a crude converging lens in air. If the bag is filled with air and placed under water, is the effective lens (a) converging or (b) diverging?

23.5 In Figure 23.25a the blue object arrow is replaced by one that is much taller than the lens. How many rays from the object will strike the lens?

23.6 An object is placed to the left of a converging lens. Which of the following statements are true, and which are false? (a) The image is always to the right of the lens. (b) The image can be upright or inverted. (c) The image is always smaller or the same size as the object.

Your success in working lens or mirror problems will be determined largely by whether you make sign errors when substituting into the lens or mirror equations. The only way to ensure you don't make sign errors is to become adept at using the sign conventions. The best way to do so is to work a multitude of problems on your own and construct confirming ray diagrams. Watching an instructor or reading the example problems is no substitute for practice.

■ APPLYING PHYSICS 23.5 | Vision and Diving Masks BIO

Diving masks often have a lens built into the glass faceplate for divers who don't have perfect vision. This lens allows the individual to dive without the necessity of glasses because the faceplate performs the necessary refraction to produce clear vision. Normal glasses have lenses that are curved on both the front and rear surfaces. The lenses in a diving-mask faceplate often have curved surfaces only on the inside of the glass. Why is this design desirable?

SOLUTION The main reason for curving only the inner surface of the lens in the diving-mask faceplate is to enable the diver to see clearly while underwater and in the air. If there were curved surfaces on both the front and the back of the diving lens, there would be two refractions. The lens could be designed so that these two refractions would give clear vision while the diver is in air. When the diver went underwater, however, the refraction between the water and the glass at the first interface would differ because the index of refraction of water is different from that of air. Consequently, the diver's vision wouldn't be clear underwater. ∎

EXAMPLE 23.7 | Images Formed by a Converging Lens

GOAL Calculate geometric quantities associated with a converging lens.

PROBLEM A converging lens of focal length 10.0 cm forms images of an object situated at various distances. **(a)** If the object is placed 30.0 cm from the lens, locate the image, state whether it's real or virtual, and find its magnification. **(b)** Repeat the problem when the object is at 10.0 cm and **(c)** again when the object is 5.00 cm from the lens.

STRATEGY All three problems require only substitution into the thin-lens equation and the associated magnification equation, Equations 23.10 and 23.11, respectively. The conventions of Table 23.3 must be followed.

Figure 23.26 (Example 23.7)

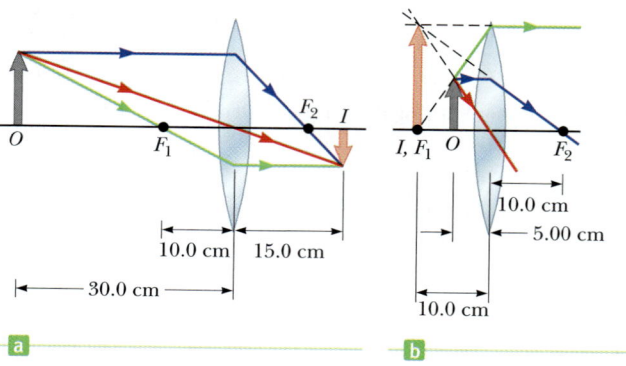

SOLUTION

(a) Find the image distance and describe the image when the object is placed at 30.0 cm.

The ray diagram is shown in Figure 23.26a. Substitute values into the thin-lens equation to locate the image:

$$\frac{1}{p} + \frac{1}{q} = \frac{1}{f}$$

$$\frac{1}{30.0 \text{ cm}} + \frac{1}{q} = \frac{1}{10.0 \text{ cm}}$$

Solve for q, the image distance. It's positive, so the image is real and on the far side of the lens:

$$q = \boxed{+15.0 \text{ cm}}$$

The magnification of the lens is obtained from Equation 23.10. M is negative and less than 1 in absolute value, so the image is inverted and smaller than the object:

$$M = -\frac{q}{p} = -\frac{15.0 \text{ cm}}{30.0 \text{ cm}} = \boxed{-0.500}$$

(b) Repeat the problem, when the object is placed at 10.0 cm.

Locate the image by substituting into the thin-lens equation:

$$\frac{1}{10.0 \text{ cm}} + \frac{1}{q} = \frac{1}{10.0 \text{ cm}} \quad \rightarrow \quad \frac{1}{q} = 0$$

This equation is satisfied only in the limit as q becomes infinite. Similarly, M becomes infinite, as well.

$$q \rightarrow \boxed{\infty}$$

(c) Repeat the problem when the object is placed 5.00 cm from the lens.

See the ray diagram shown in Figure 23.26b. Substitute into the thin-lens equation to locate the image:

$$\frac{1}{5.00 \text{ cm}} + \frac{1}{q} = \frac{1}{10.0 \text{ cm}}$$

Solve for q, which is negative, meaning the image is on the same side as the object and is virtual:

$$q = \boxed{-10.0 \text{ cm}}$$

Substitute the values of p and q into the magnification equation. M is positive and larger than 1, so the image is upright and double the object size:

$$M = -\frac{q}{p} = -\left(\frac{-10.0 \text{ cm}}{5.00 \text{ cm}}\right) = \boxed{+2.00}$$

REMARKS The ability of a lens to magnify objects led to the inventions of reading glasses, microscopes, and telescopes.

QUESTION 23.7 If the lens is used to form an image of the Sun on a screen, how far from the lens should the screen be located?

EXERCISE 23.7 Suppose the image of an object is upright and magnified 1.75 times when the object is placed 15.0 cm from a lens. Find (a) the location of the image and (b) the focal length of the lens.

ANSWERS (a) −26.3 cm (virtual, on the same side as the object) (b) 34.9 cm

---

■ EXAMPLE 23.8 | The Case of a Diverging Lens

GOAL Calculate geometric quantities associated with a diverging lens.

PROBLEM Repeat the problem of Example 23.7 for a *diverging* lens having a focal length of magnitude 10.0 cm.

STRATEGY Once again, substitution into the thin-lens equation and the associated magnification equation, together with the conventions in Table 23.3, solve the various parts. The only difference is the negative focal length.

Figure 23.27 (Example 23.8)

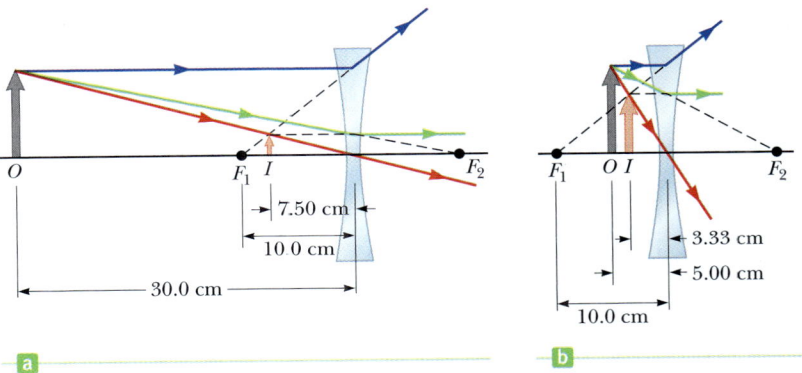

SOLUTION

(a) Locate the image and its magnification if the object is at 30.0 cm.

The ray diagram is given in Figure 23.27a. Apply the thin-lens equation with $p = 30.0$ cm to locate the image:

$$\frac{1}{p} + \frac{1}{q} = \frac{1}{f}$$

$$\frac{1}{30.0 \text{ cm}} + \frac{1}{q} = -\frac{1}{10.0 \text{ cm}}$$

Solve for q, which is negative and hence virtual:

$q = \boxed{-7.50 \text{ cm}}$

Substitute into Equation 23.10 to get the magnification. Because M is positive and has absolute value less than 1, the image is upright and smaller than the object:

$$M = -\frac{q}{p} = -\left(\frac{-7.50 \text{ cm}}{30.0 \text{ cm}}\right) = \boxed{+0.250}$$

(b) Locate the image and find its magnification if the object is 10.0 cm from the lens.

Apply the thin-lens equation, taking $p = 10.0$ cm:

$$\frac{1}{10.0 \text{ cm}} + \frac{1}{q} = -\frac{1}{10.0 \text{ cm}}$$

Solve for q (once again, the result is negative, so the image is virtual):

$q = \boxed{-5.00 \text{ cm}}$

Calculate the magnification. Because M is positive and has absolute value less than 1, the image is upright and smaller than the object:

$$M = -\frac{q}{p} = -\left(\frac{-5.00 \text{ cm}}{10.0 \text{ cm}}\right) = \boxed{+0.500}$$

(Continued)

(c) Locate the image and find its magnification when the object is at 5.00 cm.

The ray diagram is given in Figure 23.27b. Substitute $p = 5.00$ cm into the thin-lens equation to locate the image:

$$\frac{1}{5.00 \text{ cm}} + \frac{1}{q} = -\frac{1}{10.0 \text{ cm}}$$

Solve for q. The answer is negative, so once again the image is virtual:

$$q = \boxed{-3.33 \text{ cm}}$$

Calculate the magnification. Because M is positive and less than 1, the image is upright and smaller than the object:

$$M = -\left(\frac{-3.33 \text{ cm}}{5.00 \text{ cm}}\right) = \boxed{+0.666}$$

REMARKS Notice that in every case the image is virtual, hence on the same side of the lens as the object. Further, the image is smaller than the object. For a diverging lens and a real object, this is *always* the case, as can be proven mathematically.

QUESTION 23.8 Can a diverging lens be used as a magnifying glass? Explain.

EXERCISE 23.8 Repeat the calculation, finding the position of the image and the magnification if the object is 20.0 cm from the lens.

ANSWERS $q = -6.67$ cm, $M = 0.334$

Combinations of Thin Lenses

Many useful optical devices require two lenses. Handling problems involving two lenses is not much different from dealing with a single-lens problem twice. First, the image produced by the first lens is calculated as though the second lens were not present. The light then approaches the second lens *as if* it had come from the image formed by the first lens. Hence, **the image formed by the first lens is treated as the object for the second lens**. The image formed by the second lens is the final image of the system. If the image formed by the first lens lies on the back side of the second lens, the image is treated as a virtual object for the second lens, so p is negative. The same procedure can be extended to a system of three or more lenses. The overall magnification of a system of thin lenses is the *product* of the magnifications of the separate lenses. It's also possible to combine thin lenses and mirrors as shown in Example 23.10.

EXAMPLE 23.9 | Two Lenses in a Row

GOAL Calculate geometric quantities for a sequential pair of lenses.

PROBLEM Two converging lenses are placed 20.0 cm apart, as shown in Figure 23.28a, with an object 30.0 cm in front of lens 1 on the left. **(a)** If lens 1 has a focal length of 10.0 cm, locate the image formed by this lens and determine its magnification. **(b)** If lens 2 on the right has a focal length of 20.0 cm, locate the final image formed and find the total magnification of the system.

STRATEGY We apply the thin-lens equation to each lens. The image formed by lens 1 is treated as the object for lens 2. Also, we use the fact that the total magnification of the system is the product of the magnifications produced by the separate lenses.

SOLUTION

(a) Locate the image and determine the magnification of lens 1.

See the ray diagram, Figure 23.28b. Apply the thin-lens equation to lens 1:

$$\frac{1}{30.0 \text{ cm}} + \frac{1}{q} = \frac{1}{10.0 \text{ cm}}$$

Figure 23.28 (Example 23.9)

Solve for q, which is positive and hence to the right of the first lens:	$q = +15.0$ cm
Compute the magnification of lens 1:	$M_1 = -\dfrac{q}{p} = -\dfrac{15.0 \text{ cm}}{30.0 \text{ cm}} = -0.500$

(b) Locate the final image and find the total magnification.

The image formed by lens 1 becomes the object for lens 2. Compute the object distance for lens 2:

$$p = 20.0 \text{ cm} - 15.0 \text{ cm} = 5.00 \text{ cm}$$

Once again apply the thin-lens equation to lens 2 to locate the final image:

$$\dfrac{1}{5.00 \text{ cm}} + \dfrac{1}{q} = \dfrac{1}{20.0 \text{ cm}}$$

$$q = -6.67 \text{ cm}$$

Calculate the magnification of lens 2:

$$M_2 = -\dfrac{q}{p} = -\dfrac{(-6.67 \text{ cm})}{5.00 \text{ cm}} = +1.33$$

Multiply the two magnifications to get the overall magnification of the system:

$$M = M_1 M_2 = (-0.500)(1.33) = -0.665$$

REMARKS The negative sign for M indicates that the final image is inverted and smaller than the object because the absolute value of M is less than 1. Because q is negative, the final image is virtual.

QUESTION 23.9 If lens 2 is moved so it is 40 cm away from lens 1, would the final image be upright or inverted?

EXERCISE 23.9 If the two lenses in Figure 23.28 are separated by 10.0 cm, locate the final image and find the magnification of the system. *Hint:* The object for the second lens is virtual!

ANSWERS 4.00 cm behind the second lens, $M = -0.400$

EXAMPLE 23.10 | Thin Lens and a Concave Mirror

GOAL Solve a problem involving both a lens and a mirror.

PROBLEM An object is placed 20.0 cm to the right of a concave mirror with focal length 12.0 cm and 30.0 cm to the left of a converging lens with focal length 10.0 cm, as in Figure 23.29. Locate **(a)** the image formed by the lens alone, **(b)** the image created by the mirror alone, and **(c)** the image created by both the mirror and lens. **(d)** The mirror is moved so that it is 6.00 cm away from the object. Locate the image formed by the mirror and lens.

STRATEGY Part (a) is a simple application of the thin-lens equation, Equation 23.11. Part (b) can be calculated from Equation 23.6. Using the image formed by the mirror as the object for the lens, find the image location asked for in part (c), for light that first reflects off the mirror before passing through the lens.

Figure 23.29 (Example 23.10)

SOLUTION

(a) Locate the image formed by the lens alone.

Apply Equation 23.11:

$$\frac{1}{p_2} + \frac{1}{q_2} = \frac{1}{f_2}$$

Substitute values and solve for the image position, q_2:

$$\frac{1}{q_2} = \frac{1}{f_2} - \frac{1}{p_2} = \frac{1}{10.0 \text{ cm}} - \frac{1}{30.0 \text{ cm}} = \frac{1}{15.0 \text{ cm}}$$

$$q_2 = \boxed{15.0 \text{ cm}}$$

(b) Locate the image created by the mirror alone.

Apply Equation 23.6:

$$\frac{1}{p_1} + \frac{1}{q_1} = \frac{1}{f_1}$$

Substitute values for f_1 and p_1, which are both positive, and solve for q_1:

$$\frac{1}{q_1} = \frac{1}{f_1} - \frac{1}{p_1} = \frac{1}{12.0 \text{ cm}} - \frac{1}{20.0 \text{ cm}} = \frac{1}{30.0 \text{ cm}}$$

$$q_1 = \boxed{30.0 \text{ cm}}$$

(c) Locate the image created by both the mirror and lens.

Apply Equation 23.11 to the image found in part (b), which becomes a real object for the lens, noticing that the image formed by the mirror is 20.0 cm from the lens:

$$\frac{1}{q_{2f}} = \frac{1}{f_2} - \frac{1}{p_{2f}} = \frac{1}{10.0 \text{ cm}} - \frac{1}{20.0 \text{ cm}} = \frac{1}{20.0 \text{ cm}}$$

$$q_{2f} = \boxed{20.0 \text{ cm}}$$

(d) The mirror is moved so that it is 6.00 cm to the left of the object. Locate the image formed by the mirror and lens.

The object is now much closer to the mirror. Find the new location of the image created by the mirror:

$$\frac{1}{q_1} = \frac{1}{f_1} - \frac{1}{p_1} = \frac{1}{12.0 \text{ cm}} - \frac{1}{6.0 \text{ cm}} = -\frac{1}{12.0 \text{ cm}}$$

$$q_1 = -12.0 \text{ cm}$$

The image created by the mirror is virtual and therefore behind the mirror. However, it acts like a real object for the lens. Apply Equation 23.11 with $p_2 = 30.0 \text{ cm} + 18.0 \text{ cm} = 48.0 \text{ cm}$:

$$\frac{1}{q_2} = \frac{1}{f_2} - \frac{1}{p_2} = \frac{1}{10.0 \text{ cm}} - \frac{1}{48.0 \text{ cm}} = \frac{19}{2.40 \times 10^2 \text{ cm}}$$

$$q_2 = \boxed{12.6 \text{ cm}}$$

REMARKS There are two final images created to the right of the lens, as expected. As the mirror is moved closer to the object, the final image due to both the mirror and lens moves closer to the lens. The image of part (a) is inverted; however, the image of part (c) goes through two inversions, hence is upright. The virtual mirror image of part (d) is upright, so its lens image is inverted.

QUESTION 23.10 Is it possible to have a virtual object for a mirror? Explain, giving an example.

EXERCISE 23.10 The same mirror and lens are repositioned so that the mirror is 24.0 cm to the left of the lens and the object is 20.0 cm to the right of the lens. Locate the image of (a) the lens alone, (b) the first image formed by the mirror, and (c) the final, second image formed by the lens.

ANSWERS (a) 20.0 cm to the left of the lens (b) 6.00 cm behind the mirror (c) 15.0 cm to the right of the lens

23.7 Lens and Mirror Aberrations

LEARNING OBJECTIVE

1. Discuss defects in lenses and mirrors and their consequent aberrations.

One of the basic problems of systems containing mirrors and lenses is the imperfect quality of the images, which is largely the result of defects in shape and form. The simple theory of mirrors and lenses assumes rays make small angles with the principal axis and all rays reaching the lens or mirror from a point source are focused at a single point, producing a sharp image. This is not always true in the real world. Where the approximations used in this theory do not hold, imperfect images are formed.

If one wishes to analyze image formation precisely, it is necessary to trace each ray, using Snell's law, at each refracting surface. This procedure shows that there is no single point image; instead, the image is blurred. The departures of real (imperfect) images from the ideal predicted by the simple theory are called **aberrations.** Two common types of aberrations are spherical aberration and chromatic aberration.

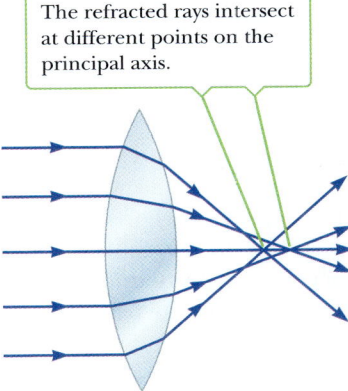

Figure 23.30 Spherical aberration used by a converging lens. Does a diverging lens use spherical aberration?

The refracted rays intersect at different points on the principal axis.

Spherical Aberration

Spherical aberration results from the fact that the focal points of light rays passing far from the principal axis of a spherical lens (or mirror) are different from the focal points of rays with the same wavelength passing near the axis. Figure 23.30 illustrates spherical aberration for parallel rays passing through a converging lens. Rays near the middle of the lens are imaged farther from the lens than rays at the edges. Hence, there is no single focal length for a spherical lens.

Most cameras are equipped with an adjustable aperture to control the light intensity and, when possible, reduce spherical aberration. (An aperture is an opening that controls the amount of light transmitted through the lens.) As the aperture size is reduced, sharper images are produced because only the central portion of the lens is exposed to the incident light when the aperture is very small. At the same time, however, progressively less light is imaged. To compensate for this loss, a longer exposure time is used. An example of the results obtained with small apertures is the sharp image produced by a pinhole camera, with an aperture size of approximately 0.1 mm.

In the case of mirrors used for very distant objects, one can eliminate, or at least minimize, spherical aberration by employing a parabolic rather than spherical surface. Parabolic surfaces are not used in many applications, however, because they are very expensive to make with high-quality optics. Parallel light rays incident on such a surface focus at a common point. Parabolic reflecting surfaces are used in many astronomical telescopes to enhance the image quality. They are also used in flashlights, in which a nearly parallel light beam is produced from a small lamp placed at the focus of the reflecting surface.

Chromatic Aberration

Different wavelengths of light refracted by a lens focus at different points, which gives rise to chromatic aberration. In Chapter 22 we described how the index of refraction of a material varies with wavelength. When white light passes through a lens, for example, violet light rays are refracted more than red light rays (see Fig. 23.31), so the focal length for red light is greater than for violet light. Other wavelengths (not shown in the figure) would have intermediate focal points. Chromatic aberration for a diverging lens is opposite that for a converging lens. Chromatic aberration can be greatly reduced by a combination of converging and diverging lenses. Chromatic aberration isn't a problem with mirrors, because all wavelengths of light are reflected at the same angle.

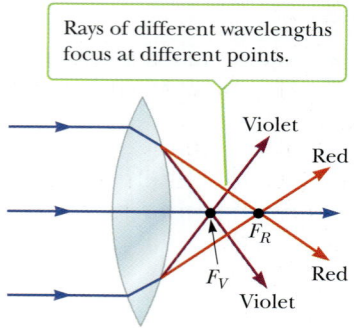

Figure 23.31 Chromatic aberration produced by a converging lens.

SUMMARY

23.1 Flat Mirrors

Images are formed where rays of light intersect or where they appear to originate. A **real image** is formed when light intersects, or passes through, an image point. In a **virtual image** the light doesn't pass through the image point, but appears to diverge from it.

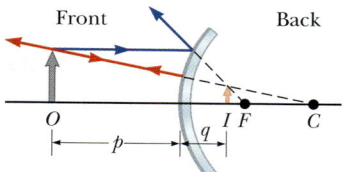

The image of a convex mirror is virtual, upright, and behind the mirror.

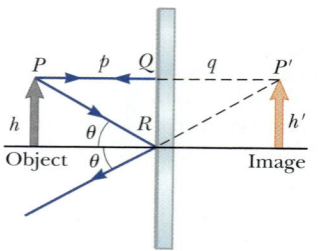

A geometric construction to locate the image of an object placed in front of a flat mirror. Because the triangles PQR and $P'QR$ are identical, $p = |q|$ and $h = h'$.

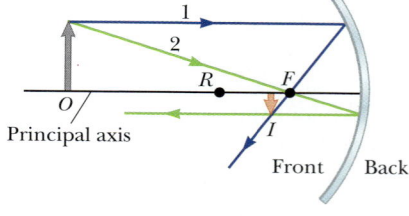

The image formed by a flat mirror has the following properties: **1.** The image is as far behind the mirror as the object is in front of it. **2.** The image is unmagnified, virtual, and upright.

23.2 Images Formed by Concave Mirrors

23.3 Convex Mirrors and Sign Conventions

The **magnification** M of a spherical mirror is defined as the ratio of the **image height** h' to the **object height** h, which is the negative of the ratio of the image distance q to the object distance p:

$$M = \frac{h'}{h} = -\frac{q}{p} \quad [23.2]$$

The **object distance** and **image distance** for a spherical mirror of radius R are related by the **mirror equation**:

$$\frac{1}{p} + \frac{1}{q} = \frac{1}{f} \quad [23.6]$$

where $f = R/2$ is the **focal length** of the mirror.

Equations 23.2 and 23.6 hold for both concave and convex mirrors, subject to the sign conventions given in Table 23.1.

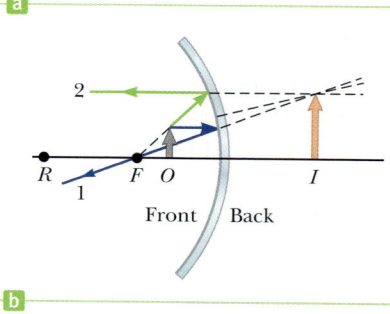

(a) The image of a concave mirror is real and inverted when the object is outside the focal point, i.e. $p > f$. The image is larger than the object when $f < p < R$, and smaller than the object when $p > R$. (b) The image of a concave mirror is virtual, upright, and larger than the object when $p < f$.

23.4 Images Formed by Refraction

An image can be formed by refraction at a spherical surface of radius R. The object and image distances for refraction from such a surface are related by

$$\frac{n_1}{p} + \frac{n_2}{q} = \frac{n_2 - n_1}{R} \quad [23.7]$$

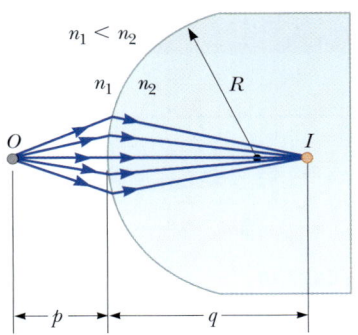

An image formed by refraction at a spherical surface. Rays making small angles with the principal axis diverge from a point object at O and pass through the image point I.

23.7 Lens and Mirror Aberrations

Aberrations are responsible for the formation of imperfect images by lenses and mirrors. **Spherical aberration** results from the focal points of light rays far from the principal axis of a spherical lens or mirror being different from those of rays passing through the center. **Chromatic aberration** arises because light rays of different wavelengths focus at different points when refracted by a lens.

The **magnification of a refracting surface** is

$$M = \frac{h'}{h} = -\frac{n_1 q}{n_2 p} \quad [23.8]$$

where the object is located in the medium with index of refraction n_1 and the image is formed in the medium with index of refraction n_2. Equations 23.7 and 23.8 are subject to the sign conventions of Table 23.2.

23.6 Thin Lenses

The **magnification of a thin lens** is

$$M = \frac{h'}{h} = -\frac{q}{p} \quad [23.10]$$

The object and image distances of a thin lens are related by the **thin-lens equation**:

$$\frac{1}{p} + \frac{1}{q} = \frac{1}{f} \quad [23.11]$$

Equations 23.10 and 23.11 are subject to the sign conventions of Table 23.3.

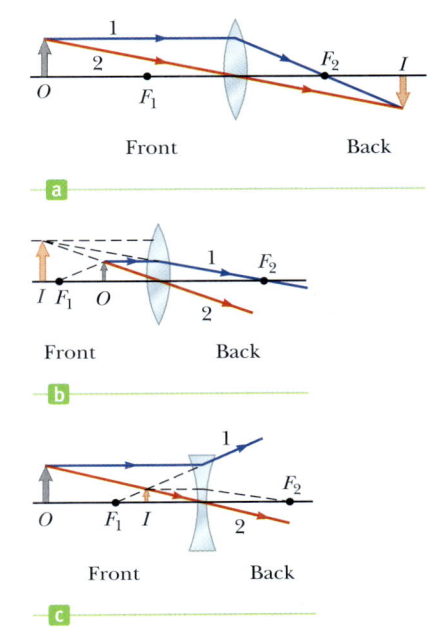

Ray diagrams for locating the image of an object. (a) The object is outside the focal point of a converging lens. (b) The object is inside the focal point of a converging lens. (c) The object is outside the focal point of a diverging lens.

■ WARM-UP EXERCISES

The warm-up exercises in this chapter may be assigned online in Enhanced WebAssign.

1. A woman stands 1.50 m in front of a full-length flat mirror. Find (a) the position of her image, with proper sign, (b) the magnification of her image, and (c) whether it is real or virtual. (See Section 23.1.)

2. An object is 20.0 cm away from a concave mirror with focal length 15.0 cm. Find (a) the image distance, q, and (b) its magnification. State whether the image is (c) real or virtual, (d) upright or inverted. (See Section 23.3.)

3. An object is placed 16.0 cm away from a convex mirror with a focal length of magnitude 6.00 cm. (a) Is the sign of the focal length negative or positive? Find (b) the image distance q, and (c) the magnification. State whether the image is (d) real or virtual, (e) upright or inverted. (See Section 23.3.)

4. A coin is at the bottom of a pool of water 1.20 m deep. Find (a) the object distance, (b) the image distance, and (c) the magnification of the image. (d) Is the image real or virtual? (See Section 23.4.)

5. A gold coin is embedded in solid ball of clear plastic of radius 25.0 cm, 10.0 cm from the surface. A person looks directly at the coin along a line going through the coin and to the center of the ball, as in Example 23.5. If the plastic has index of refraction 1.70, find (a) the image distance q, and (b) the coin's magnification. (See Section 23.4.)

6. A thin, convergent lens has a focal length of 8.00 cm. If the object distance is 24.0 cm, find (a) the image distance, and (b) the magnification. State whether the image is (c) real or virtual, (d) upright or inverted. (See Section 23.6.)

7. A real object is 10.0 cm to the left of a thin, diverging lens having a focal length of magnitude 16.0 cm. (a) Is the sign of the focal length negative or positive? Find (b) the image distance, and (c) the magnification. State whether the image is (d) real or virtual, (e) upright or inverted. (See Section 23.6.)

CONCEPTUAL QUESTIONS

1. Tape a picture of yourself on a bathroom mirror. Stand several centimeters away from the mirror. Can you focus your eyes on *both* the picture taped to the mirror *and* your image in the mirror *at the same time*? So where is the image of yourself?

2. Why does a clear stream always appear to be shallower than it actually is?

3. A flat mirror creates a virtual image of your face. Suppose the flat mirror is combined with another optical element. Can the mirror form a real image in such a combination?

4. Explain why a mirror cannot give rise to chromatic aberration.

5. A common mirage is formed when the air gets gradually cooler as the height above the ground increases. What might happen if the air grows gradually warmer as the height increases? This often happens over bodies of water or snow-covered ground; the effect is called *looming*.

6. A virtual image is often described as an image through which light rays don't actually travel, as they do for a real image. Can a virtual image be photographed?

7. Suppose you want to use a converging lens to project the image of two trees onto a screen. One tree is a distance x from the lens; the other is at $2x$, as in Figure CQ23.7. You adjust the screen so that the near tree is in focus. If you now want the far tree to be in focus, do you move the screen toward or away from the lens?

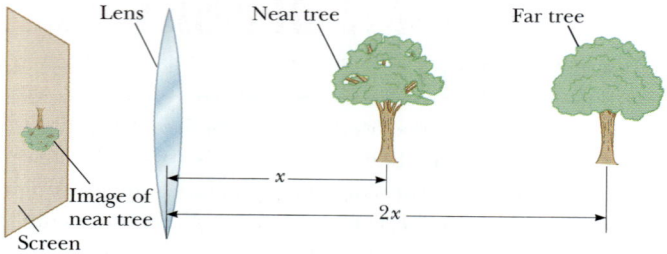

Figure CQ23.7

8. Lenses used in eyeglasses, whether converging or diverging, are always designed such that the middle of the lens curves away from the eye. Why?

9. In a Jules Verne novel, a piece of ice is shaped into a magnifying lens to focus sunlight to start a fire. Is that possible?

10. If a cylinder of solid glass or clear plastic is placed above the words LEAD OXIDE and viewed from the side, as shown in Figure CQ23.10, the word LEAD appears inverted, but the word OXIDE does not. Explain.

Figure CQ23.10

11. Can a converging lens be made to diverge light if placed in a liquid? How about a converging mirror?

12. Light from an object passes through a lens and forms a visible image on a screen. If the screen is removed, would you be able to see the image (a) if you remained in your present position and (b) if you could look at the lens along its axis, beyond the original position of the screen?

13. Why does the focal length of a mirror not depend on the mirror material when the focal length of a lens does depend on the lens material?

14. An inverted image of an object is viewed on a screen from the side facing a converging lens. An opaque card is then introduced covering only the upper half of the lens. What happens to the image on the screen? (a) Half the image would disappear. (b) The entire image would appear and remain unchanged. (c) Half the image would disappear and be dimmer. (d) The entire image would appear, but would be dimmer.

15. Why do some emergency vehicles have the symbol AMBULANCE written on the front?

16. A person spear fishing from a boat sees a stationary fish a few meters away in a direction about 30° below the horizontal. To spear the fish, and assuming the spear does not change direction when it enters the water, should the person (a) aim above where he sees the fish, (b) aim below the fish, or (c) aim precisely at the fish?

17. An object, represented by a gray arrow, is placed in front of a plane mirror. Which of the diagrams in Figure CQ23.17 best describes the image, represented by the pink arrow?

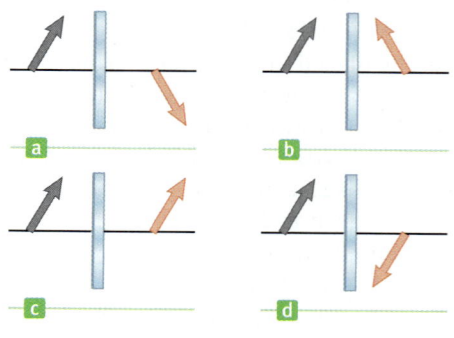

Figure CQ23.17

PROBLEMS

WebAssign The problems in this chapter may be assigned online in Enhanced WebAssign.

1. denotes straightforward problem; 2. denotes intermediate problem;
3. denotes challenging problem
[1.] denotes full solution available in *Student Solutions Manual/Study Guide*
1. denotes problems most often assigned in Enhanced WebAssign

BIO denotes biomedical problems
GP denotes guided problems
M denotes Master It tutorial available in Enhanced WebAssign
Q|C denotes asking for quantitative and conceptual reasoning
S denotes symbolic reasoning problem
W denotes Watch It video solution available in Enhanced WebAssign

23.1 Flat Mirrors

1. (a) Does your bathroom mirror show you older or younger than your actual age? (b) Compute an order-of-magnitude estimate for the age difference, based on data you specify.

2. **Q|C** Two plane mirrors stand facing each other, 3.00 m apart, and a woman stands between them. The woman faces one of the mirrors from a distance of 1.00 m and holds her left arm out to the side of her body with the palm of her left hand facing the closer mirror. (a) What is the apparent position of the closest image of her left hand, measured perpendicularly from the surface of the mirror in front of her? (b) Does it show the palm of her hand or the back of her hand? (c) What is the position of the next image? (d) Does it show the palm of her hand or the back of her hand? (e) What is the position of the third closest image? (f) Does it show the palm of her hand or the back of her hand? (g) Which of the images are real and which are virtual?

3. A person walks into a room that has, on opposite walls, two plane mirrors producing multiple images. Find the distances from the person to the first three images seen in the left-hand mirror when the person is 5.00 ft from the mirror on the left wall and 10.0 ft from the mirror on the right wall.

4. In a church choir loft, two parallel walls are 5.30 m apart. The singers stand against the north wall. The organist faces the south wall, sitting 0.800 m away from it. So that she can see the choir, a flat mirror 0.600 m wide is mounted on the south wall, straight in front of the organist. What width of the north wall can she see? *Hint:* Draw a top-view diagram to justify your answer.

5. A periscope (Fig. P23.5) is useful for viewing objects that cannot be seen directly. It can be used in submarines and when watching golf matches or parades from behind a crowd of people. Suppose the object is a distance p_1 from the upper mirror and the centers of the two flat mirrors are separated by a distance h. (a) What is the distance of the final image from the lower mirror? (b) Is the final image real or virtual? (c) Is it upright or inverted? (d) What is its magnification? (e) Does it appear to be left–right reversed?

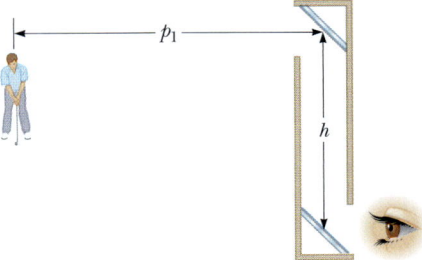

Figure P23.5

23.2 Images Formed by Concave Mirrors
23.3 Convex Mirrors and Sign Conventions

> In the following problems, algebraic signs are not given. We leave it to you to determine the correct sign to use with each quantity, based on an analysis of the problem and the sign conventions in Table 23.1.

6. A dentist uses a mirror to examine a tooth that is 1.00 cm in front of the mirror. The image of the tooth is formed 10.0 cm behind the mirror. Determine (a) the mirror's radius of curvature and (b) the magnification of the image.

7. A convex spherical mirror, whose focal length has a magnitude of 15.0 cm, is to form an image 10.0 cm behind the mirror. (a) Where should the object be placed? (b) What is the magnification of the mirror?

8. **BIO** To fit a contact lens to a patient's eye, a *keratometer* can be used to measure the curvature of the cornea—the front surface of the eye. This instrument places an illuminated object of known size at a known distance p from the cornea, which then reflects some light from the object, forming an image of it. The magnification M of the image is measured by using a small viewing telescope that allows a comparison of the image formed by the cornea with a second calibrated image projected into the field of view by a prism arrangement. Determine the radius of curvature of the cornea when $p = 30.0$ cm and $M = 0.013\ 0$.

9. A virtual image is formed 20.0 cm from a concave mirror having a radius of curvature of 40.0 cm. (a) Find the position of the object. (b) What is the magnification of the mirror?

10. **QC** While looking at her image in a cosmetic mirror, Dina notes that her face is highly magnified when she is close to the mirror, but as she backs away from the mirror, her image first becomes blurry, then disappears when she is about 30 cm from the mirror, and then inverts when she is beyond 30 cm. Based on these observations, what can she conclude about the properties of the mirror?

11. **W** A 2.00-cm-high object is placed 3.00 cm in front of a concave mirror. If the image is 5.00 cm high and virtual, what is the focal length of the mirror?

12. A dedicated sports car enthusiast polishes the inside and outside surfaces of a hubcap that is a section of a sphere. When he looks into one side of the hubcap, he sees an image of his face 30.0 cm in back of it. He then turns the hubcap over, keeping it the same distance from his face. He now sees an image of his face 10.0 cm in back of the hubcap. (a) How far is his face from the hubcap? (b) What is the magnitude of the radius of curvature of the hubcap?

13. A concave makeup mirror is designed so that a person 25 cm in front of it sees an upright image magnified by a factor of two. What is the radius of curvature of the mirror?

14. A concave mirror has a focal length of 30.0 cm. (a) What is its radius of curvature? Locate and describe the properties of the image when the object distance is (b) 100 cm and (c) 10.0 cm.

15. **M** A man standing 1.52 m in front of a shaving mirror produces an inverted image 18.0 cm in front of it. How close to the mirror should he stand if he wants to form an upright image of his chin that is twice the chin's actual size?

16. A convex mirror has a focal length of magnitude 8.0 cm. (a) If the image is virtual, what is the object location for which the magnitude of the image distance is one third the magnitude of the object distance? (b) Find the magnification of the image and state whether it is upright or inverted.

17. **M** At an intersection of hospital hallways, a convex spherical mirror is mounted high on a wall to help people avoid collisions. The magnitude of the mirror's radius of curvature is 0.550 m. (a) Locate the image of a patient located 10.0 m from the mirror. (b) Indicate whether the image is upright or inverted. (c) Determine the magnification of the image.

18. **QC** A concave mirror has a radius of curvature of 24.0 cm. (a) Determine the object position for which the resulting image is upright and larger than the object by a factor of 3.00. (b) Draw a ray diagram to determine the position of the image. (c) Is the image real or virtual?

19. **QC** A spherical mirror is to be used to form an image, five times as tall as an object, on a screen positioned 5.0 m from the mirror. (a) Describe the type of mirror required. (b) Where should the mirror be positioned relative to the object?

20. **QC** A ball is dropped from rest 3.00 m directly above the vertex of a concave mirror having a radius of 1.00 m and lying in a horizontal plane. (a) Describe the motion of the ball's image in the mirror. (b) At what time do the ball and its image coincide?

23.4 Images Formed by Refraction

21. A cubical block of ice 50.0 cm on an edge is placed on a level floor over a speck of dust. Locate the image of the speck, when viewed from directly above, if the index of refraction of ice is 1.309.

22. A goldfish is swimming inside a spherical bowl of water having an index of refraction $n = 1.333$. Suppose the goldfish is $p = 10.0$ cm from the wall of a bowl of radius $|R| = 15.0$ cm, as in Figure P23.22. Neglecting the refraction of light caused by the wall of the bowl, determine the apparent distance of the

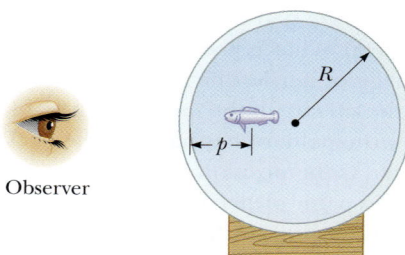

Figure P23.22

goldfish from the wall according to an observer outside the bowl.

23. A paperweight is made of a solid glass hemisphere with index of refraction 1.50. The radius of the circular cross section is 4.0 cm. The hemisphere is placed on its flat surface, with the center directly over a 2.5-mm-long line drawn on a sheet of paper. What length of line is seen by someone looking vertically down on the hemisphere?

24. **W** The top of a swimming pool is at ground level. If the pool is 2.00 m deep, how far below ground level does the bottom of the pool appear to be located when (a) the pool is completely filled with water? (b) When it is filled halfway with water?

25. A transparent sphere of unknown composition is observed to form an image of the Sun on its surface opposite the Sun. What is the refractive index of the sphere material?

26. A flint glass plate ($n = 1.66$) rests on the bottom of an aquarium tank. The plate is 8.00 cm thick (vertical dimension) and covered with water ($n = 1.33$) to a depth of 12.0 cm. Calculate the apparent thickness of the plate as viewed from above the water. (Assume nearly normal incidence of light rays.)

27. **QC** A jellyfish is floating in a water-filled aquarium 1.00 m behind a flat pane of glass 6.00 cm thick and having an index of refraction of 1.50. (a) Where is the image of the jellyfish located? (b) Repeat the problem when the glass is so thin that its thickness can be neglected. (c) How does the thickness of the glass affect the answer to part (a)?

28. **S** Figure P23.28 shows a curved surface separating a material with index of refraction n_1 from a material with index n_2. The surface forms an image I of object O. The ray shown in red passes through the surface along a radial line. Its angles of incidence and refraction are both zero, so its direction does not change at the surface. For the ray shown in blue, the direction changes according to $n_1 \sin \theta_1 = n_2 \sin \theta_2$. For paraxial rays, we assume θ_1 and θ_2 are small, so we may write $n_1 \tan \theta_1 = n_2 \tan \theta_2$. The magnification is defined as $M = h'/h$. Prove that the magnification is given by $M = -n_1 q/n_2 p$.

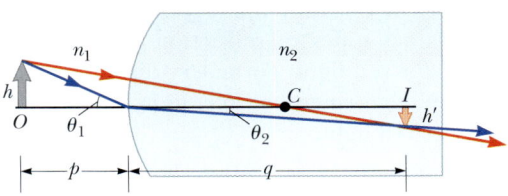

Figure P23.28

23.6 Thin Lenses

29. **BIO** A contact lens is made of plastic with an index of refraction of 1.50. The lens has an outer radius of curvature of +2.00 cm and an inner radius of curvature of +2.50 cm. What is the focal length of the lens?

30. An object is placed 50.0 cm from a screen. (a) Where should a converging lens of focal length 10.0 cm be placed to form an image on the screen? (b) Find the magnification of the lens.

31. A converging lens has a focal length of 10.0 cm. Locate the images for object distances of (a) 20.0 cm, (b) 10.0 cm, and (c) 5.00 cm, if they exist. For each case, state whether the image is real or virtual, upright or inverted, and find the magnification.

32. An object is placed 20.0 cm from a concave spherical mirror having a focal length of magnitude 40.0 cm. (a) Use graph paper to construct an accurate ray diagram for this situation. (b) From your ray diagram, determine the location of the image. (c) What is the magnification of the image? (d) Check your answers to parts (b) and (c) using the mirror equation.

33. A diverging lens has a focal length of magnitude 20.0 cm. (a) Locate the images for object distances of (i) 40.0 cm, (ii) 20.0 cm, and (iii) 10.0 cm. For each case, state whether the image is (b) real or virtual and (c) upright or inverted. (d) For each case, find the magnification.

34. **QC** A diverging lens has a focal length of 20.0 cm. Use graph paper to construct accurate ray diagrams for object distances of (a) 40.0 cm and (b) 10.0 cm. In each case determine the location of the image from the diagram and the image magnification, and state whether the image is upright or inverted. (c) Estimate the magnitude of uncertainty in locating the points in the graph. Are your answers and the uncertainty consistent with the algebraic answers found in Problem 33?

35. A transparent photographic slide is placed in front of a converging lens with a focal length of 2.44 cm. An image of the slide is formed 12.9 cm from the slide. How far is the lens from the slide if the image is (a) real? (b) Virtual?

36. **W** The nickel's image in Figure P23.36 has twice the diameter of the nickel when the lens is 2.84 cm from the nickel. Determine the focal length of the lens.

Figure P23.36

37. The projection lens in a certain slide projector is a single thin lens. A slide 24.0 mm high is to be projected so that its image fills a screen 1.80 m high. The slide-to-screen distance is 3.00 m. (a) Determine the focal length of the projection lens. (b) How far from the slide should the lens of the projector be placed to form the image on the screen?

38. **M** An object is located 20.0 cm to the left of a diverging lens having a focal length $f = -32.0$ cm. Determine (a) the location and (b) the magnification of the image. (c) Construct a ray diagram for this arrangement.

39. A converging lens is placed 30.0 cm to the right of a diverging lens of focal length 10.0 cm. A beam of parallel light enters the diverging lens from the left, and the beam is again parallel when it emerges from the converging lens. Calculate the focal length of the converging lens.

40. **QC S** (a) Use the thin-lens equation to derive an expression for q in terms of f and p. (b) Prove that for a real object and a diverging lens, the image must always be virtual. *Hint:* Set $f = -|f|$ and show that q must be less than zero under the given conditions. (c) For a real object and converging lens, what inequality involving p and f must hold if the image is to be real?

41. **W** Two converging lenses, each of focal length 15.0 cm, are placed 40.0 cm apart, and an object is placed 30.0 cm in front of the first lens. Where is the final image formed, and what is the magnification of the system?

42. Object O_1 is 15.0 cm to the left of a converging lens with a 10.0-cm focal length. A second lens is positioned 10.0 cm to the right of the first lens and is observed to form a virtual image at the position of the original object O_1. (a) What is the focal length of the second lens? (b) What is the overall magnification of this system? (c) What is the nature (i.e., real or virtual, upright or inverted) of the final image?

43. M A 1.00-cm-high object is placed 4.00 cm to the left of a converging lens of focal length 8.00 cm. A diverging lens of focal length −16.00 cm is 6.00 cm to the right of the converging lens. Find the position and height of the final image. Is the image inverted or upright? Real or virtual?

44. Two converging lenses having focal lengths of $f_1 = 10.0$ cm and $f_2 = 20.0$ cm are placed $d = 50.0$ cm apart, as shown in Figure P23.44. The final image is to be located between the lenses, at the position $x = 31.0$ cm indicated. (a) How far to the left of the first lens should the object be positioned? (b) What is the overall magnification of the system? (c) Is the final image upright or inverted?

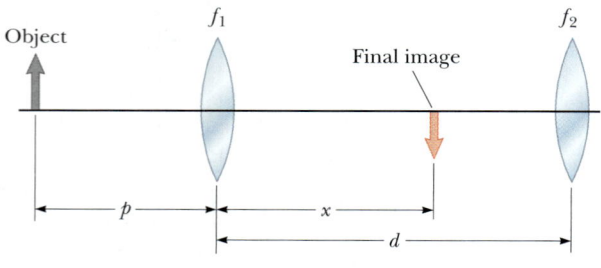

Figure P23.44

45. Lens L_1 in Figure P23.45 has a focal length of 15.0 cm and is located a fixed distance in front of the film plane of a camera. Lens L_2 has a focal length of 13.0 cm, and its distance d from the film plane can be varied from 5.00 cm to 10.0 cm. Determine the range of distances for which objects can be focused on the film.

Figure P23.45

46. GP An object is placed 15.0 cm from a first converging lens of focal length 10.0 cm. A second converging lens with focal length 5.00 cm is placed 10.0 cm to the right of the first converging lens. (a) Find the position q_1 of the image formed by the first converging lens. (b) How far from the second lens is the image of the first lens? (c) What is the value of p_2, the object position for the second lens? (d) Find the position q_2 of the image formed by the second lens. (e) Calculate the magnification of the first lens. (f) Calculate the magnification of the second lens. (g) What is the total magnification for the system? (h) Is the final image real or virtual? Is it upright or inverted (compared to the original object for the lens system)?

Additional Problems

47. An object placed 10.0 cm from a concave spherical mirror produces a real image 8.00 cm from the mirror. If the object is moved to a new position 20.0 cm from the mirror, what is the position of the image? Is the final image real or virtual?

48. S A real object's distance from a converging lens is five times the focal length. (a) Determine the location of the image q in terms of the focal length f. (b) Find the magnification of the image. (c) Is the image real or virtual? Is it upright or inverted? Is the image on the same side of the lens as the object or on the opposite side?

49. The magnitudes of the radii of curvature are 32.5 cm and 42.5 cm for the two faces of a biconcave lens. The glass has index of refraction 1.53 for violet light and 1.51 for red light. For a very distant object, locate (a) the image formed by violet light and (b) the image formed by red light.

50. A diverging lens ($n = 1.50$) is shaped like that in Figure 23.25c. The radius of the first surface is 15.0 cm, and that of the second surface is 10.0 cm. (a) Find the focal length of the lens. Determine the positions of the images for object distances of (b) infinity, (c) $3|f|$, (d) $|f|$, and (e) $|f|/2$.

51. The lens and the mirror in Figure P23.51 are separated by 1.00 m and have focal lengths of +80.0 cm and −50.0 cm, respectively. If an object is placed 1.00 m to the left of the lens, where will the final image be located? State whether the image is upright or inverted, and determine the overall magnification.

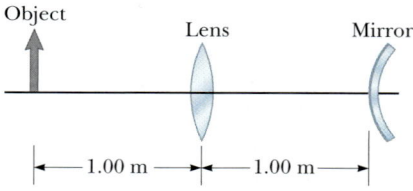

Figure P23.51

52. The object in Figure P23.52 is midway between the lens and the mirror, which are separated by a distance $d = 25.0$ cm. The magnitude of the mirror's radius of curvature is 20.0 cm, and the lens has a focal length of −16.7 cm. (a) Considering only the light that leaves the object and travels first toward the mirror, locate the final image formed by this system. (b) Is the image

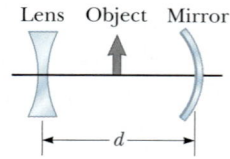

Figure P23.52

real or virtual? (c) Is it upright or inverted? (d) What is the overall magnification of the image?

53. A parallel beam of light enters a glass hemisphere perpendicular to the flat face, as shown in Figure P23.53. The radius of the hemisphere is $R = 6.00$ cm, and the index of refraction is $n = 1.56$. Determine the point at which the beam is focused. (Assume paraxial rays; i.e., assume all rays are located close to the principal axis.)

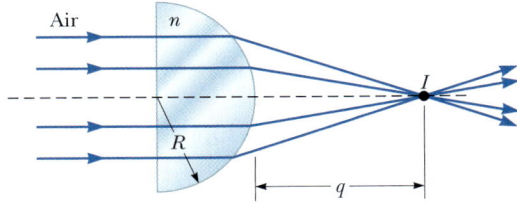

Figure P23.53

54. Two rays traveling parallel to the principal axis strike a large plano-convex lens having a refractive index of 1.60 (Fig. P23.54). If the convex face is spherical, a ray near the edge does not pass through the focal point (spherical aberration occurs). Assume this face has a radius of curvature of $R = 20.0$ cm and the two rays are at distances $h_1 = 0.500$ cm and $h_2 = 12.0$ cm from the principal axis. Find the difference Δx in the positions where each crosses the principal axis.

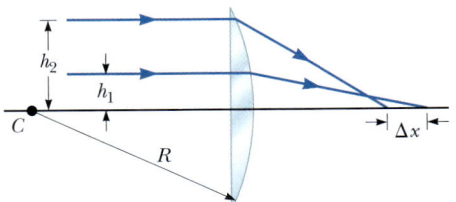

Figure P23.54

55. To work this problem, use the fact that the image formed by the first surface becomes the object for the second surface. Figure P23.55 shows a piece of glass with index of refraction $n = 1.50$ surrounded by air. The ends are hemispheres with radii $R_1 = 2.00$ cm and $R_2 = 4.00$ cm, and the centers of the hemispherical ends are separated by a distance of $d = 8.00$ cm. A point object is in air, a distance $p = 1.00$ cm from the left end of the glass. (a) Locate the image of the object due to refraction at the two spherical surfaces. (b) Is the image real or virtual?

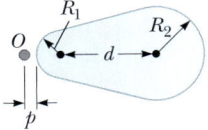

Figure P23.55

56. **S** Consider two thin lenses, one of focal length f_1 and the other of focal length f_2, placed in contact with each other, as shown in Figure P23.56. Apply the thin-lens equation to each of these lenses and combine the results to show that this combination of lenses behaves like a thin lens having a focal length f given by $1/f = 1/f_1 + 1/f_2$. Assume the thicknesses of the lenses can be ignored in comparison to the other distances involved.

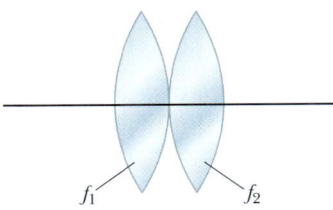

Figure P23.56

57. An object 2.00 cm high is placed 40.0 cm to the left of a converging lens having a focal length of 30.0 cm. A diverging lens having a focal length of -20.0 cm is placed 110 cm to the right of the converging lens. (a) Determine the final position and magnification of the final image. (b) Is the image upright or inverted? (c) Repeat parts (a) and (b) for the case in which the second lens is a converging lens having a focal length of $+20.0$ cm.

58. A "floating strawberry" illusion can be produced by two parabolic mirrors, each with a focal length of 7.5 cm, facing each other so that their centers are 7.5 cm apart (Fig. P23.58). If a strawberry is placed on the bottom mirror, an image of the strawberry forms at the small opening at the center of the top mirror. Show that the final image forms at that location and describe its characteristics. *Note:* A flashlight beam shone on these *images* has a very startling effect: Even at a glancing angle, the incoming light beam is seemingly reflected off the *images* of the strawberry! Do you understand why?

Figure P23.58

59. Figure P23.59 shows a converging lens with radii $R_1 = 9.00$ cm and $R_2 = -11.00$ cm, in front of a concave spherical mirror of radius $R = 8.00$ cm. The focal points (F_1 and F_2) for the thin lens and the center of

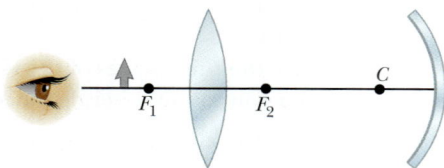

Figure P23.59

curvature (C) of the mirror are also shown. (a) If the focal points F_1 and F_2 are 5.00 cm from the vertex of the thin lens, what is the index of refraction of the lens? (b) If the lens and mirror are 20.0 cm apart and an object is placed 8.00 cm to the left of the lens, what is the position of the final image and its magnification as seen by the eye in the figure? (c) Is the final image inverted or upright? Explain.

60. **S** Find the object distances (in terms of f) for a thin converging lens of focal length f if (a) the image is real and the image distance is four times the focal length and (b) the image is virtual and the absolute value of the image distance is three times the focal length. (c) Calculate the magnification of the lens for cases (a) and (b).

61. The lens-maker's equation for a lens with index n_1 immersed in a medium with index n_2 takes the form

$$\frac{1}{f} = \left(\frac{n_1}{n_2} - 1\right)\left(\frac{1}{R_1} - \frac{1}{R_2}\right)$$

A thin diverging glass (index = 1.50) lens with $R_1 = -3.00$ m and $R_2 = -6.00$ m is surrounded by air. An arrow is placed 10.0 m to the left of the lens. (a) Determine the position of the image. Repeat part (a) with the arrow and lens immersed in (b) water (index = 1.33) (c) a medium with an index of refraction of 2.00. (d) How can a lens that is diverging in air be changed into a converging lens?

62. An observer to the right of the mirror–lens combination shown in Figure P23.62 sees two real images that are the same size and in the same location. One image is upright, and the other is inverted. Both images are 1.50 times larger than the object. The lens has a focal length of 10.0 cm. The lens and mirror are separated by 40.0 cm. Determine the focal length of the mirror. (Don't assume the figure is drawn to scale.)

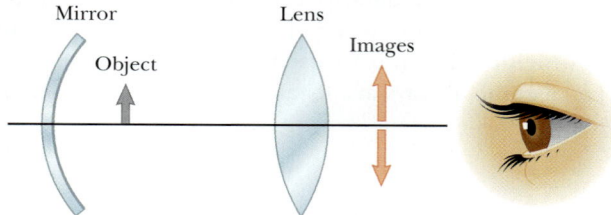

Figure P23.62

63. The lens-maker's equation applies to a lens immersed in a liquid if n in the equation is replaced by n_1/n_2. Here n_1 refers to the refractive index of the lens material and n_2 is that of the medium surrounding the lens. (a) A certain lens has focal length of 79.0 cm in air and a refractive index of 1.55. Find its focal length in water. (b) A certain mirror has focal length of 79.0 cm in air. Find its focal length in water.

64. A certain Christmas tree ornament is a silver sphere having a diameter of 8.50 cm. (a) If the size of an image created by reflection in the ornament is three-fourth's the reflected object's actual size, determine the object's location. (b) Use a principal-ray diagram to determine whether the image is upright or inverted.

65. **M** A glass sphere ($n = 1.50$) with a radius of 15.0 cm has a tiny air bubble 5.00 cm above its center. The sphere is viewed looking down along the extended radius containing the bubble. What is the apparent depth of the bubble below the surface of the sphere?

66. An object 10.0 cm tall is placed at the zero mark of a meterstick. A spherical mirror located at some point on the meterstick creates an image of the object that is upright, 4.00 cm tall, and located at the 42.0-cm mark of the meterstick. (a) Is the mirror convex or concave? (b) Where is the mirror? (c) What is the mirror's focal length?

The colors in many of a hummingbird's feathers are not due to pigment. The *iridescence* that makes the brilliant colors that often appear on the bird's throat and belly is due to an interference effect caused by structures in the feathers. The colors vary with the viewing angle.

Wave Optics 24

Colors swirl on a soap bubble as it drifts through the air on a summer day, and vivid rainbows reflect from the filth of oil films in the puddles of a dirty city street. Beachgoers, covered with thin layers of oil, wear their coated sunglasses that absorb half the incoming light. In laboratories scientists determine the precise composition of materials by analyzing the light they give off when hot, and in observatories around the world, telescopes gather light from distant galaxies, filtering out individual wavelengths in bands and thereby determining the speed of expansion of the Universe.

Understanding how these rainbows are made and how certain scientific instruments can determine wavelengths is the domain of *wave optics*. Light can be viewed as either a particle or a wave. Geometric optics, the subject of the previous chapter, depends on the particle nature of light. Wave optics depends on the wave nature of light. The three primary topics we examine in this chapter are interference, diffraction, and polarization. These phenomena can't be adequately explained with ray optics, but can be understood if light is viewed as a wave.

- **24.1** Conditions for Interference
- **24.2** Young's Double-Slit Experiment
- **24.3** Change of Phase Due to Reflection
- **24.4** Interference in Thin Films
- **24.5** Using Interference to Read CDs and DVDs
- **24.6** Diffraction
- **24.7** Single-Slit Diffraction
- **24.8** The Diffraction Grating
- **24.9** Polarization of Light Waves

24.1 Conditions for Interference

LEARNING OBJECTIVES

1. Discuss the two conditions that facilitate observations of interference in light waves.
2. Define coherent and incoherent light.

In our discussion of interference of mechanical waves in Chapter 13, we found that two waves could add together either constructively or destructively. In constructive interference the amplitude of the resultant wave is greater than that of either of the individual waves, whereas in destructive interference, the resultant amplitude is less than that of either individual wave. Light waves also interfere with one another. Fundamentally, all interference associated with light waves arises when the electromagnetic fields that constitute the individual waves combine.

Interference effects in light waves aren't easy to observe because of the short wavelengths involved (about 4×10^{-7} m to about 7×10^{-7} m). The following two conditions, however, facilitate the observation of interference between two sources of light:

◀ Conditions facilitating the observation of interference

1. The sources are **coherent**, which means that the waves they emit must maintain a constant phase with respect to one another.
2. The waves have identical wavelengths.

Two sources (producing two traveling waves) are needed to create interference. To produce a stable interference pattern, the individual waves must maintain a constant phase with one another. When this situation prevails, the sources are said to be coherent. The sound waves emitted by two side-by-side loudspeakers driven by a single amplifier can produce interference because the two speakers respond to the amplifier in the same way at the same time: they are in phase.

If two light sources are placed side by side, however, no interference effects are observed because the light waves from one source are emitted independently of the waves from the other source; hence, the emissions from the two sources don't maintain a constant phase relationship with each other during the time of observation. An ordinary light source undergoes random changes about once every 10^{-8} s. Therefore, the conditions for constructive interference, destructive interference, and intermediate states have durations on the order of 10^{-8} s. The result is that no interference effects are observed because the eye can't follow such short-term changes. Ordinary light sources are said to be **incoherent**.

An older method for producing two coherent light sources is to pass light from a single wavelength (monochromatic) source through a narrow slit and then allow the light to fall on a screen containing two other narrow slits. The first slit is needed to create a single wave front that illuminates both slits coherently. The light emerging from the two slits is coherent because a single source produces the original light beam and the slits serve only to separate the original beam into two parts. Any random change in the light emitted by the source will occur in the two separate beams at the same time, and interference effects can be observed.

Currently it's much more common to use a laser as a coherent source to demonstrate interference. A laser produces an intense, coherent, monochromatic beam over a width of several millimeters. The laser may therefore be used to illuminate multiple slits directly, and interference effects can be easily observed in a fully lighted room. The principles of operation of a laser are explained in Chapter 28.

24.2 Young's Double-Slit Experiment

LEARNING OBJECTIVES

1. Describe Young's double-slit experiment. Explain how interference creates the observed fringes.
2. Apply the conditions for constructive and destructive interferences to a Young's experiment.

Thomas Young first demonstrated interference in light waves from two sources in 1801. Figure 24.1a (page 837) is a schematic diagram of the apparatus used in this experiment. (Young used pinholes rather than slits in his original experiments.) Light is incident on a screen containing a narrow slit S_0. The light waves emerging from this slit arrive at a second screen that contains two narrow, parallel slits S_1 and S_2. These slits serve as a pair of coherent light sources because waves emerging from them originate from the same wave front and therefore are always in

24.2 | Young's Double-Slit Experiment

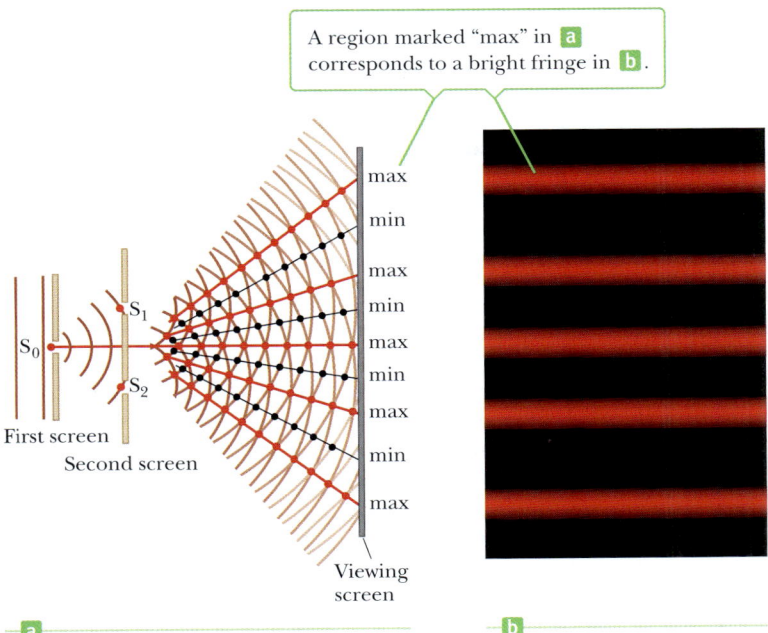

A region marked "max" in **a** corresponds to a bright fringe in **b**.

Figure 24.1 (a) A diagram of Young's double-slit experiment. The narrow slits act as sources of waves. Slits S_1 and S_2 behave as coherent sources that produce an interference pattern on screen C. (The drawing is not to scale.) (b) The fringe pattern formed on screen C could look like this.

phase. The light from the two slits produces a visible pattern on screen C consisting of a series of bright and dark parallel bands called **fringes** (Fig. 24.1b). When the light from slits S_1 and S_2 arrives at a point on the screen so that constructive interference occurs at that location, a bright fringe appears. When the light from the two slits combines destructively at any location on the screen, a dark fringe results. Figure 24.2 is a photograph of an interference pattern produced by two coherent vibrating sources in a water tank.

Figure 24.3 is a schematic diagram of some of the ways in which the two waves can combine at screen C of Figure 24.1. In Figure 24.3a two waves, which leave the two slits in phase, strike the screen at the central point P. Because these waves travel equal distances, they arrive in phase at P, and as a result, constructive interference occurs there and a bright fringe is observed. In Figure 24.3b the two light waves again start in phase, but the upper wave has to travel one wavelength farther to reach point Q on the screen. Because the upper wave falls behind the lower one by exactly one wavelength, the two waves still arrive in phase at Q, so a second

Figure 24.2 An interference pattern involving water waves is produced by two vibrating sources at the water's surface. The pattern is analogous to that observed in Young's double-slit experiment. Note the regions of constructive and destructive interference.

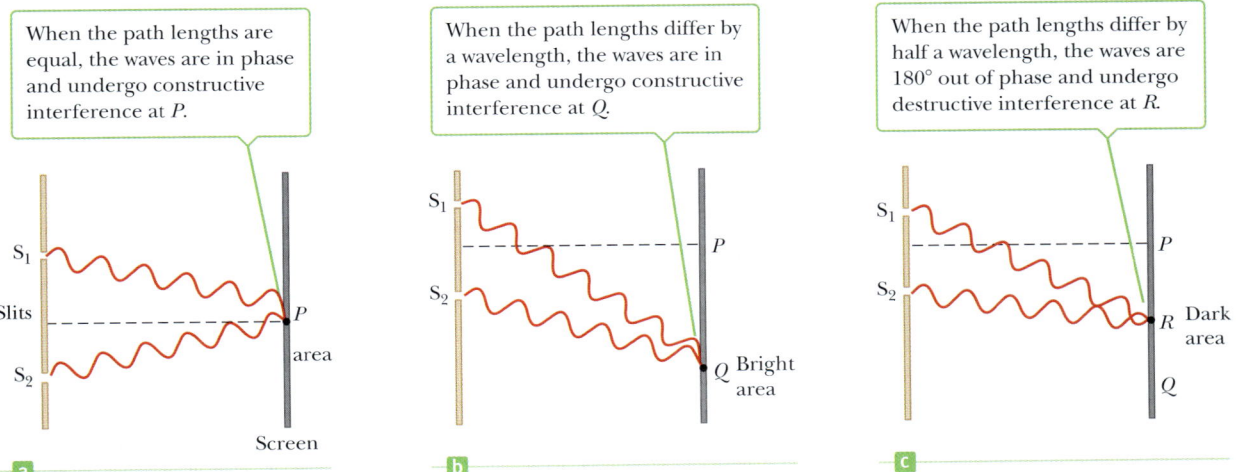

Figure 24.3 Waves leave the slits and combine at various points on the viewing screen. (All figures not to scale.)

bright fringe appears at that location. Now consider point R, midway between P and Q, in Figure 24.3c. At R, the upper wave has fallen half a wavelength behind the lower wave. This means that the trough of the bottom wave overlaps the crest of the upper wave, giving rise to destructive interference. As a result, a dark fringe can be observed at R.

We can describe Young's experiment quantitatively with the help of Figure 24.4. Consider point P on the viewing screen; the screen is positioned a perpendicular distance L from the screen containing slits S_1 and S_2, which are separated by distance d, and r_1 and r_2 are the distances the secondary waves travel from slit to screen. We assume the waves emerging from S_1 and S_2 have the same constant frequency, have the same amplitude, and start out in phase. The light intensity on the screen at P is the result of light from both slits. A wave from the lower slit, however, travels farther than a wave from the upper slit by the amount $d \sin \theta$. This distance is called the **path difference** δ (lowercase Greek delta), where

◀ Path difference

$$\delta = r_2 - r_1 = d \sin \theta \qquad [24.1]$$

Equation 24.1 assumes the two waves travel in parallel lines, which is approximately true because L is much greater than d. As noted earlier, the value of this path difference determines whether the two waves are in phase when they arrive at P. If the path difference is either zero or some integral multiple of the wavelength, the two waves are in phase at P and constructive interference results. Therefore, the condition for bright fringes, or **constructive interference**, at P is

◀ Condition for constructive interference (two slits)

$$\delta = d \sin \theta_{bright} = m\lambda \qquad m = 0, \pm 1, \pm 2, \ldots \qquad [24.2]$$

The number m is called the **order number**. The central bright fringe that appears at $\theta_{bright} = 0$ ($m = 0$) is called the *zeroth-order maximum*. The first maximum on either side, where $m = \pm 1$, is called the *first-order maximum*, and so forth.

When δ is an odd multiple of $\lambda/2$, the two waves arriving at P are 180° out of phase and give rise to destructive interference. Therefore, the condition for dark fringes, or **destructive interference**, at P is

◀ Condition for destructive interference (two slits)

$$\delta = d \sin \theta_{dark} = (m + \tfrac{1}{2})\lambda \qquad m = 0, \pm 1, \pm 2, \ldots \qquad [24.3]$$

If $m = 0$ in this equation, the path difference is $\delta = \lambda/2$, which is the condition for the location of the first dark fringe on either side of the central (bright) maximum. Likewise, if $m = 1$, the path difference is $\delta = 3\lambda/2$, which is the condition for the second dark fringe on each side, and so forth.

It's useful to obtain expressions for the positions of the bright and dark fringes measured vertically from O to P. In addition to our assumption that $L \gg d$, we assume $d \gg \lambda$. These assumptions can be valid because, in practice,

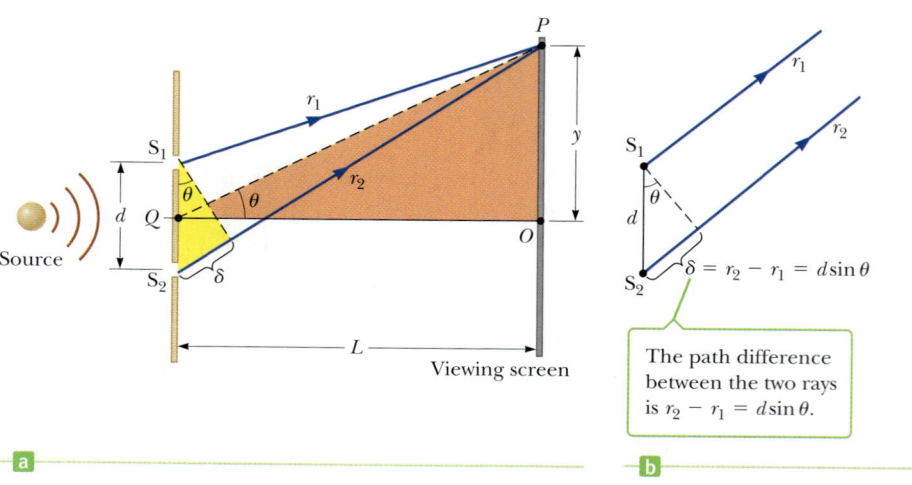

Figure 24.4 A geometric construction that describes Young's double-slit experiment. (This figure is not drawn to scale.)

L is often on the order of 1 m, d is a fraction of a millimeter, and λ is a fraction of a micrometer for visible light. Under these conditions θ is small, so we can use the approximation $\sin\theta \cong \tan\theta$. Then, from triangle OPQ in Figure 24.4, we see that

$$y = L\tan\theta \approx L\sin\theta \qquad [24.4]$$

Solving Equation 24.2 for $\sin\theta$ and substituting the result into Equation 24.4, we find that the positions of the *bright fringes*, measured from O, are

$$y_{\text{bright}} = \frac{\lambda L}{d}m \qquad m = 0, \pm 1, \pm 2, \ldots \qquad [24.5]$$

Using Equations 24.3 and 24.4, we find that the *dark fringes* are located at

$$y_{\text{dark}} = \frac{\lambda L}{d}(m + \tfrac{1}{2}) \qquad m = 0, \pm 1, \pm 2, \ldots \qquad [24.6]$$

As we will show in Example 24.1, Young's double-slit experiment provides a method for measuring the wavelength of light. In fact, Young used this technique to do just that. In addition, his experiment gave the wave model of light a great deal of credibility. It was inconceivable that particles of light coming through the slits could cancel each other in a way that would explain the dark fringes.

> **Tip 24.1** Small-Angle Approximation: Size Matters!
>
> The small-angle approximation $\sin\theta \cong \tan\theta$ is true to three-digit precision only for angles less than about 4°.

Reflection, interference, and diffraction can be seen in this aerial photograph of waves in the sea.

■ APPLYING PHYSICS 24.1 | A Smoky Young's Experiment

Consider a double-slit experiment in which a laser beam is passed through a pair of very closely spaced slits and a clear interference pattern is displayed on a distant screen. Now suppose you place smoke particles between the double slit and the screen. With the presence of the smoke particles, will you see the effects of interference in the space between the slits and the screen, or will you see only the effects on the screen?

EXPLANATION You will see the interference pattern both on the screen and in the area filled with smoke between the slits and the screen. There will be bright lines directed toward the bright areas on the screen and dark lines directed toward the dark areas on the screen. This is because Equations 24.5 and 24.6 depend on the distance to the screen, L, which can take any value. ■

■ APPLYING PHYSICS 24.2 | Television Signal Interference

Suppose you are watching television by means of an antenna rather than a cable system. If an airplane flies near your location, you may notice wavering ghost images in the television picture. What might cause this phenomenon?

EXPLANATION Your television antenna receives two signals: the direct signal from the transmitting antenna and a signal reflected from the surface of the airplane. As the airplane changes position, there are some times when these two signals are in phase and other times when they are out of phase. As a result, the intensity of the combined signal received at your antenna will vary. The wavering of the ghost images of the picture is evidence of this variation. ■

■ Quick Quiz

24.1 In a two-slit interference pattern projected on a screen, are the fringes equally spaced on the screen (a) everywhere, (b) only for large angles, or (c) only for small angles?

24.2 If the distance between the slits is doubled in Young's experiment, what happens to the width of the central maximum? (a) The width is doubled. (b) The width is unchanged. (c) The width is halved.

24.3 A Young's double-slit experiment is performed with three different colors of light: red, green, and blue. Rank the colors by the distance between adjacent bright fringes, from smallest to largest. (a) red, green, blue (b) green, blue, red (c) blue, green, red

EXAMPLE 24.1 Measuring the Wavelength of a Light Source

GOAL Show how Young's experiment can be used to measure the wavelength of coherent light.

PROBLEM A screen is separated from a double-slit source by 1.20 m. The distance between the two slits is 0.030 0 mm. The second-order bright fringe ($m = 2$) is measured to be 4.50 cm from the centerline. Determine **(a)** the wavelength of the light and **(b)** the distance between adjacent bright fringes.

STRATEGY Equation 24.5 relates the positions of the bright fringes to the other variables, including the wavelength of the light. Substitute into this equation and solve for λ. Taking the difference between y_{m+1} and y_m results in a general expression for the distance between bright fringes.

SOLUTION

(a) Determine the wavelength of the light.

Solve Equation 24.5 for the wavelength and substitute the values $m = 2$, $y_2 = 4.50 \times 10^{-2}$ m, $L = 1.20$ m, and $d = 3.00 \times 10^{-5}$ m:

$$\lambda = \frac{y_2 d}{mL} = \frac{(4.50 \times 10^{-2}\text{ m})(3.00 \times 10^{-5}\text{ m})}{2(1.20\text{ m})}$$

$$= 5.63 \times 10^{-7}\text{ m} = \boxed{563 \text{ nm}}$$

(b) Determine the distance between adjacent bright fringes.

Use Equation 24.5 to find the distance between *any* adjacent bright fringes (here, those characterized by m and $m + 1$):

$$\Delta y = y_{m+1} - y_m = \frac{\lambda L}{d}(m+1) - \frac{\lambda L}{d}m = \frac{\lambda L}{d}$$

$$= \frac{(5.63 \times 10^{-7}\text{ m})(1.20\text{ m})}{3.00 \times 10^{-5}\text{ m}} = \boxed{2.25 \text{ cm}}$$

REMARKS This calculation depends on the angle θ being small because the small-angle approximation was implicitly used. The measurement of the position of the bright fringes yields the wavelength of light, which in turn is a signature of atomic processes, as is discussed in the chapters on modern physics. This kind of measurement therefore helped open the world of the atom.

QUESTION 24.1 True or False: A larger slit creates a larger separation between interference fringes.

EXERCISE 24.1 Suppose the same experiment is run with a different light source. If the first-order maximum is found at 1.85 cm from the centerline, what is the wavelength of the light?

ANSWER 463 nm

24.3 Change of Phase Due to Reflection

LEARNING OBJECTIVE

1. Describe and discuss the conditions for which a reflected electromagnetic wave undergoes a 180° change of phase.

Young's method of producing two coherent light sources involves illuminating a pair of slits with a single source. Another simple, yet ingenious, arrangement for producing an interference pattern with a single light source is known as *Lloyd's mirror*. A point source of light is placed at point S, close to a mirror, as illustrated in Figure 24.5. Light waves can reach the viewing point P either by the direct path SP or by the path involving reflection from the mirror. The reflected ray can be treated as a ray originating at the source S' behind the mirror. Source S', which is the image of S, can be considered a virtual source.

At points far from the source, an interference pattern due to waves from S and S' is observed, just as for two real coherent sources. The positions of the dark and bright fringes, however, are *reversed* relative to the pattern obtained from two real coherent sources (Young's experiment). This is because the coherent sources S and S' differ in phase by 180°, a phase change produced by reflection.

An interference pattern is produced on a screen at P as a result of the combination of the direct ray (red) and the reflected ray (blue). The reflected ray undergoes a phase change of 180°.

Figure 24.5 Lloyd's mirror.

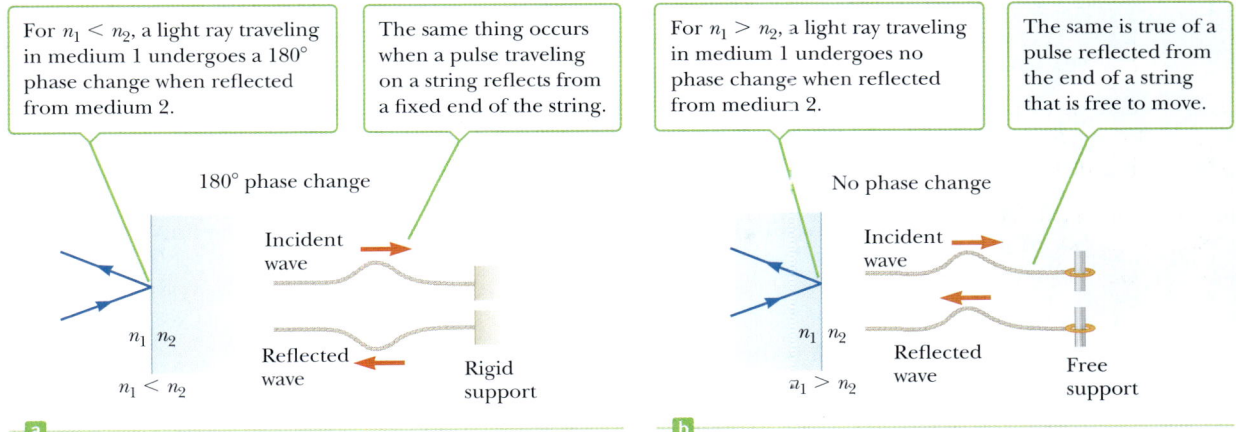

For $n_1 < n_2$, a light ray traveling in medium 1 undergoes a 180° phase change when reflected from medium 2.

The same thing occurs when a pulse traveling on a string reflects from a fixed end of the string.

For $n_1 > n_2$, a light ray traveling in medium 1 undergoes no phase change when reflected from medium 2.

The same is true of a pulse reflected from the end of a string that is free to move.

Figure 24.6 Comparisons of reflections of light waves and waves on strings.

To illustrate the point further, consider P', the point where the mirror intersects the screen. This point is equidistant from S and S'. If path difference alone were responsible for the phase difference, a bright fringe would be observed at P' (because the path difference is zero for this point), corresponding to the central fringe of the two-slit interference pattern. Instead, we observe a *dark* fringe at P, from which we conclude that a 180° phase change must be produced by reflection from the mirror. In general, **an electromagnetic wave undergoes a phase change of 180° upon reflection from a medium that has an index of refraction higher than the one in which the wave was traveling.**

An analogy can be drawn between reflected light waves and the reflections of a transverse wave on a stretched string when the wave meets a boundary, as in Figure 24.6. The pulse on a string undergoes a phase change of 180° when it is reflected from the boundary of a denser string or from a rigid support and undergoes no phase change when it is reflected from the boundary of a less dense string or free support. Similarly, an electromagnetic wave undergoes a 180° phase change when reflected from the boundary of a medium with index of refraction higher than the one in which it has been traveling. There is no phase change when the wave is reflected from a boundary leading to a medium of lower index of refraction. The transmitted wave that crosses the boundary also undergoes no phase change.

24.4 Interference in Thin Films

LEARNING OBJECTIVES

1. State the conditions for constructive and destructive interference in systems involving thin films.
2. Apply the thin film interference conditions to systems involving zero, one- or two-phase reversals.
3. Describe the phenomenon of Newton's rings.

Interference effects are commonly observed in thin films, such as the thin surface of a soap bubble or thin layers of oil on water. The varied colors observed when incoherent white light is incident on such films result from the interference of waves reflected from the two surfaces of the film.

Consider a film of uniform thickness t and index of refraction n, as in Figure 24.7. Assume the light rays traveling in air are nearly normal to the two surfaces of the

Interference in light reflected from a thin film is due to a combination of rays 1 and 2 reflected from the upper and lower surfaces of the film.

Figure 24.7 Light passes through a thin film.

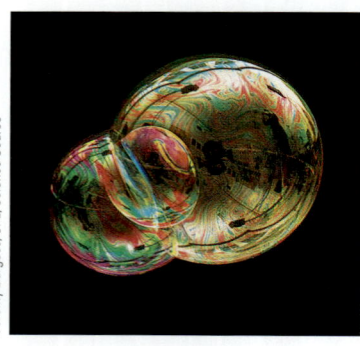

The colors observed in soap bubbles are due to interference between light rays reflected from the front and back of the thin film of soap making up the bubble. The color depends on the thickness of the film, ranging from black where the film is at its thinnest to magenta where it is thickest.

A thin film of oil on water displays interference, evidenced by the pattern of colors when white light is incident on the film. Variations in the film's thickness produce the intersecting color pattern. The razor blade gives you an idea of the size of the colored bands.

film. To determine whether the reflected rays interfere constructively or destructively, we first note the following facts:

1. An electromagnetic wave traveling from a medium of index of refraction n_1 toward a medium of index of refraction n_2 undergoes a 180° phase change on reflection when $n_2 > n_1$. There is no phase change in the reflected wave if $n_2 < n_1$.
2. The wavelength of light λ_n in a medium with index of refraction n is

$$\lambda_n = \frac{\lambda}{n} \qquad [24.7]$$

where λ is the wavelength of light in vacuum.

We apply these rules to the film of Figure 24.7. According to the first rule, ray 1, which is reflected from the upper surface A, undergoes a phase change of 180° with respect to the incident wave. Ray 2, which is reflected from the lower surface B, undergoes no phase change with respect to the incident wave. Therefore, ray 1 is 180° out of phase with respect to ray 2, which is equivalent to a path difference of $\lambda_n/2$. We must also consider, though, that ray 2 travels an extra distance of $2t$ before the waves recombine in the air above the surface. For example, if $2t = \lambda_n/2$, rays 1 and 2 recombine in phase and constructive interference results. In general, the condition for *constructive interference* in thin films is

$$2t = (m + \tfrac{1}{2})\lambda_n \qquad m = 0, 1, 2, \ldots \qquad [24.8]$$

This condition takes into account two factors: (1) the difference in path length for the two rays (the term $m\lambda_n$) and (2) the 180° phase change upon reflection (the term $\lambda_n/2$). Because $\lambda_n = \lambda/n$, we can write Equation 24.8 in the form

$$2nt = (m + \tfrac{1}{2})\lambda \qquad m = 0, 1, 2, \ldots \qquad [24.9]$$

If the extra distance $2t$ traveled by ray 2 is a multiple of λ_n, the two waves combine out of phase and the result is destructive interference. The general equation for *destructive interference* in thin films is

$$2nt = m\lambda \qquad m = 0, 1, 2, \ldots \qquad [24.10]$$

Equations 24.9 and 24.10 for constructive and destructive interferences are valid when there is only one phase reversal. This will occur when the media above and below the thin film both have indices of refraction greater than the film or when both have indices of refraction less than the film. Figure 24.7 is a case in point: the air ($n = 1$) that is both above and below the film has an index of refraction less than that of the film. As a result, there is a phase reversal on reflection off the top layer of the film but not the bottom, and Equations 24.9 and 24.10 apply. **If the film is placed between two different media, one of lower refractive index than the film and one of higher refractive index, Equations 24.9 and 24.10 are reversed: Equation 24.9 is used for destructive interference and Equation 24.10 for constructive interference.** In this case either there is a phase change of 180° for both ray 1 reflecting from surface A and ray 2 reflecting from surface B, as in Figure 24.9 of Example 24.3, or there is no phase change for either ray, which would be the case if the incident ray came from underneath the film. Hence, the net change in relative phase due to the reflections is *zero*.

■ Quick Quiz

24.4 Suppose Young's experiment is carried out in air, and then, in a second experiment, the apparatus is immersed in water. In what way does the distance between bright fringes change? (a) They move farther apart. (b) They move closer together. (c) There is no change.

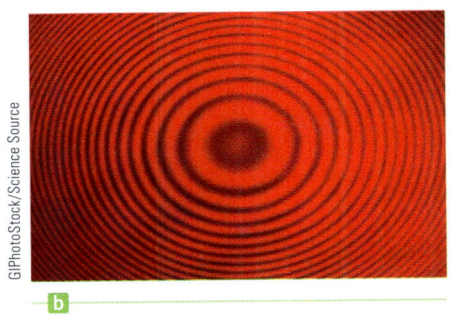

Figure 24.8 (a) The combination of rays reflected from the glass plate and the curved surface of the lens gives rise to an interference pattern known as Newton's rings. (b) A photograph of Newton's rings.

Newton's Rings

Another method for observing interference in light waves is to place a plano-convex lens on top of a flat glass surface, as in Figure 24.8a. With this arrangement, the air film between the glass surfaces varies in thickness from zero at the point of contact to some value t at P. If the radius of curvature R of the lens is much greater than the distance r and the system is viewed from above using light of wavelength λ, a pattern of light and dark rings is observed (Fig. 24.8b). These circular fringes, discovered by Newton, are called **Newton's rings**. The interference is due to the combination of ray 1, reflected from the plate, with ray 2, reflected from the lower surface of the lens. Ray 1 undergoes a phase change of 180° on reflection because it is reflected from a boundary leading into a medium of higher refractive index, whereas ray 2 undergoes no phase change because it is reflected from a medium of lower refractive index. Hence, the conditions for constructive and destructive interference are given by Equations 24.9 and 24.10, respectively, with $n = 1$ because the "film" is air. The contact point at O is dark, as seen in Figure 24.8b, because there is no path difference and the total phase change is due only to the 180° phase change upon reflection. Using the geometry shown in Figure 24.8a, we can obtain expressions for the radii of the bright and dark bands in terms of the radius of curvature R and vacuum wavelength λ. For example, the dark rings have radii of $r \approx \sqrt{m\lambda R/n}$.

One important use of Newton's rings is in the testing of optical lenses. A circular pattern like that in Figure 24.8b is achieved only when the lens is ground to a perfectly spherical curvature. Variations from such symmetry produce distorted patterns that also give an indication of how the lens must be reground and repolished to remove imperfections.

Tip 24.2 The Two Tricks of Thin Films
Be sure to include *both* effects—path length and phase change—when you analyze an interference pattern from a thin film.

APPLICATION
Checking for Imperfections in Optical Lenses

■ PROBLEM-SOLVING STRATEGY

Thin-Film Interference
The following steps are recommended in addressing thin-film interference problems:

1. Identify the thin film causing the interference, and the indices of refraction in the film and in the media on either side of it.
2. Determine the number of phase reversals: zero, one, or two.
3. Consult the following table, which contains Equations 24.9 and 24.10, and select the correct column for the problem in question:

Equation ($m = 0, 1, \ldots$)	1 Phase Reversal	0 or 2 Phase Reversals
$2nt = (m + \tfrac{1}{2})\lambda$ [24.9]	Constructive	Destructive
$2nt = m\lambda$ [24.10]	Destructive	Constructive

4. Substitute values in the appropriate equations, as selected in the previous step.

EXAMPLE 24.2 | Interference in a Soap Film

GOAL Study constructive interference effects in a thin film.

PROBLEM (a) Calculate the minimum thickness of a soap-bubble film ($n = 1.33$) that will result in constructive interference in the reflected light if the film is illuminated by light with wavelength 602 nm in free space. (b) Recalculate the minimum thickness for constructive interference when the soap-bubble film is on top of a glass slide with $n = 1.50$.

STRATEGY In part (a) there is only one inversion, so the condition for constructive interference is $2nt = (m + \frac{1}{2})\lambda$. The minimum film thickness for constructive interference corresponds to $m = 0$ in this equation. Part (b) involves two inversions, so $2nt = m\lambda$ is required.

SOLUTION

(a) Calculate the minimum thickness of the soap-bubble film that will result in constructive interference.

Solve $2nt = \lambda/2$ for the thickness t and substitute:

$$t = \frac{\lambda}{4n} = \frac{602 \text{ nm}}{4(1.33)} = \boxed{113 \text{ nm}}$$

(b) Find the minimum soap-film thickness when the film is on top of a glass slide with $n = 1.50$.

Write the condition for constructive interference, when two inversions take place:

$$2nt = m\lambda$$

Solve for t and substitute:

$$t = \frac{m\lambda}{2n} = \frac{1 \cdot (602 \text{ nm})}{2(1.33)} = \boxed{226 \text{ nm}}$$

REMARKS The different colors in a soap bubble result from the thickness of the soap layer varying from one place to another. The swirling is caused by the changing thickness of the layer with time.

QUESTION 24.2 A soap film looks red in one area and violet in a nearby area. In which area is the soap film thicker?

EXERCISE 24.2 What other film thicknesses in part (a) will produce constructive interference?

ANSWERS 339 nm, 566 nm, 792 nm, and so on

EXAMPLE 24.3 | Nonreflective Coatings for Solar Cells and Optical Lenses

GOAL Study destructive interference effects in a thin film when there are two inversions.

PROBLEM Semiconductors such as silicon are used to fabricate solar cells, devices that generate electric energy when exposed to sunlight. Solar cells are often coated with a transparent thin film, such as silicon monoxide (SiO; $n = 1.45$), to minimize reflective losses (Fig. 24.9). A silicon solar cell ($n = 3.50$) is coated with a thin film of silicon monoxide for this purpose. Assuming normal incidence, determine the minimum thickness of the film that will produce the least reflection at a wavelength of 552 nm.

Figure 24.9 (Example 24.3) Reflective losses from a silicon solar cell are minimized by coating it with a thin film of silicon monoxide (SiO).

STRATEGY Reflection is least when rays 1 and 2 in Figure 24.9 meet the condition for destructive interference. Note that *both* rays undergo 180° phase changes on reflection. The condition for a reflection *minimum* is therefore $2nt = \lambda/2$.

SOLUTION

Solve $2nt = \lambda/2$ for t, the required thickness:

$$t = \frac{\lambda}{4n} = \frac{552 \text{ nm}}{4(1.45)} = \boxed{95.2 \text{ nm}}$$

REMARKS Typically, such coatings reduce the reflective loss from 30% (with no coating) to 10% (with a coating), thereby increasing the cell's efficiency because more light is available to create charge carriers in the cell. In reality the coating is never perfectly nonreflecting because the required thickness is wavelength dependent and the incident light covers a wide range of wavelengths.

QUESTION 24.3 To minimize reflection of a smaller wavelength, should the thickness of the coating be thicker or thinner?

EXERCISE 24.3 Glass lenses used in cameras and other optical instruments are usually coated with one or more transparent thin films, such as magnesium fluoride (MgF_2), to reduce or eliminate unwanted reflection. Carl Zeiss developed this method; his first coating was 1.00×10^2 nm thick, on glass. Using $n = 1.38$ for MgF_2, what visible wavelength would be eliminated by destructive interference in the reflected light?

ANSWER 552 nm

■ EXAMPLE 24.4 | Interference in a Wedge-Shaped Film

GOAL Calculate interference effects when the film has variable thickness.

PROBLEM A pair of glass slides 10.0 cm long and with $n = 1.52$ are separated on one end by a hair, forming a triangular wedge of air, as illustrated in Figure 24.10. When coherent light from a helium–neon laser with wavelength 633 nm is incident on the film from above, 15.0 dark fringes per centimeter are observed. How thick is the hair?

STRATEGY The interference pattern is created by the thin film of air having variable thickness. The pattern is a series of alternating bright and dark parallel bands. A dark band corresponds to destructive interference, and there is one phase reversal, so $2nt = m\lambda$ should be used. We can also use the similar triangles in Figure 24.10 to obtain the relation $t/x = D/L$. We can find the thickness for any m, and if the position x can also be found, this last equation gives the diameter of the hair, D.

Figure 24.10 (Example 24.4) Interference bands in reflected light can be observed by illuminating a wedge-shaped film with monochromatic light. The dark areas in the interference pattern correspond to positions of destructive interference.

SOLUTION

Solve the destructive-interference equation for the thickness of the film, t, with $n = 1$ for air:

$$t = \frac{m\lambda}{2}$$

If d is the distance from one dark band to the next, then the x-coordinate of the mth band is a multiple of d:

$$x = md$$

By dimensional analysis, d is just the inverse of the number of bands per centimeter.

$$d = \left(15.0 \frac{\text{bands}}{\text{cm}}\right)^{-1} = 6.67 \times 10^{-2} \frac{\text{cm}}{\text{band}}$$

Now use similar triangles and substitute all the information:

$$\frac{t}{x} = \frac{m\lambda/2}{md} = \frac{\lambda}{2d} = \frac{D}{L}$$

Solve for D and substitute given values:

$$D = \frac{\lambda L}{2d} = \frac{(633 \times 10^{-9}\,\text{m})(0.100\,\text{m})}{2(6.67 \times 10^{-4}\,\text{m})} = \boxed{4.75 \times 10^{-5}\,\text{m}}$$

REMARKS Some may be concerned about interference caused by light bouncing off the top and bottom of, say, the upper glass slide. It's unlikely, however, that the thickness of the slide will be half an integer multiple of the wavelength of the helium–neon laser (for some very large value of m). In addition, in contrast to the air wedge, the thickness of the glass doesn't vary.

QUESTION 24.4 If the air wedge is filled with water, how is the distance between dark bands affected? Explain.

EXERCISE 24.4 The air wedge is replaced with water, with $n = 1.33$. Find the distance between dark bands when the helium–neon laser light hits the glass slides.

ANSWER 5.02×10^{-4} m

APPLYING PHYSICS 24.3 | Perfect Mirrors

When light hits a metallic mirror, electrons in the metal move in response to the electromagnetic fields, absorbing some of the light's energy. For many applications, such as directing high-intensity laser light, that reduction in intensity is undesirable. A dielectric mirror, on the other hand, is made of glass or plastic and doesn't conduct electricity. To improve reflectance, thin layers of different dielectric materials are stacked on the glass surface. If the thicknesses and dielectric constants are chosen properly, light reflected off one layer combines constructively with light reflected from the layer underneath, increasing the mirror's reflectance. Nearly perfect mirrors can be constructed using several thin dielectric layers. The mirrors can be designed to reflect a particular wavelength or a range of wavelengths.

Dielectric mirrors, first developed at MIT in 1998, have an enormous number of applications. One of the most important is the OmniGuide fiber, a hollow tube the size of a spaghetti noodle that can guide light without any significant loss of intensity. (Ordinary optical fibers tend to heat up.) Such fibers have up to forty concentric layers of plastic and glass and are highly flexible. Using an OmniGuide fiber, an intense laser light can be safely guided into the human body during surgery to remove tumors and other diseased tissues without harming the healthy surrounding tissue. ■

24.5 Using Interference to Read CDs and DVDs

LEARNING OBJECTIVES

1. Discuss the application of interference principles to the reading of CDs and DVDs.
2. Apply interference principles to CDs and DVDs.

APPLICATION
The Physics of CDs and DVDs

Compact discs (CDs) and digital videodiscs (DVDs) provide high-density storage of text, graphics, and movies; and high-quality sound recordings. The data on these discs are stored digitally as a series of zeros and ones, and these zeros and ones are read by laser light reflected from the disc. Strong reflections (constructive interference) from the disc are chosen to represent zeros, and weak reflections (destructive interference) represent ones.

To see in more detail how thin-film interference plays a crucial role in reading CDs and DVDs, consider Figure 24.11. This figure shows a photomicrograph of several DVD tracks, which consist of a sequence of pits (when viewed from the top or label side of the disc) of varying length formed in a reflecting-metal information layer. A cross-sectional view of a CD as shown in Figure 24.12 reveals that the pits appear as bumps to the laser beam, which shines on the metallic layer through a clear plastic coating from below.

As the disk rotates, the laser beam reflects off the sequence of bumps and lower areas into a photodetector, which converts the fluctuating reflected light intensity into an electrical string of zeros and ones. To make the light fluctuations more pronounced and easier to detect, the pit depth t is made equal to one-quarter of a

Figure 24.11 A photomicrograph of adjacent tracks on a digital video disc (DVD). The information encoded in these pits and smooth areas is read by a laser beam.

Figure 24.12 Cross section of a CD showing metallic pits of depth t and a laser beam detecting the edge of a pit.

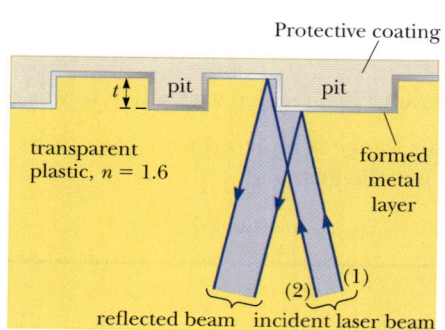

wavelength of the laser light in the plastic. When the beam hits a rising or falling bump edge, part of the beam reflects from the top of the bump and part from the lower adjacent area, ensuring destructive interference and very low intensity when the reflected beams combine at the detector. Bump edges are read as ones, and flat bump tops and intervening flat plains are read as zeros.

In Example 24.5 the pit depth for a standard CD, using an infrared laser of wavelength 780 nm, is calculated. DVDs use shorter wavelength lasers of 635 nm, so the track separation, pit depth, and minimum pit length are all smaller. These differences allow a DVD to store about 30 times more information than a CD.

■ EXAMPLE 24.5 | Pit Depth in a CD

GOAL Apply interference principles to a CD.

PROBLEM Find the pit depth in a CD that has a plastic transparent layer with index of refraction of 1.60 and is designed for use in a CD player using a laser with a wavelength of 7.80×10^2 nm in air.

STRATEGY (See Fig. 24.12.) Rays 1 and 2 both reflect from the metal layer, which acts like a mirror, so there is no phase difference due to reflection between those rays. There is, however, the usual phase difference caused by the extra distance $2t$ traveled by ray 2. The wavelength is λ/n, where n is the index of refraction in the substance.

SOLUTION

Use the appropriate condition for destructive interference in a thin film:

$$2t = \frac{\lambda}{2n}$$

Solve for the thickness t and substitute:

$$t = \frac{\lambda}{4n} = \frac{7.80 \times 10^2 \text{ nm}}{(4)(1.60)} = 1.22 \times 10^2 \text{ nm}$$

REMARKS Different CD systems have different tolerances for scratches. Anything that changes the reflective properties of the disk can affect the readability of the disk.

QUESTION 24.5 True or False: Given two plastics with different indices of refraction, the material with the larger index of refraction will have a larger pit depth.

EXERCISE 24.5 Repeat the example for a laser with wavelength 635 nm.

ANSWER 99.2 nm

24.6 Diffraction

LEARNING OBJECTIVES

1. Describe the physical origins of diffraction.
2. Discuss the appearance and ordering of maxima and minima in a diffraction pattern.

Suppose a light beam is incident on two slits, as in Young's double-slit experiment. If the light truly traveled in straight-line paths after passing through the slits, as in Figure 24.13a (page 848), the waves wouldn't overlap and no interference pattern would be seen. Instead, Huygens' principle requires that the waves spread out from the slits, as shown in Figure 24.13b. In other words, the light bends from a straight-line path and enters the region that would otherwise be shadowed. This spreading out of light from its initial line of travel is called **diffraction**.

In general, diffraction occurs when waves pass through small openings, around obstacles, or by sharp edges. For example, when a single narrow slit is placed between a distant light source (or a laser beam) and a screen, the light produces

848 CHAPTER 24 | Wave Optics

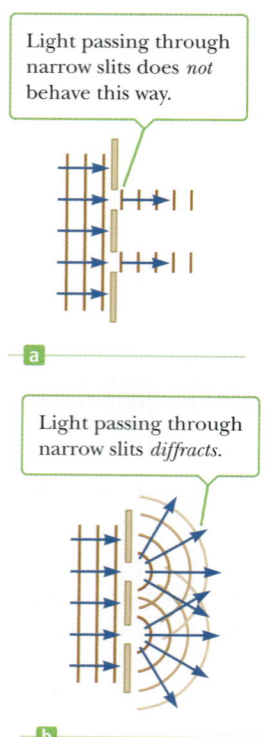

Figure 24.13 (a) If light did not spread out after passing through the slits, no interference would occur. (b) The light from the two slits overlaps as it spreads out, filling the expected shadowed regions with light and producing interference fringes.

Figure 24.16 (a) The Fraunhofer diffraction pattern of a single slit. The parallel rays are brought into focus on the screen with a converging lens. The pattern consists of a central bright region flanked by much weaker maxima. (This drawing is not to scale.) (b) A photograph of a single-slit Fraunhofer diffraction pattern.

Figure 24.14 The diffraction pattern that appears on a screen when light passes through a narrow vertical slit. The pattern consists of a broad central band and a series of less intense and narrower side bands.

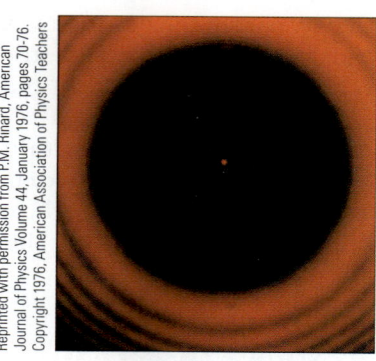

Figure 24.15 The diffraction pattern of a penny placed midway between the screen and the source. Notice the bright spot at the center.

a diffraction pattern like that in Figure 24.14. The pattern consists of a broad, intense central band flanked by a series of narrower, less intense secondary bands (called **secondary maxima**) and a series of dark bands, or **minima**. This phenomenon can't be explained within the framework of geometric optics, which says that light rays traveling in straight lines should cast a sharp image of the slit on the screen.

Figure 24.15 shows the diffraction pattern and shadow of a penny. The pattern consists of the shadow, a bright spot at its center, and a series of bright and dark circular bands of light near the edge of the shadow. The bright spot at the center (called the *Fresnel bright spot*) is explained by Augustin Fresnel's wave theory of light, which predicts constructive interference at this point for certain locations of the penny. From the viewpoint of geometric optics, there shouldn't be any bright spot: the center of the pattern would be completely screened by the penny.

One type of diffraction, called **Fraunhofer diffraction,** occurs when the rays leave the diffracting object in parallel directions. Fraunhofer diffraction can be achieved experimentally either by placing the observing screen far from the slit or by using a converging lens to focus the parallel rays on a nearby screen, as in Figure 24.16a. A bright fringe is observed along the axis at $\theta = 0$, with alternating dark and bright fringes on each side of the central bright fringe. Figure 24.16b is a photograph of a single-slit Fraunhofer diffraction pattern.

24.7 Single-Slit Diffraction

LEARNING OBJECTIVES

1. State the condition for destructive interference in a single slit experiment.
2. Apply the single-slit destructive interference condition to a single slit experiment.

Until now we have assumed slits have negligible width, acting as line sources of light. In this section we determine how their nonzero widths are the basis for understanding the nature of the Fraunhofer diffraction pattern produced by a single slit.

We can deduce some important features of this problem by examining waves coming from various portions of the slit, as shown in Figure 24.17. According to Huygens' principle, **each portion of the slit acts as a source of waves. Hence, light from one portion of the slit can interfere with light from another portion,** and the resultant intensity on the screen depends on the direction θ.

To analyze the diffraction pattern, it's convenient to divide the slit into halves, as in Figure 24.17. All the waves that originate at the slit are in phase. Consider waves 1 and 3, which originate at the bottom and center of the slit, respectively. Wave 1 travels farther than wave 3 by an amount equal to the path difference $(a/2)\sin\theta$, where a is the width of the slit. Similarly, the path difference between waves 3 and 5 is $(a/2)\sin\theta$. If this path difference is exactly half of a wavelength (corresponding to a phase difference of 180°), the two waves cancel each other, and destructive interference results. This is true, in fact, for any two waves that originate at points separated by half the slit width because the phase difference between two such points is 180°. Therefore, waves from the upper half of the slit interfere *destructively* with waves from the lower half of the slit when

$$\frac{a}{2}\sin\theta = \frac{\lambda}{2}$$

or when

$$\sin\theta = \frac{\lambda}{a}$$

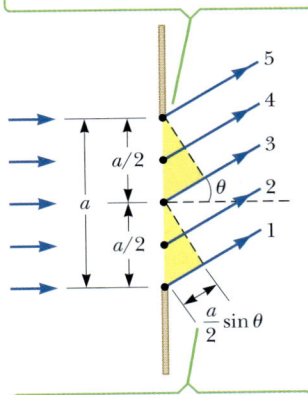

Each portion of the slit acts as a point source of light waves.

The path difference between rays 1 and 3, rays 2 and 4, or rays 3 and 5 is $(a/2)\sin\theta$.

Figure 24.17 Diffraction of light by a narrow slit of width a. (This drawing is not to scale, and the waves are assumed to converge at a distant point.)

If we divide the slit into four parts rather than two and use similar reasoning, we find that the screen is also dark when

$$\sin\theta = \frac{2\lambda}{a}$$

Continuing in this way, we can divide the slit into six parts and show that darkness occurs on the screen when

$$\sin\theta = \frac{3\lambda}{a}$$

Therefore, the general condition for **destructive interference** for a single slit of width a is

$$\sin\theta_{dark} = m\frac{\lambda}{a} \qquad m = \pm 1, \pm 2, \pm 3, \ldots \qquad [24.11]$$

Equation 24.11 gives the values of θ for which the diffraction pattern has zero intensity, where a dark fringe forms. The equation tells us nothing about the variation in intensity along the screen, however. The general features of the intensity distribution along the screen are shown in Figure 24.18. A broad central bright fringe is flanked by much weaker bright fringes alternating with dark fringes. The various dark fringes (points of zero intensity) occur at the values of θ that satisfy Equation 24.11. The points of constructive interference lie approximately halfway between the dark fringes. Note that the central bright fringe is twice as wide as the weaker maxima having $m > 1$.

Tip 24.3 The Same, but Different

Although Equations 24.2 and 24.11 have the same form, they have different meanings. Equation 24.2 describes the *bright* regions in a two-slit interference pattern, whereas Equation 24.11 describes the *dark* regions in a single-slit interference pattern.

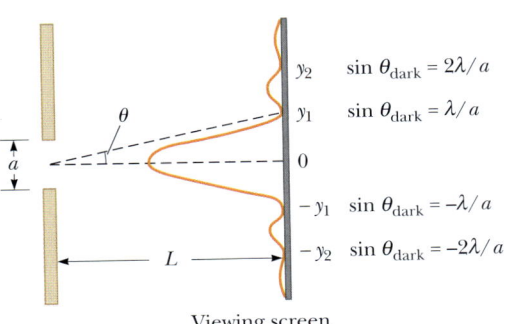

Figure 24.18 Positions of the minima for the Fraunhofer diffraction pattern of a single slit of width a. (This drawing is not to scale.)

Quick Quiz

24.5 In a single-slit diffraction experiment, as the width of the slit is made smaller, does the width of the central maximum of the diffraction pattern (a) becomes smaller, (b) become larger, or (c) remain the same?

APPLYING PHYSICS 24.4 | Diffraction of Sound Waves

If a classroom door is open even a small amount, you can hear sounds coming from the hallway, yet you can't see what is going on in the hallway. How can this difference be explained?

EXPLANATION The space between the slightly open door and the wall is acting as a single slit for waves. Sound waves have wavelengths larger than the width of the slit, so sound is effectively diffracted by the opening and the central maximum spreads throughout the room. Light wavelengths are much smaller than the slit width, so there is virtually no diffraction for the light. You must have a direct line of sight to detect the light waves.

EXAMPLE 24.6 | A Single-Slit Experiment

GOAL Find the positions of the dark fringes in single-slit diffraction.

PROBLEM Light of wavelength 5.80×10^2 nm is incident on a slit of width 0.300 mm. The observing screen is placed 2.00 m from the slit. Find the positions of the first dark fringes and the width of the central bright fringe.

STRATEGY This problem requires substitution into Equation 24.11 to find the sines of the angles of the first dark fringes. The positions can then be found with the tangent function because for small angles $\sin \theta \approx \tan \theta$. The extent of the central maximum is defined by these two dark fringes.

SOLUTION

The first dark fringes that flank the central bright fringe correspond to $m = \pm 1$ in Equation 24.11:

$$\sin \theta = \pm \frac{\lambda}{a} = \pm \frac{5.80 \times 10^{-7} \text{ m}}{0.300 \times 10^{-3} \text{ m}} = \pm 1.93 \times 10^{-3}$$

Use the triangle in Figure 24.18 to relate the position of the fringe to the tangent function:

$$\tan \theta = \frac{y_1}{L}$$

Because θ is very small, we can use the approximation $\sin \theta \approx \tan \theta$ and then solve for y_1:

$$\sin \theta \approx \tan \theta \approx \frac{y_1}{L}$$

$$y_1 \approx L \sin \theta = (2.00 \text{ m})(\pm 1.93 \times 10^{-3}) = \boxed{\pm 3.86 \times 10^{-3} \text{ m}}$$

Compute the distance between the positive and negative first-order maxima, which is the width w of the central maximum:

$$w = +3.86 \times 10^{-3} \text{ m} - (-3.86 \times 10^{-3} \text{ m}) = \boxed{7.72 \times 10^{-3} \text{ m}}$$

REMARKS Note that this value of w is much greater than the width of the slit. As the width of the slit is *increased*, however, the diffraction pattern *narrows*, corresponding to smaller values of θ. In fact, for large values of a, the maxima and minima are so closely spaced that the only observable pattern is a large central bright area resembling the geometric image of the slit. Because the width of the geometric image increases as the slit width increases, the narrowest image occurs when the geometric and diffraction widths are equal.

QUESTION 24.6 Suppose the entire apparatus is immersed in water. If the same wavelength of light (in air) is incident on the slit immersed in water, is the resulting central maximum larger or smaller? Explain.

EXERCISE 24.6 Determine the width of the first-order bright fringe in the example, when the apparatus is in air.

ANSWER 3.86 mm

24.8 The Diffraction Grating

LEARNING OBJECTIVES

1. Describe the effects of a diffraction grating on incident plane waves.
2. Discuss the order number of maxima in a diffraction pattern.
3. Apply the condition for maxima in the interference pattern of a diffraction grating to optical systems.

The diffraction grating, a useful device for analyzing light sources, consists of a large number of equally spaced parallel slits. A grating can be made by scratching parallel lines on a glass plate with a precision machining technique. The clear panes between scratches act like slits. A typical grating contains several thousand lines per centimeter. For example, a grating ruled with 5 000 lines/cm has a slit spacing d equal to the reciprocal of that number; hence, $d = (1/5\ 000)$ cm $= 2 \times 10^{-4}$ cm.

Figure 24.19 is a schematic diagram of a section of a plane diffraction grating. A plane wave is incident from the left, normal to the plane of the grating. The intensity of the pattern on the screen is the result of the combined effects of interference and diffraction. Each slit causes diffraction, and the diffracted beams in turn interfere with one another to produce the pattern. Moreover, each slit acts as a source of waves, and all waves start in phase at the slits. For some arbitrary direction θ measured from the horizontal, however, the waves must travel *different* path lengths before reaching a particular point P on the screen. In Figure 24.19, note that the path difference between waves from any two adjacent slits is $d \sin \theta$. If this path difference equals one wavelength or some integral multiple of a wavelength, waves from all slits will be in phase at P and a bright line will be observed at that point. Therefore, the condition for **maxima** in the interference pattern at the angle θ is

$$d \sin \theta_{bright} = m\lambda \qquad m = 0, \pm 1, \pm 2, \ldots \qquad [24.12]$$

◀ Condition for maxima in the interference pattern of a diffraction grating

Light emerging from a slit at an angle other than that for a maximum interferes nearly completely destructively with light from some other slit on the grating. All such pairs will result in little or no transmission in that direction, as illustrated in Figure 24.20 (page 852).

Figure 24.19 A side view of a diffraction grating. The slit separation is d, and the path difference between adjacent slits is $d \sin \theta$.

Figure 24.20 Intensity versus $\sin\theta$ for the diffraction grating. The zeroth-, first-, and second-order principal maxima are shown.

Equation 24.12 can be used to calculate the wavelength from the grating spacing and the angle of deviation, θ. The integer m is the **order number** of the diffraction pattern. If the incident radiation contains several wavelengths, each wavelength deviates through a specific angle, which can be found from Equation 24.12. All wavelengths are focused at $\theta = 0$, corresponding to $m = 0$. This point is called the *zeroth-order maximum*. The *first-order maximum*, corresponding to $m = 1$, is observed at an angle that satisfies the relationship $\sin\theta = \lambda/d$; the *second-order maximum*, corresponding to $m = 2$, is observed at a larger angle θ, and so on. Figure 24.20 is a sketch of the intensity distribution for some of the orders produced by a diffraction grating. Note the sharpness of the principal maxima and the broad range of the dark areas, a pattern in direct contrast to the broad bright fringes characteristic of the two-slit interference pattern.

A simple arrangement that can be used to measure the angles in a diffraction pattern is shown in Figure 24.21. This setup is a form of a diffraction-grating spectrometer. The light to be analyzed passes through a slit and is formed into a parallel beam by a lens. The light then strikes the grating at a 90° angle. The diffracted light leaves the grating at angles that satisfy Equation 24.12. A telescope is used to view the image of the slit. The wavelength can be determined by measuring the angles at which the images of the slit appear for the various orders.

Quick Quiz

24.6 If laser light is reflected from a phonograph record or a compact disc, a diffraction pattern appears. The pattern arises because both devices contain parallel tracks of information that act as a reflection diffraction grating. Which device, record or compact disc, results in diffraction maxima that are farther apart?

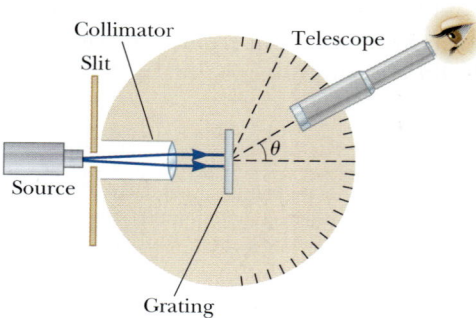

Figure 24.21 A diagram of a diffraction grating spectrometer. The collimated beam incident on the grating is diffracted into the various orders at the angles θ that satisfy the equation $d\sin\theta = m\lambda$, where $m = 0$, $\pm 1, \pm 2, \ldots$

■ APPLYING PHYSICS 24.5 | Prism vs. Grating

When white light enters through an opening in an opaque box and exits through an opening on the other side of the box, a spectrum of colors appears on the wall. From this observation, how would you be able to determine whether the box contains a prism or a diffraction grating?

EXPLANATION The determination could be made by noticing the order of the colors in the spectrum relative to the direction of the original beam of white light. For a prism, in which the separation of light is a result of dispersion, the violet light will be refracted more than the red light. Hence, the order of the spectrum from a prism will be from red, closest to the original direction, to violet. For a diffraction grating, the angle of diffraction increases with wavelength, so the spectrum from the diffraction grating will have colors in the order from violet, closest to the original direction, to red. Further, the diffraction grating will produce *two* first-order spectra on either side of the grating, whereas the prism will produce only a single spectrum. ■

24.8 | The Diffraction Grating

APPLYING PHYSICS 24.6 | Rainbows from a Compact Disc

White light reflected from the surface of a CD has a multicolored appearance, as shown in Figure 24.22. The observation depends on the orientation of the disc relative to the eye and the position of the light source. Explain how all this works.

Figure 24.22 (Applying Physics 24.5) Compact discs act as diffraction gratings when observed under white light.

EXPLANATION The surface of a CD has a spiral-shaped track (with a spacing of approximately 1 μm) that acts as a reflection grating. The light scattered by these closely spaced parallel tracks interferes constructively in certain directions that depend on both the wavelength and the direction of the incident light. Any one section of the disc serves as a diffraction grating for white light, sending beams of constructive interference for different colors in different directions. The different colors you see when viewing one section of the disc change as the light source, the disc, or you move to change the angles of incidence or diffraction. ■

Use of a Diffraction Grating in CD Tracking

If a CD player is to reproduce sound faithfully, the laser beam must follow the spiral track of information perfectly. Sometimes the laser beam can drift off track, however, and without a feedback procedure to let the player know that is happening, the fidelity of the music can be greatly reduced.

Figure 24.23 shows how a diffraction grating is used in a three-beam method to keep the beam on track. The central maximum of the diffraction pattern reads the information on the CD track, and the two first-order maxima steer the beam. The grating is designed so that the first-order maxima fall on the smooth surfaces on either side of the information track. Both of these reflected beams have their own detectors, and because both beams are reflected from smooth surfaces, they should have the same strong intensity when they are detected. If the central beam wanders off the track, however, one of the steering beams will begin to strike bumps on the information track and the amount of light reflected will decrease. This information is then used by electronic circuits to drive the main beam back to its desired location.

APPLICATION
Tracking Information on a CD

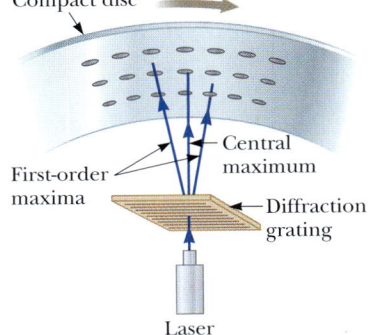

Figure 24.23 The laser beam in a CD player is able to follow the spiral track by using three beams produced with a diffraction grating.

■ EXAMPLE 24.7 | A Diffraction Grating

GOAL Calculate different-order principal maxima for a diffraction grating.

PROBLEM Monochromatic light from a helium–neon laser ($\lambda = 632.8$ nm) is incident normally on a diffraction grating containing 6.00×10^3 lines/cm. Find the angles at which one would observe the first-order maximum, the second-order maximum, and so forth.

(Continued)

STRATEGY Find the slit separation by inverting the number of lines per centimeter, then substitute values into Equation 24.12.

SOLUTION

Invert the number of lines per centimeter to obtain the slit separation:

$$d = \frac{1}{6.00 \times 10^3 \text{ cm}^{-1}} = 1.67 \times 10^{-4} \text{ cm} = 1.67 \times 10^3 \text{ nm}$$

Substitute $m = 1$ into Equation 24.12 to find the sine of the angle corresponding to the first-order maximum:

$$\sin \theta_1 = \frac{\lambda}{d} = \frac{632.8 \text{ nm}}{1.67 \times 10^3 \text{ nm}} = 0.379$$

Take the inverse sine of the preceding result to find θ_1:

$$\theta_1 = \sin^{-1} 0.379 = \boxed{22.3°}$$

Repeat the calculation for $m = 2$:

$$\sin \theta_2 = \frac{2\lambda}{d} = \frac{2(632.8 \text{ nm})}{1.67 \times 10^3 \text{ nm}} = 0.758$$

$$\theta_2 = \boxed{49.3°}$$

Repeat the calculation for $m = 3$:

$$\sin \theta_3 = \frac{3\lambda}{d} = \frac{3(632.8 \text{ nm})}{1.67 \times 10^3 \text{ nm}} = 1.14$$

Because $\sin \theta$ can't exceed 1, there is no solution for θ_3.

REMARKS The foregoing calculation shows that there can only be a finite number of principal maxima. In this case only zeroth-, first-, and second-order maxima would be observed.

QUESTION 24.7 Does a diffraction grating with more lines have a smaller or larger separation between adjacent principal maxima?

EXERCISE 24.7 Suppose light with wavelength 7.80×10^2 nm is used instead and the diffraction grating has 3.30×10^3 lines per centimeter. Find the angles of all the principal maxima.

ANSWERS 0°, 14.9°, 31.0°, 50.6°

24.9 Polarization of Light Waves

LEARNING OBJECTIVES

1. Discuss the concept of polarization of light waves.
2. Describe three processes for obtaining linearly polarized light from an unpolarized source.
3. Apply Malus's and Brewster's Laws to systems involving unpolarized light.
4. Discuss the concept of optical activity and the practical applications of liquid crystals.

In Chapter 21 we described the transverse nature of electromagnetic waves. Figure 24.24 shows that the electric and magnetic field vectors associated with an electromagnetic wave are at right angles to each other and also to the direction of wave propagation. The phenomenon of polarization, described in this section, is firm evidence of the transverse nature of electromagnetic waves.

An ordinary beam of light consists of a large number of electromagnetic waves emitted by the atoms or molecules of the light source. The vibrating charges associated with the atoms act as tiny antennas. Each atom produces a wave with its own orientation of $\vec{E}$, as in Figure 24.24, corresponding to the direction of atomic vibration. Because all directions of vibration are possible, however, the resultant electromagnetic wave is a superposition of waves produced by the individual atomic sources. The result is an **unpolarized** light wave, represented schematically in Figure 24.25a. The direction of wave propagation shown in the figure is

Figure 24.24 A schematic diagram of a polarized electromagnetic wave propagating in the x-direction. The electric field vector $\vec{E}$ vibrates in the xy-plane, whereas the magnetic field vector $\vec{B}$ vibrates in the xz-plane.

perpendicular to the page. Note that *all* directions of the electric field vector are equally probable and lie in a plane (such as the plane of this page) perpendicular to the direction of propagation.

A wave is said to be **linearly polarized** if the resultant electric field $\vec{E}$ vibrates in the same direction *at all times* at a particular point, as in Figure 24.25b. (Sometimes such a wave is described as *plane polarized* or simply *polarized*.) The wave in Figure 24.24 is an example of a wave that is linearly polarized in the y-direction. As the wave propagates in the x-direction, $\vec{E}$ is always in the y-direction. The plane formed by $\vec{E}$ and the direction of propagation is called the *plane of polarization* of the wave. In Figure 24.24 the plane of polarization is the *xy*-plane.

It's possible to obtain a linearly polarized beam from an unpolarized beam by removing all waves from the beam except those with electric field vectors that oscillate in a single plane. We now discuss three processes for doing this: (1) selective absorption, (2) reflection, and (3) scattering.

Polarization by Selective Absorption

The most common technique for polarizing light is to use a material that transmits waves having electric field vectors that vibrate in a plane parallel to a certain direction and absorbs those waves with electric field vectors vibrating in directions perpendicular to that direction.

In 1932 E. H. Land discovered a material, which he called **Polaroid**, that polarizes light through selective absorption by oriented molecules. This material is fabricated in thin sheets of long-chain hydrocarbons, which are stretched during manufacture so that the molecules align. After a sheet is dipped into a solution containing iodine, the molecules become good electrical conductors. Conduction takes place primarily along the hydrocarbon chains, however, because the valence electrons of the molecules can move easily only along those chains. (Recall that valence electrons are "free" electrons that can move easily through the conductor.) As a result, the molecules readily *absorb* light having an electric field vector parallel to their lengths and *transmit* light with an electric field vector perpendicular to their lengths. It's common to refer to the direction perpendicular to the molecular chains as the **transmission axis**. In an ideal polarizer all light with $\vec{E}$ parallel to the transmission axis is transmitted and all light with $\vec{E}$ perpendicular to the transmission axis is absorbed.

Polarizing material reduces the intensity of light passing through it. In Figure 24.26 an unpolarized light beam is incident on the first polarizing sheet, called the **polarizer**; the transmission axis is as indicated. The light that passes through this sheet is polarized vertically, and the transmitted electric field vector is $\vec{E}_0$. A second polarizing sheet, called the **analyzer**, intercepts this beam with its transmission axis at an angle of θ to the axis of the polarizer. The component of $\vec{E}_0$ that is perpendicular to the axis of the analyzer is completely absorbed. The component of $\vec{E}_0$ that is parallel to the analyzer axis, $E_0 \cos \theta$, is allowed to pass through the

Figure 24.25 (a) An unpolarized light beam viewed along the direction of propagation. The transverse electric field vector can vibrate in any direction with equal probability. (b) A linearly polarized light beam with the electric field vector vibrating in the vertical direction.

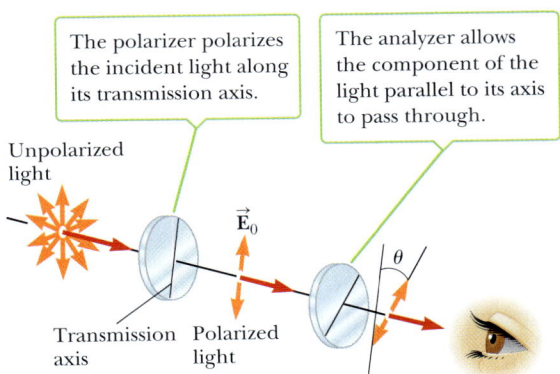

Figure 24.26 Two polarizing sheets whose transmission axes make an angle θ with each other. Only a fraction of the polarized light incident on the analyzer is transmitted.

Figure 24.27 The intensity of light transmitted through two polarizers depends on the relative orientations of their transmission axes.

The transmitted light has *maximum* intensity when the transmission axes are *aligned* with each other.

The transmitted light intensity diminishes when the transmission axes are at an angle of 45° with each other.

The transmitted light intensity is a *minimum* when the transmission axes are at *right angles* to each other.

analyzer. Because the intensity of the transmitted beam varies as the *square* of its amplitude E, we conclude that the intensity of the (polarized) beam transmitted through the analyzer varies as

Malus's law ▶

$$I = I_0 \cos^2 \theta \qquad [24.13]$$

where I_0 is the intensity of the polarized wave incident on the analyzer. This expression, known as **Malus's law**, applies to any two polarizing materials having transmission axes at an angle of θ to each other. Note from Equation 24.13 that the transmitted intensity is a maximum when the transmission axes are parallel ($\theta = 0$ or $180°$) and is a minimum (complete absorption by the analyzer) when the transmission axes are perpendicular to each other. This variation in transmitted intensity through a pair of polarizing sheets is illustrated in Figure 24.27.

When unpolarized light of intensity I_0 is sent through a single ideal polarizer, the transmitted linearly polarized light has intensity $I_0/2$. This fact follows from Malus's law because the average value of $\cos^2 \theta$ is one-half.

■ APPLYING PHYSICS 24.7 | Polarizing Microwaves

A polarizer for microwaves can be made as a grid of parallel metal wires about 1 cm apart. Is the electric field vector for microwaves transmitted through this polarizer parallel or perpendicular to the metal wires?

EXPLANATION Electric field vectors parallel to the metal wires cause electrons in the metal to oscillate parallel to the wires. Thus, the energy from the waves with these electric field vectors is transferred to the metal by accelerating the electrons and is eventually transformed to internal energy through the resistance of the metal. Waves with electric field vectors perpendicular to the metal wires are not able to accelerate electrons and pass through the wires. Consequently, the electric field polarization is perpendicular to the metal wires. ■

■ EXAMPLE 24.8 | Polarizer

GOAL Understand how polarizing materials affect light intensity.

PROBLEM Unpolarized light is incident upon three polarizers. The first polarizer has a vertical transmission axis, the second has a transmission axis rotated 30.0° with respect to the first, and the third has a transmission axis rotated 75.0° relative to the first. If the initial light intensity of the beam is I_b, calculate the light intensity after the beam passes through **(a)** the second polarizer and **(b)** the third polarizer.

STRATEGY After the beam passes through the first polarizer, it is polarized and its intensity is cut in half. Malus's law can then be applied to the second and third polarizers. The angle used in Malus's law must be relative to the immediately preceding transmission axis.

SOLUTION

(a) Calculate the intensity of the beam after it passes through the second polarizer.

The incident intensity is $I_b/2$. Apply Malus's law to the second polarizer:

$$I_2 = I_0 \cos^2 \theta = \frac{I_b}{2} \cos^2 (30.0°) = \frac{I_b}{2} \left(\frac{\sqrt{3}}{2} \right)^2 = \frac{3}{8} I_b$$

(b) Calculate the intensity of the beam after it passes through the third polarizer.

The incident intensity is now $3I_b/8$. Apply Malus's law to the third polarizer.

$$I_3 = I_2 \cos^2 \theta = \frac{3}{8} I_b \cos^2 (45.0°) = \frac{3}{8} I_b \left(\frac{\sqrt{2}}{2} \right)^2 = \frac{3}{16} I_b$$

REMARKS Notice that the angle used in part (b) was not 75.0°, but 75.0° − 30.0° = 45.0°. The angle is always with respect to the previous polarizer's transmission axis because the polarizing material physically determines what direction the transmitted electric fields can have.

QUESTION 24.8 At what angle relative to the previous polarizer must an additional polarizer be placed so as to completely block the light?

EXERCISE 24.8 The polarizers are rotated so that the second polarizer has a transmission axis of 40.0° with respect to the first polarizer, and the third polarizer has an angle of 90.0° with respect to the first. If I_b is the intensity of the original unpolarized light, what is the intensity of the beam after it passes through (a) the second polarizer and (b) the third polarizer? (c) What is the final transmitted intensity if the second polarizer is removed?

ANSWERS (a) $0.293 I_b$ (b) $0.121 I_b$ (c) 0

Polarization by Reflection

When an unpolarized light beam is reflected from a surface, the reflected light is completely polarized, partially polarized, or unpolarized, depending on the angle of incidence. If the angle of incidence is either 0° or 90° (a normal or grazing angle), the reflected beam is unpolarized. For angles of incidence between 0° and 90°, however, the reflected light is polarized to some extent. For one particular angle of incidence the reflected beam is completely polarized.

Suppose an unpolarized light beam is incident on a surface, as in Figure 24.28a. The beam can be described by two electric field components, one parallel to the surface (represented by dots) and the other perpendicular to the first component and to the direction of propagation (represented by orange arrows). It is found

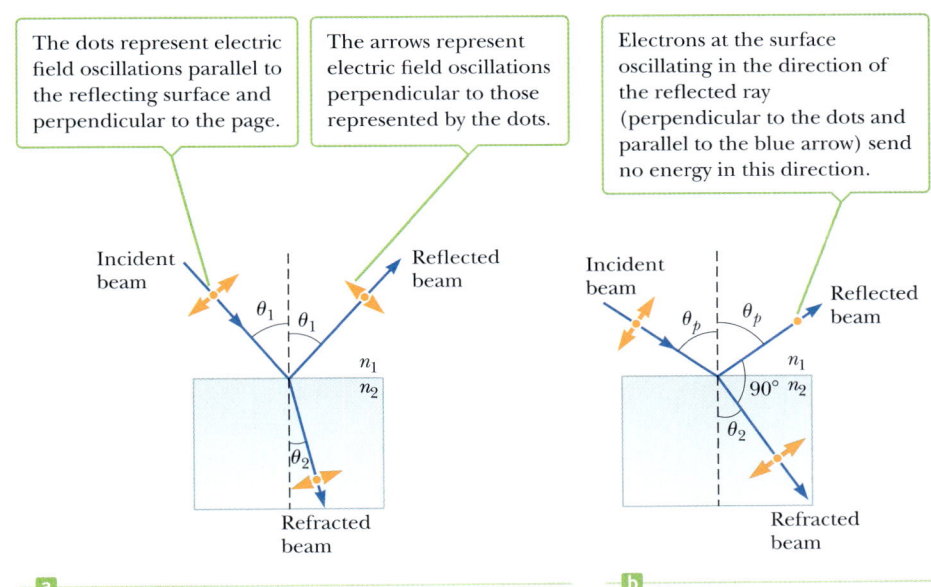

Figure 24.28 (a) When unpolarized light is incident on a reflecting surface, the reflected and refracted beams are partially polarized. (b) The reflected beam is completely polarized when the angle of incidence equals the polarizing angle θ_p, satisfying the equation $n = \tan \theta_p$.

that the parallel component reflects more strongly than the other components, and the result is a partially polarized beam. In addition, the refracted beam is also partially polarized.

Now suppose the angle of incidence, θ_1, is varied until the angle between the reflected and refracted beams is 90° (Fig. 24.28b). At this particular angle of incidence, called the **polarizing angle** θ_p, the reflected beam is completely polarized, with its electric field vector parallel to the surface, while the refracted beam is partially polarized.

An expression relating the polarizing angle to the index of refraction of the reflecting surface can be obtained by the use of Figure 24.28b. From this figure, we see that at the polarizing angle, $\theta_p + 90° + \theta_2 = 180°$, so $\theta_2 = 90° - \theta_p$. Using Snell's law and taking $n_1 = n_{air} = 1.00$ and $n_2 = n$ yields

$$n = \frac{\sin \theta_1}{\sin \theta_2} = \frac{\sin \theta_p}{\sin \theta_2}$$

Because $\sin \theta_2 = \sin(90° - \theta_p) = \cos \theta_p$, the expression for n can be written

◀ Brewster's law

$$n = \frac{\sin \theta_p}{\cos \theta_p} = \tan \theta_p \qquad [24.14]$$

Equation 24.14 is called **Brewster's law**, and the polarizing angle θ_p is sometimes called **Brewster's angle** after its discoverer, Sir David Brewster (1781–1868). For example, Brewster's angle for crown glass (where $n = 1.52$) has the value $\theta_p = \tan^{-1}(1.52) = 56.7°$. Because n varies with wavelength for a given substance, Brewster's angle is also a function of wavelength.

APPLICATION
Polaroid Sunglasses

Polarization by reflection is a common phenomenon. Sunlight reflected from water, glass, or snow is partially polarized. If the surface is horizontal, the electric field vector of the reflected light has a strong horizontal component. Sunglasses made of polarizing material reduce the glare, which *is* the reflected light. The transmission axes of the lenses are oriented vertically to absorb the strong horizontal component of the reflected light. Because the reflected light is mostly polarized, most of the glare can be eliminated without removing most of the normal light.

Polarization by Scattering

When light is incident on a system of particles, such as a gas, the electrons in the medium can absorb and reradiate part of the light. The absorption and reradiation of light by the medium, called **scattering**, is what causes sunlight reaching an observer on Earth from straight overhead to be polarized. You can observe this effect by looking directly up through a pair of sunglasses made of polarizing glass. Less light passes through at certain orientations of the lenses than at others.

Figure 24.29 illustrates how the sunlight becomes polarized. The left side of the figure shows an incident unpolarized beam of sunlight on the verge of striking an air molecule. When the beam strikes the air molecule, it sets the electrons of the molecule into vibration. These vibrating charges act like those in an antenna except that they vibrate in a complicated pattern. The horizontal part of the electric field vector in the incident wave causes the charges to vibrate horizontally, and the vertical part of the vector simultaneously causes them to vibrate vertically. A horizontally polarized wave is emitted by the electrons as a result of their horizontal motion, and a vertically polarized wave is emitted parallel to Earth as a result of their vertical motion.

Scientists have found that bees and homing pigeons use the polarization of sunlight as a navigational aid.

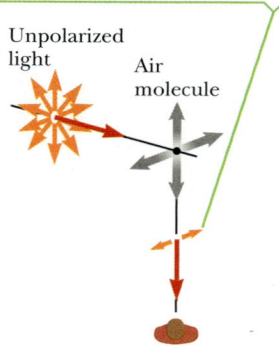

Figure 24.29 The scattering of unpolarized sunlight by air molecules.

The scattered light traveling perpendicular to the incident light is plane-polarized because the vertical vibrations of the charges in the air molecule send no light in this direction.

Optical Activity

Many important practical applications of polarized light involve the use of certain materials that display the property of **optical activity**. A substance is said to be optically active if it rotates the plane of polarization of transmitted light. Suppose

24.9 | Polarization of Light Waves

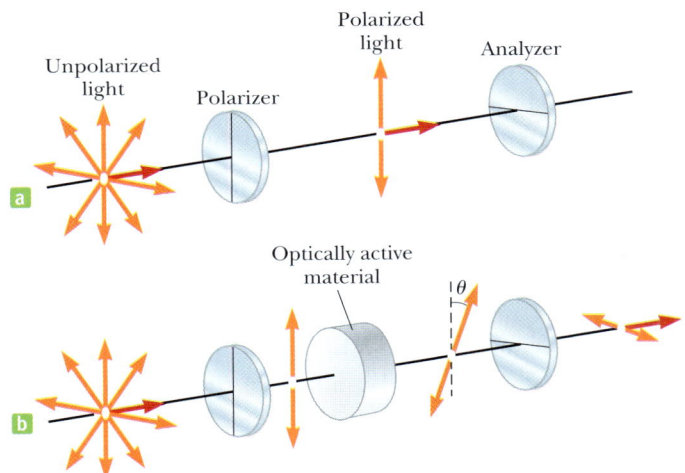

Figure 24.30 (a) When crossed polarizers are used, none of the polarized light can pass through the analyzer. (b) An optically active material rotates the direction of polarization through the angle θ, enabling some of the polarized light to pass through the analyzer.

unpolarized light is incident on a polarizer from the left, as in Figure 24.30a. The transmitted light is polarized vertically, as shown. If this light is then incident on an analyzer with its axis perpendicular to that of the polarizer, no light emerges from it. If an optically active material is placed between the polarizer and analyzer, as in Figure 24.30b, the material causes the direction of the polarized beam to rotate through the angle θ. As a result, some light is able to pass through the analyzer. The angle through which the light is rotated by the material can be found by rotating the polarizer until the light is again extinguished. It is found that the angle of rotation depends on the length of the sample and, if the substance is in solution, on the concentration. One optically active material is a solution of common sugar, dextrose. A standard method for determining the concentration of a sugar solution is to measure the rotation produced by a fixed length of the solution.

Optical activity occurs in a material because of an asymmetry in the shape of its constituent molecules. For example, some proteins are optically active because of their spiral shapes. Other materials, such as glass and plastic, become optically active when placed under stress. If polarized light is passed through an unstressed piece of plastic and then through an analyzer with an axis perpendicular to that of the polarizer, none of the polarized light is transmitted. If the plastic is placed under stress, however, the regions of greatest stress produce the largest angles of rotation of polarized light, and a series of light and dark bands are observed in the transmitted light. Engineers often use this property in the design of structures ranging from bridges to small tools. A plastic model is built and analyzed under different load conditions to determine positions of potential weakness and failure under stress. If the design is poor, patterns of light and dark bands will indicate the points of greatest weakness, and the design can be corrected at an early stage. Figure 24.31 shows examples of stress patterns in plastic.

APPLICATION
Finding the Concentrations of Solutions by Means of Their Optical Activity

Figure 24.31 A plastic model of an arch structure under load conditions observed between perpendicular polarizers. Such patterns are useful in the optimum design of architectural components.

APPLICATION
Liquid Crystal Displays (LCDs)

Liquid Crystals

An effect similar to rotation of the plane of polarization is used to create the familiar displays on pocket calculators, wristwatches, notebook computers, and so forth. The properties of a unique substance called a liquid crystal make these displays (called LCDs, for *liquid crystal displays*) possible. As its name implies, a **liquid crystal** is a substance with properties intermediate between those of a crystalline solid and those of a liquid; that is, the molecules of the substance are more orderly than those in a liquid, but less orderly than those in a pure crystalline solid. The forces that hold the molecules together in such a state are just barely strong enough to enable the substance to maintain a definite shape, so it is reasonable to call it a solid. Small inputs of mechanical or electrical energy, however, can disrupt these weak bonds and make the substance flow, rotate, or twist.

To see how liquid crystals can be used to create a display, consider Figure 24.32a. The liquid crystal is placed between two glass plates in the pattern shown, and electrical contacts, indicated by the thin lines, are made. When a voltage is applied across any segment in the display, that segment turns dark. In this fashion any number between 0 and 9 can be formed by the pattern, depending on the voltages applied to the seven segments.

To see why a segment can be changed from dark to light by the application of a voltage, consider Figure 24.32b, which shows the basic construction of a portion of the display. The liquid crystal is placed between two glass substrates that are packaged between two pieces of Polaroid material with their transmission axes perpendicular. A reflecting surface is placed behind one of the pieces of Polaroid. First consider what happens when light falls on this package and no voltages are applied to the liquid crystal, as shown in Figure 24.32b. Incoming light is polarized by the polarizer on the left and then falls on the liquid crystal. As the light passes through the crystal, its plane of polarization is rotated by 90°, allowing it to pass through the polarizer on the right. It reflects from the reflecting surface and

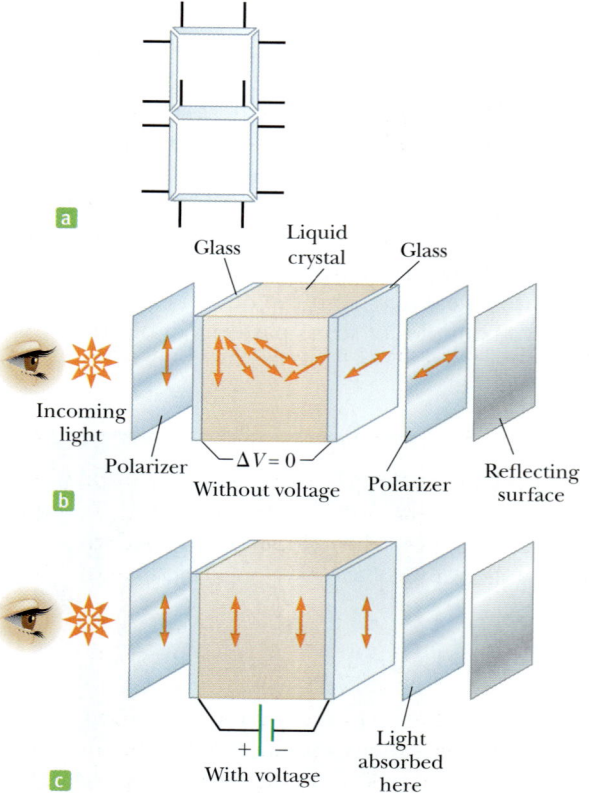

Figure 24.32 (a) The light-segment pattern of a liquid crystal display. (b) Rotation of a polarized light beam by a liquid crystal when the applied voltage is zero. (c) Molecules of the liquid crystal align with the electric field when a voltage is applied.

retraces its path through the crystal. Thus, an observer to the left of the crystal sees the segment as being bright. When a voltage is applied as in Figure 24.32c, the molecules of the liquid crystal don't rotate the plane of polarization of the light. In this case the light is absorbed by the polarizer on the right, and none is reflected back to the observer to the left of the crystal. As a result, the observer sees this segment as black. Changing the applied voltage to the crystal in a precise pattern at precise times can make the pattern tick off the seconds on a watch, display a letter on a computer display, and so forth.

SUMMARY

24.1 Conditions for Interference

Interference occurs when two or more light waves overlap at a given point. A sustained interference pattern is observed if (1) the sources are coherent (that is, they maintain a constant phase relationship with one another), (2) the sources have identical wavelengths, and (3) the superposition principle is applicable.

24.2 Young's Double-Slit Experiment

In **Young's double-slit experiment** two slits separated by distance d are illuminated by a single-wavelength light source. An interference pattern consisting of bright and dark fringes is observed on a screen a distance L from the slits. The condition for **bright fringes** (constructive interference) is

$$d \sin \theta_{bright} = m\lambda \quad m = 0, \pm 1, \pm 2, \ldots \quad [24.2]$$

The number m is called the **order number** of the fringe.

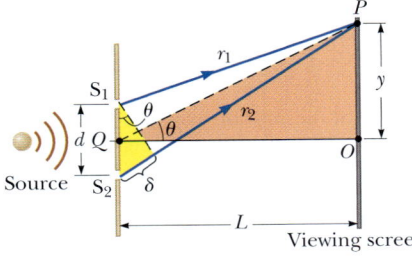

A geometric construction that describes Young's double-slit experiment. (This figure is not drawn to scale.)

The condition for **dark fringes** (destructive interference) is

$$d \sin \theta_{dark} = (m + \tfrac{1}{2})\lambda \quad m = 0, \pm 1, \pm 2, \ldots \quad [24.3]$$

The position y_m of the bright fringes on the screen can be determined by using the relation $\sin \theta \approx \tan \theta = y_m/L$, which is true for small angles. This relation can be substituted into Equations 24.2 and 24.3, yielding the location of the bright fringes:

$$y_{bright} = \frac{\lambda L}{d} m \quad m = 0, \pm 1, \pm 2, \ldots \quad [24.5]$$

A similar expression can be derived for the dark fringes. This equation can be used either to locate the maxima or to determine the wavelength of light by measuring y_m.

24.3 Change of Phase Due to Reflection

24.4 Interference in Thin Films

An electromagnetic wave undergoes a phase change of 180° on reflection from a medium with an index of refraction higher than that of the medium in which the wave is traveling. There is no change when the wave, traveling in a medium with higher index of refraction, reflects from a medium with a lower index of refraction.

Interference in light reflected from a thin film is due to a combination of rays 1 and 2 reflected from the upper and lower surfaces of the film.

The wavelength λ_n of light in a medium with index of refraction n is

$$\lambda_n = \frac{\lambda}{n} \quad [24.7]$$

where λ is the wavelength of the light in free space. Light encountering a thin film of thickness t will reflect off the top and bottom of the film, each ray undergoing a possible phase change as described above. The two rays recombine, and bright and dark fringes will be observed, with the conditions of interference given by the following table:

Equation ($m = 0, 1, \ldots$)		1 Phase Reversal	0 or 2 Phase Reversals
$2nt = (m + \tfrac{1}{2})\lambda$	[24.9]	Constructive	Destructive
$2nt = m\lambda$	[24.10]	Destructive	Constructive

24.6 Diffraction

24.7 Single-Slit Diffraction

Diffraction occurs when waves pass through small openings, around obstacles, or by sharp edges. The **diffraction pattern** produced by a single slit on a distant screen consists of a central bright maximum flanked by less bright fringes alternating with dark regions. The angles θ at which the diffraction pattern has zero intensity (regions of destructive interference) are described by

$$\sin \theta_{dark} = m\frac{\lambda}{a} \quad m = \pm 1, \pm 2, \pm 3, \ldots \quad [24.11]$$

where λ is the wavelength of the light incident on the slit and a is the width of the slit.

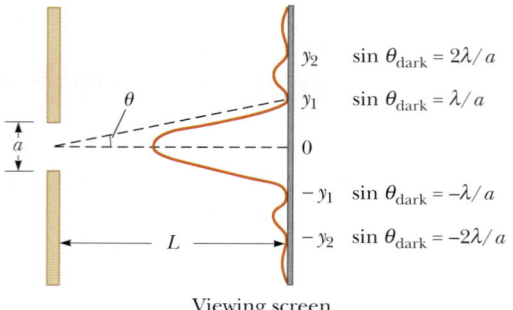

Positions of the minima for the Fraunhofer diffraction pattern of a single slit of width a. (This drawing is not to scale.)

24.8 The Diffraction Grating

A **diffraction grating** consists of many equally spaced, identical slits. The condition for **maximum intensity** in the interference pattern of a diffraction grating is

$$d \sin \theta_{bright} = m\lambda \quad m = 0, \pm 1, \pm 2, \ldots \quad [24.12]$$

where d is the spacing between adjacent slits and m is the order number of the diffraction pattern. A diffraction grating can be made by putting a large number of evenly spaced scratches on a glass slide. The number of such lines per centimeter is the inverse of the spacing d.

24.9 Polarization of Light Waves

Unpolarized light can be polarized by selective absorption, reflection, or scattering. A material can polarize light if it transmits waves having electric field vectors that vibrate in a plane parallel to a certain direction and absorbs waves with electric field vectors vibrating in directions perpendicular to that direction. When unpolarized light passes through a polarizing sheet, its intensity is reduced by half and the light becomes polarized. When this light passes through a second polarizing sheet with transmission axis at an angle of θ with respect to the transmission axis of the first sheet, the transmitted intensity is given by

$$I = I_0 \cos^2 \theta \quad [24.13]$$

where I_0 is the intensity of the light after passing through the first polarizing sheet.

In general, light reflected from an amorphous material, such as glass, is partially polarized. Reflected light is completely polarized, with its electric field parallel to the surface, when the angle of incidence produces a 90° angle between the reflected and refracted beams. This angle of incidence, called the **polarizing angle** θ_p, satisfies **Brewster's law,** given by

$$n = \tan \theta_p \quad [24.14]$$

where n is the index of refraction of the reflecting medium.

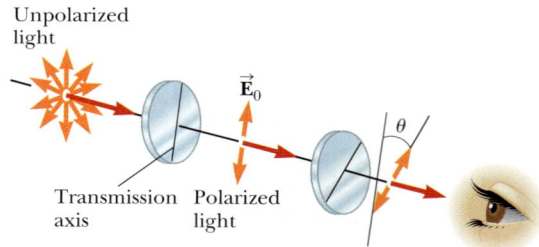

Two polarizing sheets, with transmission axes at angle θ, transmit only a fraction of the incident light.

■ WARM-UP EXERCISES

WebAssign The warm-up exercises in this chapter may be assigned online in Enhanced WebAssign.

1. **Physics Review** Two speakers, one at the origin and the other facing it at $x = 1.00$ m, are driven by the same oscillator at a frequency of 655 Hz. If the speed of sound is 343 m/s, find (a) the wavelength of sound waves produced by the speakers and (b) the location of a point of destructive interference closest to 0.500 m on the interval [0, 0.500 m]. (See Section 14.7.)

2. **Physics Review** (a) Calculate the frequency of monochromatic light having a wavelength in air of 5.30×10^2 nm. Find (b) the wavelength and (c) the frequency of the same light when inside diamond. (See Sections 21.12 and 22.3.)

3. In a Young's double-slit experiment, 633-nm laser light illuminates two slits separated by a distance of 3.50×10^{-5} m. (a) Find the angle at which the first order maximum occurs. (b) If the interference pattern is displayed on a screen 1.50 m away, determine the distance from the centerline to the first order maximum. Find also

(c) the angle at which the second-order ($m = 2$) minimum occurs and (d) the distance of that minimum from the centerline. (See Section 24.2.)

4. For a 4.76×10^{14}-Hz light wave, determine the wavelength (a) in space where the index of refraction is 1.00 and (b) in water where the index of refraction is 1.33. (See Section 24.4.)

5. Suppose the thin-film coating ($n = 1.17$) on an eyeglass lens ($n = 1.33$) is designed to eliminate reflection of 535-nm light. Determine (a) the number of phase inversions that occur, (b) the wavelength of light in the thin-film coating and (c) the minimum required coating thickness. Take the index of refraction for air to be 1.00. (See Section 24.4.)

6. Suppose 633-nm laser light illuminates a single slit of width 2.50×10^{-4} m. Determine the angle from the laser beam's path to the third dark fringe. (See Section 24.7.)

7. Light from a 543-nm laser beam is normally incident on a diffraction grating with 4.50×10^5 lines/m. Determine (a) the spacing between slits in the grating and (b) the angle to the third-order maximum. (See Section 24.8.)

8. An unpolarized beam of light with intensity of 50.0 W/m² passes through a polarizer with a vertical transmission axis. Determine the intensity of the light beam after passing through the polarizer. (See Section 24.9.)

9. A beam of light has an intensity of 70.0 W/m² and is polarized along the horizontal axis. Determine the intensity of the light beam after passing through a polarizer with a transmission axis oriented 30.0° from the horizontal. (See Section 24.9.)

10. Determine Brewster's angle for light reflected from water with an index of refraction equal to 1.33. (See Section 24.9.)

CONCEPTUAL QUESTIONS

The conceptual questions in this chapter may be assigned online in Enhanced WebAssign.

1. Your automobile has two headlights. What sort of interference pattern do you expect to see from them? Why?

2. Holding your hand at arm's length, you can readily block direct sunlight from your eyes. Why can you not block sound from your ears this way?

3. Consider a dark fringe in an interference pattern at which almost no light energy is arriving. Light from both slits is arriving at this point, but the waves cancel. Where does the energy go?

4. If Young's double-slit experiment were performed under water, how would the observed interference pattern be affected?

5. In a laboratory accident, you spill two liquids onto water, neither of which mixes with the water. They both form thin films on the water surface. As the films spread and become very thin, you notice that one film becomes bright and the other black in reflected light. Why might that be?

6. If white light is used in Young's double-slit experiment rather than monochromatic light, how does the interference pattern change?

7. A lens with outer radius of curvature R and index of refraction n rests on a flat glass plate, and the combination is illuminated from white light from above. Is there a dark spot or a light spot at the center of the lens? What does it mean if the observed rings are noncircular?

8. Fingerprints left on a piece of glass such as a windowpane can show colored spectra like that from a diffraction grating. Why?

9. In everyday experience, why are radio waves polarized, whereas light is not?

10. Suppose reflected white light is used to observe a thin, transparent coating on glass as the coating material is gradually deposited by evaporation in a vacuum. Describe some color changes that might occur during the process of building up the thickness of the coating.

11. Would it be possible to place a nonreflective coating on an airplane to cancel radar waves of wavelength 3 cm?

12. Certain sunglasses use a polarizing material to reduce the intensity of light reflected from shiny surfaces, such as water or the hood of a car. What orientation of the transmission axis should the material have to be most effective?

13. Why is it so much easier to perform interference experiments with a laser than with an ordinary light source?

14. A soap film is held vertically in air and is viewed in reflected light as in Figure CQ24.14. Explain why the film appears to be dark at the top.

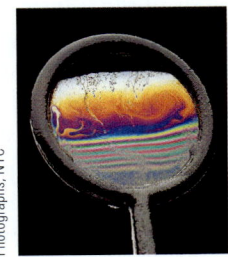

Figure CQ24.14

15. A plane monochromatic light wave is incident on a double-slit as illustrated in Figure 24.4. As the slit separation decreases, what happens to the separation between the interference fringes on the screen? (a) It decreases. (b) It increases. (c) It remains the same.

(d) It may increase or decrease, depending on the wavelength of the light. (e) More information is required.

16. A plane monochromatic light wave is incident on a double-slit as illustrated in Figure 24.4. If the viewing screen is moved away from the double slit, what happens to the separation between the interference fringes on the screen? (a) It increases. (b) It decreases. (c) It remains the same. (d) It may increase or decrease, depending on the wavelength of the light. (e) More information is required.

PROBLEMS

WebAssign The problems in this chapter may be assigned online in Enhanced WebAssign.

1. denotes straightforward problem; 2. denotes intermediate problem; 3. denotes challenging problem

1. denotes full solution available in *Student Solutions Manual/Study Guide*

1. denotes problems most often assigned in Enhanced WebAssign

BIO denotes biomedical problems
GP denotes guided problems
M denotes Master It tutorial available in Enhanced WebAssign
Q|C denotes asking for quantitative and conceptual reasoning
S denotes symbolic reasoning problem
W denotes Watch It video solution available in Enhanced WebAssign

24.2 Young's Double-Slit Experiment

1. **W** A laser beam is incident on two slits with a separation of 0.200 mm, and a screen is placed 5.00 m from the slits. If the bright interference fringes on the screen are separated by 1.58 cm, what is the wavelength of the laser light?

2. In a Young's double-slit experiment, a set of parallel slits with a separation of 0.100 mm is illuminated by light having a wavelength of 589 nm, and the interference pattern is observed on a screen 4.00 m from the slits. (a) What is the difference in path lengths from each of the slits to the location of a third-order bright fringe on the screen? (b) What is the difference in path lengths from the two slits to the location of the third dark fringe on the screen, away from the center of the pattern?

3. A pair of narrow, parallel slits separated by 0.250 mm is illuminated by the green component from a mercury vapor lamp (λ = 546.1 nm). The interference pattern is observed on a screen 1.20 m from the plane of the parallel slits. Calculate the distance (a) from the central maximum to the first bright region on either side of the central maximum and (b) between the first and second dark bands in the interference pattern.

4. Light of wavelength 620 nm falls on a double slit, and the first bright fringe of the interference pattern is seen at an angle of 15.0° from the central maximum. Find the separation between the slits.

5. In a location where the speed of sound is 354 m/s, a 2 000-Hz sound wave impinges on two slits 30.0 cm apart. (a) At what angle is the first maximum located? (b) If the sound wave is replaced by 3.00-cm microwaves, what slit separation gives the same angle for the first maximum? (c) If the slit separation is 1.00 μm, what frequency of light gives the same first maximum angle?

6. A double slit separated by 0.058 0 mm is placed 1.50 m from a screen. (a) If yellow light of wavelength 588 nm strikes the double slit, what is the separation between the zeroth-order and first-order maxima on the screen? (b) If blue light of wavelength 412 nm strikes the double slit, what is the separation between the second-order and fourth-order maxima?

7. **M** Two radio antennas separated by d = 300 m, as shown in Figure P24.7, simultaneously broadcast identical signals at the same wavelength. A car travels due north along a straight line at position x = 1 000 m from the center point between the antennas, and its radio receives the signals. (a) If the car is at the position of the second maximum after that at point O when it has traveled a distance of y = 400 m northward, what is the wavelength of the signals? (b) How much farther must the car travel from this position to encounter the next minimum in reception? *Hint:* Do not use the small-angle approximation in this problem.

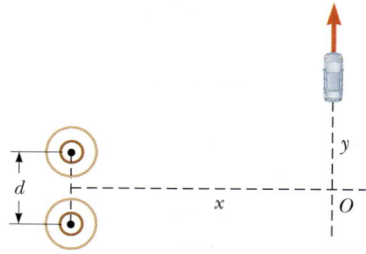

Figure P24.7

8. Light of wavelength 6.0×10^2 nm falls on a double slit, and the first bright fringe of the interference pattern is observed to make an angle of 12° with the horizontal. Find the separation between the slits.

9. Monochromatic light falls on a screen 1.75 m from two slits separated by 2.10 mm. The first- and second-order bright fringes are separated by 0.552 mm. What is the wavelength of the light?

10. A pair of slits, separated by 0.150 mm, is illuminated by light having a wavelength of λ = 643 nm. An

interference pattern is observed on a screen 140 cm from the slits. Consider a point on the screen located at y = 1.80 cm from the central maximum of this pattern. (a) What is the path difference δ for the two slits at the location y? (b) Express this path difference in terms of the wavelength. (c) Will the interference correspond to a maximum, a minimum, or an intermediate condition?

11. A riverside warehouse has two open doors, as in Figure P24.11. Its interior is lined with a sound-absorbing material. A boat on the river sounds its horn. To person A, the sound is loud and clear. To person B, the sound is barely audible. The principal wavelength of the sound waves is 3.00 m. Assuming person B is at the position of the first minimum, determine the distance between the doors, center to center.

Figure P24.11

12. **Q|C** A student sets up a double-slit experiment using monochromatic light of wavelength λ. The distance between the slits is equal to 25λ. (a) Find the angles at which the m = 1, 2, and 3 maxima occur on the viewing screen. (b) At what angles do the first three dark fringes occur? (c) Why are the answers so evenly spaced? Is the spacing even for all orders? Explain.

13. Radio waves from a star, of wavelength 250 m, reach a radio telescope by two separate paths, as shown in Figure P24.13. One is a direct path to the receiver, which is situated on the edge of a cliff by the ocean. The second is by reflection off the water. The first minimum of destructive interference occurs when the star is θ = 25.0° above the horizon. Find the height of the cliff. (Assume no phase change on reflection.)

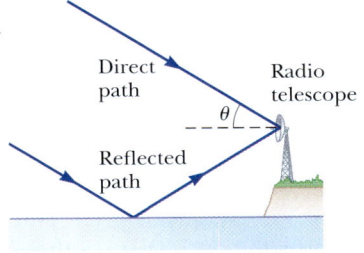

Figure P24.13

14. **GP** Monochromatic light of wavelength λ is incident on a pair of slits separated by 2.40×10^{-4} m, and forms an interference pattern on a screen placed 1.80 m away from the slits. The first-order bright fringe is 4.52 mm from the center of the central maximum. (a) Draw a picture, labeling the angle θ and the legs of the right triangle associated with the first-order bright fringe. (b) Compute the tangent of the angle θ associated with the first-order bright fringe. (c) Find the angle corresponding to the first-order bright fringe and compute the sine of that angle. Are the sine and tangent of the angle comparable in value? Does your answer always hold true? (d) Calculate the wavelength of the light. (e) Compute the angle of the fifth-order bright fringe. (f) Find its position on the screen.

15. Waves from a radio station have a wavelength of 300 m. They travel by two paths to a home receiver 20.0 km from the transmitter. One path is a direct path, and the second is by reflection from a mountain directly behind the home receiver. What is the minimum distance from the mountain to the receiver that produces destructive interference at the receiver? (Assume that no phase change occurs on reflection from the mountain.)

24.3 Change of Phase Due to Reflection
24.4 Interference in Thin Films

16. **Q|C** A soap bubble (n = 1.33) having a wall thickness of 120 nm is floating in air. (a) What is the wavelength of the visible light that is most strongly reflected? (b) Explain how a bubble of different thickness could also strongly reflect light of this same wavelength. (c) Find the two smallest film thicknesses larger than the one given that can produce strongly reflected light of this same wavelength.

17. A thin layer of liquid methylene iodide (n = 1.756) is sandwiched between two flat, parallel plates of glass (n = 1.50). What is the minimum thickness of the liquid layer if normally incident light with $\lambda = 6.00 \times 10^2$ nm in air is to be strongly reflected?

18. **S** A thin film of oil (n = 1.25) is located on smooth, wet pavement. When viewed from a direction perpendicular to the pavement, the film reflects most strongly red light at 640 nm and reflects no green light at 512 nm. (a) What is the minimum thickness of the oil film? (b) Let m_1 correspond to the order of the constructive interference and m_2 to the order of the destructive interference. Obtain a relationship between m_1 and m_2 that is consistent with the given data.

19. A thin film of glass (n = 1.52) of thickness 0.420 μm is viewed under white light at near normal incidence. What wavelength of *visible* light is most strongly reflected by the film when surrounded by air?

20. **W** A transparent oil with index of refraction 1.29 spills on the surface of water (index of refraction 1.33),

producing a maximum of reflection with normally incident orange light (wavelength 600 nm in air). Assuming the maximum occurs in the first order, determine the thickness of the oil slick.

21. **M** A possible means for making an airplane invisible to radar is to coat the plane with an antireflective polymer. If radar waves have a wavelength of 3.00 cm and the index of refraction of the polymer is $n = 1.50$, how thick would you make the coating?

22. **QC** An oil film ($n = 1.45$) floating on water is illuminated by white light at normal incidence. The film is 2.80×10^2 nm thick. Find (a) the wavelength and color of the light in the visible spectrum most strongly reflected and (b) the wavelength and color of the light in the visible spectrum most strongly transmitted. Explain your reasoning.

23. **QC** Astronomers observe the chromosphere of the Sun with a filter that passes the red hydrogen spectral line of wavelength 656.3 nm, called the H_α line. The filter consists of a transparent dielectric of thickness d held between two partially aluminized glass plates. The filter is kept at a constant temperature. (a) Find the minimum value of d that will produce maximum transmission of perpendicular H_α light if the dielectric has an index of refraction of 1.378. (b) If the temperature of the filter increases above the normal value increasing its thickness, what happens to the transmitted wavelength? (c) The dielectric will also pass what near-visible wavelength? One of the glass plates is colored red to absorb this light.

24. Two rectangular optically flat plates ($n = 1.52$) are in contact along one end and are separated along the other end by a 2.00-μm-thick spacer (Fig. P24.24). The top plate is illuminated by monochromatic light of wavelength 546.1 nm. Calculate the number of dark parallel bands crossing the top plate (including the dark band at zero thickness along the edge of contact between the plates).

Figure P24.24
Problems 24 and 25.

25. An investigator finds a fiber at a crime scene that he wishes to use as evidence against a suspect. He gives the fiber to a technician to test the properties of the fiber. To measure the diameter of the fiber, the technician places it between two flat glass plates at their ends as in Figure P24.24. When the plates, of length 14.0 cm, are illuminated from above with light of wavelength 650 nm, she observes bright interference bands separated by 0.580 mm. What is the diameter of the fiber?

26. A plano-convex lens with radius of curvature $R = 3.0$ m is in contact with a flat plate of glass. A light source and the observer's eye are both close to the normal, as shown in Figure 24.8a. The radius of the 50th bright Newton's ring is found to be 9.8 mm. What is the wavelength of the light produced by the source?

27. A plano-convex lens rests with its curved side on a flat glass surface and is illuminated from above by light of wavelength 500 nm. (See Fig. 24.8.) A dark spot is observed at the center, surrounded by 19 concentric dark rings (with bright rings in between). How much thicker is the air wedge at the position of the 19th dark ring than at the center?

28. Nonreflective coatings on camera lenses reduce the loss of light at the surfaces of multilens systems and prevent internal reflections that might mar the image. Find the minimum thickness of a layer of magnesium fluoride ($n = 1.38$) on flint glass ($n = 1.66$) that will cause destructive interference of reflected light of wavelength 550 nm near the middle of the visible spectrum.

29. A thin film of glycerin ($n = 1.473$) of thickness 524 nm with air on both sides is illuminated with white light at near normal incidence. What wavelengths will be strongly reflected in the range 300 nm to 700 nm?

30. **QC** A lens made of glass ($n_g = 1.52$) is coated with a thin film of MgF_2 ($n_s = 1.38$) of thickness t. Visible light is incident normally on the coated lens as in Figure P24.30. (a) For what minimum value of t will the reflected light of wavelength 540 nm (in air) be missing? (b) Are there other values of t that will minimize the reflected light at this wavelength? Explain.

Figure P24.30

24.7 Single-Slit Diffraction

31. Light of wavelength 5.40×10^2 nm passes through a slit of width 0.200 mm. (a) Find the width of the central maximum on a screen located 1.50 m from the slit. (b) Determine the width of the first-order bright fringe.

32. **W** Light of wavelength 600 nm falls on a 0.40-mm-wide slit and forms a diffraction pattern on a screen 1.5 m away. (a) Find the position of the first dark band on each side of the central maximum. (b) Find the width of the central maximum.

33. **M** Light of wavelength 587.5 nm illuminates a slit of width 0.75 mm. (a) At what distance from the slit should a screen be placed if the first minimum in the diffraction pattern is to be 0.85 mm from the central maximum? (b) Calculate the width of the central maximum.

34. Microwaves of wavelength 5.00 cm enter a long, narrow window in a building that is otherwise essentially opaque to the incoming waves. If the window is 36.0 cm

wide, what is the distance from the central maximum to the first-order minimum along a wall 6.50 m from the window?

35. A beam of monochromatic light is diffracted by a slit of width 0.600 mm. The diffraction pattern forms on a wall 1.30 m beyond the slit. The width of the central maximum is 2.00 mm. Calculate the wavelength of the light.

36. A screen is placed 50.0 cm from a single slit that is illuminated with light of wavelength 680 nm. If the distance between the first and third minima in the diffraction pattern is 3.00 mm, what is the width of the slit?

37. A slit of width 0.50 mm is illuminated with light of wavelength 500 nm, and a screen is placed 120 cm in front of the slit. Find the widths of the first and second maxima on each side of the central maximum.

38. The second-order dark fringe in a single-slit diffraction pattern is 1.40 mm from the center of the central maximum. Assuming the screen is 85.0 cm from a slit of width 0.800 mm and assuming monochromatic incident light, calculate the wavelength of the incident light.

24.8 The Diffraction Grating

39. Three discrete spectral lines occur at angles of 10.1°, 13.7°, and 14.8°, respectively, in the first-order spectrum of a diffraction-grating spectrometer. (a) If the grating has 3 660 slits/cm, what are the wavelengths of the light? (b) At what angles are these lines found in the second-order spectra?

40. Intense white light is incident on a diffraction grating that has 600 lines/mm. (a) What is the highest order in which the complete visible spectrum can be seen with this grating? (b) What is the angular separation between the violet edge (400 nm) and the red edge (700 nm) of the first-order spectrum produced by the grating?

41. M The hydrogen spectrum has a red line at 656 nm and a violet line at 434 nm. What angular separation between these two spectral lines is obtained with a diffraction grating that has 4 500 lines/cm?

42. Consider an array of parallel wires with uniform spacing of 1.30 cm between centers. In air at 20.0°C, ultrasound with a frequency of 37.2 kHz from a distant source is incident perpendicular to the array. (Take the speed of sound to be 343 m/s.) (a) Find the number of directions on the other side of the array in which there is a maximum of intensity. (b) Find the angle for each of these directions relative to the direction of the incident beam.

43. A helium–neon laser ($\lambda = 632.8$ nm) is used to calibrate a diffraction grating. If the first-order maximum occurs at 20.5°, what is the spacing between adjacent grooves in the grating?

44. W White light is spread out into its spectral components by a diffraction grating. If the grating has 2 000 lines per centimeter, at what angle does red light of wavelength 640 nm appear in the first-order spectrum?

45. Light from an argon laser strikes a diffraction grating that has 5 310 grooves per centimeter. The central and first-order principal maxima are separated by 0.488 m on a wall 1.72 m from the grating. Determine the wavelength of the laser light.

46. S Take red light at 700 nm and violet at 400 nm as the ends of the visible spectrum and consider the continuous spectrum of white light formed by a diffraction grating with a spacing of d meters between adjacent lines. Show that the interval $\theta_{v2} \leq \theta \leq \theta_{r2}$ of the continuous spectrum in second order must overlap the interval $\theta_{v3} \leq \theta \leq \theta_{r3}$ of the third-order spectrum. Note: θ_{v2} is the angle of the violet light in second order, and θ_{r2} is the angle made by red light in second order.

47. Sunlight is incident on a diffraction grating that has 2 750 lines/cm. The second-order spectrum over the visible range (400–700 nm) is to be limited to 1.75 cm along a screen that is a distance L from the grating. What is the required value of L?

48. GP A diffraction grating has 4.200×10^3 rulings per centimeter. The screen is 2.000 m from the grating. In parts (a) through (e), round each result to four digits, using the rounded values for subsequent calculations. (a) Compute the value of d, the distance between adjacent rulings. Express the answer in meters. (b) Calculate the angle of the second-order maximum made by the 589.0-nm wavelength of sodium. (c) Find the position of this second-order maximum on the screen. (d) Repeat parts (b) and (c) for the second-order maximum of the 589.6-nm wavelength of sodium, finding the angle and position on the screen. (e) What is the distance between the two second-order maxima on the screen? (f) Now find the distance between the two second-order maxima without rounding, carrying all digits in calculator memory. How many significant digits are in agreement? What can you conclude about rounding and the use of intermediate answers in this case?

49. Light of wavelength 500 nm is incident normally on a diffraction grating. If the third-order maximum of the diffraction pattern is observed at 32.0°, (a) what is the number of rulings per centimeter for the grating? (b) Determine the total number of primary maxima that can be observed in this situation.

50. Light containing two different wavelengths passes through a diffraction grating with 1 200 slits/cm. On a screen 15.0 cm from the grating, the third-order maximum of the shorter wavelength falls midway between the central maximum and the first side maximum for the longer wavelength. If the neighboring maxima of the longer wavelength are 8.44 mm apart on the screen, what are the wavelengths in the light? *Hint:* Use the small-angle approximation.

24.9 Polarization of Light Waves

51. The angle of incidence of a light beam in air onto a reflecting surface is continuously variable. The reflected ray is found to be completely polarized when the angle of incidence is 48.0°. (a) What is the index of refraction of the reflecting material? (b) If some of the incident light (at an angle of 48.0°) passes into the material below the surface, what is the angle of refraction?

52. Unpolarized light passes through two Polaroid sheets. The transmission axis of the analyzer makes an angle of 35.0° with the axis of the polarizer. (a) What fraction of the original unpolarized light is transmitted through the analyzer? (b) What fraction of the original light is absorbed by the analyzer?

53. The index of refraction of a glass plate is 1.52. What is the Brewster's angle when the plate is (a) in air and (b) in water? (See Problem 57.)

54. At what angle above the horizon is the Sun if light from it is completely polarized upon reflection from water?

55. A light beam is incident on a piece of fused quartz ($n = 1.458$) at the Brewster's angle. Find (a) the value of Brewster's angle and (b) the angle of refraction for the transmitted ray.

56. The critical angle for total internal reflection for sapphire surrounded by air is 34.4°. Calculate the Brewster's angle for sapphire if the light is incident from the air.

57. Equation 24.14 assumes the incident light is in air. If the light is incident from a medium of index n_1 onto a medium of index n_2, follow the procedure used to derive Equation 24.14 to show that $\tan \theta_p = n_2/n_1$.

58. Plane-polarized light is incident on a single polarizing disk, with the direction of E_0 parallel to the direction of the transmission axis. Through what angle should the disk be rotated so that the intensity in the transmitted beam is reduced by a factor of (a) 2.00, (b) 4.00, and (c) 6.00?

59. **M** Three polarizing plates whose planes are parallel are centered on a common axis. The directions of the transmission axes relative to the common vertical direction are shown in Figure P24.59. A linearly polarized beam of light with plane of polarization parallel to the vertical reference direction is incident from the left onto the first disk with intensity $I_i = 10.0$ units (arbitrary). Calculate the transmitted intensity I_f when $\theta_1 = 20.0°$, $\theta_2 = 40.0°$, and $\theta_3 = 60.0°$. *Hint:* Make repeated use of Malus's law.

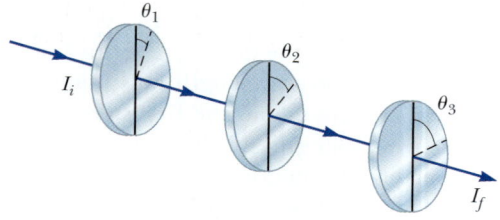

Figure P24.59 Problems 59 and 70.

60. Light of intensity I_0 and polarized parallel to the transmission axis of a polarizer is incident on an analyzer. (a) If the transmission axis of the analyzer makes an angle of 45° with the axis of the polarizer, what is the intensity of the transmitted light? (b) What should the angle between the transmission axes be to make $I/I_0 = 1/3$?

61. **BIO** Light with a wavelength in vacuum of 546.1 nm falls perpendicularly on a biological specimen that is 1.000 μm thick. The light splits into two beams polarized at right angles, for which the indices of refraction are 1.320 and 1.333, respectively. (a) Calculate the wavelength of each component of the light while it is traversing the specimen. (b) Calculate the phase difference between the two beams when they emerge from the specimen.

Additional Problems

62. Light from a helium–neon laser ($\lambda = 632.8$ nm) is incident on a single slit. What is the maximum width of the slit for which no diffraction minima are observed?

63. **Q|C** Laser light with a wavelength of 632.8 nm is directed through one slit or two slits and allowed to fall on a screen 2.60 m beyond. Figure P24.63 shows the pattern on the screen, with a centimeter ruler below it. Did the light pass through one slit or two slits? Explain how you can tell. If the answer is one slit, find its width. If the answer is two slits, find the distance between their centers.

Figure P24.63

64. **S** In a Young's interference experiment, the two slits are separated by 0.150 mm and the incident light includes two wavelengths: $\lambda_1 = 540$ nm (green) and $\lambda_2 = 450$ nm (blue). The overlapping interference patterns are observed on a screen 1.40 m from the slits. (a) Find a relationship between the orders m_1 and m_2 that determines where a bright fringe of the green light coincides with a bright fringe of the blue light. (The order m_1 is associated with λ_1, and m_2 is associated with λ_2.) (b) Find the minimum values of m_1 and m_2 such that the overlapping of the bright fringes will occur and find the position of the overlap on the screen.

65. Light of wavelength 546 nm (the intense green line from a mercury source) produces a Young's interference pattern in which the second minimum from the central maximum is along a direction that makes an angle of 18.0 min of arc with the axis through the central maximum. What is the distance between the parallel slits?

66. The two speakers are placed 35.0 cm apart. A single oscillator makes the speakers vibrate in phase at a frequency of 2.00 kHz. At what angles, measured from the perpendicular bisector of the line joining the speakers, would a distant observer hear maximum sound intensity? Minimum sound intensity? (Take the speed of sound to be 340 m/s.)

67. Interference effects are produced at point P on a screen as a result of direct rays from a 500-nm source and reflected rays off a mirror, as shown in Figure P24.67. If the source is $L = 100$ m to the left of the screen and $h = 1.00$ cm above the mirror, find the distance y (in millimeters) to the first dark band above the mirror.

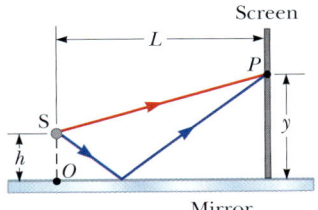

Figure P24.67

68. **BIO** Many cells are transparent and colorless. Structures of great interest in biology and medicine can be practically invisible to ordinary microscopy. An *interference microscope* reveals a difference in refractive index as a shift in interference fringes to indicate the size and shape of cell structures. The idea is exemplified in the following problem: An air wedge is formed between two glass plates in contact along one edge and slightly separated at the opposite edge. When the plates are illuminated with monochromatic light from above, the reflected light has 85 dark fringes. Calculate the number of dark fringes that appear if water ($n = 1.33$) replaces the air between the plates.

69. Figure P24.69 shows a radio-wave transmitter and a receiver, both $h = 50.0$ m above the ground and $d = 600$ m apart. The receiver can receive signals directly from the transmitter and indirectly from signals that bounce off the ground. If the ground is level between the transmitter and receiver and a $\lambda/2$ phase shift occurs upon reflection, determine the longest wavelengths that interfere (a) constructively and (b) destructively.

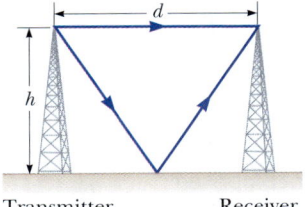

Figure P24.69

70. Three polarizers, centered on a common axis and with their planes parallel to one another, have transmission axes oriented at angles of θ_1, θ_2, and θ_3 from the vertical, as shown in Figure P24.59. Light of intensity I_i, polarized with its plane of polarization oriented vertically, is incident from the left onto the first polarizer. What is the ratio I_f/I_i of the final transmitted intensity to the incident intensity if (a) $\theta_1 = 45°$, $\theta_2 = 90°$, and $\theta_3 = 0°$? (b) $\theta_1 = 0°$, $\theta_2 = 45°$, and $\theta_3 = 90°$?

71. The transmitting antenna on a submarine is 5.00 m above the water when the ship surfaces. The captain wishes to transmit a message to a receiver on a 90.0-m-tall cliff at the ocean shore. If the signal is to be completely polarized by reflection off the ocean surface, how far must the ship be from the shore?

72. **S** A plano-convex lens (flat on one side, convex on the other) with index of refraction n rests with its curved side (radius of curvature R) on a flat glass surface of the same index of refraction with a film of index n_{film} between them. The lens is illuminated from above by light of wavelength λ. Show that the dark Newton rings that appear have radii of

$$r \approx \sqrt{m\lambda R/n_{\text{film}}}$$

where m is an integer.

73. A diffraction pattern is produced on a screen 140 cm from a single slit, using monochromatic light of wavelength 500 nm. The distance from the center of the central maximum to the first-order maximum is 3.00 mm. Calculate the slit width. *Hint:* Assume that the first-order maximum is halfway between the first- and second-order minima.

74. A flat piece of glass is supported horizontally above the flat end of a 10.0-cm-long metal rod that has its lower end rigidly fixed. The thin film of air between the rod and the glass is observed to be bright when illuminated by light of wavelength 500 nm. As the temperature is slowly increased by 25.0°C, the film changes from bright to dark and back to bright 200 times. What is the coefficient of linear expansion of the metal?

The twin Keck telescopes atop the summit of Hawaii's Mauna Kea volcano are the world's largest optical and infrared telescopes. Each telescope has a primary mirror 10 m in diameter comprised of 36 hexagonal segments that function together as a single piece of reflective glass. Far from city lights, high on a dormant volcano in dry, unpolluted air, the telescopes are ideally placed to probe the mysteries of the Universe.

25 Optical Instruments

25.1 The Camera

25.2 The Eye

25.3 The Simple Magnifier

25.4 The Compound Microscope

25.5 The Telescope

25.6 Resolution of Single-Slit and Circular Apertures

25.7 The Michelson Interferometer

We use devices made from lenses, mirrors, and other optical components every time we put on a pair of eyeglasses or contact lenses, take a photograph, look at the sky through a telescope, and so on. In this chapter we examine how optical instruments work. For the most part, our analyses involve the laws of reflection and refraction and the procedures of geometric optics. To explain certain phenomena, however, we must use the wave nature of light.

25.1 The Camera

LEARNING OBJECTIVES

1. Describe the components and operation of a single-lens camera.
2. Define the *f*-number and discuss its effects on a camera's image.

The single-lens photographic camera is a simple optical instrument having the features shown in Figure 25.1. It consists of an opaque box, a converging lens that produces a real image, and a photographic film behind the lens to receive the image. Digital cameras differ in that the image is formed on a charge-coupled device (CCD) or a complementary metal-oxide semiconductor (CMOS) sensor instead of on film. Both the CCD and the CMOS image sensors convert the image into digital form, which can then be stored in the camera's memory.

Focusing a camera is accomplished by varying the distance between the lens and sensor, with an adjustable bellows in antique cameras and other mechanisms in contemporary models. For proper focusing, which leads to sharp images, the lens-to-sensor distance depends on the object distance as well as on the focal length of the lens. The shutter, located behind the lens, is a mechanical device that is opened for selected time intervals. With this arrangement, moving objects can be photographed by using short exposure times and dark scenes (with low light levels) by using long exposure times. If this adjustment were not available, it would be impossible to take stop-action photographs. A rapidly moving vehicle, for example, could move far enough while the shutter was open to produce a blurred image. Another major cause of blurred images is movement of the *camera* while the shutter is open. To prevent such movement, you should mount the camera on a tripod or use short exposure times. Typical shutter speeds (that is, exposure times) are 1/30 s, 1/60 s, 1/125 s, and 1/250 s. Stationary objects are often shot with a shutter speed of 1/60 s.

Most cameras also have an aperture of adjustable diameter to further control the intensity of the light reaching the sensor. When an aperture of small diameter is used, only light from the central portion of the lens reaches the sensor, reducing spherical aberration.

The intensity I of the light reaching the sensor is proportional to the area of the lens. Because this area in turn is proportional to the square of the lens diameter D, the intensity is also proportional to D^2. Light intensity is a measure of the rate at which energy is received by the sensor per unit area of the image. Because the area of the image is proportional to q^2 in Figure 25.1 and $q \approx f$ (when $p \gg f$, so that p can be approximated as infinite), we conclude that the intensity is also proportional to $1/f^2$. Therefore, $I \propto D^2/f^2$. The brightness of the image formed on the sensor depends on the light intensity, so we see that it ultimately depends on both the focal length f and diameter D of the lens. The ratio f/D is called the **f-number** (or focal ratio) of a lens:

$$f\text{-number} \equiv \frac{f}{D} \qquad [25.1]$$

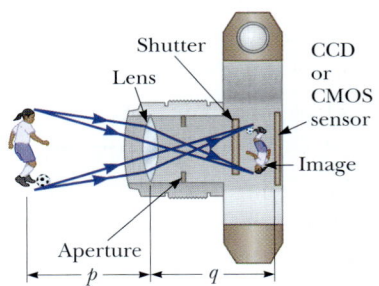

Figure 25.1 Cross-sectional view of a simple digital camera. The CCD or CMOS image sensor is the light-sensitive component of the camera. In a nondigital camera the light from the lens falls onto photographic film. In reality $p \gg q$.

The f-number is often given as a description of the lens "speed." A lens with a low f-number is a "fast" lens. Extremely fast lenses, which have an f-number as low as approximately 1.2, are expensive because of the difficulty of keeping aberrations acceptably small with light rays passing through a large area of the lens. Camera lenses are often marked with a range of f-numbers, such as 1.4, 2, 2.8, 4, 5.6, 8, and 11. Any one of these settings can be selected by adjusting the aperture, which changes the value of D. Increasing the setting from one f-number to the next-higher value (for example, from 2.8 to 4) decreases the area of the aperture by a factor of 2. The lowest f-number setting on a camera corresponds to a wide-open aperture and the use of the maximum possible lens area.

Simple cameras usually have a fixed focal length and fixed aperture size, with an f-number of about 11. This high value for the f-number allows for a large **depth of field** and means that objects at a wide range of distances from the lens form reasonably sharp images on the sensor. In other words, the camera doesn't have to be focused. Most cameras with variable f-numbers adjust them automatically.

25.2 The Eye BIO

LEARNING OBJECTIVES

1. Identify the essential parts of the eye and discuss the eye's ability to adjust to varying light conditions.
2. Define near point and far point and discuss the focusing process of accommodation.
3. Describe several focusing conditions of the eye and the characteristics of optical lenses used to correct them.
4. Define the power of a lens.
5. Calculate the required power of corrective optical lenses.

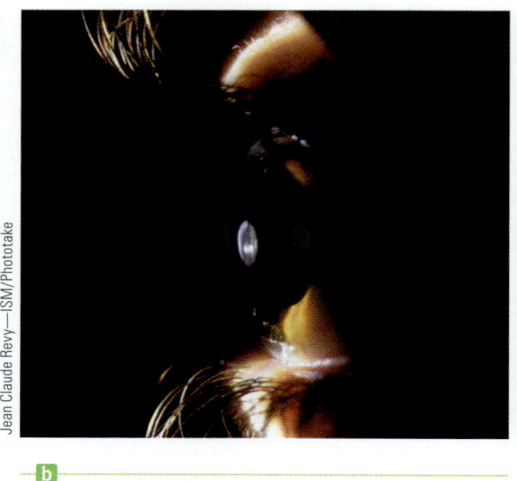

Figure 25.2 (a) Essential parts of the eye. Can you correlate the essential parts of the eye with those of the simple camera in Figure 25.1? (b) Close-up photograph of the human cornea.

Like a camera, a normal eye focuses light and produces a sharp image. The mechanisms by which the eye controls the amount of light admitted and adjusts to produce correctly focused images, however, are far more complex, intricate, and effective than those in even the most sophisticated camera. In all respects the eye is a physiological wonder.

Figure 25.2a shows the essential parts of the eye. Light entering the eye passes through a transparent structure called the *cornea*, behind which are a clear liquid (the *aqueous humor*), a variable aperture (the *pupil*, which is an opening in the *iris*), and the *crystalline lens*. Most of the refraction occurs at the outer surface of the eye, where the cornea is covered with a film of tears. Relatively little refraction occurs in the crystalline lens because the aqueous humor in contact with the lens has an average index of refraction close to that of the lens. The iris, which is the colored portion of the eye, is a muscular diaphragm that controls pupil size. The iris regulates the amount of light entering the eye by dilating the pupil in low-light conditions and contracting the pupil under conditions of bright light. The *f*-number of the eye ranges from about 2.8 to 16.

The cornea–lens system focuses light onto the back surface of the eye—the *retina*—which consists of millions of sensitive receptors called *rods* and *cones*. When stimulated by light, these structures send impulses to the brain via the optic nerve, converting them into our conscious view of the world. The process by which the brain performs this conversion is not well understood and is the subject of much speculation and research. Unlike film in a camera, the rods and cones chemically adjust their sensitivity according to the prevailing light conditions. This adjustment, which takes about 15 minutes, is responsible for the experience of "getting used to the dark" in such places as movie theaters. Iris aperture control, which takes less than a second, helps protect the retina from overload in the adjustment process.

The eye focuses on an object by varying the shape of the pliable crystalline lens through an amazing process called **accommodation**. An important component in accommodation is the *ciliary muscle*, which is situated in a circle around the rim of the lens. Thin filaments, called *zonules*, run from this muscle to the edge of the lens. When the eye is focused on a distant object, the ciliary muscle is relaxed, tightening the zonules that attach the ciliary muscle to the edge of the lens. The force of the zonules causes the lens to flatten, increasing its focal length. For an object distance of infinity, the focal length of the eye is equal to the fixed distance between lens and retina, about 1.7 cm. The eye focuses on nearby objects by tensing the ciliary muscle, which relaxes the zonules. This action allows the lens to bulge a bit and its focal length decreases, resulting in the image being focused on the retina. All these lens adjustments take place so swiftly that we are not even aware of the change. In this respect even the finest electronic camera is a toy compared with the eye.

There is a limit to accommodation because objects that are very close to the eye produce blurred images. The **near point** is the closest distance for which the lens can accommodate to focus light on the retina. This distance usually increases with age and has an average value of 25 cm. Typically, at age 10 the near point of the eye is about 18 cm. This increases to about 25 cm at age 20, 50 cm at age 40, and 500 cm or greater at age 60. The **far point** of the eye represents the farthest distance for which the lens of the relaxed eye can focus light on the retina. A person with normal vision is able to see very distant objects, such as the Moon, and so has a far point at infinity.

Conditions of the Eye

When the eye suffers a mismatch between the focusing power of the lens–cornea system and the length of the eye, so that light rays reach the retina before they converge to form an image, as in Figure 25.3a, the condition is known as **farsightedness** (or *hyperopia*). A farsighted person can usually see faraway objects clearly but not nearby objects. Although the near point of a normal eye is approximately 25 cm, the near point of a farsighted person is much farther than that. The eye of a farsighted person tries to focus by accommodation, by shortening its focal length. Accommodation works for distant objects, but because the focal length of the farsighted eye is longer than normal, the light from nearby objects can't be brought to a sharp focus before it reaches the retina, causing a blurred image. The condition can be corrected by placing a converging lens in front of the eye, as in Figure 25.3b. The lens refracts the incoming rays more toward the principal axis before entering the eye, allowing them to converge and focus on the retina.

Nearsightedness (or *myopia*) is another mismatch condition in which a person is able to focus on nearby objects, but not faraway objects. In the case of *axial myopia*, nearsightedness is caused by the lens being too far from the retina. It is also possible to have *refractive myopia*, in which the lens–cornea system is too powerful for the normal length of the eye. The far point of the nearsighted eye is not at infinity and may be less than 1 meter. The maximum focal length of the nearsighted eye is insufficient to produce a sharp image on the retina, and rays from a distant object converge to a focus in front of the retina. They then continue past that point, diverging before they finally reach the retina and produce a blurred image (Fig. 25.4a, page 874).

Nearsightedness can be corrected with a diverging lens, as shown in Figure 25.4b. The lens refracts the rays away from the principal axis before they enter the eye, allowing them to focus on the retina.

Beginning with middle age, most people lose some of their accommodation ability as the ciliary muscle weakens and the lens hardens. Unlike farsightedness, which is a

BIO APPLICATION
Using Optical Lenses to Correct for Defects

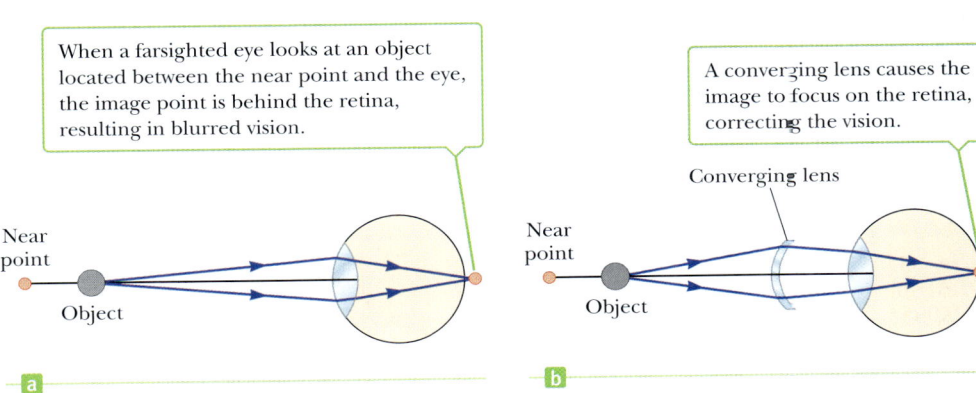

Figure 25.3 (a) An uncorrected farsighted eye. (b) A farsighted eye corrected with a converging lens. (The object is assumed to be very small in these figures.)

Figure 25.4 (a) An uncorrected nearsighted eye. (b) A nearsighted eye corrected with a diverging lens. (The object is assumed to be very small in these figures.)

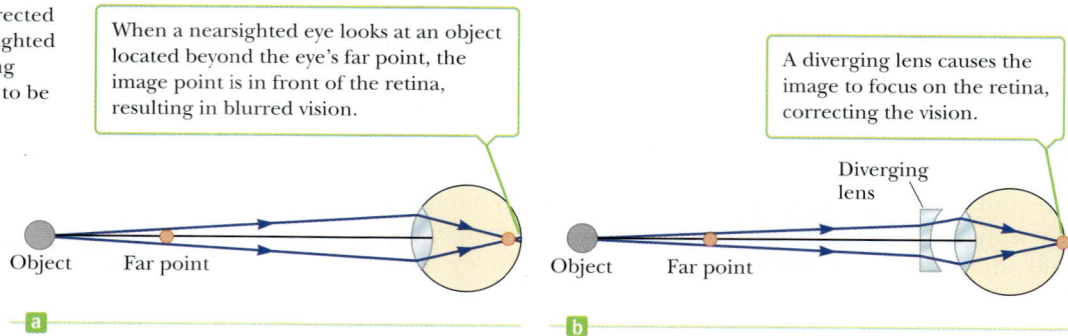

mismatch of focusing power and eye length, **presbyopia** (literally, "old-age vision") is due to a reduction in accommodation ability. This means the cornea and lens aren't able to bring nearby objects into focus on the retina. The symptoms are the same as with farsightedness, and the condition can be corrected with converging lenses.

In the eye defect known as **astigmatism**, light from a point source produces a line image on the retina. This condition arises when either the cornea or the lens (or both) is not perfectly symmetric. Astigmatism can be corrected with lenses having different curvatures in two mutually perpendicular directions.

Optometrists and ophthalmologists usually prescribe lenses measured in **diopters**:

> The **power** P of a lens in diopters equals the inverse of the focal length in meters: $P = 1/f$.

For example, a converging lens with a focal length of +20 cm has a power of +5.0 diopters, and a diverging lens with a focal length of −40 cm has a power of −2.5 diopters. (Although the symbol P is the same as for mechanical power, there is no relationship between the two concepts.)

The position of the lens relative to the eye causes differences in power, but they usually amount to less than one-quarter diopter, which isn't noticeable to most patients. As a result, practicing optometrists deal in increments of one-quarter diopter. Neglecting the eye–lens distance is equivalent to doing the calculation for a contact lens, which rests directly on the eye.

EXAMPLE 25.1 | Prescribing a Corrective Lens for a Farsighted Patient BIO

GOAL Apply geometric optics to correct farsightedness.

PROBLEM The near point of a patient's eye is 50.0 cm. **(a)** What focal length must a corrective lens have to enable the eye to clearly see an object 25.0 cm away? Neglect the eye–lens distance. **(b)** What is the power of this lens? **(c)** Repeat the problem, taking into account that, for typical eyeglasses, the corrective lens is 2.00 cm in front of the eye.

STRATEGY This problem requires substitution into the thin-lens equation (Eq. 23.11) and then using the definition of lens power in terms of diopters. The object is at 25.0 cm, but the lens must form an image at the patient's near point, 50.0 cm, the closest point at which the patient's eye can see clearly. In part (c) 2.00 cm must be subtracted from both the object distance and the image distance to account for the position of the lens.

SOLUTION

(a) Find the focal length of the corrective lens, neglecting its distance from the eye.

Apply the thin-lens equation:

$$\frac{1}{p} + \frac{1}{q} = \frac{1}{f}$$

Substitute $p = 25.0$ cm and $q = -50.0$ cm (the latter is negative because the image must be virtual) on the same side of the lens as the object:

$$\frac{1}{25.0 \text{ cm}} + \frac{1}{-50.0 \text{ cm}} = \frac{1}{f}$$

Solve for f. The focal length is positive, corresponding to a converging lens.

$$f = 50.0 \text{ cm}$$

(b) What is the power of this lens?

The power is the reciprocal of the focal length in meters:

$$P = \frac{1}{f} = \frac{1}{0.500 \text{ m}} = +2.00 \text{ diopters}$$

(c) Repeat the problem, noting that the corrective lens is actually 2.00 cm in front of the eye.

Substitute the corrected values of p and q into the thin-lens equation:

$$\frac{1}{p} + \frac{1}{q} = \frac{1}{23.0 \text{ cm}} + \frac{1}{(-48.0 \text{ cm})} = \frac{1}{f}$$

$$f = 44.2 \text{ cm}$$

Compute the power:

$$P = \frac{1}{f} = \frac{1}{0.442 \text{ m}} = +2.26 \text{ diopters}$$

REMARKS Notice that the calculation in part (c), which doesn't neglect the eye–lens distance, results in a difference of 0.26 diopter.

QUESTION 25.1 True or False: The larger the distance to a near point, the larger the power of the required corrective lens.

EXERCISE 25.1 Suppose a lens is placed in a device that determines its power as 2.75 diopters. Find (a) the focal length of the lens and (b) the minimum distance at which a patient will be able to focus on an object if the patient's near point is 60.0 cm. Neglect the eye–lens distance.

ANSWERS (a) 36.4 cm (b) 22.7 cm

EXAMPLE 25.2 | A Corrective Lens for Nearsightedness

GOAL Apply geometric optics to correct nearsightedness.

PROBLEM A particular nearsighted patient can't see objects clearly when they are beyond 25 cm (the far point of the eye). **(a)** What focal length should the prescribed contact lens have to correct this problem? **(b)** Find the power of the lens, in diopters. Neglect the distance between the eye and the corrective lens.

STRATEGY The purpose of the lens in this instance is to take objects at infinity and create an image of them at the patient's far point. Apply the thin-lens equation.

SOLUTION

(a) Find the focal length of the corrective lens.

Apply the thin-lens equation for an object at infinity and image at 25.0 cm:

$$\frac{1}{p} + \frac{1}{q} = \frac{1}{\infty} + \frac{1}{(-25.0 \text{ cm})} = \frac{1}{f}$$

$$f = -25.0 \text{ cm}$$

(b) Find the power of the lens in diopters.

$$P = \frac{1}{f} = \frac{1}{-0.250 \text{ m}} = -4.00 \text{ diopters}$$

REMARKS The focal length is negative, consistent with a diverging lens. Notice that the power is also negative and has the same numeric value as the sum on the left side of the thin-lens equation.

QUESTION 25.2 True or False: The shorter the distance to a patient's far point, the more negative the power of the required corrective lens.

EXERCISE 25.2 (a) What power lens would you prescribe for a patient with a far point of 35.0 cm? Neglect the eye–lens distance. (b) Repeat, assuming an eye-corrective lens distance of 2.00 cm.

ANSWERS (a) −2.86 diopters (b) −3.03 diopters

APPLYING PHYSICS 25.1 | Vision of the Invisible Man

A classic science fiction story, *The Invisible Man* by H. G. Wells, tells of a man who becomes invisible by changing the index of refraction of his body to that of air. Students who know how the eye works have criticized this story; they claim that the invisible man would be unable to see. On the basis of your knowledge of the eye, would he be able to see?

EXPLANATION He wouldn't be able to see. For the eye to see an object, incoming light must be refracted at the cornea and lens to form an image on the retina. If the cornea and lens have the same index of refraction as air, refraction can't occur and an image wouldn't be formed. ∎

Quick Quiz

25.1 Two campers wish to start a fire during the day. One camper is nearsighted and one is farsighted. Whose glasses should be used to focus the Sun's rays onto some paper to start the fire? (a) either camper's (b) the nearsighted camper's (c) the farsighted camper's

25.3 The Simple Magnifier

LEARNING OBJECTIVES
1. Define the angular magnification of a simple magnifier.
2. Apply geometric optics to systems involving a simple magnifier.

The **simple magnifier** is one of the most basic of all optical instruments because it consists only of a single converging lens. As the name implies, this device is used to increase the apparent size of an object. Suppose an object is viewed at some distance p from the eye, as in Figure 25.5. Clearly, the size of the image formed at the retina depends on the angle θ subtended by the object at the eye. As the object moves closer to the eye, θ increases and a larger image is observed. A normal eye, however, can't focus on an object closer than about 25 cm, the near point (Fig. 25.6a). (Try it!) Therefore, θ is a maximum at the near point.

To further increase the apparent angular size of an object, a converging lens can be placed in front of the eye with the object positioned at point O, just inside the focal point of the lens, as in Figure 25.6b. At this location, the lens forms a virtual, upright, and enlarged image, as shown. The lens allows the object to be viewed closer to the eye than is otherwise possible. We define the **angular magnification** m as the ratio of the angle subtended by a small object when the lens is in use (angle θ in Fig. 25.6b) to the angle subtended by the object placed at the near point with no lens in use (angle θ_0 in Fig. 25.6a):

$$m \equiv \frac{\theta}{\theta_0} \qquad [25.2]$$

◀ Angular magnification with the object at the near point

For the case in which the lens is held close to the eye, the angular magnification is a maximum when the image formed by the lens is at the near point of

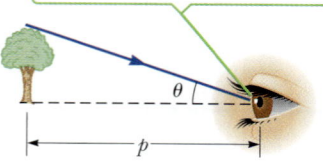

Figure 25.5 An observer looks at an object at distance p.

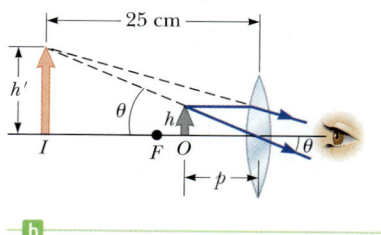

Figure 25.6 (a) An object placed at the near point ($p = 25$ cm) subtends an angle of $\theta_0 \approx h/25$ at the eye. (b) An object placed near the focal point of a converging lens produces a magnified image, which subtends an angle of $\theta \approx h'/25$ at the eye. Note that in this situation $q = -25$ cm.

the eye, which corresponds to $q = -25$ cm (see Fig. 25.6b). The object distance corresponding to this image distance can be calculated from the thin-lens equation:

$$\frac{1}{p} + \frac{1}{-25 \text{ cm}} = \frac{1}{f} \quad [25.3]$$

$$p = \frac{25f}{25 + f}$$

Here, f is the focal length of the magnifier in centimeters. From Figures 25.6a and 25.6b, the small-angle approximation gives

$$\tan \theta_0 \approx \theta_0 \approx \frac{h}{25} \quad \text{and} \quad \tan \theta \approx \theta \approx \frac{h}{p} \quad [25.4]$$

Equation 25.2 therefore becomes

$$m_{max} = \frac{\theta}{\theta_0} = \frac{h/p}{h/25} = \frac{25}{p} = \frac{25}{25f/(25+f)}$$

so that

$$m_{max} = 1 + \frac{25 \text{ cm}}{f} \quad [25.5]$$

The maximum angular magnification given by Equation 25.5 is the ratio of the angular size seen with the lens to the angular size seen without the lens, with the object at the near point of the eye. Although the normal eye can focus on an image formed anywhere between the near point and infinity, it's most relaxed when the image is at infinity (Sec. 25.2). For the image formed by the magnifying lens to appear at infinity, the object must be placed at the focal point of the lens so that $p = f$. In this case Equation 25.4 becomes

$$\theta_0 \approx \frac{h}{25} \quad \text{and} \quad \theta \approx \frac{h}{f}$$

and the angular magnification is

$$m = \frac{\theta}{\theta_0} = \frac{25 \text{ cm}}{f} \quad [25.6]$$

With a single lens, it's possible to achieve angular magnifications up to about 4 without serious aberrations. Magnifications up to about 20 can be achieved by using one or two additional lenses to correct for aberrations.

EXAMPLE 25.3 | Magnification of a Lens

GOAL Compute magnifications of a lens when the image is at the near point and when it's at infinity.

PROBLEM (a) What is the maximum angular magnification of a lens with a focal length of 10.0 cm? (b) What is the angular magnification of this lens when the eye is relaxed? Assume an eye–lens distance of zero.

STRATEGY The maximum angular magnification occurs when the image formed by the lens is at the near point of the eye. Under these circumstances, Equation 25.5 gives us the maximum angular magnification. In part (b) the eye is relaxed only if the image is at infinity, so Equation 25.6 applies.

SOLUTION

(a) Find the maximum angular magnification of the lens.

Substitute into Equation 25.5:

$$m_{max} = 1 + \frac{25 \text{ cm}}{f} = 1 + \frac{25 \text{ cm}}{10.0 \text{ cm}} = \boxed{3.5}$$

(Continued)

(b) Find the magnification of the lens when the eye is relaxed.

When the eye is relaxed, the image is at infinity, so substitute into Equation 25.6:

$$m = \frac{25 \text{ cm}}{f} = \frac{25 \text{ cm}}{10.0 \text{ cm}} = \boxed{2.5}$$

QUESTION 25.3 For greater magnification, should a lens with a larger or smaller focal length be selected?

EXERCISE 25.3 What focal length would be necessary if the lens were to have a maximum angular magnification of 4.0?

ANSWER 8.3 cm

25.4 The Compound Microscope

LEARNING OBJECTIVES
1. Discuss the optical components, characteristics, and limitations of a compound microscope.
2. Evaluate the magnification of a compound microscope.

A simple magnifier provides only limited assistance with inspection of the minute details of an object. Greater magnification can be achieved by combining two lenses in a device called a compound microscope, a schematic diagram of which is shown in Figure 25.7a. The instrument consists of two lenses: an objective with a very short focal length f_o (where $f_o < 1$ cm) and an ocular lens, or eyepiece, with a focal length f_e of a few centimeters. The two lenses are separated by a distance L that is much greater than either f_o or f_e.

The basic approach used to analyze the image formation properties of a microscope is that of two lenses in a row: the image formed by the first becomes the object for the second. The object O placed just outside the focal length of the objective forms a real, inverted image at I_1 that is at or just inside the focal point of the eyepiece. This image is much enlarged. (For clarity, the enlargement of I_1 is not shown in Fig. 25.7a.) The eyepiece, which serves as a simple magnifier, uses the

Figure 25.7 (a) A diagram of a compound microscope, which consists of an objective and an eyepiece, or ocular lens. (b) A compound microscope. Combinations of eyepieces with different focal lengths and different objectives can produce a wide range of magnifications.

image at I_1 as its object and produces an image at I_2. The image seen by the eye at I_2 is virtual, inverted, and very much enlarged.

The lateral magnification M_1 of the first image is $-q_1/p_1$. Note that q_1 is approximately equal to L because the object is placed close to the focal point of the objective lens, which ensures that the image formed will be far from the objective lens. Further, because the object is very close to the focal point of the objective lens, $p_1 \approx f_o$. Therefore, the lateral magnification of the objective is

$$M_1 = -\frac{q_1}{p_1} \approx -\frac{L}{f_o}$$

From Equation 25.6, the angular magnification of the eyepiece for an object (corresponding to the image at I_1) placed at the focal point is found to be

$$m_e = \frac{25 \text{ cm}}{f_e}$$

The overall magnification of the compound microscope is defined as the product of the lateral and angular magnifications:

$$m = M_1 m_e = -\frac{L}{f_o}\left(\frac{25 \text{ cm}}{f_e}\right) \qquad [25.7]$$

◀ Magnification of a microscope

The negative sign indicates that the image is inverted with respect to the object.

The microscope has extended our vision into the previously unknown realm of incredibly small objects, and the capabilities of this instrument have increased steadily with improved techniques in precision grinding of lenses. A natural question is whether there is any limit to how powerful a microscope could be. For example, could a microscope be made powerful enough to allow us to see an atom? The answer to this question is no, as long as visible light is used to illuminate the object. To be seen, the object under a microscope must be at least as large as a wavelength of light. An atom is many times smaller than the wavelength of visible light, so its mysteries must be probed via other techniques.

The wavelength dependence of the "seeing" ability of a wave can be illustrated by water waves set up in a bathtub in the following way. Imagine that you vibrate your hand in the water until waves with a wavelength of about 6 in. are moving along the surface. If you fix a small object, such as a toothpick, in the path of the waves, you will find that the waves are not appreciably disturbed by the toothpick, but continue along their path. Now suppose you fix a larger object, such as a toy sailboat, in the path of the waves. In this case the waves are considerably disturbed by the object. The toothpick was much smaller than the wavelength of the waves, and as a result the waves didn't "see" it. The toy sailboat, however, is about the same size as the wavelength of the waves and hence creates a disturbance. Light waves behave in this same general way. The ability of an optical microscope to view an object depends on the size of the object relative to the wavelength of the light used to observe it. Hence, it will never be possible to observe atoms or molecules with such a microscope because their dimensions are so small (≈ 0.1 nm) relative to the wavelength of the light (≈ 500 nm).

EXAMPLE 25.4 | Microscope Magnifications

GOAL Understand the critical factors involved in determining the magnifying power of a microscope.

PROBLEM A certain microscope has two interchangeable objectives. One has a focal length of 2.0 cm, and the other has a focal length of 0.20 cm. Also available are two eyepieces of focal lengths 2.5 cm and 5.0 cm. If the length of the microscope is 18 cm, compute the magnifications for the following combinations: the 2.0-cm objective and 5.0-cm eyepiece, the 2.0-cm objective and 2.5-cm eyepiece, and the 0.20-cm objective and 5.0-cm eyepiece.

STRATEGY The solution consists of substituting into Equation 25.7 for three different combinations of lenses.

(Continued)

SOLUTION

Apply Equation 25.7 and combine the 2.0-cm objective with the 5.0-cm eyepiece:

$$m = -\frac{L}{f_o}\left(\frac{25 \text{ cm}}{f_e}\right) = -\frac{18 \text{ cm}}{2.0 \text{ cm}}\left(\frac{25 \text{ cm}}{5.0 \text{ cm}}\right) = \boxed{-45}$$

Combine the 2.0-cm objective with the 2.5-cm eyepiece:

$$m = -\frac{18 \text{ cm}}{2.0 \text{ cm}}\left(\frac{25 \text{ cm}}{2.5 \text{ cm}}\right) = \boxed{-9.0 \times 10^1}$$

Combine the 0.20-cm objective with the 5.0-cm eyepiece:

$$m = -\frac{18 \text{ cm}}{0.20 \text{ cm}}\left(\frac{25 \text{ cm}}{5.0 \text{ cm}}\right) = \boxed{-450}$$

REMARKS Much higher magnifications can be achieved, but the resolution starts to fall, resulting in fuzzy images that don't convey any details. (See Section 25.6 for further discussion of this point.)

QUESTION 25.4 True or False: A shorter focal length for either the eyepiece or objective lens will result in greater magnification.

EXERCISE 25.4 Combine the 0.20-cm objective with the 2.5-cm eyepiece and find the magnification.

ANSWER -9.0×10^2

25.5 The Telescope

LEARNING OBJECTIVES

1. Describe the optical components of reflecting and refracting telescopes and discuss the advantages and disadvantages of each.
2. Evaluate the angular magnification of a telescope.

The Hubble Space Telescope enables us to see both further into space and further back in time than ever before.

There are two fundamentally different types of telescope, both designed to help us view distant objects such as the planets in our solar system: (1) the **refracting telescope**, which uses a combination of lenses to form an image, and (2) the **reflecting telescope**, which uses a curved mirror and a lens to form an image. Once again, we can analyze the telescope by considering it to be a system of two optical elements in a row. As before, the image formed by the first element becomes the object for the second.

In the refracting telescope two lenses are arranged so that the objective forms a real, inverted image of the distant object very near the focal point of the eyepiece (Fig. 25.8a). Further, the image at I_1 is formed at the focal point of the objective because the object is essentially at infinity. Hence, the two lenses are separated by the distance $f_o + f_e$, which corresponds to the length of the telescope's tube. Finally, at I_2, the eyepiece forms an enlarged image of the image at I_1.

The angular magnification of the telescope is given by θ/θ_o, where θ_o is the angle subtended by the object at the objective and θ is the angle subtended by the final image. From the triangles in Figure 25.8a, and for small angles, we have

$$\theta_o \approx \frac{h'}{f_e} \quad \text{and} \quad \theta_o \approx \frac{h'}{f_o}$$

Therefore, the angular magnification of the telescope can be expressed as

▶ Angular magnification of a telescope

$$m = \frac{\theta}{\theta_o} = \frac{h'/f_e}{h'/f_o} = \frac{f_o}{f_e} \quad [25.8]$$

This equation says that the angular magnification of a telescope equals the ratio of the objective focal length to the eyepiece focal length. Here again, the angular magnification is the ratio of the angular size seen with the telescope to the angular size seen with the unaided eye.

In some applications—for instance, the observation of relatively nearby objects such as the Sun, the Moon, or planets—angular magnification is important. Stars, however, are so far away that they always appear as small points of light regardless

25.5 | The Telescope

Figure 25.8 (a) A diagram of a refracting telescope, with the object at infinity. (b) A refracting telescope.

of how much angular magnification is used. The large research telescopes used to study very distant objects must have great diameters to gather as much light as possible. It's difficult and expensive to manufacture such large lenses for refracting telescopes. In addition, the heaviness of large lenses leads to sagging, which is another source of aberration.

These problems can be partially overcome by replacing the objective lens with a reflecting, concave mirror, usually having a parabolic shape so as to avoid spherical aberration. Figure 25.9 shows the design of a typical reflecting telescope. Incoming light rays pass down the barrel of the telescope and are reflected by a parabolic mirror at the base. These rays converge toward point A in the figure, where an image would be formed on a photographic plate or another detector. Before this image is formed, however, a small, flat mirror at M reflects the light toward an opening in the side of the tube that passes into an eyepiece. This design is said to have a *Newtonian focus*, after its developer. Note that in the reflecting telescope the light never passes through glass (except in the small eyepiece). As a result, problems associated with chromatic aberration are virtually eliminated.

The largest optical telescopes in the world are the two 10-m-diameter Keck reflectors on Mauna Kea in Hawaii. The largest single-mirrored reflecting telescope in the United States is the 5-m-diameter instrument on Mount Palomar in California. (See Fig. 25.10.) In contrast, the largest refracting telescope in the world, at the Yerkes Observatory in Williams Bay, Wisconsin, has a diameter of only 1 m.

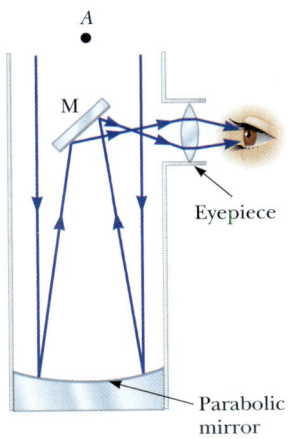

Figure 25.9 A reflecting telescope with a Newtonian focus.

Figure 25.10 The Hale telescope at Mount Palomar Observatory. Just before taking the elevator up to the prime-focus cage, a first-time observer is always told, "Good viewing! And, if you should fall, try to miss the mirror."

EXAMPLE 25.5 Hubble Power

GOAL Understand magnification in telescopes.

PROBLEM The Hubble Space Telescope is 13.2 m long, but has a secondary mirror that increases its effective focal length to 57.8 m. (See Fig. 25.11.) The telescope doesn't have an eyepiece because various instruments, not a human eye, record the collected light. It can, however, produce images several thousand times larger than they would appear with the unaided human eye. What focal-length eyepiece used with the Hubble mirror system would produce a magnification of 8.00×10^3?

STRATEGY Equation 25.8 for telescope magnification can be solved for the eyepiece focal length. The equation for finding the angular magnification of a reflector is the same as that for a refractor.

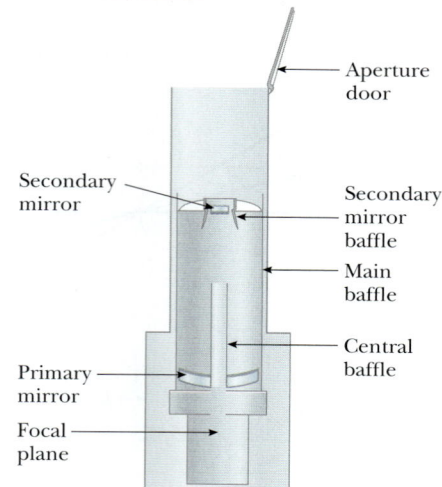

Figure 25.11 A schematic of the Hubble Space Telescope.

SOLUTION

Solve for f_e in Equation 25.8 and substitute values:

$$m = \frac{f_o}{f_e} \quad \rightarrow \quad f_e = \frac{f_o}{m} = \frac{57.8 \text{ m}}{8.00 \times 10^3} = \boxed{7.23 \times 10^{-3} \text{ m}}$$

REMARKS The light-gathering power of a telescope and the length of the baseline over which light is gathered are in fact more important than a telescope's magnification, because these two factors contribute to the resolution of the image. A high-resolution image can always be magnified so its details can be examined. A low resolution image, however, is often fuzzy when magnified. (See Section 25.6.)

QUESTION 25.5 Can greater magnification of a telescope be achieved by increasing the focal length of the mirror? What effect will increasing the focal length of the eyepiece have on the magnification?

EXERCISE 25.5 The Hale telescope on Mount Palomar has a focal length of 16.8 m. Find the magnification of the telescope in conjunction with an eyepiece having a focal length of 5.00 mm.

ANSWER 3.36×10^3

25.6 Resolution of Single-Slit and Circular Apertures

LEARNING OBJECTIVES

1. State and discuss the physical origins of Rayleigh's criterion.
2. Determine the limiting angle for slit and circular apertures.
3. Evaluate the resolution limitations of optical instruments and the resolving power of a diffraction grating.

The ability of an optical system such as the eye, a microscope, or a telescope to distinguish between closely spaced objects is limited because of the wave nature of light. To understand this difficulty, consider Figure 25.12, which shows two light sources far from a narrow slit of width a. The sources can be taken as two point

25.6 | Resolution of Single-Slit and Circular Apertures

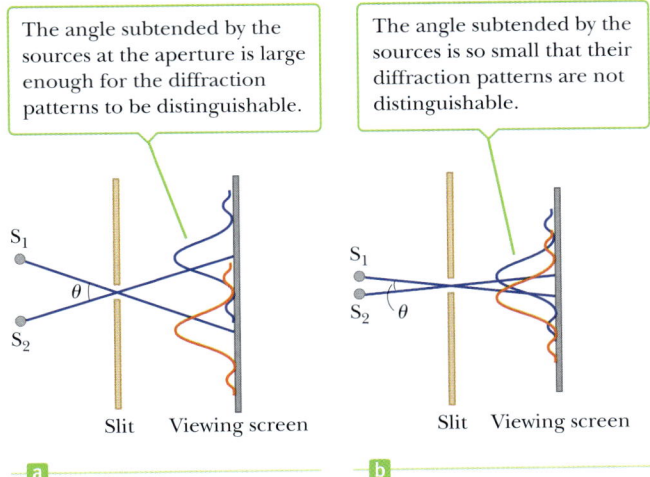

Figure 25.12 Two point sources far from a narrow slit each produce a diffraction pattern. (a) The sources are separated by a large angle. (b) The sources are separated by a small angle. (Notice that the angles are greatly exaggerated. The drawing is not to scale.)

sources S_1 and S_2 that are *not* coherent. For example, they could be two distant stars. If no diffraction occurred, two distinct bright spots (or images) would be observed on the screen at the right in the figure. Because of diffraction, however, each source is imaged as a bright central region flanked by weaker bright and dark rings. What is observed on the screen is the sum of two diffraction patterns, one from S_1 and the other from S_2.

If the two sources are separated so that their central maxima don't overlap, as in Figure 25.12a, their images can be distinguished and are said to be *resolved*. If the sources are close together, however, as in Figure 25.12b, the two central maxima may overlap and the images are *not resolved*. To decide whether two images are resolved, the following condition is often applied to their diffraction patterns:

When the central maximum of one image falls on the first minimum of another image, the images are said to be just resolved. This limiting condition of resolution is known as **Rayleigh's criterion**.

◀ Rayleigh's criterion

Figure 25.13 (page 884) shows diffraction patterns in three situations. In Fig. 25.13a, the sources are sufficiently separated, so the images are resolved. As the sources are brought closer together, as in Figure 25.13b, the central maximum of one image is centered on the first minimum of the other, so by Rayleigh's criterion, the images are just resolved. Finally, when the sources are very close to each other, their images are not resolved (Fig. 25.13c).

From Rayleigh's criterion, we can determine the minimum angular separation θ_{min} subtended by the source at the slit so that the images will be just resolved. In Chapter 24 we found that the first minimum in a single-slit diffraction pattern occurs at the angle that satisfies the relationship

$$\sin\theta = \frac{\lambda}{a}$$

where a is the width of the slit. According to Rayleigh's criterion, this expression gives the smallest angular separation for which the two images can be resolved. Because $\lambda \ll a$ in most situations, $\sin\theta$ is small and we can use the approximation $\sin\theta \approx \theta$. Therefore, the limiting angle of resolution for a slit of width a is

$$\theta_{min} \approx \frac{\lambda}{a} \qquad [25.9]$$

◀ Limiting angle for a slit

where θ_{min} is in radians. Hence, the angle subtended by the two sources at the slit must be *greater* than λ/a if the images are to be resolved.

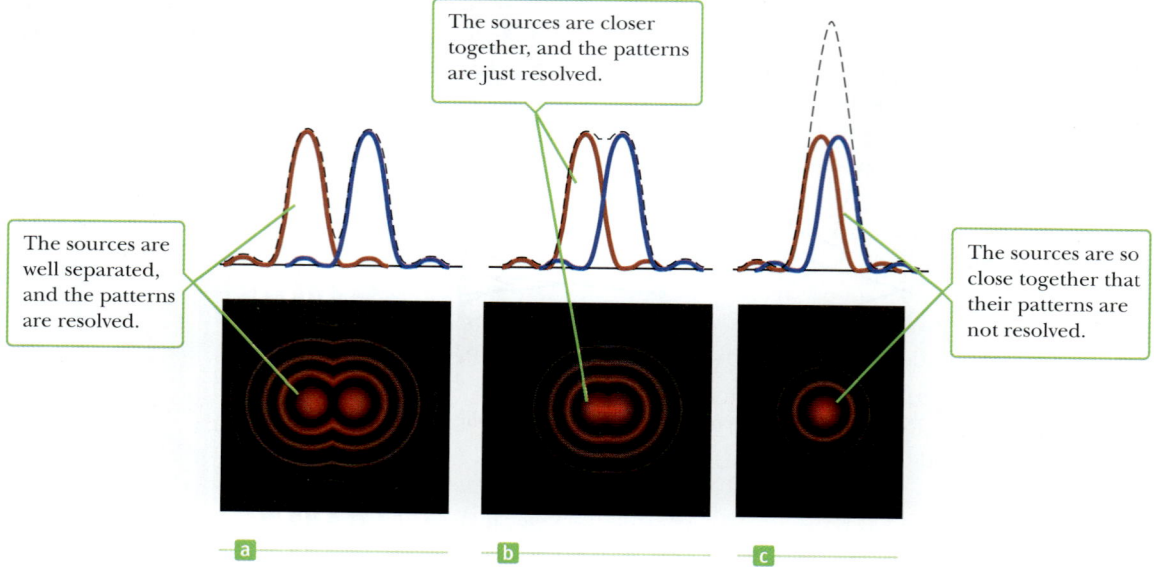

Figure 25.13 The diffraction patterns of two point sources (solid curves) and the resultant pattern (dashed curve) for three angular separations of the sources.

Many optical systems use circular apertures rather than slits. The diffraction pattern of a circular aperture (Fig. 25.14) consists of a central circular bright region surrounded by progressively fainter rings. Analysis shows that the limiting angle of resolution of the circular aperture is

$$\theta_{min} = 1.22 \frac{\lambda}{D} \qquad [25.10]$$

where D is the diameter of the aperture. Note that Equation 25.10 is similar to Equation 25.9 except for the factor 1.22, which arises from a complex mathematical analysis of diffraction from a circular aperture.

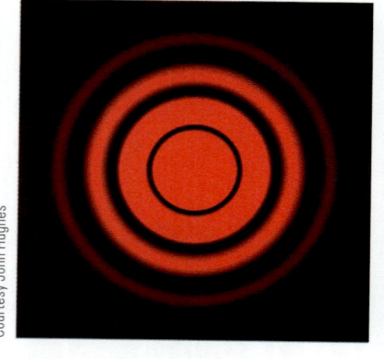

Figure 25.14 The diffraction pattern of a circular aperture consists of a central bright disk surrounded by concentric bright and dark rings.

■ **Quick Quiz**

25.2 Suppose you are observing a binary star with a telescope and are having difficulty resolving the two stars. Which color filter will better help resolve the stars? (a) blue (b) red (c) neither because colored filters have no effect on resolution

■ **APPLYING PHYSICS 25.2** | **Cat's Eyes** BIO

Cats' eyes have vertical pupils in dim light. Which would cats be most successful at resolving at night, headlights on a distant car or vertically separated running lights on a distant boat's mast having the same separation as the car's headlights?

EXPLANATION The effective slit width in the vertical direction of the cat's eye is larger than that in the horizontal direction. Thus, it has more resolving power for lights separated in the vertical direction and would be more effective at resolving the mast lights on the boat. ■

■ **EXAMPLE 25.6** | **Resolution of a Microscope**

GOAL Study limitations on the resolution of a microscope.

PROBLEM Sodium light of wavelength 589 nm is used to view an object under a microscope. The aperture of the objective has a diameter of 0.90 cm. **(a)** Find the limiting angle of resolution for this microscope. **(b)** Using visible light of any wavelength you desire, find the best limit of resolution for this microscope. **(c)** Water of index of refraction 1.33 now fills the space between the object and the objective. What effect would this water have on the resolving power of the microscope, using 589-nm light?

25.6 | Resolution of Single-Slit and Circular Apertures

STRATEGY Parts (a) and (b) require substitution into Equation 25.10. Because the wavelength appears in the numerator, violet light, with the shortest visible wavelength, gives the maximum resolution. In part (c) the only difference is that the wavelength changes to λ/n, where n is the index of refraction of water.

SOLUTION

(a) Find the limiting angle of resolution for this microscope.

Substitute into Equation 25.10 to obtain the limiting angle of resolution:

$$\theta_{min} = 1.22 \frac{\lambda}{D} = 1.22 \left(\frac{589 \times 10^{-9} \text{ m}}{0.90 \times 10^{-2} \text{ m}} \right)$$

$$= 8.0 \times 10^{-5} \text{ rad}$$

(b) Calculate the microscope's best limit of resolution.

To obtain the best resolution, substitute the shortest visible wavelength available, which is violet light, of wavelength 4.0×10^2 nm:

$$\theta_{min} = 1.22 \frac{\lambda}{D} = 1.22 \left(\frac{4.0 \times 10^{-7} \text{ m}}{0.90 \times 10^{-2} \text{ m}} \right)$$

$$= 5.4 \times 10^{-5} \text{ rad}$$

(c) What effect does water between the object and the objective lens have on the resolution, with 589-nm light?

Calculate the wavelength of the sodium light in the water:

$$\lambda_w = \frac{\lambda_a}{n} = \frac{589 \text{ nm}}{1.33} = 443 \text{ nm}$$

Substitute this wavelength into Equation 25.10 to get the resolution:

$$\theta_{min} = 1.22 \left(\frac{443 \times 10^{-9} \text{ m}}{0.90 \times 10^{-2} \text{ m}} \right) = 6.0 \times 10^{-5} \text{ rad}$$

REMARKS In each case any two points on the object subtending an angle of less than the limiting angle θ_{min} at the objective cannot be distinguished in the image. Consequently, it may be possible to see a cell but then be unable to clearly see smaller structures within the cell. Obtaining an increase in resolution is the motivation behind placing a drop of oil on the slide for certain objective lenses.

QUESTION 25.6 Does having two eyes instead of one improve the human ability to resolve distant objects? In general, would more widely spaced eyes increase visual resolving power? Explain.

EXERCISE 25.6 Suppose oil with $n = 1.50$ fills the space between the object and the objective for this microscope. Calculate the limiting angle θ_{min} for sodium light of wavelength 589 nm in air.

ANSWER 5.3×10^{-5} rad

■ EXAMPLE 25.7 | Resolving Craters on the Moon

GOAL Calculate the resolution of a telescope.

PROBLEM The Hubble Space Telescope has an aperture of diameter 2.40 m. **(a)** What is its limiting angle of resolution at a wavelength of 6.00×10^2 nm? **(b)** What's the smallest crater it could resolve on the Moon? (The Moon's distance from Earth is 3.84×10^8 m.)

STRATEGY After substituting into Equation 25.10 to find the limiting angle, use $s = r\theta$ to compute the minimum size of crater that can be resolved.

SOLUTION

(a) What is the limiting angle of resolution at a wavelength of 6.00×10^2 nm?

Substitute $D = 2.40$ m and $\lambda = 6.00 \times 10^{-7}$ m into Equation 25.10:

$$\theta_{min} = 1.22 \frac{\lambda}{D} = 1.22 \left(\frac{6.00 \times 10^{-7} \text{ m}}{2.40 \text{ m}} \right)$$

$$= 3.05 \times 10^{-7} \text{ rad}$$

(Continued)

(b) What's the smallest lunar crater the Hubble Space Telescope can resolve?

The two opposite sides of the crater must subtend the minimum angle. Use the arc length formula:

$$s = r\theta = (3.84 \times 10^8 \text{ m})(3.05 \times 10^{-7} \text{ rad}) = \boxed{117 \text{ m}}$$

REMARKS The distance is so great and the angle so small that using the arc length of a circle is justified because the circular arc is very nearly a straight line. The Hubble Space Telescope has produced several gigabytes of data every day since it first began operation.

QUESTION 25.7 Is the resolution of a telescope better at the red end of the visible spectrum or the violet end?

EXERCISE 25.7 The Hale telescope on Mount Palomar has a diameter of 5.08 m (200 in.). (a) Find the limiting angle of resolution for a wavelength of 6.00×10^2 nm. (b) Calculate the smallest crater diameter the telescope can resolve on the Moon. (c) The answers appear better than what the Hubble can achieve. Why are the answers misleading?

ANSWERS (a) 1.44×10^{-7} rad (b) 55.3 m (c) Although the numbers are better than Hubble's, the Hale telescope must contend with the effects of atmospheric turbulence, so the smaller space-based telescope actually obtains far better results.

It's interesting to compare the resolution of the Hale telescope with that of a large radio telescope, such as the system at Arecibo, Puerto Rico, which has a diameter of 1 000 ft (305 m). This telescope detects radio waves at a wavelength of 0.75 m. The corresponding minimum angle of resolution can be calculated as 3.0×10^{-3} rad (10 min 19 s of arc), which is more than 10 000 times larger than the calculated minimum angle for the Hale telescope.

With such relatively poor resolution, why is Arecibo considered a valuable astronomical instrument? Unlike its optical counterparts, Arecibo can see through clouds of dust. The center of our Milky Way galaxy is obscured by such dust clouds, which absorb and scatter visible light. Radio waves easily penetrate the clouds, so radio telescopes allow direct observations of the galactic core.

Resolving Power of the Diffraction Grating

The diffraction grating studied in Chapter 24 is most useful for making accurate wavelength measurements. Like the prism, it can be used to disperse a spectrum into its components. Of the two devices, the grating is better suited to distinguishing between two closely spaced wavelengths. We say that the grating spectrometer has a higher *resolution* than the prism spectrometer. If λ_1 and λ_2 are two nearly equal wavelengths between which the spectrometer can just barely distinguish, the **resolving power** of the grating is defined as

$$R \equiv \frac{\lambda}{\lambda_2 - \lambda_1} = \frac{\lambda}{\Delta\lambda} \quad \text{[25.11]}$$

where $\lambda \approx \lambda_1 \approx \lambda_2$ and $\Delta\lambda = \lambda_2 - \lambda_1$. From this equation, it's clear that a grating with a high resolving power can distinguish small differences in wavelength. Further, if N lines of the grating are illuminated, it can be shown that the resolving power in the mth-order diffraction is given by

▶ Resolving power of a grating

$$R = Nm \quad \text{[25.12]}$$

So, the resolving power R increases with the order number m and is large for a grating with a great number of illuminated slits. Note that for $m = 0$, $R = 0$, which signifies that *all wavelengths are indistinguishable* for the zeroth-order maximum. (All wavelengths fall at the same point on the screen.) Consider, however, the second-order diffraction pattern of a grating that has 5 000 rulings illuminated by the light source. The resolving power of such a grating in second order is $R = 5\,000 \times 2 = 10\,000$. Therefore, the *minimum* wavelength separation between two spectral lines that can be just resolved, assuming a mean wavelength of 600 nm, is calculated from Equation 25.12 to be $\Delta\lambda = \lambda/R = 6 \times 10^{-2}$ nm. For the third-order principal maximum, $R = 15\,000$ and $\Delta\lambda = 4 \times 10^{-2}$ nm, and so on.

EXAMPLE 25.8 | Light from Sodium Atoms

GOAL Find the necessary resolving power to distinguish spectral lines.

PROBLEM Two bright lines in the spectrum of sodium have wavelengths of 589.00 nm and 589.59 nm, respectively. **(a)** What must the resolving power of a grating be so as to distinguish these wavelengths? **(b)** To resolve these lines in the second-order spectrum, how many lines of the grating must be illuminated?

STRATEGY This problem requires little more than substituting into Equations 25.11 and 25.12.

SOLUTION

(a) What must the resolving power of a grating be in order to distinguish the given wavelengths?

Substitute into Equation 25.11 to find R:

$$R = \frac{\lambda}{\Delta\lambda} = \frac{589.00 \text{ nm}}{589.59 \text{ nm} - 589.00 \text{ nm}} = \frac{589 \text{ nm}}{0.59 \text{ nm}}$$

$$= \boxed{1.0 \times 10^3}$$

(b) To resolve these lines in the second-order spectrum, how many lines of the grating must be illuminated?

Solve Equation 25.12 for N and substitute:

$$N = \frac{R}{m} = \frac{1.0 \times 10^3}{2} = \boxed{5.0 \times 10^2 \text{ lines}}$$

REMARKS The ability to resolve spectral lines is particularly important in experimental atomic physics.

QUESTION 25.8 True or False: If two diffraction gratings differ only by the number of lines, the grating with the larger number of lines can yield a greater resolving power.

EXERCISE 25.8 Due to a phenomenon called electron spin, when the lines of a spectrum are examined at high resolution, each line is actually found to be two closely spaced lines called a doublet. An example is the doublet in the hydrogen spectrum having wavelengths of 656.272 nm and 656.285 nm. **(a)** What must be the resolving power of a grating so as to distinguish these wavelengths? **(b)** How many lines of the grating must be illuminated to resolve these lines in the third-order spectrum?

ANSWERS (a) 5.0×10^4 (b) 1.7×10^4 lines

25.7 The Michelson Interferometer

LEARNING OBJECTIVE

1. Describe the optical components and operating principles of the Michelson interferometer.

The Michelson interferometer is an optical instrument having great scientific importance. Invented by American physicist A. A. Michelson (1852–1931), it is an ingenious device that splits a light beam into two parts and then recombines them to form an interference pattern. The interferometer is used to make accurate length measurements.

Figure 25.15 (page 888) is a schematic diagram of an interferometer. A beam of light provided by a monochromatic source is split into two rays by a partially silvered mirror M inclined at an angle of 45° relative to the incident light beam. One ray is reflected vertically upward to mirror M_1, and the other ray is transmitted horizontally through mirror M to mirror M_2. Hence, the two rays travel separate paths, L_1 and L_2. After reflecting from mirrors M_1 and M_2, the two rays eventually recombine to produce an interference pattern, which can be viewed through a telescope. The glass plate P, equal in thickness to mirror M, is placed in the path of the horizontal ray to ensure that the two rays travel the same distance through glass.

The interference pattern for the two rays is determined by the difference in their path lengths. When the two rays are viewed as shown, the image of M_2 is at

Figure 25.15 A diagram of the Michelson interferometer.

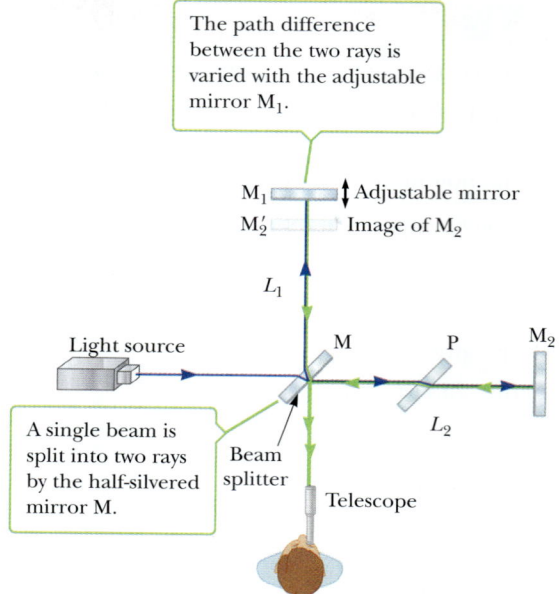

M'_2, parallel to M_1. Hence, the space between M'_2 and M_1 forms the equivalent of a parallel air film. The effective thickness of the air film is varied by using a finely threaded screw to move mirror M_1 in the direction indicated by the arrows in Figure 25.15. If one of the mirrors is tipped slightly with respect to the other, the thin film between the two is wedge shaped and an interference pattern consisting of parallel fringes is set up, as described in Example 24.4. Now suppose we focus on one of the dark lines with the crosshairs of a telescope. As mirror M_1 is moved to lengthen the path L_1, the thickness of the wedge increases. When the thickness increases by $\lambda/4$, the destructive interference that initially produced the dark fringe has changed to constructive interference, and we now observe a bright fringe at the location of the crosshairs. The term *fringe shift* is used to describe the change in a fringe from dark to light or from light to dark. Successive light and dark fringes are formed each time M_1 is moved a distance of $\lambda/4$. The wavelength of light can be measured by counting the number of fringe shifts for a measured displacement of M_1. Conversely, if the wavelength is accurately known (as with a laser beam), the mirror displacement can be determined to within a fraction of the wavelength. Because the interferometer can measure displacements precisely, it is often used to make highly accurate measurements of the dimensions of mechanical components.

If the mirrors are perfectly aligned rather than tipped with respect to each other, the path difference differs slightly for different angles of view. This arrangement results in an interference pattern that resembles Newton's rings. The pattern can be used in a fashion similar to that for tipped mirrors. An observer pays attention to the center spot in the interference pattern. For example, suppose the spot is initially dark, indicating that destructive interference is occurring. If M_1 is now moved a distance of $\lambda/4$, this central spot changes to a light region, corresponding to a fringe shift.

SUMMARY

25.1 The Camera

The light-concentrating power of a lens of focal length f and diameter D is determined by the *f*-number, defined as

$$f\text{-number} \equiv \frac{f}{D} \quad [25.1]$$

The smaller the *f*-number of a lens, the brighter the image formed.

25.2 The Eye

Hyperopia (farsightedness) is a defect of the eye that occurs either when the eyeball is too short or when the ciliary muscle cannot change the shape of the lens enough to form a properly focused image. **Myopia** (nearsightedness) occurs either when the eye is longer than normal or when the maximum focal length of the lens is insufficient to produce a clearly focused image on the retina.

25.5 The Telescope
The **angular magnification of a telescope** is

$$m = \frac{f_o}{f_e} \quad [25.8]$$

where f_o is the focal length of the objective and f_e is the focal length of the eyepiece.

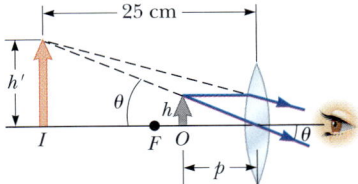

An object placed near the focal point of a converging lens produces a magnified image, which subtends an angle of $\theta \approx h'/25$ at the eye. Note that in this situation $q = -25$ cm.

The **power** of a lens in **diopters** is the inverse of the focal length in meters.

25.3 The Simple Magnifier
The **angular magnification of a lens** is defined as

$$m \equiv \frac{\theta}{\theta_0} \quad [25.2]$$

where θ is the angle subtended by an object at the eye with a lens in use and θ_0 is the angle subtended by the object when it is placed at the near point of the eye and no lens is used. The **maximum angular magnification of a lens** is

$$m_{max} = 1 + \frac{25 \text{ cm}}{f} \quad [25.5]$$

When the eye is relaxed, the angular magnification is

$$m = \frac{25 \text{ cm}}{f} \quad [25.6]$$

25.4 The Compound Microscope
The overall **magnification of a compound microscope** of length L is the product of the magnification produced by the objective, of focal length f_o, and the magnification produced by the eyepiece, of focal length f_e:

$$m = -\frac{L}{f_o}\left(\frac{25 \text{ cm}}{f_e}\right) \quad [25.7]$$

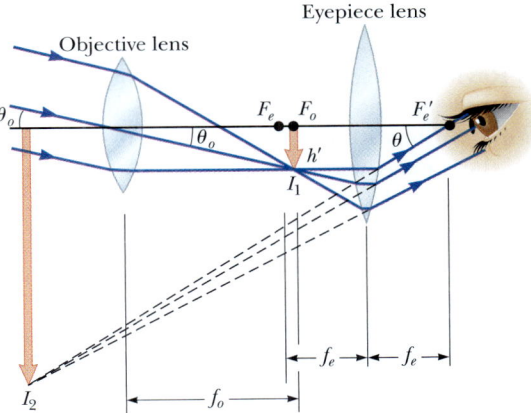

A ray diagram for a refracting telescope, with the object at infinity.

25.6 Resolution of Single-Slit and Circular Apertures
Two images are said to be **just resolved** when the central maximum of the diffraction pattern for one image falls on the first minimum of the other image. This limiting condition of resolution is known as **Rayleigh's criterion**. The limiting angle of resolution for a **slit** of width a is

$$\theta_{min} \approx \frac{\lambda}{a} \quad [25.9]$$

The limiting angle of resolution of a **circular aperture** is

$$\theta_{min} = 1.22 \frac{\lambda}{D} \quad [25.10]$$

where D is the diameter of the aperture.

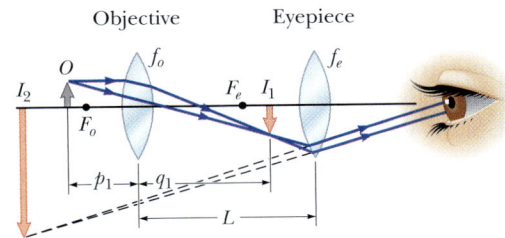

A ray diagram for a compound microscope, which consists of an objective and an eyepiece, or ocular lens.

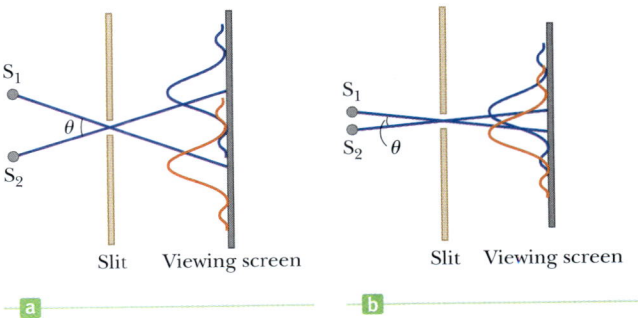

Two point sources far from a narrow slit each produce a diffraction pattern. (a) The sources are separated by a large angle. (b) The sources are separated by a small angle.

If λ_1 and λ_2 are two nearly equal wavelengths between which a grating spectrometer can just barely distinguish, the **resolving power** R of the grating is defined as

$$R \equiv \frac{\lambda}{\lambda_2 - \lambda_1} = \frac{\lambda}{\Delta\lambda} \quad [25.11]$$

where $\lambda \approx \lambda_1 \approx \lambda_2$ and $\Delta\lambda = \lambda_2 - \lambda_1$. The **resolving power** of a diffraction grating in the mth order is

$$R = Nm \quad [25.12]$$

where N is the number of illuminated rulings on the grating.

■ WARM-UP EXERCISES

The warm-up exercises in this chapter may be assigned online in Enhanced WebAssign.

1. **Physics Review** A real object is 12.0 cm in front of a thin, convergent lens with a focal length of 10.0 cm. Determine (a) the distance from the lens to the image and (b) the image magnification. (c) Is the image upright or inverted? (d) Is the image real or virtual? (See Section 23.6.)

2. **Physics Review** A thin, diverging lens has a focal length of magnitude 15.0 cm. An object is placed 10.0 cm in front of the lens. Find (a) the image distance and (b) the image magnification. (c) Is the image upright or inverted? (d) Real or virtual? (See Section 23.6.)

3. A large telephoto camera lens has a focal length of 0.825 m and an f-number of 4. Determine the lens diameter. (See Section 25.1.)

4. A prescription lens has a focal length of 25.0 cm. What is the power of the lens in diopters? (See Section 25.2.)

5. A man can see no farther than 42.0 cm without corrective eyeglasses. (a) Is the man nearsighted or farsighted? Find (b) the focal length of the appropriate corrective lens and (c) the power of the lens in diopters. (See Section 25.2.)

6. A woman can see clearly only when objects are 48.0 cm or farther away from her. (a) Is she nearsighted or farsighted? Find (b) the focal length of the appropriate corrective lens and (c) the power of the lens in diopters. (See Section 25.2.)

7. A single-lens magnifier has a maximum angular magnification of 9.33. Determine (a) the lens's focal length (in cm) and (b) the magnification when used with a relaxed eye. (See Section 25.3.)

8. A compound microscope has objective and eyepiece lenses of focal lengths 0.80 cm and 4.0 cm, respectively. If the microscope length is 15 cm, what is the magnification of the microscope? (See Section 25.4.)

9. Determine the angular magnification of a small, handheld telescope if its objective lens has a focal length of 25.0 cm and its eyepiece has a focal length of 2.20 cm. (See Section 25.5.)

10. Determine the approximate limiting angle of resolution (in rad) for 645-nm light incident on a single slit having a width of 1.12×10^{-3} m. (See Section 25.6.)

11. Two stars have an angular separation of 7.50×10^{-8} rad when viewed in the night sky from Earth. Determine the minimum diameter of a telescope's circular aperture if the two stars are to be angularly resolved at a wavelength of 532 nm. (See Section 25.6.)

12. What minimum resolving power must a diffraction grating have to distinguish between two emissions near a wavelength of 632 nm and separated by 2.00 nm? (See Section 25.6.)

■ CONCEPTUAL QUESTIONS

The conceptual questions in this chapter may be assigned online in Enhanced WebAssign.

1. A lens is used to examine an object across a room. Is the lens probably being used as a simple magnifier? Explain in terms of focal length, the image, and magnification.

2. A laser beam is incident at a shallow angle on a horizontal machinist's ruler that has a finely calibrated scale. The engraved rulings on the scale give rise to a diffraction pattern on a vertical screen. Discuss how you can use this technique to obtain a measure of the wavelength of the laser light.

3. **BIO** The optic nerve and the brain invert the image formed on the retina. Why don't we see everything upside down?

4. Suppose you are observing the interference pattern formed by a Michelson interferometer in a laboratory and a joking colleague holds a lit match in the light path of one arm of the interferometer. Will this match have an effect on the interference pattern?

5. If you want to examine the fine detail of an object with a magnifying glass with a power of +20.0 diopters, where should the object be placed so as to observe a magnified image of the object?

6. **BIO** Compare and contrast the eye and a camera. What parts of the camera correspond to the iris, the retina, and the cornea of the eye?

7. If you want to use a converging lens to set fire to a piece of paper, why should the light source be farther from the lens than its focal point?

8. Large telescopes are usually reflecting rather than refracting. List some reasons for this choice.

9. Explain why it is theoretically impossible to see an object as small as an atom regardless of the quality of the light microscope being used.

10. Which is most important in the use of a camera photoflash unit, the intensity of the light (the energy per unit area per unit time) or the product of the intensity and the time of the flash, assuming the time is less than the shutter speed?

11. BIO A patient has a near point of 1.25 m. Is she nearsighted or farsighted? Should the corrective lens be converging or diverging?

12. A lens with a certain power is used as a simple magnifier. If the power of the lens is doubled, does the angular magnification increase or decrease?

13. A laser produces a beam a few millimeters wide, with uniform intensity across its width. A hair is stretched vertically across the front of the laser to cross the beam. (a) How is the diffraction pattern it produces on a distant screen related to that of a vertical slit equal in width to the hair? (b) How could you determine the width of the hair from measurements of its diffraction pattern?

14. BIO During LASIK eye surgery (laser-assisted *in situ* keratomileusis), the shape of the cornea is modified by vaporizing some of its material. If the surgery is performed to correct for nearsightedness, how does the cornea need to be reshaped?

15. If you increase the aperture diameter of a camera by a factor of 3, how is the intensity of the light striking the film affected? (a) It increases by a factor of 3. (b) It decreases by a factor of 3. (c) It increases by a factor of 9. (d) It decreases by a factor of 9. (e) Increasing the aperture doesn't affect the intensity.

PROBLEMS

WebAssign The problems in this chapter may be assigned online in Enhanced WebAssign.

1. denotes straightforward problem; 2. denotes intermediate problem;
3. denotes challenging problem
1. denotes full solution available in *Student Solutions Manual/Study Guide*
1. denotes problems most often assigned in Enhanced WebAssign

BIO denotes biomedical problems
GP denotes guided problems
M denotes Master It tutorial available in Enhanced WebAssign
QC denotes asking for quantitative and conceptual reasoning
S denotes symbolic reasoning problem
W denotes Watch It video solution available in Enhanced WebAssign

25.1 The Camera

1. A lens has a focal length of 28 cm and a diameter of 4.0 cm. What is the f-number of the lens?

2. A certain camera has f-numbers that range from 1.2 to 22. If the focal length of the lens is 55 mm, what is the range of aperture diameters for the camera?

3. A photographic image of a building is 0.092 0 m high. The image was made with a lens with a focal length of 52.0 mm. If the lens was 100 m from the building when the photograph was made, determine the height of the building.

4. The image area of a typical 35 mm slide is 23.5 mm by 35.0 mm. If a camera's lens has a focal length of 55.0 mm and forms an image of the constellation Orion, which is 20° across, will the full image fit on a 35-mm slide?

5. A camera is being used with a correct exposure at $f/4$ and a shutter speed of $\frac{1}{15}$ s. In addition to the f-numbers listed in Section 25.1, this camera has f-numbers $f/1$, $f/1.4$, and $f/2$. To photograph a rapidly moving subject, the shutter speed is changed to $\frac{1}{125}$ s. Find the new f-number setting needed on this camera to maintain satisfactory exposure.

6. (a) Use conceptual arguments to show that the intensity of light (energy per unit area per unit time) reaching the film in a camera is proportional to the square of the reciprocal of the f-number as

$$I \propto \frac{1}{(f/D)^2}$$

(b) The correct exposure time for a camera set to $f/1.8$ is $(1/500)$ s. Calculate the correct exposure time if the f-number is changed to $f/4$ under the same lighting conditions. *Note:* "$f/4$," on a camera, means "an f-number of 4."

7. A certain type of film requires an exposure time of 0.010 s with an $f/11$ lens setting. Another type of film requires twice the light energy to produce the same level of exposure. What f-number does the second type of film need with the 0.010-s exposure time?

8. A certain camera lens has a focal length of 175 mm. Its position can be adjusted to produce images when the lens is between 180 mm and 210 mm from the plane of the film. Over what range of object distances is the lens useful?

25.2 The Eye

9. **BIO** The near point of a person's eye is 60.0 cm. To see objects clearly at a distance of 25.0 cm, what should be the (a) focal length and (b) power of the appropriate corrective lens? (Neglect the distance from the lens to the eye.)

10. **BIO** **GP** A patient can't see objects closer than 40.0 cm and wishes to clearly see objects that are 20.0 cm from his eye. (a) Is the patient nearsighted or farsighted? (b) If the eye–lens distance is 2.00 cm, what is the minimum object distance p from the lens? (c) What image position with respect to the lens will allow the patient to see the object? (d) Is the image real or virtual? Is the image distance q positive or negative? (e) Calculate the required focal length. (f) Find the power of the lens in diopters. (g) If a contact lens is to be prescribed instead, find p, q, and f, and the power of the lens.

11. **BIO** **M** The accommodation limits for Nearsighted Nick's eyes are 18.0 cm and 80.0 cm. When he wears his glasses, he is able to see faraway objects clearly. At what minimum distance is he able to see objects clearly?

12. **BIO** **W** A certain child's near point is 10.0 cm; her far point (with eyes relaxed) is 125 cm. Each eye lens is 2.00 cm from the retina. (a) Between what limits, measured in diopters, does the power of this lens–cornea combination vary? (b) Calculate the power of the eyeglass lens the child should use for relaxed distance vision. Is the lens converging or diverging?

13. **BIO** An individual is nearsighted; his near point is 13.0 cm and his far point is 50.0 cm. (a) What lens power is needed to correct his nearsightedness? (b) When the lenses are in use, what is this person's near point?

14. **BIO** **QC** A patient has a near point of 45.0 cm and far point of 85.0 cm. (a) Can a single lens correct the patient's vision? Explain the patient's options. (b) Calculate the power lens needed to correct the near point so that the patient can see objects 25.0 cm away. Neglect the eye–lens distance. (c) Calculate the power lens needed to correct the patient's far point, again neglecting the eye–lens distance.

15. **BIO** An artificial lens is implanted in a person's eye to replace a diseased lens. The distance between the artificial lens and the retina is 2.80 cm. In the absence of the lens, an image of a distant object (formed by refraction at the cornea) falls 5.33 cm behind the implanted lens. The lens is designed to put the image of the distant object on the retina. What is the power of the implanted lens? *Hint:* Consider the image formed by the cornea to be a virtual object.

16. **BIO** A person is to be fitted with bifocals. She can see clearly when the object is between 30 cm and 1.5 m from the eye. (a) The upper portions of the bifocals (Fig. P25.16) should be designed to enable her to see

Figure P25.16

distant objects clearly. What power should they have? (b) The lower portions of the bifocals should enable her to see objects located 25 cm in front of the eye. What power should they have?

17. **BIO** A nearsighted woman can't see objects clearly beyond 40.0 cm (her far point). If she has no astigmatism and contact lenses are prescribed, what power and type of lens are required to correct her vision?

18. **BIO** **QC** A person sees clearly wearing eyeglasses that have a power of −4.00 diopters when the lenses are 2.00 cm in front of the eyes. (a) What is the focal length of the lens? (b) Is the person nearsighted or farsighted? (c) If the person wants to switch to contact lenses placed directly on the eyes, what lens power should be prescribed?

25.3 The Simple Magnifier

19. A stamp collector uses a lens with 7.5-cm focal length as a simple magnifier. The virtual image is produced at the normal near point (25 cm). (a) How far from the lens should the stamp be placed? (b) What is the expected angular magnification?

20. A lens that has a focal length of 5.00 cm is used as a magnifying glass. (a) To obtain maximum magnification and an image that can be seen clearly by a normal eye, where should the object be placed? (b) What is the angular magnification?

21. **W** A biology student uses a simple magnifier to examine the structural features of an insect's wing. The wing is held 3.50 cm in front of the lens, and the image is formed 25.0 cm from the eye. (a) What is the focal length of the lens? (b) What angular magnification is achieved?

22. A jeweler's lens of focal length 5.0 cm is used as a magnifier. With the lens held near the eye, determine (a) the angular magnification when the object is at the focal point of the lens and (b) the angular magnification when the image formed by the lens is at the near point of the eye (25 cm). (c) What is the object distance giving the maximum magnification?

23. A leaf of length h is positioned 71.0 cm in front of a converging lens with a focal length of 39.0 cm. An observer views the image of the leaf from a position 1.26 m behind the lens, as shown in Figure P25.23. (a) What is the magnitude of the lateral magnification (the ratio of the image size to the object size) produced

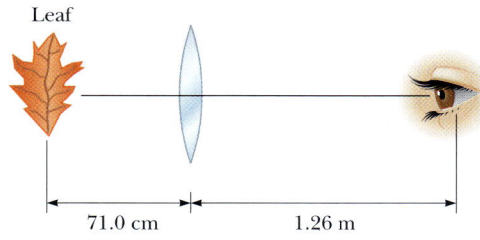

Figure P25.23

by the lens? (b) What angular magnification is achieved by viewing the image of the leaf rather than viewing the leaf directly?

24. (a) What is the maximum angular magnification of an eyeglass lens having a focal length of 18.0 cm when used as a simple magnifier? (b) What is the magnification of this lens when the eye is relaxed?

25.4 The Compound Microscope
25.5 The Telescope

25. The desired overall magnification of a compound microscope is 140×. The objective alone produces a lateral magnification of 12×. Determine the required focal length of the eyepiece.

26. The distance between the eyepiece and the objective lens in a certain compound microscope is 20.0 cm. The focal length of the objective is 0.500 cm, and that of the eyepiece is 1.70 cm. Find the overall magnification of the microscope.

27. Find the magnification of a telescope that uses a 2.75-diopter objective lens and a 35.0-diopter eyepiece.

28. **BIO** A microscope has an objective lens with a focal length of 16.22 mm and an eyepiece with a focal length of 9.50 mm. With the length of the barrel set at 29.0 cm, the diameter of a red blood cell's image subtends an angle of 1.43 mrad with the eye. If the final image distance is 29.0 cm from the eyepiece, what is the actual diameter of the red blood cell? *Hint:* To solve this question, go back to basics and use the thin-lens equation.

29. A certain telescope has an objective mirror with an aperture diameter of 200 mm and a focal length of 2 000 mm. It captures the image of a nebula on photographic film at its prime focus with an exposure time of 1.50 min. To produce the same light energy per unit area on the film, what is the required exposure time to photograph the same nebula with a smaller telescope that has an objective with a 60.0-mm diameter and a 900-mm focal length?

30. **S** (a) Find an equation for the length L of a refracting telescope in terms of the focal length of the objective f_o and the magnification m. (b) A knob adjusts the eyepiece forward and backward. Suppose the telescope is in focus with an eyepiece giving a magnification of 50.0. By what distance must the eyepiece be adjusted when the eyepiece is replaced, with a resulting magnification of 1.00×10^2? Must the eyepiece be adjusted backward or forward? Assume the objective lens has a focal length of 2.00 m.

31. Suppose an astronomical telescope is being designed to have an angular magnification of 34.0. If the focal length of the objective lens being used is 86.0 cm, find (a) the required focal length of the eyepiece and (b) the distance between the two lenses for a relaxed eye. *Hint:* For a relaxed eye, the image formed by the objective lens is at the focal point of the eyepiece.

32. A certain telescope has an objective of focal length 1 500 cm. If the Moon is used as an object, a 1.0-cm-long image formed by the objective corresponds to what distance, in miles, on the Moon? Assume 3.8×10^8 m for the Earth–Moon distance.

33. **S** Astronomers often take photographs with the objective lens or mirror of a telescope alone, without an eyepiece. (a) Show that the image size h' for a telescope used in this manner is given by $h' = fh/(f - p)$, where h is the object size, f is the objective focal length, and p is the object distance. (b) Simplify the expression in part (a) if the object distance is much greater than the objective focal length. (c) The "wingspan" of the International Space Station is 108.6 m, the overall width of its solar panel configuration. When it is orbiting at an altitude of 407 km, find the width of the image formed by a telescope objective of focal length 4.00 m.

34. **BIO W** An elderly sailor is shipwrecked on a desert island, but manages to save his eyeglasses. The lens for one eye has a power of +1.20 diopters, and the other lens has a power of +9.00 diopters. (a) What is the magnifying power of the telescope he can construct with these lenses? (b) How far apart are the lenses when the telescope is adjusted for minimum eyestrain?

35. **M** A person decides to use an old pair of eyeglasses to make some optical instruments. He knows that the near point in his left eye is 50.0 cm and the near point in his right eye is 100 cm. (a) What is the maximum angular magnification he can produce in a telescope? (b) If he places the lenses 10.0 cm apart, what is the maximum overall magnification he can produce in a microscope? *Hint:* Go back to basics and use the thin-lens equation to solve part (b).

36. **QC** Galileo devised a simple terrestrial telescope that produces an upright image. It consists of a converging objective lens and a diverging eyepiece at opposite ends of the telescope tube. For distant objects, the tube length is the objective focal length less the absolute value of the eyepiece focal length. (a) Does the user of the telescope see a real or virtual image? (b) Where is the final image? (c) If a telescope is to be constructed with a tube of length 10.0 cm and a magnification of 3.00, what are the focal lengths of the objective and eyepiece?

25.6 Resolution of Single-Slit and Circular Apertures

37. A converging lens with a diameter of 30.0 cm forms an image of a satellite passing overhead. The satellite has two green lights (wavelength 500 nm) spaced 1.00 m apart. If the lights can just be resolved according to the Rayleigh criterion, what is the altitude of the satellite?

38. While flying at an altitude of 9.50 km, you look out the window at various objects on the ground. If your ability to distinguish two objects is limited only by diffraction, find the smallest separation between two objects on the ground that are distinguishable. Assume your pupil has a diameter of 4.0 mm and take $\lambda = 575$ nm.

39. **M** To increase the resolving power of a microscope, the object and the objective are immersed in oil ($n = 1.5$). If the limiting angle of resolution without the oil is 0.60 μrad, what is the limiting angle of resolution with the oil? *Hint:* The oil changes the wavelength of the light.

40. **BIO** (a) Calculate the limiting angle of resolution for the eye, assuming a pupil diameter of 2.00 mm, a wavelength of 500 nm *in air*, and an index of refraction for the eye of 1.33. (b) What is the maximum distance from the eye at which two points separated by 1.00 cm could be resolved?

41. A vehicle with headlights separated by 2.00 m approaches an observer holding an infrared detector sensitive to radiation of wavelength 885 nm. What aperture diameter is required in the detector if the two headlights are to be resolved at a distance of 10.0 km?

42. Two stars located 23 light-years from Earth are barely resolved using a reflecting telescope having a mirror of diameter 68 cm. Assuming $\lambda = 575$ nm and assuming that the resolution is limited only by diffraction, find the separation between the stars.

43. Suppose a 5.00-m-diameter telescope were constructed on the Moon, where the absence of atmospheric distortion would permit excellent viewing. If observations were made using 500-nm light, what minimum separation between two objects could just be resolved on Mars at closest approach (when Mars is 8.0×10^7 km from the Moon)?

44. **W** A spy satellite circles Earth at an altitude of 200 km and carries out surveillance with a special high-resolution telescopic camera having a lens diameter of 35 cm. If the angular resolution of this camera is limited by diffraction, estimate the separation of two small objects on Earth's surface that are just resolved in yellow-green light ($\lambda = 550$ nm).

45. A 15.0-cm-long grating has 6 000 slits per centimeter. Can two lines of wavelengths 600.000 nm and 600.003 nm be separated with this grating? Explain.

46. The H_α line in hydrogen has a wavelength of 656.20 nm. This line differs in wavelength from the corresponding spectral line in deuterium (the heavy stable isotope of hydrogen) by 0.18 nm. (a) Determine the minimum number of lines a grating must have to resolve these two wavelengths in the first order. (b) Repeat part (a) for the second order.

25.7 The Michelson Interferometer

47. Light of wavelength 550 nm is used to calibrate a Michelson interferometer. With the use of a micrometer screw, the platform on which one mirror is mounted is moved 0.180 mm. How many fringe shifts are counted?

48. **Q|C** Monochromatic light is beamed into a Michelson interferometer. The movable mirror is displaced 0.382 mm, causing the central spot in the interferometer pattern to change from bright to dark and back to bright $N = 1\,700$ times. (a) Determine the wavelength of the light. What color is it? (b) If monochromatic red light is used instead and the mirror is moved the same distance, would N be larger or smaller? Explain.

49. **BIO** An interferometer is used to measure the length of a bacterium. The wavelength of the light used is 650 nm. As one arm of the interferometer is moved from one end of the cell to the other, 310 fringe shifts are counted. How long is the bacterium?

50. Mirror M_1 in Figure 25.15 is displaced a distance ΔL. During this displacement, 250 fringe shifts are counted. The light being used has a wavelength of 632.8 nm. Calculate the displacement ΔL.

51. A thin sheet of transparent material has an index of refraction of 1.40 and is 15.0 μm thick. When it is inserted in the light path along one arm of an interferometer, how many fringe shifts occur in the pattern? Assume the wavelength (in a vacuum) of the light used is 600 nm. *Hint:* The wavelength will change within the material.

52. **M** The Michelson interferometer can be used to measure the index of refraction of a gas by placing an evacuated transparent tube in the light path along one arm of the device. Fringe shifts occur as the gas is slowly added to the tube. Assume 600-nm light is used, the tube is 5.00 cm long, and 160 fringe shifts occur as the pressure of the gas in the tube increases to atmospheric pressure. What is the index of refraction of the gas? *Hint:* The fringe shifts occur because the wavelength of the light changes inside the gas-filled tube.

Additional Problems

53. The Yerkes refracting telescope has a 1.00-m-diameter objective lens of focal length 20.0 m. Assume it is used with an eyepiece of focal length 2.50 cm. (a) Determine the magnification of the planet Mars as seen through the telescope. (b) Are the observed Martian polar caps right side up or upside down?

54. Estimate the minimum angle subtended at the eye of a hawk flying at an altitude of 50 m necessary to recognize a mouse on the ground.

55. An American standard analog television picture (non-HDTV), also known as NTSC, is composed of approximately 485 visible horizontal lines of varying light intensity. Assume your ability to resolve the lines is limited only by the Rayleigh criterion, the pupils of your eyes are 5.00 mm in diameter, and the average wavelength of the light coming from the screen is 550 nm. Calculate the ratio of the minimum viewing distance to the vertical dimension of the picture such that you will not be able to resolve the lines.

56. BIO A person with a nearsighted eye has near and far points of 16 cm and 25 cm, respectively. (a) Assuming a lens is placed 2.0 cm from the eye, what power must the lens have to correct this condition? (b) Suppose contact lenses placed directly on the cornea are used to correct the person's eyesight. What is the power of the lens required in this case, and what is the new near point? *Hint:* The contact lens and the eyeglass lens require slightly different powers because they are at different distances from the eye.

57. BIO The near point of an eye is 75.0 cm. (a) What should be the power of a corrective lens prescribed to enable the eye to see an object clearly at 25.0 cm? (b) If, using the corrective lens, the person can see an object clearly at 26.0 cm but not at 25.0 cm, by how many diopters did the lens grinder miss the prescription?

58. BIO If a typical eyeball is 2.00 cm long and has a pupil opening that can range from about 2.00 mm to 6.00 mm, what are (a) the focal length of the eye when it is focused on objects 1.00 m away, (b) the smallest f-number of the eye when it is focused on objects 1.00 m away, and (c) the largest f-number of the eye when it is focused on objects 1.00 m away?

59. BIO M A cataract-impaired lens in an eye may be surgically removed and replaced by a manufactured lens. The focal length required for the new lens is determined by the lens-to-retina distance, which is measured by a sonarlike device, and by the requirement that the implant provide for correct distance vision. (a) If the distance from lens to retina is 22.4 mm, calculate the power of the implanted lens in diopters. (b) Since there is no accommodation and the implant allows for correct distance vision, a corrective lens for close work or reading must be used. Assume a reading distance of 33.0 cm, and calculate the power of the lens in the reading glasses.

60. BIO If the aqueous humor of the eye has an index of refraction of 1.34 and the distance from the vertex of the cornea to the retina is 2.00 cm, what is the radius of curvature of the cornea for which distant objects will be focused on the retina? (For simplicity, assume all refraction occurs in the aqueous humor.)

61. A Boy Scout starts a fire by using a lens from his eyeglasses to focus sunlight on kindling 5.0 cm from the lens. The Boy Scout has a near point of 15 cm. When the lens is used as a simple magnifier, (a) what is the maximum magnification that can be achieved and (b) what is the magnification when the eye is relaxed? *Caution:* The equations derived in the text for a simple magnifier assume a "normal" eye.

62. A laboratory (astronomical) telescope is used to view a scale that is 300 cm from the objective, which has a focal length of 20.0 cm; the eyepiece has a focal length of 2.00 cm. Calculate the angular magnification when the telescope is adjusted for minimum eyestrain. *Note:* The object is not at infinity, so the simple expression $m = f_o/f_e$ is not sufficiently accurate for this problem. Also, assume small angles, so that $\tan \theta \approx \theta$.

Mathematics Review

■ Appendix A

A.1 Mathematical Notation

Many mathematical symbols are used throughout this book. These symbols are described here, with examples illustrating their use.

Equals Sign: =

The symbol = denotes the mathematical equality of two quantities. In physics, it also makes a statement about the relationship of different physical concepts. An example is the equation $E = mc^2$. This famous equation says that a given mass m, *measured in kilograms*, is equivalent to a certain amount of energy, E, *measured in joules*. The speed of light squared, c^2, can be considered a constant of proportionality, neccessary because the units chosen for given quantities are rather arbitrary, based on historical circumstances.

Proportionality: ∝

The symbol ∝ denotes a proportionality. This symbol might be used when focusing on relationships rather than an exact mathematical equality. For example, we could write $E \propto m$, which says "the energy E associated with an object is proportional to the mass m of the object." Another example is found in kinetic energy, which is the energy associated with an object's motion, defined by $KE = \frac{1}{2}mv^2$, where m is again the mass and v is the speed. Both m and v are variables in this expression. Hence, the kinetic energy KE is proportional to m, $KE \propto m$, and at the same time KE is proportional to the speed squared, $KE \propto v^2$. Another term used here is "directly proportional." The density ρ of an object is related to its mass and volume by $\rho = m/V$. Consequently, the density is said to be directly proportional to mass and inversely proportional to volume.

Inequalities

The symbol < means "is less than," and > means "is greater than." For example, $\rho_{Fe} > \rho_{Al}$ means that the density of iron, ρ_{Fe}, is greater than the density of aluminum, ρ_{Al}. If there is a line underneath the symbol, there is the possibility of equality: ≤ means "less than or equal to," whereas ≥ means "greater than or equal to." Any particle's speed v, for example, is less than or equal to the speed of light, c: $v \leq c$.

Sometimes the size of a given quantity greatly differs from the size of another quantity. Simple inequality doesn't convey vast differences. For such cases, the symbol ≪ means "is much less than" and ≫ means "is much greater than." The mass of the Sun, M_{Sun}, is much greater than the mass of the Earth, M_E: $M_{Sun} \gg M_E$. The mass of an electron, m_e, is much less than the mass of a proton, m_p: $m_e \ll m_p$.

Approximately Equal: ≈

The symbol ≈ indicates that two quantities are approximately equal to each other. The mass of a proton, m_p, is approximately the same as the mass of a neutron, m_n. This relationship can be written $m_p \approx m_n$.

Equivalence: ≡

The symbol ≡ means "is defined as," which is a different statement than a simple =. It means that the quantity on the left—usually a single quantity—is another way

A.1

of expressing the quantity or quantities on the right. The classical momentum of an object, p, is defined to be the mass of the object m times its velocity v, hence $p \equiv mv$. Because this equivalence is by definition, there is no possibility of p being equal to something else. Contrast this case with that of the expression for the velocity v of an object under constant acceleration, which is $v = at + v_0$. This equation would never be written with an equivalence sign because v in this context is not a defined quantity; rather it is an equality that holds true only under the assumption of constant acceleration. The expression for the classical momentum, however, is always true by definition, so it would be appropriate to write $p \equiv mv$ the first time the concept is introduced. After the introduction of a term, an ordinary equals sign generally suffices.

Differences: Δ

The Greek letter Δ (capital delta) is the symbol used to indicate the difference in a measured physical quantity, usually at two different times. The best example is a displacement along the x-axis, indicated by Δx (read as "delta x"). Note that Δx doesn't mean "the product of Δ and x." Suppose a person out for a morning stroll starts measuring her distance away from home when she is 10 m from her doorway. She then continues along a straight-line path and stops strolling 50 m from the door. Her change in position during the walk is $\Delta x = 50 \text{ m} - 10 \text{ m} = 40 \text{ m}$. In symbolic form, such displacements can be written

$$\Delta x = x_f - x_i$$

In this equation, x_f is the final position and x_i is the initial position. There are numerous other examples of differences in physics, such as the difference (or change) in momentum, $\Delta p = p_f - p_i$; the change in kinetic energy, $\Delta K = K_f - K_i$; and the change in temperature, $\Delta T = T_f - T_i$.

Summation: Σ

In physics there are often contexts in which it's necessary to add several quantities. A useful abbreviation for representing such a sum is the Greek letter Σ (capital sigma). Suppose we wish to add a set of five numbers represented by x_1, x_2, x_3, x_4, and x_5. In the abbreviated notation, we would write the sum as

$$x_1 + x_2 + x_3 + x_4 + x_5 = \sum_{i=1}^{5} x_i$$

where the subscript i on x represents any one of the numbers in the set. For example, if there are five masses in a system, m_1, m_2, m_3, m_4, and m_5, the total mass of the system $M = m_1 + m_2 + m_3 + m_4 + m_5$ could be expressed as

$$M = \sum_{i=1}^{5} m_i$$

The x-coordinate of the center of mass of the five masses, meanwhile, could be written

$$x_{CM} = \frac{\sum_{i=1}^{5} m_i x_i}{M}$$

with similar expressions for the y- and z-coordinates of the center of mass.

Absolute Value: | |

The magnitude of a quantity x, written $|x|$, is simply the absolute value of that quantity. The sign of $|x|$ is always positive, regardless of the sign of x. For example, if $x = -5$, then $|x| = 5$; if $x = 8$, then $|x| = 8$. In physics this sign is useful whenever the magnitude of a quantity is more important than any direction that might be implied by a sign.

A.2 Scientific Notation

Many quantities in science have very large or very small values. The speed of light is about 300 000 000 m/s, and the ink required to make the dot over an *i* in this textbook has a mass of about 0.000 000 001 kg. It's very cumbersome to read, write, and keep track of such numbers because the decimal places have to be counted and because a number with one significant digit may require a large number of zeros. Scientific notation is a way of representing these numbers without having to write out so many zeros, which in general are only used to establish the magnitude of the number, not its accuracy. The key is to use powers of 10. The nonnegative powers of 10 are

$$10^0 = 1$$
$$10^1 = 10$$
$$10^2 = 10 \times 10 = 100$$
$$10^3 = 10 \times 10 \times 10 = 1\,000$$
$$10^4 = 10 \times 10 \times 10 \times 10 = 10\,000$$
$$10^5 = 10 \times 10 \times 10 \times 10 \times 10 = 100\,000$$

and so on. The number of decimal places following the first digit in the number and to the left of the decimal point corresponds to the power to which 10 is raised, called the **exponent** of 10. The speed of light, 300 000 000 m/s, can then be expressed as 3×10^8 m/s. Notice there are eight decimal places to the right of the leading digit, 3, and to the left of where the decimal point would be placed. Some representative numbers smaller than 1 are

$$10^{-1} = \frac{1}{10} = 0.1$$
$$10^{-2} = \frac{1}{10 \times 10} = 0.01$$
$$10^{-3} = \frac{1}{10 \times 10 \times 10} = 0.001$$
$$10^{-4} = \frac{1}{10 \times 10 \times 10 \times 10} = 0.000\,1$$
$$10^{-5} = \frac{1}{10 \times 10 \times 10 \times 10 \times 10} = 0.000\,01$$

In these cases, the number of decimal places to the right of the decimal point up to and including only the first nonzero digit equals the value of the (negative) exponent.

Numbers expressed as some power of 10 multiplied by another number between 1 and 10 are said to be in **scientific notation**. For example, Coulomb's constant, which is associated with electric forces, is given by 8 987 551 789 N·m²/C² and is written in scientific notation as $8.987\,551\,789 \times 10^9$ N·m²/C². Newton's constant of gravitation is given by 0.000 000 000 066 731 N·m²/kg², written in scientific notation as $6.673\,1 \times 10^{-11}$ N·m²/kg².

When numbers expressed in scientific notation are being multiplied, the following general rule is very useful:

$$10^n \times 10^m = 10^{n+m} \qquad [\text{A.1}]$$

where n and m can be any numbers (not necessarily integers). For example, $10^2 \times 10^5 = 10^7$. The rule also applies if one of the exponents is negative: $10^3 \times 10^{-8} = 10^{-5}$.

When dividing numbers expressed in scientific notation, note that

$$\frac{10^n}{10^m} = 10^n \times 10^{-m} = 10^{n-m}$$ [A.2]

Exercises
With help from the above rules, verify the answers to the following:

1. $86\,400 = 8.64 \times 10^4$
2. $9\,816\,762.5 = 9.816\,762\,5 \times 10^6$
3. $0.000\,000\,039\,8 = 3.98 \times 10^{-8}$
4. $(4 \times 10^8)(9 \times 10^9) = 3.6 \times 10^{18}$
5. $(3 \times 10^7)(6 \times 10^{-12}) = 1.8 \times 10^{-4}$
6. $\dfrac{75 \times 10^{-11}}{5 \times 10^{-3}} = 1.5 \times 10^{-7}$
7. $\dfrac{(3 \times 10^6)(8 \times 10^{-2})}{(2 \times 10^{17})(6 \times 10^5)} = 2 \times 10^{-18}$

A.3 Algebra

A. Some Basic Rules

When algebraic operations are performed, the laws of arithmetic apply. Symbols such as x, y, and z are usually used to represent quantities that are not specified, what are called the **unknowns**.

First, consider the equation

$$8x = 32$$

If we wish to solve for x, we can divide (or multiply) each side of the equation by the same factor without destroying the equality. In this case, if we divide both sides by 8, we have

$$\frac{8x}{8} = \frac{32}{8}$$

$$x = 4$$

Next consider the equation

$$x + 2 = 8$$

In this type of expression, we can add or subtract the same quantity from each side. If we subtract 2 from each side, we obtain

$$x + 2 - 2 = 8 - 2$$

$$x = 6$$

In general, if $x + a = b$, then $x = b - a$.

Now consider the equation

$$\frac{x}{5} = 9$$

If we multiply each side by 5, we are left with x on the left by itself and 45 on the right:

$$\left(\frac{x}{5}\right)(5) = 9 \times 5$$

$$x = 45$$

In all cases, **whatever operation is performed on the left side of the equality must also be performed on the right side.**

The following rules for multiplying, dividing, adding, and subtracting fractions should be recalled, where a, b, and c are three numbers:

	Rule	Example
Multiplying	$\left(\dfrac{a}{b}\right)\left(\dfrac{c}{d}\right) = \dfrac{ac}{bd}$	$\left(\dfrac{2}{3}\right)\left(\dfrac{4}{5}\right) = \dfrac{8}{15}$
Dividing	$\dfrac{(a/b)}{(c/d)} = \dfrac{ad}{bc}$	$\dfrac{2/3}{4/5} = \dfrac{(2)(5)}{(4)(3)} = \dfrac{10}{12} = \dfrac{5}{6}$
Adding	$\dfrac{a}{b} \pm \dfrac{c}{d} = \dfrac{ad \pm bc}{bd}$	$\dfrac{2}{3} - \dfrac{4}{5} = \dfrac{(2)(5) - (4)(3)}{(3)(5)} = -\dfrac{2}{15}$

Very often in physics we are called upon to manipulate symbolic expressions algebraically, a process most students find unfamiliar. It's very important, however, because substituting numbers into an equation too early can often obscure meaning. The following two examples illustrate how these kinds of algebraic manipulations are carried out.

■ EXAMPLE

A ball is dropped from the top of a building 50.0 m tall. How long does it take the ball to fall to a height of 25.0 m?

SOLUTION First, write the general ballistics equation for this situation:

$$x = \tfrac{1}{2}at^2 + v_0 t + x_0$$

Here, $a = -9.80$ m/s² is the acceleration of gravity that causes the ball to fall, $v_0 = 0$ is the initial velocity, and $x_0 = 50.0$ m is the initial position. Substitute only the initial velocity, $v_0 = 0$, obtaining the following equation:

$$x = \tfrac{1}{2}at^2 + x_0$$

This equation must be solved for t. Subtract x_0 from both sides:

$$x - x_0 = \tfrac{1}{2}at^2 + x_0 - x_0 = \tfrac{1}{2}at^2$$

Multiply both sides by $2/a$:

$$\left(\dfrac{2}{a}\right)(x - x_0) = \left(\dfrac{2}{a}\right)\tfrac{1}{2}at^2 = t^2$$

It's customary to have the desired value on the left, so switch the equation around and take the square root of both sides:

$$t = \pm\sqrt{\left(\dfrac{2}{a}\right)(x - x_0)}$$

Only the positive root makes sense. Values could now be substituted to obtain a final answer.

■ EXAMPLE

A block of mass m slides over a frictionless surface in the positive x-direction. It encounters a patch of roughness having coefficient of kinetic friction μ_k. If the rough patch has length Δx, find the speed of the block after leaving the patch.

SOLUTION Using the work-energy theorem, we have

$$\tfrac{1}{2}mv^2 - \tfrac{1}{2}mv_0^2 = -\mu_k mg\,\Delta x$$

Add $\tfrac{1}{2}mv_0^2$ to both sides:

$$\tfrac{1}{2}mv^2 = \tfrac{1}{2}mv_0^2 - \mu_k mg\,\Delta x$$

(Continued)

Multiply both sides by $2/m$:

$$v^2 = v_0^2 - 2\mu_k g \Delta x$$

Finally, take the square root of both sides. Because the block is sliding in the positive x-direction, the positive square root is selected.

$$v = \sqrt{v_0^2 - 2\mu_k g \Delta x}$$

Exercises

In Exercises 1–4, solve for x:

Answers

1. $a = \dfrac{1}{1+x}$ $x = \dfrac{1-a}{a}$

2. $3x - 5 = 13$ $x = 6$

3. $ax - 5 = bx + 2$ $x = \dfrac{7}{a-b}$

4. $\dfrac{5}{2x+6} = \dfrac{3}{4x+8}$ $x = -\dfrac{11}{7}$

5. Solve the following equation for v_1:

$$P_1 + \tfrac{1}{2}\rho v_1^2 = P_2 + \tfrac{1}{2}\rho v_2^2$$

Answer: $v_1 = \pm\sqrt{\dfrac{2}{\rho}(P_2 - P_1) + v_2^2}$

B. Powers

When powers of a given quantity x are multiplied, the following rule applies:

$$x^n x^m = x^{n+m} \qquad [A.3]$$

For example, $x^2 x^4 = x^{2+4} = x^6$.

When dividing the powers of a given quantity, the rule is

$$\dfrac{x^n}{x^m} = x^{n-m} \qquad [A.4]$$

For example, $x^8/x^2 = x^{8-2} = x^6$.

A power that is a fraction, such as $\tfrac{1}{3}$, corresponds to a root as follows:

$$x^{1/n} = \sqrt[n]{x} \qquad [A.5]$$

For example, $4^{1/3} = \sqrt[3]{4} = 1.587\,4$. (A scientific calculator is useful for such calculations.)

Finally, any quantity x^n raised to the mth power is

$$(x^n)^m = x^{nm} \qquad [A.6]$$

Table A.1 summarizes the rules of exponents.

Table A.1 Rules of Exponents

$x^0 = 1$
$x^1 = x$
$x^n x^m = x^{n+m}$
$x^n/x^m = x^{n-m}$
$x^{1/n} = \sqrt[n]{x}$
$(x^n)^m = x^{nm}$

Exercises

Verify the following:

1. $3^2 \times 3^3 = 243$
2. $x^5 x^{-8} = x^{-3}$
3. $x^{10}/x^{-5} = x^{15}$
4. $5^{1/3} = 1.709\,975$ (Use your calculator.)
5. $60^{1/4} = 2.783\,158$ (Use your calculator.)
6. $(x^4)^3 = x^{12}$

C. Factoring

The following are some useful formulas for factoring an equation:

$$ax + ay + az = a(x + y + z) \quad \text{common factor}$$
$$a^2 + 2ab + b^2 = (a + b)^2 \quad \text{perfect square}$$
$$a^2 - b^2 = (a + b)(a - b) \quad \text{difference of squares}$$

D. Quadratic Equations

The general form of a quadratic equation is

$$ax^2 + bx + c = 0 \quad \text{[A.7]}$$

where x is the unknown quantity and a, b, and c are numerical factors referred to as coefficients of the equation. This equation has two roots, given by

$$x = \frac{-b \pm \sqrt{b^2 - 4ac}}{2a} \quad \text{[A.8]}$$

If $b^2 - 4ac > 0$, the roots are real.

■ EXAMPLE

The equation $x^2 + 5x + 4 = 0$ has the following roots corresponding to the two signs of the square-root term:

$$x = \frac{-5 \pm \sqrt{5^2 - (4)(1)(4)}}{2(1)} = \frac{-5 \pm \sqrt{9}}{2} = \frac{-5 \pm 3}{2}$$

$$x_1 = \frac{-5 + 3}{2} = \boxed{-1} \qquad x_2 = \frac{-5 - 3}{2} = \boxed{-4}$$

where x_1 refers to the root corresponding to the positive sign and x_2 refers to the root corresponding to the negative sign.

■ EXAMPLE

A ball is projected upwards at 16.0 m/s. Use the quadratic formula to determine the time necessary for it to reach a height of 8.00 m above the point of release.

SOLUTION From the discussion of ballistics in Chapter 2, we can write

$$(1) \quad x = \tfrac{1}{2}at^2 + v_0 t + x_0$$

The acceleration is due to gravity, given by $a = -9.80$ m/s²; the initial velocity is $v_0 = 16.0$ m/s; and the initial position is the point of release, taken to be $x_0 = 0$. Substitute these values into Equation (1) and set $x = 8.00$ m, arriving at

$$x = -4.90t^2 + 16.00t = 8.00$$

where units have been suppressed for mathematical clarity. Rearrange this expression into the standard form of Equation A.7:

$$-4.90t^2 + 16.00t - 8.00 = 0$$

The equation is quadratic in the time, t. We have $a = -4.9$, $b = 16$, and $c = -8.00$. Substitute these values into Equation A.8:

$$t = \frac{-16.0 \pm \sqrt{16^2 - 4(-4.90)(-8.00)}}{2(-4.90)} = \frac{-16.0 \pm \sqrt{99.2}}{-9.80}$$

$$= 1.63 \mp \frac{\sqrt{99.2}}{9.80} = \boxed{0.614 \text{ s}, 2.65 \text{ s}}$$

Both solutions are valid in this case, one corresponding to reaching the point of interest on the way up and the other to reaching it on the way back down.

Exercises

Solve the following quadratic equations:

	Answers	
1. $x^2 + 2x - 3 = 0$	$x_1 = 1$	$x_2 = -3$
2. $2x^2 - 5x + 2 = 0$	$x_1 = 2$	$x_2 = \frac{1}{2}$
3. $2x^2 - 4x - 9 = 0$	$x_1 = 1 + \sqrt{22}/2$	$x_2 = 1 - \sqrt{22}/2$

4. Repeat the ballistics example for a height of 10.0 m above the point of release.
   Answer: $t_1 = 0.842$ s $t_2 = 2.42$ s

E. Linear Equations

A linear equation has the general form

$$y = mx + b \qquad [A.9]$$

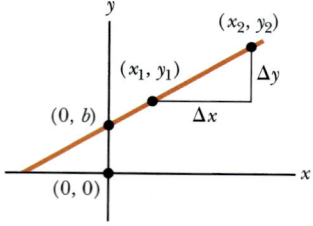

Figure A.1

where m and b are constants. This kind of equation is called linear because the graph of y versus x is a straight line, as shown in Figure A.1. The constant b, called the y-intercept, represents the value of y at which the straight line intersects the y-axis. The constant m is equal to the slope of the straight line. If any two points on the straight line are specified by the coordinates (x_1, y_1) and (x_2, y_2), as in Figure A.1, the slope of the straight line can be expressed as

$$\text{Slope} = \frac{y_2 - y_1}{x_2 - x_1} = \frac{\Delta y}{\Delta x} \qquad [A.10]$$

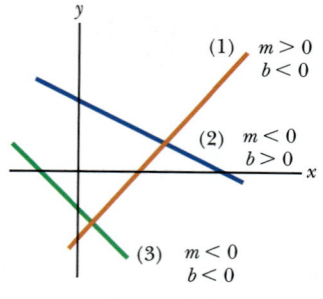

Figure A.2

Note that m and b can have either positive or negative values. If $m > 0$, the straight line has a positive slope, as in Figure A.1. If $m < 0$, the straight line has a negative slope. In Figure A.1, both m and b are positive. Three other possible situations are shown in Figure A.2.

■ EXAMPLE

Suppose the electrical resistance of a metal wire is 5.00 Ω at a temperature of 20.0°C and 6.14 Ω at 80.0°C. Assuming the resistance changes linearly, what is the resistance of the wire at 60.0°C?

SOLUTION Find the equation of the line describing the resistance R and then substitute the new temperature into it. Two points on the graph of resistance versus temperature, (20.0°C, 5.00 Ω) and (80.0°C, 6.14 Ω), allow computation of the slope:

$$(1) \quad m = \frac{\Delta R}{\Delta T} = \frac{6.14\ \Omega - 5.00\ \Omega}{80.0°\text{C} - 20.0°\text{C}} = 1.90 \times 10^{-2}\ \Omega/°\text{C}$$

Now use the point-slope formulation of a line, with this slope and (20.0°C, 5.00 Ω):

$$(2) \quad R - R_0 = m(T - T_0)$$

$$(3) \quad R - 5.00\ \Omega = (1.90 \times 10^{-2}\ \Omega/°\text{C})(T - 20.0°\text{C})$$

Finally, substitute $T = 60.0°$ into Equation (3) and solve for R, getting $R = 5.76$ Ω.

Exercises

1. Draw graphs of the following straight lines:
   (a) $y = 5x + 3$ (b) $y = -2x + 4$ (c) $y = -3x - 6$
2. Find the slopes of the straight lines described in Exercise 1.
   Answers: (a) 5 (b) −2 (c) −3
3. Find the slopes of the straight lines that pass through the following sets of points: (a) (0, −4) and (4, 2) (b) (0, 0) and (2, −5) (c) (−5, 2) and (4, −2)
   Answers: (a) 3/2 (b) −5/2 (c) −4/9

4. Suppose an experiment measures the following displacements (in meters) from equilibrium of a vertical spring due to attaching weights (in Newtons): $(0.025\ 0\ \text{m}, 22.0\ \text{N})$, $(0.075\ 0\ \text{m}, 66.0\ \text{N})$. Find the spring constant, which is the slope of the line in the graph of weight versus displacement.
Answer: 880 N/m

F. Solving Simultaneous Linear Equations

Consider the equation $3x + 5y = 15$, which has two unknowns, x and y. Such an equation doesn't have a unique solution. For example, note that $(x = 0, y = 3)$, $(x = 5, y = 0)$, and $(x = 2, y = 9/5)$ are all solutions to this equation.

If a problem has two unknowns, a unique solution is possible only if we have two equations. In general, if a problem has n unknowns, its solution requires n equations. To solve two simultaneous equations involving two unknowns, x and y, we solve one of the equations for x in terms of y and substitute this expression into the other equation.

■ EXAMPLE

Solve the following two simultaneous equations:

$$(1)\quad 5x + y = -8 \qquad (2)\quad 2x - 2y = 4$$

SOLUTION From Equation (2), we find that $x = y + 2$. Substitution of this value into Equation (1) gives

$$5(y + 2) + y = -8$$
$$6y = -18$$
$$y = \boxed{-3}$$
$$x = y + 2 = \boxed{-1}$$

ALTERNATE SOLUTION Multiply each term in Equation (1) by the factor 2 and add the result to Equation (2):

$$10x + 2y = -16$$
$$\underline{2x - 2y = 4}$$
$$12x = -12$$
$$x = \boxed{-1}$$
$$y = x - 2 = \boxed{-3}$$

Two linear equations containing two unknowns can also be solved by a graphical method. If the straight lines corresponding to the two equations are plotted in a conventional coordinate system, the intersection of the two lines represents the solution. For example, consider the two equations

$$x - y = 2$$
$$x - 2y = -1$$

These equations are plotted in Figure A.3. The intersection of the two lines has the coordinates $x = 5, y = 3$, which represents the solution to the equations. You should check this solution by the analytical technique discussed above.

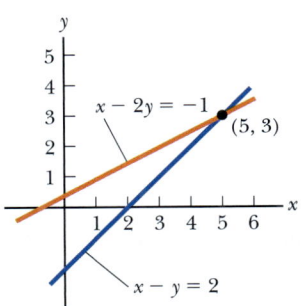

Figure A.3

■ EXAMPLE

A block of mass $m = 2.00$ kg travels in the positive x-direction at $v_i = 5.00$ m/s, while a second block, of mass $M = 4.00$ kg and leading the first block, travels in the positive x-direction at 2.00 m/s. The surface is frictionless. What are the blocks' final velocities if the collision is perfectly elastic?

(Continued)

SOLUTION As can be seen in Chapter 6, a perfectly elastic collision involves equations for the momentum and energy. With algebra, the energy equation, which is quadratic in v, can be recast as a linear equation. The two equations are given by

$$(1) \quad mv_i + MV_i = mv_f + MV_f$$
$$(2) \quad v_i - V_i = -(v_f - V_f)$$

Substitute the known quantities $v_i = 5.00$ m/s and $V_i = 2.00$ m/s into Equations (1) and (2):

$$(3) \quad 18 = 2v_f + 4V_f$$
$$(4) \quad 3 = -v_f + V_f$$

Multiply Equation (4) by 2 and add to Equation (3):

$$18 = 2v_f + 4V_f$$
$$6 = -2v_f + 2V_f$$
$$24 = 6V_f \quad \rightarrow \quad V_f = \boxed{4.00 \text{ m/s}}$$

Substituting the solution for V_f back into Equation (4) yields $v_f = \boxed{1.00 \text{ m/s}}$.

Exercises

Solve the following pairs of simultaneous equations involving two unknowns:

Answers

1. $x + y = 8$
   $x - y = 2$ $x = 5, y = 3$

2. $98 - T = 10a$
   $T - 49 = 5a$ $T = 65.3, a = 3.27$

3. $6x + 2y = 6$
   $8x - 4y = 28$ $x = 2, y = -3$

G. Logarithms and Exponentials

Suppose a quantity x is expressed as a power of some quantity a:

$$x = a^y \quad \text{[A.11]}$$

The number a is called the **base** number. The **logarithm** of x with respect to the base a is equal to the exponent to which the base must be raised so as to satisfy the expression $x = a^y$:

$$y = \log_a x \quad \text{[A.12]}$$

Conversely, the **antilogarithm** of y is the number x:

$$x = \text{antilog}_a y \quad \text{[A.13]}$$

The antilog expression is in fact identical to the exponential expression in Equation A.11, which is preferable for practical purposes.

In practice, the two bases most often used are base 10, called the *common* logarithm base, and base $e = 2.718\ldots$, called the *natural* logarithm base. When common logarithms are used,

$$y = \log_{10} x \quad (\text{or } x = 10^y) \quad \text{[A.14]}$$

When natural logarithms are used,

$$y = \ln x \quad (\text{or } x = e^y) \quad \text{[A.15]}$$

For example, $\log_{10} 52 = 1.716$, so $\text{antilog}_{10} 1.716 = 10^{1.716} = 52$. Likewise, $\ln_e 52 = 3.951$, so $\text{antiln}_e 3.951 = e^{3.951} = 52$.

In general, note that you can convert between base 10 and base e with the equality

$$\ln x = (2.302\,585)\log_{10} x \quad \text{[A.16]}$$

Finally, some useful properties of logarithms are

$$\log(ab) = \log a + \log b \qquad \ln e = 1$$
$$\log(a/b) = \log a - \log b \qquad \ln e^a = a$$
$$\log(a^n) = n \log a \qquad \ln\left(\frac{1}{a}\right) = -\ln a$$

Logarithms in college physics are used most notably in the definition of decibel level. Sound intensity varies across several orders of magnitude, making it awkward to compare different intensities. Decibel level converts these intensities to a more manageable logarithmic scale.

■ EXAMPLE (Logs)

Suppose a jet testing its engines produces a sound intensity of $I = 0.750$ W at a given location in an airplane hangar. What decibel level corresponds to this sound intensity?

SOLUTION Decibel level β is defined by

$$\beta = 10 \log\left(\frac{I}{I_0}\right)$$

where $I_0 = 1 \times 10^{-12}$ W/m² is the standard reference intensity. Substitute the given information:

$$\beta = 10 \log\left(\frac{0.750 \text{ W/m}^2}{10^{-12} \text{ W/m}^2}\right) = 119 \text{ dB}$$

■ EXAMPLE (Antilogs)

A collection of four identical machines creates a decibel level of $\beta = 87.0$ dB in a machine shop. What sound intensity would be created by only one such machine?

SOLUTION We use the equation of decibel level to find the total sound intensity of the four machines, and then we divide by 4. From Equation (1):

$$87.0 \text{ dB} = 10 \log\left(\frac{I}{10^{-12} \text{ W/m}^2}\right)$$

Divide both sides by 10 and take the antilog of both sides, which means, equivalently, to exponentiate:

$$10^{8.7} = 10^{\log(I/10^{-12})} = \frac{I}{10^{-12}}$$

$$I = 10^{-12} \cdot 10^{8.7} = 10^{-3.3} = 5.01 \times 10^{-4} \text{ W/m}^2$$

There are four machines, so this result must be divided by 4 to get the intensity of one machine:

$$I = 1.25 \times 10^{-4} \text{ W/m}^2$$

■ EXAMPLE (Exponentials)

The half-life of tritium is 12.33 years. (Tritium is the heaviest isotope of hydrogen, with a nucleus consisting of a proton and two neutrons.) If a sample contains 3.0 g of tritium initially, how much remains after 20.0 years?

SOLUTION The equation giving the number of nuclei of a radioactive substance as a function of time is

$$N = N_0 \left(\frac{1}{2}\right)^n$$

(Continued)

where N is the number of nuclei remaining, N_0 is the initial number of nuclei, and n is the number of half-lives. Note that this equation is an exponential expression with a base of $\frac{1}{2}$. The number of half-lives is given by $n = t/T_{1/2} =$ 20.0 yr/12.33 yr = 1.62. The fractional amount of tritium that remains after 20.0 yr is therefore

$$\frac{N}{N_0} = \left(\frac{1}{2}\right)^{1.62} = 0.325$$

Hence, of the original 3.00 g of tritium, 0.325×3.00 g = 0.975 g remains.

A.4 Geometry

Table A.2 gives the areas and volumes for several geometric shapes used throughout this text. These areas and volumes are important in numerous physics applications. A good example is the concept of pressure P, which is the force per unit area. As an equation, it is written $P = F/A$. Areas must also be calculated in problems involving the volume rate of fluid flow through a pipe using the equation of continuity, the tensile stress exerted on a cable by a weight, the rate of thermal energy transfer through a barrier, and the density of current through a wire. There are numerous other applications. Volumes are important in computing the buoyant force exerted by water on a submerged object, in calculating densities, and in determining the bulk stress of fluid or gas on an object, which affects its volume. Again, there are numerous other applications.

Table A.2 Useful Information for Geometry

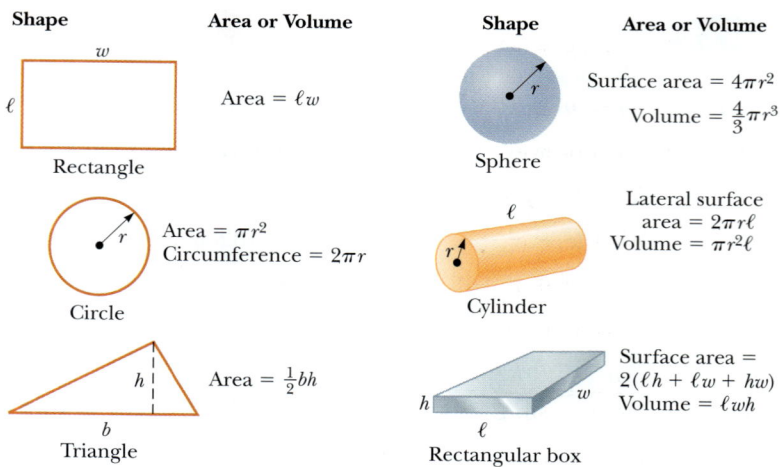

A.5 Trigonometry

Some of the most basic facts concerning trigonometry are presented in Chapter 1, and we encourage you to study the material presented there if you are having trouble with this branch of mathematics. The most important trigonometric concepts include the Pythagorean theorem:

$$\Delta s^2 = \Delta x^2 + \Delta y^2 \qquad [\text{A.17}]$$

This equation states that the square distance along the hypotenuse of a right triangle equals the sum of the squares of the legs. It can also be used to find distances between points in Cartesian coordinates and the length of a vector, where Δx is replaced by the *x*-component of the vector and Δy is replaced by the *y*-component of the vector. If the vector $\vec{A}$ has components A_x and A_y, the magnitude A of the vector satisfies

$$A^2 = A_x^2 + A_y^2 \qquad [\text{A.18}]$$

which has a form completely analogous to the form of the Pythagorean theorem. Also highly useful are the cosine and sine functions because they relate the length of a vector to its *x*- and *y*-components:

$$A_x = A \cos \theta \qquad [\text{A.19}]$$
$$A_y = A \sin \theta \qquad [\text{A.20}]$$

The direction θ of a vector in a plane can be determined by use of the tangent function:

$$\tan \theta = \frac{A_y}{A_x} \qquad [\text{A.21}]$$

A relative of the Pythagorean theorem is also frequently useful:

$$\sin^2 \theta + \cos^2 \theta = 1 \qquad [\text{A.22}]$$

Details on the above concepts can be found in the extensive discussions in Chapters 1 and 3. The following are some other useful trigonometric identities:

$$\sin \theta = \cos(90° - \theta)$$

$$\cos \theta = \sin(90° - \theta)$$

$$\sin 2\theta = 2 \sin \theta \cos \theta$$

$$\cos 2\theta = \cos^2 \theta - \sin^2 \theta$$

$$\sin(\theta \pm \phi) = \sin \theta \cos \phi \pm \cos \theta \sin \phi$$

$$\cos(\theta \pm \phi) = \cos \theta \cos \phi \mp \sin \theta \sin \phi$$

The following relationships apply to any triangle, as shown in Figure A.4:

$$\alpha + \beta + \gamma = 180°$$

$$a^2 = b^2 + c^2 - 2bc \cos \alpha$$
$$b^2 = a^2 + c^2 - 2ac \cos \beta \qquad \text{law of cosines}$$
$$c^2 = a^2 + b^2 - 2ab \cos \gamma$$

$$\frac{a}{\sin \alpha} = \frac{b}{\sin \beta} = \frac{c}{\sin \gamma} \qquad \text{law of sines}$$

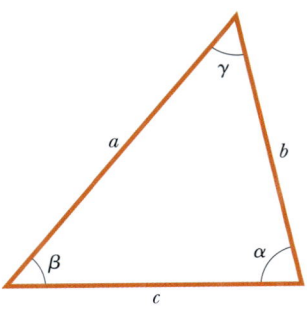

Figure A.4

Appendix B
An Abbreviated Table of Isotopes

Atomic Number Z	Element	Symbol	Chemical Atomic Mass (u)	Mass Number (* Indicates Radioactive) A	Atomic Mass (u)	Percent Abundance	Half-Life (If Radioactive) $T_{1/2}$
0	(Neutron)	n		1*	1.008 665		10.4 min
1	Hydrogen	H	1.007 94	1	1.007 825	99.988 5	
	Deuterium	D		2	2.014 102	0.011 5	
	Tritium	T		3*	3.016 049		12.33 yr
2	Helium	He	4.002 602	3	3.016 029	0.000 137	
				4	4.002 603	99.999 863	
3	Lithium	Li	6.941	6	6.015 122	7.5	
				7	7.016 004	92.5	
4	Beryllium	Be	9.012 182	7*	7.016 929		53.3 days
				9	9.012 182	100	
5	Boron	B	10.811	10	10.012 937	19.9	
				11	11.009 306	80.1	
6	Carbon	C	12.010 7	10*	10.016 853		19.3 s
				11*	11.011 434		20.4 min
				12	12.000 000	98.93	
				13	13.003 355	1.07	
				14*	14.003 242		5 730 yr
7	Nitrogen	N	14.006 7	13*	13.005 739		9.96 min
				14	14.003 074	99.632	
				15	15.000 109	0.368	
8	Oxygen	O	15.999 4	15*	15.003 065		122 s
				16	15.994 915	99.757	
				18	17.999 160	0.205	
9	Fluorine	F	18.998 403 2	19	18.998 403	100	
10	Neon	Ne	20.179 7	20	19.992 440	90.48	
				22	21.991 385	9.25	
11	Sodium	Na	22.989 77	22*	21.994 437		2.61 yr
				23	22.989 770	100	
				24*	23.990 963		14.96 h
12	Magnesium	Mg	24.305 0	24	23.985 042	78.99	
				25	24.985 837	10.00	
				26	25.982 593	11.01	
13	Aluminum	Al	26.981 538	27	26.981 539	100	
14	Silicon	Si	28.085 5	28	27.976 926	92.229 7	
15	Phosphorus	P	30.973 761	31	30.973 762	100	
				32*	31.973 907		14.26 days
16	Sulfur	S	32.066	32	31.972 071	94.93	
				35*	34.969 032		87.5 days
17	Chlorine	Cl	35.452 7	35	34.968 853	75.78	
				37	36.965 903	24.22	
18	Argon	Ar	39.948	40	39.962 383	99.600 3	
19	Potassium	K	39.098 3	39	38.963 707	93.258 1	
				40*	39.963 999	0.011 7	1.28×10^9 yr
20	Calcium	Ca	40.078	40	39.962 591	96.941	
21	Scandium	Sc	44.955 910	45	44.955 910	100	
22	Titanium	Ti	47.867	48	47.947 947	73.72	

Atomic Number Z	Element	Symbol	Chemical Atomic Mass (u)	Mass Number (* Indicates Radioactive) A	Atomic Mass (u)	Percent Abundance	Half-Life (If Radioactive) $T_{1/2}$
23	Vanadium	V	50.941 5	51	50.943 964	99.750	
24	Chromium	Cr	51.996 1	52	51.940 512	83.789	
25	Manganese	Mn	54.938 049	55	54.938 050	100	
26	Iron	Fe	55.845	56	55.934 942	91.754	
27	Cobalt	Co	58.933 200	59	58.933 200	100	
				60*	59.933 822		5.27 yr
28	Nickel	Ni	58.693 4	58	57.935 348	68.076 9	
				60	59.930 790	26.223 1	
29	Copper	Cu	63.546	63	62.929 601	69.17	
				65	64.927 794	30.83	
30	Zinc	Zn	65.39	64	63.929 147	48.63	
				66	65.926 037	27.90	
				68	67.924 848	18.75	
31	Gallium	Ga	69.723	69	68.925 581	60.108	
				71	70.924 705	39.892	
32	Germanium	Ge	72.61	70	69.924 250	20.84	
				72	71.922 076	27.54	
				74	73.921 178	36.28	
33	Arsenic	As	74.921 60	75	74.921 596	100	
34	Selenium	Se	78.96	78	77.917 310	23.77	
				80	79.916 522	49.61	
35	Bromine	Br	79.904	79	78.918 338	50.69	
				81	80.916 291	49.31	
36	Krypton	Kr	83.80	82	81.913 485	11.58	
				83	82.914 136	11.49	
				84	83.911 507	57.00	
				86	85.910 610	17.30	
37	Rubidium	Rb	85.467 8	85	84.911 789	72.17	
				87*	86.909 184	27.83	4.75×10^{10} yr
38	Strontium	Sr	87.62	86	85.909 262	9.86	
				88	87.905 614	82.58	
				90*	89.907 738		29.1 yr
39	Yttrium	Y	88.905 85	89	88.905 848	100	
40	Zirconium	Zr	91.224	90	89.904 704	51.45	
				91	90.905 645	11.22	
				92	91.905 040	17.15	
				94	93.906 316	17.38	
41	Niobium	Nb	92.906 38	93	92.906 378	100	
42	Molybdenum	Mo	95.94	92	91.906 810	14.84	
				95	94.905 842	15.92	
				96	95.904 679	16.68	
				98	97.905 408	24.13	

(*Continued*)

APPENDIX B | An Abbreviated Table of Isotopes

Atomic Number Z	Element	Symbol	Chemical Atomic Mass (u)	Mass Number (* Indicates Radioactive) A	Atomic Mass (u)	Percent Abundance	Half-Life (If Radioactive) $T_{1/2}$
43	Technetium	Tc		98*	97.907 216		4.2×10^6 yr
				99*	98.906 255		2.1×10^5 yr
44	Ruthenium	Ru	101.07	99	98.905 939	12.76	
				100	99.904 220	12.60	
				101	100.905 582	17.06	
				102	101.904 350	31.55	
				104	103.905 430	18.62	
45	Rhodium	Rh	102.905 50	103	102.905 504	100	
46	Palladium	Pd	106.42	104	103.904 035	11.14	
				105	104.905 084	22.33	
				106	105.903 483	27.33	
				108	107.903 894	26.46	
				110	109.905 152	11.72	
47	Silver	Ag	107.868 2	107	106.905 093	51.839	
				109	108.904 756	48.161	
48	Cadmium	Cd	112.411	110	109.903 006	12.49	
				111	110.904 182	12.80	
				112	111.902 757	24.13	
				113*	112.904 401	12.22	9.3×10^{15} yr
				114	113.903 358	28.73	
49	Indium	In	114.818	115*	114.903 878	95.71	4.4×10^{14} yr
50	Tin	Sn	118.710	116	115.901 744	14.54	
				118	117.901 606	24.22	
				120	119.902 197	32.58	
51	Antimony	Sb	121.760	121	120.903 818	57.21	
				123	122.904 216	42.79	
52	Tellurium	Te	127.60	126	125.903 306	18.84	
				128*	127.904 461	31.74	$> 8 \times 10^{24}$ yr
				130*	129.906 223	34.08	$\leq 1.25 \times 10^{21}$ yr
53	Iodine	I	126.904 47	127	126.904 468	100	
				129*	128.904 988		1.6×10^7 yr
54	Xenon	Xe	131.29	129	128.904 780	26.44	
				131	130.905 082	21.18	
				132	131.904 145	26.89	
				134	133.905 394	10.44	
				136*	135.907 220	8.87	$\geq 2.36 \times 10^{21}$ yr
55	Cesium	Cs	132.905 45	133	132.905 447	100	
56	Barium	Ba	137.327	137	136.905 821	11.232	
				138	137.905 241	71.698	
57	Lanthanum	La	138.905 5	139	138.906 349	99.910	
58	Cerium	Ce	140.116	140	139.905 434	88.450	
				142*	141.909 240	11.114	$> 5 \times 10^{16}$ yr
59	Praseodymium	Pr	140.907 65	141	140.907 648	100	
60	Neodymium	Nd	144.24	142	141.907 719	27.2	
				144*	143.910 083	23.8	2.3×10^{15} yr
				146	145.913 112	17.2	

Atomic Number Z	Element	Symbol	Chemical Atomic Mass (u)	Mass Number (* Indicates Radioactive) A	Atomic Mass (u)	Percent Abundance	Half-Life (If Radioactive) $T_{1/2}$
61	Promethium	Pm		145*	144.912 744		17.7 yr
62	Samarium	Sm	150.36	147*	146.914 893	14.99	1.06×10^{11} yr
				149*	148.917 180	13.82	$> 2 \times 10^{15}$ yr
				152	151.919 728	26.75	
				154	153.922 205	22.75	
63	Europium	Eu	151.964	151	150.919 846	47.81	
				153	152.921 226	52.19	
64	Gadolinium	Gd	157.25	156	155.922 120	20.47	
				158	157.924 100	24.84	
				160	159.927 051	21.86	
65	Terbium	Tb	158.925 34	159	158.925 343	100	
66	Dysprosium	Dy	162.50	162	161.926 796	25.51	
				163	162.928 728	24.90	
				164	163.929 171	28.18	
67	Holmium	Ho	164.930 32	165	164.930 320	100	
68	Erbium	Er	167.6	166	165.930 290	33.61	
				167	166.932 045	22.93	
				168	167.932 368	26.78	
69	Thulium	Tm	168.934 21	169	168.934 211	100	
70	Ytterbium	Yb	173.04	172	171.936 378	21.83	
				173	172.938 207	16.13	
				174	173.938 858	31.83	
71	Lutecium	Lu	174.967	175	174.940 768	97.41	
72	Hafnium	Hf	178.49	177	176.943 220	18.60	
				178	177.943 698	27.28	
				179	178.945 815	13.62	
				180	179.946 549	35.08	
73	Tantalum	Ta	180.947 9	181	180.947 996	99.988	
74	Tungsten (Wolfram)	W	183.84	182	181.948 206	26.50	
				183	182.950 224	14.31	
				184*	183.950 933	30.64	$> 3 \times 10^{17}$ yr
				186	185.954 362	28.43	
75	Rhenium	Re	186.207	185	184.952 956	37.40	
				187*	186.955 751	62.60	4.4×10^{10} yr
76	Osmium	Os	190.23	188	187.955 836	13.24	
				189	188.958 145	16.15	
				190	189.958 445	26.26	
				192	191.961 479	40.78	
77	Iridium	Ir	192.217	191	190.960 591	37.3	
				193	192.962 924	62.7	
78	Platinum	Pt	195.078	194	193.962 664	32.967	
				195	194.964 774	33.832	
				196	195.964 935	25.242	
79	Gold	Au	196.966 55	197	196.966 552	100	
80	Mercury	Hg	200.59	199	198.968 262	16.87	
				200	199.968 309	23.10	
				201	200.970 285	13.18	
				202	201.970 626	29.86	

(Continued)

APPENDIX B | An Abbreviated Table of Isotopes

Atomic Number Z	Element	Symbol	Chemical Atomic Mass (u)	Mass Number (* Indicates Radioactive) A	Atomic Mass (u)	Percent Abundance	Half-Life (If Radioactive) $T_{1/2}$
81	Thallium	Tl	204.383 3	203	202.972 329	29.524	
				205	204.974 412	70.476	
		(Th C″)		208*	207.982 005		3.053 min
		(Ra C″)		210*	209.990 066		1.30 min
82	Lead	Pb	207.2	204*	203.973 029	1.4	$\geq 1.4 \times 10^{17}$ yr
				206	205.974 449	24.1	
				207	206.975 881	22.1	
				208	207.976 636	52.4	
		(Ra D)		210*	209.984 173		22.3 yr
		(Ac B)		211*	210.988 732		36.1 min
		(Th B)		212*	211.991 888		10.64 h
		(Ra B)		214*	213.999 798		26.8 min
83	Bismuth	Bi	208.980 38	209	208.980 383	100	
		(Th C)		211*	210.987 258		2.14 min
84	Polonium	Po					
		(Ra F)		210*	209.982 857		138.38 days
		(Ra C′)		214*	213.995 186		164 µs
85	Astatine	At		218*	218.008 682		1.6 s
86	Radon	Rn		222*	222.017 570		3.823 days
87	Francium	Fr					
		(Ac K)		223*	223.019 731		22 min
88	Radium	Ra		226*	226.025 403		1 600 yr
		(Ms Th$_1$)		228*	228.031 064		5.75 yr
89	Actinium	Ac		227*	227.027 747		21.77 yr
90	Thorium	Th	232.038 1				
		(Rd Th)		228*	228.028 731		1.913 yr
		(Th)		232*	232.038 050	100	1.40×10^{10} yr
91	Protactinium	Pa	231.035 88	231*	231.035 879		32.760 yr
92	Uranium	U	238.028 9	232*	232.037 146		69 yr
				233*	233.039 628		1.59×10^5 yr
		(Ac U)		235*	235.043 923	0.720 0	7.04×10^8 yr
				236*	236.045 562		2.34×10^7 yr
		(UI)		238*	238.050 783	99.274 5	4.47×10^9 yr
93	Neptunium	Np		237*	237.048 167		2.14×10^6 yr
94	Plutonium	Pu		239*	239.052 156		2.412×10^4 yr
				242*	242.058 737		3.73×10^6 yr
				244*	244.064 198		8.1×10^7 yr

Sources: Chemical atomic masses are from T. B. Coplen, "Atomic Weights of the Elements 1999," a technical report to the International Union of Pure and Applied Chemistry, and published in *Pure and Applied Chemistry,* 73(4), 667–683, 2001. Atomic masses of the isotopes are from G. Audi and A. H. Wapstra, "The 1995 Update to the Atomic Mass Evaluation," *Nuclear Physics,* A595, vol. 4, 409–480, December 25, 1995. Percent abundance values are from K. J. R. Rosman and P. D. P. Taylor, "Isotopic Compositions of the Elements 1999," a technical report to the International Union of Pure and Applied Chemistry, and published in *Pure and Applied Chemistry,* 70(1), 217–236, 1998.

Some Useful Tables

Appendix C

Table C.1 Mathematical Symbols Used in the Text and Their Meaning

Symbol	Meaning
$=$	is equal to
$\neq$	is not equal to
$\equiv$	is defined as
$\propto$	is proportional to
$>$	is greater than
$<$	is less than
$\gg$	is much greater than
$\ll$	is much less than
$\approx$	is approximately equal to
$\sim$	is on the order of magnitude of
Δx	change in x or uncertainty in x
Σx_i	sum of all quantities x_i
$\|x\|$	absolute value of x (always a positive quantity)

Table C.2 Standard Symbols for Units

Symbol	Unit	Symbol	Unit
A	ampere	kcal	kilocalorie
Å	angstrom	kg	kilogram
atm	atmosphere	km	kilometer
Bq	bequerel	kmol	kilomole
Btu	British thermal unit	L	liter
C	coulomb	lb	pound
°C	degree Celsius	ly	lightyear
cal	calorie	m	meter
cm	centimeter	min	minute
Ci	curie	mol	mole
d	day	N	newton
deg	degree (angle)	nm	nanometer
eV	electronvolt	Pa	pascal
°F	degree Fahrenheit	rad	radian
F	farad	rev	revolution
ft	foot	s	second
G	Gauss	T	tesla
g	gram	u	atomic mass unit
H	henry	V	volt
h	hour	W	watt
hp	horsepower	Wb	weber
Hz	hertz	yr	year
in.	inch	μm	micrometer
J	joule	Ω	ohm
K	kelvin		

Table C.3 The Greek Alphabet

Alpha	A	α	Nu	N	ν
Beta	B	β	Xi	Ξ	ξ
Gamma	Γ	γ	Omicron	O	o
Delta	Δ	δ	Pi	Π	π
Epsilon	E	ϵ	Rho	P	ρ
Zeta	Z	ζ	Sigma	Σ	σ
Eta	H	η	Tau	T	τ
Theta	Θ	θ	Upsilon	Υ	υ
Iota	I	ι	Phi	Φ	ϕ
Kappa	K	κ	Chi	X	χ
Lambda	Λ	λ	Psi	Ψ	ψ
Mu	M	μ	Omega	Ω	ω

Table C.4 Physical Data Often Used[a]

Average Earth–Moon distance	3.84×10^8 m
Average Earth–Sun distance	1.496×10^{11} m
Equatorial radius of Earth	6.38×10^6 m
Density of air (20°C and 1 atm)	1.20 kg/m^3
Density of water (20°C and 1 atm)	1.00×10^3 kg/m^3
Free-fall acceleration	9.80 m/s^2
Mass of Earth	5.98×10^{24} kg
Mass of Moon	7.36×10^{22} kg
Mass of Sun	1.99×10^{30} kg
Standard atmospheric pressure	1.013×10^5 Pa

[a] These are the values of the constants as used in the text.

Table C.5 Some Fundamental Constants

Quantity	Symbol	Value[a]
Atomic mass unit	u	$1.660\,538\,782\,(83) \times 10^{-27}$ kg
		$931.494\,028\,(23)$ MeV/c^2
Avogadro's number	N_A	$6.022\,141\,79\,(30) \times 10^{23}$ particles/mol
Bohr magneton	$\mu_B = \dfrac{e\hbar}{2m_e}$	$9.274\,009\,15\,(23) \times 10^{-24}$ J/T
Bohr radius	$a_0 = \dfrac{\hbar^2}{m_e e^2 k_e}$	$5.291\,772\,085\,9\,(36) \times 10^{-11}$ m
Boltzmann's constant	$k_B = \dfrac{R}{N_A}$	$1.380\,650\,4\,(24) \times 10^{-23}$ J/K
Compton wavelength	$\lambda_C = \dfrac{h}{m_e c}$	$2.426\,310\,217\,5\,(33) \times 10^{-12}$ m
Coulomb constant	$k_e = \dfrac{1}{4\pi\epsilon_0}$	$8.987\,551\,788\ldots \times 10^9$ N·m²/C² (exact)
Deuteron mass	m_d	$3.343\,583\,20\,(17) \times 10^{-27}$ kg
		$2.013\,553\,212\,724\,(78)$ u
Electron mass	m_e	$9.109\,382\,15\,(45) \times 10^{-31}$ kg
		$5.485\,799\,094\,3\,(23) \times 10^{-4}$ u
		$0.510\,998\,910\,(13)$ MeV/c^2
Electron volt	eV	$1.602\,176\,487\,(40) \times 10^{-19}$ J
Elementary charge	e	$1.602\,176\,487\,(40) \times 10^{-19}$ C
Gas constant	R	$8.314\,472\,(15)$ J/mol·K
Gravitational constant	G	$6.674\,28\,(67) \times 10^{-11}$ N·m²/kg²
Neutron mass	m_n	$1.674\,927\,211\,(84) \times 10^{-27}$ kg
		$1.008\,664\,915\,97\,(43)$ u
		$939.565\,346\,(23)$ MeV/c^2
Nuclear magneton	$\mu_n = \dfrac{e\hbar}{2m_p}$	$5.050\,783\,24\,(13) \times 10^{-27}$ J/T
Permeability of free space	μ_0	$4\pi \times 10^{-7}$ T·m/A (exact)
Permittivity of free space	$\epsilon_0 = \dfrac{1}{\mu_0 c^2}$	$8.854\,187\,817\ldots \times 10^{-12}$ C²/N·m² (exact)
Planck's constant	h	$6.626\,068\,96\,(33) \times 10^{-34}$ J·s
	$\hbar = \dfrac{h}{2\pi}$	$1.054\,571\,628\,(53) \times 10^{-34}$ J·s
Proton mass	m_p	$1.672\,621\,637\,(83) \times 10^{-27}$ kg
		$1.007\,276\,466\,77\,(10)$ u
		$938.272\,013\,(23)$ MeV/c^2
Rydberg constant	R_H	$1.097\,373\,156\,852\,7\,(73) \times 10^7$ m⁻¹
Speed of light in vacuum	c	$2.997\,924\,58 \times 10^8$ m/s (exact)

Note: These constants are the values recommended in 2006 by CODATA, based on a least-squares adjustment of data from different measurements. For a more complete list, see P. J. Mohr, B. N. Taylor, and D. B. Newell, "CODATA Recommended Values of the Fundamental Physical Constants: 2006." *Rev. Mod. Phys.* **80:**2, 633–730, 2008.

[a] The numbers in parentheses represent the uncertainties of the last two digits.

SI Units

Appendix D

Table D.1 SI Base Units

Base Quantity	SI Base Unit Name	Symbol
Length	meter	m
Mass	kilogram	kg
Time	second	s
Electric current	ampere	A
Temperature	kelvin	K
Amount of substance	mole	mol
Luminous intensity	candela	cd

Table D.2 Derived SI Units

Quantity	Name	Symbol	Expression in Terms of Base Units	Expression in Terms of Other SI Units
Plane angle	radian	rad	m/m	
Frequency	hertz	Hz	s^{-1}	
Force	newton	N	$kg \cdot m/s^2$	J/m
Pressure	pascal	Pa	$kg/m \cdot s^2$	N/m^2
Energy: work	joule	J	$kg \cdot m^2/s^2$	$N \cdot m$
Power	watt	W	$kg \cdot m^2/s^3$	J/s
Electric charge	coulomb	C	$A \cdot s$	
Electric potential (emf)	volt	V	$kg \cdot m^2/A \cdot s^3$	W/A, J/C
Capacitance	farad	F	$A^2 \cdot s^4/kg \cdot m^2$	C/V
Electric resistance	ohm	Ω	$kg \cdot m^2/A^2 \cdot s^3$	V/A
Magnetic flux	weber	Wb	$kg \cdot m^2/A \cdot s^2$	$V \cdot s, T \cdot m^2$
Magnetic field intensity	tesla	T	$kg/A \cdot s^2$	Wb/m^2
Inductance	henry	H	$kg \cdot m^2/A^2 \cdot s^2$	Wb/A

Answers to Quick Quizzes, Example Questions, Odd-Numbered Warm-Up Exercises, Conceptual Questions, and Problems

CHAPTER 1

Example Questions
1. False
2. True
3. 2.6×10^2 m^2
4. 28.0 m/s $= \left(28.0 \dfrac{\text{m}}{\text{s}}\right)\left(\dfrac{2.24 \text{ mi/h}}{1.00 \text{ m/s}}\right) = 62.7$ mi/h

   The answer is slightly different because the different conversion factors were rounded, leading to small, unpredictable differences in the final answers.
5. $\left(\dfrac{60.0 \text{ min}}{1.00 \text{ h}}\right)^2$
6. An answer of 10^{12} cells is within an order of magnitude of the given answer, corresponding to slightly different choices in the volume estimations. Consequently, 10^{12} cells is also a reasonable estimate. An estimate of 10^9 cells would be suspect because (working backwards) it would imply cells that could be seen with the naked eye!
7. $\sim 4 \times 10^{11}$
8. $\sim 10^{12}$
9. Working backwards, $r = 4.998$, which further rounds to 5.00, whereas $\theta = 37.03°$, which further rounds to 37.0°. The slight differences are caused by rounding.
10. Yes. The cosine function divided into the distance to the building will give the length of the hypotenuse of the triangle in question.
11. 540 km

Warm-Up Exercises
1. (a) $5.680\ 17 \times 10^5$ (b) 3.09×10^{-4}
3. $25.2 \dfrac{\text{km}}{\text{min}^2}$
5. 132 m^2
7. 58
9. 22 m

Conceptual Questions
1. (a) ~ 0.1 m   (b) ~ 1 m   (c) Between 10 m and 100 m
   (d) ~ 10 m   (e) ~ 100 m
3. $\sim 10^9$ s
5. (a) $\sim 10^6$ beats   (b) $\sim 10^9$ beats
7. The length of a hand varies from person to person, so it isn't a useful standard of length.
9. (a) A dimensionally correct equation isn't necessarily true. For example, the equation 2 dogs = 5 dogs is dimensionally correct, but isn't true. (b) If an equation is not dimensionally correct, it cannot be true.
11. An estimate, even if imprecise by an order of magnitude, greatly reduces the range of plausible answers to a given question. The estimate gives guidance as to what corrective measures might be feasible. For example, if you estimate that 40 000 people in a country will die unless they have food assistance and if this number is reliable within an order of magnitude, you know that at most 400 000 people will need provisions.
13. (a) yes   (b) no   (c) yes   (d) no   (e) yes

Problems
3. (b) $A_{\text{cylinder}} = \pi R^2$, $A_{\text{rectangular plate}} = \text{length} \times \text{width}$
5. m^3/(kg · s^2)
7. 52 m^2
9. (a) three significant figures (b) four significant figures (c) three significant figures (d) two significant figures
11. (a) 35.7 cm^2 (b) 99.2 cm^3 (c) 17.7 cm^2, 99.5 cm^3 (d) In the rounding process, small amounts are either added to or subtracted from an answer to satisfy the rules of significant figures. For a given rounding, different small adjustments are made, introducing a certain amount of randomness in the last significant digit of the final answer.
13. 5.9×10^3 cm^3
15. 2×10^8 fathoms
17. 0.204 m^3
19. Yes. The speed of 38.0 m/s converts to a speed of 85.0 mi/h, so the driver exceeds the speed limit by 10 mi/h.
21. (a) 6.81 cm   (b) 5.83×10^2 cm^2 (c) 1.32×10^3 cm^3
23. 6.71×10^8 mi/h
25. 3.08×10^4 m^3
27. 9.82 cm
29. $\sim 10^8$ breaths
31. (a) $\sim 10^{14}$. Other answers are possible. This answer assumes an intestinal length of about 7 m and average diameter of about 4 cm. There are probably more bacteria in your intestinal tract than there are cells in your body! (b) The large number means that they are probably not dangerous. Intestinal bacteria help digest food and provide important nutrients. Humans and bacteria enjoy a mutually beneficial symbiotic relationship.
33. $\sim 10^7$ rev
35. (2.0 m, 1.4 m)
37. $r = 2.2$ m, $\theta = 27°$
39. 5.69 m
41. (a) 6.71 m   (b) 0.894   (c) 0.746
43. 3.41 m
45. (a) 3.00   (b) 3.00   (c) 0.800   (d) 0.800   (e) 1.33
47. 5.00/7.00; the angle itself is 35.5°
49. 70.0 m
51. (a) MLT^{-2} (b) kg · m/s^2
53. (a) 1.0×10^3 kg   (b) $m_{\text{cell}} = 5.2 \times 10^{-16}$ kg
    (c) $m_{\text{kidney}} = 0.27$ kg   (d) $m_{\text{fly}} = 1.3 \times 10^{-5}$ kg
55. The value of k, a dimensionless constant, cannot be found by dimensional analysis.
57. 152 μm
59. 1×10^{10} gal/yr
61. (a) 3.16×10^7 s   (b) Between 10^{10} yr and 10^{11} yr
63. 1×10^5

CHAPTER 2

Quick Quizzes
1. (a) 200 yd   (b) 0   (c) 0   (d) 8 yd/s
2. (a) False   (b) True   (c) True
3. The velocity vs. time graph (a) has a constant slope, indicating a constant acceleration, which is represented by the acceleration vs. time graph (e).

A.23

Graph (b) represents an object with increasing speed, but as time progresses, the lines drawn tangent to the curve have increasing slopes. Since the acceleration is equal to the slope of the tangent line, the acceleration must be increasing, and the acceleration vs. time graph that best indicates this behavior is (d).

Graph (c) depicts an object which first has a velocity that increases at a constant rate, which means that the object's acceleration is constant. The velocity then stops changing, which means that the acceleration of the object is zero. This behavior is best matched by graph (f).

4. (b)
5. (a) blue graph   (b) red graph   (c) green graph
6. (e)
7. (c)
8. (a) and (f)

Example Questions

1. No. The object may not be traveling in a straight line. If the initial and final positions are in the same place, for example, the displacement is zero regardless of the total distance traveled during the given time.
2. No. A vertical line in a position vs. time graph would mean that an object had somehow traversed all points along the given path instantaneously, which is physically impossible.
3. No. A vertical tangent line would correspond to an infinite acceleration, which is physically impossible.
4. 35.0 m/s
5. The graphical solution is the intersection of a straight line and a parabola.
6. The coasting displacement would double to 143 m, with a total displacement of 715 m.
7. The acceleration is zero wherever the tangent to the velocity versus time graph is horizontal. Visually, that occurs from -50 s to 0 s and then again at approximately 180 s, 320 s, and 330 s.
8. The upward jump would slightly increase the ball's initial velocity, slightly increasing the maximum height.
9. 6
10. The engine should be fired again at 235 m.

Warm-Up Exercises

1. $t = 4.10, -1.10$
3. (a) 10.3 s   (b) 3.20×10^2 m
5. 30.5 m/s
7. 1.00 s

Conceptual Questions

1. Yes. If the velocity of the particle is nonzero, the particle is in motion. If the acceleration is zero, the velocity of the particle is unchanging or is constant.
3. Yes. If this occurs, the acceleration of the car is opposite to the direction of motion, and the car will be slowing down.
5. No. Car B may be traveling at a lower velocity but have a greater acceleration at that instant.
7. (a) Yes.   (b) Yes.
9. (a) The car is moving to the east and speeding up.
   (b) The car is moving to the east but slowing down.
   (c) The car is moving to the east at constant speed.
   (d) The car is moving to the west but slowing down.
   (e) The car is moving to the west and speeding up.
   (f) The car is moving to the west at constant speed.
   (g) The car starts from rest and begins to speed up toward the east.
   (h) The car starts from rest and begins to speed up toward the west.
11. (b)
13. (d)

Problems

1. ≈ 0.02 s
3. (a) 52.9 km/h   (b) 90.0 km
5. (a) Boat A wins by 60 km   (b) 0
7. (a) 180 km   (b) 63.4 km/h
9. Yes. Taking the initial speed to be zero, the final speed as 75 m/s, and $a = 1.3$ m/s^2, we find that the distance the plane travels before takeoff is $x = v_{to}^2/2a = 2.2$ km, which is less than the length of the runway.
11. (a) 4.4 m/s^2 (b) 27 m
13. (a) 2.80 h   (b) 218 km
15. 274 km/h
17. (a) 5.00 m/s   (b) -2.50 m/s   (c) 0   (d) 5.00 m/s
19. 0.18 mi west of the flagpole
21. $|\bar{a}| = 1.34 \times 10^4$ m/s^2
23. 3.7 s
25. (a) 70.0 mi/h · s = 31.3 m/s^2 = $3.19g$ (b) 321 ft = 97.8 m
27. -16.0 cm/s^2
29. (a) 6.61 m/s   (b) -0.448 m/s^2
31. (a) 2.32 m/s^2 (b) 14.4 s
33. (a) 8.14 m/s^2 (b) 1.23 s   (c) Yes. For uniform acceleration, the velocity is a linear function of time.
35. There will be a collision only if the car and the van meet at the same place at some time. Writing expressions for the position versus time for each vehicle and equating the two gives a quadratic equation in t whose solution is either 11.4 s or 13.6 s. The first solution, 11.4 s, is the time of the collision. The collision occurs when the van is 212 m from the original position of Sue's car.
37. 200 m
39. (a) 1.5 m/s   (b) 32 m
41. 958 m
43. (a) 8.2 s   (b) 1.3×10^2 m
45. (a) 31.9 m   (b) 2.55 s   (c) 2.55 s   (d) -25.0 m/s
47. (a) -12.7 m/s   (b) 38.2 m
49. Hardwood floor: $a = 2.0 \times 10^3$ m/s^2, $\Delta t = 1.4$ ms; carpeted floor: $a = 3.9 \times 10^2$ m/s^2, $\Delta t = 7.1$ ms
51. (a) It is a freely falling object, so its acceleration is -9.80 m/s^2 (downward) while in flight.   (b) 0   (c) 9.80 m/s   (d) 4.90 m
53. (a) It continues upward, slowing under the influence of gravity until it reaches a maximum height, and then it falls to Earth.   (b) 308 m   (c) 8.51 s   (d) 16.4 s
55. 15.0 s
57. (a) -3.50×10^5 m/s^2 (b) 2.86×10^{-4} s
59. (a) 10.0 m/s upward   (b) 4.68 m/s downward
61. (a) 4.0 m/s   (b) 1.0 ms   (c) 0.82 m
63. 0.60 s
65. (a) 0.231 m   (b) 0.364 m   (c) 0.399 m   (d) 0.175 m
67. (a) 1.21 s after the first ball is dropped   (b) 7.18 m below the window
69. No. In the time interval equal to David's reaction time, the \$1 bill (a freely falling object) falls a distance of $gt^2/2 \cong$ 20 cm, which is about twice the distance between the top of the bill and its center.
71. (a) $t_1 = 5.0$ s, $t_2 = 85$ s   (b) 200 ft/s   (c) 18 500 ft from starting point   (d) 10 s after starting to slow down (total trip time = 100 s)
73. (a) 5.5×10^3 ft   (b) 3.7×10^2 ft/s   (c) The plane would travel only 0.002 ft in the time it takes the light from the bolt to reach the eye.
75. (a) 7.82 m   (b) 0.782 s

CHAPTER 3

Quick Quizzes

1. (c)

2.

Vector	x-component	y-component
$\vec{A}$	−	+
$\vec{B}$	+	−
$\vec{A} + \vec{B}$	−	−

3. Vector $\vec{B}$
4. (b)
5. (a)
6. (c)
7. (b)

Example Questions

1. If the vectors point in the same direction, the sum of the magnitudes of the two vectors equals the magnitude of the resultant vector.
2. Because B_x, B_y, and B are all known, any of the trigonometric functions can be used to find the angle.
3. The hikers' displacement vectors are the same.
4. The initial and final velocity vectors are equal in magnitude because the x-component doesn't change and the y-component changes only by a sign.
5. To the pilot, the package appears to drop straight down because the x-components of velocity for the plane and package are the same.
6. False
7. 45°
8. False
9. False
10. To an observer on the ground, the ball drops straight down.
11. The angle decreases with increasing speed.
12. The angle is different because relative velocity depends on both the magnitude and the direction of the velocity vectors. In Example 3.12, the boat's velocity vector forms the hypotenuse of a right triangle, whereas in Example 3.11, that vector forms a leg of a right triangle.

Warm-Up Exercises

1. (a) 10.3 m (b) 119°
3. (a) 6.0 m (b) 16.0 m
5. (a) 4.20 km/h (b) 3.00 km/h
7. (a) 0.808 s (b) 17.8 m
9. (a) 3.68 m/s (b) 0.82 m/s

Conceptual Questions

1. The magnitudes add when $\vec{A}$ and $\vec{B}$ are in the same direction. The resultant will be zero when the two vectors are equal in magnitude and opposite in direction.
3. (a) At the top of the projectile's flight, its velocity is horizontal and its acceleration is downward. This is the only point at which the velocity and acceleration vectors are perpendicular. (b) If the projectile is thrown straight up or down, then the velocity and acceleration will be parallel throughout the motion. For any other kind of projectile motion, the velocity and acceleration vectors are never parallel.
5. (a) The acceleration is zero, since both the magnitude and direction of the velocity remain constant. (b) The particle has an acceleration because the direction of $\vec{v}$ changes.
7. The spacecraft will follow a parabolic path equivalent to that of a projectile thrown off a cliff with a horizontal velocity. As regards the projectile, gravity provides an acceleration that is always perpendicular to the initial velocity, resulting in a parabolic path. As regards the spacecraft, the initial velocity plays the role of the horizontal velocity of the projectile, and the leaking gas plays the role that gravity plays in the case of the projectile. If the orientation of the spacecraft were to change in response to the gas leak (which is by far the more likely result), then the acceleration would change direction and the motion could become very complicated.
9. For angles $\theta < 45°$, the projectile thrown at angle θ will be in the air for a shorter interval. For the smaller angle, the vertical component of the initial velocity is smaller than that for the larger angle. Thus, the projectile thrown at the smaller angle will not go as high into the air and will spend less time in the air before landing.
11. (a) Yes, the projectile is in free fall. (b) Its vertical component of acceleration is the downward acceleration of gravity. (c) Its horizontal component of acceleration is zero.
13. (b)
15. (i) (a) (ii) (b)

Problems

1. 43 units in the negative y-direction
3. (a) Approximately 5.0 units at −53° (b) Approximately 5.0 units at +53°
5. Approximately 421 ft at 3° below the horizontal
7. Approximately 310 km at 57° S of W
9. Approximately 15 m at 58° S of E
11. 28.7 units, −20.1 units
13. (a) 5.00 blocks at 53.1° N of E (b) 13.0 blocks
15. 47.2 units at 122° from the positive x-axis
17. 157 km
19. 245 km at 21.4° W of N
21. 4.64 m at 78.6° N of E
23. (a) $(x, y) = (0, 50.0 \text{ m})$ (b) $v_{0x} = 18.0$ m/s, $v_{0y} = 0$
    (c) $v_x = 18.0$ m/s, $v_y = -(9.80 \text{ m/s}^2)t$
    (d) $x = (18.0 \text{ m/s})t$, $y = -(4.90 \text{ m/s}^2)t^2 + 50.0$ m
    (e) 3.19 s (f) 36.1 m/s, −60.1°
25. 12 m/s
27. (a) The ball clears by 0.89 m (b) while descending
29. 25 m
31. (a) 32.5 m from the base of the cliff (b) 1.78 s
33. (a) 52.0 m/s horizontally (b) 212 m
35. (a) $(\vec{v}_{JA})_x = 3.00 \times 10^2$ mi/h, $(\vec{v}_{JA})_y = 0$,
    (b) $(\vec{v}_{AE})_x = 86.6$ mi/h, $(\vec{v}_{AE})_y = 50.0$ mi/h
    (c) $\vec{v}_{JA} = \vec{v}_{JE} - \vec{v}_{AE}$
    (d) $\vec{v}_{JE} = 3.90 \times 10^2$ mi/h, 7.36° N of E
37. (a) 9.80 m/s² down and 2.50 m/s² south (b) 9.80 m/s² down
    (c) The bolt moves on a parabola with a vertical axis.
39. (a) 1.26 h (b) 1.13 h (c) 1.19 h
41. (a) 2.02×10^3 s (b) 1.67×10^3 s (c) The time savings going downstream with the current is always less than the extra time required to go the same distance against the current.
43. (a) 6.80 km (b) 3.00 km vertically above the impact point
    (c) 66.2°
45. 18.0 s
47. (a) 40.5 m/s (b) Rather than falling like a rock, the skier glides through the air like a bird, prolonging the jump.
49. 68.6 km/h
51. (a) 1.53×10^3 m (b) 36.2 s (c) 4.04×10^3 m
53. (a) $R_{\text{moon}} = 18$ m (b) $R_{\text{Mars}} = 7.9$ m
55. (a) 42 m/s (b) 3.8 s (c) $v_x = 34$ m/s, $v_y = -13$ m/s; $v = 37$ m/s
57. (a) in the direction the ball was thrown (b) 7.5 m/s
61. $R/2$
63. 7.5 min
65. 10.8 m
67. (b) $y = Ax^2$ with $A = \dfrac{g}{2v_i^2}$ where v_i is the muzzle velocity
    (c) 14.5 m/s
69. (a) 35.1° or 54.9° (b) 42.2 m or 85.4 m, respectively
71. (a) 20.0° above the horizontal (b) 3.05 s
73. (a) 23 m/s (b) 360 m horizontally from the base of cliff

CHAPTER 4

Quick Quizzes

1. (a), (c), and (d) are true.
2. (b)
3. (a) False
   (b) True
   (c) False
   (d) False
4. (c); (d)
5. (c)
6. (c)
7. (b)
8. (b)
9. (b) By exerting an upward force component on the sled, you reduce the normal force on the ground and so reduce the force of kinetic friction.

Example Questions

1. Other than the forces mentioned in the problem, the force of gravity pulls downwards on the boat. Because the boat doesn't sink, a force exerted by the water on the boat must oppose the gravity force. (In Chapter 9 this force will be identified as the buoyancy force.)
2. False. The angles at which the forces are applied are also important in determining the magnitude of the acceleration vector.
3. 0.2 N
4. $3g$
5. The gravitational force of the Earth acts on the man, and an equal and opposite gravitational force of the man acts on the Earth. The normal force acts on the man, and the reaction force consists of the man pressing against the surface.
6. The tensions would double.
7. The magnitude of the tension force would be greater, and the magnitude of the normal force would be smaller.
8. Doubling the weight doubles the mass, which halves both the acceleration and displacement.
9. A gentler slope means a smaller angle and hence a smaller acceleration down the slope. Consequently, the car would take longer to reach the bottom of the hill.
10. The scale reading is greater than the weight of the fish during the first acceleration phase. When the velocity becomes constant, the scale reading is equal to the weight. When the elevator slows down, the scale reading is less than the weight.
11. Attach one end of the cable to the object to be lifted and the other end to a platform. Place lighter weights on the platform until the total mass of the weights and platform exceeds the mass of the heavy object.
12. A larger static friction coefficient would increase the maximum angle.
13. The coefficient of kinetic friction would be larger than in the example.
14. Both the acceleration and the tension increase when m_2 is increased.
15. The top block would slide off the back end of the lower block.

Warm-Up Exercises

1. (a) -0.800 m/s^2 (b) 8.4 m/s   (c) 5.00 s
3. (a) 3.75 m/s^2 (b) 22.5 m/s
5. 3.00
7. (a) 147 N   (b) 127 N   (c) 192 N   (d) 84.5 N
9. (a) -5.36 m/s^2 (b) 4 690 N   (c) 84.0 m
11. 0.272
13. (a) 3.0 m/s^2 (b) 48 N

Conceptual Questions

1. The inertia of the suitcase would keep it moving forward as the bus stops. There would be no tendency for the suitcase to be thrown backward toward the passenger. The case should be dismissed.
3. (a) $w = mg$ and g decreases with altitude. Thus, to get a good buy, purchase it in Denver. (b) If gold were sold by mass, it would not matter where you bought it.
5. (a) Two external forces act on the ball. (i) One is a downward gravitational force exerted by Earth. (ii) The second force on the ball is an upward normal force exerted by the hand. The reactions to these forces are (i) an upward gravitational force exerted by the ball on Earth and (ii) a downward force exerted by the ball on the hand. (b) After the ball leaves the hand, the only external force acting on the ball is the gravitational force exerted by Earth. The reaction is an upward gravitational force exerted by the ball on Earth.
7. The force causing an automobile to move is the friction between the tires and the roadway as the automobile attempts to push the roadway backward. The force driving a propeller airplane forward is the reaction force exerted by the air on the propeller as the rotating propeller pushes the air backward (the action). In a rowboat, the rower pushes the water backward with the oars (the action). The water pushes forward on the oars and hence the boat (the reaction).
9. When the bus starts moving, Claudette's mass is accelerated by the force exerted by the back of the seat on her body. Clark is standing, however, and the only force acting on him is the friction between his shoes and the floor of the bus. Thus, when the bus starts moving, his feet accelerate forward, but the rest of his body experiences almost no accelerating force (only that due to his being attached to his accelerating feet!). As a consequence, his body tends to stay almost at rest, according to Newton's first law, relative to the ground. Relative to Claudette, however, he is moving toward her and falls into her lap. Both performers won Academy Awards.
11. (a) As the man takes the step, the action is the force his foot exerts on Earth; the reaction is the force exerted by Earth on his foot. (b) Here, the action is the force exerted by the snowball on the girl's back; the reaction is the force exerted by the girl's back on the snowball. (c) This action is the force exerted by the glove on the ball; the reaction is the force exerted by the ball on the glove. (d) This action is the force exerted by the air molecules on the window; the reaction is the force exerted by the window on the air molecules. In each case, we could equally well interchange the terms "action" and "reaction."
13. The tension in the rope is the maximum force that occurs in *both* directions. In this case, then, since both are pulling with a force of magnitude 200 N, the tension is 200 N. If the rope does not move, then the force on each athlete must equal zero. Therefore, each athlete exerts 200 N against the ground.
15. (c)
17. (b)
19. (b)
21. (b)

Problems

1. 2×10^4 N
3. (a) 12 N   (b) 3.0 m/s^2
5. 3.71 N, 58.7 N, 2.27 kg
7. 1.2 m/s^2 upward
9. (a) 13.5 m/s^2 (b) 1.42×10^3 N
11. (a) 0.200 m/s^2 (b) 10.0 m   (c) 2.00 m/s
13. 4.85 kN eastward
15. 1.1×10^4 N
17. (a) 600 N in vertical cable, 997 N in inclined cable, 796 N in horizontal cable (b) If the point of attachment were

moved higher up on the wall, the left cable would have a y-component that would help support the cat burglar, thus reducing the tension in the cable on the right.
19. 150 N in vertical cable, 75 N in right-side cable, 130 N in left-side cable
21. (a) $T_1 = 79.8$ N, $T_2 = 39.9$ N (b) 2.34 m/s^2
23. 613 N
25. 64 N
27. (a) $T_1 = 3mg \sin \theta$ (b) $T_2 = 2mg \sin \theta$
29. (a) 7.0 m/s^2 horizontal and to the right (b) 21 N (c) 14 N horizontal and to the right
31. (a) 78.4 N (b) 105 N
33. (a) 4.90 kN (b) 607 N
35. (a) $T > w$ (b) $T = w$ (c) $T < w$ (d) 1.85×10^4 N; yes (e) 1.47×10^4 N; yes (f) 1.25×10^4 N; yes
37. (a) 2.15×10^3 N forward (b) 645 N forward (c) 645 N rearward (d) 1.02×10^4 N at 74.1° below horizontal and rearward
39. $\mu_s = 0.38$, $\mu_k = 0.31$
41. (a) 0.256 (b) 0.509 m/s^2
43. (a) 14.7 m (b) Neither mass is necessary
45. $\mu_k = 0.287$
47. 32.1 N
49. (a) 33 m/s (b) No. The object will speed up to 33 m/s from any lower speed and will slow down to 33 m/s from any higher speed.
51. (a) 1.78 m/s^2 (b) 0.368 (c) 9.37 N (d) 2.67 m/s
53. (a) 1.11 s (b) 0.875 s
55. (a) 0.404 (b) 45.8 lb
57. (a)

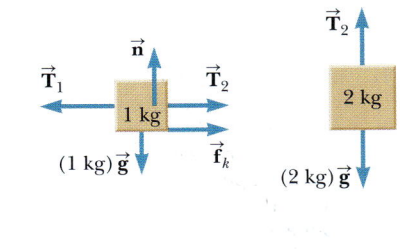

(b) 2.31 m/s^2, down for the 4.00-kg object, left for the 1.00-kg object, up for the 2.00-kg object (c) 30.0 N in the left cord, 24.2 N in the right cord (d) Without friction, the 4-kg block falls more freely, so the tension T_1 in the string attached to it is reduced. The 2-kg block is accelerated upwards at a higher rate, hence the tension force T_2 acting on it must be greater.
59. (a) 84.9 N upward (b) 84.9 N downward
61. 50 m
63. (a) friction between box and truck (b) 2.94 m/s^2
65. (a) 2.22 m (b) 8.74 m/s down the incline
67. (a) 0.23 m/s^2 (b) 9.7 N
69. (a) 1.7 m/s^2, 17 N (b) 0.69 m/s^2, 17 N
71. (a) 3.43 kN (b) 0.967 m/s horizontally forward (c) Both would increase.
73. (a) 30.7° (b) 0.843 N
75. 5.5×10^2 N
77. 72.0 N
79. (a) 7.1×10^2 N (b) 8.1×10^2 N (c) 7.1×10^2 N (d) 6.5×10^2 N
81. (a) 0.408 m/s^2 upward (b) 83.3 N
83. (b) If θ is greater than $\tan^{-1}(1/\mu_s)$, motion is impossible.
85. $(M + m_1 + m_2)(m_1 g/m_2)$

CHAPTER 5

Quick Quizzes
1. (c)
2. (d)
3. (c)
4. (c)

Example Questions
1. As long as the same displacement is produced by the same force, doubling the load will not change the amount of work done by the applied force.
2. Doubling the displacement doubles the amount of work done in each case.
3. The wet road would reduce the coefficient of kinetic friction, so the final velocity would be greater.
4. (c)
5. In each case the velocity would have an additional horizontal component, meaning that the overall speed would be greater.
6. A smaller angle means that a larger initial speed would be required to allow the grasshopper to reach the indicated height.
7. In the presence of friction a different shape slide would result in different amounts of mechanical energy lost through friction, so the final answer would depend on the slide's shape.
8. In the crouching position there is less wind resistance. Crouching also lowers the skier's center of mass, making it easier to balance.
9. 73.5%
10. If the acrobat bends her legs and crouches immediately after contacting the springboard and then jumps as the platform pushes her back up, she can rebound to a height greater than her initial height.
11. The continuing vibration of the spring means that some energy wasn't transferred to the block. As a result, the block will go a slightly smaller distance up the ramp.
12. (a)
13. The work required would quadruple but time would double, so overall the average power would double.
14. The instantaneous power is 9.00×10^4 W, which is twice the average power.
15. False. The correct answer is one-quarter.
16. No. Using the same-size boxes is simply a matter of convenience.

Warm-Up Exercises
1. 1.79×10^4 N
3. (a) 375 J (b) 307 J
5. (a) 4.90 m/s (b) 5.66 m/s (c) 6.32 m/s
7. 11.7 m/s
9. 82.6 W

Conceptual Questions
1. Because no motion is taking place, the rope undergoes no displacement and no work is done on it. For the same reason, no work is being done on the pullers or the ground. Work is being done only within the bodies of the pullers. For example, the heart of each puller is applying forces on the blood to move blood through the body.
3. (a) When the slide is frictionless, changing the length or shape of the slide will not make any difference in the final speed of the child, as long as the difference in the heights of the upper and lower ends of the slide is kept constant. (b) If friction must be considered, the path length along which the friction force does negative work will be greater when the slide is made longer or given bumps. Thus, the child will arrive at the lower end with less kinetic energy (and hence less speed).
5. If we ignore any effects due to rolling friction on the tires of the car, we find that the same amount of work would be done in driving up the switchback and in driving straight

up the mountain because the weight of the car is moved upwards against gravity by the same vertical distance in each case. If we include friction, there is more work done in driving the switchback because the distance over which the friction force acts is much longer. So why do we use switchbacks? The answer lies in the force required, not the work. The force required from the engine to follow a gentle rise is much smaller than that required to drive straight up the hill. To negotiate roadways running straight uphill, engines would have to be redesigned to enable them to apply much larger forces. (It is for much the same reason that ramps are designed to move heavy objects into trucks, as opposed to lifting the objects vertically.)

7. (a) The tension in the supporting cord does no work, because the motion of the pendulum is always perpendicular to the cord and therefore to the tension force. (b) The air resistance does negative work at all times, because the air resistance is always acting in a direction opposite that of the motion. (c) The force of gravity always acts downwards; therefore, the work done by gravity is positive on the downswing and negative on the upswing.

9. During the time that the toe is in contact with the ball, the work done by the toe on the ball is given by

$$W_{\text{toe}} = \tfrac{1}{2} m_{\text{ball}} v^2 - 0 = \tfrac{1}{2} m_{\text{ball}} v^2$$

where v is the speed of the ball as it leaves the toe. After the ball loses contact with the toe, only the gravitational force and the retarding force due to air resistance continue to do work on the ball throughout its flight.

11. (a) Yes, the total mechanical energy of the system is conserved because the only forces acting are conservative: the force of gravity and the spring force. (b) There are two forms of potential energy in this case: gravitational potential energy and elastic potential energy stored in the spring.

13. Let's assume you lift the book slowly. In this case, there are two forces on the book that are almost equal in magnitude: the lifting force and the force of gravity. Thus, the positive work done by you and the negative work done by gravity cancel. There is no net work performed and no net change in the kinetic energy, so the work–energy theorem is satisfied.

15. As the satellite moves in a circular orbit about the Earth, its displacement during any small time interval is perpendicular to the gravitational force, which always acts toward the center of the Earth. Therefore, the work done by the gravitational force during any displacement is zero. (Recall that the work done by a force is defined to be $F\Delta x \cos\theta$, where θ is the angle between the force and the displacement. In this case, the angle is 90°, so the work done is zero.) Because the work–energy theorem says that the net work done on an object during any displacement is equal to the change in its kinetic energy, and the work done in this case is zero, the change in the satellite's kinetic energy is zero: hence, its speed remains constant.

17. (a)
19. (d)

Problems

1. 700 J
3. 2×10^4 J
5. (a) 61.3 J (b) −46.3 J (c) 0 (d) The work done by gravity would not change, the work done by the friction force would decrease, and the work done by the normal force would not change.
7. (a) 79.4 N (b) 1.49 kJ (c) −1.49 kJ
9. (a) 2.00 m/s (b) 200 N
11. (a) 879 J (b) His kinetic energy increases by a factor of 4.
13. (a) -5.6×10^2 J (b) 1.2 m
15. (a) 2.34×10^4 N (b) 1.91×10^{-4} s
17. (a) 4.68×10^9 J (b) -4.68×10^9 J (c) 1.87×10^6 N
19. (a) 2.5 J (b) −9.8 J (c) −12.3 J
21. 7.37 N/m
23. 878 kN up
25. $h = 6.94$ m
27. $W_{\text{biceps}} = 120$ J, $W_{\text{chin-up}} = 290$ J, additional muscles must be involved
29. (a) 4.30×10^5 J (b) -3.97×10^4 J (c) 115 m/s
31. (a) The mass, spring, and Earth (including the wall) constitute the system. The mass and Earth interact through the spring force, gravity, and the normal force. (b) the point of maximum extension, $x = 0.060\,0$ m, and the equilibrium point, $x = 0$ (c) 1.53 J at $x = 6.00$ cm; 0 J at $x = 0$ (d) $\tfrac{1}{2}mv_1^2 + \tfrac{1}{2}kx_1^2 = \tfrac{1}{2}mv_2^2 + \tfrac{1}{2}kx_2^2$

$$\rightarrow v_2 = \sqrt{v_1^2 + \frac{k}{m}(x_1^2 - x_2^2)},$$

1.75 m/s (e) 1.51 m/s. This answer is not one-half the first answer because the equation is not linear.
33. 0.459 m
35. (a) 0.350 m (b) The result would be less than 0.350 m because some of the mechanical energy is lost as a result of the force of friction between the block and track.
37. (a) 10.9 m/s (b) 11.6 m/s
39. (a) Initially, all the energy is stored in the compressed spring. After the gun is fired and the projectile leaves the gun, the energy is transferred to the kinetic energy of the projectile, resulting in a small increase in gravitational potential energy. Once the projectile reaches its maximum height, the energy is all associated with its gravitational potential energy. (b) 544 N/m (c) 19.7 m/s
41. (a) Yes. There are no nonconservative forces acting on the child, so the total mechanical energy is conserved. (b) No. In the expression for conservation of mechanical energy, the mass of the child is included in every term and therefore cancels out. (c) The answer is the same in each case. (d) The expression would have to be modified to include the work done by the force of friction. (e) 15.3 m/s.
43. (a) 372 N (b) $T_1 = 372$ N, $T_2 = T_3 = 745$ N (c) 1.34 kJ
45. 3.8 m/s
47. 289 m
49. (a) 24.5 m/s (b) Yes (c) 206 m (d) Unrealistic; the actual retarding force will vary with speed.
51. 236 s or 3.93 min
53. 8.01 W
55. The power of the sports car is four times that of the older-model car.
57. (a) 2.38×10^4 W = 32.0 hp (b) 4.77×10^4 W = 63.9 hp
59. (a) 24.0 J (b) −3.00 J (c) 21.0 J
61. (a) The graph is a straight line passing through the points (0 m, −16 N), (2 m, 0 N), and (3 m, 8 N). (b) −12.0 J
63. 0.265 m/s
65. (a) $PE_{\text{Ⓐ}} = 3.94 \times 10^5$ J, $PE_{\text{Ⓑ}} = 0$, $\Delta PE = -3.94 \times 10^5$ J (b) $PE_{\text{Ⓐ}} = 5.63 \times 10^5$ J, $PE_{\text{Ⓑ}} = 1.69 \times 10^5$ J, $\Delta PE = -3.94 \times 10^5$ J
67. (a) 575 N/m (b) 46.0 J
69. (a) 4.4 m/s (b) 1.5×10^5 N
71. (a) 3.13 m/s (b) 4.43 m/s (c) 1.00 m
73. (a) 0.588 J (b) 0.588 J (c) 2.42 m/s (d) $PE_C = 0.392$ J (e) $KE_C = 0.196$ J
75. (a) 0.020 4 m (b) 7.20×10^2 N/m
77. (a) 423 mi/gal (b) 776 mi/gal
79. (a) 28.0 m/s (b) 30.0 m (c) 88.9 m beyond the end of the track
81. (a) 101 J (b) 0.412 m (c) 2.84 m/s (d) −9.80 mm (e) 2.86 m/s
83. 914 N/m

85. (a) $W_{net} = 0$ (b) $W_{grav} = -2.0 \times 10^4$ J (c) $W_{normal} = 0$,
    (d) $W_{friction} = 2.0 \times 10^4$ J
87. (a) 10.2 kW (b) 10.6 kW (c) 5.82×10^6 J
89. 4.3 m/s
91. between 25.2 km/h and 27.0 km/h

CHAPTER 6

Quick Quizzes

1. (b)
2. (c)
3. (c)
4. (a)
5. (a) Perfectly inelastic (b) Inelastic (c) Inelastic
6. (a)

Example Questions

1. 44 m/s
2. When one car is overtaking another, the relative velocity is small, so on impact the change in momentum is also small. In a head-on collision, however, the relative velocity is large because the cars are traveling in opposite directions. Consequently, the change in momentum of a passenger in a head-on collision is greater than when hit from behind, which implies a larger average force.
3. Assuming the kinetic energies of the two arrows are identical, the heavier arrow would have a greater momentum, because $p^2 = 2mK$. Greater arrow momentum means greater recoil speed for the archer.
4. The final velocity would be unaffected, but the change in kinetic energy would be doubled.
5. Energy can be lost due to friction during the impact, work done in deforming the bullet and block, friction in the physical mechanisms, air drag, and the creation of sound waves.
6. No. If that were the case, energy could not be conserved.
7. The blocks cannot both come to rest at the same time because then by Equation (1) momentum would not be conserved.
8. 45°
9. $m(a + g)$

Warm-Up Exercises

1. (a) -0.800 m/s (b) 3.20 m/s
3. (a) 5.55 (b) 4.77×10^3 kg
5. (a) 2.62 kg · m/s (b) 43.7 N
7. (a) 0.369 J (b) 0.640 J (c) 0.184 J
9. (a) 15.0 m/s (b) 26.0 m/s
11. (a) 4.54×10^5 kg (b) 2.16×10^5 kg

Conceptual Questions

1. (a) No. It cannot carry more kinetic energy than it possesses. That would violate the law of energy conservation. (b) Yes. By bouncing from the object it strikes, it can deliver more momentum in a collision than it possesses in its flight.
3. If all the kinetic energy disappears, there must be no motion of either of the objects after the collision. If neither is moving, the final momentum of the system is zero, and the initial momentum of the system must also have been zero. A situation in which this could be true would be the head-on collision of two objects having momenta of equal magnitude but opposite direction.
5. Initially, the clay has momentum directed toward the wall. When it collides and sticks to the wall, neither the clay nor the wall appears to have any momentum. It is therefore tempting to (wrongfully) conclude that momentum is not conserved. The "lost" momentum, however, is actually imparted to the wall and the Earth, causing both to move. Because of the Earth's enormous mass, its recoil speed is too small to detect.
7. As the water is forced out of the holes in the arm, the arm imparts a horizontal impulse to the water. The water then exerts an equal and opposite impulse on the spray arm, causing the spray arm to rotate in the direction opposite that of the spray.
9. It will be easiest to catch the medicine ball when its speed (and kinetic energy) is lowest. The first option—throwing the medicine ball at the same velocity—will be the most difficult, because the speed will not be reduced at all. The second option, throwing the medicine ball with the same momentum, will reduce the velocity by the ratio of the masses. Since $m_t v_t = m_m v_m$, it follows that

$$v_m = v_t \left(\frac{m_t}{m_m} \right)$$

The third option, throwing the medicine ball with the same kinetic energy, will also reduce the velocity, but only by the square root of the ratio of the masses. Since

$$\tfrac{1}{2} m_t v_t^2 = \tfrac{1}{2} m_m v_m^2$$

it follows that

$$v_m = v_t \sqrt{\frac{m_t}{m_m}}$$

The slowest and easiest throw will be made when the momentum is held constant. If you wish to check this answer, try substituting in values of $v_t = 1$ m/s, $m_t = 1$ kg, and $m_m = 100$ kg. Then the same-momentum throw will be caught at 1 cm/s, while the same-energy throw will be caught at 10 cm/s.
11. It is the product mv that is the same for both the bullet and the gun. The bullet has a large velocity and a small mass, while the gun has a small velocity and a large mass. Furthermore, the bullet carries much more kinetic energy than the gun.
13. (a) The follow-through keeps the club in contact with the ball as long as possible, maximizing the impulse. Thus, the ball accrues a larger change in momentum than without the follow-through, and it leaves the club with a higher velocity and travels farther. (b) With a short shot to the green, the primary factor is control, not distance. Hence, there is little or no follow-through, allowing the golfer to have a better feel for how hard he or she is striking the ball.
15. No. Impulse, $\vec{F}\Delta t$, depends on the force and the time interval during which it is applied.
17. (c)

Problems

1. (a) 8.35×10^{-21} kg · m/s (b) 4.50 kg · m/s (c) 750 kg · m/s
   (d) 1.78×10^{29} kg · m/s
3. (a) 3.0 m/s (b) the bullet, 3.38×10^3 J versus 69.7 J
5. (a) 0.42 N downward (b) The hailstones would exert a larger average force on the roof because they would bounce off the roof, and the impulse acting on the hailstones would be greater than the impulse acting on the raindrops. Newton's third law then tells us that the hailstones exert a greater force on the roof.
7. (a) 22.0 m/s (b) 1.14 kg
9. (a) 10.4 kg · m/s in the direction of the final velocity of the ball (b) 173 N
11. 1.39 N · s up
13. (a) 364 kg · m/s forward (b) 438 N forward
15. (a) 8.0 N · s (b) 5.3 m/s (c) 3.3 m/s
17. (a) 12 N · s (b) 8.0 N · s (c) 8.0 m/s (d) 5.3 m/s
19. (a) 9.60×10^{-2} s (b) 3.65×10^5 N (c) 26.6g

21. 65 m/s
23. (a) 1.15 m/s (b) 0.346 m/s directed opposite to girl's motion
25. (a) 154 m (b) By Newton's third law, when the astronaut exerts a force on the tank, the tank exerts a force back on the astronaut. This reaction force accelerates the astronaut towards the spacecraft.
27. $v_{thrower} = 2.48$ m/s, $v_{catcher} = 2.25 \times 10^{-2}$ m/s
29. (a)

(b) The collision is best described as perfectly inelastic because the skaters remain in contact after the collision.
(c) $m_1 v_1 + m_2 v_2 = (m_1 + m_2) v_f$
(d) $v_f = (m_1 v_1 + m_2 v_2)/(m_1 + m_2)$ (e) 6.33 m/s
31. 15.6 m/s
33. (a) 1.80 m/s (b) 2.16×10^4 J
35. (a) 1
37. 1.32 m
39. 57 m
41. 273 m/s
43. (a) He moves at a velocity of $\left(\dfrac{m_g}{m_b}\right) v_g$ toward the west.

(b) $KE_g = \frac{1}{2} m_g v_g^2$; $KE_b = \left(\dfrac{m_g^2}{2 m_b}\right) v_g^2$ The ratio $\dfrac{KE_g}{KE_b} = \dfrac{m_b}{m_g}$ is greater than 1 because $m_b > m_g$. Hence, the girl has more kinetic energy. (c) Work is done by both the boy and girl as they push each other apart and the origin of this work is chemical energy in their bodies.
45. 17.1 cm/s (25.0-g object), 22.1 cm/s (10.0-g object)
47. (a) Over a very short time interval, external forces have no time to impart significant impulse to the players during the collision. The two players move together after the tackle, so the collision is completely inelastic. (b) 2.88 m/s at 32.3° (c) 785 J. (d) The lost kinetic energy is transformed into other forms of energy such as thermal energy and sound.
49. 5.59 m/s north
51. (a) 2.50 m/s at $-60°$ (b) elastic collision
53. 1.78×10^3 N on truck driver, 8.89×10^3 N on car driver
55. (a) 8/3 m/s (incident particle), 32/3 m/s (target particle)
    (b) $-16/3$ m/s (incident particle), 8/3 m/s (target particle)
    (c) 7.1×10^{-2} J in case (a), and 2.8×10^{-3} J in case (b). The incident particle loses more kinetic energy in case (a), in which the target mass is 1.0 g.
57. $v_1 = -1.7$ m/s (to the left); $v_2 = 1.1$ m/s (to the right)
59. 62 s
61. (a) 3 (b) 2
63. (a) -2.33 m/s, 4.67 m/s (b) 0.277 m (c) 2.98 m (d) 1.49 m
65. (a) -0.667 m/s (b) 0.952 m
67. (a) 3.54 m/s (b) 1.77 m (c) 3.54×10^4 N (d) No, the normal force exerted by the rail contributes upward momentum to the system.
69. (a) 0.28 or 28% (b) 1.1×10^{-13} J for the neutron, 4.5×10^{-14} J for carbon
71. (a) No. After colliding, the cars, moving as a unit, would travel northeast, so they couldn't damage property on the southeast corner. (b) x-component: 16.3 km/h, y-component: 9.17 km/h, angle: the final velocity of the car is 18.7 km/h at 29.4° north of east, consistent with part (a).
73. (a) 4.85 m/s (b) 8.41 m
75. $v_0 = \left(\dfrac{M + m}{m}\right) \sqrt{2 \mu g d}$
77. (a) 1.1 m/s at 30° below the positive x-axis (b) 0.32 or 32%
79. (a) The momentum of the bullet-block system is conserved in the collision, so you can relate the speed of the block and bullet immediately after the collision to the initial speed of the bullet. Then, you can use conservation of mechanical energy for the bullet–block–Earth system to relate the speed after the collision to the maximum height. (b) 521 m/s upward

CHAPTER 7

Quick Quizzes

1. (c)
2. (b)
3. (b)
4. (b)
5. (a)
6. 1. (e) 2. (a) 3. (b)
7. (c)
8. (b), (c)
9. (e)
10. (d)

Example Questions

1. Yes. The conversion factor is $(180°/\pi$ rad) or 57.3° s^{-1}.
2. All given quantities and answers are in angular units, so altering the radius of the wheel has no effect on the answers.
3. In this case, doubling the angular acceleration doubles the angular displacement. That is true here because the initial angular speed is zero.
4. The angular acceleration of a record player during play is zero. A CD player must have nonzero acceleration because the angular speed must change.
5. (b)
6. It would be increased.
7. The angle of the bank, the coefficient of friction, and the radius of the circle determine the minimum and maximum safe speeds.
8. The normal force is still zero.
9. Yes. The force of gravity acting on each billiard ball holds the balls against the table and assists in creating friction forces that allow the balls to roll. The gravity forces between the balls are insignificant, however.
10. First, most asteroids are irregular in shape, so Equation (1) will not apply because the acceleration may not be uniform. Second, the asteroid may be so small that there will be no significant or useful region where the acceleration is uniform. In that case, Newton's more general law of gravitation would be required.
11. (b)
12. Mechanical energy is conserved in this system. Because the potential energy at perigee is lower, the kinetic energy must be higher.
13. 5 days

Warm-Up Exercises

1. (a) 785 m (b) 2.20 radians
3. (a) 1.26 rad/s (b) 24.5 rev/min
5. 1.99×10^{-7} rad/s
7. (a) 0.608 m/s^2 (b) 28.9 rad/s (c) 261 rad (d) 99.2 m
9. (a) 4.44 N (b) 8.36 N
11. $0.5g$
13. 1.34×10^4 m/s
15. (a) 18.0 AU (b) 35.4 AU

Conceptual Questions

1. (a) The head will tend to lean toward the right shoulder (that is, toward the outside of the curve). (b) When there is no strap, tension in the neck muscles must produce the centripetal acceleration. (c) With a strap, the tension in the strap performs this function, allowing the neck muscles to remain relaxed.

3. The speedometer will be inaccurate. The speedometer measures the number of tire revolutions per second, so its readings will be too low.
5. (a) Point C. The total acceleration here is centripetal acceleration, straight up. (b) Point A. The speed at A is zero where the bob is reversing direction. The total acceleration here is tangential acceleration, to the right and downward perpendicular to the cord. (c) Point B. The total acceleration here is to the right and pointing in a direction somewhere in between the tangential and radial directions, depending on their relative magnitudes.
7. Consider an individual standing against the inside wall of the cylinder with her head pointed toward the axis of the cylinder. As the cylinder rotates, the person tends to move in a straight-line path tangent to the circular path followed by the cylinder wall. As a result, the person presses against the wall, and the normal force exerted on her provides the radial force required to keep her moving in a circular path. If the rotational speed is adjusted such that this normal force is equal in magnitude to her weight on Earth, she will not be able to distinguish between the artificial gravity of the colony and ordinary gravity.
9. The tendency of the water is to move in a straight-line path tangent to the circular path followed by the container. As a result, at the top of the circular path, the water presses against the bottom of the pail, and the normal force exerted by the pail on the water provides the radial force required to keep the water moving in its circular path.
11. Any object that moves such that the *direction* of its velocity changes has an acceleration. A car moving in a circular path will always have a centripetal acceleration.
13. The speed changes. The component of force tangential to the path causes a tangential acceleration.

Problems

1. (a) 7.27×10^{-5} rad/s (b) Because of its rotation about its axis, Earth bulges at the equator.
3. (a) 3.2×10^8 rad (b) 5.0×10^7 rev
5. (a) 821 rad/s^2 (b) 4.21×10^3 rad
7. (a) 3.5 rad (b) The angular displacement increases by a factor of 4 because Equation 7.9 is quadratic in the angular velocities.
9. Main rotor: 179 m/s = $0.522 v_{sound}$
   Tail rotor: 221 m/s = $0.644 v_{sound}$
11. (a) 116 rev (b) 62.1 rad/s
13. 13.7 rad/s^2
15. (a) 6.53 m/s (b) 0.285 m/s^2 directed toward the center of the circular arc
17. (a) 0.346 m/s^2 (b) 1.04 m/s (c) 0.346 m/s^2 (d) 0.943 m/s^2 (e) 1.00 m/s^2 at 20.1° forward with respect to the direction of $\vec{a}_c$
19. (a) 20.6 N (b) 3.35 m/s^2 downward tangent to the circle; 32.0 m/s^2 radially inward (c) 32.2 m/s^2 at 5.98° to the cord, pointing toward a location below the center of the circle. (d) No change. (e) If the object is swinging down it is gaining speed. If it is swinging up it is losing speed but its acceleration has the same magnitude and its direction can be described in the same terms.
21. (a) 1.10 kN (b) 2.04 times her weight
23. 22.6 m/s
25. (a) 18.0 m/s^2 (b) 900 N (c) 1.84; this large coefficient is unrealistic, and she will not be able to stay on the merry-go-round.
27. (a) 9.8 N (b) 9.8 N (c) 6.3 m/s
29. (a) The force of static friction acting toward the road's center of curvature causes the briefcase's centripetal acceleration. When the necessary centripetal force exceeds the maximum value of the static friction force, $\mu_s n$, the briefcase begins to slide. (b) 0.370

31. (a) 1.58 m/s^2 (b) 455 N upward (c) 329 N upward (d) 397 N directed inward and 80.8° above horizontal
33. 321 N toward Earth
35. 1.1×10^{-10} N at 72° above the $+x$-axis
37. (a) 2.50×10^{-5} N toward the 500-kg object (b) between the two objects and 0.245 m from the 500-kg object
39. (a) $r = \dfrac{9}{8} R_E = 7.18 \times 10^6$ m (b) 7.98×10^5 m
41. (a) 2.43 h (b) 6.59 km/s (c) 4.73 m/s^2 toward the Earth
43. 6.3×10^{23} kg
45. $\dfrac{T_B}{T_A} = \sqrt{\dfrac{27}{8}} = 1.84$
47. (a) 1.90×10^{27} kg (b) 1.89×10^{27} kg (c) yes
49. (a) 9.40 rev/s (b) 44.2 rev/s^2; $a_r = 2\,590$ m/s^2; $a_t = 206$ m/s^2 (c) $F_r = 515$ N; $F_t = 40.7$ N
51. (a) 2.51 m/s (b) 7.90 m/s^2 (c) 4.00 m/s
53. (a) 7.76×10^3 m/s (b) 89.3 min
55. (a) $n = m\left(g - \dfrac{v^2}{r}\right)$ (b) 17.1 m/s
57. (a) $F_{g,\text{true}} = F_{g,\text{apparent}} + mR_E\omega^2$ (b) 732 N (equator), 735 N (either pole)
59. 11.8 km/s
61. 0.835 rev/s
65. (a) 10.6 kN (b) 14.1 m/s
67. (a) 0.71 yr (b) The departure must be timed so that the spacecraft arrives at aphelion when the target planet is there.
69. (a) $t = \dfrac{2R\omega}{g}$ (b) $\omega = \sqrt{\dfrac{\pi g}{R}}$
71. (a) 109 N (b) 56.4 N
73. (a) 106 N (b) 0.396
75. 0.131

CHAPTER 8

Quick Quizzes

1. (d)
2. (b)
3. (b)
4. (a)
5. (c)
6. (a)

Example Questions

1. The revolving door begins to move counterclockwise instead of clockwise.
2. Placing the wedge closer to the doorknob increases its effectiveness.
3. If the woman leans backwards, the torque she exerts on the seesaw increases and she begins to descend.
4. (b)
5. The system would begin to rotate clockwise.
6. The angle made by the biceps force would still not vary much from 90° but the length of the moment arm would be doubled, so the required biceps force would be reduced by nearly half.
7. (c)
8. The tension in the cable would increase.
9. Lengthening the rod between balls 2 and 4 would create the larger change in the moment of inertia.
10. Stepping forward transfers the momentum of the pitcher's body to the ball. Without proper timing, the transfer will not take place or will have less effect.
11. The magnitude of the acceleration would decrease; that of the tension would increase.
12. Block, ball, cylinder
13. The final answer wouldn't change.

14. His angular speed remains the same.
15. Energy conservation is not violated. The positive net change occurs because the student is doing work on the system.

Warm-Up Exercises

1. (a) 36.0 N and (b) 24.0 N
3. (a) -0.942 rad/s^2 (b) 78.6 rad
5. (a) 25.0 N · m (b) 50.0 N · m
7. (1.33 m, 2.00 m)
9. (a) 8.57 rad/s (b) 0 N · m
11. (a) 0.331 kg · m^2 (b) 0.248 N · m (c) 0.709 N
13. (a) 1.04 J (b) 0.110 kg · m^2/s

Conceptual Questions

1. In order for you to remain in equilibrium, your center of gravity must always be over your point of support, the feet. If your heels are against a wall, your center of gravity cannot remain above your feet when you bend forward, so you lose your balance.
3. No, only if its angular velocity changes.
5. The long pole has a large moment of inertia about an axis along the rope. An unbalanced torque will then produce only a small angular acceleration of the performer–pole system, to extend the time available for getting back in balance. To keep the center of mass above the rope, the performer can shift the pole left or right, instead of having to bend his body around. The pole sags down at the ends to lower the system's center of gravity, increasing the relative stability of the system.
7. As the motorcycle leaves the ground, the friction between the tire and the ground suddenly disappears. If the motorcycle driver keeps the throttle open while leaving the ground, the rear tire will increase its angular speed and, hence, its angular momentum. The airborne motorcycle is now an isolated system, and its angular momentum must be conserved. The increase in angular momentum of the tire directed, say, clockwise must be compensated for by an increase in angular momentum of the entire motorcycle counterclockwise. This rotation results in the nose of the motorcycle rising and the tail dropping.
9. The angular momentum of the gas cloud is conserved. Thus, the product $I\omega$ remains constant. As the cloud shrinks in size, its moment of inertia decreases, so its angular speed ω must increase.
11. We can assume fairly accurately that the driving motor will run at a constant angular speed and at a constant torque. (a) As the radius of the take-up reel increases, the tension in the tape will decrease, in accordance with the equation.

$$T = \tau_{\text{const}}/R_{\text{take-up}} \quad (1)$$

As the radius of the source reel decreases, given a decreasing tension, the torque in the source reel will decrease even faster, as the following equation shows:

$$\tau_{\text{source}} = TR_{\text{source}} = \tau_{\text{const}} R_{\text{source}}/R_{\text{take-up}} \quad (2)$$

(b) In the case of a sudden jerk on the tape, the changing angular speed of the source reel becomes important. If the source reel is full, then the moment of inertia will be large and the tension in the tape will be large. If the source reel is nearly empty, then the angular acceleration will be large instead. Thus, the tape will be more likely to break when the source reel is nearly full. One sees the same effect in the case of paper towels: It is easier to snap a towel free when the roll is new than when it is nearly empty.
13. When a ladder leans against a wall, both the wall and the floor exert forces of friction on the ladder. If the floor is perfectly smooth, it can exert no frictional force in the horizontal direction to counterbalance the wall's normal force. Therefore, a ladder on a smooth floor cannot stand in equilibrium. However, a smooth wall can still exert a normal force to hold the ladder in equilibrium against horizontal motion. The counterclockwise torque of this force prevents rotation about the foot of the ladder. So you should choose a rough floor.
15. (d)
17. (e)

Problems

1. 168 N · m
3. (a) 30 N · m (counterclockwise)
   (b) 36 N · m (counterclockwise)
5. (a) 5.1 N · m (b) The torque increases, because the torque is proportional to the moment arm, $L \sin\theta$, and this factor increases as θ increases.
7. $F_t = 724$ N, $F_s = 716$ N
9. 312 N
11. $x_{cg} = 3.33$ ft, $y_{cg} = 1.67$ ft
13. $(-1.5$ m, -1.5 m$)$
15. 1.01 m in Figure P8.15b; 0.015 m toward the head in Figure P8.15c.
17. (a) $T = 2.71$ kN (b) $R_x = 2.65$ kN
19. (a) 443 N, (b) 222 N (to the right), 216 N (upward)
21. $T_1 = 501$ N, $T_2 = 672$ N, $T_3 = 384$ N
23. (a) $d = \dfrac{mg}{2k \tan\theta}$ (b) $R_x = \dfrac{mg}{2\tan\theta}$; $R_y = mg$
25. $\theta = \tan^{-1}\left(\dfrac{w}{h}\right)$
27. $R = 107$ N, $T = 157$ N
29. 209 N
31. (a) 99.0 kg · m^2 (b) 44.0 kg · m^2 (c) 143 kg · m^2
33. (a) 87.8 kg · m^2 (b) 1.61×10^3 kg (c) 4.70 rad/s
35. (a) 3.92 m/s^2 (b) 9.80 rad/s^2 (c) As the rope leaves the cylinder, the mass of the cylinder decreases, also decreasing the cylinder's moment of inertia. At the same time, the mass of rope leaving the cylinder would increase the torque exerted on the cylinder. Hence, the acceleration of the system would increase in magnitude with time. (The increase would be slight in this case, given the large mass of the cylinder.)
37. (a) 24.0 N · m (b) 0.035 6 rad/s^2 (c) 1.07 m/s^2
39. 177 N
41. 0.524
43. 276 J
45. (a) 5.47 J (b) 5.99 J
47. (a) 3.90 m/s (b) 15.6 rad/s
49. 149 rad/s
51. (a) 500 J (b) 250 J (c) 750 J
53. (a) 5.00 m (b) 2.31×10^4 J (c) 1.79×10^4 J (d) 9.27 m/s
55. (a) 7.08×10^{33} J · s (b) 2.66×10^{40} J · s
57. 17.5 J · s counterclockwise
59. (a) $7md^2/3$ (b) mgd counterclockwise (c) $3g/7d$ counterclockwise (d) $2g/7$ upward (e) mgd (f) $\sqrt{6g/7d}$ (g) $m\sqrt{14gd^3/3}$ (h) $\sqrt{2gd/21}$
61. 6.73 rad/s
63. 5.99×10^{-2} J
65. (a) $\omega = \left(\dfrac{I_1}{I_1 + I_2}\right)\omega_0$ (b) $\dfrac{KE_f}{KE_i} = \dfrac{I_1}{I_1 + I_2} < 1$
67. (a) 2.6 rad/s (b) 5.1×10^5 kg · m^2 (c) 1.7×10^6 J
69. (a) As the child walks to the right end of the boat, the boat moves left (toward the pier). (b) The boat moves 1.45 m closer to the pier, so the child will be 5.55 m from the pier. (c) No. He will be short of reaching the turtle by 0.45 m.
71. 36.9°
73. (a) Mvd (b) Mv^2 (c) Mvd (d) $2v$ (e) $4Mv^2$ (f) $3Mv^2$
75. (a) 6.73 N upward (b) $x = 0.389$ m

77. (a) 0 (b) 0; Because the monkey and the bananas are both moving upward with the same speed, the angular momentum of the system is zero. (c) The monkey won't reach the bananas until they become caught in the pulley.
79. 1.09 m
81. (a) $T = \dfrac{Mmg}{M + 4m}$ (b) $a_t = \dfrac{4mg}{M + 4m}$
83. 24.5 m/s
85. 9.3 kN
87. (a) 3.12 m/s² (b) $T_1 = 26.7$ N, $T_2 = 9.36$ N
89. (a) 0.73 m (b) 1.6 m/s

CHAPTER 9

Quick Quizzes

1. (c)
2. (a)
3. (c)
4. (b)
5. (c)
6. (b)
7. (a)

Example Questions

1. Because water is more dense than oil, the pressure exerted by a column of water is greater than the pressure exerted by a column of oil.
2. Tungsten, steel, aluminum, rubber
3. The lineman's skull and neck would undergo compressional stress.
4. Steel, copper, mercury, water
5. At higher altitude, the column of air above a given area is progressively shorter and less dense, so the weight of the air column is reduced. Pressure is caused by the weight of the air column, so the pressure is also reduced.
6. As fluid pours out through a single opening, the air inside the can above the fluid expands into a larger volume, reducing the pressure to below atmospheric pressure. Air must then enter the same opening going the opposite direction, resulting in disrupted fluid flow. A separate opening for air intake maintains air pressure inside the can without disrupting the flow of the fluid.
7. True
8. False
9. (a)
10. The aluminum cube would float free of the bottom.
11. The speed of the blood in the narrowed region increases.
12. A factor of 2
13. The speed decreases with time.
14. The limit is $v_1 = \sqrt{2gh}$, called Torricelli's Law. (See Example 9.13).
15. The pressure difference across the wings depends linearly on the density of air. At higher altitude, the air's density decreases, so the lift force decreases as well.
16. False
17. No. There are many plants taller than 0.3 m, so there must be some additional explanation.
18. A factor of 16
19. False

Warm-Up Exercises

1. 1.96×10^{-3} N
3. 5.35×10^5 Pa
5. 8.9×10^{-6} m
7. 2.35×10^6 Pa
9. (a) 1.77×10^{-3} m³ (b) 17.3 N (c) 196 N (d) 179 N
11. 6.57 m/s

Conceptual Questions

1. She exerts enough pressure on the floor to dent or puncture the floor covering. The large pressure is caused by the fact that her weight is distributed over the very small cross-sectional area of her high heels. If you are the homeowner, you might want to suggest that she remove her high heels and put on some slippers.
3. The density of air is lower in the mile-high city of Denver than it is at lower altitudes, so the effect of air drag is less in Denver than it would be in a city such as New York. The reduced air drag means a well-hit ball will go farther, benefiting home-run hitters. On the other hand, curve ball pitchers prefer to throw at lower altitudes where the higher-density air produces greater deflecting forces on a spinning ball.
5. If you think of the grain stored in the silo as a fluid, the pressure the grain exerts on the walls of the silo increases with increasing depth, just as water pressure in a lake increases with increasing depth. Thus, the spacing between bands is made smaller at the lower portions to counterbalance the larger outward forces on the walls in these regions.
7. In the ocean, the ship floats due to the buoyant force from *salt water*, which is denser than fresh water. As the ship is pulled up the river, the buoyant force from the fresh water in the river is not sufficient to support the weight of the ship, and it sinks.
9. At lower elevation the water pressure is greater because pressure increases with increasing depth below the water surface in the reservoir (or water tower). The penthouse apartment is not so far below the water's surface. The pressure behind a closed faucet is weaker there and the flow weaker from an open faucet. Your fire department likely has a record of the precise elevation of every fire hydrant.
11. The two cans displace the same volume of water and hence are acted upon by buoyant forces of equal magnitude. The total weight of the can of diet cola must be less than this buoyant force, whereas the total weight of the can of regular cola is greater than the buoyant force. This is possible even though the two containers are identical and contain the same volume of liquid. Because of the difference in the quantities and densities of the sweeteners used, the volume V of the diet mixture will have less mass than an equal volume of the regular mixture.
13. Opening the windows results in a smaller pressure difference between the exterior and interior of the house and, therefore, less tendency for severe damage to the structure due to the Bernoulli effect.
15. (b)

Problems

1. $4.74\rho_0$
3. (a) 2 550 N (b) 3.00×10^4 Pa (c) No. The pressure will be the same only if the acrobats all wear the same-size shoe.
5. (a) $\sim 4 \times 10^{17}$ kg/m³ (b) The density of an atom is about 10^{14} times greater than the density of iron and other common solids and liquids. This shows that an atom is mostly empty space. Liquids and solids, as well as gases, are mostly empty space.
7. (a) 1.01×10^6 N (b) 3.88×10^5 N (c) 1.11×10^5 Pa
9. 4.90 mm
11. 1.05×10^7 Pa
13. 3.5×10^8 Pa
15. 4.4 mm
17. 8.6×10^{-4} m
19. 1.9 cm

21. (a) 3.71×10^5 Pa (b) 3.57×10^4 N
23. 0.133 m
25. 1.05×10^5 Pa
27. 27 N · m
29. 0.258 N down
31. 9.41 kN
33. (a) 408 kg/m³ (b) If the steel object's mass is just slightly greater than 0.310 kg, then the block is submerged but the steel object and the block do not sink to the bottom, since the steel object starts to displace water and the buoyant force is increasing. If the steel object is solid, it will displace a little more than 0.039 kg of water when fully submerged. The steel object and the block will sink to the bottom when the steel object's mass exceeds about 0.350. For values less than this, but greater than 0.310 kg, the block will be submerged and the steel object partially submerged.
35. (a) 1.43 kN upward (b) 1.28 kN upward (c) The balloon expands because the external pressure declines with increasing altitude.
37. (a) 4.0 kN (b) 2.2 kN (c) The air pressure at this high altitude is much lower than atmospheric pressure at the surface of Earth, so the balloons expanded and eventually burst.
39. (a) 7.00 cm (b) 2.80 kg
41. (a) 1.46×10^{-2} m³ (b) 2.10×10^3 kg/m³
43. 17.3 N (upper scale), 31.7 N (lower scale)
45. (a) 80 g/s (b) 0.27 mm/s
47. 12.6 m/s
49. (a) 9.43×10^3 Pa (b) 255 m/s (c) The density of air decreases with increasing height, resulting in a smaller pressure difference. Beyond the maximum operational altitude, the pressure difference can no longer support the aircraft.
51. (a) 0.553 s (b) 14.5 m/s (c) 0.145 m/s (d) 1.013×10^5 Pa (e) 2.06×10^5 Pa; gravity terms can be neglected. (f) 33.0 N
53. 9.00 cm
55. 1.47 cm
57. (a) 28.0 m/s (b) 28.0 m/s (c) 2.11 MPa
59. 8.3×10^{-2} N/m
61. 5.6×10^{-2} N/m
63. 8.6 N
65. 2.1 MPa
67. 2.8 μm
69. 0.21 mm
71. $RN = 4.3 \times 10^3$; turbulent flow
73. 1.8×10^{-3} kg/m³
75. 1.4×10^{-5} N · s/m²
77. (a) The buoyant forces are the same because the two blocks displace equal amounts of water. (b) The spring scale reads largest value for the iron block. (c) $B = 2.0 \times 10^3$ N for both blocks, $T_{iron} = 13 \times 10^3$ N, $T_{aluminum} = 3.3 \times 10^3$ N.
79. (b) 5.9 km
81. 2.5×10^7 capillaries
83. (a) 1.57 kPa, 1.55×10^{-2} atm, 11.8 mm of Hg (b) The fluid level in the tap should rise. (c) Blockage of flow of the cerebrospinal fluid
85. 2.25 m above the level of point B
87. 0.605 m
89. $F = \pi R^2 (P_0 - P)$
91. 17.0 cm above the floor

CHAPTER 10

Quick Quizzes

1. (c)
2. (b)
3. (c)
4. (c)
5. Unlike land-based ice, ocean-based ice already displaces water, so when it melts ocean levels won't change much.
6. (b)

Example Questions

1. A Celsius degree
2. True
3. When the temperature decreases, the tension in the wire increases.
4. The magnitude of the required temperature change would be larger because the linear expansion coefficient of steel is less than that of copper.
5. Glass, aluminum, ethyl alcohol, mercury
6. The balloon expands.
7. The pressure is slightly reduced.
8. The volume of air decreases.
9. The pressure would increase, up to double if the reflections were all elastic.
10. True

Warm-Up Exercises

1. 15.6
3. (a) $-13.0°$F (b) 248 K
5. (a) 3.4×10^{-5} (°C)$^{-1}$ (b) 1.2×10^{-4} m²
7. (a) 298 K (b) 191 K
9. (a) 425 K (b) 1.29 mol (c) 6 830 J

Conceptual Questions

1. (a) An ordinary glass dish will usually break because of stresses that build up as the glass expands when heated. (b) The expansion coefficient for Pyrex glass is much lower than that of ordinary glass. Thus, the Pyrex dish will expand much less than the dish of ordinary glass and does not normally develop sufficient stress to cause breakage.
3. Mercury must have the larger coefficient of expansion. As the temperature of a thermometer rises, both the mercury and the glass expand. If they both had the same coefficient of linear expansion, the mercury and the cavity in the glass would expand by the same amount, and there would be no apparent movement of the end of the mercury column relative to the calibration scale on the glass. If the glass expanded more than the mercury, the reading would go down as the temperature went up! Now that we have argued this conceptually, we can look in a table and find that the coefficient for mercury is about 20 times as large as that for glass, so that the expansion of the glass can sometimes be ignored.
5. We can think of each bacterium as being a small bag of liquid containing bubbles of gas at a very high pressure. The ideal gas law indicates that if the bacterium is raised rapidly to the surface, then its volume must increase dramatically. In fact, the increase in volume is sufficient to rupture the bacterium.
7. Additional water vaporizes into the bubble, so that the number of moles n increases.
9. The bags of chips contain a sealed sample of air. When the bags are taken up the mountain, the external atmospheric pressure on them is reduced. As a result, the difference between the pressure of the air inside the bags and the reduced pressure outside results in a net force pushing the plastic of the bag outward.
11. The coefficient of expansion for metal is generally greater than that of glass; hence, the metal lid loosens because it expands more than the glass.
13. As the water rises in temperature, it expands or rises in pressure or both. The excess volume would spill out of the cooling system, or else the pressure would rise very high indeed. The expansion of the radiator itself provides only a little relief, because in general solids expand far less than liquids for a given positive change in temperature.

Problems

1. (a) $-460°C$ (b) $37.0°C$ (c) $-280°C$
3. (a) $-253°C$ (b) $-423°F$
9. (a) $107°F$ (b) Yes; the normal body temperature is $98.6°F$, so the patient has a high fever that needs immediate attention.
11. 31 cm
13. $55.0°C$
15. (a) $-179°C$ (attainable) (b) $-376°C$ (below 0 K, unattainable)
17. (a) 11.2×10^3 kg/m³ (b) No. Although the density of gold would be less on a warm day, the mass of the bar would be the same regardless of its temperature, and that is what you are paying for. (Note that the volume of the bar increases with increasing temperature, whereas its density decreases. Its mass, however, remains constant.)
19. 1.02×10^3 gallons
21. (a) 0.10 L (b) 2.009 L (c) 1.0 cm
23. 2.7×10^2 N
25. 0.548 gal
27. (a) increases (b) 1.603 cm
29. (a) $627°C$ (b) $927°C$
31. (a) 2.5×10^{19} molecules (b) 4.1×10^{-21} mol
33. 4.28 atm
35. 7.1 m
37. 16.0 cm³
39. 6.21×10^{-21} J
41. 6.64×10^{-27} kg
43. (a) 2.01×10^4 K (b) 901 K
45. 16 N
47. 0.66 mm to the right at 78° below the horizontal
49. 3.55 L
51. 35.016 m
53. 6.57 MPa
55. (a) 99.8 mL (b) The change in volume of the flask is far smaller because Pyrex has a much smaller coefficient of expansion than acetone. Hence, the change in volume of the flask has a negligible effect on the answer.
57. (b) The expansion of the mercury is almost 20 times that of the flask (assuming Pyrex glass).
59. 2.7 m
61. (a) $\theta = \dfrac{(\alpha_2 - \alpha_1) L_0 (\Delta T)}{\Delta r}$

    (c) The bar bends in the opposite direction.
63. (a) No torque acts on the disk so its angular momentum is constant. Yes, the angular speed increases. As the disk cools, its radius, and hence its moment of inertia, decreases. Conservation of angular momentum then requires that its angular speed increase. (b) 25.7 rad/s

CHAPTER 11

Quick Quizzes

1. (a) Water, glass, iron. (b) Iron, glass, water.
2. (b) The slopes are proportional to the reciprocal of the specific heat, so a larger specific heat results in a smaller slope, meaning more energy is required to achieve a given temperature change.
3. (c)
4. (b)
5. (a) 4 (b) 16 (c) 64

Example Questions

1. From the point of view of physics, faster repetitions don't affect the final answer; physiologically, however, the weightlifter's metabolic rate would increase.
2. (c)
3. No
4. (c)
5. No
6. The mass of ice melted would double.
7. Nickel–iron asteroids have a higher density and therefore a greater mass, which means they can deliver more energy on impact for a given speed.
8. A runner's metabolism is much higher when he is running than when he is at rest, and because muscles are only about 20% efficient, a great amount of random thermal energy is created through muscular exertion. Consequently, the runner needs to eliminate far more thermal energy when running than when resting. Once the run is over, muscular exertions cease and the metabolism starts to return to normal, so the runner begins to feel chilled.
9. (a)
10. (a)
11. If the planet doesn't reemit all the energy that it absorbs from its star, it will increase in temperature. As the temperature increases, the planet will radiate at a greater and greater rate until it reaches thermal equilibrium, when it emits as much as it absorbs.

Warm-Up Exercises

1. $8.09°C$
3. 6.8×10^{-4} m
5. 14.1 J
7. $29.4°C$
9. (a) 408 K (b) 50.3 m² (c) 2.54×10^4 W

Conceptual Questions

1. When you rub the surface, you increase the temperature of the rubbed region. With the metal surface, some of this energy is transferred away from the rubbed site by conduction. Consequently, the temperature in the rubbed area is not as high for the metal as it is for the wood, and it feels relatively cooler than the wood.
3. The fruit loses energy into the air by radiation and convection from its surface. Before ice crystals can form inside the fruit to rupture cell walls, all of the liquid water on the skin will have to freeze. The resulting time delay may prevent damage within the fruit throughout a frosty night. Further, a surface film of ice provides some insulation to slow subsequent energy loss by conduction from within the fruit.
5. One of the ways that objects transfer energy is by radiation. The top of the mailbox is oriented toward the clear sky. Radiation emitted by the top of the mailbox goes upward and into space. There is little radiation coming down from space to the top of the mailbox. Radiation leaving the sides of the mailbox is absorbed by the environment. Radiation from the environment (tree, houses, cars, etc.), however, can enter the sides of the mailbox, keeping them warmer than the top. As a result, the top is the coldest portion and frost forms there first.
7. (a) The operation of an immersion coil depends on the convection of water to maintain a safe temperature. As the water near a coil warms up, the warmed water floats to the top due to Archimedes' principle. The temperature of the coil cannot go higher than the boiling temperature of water, $100°C$. If the coil is operated in air, convection is reduced, and the upper limit of $100°C$ is removed. As a result, the coil can become hot enough to be damaged. (b) If the coil is used in an attempt to warm a thick liquid like stew, convection cannot occur fast enough to carry energy away from the coil, so that it again may become hot enough to be damaged.
9. Tile is a better conductor of energy than carpet, so the tile conducts energy away from your feet more rapidly than does the carpeted floor.

Problems

1. 16.9°C
3. (a) 1.67×10^{18} J (b) 52.9 yr
5. (a) 25.8°C (b) No. Gravitational potential energy is proportional to the mass, just like the temperature change.
7. (a) 4.5×10^3 J (b) 910 W (c) 0.87 Cal/s (d) The excess thermal energy is transported by conduction and convection to the surface of the skin and disposed of through the evaporation of sweat.
9. 176°C
11. 88 W
13. 4.2×10^6 J
15. 0.845 kg
17. 80 g
19. (a) 1.82×10^3 J/kg·°C (b) We cannot make a definite identification. It might be beryllium. (c) The material might be an unknown alloy or a material not listed in the table.
21. 0.26 kg
23. (a) 21.3°C (b) 178 J/kg·K (c) $N_{Sn} = 2.03 \times 10^{24}$ atoms; $N_{Pb} = 1.16 \times 10^{24}$ atoms (d) $N_{Sn}/N_{Pb} = 1.75$; $c_{Sn}/c_{Pb} = 1.77$. The specific heat of an element is proportional to the number of atoms per unit mass of that element.
25. 16°C
27. 65°C
29. 2.3 km
31. 16°C
33. (a) $t_{boil} = 2.8$ min (b) $t_{evaporate} = 18$ min
35. (a) all ice melts, $T_f = 40°C$ (b) 8.0 g melts, $T_f = 0°C$
37. (a) The bullet loses all its kinetic energy as it is stopped by the ice. Also, thermal energy must be removed from the bullet to cool it from 30.0°C to 0°C. The sum of these two energies equals the energy it takes to melt part of the ice. The final temperature of the bullet is 0°C because not all the ice melts. (b) 0.294 g
39. 3×10^3 W
41. 402 MW
43. 709 s
45. 9.0 cm
47. 2.7×10^7 J
49. 16:1
51. 12 kW
53. 2.3 kg
55. 29°C
57. 8.00×10^2 J/kg·°C. This value differs from the tabulated value by 11%, so they agree within 15%.
59. 66.7 min
61. 51.2°C
63. (a) seven times (b) As the car stops, it transforms part of its kinetic energy into internal energy due to air resistance. As soon as the brakes rise above the air temperature, they transfer energy by heat into the air. If they reach a high temperature, they transfer energy very quickly.
65. (b) 2.7×10^3 J/kg·°C
67. 12 h
69. 3.85×10^{26} J
71. 1.4 kg
73. (a) 75.0°C (b) 36 kJ

CHAPTER 12

Quick Quizzes

1. (b)
2. A is isovolumetric, B is adiabatic, C is isothermal, D is isobaric.
3. (c)
4. (b)
5. The number 7 is the most probable outcome. The numbers 2 and 12 are the least probable outcomes.

Example Questions

1. No
2. No
3. True
4. True
5. The change in temperature must always be negative because the system does work on the environment at the expense of its internal energy and no thermal energy can be supplied to the system to compensate for the loss.
6. A diatomic gas does more work under these assumptions.
7. The carbon dioxide gas would have a final temperature lower than 380 K.
8. False
9. (a)
10. No. The efficiency improves only if the ratio $|Q_c/Q_h|$ becomes smaller. Further, too large an increase in Q_h will damage the engine, so there is a limit even if Q_c remains fixed.
11. If the path from B to C were a straight line, more work would be done per cycle.
12. No. The compressor does work and warms the kitchen. With the refrigerator door open, the compressor would run continuously.
13. False
14. Silver, lead, ice
15. False
16. The thermal energy created by your body during the exertion would be dissipated into the environment, increasing the entropy of the Universe.
17. Skipping meals can lower the basal metabolism, reducing the rate at which energy is used. When a large meal is eaten later, the lower metabolism means more food energy will be stored, and weight will be gained even if the same number of total calories is consumed in a day.

Warm-Up Exercises

1.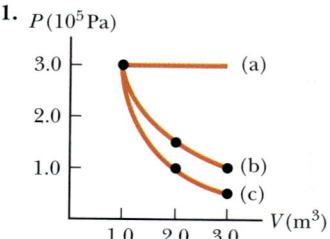

3. 7.54×10^4 J
5. (a) 9.19×10^3 J (b) 5.12×10^3 J
7. (a) 0 J (b) 1.00×10^2 J (c) 4.01 K (d) 2.41 K
9. (a) 1.40 (b) 1.73×10^4 Pa
11. (a) 0.200 (b) 60.0 kJ
13. (a) 0.333 (b) 15.0 kJ

Conceptual Questions

1. First, the efficiency of the automobile engine cannot exceed the Carnot efficiency: it is limited by the temperature of the burning fuel and the temperature of the environment into

which the exhaust is dumped. Second, the engine block cannot be allowed to exceed a certain temperature. Third, any practical engine has friction, incomplete burning of fuel, and limits set by timing and energy transfer by heat.

3. The energy that is leaving the body by work and heat is replaced by means of biological processes that transform chemical energy in the food that the individual ate into internal energy. Thus, the temperature of the body can be maintained.

5. If there is no change in internal energy, then, according to the first law of thermodynamics, the heat is equal to the negative of the work done on the gas (and thus equal to the work done *by* the gas). Thus, $Q = -W = W_{\text{by gas}}$.

7. Practically speaking, it isn't possible to create a heat engine that creates no thermal pollution, because there must be both a hot heat source (energy reservoir) and a cold heat sink (low-temperature energy reservoir). The heat engine will warm the cold heat sink and will cool down the heat source. If either of those two events is undesirable, then there will be thermal pollution.

    Under some circumstances, the thermal pollution would be negligible. For example, suppose a satellite in space were to run a heat pump between its sunny side and its dark side. The satellite would intercept some of the energy that gathered on one side and would 'dump' it to the dark side. Since neither of those effects would be particularly undesirable, it could be said that such a heat pump produced no thermal pollution.

9. Although no energy is transferred into or out of the system by heat, work is done on the system as the result of the agitation. Consequently, both the temperature and the internal energy of the coffee increase.

11. The first law is a statement of conservation of energy that says that we cannot devise a cyclic process that produces more energy than we put into it. If the cyclic process takes in energy by heat and puts out work, we call the device a heat engine. In addition to the first law's limitation, the second law says that, during the operation of a heat engine, some energy must be ejected to the environment by heat. As a result, it is theoretically impossible to construct a heat engine that will work with 100% efficiency.

13. If the system is isolated, no energy enters or leaves the system by heat, work, or other transfer processes. Within the system energy can change from one form to another, but since energy is conserved these transformations cannot affect the total amount of energy. The total energy is constant.

15. (b)
17. (a)

Problems

1. (a) -465 J (b) The negative sign for work done on the gas indicates that the expanding gas does positive work on the surroundings.
3. (a) -6.1×10^5 J (b) 4.6×10^5 J
5. (a) -810 J (b) -507 J (c) -203 J
7. 96.3 mg
9. (a) 1.09×10^3 K (b) -6.81 kJ
13. (a) -88.5 J (b) 722 J
15. (a) 567 J (b) 167 J
17. (a) -180 J (b) $+188$ J
19. (a) 3.25 kJ (b) 0 (c) -3.25 kJ (d) The internal energy would increase, resulting in an increase in temperature of the gas.
21. (a) -4.58×10^4 J (b) 4.58×10^4 J (c) 0
23. (a) 40.6 moles (b) 506 J/K (c) $W = 0$ (d) 16.0 kJ
    (e) 31.6 K (f) 332 K (g) 5.53 atm
25. (a) 0.95 J (b) 3.2×10^5 J (c) 3.2×10^5 J
27. (b) 1.3×10^{56} K
29. 405 kJ
31. 0.540 (or 54.0%)
33. (a) 0.25 (or 25%) (b) 3/4
35. (a) 0.672 (or 67.2%) (b) 58.8 kW
37. (a) 0.294 (or 29.4%) (b) 5.00×10^2 J (c) 1.67 kW
39. (a) 4.50×10^6 J (b) 2.84×10^7 J (c) 68.2 kg
41. 1/3
43. (a) 30.6% (b) 985 MW
45. 143 J/K
47. (a) -1.2 kJ/K (b) 1.2 kJ/K
49. 57.2 J/K
51. 3.27 J/K
53. (a)

End Result	Possible Tosses	Total Number of Same Result
All H	HHHH	1
1T, 3H	HHHT, HHTH, HTHH, THHH	4
2T, 2H	HHTT, HTHT, THHT, HTTH, THTH, TTHH	6
3T, 1H	TTTH, TTHT, THTT, HTTT	4
All T	TTTT	1

    (b) all H or all T (c) 2H and 2T
55. -6.5 MJ
57. 1 300 W
59. 18°C
61. (a) 12.2 kJ (b) 4.05 kJ (c) 8.15 kJ
63. (a) 26 J (b) 9.0×10^5 J (c) 9.0×10^5 J
65. (a) 2.49 kJ (b) 1.50 kJ (c) -990 J
67. (a) $T_Ⓐ = 1.20 \times 10^2$ K; $T_Ⓑ = 722$ K
    (b) 1.10×10^5 J (c) 7.50×10^4 J (d) 1.85×10^5 J
69. 0.146; 486 kcal
71. (a) 2.6×10^3 metric tons/day (b) \$$7.6 \times 10^6$/yr
    (c) 4.1×10^4 kg/s

CHAPTER 13

Quick Quizzes

1. (d)
2. (c)
3. (b)
4. (a)
5. (c)
6. (d)
7. (c), (b)
8. (a)
9. (b)

Example Questions

1. No. If a spring is stretched too far, it no longer satisfies Hooke's law and can become permanently deformed.
2. $k_{\text{eq}} = k_1 + k_2$
3. False
4. True
5. False
6. (b)
7. True
8. No
9. (a), (c)
10. The speed is doubled.

Warm-Up Exercises

1. (a) 5.00 m (b) -5.00 m (c) 0.250 s
3. 0.123 m
5. (a) 98.0 J (b) 0.327 m
7. (a) 18.8 rad/s (b) 3.00 Hz (c) 0.333 s (d) 1.78×10^3 N/m

9. (a) 0.400 Hz (b) 2.51 rad/s (c) 1.56 m
11. (a) 0.006 25 kg/m (b) 2.10 × 10² m/s

Conceptual Questions

1. No. Because the total energy is $E = \frac{1}{2}kA^2$, changing the mass of the object while keeping A constant has no effect on the total energy. When the object is at a displacement x from equilibrium, the potential energy is $\frac{1}{2}kx^2$, independent of the mass, and the kinetic energy is $KE = E - \frac{1}{2}kx^2$, also independent of the mass.

3. When the spring with two objects on opposite ends is set into oscillation in space, the coil at the exact center of the spring does not move. Thus, we can imagine clamping the center coil in place without affecting the motion. If we do this, we have two separate oscillating systems, one on each side of the clamp. The half-spring on each side of the clamp has twice the spring constant of the full spring, as shown by the following argument: The force exerted by a spring is proportional to the separation of the coils as the spring is extended. Imagine that we extend a spring by a given distance and measure the distance between coils. We then cut the spring in half. If one of the half-springs is now extended by the same distance, the coils will be twice as far apart as they were in the complete spring. Thus, it takes twice as much force to stretch the half-spring, from which we conclude that the half-spring has a spring constant which is twice that of the complete spring. Hence, our clamped system of objects on two half-springs will vibrate with a frequency that is higher than f by a factor of the square root of two.

5. We assume that the buoyant force acting on the sphere is negligible in comparison to its weight, even when the sphere is empty. We also assume that the bob is small compared with the pendulum length. Then, the frequency of the pendulum is $f = 1/T = (1/2\pi)\sqrt{g/L}$, which is independent of mass. Thus, the frequency will not change as the water leaks out.

7. (a) The bouncing ball is not an example of simple harmonic motion. The ball does not follow a sinusoidal function for its position as a function of time. (b) The daily movement of a student is also not simple harmonic motion, because the student stays at a fixed location, school, for a long time. If this motion were sinusoidal, the student would move more and more slowly as she approached her desk, and as soon as she sat down at the desk, she would start to move back toward home again.

9. The speed of a wave on a string is given by $v = \sqrt{F/\mu}$. This says the speed is independent of the frequency of the wave. Thus, doubling the frequency leaves the speed unaffected.

11. The kinetic energy is proportional to the square of the speed, and the potential energy is proportional to the square of the displacement. Therefore, both must be positive quantities.

Problems

1. (a) 17 N to the left (b) 28 m/s² to the left
3. (a) 6.58 N (b) 10.1 N
5. 0.242 kg
7. (a) 0.206 m (b) −0.042 1 m (c) The block oscillates around the unstretched position of the spring with an amplitude of 0.248 m.
9. (a) 60 J (b) 49 m/s
11. 2.94 × 10³ N/m
13. 0.478 m
15. (a) 1 630 N/m (b) 47.0 J (c) 7.90 kg (d) 2.57 m/s (e) 26.1 J (f) 20.9 J (g) −0.201 m
17. (a) 0.28 m/s (b) 0.26 m/s (c) 0.26 m/s (d) 3.5 cm
19. 39.2 N
21. (a) 1.53 J (b) 1.75 m/s (c) 1.51 m/s
23. The horizontal displacement is described by $x(t) = A \cos \omega t$, where A is the distance from the center of the wheel to the crankpin.
25. 0.63 s
27. (a) 1.0 s (b) 0.28 m/s (c) 0.25 m/s
29. (a) 5.98 m/s (b) 206 N/m (c) 0.238 m
31. (a) 11.0 N toward the left (b) 0.881 oscillations
33. $v = \pm \omega A \sin \omega t$, $a = -\omega^2 A \cos \omega t$
35. (a) 1.46 s (b) 9.59 m/s²
37. (a) slow (b) 9:47
39. (a) $L_{\text{Earth}} = 25$ cm (b) $L_{\text{Mars}} = 9.4$ cm (c) $m_{\text{Earth}} = 0.25$ kg (d) $m_{\text{Mars}} = 0.25$ kg
41. (a) 4.13 cm (b) 10.4 cm (c) 5.56 ×10⁻² s (d) 187 cm/s
43. (a) 5.45 × 10¹⁴ Hz (b) 1.83 × 10⁻¹⁵ s
45. 31.9 cm
47. 58.8 s
49. 80.0 N
51. 5.20 × 10² m/s
53. (a) 30.0 N (b) 25.8 m/s
55. 28.5 m/s
57. (a) 0.051 0 kg/m (b) 19.6 m/s
59. (a) 13.4 m/s (b) The worker could throw an object such as a snowball at one end of the line to set up a pulse and then use a stopwatch to measure the time it takes the pulse to travel the length of the line. From this measurement, the worker would have an estimate of the wave speed, which in turn can be used to estimate the tension.
61. (a) Constructive interference gives A = 0.50 m (b) Destructive interference gives A = 0.10 m
63. (a) 219 N/m (b) 6.12 kg
65. (a) 1.68 s (b) 16.8 N/m
67. (a) 588 N/m (b) 0.700 m/s
69. 6.62 cm
71. (a) Using s for the displacement from equilibrium along the arc, the restoring force on the balloon takes the form of Hooke's law: $F_{\text{tangential}} = -[(\rho_{\text{air}} - \rho_{\text{He}})Vg/L]s$ (b) $T = 1.40$ s
75. (a) 15.8 rad/s (b) 5.23 cm

CHAPTER 14

Quick Quizzes

1. (c)
2. (c)
3. (b)
4. (b), (e)
5. (d)
6. (a)
7. (b)

Example Questions

1. Rubber is easier to compress than solid aluminum, so aluminum must have the larger bulk modulus and by Equation 14.1, a higher sound speed.
2. 3.0 dB
3. You should increase your distance from the sound source by a factor of 5.
4. Yes. It changes because the speed of sound changes with temperature. Answer (b) is correct.
5. No
6. True
7. True
8. (b)
9. True
10. True
11. The notes are so different from each other in frequency that the beat frequency is very high and cannot be distinguished.

Warm-Up Exercises

1. (a) 0.699 (b) 33.3
3. 0.780 m
5. 364 m/s
7. 3.16×10^{-2} W/m^2
9. (a) 352 m/s (b) 267 Hz
11. (a) 1.47 m (b) 0.735 m
13. (a) 148 Hz (b) 444 Hz (c) 296 Hz

Conceptual Questions

1. (a) higher (b) lower
3. Yes. The speed of sound in air is proportional to the square root of the absolute temperature, $\sqrt{T}$. The speed of sound is greater in warmer air, so the pulse from the camera would return sooner than it would on a cooler day from an object at the same distance. The camera would interpret an object as being closer than it actually is on a hot day.
5. Sophisticated electronic devices break the frequency range used in telephone conversations into several frequency bands and then mix them in a predetermined pattern so that they become unintelligible. The descrambler moves the bands back into their proper order.
7. (a) The echo is Doppler shifted, and the shift is like both a moving source and a moving observer. The sound that leaves your horn in the forward direction is Doppler shifted to a higher frequency, because it is coming from a moving source. As the sound reflects back and comes toward you, you are a moving observer, so there is a second Doppler shift to an even higher frequency. (b) If the sound reflects from the spacecraft coming toward you, there is a different moving-source shift to an even higher frequency. The reflecting surface of the spacecraft acts as a moving source.
9. The bowstring is pulled away from equilibrium and released, in a manner similar to the way a guitar string is pulled and released when it is plucked. Thus, standing waves will be excited in the bowstring. If the arrow leaves from the exact center of the string, then a series of odd harmonics will be excited. Even harmonics will not be excited, because they have a node at the point where the string exhibits its maximum displacement.
11. The two engines are running at slightly different frequencies, thus producing a beat frequency between them.

Problems

1. (a) 5.56 km (b) No. The speed of light is much greater than the speed of sound, so the time interval required for the light to reach you is negligible compared to the time interval for the sound.
3. 358 m/s
5. 515 m
7. The wavelength decreases by 2.5 mm.
9. (a) The pulse that travels through the rail (b) 23.4 ms
11. 150 dB
13. 3.0×10^{-8} W/m^2
15. (a) 1.00×10^{-2} W/m^2 (b) 105 dB
17. 37 dB
19. (a) 1.3×10^2 W (b) 96 dB
23. (a) 75 Hz drop (b) 0.948 m
25. 596 Hz
27. 9.09 m/s
29. 19.7 m
31. (a) 56.3 s (b) 56.6 km farther along
33. At distances of 0.089 m, 0.303 m, 0.518 m, 0.732 m, 0.947 m, and 1.16 m from either speaker
35. 800 m
37. (a) 0.240 m (b) 0.855 m
39. (a) Nodes at 0, 2.67 m, 5.33 m, and 8.00 m; antinodes at 1.33 m, 4.00 m, and 6.67 m (b) 18.6 Hz
41. 378 Hz
43. (a) 1.85×10^{-2} kg/m (b) 90.6 m/s (c) 152 N (d) 2.20 m (e) 8.33 m
45. (a) 73.9 N (b) 211 Hz
47. 19.976 kHz
49. 58 Hz
51. 3.1 kHz
53. (a) 0.552 m (b) 316 Hz
55. 5.64 beats/s
57. 3.85 m/s away from the station or 3.77 m/s toward the station
59. (a) 1.99 beats/s (b) 3.38 m/s
61. 1.76 cm
63. (a) 0.642 W (b) 0.43%
65. 67.0 dB
67. $r_1 = 3.3$ m and $r_2 = 32.7$ m
69. 262 kHz
71. 64 dB
73. (a) 439 Hz (b) 441 Hz
75. 1.34×10^4 N

CHAPTER 15

Quick Quizzes

1. (b)
2. (b)
3. (c)
4. (a)
5. (c) and (d)
6. (a)
7. (c)
8. (b)
9. (d)
10. (b) and (d)

Example Questions

1. $\frac{1}{4}$
2. (a)
3. Fourth quadrant
4. The suspended droplet would accelerate downward at twice the acceleration of gravity.
5. The angle ϕ would increase.
6. (c)
7. The charge on the inner surface would be negative.
8. Zero

Warm-Up Exercises

1. (a) 8.32 N (b) 147°
3. (a) 7.50 N (b) −16.0 N (c) 17.7 N (d) −64.9°
5. (a) 2.97 N (b) 34.3°
7. (a) 2.85 N (b) 5.70 m/s^2
9. (a) 5.00 m (b) 53.1° (c) 5.75 N/C (d) 3.45 N/C (e) 4.60 N/C
11. (a) -1.47×10^{-14} N (b) 1.44×10^{-14} N (c) -0.200 m/s^2
13. 198 N·m^2/C

Conceptual Questions

1. Electrons have been removed from the glass object. Negative charge has been removed from the initially neutral rod, resulting in a net positive charge on the rod. The protons cannot be removed from the rod; protons are not mobile because they are within the nuclei of the atoms of the rod.
3. No. The charge on the metallic sphere resides on its outer surface, so the person is able to touch the surface without causing any charge transfer.
5. No. Life would be no different if electrons were + charged and protons were − charged. Opposite charges would still attract, and like charges would repel. The naming of + and − charge is merely a convention.

7. Move an object A with a net positive charge so it is near, but not touching, a neutral metallic object B that is insulated from the ground. The presence of A will polarize B, causing an excess negative charge to exist on the side nearest A and an excess positive charge of equal magnitude to exist on the side farthest from A. While A is still near B, touch B with your hand. Additional electrons will then flow from ground, through your body, and onto B. With A continuing to be near but not in contact with B, remove your hand from B, thus trapping the excess electrons on B. When A is now removed, B is left with excess electrons, or a net negative charge. By means of mutual repulsion, this negative charge will now spread uniformly over the entire surface of B.

9. She is not shocked. She becomes part of the dome of the Van de Graaff, and charges flow onto her body. They do not jump to her body via a spark, however, so she is not shocked.

11. The dry paper is initially neutral. The comb attracts the paper because its electric field causes the molecules of the paper to become polarized—the paper as a whole cannot be polarized because it is an insulator. Each molecule is polarized so that its unlike-charged side is closer to the charged comb than its like-charged side, so the molecule experiences a net attractive force toward the comb. Once the paper comes in contact with the comb, like charge can be transferred from the comb to the paper, and if enough of this charge is transferred, the like-charged paper is then repelled by the like-charged comb.

13. The electric shielding effect of conductors depends on the fact that there are two kinds of charge: positive and negative. As a result, charges can move within the conductor so that the combination of positive and negative charges establishes an electric field that exactly cancels the external field within the conductor and any cavities inside the conductor. There is only one type of gravitation charge, however, because there is no negative mass. As a result, gravitational shielding is not possible. A room cannot be gravitationally shielded because mass is always positive or zero, never negative.

15. You can only conclude that the net charge inside the Gaussian surface is positive.

17. (b)

Problems

1. (a) 8.74×10^{-8} N (b) repulsive
3. (a) 2.36×10^{-5} C (b) The charges induce opposite charges in the bulkheads, but the induced charge in the bulkhead near ball B is greater because of ball B's greater charge. The system therefore moves slowly toward the bulkhead closer to ball B.
5. (a) 36.8 N (b) 5.54×10^{27} m/s^2
7. 49.3 N/m
9. (a) 2.2×10^{-5} N (attraction) (b) 9.0×10^{-7} N (repulsion)
11. 1.38×10^{-5} N at 77.5° below the negative x-axis
13. 0.437 N at −85.3° from the +x-axis.
15. 7.2 nC
17. 2.07×10^3 N/C down
19. 3.15 N due north
21. (a) $\dfrac{mg}{|Q|}\sin\theta$ (b) 3.19×10^3 N/C down the incline
23. (a) 6.12×10^{10} m/s^2 (b) 19.6 μs (c) 11.8 m (d) 1.20×10^{-15} J
25. $E_x = (1 - \sqrt{2})k_e\dfrac{Q}{d^2}$, $E_y = \sqrt{2}k_e\dfrac{Q}{d^2}$
27. 1.8 m to the left of the −2.5-μC charge
29. zero
35. (a) 0 (b) 5 μC inside, −5 μC outside (c) 0 inside, −5 μC outside (d) 0 inside, −5 μC outside
37. 1.3×10^{-3} C
39. (a) 2.54×10^{-15} N (b) 1.59×10^4 N/C radially outward
41. (a) 858 N·m^2/C (b) 0 (c) 657 N·m^2/C
43. $-Q/\epsilon_0$ for S_1; 0 for S_2; $-2Q/\epsilon_0$ for S_3; 0 for S_4
45. (a) 0 (b) k_eq/r^2 outward
47. (a) −7.99 N/C (b) 0 (c) 1.44 N/C (d) 2.00 nC on the inner surface; 1.00 nC on the outer surface
49. 115 N
51. 24 N/C in the positive x-direction
53. (a) $E = 2k_eqb\,(a^2 + b^2)^{-3/2}$ in the positive x-direction (b) $E = k_eQb(a^2 + b^2)^{-3/2}$ in the positive x-direction
55. $F_x = (0.354)k_e\dfrac{Q^2}{d^2}$; $F_y = (1.65)k_e\dfrac{Q^2}{d^2}$
57. 4.4×10^5 N/C
59. 1.14×10^{-7} C on one sphere and 5.69×10^{-8} C on the other
61. (a) 0 (b) 7.99×10^7 N/C (outward) (c) 0 (d) 7.34×10^6 N/C (outward)
63. (a) 1.00×10^3 N/C (b) 3.37×10^{-8} s (c) accelerate at 1.76×10^{14} m/s^2 in the direction opposite that of the electric field

CHAPTER 16

Quick Quizzes

1. (b)
2. (a)
3. (b)
4. (d)
5. (d)
6. (c)
7. (a)
8. (c)
9. (a) C decreases. (b) Q stays the same. (c) E stays the same. (d) ΔV increases. (e) The energy stored increases.
10. (a) C increases. (b) Q increases. (c) E stays the same. (d) ΔV remains the same. (e) The energy stored increases.
11. (a)

Example Questions

1. True
2. True
3. False
4. (a)
5. Three
6. Each answer would be reduced by a factor of one-half.
7. One-quarter
8. The voltage drop is the smallest across the 24-μF capacitor and largest across the 3.0-μF capacitor.
9. The 3.0-μF capacitor
10. (c)
11. (b)
12. $C = \dfrac{\epsilon_0 A}{d}$

Warm-Up Exercises

1. 1.96 Ω
3. (a) 9.00×10^{-6} J (b) -9.00×10^{-6} J (c) −4.50 V
5. (a) 2.40×10^{-9} V (b) 3.84×10^{-28} J
7. (a) 2.70×10^{-5} C (b) 24.0 V
9. (a) 5.00 μF (b) 1.20 μF
11. (a) 8.11 μF (b) 9.73×10^{-5} C (c) 8.11 V (d) 3.89 V

Conceptual Questions

1. (a) The proton moves in a straight line with constant acceleration in the direction of the electric field. (b) As its velocity increases, its kinetic energy increases, and the electric potential energy associated with the proton decreases.
3. The work done in pulling the capacitor plates farther apart is transferred into additional electric energy stored in the

capacitor. The charge is constant, and the capacitance decreases, but the potential difference between the plates increases, which results in an increase in the stored electric energy.

5. If the power line makes electrical contact with the metal of the car, it will raise the potential of the car to 20 kV. It will also raise the potential of your body to 20 kV, because you are in contact with the car. In itself, this is not a problem. If you step out of the car, however, your body at 20 kV will make contact with the ground, which is at zero volts. As a result, a current will pass through your body, and you will likely be injured. Thus, it is best to stay in the car until help arrives.

7. If two points on a conducting object were at different potentials, then free charges in the object would move, and we would not have static conditions, in contradiction to the initial assumption. (Free positive charges would migrate from locations of higher to locations of lower potential. Free electrons would rapidly move from locations of lower to locations of higher potential.) All of the charges would continue to move until the potential became equal everywhere in the conductor.

9. (a) The capacitor often remains charged long after the voltage source is disconnected. This residual charge can be lethal. (b) The capacitor can be safely handled after discharging the plates by short-circuiting the device with a conductor, such as a screwdriver with an insulating handle.

11. $D > C > B > A$

13. Not all connections are simple combinations of series and parallel circuits. As an example of such a complex circuit, consider the network of five capacitors, C_1, C_2, C_3, C_4, and C_5 shown below

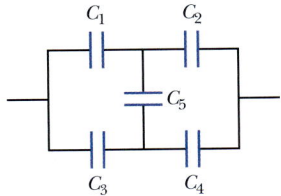

This combination cannot be reduced to a simple equivalent by the techniques of combining series and parallel capacitors.

Problems

1. (a) 1.92×10^{-18} J (b) -1.92×10^{-18} J (c) 2.05×10^6 m/s in the negative x-direction
3. 1.4×10^{-20} J
5. 1.7×10^6 N/C
7. (a) 1.13×10^5 N/C (b) 1.80×10^{-14} N (c) 4.37×10^{-17} J
9. (a) The spring stretches by 4.36 cm.
   (b) Equilibrium: $x = 2.18$ cm; $A = 2.18$ cm (c) $\Delta V = -\dfrac{2kA^2}{Q}$
11. (a) -5.75×10^{-7} V (b) -1.92×10^{-7} V, $\Delta V = 3.84 \times 10^{-7}$ V
    (c) No. Unless fixed in place, the electron would move in the opposite direction, increasing its distance from points A and B and lowering the potential difference between them.
13. (a) 2.67×10^6 V (b) 2.13×10^6 V
15. (a) 103 V (b) -3.85×10^{-7} J; positive work must be done to separate the charges.
17. -11.0 kV
19. (a) 3.84×10^{-14} J (b) 2.55×10^{-13} J (c) -2.17×10^{-13} J
    (d) 8.08×10^6 m/s (e) 1.24×10^7 m/s
21. (a) 0.294 J (b) 271 m/s
23. 2.74×10^{-14} m
25. (a) 1.1×10^{-8} F (b) 27 C
27. (a) 1.36 pF (b) 16.3 pC (c) 8.00×10^3 V/m
29. (a) 11.1 kV/m toward the negative plate (b) 3.74 pF
    (c) 74.8 pC and -74.8 pC
31. (a) 5.90×10^{-10} F (b) 3.54×10^{-9} C (c) 2.00×10^3 N/C
    (d) 1.77×10^{-8} C/m² (e) All the answers are reduced.
33. (a) 10.7 μC on each capacitor (b) 15.0 μC on the 2.50-μF capacitor and 37.5 μC on the 6.25-μF capacitor
35. (a) 2.67 μF (b) 24.0 μC on each 8.00-μF capacitor, 18.0 μC on the 6.00-μF capacitor, 6.00 μC on the 2.00-μF capacitor
    (c) 3.00 V across each capacitor
37. (a) 3.33 μF (b) 180 μC on the 3-μF and the 6-μF capacitors, 120 μC on the 2.00-μF and 4.00-μF capacitors (c) 60.0 V across the 3-μF and the 2-μF capacitors, 30.0 V across the 6-μF and the 4-μF capacitors
39. $Q_1 = 16.0$ μC, $Q_5 = 80.0$ μC, $Q_8 = 64.0$ μC, $Q_4 = 32.0$ μC
41. (a) $Q_{25} = 1.25$ mC, $Q_{40} = 2.00$ mC (b) $Q'_{25} = 288$ μC, $Q'_{40} = 462$ μC (c) $\Delta V = 11.5$ V
43. $Q'_1 = 3.33$ μC, $Q'_2 = 6.67$ μC
45. 3.24×10^{-4} J
47. (a) 54.0 μJ (b) 108 μJ (c) 27.0 μJ
49. (a) $\kappa = 3.4$. The material is probably nylon (see Table 16.1).
    (b) The voltage would lie somewhere between 25.0 V and 85.0 V.
51. (a) 8.1 nF (b) 2.4 kV
53. (a) volume 9.09×10^{-16} m³, area 4.54×10^{-10} m²
    (b) 2.01×10^{-13} F (c) 2.01×10^{-14} C, 1.26×10^5 electronic charges
59. 0.188 m²
61. 6.25 μF
63. 4.47 kV
65. 0.75 mC on C_1, 0.25 mC on C_2
67. Sphere A: 0.800 μC; sphere B: 1.20 μC

CHAPTER 17

Quick Quizzes

1. (d)
2. (b)
3. (c), (d)
4. (b)
5. (b)
6. (b)
7. (a)
8. (b)
9. (a)
10. (c)

Example Questions

1. No. Such a current corresponds to the passage of one electron every 2 seconds. The average current, however, can have any value.
2. True
3. Higher
4. (b)
5. (a)
6. (c)

Warm-Up Exercises

1. 10.4 W
3. (a) 3.39×10^4 m/s (b) 3.59×10^4 m/s
5. (a) 3.30×10^{-6} m² (b) 2.24×10^{-4} m/s
7. 3.9×10^2 m
9. (a) 1.20×10^{-2} A (b) 0.144 W (c) 518 J

Conceptual Questions

1. In the electrostatic case in which charges are stationary, the electric field inside a conductor must be zero. A non-zero field would produce a current (by interacting with the free electrons in the conductor), which would violate the condition of static equilibrium. In this chapter we deal with conductors that carry current, a nonelectrostatic

situation. The current arises because of a potential difference applied between the ends of the conductor, which produces an internal electric field.

3. Because there are so many electrons in a conductor (approximately 10^{28} electrons/m^3), the average velocity of charges is very slow. When you connect a wire to a potential difference, you establish an electric field everywhere in the wire nearly instantaneously, to make electrons start drifting everywhere all at once.

5. We would conclude that the conductor is nonohmic.

7. A voltage is not something that "surges through" a completed circuit. A voltage is a potential difference that is applied across a device or a circuit. It would be more correct to say "1 ampere of electricity surged through the victim's body." Although this amount of current would have disastrous results on the human body, a value of 1 (ampere) doesn't sound as exciting for a newspaper article as 10 000 (volts). Another possibility is to write "10 000 volts of electricity were applied across the victim's body," which still doesn't sound quite as exciting.

9. The drift velocity might increase steadily as time goes on, because collisions between electrons and atoms in the wire would be essentially nonexistent and the conduction electrons would move with constant acceleration. The current would rise steadily without bound also, because I is proportional to the drift velocity.

11. Once the switch is closed, the line voltage is applied across the bulb. As the voltage is applied across the cold filament when it is first turned on, the resistance of the filament is low, the current is high, and a relatively large amount of power is delivered to the bulb. As the filament warms, its resistance rises and the current decreases. As a result, the power delivered to the bulb decreases. The large current spike at the beginning of the bulb's operation is the reason that lightbulbs often fail just after they are turned on.

Problems

1. (a) 3.00×10^{20} electrons (b) the direction opposite to the current
3. 1.05 mA
5. 0.678 mm
7. 27 yr
9. (a) 55.85×10^{-3} kg/mol (b) 1.41×10^5 mol/m^3 (c) 8.49×10^{28} iron atoms/m^3 (d) 1.70×10^{29} conduction electrons/m^3 (e) 2.21×10^{-4} m/s
11. 32 V is 200 times larger than 0.16 V
13. (a) 13.0 Ω (b) 17.0 m
15. (a) 30 Ω (b) 4.7×10^{-4} Ω · m
17. silver ($\rho = 1.59 \times 10^{-8}$ Ω · m)
19. 256 Ω
23. 2.0 A
25. 2.71 MΩ
27. 26 mA
29. (a) 3.0 A (b) 2.9 A
31. (a) 1.2 Ω (b) 8.0×10^{-4} (a 0.080% increase)
33. (a) 8.33 A (b) 14.4 Ω
35. (a) 2.1 W (b) 3.42 W (c) The aluminum wire would not be as safe. If surrounded by thermal insulation, it would get much hotter than a copper wire.
37. 11.2 min
39. (a) $R_A = 576$ Ω; $R_B = 144$ Ω (b) 4.80 s. (c) The charge is the same. It is at a location that is lower in potential. (d) 0.040 0 s (e) Energy enters the lightbulb by electric transmission and leaves by heat and electromagnetic radiation. (f) $1.98
41. 1.6 cm
43. 15.0 μW
45. (a) 116 V (b) 12.8 kW
47. $\dfrac{d_A}{d_B} = \sqrt{3}$
49. 2.16×10^5 C
51. (a) 1.50 Ω (b) 6.00 A
53. 1.1 km
55. 1.47×10^{-6} Ω · m; differs by 2.0% from value in Table 17.1
57. (a) $3.06 (b) No. The circuit must be able to handle at least 26 A.
59. (a) 17.3 A (b) 22.4 MJ (c) $0.684
61. 3.77×10^{28}/m^3
63. 37 MΩ
65. 0.48 kg/s

CHAPTER 18

Quick Quizzes

1. True
2. Because of the battery's internal resistance, power is delivered to the battery material, raising its temperature.
3. (b)
4. (a)
5. (a)
6. (b)
7. *Parallel:* (a) unchanged (b) unchanged (c) increase (d) decrease
8. *Series:* (a) decrease (b) decrease (c) decrease (d) increase
9. (c)

Example Questions

1. As the resistors get warmer, their resistance increases, reducing the current.
2. (b)
3. The 8.00-Ω resistor
4. The answers would be negative, but they would have the same magnitude as before.
5. Yes
6. (a)
7. (a)

Warm-Up Exercises

1. (a) 9.00 A (b) 18.0 A
3. (a) 2.46 C (b) 0.183 s
5. (a) 2.50 μF (b) 0.400 μF
7. (a) 2.71 A (b) 8.12 V (c) 24.9 W
9. (a) 1.56 Ω (b) 37.0 Ω
11. (a) 1.94 Ω (b) 3.94 Ω (c) 3.05 A (d) 6.10 V (e) 5.90 V (f) 0.590 A
13. (a) 9.90×10^{-3} s (b) 4.32×10^{-4} C (c) 1.71×10^{-4} C

Conceptual Questions

1. No. When a battery serves as a source and supplies current to a circuit, the direction of the current is from the negative terminal of the battery to the positive one. However, when a source having a larger emf than the battery is used to charge the battery, the direction of the current is from the positive terminal of the battery to the negative one.

3. The total amount of energy delivered by the battery will be less than W. Recall that a battery can be considered an ideal, resistanceless battery in series with the internal resistance. When the battery is being charged, the energy delivered to it includes the energy necessary to charge the ideal battery, plus the energy that goes into raising the temperature of the battery due to I^2r heating in the internal resistance. This latter energy is not available during discharge of the battery, when part of the reduced available energy again transforms into internal energy in the internal resistance, further reducing the available energy below W.

5. Connecting batteries in parallel does not increase the emf. A high-current device connected to two batteries in parallel can draw currents from both batteries. Thus, connecting the

batteries in parallel increases the possible current output and, therefore, the possible power output.

7. The starter in the automobile draws a relatively large current from the battery. This large current causes a significant voltage drop across the internal resistance of the battery. As a result, the terminal voltage of the battery is reduced, and the headlights dim accordingly.

9. The bird is resting on a wire of fixed potential. In order to be electrocuted, a large potential difference is required between the bird's feet. The potential difference between the bird's feet is too small to harm the bird.

11. She will not be electrocuted if she holds onto only one high-voltage wire, because she is not completing a circuit. There is no potential difference across her body as long as she clings to only one wire. However, she should release the wire immediately once it breaks, because she will become part of a closed circuit when she reaches the ground or comes into contact with another object.

13. The junction rule is a statement of conservation of charge. It says that the amount of charge that enters a junction in some time interval must equal the charge that leaves the junction in that time interval. The loop rule is a statement of conservation of energy. It says that the increases and decreases in potential around a closed loop in a circuit must add to zero.

Problems

1. $4.92\ \Omega$
3. (a) Your circuit diagram will consist of two $0.800\text{-}\Omega$ resistors in series with the $192\text{-}\Omega$ resistance of the bulb. (b) 73.8 W
5. (a) $17.1\ \Omega$ (b) 1.99 A for $4.00\ \Omega$ and $9.00\ \Omega$, 1.17 A for $7.00\ \Omega$, 0.820 A for $10.0\ \Omega$
7. $470\ \Omega$ and $220\ \Omega$
9. (a) 5.68 V (b) 0.227 A
11. $55\ \Omega$
13. 0.43 A
15. (a) Connect two $50\text{-}\Omega$ resistors in parallel, and then connect this combination in series with a $20\text{-}\Omega$ resistor. (b) Connect two $50\text{-}\Omega$ resistors in parallel, connect two 20-V resistors in parallel, and then connect these two combinations in series with each other.
17. 0.714 A, 1.29 A, 12.6 V
19. 50.0 mA from a to e
21. (a) $I_1 = 1.00$ A upward in $200\ \Omega$; $I_4 = 4.00$ A upward in $70.0\ \Omega$; $I_2 = 3.00$ A upward in $80.0\ \Omega$; $I_3 = 8.00$ A downward in $20.0\ \Omega$; (b) 200 V
23. (a) 0.385 mA, 3.08 mA, 2.69 mA (b) 69.2 V, with c at the higher potential
25. (a) No. The only simplification is to note that the $2.0\text{-}\Omega$ and $4.0\text{-}\Omega$ resistors are in series and add to a resistance of $6.0\ \Omega$. Likewise, the $5.0\text{-}\Omega$ and $1.0\text{-}\Omega$ resistors are in series and add to a resistance of $6.0\ \Omega$. The circuit cannot be simplified any further. Kirchhoff's rules must be used to analyze the circuit. (b) $I_1 = 3.5$ A, $I_2 = 2.5$ A, $I_3 = 1.0$ A
27. (a) No. The multiloop circuit cannot be simplified any further. Kirchhoff's rules must be used to analyze the circuit. (b) $I_{30} = 0.353$ A directed to the right, $I_5 = 0.118$ A directed to the right, $I_{20} = 0.471$ A directed to the left
29. $\Delta V_2 = 3.05$ V, $\Delta V_3 = 4.57$ V, $\Delta V_4 = 7.38$ V, $\Delta V_5 = 1.62$ V
31. (a) 1.88 s (b) 1.90×10^{-4} C
33. (a) 5.00 s (b) 150 μC (c) 4.06 μA
35. (a) 0.432 s (b) 6.00 μF
37. 48 lightbulbs
39. (a) 6.25 A (b) 750 W
41. (a) 1.2×10^{-9} C, 7.3×10^9 K^+ ions. Not large, only $1e/(290\ \text{Å})^2$. (b) 1.7×10^{-9} C, 1.0×10^{10} Na^+ ions (c) 0.83 mA (d) 7.5×10^{-12} J
43. 11 nW
45. (a) 4.00 V (b) Point a is at the higher potential.

47. (a) $15\ \Omega$ (b) $I_1 = 1.0$ A, $I_2 = I_3 = 0.50$ A, $I_4 = 0.30$ A, and $I_5 = 0.20$ A (c) $(\Delta V)_{ac} = 6.0$ V, $(\Delta V)_{ce} = 1.2$ V, $(\Delta V)_{ed} = (\Delta V)_{fd} = 1.8$ V, $(\Delta V)_{cd} = 3.0$ V, $(\Delta V)_{db} = 6.0$ V (d) $P_{ac} = 6.0$ W, $P_{ce} = 0.60$ W, $P_{ed} = 0.54$ W, $P_{fd} = 0.36$ W, $P_{cd} = 1.5$ W, $P_{db} = 6.0$ W
49. $6.00\ \Omega$, $3.00\ \Omega$
51. (a) $R_{eq}^{open} = 3R$, $R_{eq}^{closed} = 2R$ (b) $P = \dfrac{\mathcal{E}^2}{3R}$, $P = \dfrac{\mathcal{E}^2}{2R}$
(c) Lamps A and B increase in brightness, lamp C goes out.
53. (a) 1.02 A down (b) 0.364 A down (c) 1.38 A up (d) 0 (e) 66.0 μC
55. (a) $R_x = R_2 - \tfrac{1}{4}R_1$ (b) $R_x = 2.8\ \Omega$ (inadequate grounding)
59. $P = \dfrac{(144\ \text{V}^2)R}{(R + 10.0\ \Omega)^2}$

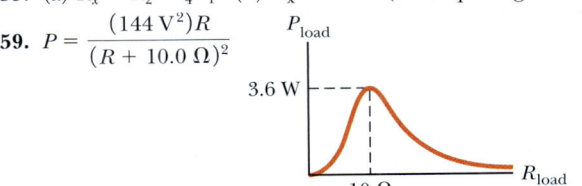

61. (a) 6.0×10^2 W (b) 1.2 J (c) 37 m
63. 0.395 A; 1.50 V
65. (a) 2.41×10^{-5} C (b) 1.61×10^{-5} C (c) 1.61×10^{-2} A

CHAPTER 19

Quick Quizzes

1. (b)
2. (c)
3. (c)
4. (a)
5. (a), (c)
6. (b)

Example Questions

1. The force on the electron is opposite the direction of the force on the proton, and the acceleration of the electron is far greater than the acceleration on the proton due to the electron's lower mass.
2. A magnetic field always exerts a force perpendicular to the velocity of a charged particle, so it can change the particle's direction but not its speed.
3. Zero
4. As the angle approaches 90°, the magnitude of the force increases. After going beyond 90°, it decreases.
5. The magnitude of the momentum remains constant. The direction of momentum changes, unless the particle's velocity is parallel or antiparallel to the magnetic field.
6. 20 m
7. (a) to the left (b) to the right
8. Earth's magnetic field is extremely weak, so the current carried by the car would have to be correspondingly very large. Such a current would be impractical to generate, and if generated, it would heat and melt the wires carrying it.
9. The proton would accelerate downward due to gravity while circling.

Warm-Up Exercises

1. (a) 1.19×10^3 N·m (b) 1.37×10^3 N·m
3. 0.200 T
5. 6.00×10^{-3} N·m
7. 4.79×10^3 m/s
9. 6.03×10^{-4} T

Conceptual Questions

1. The set should be oriented such that the beam is moving either toward the east or toward the west.

3. The magnetic force on a moving charged particle is always perpendicular to the particle's direction of motion. There is no magnetic force on the charge when it moves parallel to the direction of the magnetic field. However, the force on a charged particle moving in an electric field is never zero and is always parallel to the direction of the field. Therefore, by projecting the charged particle in different directions, it is possible to determine the nature of the field.

5. In the figure, the magnetic field created by wire 1 at the position of wire 2 is into the paper. Hence, the magnetic force on wire 2 is in direction (current down) × (field into the paper) = (force to the right), away from wire 1. Now wire 2 creates a magnetic field into the page at the location of wire 1, so wire 1 feels force (current up) × (field into the paper) = (force to the left), away from wire 2.

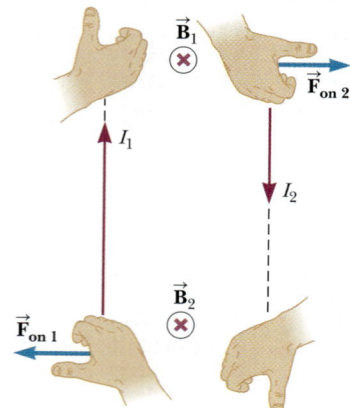

7. Such levitation could never occur. At the North Pole, where Earth's magnetic field is directed downward, toward the equivalent of a buried south pole, a coffin would be repelled if its south magnetic pole were directed downward. However, equilibrium would be only transitory, as any slight disturbance would upset the balance between the magnetic force and the gravitational force.

9. A compass does not detect currents in wires near light switches, for two reasons. The first is that, because the cable to the light switch contains two wires, one carrying current to the switch and the other carrying it away from the switch, the net magnetic field would be very small and would fall off rapidly with increasing distance. The second reason is that the current is alternating at 60 Hz. As a result, the magnetic field is oscillating at 60 Hz also. This frequency would be too fast for the compass to follow, so the effect on the compass reading would average to zero.

11. If you were moving along with the electrons, you would measure a zero current for the electrons, so they would not produce a magnetic field according to your observations. However, the fixed positive charges in the metal would now be moving backward relative to you, creating a current equivalent to the forward motion of the electrons when you were stationary. Thus, you would measure the same magnetic field as when you were stationary, but it would be due to the positive charges presumed to be moving from your point of view.

13. Each coil of the Slinky® will become a magnet, because a coil acts as a current loop. The sense of rotation of the current is the same in all coils, so each coil becomes a magnet with the same orientation of poles. Thus, all of the coils attract, and the Slinky® will compress.

15. There is no net force on the wires, but there is a torque. To understand this distinction, imagine a fixed vertical wire and a free horizontal wire (see the figure below). The vertical wire carries an upward current and creates a magnetic field that circles the vertical wire itself. To the right, the magnetic field of the vertical wire points into the page, while on the left side it points out of the page, as indicated. Each segment of the horizontal wire (of length ℓ) carries current that interacts with the magnetic field according to the equation $F = BI\ell \sin \theta$. Apply the right-hand rule on the right side: point the fingers of your right hand in the direction of the horizontal current and curl them into the page in the direction of the magnetic field. Your thumb points downward, the direction of the force on the right side of the wire. Repeating the process on the left side gives a force upward on the left side of the wire. The two forces are equal in magnitude and opposite in direction, so the net force is zero, but they create a net torque around the point where the wires cross.

17. If they are projected in the same direction into the same magnetic field, the charges are of opposite sign.
19. (b)

Problems

1. (a) west   (b) zero deflection   (c) up   (d) down
3. (a) into the page   (b) toward the right   (c) toward the bottom of the page
5. (a) 1.44×10^{-12} N   (b) 8.62×10^{14} m/s²   (c) A force would be exerted on the electron that had the same magnitude as the force on a proton, but in the opposite direction because of its negative charge. (d) The magnitude of the acceleration of the electron would be much greater than that of the proton because the mass of the electron is much smaller. The electron's acceleration would also be in the opposite direction.
7. 2.83×10^7 m/s west
9. 48.9° or 131°
11. $F_g = 8.93 \times 10^{-30}$ N (downward), $F_e = 1.60 \times 10^{-17}$ N (upward), $F_m = 4.80 \times 10^{-17}$ N (downward)
13. 8.0×10^{-3} T in the +z-direction
15. (a) into the page   (b) toward the right   (c) toward the bottom of the page
17. 7.50 N
19. 0.131 T (downward)
21. (a) The magnetic force and the force of gravity both act on the wire. When the magnetic force is upward and balances the downward force of gravity, the net force on the wire is zero, and the wire can move upward at constant velocity. (b) 0.20 T out of the page (c) If the field exceeds 0.20 T, the upward magnetic force exceeds the downward force of gravity, so the wire accelerates upward.
23. ab: 0, bc: 0.040 0 N in −x-direction, cd: 0.040 0 N in the −z-direction, da: 0.056 6 N parallel to the xz-plane and at 45° to both the +x- and the +z-directions
25. 118 µN · m
27. 9.05×10^{-4} N · m, tending to make the left-hand side of the loop move toward you and the right-hand side move away.
29. (a) 0.56 A   (b) 0.064 N · m
31. (a) 3.97°   (b) 3.39×10^{-3} N·m
33. (a) 8.51×10^5 m/s   (b) 2.06×10^{-5} m
37. 1.77 cm
39. $r = 3R/4$
41. (a) 2.08×10^{-7} kg/C   (b) 6.66×10^{-26} kg   (c) Calcium

43. 20.0 μT
45. 2.0×10^{-10} A
47. 2.4 mm
49. 20.0 μT toward the bottom of page
51. 0.167 mT out of the page
53. (a) 4.00 m (b) 7.50 nT (c) 1.26 m (d) zero
55. (a) 3.50×10^{-5} N/m (b) repulsive
57. 4.5 mm
59. 31.8 mA
61. (a) 920 turns (b) 12 cm
63. (a) 2.8 μT (b) 0.89 mA
65. (a) 5.24 μT (b) into the page (c) 7.20 cm
67. (a) 0.500 μT out of the page (b) 3.88 μT parallel to xy-plane and at 59.0° clockwise from $+x$-direction
69. (a) 1.33 m/s (b) The sign of the emf is independent of the charge.
71. 53 μT toward the bottom of the page, 20 μT toward the bottom of the page, and 0
73. (a) -8.00×10^{-21} kg·m/s (b) 8.90°
75. 1.41×10^{-6} N

CHAPTER 20

Quick Quizzes

1. b, c, a
2. (c)
3. (b)
4. (a)
5. (b)
6. (b)

Example Questions

1. False
2. 5.06×10^{-3} V; the current would be in the opposite direction.
3. (a), (e)
4. A magnetic force directed to the right will be exerted on the bar.
5. Doubling the frequency doubles the maximum induced emf.
6. (b)
7. (a)
8. 72.6 A
9. (b)
10. False

Warm-Up Exercises

1. (a) 0.233 Hz (b) 1.46 rad/s
3. 2.78×10^{-2} s
5. (a) 6.40×10^{-4} V (b) 0.533 A
7. (a) 135 V (b) 13.5 A
9. 1.60×10^{-6} s

Conceptual Questions

1. As the spacecraft moves through space, it is apparently moving from a region of one magnetic field strength to a region of a different magnetic field strength. The changing magnetic field through the coil induces an emf and a corresponding current in the coil.
3. According to Faraday's law, an emf is induced in a wire loop if the magnetic flux through the loop changes with time. In this situation, an emf can be induced either by rotating the loop around an arbitrary axis or by changing the shape of the loop.
5. If the bar were moving to the left, the magnetic force on the negative charges in the bar would be upward, causing an accumulation of negative charges on the top and positive charges at the bottom. Hence, the electric field in the bar would be upward, as well.
7. If, for any reason, the magnetic field should change rapidly, a large emf could be induced in the bracelet. If the bracelet were not a continuous band, this emf would cause high-voltage arcs to occur at any gap in the band. If the bracelet were a continuous band, the induced emf would produce a large induced current and result in resistance heating of the bracelet.
11. As the aluminum plate moves into the field, eddy currents are induced in the metal by the changing magnetic field at the plate. The magnetic field of the electromagnet interacts with this current, producing a retarding force on the plate that slows it down. In a similar fashion, as the plate leaves the magnetic field, a current is induced, and once again there is an upward force to slow the plate.
13. Oscillating current in the solenoid produces an always-changing magnetic field. Vertical flux through the ring, alternately increasing and decreasing, produces current in it with a direction that is alternately clockwise and counterclockwise. The current through the ring's resistance converts electrically transmitted energy into internal energy at the rate I^2R.

Problems

1. 4.8×10^{-3} T · m²
3. (a) 0.177 T (b) 0
5. (a) $\Phi_{B,\text{net}} = 0$ (b) 0
7. (a) 3.1×10^{-3} T · m² (b) $\Phi_{B,\text{net}} = 0$
9. (a) zero, since the flux through the loop A doesn't change (b) counterclockwise for loop B (c) clockwise for loop C
11. 94 mV
13. 33 mV
15. (a) from left to right (b) from right to left
17. (a) The current is zero, because the magnetic flux due to the current through the right side of the loop (into the page) is opposite the flux through the left side of the loop (out of the page), so net flux through the loop is always zero. (b) Clockwise, because the flux due to the long wire through the loop is out of the page, and the induced current in the loop must be clockwise to compensate for the increasing external flux.
19. (a) 3.77×10^{-3} T (b) 9.42×10^{-3} T (c) 7.07×10^{-4} m² (d) 3.99×10^{-6} Wb (e) 1.77×10^{-5} V; the average induced emf is equal to the instantaneous in this case because the current increases steadily. (f) The induced emf is small, so the current in the 4-turn coil and its magnetic field will also be small.
21. 10.2 μV
23. 13.1 mV
25. 2.6 mV
27. (a) toward the east (b) 4.58×10^{-4} V
29. 2.80 m/s
31. (a) 81.9 mV (b) 105 mA
33. 1.9×10^{-11} V
35. (a) 60 V (b) 57 V (c) 0.13 s
37. (a) 18.1 μV (b) 0
39. 4.0 mH
41. (a) 2.0 mH (b) 38 A/s
43. (a) 5.33 A (b) 2.67 s (c) 8.00 s
45. (a) 2.4 V (b) 75 mH (c) 5.1 A (d) $\Delta V_R = -1.5$ V (e) $\Delta V_L = -0.9$ V
47. 1.92 Ω
49. (a) 0.208 mH (b) 0.936 mJ
51. (a) 18 J (b) 7.2 J
53. counterclockwise
55. (a) $F = N^2B^2w^2v/R$ to the left (b) 0 (c) $F = N^2B^2w^2v/R$ to the left
57. (a) 20.0 ms (b) 37.9 V (c) 1.52 mV (d) 51.8 mA
59. (a) 0.500 A (b) 2.00 W (c) 2.00 W
61. 115 kV

63. (a) 0.157 mV (end B is positive) (b) 5.89 mV (end A is positive)
65. (a) 9.00 A (b) 10.8 N (c) b is at the higher potential (d) No
67. (b) A larger resistance would make the current smaller, so the loop must reach higher speed before the magnitude of the magnetic force will equal the gravitational force. (c) The magnetic force is proportional to the product of the field and the current, while the current itself is proportional to the field. If B is cut in half, the speed must become four times larger to compensate and yield a magnetic force with magnitude equal to that of the gravitational force.

CHAPTER 21

Quick Quizzes

1. (c)
2. (b)
3. (b)
4. (a)
5. (a)
6. (b)
7. (b), (c)
8. (b), (d)

Example Questions

1. False
2. False
3. True
4. True
5. The average power will be zero if the phase angle is 90° or −90° (i.e., when $R = 0$).
6. False
7. (d)
8. Yes. When the roof is perpendicular to the Sun's rays, the amount of energy intercepted by the roof is a maximum. At other angles the amount is smaller.
9. (c)

Warm-Up Exercises

1. (a) 2.92 m (b) 59.0°
3. 229 Hz
5. (a) $5.64 \times 10^3 \, \Omega$ (b) 2.13×10^{-2} A
7. (a) $3.12 \times 10^3 \, \Omega$ (b) $4.90 \times 10^3 \, \Omega$ (c) $2.33 \times 10^3 \, \Omega$ (d) 7.28×10^{-2} A
9. 5.32×10^{-9} F
11. 6.67×10^{-6} N/m^2

Conceptual Questions

1. Different stations have transmitting antennas at different locations. For best reception align your rabbit ears perpendicular to the straight-line path from your TV to the transmitting antenna. The transmitted signals are also polarized. The polarization direction of the wave can be changed by reflection from surfaces—including the atmosphere—and through Kerr rotation—a change in polarization axis when passing through an organic substance. In your home, the plane of polarization is determined by your surroundings, so antennas need to be adjusted to align with the polarization of the wave.
3. The fundamental source of an electromagnetic wave is an accelerating charge. For example, in a transmitting antenna of a radio station, charges are caused to move up and down at the frequency of the radio station. These moving charges set up electric and magnetic fields, the electromagnetic wave, in the space around the antenna.
5. (a) The flashing of the light according to Morse code is a drastic amplitude modulation (AM)—the amplitude is changing from a maximum to zero. In this sense, it is similar to the on-and-off binary code used in computers and compact disks. (b) The carrier frequency is that of the light, on the order of 10^{14} Hz. (c) The frequency of the signal depends on the skill of the signal operator, but it is on the order of a single hertz, as the light is flashed on and off. (d) The broadcasting antenna for this modulated signal is the filament of the lightbulb in the signal source. (e) The receiving antenna is the eye.
7. The sail should be as reflective as possible, so that the maximum momentum is transferred to the sail from the reflection of sunlight.
9. No. The wire will emit electromagnetic waves only if the current varies in time. The radiation is the result of accelerating charges, which can occur only when the current is not constant.
11. Photographs created using infrared light make warmer areas of the face brighter; visible light photographs capture mainly reflected ambient light, and are not affected by temperature. Consequently, bright areas in infrared light would not necessarily be bright in visible light. Pupils, for example, are very bright in the infrared but dark in visible light.
13. It is far more economical to transmit power at a high voltage than at a low voltage because the I^2R loss on the transmission line is significantly lower at high voltage. Transmitting power at high voltage permits the use of step-down transformers to make "low" voltages and high currents available to the end user.
15. The resonance frequency is determined by the inductance and the capacitance in the circuit. If both L and C are doubled, the resonance frequency is reduced by a factor of two.
17. (a)
19. (c) and (d)

Problems

1. (a) 193 Ω (b) 145 Ω
3. 14.6 Hz
5. 6.76 W
7. (a) $f > 41.3$ Hz (b) $X_C < 87.5 \, \Omega$
9. 4.0×10^2 Hz
11. 100 mA
13. 3.14 A
15. (a) 0.042 4 H (b) 942 rad/s
17. 0.450 T · m^2
19. (a) 0.361 A (b) 18.1 V (c) 23.9 V (d) −53.0°
21. (a) 1.43 kΩ (b) 0.097 9 A (c) 51.1° (d) voltage leads current
23. (a) 17.4° (b) the voltage
25. 1.88 V
27. (a) 47.1 Ω (b) 637 Ω (c) 2.40 kΩ (d) 2.33 kΩ (e) −14.2°
29. (a) 103 V (b) 150 V (c) 127 V (d) 23.6 V
31. (a) 208 Ω (b) 40.0 Ω (c) 0.541 H
33. 8.00 W
35. (a) $1.8 \times 10^2 \, \Omega$ (b) 0.71 H
37. 1.82 pF
39. $C_{min} = 4.9$ nF, $C_{max} = 51$ nF
41. 3.60 kHz
43. 721 V
45. 0.18% is lost
47. (a) 1.1×10^3 kW (b) 3.1×10^2 A (c) 8.3×10^3 A
49. 1 000 km; there will always be better use for tax money.
51. (a) 6.30×10^{-6} Pa (b) Atmospheric pressure is 1.60×10^{10} times larger than the radiation pressure.
53. $f_{red} = 4.55 \times 10^{14}$ Hz, $f_{IR} = 3.19 \times 10^{14}$ Hz, $E_{max,f}/E_{max,i} = 0.57$
55. $B_{max} = 3.39 \times 10^{-6}$ T, $E_{max} = 1.02 \times 10^3$ V/m
57. 2.94×10^8 m/s
59. (a) 6.00 pm (b) 7.50 cm
61. (a) 188 m to 556 m (b) 2.78 m to 3.4 m
63. 5.2×10^{13} Hz, 5.8 μm

65. $4.299\,999\,84 \times 10^{14}$ Hz; The frequency decreases by 1.6×10^7 Hz.
67. 5.31×10^{-5} N/m^2
69. 1.7 cents
71. 99.6 mH
73. (a) 6.7×10^{-16} T (b) 5.3×10^{-17} W/m^2 (c) 1.7×10^{-14} W
75. (a) 0.536 N (b) 8.93×10^{-5} m/s^2 (c) 33.9 days

CHAPTER 22

Quick Quizzes
1. (a)
2. Beams 2 and 4 are reflected; beams 3 and 5 are refracted.
3. (b)
4. (c)

Example Questions
1. (a)
2. (b)
3. False
4. (c)
5. (b)
6. (b)

Warm-Up Exercises
1. 16
3. (a) 16.8° (b) 326 nm
5. 68.8°

Conceptual Questions
1. The spectrum of the light sent back to you from a drop at the top of the rainbow arrives such that the red light (deviated by an angle of 42°) strikes the eye while the violet light (deviated by 40°) passes over your head. Thus, the top of the rainbow looks red. At the bottom of the rainbow, violet arrives at your eye and red light is deviated toward the ground. Thus, the bottom part of the rainbow appears violet.
3. $n_a > n_c > n_b$
5. On the one hand, a ball covered with mirrors sparkles by reflecting light from its surface. On the other hand, a faceted diamond lets in light at the top, reflects it by total internal reflection in the bottom half, and sends the light out through the top again. Because of its high index of refraction, the critical angle for diamond in air for total internal reflection, namely $\theta_c = \sin^{-1}(n_{air}/n_{diamond})$, is small. Thus, light rays enter through a large area and exit through a very small area with a much higher intensity. When a diamond is immersed in carbon disulfide, the critical angle is increased to $\theta_c = \sin^{-1}(n_{carbon\ disulfide}/n_{diamond})$. As a result, the light is emitted from the diamond over a larger area and appears less intense.
7. There is no dependence of the angle of reflection on wavelength, because the light does not enter deeply into the material during reflection—it reflects from the surface.
9. The color traveling slowest is bent the most. Thus, X travels more slowly in the glass prism.
11. Light rays coming from parts of the pencil under water are bent away from the normal as they emerge into the air above. The rays enter the eye (or camera) at angles closer to the horizontal, thus the parts of the pencil under water appear closer to the surface than they actually are, so the pencil appears bent.
13. Light travels through a vacuum at a speed of 3×10^8 m/s. Thus, an image we see from a distant star or galaxy must have been generated some time ago. For example, the star Altair is 16 light-years away; if we look at an image of Altair today, we know only what Altair looked like 16 years ago. This may not initially seem significant; however, astronomers who look at other galaxies can get an idea of what galaxies looked like when they were much younger. Thus, it does make sense to speak of "looking backward in time."
15. (c)

Problems
1. 3.00×10^8 m/s
3. (a) 2.07×10^3 eV (b) 4.14 eV
5. (a) 2.25×10^8 m/s (b) 1.97×10^8 m/s (c) 1.24×10^8 m/s
7. (a) 29.0° (b) 25.7° (c) 32.0°
9. 19.5° above the horizontal
11. (a) 1.52 (b) 416 nm (c) 4.74×10^{14} Hz (d) 1.97×10^8 m/s
13. 110.6°
15. (a) 1.56×10^8 m/s (b) 329.1 nm (c) 4.74×10^{14} Hz
17. five times from the right-hand mirror and six times from the left
19. $\theta = 30.4°, \theta' = 22.3°$
21. (a) 4.84 cm (b) 67.5°
23. (b) 4.73 cm
25. $\theta = \tan^{-1}(n_g)$
27. 3.39 m
29. (a) $\theta_{blue} = 47.79°$ (b) $\theta_{red} = 48.22°$
31. $\theta_{red} - \theta_{violet} = 0.32°$
33. (a) $\theta_{1i} = 30°, \theta_{1r} = 19°, \theta_{2i} = 41°, \theta_{2r} = 77°$ (b) First surface: $\theta_{reflection} = 30°$; second surface: $\theta_{reflection} = 41°$
35. (a) 66.1° (b) 63.5° (c) 59.7°
37. (a) 40.8° (b) 60.6°
39. 1.414
41. (a) 10.7° (b) air (c) Sound falling on the wall from most directions is 100% reflected.
43. 27.5°
45. 22.0°
47. (a) 1.07 m (b) 1.65 m
49. (a) 38.5° (b) 1.44
53. 24.7°
55. 1.93
57. (a) 1.20 (b) 3.39 ns
59. (a) 53.1° (b) $\theta_1 \geq 38.7°$
61. 15.9°

CHAPTER 23

Quick Quizzes
1. At C
2. (c)
3. (a) False (b) False (c) True
4. (b)
5. An infinite number
6. (a) False (b) True (c) False

Example Questions
1. No
2. (b)
3. True
4. (b)
5. (a)
6. (c)
7. The screen should be placed one focal length away from the lens.
8. No. The largest magnification would be 1, so a diverging lens would not make a good magnifying glass.
9. Upright
10. Yes, it's possible for a mirror to have a virtual object. Using the present system of mirror and lens, place the object to the right of the lens just outside the focal length. The image distance can be made as large as desired by moving the object toward the focal point, so at some point the lens's image will

Warm-Up Exercises

1. (a) −1.50 m (b) 1.00 (c) virtual
3. (a) negative (b) −4.36 cm (c) 0.273 (d) virtual (e) upright
5. (a) −7.04 cm (b) 1.20
7. (a) negative (b) −6.15 cm (c) 0.615 (d) virtual (e) upright

Conceptual Questions

1. You will not be able to focus your eyes on both the picture and your image at the same time. To focus on the picture, you must adjust your eyes so that an object several centimeters away (the picture) is in focus. Thus, you are focusing on the mirror surface. But your image in the mirror is as far behind the mirror as you are in front of it. Thus, you must focus your eyes beyond the mirror, twice as far away as the picture to bring the image into focus.
3. A single flat mirror forms a virtual image of an object due to two factors. First, the light rays from the object are necessarily diverging from the object, and second, the lack of curvature of the flat mirror cannot convert diverging rays to converging rays. If another optical element is first used to cause light rays to converge, then the flat mirror can be placed in the region in which the converging rays are present, and it will change the direction of the rays so that the real image is formed at a different location. For example, if a real image is formed by a convex lens, and the flat mirror is placed between the lens and the image position, the image formed by the mirror will be real.
5. An effect similar to a mirage is produced except the "mirage" is seen hovering in the air. Ghost lighthouses in the sky have been seen over bodies of water by this effect.
7. We consider the two trees to be two separate objects. The far tree is an object that is farther from the lens than the near tree. Thus, the image of the far tree will be closer to the lens than the image of the near tree. The screen must be moved closer to the lens to put the far tree in focus.
9. This is a possible scenario. When light crosses a boundary between air and ice, it will refract in the same manner as it does when crossing a boundary of the same shape between air and glass. Thus, a converging lens may be made from ice as well as glass. However, ice is such a strong absorber of infrared radiation that it is unlikely you will be able to start a fire with a small ice lens.
11. If a converging lens is placed in a liquid having an index of refraction larger than that of the lens material, the direction of refractions at the lens surfaces will be reversed, and the lens will diverge light. A mirror depends only on reflection that is independent of the surrounding material, so a converging mirror will be converging in any liquid.
13. The focal length for a mirror is determined by the law of reflection from the mirror surface. The law of reflection is independent of the material of which the mirror is made and of the surrounding medium. Thus, the focal length depends only on the radius of curvature and not on the material. The focal length of a lens depends on the indices of refraction of the lens material and surrounding medium. Thus, the focal length of a lens depends on the lens material.
15. Because when you look at the ƎƆИAJUᗺMA in your rear view mirror, the apparent left-right inversion clearly displays the name of the AMBULANCE behind you.
17. (b)

Problems

1. (a) younger (b) on the order of 10^{-9} s
3. 10.0 ft, 30.0 ft, 40.0 ft
5. (a) $p_1 + h$, behind the lower mirror (b) virtual (c) upright (d) 1.00 (e) no
7. (a) The object should be placed 30.0 cm in front of the mirror. (b) 0.333
9. (a) The object is 10.0 cm in front of the mirror. (b) 2.00
11. 5.00 cm
13. 1.0 m
15. 8.05 cm
17. (a) 26.8 cm behind the mirror (b) upright (c) 0.026 8
19. (a) concave with focal length $f = 0.83$ m (b) Object must be 1.0 m in front of the mirror.
21. 38.2 cm below the upper surface of the ice
23. 3.8 mm
25. $n = 2.00$
27. (a) The image is 0.790 m from the outer surface of the glass pane, inside the water tank. (b) With a glass pane of negligible thickness, the image is 0.750 m inside the tank. (c) The thicker the glass, the greater the distance between the final image and the outer surface of the glass.
29. 20.0 cm
31. (a) 20.0 cm beyond the lens; real, inverted, $M = -1.00$ (b) No image is formed. Parallel rays leave the lens. (c) 10.0 cm in front of the lens; virtual, upright, $M = +2.00$
33. (i) (a) 13.3 cm in front of the lens (b) virtual (c) upright (d) +0.333 (ii) (a) 10.0 cm in front of the lens (b) virtual (c) upright (d) +0.500 (iii) (a) 6.67 cm in front of the lens (b) virtual (c) upright (d) +0.667
35. (a) either 9.63 cm or 3.27 cm (b) 2.10 cm
37. (a) 39.0 mm (b) 39.5 mm
39. 40.0 cm
41. 30.0 cm to the left of the second lens, $M = -3.00$
43. 7.47 cm in front of the second lens, 1.07 cm, virtual, upright
45. from 0.224 m to 18.2 m
47. real image, 5.71 cm in front of the mirror
49. (a) −34.7 cm (b) −36.1 cm
51. 160 cm to the left of the lens, inverted, $M = -0.800$
53. $q = 10.7$ cm
55. (a) 32.0 cm to the right of the second surface (b) real
57. (a) 20.0 cm to the right of the second lens, $M = -6.00$ (b) inverted (c) 6.67 cm to the right of the second lens, $M = -2.00$, inverted
59. (a) 1.99 (b) 10.0 cm to the left of the lens (c) inverted
61. (a) 5.45 m to the left of the lens (b) 8.24 m to the left of the lens (c) 17.1 m to the left of the lens (d) by surrounding the lens with a medium having a refractive index greater than that of the lens material
63. (a) 263 cm (b) 79.0 cm
65. 8.57 cm

CHAPTER 24

Quick Quizzes

1. (c)
2. (c)
3. (c)
4. (b)
5. (b)
6. The compact disc

Example Questions

1. False
2. The soap film is thicker in the region that reflects red light.
3. The coating should be thinner.
4. The wavelength is smaller in water than in air, so the distance between dark bands is also smaller.
5. False

6. Because the wavelength becomes smaller in water, the angles to the first maxima become smaller, resulting in a smaller central maximum.
7. The separation between principal maxima will be larger.
8. The additional polarizer must make an angle of 90° with respect to the previous polarizer.

Warm-Up Exercises

1. (a) 0.524 m (b) 0.369 m
3. (a) 1.04° (b) 2.71×10^{-2} m (c) 2.59° (d) 6.79×10^{-2} m
5. (a) two (b) 4.57×10^{-7} m (c) 1.14×10^{-7} m
7. (a) 2.22×10^{-6} m (b) 47.1°
9. 52.5 W/m²

Conceptual Questions

1. You will *not* see an interference pattern from the automobile headlights, for two reasons. The first is that the headlights are not coherent sources and are therefore incapable of producing sustained interference. Also, the headlights are so far apart in comparison to the wavelengths emitted that, even if they were made into coherent sources, the interference maxima and minima would be too closely spaced to be observable.
3. The result of the double slit is to redistribute the energy arriving at the screen. Although there is no energy at the location of a dark fringe, there is four times as much energy at the location of a bright fringe as there would be with only a single narrow slit. The total amount of energy arriving at the screen is twice as much as with a single slit, as it must be according to the law of conservation of energy.
5. One of the materials has a higher index of refraction than water, and the other has a lower index. The material with the higher index will appear black as it approaches zero thickness. There will be a 180° phase change for the light reflected from the upper surface, but no such phase change for the light reflected from the lower surface, because the index of refraction for water on the other side is lower than that of the film. Thus, the two reflections will be out of phase and will interfere destructively. The material with index of refraction lower than water will have a phase change for the light reflected from both the upper and the lower surface, so that the reflections from the zero-thickness film will be back in phase and the film will appear bright.
7. Because the light reflecting at the lower surface of the film undergoes a 180° phase change, while light reflecting from the upper surface of the film does not undergo such a change, the central spot (where the film has near zero thickness) will be dark. If the observed rings are not circular, the curved surface of the lens does not have a true spherical shape.
9. For regional communication at the Earth's surface, radio waves are typically broadcast from currents oscillating in tall vertical towers. These waves have vertical planes of polarization. Light originates from the vibrations of atoms or electronic transitions within atoms, which represent oscillations in all possible directions. Thus, light generally is not polarized.
11. Yes. In order to do this, first measure the radar reflectivity of the metal of your airplane. Then choose a light, durable material that has approximately half the radar reflectivity of the metal in your plane. Measure its index of refraction, and place onto the metal a coating equal in thickness to one-quarter of 3 cm, divided by that index. Sell the plane quickly, and then you can sell the supposed enemy new radars operating at 1.5 cm, which the coated metal will reflect with extra-high efficiency.
13. If you wish to perform an interference experiment, you need monochromatic coherent light. To obtain it, you must first pass light from an ordinary source through a prism or diffraction grating to disperse different colors into different directions. Using a single narrow slit, select a single color and make that light diffract to cover both slits for a Young's experiment. The procedure is much simpler with a laser because its output is already monochromatic and coherent.
15. (b)

Problems

1. 632 nm
3. (a) 2.6 mm (b) 2.62 mm
5. (a) 36.2° (b) 5.08 cm (c) 5.08×10^{14} Hz
7. (a) 55.7 m (b) 124 m
9. 662 nm
11. 11.3 m
13. 148 m
15. 75.0 m
17. 85.4 nm
19. 511 nm
21. 0.500 cm
23. (a) 238 nm (b) λ will increase (c) 328 nm
25. 78.4 μm
27. 4.75 μm
29. 617 nm, 441 nm, 343 nm
31. (a) 8.10 mm (b) 4.05 mm
33. (a) 1.1 m (b) 1.7 mm
35. 462 nm
37. 1.20 mm, 1.20 mm
39. (a) 479 nm, 647 nm, 698 nm (b) 20.5°, 28.3°, 30.7°
41. 5.91° in first order; 13.2° in second order; and 26.5° in third order
43. 1.81 μm
45. 514 nm
47. 9.17 cm
49. (a) 3.53×10^3 grooves/cm (b) 11 maxima
51. (a) 1.11 (b) 42.0°
53. (a) 56.7° (b) 48.8°
55. (a) 55.6° (b) 34.4°
59. 6.89 units
61. (a) 413.7 nm, 409.7 nm (b) 8.6°
63. One slit 0.12 mm wide. The central maximum is twice as wide as the other maxima.
65. 0.156 mm
67. 2.50 mm
69. (a) 16.6 m (b) 8.28 m
71. 127 m
73. 0.350 mm

CHAPTER 25

Quick Quizzes

1. (c)
2. (a)

Example Questions

1. True
2. True
3. A smaller focal length gives a greater magnification and should be selected.
4. True
5. Yes. Increasing the focal length of the mirror increases the magnification. Increasing the focal length of the eyepiece decreases the magnification.
6. More widely spaced eyes increase visual resolving power by effectively increasing the aperture size, D, in Equation 25.10. The limiting angle of resolution is thereby decreased, meaning finer details of distant objects can be resolved.

7. Resolution is better at the violet end of the visible spectrum.
8. True

Warm-Up Exercises
1. (a) 60.0 cm (b) −5.00 (c) inverted (d) real
3. 0.206 m
5. (a) nearsighted (b) −42.0 cm (c) −2.38 diopters
7. (a) 3.00 cm (b) 8.33
9. 11.4
11. 8.65 m

Conceptual Questions
1. The observer is *not* using the lens as a simple magnifier. For a lens to be used as a simple magnifier, the object distance must be less than the focal length of the lens. Also, a simple magnifier produces a virtual image at the normal near point of the eye, or at an image distance of about $q = -25$ cm. With a large object distance and a relatively short image distance, the magnitude of the magnification by the lens would be considerably less than one. Most likely, the lens in this example is part of a lens combination being used as a telescope.
3. The image formed on the retina by the lens and cornea is already inverted.
5. For a lens to operate as a simple magnifier, the object should be located just inside the focal point of the lens. If the power of the lens is +20.0 diopters, its focal length is
$f = (1.00 \text{ m})/P = (1.00 \text{ m})/+20.0 = 0.050\ 0$ m $= 5.00$ cm
The object should be placed slightly less than 5.00 cm in front of the lens.
7. You want a real image formed at the location of the paper. To form such an image, the object distance must be greater than the focal length of the lens.
9. In order for someone to see an object through a microscope, the wavelength of the light in the microscope must be smaller than the size of the object. An atom is much smaller than the wavelength of light in the visible spectrum, so an atom can never be seen with the use of visible light.
11. farsighted; converging
13. (a) The diffraction pattern of a hair is the same as the diffraction pattern produced by a single slit of the same width. (b) The central maximum is flanked by minima. Measure the width 2y of the central maximum between the minima bracketing it. Because the angle is small, you can use

$$\sin \theta_{dark} \approx \tan \theta_{dark}$$
$$m\lambda/a \approx y/L$$

to find the width a of the hair.
15. (c)

Problems
1. 7.0
3. 177 m
5. $f/1.4$
7. 8.0
9. (a) 42.9 cm (b) +2.33 diopters
11. 23.2 cm
13. (a) −2.00 diopters (b) 17.6 cm
15. +17.0 diopters
17. −2.50 diopters, a diverging lens
19. (a) 5.8 cm (b) $m = 4.3$
21. (a) 4.07 cm (b) $m = +7.14$
23. (a) $|M| = 1.22$ (b) $\theta/\theta_0 = 6.08$
25. 2.1 cm
27. 12.7
29. 3.38 min
31. (a) 2.53 cm (b) 88.5 cm
33. (b) $-fh/p$ (c) −1.07 mm
35. (a) 1.50 (b) 1.9
37. 492 km
39. 0.40 μrad
41. 5.40 mm
43. 9.8 km
45. No. A resolving power of 2.0×10^5 is needed, and that available is only 1.8×10^5.
47. 1.31×10^3
49. 50.4 μm
51. 40
53. (a) 8.00×10^2 (b) The image is inverted.
55. 15.4
57. (a) +2.67 diopters (b) 0.16 diopter too low
59. (a) +44.6 diopters (b) 3.03 diopters
61. (a) $m = 4.0$ (b) $m = 3.0$

CHAPTER 26

Quick Quizzes
1. False: the speed of light is c for all observers.
2. (a)
3. False
4. No. From your perspective you're at rest with respect to the cabin, so you will measure yourself as having your normal length, and will require a normal-sized cabin.
5. (i) (a), (e) (ii) (a), (e)
6. False
7. (a) False (b) False (c) True (d) False

Example Questions
1. 9.61 s
2. (c)
3. (a)
4. $-0.946c$
5. (a)
6. Very little of the mass is converted to other forms of energy in these reactions because the total number of neutrons and protons doesn't change. The energy liberated is only the energy associated with their interactions.

Warm-Up Exercises
1. −122 m/s
3. 5.74 m/s
5. (a) 1.36 (b) 2.82×10^5 km
7. (a) $0.750c$ (b) $0.667c$
9. (a) 0.334 MeV (b) 6.02×10^{-13} J

Conceptual Questions
1. An ellipsoid. The dimension in the direction of motion would be measured to be less than D.
3. No. The principle of relativity implies that nothing can travel faster than the speed of light in a *vacuum*, which is equal to 3.00×10^8 m/s.
5. As the object approaches the speed of light, its kinetic energy grows without limit. It would take an infinite investment of work to accelerate the object to the speed of light.
7. For a wonderful fictional exploration of this question, get a "Mr. Tompkins" book by George Gamow. All of the relativity effects would be obvious in our lives. Time dilation and length contraction would both occur. Driving home in a hurry, you would push on the gas pedal not to increase your speed very much, but to make the blocks shorter. Big Doppler shifts in wave frequencies would make red lights look green as you approached and make car horns and radios useless. High-speed transportation would be very expensive, requiring huge fuel purchases, as well as dangerous, since a speeding car could knock down a building. When you got home, hungry for lunch, you would find that you had missed

dinner; there would be a five-day delay in transit when you watch a live TV program originating in Australia. Finally, we would not be able to see the Milky Way, since the fireball of the Big Bang would surround us at the distance of Rigel or Deneb.

9. A photon transports energy. The relativistic equivalence of mass and energy is enough to give it momentum.
11. Suppose a railroad train is moving past you. One way to measure its length is this: you mark the tracks at the cowcatcher forming the front of the moving engine at 9:00:00 AM, while your assistant marks the tracks at the back of the caboose at the same time. Then you find the distance between the marks on the tracks with a tape measure. You and your assistant must make the marks simultaneously in your frame of reference, for otherwise the motion of the train would make its length different from the distance between marks.

Problems

1. (a) 1.38 yr (b) 1.31 light-years
3. $0.103c$
5. (a) 1.3×10^{-7} s (b) 38 m (c) 7.6 m
7. (a) 1.55×10^{-5} s (b) 7.09 (c) 2.19×10^{-6} s (d) 649 m (e) From the third observer's point of view, the muon is traveling faster, so according to the third observer, the muon's lifetime is longer than that measured by the observer at rest with respect to Earth.
9. $0.950c$
11. Yes, with 19 m to spare
13. (a) $\dfrac{2\sqrt{2}}{3}c = 0.943c = 2.83 \times 10^8$ m/s (b) The result would be the same.
15. $0.285c$
17. $0.54c$ to the right
19. $0.357c$
21. $0.998c$ toward the right
23. (a) $0.999\,997c$ (b) 3.74×10^5 MeV
25. $0.786c$
27. (a) 8.65×10^{22} J (b) 9.61×10^5 kg
31. (a) 3.91×10^4 (b) 7.67 cm
33. (a) $0.979c$ (b) $0.065\,2c$ (c) $0.914c$
35. $0.917c$
37. (a) $v/c = 1 - 1.13 \times 10^{-10}$ (b) 5.99×10^{27} J (c) $\$2.16 \times 10^{20}$
39. (a) $0.800c$ (b) 7.50×10^3 s (c) 1.44×10^{12} m (d) $0.385c$
41. $0.80c$
43. (a) $v = 0.141c$ (b) $v = 0.436c$
45. (a) 7.0 μs (b) 1.1×10^4 muons
47. 5.5 yr; Goslo is older
49. (a) 17.4 m (b) 3.30° with respect to the direction of motion

CHAPTER 27

Quick Quizzes

1. True
2. (b)
3. (c)
4. False
5. (c)

Example Questions

1. False
2. False
3. Some of the photon's energy is transferred to the electron.
4. Doubling the speed of a particle doubles its momentum, reducing the particle's wavelength by a factor of one-half. This answer is no longer true when the doubled speed is relativistic.
5. True

Warm-Up Exercises

1. (a) 7.69×10^{14} Hz (b) 4.29×10^{14} Hz
3. 1.96×10^6 m/s
5. 2.46 eV
7. 0.124 nm
9. 4.85×10^{-3} nm
11. 2.32×10^4 m/s

Conceptual Questions

1. The shape of an object is normally determined by observing the light reflecting from its surface. In a kiln, the object will be very hot and will be glowing red. The emitted radiation is far stronger than the reflected radiation, and the thermal radiation emitted is only slightly dependent on the material from which the object is made. Unlike reflected light, the emitted light comes from all surfaces with equal intensity, so contrast is lost and the shape of the object is harder to discern.
3. The "blackness" of a blackbody refers to its ideal property of absorbing all radiation incident on it. If an observed room temperature object in everyday life absorbs all radiation, we describe it as (visibly) black. The black appearance, however, is due to the fact that our eyes are sensitive only to visible light. If we could detect infrared light with our eyes, we would see the object emitting radiation. If the temperature of the blackbody is raised, Wien's law tells us that the emitted radiation will move into the visible range of the spectrum. Thus, the blackbody could appear as red, white, or blue, depending on its temperature.
5. All objects do radiate energy, but at room temperature this energy is primarily in the infrared region of the electromagnetic spectrum, which our eyes cannot detect. (Pit vipers have sensory organs that *are* sensitive to infrared radiation; thus, they can seek out their warm-blooded prey in what we would consider absolute darkness.)
7. We can picture higher frequency light as a stream of photons of higher energy. In a collision, one photon can give all of its energy to a single electron. The kinetic energy of such an electron is measured by the stopping potential. The reverse voltage (stopping voltage) required to stop the current is proportional to the frequency of the incoming light. More intense light consists of more photons striking a unit area each second, but atoms are so small that one emitted electron never gets a "kick" from more than one photon. Increasing the intensity of the light will generally increase the size of the current, but it will not change the energy of the individual electrons that are ejected. Thus, the stopping potential remains constant.
9. Wave theory predicts that the photoelectric effect should occur at any frequency, provided that the light intensity is high enough. However, as seen in photoelectric experiments, the light must have sufficiently high frequency for the effect to occur.
11. (a) Electrons are emitted only if the photon frequency is greater than the cutoff frequency.
13. No. Suppose that the incident light frequency at which you first observed the photoelectric effect is above the cutoff frequency of the first metal, but less than the cutoff frequency of the second metal. In that case, the photoelectric effect would not be observed at all in the second metal.
15. An electron has both classical-wave and classical-particle characteristics. In single- and double-slit diffraction and interference experiments, electrons behave like classical waves. An electron has mass and charge. It carries kinetic energy and momentum in parcels of definite size, as classical particles do. At the same time it has a particular wavelength and frequency. Since an electron displays characteristics of both classical waves and classical particles, it is neither a

classical wave nor a classical particle. It is customary to call it a *quantum particle*, but another invented term, such as "wavicle," could serve equally well.

Problems

1. (a) ≈ 3 000 K (b) ≈ 20 000 K
3. 9.47 μm, which is in the infrared region of the spectrum
5. (a) 2.49×10^{-5} eV (b) 2.49 eV (c) 249 eV
7. (a) 2.57 eV (b) 1.28×10^{-5} eV (c) 1.91×10^{-7} eV
9. (a) 2.24 eV (b) 555 nm (c) 5.41×10^{14} Hz
11. (a) 1.02×10^{-18} J (b) 1.53×10^{15} Hz (c) 196 nm (d) 2.15 eV (e) 2.15 V
13. 4.8×10^{14} Hz, 2.0 eV
15. 1.2×10^{2} V and 1.2×10^{7} V, respectively
17. 17.8 kV
19. 0.078 nm
21. 0.281 nm
23. 70.0°
25. (a) 1.18×10^{-23} kg·m/s (b) 478 eV
27. (a) 1.46 km/s (b) 7.28×10^{-11} m
29. 3.58×10^{-13} m
31. (a) 15 keV (b) 1.2×10^{2} keV
33. ~10^{6} m/s
35. Within 1.16 mm for the electron, 5.28×10^{-32} m for the bullet
37. 3×10^{-29} J
39. (a) 2.21×10^{-32} kg·m/s (b) 1.00×10^{10} Hz (c) 4.14×10^{-5} eV
41. 2.5×10^{20} photons
43. (a) 5 200 K (b) Clearly, a firefly is not at this temperature, so this is not blackbody radiation.
45. 1.36 eV
47. 2.00 eV
49. (a) $0.022\,0c$ (b) $0.999\,2c$

CHAPTER 28

Quick Quizzes

1. (b)
2. (a) 5 (b) 9 (c) 25
3. (d)

Example Questions

1. The energy associated with the quantum number n increases with increasing quantum number n, going to zero in the limit of arbitrarily large n. A transition from a very high energy level to the ground state therefore results in the emission of photons approaching an energy of 13.6 eV, the same as the ionization energy.
2. The energy difference in the helium atom will be four times that of the same transition in hydrogen. Energy levels in hydrogen-like atoms are proportional to Z^2, where Z is the atomic number.
3. The quantum numbers n and ℓ are never negative.
4. No. The M shell is at a higher energy; hence, transitions from the M shell to the K shell will always result in more energetic photons than any transition from the L shell to the K shell.

Warm-Up Exercises

1. (a) 6.48×10^{-19} J (b) 307 nm
3. (a) 0.967 eV (b) 1.55×10^{-19} J (c) 2.33×10^{14} Hz
5. (a) 0.212 nm (b) 0.846 nm
7. (a) 2 (b) 8 (c) 18
9. (a) 6 (b) 3

Conceptual Questions

1. If the energy of the hydrogen atom were proportional to n (or any power of n), then the energy would become infinite as n grew to infinity. But the energy of the atom is inversely proportional to n^2. Thus, as n grows to infinity, the energy of the atom approaches a value that is above the ground state by a finite amount, namely, the ionization energy 13.6 eV. As the electron falls from one bound state to another, its energy loss is always less than the ionization energy. The energy and frequency of any emitted photon are finite.

3. The characteristic x-rays originate from transitions within the atoms of the target, such as an electron from the L shell making a transition to a vacancy in the K shell. The vacancy is caused when an accelerated electron in the x-ray tube supplies energy to the K shell electron to eject it from the atom. If the energy of the bombarding electrons were to be increased, the K shell electron will be ejected from the atom with more remaining kinetic energy. But the energy difference between the K and L shell has not changed, so the emitted x-ray has exactly the same wavelength.

5. A continuous spectrum without characteristic x-rays is possible. At a low accelerating potential difference for the electron, the electron may not have enough energy to eject an electron from a target atom. As a result, there will be no characteristic x-rays. The change in speed of the electron as it enters the target will result in the continuous spectrum.

7. The hologram is an interference pattern between light scattered from the object and the reference beam. If anything moves by a distance comparable to the wavelength of the light (or more), the pattern will wash out. The effect is just like making the slits vibrate in Young's experiment, to make the interference fringes vibrate wildly so that a photograph of the screen displays only the average intensity everywhere.

9. If the Pauli exclusion principle were not valid, the elements and their chemical behavior would be grossly different because every electron would end up in the lowest energy level of the atom. All matter would therefore be nearly alike in its chemistry and composition, since the shell structures of each element would be identical. Most materials would have a much higher density, and the spectra of atoms and molecules would be very simple, resulting in the existence of less color in the world.

11. The three elements have similar electronic configurations, with filled inner shells plus a single electron in an s orbital. Because atoms typically interact through their unfilled outer shells, and since the outer shells of these atoms are similar, the chemical interactions of the three atoms are also similar.

13. Each of the eight electrons must have at least one quantum number different from each of the others. They can differ (in m_s) by being spin-up or spin-down. They can differ (in ℓ) in angular momentum and in the general shape of the wave function. Those electrons with $\ell = 1$ can differ (in m_ℓ) in orientation of angular momentum.

15. Stimulated emission is the reason laser light is coherent and tends to travel in a well-defined parallel beam. When a photon passing by an excited atom stimulates that atom to emit a photon, the emitted photon is in phase with the original photon and travels in the same direction. As this process is repeated many times, an intense, parallel beam of coherent light is produced. Without stimulated emission, the excited atoms would return to the ground state by emitting photons at random times and in random directions. The resulting light would not have the useful properties of laser light.

Problems

1. (a) 121.5 nm, 102.5 nm, 97.20 nm (b) Far ultraviolet
3. (a) 2.3×10^{-8} N (b) −14 eV

5. (a) 1.6×10^6 m/s (b) No. $v/c = 5.3 \times 10^{-3} \ll 1$
    (c) 0.45 nm (d) Yes. The wavelength is roughly the same size as the atom.
7. (a) 0.212 nm (b) 9.95×10^{-25} kg·m/s (c) 2.11×10^{-34} J·s
    (d) 3.40 eV (e) -6.80 eV (f) -3.40 eV
11. $E = -1.51$ eV ($n = 3$) to $E = -3.40$ eV ($n = 2$)
13. (a) 2.86 eV (b) 0.472 eV
15. (a) 1.89 eV (b) 658 nm (c) 3.02 eV (d) 412 nm
    (e) 366 nm
17. (a) 488 nm (b) 0.814 m/s
19. (a) 3 (b) 520 km/s
21. (a) 0.476 nm (b) 0.997 nm
23. (a) 0.026 5 nm (b) 0.017 6 nm (c) 0.013 2 nm
25. (a) 1.31 μm (b) 164 nm
27.

n	ℓ	m_ℓ	m_s
3	2	2	$\frac{1}{2}$
3	2	2	$-\frac{1}{2}$
3	2	1	$\frac{1}{2}$
3	2	1	$-\frac{1}{2}$
3	2	0	$\frac{1}{2}$
3	2	0	$-\frac{1}{2}$
3	2	-1	$\frac{1}{2}$
3	2	-1	$-\frac{1}{2}$
3	2	-2	$\frac{1}{2}$
3	2	-2	$-\frac{1}{2}$

29. Fifteen possible states, as summarized in the following table:

n	3	3	3	3	3	3	3	3	3	3	3	3	3	3	3
ℓ	2	2	2	2	2	2	2	2	2	2	2	2	2	2	2
m_ℓ	2	2	2	1	1	1	0	0	0	-1	-1	-1	-2	-2	-2
m_s	1	0	-1	1	0	-1	1	0	-1	1	0	-1	1	0	-1

31. aluminum
33. (a) $n = 4$ and $\ell = 2$ (b) $m_\ell = (0, \pm 1, \pm 2)$, $m_s = \pm 1/2$
    (c) $1s^2 2s^2 2p^6 3s^2 3p^6 3d^{10} 4s^2 4p^6 4d^2 5s^2$ = [Kr] $4d^2 5s^2$
35. (a) 14 keV (b) 8.9×10^{-11} m
37. L shell: 11.7 keV; M shell: 10.0 keV; N shell: 2.30 keV
39. (a) 10.2 eV (b) 7.88×10^4 K
41. (a) 5.40 keV (b) 3.40 eV (c) 5.40 keV
43. (a) 4.24×10^{15} W/m^2 (b) 1.20×10^{-12} J

CHAPTER 29

Quick Quizzes

1. False
2. (c)
3. (c)
4. (a) and (b)
5. (b)

Example Questions

1. Tritium has the greater binding energy. Unlike tritium, helium has two protons that exert a repulsive electrostatic force on each other. The helium-3 nucleus is therefore not as tightly bound as the tritium nucleus.
2. Doubling the initial mass of radioactive material doubles the initial activity. Doubling the mass has no effect on the half-life.
3. 7.794×10^{-13} J, 4.69×10^{11} J
4. The binding energy of carbon-14 should be greater than nitrogen-14 because nitrogen has more protons in its nucleus. The mutual repulsion of the protons means that the nitrogen nucleus would require less energy to break up than the carbon nucleus.
5. No
6. Reactions between helium and beryllium can be found in the Sun.

Warm-Up Exercises

1. (a) 0.657 J (b) 4.11×10^{12} MeV
3. 2.7×10^{-15} m
5. 1.49×10^4 MeV/c^2
7. 3.12×10^{-17} s^{-1}
9. (a) 230 (b) 88

Conceptual Questions

1. An alpha particle contains two protons and two neutrons. Because a hydrogen nucleus contains only one proton, it cannot emit an alpha particle.
3. Isotopes of a given element correspond to nuclei with different numbers of neutrons. This will result in a variety of different physical properties for the nuclei, including the obvious one of mass. The chemical behavior, however, is governed by the element's electrons. All isotopes of a given element have the same number of electrons and, therefore, the same chemical behavior.
5. Nucleus Y will be more unstable. The nucleus with the higher binding energy requires more energy to be disassembled into its constituent parts and has less available energy to release in a decay.
7. In alpha decay, there are only two final particles: the alpha particle and the daughter nucleus. There are also two conservation principles: of energy and of momentum. As a result, the alpha particle must be ejected with a discrete energy to satisfy both conservation principles. However, beta decay is a three-particle decay: the beta particle, the neutrino (or antineutrino), and the daughter nucleus. As a result, the energy and momentum can be shared in a variety of ways among the three particles while still satisfying the two conservation principles. This allows a continuous range of energies for the beta particle.
11. The larger rest energy of the neutron means that a free proton in space will not spontaneously decay into a neutron and a positron. When the proton is in the nucleus, however, the important question is that of the total rest energy of the nucleus. If it is energetically favorable for the nucleus to have one less proton and one more neutron, then the decay process will occur to achieve this lower energy.
13. Carbon dating cannot generally be used to estimate the age of a rock, because the rock was not alive to receive carbon, and hence radioactive carbon-14, from the environment. Only the ages of objects that were once alive can be estimated with carbon dating.

Problems

1. (a) 1.5 fm (b) 4.7 fm (c) 7.0 fm (d) 7.4 fm
3. 1.8×10^2 m
5. (a) 27.6 N (b) 4.16×10^{27} m/s^2 (c) 1.73 MeV
7. (a) 1.9×10^7 m/s (b) 7.1 MeV
9. (a) 8.26 MeV/nucleon (b) 8.70 MeV/nucleon
11. 3.54 MeV
13. (a) 0.210 MeV/nucleon greater for $^{23}_{11}$Na (b) attributable to less proton repulsion
15. 2.29 g
17. (a) 6.95×10^5 s (b) 9.97×10^{-7} s^{-1} (c) 1.9×10^4 decays/s
    (d) 1.9×10^{10} nuclei (e) 5, 0.200 mCi
19. 8.06 days
21. (a) 5.58×10^{-2} h^{-1}, 12.4 h (b) 2.39×10^{13} nuclei
    (c) 1.9 mCi
23. (a) $^{203}_{81}$Tl (b) $^{95}_{37}$Rb (c) $^{144}_{60}$Nd
25. e^+ decay, $^{56}_{27}$Co $\rightarrow$ $^{56}_{26}$Fe $+ e^+ + \nu$
27. (a) cannot occur spontaneously (b) can occur spontaneously
29. 18.6 keV
31. 4.22×10^3 yr
33. (a) $^{21}_{10}$Ne (b) $^{144}_{54}$Xe (c) X = e^+, X' = ν
35. (a) $^{13}_{6}$C (b) $^{10}_{5}$B

37. (a) $^{197}_{79}\text{Au} + \text{n} \rightarrow {}^{198}_{80}\text{Hg} + e^- + \bar{\nu}$ (b) 7.89 MeV
39. (a) The product nucleus is $^{10}_{5}\text{B}$. (b) -2.79 MeV
41. 18.8 J
43. 24 d
45. (a) 8.97×10^{11} electrons (b) 0.100 J (c) 100 rad
47. (a) 3.18×10^{-7} mol (b) 1.91×10^{17} nuclei (c) 1.08×10^{14} Bq (d) 8.92×10^6 Bq
49. 46.5 d
51. (a) 4.0×10^9 yr (b) It could be no older. (c) The rock could be younger if some ^{87}Sr were initially present.
53. 54 μCi
55. 5.9 billion years

CHAPTER 30

Quick Quizzes

1. (a)
2. (b)

Example Questions

1. 1 mg/yr
2. 1.77×10^{12} J
3. True
4. No. That reaction violates conservation of baryon number.

Warm-Up Exercises

1. (a) 7.11×10^{13} J (b) 7.90×10^{-4} kg
3. (a) 3.33×10^{-11} J (b) 3.70×10^{-28} kg
5. 6.67×10^{15} cm^{-3}
7. 0
9. 1

Conceptual Questions

1. The experiment described is a nice analogy to the Rutherford scattering experiment. In the Rutherford experiment, alpha particles were scattered from atoms and the scattering was consistent with a small structure in the atom containing the positive charge.
3. The largest quark charge is $2e/3$, so a combination of only two particles, a quark and an antiquark forming a meson, could not have an electric charge of $+2e$. Only particles containing three quarks, each with a charge of $2e/3$, can combine to produce a total charge of $2e$.
5. Until about 700 000 years after the Big Bang, the temperature of the Universe was high enough for any atoms that formed to be ionized by ambient radiation. Once the average radiation energy dropped below the hydrogen ionization energy of 13.6 eV, hydrogen atoms could form and remain as neutral atoms for relatively long period of time.
7. In the quark model, all hadrons are composed of smaller units called quarks. Quarks have a fractional electric charge and a baryon number of $\frac{1}{3}$. There are six flavors of quarks: up (u), down (d), strange (s), charmed (c), top (t), and bottom (b). All baryons contain three quarks, and all mesons contain one quark and one antiquark. Section 30.8 has a more detailed discussion of the quark model.
9. Baryons and mesons are hadrons, interacting primarily through the strong force. They are not elementary particles, being composed of either three quarks (baryons) or a quark and an antiquark (mesons). Baryons have a nonzero baryon number with a spin of either $\frac{1}{2}$ or $\frac{3}{2}$. Mesons have a baryon number of zero and a spin of either 0 or 1.
11. All stable particles other than protons and neutrons have baryon number zero. Since the baryon number must be conserved, and the final states of the kaon decay contain no protons or neutrons, the baryon number of all kaons must be *zero*.

Problems

1. 1.1×10^{16} fissions
3. 126 MeV
5. (a) 16.2 kg (b) 117 g
7. (a) $3/a$ (b) $3.72/a$ (c) The surface-to-volume ratio is lowest for the sphere; therefore, the sphere has the better shape for minimum leakage.
9. 1.01 g
11. (a) $^{8}_{4}\text{Be}$ (b) $^{12}_{6}\text{C}$ (c) 7.27 MeV
13. 4.03 MeV
15. 14.1 MeV
17. (a) 4.53×10^{23} Hz (b) 0.662 fm
19. 67.5 MeV, 67.5 MeV/c, 1.63×10^{22} Hz
21. (a) conservation of electron-lepton number and conservation of muon-lepton number (b) conservation of charge (c) conservation of baryon number (d) conservation of baryon number (e) conservation of charge
23. (a) Violates conservation of baryon number (not allowed). (b) This reaction may occur via the strong interaction. (c) This reaction may occur via the weak interaction only. (d) This reaction may occur by the weak interaction only. (e) This reaction may occur via the electromagnetic interaction.
25. (a) not conserved (b) conserved (c) conserved (d) not conserved (e) not conserved (f) not conserved
27. 3.34×10^{26} electrons, 9.36×10^{26} up quarks, 8.70×10^{26} down quarks
29. (a) Σ^+ (b) π^- (c) K^0 (d) Ξ^-
31. a neutron, udd
33. 18.8 MeV
35. (a) K^+ (b) Ξ^0 (c) π^0
37. 26
39. (b) 12 days
41. 29.8 MeV
43. 3.60×10^{38} protons/s

Index

Note: Page numbers followed by *f* and *t* refer to figures and tables respectively.

A

Aberration
   chromatic, 826, 826*f*, 881
   defined, 825
   spherical, 805, 805*f*, 825, 825*f*
Absolute pressure, 297
Absolute value (| |), A.2–A.3
Absolute temperature, 339–340
Absolute zero, 339–340
Absorber, ideal, 388
Absorption spectrum, 949*f*, 950, 950*f*
AC. *See entries under* Alternating-current
Acceleration (*a*)
   angular, 205–206, 205*f*
      average, 205–206
      constant, rotational motion under, 206–208
      instantaneous, 206
      rotational analog of Newton's second law, 254, 256–259
      tangential acceleration relationship to, 209
      torque and, 241, 252–259
   average ($\bar{a}$)
      defined, 64
      in one dimension, 34–35, 34*f*
      SI (Système International) units, 34, 64
      in two dimensions, 64
   centripetal, 7, 211–219, 673
   constant
      angular, rotational motion under, 206–208
      equations for, 40*t*
      in one dimension, 38–44, 38*f*, 40*t*
      problem solving strategy, 40
      in two dimensions, 72–73, 72*f*
   deceleration, 35
   defined, 34
   dimensions of, 5–6, 5*t*
   free-fall, 45, 220, 220*t*, 461–462, A.19*t*
   of gravity, 95
   instantaneous
      defined, 64
      in one dimension, 35–37, 36*f*
      SI (Système International) units, 35, 64
      in two dimensions, 63
      from velocity-time graph, 36–37, 36*f*
   mass relationship to, 92
   in motion diagrams, 37–38, 37*f*
   negative, 35, 38
   Newton's second law of motion and, 92, 100, 104–107
   in one dimension, 34–37, 34*f*, 36*f*
   peak, 43–44, 43*f*
   positive, 38
   projectile motion and, 65–68, 65*f*
   in simple harmonic motion
      as function of position, 447–448
      as function of time, 457–459, 458*f*
   tangential, 209, 213
   total, 213
   translational, 245
   in two dimensions, 63–65
   uniform, 36
   as vector quantity, 35
   velocity vs., 34–35, 34*f*
Acceleration-time graph, 36, 36*f*
   for motion with constant acceleration, 38, 38*f*

Accommodation, 872–874
Accuracy in measurement, 7–11
Acetone, coefficient of volume expansion for, 344*t*
Acrobatics, 146–147, 146*f*, 264
Actinium (Ac), radioactive series, 985*t*
Action potential, 647–648, 649*f*
Activity (decay rate) (*R*), 977–979
Addition
   commutative law of, 58
   of fractions, A.5
   scalar, 58
   significant figures and, 9
   triangle method of, 58
   of vectors
      algebraically, 62–63, 62*f*
      geometrical, 58, 58*f*, 59, 59*f*
Adenosine triphosphate (ATP), 760
Adhesive forces, 316–319, 316*f*–318*f*
Adiabatic index of a gas (γ), 411
Adiabatic processes, 411–413, 417*t*
Air. *See also* Air flow; Air resistance; Atmosphere
   coefficient of volume expansion, 344*t*
   density of, 284*t*, A.19*t*
   dielectric constant and dielectric strength of, 585*t*
   index of refraction, 778*t*
   R-value of stagnant, 383*t*, 384
   speed of sound in, 484–486, 485*t*
   thermal conductivity, 381*t*
Air bags, automobile, 175
Air circulation at the beach, 370, 370*f*, 385
Air cleaner, electrostatic, 572
Air columns, standing waves in, 504–508, 505*f*
Air conditioner, 421
Air drag. *See* Air resistance
Air flow
   around airplane wing, 311–312, 311*f*
   around spinning golf or tennis ball, 311
Airplanes
   airport runways, labeling, 663
   coasting and braking distances for, 42–43, 42*f*
   lift on airplane wing, 311–312, 311*f*
   potential difference induced across wing, 710–711
   runway length problem, 42–43, 42*f*
   sonic boom produced by, 496
   sound intensity levels of, 488, 488*t*
   vertical loop, 218, 218*f*
Airport runways, labeling, 663
Air resistance (air drag), 114–115, 114*f*
   on airplane wing, 311, 311*f*
   on moving automobile, 114, 114*f*
   as nonconservative force, 135
   skydiving and, 114–115, 114*f*
   terminal speed, 114
Alcohol thermometer, 338, 339
Algal blooms in ponds and lakes, 386–387, 386*f*
Algebra
   basic rules, A.4–A.6
   factoring, A.7
   linear equations, A.8–A.9
   logarithms and exponentials, A.10–A.12
   order of calculations, 10
   powers, A.6
   quadratic equations, A.7–A.8
   solving simultaneous linear equations, A.9–A.10
Alkali metals, 961
Alphabet, Greek, A.19*t*
Alpha (α) decay, 979–981, 980*f*

Alpha (α) particles, 947–948, 948*f*
   decay process, 979–981, 980*f*
   formation in proton-proton cycle, 1001
   in magnetic field ($\vec{\mathbf{B}}$), 976–977, 976*f*
   in nuclear reaction, 985–987
   penetrating power of, 977
   RBE (relative biological effectiveness) factor, 988, 989*t*
   Rutherford's experiments with, 972–973, 972*f*
Alpher, Ralph, 1014*f*
Alternating-current (AC) circuits, 733–747
   capacitors in, 737–738, 737*f*
   impedance values and phase angle, 742*t*
   inductors in, 738–740, 738*f*, 739*f*
   phasor diagram, 741, 741*f*
   power in, 744–745
   purely capacitive, 738
   purely inductive, 739–740
   resistors in, 733–737, 734*f*, 735*f*
   RLC series circuit, 740–744, 740*f*, 741*f*
      average power ($\bar{P}$) in, 745
      energy stored in, 744
      impedance values and phase angle, 742*t*
      phase angle, 742*t*, 743–744
      problem-solving strategies, 743
      resonance in, 746–747, 746*f*
   rms current, 734–736
Alternating-current (AC) generator, 712–713, 712*f*, 714–715
Alternating-current (AC) motor, 673
Alternating-current (AC) transformer, 747–749, 747*f*
Aluminized Mylar film, 756
Aluminum (Al)
   boiling point, 375*t*
   coefficient of linear expansion, 344*t*
   density, 284*t*
   elastic modulus, 288*t*
   electron configuration of, 960*t*
   latent heat of fusion, 375*t*
   latent heat of vaporization, 375*t*
   melting point, 375*t*
   resistivity, 608*t*
   specific heat, 370*t*
   superconductor, 615, 615*t*
   temperature coefficient of resistivity, 608*t*
   thermal conductivity, 381*t*
   ultimate strength, 290*t*
   work function of, 927*t*
Ammeter, 605*f*, 606, 700, 700*f*
Amorphous solid, 283, 283*f*
Amount of substance, SI units for, A.21*t*
Ampere (A), 601, 680
Ampère, André-Marie, 677, 678
Ampère's circuital law, 677–679, 678*f*, 750
   applied to solenoids, 683, 684–685, 685*f*
Amplitude
   interference of waves and, 470, 470*f*, 471*f*
   in periodic motion, 446
   of waves, 467, 467*f*
Anaerobic bacteria, 663
Analyzers, 855–856, 855*f*, 859, 859*f*
Anderson, Carl, 1005
Aneurysms, 311
Angle(s)
   Brewster's, 858
   contact, 317, 317*f*
   critical in total internal reflection, 788–790, 789*f*, 791*f*, 792

I.1

I.2 | Index

Angle(s) (*Continued*)
  radian measure of
    converting degrees to/from, 203, 204
    defined, 203, 203f
  trigonometric functions of, in a right triangle, 15–16
Angle of deviation (δ), 782, 783f
Angle of incidence
  critical angle in total internal reflection, 788–790, 789f, 791f, 792
  in reflection, 775, 775f, 857
  in refraction, 777, 777f
Angle of polarization (θ_p), 857f, 858
Angle of reflection, 775, 775f
Angle of refraction, 777, 777f, 778f, 780, 789f
Angle of resolution, limiting (θ_{min}), 883–885
Angstrom (Å), 758
Angular acceleration, 205–206, 205f
  average (α_{av}), 205–206
  constant, rotational motion under, 206–208
  instantaneous (α), 206
  rotational analog of Newton's second law, 254, 256–259
  tangential acceleration relationship to, 209
  torque and, 241, 252–259
Angular displacement (Δθ), 203–204, 203f
Angular frequency (ω)
  frequency and, 455
  in pendulums, 460–461
  in simple harmonic motion, 455–457
Angular magnification (*m*)
  of compound microscope, 879
  of simple magnifier, 876–878
  of telescope, 880–882
Angular momentum (L), 262–267
  conservation of, 227, 263–267
  defined, 263
  torque and, 262–263, 262f
Angular position (θ), 203, 203f
Angular quantities
  relation to linear quantities, 208–211
  as vectors, 214, 214f
Angular speed
  average (ω_{av}), 204
  helicopter, 204–205
  instantaneous (ω), 204
  moment of inertia and, 263–266, 263f, 265f, 266f
  of phonograph records and compact discs, 210–211
  rotational kinetic energy and, 259–260, 260f
  tangential speed relationship to, 208
Angular velocity ($\vec{\omega}$)
  right-hand rule for, 214, 214f
  as vector quantity, 214
Antenna
  electromagnetic waves production by, 752–753, 752f
  microwave horn, 1016, 1016f
Antibaryon, 1007
Antifreeze, 301, 301f
Antileptons, 1008
Antilogarithm, A.10, A.11
Antimuons (μ^+), 1008
Antineutrino ($\bar{\nu}$), 982, 1005
Antinode, of standing wave, 498–500, 499f, 504–505
Antiparticles, 976, 982, 1004–1005, 1006t, 1008, 1010
Antiproton ($\bar{p}$), 1005
Antiquarks, 1011–1012, 1011t, 1012f
Antitau (τ^+), 1006
Aperture
  camera, 871, 871f
  diffraction pattern of circular, 884, 884f
  diffraction pattern of single-slit, 883, 883f, 884f
Apnea monitor, 707–708, 708f
Appliances
  circuits for household, 645
  third wire on consumer, 646

Approximately equal (≈), A.1, A19t
Aqueous humor, 872, 872f
Archery, 177–179, 177f, 450, 450f
Arches, 292–293, 293f
Archimedes, 299
Archimedes' principle, 299–304, 324
Architecture
  arches, 292–293, 293f
  maximum load of steel beams, 290
  post-and-beam construction, 292, 293f
  pyramids, level construction of, 297
  ultimate strength of building materials, 290t
Area
  dimension of, 5t
  expansion (γ), 345, 346
Area expansion coefficient (γ), 345, 346
Arecibo radio telescope, 886
Argon (Ar)
  electron configuration of, 960t
  molar specific heat of, 409t
  as noble gas, 960
Aristarchus, 26
Aristotle, 26, 45
Arsenic (As), electron configuration of, 960t
Arteriosclerosis, 311, 311f
Artificial gravity, 215
Artificial heart, 669, 669f
Artificial radioactivity, 985
Asbestos, thermal conductivity of, 381, 381t
ASDEX Upgrade Fusion Reactor, 971f
Asteroids
  Ceres, 221–222
  impact of, 150
  map of inner solar system, 150f
  work done by falling near-earth asteroid, 223–224
Astigmatism, 874
Astronomy and astrophysics. *See also* Cosmology; Moon; Planets; Sun; Universe; *specific planets*
  asteroids, 150, 150f, 221–222, 223–224
  atmospheric refraction, 815–816, 815f
  Big Bang theory, 1014–1017, 1015f
  black holes, 88
  cosmological red shift, 761
  Crab Nebula, 759, 760f
  curvature of spacetime, 916, 917
  dark energy, 284, 1017, 1017f
  dark matter, 283–284, 1016–1017, 1017f
  dust clouds, 886
  escape speed, 225–226, 225t, 358
  galaxies, estimated number in universe, 14
  Kepler's laws, 226–229
  pulsars, 264, 264f
  solar system
    age of, 985
    asteroid map of inner, 150f
    dust, 755
    geocentric model of, 26, 226
    heliocentric model of, 26, 45, 226
  stars
    color of, 924–925
    dark, 88
    fusion in, 1000–1001
    neutron, 264, 264f
    radiation and, 390–391, 390f
    supernova, 264, 264f, 1017
    velocity of, 1016–1017, 1016f
  telescope (*See* Telescope)
Atmosphere. *See also* Air
  carbon dioxide concentration in, 392–393, 392f
  greenhouse effect, 392
  greenhouse gases, 392–393, 392f
  rainbows, 785–786, 786f
  refraction of light by, 815–816, 815f
Atmosphere (atm), 298
Atmospheric electric field, measuring, 538, 538f

Atmospheric pressure, 285, 294, 297–298
  standard, A.19t
Atmospheric refraction, 815–816, 815f
Atom(s). *See also* Atomic nucleus
  characteristic x-rays, 961–962, 961f, 962f
  excited states, 963–964, 963f, 964f
  ground state, 963, 963f, 964f
  models of, 947–948, 948f
    Bohr's, 950–955, 951f, 952f
    Geiger and Marsden's, 947, 948f
    Rutherford's, 948, 948f
    Thomson's, 947, 948f
  origin of term, 4
  spontaneous emission, 963, 963f
  stimulated absorption process, 963, 963f
  stimulated emission, 963–964, 964f
  structure of, 4–5, 5
  vibrating, 283
Atomic clock, 2f, 3, 4, 917
Atomic mass, 349, 973
Atomic mass unit (u), 349–350, 972, 972t, A.20
Atomic nucleus, 971–976
  atomic number, 971–972
  binding energy, 974–976, 975f
  charge, 972, 972t
  daughter nucleus, 980–982.997
  density, 973
  discovery of, 948, 973
  effective nuclear charge (Z_{eff}), 962
  mass, 972
  mass number, 971, 972
  neutron number, 971
  nuclear fission, 996–1000, 997f, 999f
  nuclear fusion, 1000–1003
  parent nucleus, 980–982
  size, 972–973
  stability, 973, 974f
Atomic number (Z), 961–962, 962f, 971–972, 973, 974f
Atomic physics, 947–964
Atomic spectra, 948–950, 949f
  absorption spectrum, 949f, 950, 950f
  emission spectrum, 949, 949f
Atomizers, 311, 311f
Atom smashers, 564
Atrioventricular (AV) node, 616f, 617
Atwood's machine, 107, 107f
Audible waves, 482
Auroras, 950
Automobile(s)
  accident reconstruction, 148
  air bags, 175
  air resistance on moving, 114, 114f
  antifreeze, 301, 301f
  battery
    charge, checking, 301, 301f
    potential difference in, 564
  bumpers, 174
  car lift, 296, 296f
  collisions
    accident reconstruction, 148
    analysis of, 134
    crash test, 174, 174f
    force on car *versus* time, 175, 175f
    with guard rail, spring-loaded, 450–451, 451f
    injury in, 148, 174–175
    Newton's third law and, 99
    perfectly inelastic, 180–181
    two-dimensional inelastic collision, 187–188, 187f
  crash test, 174, 174f
  Doppler effect in sirens, 495
  engines
    conversion of simple harmonic to circular motion in, 454
    cooling, 386
    fluid dynamics in, 311
    as heat engine, 418
    work done in adiabatic expansion, 412

friction and, 114
motion and
  acceleration, 34–36, 34f, 36f
  average speed, 29–30
  average velocity ($\bar{v}$), 30, 31
  on banked, curved track, 217–218, 217f
  car chase problem, 41–42, 41f
  displacement, 27–28, 28f
  friction, 114
  instantaneous velocity, 32, 32t, 33f, 33f, 64
  motion diagrams, 37–38, 37f
  rearview mirrors, 803–804, 803f
  seat belts, 91, 91f, 175
  shock absorbers, 463, 463f
  side-view mirrors, 809, 809f
  skid marks, 133
  spring constants of, 148
  streamline flow around, 305f
  windshield wipers, 641
Avalanche, 465
Average acceleration
  defined, 34, 64
  in one dimension, 34–35, 34f
  SI (Système International) units, 34, 64
  in two dimensions, 64
Average angular acceleration (α_{av}), 205–206
Average angular speed (ω_{av}), 204
Average current (I_{av}), 601
Average force ($\vec{F}_{av}$), 172, 173, 174
Average power ($\bar{P}$), 150–151, 152
  in alternating-current (AC) circuits, 745
Average speed, 29–30
  average velocity ($\bar{v}$) vs., 31
Average velocity ($\bar{v}$)
  average speed vs., 31
  in constant acceleration, 39, 40t
  defined, 30, 64
  in one dimension, 30–31, 30f–31f
  from position vs. time graph, 31, 31f
  SI (Système International) units, 30, 64
  in two dimensions, 64
Avogadro's number (N_A), 349–350, A.20
Axis of rotation
  selecting for problem solving, 247
  torque and, 243–244, 253
Axis of symmetry, 247
Axons, 647–648, 647f–649f

B

Back emf, 715–716, 715f, 739
Bacteria
  anaerobic, 663
  bioluminescence in, 149
  growth of, 641
  magnetic, 663
  mass of, 3t
Bakelite®, dielectric constant and dielectric strength of, 585t
Ballast, lead, 291–292
Ballistic pendulum, 181–182, 182f
Balloons, hot air, 282, 300
Ballpoint pens, 298
Balmer, Johann, 949
Balmer series, 949, 950f, 952f, 953–954
Barometer, 297, 297f
Baryonic charge, 1007
Baryon number, conservation of, 1007–1008
Baryons, 1006, 1006t, 1007–1008, 1010, 1010f, 1011–1012, 1011f, 1012f
Base number, A.10
Basilar membrane, ear, 512
Baton twirling, 254–256, 254f–256f
Battery
  automobile
    charge, checking, 301, 301f
    potential difference in, 564
  in defibrillators, 583

electric circuit symbol for, 576
emf source, 626–627, 627f
invention of, 600
lithium, 618
Beach, air circulation at, 370, 370f, 385
Beat(s), 508–510, 508f
Beat frequency, 509–510
Becquerel (Bq), 978
Becquerel, Antoine Henri, 976
Bednorz, J. Georg, 615
Bed of nails, 286, 286f
Bell, Alexander Graham, 487
Bell Laboratories, 1015, 1016f
Benzene
  coefficient of volume expansion, 344t
  density, 284t
  index of refraction, 778t
Bernoulli, Daniel, 307, 308f
Bernoulli's equation, 307–313
  applications of, 309–313
    aneurysms, 311
    atomizers, 311, 311f
    finding speed of a fluid, 309–310
    golf ball flight, 311
    home plumbing, 313
    lift on airplane wing, 311–312, 311f
    rocket engines, 313
    sailing upwind, 312
    vascular flutter, 311, 311f
  Venturi tube and, 308, 309f
Beryllium (Be)
  electron configuration of, 960t
  specific heat of, 370t
Beta (β) decay, 981–982, 982f
Beta (β) particles
  beta decay process, 981–982, 982f
  in magnetic field ($\vec{B}$), 976–977, 976f
  penetrating power of, 977
  RBE (relative biological effectiveness) factor, 989t
Biconcave lens, 816f
Biconvex lens, 816f
Bicycle gears, 254, 254f
Big Bang theory, 1014–1017, 1015f
Bimetallic strips, 345, 345f
Binding energy of atomic nucleus, 974–976, 975f
Biology. See Medicine and biology
Blackbody, 388, 1016
Blackbody radiation, 923–925, 924f, 925f, 1016, 1016f
Black holes, 88
Blood
  dialysis, 323, 323f
  flow
    Poiseuille's law and, 320
    turbulent, 321
  skin temperature and, 341
  viscosity of, 319t, 320
Blood pressure measurement, 298, 298f
Blue-shift, 910
Body temperature, maintaining, 387
Bohr, Niels, 4, 948, 950–953, 951f
Bohr magneton (μ_B), A.20
Bohr radius (a_0), 952, A.20
Bohr's correspondence principle, 954
Bohr's model of the atom, 950–955, 951f, 952f
Boiling point of water, 339
Boltzmann's constant (k_B), 354, 430, A.20
Bond energy, 368
Bone
  elastic modulus of, 288t
  fractures, in automobile collisions, 174–175
  ultimate strength of, 290t
Boron (B), electron configuration of, 960t
Bose, Satyendra Nath, 896
Bose-Einstein condensate, 896
Bosons, 1004, 1004f, 1010, 1013–1014, 1013f
  Higgs, 1014

W, 1004, 1004t, 1013–1014, 1013f
Z, 1004, 1004t, 1013–1014, 1013f
Bottom quark (b), 1011–1012, 1011t
Boxing and brain injury, 172–173
Boyle's law, 350
Brache, Tycho, 226
Brachial artery, 298
Bragg, W. L., 931
Bragg's law, 931–932
Brain
  boxing injury to, 172–173
  cerebrospinal fluid, 301
  estimated number of cells, 13
  high-intensity focused ultrasound (HIFU) use on, 484
Brass
  coefficient of linear expansion, 344t
  elastic modulus, 288t
Breaking point, 288
Breeder reactors, 1000
Bremsstrahlung, 930, 961f
Brewster, David, 858
Brewster's angle, 858
Brewster's law, 858
Brick, ultimate strength of, 290t
Bridges, thermal expansion joints of, 343, 343f
Bright fringes, 837–841, 837f, 848–849
British thermal unit (Btu), 368
Bromine (Br), electron configuration of, 960t
Bronze, coefficient of linear expansion for, 344t
Brushes, of electric motors, 673, 673f
Bugles, 506
Buildings
  arch structures in, 292–293, 293f
  bursting of water pipes in winter, 348–349
  home insulation, 383–385, 383t
  radiator heating of, 385–386, 386f
  R-values for common building materials, 383t
  thermal expansion joints, 343, 343f
  thermogram, 388, 388f
Bulk modulus (B), 289–290, 484
Bulk strain, 291–292
Bulk stress, 289, 289f, 291–292
Bungee jumping, 456, 456f
Buoyant forces (B), 299–304, 299f–304f
  applications of, 301
  defined, 299
  on floating object, 300–301, 301f, 303–304, 304f
  on submerged object, 300, 300f, 302–303, 302f, 352–353
Bursting of water pipes in winter, 348–349

C

Cadmium (Cd)
  in nuclear reactor control rods, 999
  specific heat of, 370t
Calcite, x-ray diffraction of, 932–933
Calcium (Ca), electron configuration of, 960t
Calculations. See also Measurement
  order-of-magnitude, 12–14
Calculators
  inverse tangent function on, 60
  significant figures on, 7, 8
Calculus, invention of, 88
Calorie (C), 155, 368
Calorie (cal), 368
Calorimeters, 372
Calorimetry, 372–374
  problem-solving strategy, 377
Camera, 870–871, 871f
  digital, 870
  flash attachments, capacitor in, 574
  f-number, 871
  red eye in flash photographs, 775
  spherical aberration, 825
  ultrasonic ranging unit, 484
Cancer treatment, 737

Capacitance (C)
  change in computer keyboards, 574
  defined, 573
  equivalent
    for parallel capacitors, 576f, 577–578
    for series capacitors, 579–580, 579f
  in pair of conductors, 573
  of parallel-plate capacitors, 574, 576f, 577
  SI units of, 573
Capacitive reactance (X_C), 738
Capacitor(s)
  in alternating-current (AC) circuits, 737–738, 737f
  applications
    camera flash attachments, 574
    computer keyboards, 574
    defibrillators, 583
    electrostatic confinement, 575
    maximum energy design, 584
    stud finders, 586, 586f
  in cardiac pacemaker, 618
  combinations of
    in parallel, 576–578, 576f, 577f
    problem-solving strategy, 580
    in series, 578–580, 579f
  commercially available, 585–586, 586f
  defined, 573
  dielectrics, 584–590, 585f, 586f
  electrolytic, 586, 586f
  energy stored in, 582–584, 582f, 722
  impedance values and phase angle, 742t
  instantaneous voltage across, 740, 740f
  LC circuits, 750–751
  in parallel, 576–578, 576f, 577f
  parallel-plate, 548, 548f, 573–578, 573f, 574f, 576f, 577f
    dielectric in, 584–585, 585f
  power and, 583
  pure, 744
  RC circuits, 640–644, 640f–641f
    charging capacitor in, 642–643
    discharging capacitor in, 643–644
  in RLC series circuit, 740–745, 740f
  in series, 578–580, 579f
  symbol, 575
  voltage across a, 738, 740–741, 740f
Car. See Automobile
Carbon (C)
  atomic mass of, 349, 350
  beta decay of carbon-14, 981–982, 983, 984–985
  electron configuration of, 960t
  isotopes, 972, 981–982
  resistivity, 608t
  temperature coefficient of resistivity, 608t
Carbon dating, 983, 984–985
Carbon dioxide (CO_2)
  atmospheric concentration of, 392–393, 392f
  dry ice, 355, 375
  index of refraction, 778t
  molar specific heat, 409t
  root-mean-square (rms) speed, 358t
Carbon disulfide, index of refraction, 778t
Carbon monoxide (CO), molar specific heat of, 409t
Carbon tetrachloride, index of refraction, 778t
Carburetor, 311
Cardiac pacemakers, 617–618
Carlson, Chester, 572
Carnot, Jadi, 424, 425f
Carnot cycle, 424–425
Carnot engine, 424–426, 424f, 425f
Carnot theorem, 424
Cartesian coordinate system, 15, 15f. See also Rectangular coordinate system
  conversion to/from polar coordinates, 16–17, 16f
Case ground, 646
Catapult, space, 709–710

Cathode ray
  tubes, 683f
  TV sets, 683
Cavendish, Henry, 220
Cavitron ultrasonic surgical aspirator (CUSA), 484
CCD (charge-coupled device), 870
CDs. See Compact discs
Cells
  osmosis and, 322–323
  size of, 3, 322
Celsius temperature scale, 338–339, 340–342, 341f, 373
Center of gravity, 246–249, 246f–248f
Center of mass (CM), 154
Centigrade scale. See Celsius temperature scale
Centimeter (cm), 4
Centipoise (cp), 319
Centrifugal force, as fictitious force, 219
Centrifugation, 324–325, 325f
Centripetal acceleration (a_c), 7, 211–219
  defined, 212
  forces causing, 214–219, 215f
  magnetic force and, 673
  problem-solving strategy, 215–216
Centripetal force (F_c), 214–219, 215f
Cerebrospinal fluid, 301
Ceres (asteroid), 221–222
CERN (European Council for Nuclear Research), 904, 1013f, 1014
Cesium atomic clock, 2f, 3
Cgs. See Gaussian (cgs) system of units
Change in, symbol for (Δ), A.2, A.19t
Chain reaction, nuclear, 998–999, 999f
Challenger disaster, 336
Characteristic x-rays, 961–962, 961f, 962f
Charge. See also Electric charge
  baryonic, 1007
  of quarks, 1011, 1011f
Charge carriers, 601
Charge-coupled device (CCD), 870
Charged particles
  moving through magnetic field, 663–667, 665f, 673–676, 674f
  storing (trapping), 674
Charge exchange, 1013
Charge pump, 627
Charles's law, 350
Charmed quark (c), 1011–1012, 1011t
Cheney, Dick, 618
Chlorine (Cl)
  electron configuration of, 960t
  molar specific heat of, 409t
Christmas lights in series, 629, 629f
Chromatic aberration, 826, 826f, 881
Chromium (Cr)
  electron configuration of, 960t
  radioactive tracers, 989
Ciliary muscle, 872, 872f
Circle, reference, 457, 457f
Circuit breakers, 634, 644–645, 645f
Circuit diagrams, 576, 576f, 605, 605f
  for photoelectric effect, 926f
Circuits. See Alternating-current circuits; Direct-current circuits; Electric circuits
Circular motion, uniform compared to simple harmonic motion, 453–457, 454f
Clarinet, 510, 511f
Clausius, Rudolf, 427
Clock
  atomic, 2f, 3, 917
  pendulum, 461–462
  time dilation and, 903–94
Clothing, color of, 388
Cloud chamber, 1005
CMOS (complementary metal-oxide semiconductor), 870
Coal, 1000

Coastal flooding, 347
Coaxial cable, magnetic field of, 678–679, 678f
Cobalt (Co)
  electron configuration of, 960t
  radioisotope use for irradiation of materials, 989
Cochlea, 512–513, 512f
Cochlear implant, 513
Coefficient of area expansion (γ), 345, 346
Coefficient of kinetic friction (μ_k), 109, 109t
Coefficient of linear expansion (α), 343, 344t, 347
Coefficient of performance (COP)
  heat pumps, 421–422
  refrigerators, 421–422
Coefficient of static friction (μ_s), 109, 109t
Coefficient of viscosity (η), 319, 319t
Coefficient of volume expansion (β), 344t, 346–347
Coherent light sources, 836
Cohesive forces, 316–319, 316f–318f
Coil
  induction, 751, 751f
  magnetic moment of, 671
  pickup, 707, 707f
  primary, 699–700, 700f, 747–748
  secondary, 700, 747–748
  sensing, 706, 706f
Collisions. See also Automobile(s), collisions
  comet impacting Earth, 378–380
  conservation of momentum and, 176, 176f, 179–188
  elastic
    defined, 179, 180
    in one dimension, 183–185, 183f
    in two dimensions, 187–188
  glancing, 186–188, 186f
  inelastic
    defined, 179, 180
    glaucoma testing, 179–180
    perfectly inelastic compared to, 179
    in two dimensions, 186–188, 187f
  kinetic energy in, 179–181, 183, 378–379
  in one dimension
    elastic, 183–185, 183f
    perfectly inelastic, 180–183, 180f, 182f
    problem-solving strategy, 184
  perfectly inelastic
    defined, 179, 180
    inelastic compared to, 179
    in one dimension, 180–183, 180f, 182f
  in two dimensions, 186–188, 186f
    elastic, 187–188
    inelastic, 186–188, 187f
    problem-solving strategy, 187
Color
  interference in thin films and, 841, 842
  iridescence, 835f
  light dispersion and, 782–786, 783f
  quarks, 1012, 1012f
  rainbows, 785–786, 786f
  of stars, 924–925
  of water ripples at sunset, 776
Color charge of quarks, 1012, 1012f
Color force, 1012, 1014
Comet impacting Earth, 378–380
Common logarithm base, A.10
Commutative law of addition, 58
Compact discs (CDs)
  angular and tangential speeds of, 210–211
  reading, 846–847, 846f
  tracking using diffraction grating, 853, 853f
Compass, 662, 663, 676–677, 676f
Complementary metal-oxide semiconductor (CMOS), 870
Components, vector, 60–63, 60f–62f
Compound microscope, 878–880, 878f
Compressibility of materials, 290
Compression
  due to tensile stress and maximum load, calculating, 290

in sound waves, 482, 482f
Young's modulus and, 288, 290
Compressive strain, 290
Compressive strength, 290t
Compressive stress, 290, 371
Compton, Arthur H., 933, 933f
Compton effect, 933–935, 933f
Compton shift, 933–934
Compton wavelength (h/m_ec), 934, A.20
Computer keyboard, capacitors in, 574
Concave mirrors, 804–806, 804f–806f, 809–810
  with thin lenses, 824–825, 824f
Concentration
  diffusion and, 321–322, 321f
  gradient, 322
Concrete
  coefficient of linear expansion, 344t
  thermal conductivity, 381t
  ultimate strength of, 290t
Conduction. *See also* Thermal conduction
  charging by, 526, 526f
  electrical
    charging by, 526, 526f
    in heart, 616–617, 616f
    by neurons, 647–648, 647f–649f
  in Polaroid, 855
Conductors. *See* Electric conductors
Conservation
  of angular momentum, 227, 263–267
  of baryon number, 1007–1008
  defined, 138
  of electric charge, 525, 636
  of electron-lepton number, 1008
  elementary particles and, 1007–1009
  of energy, 149–150, 158
    applied to ideal fluid, 307
    in biological systems, 149
    calorimetry and, 372
    defined, 149
    Earth escape speed and, 225
    in electric potential energy problems, 562, 568
    energy transfer to/from system and, 148–149
    loop rule and, 637
    Sun formation and, 224
  in isolated system, 176–177, 176f
  of lepton number, 1008
  of mechanical energy
    gravity and, 138–141
    problem examples, 140–141, 140f, 145–148, 145f–147f
    problem-solving strategy, 139
    in rotational motion, 260–261
    in springs, 145–148
    in vertical jump, 154
  of momentum, 170, 176–179, 176f, 177f
    in collisions, 79–188, 176, 176f
    defined, 176
    recoil and, 177–179
    in rocket propulsion, 189
    squid propulsion and, 177, 177f
  of strangeness, 1009, 1009f
Conservative force(s), 134–135, 135f, 148, 158
  Coulomb force as, 449
Constant acceleration
  in one dimension, 38–44, 38f, 40t
  problem solving strategy, 40
  rotational motion under constant angular acceleration, 206–208
  in two dimensions, 72–73, 72f
Constant of gravitation (G), 95, 219, 220, 220f, 915, A.20
Constants, table of fundamental, A.20
Constant-volume gas thermometer, 339–340, 339f, 340f
Constructive interference
  of light waves, 888
    conditions for, 835–836, 837f, 838, 842–844

equation for, 842
position of, 838–839, 849
in single-slit diffraction, 849
in thin films, 842–843
of water waves, 837, 837f
of waves, generally, 470–471, 470f–471f, 497, 509
in x-ray diffraction, 931
Consumer appliances, third wire on, 646
Contact angle, 317, 317f
Continuity, equation of, 305–307
Control rods, of nuclear reactor, 999, 999f
Convection, 385–387, 385f–386f
Converging lens, 816–818, 816f, 820–821, 820f, 822–824, 825, 825f, 826f
  for correcting farsightedness, 873, 873f, 875
  for correcting presbyopia, 874
  of simple magnifier, 876–878
  thin lenses, 816–818, 816f, 820–821, 820f, 822–824
Conversion
  Calories to joules, 369
  of Cartesian coordinates to/from polar coordinates, 16–17, 16f
  of force units, 92
  radians to/from degrees, 203, 204
  SI (Système International) units to/from U.S. customary units, 11–12
  of temperature scales, 341–342
Convex-concave lens, 816f
Convex mirrors, 806–812, 807f–809f
Coordinate system
  Cartesian, 15, 15f
  components of, 15
  conversion between, 16–17, 16f
  frame of reference, 27
  plane polar, 15, 15f
  vector components on, 60–61
COP (coefficient of performance)
  heat pumps, 421–422
  refrigerators, 421–422
Copernicus, Nicolas, 27, 45, 226
Copper (Cu)
  boiling point, 375t
  coefficient of linear expansion, 344t
  density, 284t
  as ductile metal, 288
  elastic modulus, 288t
  electron configuration of, 960t
  latent heat of fusion, 375t
  latent heat of vaporization, 375t
  melting point, 375t
  resistivity, 608t
  specific heat, 370t
  temperature coefficient of resistivity, 608t
  thermal conductivity, 381t
  work function of, 927t
Cornea, 872, 872f
Correspondence principle, 954
Cosine (cos), 15f, 16
Cosmic Background Explorer (COBE), 1016, 1016f
Cosmic background radiation, 1015–1016, 1016f, 1017–1018
Cosmological constant, Einstein's, 1017
Cosmological red shift, 761
Cosmology
  background radiation, 1015–1016, 1016f
  Big Bang theory, 1014–1017, 1015f
  dark energy, 1017, 1017f
  dark matter, 1016–1017, 1017f
  evolution and fate of the Universe, 1017–1019, 1018f
  horizon problem, 1018
  monopole problem, 1018
  unanswered questions in, 1016–1019
Coulomb (C), 525, 528, 680
Coulomb, Charles, 527, 528
Coulomb constant (k_e), 527, 528, 951, A.20
Coulomb force, 528–529, 559
  in Bohr's model of the atom, 951

as conservative force, 449
repulsion by, 972, 973
Coulomb per second, 601
Coulomb per volt (C/V), 573
Coulomb's law, 527–532, 528f, 750
Crab Nebula, 759, 760f
Crash test, automobile, 174, 174f
Crick, Francis, 932
Critical angle in total internal reflection, 788–790, 789f, 791f, 792
Critical condition, of nuclear reactor, 999
Critical temperature, 615, 615f, 615t
Crystal(s)
  liquid, 860–861, 860f
  piezoelectric, 483, 483f
  x-ray diffraction by, 930–933, 931f, 932f
Crystal ball, gazing into, 813–814, 813f
Crystalline lens, 872
Crystalline solids, 283, 283f
Curie (Ci), 978
Curie, Marie, 976, 976f
Curie, Pierre, 976
Current. *See also* Electric current
  induced
    direction of, 705–708, 705f–707f
  torque on current loop in magnetic field, 670–673, 670f, 672f
Curvature of spacetime, 916, 917
CUSA (cavitron ultrasonic surgical aspirator), 484
Cut-off frequency (f_c), 927–928
Cutoff wavelength (λ_c), 928
Cyclotron equation
  applying, 675–676
  described, 674
Cyclotrons, 564, 659

D

Damped oscillations, 463–464, 464f
Dark energy, 284, 1017, 1017f
Dark fringes, 837–839, 837f, 840–841, 848–850
Dark matter, 283–284, 1016–1017, 1017f
Daughter nucleus, 980–982, 997
Davisson-Germer experiment, 935
Day, time interval for one, 3t
Daytona International Speedway, banked curves at, 214
DC. *see entries under* Direct-current
Deafness, 513
de Broglie, Louis, 935, 936f
de Broglie hypothesis, 935–936
de Broglie wavelength, 935–936
Decay constant (λ), 977–979
Decay curve, 977, 977f
Decay rate (activity) (R), 977–979
Deceleration, 35
Decibels, intensity of sound waves in, 487–489, 488t
Decompression, injury from, 298
Defibrillators, 583
Deformation
  of objects by force, 90
  of solids, 287–293
Degree of freedom, 408
Degrees
  converting radians to/from, 203, 204
  setting calculator to, 16
Delta (Δ), A.2, A.19t
Democritus, 4
Dendrites, 647, 647f
Density (ρ), 284, 286–287
  of atomic nuclei, 973
  of common substances, 284t
  defined, 284
  specific gravity and, 284
  units of, 284
Density waves, 466
Depolarization wave, in heart, 616–617, 616f
Depth, pressure variation with, 293–297, 293f, 294f

Depth of field, camera, 871
Destructive interference
  of light waves, 773, 888
    conditions for, 835–836, 837f, 838, 842–843, 849
    equation for, 842
    position of, 838–839, 849
    in single-slit diffraction, 849
    in thin films, 842–843
  of water waves, 837, 837f
  of waves, generally, 471, 471f, 497, 509
Detergents, 317
Deuterium, 4
  binding energy of nucleus, 974–975
  as isotope, 972
  mass (m_d), A.20
  nuclear fusion and, 1001–1003
Deuteron
  binding energy of, 974–975
  mass, A.20
Dewar flask, 391, 391f
Dextrose, as optically active material, 859
Dialysis, kidney, 323, 323f
Diamond, index of refraction, 778t
Diatomic gases, molar specific heat of, 408, 409t
Dielectric constant, 584, 585t
Dielectric mirrors, 846
Dielectrics
  atomic description of, 589–590, 589f
  capacitors and, 584–590, 585f, 586f
  defined, 584
  of example materials, 585t
Dielectric strength, 585, 585t
Diet vs. exercise in weight loss programs, 154–155
Difference in, symbol for (Δ), A.2
Diffraction
  defined, 773
  of light waves, 847–854
    defined, 847
    Fraunhofer, 848, 848f, 849f
    single-slit, 848–850, 849f
  resolution and, 883–884
  of sound waves, 850
  of x-rays by crystals, 930–933, 931f, 932f
Diffraction grating, 851–854, 851f–853f
  CD tracking, use in, 853, 853f
  effect on plane waves, 851, 851f
  maxima for, 851–852, 851f, 853–854
  order number of diffraction pattern, 852, 852f
  prism compared, 852
  resolution of, 886–887
Diffraction-grating spectrometer, 852, 852f
Diffraction pattern
  of circular aperture, 884, 884f
  Fraunhofer, 848, 848f, 849f
  order number of, 852, 852f
  of single-slit, 883, 883f, 884f
  x-ray, 929, 929f, 931, 932, 932f
Diffuse reflection, 774, 775f
Diffusion, 321–322, 321f, 322t
  coefficient, 322, 322t
  concentration and, 321–322, 321f
  Fick's law, 322
  osmosis, 322–323, 323f
Diffusion rate, 322
Digital videodisc (DVD), reading of, 781–782, 781f, 846–847, 846f
Diamagnetic materials, 687, 687f
Dimensional analysis, 5–7, 5t
Dimensions as algebraic quantities, 5
Diode, as nonohmic device, 607, 608f
Diopters, 874
Dipole
  electric potential, 566, 566f
  water molecules and, 589
Dirac, Paul Adrien Maurice, 957, 1004–1005, 1005f
Direct-current (DC) circuits, 626–649
  electrical safety, 645–646, 646f

  household circuits, 634–635, 644–645, 645f
  Kirchhoff's rules, 636–640, 636f–637f
  problem-solving strategies
    Kirchhoff's rules, 637–638
    resistors, 635
  RC circuits, 640–644, 640f–641f
  resistors in parallel, 631–636, 631f
  resistors in series, 628–630, 628f–630f
  RL circuits, 718–721, 719f, 720f
  sources of emf, 626–627, 627f
Direct-current (DC) generator, 713–714, 713f
Direct-current (DC) motor, 673, 673f
Disorder and entropy, 429–431
Dispersion of light, 782–786, 782f–784f, 786f, 790
Displacement. See also Linear displacement; Position
  angular (Δθ), 203–204, 203f
  defined, 27, 64
  distance distinct from, 28
  as function of time, 39, 40t
  Hooke's law and, 446, 446f
  in one dimension, 27–28, 28f, 39, 40t
  resultant, 59, 59f
  runway length problem, 42–43, 42f
  SI (Système International) units, 27, 64
  in two dimensions, 63–64, 63f
  as vector quantity, 28
  velocity as function of, 39, 40t
  work and, 128–130f, 128f–130f
  work done by a varying force and, 155–156, 155f
Dissipative forces
  nonconservative force as, 135
  work and, 131–132
Distance (d)
  dimension of, 5–6
  displacement distinct from, 28
Diverging lens, 816–819, 816f, 821–822, 822f, 826
  for correcting nearsightedness, 873, 874f, 875
  thin lenses, 816–819, 816f, 821–822, 822f
Diverging mirror, 806. See also Convex mirrors
Diving, 135, 135f, 140, 140f
Diving masks, 819
Division
  of fractions, A.5
  of powers, A.6
  and scientific notation, A.4
  significant figures and, 8
  of vector by a scalar, 59
DNA
  structure of, 932, 932f
  x-ray diffraction, 932, 932f
Domains, magnetic, 686, 686f
Doppler, Christian, 491
Doppler effect
  of electromagnetic waves, 760–761
  general equation for, 493
  independence from distance, 493
  problem-solving strategies, 493–495
  sound waves, 491–496, 492f, 509–510
    observer moving relative to stationary source, 492, 492f
    source moving relative to stationary observer, 492–493, 492f
  ultrasound and, 483
Doppler shift, 493–495
Double images, 776
Down quark (d), 1011, 1011t
Drift speed, 603–605, 603f
Drive wheel mechanism of locomotives, 454, 454f
Dry ice, 355, 375
Dust, solar system, 755
Dust clouds, 886
DVD (digital videodisc), reading of, 781–782, 781f, 846–847, 846f
Dynamics. See also Thermodynamics
  defined, 26
  fluid, applications of, 311–313

  kinematics, 26
  rotational, 240

E

e (fundamental unit of electric charge), 525
Ear anatomy and function, 511–513, 512f
Ear canal, 511–512, 512f
Eardrum, 512–513, 512f
  pressure on while swimming, 295
Earplugs, 488
Earth
  age of, 3t
  atmospheric pressure, 285, 294, 297–298
  atmospheric refraction, 815–816, 815f
  comet impacting, 378–380
  distance to Moon, A.19t
  distance to Sun, A.19t
  equatorial radius of, A.19t
  escape speed, 225, 225t, 358
  geocentric model of universe, 26, 226
  geosynchronous orbits of telecommunications satellites, 229
  global warming, 391–393, 392f, 393f
    coastal flooding, 347
    greenhouse gases, 392–393, 392f
  heliocentric model of solar system, 26, 45
  magnetic field of, 661–663, 662f, 701
    current-carrying wire in, 669–670
    lightning strike deflection by, 669
    magnetic force on proton moving through, 665–666
  reversal of, 662–663
  mass of, 3t, 228t, A.19t
  planetary data, 228t
  tangential speed of rotating surface, 209
Earthquake waves, 482
Eddington, Arthur, 916
Eddy currents, 305
Effective nuclear charge (Z_{eff}), 962
Efficiency. See also Thermal efficiency
  of heat engine, 418–419, 423, 424–425
  of human body, 434–435, 434t
Eightfold way, 1010, 1010f
Einstein, Albert, 896f, 898f, 899f
  cosmological constant, 1017
  general relativity and, 915–916, 1017, 1018
  kinetic energy expression, 911
  mass-energy equivalence equation, 912
  photoelectric effect and, 773, 927
  photons and, 773
  special relativity and, 900, 901
  stimulated emission, prediction of, 963
EKG (electrocardiogram), 616–617, 617f
Elastic collisions
  defined, 179, 180
  in one dimension, 183–185, 183f
  in two dimensions, 187–188
Elasticity of solids, 283, 283f
Elastic limit, 288, 288f
Elastic modulus, 287–288, 288t
Elastic potential energy (PE_s), 449–453, 449f–452f
  defined, 144
  spring potential energy, 143–148
  work done by, 144
  work-energy theorem, 144–145
Elastic property of medium, wave speed and, 485
Electrical conduction
  charging by, 526, 526f
  in heart, 616–617, 616f
  by neurons, 647–648, 647f–649f
Electrical energy, 582–584, 582f, 611–615
Electrical resistance. See Resistance
Electrical safety, 645–646, 646f
Electrical transmission and energy transfer to/from system, 149
Electric charge
  attraction of unlike, 524, 524f

charging by conduction, 526, 526f
charging by induction, 526, 527f
conservation of, 525, 636
of electrons, 324, 525, 528, 528t, 541–542
fundamental unit of, 525
Millikan oil-drop experiment, 541–542, 541f–542f
negative, 524–525, 524f, 525f
polarization of, 526
positive, 524–525, 524f, 525f
properties of, 524–525, 524f, 525f
of protons, 524–525, 528, 528t
quantization of, 525, 542
repelling of like, 524, 524f
SI units, 525, 528, 680, A.21t
transfer of, 525, 525f
Van de Graaff generator and, 540, 542–543, 542f
Electric circuits. *See also* Alternating-current (AC) circuits; Direct-current (DC) circuits
defined, 605
direct-current, 626–649
energy transfer in, 611–612
household circuits, 634–635, 644–645, 645f
impedance of, 742
junction rule, 636–639, 636f
Kirchhoff's rules, 636–640, 636f–637f
LC circuits, 750–751, 751f
loop rule, 636–639, 637f
measuring current and voltage in, 605–606, 605f
radio tuning circuit, 746
resistance in, 607
secondary, 700, 700f
short circuit, 707, 707f
sources of emf, 626–627, 627f
sparks in, 717–718
symbols, 575–576
battery, 576
capacitor, 575
ground, 612
inductor, 719
resistor, 576, 607
series circuit, 734, 734f
twisted wires in, 682
Electric conductors
capacitance in pair of, 573
current-carrying
drift speed, 603–605, 603f
electric potential in, 601
magnetic force on, 667–670, 667f–669f
defined, 525
electric potential on charged conductor in electrostatic equilibrium, 569, 569f
in electrostatic equilibrium, 539–541, 539f, 540f
magnetic force
on current-carrying, 667–670, 667f–669f
between two parallel conductors, 679–681, 679f
resistance, 606–608, 606f
resistivity of, 608, 608f
superconductors, 615–616, 615f, 615t, 616f
Electric current, 601–603
average (I_{av}), 601
from changing magnetic field, 700
defined, 601
direction of, 601, 601f
electric shock, 645–646, 646f
instantaneous (I), 601
in resistor, 734
junction rule, 636–639, 636f
magnetic fields of current loops and solenoids, 681–685, 681f–683f, 685f
magnetic force on current-carrying conductor, 667–670, 667f–669f
measuring in electrical circuit, 605–606, 605f
microscopic view of, 603–605, 603f
misconceptions concerning, 612
motors and back emf, 715–716
in Ohm's law, 607
path of least resistance, 639

rms, 734–736
SI units, 601, 680, A.21t
in superconductor, 615
Electric dipole
electric field lines and, 537, 537f
electric potential, 566, 566f
Electric field, 532–536
atmospheric, 538, 538f
of charged spherical surface, 546–547, 546f
defined, 532
direction of, 532–533, 532f, 533f
electric flux and, 543–544
in electrostatic precipitator, 571
Gauss's law and, 543, 545–548
magnitude, 533
Maxwell's predictions, 749–750
of parallel-plate capacitors, 574, 574f
potential energy differences in, 561
problem-solving strategy, 535
SI unit of, 532
superposition principle and, 533, 535–536
Electric field lines, 536–538, 537f, 538f
electric flux and, 543–544, 543f, 544f
Electric field therapy, 737
Electric field vector ($\vec{E}$), 854–855, 854f, 855f
Electric flux (Φ_E), 543–544
calculating through a closed surface, 544–545, 544f
defined, 543
expression for, 544
Gauss's law and, 543, 545–548
Electric force (F_e)
Coulomb's law, 527–532, 528f
as field force, 89
gravitational force compared to, 529
magnetic force compared to, 664
problem-solving strategy, 535
properties of, 527
superposition principle, 529–532
Electric generator, 700
Electric guitar, 707, 707f
Electric heater, power converted by, 614–615
Electricity
early studies of, 523, 524
generation by heat engine, 418
Maxwell's predictions, 749–750
Electric monopoles, 750
Electric motors. *See* Motors, electric
Electric potential (voltage; V), 562–565
on charged conductor, 569, 569f
current and, 601–602
difference (ΔV), 563–564
due to point charge, 565–568, 565f, 566f
electric dipole associated, 566, 566f
equipotential surfaces, 570–571, 570f
problem-solving strategy, 567
relationship to electric potential energy, 559
as scalar quantity, 563
SI units, A.21t
superposition principle, 565–566
Electric potential energy (PE_{el})
change in, 560
in conservation of energy problems, 562
gravitational potential energy compared, 560, 560f
potential energy differences in electric field, 561
relationship to electric potential, 559
work and, 559–560, 559f
Electric resistance. *See* Resistance
Electric shock, 645–646, 646f
Electrocardiogram (EKG), 616–617, 617f
Electrocardiograph, 616
Electromagnet, 682–683, 683f, 686. *See also* Solenoid
Electromagnetic force
Big Bang and, 1015, 1015f
as field force, 89

as fundamental force, 1003–1004, 1004t, 1015, 1015f
photons and, 1004, 1004f, 1004t, 1013–1014, 1013f
Electromagnetic induction, 583
Electromagnetic pump, 669, 669f
Electromagnetic radiation, 387
energy transfer to/from system, 149
Electromagnetic theory, speed of light and, 898
Electromagnetic waves, 733, 750. *See also* Light and light waves
Doppler effect of, 760–761
energy in, 754
Hertz and, 750–751, 751f
intensity of, 754
light as, 754–755, 899
momentum of, 754–755
phase change due to reflection, 840–841, 840f, 841f
plane wave, 753
polarized, 854–855, 854f
production by an antenna, 752–753, 752f
properties of, 753–757, 753f, 755f
radio-wave transmission, 752
spectrum of, 757–760, 758f, 760f
speed of, 754
as transverse wave, 753, 754
types, 759
gamma rays, 758f, 759
infrared waves, 758f, 759
microwaves, 758f, 759
radio waves, 758f, 759
ultraviolet (UV) light, 758f, 759
visible light, 758f, 759
X-rays, 758f, 759
Electrometer, 540
Electron(s)
antiparticle of, 976, 982
in atomic structure, 4
in beta decay, 976, 981
binding, 975–976
charge of, 324, 525, 528, 528t
Millikan oil-drop experiment and, 541–542, 541f–542f
Compton effect, 933–934, 933f
deflection by electromagnets, 683, 683f
drift speed, 603–605, 603f
in electric field, 605
energy of, 913
as leptons, 1006, 1006t
mass of, 3t, 528t, 982, A.20
notations for, 982
photoelectric effect, 773
photoelectrons, 926–928, 926f, 928f
photon production and, 930, 930f
relativistic momentum of, 909
spin of, 685, 685f, 686, 956–957, 957f
uncertainty principle and, 939–940, 939f
x-ray production and, 930, 961–962
Electron clouds, 958–959, 959f
Electron configuration, 959–961, 960t
Electron-lepton number, conservation of, 1008
Electron microscope, 937–938, 937f
Electron neutrino (ν_e), 1006, 1006t, 1007, 1008
Electron particle beam, pressure of, 356–357
Electron-positron annihilation, 1005
Electron volt (eV), 569–570, A.20
Electrostatic air cleaner, 572
Electrostatic confinement, capacitors for, 575
Electrostatic equilibrium
conductors in
electric potential on charged conductor, 569, 569f
properties of, 539–541, 539f, 540f
defined, 539
Electrostatic forces, 528, 532, 572, 750, 973
Electrostatic generator, 542–543, 542f

Index

Electrostatic precipitator, 571, 571f
Electroweak force, 1014–1015, 1015f
Electroweak theory, 1013–1014
Elementary charge (e), 541–542, A.20
Elementary particles
  classification of, 1005–1007, 1006t
  conservation laws, 1007–1009
    baryon number, 1007–1008
    lepton number, 1008
    strangeness, 1009, 1009f
  eightfold way, 1010, 1010f
  hadrons, 1005–1006, 1006t, 1011–1012, 1011f, 1012f
    baryons, 1006, 1006t, 1007–1008, 1010, 1010f, 1011–1012, 1011f, 1012f
    mesons, 1005, 1006t, 1010, 1010f, 1011–1012, 1011f, 1012, 1012f
    quarks, 1011–1012, 1011f, 1011t, 1012f, 1013–1014, 1013f
  leptons, 1003, 1006–1007, 1006t, 1008, 1013–1014, 1013f
Elements
  atomic spectra, 948–950, 949f
    absorption spectrum, 949f, 950, 950f
    emission spectrum, 949, 949f
  electron configuration of, 959–961, 960t
  periodic table of, 349, 960–961
  table of isotopes, A.14–A.18
Elevator, power delivered by motor, 151–152, 151f
Elliptical orbits, 227–228, 227f
Emf
  induced, 699–702
    back emf, 715–716, 715f, 739
    calculating, 704
    direction of induced current, 705–708, 705f–707f
    Faraday's experiment, 700, 700f
    Faraday's law of induction, 703, 704, 707, 709, 716, 748, 750
    in generators, 712–716, 712f–713f
    Lenz's law, 703–709, 705f–706f, 716
    magnetic flux and, 700–702
    motional, 708–712, 708f–709f
    by motion through a magnetic field ($\vec{B}$), 710–711
    self-induced, 716–718, 716f
  motional, 708–712, 708f–709f
  sources of, 626–627, 627f
Emission spectrum, 949, 949f
Emissivity, 387
Endothermic nuclear reaction, 987
Energy, 127–157. *See also* Elastic potential energy; Gravitational potential energy; Heat; Internal energy; Kinetic energy; Mechanical energy; Potential energy
  binding energy of atomic nucleus, 974–976, 975f
  bond, 368
  of characteristic x-rays, 961, 962
  chemical, 154–155
  conservation of, 149–150, 158
    applied to ideal fluid, 307
    in biological systems, 149
    calorimetry and, 372
    defined, 149
    Earth escape speed and, 225
    in electric potential energy problems, 562, 568
    energy transfer to/from system and, 148–149
    loop rule and, 637
    Sun formation and, 224
  dark, 284, 1017, 1017f
  degradation of, 431
  elastic potential energy, 144, 449–453, 449f–452f
  electrical, 582–584, 582f, 611–615
  in electromagnetic wave, 754
  electron volt unit of, 569–570
  from fission, 997–1000
  in food, 154–155
  forms of, 127–157
  ionization, 952
  mass-energy equivalence equation, 912
  from nuclear fusion, 1001–1003
  of photons, 773, 927
  relativistic, 911–915
  rest, 911
  SI units, A.21t
  of sound waves, 486–487
  spring potential energy, 143–148
  stored in alternating-current (AC) circuits, 744
  stored in capacitors, 582–584, 582f
  stored in magnetic field ($\vec{B}$), 722–723
  in thermal processes, 367–393
  total energy, relativity and, 911–914
  transfer
    in electrical circuit, 611–612
    heat engine, 417–426
    minimizing with Dewar flask (Thermos bottle), 391
    power, 150–155, 158
    to/from system, 148–149, 368, 406–407
    thermal energy, 372–374, 380–391, 1000
  in vertical jump, 153–155, 154f
  zero-point, 340
Energy levels of hydrogen, 952–953, 953f
Engine(s)
  automobile
    cooling, 386
    fluid dynamics in, 311
    as heat engine, 418
    work done in adiabatic expansion, 412
  heat engine, 417–426
    applications of, 418
    Carnot engine, 424–426, 424f, 425f
    cyclic process in, 418, 419–421
    defined, 418
    PV diagram for, 418, 418f
    refrigerators and heat pumps as, 421–423, 421f–422f
    second law of thermodynamics and, 423
    thermal efficiency of, 418–419, 423, 424–425
    work done by, 418–419, 423
  rocket, 313
Entropy (S), 426–431
  change
    calculating, 427–429
    with phase change, 427–429
  defined, 426–427
  disorder and, 429–431
  of a system, 427
  of Universe, 427, 431
Environment
  energy transfer to/from system, 148–149
  work done by gas on, 403, 404, 407, 409, 410
Equal sign (=), A.1, A.19t
Equation of continuity, 305–307
Equation of hydrostatic equilibrium, 294, 352–353
Equation of state, 349
  for ideal gas, 350
Equations, 5–7
  factoring, A.7
  linear, A.8–A.9
    solving simultaneous, A.9–A.10
  quadratic, A.7–A.8
Equilibrium
  electrostatic
    conductors in, 539–541, 539f, 540f
    defined, 539
    electric potential on charged conductor in, 569, 569f
    equation of hydrostatic equilibrium, 294, 352–353
  mechanical
    conditions for, 245–246, 249
    examples of objects in, 249–252, 250f
    problem-solving strategy, 249
  Newton's second law and, 101–104
  spring position, 143–145
  static in fluids, 293, 293f
  temperature, 373–374
  thermal, 337–338, 337f
    zeroth law of thermodynamics, 337f, 338
  translational, 245
Equipotentials, 570–571, 570f
Equipotential surfaces, 570–571, 570f
Equivalence, principle of, 915
Equivalence symbol (≡), A.1–A.2, A.19t
Equivalent capacitance
  for parallel capacitors, 576f, 577–578
  for series capacitors, 579–580, 579f
Equivalent resistance. *See* Resistance, equivalent
Escape speed, 225–226, 225t, 358
Estimates, 12–14
Eta (η) particle, 1006
Ether, luminiferous, 898–899, 898f, 899f
Ether wind, 898–899, 898f, 899f
Ethyl alcohol
  boiling point, 375t
  coefficient of volume expansion, 344t
  density, 284t
  index of refraction, 778t
  latent heat of fusion, 375t
  latent heat of vaporization, 375t
  melting point, 375t
  specific heat, 370t
  surface tension of, 314t
Euler's constant (e), 640, 977
European Council for Nuclear Research (CERN), 904, 1013f, 1014
European Space Agency launch site, 209
Excited states, 963–964, 963f, 964f
Exclusion principle, 957, 959–960, 959t, 1012
Exercise
  calories burned in, 154–155
  diet vs. exercise in weight loss programs, 154–155
  physiology of, 369
  power output, 154–155, 155t
Exhaust speed, 189, 313
Exothermic nuclear reaction, 987
Expansion, thermal. *See* Thermal expansion
Explosion
  over an ice sheet, 486
  sonic boom, 496
Exponential decay, 977
Exponentials, A.10–A.12
Exponents, rules of, A.6, A.6t
Eye, 871–876, 872f–874f
  accommodation, 872–874
  conditions, 873–875
    astigmatism, 874
    farsightedness (hyperopia), 873, 873f, 874–875
    nearsightedness (myopia), 873–874, 874f, 875
    presbyopia, 874
  far point, 873
  glaucoma testing of, 179–180
  near point, 873
  red in flash photographs, 775
  resolution of cat's, 884
  structure, 872, 872f
  Sun and evolution of, 760

F

Factoring, A.7
Fahrenheit temperature scale, 340–342, 341f
Farad (F), 573
Faraday, Michael, 89, 532, 536, 540, 700, 703
Faraday's ice pail experiment, 540, 540f
Faraday's law of magnetic induction, 703, 704, 707, 709, 716, 748, 750
Farnsworth, Philo, 1002
Far point, eye, 873
Farsightedness (hyperopia), 873, 873f, 874–875

Femtometer (fm), 973
Fermi (fm), 973
Fermi, Enrico, 982, 983f, 999
Fermions, 959, 1010
Ferromagnetic materials, 686–687
Fetus, ultrasound imaging of, 483, 483f
Feynman, Richard P., 1003f, 1004
Feynman diagram, 1004, 1004f
Fiberglass insulation, 383, 383t
Fiber optics, 791–792, 791f, 846
Fibrillation, heart, 617, 617f
Fick's law, 322
Fictitious force(s), 219
Field, introduction of concept, 89
Field forces, 89–90, 89f. See also Electric force; Gravitational Force; Magnetic Force
Field mill, 538
Field quanta, 1004
Fifth harmonic, 505f
Figure skating, 263–264, 263f
Fine structure, 956
First harmonic (fundamental frequency), 499f, 500–503, 505–508, 505f
First law of motion (Newton's), 88, 90–91, 90f–91f, 897
   application of, 91, 91f
   mass and inertia, 91
First law of thermodynamics, 402, 406–408, 406f
   in adiabatic processes, 411–413, 417t
   in general processes, 416–417
   human metabolism and, 432
   in isobaric processes, 409–411, 417t
   in isothermal processes, 414–416, 417t
   in isovolumetric processes, 413–414, 417t
First-order maximum, 838, 851f, 852, 853
First overtone (second harmonic), 499f, 500–503, 505, 505f
Fish
   apparent depth of, 814–815, 814f
   bioluminescent, 149
   buoyancy control in, 301
   view from eye of, 790, 790f
Fission, nuclear, 912, 914, 996–1000, 997f, 999f
Flagella, 149
Flash attachments, camera, 574
Flashers, roadway, 642
Floating object, buoyant forces on, 300–301, 301f, 303–304, 304f
Flow rate, 306, 307
Fluid(s). See also Fluids in motion; Liquids
   buoyant forces, 299–304, 299f–304f
   compressibility of, 484
   damped oscillations and, 463–464, 463f
   ideal, 305
   inertia of, 484
   pressure and, 285, 285f, 286–287
      Bernoulli's equation, 307–310
      Poiseuille's law, 319–320, 319f
      variation with depth, 293–297, 293f, 294f
   speed of sound in, 484–486, 485t
   static equilibrium, 293, 293f
   transport phenomena, 321–325
      diffusion, 321–322, 321f, 322t
      motion through a viscous medium, 323–324, 324f
      osmosis, 322–323, 323f
      sedimentation and centrifugation, 324–325
   viscosity of, 318–320, 319f, 319t
Fluid dynamics, applications of, 311–313. See also Fluids in motion
   aneurysms, 311
   atomizers, 311, 311f
   golf ball flight, 311
   home plumbing, 313
   lift on airplane wing, 311–312, 311f
   rocket engines, 313

sailing upwind, 312
vascular flutter, 311, 311f
Fluids in motion, 304–310
   Bernoulli's equation, 307–313
      applications of, 309–313
      Venturi tube and, 308, 309f
   equation of continuity, 305–307
   flow rate, 306, 307
   ideal fluid, characteristics of, 305
   Poiseuille's law, 319–320, 319f
   streamline (laminar) flow, 305, 305f
   turbulent flow, 305, 305f
      Reynolds number, 320–321
   viscous flow, 318–320, 319f, 319t
Fluorine (Fl)
   absorption by teeth, 989
   electron configuration of, 960t
Fluorite, index of refraction, 778t
Flute, 510, 511f
Flying buttress, 293, 293f
f-number
   camera, 871
   eye, 872
Focal length (f)
   in microscopes, 878–879, 878f
   mirror, 806, 807t, 811
   in telescopes, 880–882, 881f
   thin lens, 817, 818, 818t
Focal point (F)
   mirror, 806, 807–808
   thin lens, 816–817, 816f, 819, 819f
Focal points (foci), ellipses and, 227, 227f
Focus, Newtonian, 881, 881f
Food
   chemical energy in, 154–155
   diet vs. exercise in weight loss programs, 154–155
   irradiation of, 989
Foot (ft), 4
Football injuries, 291
Foot-pound, 129
Force(s). See also Air resistance; Buoyant forces; Dissipative forces; Electric force; Friction; Gravitational force; Newton's laws
   action, 98–100
   adhesive, 316–319, 316f–318f
   in automobile collisions, 174–175, 175f
   average ($\vec{F}_{av}$), 172, 173, 174
   centrifugal, as fictitious force, 219
   centripetal, 214–219, 215f
   cohesive, 316–318, 316f–318f
   color, 1012, 1014
   conservative, 134–135, 135f, 148, 158
   contact, 89, 89f
   Coulomb, 449, 528–529, 559, 951, 972, 973
   defined, 89
   deformation of objects by, 90
   electromagnetic, 89, 1003–1004, 1004t, 1013–1014, 1013f, 1015, 1015f
   electrostatic, 528
   electroweak, 1014–1015, 1015f
   fictitious, 219
   field, 89–90, 89f
   fundamental, 89, 95, 1003–1004, 1004f, 1014–1015, 1015f
   grand unified, 1014
   Hooke's law, 446–448, 446f
   Newton's second law of motion and, 88, 91–94, 92f, 673
   nonconservative, 135, 135f
      work done by, 131–132, 141–143, 141f, 142f, 148
   normal ($\vec{n}$), 99
   nuclear, 973, 976
   range of, 89–90
   reaction
      Newton's third law and, 98–100, 98f, 99f
      rocket propulsion and, 188, 188f

restoring force
   of pendulum, 460, 460f
   of spring, 143, 446
   SI units, A.21t
   spring, 143–148
   strong force, 4, 89, 90, 1003–1004, 1004t, 1005, 1009, 1012, 1014, 1015f
   torque and, 241–246, 241f–244f
   as vector quantity, 285
   weak force, 89, 90, 1003–1004, 1004t, 1013–1014, 1013f, 1015, 1015f
   work done by
      constant force
         at angle to displacement, 129, 129f, 130–131
         in linear displacement, 128–129, 129f, 130–131
      dissipative forces, 131–132
      friction, 131–132, 142–143, 142f
      gravity, 136–137, 136f
      nonconservative forces, 131–132, 141–143, 141f, 142f, 148
      spring force, 144
      varying force, 155–157, 155f–157f
Forced convection, 385
Forced vibration, 503
Fossil fuels, burning of, 392
Fourth harmonic, 501–502, 505f
Fractions
   mathematical operations with, A.5
   powers as, A.6
Fractures, in automobile collisions, 174–175
Frame of reference, 27, 27f, 73, 897
   absolute, 898
   accelerated, 915–916, 916f
   inertial, 897, 902, 905–906
   laws of mechanics and, 897
   moving, 73
   speed of light and, 898
   stationary, 73
Franklin, Benjamin, 524, 541
Franklin, Rosalind, 932
Fraunhofer diffraction, 848, 848f, 849f
Free-body diagram, 100f, 101, 216, 216f
Free fall, 44–49
   acceleration, 45, 461–462, A.19t
      variation with altitude, 220, 220t
   defined, 45
   example problems, 46–49, 46f, 48f
   Galileo Galilei and, 45
Freezing point of water, 339
Frequency (f)
   angular, 455
      in pendulums, 460–461
      in simple harmonic motion, 455–457
   beat, 509
   computation using interference, 497–498, 497f
   fundamental (first harmonic), 499f, 500–503
   interference and, 470, 470f, 471f
   in pendulums, 461
   in periodic motion, 446
   pitch distinct from, 510
   resonant, 503
   of simple harmonic motion, 455–457
   SI units, A.21t
   of waves, 466–468
Fresnel, Augustin, 848
Fresnel bright spot, 848
Friction, 108–115
   air drag, 114–115, 114f
   coefficient of kinetic friction (μ_k), 109, 109t
   coefficient of static friction (μ_s), 109, 109t, 250–251
   defined, 108
   kinetic ($\vec{f}_k$), 108f, 109, 111
   as nonconservative force, 135, 135f
   static ($\vec{f}_s$), 108–109, 108f, 110, 113
   system approach, 111–113

Friction (Continued)
thermal energy created by, 149
viscosity and, 318
work done by, 131–132, 142–143, 142f
Fringes
bright, 837–841, 837f, 848–849
conditions for, 838
dark, 837–839, 837f, 840–841, 848–850
defined, 837
Fraunhofer diffraction and, 848
location of, 838–839, 840–842, 849
in single-slit diffraction, 849, 850
in Young's double-slit experiment, 837–840, 837f
Fringe shift, 888, 899
Fuel elements, of nuclear reactor, 999, 999f
Fundamental forces, 89, 95, 1003–1004
Big Bang and, 1014–1015, 1015f
Fundamental frequency (first harmonic), 499f, 500–503, 505–508, 505f
Fundamental particles, 1003
Fuses, 634, 644–645
Fusion, latent heat of, 375–377, 375t
Fusion, nuclear, 1000–1003
Fusion bomb, 1001
Fusion reactors, 971f, 1001–1002

G

Galaxy
estimated number in universe, 14
Milky Way
dust clouds and, 886
mass of, 3t
velocity of stars in, 1016–1017, 1016f
Galilean relativity, 897, 897f, 898, 899, 910
Galileo Galilei, 27, 45, 90, 461
Gallium (Ga), electron configuration of, 960t
Gamma (γ) decay, 982–983
Gamma (γ) rays, 758f, 759, 1005
gamma decay process, 982–983
in magnetic field ($\vec{B}$), 976–977, 976f
nuclear fusion and, 1001
penetrating power of, 977
RBE (relative biological effectiveness) factor, 988, 989t
Gamow, George, 1014f
Gas(es). See also Ideal gas
adiabatic index of, 411
Bernoulli's principle for, 308
Boyle's law, 350
characteristics of, 282–283
Charles's law, 350
density, 284
electric discharge in, 571
Gay-Lussac's law, 350
greenhouse, 392–393, 392f
identifying with spectrometer, 784
internal energy for monoatomic, 357
kinetic theory of gases, 336, 354–359
molar specific heat, 408, 409t
phase change, 374–380
root-mean-square (rms) speed, 358, 358t
speed of sound in, 485–486, 485t
storage of energy as motion, 48
thermal conductivity, 381, 381t
universal gas constant, 350
work done by gas on its environment, 403, 404, 407, 409, 410
work done on
in adiabatic processes, 411–413, 417t
calculating from PV diagram, 404, 405
in general thermodynamic processes, 416–417
in isobaric processes, 409–411, 417t
in isothermal processes, 414–416, 417t
in isovolumetric processes, 413–414, 417t
work done by a gas compared, 403
Gas constant (R), 350, A.20
Gasoline, coefficient of volume expansion for, 344t

Gas thermometer, 339–340, 339f, 340f
Gauge pressure, 297
Gauss (G), 664
Gauss, Karl Fredrich, 200, 543
Gaussian surface, 546, 548f
Gaussian (cgs) system of units
acceleration, 5t
area, 5t
average velocity ($\overline{v}$), 30
density, 284
length, 4
magnetic field ($\vec{B}$), 664
mass, 4
prefixes for, 4, 4t
time, 4
velocity, 5t
volume, 5t
Gauss's law
electric flux and, 543, 545–548
for gravitational force, 200
Gay-Lussac's law, 350
Gears, bicycle, 254, 254f
Geiger, Hans, 947, 948f
Gell-Mann, Murray, 1010, 1010f
General relativity, 899f, 915–917, 916f, 917f, 1017, 1018
Generator, 700, 712–716
alternating-current (AC), 712–713, 714–715
direct-current (DC), 713–714, 713f
electrostatic, 542–543, 542f
motors and back emf, 715–716, 715f
Van de Graaff, 540, 542–543, 542f
Genetic damage, from radiation, 988
Geocentric model of universe, 26, 226
Geometry, A.12, A.12t
Germanium (Ge)
electron configuration of, 960t
resistivity, 608t
temperature coefficient of resistivity, 608t
GFIs (ground-fault interrupters), 646, 706–707, 706f–707f
Ghost images, 839
Glaciers, 392–393
Glancing collisions, 186–188, 186f
Glass
elastic modulus, 288t
index of refraction, 778t
optically active, 859
resistivity, 608t
specific heat, 370t
thermal conductivity, 381t
thermal expansion, 343, 344f
Glaucoma testing, 179–180
Global warming, 347, 391–393, 392f, 393f
Glomerulus, 323, 323f
Glucose, in positron-emission tomography (PET), 1005
Gluons, 1004, 1004f, 1012, 1013–1014, 1013t
Glycerin
coefficient of volume expansion, 344t
density, 284t
index of refraction, 778t
viscosity of, 319t
Goddard, Robert, 188, 313
Goeppert-Mayer, Maria, 973f
Gold (Au)
boiling point, 375t
density, 284t
as ductile metal, 288
latent heat of fusion, 375t
latent heat of vaporization, 375t
melting point, 375t
resistivity, 608t
specific heat, 370t
temperature coefficient of resistivity, 608t
thermal conductivity, 381t
Golf, 173, 311

Goudsmit, Samuel, 956–957
Gram (gm), 4
Grand unification theory (GUT), 1014, 1015
Grand unified force, 1014
Graphite, as moderator in nuclear reactor, 999
Grating spectrometer, 886
Gravitation, Newton's law of, 88, 89, 95, 96–97, 219–226, 227
Gravitational constant (G), 95, 219, 220, 220f, 915, A.20
Gravitational force (F_g)
Big Bang and, 1014, 1015f
described, 95, 95f
Einstein and, 915–916
electric force compared to, 529
escape speed and, 225–226, 225t
as field force, 89, 158
as fundamental force, 1003–1004, 1004t, 1014, 1015f
Gauss's law and, 200
general relativity and, 915–916, 916f
gravitational constant, 95, 219, 220, 220f, 915, A.20
measurement of, 220, 220f
law of universal gravitation (Newton), 88, 89, 95, 96–97, 219–226, 227
problem examples, 220–222, 220f, 223–224, 225–226
projectile motion and, 65
torque and, 245–249, 246f
variation with altitude, 220, 220f
weight and, 95, 95f, 97
work done by, 136–137, 136f, 158, 223–224
Gravitational mass (m_g), 915
Gravitational potential energy (PE_g), 135–143, 158
conservation of mechanical energy and, 138–141
defined, 136
electric potential energy compared, 560, 560f
fluid flow and, 308
gravitational work and, 136–137, 136f, 158, 223–224
for high objects, 222–224, 222f
nonconservative forces and, 141–143
reference levels for, 137–138, 137f
relating to mgh, 224, 224f
in rotational motion, 260–262
SI (Système International) units, 136, 222
Sun formation and, 224
as system property, 222–223
work-energy theorem and, 136–137, 158
with elastic (spring) potential energy component, 144–148
Gravitational work (W_g), 136–137, 136f, 158, 223–224
Gravitons, 1004, 1004f
Gravity
artificial, 215
center of, 246–249, 246f–248f
conservation of mechanical energy and, 138–141
as conservative force, 134–135, 135f, 136
nonconservative forces and, 141–143
work done by, 136–137, 136f
Greater than symbol (>), A.1, A.19t
Greek alphabet, A.19t
Green, Andy, 26
Greenhouse effect, 392
Greenhouse gases, 392–393, 392f
Grimaldi, Francesco, 773
Grinding machine, sound intensity of, 488–489
Ground, electrical, 565
case, 646
defined, 526
for electric shock prevention, 645–646, 646f
symbol for, 612
Ground-fault interrupters (GFIs), 646, 706–707, 706f–707f

Ground state, 952
Guitar, 469, 501–502, 501*f*, 509
Guitar, electric, 707, 707*f*
GUT (grand unification theory), 1014, 1015

H

Hadrons, 1005–1006, 1006*t*, 1011–1012, 1011*f*, 1012*f*
   baryons, 1006, 1006*t*, 1007–1008, 1010, 1010*f*, 1011–1012, 1011*f*, 1012*f*
   mesons, 1005, 1006*t*, 1010, 1010*f*, 1011–1012, 1011*f*, 1012, 1012*f*
   quarks, 1011–1012, 1011*f*, 1011*t*, 1012*f*, 1013–1014, 1013*f*
Hahn, Otto, 914
Hale telescope, 881
Half-life ($T_{1/2}$), 977–979, 977*f*
   energy of decay and, 980
   of selected elements, A.14–A.18
Halogens, 961
Harmonic oscillator equation, 447
Harmonics
   in air column waves, 505–507, 505*f*
   quality of sound and, 510–511, 511*f*
   in string waves, 499*f*, 500–503
Harmonic series, 500
Hearing, human
   damage from high intensity sound, 488–489
   ear anatomy and physiology, 511–513, 512*f*
   frequency response curves, 512–513, 512*f*
   sensitivity range of, 482
   threshold of hearing, 487–488, 512–513, 512*f*
   threshold of pain, 487–488, 512*f*
Heart
   artificial, 669, 669*f*
   cardiac pacemakers for, 617–618
   defibrillators, 583
   electrical activity in, 616–618, 616*f*–618*f*
   fibrillation, 583, 617, 617*f*
   implanted cardioverter defibrillator (ICD), 618, 618*f*, 619*t*
Heart beats, time between, 3*t*
Heat (*Q*)
   convection, 385–387, 385*f*–386*f*
   defined, 337, 368, 406
   energy transfer to/from system, 149
   in first law of thermodynamics, 406–407
   ideal absorber, 388
   ideal reflector, 388
   internal energy distinct from, 367–368
   latent (*L*)
      defined, 374
      of fusion, 375–377, 375*t*
      of sublimation, 375
      units of, 375
      of vaporization, 375–377, 375*t*
   mechanical equivalent of, 368
   molar specific heat, 408, 409*t*
   radiation, 387–391, 387*f*–390*f*
   specific (*c*), 370–372
      air circulation/flow and, 370–371, 370*f*
      defined, 370
      finding with calorimetry, 372
      values for selected substances, 370*t*
   thermal conduction, 380–385, 380*f*–381*f*
      conductivity of selected substances, 381*t*
      defined, 380
      home insulation and, 383–385, 383*t*
      losses from human body, 382
   units of, 368
Heat capacity and vibration, 408
Heat engine, 417–426
   applications of, 418
   Carnot engine, 424–426, 424*f*, 425*f*
   cyclic process in, 418, 419–421
   defined, 418
   PV diagram for, 418, 418*f*
   refrigerators and heat pumps as, 421–423, 421*f*–422*f*
   second law of thermodynamics and, 423
   thermal efficiency of, 418–419, 423, 424–425
   work done by, 418–419, 423
Heat pumps, 421–423, 421*f*
Height of building, measuring, 17, 17*f*
Heisenberg, Werner, 939–940, 939*f*, 955
Helicopter, 204–205
   flight and Newton's third law, 98–99
Heliocentric model of solar system, 26, 45, 226
Helium (He)
   affect on voice, 511
   boiling point, 375*t*
   coefficient of volume expansion, 344*t*
   density, 284*t*
   electron configuration of, 960, 960*t*
   fusion and, 1000–1002
   helium-neon gas laser, 964
   latent heat of fusion, 375*t*
   latent heat of vaporization, 375*t*
   liquid, 391
   melting point, 375*t*
   molar specific heat, 409*t*
   as noble gas, 960
   nucleus, 948
   root-mean-square (rms) speed, 358*t*
   singly ionized, 954–955
   storage of energy as motion, 408
   thermal conductivity, 381*t*
Hemoglobin, diffusion coefficient of, 522*t*
Henry (H), 717
Henry, Joseph, 717
Herman, Robert, 1014*f*
Herschel, William, 758
Hertz, Heinrich Rudolf, 750–751, 757–758, 773
Higgs boson, 1014
High-intensity focused ultrasound (HIFU), 484
Hilbert, David, 899
Hockey, 111, 111*f*
Home plumbing, use of vent in, 313
Homes
   insulation, 383–385, 383*t*
   solar-powered, 756
Hooke, Robert, 143, 446
Hooke's law, 445–449
   pendulum motion and, 460
   springs and, 143, 445–449, 446*f*
Horizon problem, 1018
Horsepower (hp), 151
Hot air balloons, 282, 300
"Hot" wire, 646, 646*f*
Household circuits, 634–635, 644–645, 645*f*
Hubble, Edwin, 1017
Hubble Space Telescope, 880*f*, 882, 882*f*
Human body. *See* Medicine and biology
Human metabolism, 432–435
   basal metabolic rate, 433
   estimating energy daily energy usage, 433–434
   first law of thermodynamics and, 432
   metabolic rate measurement, 432–433, 432*f*, 433*t*
   oxygen consumption
      metabolic rate and, 432–433, 432*f*, 433*t*
      physical fitness and, 434–435
   weight gain and, 433
   work done by body on its surroundings, 432
Human tissue
   specific heat of, 370*t*
   thermal conductivity, 382
Hummingbird, feather color in, 835*f*
Huygens, Christian, 773, 786
Huygen's principle, 786–788, 787*f*, 788*f*, 847
   applied to reflection of light, 787, 787*f*
   applied to refraction of light, 787–788, 788*f*
Hydraulic press, 295–296, 296*f*
Hydroelectric plant, 712
Hydrogen atom(s)
   atomic spectra, 949, 949*f*, 950*f*, 953
   Balmer series of, 949, 950*f*, 953
   Bohr's model of atom, 951–953, 951*f*, 952*f*
   density, 284*t*
   diameter of, 3*t*
   electron clouds for, 958–959, 959*f*
   electron configuration of, 959–960, 960*t*
   energy levels of, 952–953, 953*f*
   fusion, 1000–1002
   isotopes, 4, 972, 1001–1003 (*See also* Deuterium; Tritium)
   mass of, 3*t*
   molar specific heat, 409*t*
   molecular mass, 349–350
   quantum mechanics and, 955–959
   quantum numbers, 952–953, 952*f*, 956*t*
   root-mean-square (rms) speed, 358*t*
   size of, 3*t*, 4
   storage of energy as motion, 408
   thermal conductivity, 381*t*
   wave function in ground state, 958
Hydrogen bomb, 1001
Hydrogen-like atoms, 954
Hydrostatic equilibrium, equation of, 294, 352–353
Hyperopia (farsightedness), 873, 873*f*, 874–875

I

ICD (implanted cardioverter defibrillator), 618, 618*f*, 619*t*
Ice
   density, 284*t*
   index of refraction, 778*t*
   as insulating layer over water, 348
   phase change, 375–379, 376*f*, 428–429
   specific heat, 370*t*
   thermal conductivity, 381*t*
Iceberg, 300
Ice point of water, 339
Ice sheet, explosion over an, 486
Ideal absorber, 388
Ideal gas, 336
   defined, 349
   equation of state for, 350
   first law and thermodynamic processes, 417*t*
   internal energy of, 408
   kinetic theory of gases, 354–359
   macroscopic description of, 349–354
   molar specific heat at constant volume for, 408
   molecular models for, 354–359
   pressure of, 354–357
Ideal gas law
   applying in example problems, 351–353
   Boltzmann's constant, 354
   defined, 350
   kinetic theory of gases, 354–359
Ideal reflector, 388
Ideal springs, 447
Illusionist's trick with mirrors, 804, 804*f*
Image(s)
   apparent left-right reversal, 802
   defined, 801
   double images and light reflection, 776
   formed by atmospheric refraction, 815–816, 815*f*
   formed by reflection from flat mirror, 801–802, 802*f*
   formed by refraction, 812–815, 812*f*–814*f*
   ghost, 839
   real, 802, 805, 808, 808*f*, 818–819, 819*f*
   virtual, 802, 808–809, 808*f*
Image distance, 801, 802*f*, 806
Image point, 805, 806, 807–808
Imaging, ultrasound, 483–484, 483*f*
Impedance (*Z*) of electric circuits, 742, 742*t*
Impending motion, 109
Implanted cardioverter defibrillator (ICD), 618, 618*f*, 619*t*

Impulse ($\vec{\mathbf{I}}$), 172–174
Impulse-momentum theorem, 172–174, 176
Incoherent light sources, 836
Index of refraction (n), 778–782, 778f, 778t, 779f, 782f, 788–789, 789f
 of thin film, 841–845, 841f
Indium, as superconductor, 615, 615t
Inductance (L)
 defined, 717
 in RL circuits, 718–721, 719f
 self-inductance, 716–718, 716f
 SI units, 717, A.21t
Induction
 electromagnetic, 583
 Faraday's law of magnetic induction, 703, 704, 707, 709, 716, 748, 750
Induction, charging by, 526, 527f
Inductive reactance (X_L), 739
Inductor
 in alternating-current (AC) circuits, 738–740, 738f, 739f
 circuit symbol for, 719
 defined, 719
 energy stored in, 722
 impedance values and phase angle, 742t
 instantaneous voltage across, 740, 740f
 LC circuits, 750–751, 751f
 pure, 744
 in RL circuits, 719–721, 719f
 in RLC series circuit, 740–745, 740f
 voltage across, 739, 740–741, 740f
Inelastic collisions
 defined, 179, 180
 glaucoma testing, 179–180
 perfectly inelastic
  defined, 179, 180
  inelastic compared to, 179
  in one dimension, 180–183, 180f, 182f
  in two dimensions, 186–188, 187f
Inequalities, A.1, A.19t
Inertia, 91, 897
Inertial confinement fusion, 1002
Inertial mass (m_i), 915
Inertia property of medium, wave speed and, 485
Infrared waves, 758f, 759
Infrasonic waves, 482
Inner ear, 512–513, 512f
Insects, walking on water by, 315–316
Instantaneous acceleration (a)
 defined, 35, 64
 in one dimension, 35–37, 36f
 SI (Système International) units, 35, 64
 in two dimensions, 63
 from velocity-time graph, 36–37, 36f
Instantaneous angular acceleration (α), 206
Instantaneous angular speed (ω), 204
Instantaneous current (I), 601
Instantaneous power (P), 150
Instantaneous speed (v), 33
Instantaneous thrust, 189
Instantaneous velocity
 defined, 32, 64
 in one dimension (v), 32–34, 32t, 33f
 from position-time graph, 32–34, 33f
 SI (Système International) units, 32, 64
 in two dimensions ($\vec{\mathbf{v}}$), 64
Insulation (R-value), home, 383–385, 383t
Insulators, 525–526
 defined, 525
 induced charge on, 526, 527f
Intensity (I)
 of electromagnetic wave, 754
 of sound waves, 487–490, 488t
  hearing damage from, 488–489
  intensity level in decibels, 487–489, 488t
  threshold of hearing, 487–488
  threshold of pain, 487–488
  variation with distance, 489–491, 489f
 of wave defined, 487
Intensity level, in decibels, 487–489, 488t
Interference
 constructive
  in x-ray diffraction, 931
 frequency computation using, 497–498, 497f
 of light waves (See also Diffraction)
  conditions for, 835–836, 838
  constructive
   conditions for, 835–836, 837f, 838, 842–844
   equation for, 842
   position of, 838–839, 849
   in single-slit diffraction, 849
   in thin films, 842–844
  destructive, 773, 888
   conditions for, 835–836, 837f, 838, 842–843, 849
   equation for, 842
   position of, 838–839, 849
   in single-slit diffraction, 849
   in thin films, 842–843
  iridescence and, 835f
  in Michelson interferometer, 887–888, 888f, 899, 899f
  Newton's rings, 843, 843f
  phase change due to reflection, 840–841, 840f, 841f
  in thin films, 841–846, 841f–845f
   Newton's ring, 843, 843f
   problem-solving strategy, 843
  use in reading CDs and DVDs, 846–847, 846f
  in Young's double-slit experiment, 836–840, 837f, 838f
 of sound waves, 496–498, 497f
  beats, 508–510, 508f
 with television signal, 839
 of water waves, 837, 837f
 of waves, generally, 470–471, 470f–471f
  beats, 508–510, 508f
  constructive, 470–471, 470f–471f, 497, 509
  destructive, 471, 471f, 497, 509
Interferometer, Michelson, 887–888, 888f, 899, 899f
Internal energy (U)
 change in
  adiabatic processes, 411–413, 417t
  first law of thermodynamics, 406–407
  general thermodynamic processes, 416–417
  isobaric processes, 409–411, 417t
  isothermal processes, 414–416, 417t
  isovolumetric processes, 413–414, 417t
 conservation of energy and, 149
 defined, 368
 heat distinct from, 367–368
 in human body, 432
 of ideal gas, 408
 for monoatomic gas, 357
 of system, calculating, 359
 total translational kinetic energy as, 357, 368
International Bureau of Weights and Measures, 2
International Space Station, 202
Interneurons, 647
Invar, coefficient of linear expansion of, 344t
Inverse sine function, 16
Inverse-square law, 219–220
Inverse tangent function, 16, 60
Inverse trigonometric functions, 16
Iodine (I), radioisotope of, 989
Ionization chamber, in smoke detectors, 983, 983f
Ionization energy, 952
Ionizing radiation, 988
Ions, RBE (relative biological effectiveness) factor for heavy, 989t
Iridescence, 835f
Iris, 872, 872f

Iron (Fe)
 density, 284t
 electron configuration of, 960t
 resistivity, 608t
 specific heat, 370t
 temperature coefficient of resistivity, 608t
 thermal conductivity, 381t
 ultimate strength, 290t
 work function of, 927t
Iron, magnetizing, 660
Irradiation of food and medical equipment, 989
Irreversible process, 423
Isobaric (constant pressure) process, 403–404, 407–408, 409–411, 417t
Isochoric (isovolumetric) processes, 413–414, 417t
Isothermal processes, 414–416, 417t, 424, 424f
Isotopes
 defined, 972
 of hydrogen, 4
 natural abundance of, 972
 separating with mass spectrometer, 675–676, 675f
 table of, A.14–A.18
Isovolumetric processes, 413–414, 417t

J

Jensen, Hans, 973
Jet airplanes
 coasting and braking distances for, 42–43, 42f
 runway length problem, 42–43, 42f
 sonic boom produced by, 496
 sound intensity levels of, 488, 488t
 vertical loop, 218, 218f
Joule (J), 128, 129
Joule, James Prescott, 367, 368
Joule per kilogram, 375
Jump, energy and power in vertical, 153–155, 154f
Junction rule, 636–639, 636f
Jupiter
 escape speed, 225t
 magnetic field of, 661
 planetary data, 228t

K

Kaon (K) particle, 1006t, 1009, 1009f
Keck telescopes, 870, 881
Kelvin (K), 340
Kelvin, Lord (William Thomson), 423
Kelvin temperature scale, 340–342, 341f, 350, 373
Kepler, Johannes, 226
Kepler's laws, 226–229
 first, 227, 227f
 second, 227, 227f
 third, 227–229
Keyboard, capacitors in computer, 574
Kidneys
 dialysis, 323, 323f
 structure and function, 323, 323f
 ultrasonic disruption of stones, 484
Killer whale, power generated by, 152
Kilogram (kg), 2, 2f, 3, 91, 93t
Kilogram-meter per second, 171
Kilogram per meter cubed, 284
Kilowatt-hour (kWh), 151, 612
Kinematics, 26. See also Motion
Kinetic energy (KE), 157–158
 in alpha decay, 980
 in Bernoulli's equation, 308
 in beta decay, 982, 982f
 in collisions, 179–181, 183, 378–379
 conservation of mechanical energy and, 138–141
 conversion of mass to in uranium fission, 914
 defined, 133
 fluid flow and, 308
 momentum and, 171
 in nuclear fission of uranium, 997
 in nuclear reactions, 987
 photoelectrons, 926–928, 928f

relativistic, 911–915
rotational (KE_r), 259–262
temperature of the Sun and, 224
translational, 357, 368
in vertical jump, 154–155
Kinetic friction ($\vec{f}_k$), 108f, 109, 111
  coefficient of (μ_k), 109, 109t
Kinetic theory of gases, 336, 354–359
Kirchhoff, Gustav, 637
Kirchhoff's rules, 636–640, 636f–637f
Knievel, Evel, 73
Krypton (Kr)
  electron configuration of, 960t
  molar specific heat of, 409t
  as noble gas, 960

L

Lambda ($\overline{\Lambda}^0$), antiparticle, 1006
Lambda (Λ^0) particle, 1006t, 1009, 1009f
Laminar flow, 305, 305f
Land, E. H., 855
Laser light
  as coherent, monochromatic beam, 836
  OmniGuide fiber and, 846
  reading CDs and DVDs, 781–782, 781f, 846–847, 846f
  refraction of in digital videodisc (DVD), 781–782, 781f
Laser printers, 572–573, 572f
Lasers
  applications of, 964
  helium-neon gas laser, 964
  ruby, 964f
  stimulated emission and, 963–964, 964f
Latent heat (L)
  defined, 374
  of fusion, 375–377, 375t
  of sublimation, 375
  units of, 375
  of vaporization, 375–377, 375t
Lateral magnification
  of compound microscope, 879
  of mirror, 802
Laue, Max von, 929
Laue pattern, 931
Laws of motion (Newton's)
  applications of, 100–107, 100f, 102f–107f
  first, 88, 90–91, 90f–91f, 897
    application of, 91, 91f
    mass and inertia, 91
  second, 88, 91–94, 92f, 673, 915
    accelerating objects, 104–107
    applications of, 100–107, 100f, 102f–107f
    described, 88, 91–94, 92f
    example problems, 93–94, 93f–94f, 102–107, 102f–107f
    free-body diagram and, 100f, 101
    momentum and, 171
    objects in equilibrium, 101–104
    problem-solving strategy for, 101
    rotational analog of, 254, 256–259
    as vector equation, 92, 93
  third, 88, 97–100, 98f–99f, 176
    applications of, 98–99
    gravitational force and, 219–220, 219f
    snowshoes and, 285
Lawson's criterion, 1002
LC circuits, 750–751, 751f
LCD (liquid crystal display), 860–861, 860f
Lead (Pb)
  ballast, 291–292
  boiling point, 375t
  coefficient of linear expansion, 344t
  density, 284t
  latent heat of fusion, 375t
  latent heat of vaporization, 375t
  melting and entropy change, 427–428

melting point, 375t
resistivity, 608t
specific heat, 370t
superconductor, 615, 615t
temperature coefficient of resistivity, 608t
thermal conductivity, 381t
work function of, 927t
Length (L)
  approximate values of examples, 3, 3t
  dimensions, 5–6, 5t
  relativity and, 901, 906–908, 907f
  units of, 1, 2, 3–4, A.21t
Length contraction, 906–908, 907f
Length elasticity in solids, 287–288, 287f, 288f, 288t
Lens(es). See also Thin lenses
  camera, 870–871, 871f
  checking for imperfections in optical, 843
  chromatic aberration, 826, 826f
  converging, 816–818, 816f, 820–821, 820f, 822–824, 825, 825f, 826f
    for correcting farsightedness, 873, 873f, 875
    for correcting presbyopia, 874
    of simple magnifier, 876–878
  crystalline, 872
  diopter measurements, 874
  diverging, 816–819, 816f, 821–822, 822f, 826
    for correcting nearsightedness, 873, 874f, 875
    thin lenses, 816–819, 816f, 821–822, 822f
  diving masks, 819
  eye, 872–874, 872f
  f-number
    camera, 871
    eye, 872
  lens-maker's equation, 818
  microscope
    objective, 878–880, 878f
    ocular, 878
  nonreflective coatings for, 844–845
  power, 874
  shapes, 816, 816f
  spherical aberration in, 825, 825f
  telescope, 880–881, 881f
Lens-maker's equation, 818
Lenz's law, 703–709, 705f–706f, 716
Lepton number, conservation of, 1008
Leptons, 1003, 1006–1007, 1006t, 1008, 1013–1014, 1013f
Less than symbol (<) A.1, A.19t
Leucippus, 4
Lever arm, 243
Lift on airplane wing, 311–312, 311f
Light and light waves
  blue-shift, 910
  coherent light sources, 836
  color, 782–786, 783f, 835f
  diffraction (See Diffraction)
  dispersion, 782–786, 782f–784f, 786f, 790
  dual nature of light and matter, 935–938
  as electromagnetic wave, 754–755
  Huygen's principle, 786–788, 787f, 788f, 847
  incoherent light sources, 836
  interference (See Interference)
  momentum of, 754–755
  nature of, 772–773
  photoelectric effect, 773
  polarization (See Polarization)
  ray approximation, 774, 774f
  red-shift, 910
  reflection
    angle of reflection, 775, 775f
    applications of, 774–776, 775f
    complete, 755
    defined, 774
    diffuse, 774, 775f
    double images, 776
    Huygen's principle applied to, 787, 787f
    phase change due to, 840–841, 840f, 841f

polarization by, 857–858, 857f
    retroreflection, 775
    specular, 774, 775f
  light, speed of (c), 2, A.20
    total internal reflection, 788–792, 788f–791f
    wet-road example, 774–775, 775f
  refraction (See Refraction)
  speed of (c), 2, 754, 773, 778
    constancy of, 900
    electromagnetic theory and, 898
    ether wind theory and, 898–899, 898f, 899f
    frames of reference and, 898
    Galilean relativity and, 898, 898f
    index of refraction and, 778
    Michelson-Morley experiment, 899, 899f
  unpolarized, 854–855, 855f
  wave fronts, 774, 774f, 786–787, 787f
  wavelength (See Wavelength)
  wound treatment, 760, 760f
Lightbulbs
  circuit diagram symbol for, 576, 576f
  dimming of aging, 608
  failure of, 613
  three-way, 634, 634f
Lightning
  strike deflection by Earth's magnetic field, 669
  thunder and, 485
Lightning rods, 540–541
Light year, 3t
Limiting angle of resolution (θ_{min}), 883–885
Linear accelerators, 564
Linear density, 469
Linear displacement (Δx)
  defined, 27, 64
  distance distinct from, 28
  as function of time, 39, 40t
  Hooke's law and, 446, 446f
  in one dimension, 27–28, 28f, 39, 40t
  resultant, 59, 59f
  runway length problem, 42–43, 42f
  SI (Système International) units, 27, 64
  in two dimensions, 63–64, 63f
  as vector quantity, 28
  velocity as function of, 39, 40t
  work and, 128–130f, 128f–130f
  work done by a varying force and, 155–156, 155f
Linear equations, A.8–A.9
  solving simultaneous, A.9–A.10
Linear expansion, 343–344, 344t, 347, 371
Linear expansion coefficient (α), 343, 344t, 347
Linearly polarized light, 855, 855f. See also Polarization
Linear momentum ($\vec{p}$), 170–179
  conservation of, 170, 176–179, 176f, 177f
    in collisions, 79–188, 176, 176f
    defined, 176
    recoil and, 177–179
    in rocket propulsion, 189
    squid propulsion and, 177, 177f
  defined, 171
  impulse and, 172–174
  kinetic energy and, 171
  Newton' second law and, 171
  in rocket propulsion, 188–191
  SI (Système International) units, 171
Linear quantities, relation to angular quantities, 208–211
Line spectra, 949, 949f
Liquid crystals, 860–861, 860f
Liquids. See also Fluid(s); Fluids in motion
  adhesive forces, 316–319, 316f–318f
  capillary action of, 317–318
  characteristics of, 282–283
  cohesive forces, 316–318, 316f–318f
  density, 284
  fluid dynamics, applications of, 311–313
  phase change, 374–380

speed of sound in, 484–486, 485t
surface of, 316–317, 316f–317f
surface tension of (γ), 314–316, 314f–315f, 314t
volume expansion of, 344t, 347, 348–349
Lithium (Li), electron configuration of, 960, 960t
Lithium battery, 618
Lloyd's mirror, 840, 840f
Load resistance, 627
Logarithms, A.10–A.12
Loma Prieta earthquake, 504
Longitudinal waves
 described, 465–466, 465f–466f
 motion of elements in, 482
 sound wave as, 481, 482–483
 speed in a solid rod, 485
Long jump, 69–70, 69f
Loop rule, 636–639, 637f
Loudspeakers, 668–669, 668f
Luminiferous ether, 898–899, 898f, 899f
Luminous intensity, SI units of, A.21t
Lungs
 injury from decompression, 298
 surface tension in air sacs, 315
Lyman series, 949, 952f

M
Machinery, sound intensity of, 488–489
Mach number, 496
Magnesium (Mg), electron configuration of, 960t
Magnetic declination, 662
Magnetic domains, 686, 686f
Magnetic field(s) ($\vec{B}$), 663–667
 Ampère's circuital law and, 677–679, 678f, 750
 at atomic level, 685
 of bar magnet, 661
 charged particle moving through, 663–667, 665f, 673–676, 674f
 of coaxial cable, 678–679, 678f
 compasses and, 662, 663, 676–677, 676f
 of current loops, 681–682, 681f–682f
 current produced by changing, 700
 direction of induced current, 705–708, 705f–707f
 direction of magnetic force, 664–665, 665f, 668–669, 669f, 673–674, 674f, 676–677, 676f–677f
 Earth's, 661–663, 662f, 701
  current-carrying wire in, 669–670
  lightning strike deflection by, 669
  magnetic force on proton moving through, 665–666
  reversal of, 662–663
 emf induced by motion through, 710–711
 energy stored in, 722–723
 inductor's, 721
 of long, straight wire, 676–679, 676f, 678f
 magnetic domains, 686, 686f
 in magnetic resonance imaging (MRI), 989–990
 Maxwell's predictions, 749–750
 of planets, 661
 solenoids, 682–685, 682f, 683f, 685f, 705–706, 705f–706f
 storing charged particles, 674
 in tokamak, 1002, 1002f
 torque on current loop in, 670–673, 670f, 672f
 units of measurement, 664
Magnetic field intensity, SI unit of, A.21t
Magnetic field lines, 661, 661f, 752–753, 753f
 for solenoids
  loosely wound, 682, 682f
  tightly wound, 682, 683f
Magnetic field pattern, of bar magnet, 661, 661f
Magnetic flux (Φ_B)
 in AC transformer, 747–748
 calculating, 701–702
 changing, 700
 defined, 700
 of Earth's magnetic field, 701

induced emf, 699–702
 calculating, 704
 direction of induced current, 705–708, 705f–707f
 Faraday's experiment, 700, 700f
 Faraday's law of induction, 703, 704, 707, 709, 716, 748, 750
 in generators, 712–716, 712f–713f
 Lenz's law, 703–709, 705f–706f, 716
 magnetic flux and, 700–702
 motional, 708–712, 708f–709f
 self-induced, 716–718, 716f
 SI units, 700, A.21t
 value of, 701
Magnetic flux quantum (Φ_0), A.20
Magnetic force
 centripetal acceleration and, 673
 charged particle moving through magnetic field, 663–667, 665f, 673–676, 674f
 on current-carrying conductor, 667–670, 667f–669f
 direction of, 664–665, 665f, 668–669, 669f, 673–674, 674f, 676–677, 676f–677f
 electric force compared to, 664
 of proton traveling through magnetic field, 665–667
 rail guns and, 659, 660f
 right-hand rule, 664–665, 665f, 668–669, 669f, 673, 677, 677f
 between two parallel conductors, 679–681, 679f
Magnetic materials, 686–687
 diamagnetic, 687, 687f
 ferromagnetic, 685, 686
 paramagnetic, 686–687
Magnetic moment ($\vec{\mu}$), 671
Magnetic monopoles, 750, 1018
Magnetic north pole, 662, 662f
Magnetic poles, 660
Magnetic resonance imaging (MRI), 989–990, 990f
Magnetic south pole, 661–662, 662f
Magnetism, 659–687
 hard magnetic materials, 660
 magnets, 659–661
  magnetic field of bar magnet, 661
  permanent, 660, 686, 686f
  poles of, 660
  superconducting, 616, 616f, 659, 664
 Maxwell's predictions, 749–750
 soft magnetic materials, 660
 torque on current loop, 670–673, 670f, 672f
 types of magnetic materials, 686–687
Magnetometers, 616
Magnets
 hard and soft materials, 660
 magnetic field of bar magnet, 661
 magnetizing iron with, 660
 permanent, 660, 686, 686f
 poles of, 660
 superconducting, 616, 616f, 659, 664
Magnification
 angular (m)
  of compound microscope, 879
  of simple magnifier, 876–878
  of telescope, 880–882
 of compound microscope, 879
 enlargement distinct from, 802
 lateral, 802
  of compound microscope, 879
  of mirror, 802
Magnifier, 876–878, 876f
Malus's law, 856
Manganese (Mn), electron configuration of, 960t
Manometer
 open-tube, 297, 297f
 sphygmomanometer, 298, 298f
Marble, ultimate strength of, 290t
Mars
 escape speed, 225t

orbit of, 226
 planetary data, 228t
Marsden, Ernest, 947, 948f
Mass (m)
 acceleration relationship to, 92
 approximate values of examples, 3, 3t
 center of (CM), 154
 conversion to kinetic energy in uranium fission, 914
 energy equivalent, 972
 general relativity, 761, 896, 915–917, 938, 1017–1018
 gravitational force, 529, 532, 665–666, 669–680, 755, 916, 1003–1004, 1014–1015, 1017
 inertial, 91, 915
 linear density, 469
 mass-energy equivalence equation, 912
 moment of inertia vs., 256
 unified mass unit (u), 972
 units of, 1, 2, 2f, 3–4, 92, 93t, 972, 972t, A.21t
Mass-energy equivalence equation, 912
Mass number (A), 971, 972
Mass spectrometer
 identifying particles, 675
 separating isotopes, 675–676, 675f
Mathematical notation, A.1–A.3, A.19t
Matter
 dark, 283–284, 1016–1017, 1017f
 dual nature of light and matter, 935–938
 levels of organization, 4–5, 5f
 phase change, 374–380
 radiation damage to, 988–989, 989t
 states of, 282–284
 vibrations in, 283
Maxima
 for diffraction grating, 851–852, 851f, 853–854
 secondary, 848
 in single-slit diffraction, 848f, 850
Maximum load, 290
Maxwell, James Clerk, 758
 Hertz's confirmation of Maxwell's predictions, 750–751, 751f
 predictions of, 749–750, 773
Maxwell velocity distribution, 358, 358f
Measurement. See also Gaussian (cgs) system of units; SI (Système International) units; U.S. customary units
 dimensional analysis, 5–7, 5t
 significant figures, 7–11
 uncertainty in, 7–11
Mechanical energy (E), 127
 calorie energy related to, 369
 change in system, 148, 158
 conservation of, 138–141
  in rotational motion, 260–261
 entropy and, 429
Mechanical equilibrium
 conditions for, 245–246
 examples of objects in, 249–252, 250f
 rigid bodies in, 249–252, 250f
Mechanical equivalent of heat, 368
Mechanical power. See Power
Mechanical waves
 energy transfer to/from system, 149
 requirements for, 464
Mechanics
 frames of reference and, 897
 laws of, 897
Medical equipment, irradiation of, 989
Medicine and biology
 algal blooms in ponds and lakes, 386–387, 386f
 aneurysms, 311
 apnea monitor, 707–708, 708f
 arteriosclerosis, 311, 311f
 bacteria
  anaerobic, 663

growth of, 641
magnetic, 663
bioluminescence, 149
blood
   dialysis, 323, 323f
   flow, 320, 321
   skin temperature and, 341
   viscosity, 319t, 320
blood pressure measurement, 298, 298f
body temperature, maintaining, 387
boxing and brain injury, 172–173
brain
   cerebrospinal fluid, 301
   estimated number of cells, 13
   high-intensity focused ultrasound (HIFU) use on, 484
   injury, 172–173, 174
buoyancy control in fish, 301
calorie energy related to mechanical energy, 369
cancer treatment, 737
capillary action in, 318
cavitron ultrasonic surgical aspirator (CUSA), 484
cell size, 322
centrifugation, 325, 325f
cerebrospinal fluid, 301
cochlear implant, 513
conductive losses from human body, 382
deafness, 513
defibrillators
   external, 583
   implanted cardioverter defibrillator (ICD), 618, 618f, 619t
diet vs. exercise in weight loss programs, 154–155
DNA, structure of, 932, 932f
ear anatomy and physiology, 511–513, 512f
eardrum, pressure on while swimming, 295
electrical conduction by neurons, 647–648, 647f–649f
electric field therapy, 737
electric shock, 645–646
electrocardiogram (EKG), 616–617, 617f
exercise
   calories burned in, 154–155
   diet vs. exercise in weight loss programs, 154–155
   physiology of, 369
   power output, 154–155, 155t
eye, 871–876, 872f–874f
   accommodation, 872–874
   astigmatism, 874
   conditions, 873–875
   far point, 873
   farsightedness (hyperopia), 873, 873f, 874–875
   glaucoma testing of, 179–180
   near point, 873
   nearsightedness (myopia), 873–874, 874f, 875
   presbyopia, 874
   red in flash photographs, 775
   resolution of cat's, 884
   structure, 872, 872f
   Sun and evolution of, 760
fetus, ultrasound imaging of, 483, 483f
fiber optic devices, 791
fish
   apparent depth of, 814–815, 814f
   bioluminescent, 149
   buoyancy control in, 301
   view from eye of, 790, 790f
flagellar movement, 149
food
   chemical energy in, 154–155
   diet vs. exercise in weight loss programs, 154–155
   irradiation of, 989
fractures, in automobile collisions, 174–175
glaucoma testing, 179–180

global warming
   coastal flooding and, 347
   greenhouse gases and, 391–393, 392f, 393f
hearing, human
   damage from high intensity sound, 488–489
   ear anatomy and physiology, 511–513, 512f
   sensitivity range of, 482
   threshold of hearing, 487–488, 512–513, 512f
   threshold of pain, 487–488, 512f
heart
   artificial, 669, 669f
   cardiac pacemakers for, 617–618
   defibrillators, 583
   electrical activity in, 616–618, 616f–618f
   fibrillation, 583, 617, 617f
   implanted cardioverter defibrillator (ICD), 618, 618f, 619t
heart beats, time between, 3t
high-intensity focused ultrasound (HIFU), 484
injury
   in automobile collisions, 148, 174–175
   boxing and brain injury, 172–173
   football, 291
   to lungs with decompression, 298
insects, walking on water by, 315–316
irradiation of food and medical equipment, 989
kidneys
   dialysis, 323, 323f
   structure and function, 323, 323f
   ultrasonic disruption of stones, 484
killer whale, power generated by, 152
lungs
   injury from decompression, 298
   surface tension in air sacs, 315
magnetic resonance imaging (MRI), 989–990, 990f
mechanical equilibrium conditions applied to human body, 249–250, 259f
metabolism, human, 432–435
   basal metabolic rate, 433
   estimating energy daily energy usage, 433–434
   first law of thermodynamics and, 432
   metabolic rate measurement, 432–433, 432f, 433t
   oxygen consumption and metabolic rate, 432–433, 432f, 433t
   physical fitness and oxygen consumption, 434–435, 434t
   weight gain and, 433
   work done by body on its surroundings, 432
neurons, electrical conduction by, 647–648, 647f–649f
osmosis, 322–323, 323f
positron-emission tomography (PET), 1005
power, human output of, 154–155, 155t
radiant thermometer, 389, 389f
radiation
   damage to matter, 988–989, 989t
   medical applications of, 988–990
radioactive tracers, 989
squid propulsion, 177, 177f
sudden infant death syndrome (SIDS), 707–708, 708f
surface tension in air sacs, 315
thermal protection of ice over water, 348
thermogram, 388, 388f
ultrasound applications, 483–484, 483f
underwater vision, 813
vascular flutter, 311, 311f
vertical jump, energy and power in, 153–155
vision
   diving masks and, 819
   underwater, 813
voice, shattering of glass, 503–504
whale, power generated by, 152
wound treatment, 760, 760f
Meissner effect, 616f
Meitner, Lise, 914

Membrane, selectively permeable, 322
Mendeleyev, Dimitry, 960
Mercury (Hg)
   coefficient of volume expansion, 344t
   density, 284t
   elastic modulus, 288t
   in pressure measurement devices, 297, 297f
   specific heat, 370t
   spectral lines, 784
   as superconductor, 615, 615f, 615t
   surface of, in glass container, 316, 316f
   surface tension of, 314t
Mercury (planet)
   escape speed, 225, 225t
   orbit of, 916
   planetary data, 228t
Merry-go-round, 219, 219f, 241, 265–267, 266f
Mesons, 1005, 1006t, 1010, 1010f, 1011–1012, 1011f, 1012, 1012f
Metabolic rate, 369
   basal, 433
   equation, 432
   measurement, 432–433, 432f, 433t
   oxygen consumption and, 432–433, 432f, 433t
   physical activities and, 434t
Metabolism, human, 432–435
   basal metabolic rate, 433
   estimating energy daily energy usage, 433–434
   first law of thermodynamics and, 432
   metabolic rate measurement, 432–433, 432f, 433t
   oxygen consumption
      metabolic rate and, 432–433, 432f, 433t
      physical fitness and, 434–435
   weight gain and, 433
   work done by body on its surroundings, 432
Metal(s)
   alkali metals, 961
   thermal conductivity of, 381, 381t
   work function of, 927–928, 927t
Metal detectors, 746
Metastable state, 963
Meter (m), 2, 3, 27
Methane (CH_4)
   as greenhouse gas, 392
   molar specific heat, 409t
Michelson, Albert A., 887, 899
Michelson interferometer, 887–888, 888f, 899, 899f
Michelson-Morley experiment, 899, 899f
Microscope
   compound, 878–880, 878f
   electron, 937–938, 937f
   limitation of light microscope, 879
   resolution of, 884–885
   x-ray, 938
Microwaves, 758f, 759
   background radiation, 1015–1016, 1017–1018
   polarizing, 856
Middle ear, 512–513, 512f
Milky Way galaxy
   dust clouds and, 886
   mass of, 3t
   velocity of stars in, 1016–1017, 1016f
Millikan, Robert, 324, 525, 541–542
Millikan oil-drop experiment, 541–542, 541f–542f
Mineral oil, use in ultrasound imaging, 483
Minima, 848
   in single-slit diffraction, 849f
Mirage, 815–816, 815f
Mirror equation, 806
Mirror reflection, 1013
Mirrors, 801–812
   concave, 804–806, 804f–806f, 809–810
   with thin lenses, 824–825, 824f
   convex, 806–812, 807f–809f
   dielectric, 846
   diverging, 806
   flat, 801–804, 802f, 803f

Mirrors (Continued)
  illusionist's trick with, 804, 804f
  lateral magnification, 802
  Lloyd's, 840, 840f
  in Michelson interferometer, 887–888, 888f
  mirror equation, 806
  perfect, 846
  principle axis, 805, 805f, 808f
  ray diagrams for, 807–809, 808f
  rearview, 803–804, 803f
  side-view, 809, 809f
  sign conventions for, 807, 807f, 807t
  spherical
    defined, 804
    ray diagrams for, 808f
    spherical aberration in, 805, 805f, 825, 825f
  telescope, 880–881, 881f, 882f
Moderator, of nuclear reactor, 999, 999f
Modified Newtonian Dynamics (MOND), 1017
Molar specific heat, 408, 409t
Molecular mass, 349–350
Moles, 349–350
Moment of inertia (I), 253–258, 254f
  angular speed and, 263–266, 263f, 265f, 266f
  of composite object, 254–256, 255f
  defined, 253
  for extended objects, 256–259, 256f, 257f
  mass vs., 256
  rotational analog of Newton's second law, 254, 256–259
  SI units, 254
  of system, 255
Momentum, 170–179
  alpha decay and, 980
  angular (L), 262–267
    conservation of, 227, 263–267
    defined, 263
    torque and, 262–263, 262f
  conservation of, 170, 176–179, 176f, 177f
    in collisions, 79–188, 176, 176f
    defined, 176
    recoil and, 177–179
    in rocket propulsion, 189
    squid propulsion and, 177, 177f
  defined, 171
  of electromagnetic wave, 754–755
  impulse and, 172–174
  kinetic energy and, 171
  linear (See Linear momentum)
  Newton' second law and, 171
  of photon, 935
  position-momentum uncertainty principle, 939–940, 939f
  relativistic, 908–909, 912–913
  in rocket propulsion, 188–191
  SI (Système International) units, 171
  of x-ray photon, 933
MOND (Modified Newtonian Dynamics), 1017
Monoatomic gas
  internal energy of, 357
  molar specific heat of, 408, 409t
Monopole problem, 1018
Moon
  distance from Earth, 3t, A.19t
  escape speed, 225t
  mass of, 3t, 228t, A.19t
Morley, Edward W., 899
Moseley, G. J., 962
Moseley plot, 962, 962f
Motion. See also Acceleration; Displacement; Laws of motion; Projectile motion; Rotational motion; Simple harmonic motion; Speed; Velocity
  energy stored as, in gases, 408
  history of study, 26–27
  impending, 109

  in one dimension, 36–49
    acceleration, 34–37, 34f, 36f
    with constant acceleration, 38–44, 38f, 40t
    displacement, 27–28, 27f–28f, 39, 40t
    free fall, 44–49
    motion diagrams, 37–38, 37f
    velocity, 28–34, 30f–31f, 30t, 32t, 33f
  through a viscous medium, 323–324, 324f
  in two dimensions, 63–77
    constant acceleration, 72–73, 72f
    projectile motion, 65–72, 65f–69f, 71f
    relative velocity, 73–77, 74f–76f
    uniform circular motion, 453–457, 454f
Motional emf, 708–712, 708f–709f
Motion diagram
  described, 37–38, 37f
  projectile motion in, 67–68, 67f
Motor oil, viscosity of, 319t
Motors, electric
  back emf, 715–716, 715f
  induced current in, 715–716
  structure and functioning of, 673, 673f
Mount Palomar Observatory, 881
Moving frame of reference, 73
MRI (magnetic resonance imaging), 989–990, 990f
Müller, K. Alex, 615
Multimeter, 605f, 606
Multiplication
  of fractions, A.5
  of powers, A.6
  significant figures and, 8
  of vector by a scalar, 59
Muon (μ^-), 1005, 1006, 1006t, 1008
Muon-lepton number, conservation of, 1008
Musical instruments. See also Sound waves
  bugle, 506
  clarinet, 510, 511f
  flute, 510, 511f
  organ pipe, standing waves in, 504–506, 505f
  piano, 481, 509–510
  quality (timbre) of sound, 510–511, 511f
  strings
    guitar, 469, 501–502, 501f, 509
    sound production in, 481
    tuning of, 500–501, 509
  tuning of, 500–501, 509
  warming up, 506
Mylar film, aluminized, 756
Myopia (nearsightedness), 873–874, 874f, 875

N

Natural logarithm base, A.10
Natural radioactivity, 985, 985f, 985t
Near point, eye, 873
Nearsightedness (myopia), 873–874, 874f, 875
Ne'eman, Yuval, 1010
Negative acceleration, 35, 38
Neon (Ne)
  electron configuration of, 960t, 961
  helium-neon gas laser, 964
  molar specific heat, 409t
  as noble gas, 960
  root-mean-square (rms) speed, 358t
Neoprene rubber, dielectric constant and dielectric strength of, 585t
Neptune
  escape speed, 225t
  planetary data, 228t
Neptunium (Np), radioactive series, 985, 985t
Neurons, electrical conduction by, 647–648, 647f–649f
Neutrino (ν), 982, 1001, 1006–1007, 1008, 1017
Neutrino oscillation, 1017
Neutron(s)
  in atomic structure, 4, 5f
  as baryons, 1006, 1006t
  decay of, 1008

  discovery of, 986
  in isotopes, 4
  mass of, 528t, 972, A.20
  in nuclear fission of uranium, 996–999, 997f, 999f
  quarks in, 5, 5f
  RBE (relative biological effectiveness) factor, 989t
Neutron number (N), 971, 973, 974f
Neutron star, 264, 264f
Newton (N), 92, 93t, 95
Newton, Isaac, 92, 772, 773
Newtonian focus, 881, 881f
Newton-meter, 128, 129, 241, 243, 314
Newton per coulomb (N/C), 532
Newton's law of universal gravitation, 88, 89, 95, 96–97, 219–226, 227, 915
Newton's laws of motion
  applications of, 100–107, 100f, 102f–107f
  first, 88, 90–91, 90f–91f, 897
    application of, 91, 91f
    mass and inertia, 91
  second, 88, 91–94, 92f, 673, 915
    accelerating objects, 104–107
    applications of, 100–107, 100f, 102f–107f
    described, 88, 91–94, 92f
    example problems, 93–94, 93f–94f, 102–107, 102f–107f
    free-body diagram and, 100f, 101
    momentum and, 171
    objects in equilibrium, 101–104
    problem-solving strategy for, 101
    rotational analog of, 254, 256–259
    as vector equation, 92, 93
  third, 88, 97–100, 98f–99f, 176
    applications of, 98–99
    gravitational force and, 219–220, 219f
    snowshoes and, 285
Newton's particle theory of light, 772, 773
Newton's rings, 843, 843f
Niagara Falls, 306
Nichrome
  resistance of a wire, 609
  resistivity, 608t
  temperature coefficient of resistivity, 608t
Nickel (Ni), electron configuration of, 960t
Night vision, thermal radiation and, 389
Nitrogen (N)
  beta decay, 982
  boiling point, 375t
  electron configuration of, 960t
  isotopes, 982
  latent heat of fusion, 375t
  latent heat of vaporization, 375t
  liquid, 391
  Maxwell speed distribution for, 358f
  melting point, 375t
  molar specific heat, 409t
  root-mean-square (rms) speed, 358t
  thermal conductivity, 381t
Nitrous oxide, as greenhouse gas, 392
Noble gases, 960–961
Node, of standing wave, 498–502, 499f, 505
Noise-level regulations, OSHA, 489
Noise pollution, 488
Nonconducting plane sheet of charge, 547–548, 548f
Nonconservative force, 135, 135f, 158
  work done by, 131–132, 141–143, 141f, 142f, 148
Nonmetals, thermal conductivity of, 381, 381t
Nonohmic materials, 607, 608f
Nonreflective coatings for solar cells and optical lens, 844–845
Normal force $(\vec{n})$, 99
North pole, magnetic, 660
North Pole of Earth, geographic, 661–662, 662f

Notation
  for instantaneous acceleration, 35
  for instantaneous velocity, 32
  for many-digit numbers, 2
  mathematical, A.1–A.3, A.19t
  powers of 10, 3, 4, 4t
  scientific, 8–9, A.3–A.4
  shell and subshell, 956t
  significant figures, 7–11
  for vector, 28, 58
Nuclear bomb, 999
Nuclear fission, 996–1000, 997f, 999f
Nuclear force, 973, 976
Nuclear fusion, 1000–1003
Nuclear magnetic resonance, 990
Nuclear magneton (μ_n), A.20
Nuclear physics, 971–990
  decay process, 979–985
    alpha decay, 979–981, 980f
    beta decay, 981–982, 982f
    gamma decay, 982–983
  medical applications of radiation, 988–990
  nuclear reactions, 985–987
  properties of nuclei, 971–976
    atomic number, 971–972
    binding energy, 974–976, 975f
    charge, 972, 972t
    density, 973
    mass, 972
    mass number, 971, 972
    neutron number, 971
    size, 972–973
    stability, 973, 974f
  radioactivity, 976–979, 976f, 977f
    artificial, 985
    decay constant and half-life, 977–979
    natural, 985, 985f, 985t
    practical uses of, 983–985
Nuclear reaction
  endothermic, 987
  exothermic, 987
  Q values, 986–987
  threshold energy, 987
Nuclear reactions
  fission, 996–1000, 997f, 999f
  fusion, 1000–1003
Nuclear reactors, 998–1000, 999f
Nucleons, 975–976. See also Neutron(s); Proton(s)
Nucleus, atomic, 971–976
  atomic number, 971–972
  binding energy, 974–976, 975f
  charge, 972, 972t
  daughter nucleus, 980–982.997
  density, 973
  diameter of, 3t
  discovery of, 4, 948, 973
  effective nuclear charge (Z_{eff}), 962
  mass, 972
  mass number, 971, 972
  neutron number, 971
  nuclear fission, 996–1000, 997f, 999f
  nuclear fusion, 1000–1003
  parent nucleus, 980–982
  size, 972–973
  stability, 973, 974f
  strong nuclear force, 4, 89, 90
  structure of, 4–5, 5f
  weak nuclear force, 89, 90
Nylon, dielectric constant and dielectric strength of, 585t

O

Object distance, 801
Ocean sea levels, rising, 347
Ocean waves, 445, 464, 466, 506
Oersted, Hans Christian, 676, 677
Ohm (Ω), 607

Ohm, Georg Simon, 607
Ohmic materials, 607, 608f
Ohm's law, 607, 609, 736, 739
Oil film on water, interference and, 841, 842
Omega minus (Ω^-) particle, 1010, 1012
Omega (Ω) particle, 1006t
OmniGuide fiber, 846
Onnes, H. Kamerlingh, 615
Open-circuit voltage, 627
Open-tube manometer, 297, 297f
Optical activity, 858–859, 859f
Optical fiber, 791–792, 791f
Optical instruments, 870–888
  camera, 870–871, 871f
  compound microscope, 878–880, 873f
  eye, 871–876, 872f–874f
  magnifier, 876–878, 876f
  Michelson interferometer, 887–888, 888f
  resolution of single-slit and circular apertures, 882–887
  telescope, 880–882, 880f–882f
Optics
  geometric, 835 (See also Lens(es); Mirrors)
  wave (See Wave optics)
Orbital magnetic quantum number (m_ℓ), 955–959, 956t
Orbital quantum number (ℓ), 955–959, 956t
Orbits
  geosynchronous, 229
  Kepler's laws and, 226–229
Order number
  of diffraction pattern, 852, 852f
  fringe, 838
Order-of-magnitude calculations, 12–14
Organ pipe, standing waves in, 504–506, 505f
Oscillating charges, 752–753, 752f, 753f
Oscillations. See also Waves
  damped, 463–464, 464f
  standing waves
    in air columns, 504–508, 505f
    on strings, 498–503, 498f, 499f
Oscillator, 927
OSHA noise-level regulations, 489
Osmosis, 322–323, 323f
Outer ear, 511–513, 512f
Overtones, 500
Oxygen (O)
  atomic mass of, 349
  boiling point, 375t
  consumption and human metabolism
    metabolic rate and, 432–433, 432f, 433t
    physical fitness and, 434–435
  density, 284t
  diffusion coefficient, 322t
  electron configuration of, 960t
  latent heat of fusion, 375t
  latent heat of vaporization, 375t
  melting point, 375t
  molar specific heat, 408, 409t
  root-mean-square (rms) speed, 358t
  thermal conductivity, 381t

P

Pacemakers, cardiac, 617–618
Pain, threshold of, 487–488, 512f
Paintings, using x-rays to study, 930
Paper, dielectric constant and dielectric strength of, 585t
Parabolic surfaces, 825
Parallel-plate capacitor(s), 548, 548f, 573–578, 573f, 574f, 576f, 577f
  dielectric in, 584–585, 585f
Paramagnetic materials, 686–687
Parent nucleus, 980–982
Particle(s). See also specific particles
  antiparticles, 976, 982, 1004–1005 1006t, 1008, 1010

charged
  moving through magnetic field, 663–667, 665f, 673–676, 674f
  storing (trapping), 674
classification of, 1005–1007, 1006t
conservation laws, 1007–1009
eightfold way, 1010, 1010f
fundamental, 1003
hadrons, 1005–1006, 1006t
  baryons, 1006, 1006t
  mesons, 1005, 1006t
leptons, 1006–1007, 1006t
Particle accelerators, 683, 683f, 986
Particle-antiparticle annihilation, 1005
Particle physics
  Standard Model, 1013–1014, 1013f
  unanswered questions in, 1019
Particle theory of light, 772, 773
Pascal (Pa), 285, 288
Pascal, Blaise, 295
Pascal's principle, 295–297
Paschen series, 949, 952f
Path difference (δ), 838
Pauli, Wolfgang, 957, 959, 959f, 982
Pauli exclusion principle, 957, 959–960, 959t, 1012
Pendulums
  applications, 461–462
  ballistic pendulum, 181–182, 182f
  physical, 463, 463f
  resonance in, 503
  simple, 460–462, 460f, 461f
  simple harmonic motion compared to, 460–461, 461f
  time dilation and period of, 905
Penning trap, 674
Pens, ballpoint, 298
Penzias, Arno A., 1015, 1016f
Perfectly inelastic collisions
  defined, 179, 180
  inelastic compared to, 179
  in one dimension, 180–183, 180f, 182f
Perihelion, 916
Period (T)
  in pendulums, 461–463
  in periodic motion, 446
  of planet, 227, 228t
  of simple harmonic motion, 454–457
Periodic motion, 445–446. See also Simple harmonic motion
  pendulums, 460–463, 460f, 461f, 463f
  terminology, 446
  velocity as function of position, 452
Periodic table of elements, 349, 960–961
Periscopes, submarine, 789
Perlmutter, Saul, 1017
Permeability of free space (μ_0), 677, A.20
Permittivity of free space (ϵ_0), 545
Perpetual motion machines, 427
PET (positron-emission tomography), 1005
Phase angle (ϕ), 742, 742t, 743–744
Phase change, 374–380
  defined, 374
  due to reflection, 840–841, 840f, 841f
  entropy change with, 427–429
  incomplete, 378
  latent heat of fusion, 375–377, 375t
  latent heat of sublimation, 375
  latent heat of vaporization, 375–377, 375t
  problem-solving strategy, 377
  in water, 375–378, 376f
  of waves on strings, 841, 841f
Phasor diagram, 741, 741f
Phasors, 741
Phonograph record, angular and tangential speeds, 210
Phosphorus (P), electron configuration of, 960t
Photocells, 929

Index

Photoconductor, 572, 572f
Photocopiers, 572–573, 572f
Photoelectric effect, 773, 926–929, 926f, 928f
Photoelectrons, 926–928, 926f, 928f
Photography. *See also* Camera
   red eye in flash photographs, 775
Photons
   Compton effect, 933–935, 933f
   Einstein and, 773
   electromagnetic force and, 1004, 1004f, 1004t, 1013–1014, 1013f
   electrons and production of, 930, 930f
   energy of, 773, 927
   gamma rays, 758f, 759, 976–977, 976f, 982–983
   history of concept, 773
   momentum of, 935
   from particle-antiparticle annihilation, 1005
   stimulated absorption process, 963, 963f
   stimulated emission, 963–964, 964f
   uncertainty principle and, 939–940, 939f
   virtual, 1004
   x-ray, 933–935, 933f, 962–963
Physical fitness, 434–435, 434t
Physical pendulum, 463, 463f
Physical quantities
   conservation of, 138
   conversion of units, 11–12
   dimensional analysis of, 5–7, 5t
   estimates, 12–14
   uncertainty in measurement, 7–11
Physiology of exercise, 369
Piano, 481, 509–510
Piezoelectric effect, 483, 483f
Pion (π), 1005, 1006t, 1007, 1009, 1009f
Pipe organ, standing waves in, 504–506, 505f
Pipes
   harmonics of, 506–507
   resonance in, 507–508, 507f
   standing waves in, 504–506, 505f
Pistons, simple harmonic motion and, 454
Pitch, 510
Pitchblende, 976
Planck, Max, 924f, 925
Planck's constant (h), 773, 925, 927, A.20
Plane angle, SI units, A.21
Plane diffraction grating, 851, 851f
Plane polar coordinates (r, θ), 15, 15f
   conversion to/from Cartesian coordinates, 16–17, 16f
Plane polarized light wave, 855. *See also* Polarization
Planets. *See also specific planets*
   escape speeds, 225, 225t
   magnetic fields of, 661
   orbits of, 227–228, 227f
   periods of, 227, 228t
   planetary data, 228t
   radiation and, 390–391, 390f
Plane wave, 490, 490f
   electromagnetic wave as, 753
   of light, 774f
   plane diffraction grating and, 851, 851f
Plano-concave lens, 816f
Plano-convex lens, 816f
Plants, capillary action in, 318
Plaque, arterial, 311, 311f
Plasma, 282, 283, 1002
Plasma confinement time (τ), 1002
Plasma ion density (n), 1002
Plastic, optically active, 859
Platform diving, 140, 140f
Platinum
   resistance thermometer, 610–611
   resistivity, 608t
   temperature coefficient of resistivity, 608t
   work function of, 927t

Platinum (Pt), density of, 284t
Plumb bob, 247
Pluto
   escape speed, 225t
   planetary data, 228t
Pogo-stick, 127
Point charge, electric potential due to, 565–568, 565f, 566f
Pointed (Gothic) arch, 293, 293f
Point source, sound waves from, 489–491, 490f
Poise (unit), 319
Poiseuille, J. L., 319
Poiseuille's law, 319–320, 319f
Poker hands, 431
Polarization
   of electric charge, 526
   induced, 589
   of light
     angle of polarization (θ_p), 857f, 858
     sunglasses and, 858
     transmission axis of polarizing material, 855–856, 855f, 856f
   of light waves, 854–861
     concept described, 854–855, 854f, 855f
     liquid crystals, 860–861, 860f
     optical activity, 858–859, 859f
     plane of, 855
     by reflection, 857–858, 857f
     by scattering, 858, 858f
     by selective absorption, 855–857, 855f, 856f
   of molecules in dielectric, 589–590, 589f
Polarizers, 855–857, 855f, 859, 859f
Polarizing angle (θ_p), 857f, 858
Polaroid, 855
Polaroid sunglasses, 858
Poles, magnetic field, 660
Pollution, radon, 984
Polyatomic gases, molar specific heat of, 409t
Polystyrene
   dielectric constant and dielectric strength of, 585t
   index of refraction, 778t
Pool/billiards, 179, 184–185, 220–221
Population inversion, 963
Position. *See also* Displacement
   acceleration as function of, 447–448
   angular, 203, 203f
   equilibrium position of springs, 143–145
   as function of time, 457–459, 458f
   velocity as function of, 452, 454
Position-momentum uncertainty principle, 939–940, 939f
Position-time graph, 38
   average velocity from, 31, 31f
   instantaneous velocity from, 32–34, 33f
   for motion with constant acceleration, 38f, 39
   slope, 31
Positron (e^+), 976–977, 982, 1001, 1004–1005, 1007
Positron-emission tomography (PET), 1005
Post-and-beam construction, 292, 293f
Potassium (K)
   electron configuration of, 960t
   ions, electric signal conduction in neurons and, 648
Potential difference. *See also* Voltage
   equivalent to voltage, 577
   in household circuits, 644
   induced across airplane wing, 710–711
   loop rule, 636–639, 637f
   photoelectric effect *versus*, 926, 926f
Potential energy (PE). *See also* Elastic potential energy; Electric potential energy; Gravitational potential energy
   conservative forces and, 135
   defined, 135
   elastic, 449–453, 449f–452f
   as property of a system, 135–136

   as scalar quantity, 136
   of springs, 143–148
Potential energy function, 136
Pound (lb), 92, 93t
Pound per inch squared, 285
Power (P), 150–155, 158
   in alternating-current (AC) circuits, 744–745
   average ($\bar{P}$), 150–151, 152
     in alternating-current (AC) circuits, 745
   capacitors and, 583
   commercial units of (kilowatt-hour), 151
   converted by electric heater, 614–615
   defined, 150
   delivered by electric motor, 151–152
   delivered to a resistor, 612
   electrical energy and power, 611–615
   human output of, 154–155, 155t
   instantaneous, 150
   of lens, 874
   output of human body, 434–435, 434t
   problem examples, 151–153
   in *RLC* series circuits, 745
   SI (Système International) units, 150–151, A.21t
   sound wave, 487
   usage in kilowatt-hours, 612, 613–615
   in vertical jump, 153–155, 154f
Power factor, 745
Powers of ten
   rules for, A.6
Powers, rules of, A.6, A.6t
Precipitator, electrostatic, 571, 571f
Presbyopia, 874
Pressure (P)
   absolute, 297
   atmospheric, 285, 294, 297–298
   defined, 285
   of fluid
     Bernoulli's equation, 307–313
     Pascal's principle, 295–297
     Poiseuille's law, 319–320, 319f
     variation with depth, 293–297, 293f, 294f
     of water, 286–287, 286f
   force and, 285
   of gas
     Boyle's law, 350
     Gay-Lussac's law, 350
     ideal gas law, 350
     molecular model for ideal gas, 354–357
   gauge, 297
   isobaric (constant pressure) process, 403–404, 407–408, 409–411, 417t
   measurement of, 297–298
     with barometer, 297, 297f
     blood pressure, 298, 298f
     with open-tube manometer, 297, 297f
   one atmosphere, definition of, 298
   as scalar quantity, 285
   standard temperature and pressure (STP), 350
   units of, 285, A.21t
Pressure waves, 466
Primary coil, 699–700, 700f, 747–748
Principle axis, mirror, 805, 805f, 808f
Principle of equivalence, 915
Principle quantum number (n), 955–956, 956t
Prisms
   alteration of direction of light beam by, 789, 789f
   diffraction grating compared, 852
   dispersion of light and, 782–786, 782f–784f, 786f
   prism spectrometer, 783–784, 783f, 886
Prism spectrometer, 783–784, 783f, 886
Probability
   electron clouds and, 958–959, 959f
   wave function and, 938
Problem-solving strategies
   algebraic order of calculations, 10
   calorimetry with phase change, 377
   capacitor combinations, 580

centripetal acceleration, forces causing, 215–216
collisions in one dimension, 184
collisions in two dimensions, 187
conservation of mechanical energy, 139
constant acceleration in one dimension, 40
conversion of units, 11–12
direct-current circuits
    Kirchhoff's rules, 637–638
    resistors, 635
Doppler effect, 493–495
electric forces and fields, 535
electric potential, 567
energy methods and rotation, 260
general strategy, 18–19
interference in thin films, 843
Newton's second law, 101
objects in equilibrium, 249
projectile motion, 68
relative velocity, 74
RLC series circuit, 743
rotational motion, 260
rounding, 8, 10–11
spring potential energy, 145
symbol use, 19
work-energy theorem, 157–158
Projectile motion, 65–72
    described, 65–67, 65f, 66f
    example problems, 47–49, 67–72
    height maximum, 47–48
    independence of horizontal and vertical motions, 65
    in motion diagram, 67–68, 67f
    parabolic path of, 65, 65f
    powered ascent followed by free fall, 48–49, 48f
    problem-solving strategy, 68
    projection angle, 65f, 66
    range, 70–71
    summary of important facts, 67
Projection angle, 65f, 66
Proportionality ($\propto$), A.1, A.19t
Propulsive force, as nonconservative force, 135
Prospecting, pendulum use in, 462
Proteins
    as optically active material, 859
    x-ray diffraction of, 932
Proton(s)
    in atomic structure, 4, 5f
    as baryons, 1006, 1006t
    charge of, 524–525, 528, 528t, 972
    diameter of, 3t
    in electrostatic generator, 543
    half-life of, 1007
    magnetic force of proton traveling through magnetic field, 665–667
    mass of, 528t, 972, 972f, A.20
    quarks in, 5, 5f
    RBE (relative biological effectiveness) factor, 989t
Proton-proton cycle, 1001
Ptolemy, Claudius, 26, 226
Pulsars, 264, 264f
Pump
    electromagnetic, 669, 669f
    heat, 421–423, 421f
Pupil, eye, 872, 872f
Purkinje fibers, 616f, 617
PV diagram, 404–405, 404f–405f
    for adiabatic expansion, 412–413, 412f
    for Carnot engine, 425, 425f
    for general thermodynamic process, 416, 416f
    for heat engine, 418, 418f
    for isothermal expansion, 414–415, 415f
Pyramids, level construction of, 297
Pyrex®, 343, 344f, 391
    dielectric constant and dielectric strength of, 585t
Pythagorean theorem, 16–17, 60, 213, 355, A.12–A.13

Q

QCD (quantum chromodynamics), 1012
QRS pulse, 617, 617f
Quadratic equations, A.7–A.8
Quality (timbre) of sound, 510–511, 511f
Quantization of electric charge, 525, 542
Quantum chromodynamics (QCD), 1012
Quantum mechanics
    Dirac's theory, 1004–1005
    hydrogen atom and, 955–959
Quantum number
    defined, 925
    hydrogen atom, 952–953, 952f, 956t
    orbital (ℓ), 955–959, 956t
    orbital magnetic (m_ℓ), 955–959, 956t
    Pauli exclusion principle and, 957, 959–960, 959t
    principle (n), 955–956, 956t
    of quarks, 1011t, 1012
    spin magnetic (m_s), 957–958, 1010
Quantum physics, 923–940
    blackbody radiation and, 923–925, 924f, 925f
    Compton effect, 933–935, 933f
    dual nature of light and matter, 935–938
    photoelectric effect, 926–929, 926f, 928f
    uncertainty principle, 939–940, 939f
    wave function, 938
    x-rays
        diffraction by crystals, 930–933, 931f, 932f
        spectrum emitted by x-ray tube, 930, 930f
Quantum state, 925
    energies of, 956–958
Quantum tunneling, 1001
Quarks, 5, 5f, 1003, 1011–1012, 1011f, 1011t, 1012f, 1013–1014, 1013f, 1013t
Quartz, 283f
    dielectric constant and dielectric strength of, 585t
    elastic modulus, 288t
    index of refraction, 778t, 780–781
    resistivity, 608t
Quintessence theory, 1019
Q values, 986–987

R

Racetrack, banked, 217–218, 217f
Rad (radiation absorbed dose), 988
Radians
    converting degrees to/from, 203, 204
    defined, 203, 203f
    setting calculator to, 16
    as unit of angular displacement, 204
Radiant thermometer, 389, 389f
Radiation. *See also* Radioactivity
    alpha (α) particles, 947–948, 948f
        decay process, 979–981, 980f
        formation in proton-proton cycle, 1001
        in magnetic field ($\vec{B}$), 976–977, 976f
        in nuclear reaction, 985–987
        penetrating power of, 977
        RBE (relative biological effectiveness) factor, 988, 989t
        Rutherford's experiments with, 972–973, 972f
    beta (β) particles
        beta decay process, 981–982, 982f
        in magnetic field ($\vec{B}$), 976–977, 976f
        penetrating power of, 977
        RBE (relative biological effectiveness) factor, 989t
    Big Bang and, 1015
    blackbody, 923–925, 924f, 925f, 1016, 1016f
    bremsstrahlung, 930
    cosmic background, 1015–1016, 1016f, 1017–1018
    damage to matter, 988–989, 989t
    electromagnetic, 149, 387
    gamma (γ) rays, 758f, 759, 1005
        gamma decay process, 982–983
        in magnetic field ($\vec{B}$), 976–977, 976f
        nuclear fusion and, 1001
        penetrating power of, 977
        RBE (relative biological effectiveness) factor, 988, 989t
    ionizing, 988
    medical applications of, 988–990, 990f
    occupational exposure limits, 988–989
    penetrating powers, 977
    RBE (relative biological effectiveness) factor, 988, 989t
    thermal, 387–391, 387f–390f
        quantum physics and, 924–925
        solar-powered homes, 756
Radiation absorbed dose (rad), 988
Radiation damage to matter, 988–989, 989t
Radiator
    automobile, 386
    home heating, 385–386, 386f
Radio, tuning, 746
Radioactive series, 985, 985f, 985t
Radioactive tracers, 989
Radioactivity, 976–979, 976f, 977f
    artificial, 985
    damage to matter, 988–989, 989t
    decay constant (λ), 977–979
    decay curve, 977, 977f
    decay process, 979–985
        alpha decay, 979–981, 980f
        beta decay, 981–982, 982f
        gamma decay, 982–983
    decay rate (activity) (R), 977–979
    discovery of, 976
    half-life ($T_{1/2}$), 977–979, 977f
        energy of decay and, 980
        of selected elements, A.14–A.18
    medical applications of radiation, 988–990
        irradiation of food and medical equipment, 989
        magnetic resonance imaging (MRI), 989–990, 990f
        tracing, 989
    natural, 985, 985f, 985t
    practical uses of, 983–985
        carbon dating, 983, 984–985
        radon detection, 983–984
        smoke detectors, 983, 983f
    spontaneous decay (transmutation), 980
Radio telescope, 886
Radio waves, 758f, 759, 886
Radio-wave transmission, 752
Radium (Ra)
    activity of, 978–979
    alpha decay of, 980–981, 980f
    radon from, 983–984
Radon (Rn), as noble gas, 960
Radon detection, 983–984
Raft, buoyancy of, 303, 303f
Rail guns, 659, 660f
Railroads
    Acela train, 43–44, 44f
    Doppler effect and, 494
    drive wheel mechanism of locomotives, 454, 454f
    track, expansion of, 344, 344f
Rainbows, 785–786, 786f
Rankine scale, 341
Rarefaction, in sound waves, 482, 482f
Ray approximation, in geometric objects, 774, 774f
Ray diagrams
    for mirrors, 807–809, 808f
    for thin lenses, 818–819, 819f
Ray diagrams for mirrors, 807–809, 808f
Rayleigh's criterion, 883
Razor blades, 131
RBE (relative biological effectiveness) factor, 988, 989t

RC circuits, 640–644, 640f–641f
  charging capacitor in, 642–643
  discharging capacitor in, 643–644
  impedance values and phase angle, 742t
Reactance
  capacitive, 738
  inductive, 739
Reaction force
  Newton's third law and, 98–100, 98f, 99f
  rocket propulsion and, 188, 188f
Real image, 802, 805, 808, 808f, 818–819, 819f
Rearview mirrors, 803–804, 803f
Recoil, 177–179
Rectangular coordinate system, 15, 15f
  conversion to/from polar coordinates, 16–17, 16f
  vector components on, 60–61
Red shift, 761, 910, 1017
Reference circle, 457, 457f
Reference frame, 27, 27f, 73, 897
  absolute, 898
  accelerated, 915–916, 916f
  inertial, 897, 902, 905–906
  laws of mechanics and, 897
  moving, 73
  speed of light and, 898
  stationary, 73
Reference levels for gravitational potential energy, 137–138, 137f
Reflecting telescope, 880–882, 881f
Reflection
  of light
    angle of reflection, 775, 775f, 857
    applications of, 774–776, 775f
    complete, 755
    defined, 774
    diffuse, 774, 775f
    double images, 776
    Huygen's principle applied to, 787, 787f
    nonreflective coatings for solar cells and optical lens, 844–845
    phase change due to, 840–841, 840f, 841f
    retroreflection, 775
    specular, 774, 775f
    total internal reflection, 788–792, 788f–791f
    wet-road example, 774–775, 775f
  of sound waves, 483
  of waves, generally, 471–472, 472f
  waves on strings, 471–472, 472f, 841, 841f
Reflector, ideal, 388
Refracting telescope, 880–881, 881f
Refraction
  angle of deviation, 782, 783f
  angle of refraction, 777, 777f, 778f, 780, 789f
  atmospheric, 815–816, 815f
  defined, 774
  dispersion, 782–786, 782f–784f, 786f
  Huygen's principle applied to, 787–788, 788f
  images formed by, 812–815, 812f–814f
    flat surface, 813, 814
    sign conventions, 812, 812t
    spherical lens, 813–814, 813f
  index of refraction, 778–782, 778f, 778t, 779f, 782f, 788–789, 789f
  of laser light in digital videodisc (DVD), 781–782, 781f
  Snell's law, 779–782, 789, 825
Refrigerators, 421–423, 422f
Relative biological effectiveness (RBE) factor, 988, 989t
Relative velocity, 73–77, 74f–76f, 909–911
  problem-solving strategy, 74
Relativistic momentum, 908–909, 912–913
Relativity, 896–917
  Galilean, 897, 897f, 898, 899, 910
  general, 899f, 915–917, 916f, 917f, 1017, 1018

  special, 900–915
    consequences of, 900–908
    length contraction, 906–908, 907f
    mass-energy equivalence, 912
    postulates of, 900
    relative velocity, 909–911
    relativistic energy, 911–915
    relativistic momentum, 908–909, 912–913
    relativity of time, 901–902, 901f
    simultaneity and, 901–902, 901f
    time dilation, 902–905, 902f, 904f
    twin paradox, 905–906, 906f
  speed of light and, 898–899, 898f, 899f
Rem (roentgen equivalent in man), 988
Resistance (R)
  of conductors, 606–60, 606*8*
  defined, 607
  equivalent
    for parallel resistors, 631–636, 631f
    for series resistors, 628–630, 628f
  load, 627
  of nichrome wire, 609
  ohmic and nonohmic materials, 607, 608f
  Ohm's law, 607, 609
  in RL circuits, 719
  SI units, 607, A.21t
  superconductors and, 615–616, 615f, 616f
  temperature variation of, 610–611, 610f
Resistance thermometer, 610–611
Resistivity (ρ), 607–608, 608t, 609
  temperature coefficient of, 608t, 610
Resistors, 607f
  in alternating-current (AC) circuits, 733–737, 734f, 735f
  defined, 607
  direct-current circuits
    in parallel, 631–636, 631f
    problem-solving strategies, 635
    in series, 628–630, 628f–630f
  impedance values and phase angle, 742t
  instantaneous voltage across, 740, 740f
  power delivered to, 612
  in RL circuits, 719, 719f
  in RLC series circuit, 740–745, 740f
  rms voltage across, 735–736
  symbol for, 576, 607
  voltage across, 739, 740–741, 740f
Resolution
  of cat's eye, 884
  of diffraction grating, 886–887
  limiting angle of resolution (θ_{min}), 883–885
  of microscope, 884–885
  Rayleigh's criterion and, 883
  of single-slit and circular apertures, 882–887
  of telescope, 885–886
Resolving power (R) of diffraction grating, 886–887
Resonance, 503–504, 503f, 504f, 507–508, 507f
  in pendulum, 503, 503f
  in RLC series circuit, 746–747, 746f
  singers and shattered glass, 503–504
  structural, 504, 504f
  tube of variable length, 507–508, 507f
Resonance frequency, 746–747, 751
Resonators, 925
Rest energy, 911
Restoring force
  of pendulum, 460, 460f
  of springs, 143, 446
Resultant vector, 58, 58f, 59, 59f, 62–63, 62f
Retina, 872, 872f
Retroreflection, 775
Reversible process, 423–424, 424f
Revolving door, torque and, 242, 242f
Reynolds number (RN), 320–321
Rib cartilage, elastic modulus of, 288t
Riess, Adam, 1017

Right-hand rule
  for angular velocity ($\vec{\omega}$), 214, 214f
  of magnetic force
    rule number 1, 664–665, 665f, 668–669, 669f, 673
    rule number 2, 667, 667f
  for torque, 243–244, 243f
Right triangle, trigonometric functions of, 15–16, 15f
RL circuits, 718–721, 719f, 720f
  energy stored in, 722
  impedance values and phase angle, 742t
RLC series circuit, 740–744, 740f, 741f
  average power in, 745
  energy stored in, 744
  impedance values and phase angle, 742t
  phase relations in, 740–741, 740f
  problem-solving strategies, 743
  resonance in, 746–747, 746f
rms current, 734–736
rms speed, 358, 358t
rms voltage, 735–736
Roadway flashers, 642
Rock concert, sound intensity levels of, 488, 488t
Rockets
  with acceleration in two dimensions, 72–73, 72f
  engines, 313
  European Space Agency launch site, 209
  exhaust speed, 189
  flight, analysis of, 48–49, 48f
  instantaneous thrust, 189
  multistage, 190
  propulsion, 188–191, 188f, 189f
  single stage to orbit (SSTO), 190–191
Roentgen (R), 988
Roentgen equivalent in man (rem), 988
Roller-coaster, circular loop in, 218, 218f
Röntgen, Wilhelm, 929
Root-mean-square (rms) current, 734–736
Root-mean-square (rms) speed, 358, 358t
Root-mean-square (rms) voltage, 735–736
Rotational dynamics, applications of, 240
Rotational equilibrium
  applications of, 240
  conditions for, 245–246
Rotational kinetic energy (KE_r), 259–262, 260f
Rotational motion, 202–219. See also Centripetal acceleration; Torque
  basic concepts, 203–206
  conservation of mechanical energy, 260–261
  under constant angular acceleration, 206–208
  kinematic equations for, 206–207
  problem-solving strategy, 260
  relations between angular and linear quantities, 208–211
  work-energy theorem and, 260, 261–262
  under zero torque, 249
Rounding, 8, 10–11
Rubber
  dielectric constant and dielectric strength of, 585t
  elastic modulus, 288t
  resistivity, 608t
  thermal conductivity, 381t
Runway length problem, 42–43, 42f
Russell, John Scott, 466
Rutherford, Ernest, 947–948, 948f, 972–973, 972f, 983, 985–986
R-values, for insulation materials, 383–385, 383t
Rydberg constant (R_H), 949, 952, A.20
Rydberg equation, 949

S

Safety, electrical, 645–646, 646f
Sailing upwind, 312
Satellites
  altitude of orbiting, 3t
  European Space Agency launch site, 209

geosynchronous orbit, 229
protection from thermal radiation, 391
Saturn
   escape speed, 225t
   planetary data, 228t
Scalar quantity
   instantaneous speed, 33
   speed as, 28
   vector quantity compared to, 28, 57
Scandium (Sc), electron configuration of, 960t
Scattering
   polarization of light waves by, 858, 858f
   of x-rays, 934
Schmidt, Brian, 1017
Schmitt, Harrison, 95
Schrödinger, Erwin, 938, 938f, 955
Schrödinger wave equation, 938
Schwinger, Julian S., 1003f
Scientific notation, 8–9, A.3–A.4
Scott, David, 45
Seat belts, automobile, 91, 91f, 175
Second(s), 3, 4
Secondary circuit, 700, 700f
Secondary coil, 700, 747–748
Secondary maxima, 848
Second harmonic (first overtone), 499f, 500–503, 505, 505f
Second law of motion (Newton's), 88, 91–94, 92f, 673, 915
   accelerating objects, 104–107
   applications of, 100–107, 100f, 102f–107f
   described, 88, 91–94, 92f
   example problems, 93–94, 93f–94f, 102–107, 102f–107f
   free-body diagram and, 100f, 101
   momentum and, 171
   objects in equilibrium, 101–104
   problem-solving strategy for, 101
   rotational analog of, 254, 256–259
   as vector equation, 92, 93
Second law of thermodynamics, 402
   direction of time and, 430–431
   entropy and, 426, 427, 430
   heat engines and, 423
   Lord Kelvin and, 423
   as statement of the most probable, 430–431
Second-order maximum, 852
Second overtone (third harmonic), 499f, 500–503, 505f
Sedimentation rate, 325
See-saw, 245–246
Seiche, 506
Selective absorption, polarization of light waves by, 855–857, 855f, 856f
Selectively permeable membrane, 322
Selenium (Se)
   electron configuration of, 960t
   in laser printers, 573
Self-inductance, 716–718, 716f
Self-sustained chain reaction, 999
Semicircular arch, 292–293, 293f
Semiconductors
   complementary metal-oxide semiconductor (CMOS), 870
   defined, 526
Sensing coil, 706, 706f
Shape
   deformation of objects by force, 90
   elasticity in solids, 288t, 289, 289f
Shear modulus (S), 288t, 289
Shear strain, 289
Shear stress, 289, 289f, 291
Shells, atomic, 956–957
   notation for, 956, 956t
   Pauli exclusion principle and, 959–960, 959t
Shock, electric, 645–646, 646f
Shock absorbers, 463, 463f

Shock waves, 495–496, 496f
Short circuit, 707, 707f
Shutter, camera, 870–871, 871f
Side-view mirrors, 809, 809f
SIDS (sudden infant death syndrome), 707–708, 708f
Sigma (Σ) particle, 1006t, 1009
Significant figures, 7–11
Silicon (Si)
   electron configuration of, 960t
   resistivity, 608t
   specific heat of, 370t
   temperature coefficient of resistivity, 608t
Silicone oil, dielectric constant and dielectric strength of, 585t
Silicon monoxide, nonreflective coatings for solar cells and optical lens, 844–845
Silver (Ag)
   boiling point, 375t
   density, 284t
   latent heat of fusion, 375t
   latent heat of vaporization, 375t
   melting point, 375t
   resistivity, 608t
   specific heat, 370t
   temperature coefficient of resistivity, 608t
   thermal conductivity, 381t
   work function of, 927t
Simple harmonic motion, 446–447
   acceleration
      as function of position, 447–448
      as function of time, 457–459, 453f
   angular frequency of, 455–457
   bungee jumping, 456, 456f
   defined, 446
   frequency of, 455–457
   on frictionless surface, 447–448
   Hooke's law and, 446–447
   pendulum motion compared to, 450–461, 461f
   period of, 454–457
   pistons and drive wheels, 454, 454f
   position as function of time, 457–459, 458f
   sinusoidal nature of, 457–459, 458f
   uniform circular motion compared to, 453–457, 454f
   velocity
      as function of displacement, 454
      as function of position, 452, 454
      as function of time, 457–459, 458f
Simple magnifier, 876–878, 876f
Simultaneity, special relativity and, 901–902, 901f
Simultaneous linear equations, solving, A.9–A.10
Sine (sin), 15–16, 15f
Singers and shattered glass, 503–504
Single-slit aperture, diffraction pattern from, 883, 883f, 884f
Single-slit diffraction of light waves, 848–850, 849f
Singularity, 1014
Sinoatrial (SA) node, 616–617, 616f
Sinusoidal motion
   simple harmonic motion as, 457–459, 458f
   sound wave as, 482, 482f
   waves as, 466, 466f
Siren, Doppler shift and, 495
SI (Système International) units
   acceleration, 5t, 92, 93t
   amount of substance, A.21t
   angular displacement, 204
   area, 5t
   average acceleration, 34, 64
   average angular acceleration, 205
   average angular speed, 204
   average speed, 29
   average velocity, 30, 64
   base units, A.21t

   capacitance, 573, A.21t
   conversion of units, 11–12
   current, 680
   density, 284
   derived units, A.21t
   dimensions in, 5t
   displacement, 27, 64
   electric charge, 525, 528, 680, A.21t
   electric current, 601, A.21t
   electric field, 532
   electric potential (emf), A.21t
   electric resistance, A.21t
   energy, A.21t
   force, 92, 93t, 95, A.21t
   frequency, A.21t
   gravitational potential energy, 136, 222
   heat, 368
   impulse, 172
   inductance, 717, A.21t
   instantaneous acceleration, 35, 64
   instantaneous angular acceleration, 206
   instantaneous angular speed, 204
   instantaneous velocity, 32, 64
   latent heat, 375
   length, 2, 3–4, A.21t
   luminous intensity, A.21t
   magnetic field, 664
   magnetic field intensity, A.21t
   magnetic flux, 700, A.21t
   mass, 2, 2f, 3–4, 91, 92, 93t, A.21t
   moment of inertia, 254
   momentum, 171
   plane angle, A.21t
   power, 150–151, A.21t
   prefixes for, 4, 4t
   pressure, 285, A.21t
   resistance, 607
   specific heat, 370
   stress, 288
   surface tension, 314
   temperature, A.21t
   time, 2–4, A.21t
   torque, 241, 243
   velocity, 5t
   viscosity, 319
   volume, 5t
   watt per meter squared (W/m^2), 487
   work, 128, 129, 136, A.21t
Skid marks, automobile, 133
Skiing, 137–138, 137f, 142–143, 142f
Skin temperature, 341
Skydiving, 114–115, 114f
Slope
   defined, 31
   from position vs. time graph, 31
Slug (unit of mass), 4, 93t
Smith, Dave, 57
Smoke detectors, 983, 983f
Snell's law of refraction, 779–782, 789, 825
Snowshoes, 285, 285f
Soap film, interference and, 841, 842, 844
Soddy, Frederick, 983
Sodium (Na)
   electron configuration of, 960t
   ions, electric signal conduction in neurons and, 648
   in nuclear reactors, 1000
   photoelectrons from, 928
   spectral lines, 784, 887, 956
   work function of, 927t
Sodium chloride (NaCl), 283, 283f
   crystalline structure of, 931, 931f
   index of refraction, 778t
Solar cells, nonreflective coatings for, 844–845
Solar day, 2
Solar panels, 1000
Solar-powered homes, 756

Solar system. *See also specific planets*
  age of, 985
  asteroid map of inner, 150f
  dust, 755
  geocentric model of, 26, 226
  heliocentric model of, 26, 45, 226
  Kepler's laws and, 226–229
Solenoid
  Ampère's law applied to, 683, 684–685, 685f
  calculating inductance and self-induced emf in, 718
  direction of induced current, 705–706, 705f–706f
  magnetic field of, 682–685, 682f, 683f, 685f
Solid(s)
  amorphous, 283, 283f
  characteristics of, 282–283, 283f
  crystalline, 283, 283f
  deformation of, 287–293
  density, 284
  elasticity of, 283, 283f, 287–290
    in length, 287–288, 287f, 288f, 288t
    of shape, 288t, 289, 289f
    volume, 288t, 289–290, 289f
  phase change, 374–380
  speed of sound in, 485–486, 485t
  thermal conductivity, 381, 381t
Soliton, 466
Somatic damage, from radiation, 988
Sonic boom, 496
Sound waves, 481–513
  audible, 482
  beats, 508–510, 508f
  categories of, 482–483
  characteristics of, 482–484
  diffraction of, 850
  Doppler effect, 491–496, 492f
  energy of, 486–487
  forced vibration, 503
  human ear and, 511–513, 512f
  infrasonic, 482
  intensity of, 487–490, 488t
  interference of, 496–498, 497f
  as longitudinal wave, 481, 482–483
  OSHA noise-level regulations, 489
  pitch, 510
  plane, 490, 490f
  producing, 481–482, 482f
  quality (timbre) of, 510–511, 511f
  reflection of, 483
  resonance, 503–504, 503f, 504f, 507–508, 507f
  shock waves, 495–496, 496f
  speed of sound, 484–486, 485t
  spherical, 489–490, 489f–490f
  standing
    in air columns, 504–508, 505f
    on strings, 498–503, 498f, 499f
  thunder, 485
  ultrasound, 483
    applications of, 483–484, 483f
    production of, 483, 483f
South pole, magnetic, 660
South Pole of Earth, geographic, 661–662, 662f
Space catapult, 709–710
Spacecraft, sails for, 756–757
Space shuttle, sonic boom produced by, 496
Spacetime, curvature of, 916, 917
Space travel
  artificial gravity, 215
  International Space Station, 202
  length contraction and, 907–908
  Newton's first law of motion and, 91
  sails for spacecraft, 756–757
  twin paradox and special relativity, 905–906, 906f
Sparks in circuits, 717–718
Speakers

Doppler shift and, 494
interference and, 497–498, 497f
magnetic force and, 668–669, 668f
Special relativity, 900–915
  consequences of, 900–908
  length contraction, 906–908, 907f
  mass-energy equivalence, 912
  postulates of, 900
  relative velocity, 909–911
  relativistic energy, 911–915
  relativistic momentum, 908–909, 912–913
  relativity of time, 901–902, 901f
  simultaneity and, 901–902, 901f
  time dilation, 902–905, 902f, 904f
  twin paradox, 905–906, 906f
Specific gravity, 284
Specific heat (c), 370–372
  air circulation/flow and, 370–371, 370f
  defined, 370
  finding with calorimetry, 372
  molar specific heat, 408
  values for selected substances, 370t
Spectral analysis, 784
Spectral lines, 784, 887
  splitting of, 955–956, 955f
Spectrometer
  diffraction-grating, 852, 852f
  grating, 886
  identifying gases with, 784
  mass
    identifying particles, 675
    separating isotopes, 675–676, 675f
  prism, 783–784, 783f, 886
Spectrum
  atomic, 948–950, 949f
    absorption spectrum, 949f, 950, 950f
    emission spectrum, 949, 949f
  of electromagnetic wave, 757–760, 758f, 760f
  identifying gases with spectrometer, 784
  visible, 782, 782f, 784, 949f
  x-ray, 929–930, 929f, 961f, 961f
Specular reflection, 774, 775f
Speed
  angular
    average, 204
    helicopter, 204–205
    instantaneous, 204
    moment of inertia and, 263–266, 263f, 265f, 266f
    of phonograph records and compact discs, 210–211
    rotational kinetic energy and, 259–260, 260f
    tangential speed relationship to, 208
  average, 29–30
  changes in, 35
  drift speed, 603–605, 603f
  escape, 225–226, 225t, 358
  exhaust speed of rockets, 189, 313
  of fluid, finding with Bernoulli's equation, 309–310
  instantaneous, 33
  Maxwell speed distribution, 358, 358f
  root-mean-square (rms), 358, 358t
  tangential
    angular speed relationship to, 208
    centripetal acceleration and, 212, 213
    defined, 208
    of phonograph records and compact discs, 210–211
    of rotating Earth surface, 209
  terminal (v_t), 114, 324, 325
  velocity versus, 28–30
Speed of light (c), 2, 754, 773
  constancy of, 900
  electromagnetic theory and, 898
  ether wind theory and, 898–899, 898f, 899f
  frames of reference and, 898

Galilean relativity and, 898, 898f
index of refraction and, 778
Michelson-Morley experiment, 899, 899f
in vacuum, A.20
Speed of sound, 484–486, 485t
Spherical aberration, 805, 805f, 825, 825f
Spherical mirror. *See also* Concave mirrors; Convex mirrors
  defined, 804
  ray diagrams for, 808f
  spherical aberration in, 805, 805f, 825, 825f
Spherical sound waves, 489–490, 489f–490f
Sphygmomanometer, 298, 298f
Spin magnetic quantum number (m_s), 957–958, 1010
Spontaneous decay, 980
Spontaneous emission, 963, 963f
Sports
  acrobatics, 146–147, 146f, 264
  archery, 177–179, 177f, 450, 450f
  bicycling, 254, 254f
  boxing, 172–173
  bungee jumping, 456, 456f
  diving, 135, 135f, 140, 140f, 263f, 264
  figure skating, 263–264, 263f
  football, 291
  golf, 173, 311
  hockey, 111, 111f
  long jump, 69–70, 69f
  pool/billiards, 179, 184–185, 220–221
  skiing, 137–138, 137f, 142–143, 142f
  skydiving, 114–115, 114f
  swimming, 295
  tennis, 240, 311
Spring(s)
  bungee jumping modeled as, 456, 456f
  elastic potential energy, 449–453, 449f–452f
  equilibrium position, 143–145
  Hooke's law and, 143, 445–449, 446f
  ideal, 447
  longitudinal wave on, 466, 466f
  mass on vertical, 448–449, 448f
  restoring force of, 143, 446
  simple harmonic motion on a frictionless surface, 447–448
  work to stretch a spring, 156–157, 156f
Spring constant (k), 143, 148, 446, 448
Spring force, 143–148
Spring potential energy, 143–148, 158
  problem examples, 145–148
  problem-solving strategy for, 145
  work-energy theorem, 144–145, 158
Squid propulsion, 177, 177f
Standard atmospheric pressure, A.19t
Standard Model, 1013–1014, 1013f
Standard temperature and pressure (STP), 350
Standing waves
  in air columns, 504–508, 505f
  seiche, 506
  on strings, 498–503, 498f, 499f
Stars
  color of, 924–925
  dark, 88
  fusion in, 1000–1001
  neutron, 264, 264f
  radiation and, 390–391, 390f
  supernova, 264, 264f, 1017
  velocity of, 1016–1017, 1016f
Static equilibrium in fluids, 293, 293f
Static friction $(\vec{f}_s)$, 108–109, 108f, 110, 113
  coefficient of (μ_s), 109, 109t, 250–251
Stationary frame of reference, 73
Steam
  phase change in water, 375–376, 376f
  specific heat of, 370t
  superheated, 379

Steam point of water, 339
Steel
 coefficient of linear expansion, 344t
 elastic modulus, 288t
 maximum load of steel beams, 290
 ultimate strength, 290t
Stefan-Boltzmann constant, 387
Stefan's law, 387, 389–391, 924
Step-down transformer, 748, 749
Step-up transformer, 748
Stethoscope, 298, 298f
Stimulated absorption process, 963, 963f
Stimulated emission, 963–964, 964f
Stokes, George, 323
Stoke's law, 324
Stopping potential, 926
Strain
 bulk, 291–292
 compressive, 290
 defined, 287
 shear, 289
 tensile, 288
 volume, 289
Strangeness (S)
 conservation of, 1009, 1009f
 eightfold way and, 1010, 1010f
 of quarks, 1011, 1011t
Strange particles, 1009, 1009f
Strange quark (s), 1011, 1011t
Streamline (laminar) flow, 305, 305f
Stress
 bulk, 289, 289f, 291–292
 compressive, 290, 371
 defined, 287
 as proportional to strain, 287
 shear, 289, 289f, 291
 SI units, 288
 tensile, 288
 thermal, 343
 volume, 289
Stress-strain curve, 288, 288f
Stretching, Young's modulus and, 288
Strings, waves on, 465–467, 467
 interference, 471, 471f
 phase change, 841, 841f
 reflection of, 471–472, 472f, 841, 841f
 speed of, 468–470, 469f
 standing waves, 498–503, 498f, 499f
Strong force, 4, 89, 90, 1003–1004, 1004t, 1005, 1009, 1012, 1014, 1015f
Strontium, radioisotope of, 989
Strontium titanate, dielectric constant and dielectric strength of, 585t
Stud finders, 586, 586f
Subcritical condition, of nuclear reactor, 999
Sublimation, 375
Submarine periscopes, 789
Submerged object, buoyant forces on, 300, 300f, 302–303, 302f, 352–353
Subshells, atomic, 956–957
 notation for, 956, 956t
 Pauli exclusion principle and, 959–960, 959t
Subtraction
 significant figures and, 9
 vector, 59, 59f
Sucrose, diffusion coefficient of, 322t
Sudden infant death syndrome (SIDS), 707–708, 708f
Sugar, as optically active material, 859
Sulfur (S)
 boiling point, 375t
 electron configuration of, 960t
 latent heat of fusion, 375t
 latent heat of vaporization, 375t
 melting point, 375t
 resistivity, 608t

Sulfur dioxide (SO_2)
 as greenhouse gas, 392
 molar specific heat, 409t
 root-mean-square (rms) speed of, 358t
Summation (Σ), symbol for, A.2
Sun
 distance from Earth, A.19t
 formation of, 224
 fusion in, 1000–1001
 geocentric model of solar system, 26, 226
 gravitational force, 89, 96–97
 heliocentric model of solar system, 26, 45, 226
 Kepler's laws and orbits around, 226–228, 227f
 mass of, 3t, 228, 228t, A.19t
 radiant energy of, 387
 sunlight
  polarization of, 858
  scattering by air molecules, 858, 858f
 temperature of, 224
Sunglasses
 Polaroid, 858
 with UV protection, 758
Sunset, color of water ripples at, 776
Superconducting magnets, 616, 616f, 659, 664
Superconductors, 615–616, 615f, 615t, 616f
Superimposition principle
 electric force and, 529–532
Supernova, 264, 264f, 1017
Superposition principle
 beats and, 508–510, 508f
 defined, 470
 electric field and, 533, 535–536
 electric potential and, 565–566
 interference and, 470–471
 quality of sound and, 510–511
 standing waves on string, 498
Supersonic speeds, 496
Surface tension (γ), 314–316, 314f–315f, 314t
 in air sacs, 315
 defined, 314
 insects walking on water and, 315–316
 SI units, 314
 surfactant effect on, 315
 temperature and, 315
 values for common liquids, 314t
Surfactants, 315
Surgery
 cavitron ultrasonic surgical aspirator (CUSA), 484
 high-intensity focused ultrasound (HIFU), 484
Swim bladder, in fish, 301
Swimming, 295
Symbols
 of fundamental constants, A.20
 mathematical notation, A.1–A.3, A.19t
 SI base units, A.21
 SI derived units, A.21
 for units, A.19t
 use in problem solving, 19
Symmetry breaking, 1014
System
 defined, 135–136
 energy transfer to/from, 148–149, 368 (See also Heat)
 entropy of, 427
 internal energy of, calculating, 359
 moment of inertia of, 255
 translational kinetic energy of, 357
System approach, 111–113
Système International. See SI (Système International) units

T

Tacking, by sailboats, 312, 312f
Tacoma Narrows bridge, 504, 504f
Tangent (tan), 15f, 16
Tangential acceleration, 209, 213

Tangential speed
 angular speed relationship to, 208
 centripetal acceleration and, 212, 213
 defined, 208
 of phonograph records and compact discs, 210–211
 of rotating Earth surface, 209
Tau (τ^-), 1006, 1006t, 1007, 1008
Tau-lepton number, conservation of, 1008
Tau-neutrino (ν_τ), 1006, 1006t
Teflon®, dielectric constant and dielectric strength of, 585t
Tekalli, Jamila, 481f
Telecommunications, fiber optics in, 791–792, 791f
Telecommunications satellites, geosynchronous orbits of, 229
Telescope, 880–882, 880f–882f
 angular magnification (m) of, 880–882
 Hale, 881, 886
 Hubble Space Telescope, 880f, 882, 882f
 Keck, 870, 881
 mirrors in, 880–881, 881f, 882f
 Planck, 923
 prism spectrometer, 783–784, 783f
 radio, 886
 reflecting, 880–882, 881f
 refracting, 880–881, 881f
 resolution of, 885–886
Television signal interference, 839
Temperature (T)
 body temperature, maintaining, 387
 Celsius scale, 338–339, 340–342, 341f, 373
 critical, 615, 615f, 615t
 definition of, 337–338
 density of water and, 348, 348f
 equilibrium, 373–374
 expansion and, 359 (See also Thermal expansion)
 Fahrenheit scale, 340–342, 341f
 of gas
  Charles's law, 350
  Gay-Lussac's law, 350
  ideal gas law, 350
  molecular interpretation of, 357–359
 internal energy and, 149
 isothermal processes, 414–416, 417t, 424, 424f
 Kelvin scale, 340–342, 341f, 350, 373
 measurement of, 337–340, 338f, 339f
 molecular interpretation of, 357–359
 resistance, variation of, 610–611, 610f
 SI units, A.21t
 speed of sound and, 485
 standard temperature and pressure (STP), 350
 of Sun, 224
 surface tension and, 315
 volume expansion in liquids and, 348–349
Temperature coefficient of resistivity (α), 608t, 610, 611
Tendon, elastic modulus of, 288t
Tennis, 240, 311
Tensile strain, 288
Tensile strength, 290t
Tensile stress, 288
Tension (T), 100–101, 100f
 on string and speed of waves, 468–470
Terminal speed (v_t), 114, 324, 325
Terminal voltage, 627
Tesla, 664, 701
Tesla, Nikola, 742
Thermal conduction, 380–385, 380f–381f
 conductivity of selected substances, 381t
 defined, 380
 home insulation and, 383–385, 383t
 losses from human body, 382
Thermal conductivity (k), 381, 381t
Thermal contact, 337
Thermal efficiency (e)
 defined, 418

of heat engine, 418–419, 423, 424–425
of human body, 434–435, 434t
Thermal energy
  created by friction, 149
  global warming and greenhouse gases, 391–393, 392f, 393f
  transfer, 380–391 (See also Heat)
    calorimetry and, 372–374
    conduction, 380–385, 380f–381f
    convection, 385–387, 385f–386f
    minimizing, 391, 391f
    in nuclear reactor design, 100
    radiation, 387–391, 387f–390f
Thermal equilibrium, 337–338, 337f
Thermal expansion, 343–349
  area expansion, 345, 346
  bimetallic strips, 345, 345f
  coefficient of area expansion (γ), 345, 346
  coefficient of linear expansion (α), 343, 344t
  coefficient of volume expansion (β), 344t, 346-347
  defined, 343
  joints, 343, 343f
  linear expansion, 343–344, 344t, 347, 371
  volume expansion, 344t, 346–347
  of water, 347, 348–349
Thermal physics, 336–359. See also Temperature; Thermodynamics
Thermal processes, 408–417
  adiabatic processes, 411–413, 417t
  general processes, 416–417
  isobaric processes, 404, 409–411, 417t
  isothermal processes, 414–416, 417t
  isovolumetric processes, 413–414, 417t
Thermal processes, energy in, 367–393
Thermal radiation, 387–391, 387f–390f
  quantum physics and, 924–925
  solar-powered homes, 756
Thermals, 370–371
Thermal stress, 343
Thermoclines, 386–387, 386f
Thermodynamics, 402–435
  Carnot as founder of science of, 425
  first law of, 402, 406–408, 406f
    in adiabatic processes, 411–413, 417t
    in general processes, 416–417
    human metabolism and, 432
    in isobaric processes, 409–411, 417t
    in isothermal processes, 414–416, 417t
    in isovolumetric processes, 413–414, 417t
  second law of, 402
    direction of time and, 430–431
    entropy and, 426, 427, 430
    heat engines and, 423
    Lord Kelvin and, 423
    as statement of the most probable, 430–431
  third law of, 425
  work in thermodynamic processes, 402–405
  zeroth law of, 337f, 338
Thermogram, 388, 388f
Thermography, 388
Thermometers, 337–340, 337f–340
  alcohol, 338, 339
  calibration of, 338, 339
  constant-volume gas, 339–340, 339f, 340f
  defined, 337
  mercury, 338–339, 338f
  radiant, 389, 389f
  resistance, 610–611
Thermonuclear fusion reactions, 1001
Thermos bottle, 391, 391f
Thermostats, 345, 345f
Thin films, interference of light waves in, 841–846, 841f–845f
  index of refraction and, 841–845, 841f
  Newton's rings, 843, 843f
  problem-solving strategy, 843
  wedge-shaped film, 845
Thin-lens equation, 817f, 818
Thin lenses, 816–825
  combinations of, 822–825, 823f, 824f
  with concave mirror, 824–825, 824f
  converging, 816–818, 816f, 820–821, 820f, 822–824, 825, 825f, 826f
  diverging, 816–819, 816f, 821–822, 822f, 826
  focal point, 816–817, 816f, 819, 819f
  lens-maker's equation, 818
  ray diagrams for, 818–819, 819f
  thin-lens equation, 817f, 818
Third harmonic (second overtone), 499f, 500–503, 505f
Third law of motion (Newton's), 88, 97–100, 98f–99f, 176
  applications of, 98–99
  gravitational force and, 219–220, 219f
  snowshoes and, 285
Third law of thermodynamics, 425
Third wire on consumer appliances, 646
Thomson, J. J., 947, 948f
Thorium (Th)
  in breeder reactors, 1000
  radioactive series, 985, 985f, 985t
Three-way lightbulb, 634, 634f
Threshold voltage, 930
Thrust, 189
Thunder, 485
Time (t)
  approximate values of examples, 3, 3t
  dimensions, 5–6, 5t
  direction of, 430–431
  position, velocity, and acceleration as function of in simple harmonic motion, 457–459, 458f
  in special relativity
    simultaneity, 901–902, 901f
    time dilation, 902–905, 902f, 904f
  units of, 2–4, A.21t
Time constant (τ), 641, 719, 720
Time dilation, 902–905, 902f, 904f
Tin
  specific heat of, 370t
  superconductor, 615, 615t
Titanium (Ti), electron configuration of, 960t
Tokamaks, 683, 683f, 971f, 1002, 1002f
Tomonaga, Shinichiro, 1003f
Toner, photocopier, 572–573, 572f
Tonometer, 179–180
Top quark (t), 1011–1012, 1011t
Toroid, 1002
Torque ($\vec{\tau}$), 241–246, 241f
  angular acceleration and, 241, 252–259
  angular momentum and, 262–263, 262f
  axis of rotation and, 243–244, 253
  bicycle gears and, 254, 254f
  center of gravity and, 246–249
  definition
    basic, 241, 242
    general, 243, 244
  direction
    positive and negative, 241–242
    right-hand rule for, 243–244, 243f
  equilibrium conditions, 245–246
  force and, 241–246, 241f–244f
  moment of inertia, 254–258, 254f
  objects in equilibrium, 249–252, 250f
  point of application, 241
  on rotating object, 253–254, 253f–254f
  rotational analog of Newton's second law, 254, 256–259
  rotational motion under zero torque, 249
  SI units, 241, 243
Torque on current loop in magnetic field, 670–673, 670f, 672f

Torricelli, Evangelista, 297
Total acceleration, 213
Total energy, relativity and, 911–914
Total internal reflection, 788–792, 788f–791f
Tracers, radioactive, 989
Trains
  Acela, 43–44, 44f
  Doppler effect and, 494
  drive wheel mechanism of locomotives, 454, 454f
  railroad track, expansion of, 344, 344f
Transformer
  alternating-current (AC), 747–749, 747f
  role in reducing power loss, 748–749
  step-down, 748, 749
  step-up, 748
Transistor, as nonohmic device, 607
Translational acceleration, 245
Translational equilibrium, 245
Translational kinetic energy (KE_t), 357, 368
Transmission axis, of polarizing material, 855–856, 855f, 856f
Transmission electron microscope, 937, 937f
Transmutation, 980
Transport phenomena in fluids, 321–325
  diffusion, 321–322, 321f, 322f
  motion through a viscous medium, 323–324, 324f
  osmosis, 322–323, 323f
  sedimentation and centrifugation, 324–325
Transverse waves
  described, 465–466, 465f
  electromagnetic wave as, 753, 754
  motion of elements in, 482
  reflection of wave on string, 841, 841f
Traveling waves
  described, 465–468, 465f–466f
  reflection of, 472, 472f
  speed of pulse on, 469–470
Triangle, trigonometric functions of, 15–16, 15f
Triangle method of addition, 58
Trigonometry, 15–17, 15f, 17f, A.12–A–13
Triple point of water, 340
Tritium, 4
  as isotope, 972
  nuclear fusion and, 1001–1002
Tubes, resonance in, 507–508, 507f
Tungsten
  resistivity, 608t
  temperature coefficient of resistivity, 608t
Tungsten, elastic modulus of, 288t
Tuning fork, 481–482, 482f, 486–487, 509–511, 511f
Tuning musical instruments, 500–501, 509
Turbulent flow, 305, 305f
  of blood, 321
  Reynolds number, 320–321
Turntable, 453–454, 454f
Turpentine, coefficient of volume expansion for, 344t
Twin paradox, in special relativity, 905–906, 906f

U

Uhlenbeck, George, 957
Ultimate strength, 288, 290t
Ultrasonic flow meter, 483
Ultrasonic ranging unit, 484
Ultrasonic waves, 483
Ultrasound
  applications of, 483–484, 483f
  production of, 483, 483f
Ultraviolet (UV) light, 758f, 759
Uncertainty in measurement, 7–11
Uncertainty principle, 939–940, 939f
Underwater vision, 813
Unified mass unit (u), 972, 972f
Uniform circular motion, compared to simple harmonic motion, 453–457, 454f
Units. See also Gaussian (cgs) system of units; SI (Système International) units; U.S. customary units

conversion of, 11–12
dimensional, 5–7, 5t
prefixes for, 4, 4t
standard symbols for, A.19t
Universal gas constant (R), 350
Universal gravitation, Newton's law of, 88, 89, 95, 96–97, 219–226, 227
Universal gravitation constant (G), 95, 219, 220, 220f
Universe
   age of, 3t, 1007
   Big Bang theory, 1014–1017, 1015f
   composition of, 1017, 1017f
   dark energy in, 284
   dark matter in, 283–284
   distances, sample, 3t
   entropy of, 427, 431
   estimated number of galaxies in, 14
   evolution and fate of, 1017–1019, 1018f
   expansion of, 1015, 1015f, 1017, 1018, 1018f
   geocentric model, 26, 226
   geometry of, 1018, 1018f
   heliocentric model, 26, 45, 226
   mass of, 3t
Unknown, algebraic, A.4
Unpolarized light wave, 854–855, 855f
Up quark (u), 1011, 1011t
Upwelling, 348
Uranium (U)
   alpha decay of, 980
   density of, 284t
   discovery of radioactivity, 976
   fission of, 912, 914, 996–999, 997f, 999f
   radioactive series, 985, 985t
   radon gas in mines, 984
   sources of, 1000
Uranus
   escape speed, 225t
   planetary data, 228t
U.S. customary units
   acceleration, 5t, 35, 92, 93t
   area, 5t
   average velocity, 30
   conversion of units, 11–12
   force, 92, 93t
   heat, 368
   length, 4
   mass, 4, 92, 93t
   power, 151
   pressure, 285
   time, 4
   velocity, 5t
   volume, 5t
   work, 129

V

Vacuum dielectric constant, 585t
Van de Graaff generator, 540, 542–543, 542f
Vanadium (V), electron configuration of, 960t
Vaporization, latent heat of, 375–377, 375t
Vascular flutter, 311, 311f
Vector(s)
   addition of
      algebraically, 62–63, 62f
      geometrical, 58, 58f, 59, 59f
   angular quantities as, 214, 214f
   components of, 60–63, 60f–62f
   defined, 28
   displacement, velocity, and acceleration in two dimensions, 63–65
   electric field, 854–855, 854f, 855f
   equality of two, 58
   negative of, 59
   notation for, 28, 58
   properties of, 57–59
   resultant, 58, 58f, 59, 59f, 62–63, 62f
   subtraction of, 59, 59f

Vector quantity
   defined, 57
   examples of, 58
   notation for, 58
   scalar quantity compared to, 28, 57
Velocity
   acceleration vs., 34–35, 34f
   angular ($\vec{\omega}$)
      right-hand rule for, 214, 214f
      as vector quantity, 214
   average ($\bar{v}$)
      in constant acceleration, 39, 40t
      defined, 64
      in one dimension, 30–31, 30f–31f
      from position-time graph, 31, 31f
      SI (Système International) units, 30, 64
      in two dimensions ($\vec{v}_{av}$), 64
   dimensions of, 5–6, 5t
   as function of displacement in constant acceleration, 39, 40t
   as function of time in constant acceleration, 38f, 39, 40t
   graphical interpretation of, 31, 31f
   instantaneous
      defined, 32, 64
      in one dimension (v), 32–34, 32t, 33f
      from position-time graph, 32–34, 33f
      SI (Système International) units, 32, 64
      in two dimensions ($\vec{v}$), 64
   Maxwell velocity distribution, 358, 358f
   in motion diagrams, 37–38, 37f
   Newton's first law of motion and, 90–91, 90f–91f
   in one dimension, 28–34, 30f–31f, 30t, 32t, 33f
   projectile motion and, 65f, 66–69, 66f
   recoil, 177–179
   relative, 73–77, 74f–76f
      problem solving, 74–77
      in special relativity, 909–911
   in simple harmonic motion
      displacement and, 454
      as function of time, 457–459, 458f
      position and, 452, 454
   speed versus, 28–30
   of stars, 1016–1017, 1016f
   in two dimensions, 63–64
   as vector, 28–29
Velocity-time graph
   for Acela train, 43–44, 44f
   instantaneous acceleration from, 35–37, 36f
   for motion with constant acceleration, 38f, 39
Venturi tube, 308, 309f
Venus
   escape speed, 225, 225t
   magnetic field of, 661
   planetary data, 228t
Vertical jump, energy and power in, 153–155, 154f
Vibration
   damped oscillations, 463–464, 464f
   forced, 503
   heat capacity and, 408
   in matter, 283
   sound and (See Sound waves)
   standing waves on strings, 498–503, 498f, 499f
   tuning fork, 481–482, 482f, 486–487, 509–511, 511f
Virtual image, 802, 802f, 808–809, 808f
Virtual photon, 1004
Viscosity, 318–320, 319f, 319t
   coefficient of (η), 319, 319t
   Poiseuille's law and, 319–320, 319f
   SI units, 319
Visible light
   visible spectrum, 782, 782f, 784
   wavelength of, 758, 758f
Visible spectrum, 782, 782f, 784, 949f
Vision. See also Eye
   diving masks and, 819

   of invisible man, 876
   underwater, 813
Vitreous humor, 872, 872f
Voice
   helium affect on, 511
   shattering of glass, 503–504
Volta, Alessandro, 600
Voltage
   across a capacitor, 738, 740–741, 740f
   across an inductor, 739, 740–741, 740f
   across resistor, 739, 740–741, 740f
   equivalent to potential difference, 577
   in heart, 616–618, 616f
   in household circuits, 644–645, 645f
   instantaneous in RLC series circuit, 740, 740f
   measuring in electrical circuit, 605–606, 605f
   in Ohm's law, 607
   open-circuit, 627
   phasor diagram, 741, 741f
   rms, 735–736
   terminal, 627
   threshold, 930
Voltaic cell, invention of, 600
Voltmeter, 605f, 606
Volume (V)
   dimension of, 5t
   of gas
      Boyle's law, 350
      Charles's law, 350
      ideal gas law, 350
   isovolumetric processes, 413–414, 417t
Volume elasticity of solids, 288t, 289–290, 289f
Volume expansion (β), 344t, 346–347, 348–349
Volume strain, 289
Volume (bulk) stress, 289

W

Water
   adhesive forces, 316–317, 316f–317f
   algal blooms in ponds and lakes, 386–387, 386f
   boiling point (steam point) of, 339, 375t
   cohesive forces, 316–317, 316f–317f
   convection in boiling, 385
   density, 284t, 348, 348t, A.19t
   dielectric constant, 585t
   elastic modulus, 288t
   freezing point (ice point) of, 339
   heavy as moderator in nuclear reactor, 999
   index of refraction, 778t
   latent heat of fusion, 375, 375t
   latent heat of vaporization, 375, 375t
   melting point, 375t
   molar specific heat, 409t
   phase change in, 375–378, 376f, 428–429
   polarization of molecule, 589, 589f
   pressure and weight of, 286–287, 286f
   root-mean-square (rms) speed, 358t
   specific heat, 370, 370t
   speed of sound in, 485t, 486
   surface tension of, 314, 314f, 314t, 315–316, 315f
   thermal conductivity, 381t
   thermal expansion of, 347, 348–349
   triple point of, 340
   viscosity of, 319t
   waves
      interference and, 837, 837f
      ocean, 445, 464, 466, 506
Water pipes, bursting in winter, 348–349
Waterproofing agents, 317
Waterslides, 141–142, 141f
Watson, James, 932
Watt (W), 150–151
Watt, James, 151
Watt per meter squared (W/m^2), 487
Wave(s), 464–472. See also Electromagnetic waves; Light and light waves; Sound waves
   amplitude of, 467, 467f

defined, 464
density (pressure), 466
energy transfer to/from system, 149
frequency of, 466–468
graphical representations of, 466, 466f
interference (See Interference)
longitudinal, 465–466, 465f–466f
ocean, 445, 464, 466, 506
plane, 490, 490f, 753, 774f
radio, 886
reflection of, 471–472, 472f
resonance, 503–504, 503f, 504f, 507–508, 507f
shock, 495–496, 496f
soliton, 466
speed on strings, 468–470, 469f
spherical, 489–490, 489f–490f
standing
  in air columns, 504–508, 505f
  seiche, 506
  on strings, 498–503, 498f, 499f
transverse, 465–466, 465f
traveling, 465–470, 465f–466f, 472, 472f
types of, 465–466, 465f
water
  interference, 470–471, 471f
  ocean waves, 445, 464, 466, 506
  solitons, 466
  wave production, 464
Wave front
  of light, 774, 774f, 786–787, 787f
  of spherical wave, 490, 490f
Wave function (Ψ), 938
Wavelength (λ)
  Compton, 934, A.20
  cutoff, 928
  de Broglie, 935–936
  electromagnetic spectrum, 758–760, 758f
  of light
    limitations of compound microscope and, 879
    measurement with diffraction grating, 886
    measuring of coherent light source, 840
    visible light, 758, 758f
  of linear wave, 467–468, 467f
  of spherical wave, 490, 490f
Wave optics, 835–861
  diffraction, 847–854
    defined, 847
    Fraunhofer, 848, 848f
    single-slit, 848–850, 849f
  diffraction grating, 851–854, 851f–853f
  interference
    conditions for, 835–836
    Newton's rings, 843, 843f
    phase change due to reflection, 840–841, 840f, 841f
    in thin films, 841–846, 841f–845f
    use in reading CDs and DVDs, 846–847, 846f
    in Young's double-slit experiment, 836–840, 837f, 838f
  phase change due to reflection, 840–841, 840f, 841f
  polarization of light waves, 854–861
    concept described, 854–855, 854f, 855f
    liquid crystals, 860–861, 860f
    optical activity, 858–859, 859f
    by reflection, 857–858, 857f

by scattering, 858, 858f
by selective absorption, 855–857, 855f, 856f
Young's double-slit experiment, 836–840, 837f, 838f
Wave speed (v), 484–485
  equation for, 467
  shock waves, 495–496, 496f
  on strings, 468–470, 469f
Wave theory of light, 773
  Fresnel's, 848
W bosons, 1004, 1004f, 1013–1014, 1013f
Weak force, 89, 90, 1003–1004, 1004t, 1013–1014, 1013f, 1015, 1015f
Weakly interacting massive particle (WIMP), 1017
Weber (Wb), 664, 700
Weight, 89
  gravitational force and, 95, 95f, 97
  of water, 286–287, 286f
Weight gain, human, 433
Weight loss programs, diet vs. exercise in, 154–155
Wetting agents, 317
Whale, power generated by, 152
Wien's displacement law, 924
Wilkins, Maurice, 932
Wilson, Robert W., 1015, 1016f
WIMP (weakly interacting massive particle), 1017
Windshield wipers, 641
Wind turbine, 240
Wood, thermal conductivity of, 381t
Work (W), 128–132, 157
  by body on its surroundings, 432
  calculating from PV diagram, 404, 405
  coefficient of performance (COP), 421–422
  conservative and nonconservative forces, 134–135, 135f, 148, 158
  by constant force
    at angle to displacement, 129, 129f, 130–131
    in linear displacement, 128–129, 129f, 130–131
  defined, 128–129
  dissipative forces and, 131–132
  by elastic potential energy, 144
  energy transfer to/from system, 149
  estimating by counting boxes in graphical method, 157
  in first law of thermodynamics, 406–407
  fluid flow and, 308, 308f
  by friction, 131–132, 142–143, 142f
  on gas
    in adiabatic processes, 411–413, 417t
    calculating from PV diagram, 404, 405
    in general thermodynamic processes, 416–417
    in isobaric processes, 409–411, 417t
    in isothermal processes, 414–416, 417t
    in isovolumetric processes, 413–414, 417t
    work done by a gas compared, 403
  by gas on its environment (W_{env}), 403, 404, 407, 409, 410
  gravitational, 136–137, 136f, 158
  done by gravity and gravitational potential energy, 136, 144
  by heat engine, 418–419, 423
  as scalar quantity, 128, 129
  SI (Système International) units, 128, 129, 136, A.21t
  done by springs and spring potential energy, 144
  by spring force, 144

to stretch a spring, 156–157, 156f
in thermodynamic processes, 402–405
by a varying force, 155–157, 155f–157f
Work-energy theorem, 157–158
  applying, 134
  conservative and nonconservative forces, 135, 148, 158
  described, 133
  fluid flow and, 308f
  with gravitational potential energy and elastic (spring) potential energy components, 144–148, 158
  with gravitational potential energy component, 136–137, 158
  problem-solving strategy for, 157–158
  rotational motion and, 260, 261–262
Work function (ϕ), 927–928, 927t
Wound treatment, 760, 760f

X

X-component, of vector, 60–63, 60f–62f
Xenon (Xe), as noble gas, 960
Xerography, 572–573, 572f
Xi (Ξ) particle, 1006, 1006t
X-rays, 758f, 759, 929–930
  characteristic, 961–962, 961f, 962f
  diffraction by crystals, 930–933, 931f, 932f
  diffraction pattern, 929, 929f, 931, 932, 932f
  discovery of, 929
  electrons and production of, 930
  paintings studied with, 930
  RBE (relative biological effectiveness) factor, 988, 989t
  scattering of, 934
  spectrum emitted by x-ray tube, 929–930, 929f
  wavelength of, 929
X-ray tube
  spectrum emitted by, 930, 930f
  structure of, 929–930, 929f

Y

Y-component, of vector, 60–63, 60f–62f
Year, time interval for one, 3t
Yerkes Observatory, 881
Young, Thomas, 773, 836
Young's double-slit experiment, 836–840, 837f, 838f
Young's modulus (Y), 287–288, 288t, 485

Z

Z bosons, 1004, 1004f, 1013–1014, 1013f
Zeeman effect, 955–956, 955f
Zero-point energy, 340
Zeros, as significant figures, 8
Zero temperature, absolute, 339–340
Zeroth law of thermodynamics, 337f, 338
Zeroth-order maximum, 838, 851f, 852
Zinc (Zn)
  electron configuration of, 960t
  as superconductor, 615, 615t
  work function of, 927t
Zircon, index of refraction, 778t
Zonules, 872
Zwicky, Fritz, 1016

PHYSICAL CONSTANTS

Quantity	Symbol	Value	SI unit
Avogadro's number	N_A	6.02×10^{23}	particles/mol
Bohr radius	a_0	5.29×10^{-11}	m
Boltzmann's constant	k_B	1.38×10^{-23}	J/K
Coulomb constant, $1/4\pi\epsilon_0$	k_e	8.99×10^9	$N \cdot m^2/C^2$
Electron Compton wavelength	$h/m_e c$	2.43×10^{-12}	m
Electron mass	m_e	9.11×10^{-31}	kg
		5.49×10^{-4}	u
		$0.511 \text{ MeV}/c^2$	
Elementary charge	e	1.60×10^{-19}	C
Gravitational constant	G	6.67×10^{-11}	$N \cdot m^2/kg^2$
Mass of Earth	M_E	5.98×10^{24}	kg
Mass of Moon	M_M	7.36×10^{22}	kg
Molar volume of ideal gas at STP	V	22.4	L/mol
		2.24×10^{-2}	m^3/mol
Neutron mass	m_n	$1.674\,93 \times 10^{-27}$	kg
		$1.008\,665$	u
		$939.565 \text{ MeV}/c^2$	
Permeability of free space	μ_0	1.26×10^{-6} ($4\pi \times 10^{-7}$ exactly)	$T \cdot m/A$
Permittivity of free space	ϵ_0	8.85×10^{-12}	$C^2/N \cdot m^2$
Planck's constant	h	6.63×10^{-34}	$J \cdot s$
	$\hbar = h/2\pi$	1.05×10^{-34}	$J \cdot s$
Proton mass	m_p	$1.672\,62 \times 10^{-27}$	kg
		$1.007\,276$	u
		$938.272 \text{ MeV}/c^2$	
Radius of Earth (at equator)	R_E	6.38×10^6	m
Radius of Moon	R_M	1.74×10^6	m
Rydberg constant	R_H	1.10×10^7	m^{-1}
Speed of light in vacuum	c	3.00×10^8	m/s
Standard free-fall acceleration	g	9.80	m/s^2
Stefan-Boltzmann constant	σ	5.67×10^{-8}	$W/m^2 \cdot K^4$
Universal gas constant	R	8.31	$J/mol \cdot K$

The values presented in this table are those used in computations in the text. Generally, the physical constants are known to much better precision.